Floristic Divisions and Sections of the Intermountain Region

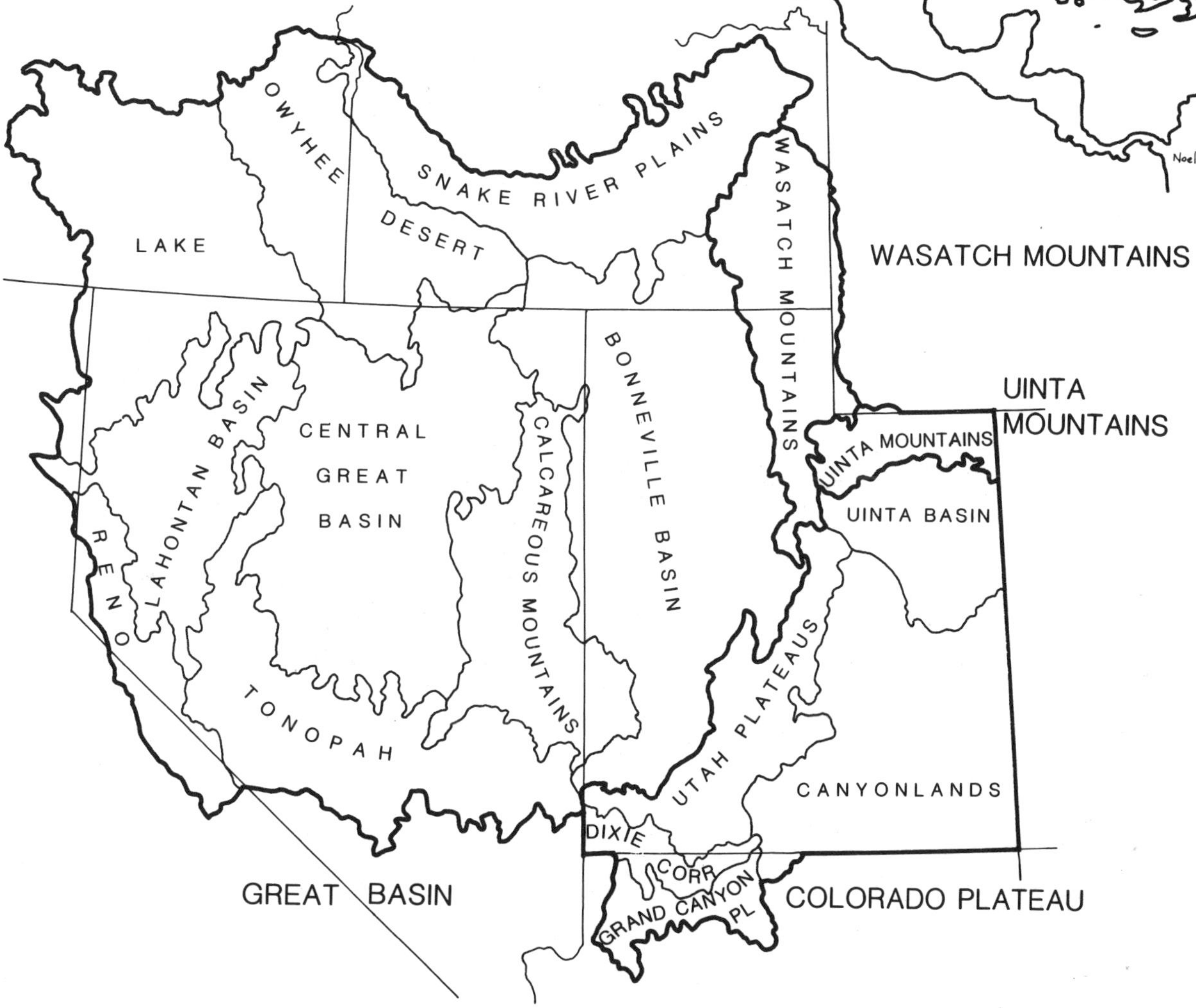

INTERMOUNTAIN FLORA

Sego Lily, *Calochortus nuttallii;* Bear River Canyon, Box Elder County, Utah. Photo by Valoie Albrecht.

INTERMOUNTAIN FLORA

Vascular Plants of the Intermountain West, U.S.A.

By

ARTHUR CRONQUIST
ARTHUR H. HOLMGREN
NOEL H. HOLMGREN
JAMES L. REVEAL
PATRICIA K. HOLMGREN

VOLUME SIX

THE MONOCOTYLEDONS

REPRINTED 1994
THE NEW YORK BOTANICAL GARDEN

ORIGINALLY PUBLISHED FOR
THE NEW YORK BOTANICAL GARDEN
BY
COLUMBIA UNIVERSITY PRESS
NEW YORK
1977

LIBRARY OF CONGRESS CATALOGING IN PUBLICATION DATA (revised)
Main entry under title:
Intermountain flora.
Includes bibliographies.
CONTENTS: v. 1. Geological and botanical history of the region, its plant geography and a glossary. The vascular cryptogams and the gymnosperms.
v. 6. The monocotyledons.
1. Botany—Great Basin. 2. Plants—Identification.
I. Cronquist, Arthur. II. New York (City).
Botanical Garden.
QK141.158 581.9′79 73-134298
ISBN 0-231-04120-9

Printed in the United States of America

Reprinted 1994
by
The New York Botanical Garden
ISBN 0-89327-386-4

The paper used in this publication meets the requirements of the American National Standard for Information Sciences—Permanence of Paper for Publications and Documents in Libraries and Archives, ANSI/NISO (Z39.48–1992).

In publishing this volume of the Intermountain Flora we take particular note of our indebtedness to the men and women of the United States Forest Service, the Bureau of Land Management, and the National Park Service, whose studies over many years have contributed so much to the understanding of grasslands, and the grasses and other plants in the Intermountain Region.

CONTENTS OF VOLUME SIX

INTERMOUNTAIN FLORA

INTRODUCTION

Volume 6 of our Flora, treating the monocotyledons, is chronologically the next volume to appear after volume 1. We thought it desirable to make the treatment of the Poaceae available to land-use managers as soon as possible. Furthermore, the treatment of the Cyperaceae had been written several years ago and needed only minor updating for publication at this time. We trust that librarians and bibliographers will not be unduly inconvenienced by this unorthodox but to us useful sequence of volumes.

In this volume, as in others, some one or two authors are responsible for each segment of the work, the other authors serving only in advisory capacity. We have added Patricia K. Holmgren to our group of authors. She serves as general editor, and she has written the treatment of the Iridaceae. The family Orchidaceae has been done by Arthur Holmgren, and the family Poaceae by Arthur Holmgren and Noel Holmgren together. Arthur Cronquist has done the Cyperaceae, the Juncaceae, and the genus *Allium,* the latter in collaboration with the late Marion Ownbey, a well-known student of the genus. The artificial key to the families and the synoptical arrangement of the subclasses are also by Cronquist. James Reveal has done all the remaining families: Agavaceae, Alismataceae, Butomaceae, Commelinaceae, Hydrocharitaceae, Juncaginaceae, Lemnaceae, Liliaceae (except *Allium*), Najadaceae, Potamogetonaceae, Ruppiaceae, Sparganiaceae, Typhaceae, and Zannichelliaceae.

We wish to emphasize that the Flora represents a cooperative effort not only by the several authors, but also by their respective institutions. Three of the authors (Cronquist, Noel Holmgren, and Patricia Holmgren) are based at the New York Botanical Garden; Arthur Holmgren is at Utah State University; Reveal is at the University of Maryland and also makes frequent use of the collections and facilities at the Smithsonian Institution.

Species that occur in the Pacific Northwest as well as in the Intermountain Region are mostly illustrated by drawings that were prepared by Jeanne R. Janish and published in Vascular Plants of the Pacific Northwest (Univ. Wash. Publ. Biol. 17. 1955–69). These drawings are reproduced by permission of the University of Washington Press. Other species in our Flora are illustrated by new drawings by Page R. Allen, Rupert C. Barneby, Charles C. Clare, Jeanne R. Janish, Melissa Marshall, William Moye III, Frances Runyan, and Kaye H. Thorne. The drawings of flowers of *Elodea* are taken from those of Harold St. John, with his kind permission.

We acknowledge with pleasure the advice and assistance received from Dr. Douglas R. Dewey on the Triticeae, Dr. Robert R. Haynes on the Potamogetonaceae, Dr. Douglass Henderson on *Sisyrinchium,* Dr. Frederick J. Hermann on the Cyperaceae and Juncaceae, Dr. Richard Spellenberg on *Panicum* and *Stipa,* and Lois A. Arnow and Dr. Stanley L. Welsh on the local distribution of numerous species. The opinions expressed herein are our own, and should not be attributed to those who helped us.

Secretaries play an insufficiently appreciated role in the production of books. We are indebted to several secretaries, especially to Lucy F. Kluska, who typed most of the final version of the manuscript, as well as interim versions of individual segments while the work was in progress.

We gratefully acknowledge the long-continued support of the National Science Foundation for the Intermountain Flora project. The following grants, dating from 1959, were used in whole or in part for work on the Intermountain Flora: G-7120 (New York

Botanical Garden, for Cronquist), GB-1299 (New York Botanical Garden, for Cronquist), GB-5361 (New York Botanical Garden, for Cronquist and Noel Holmgren), GB-30894 (New York Botanical Garden, for Cronquist, Noel Holmgren, and Patricia Holmgren), GB-36163 (New York Botanical Garden, for Noel Holmgren), and GB-22645 (University of Maryland, for Reveal).

Class **LILIOPSIDA**, Monocotyledons

Herbs (nearly all ours) or woody plants with anomalous or no secondary thickening, always without an intrafascicular cambium, very often with the vessels confined to the roots; vascular bundles of the stem generally scattered, or in two or more rings; root system wholly adventitious; leaves most commonly narrow and parallel-veined, frequently with a basal sheath rather than a petiole, but varying to broader, more net-veined, and sometimes also petiolate; floral parts, when (as usually) of definite number, typically borne in sets of 3, seldom 4, never 5 (stamens or carpels often fewer); pollen mostly uniaperturate; embryo with a single cotyledon, or undifferentiated.

The class Liliopsida consists of 4 rather poorly characterized subclasses, all of which occur in our region. The synoptical keys to subclasses, orders, and families are preceded by an artificial key to the families of monocotyledons in our flora.

Artificial Key to the Families of Monocotyledons
(as represented in our flora)

1 Plants submersed or free-floating or floating-leaved aquatics, only the individual flowers or the small inflorescences sometimes elevated a little above the surface of the water.
  2 Plants thalloid, free-floating, rootless or with 1–several simple roots hanging down in the water; flowers (seldom produced) tiny, without perianth, borne in small, marginal pockets of the thallus .... Lemnaceae, p. 471
  2 Plants differentiated into roots, stems, and leaves, often rooted to the bottom, but sometimes free-floating; flowers otherwise.
    3 Flowers in globose, unisexual clusters along the upper part of the stem, nearly or quite without perianth .... Sparganiaceae, p. 465
    3 Flowers in terminal spikes, or solitary in the axils, or in small, axillary cymes.
      4 Perianth wanting; carpels 1–several, separate.
        5 Leaves all clustered at the base, the plants acaulescent (*Lilaea*) .... Juncaginaceae, p. 18
        5 Leaves distributed along the stems, variously alternate to opposite or whorled.
          6 Flowers perfect, borne in terminal spikes; leaves alternate .... Ruppiaceae, p. 42
          6 Flowers unisexual, solitary in the axils, or in axillary cymes; leaves mostly opposite or subopposite to verticillate.
            7 Carpel solitary .... Najadaceae, p. 22
            7 Carpels 2 or more, separate .... Zannichelliaceae, p. 43
      4 Perianth evident, though not always showy.
        8 Flowers tetramerous, hypogynous, perfect, with a single set of 4 tepals to which the 4 stamens are adnate, and with 4 separate, uniovulate carpels .... Potamogetonaceae, p. 24
        8 Flowers trimerous, epigynous, mostly unisexual, with well differentiated sepals and petals, and with the 3 carpels united to form a compound pistil with intruded partial partitions .... Hydrocharitaceae, p. 14
1 Plants terrestrial, or emergent aquatics, sometimes with floating or in part submersed leaves, but the inflorescence then elevated well above the surface of the water.
  9 Perianth without petal-like parts, consisting instead of small, dry, chaffy or scale-like parts, or slender bristles, or essentially wanting.
    10 Ovary with a single locule and a single ovule.
      11 Styles or stigmas 2 or 3 (only 1 in *Zea mays*).
        12 Flowers spirally or less often distichously arranged on the axis of a spike or spikelet, usually each flower seemingly or actually subtended by only a single bract, without an evident bract between the flower and the axis; seed coat generally free from the pericarp; leaf sheath usually closed; stem usually solid, often triangular; flowers often with a perianth of evident, slender bristles; carpels 3 or less often 2 .... Cyperaceae, p. 67

12 Flowers distichously arranged on the axis of a spikelet (or only one per spikelet), each flower ordinarily subtended by a pair of bracts (lemma and palea), the palea inserted between the flower and the axis; seed coat generally adnate to the pericarp; leaf sheath usually open; stem usually hollow, never triangular; flowers without a perianth, unless the lodicules are so interpreted; carpels 2 .... POACEAE, p. 175

11 Style and stigma 1.

13 Plants acaulescent, the leaves slender, terete, all tufted at the base (*Lilaea*) .... JUNCAGINACEAE, p. 18

13 Plants more or less leafy-stemmed, the leaves flat or keeled.

14 Inflorescence of globose heads; achenes sessile or nearly so, not wind-distributed .... SPARGANIACEAE, p. 465

14 Inflorescence of dense, cylindrical spikes; achenes slenderly long-stipitate, with long hairs on the stipe, wind-distributed .... TYPHACEAE, p. 468

10 Ovary either with more than one locule, or with 3 or more ovules in a single locule.

15 Ovules and seeds as many as the locules; fruit indehiscent, or follicular.

16 Plants scapose, the leaves all clustered at the base; inflorescence an elongate, slender spike or raceme .... JUNCAGINACEAE, p. 18

16 Plants more or less leafy-stemmed; inflorescence of globose heads arranged along the upper part of the stem .... SPARGANIACEAE, p. 465

15 Ovules and seeds obviously more numerous than the locules; fruits capsular .... JUNCACEAE, p. 47

9 Perianth partly or wholly of petaloid or corolloid, more or less showy parts.

17 Carpels essentially separate; plants semi-aquatic.

18 Ovary with more or less numerous ovules; fruit a follicle; leaves not differentiated into blade and petiole .... BUTOMACEAE, p. 5

18 Ovary with a single ovule; fruit an achene; leaves often (not always) differentiated into blade and petiole .... ALISMATACEAE, p. 7

17 Carpels united to form a compound pistil.

19 Stamens 3–6, separate from the style; flowers regular or irregular; ovary superior to inferior.

20 Plants coarse, often shrubby or arborescent, with firm, perennial leaves .... AGAVACEAE, p. 526

20 Plants strictly herbaceous, the leaves generally lasting only a single season.

21 Stamens (including staminodes, if any) 6; ovary superior or inferior.

22 Corolla ephemeral; sepals herbaceous, not at all petaloid; ovules only 1 or 2 in each locule .... COMMELINACEAE, p. 46

22 Corolla lasting more than one day, not ephemeral; sepals usually showy, like the petals (but not in *Calochortus*); ovules in most species several or numerous in each locule .... LILIACEAE, p. 476

21 Stamens 3; ovary inferior .... IRIDACEAE, p. 538

19 Stamens 1–2, adnate to the thickened style; flowers strongly irregular; ovary inferior, with extraordinarily numerous and small seeds .... ORCHIDACEAE, p. 546

SYNOPTICAL ARRANGEMENT OF THE SUBCLASSES OF LILIOPSIDA
(with special reference to the species of our flora)

1 Flowers mostly apocarpous, with 1 to more often 3–many carpels, the syncarpous forms with a modified type of laminar placentation (except Juncaginaceae, of the order Najadales); plants aquatic or semi-aquatic, but not thalloid; endosperm wanting .... Subclass ALISMATIDAE, p. 5

1 Flowers syncarpous (ovary pseudomonomerous in Typhales); placentation various, but not laminar; plants variously terrestrial or less often aquatic; endosperm present or (Orchidaceae) wanting.

2 Our plants thalloid, free-floating aquatics, the flowers seldom produced, few, much reduced, without perianth; more typical (but extralimital) members of the subclass, such as palms and aroids, commonly have broad, petiolate leaves that do not have the familiar parallel venation of other monocots, the plants often woody and with well developed vessels throughout, and the flowers commonly subtended by a spathe and often forming a spadix .... Subclass ARECIDAE, p. 470

2 Plants neither thalloid nor free-floating; leaves typically narrow and parallel-veined, seldom (in extralimital species) broader and net-veined; flowers few or solitary to numerous, large or small, but seldom with a conspicuous spathe and never aggregated into a spadix.

3 Flowers either with well differentiated sepals and petals (the sepals not petaloid), or more often more or less reduced and with chaffy or bristly (or no) perianth; species with reduced or no perianth mostly wind-pollinated, or self-pollinated, or apomictic, those with well developed petals insect-pollinated but without nectar; our families with the vessels well developed in all vegetative organs; seeds mostly with a starchy endosperm (endosperm of Typhaceae with more oil and protein than starch) .... Subclass COMMELINIDAE, p. 44

3 Flowers generally with both the sepals and the petals petaloid and often much alike (sepals not petaloid in *Calochortus* and *Cypripedium*); most species nectariferous and insect-pollinated; our families with the vessels largely or wholly confined to the roots; seeds mostly with proteinaceous, oily, or cellulosic, nonstarchy endosperm, or without endosperm .... Subclass LILIIDAE, p. 476

Subclass **ALISMATIDAE**

Monocotyledonous plants with mostly apocarpous flowers, the syncarpous forms (except Juncaginaceae) with a modified type of laminar placentation; pollen trinucleate; seeds mostly (including all of our species) without endosperm; aquatic and semiaquatic herbs.

Four orders, three of which occur in our area, with the fourth (Triuridales) wholly tropical and only doubtfully included in the subclass.

SYNOPTICAL ARRANGEMENT OF THE ORDERS OF ALISMATIDAE
(as represented in our flora)

1 Perianth differentiated into evident sepals and petals; flowers often bracteate.
  2 Flowers hypogynous; carpels free or nearly so .... Order Alismatales, p. 5
  2 Flowers epigynous; ovary compound and with modified laminar placentation on intruded partial partitions .... Order Hydrocharitales, p. 14
1 Perianth, when present, not differentiated into sepals and petals; bracts usually wanting .... Order Najadales, p. 16

Order **ALISMATALES**

Flowers hypogynous, each generally subtended by a bract; perianth well developed, biseriate, generally differentiated into three sepals and three petals; stamens 6–many, free, with an equal number of distinct pistils; pollen trinucleate; seeds without endosperm; aquatic and semiaquatic herbs with slender and parallel-veined to broader and net-veined leaves, with or without distinction into a blade and petiole.

Three families, two of them represented in our range, with the third (Limnocharitaceae) wholly tropical.

KEY TO THE FAMILIES OF ALISMATALES

1 Fruits follicles with numerous seeds; leaves not differentiated into a blade and petiole; tall, herbaceous perennials .... BUTOMACEAE, p. 5
1 Fruits achenes with a single basal seed; leaves often differentiated into a blade and petiole; low, herbaceous perennials .... ALISMATACEAE, p. 7

Family **BUTOMACEAE,** the Flowering-rush Family

Herbaceous, aquatic or marsh-inhabiting perennials from a stout, elongated rhizome with long, basal, linear, flat leaves, these often floating in deep water; flowers perfect, regular, trimerous, borne in a terminal involucrate umbel; perianth petaloid, persistent, the sepals 3, greenish, the petals 3, reddish, usually conspicuous and showy; stamens 9, free, all fertile, hypogynous, the filaments slender, the anthers introrse, 2-celled; pistils simple, 6, connate only at the base, whorled, the ovules numerous, scattered on the inner wall, the style terminal, the stigma mostly sessile; fruits a whorl of hardened many-seeded follicles, each follicle with a persistent, short, slightly curved beak and dehiscing ventrally, with numerous exalbuminous, straight seeds borne over much of the inner surface.

As defined here a monotypic family, native of Europe, but widely introduced in N. Amer.

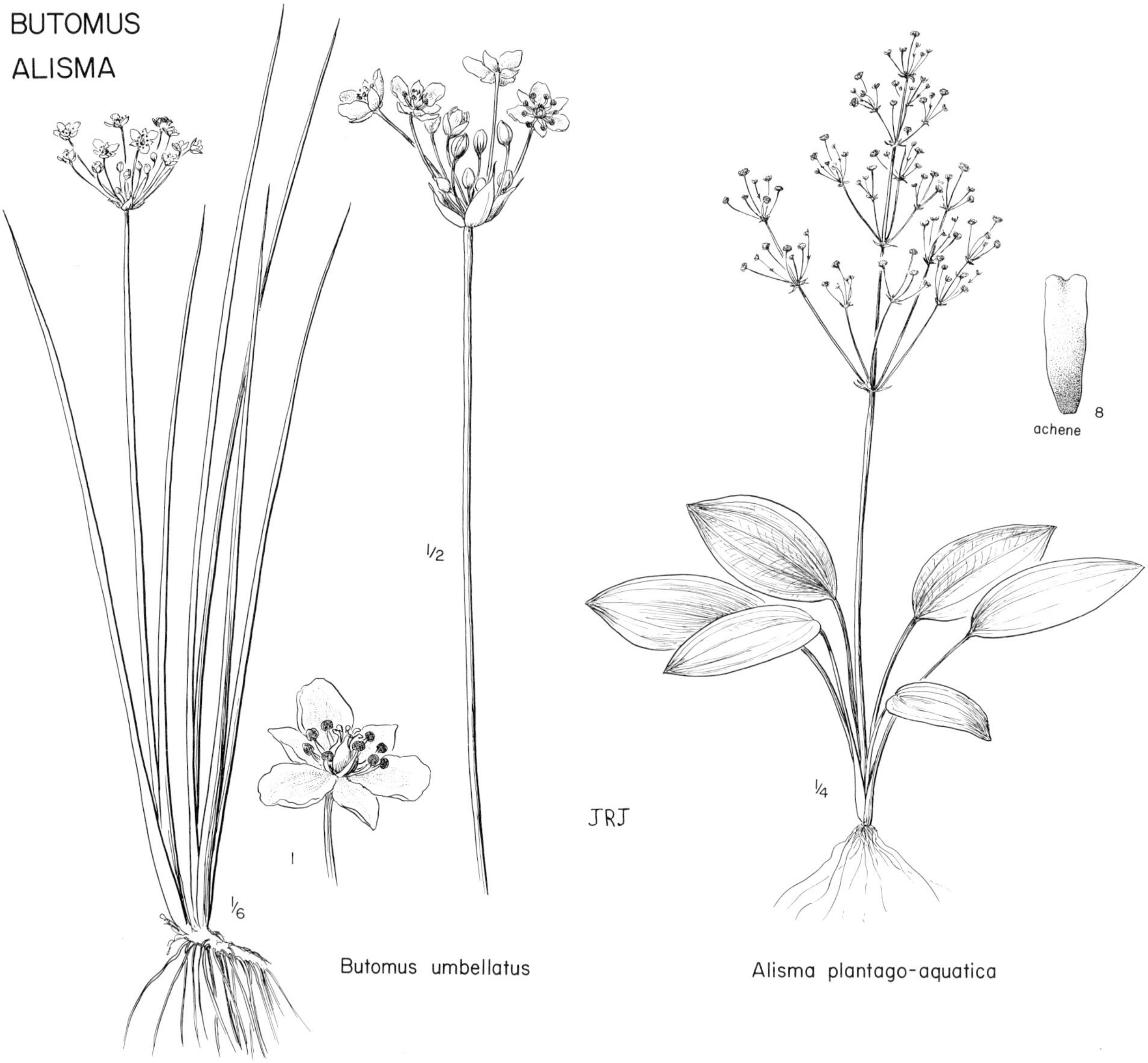

1. BUTOMUS L. Flowering-rush

A monotypic genus of temp. Eurasia, becoming established elsewhere as in parts of N. Amer. (From the Greek *bous*, ox, and *temno*, to cut, alluding to the leaves but not necessarily in the same sense as the Greek word *butomus* was originally used.)

1. Butomus umbellatus L.

Butomus umbellatus L. Sp. Pl. 372. 1753. (N. Europe.)

B. umbellatus f. *vallisneriaefolius* Sagorski ex Aschers. & Graebner, Syn. Mitteleur. Fl. 1: 396. 1897. (Ostpreussen.)

Tall perennials up to 1 m high arising from stout fleshy rhizomes which bear many freely deciduous erect grain-like tubers late in the season; leaves erect, 2-ranked, 3-angled at the base, ensiform, linear, 5–10 mm wide, about equalling the scape; scapes 1–few, erect, with a large terminal umbel, the umbel in pre-anthesis enclosed by 3 large scarious purplish involucral-like bracts, these bending back during anthesis revealing the ascending slender pedicels up to 5 cm long; flowers numerous, the perianth 2–2.5 cm across, roseate, the sepals usually greenish and less showy than the larger rose-colored petals; follicles inflated, beaked, the body 8–10 mm long, dark brown; $2n$ = 20, 26, 39.

Mud and shallow water along streams and marsh areas; Eurasia, introduced and established in many regions of e. N. Amer., especially in the Great Lakes area and along the St. Lawrence system; found in our area in the Snake River drainage, near Idaho Falls, Bonneville Co., Idaho. June–Sept.

Family **ALISMATACEAE**, the Water-plantain Family

Herbaceous, aquatic or semiaquatic lactiferous annuals or (ours) perennials with fibrous roots arising from a somewhat thickened rootstock and with a cluster of erect to floating leaves, these usually well differentiated into a sheathing petiole and at first bladeless, eventually developing a linear or sagittate blade with prominent nerves and transverse veinlets, the ligule lacking; flowers perfect or occasionally imperfect (unisexual), regular, trimerous, borne in umbels, racemes or panicles; perianth imbricate or involute in bud, differentiating into a persistent calyx and deciduous corolla, the sepals 3, greenish, the petals 3, white to pinkish or purplish, often conspicuous and showy, rarely lacking; stamens 6–many, infrequently only 3, included, free, hypogynous, the filaments slender to somewhat thickened and fleshy, the anthers extrorse, 2-celled; pistils simple, usually several to many, whorled, arranged in a ring or crowded so as to produce a headlike cluster of achenes on the receptacle, the ovules mostly 1, rarely more, restricted to the base of the ovary, the style terminal, the stigma mostly sessile; fruits usually achenes but sometimes dehiscent, with 1-several exalbuminous, straight seeds borne at the base.

A family of 10 to 12 genera and some 70 species widely distributed throughout most of the N. Hemisphere except in the Arctic.

1 Achenes arranged in a single ring on the receptacle; stamens 6.
  2 Achenes rounded on the back, not conspicuously beaked, the beak short and less than 1 mm long, the ring of achenes smooth; petals entire or only slightly erose .... 1. *Alisma*
  2 Achenes conspicuously horned on the back with a sharp beak 3–6 mm long, the ring of achenes forming a stellate pattern; petals toothed .... 2. *Machaerocarpus*
1 Achenes densely packed over the entire surface of the receptacle; stamens more than 6.
  3 Leaves rounded to truncate or subcordate at the base; flowers perfect; achenes arranged into a bur-like head; stems and petioles angled; known in our area only from Uintah Co., Utah .... 3. *Echinodorus*
  3 Leaves sagittate to hastate or infrequently tapered at the base, often ribbon-like when emersed; flowers perfect or all pistillate or all staminate; achenes usually arranged in a smooth head; stems and petioles smooth and terete; common .... 4. *Sagittaria*

1. **ALISMA** L. Water-plantain

Scapose, herbaceous perennials with erect to floating basal leaves; flowers few to several in bracteate, compound, terminal whorled panicles, rather small and inconspicuous, perfect; sepals greenish, persistent; petals white, rarely pinkish or rose, the margin entire or only slightly erose, deciduous; stamens 6–9; pistils (5) 10–25 in a smooth whorl; fruits strongly compressed achenes, ribbed on the curved back, the beak short and straight or (in ours) slightly curved.

A complex genus of 4 to 9 species, depending upon interpretation, widely distributed in temp. and trop. regions. (From the Greek *alisma*, a kind of aquatic plant whose identity is not now known; the derivation of the word is presently obscure.)

References:
Björkqvist, I. Studies in *Alisma* L. I. Distribution, variation and germination. Opera Bot. 17: 1–128. 1967. II. Chromosome studies, crossing experiments and taxonomy. Opera Bot. 19: 1–138. 1968.
Fernald, M. L. The North American representatives of *Alisma plantago-aquatica*. Rhodora 48: 86–88. 1946.
Hendricks, A. J. A revision of the genus *Alisma* (Dill.) L. Amer. Midl. Naturalist 58: 470–493. 1957.
Pogan, E. Taxonomical value of *Alisma triviale* Pursh and *Alisma subcordatum* Rafin. Canad. J. Bot. 41: 1011–1013. 1963.
Rubtzoff, P. Notes on the genus *Alisma*. Leafl. W. Bot. 10: 90–95. 1964.
Samuelsson, G. Die Arten der Gattung *Alisma* L. Ark. Bot. 24A(7): 1–46. 1932.

1 Achenes centrally grooved at the tip, the style straight or nearly so, (0.4) 0.5–0.8 mm long; leaves lanceolate to oblong or broadly elliptic to ovate, mostly 2–20 cm wide; pedicels ascending or erect; petals 3.5–6 mm long, white; inflorescences surpassing the leaves .... 1. *A. plantago-aquatica*
1 Achenes centrally ridged and with two grooves towards the tip, the style curled, 0.4–0.5 mm long; leaves linear to narrowly elliptic, mostly less than 3 cm wide; pedicels widely spreading and curved; petals 2–4 mm long, pinkish or whitish; inflorescences not always surpassing the leaves .... 2. *A. gramineum*

1. **Alisma plantago-aquatica** L.

Alisma plantago-aquatica (as *plantago*-Δ) L. Sp. Pl. 342. 1753. (Europe.)

A. subcordatum Raf. Med. Repos. II. 5: 362. 1808. ("Common almost all over the United States of America," a neotype has been chosen by Björkqvist as *Moser*, Bethlehem, Pa., collected in 1822, but no effort was made to check for an original Rafinesque collection—see comments below.) = var. *americanum?*

A. plantago sensu Michx. Fl. Boreali-Amer. 1: 218. 1803; not L. 1753. *A. trivialis* Pursh, Fl. Amer. Sept. 252. 1814. *A. plantago-aquatica* var. *americanum* Schultes & Schultes, Syst. Veg. 7(2): 1598. 1830. *A. plantago* var. *triviale* Britton, Sterns & Poggenburg, Prelim. Cat. Antho. & Pterid. within 100 Mi. New York City 58. 1888. *A. plantago-aquatica* var. *triviale* Farw. Annual Rep. Commiss. Park & Boulevards Detroit, Michigan 11: 44. 1900. (*Michaux*, Quebec, Can., 1792.)

A. brevipes Greene, Pittonia 4: 158. 1900. *A. plantago-aquatica* subsp. *brevipes* G. Samuelsson, Ark. Bot. 24A(7): 19. 1932. *A. plantago-aquatica* var. *brevipes* (attrib. to G. Samuelsson by) Marie-Vict. Fl. Laurent. 615. 1935. (*C. F. Baker 138*, Piedra, Archuleta Co., Colo., 12 July 1899.) = var. *americanum*.

Terrestrial herbaceous perennials (1) 2–6 (12) dm tall arising from short, crowded, fleshy rhizomes; leaves basal, erect or spreading, much shorter than the inflorescence, the blade linear-lanceolate to oblong or more commonly broadly elliptic to ovate, 5–20 cm long and wide, rounded to slightly cordate at the base, 5 to 7-veined, the petiole sheathing, 3–15 (20) cm long; scapes 1–few, erect, with 1 or more commonly several compound-paniculately arranged bracteate verticils of 1 or more whorls of flowers, the internodes of the main stem gradually becoming more reduced above, those of the secondary stems much more rapidly so; bracts papery, lanceolate, 2–3 subtending each branch and pedicel, greenish; pedicels ascending or erect, 0.5–4 cm long; sepals persistent, ovate to suborbicular, gibbous in most, (2.5) 3–4 mm long, greenish; petals deciduous, rhombic, 3.5–6 mm long, white, the margin entire; stamens 6, the filament glabrous, 2–3 mm long, slender, the anther ovoid, 0.6–0.8 mm long; pistils (5) 10–20 in a single, somewhat 3-sided, smooth whorl, the style (0.4) 0.5–0.7 (0.8) mm long, straight or only slightly curved and somewhat lateral on the ovary; achenes 2–2.3 mm long, usually 3-ribbed and with a central groove near the tip, the beak erect; $2n = 28$ for our plants, otherwise $n = 6, 7$.

Common in fresh water marshes usually on mud flats, rarely in deep water; a cosmopolitan species of Eurasia, e. Africa, Australia, N. and S. Amer.; s. Can., s. to s. Calif., Ariz., Colo., Mo., and ne. U.S.; in our area in se. Oregon (Lake Sect.), e. Idaho, n. Nev. and Utah, and on the Kaibab Plateau of n. Ariz. June–Sept.

Our plants are referred to var. *americanum* Schultes & Schultes as indicated below.

The *A. plantago-aquatica* complex is the subject of considerable disagreement in botanical circles, and its problems can not be fully resolved here. In spite of three recent monographs on *Alisma,* major taxonomic and nomenclatural questions remain. Most early authors referred the North American plants to Linnaeus' species (written as *A. plantago, A. plantago-*Δ or *A. plantago aquatica*), but in 1808 Rafinesque proposed *A. subcordatum* for these plants, and considered them to be specifically distinct from the European form. In 1814, Pursh proposed *A. triviale* (as *trivialis*) and *A. parviflorum* (as *parviflora*). Pursh took his description of *A. triviale* from Michaux's 1803 *Flora.* Although the type of *A. triviale* came from Canada, Pursh gave the range of his new species as from Canada to Florida, with the exception of those plants found along the coast of New Jersey and Pennsylvania; to this latter group he gave the name *A. parviflorum.* As early as 1817, Rafinesque considered his name, *A. subcordatum,* and *A. triviale* to represent the same kind of plant. Nuttall, in 1818, felt both of Pursh's species were merely minor variants of the European species. In 1824 Torrey placed *A. triviale* in synonymy under *"A. plantago,"* while *A. parviflorum* was reduced to the varietal rank.

In 1932, Samuelsson recognized *A. subcordatum* as a valid species and placed both *A. triviale* and *A. parviflorum* in synonymy under it; he distinguished the western American phase of the complex under the name *A. plantago-aquatica* subsp. *brevipes.* Fernald later commented that the two American elements were distinct species and suggested (without any supporting evidence) that the two correct names were *A. subcordatum* and *A. triviale.* Pogan agreed, and noted that the first was a diploid ($2n = 14$) and the second tetraploid ($2n = 28$). Hendricks called the two American forms varieties of *A. plantago-aquatica,* and used the name var. *americanum* for the tetraploid and var. *parviflorum* for the diploid. He also recognized var. *michaletii* (Aschers. & Graebner) Buchenau (in the sense we would use "var. *plantago-aquatica*" today) as occurring in the United States, but these specimens are almost all referable to var. *americanum.* Hendricks placed *A. subcordatum* in synonymy under var. *parviflorum,* while *A. triviale* and *A. brevipes* were referred to var. *americanum.*

Björkqvist contends that *A. triviale* and *A. subcordatum* are valid species, and outlines them much as Fernald and Pogan did before him. However, he does attempt to typify some of the names and correct some oversights made by previous authors. He correctly selects the Michaux specimen from Quebec, Canada, as the type of *A. triviale,* proclaims it to be a tetraploid, but fails to understand var. *americanum.* Schultes and Schultes state that their new name is based on *A. plantago* sensu Michaux, and place *A. triviale* in synonymy under their var. *americanum.* As a result, both names are based on the same type. They also mention *A. subcordatum,* but only questionably refer the name to var. *americanum.* Thus Björkqvist's placement of var. *americanum* under *A. subcordatum* is wholly incorrect. Likewise, his selection of a lectotype (*Schopf,* e. U.S.) for var. *americanum* has no basis. Björkqvist also attempts to neotypify Rafinesque's *A. subcordatum,* proposing a Moser collection from Pennsylvania. Such action is proper and needed if, and only if, all efforts were made to discover an original type; no such attempt was made in the herbaria where Rafinesque supposedly deposited his specimens gathered in the United States in 1803 and 1804. Until some of these nomenclatural matters are resolved, I am maintaining a conservative species definition for the *A. plantago-aquatica* complex, and using the var. *americanum* for our plants.

Two other variants occur in the United States and Canada. The var. *parviflorum* (Pursh) Farw. is east of us, and Rubtzoff has reported var. *plantago-aquatica* naturalized along the West Coast.

2. **Alisma gramineum** Lej.

Alisma gramineum Lej. Fl. Environs Spa. 175. 1811. (*Lejeune,* "dans les fossés du chateau de Bel-Oeil, sur la gauche du Pont," Belgium.)

A. plantago var. *angustissimum* DC. in Lam. & DC. Fl. Franç. ed. 3. 3: 189. 1815. *A. plantago* f. *angustissimum* Aschers. & Graebner, Syn. Mitteleur. Fl. 1: 382. 1897. *A. arcuatum* var. *angustissimum* Lunell, Bot. Gaz. 43: 211. 1907. *A. geyeri* var. *angustissimum* Lunell, Bull. Leeds Herb. 2: 5. 1908. *A. gramineum* C. C. Gmelin var. *angustissimum* Hendricks, Amer. Midl. Naturalist 58: 489. 1957. (*W. D. J. Koch,* Rheinland Pfalz, Gernersheim, Germany.)

A. graminea C. C. Gmelin, Fl. Badensis 4: 256. 1826. ("Dachsland," the Alsace-Baden region of France.), a later homonym; not Lej. 1811.

A. geyeri Torr. ex Nicollet, Rep. Upper Mississippi River 162. 1843. *A. gramineum* C. C. Gmelin var. *geyeri* G. Samuelsson, Ark. Bot. 24A(7): 43. 1932. (*Geyer,* Devil's Lake, N.D., 6 Aug. 1839.)

A. arcuatum Michalet in Fermond, Bull. Soc. Bot. France 1: 312. 1854. *A. plantago-aquatica* subsp. *arcuatum* Aschers. & Graebner, Syn. Mitteleur. Fl. 1: 384. 1897. *A. gramineum* C. C. Gmelin subsp. *arcuatum* Hylander, Uppsala Univ. Årsskr. 7: 69. 1945. (*Michalet,* Jura, France, 1854; lectotype by Björkqvist.)

A. validum Greene, Pittonia 3: 115. 1896. (*Greene,* near Palisade, Eureka Co., Nev., July 1896; lectotype by Björkqvist.)

Submerged or amphibious herbaceous perennials 0.5–3 (5) dm tall arising from short, compact, fleshy rhizomes; leaves basal, erect or ascending, much longer than the inflorescence, often floating, the blade linear-lanceolate to narrowly elliptic or rarely oblong when emergent, linear or nearly so when floating, 5–9 cm long, up to 3 cm wide, obtuse to rounded at the base, 3- to 5-veined, the petiole sheathing, mostly linear and gradually tapering into the blade; scapes 1–few, erect, 2–4 compound-paniculately arranged bracteate verticils of 1–2 whorls of flowers, the internodes of the main stem rapidly becoming reduced above, those of secondary stems very short; bracts papery, lanceolate, 2–3 subtending each branch and pedicel; pedicels recurved, 0.5–4.5 cm long; sepals persistent, ovate, not obviously gibbous, 1.5–3 mm long, greenish with purplish margins; petals deciduous, rhombic, 2–4

ALISMA

Alisma gramineum

mm long, white but becoming reddish or pinkish (rarely purplish) tinged, the margin entire or infrequently somewhat erose; stamens 6–9, the filament glabrous, 1–2 mm long, nearly as broad as long, the anther ovoid, 0.4–0.5 mm long; pistils (5) 10–20 in a single, somewhat 3-sided, smooth whorl, the style 0.4–0.5 mm long, curved and somewhat lateral on the ovary; achenes 2–2.7 mm long, with 2 shallow grooves lateral to a slightly central ridge, the beak becoming curled at maturity in most; $2n = 14$, 16.

Infrequent in fresh water marshes in mud at the water's edge or almost completely submersed; N. Temp., Eurasia and N. Amer.; sw. B.C. to Que., s. to n. Calif., Nev., Ariz., N.M. and N.Y.; se. Oregon (Lake Sect.), n. Nev. (n.-central Great Basin Sect.) and Utah. June–Aug.

This variable species is often divided into two or three varieties whose merits are dubious. Our plants have been called var. *angustissimum* (DC.) Hendricks (but the combination was made under *A. gramineum* C. C. Gmelin, not Lej.) by most authors. The name var. *graminifolium* (Wahlenb.) Hendricks has also been used for our plants, but according to Björkqvist, this is a synonym of *A. wahlenbergii* (O. R. Holmberg) Juzepczuk, a species restricted to Sweden, Finland and western Russia. However, the name, var. *graminifolium,* may prove to be a superfluous substitute for *A. natans* L., now known as *Luronium natans* (L.) Raf., and as the specific name, *A. wahlenbergii,* is based on the variety, the status of the specific name, *A. wahlenbergii,* must remain in doubt.

2. MACHAEROCARPUS Small

Scapose, herbaceous perennials with erect, spreading or floating basal leaves; flowers several in bracteate umbels or more commonly in few-whorled panicles, rather showy, perfect; sepals greenish, persistent; petals white or rarely pinkish, irregularly toothed or fimbriate, deciduous; stamens 6; pistils (6) 9–15 in a single whorl; fruits flattened achenes, strongly ribbed with an erect beak about as long as the body.

A monotypic genus of w.-central N. Amer. that is close to *Damasonium*. (From the Greek *machaira*, dagger, and *karpos*, fruit, alluding to the dagger-like beak of the achene.)

1. Machaerocarpus californicus (Torr. ex Benth.) Small

Damasonium californicum Torr. ex Benth. Pl. Hartwegianus 341. 1857. *Alisma californica* M. Mich. in A. DC. & C. DC. Monogr. Phanerog. 3: 34. 1881. *Machaerocarpus californicus* Small, N. Amer. Fl. 17: 44. 1909. (*Hartweg 2015*, "In uliginosis vallis Sacramento," near Chico, Calif., according to Jepson—Torrey actually based his description on another collection [*Williamson*, deposited at NY], but while Bentham accepted Torrey's name and gave him credit for it, the type is the Hartweg specimen.)

Damasonium californicum var. *biddlei* L. F. Henderson, Rhodora 32: 21. 1930. (*Henderson 8256*, near Burns, Harney Co., Oregon, 23 June 1927.)

Aquatic or amphibious herbaceous perennials (1) 2–4 (6) dm tall arising from short, crowded, fleshy rhizomes; leaves basal, erect, spreading or floating, the blade linear to lanceolate or elliptic, 3–6 (8) cm long, mostly less than 4 cm wide, rounded to subcuneate at the base, 3- to 5-veined, the petiole long, about equalling the inflorescence; scapes 1–several, erect or ascending, with 1–several racemosely or paniculately arranged involucrate verticils of 2–4 whorls of flowers, the internodes gradually shortening above; bracts scarious, broadly lanceolate to ovate, greenish; pedicels spreading to recurved, 2–6 (8) cm long, mostly subequal; sepals persistent, oblong to ovate, 4–5 mm long, greenish; petals deciduous, suborbicular or rhombic, 7–10 mm long, white with a yellowish patch near the base, rarely pink, the margins irregularly fimbriated or toothed; stamens 6, the filament glabrous, subulate, 2–3 mm long, the anther reddish, 1.5–2 mm long, becoming twisted after dehiscence; achenes 6–10 (15), horizontally radiating in a stellate pattern, 4–6 (12) mm long, the body ribbed on the angles, tapering to a stout, subulate beak.

Infrequent in vernal pools, on margins of intermittent streams, sloughs, and on mud flats in marshy places; w.-central N. Amer. from c. Oregon s. to c. Calif., e. to sw. Idaho and nw. Nev.; in the Lake and Owyhee Desert sects. of se. Oregon, Idaho, nw. Nev., and ne. Calif. June–Sept.

3. ECHINODORUS Rich. Bur-head

Scapose, herbaceous annuals (ours) or short-lived perennials with erect, submersed or emergent, spreading or floating basal leaves; flowers several in bracteate, verticillate panicles or racemes, rather showy, trimerous, perfect; sepals greenish, persistent; petals white or greenish-white, deciduous; stamens 6–30; pistils numerous and arranged in a dense cluster on the receptacle; fruits flattened achenes, ribbed with a stout persistent beak shorter than the body.

About 30 species widely distributed in N. and S. Amer. and Africa in temp. and trop. regions. (From the Greek *echinos*, spiny, and *doros*, utricle, alluding to the stout, beaked fruit.)

Reference: Fassett, N. C. *Echinodorus* in the American tropics. Rhodora 57: 133–156, 174–188, 202–212. 1955.

1. Echinodorus rostratus (Nutt.) Engelm. ex A. Gray

Alisma rostrata Nutt. Trans. Amer. Philos. Soc. II. 5: 159. 1837. *Echinodorus rostratus* Engelm. ex A. Gray, Manual Bot. 460. 1848. (*Nuttall*, "In the ponds of the Verdigris river of Arkansas," se. Okla., 1819.)

E. rostratus var. *lanceolatus* Engelm. ex Wats. & Coulter in Gray's Manual Bot. ed. 6. 556. 1890. *E. cordifolius* var. *lanceolatus* Mackenzie & Bush, Manual Fl. Jackson County, Missouri 10. 1902. *E. cordifolius* f. *lanceolatus* Fern. Rhodora 38: 73. 1936. *E. rostratus* f. *lanceolatus* Fern. Rhodora 49: 108. 1947. *E. berteroi* var. *lanceolatus* Fassett, Rhodora 57: 144. 1955. (*Engelmann*, American Bottom, Menard Co., Ill., Sept. 1846; lectotype selected by Fassett.)

Submerged or amphibious annuals (1) 2–4 (6) dm tall arising from short, compact, fleshy rhizomes; leaves basal, erect or ascending, occasionally floating, sheathing, the blade variable, often beginning as submersed and lanceolate to narrowly elliptic (especially in ours), becoming broadly oval or cordate when emersed, up to 1 dm long and wide (but not in ours), distinctly veined with coarse reticulate veinlets, 5- to 9-veined, the petiole long, erect and rigid, angled, shorter than the inflorescence; scapes few, erect, ribbed or angled, with 1–several bracteate verticillate panicles of few to several flowers; bracts papery, linear-lanceolate (ours) to lanceolate, 2–3 subtending each branch and pedicel, greenish with hyaline margins; pedicels ascending, unequal in length, up to 2 cm long; sepals persistent, broadly triangular, 2–4 (5) mm long, dark green; petals deciduous, rhombic (in ours) to orbicular, (5) 6–10 mm long, white to cream or greenish-white, the margin entire; stamens 12, the filament glabrous, 1–2 mm long, slender, the anther lanceolate, 0.5–1.2 mm long; pistils numerous, densely packed over the entire surface of the receptacle, the style longer than the ovary, straight or nearly so; achenes (1) 2–3 (4) mm long, with 2-winged ribs alternating with 3 nonwinged ribs, the beak stout and persistent, 0.5–2 mm long, the fruiting head bur-like at maturity.

In fresh water marshes, shallow ponds, ditches and streams, or on mud flats or almost totally submersed; N. Amer. from s. Calif., Texas, Mo. and Ill. s. to Mex. with an isolated population in the Green River Bird Refuge, Uintah Co., Utah. June–Oct.

This species is known in our area from a single location in ne. Utah, but is much more common to the south of us. Most authors consider the species to consist of two variants, var. *rostratus* and var. *lanceolatus* Engelm. ex Wats. & Coulter, but the two completely intergrade. Our plant seems to belong to var. *lanceolatus*. Past references to *E. cordifolius* (L.) Griseb. or *E. berteroi* (Sprengel) Fassett by western authors are based on specimens of *E. rostratus*.

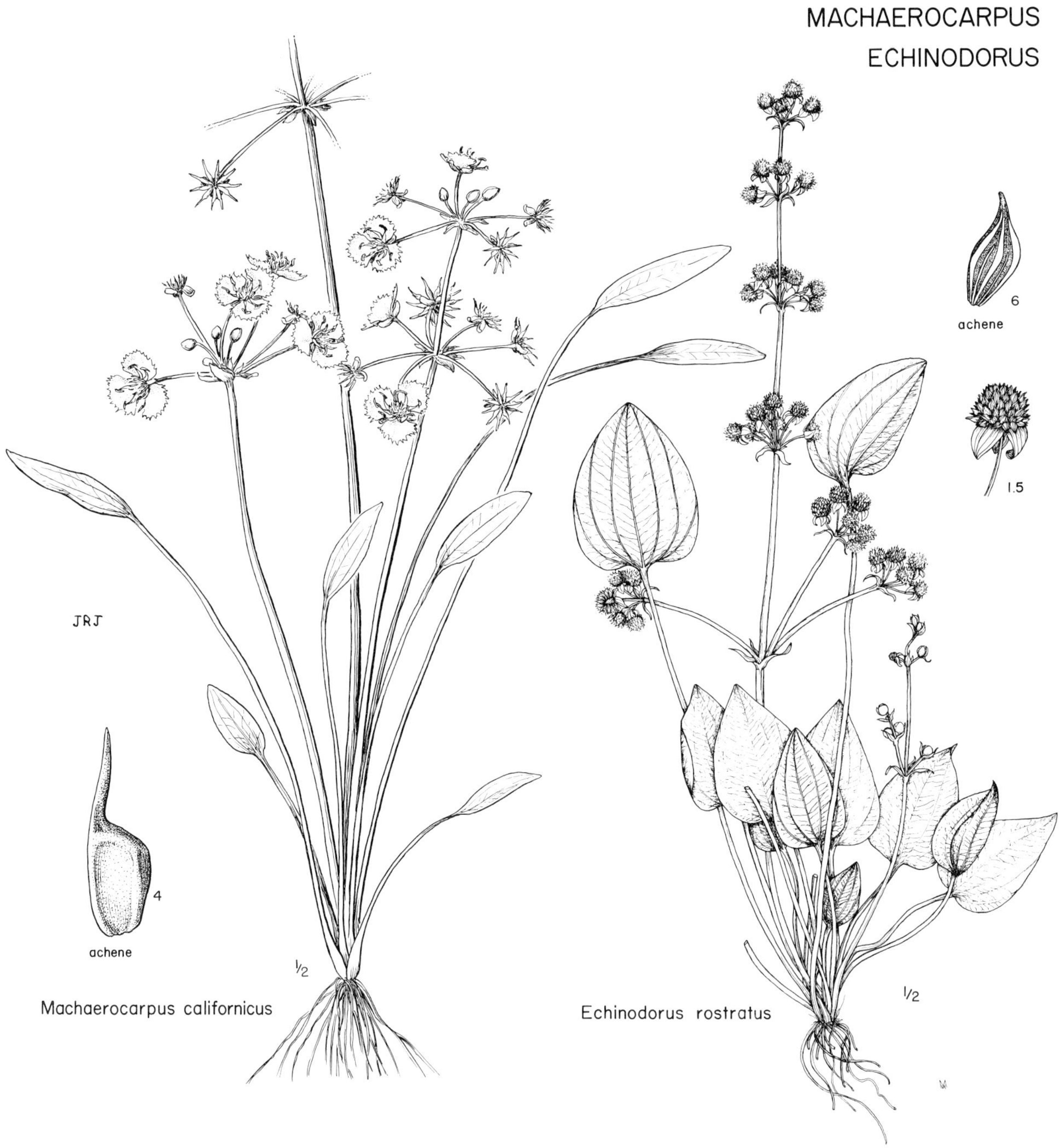

4. SAGITTARIA L. Arrowhead

Scapose, herbaceous annuals or perennials with erect or spreading to floating, submersed or emergent, basal leaves; flowers few to several in bracteate, simple or branched, verticillate inflorescences, rather showy, monoecious or occasionally dioecious or with some flowers perfect, usually the lower ones pistillate and the upper one staminate, or the lower flowers perfect and the upper ones staminate, the unisexual flowers often bearing rudimentary stamens and pistils; sepals greenish, usually reflexed and persistent although often withered and inconspicuous, occasionally erect and growing with the fruit; petals white with a basal spot or only whitish, deciduous; stamens many; pistils many and spirally arranged in a crowded head on a large globose receptacle; fruits flattened achenes, ribbed or winged with a short, erect or divergent persistent beak shorter than the body.

About 20 species of temp. and trop. regions, with the majority found in the New World. (From the Latin *sagitta,* arrow, as to the leaf shape.)

The reference to *S. montevidensis* Cham. & Schlecht. subsp. *calycina* (Engelm. in Torr.) Bogin in the "Checklist" was based on specimens found west of our area.

References:
Bogin, C. Revision of the genus *Sagittaria.* Mem. New York Bot. Gard. 9: 179–233. 1955.
Smith, J. G. A revision of the North American species of *Sagittaria* and *Lophotocarpus.* Annual Rep. Missouri Bot. Gard. 6: 27–64. 1897.

1 Achenes with beaks extending at right angles to the body, the beak 0.5–1.5 mm long; plants sometimes dioecious; lobes of the leaves about equal length; bracts of the inflorescences 5–10 (15) mm long; rare .... 1. *S. latifolia*
1 Achenes with beaks erect, the beak 0.2–0.4 mm long; plants monoecious; lobes of the leaves unequal, the terminal one longer than the two lateral lobes; bracts of the inflorescences (5) 10–30 mm long; common .... 2. *S. cuneata*

1. Sagittaria latifolia Willd.

Sagittaria latifolia Willd. Sp. Pl. 4: 409. 1805. *S. sagittifolia* var. *latifolia* Muhl. Cat. Pl. N. Amer. 86. 1813. *S. variabilis* var. *latifolia* Engelm. in A. Gray, Manual Bot. ed. 2. 439. 1856. *S. sagittaefolia* f. *latifolia* Britton, Bull. Torrey Bot. Club 17: 125. 1890. *Sagitta latifolia* Nieuwl. Amer. Midl. Naturalist 3: 21. 1913. (*Muhlenberg,* Canada; type as indicated by Bogin.)

Sagittaria obtusa Muhl. ex Willd. Sp. Pl. 4: 409. 1805. *S. variabilis* var. *obtusa* Engelm. in A. Gray, Manual Bot. ed. 2. 439. 1856. *S. sagittaefolia* f. *obtusa* Britton, Bull. Torrey Bot. Club 17: 125. 1890. *S. latifolia* f. *obtusa* B. L. Robinson, Rhodora 10: 31. 1908. *S. latifolia* var. *obtusa* Wieg. Rhodora 27: 186. 1925. ("America boreali," no specimen mentioned.)

S. latifolia var. *major* Pursh, Fl. Amer. Sept. 396. 1814. ("Canada to Carolina.")

S. hastata Pursh, Fl. Amer. Sept. 396. 1814. *S. sagittifolia* var. *hastata* Torr. Compend. Fl. N. Middle States 355. 1826. *S. variabilis* var. *hastata* Engelm. in A. Gray, Manual Bot. ed. 5. 493. 1867. *S. sagittaefolia* f. *hastata* Britton, Bull. Torrey Bot. Club 17: 125. 1890. *S. latifolia* f. *hastata* B. L. Robinson, Rhodora 10: 64. 1908. ("Pensylvania.")

S. gracilis Pursh, Fl. Amer. Sept. 396. 1814. *S. hastata* var. *gracilis* Nutt. Gen. N. Amer. Pls. 2: 213. 1818. *S. sagittifolia* var. *gracilis* Torr. Compend. Fl. N. Middle States 356. 1826. *S. variabilis* var. *gracilis* Engelm. in A. Gray, Manual Bot. ed. 2. 439. 1856. *S. sagittaefolia* f. *gracilis* Britton, Bull. Torrey Bot. Club 17: 125. 1890. *S. latifolia* f. *gracilis* B. L. Robinson, Rhodora 10: 31. 1908. ("Pensylvania to Virginia, particularly on the mountains.")

S. simplex Pursh, Fl. Amer. Sept. 397. 1814. *S. sagittifolia* var. *simplex* Hook. Fl. Boreali-Amer. 2: 167. 1838. ("New Jersey.")

S. sagittifolia var. *vulgaris* Hook. Fl. Boreali-Amer. 2: 167. 1838. [Two interpretations of this name are possible: One is that var. *vulgaris* was intended to be the typical phase of the species, much as later authors used "typica," or the second is that the name was intended to indicate a new variety. It is unclear which Hooker meant. He did cite two specimens at the end of the protologue, "Hab. Ditches and pools, and throughout Canada and the Saskatchawan; Columbia, and Walla-wallah Rivers, *Dr. Gairdner, Mr. Tolmie.*"]

S. sagittifolia var. *macrophylla* Hook. Fl. Boreali-Amer. 2: 167. 1838. (*Scouler,* "N. W. [North West] coast.")

S. variabilis Engelm. ex A. Gray, Manual Bot. 461. 1848. *S. sagittaefolia* var. *variabilis* M. Mich. in A. DC. & C. DC. Monogr. Phanerog. 3: 69. 1881. (*Engelmann s.n.,* near the Episcopal Cemetery, St. Louis, Mo., Sept. 1846; lectotypified here, deposited at MO.)

S. esculenta Howell, Fl. N.W. Amer. 679. 1903. ("Brit. Columbia to California.")

Wapato, arrowhead.

Emersed aquatic herbaceous perennials 2–15 (20) dm tall arising from stout, tuber-bearing, spreading rhizomes; leaves basal, erect, emersed, but if emersed then blade linear and ribbon-like or lacking, the emersed blade usually sagittate (rarely otherwise in ours), the lateral lobes nearly as long as the terminal lobe, linear to ovate-triangular, 2–20 cm long, 0.5–12 cm wide, widely divergent to nearly parallel, the terminal lobe varying from obtuse to deltoid, 3–25 cm long, 0.7–35 cm wide, acute and acuminate to obtuse, the petiole erect, sheathing, up to 7 dm long, often with distinct cross-ridges; scapes few, erect, 1.5–7 (9) dm long, shorter to slightly longer than the leaves, with 2–7 verticils of flowers, infrequently branched at the lower nodes, glabrous (ours) or pubescent, the internodes abruptly shortened at the apex; bracts papery, lanceolate to deltoid, 5–10 (15) mm long, usually joined at the base, gibbous in most, 2–3 subtending each flower, greenish; pedicels ascending, 0.5–6 cm long, those of the pistillate flowers about as long as those of the staminate ones; plants monoecious or dioecious, often the lower flowers pistillate and the upper ones staminate with some perfect flowers inbetween, infrequently all perfect or all staminate; sepals persistent, reflexed, ovate, 5–10 mm long, greenish; petals slowly deciduous, broadly and abruptly narrowed to a claw, 1–2 cm long, white; stamens numerous, 21 or more, the filament stout, 1–3.5 mm long, glabrous, the anther oblong, 1–2 mm long; pistils numerous; fruiting heads globose, 1.5–3 cm in diam; achenes 2.5–4 mm long, with a stout beak extending at a right angle to the body, the beak 0.5–1.5 mm long, the margins of the achene with a broad crested dorsal wing 0.4–0.8 mm wide; $n = 11$.

Ponds, ditches and swampy areas in water or on mud flats, rarely totally submersed; s. Can. s. to Mex. and C. Amer., W. I., S. Amer., and Hawaii; s. Can. s. along the Pacific Coast from Wash. to Calif., e. across s. Ariz. to Colo., becoming common in the Midwest and e. U.S.; rare in our area, known only from se. Oregon and se. Idaho. June–Sept.

Our plants are var. *latifolia.* A second variant, var. *pubescens* (Muhl.) J. G. Smith, occurs in the se. U.S.

2. Sagittaria cuneata Sheldon

Sagittaria sagittifolia var. *minor* Pursh, Fl. Amer. Sept. 395. 1814. *S. arifolia* Nutt. ex J. G. Smith, Annual Rep. Missouri Bot. Gard. 6: 32. 1895. ("Pensylvania to Carolina," certainly an error; according to Smith probably a Nuttall specimen from the "Rocky Mountains," which is also an error. Probably based on a specimen from the Upper Missouri River drainage of N. or S. Dakota collected by Nuttall in 1811.)

S. cuneata Sheldon, Bull. Torrey Bot. Club 20: 283. 1893. *S. arifolia* var. *cuneata* Lunell, Bull. Leeds Herb. 1: 3. 1907.

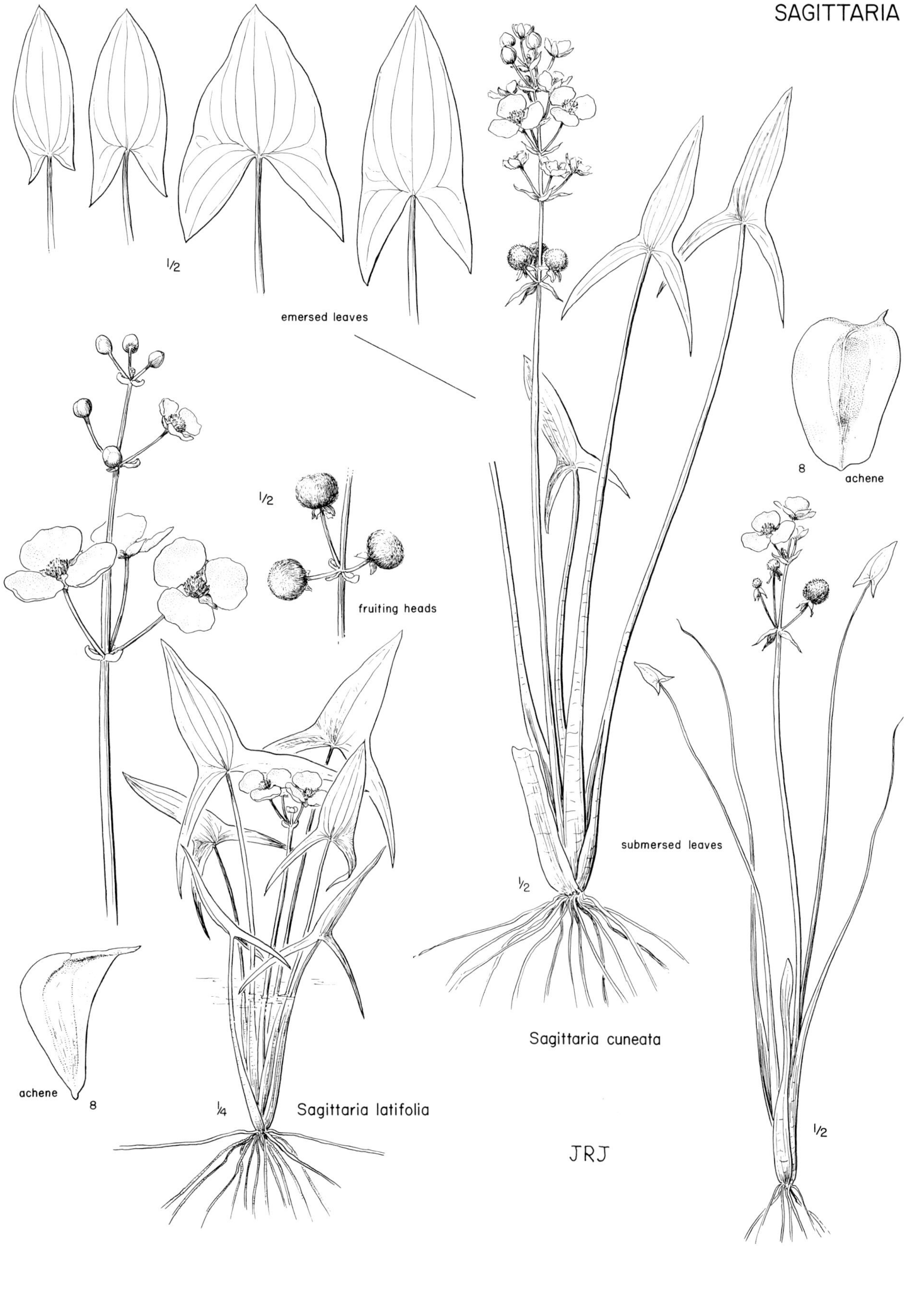
SAGITTARIA
1/2
emersed leaves
1/2
fruiting heads
8
achene
submersed leaves
1/2
achene
8
1/4
Sagittaria latifolia
Sagittaria cuneata
JRJ
1/2

(*Sheldon*, E. Battle Lake, Otter Tail Co., Minn., Aug. 1892.)
S. arifolia var. *stricta* J. G. Smith, Annual Rep. Missouri Bot. Gard. 6: 34. 1895. (*Suksdorf 1319* [*674*], Falcon Valley, Klickitat Co., Wash., 1 Aug. 1885.)
S. arifolia var. *aquatilis* J. G. Smith ex Norton, Annual Rep. Missouri Bot. Gard. 9: 156. 1898. (*Henderson 2977*, ponds at the mouth of Pack River, Lake Pend d'Oreille, Bonner Co., Idaho, 1891; lectotypified here, deposited at MO.)
S. hebetiloba A. Nels. Bull. Torrey Bot. Club 26: 6. 1899. (*A. Nelson 2763*, Platte Canyon, Laramie Co., Wyo., 27 Aug. 1896.)
S. paniculata Blankinship, Montana Coll. Agric. Sci. Stud., Bot. 1: 40. 1905. (*Blankinship*, Box Elder Cr., Valley Co., Mont., 14 July 1900, is the first of 5 collections cited.)
S. arifolia var. *tenuor* Blankinship, Montana Coll. Agric. Sci. Stud., Bot. 1: 40. 1905. (*Blankinship*, Flathead R., Mont., 20 and 23 July 1900, is the first of 3 collections cited.)
S. suksdorfii Gandoger, Bull. Soc. Bot. France 66: 294. 1919. (*Suksdorf 1319c*, Falcon Valley, Klickitat Co., Wash., June–Aug. 1896–98.)

Wapato, arrowhead.

Emersed aquatic herbaceous perennials 1–5 dm tall arising from stout, tuber-bearing, spreading rhizomes; leaves basal, erect or spreading, rarely prostrate, emersed, but if submersed then blade reduced and ribbon-like, the emersed blade sagittate, the lateral lobes shorter than the terminal lobe, deltoid, 1–8 cm long, 1–3 cm wide, divergent, the terminal lobe mostly deltoid, occasionally lanceolate to ovate, 1–10 cm long, 1–6 cm wide, acute and acuminate, the petiole erect or nearly so, sheathing, up to 5 dm long, widening at the base, often with distinct cross-ridges; scapes few, erect, 1–5 dm long, shorter than or as long as the leaves, with 2–8 verticils of flowers, rarely branched at the first node, glabrous, the internodes gradually shortening near the apex; bracts papery, lanceolate, (5) 10–35 mm long, joined at the base, not gibbous, 2–3 subtending each flower, greenish; pedicels ascending, those of the pistillate flowers 5–20 mm long, those of the staminate flowers 5–30 mm long; plants monoecious, the lower flowers pistillate, the upper ones staminate; sepals persistent, reflexed, ovate, 4–8 mm long, greenish; petals slowly deciduous, with a rounded blade and cuneate base, 6–10 mm long, white; stamens 15–24, the filament subulate, 1–2 mm long, glabrous, the anther oblong, 1–2 mm long; pistils numerous; fruiting heads globose, 1–1.5 cm in diam; achenes mostly 2–2.5 mm long, with a short erect straight beak, the beak 0.2–0.4 mm long, the margins of the achene with a rounded dorsal wing 0.5–0.8 mm wide; $n = 11$.

Ponds, ditches, slow-moving streams, lake shores and swampy areas, in the water or on mud flats, rarely totally submersed; s. Can. and the n. U.S. s. to Calif., c. N.M. and n. Texas; common throughout most of our area except in s. Nev. and n. Ariz. June–Sept.

Order **HYDROCHARITALES**

Flowers epigynous, each associated with a thin papery spathe of united bracts; perianth small, usually biseriate and well differentiated into three sepals and three petals, rarely with the petals lacking; stamens 3–12, free; ovary compound and with modified laminar placentation, 1-celled with 3–6 or rarely more parietal placentas; pollen trinucleate; seeds without endosperm; submersed aquatic herbs with narrow, tufted, alternate to opposite or whorled leaves.

A single family, the Hydrocharitaceae, as described below.

Family **HYDROCHARITACEAE,** the Frog's-bit Family

Herbaceous submersed aquatic perennials of fresh or salt waters, with anchored or floating roots and simple, tufted, alternate to opposite or whorled leaves; flowers mostly imperfect and the plants dioecious or polygamous (mostly polygamo-monoecious), or rarely perfect, regular, the pistillate ones solitary, the staminate flowers numerous, trimerous, borne sessile or on scape-like, sometimes spirally twisted peduncles, all associated with a thin papery spathe, the spathe composed of (1) 2 (3) united bracts which usually enclose the flowers, the flowers on a slender, pedicel-like perianth tube (or elongated hypanthium), sometimes long-pediceled and reaching the surface of the water, or if sessile, then soon deciduous from the plant at anthesis and floating to the surface of the water, the perianth members in 1 or 2 series of 3, free, the outer whorl (sepals) valvate, the inner whorl (petals) somewhat imbricate; stamens (1) 3–9 (12), free, the filament mostly thick, the anther extrorse, 2-celled, sometimes reduced to a staminodium; ovary inferior, with modified laminar placentation, 1-celled, the styles usually 3 or 6 and each 2- to 3-branched; fruits dry, indehiscent but ripening under water, with many exalbuminous seeds.

A family of about 14 or 16 genera and from 80 to 100 species, widely distributed in temp. and trop. waters. Only one genus, *Elodea*, is found in our area.

1. **ELODEA** Michx. Waterweed, ditchmoss

Slender submersed dioecious to polygamo-monoecious herbaceous perennials with branching and usually nodally rooting stems and 1-nerved, sessile, opposite to whorled (rarely alternate), usually denticulate leaves; flowers few, solitary or rarely in 3's, borne in sessile to pedunculate, 2-lobed, axillary tubular spathes, rather small and inconspicuous, mostly imperfect or sometimes perfect in part; sepals greenish; petals 3 or wanting, whitish, often hyaline; stamens (1) 3–9, mostly in 2 series, the outer whorl of 6, the inner whorl of 3, the staminate flowers usually 1 but sometimes 2–3, arising from a globose spathe, sessile or on a short pedicel-like perianth tube, but if sessile, then soon deciduous at anthesis and floating; ovary inferior, the stigmas 3, usually bilobed, tending to float, the style slender, the pistillate flowers in tubular spathes, the pedicel-like perianth tube usually elongated and placing the flower on the surface of the water; fruits coriaceous, cylindric, several seeded.

A variable genus of perhaps 15 species of fresh or brackish waters in temp. and subtrop. N. and S. Amer. (From the Greek *elodes,* marshy, referring to the habitat.)

The final discriminator in determining species of *Elodea* is the flower, but unfortunately they are rarely found, and thus the variable vegetative features must be used. In the key presented here, the vegetative characters have been stressed, but final determination is always best when the nature of the flower can be determined. The authorship of the genus is given as "Michx." contrary to the arguments put forth by St. John. We are following Stafleu ("Taxonomic Literature," Regnum Veg. 52: 310. 1967) who says there is no justification for crediting Richard with the new taxa in Michaux's 1803 flora.

References:

St. John, H. Monograph of the genus *Elodea* (Hydrocharitaceae). Part I. Res. Stud. State Coll. Wash. 30(2): 19–44. 1962.

———. Monograph of the genus *Elodea:* Part 4 summary. Rhodora 67: 1–35, 155–180. 1965.

1 Leaves in whorls of (4) 6, the main leaves averaging about 2.5 cm long, 2–3 (5) mm wide; staminate flowers 2–3 per spathe; petals 8–10 mm long; used in aquaria and perhaps planted or escaped, but not known to be naturalized in our area .... *Egeria densa* Planchon
1 Leaves opposite or in whorls of 3, or not at all whorled but still 2–3 per node, the main leaves less than 2.5 cm long; staminate flowers solitary in each spathe; petals less than 5 mm long.
  2 Upper and middle leaves opposite, (5) 15–25 mm long; spathe of the pistillate flowers 3–7 cm long; petals of the staminate flowers 3.5–5 mm long; n. Nev. and Utah .... 2. *E. longivaginata*
  2 Upper and middle leaves in whorls of 3.
    3 Upper and middle leaves mostly (1) 2–3 (5) mm wide, dark green.
      4 Upper and middle leaves (3) 5–8 mm long; flowers perfect; stigma entire, 0.3 mm long; just w. of our area near Truckee, Nevada Co., Calif. .... *E. brandegeae* St. John
      4 Upper and middle leaves 6–13 mm long; plants dioecious; stigma bifurcate, 4 mm long; common .... 1. *E. canadensis*
    3 Upper and middle leaves 0.3–2 mm wide, mostly pale green.
      5 Upper and middle leaves 1–2 mm wide, not flaccid; sepals of the pistillate flower 3–3.5 mm long, the staminodia acicular with a petaloid tip; originally collected in a pond at Wadsworth, Washoe Co., Nev. but probably now extinct .... 3. *E. nevadensis*
      5 Upper and middle leaves 0.3–1.5 mm wide, flaccid; sepals of the pistillate flower 2–2.2 mm long, the staminodia acicular but lacking a petaloid tip; not known from our area but to be expected in e. Calif. and se. Oregon. .... *E. nuttallii* (Planchon) St. John

1. Elodea canadensis Michx.

Elodea canadensis Michx. Fl. Boreali-Amer. 1: 20. 1803. *Serpicula occidentalis* Pursh, Fl. Amer. Sept. 33. 1814, a superfluous substitute for *E. canadensis,* but usually misapplied to specimens of *E. nuttallii. Udora canadensis* Nutt. Gen. N. Amer. Pls. 2: 242. 1818, as to basionym, but not as to concept. *Serpicula canadensis* Eat. Manual Bot. N. States, ed. 5. 390. 1829. *Udora occidentalis* Koch, Syn. Fl. German. & Helv. 669. 1837. *Anacharis canadensis* Planchon, Ann. Mag. Nat. Hist. II. 1: 86. 1848. *Philotria canadensis* Britton, Science, II. 2: 5. 1895. *E. occidentalis* St. John, Rhodora 22: 27. 1920. *Philotria occidentalis* H. D. House, New York State Mus. Bull. 243: 55. 1921. *Anacharis occidentalis* Marie-Vict. Contr. Lab. Bot. Univ. Montréal 18: 40. 1931. (*Michaux,* "in rivulis Canadae," near Montreal, Can., June or July 1792.)

E. ioensis Wylie, Bull. Lab. Nat. Hist. Iowa State Univ. 6: 48. 1913, a printer's error for *iowensis.* (*Wylie,* E. Okoboji Lake, near Spirit Lake, Dickinson Co., Iowa, 1910.)

E. bifoliata St. John, Res. Stud. State Coll. Wash. 30(2): 23. 1962. (*Hanson & Hanson A892,* "in slough near Flagstaff," Coconino Co., Ariz., 27 Aug. 1923.)

Submersed herbaceous aquatic perennials rooting in mud; stems slender, sparsely dichotomously branched; leaves opposite and lance-ovate below, in whorls of 3 and linear to lance-oblong in the middle and above, crowded, the blade 6–15 (17) mm long, (1) 1.5–2.5 (5) mm wide, firm or subflaccid, dark green, finely serrulate, the apex mostly rounded or obtuse; plants dioecious, with the flowers produced in the upper axils; spathe of the staminate flower up to 1.5 cm long, the lower portion slender, the upper portion 6–8 mm long, 3.4–5 mm broad, ellipsoid or ovoid, inflated; spathe of the pistillate flower 1–2 cm long, similar to the staminate one but cylindric and not inflated near the apex; staminate flowers with the thread-like perianth tube 3–20 cm long, the sepals elliptic, 3.5–5 mm long, 2–2.5 mm wide, the petal lanceolate, 4.5–5 mm long, 0.4–0.8 mm wide, white, the 9 stamens mostly exserted, the anther 2–3 mm long, oblong; pistillate flowers with the thread-like perianth tube 2–28 cm long, the sepal oblong-elliptic, 2–2.2 mm long, 1–1.1 mm wide, the petal spathulate, 2.3–2.6 mm long, 1.1–1.3 mm wide, white, the 3 minute staminodia acicular, 0.6–0.7 mm long, the stigma 4 mm long, 2-cleft at the apex; ovary 3 mm long with 3–4 erect seeds; fruit a capsule 5–6 mm long, the narrowly cylindric glabrous seeds 4–4.5 mm long; $2n = 24, 48$.

Ditches, streams, ponds, lakes and marshy areas, often where somewhat alkaline; s. Can. s. to c. Calif., w. and n. Nev., s. Utah and n. Ariz., N.M., Ark., Ala. and N.C.; common in our area. July–Sept.

2. Elodea longivaginata St. John

Elodea longivaginata St. John, Res. Stud. State Coll. Wash. 30(2): 38. 1962. (*Macoun 13725,* lake at the northern edge of Milk River Ridge, where the Lethbridge Trail leaves the summit, Alta., Can., 20 July 1895.)

Submersed herbaceous aquatic perennials rooting in mud; stems slender, sparsely dichotomously branched; leaves opposite throughout, rarely alternate, in 2 or rarely with an occasional whorl of 3, the blade linear, (5) 15–25 mm long, (0.5) 1–2 mm wide, flaccid, bright green, finely serrulate, the apex acute to obtuse; plants dioecious, with the flowers produced along the upper portion of the stem; spathe of the staminate flower 2–5 cm long, the lower portion slender, the upper portion 8–10 mm long, 2.5–4 mm broad, ellipsoid, inflated; spathe of the pistillate flower 3–7 cm long, linear-cylindric, the upper portion slightly inflated with two large teeth; staminate flowers with the thread-like perianth tube 4–30 cm long, the sepal elliptic, 3.5–5 mm long, 2–2.5 mm wide, the petal linear, 4.5–5 mm long, 0.5–0.6 mm wide, white, the 9 stamens exserted, the anther 2–3 mm long, oblong; pistillate flowers with the thread-like perianth tube up to 3 dm long, extending the flower to the surface of the water, the sepal elliptic, 2.5–2.8 mm long, 1–1.3 mm wide, the petal spathulate, 3.5–4 mm long, 1.1–1.3 mm wide, white, the 3 staminodia strap-shaped, the stigma 1–2 mm long, entire; ovary 3–4 mm long with 6 seeds; fruit a capsule 8–10 mm long, the narrow seeds 5–6 mm long.

Ditches, ponds and lakes; Rocky Mts. area from Alta. and N.D. s. through Mont., Wyo., Colo. to N.M., w. to n. Nev.; infrequent in our area from n. Nev. and n. Utah. July–Sept.

3. **Elodea nevadensis** St. John

Elodea nevadensis St. John, Res. Stud. State Coll. Wash. 30(2): 41. 1962. (*Tracy & Evans 475,* in ponds at Wadsworth, Washoe Co., Nev., 21 July 1887.)

Submersed herbaceous aquatic perennials probably rooting in mud; stems slender, dichotomously branched; leaves opposite and ovate below, in whorls of 3 and linear-lanceolate in the middle and above, crowded, the blade 7–15 mm long, 1–2 mm wide, firm or subflaccid, dull green, finely serrulate, the apex mostly acute; plants dioecious; staminate flowers unknown; spathe of the pistillate flower 0.9–1.1 cm long, cylindric, not inflated near the apex; pistillate flowers with the thread-like perianth tube 1.5–4 cm long, the sepal elliptic, 3–3.5 mm long, 1 mm wide, the petal spathulate, 3 mm long, 1–1.1 mm wide, the 3 minute staminodia subulate below, expanding and petaloid above, the stigma 3–3.3 mm long, 2-cleft at the apex; ovary unknown; fruit a slender capsule.

Apparently in ponds near Wadsworth, Washoe Co., Nev., but probably extinct.

This dubious species is seemingly most closely related to *Elodea nuttallii* (Planchon) St. John, and may ultimately prove to be a synonym of that species once the unknown parts of *E. nevadensis* are discovered and studied. However, because of some similar vegetative features, it is possible that this species could be referred to *E. canadensis.*

In 1971, I visited western Nevada and adjacent eastern California in hopes of rediscovering *Elodea brandegeae* and *E. nevadensis.* Neither was found.

The type area of the first is now the playground of the high school in Truckee. According to Forest Service records, a pond was built below a furniture mill in the 1890s, but the pond was not actively maintained after the 1920s. In the 1960s, the resultant "marshy area" was filled in. I searched the mill ponds of the lumber mill and the Truckee River for *E. brandegeae,* but none was found. I have seen no other specimens of this species except the holotype. The species closely resembles *E. canadensis,* and can easily be associated with that species based upon vegetative morphology; the unusual floral condition of *E. brandegeae* (the flowers are perfect), however, readily distinguishes the two. I must admit that I would like to know if the perfect condition of *Elodea* flowers is not a haphazard condition. Another perfect species, *E. schweinitzii* (Planchon) Casp. of New Jersey and Pennsylvania, is now considered extinct, having been last collected in 1834; it too is closely related to *E. canadensis.*

As for *Elodea nevadensis,* its type location is not extant either, or so it seems. The one possible site was the pond area later developed into an after-bay for a local power station which provided electricity for Wadsworth up to 1919. However, with the death of the owner, the station stopped and it was not until 1924 that power was restored, this time via a larger power line system in western Nevada. It is possible that *E. nevadensis* grew in this pond which is now a large roundish depression on the north side of the townsite. In 1973, the area around Wadsworth was again explored, with *E. canadensis* being commonly found in the area. The flowers of *E. nevadensis* differ from those of *E. canadensis,* and thus—reluctantly—I am maintaining *E. nevadensis* as a distinct species until additional information can be obtained.

Elodea nuttallii apparently does not occur in the Intermountain Region, but is found to the north of us in Idaho and to the east in Colorado, at least according to St. John. However, specimens seen in various herbaria seem to suggest that the species extends into northern California, western Oregon and in scattered locations in Washington and northern Idaho. We would expect to find *E. nuttallii* in the Lake Section of southeastern Oregon.

Order **NAJADALES**

Flowers hypogynous or naked, bractless except for a single bract-like tepal in *Lilaea;* perianth, when present, not well differentiated into sepals and petals; stamens 1–6 (many), free; ovary 1–several, free or (in Juncaginaceae) connate; pollen trinucleate; seeds without endosperm; aquatic or semiaquatic herbs with slender and parallel-veined to broader and net-veined leaves, with or without distinction into a blade and petiole.

Eight families, five of which occur in our range. Of the other three, Scheuchzeriaceae is more northern, Aponogetonaceae is largely tropical, and the last, Zosteraceae, is marine.

Key to the Families of Najadales

1 More or less emergent, rooted, marsh plants with onion-like chiefly or wholly basal leaves, scapose or subscapose; carpels and stamens usually more than 1 . . . . Juncaginaceae, p. 18

1 Submersed or floating aquatic plants with scattered linear to broad leaves, never scapose.

  2 Flowers solitary in the axils; carpel and stamen 1; leaves serrate or toothed . . . . Najadaceae, p. 22

  2 Flowers clustered, or if solitary, then pistils more than 1; leaves various.

ELODEA

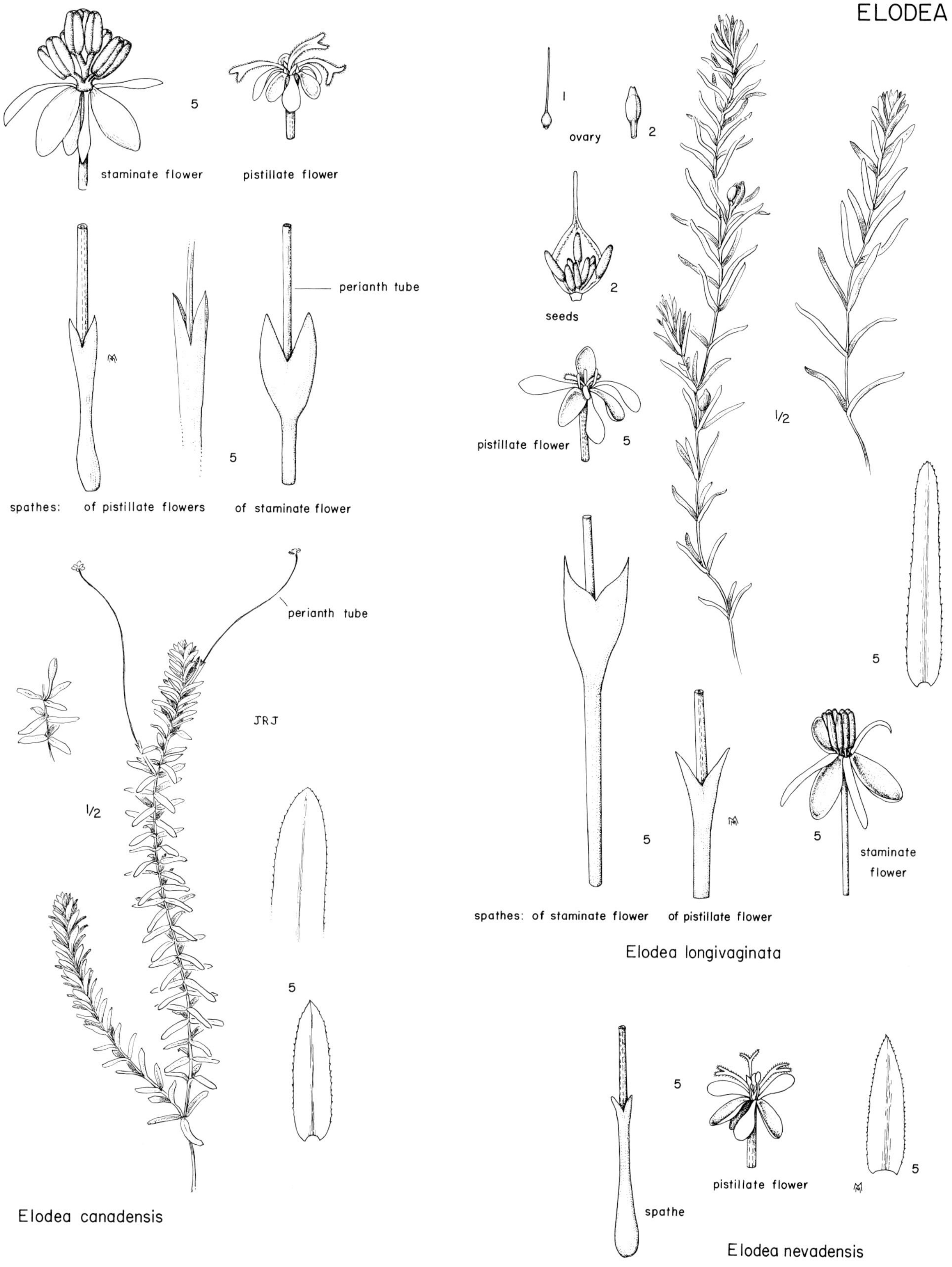
5
staminate flower
pistillate flower
perianth tube
5
spathes: of pistillate flowers
of staminate flower
perianth tube
JRJ
1/2
5
Elodea canadensis
1
ovary
2
2
seeds
pistillate flower
5
1/2
5
5
5
staminate flower
spathes: of staminate flower
of pistillate flower
Elodea longivaginata
5
spathe
pistillate flower
5
Elodea nevadensis

3 Flowers in spikes or racemes, perfect or less often unisexual; fruits mostly rounded.
4 Tepals 4; stamens 4; spikes axillary; fruiting carpels sessile; pollen ellipsoid or spheroid .... Potamogetonaceae, p. 24
4 Tepals lacking; stamens 2; spikes terminal, umbelliform; fruiting carpels stipitate; pollen bilateral .... Ruppiaceae, p. 42
3 Flowers in axillary cymes or solitary in the axils, unisexual; fruits flattened .... Zannichelliaceae, p. 43

Family **JUNCAGINACEAE**, the Arrow-grass Family

Herbaceous annual to perennial terrestrial marsh-inhabiting plants, scapose or nearly so, often rhizomatous, with tufted basal sheathing, terete or onion-like, rarely ribbon-like leaves; flowers perfect or imperfect, the plants monoecious, dioecious or polygamous, regular, bractless, or in *Lilaea* the staminate flower in axil of bracts, the pistillate flowers without bracts or borne in the axils of the leaves at the base of the plants, the small flowers sessile or borne on short peduncles in racemes or spikes; perianth parts usually 6, sometimes lacking, greenish or purplish, in two series of 3, free; stamens 1, 4 or 6, the filament thick and short or nearly lacking in the axis of the perianth segments, the anther extrorse, 2-celled; ovary 1, 3, 4 or 6, free to completely connate (ours), the style short to long or lacking, the capitate stigma as many as the carpels, the ovules solitary, basal, erect or pendulous; fruits several and simple or solitary, compound, ovoid to cylindric, ribbed in some, the seeds exalbuminous.

A family of four small genera found primarily in the S. Hemisphere.

1 Perianth 6-parted; flowers bractless, all perfect, borne in terminal racemes; ovary 3 or 6, separating at maturity; stamens 3–6 .... 1. *Triglochin*
1 Perianth lacking; flowers bracteate, of two types, the pistillate ones borne in basal leaf axils, the perfect ones in pedunculate spikes; ovary appearing solitary (or 3 and completely fused); stamen 1 .... 2. *Lilaea*

1. **TRIGLOCHIN** L. Arrow-grass

Rhizomatous herbaceous perennials with long, narrow, terete or somewhat flattened tufted basal, sheathing leaves, generally of saline or alkaline moist places; flowers ebracteate, solitary, borne racemosely on decurrent pedicels, rather small and inconspicuous, perfect; sepals 3, free, greenish or purplish; petals 3 or lacking, mostly greenish, the sepals and petals not well differentiated; stamens 6, in two series of 3, subsessile in the axils of the perianth segments; ovary 3 or 6, connate on the adaxial side, each with a plumose sessile stigma; fruits follicular, 1-seeded, the carpels tardily separating and spreading outwardly revealing a terete or winged central axis; seeds erect, narrowly fusiform, compressed or angular.

A genus of 12–15 species, widely distributed especially in S. Amer. and Australia. (From the Greek *treis*, three, and *glochin*, point, alluding to the three-pointed fruits of some species.)

In accordance with Recommendation 75A of the present Code, the name *Triglochin* is considered to be feminine instead of neuter as used by Linnaeus.

References:
Howell, J. T. Remarks on *Triglochin concinna*. Leafl. W. Bot. 5: 13–19. 1947.
Löve, A., and D. Löve. Biosystematics of *Triglochin maritimum* agg. Naturaliste Canad. 85: 156–165. 1958.

1 Fertile carpels and stigmas 3; fruits elongate, linear to clavate, 5–7 (8) mm long, the carpels subulate at the base; ligules bilobate .... 1. *T. palustris*
1 Fertile carpels and stigmas 6; fruits oblong or ovoid-prismatic, 4–5 (6) mm long, the carpels not subulate at the base.
2 Ligules entire or only slightly bilobate, (1) 1.5–5 mm long; leaf-blades obcompressed, 1.5–2.5 (4) mm wide; plants closely tufted on a proliferating rhizome, (1) 3–10 (12) dm tall .... 2. *T. maritima*
2 Ligules bilobed or emarginated, 0.5–1 mm long; leaf-blades nearly terete, 0.5–1 (1.5) mm wide; aerial stems well spaced on the rhizome, 1.5–3 (4) dm tall .... 3. *T. concinna*

1. Triglochin palustris L.

Triglochin palustre L. Sp. Pl. 338. 1753. *Juncago palustris* Moench, Meth. Pl. Horti Bot. & Agri Marburg. 644. 1794. *Abbotia palustris* Raf. New Fl. N. Amer. 1: 37. 1836. (Europe.)

Slender erect herbaceous perennials 1.5–6 dm tall, arising from a stout, ascending rhizome; leaves linear, 5–30 cm long, 1/2 to 3/4 as long as the plants, the blade 1–2 mm wide, sharp-pointed, the ligule bilobed, divided to the base, 0.5–1.5 mm long; scapes slender, 1–2 dm long, terminated by a spike-like raceme 0.5–3 dm long, the pedicel slender, erect in fruit and mostly 4–6 mm long; perianth segments 6, 1.5–2 mm long, greenish-yellow; stamens 6, the anther subsessile; stigmas 3; fruits linear to clavate,

TRIGLOCHIN

6
fruit
6
6
fruit
1/2
JRJ
6
ligule
1/2
1/2
4
ligule
Triglochin maritima
Triglochin palustris

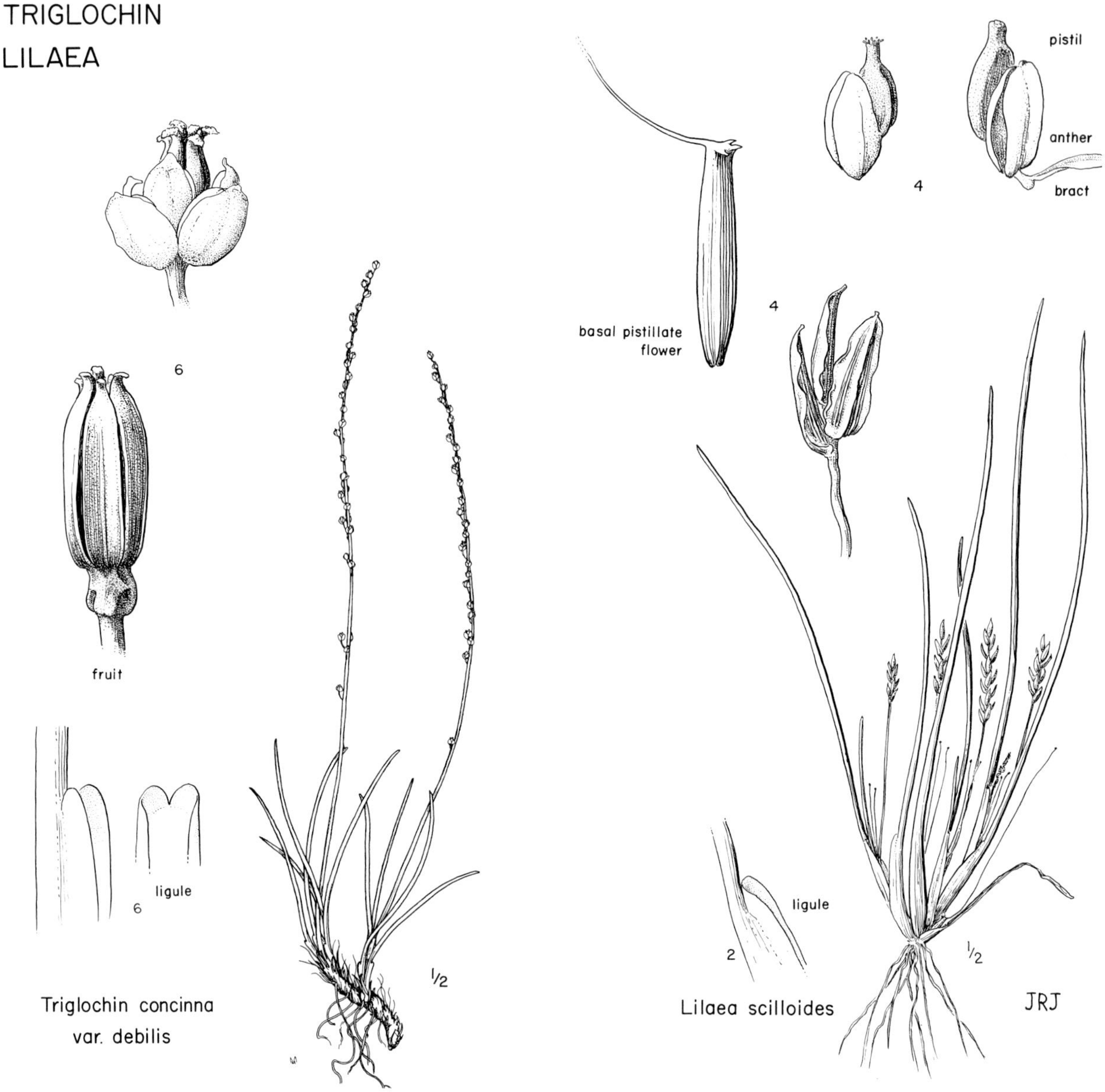

5–7 (8) mm long, the 3 carpels separating at the base and proceeding upwardly, the individual segments attached to the tips and remaining pendulous on the central axis, the base acicular; n = 12, 18, 24.

Meadows and flats, streams or lake margins, often where alkaline; Eurasia, N. and S. Amer.; Alaska to Lab. and Greenl., s. to Calif., Nev., Utah, N.M., N.D., Ill., N.Y., and Me.; s. Idaho, ne. Nev. (Ruby Mts.), n. and s.-central Utah. June–Aug.

2. Triglochin maritima L.

Triglochin maritimum L. Sp. Pl. 339. 1753. *Juncago maritima* Bubani, Fl. Pyrén. 4: 9. 1901. *Hexaglochin maritima* Nieuwl. Amer. Midl. Naturalist 3: 20. 1913. (Sweden, according to Löve & Löve.)

T. elatum Nutt. Gen. N. Amer. Pls. 1: 237. 1818. *T. maritima* var. *elata* A. Gray, Manual Bot. ed. 2. 437. 1856. ("In the state of New York.")

Coarse to slender erect herbaceous perennials (1) 3–10 (12) dm tall, arising from a stout rhizome; leaves linear, 10–80 cm long, usually 1/2 as long as the plants, the blade 1.5–2.5 (4) mm wide, strongly obcompressed, acute at the tip, the ligule entire or only shallowly lobed, (1) 1.5–4 (5) mm long; scapes slender, 1–8 dm long, terminated by a raceme 1–4 dm long, densely clothed with numerous pedicellate flowers, the pedicel slender, ascending to decurrent, 2–3 (6) mm long; perianth segments 6, 1.5–2 mm long, greenish-yellow; stamens 6, the anther subsessile; stigmas 6; fruits oblong to oblong-ovoid, 4–6 mm long, the 6 (rarely 3 fertile and 3 aborted) fertile carpels rounded at the base and united, completely deciduous from the terete axis; $2n$ = 12, 24, 30, 36, 48 and 60 for European populations, 120 for an Asiatic variety, 96 for a closely related species of e. Can., 48 for n. Alaska populations, and 144 for e. N. Amer. plants of var. *elata*.

Marsh areas and moist meadows, mostly where alkaline; Eurasia, N. and S. Amer.; Alaska to Lab., s. to Mex., Nebr., Ind. and Pa.; se. Oregon and s. Idaho, s. to Mono Co., Calif. and Lincoln Co., Nev., and through s. Utah to n. Ariz. May–Aug.

Our material is var. *elata* (Nutt.) A. Gray.

The *T. maritima* complex is in need of critical world-wide study. The typical phase occurs from western Europe across northern Asia into Alaska and northwestern Canada. These plants have a wide variety of chromosome numbers. In Japan and adjacent areas of southeast Asia, var. *maritima* (with $2n$ = 48) and var. *asiatica* (Kitagawa) Ohwi (with $2n$ = 120) are common. The most frequently encountered phase of the species in North America appears to be var. *elata*. It differs from var. *maritima* in the longer and narrower fruits; those of var. *maritima* being ovoid and 3–4 mm long, while those of var. *elata* are oblong to oblong-ovoid and 4–6 mm long. The var. *maritima* found in the northwestern part of North America has a chromosome number of $2n$ = 48, while var. *elata* (so far as known) is $2n$ = 144. Where these two varieties overlap in their ranges, if indeed they do, is not known at this time.

Members of the genus *Triglochin*, and in particular this species, are poisonous to livestock.

3. **Triglochin concinna** Davy

Triglochin concinna Davy, Erythea 3: 117. 1895. (*Davy 1116,* salt marshes around San Francisco Bay, Calif.)

T. maritimum var. *debile* M. E. Jones, Proc. Calif. Acad. Sci. II. 5: 722. 1895. *T. concinna* var. *debilis* J. T. Howell, Leafl. W. Bot. 5: 18. 1947. *T. debile* A. Löve & D. Löve, Naturaliste Canad. 85: 161. 1958. (*M. E. Jones 5289,* Johnson, Kane Co., Utah, 23 May 1894.)

Slender erect herbaceous perennials (1.5) 2–3 (4) dm tall, arising from a slender rhizome; leaves linear, 8–20 cm long, usually less than 1/2 as long as the plants, the blade 0.5–1 (1.5) mm wide, subterete, rather blunt at the tip, the ligule bilobed to the base, 0.5–1 mm long; scapes slender, 1–3 dm long, terminated by a sparsely flowered raceme 1–2 dm long, the pedicels ascending, 2–4 (5) mm long; perianth segments 6, 1–1.5 mm long, greenish; stamens 6, the anther subsessile; stigmas 6; fruits oblong-obovoid, 4–5 mm long, the 6 fertile carpels rounded at the base and united, completely deciduous from the terete axis; $2n$ = 48, 96.

Wet meadows and marshes, usually where alkaline; sw. B.C., s. along the Pacific Coast to n. Baja Calif., e. to the Dakotas, Wyo., Colo. and Ariz., and in sw. S. Amer.; se. Oregon, s. to Mono Co., Calif., e. across n. and s. Nev. to Utah and n. Ariz. May–Aug.

Only the var. *debilis* occurs in our area, the var. *concinna* being found along the Pacific Coast of North America with a third variant restricted to South America.

2. **LILAEA** Humb. & Bonpl. Flowering Quillwort

Aquatic acaulescent herbaceous annuals with fibrous roots and basal tufted linear or ribbon-like leaves; flowers bracteated or ebracteated, solitary or in clusters, variously borne, rather small and inconspicuous, perfect or imperfect, the pistillate flowers—in part—solitary in the sheathing leaf axils at the base of the plant, or on a solitary scapose spike of both perfect and staminate flowers or all perfect (infrequently with the pistillate flowers mixed with perfect ones at the lower portion of the inflorescence); sepals lacking; petals lacking; pistillate flowers basal or in spikes, the basal ones enclosed in the sheathing leaf base, consisting of a solitary sessile, erect carpel with a long filiform style usually floating on the surface of the water if plant partially submersed, this terminated by a capitate stigma, the pistillate flowers of the spike bractless, consisting of a solitary, sessile, erect carpel with a short (or lacking) style, the pistil in both types 1-celled and 1-ovuled; perfect flowers small, irregularly imbedded in the fleshy axis of the spike, consisting of a single caducous bract, a single subsessile, 2-celled extrorse anther, and a solitary, sessile carpel with a short style; staminate flowers restricted to the upper portions of the spike, small, bracteate, consisting of a single, sessile, 2-celled, extrorse anther; fruits variable but characteristic of position on the plant, those of the basal pistillate flowers indehiscent, divergent, with 25–30 longitudinal ribs, obliquely beaked by a persistent, recurved base of the style, those of the perfect flowers and of the scapose pistillate flowers similarly ribbed but with 2–4 ribs drawn into narrow, undulate wings, the beaks short and stout; ovules basal, with erect seeds.

A monotypic genus occurring in the mountainous regions of N. and S. Amer., and introduced in deep water ponds in e. Australia. (Named for the French botanist Alire Raffeneau-Delile, 1778–1850, an authority on the flora of n. Africa and Asia Minor.)

Reference: Larsen, K. Cytotaxonomical note on *Lilaea*. Bot. Not. 119: 496–497. 1966.

1. **Lilaea scilloides** (Poiret) Hauman

Phalangium scilloides Poiret in Lam. Encycl. Méth. Bot. 5: 251. 1804. *Anthericum scilloides* Schultes f. in Roemer & Schultes, Syst. Veg. 7: 477. 1829. *Lilaea scilloides* Hauman, Fac. Filos. y Letras Buenos Aires, Publ. Inst. Invest. Geogr. 10: 26. 1925. (Argentina.)

L. subulata Humb. & Bonpl. Pl. Aequinoctiales 1: 221. 1808. (Santa Fé de Bogota, Colombia.)

Heterostylus gramineus Hook. Fl. Boreali-Amer. 2: 171. 1838. (*Scouler,* "Wet plains of the Columbia.")

Tufted acaulescent annuals either in shallow water where the stigmas of the basal flowers and the spike of mixed flowers tend to float, or stranded and exposed, rooting on mud flats; leaves basal or nearly so, terete, linear, (3) 5–25 (35) cm long, 1–4 mm thick, sheathing the stem with a hyaline stipule which converges across the top to form a low, emarginate ligule; flowers various and of three possible types; basal pistillate flowers enclosed by the sheathing leaf base, more or less triquetrous and in fruit 3-lobed or -horned at the tip, 4–6 mm long, the style thread-like and 5–25 (30) cm long, arising laterally and at right angles to the top of the ovary, occasionally reflexed, bearing a capitate stigma; flowering spike on a peduncle 3–20 (30) cm long, the flowering portion of the spike 0.5–4 cm long, composed of perfect flowers, staminate and pistillate flowers, or of perfect and staminate flowers, or even strictly of pistillate flowers; staminate flowers with a caducous bract 2–3 mm long; pistillate flowers bractless; perfect flowers with small recurved, glabrous, caducous bracts; stamens sessile or nearly so, the anther 1–1.5 mm long; pistils 1-celled and 1-ovuled; fruits winged, glabrous, (2) 3–5 mm long, those at the base of the plant with beaks or horns 1–2 mm long, those of the spike essentially beakless; $2n = 12$.

Infrequent in our area in shallow water or on adjacent mud flats; N. and S. Amer., introduced in Australia; B.C. to Alta., s. to Baja Calif. and w. Mex., e. to s. Idaho and w. Mont.; in the Lake Sect. of se. Oregon, ne. Calif., and in the Salmon Falls Creek drainage of the Bonneville Basin Sect. of Elko Co., Nev. June–Aug.

Family **NAJADACEAE**, the Water-nymph Family

Submersed, monoecious or dioecious, slender branched herbaceous aquatic annuals with toothed to entire, subopposite or verticillate, sessile leaves with a dilated and sheathing base, the flowering leaves usually clustered toward the branch-tip and thus concealing the flowers, the fruit clearly obvious as the branch elongates and fruit matures; flowers imperfect, axillary, sessile or pedicelled; staminate flowers composed of a single stamen enclosed in a membranous and hyaline, saccate bract surrounded by a firmer, 2-lipped, entire to 4-lobed, perianth segment, the stamen subsessile to sessile, the anther 1- to 4-celled; pistillate flowers naked or surrounded by a membranous perianth segment, the pistil one, 1-celled with a basal ovule, the stigmas 2–4, sessile; fruits sessile and indehiscent achenes, fusiform, with a thin membranous pericarp, the seed exalbuminous.

A monotypic family.

1. **NAJAS** L. Water-nymph

Perhaps 40 species, widely distributed in fresh and brackish water in temp. and warm parts of the world. (From the Greek *Naias,* a water-nymph.)

Reference: Clausen, R. T. Studies in the genus *Najas* in the northern United States. Rhodora 38: 333–345. 1936.

1 Leaves coarsely toothed, (0.5) 1–3 mm wide, the internodes and backs of the leaves obviously spiny; plants dioecious; rare . . . . 1. *N. marina*

1 Leaves minutely toothed, 0.5–1.5 mm wide, the internodes and backs of the leaves not spiny; plants monoecious.

  2 Seed coat smooth and shining with (40) 50–70 rows of minute areolae; styles 0.7–1 mm long; leaves gradually tapering to a long slender point; plants less than 0.5 dm long; c. Utah . . . . 2. *N. caespitosa*

  2 Seed coats coarsely pitted and dull with 10–20 rows of large areolae; styles less than 0.5 mm long; leaves abruptly tapering to a rounded or obtuse but still slender point; plants 3–6 (7) dm long; n. Utah and introduced in ponds near Reno, Nev. . . . . 3. *N. guadalupensis*

1. **Najas marina** L.

Najas marina L. Sp. Pl. 1015. 1753. (Europe.)

N. major All. Fl. Pedemontana 2: 221. 1785. *Ittnera major* Reichenb. Fl. German. Excursoria 150. 1830. (Italy.)

N. major var. *gracilis* Morong, Bot. Gaz. 10: 255. 1885. *N. marina* var. *gracilis* Dudley, Cornell Univ. Sci. Bull. 2: 104. 1886. (*Curtiss 2705,* Palm Cr., near Cape Romano, Collier Co., Fla., July 1881.)

N. marina var. *recurvata* Dudley, Cornell Univ. Sci. Bull. 2: 104. 1886. (*Dudley,* Cayuga marshes, N.Y.)

N. marina var. *californica* Rendle, Trans. Linn. Soc. London, II. 5: 398. 1899. (Calif., *Orcutt* and *Coulter 188* are cited.)

Stems rather stout, branching near the base, 1–4 dm long, often armed with prickles on the internodes; leaves narrowly linear, 1–4 (4.5) cm long, (0.5) 1–3 mm wide, obviously and coarsely spiny-toothed along the margin and often spiny on the abaxial surface along the midrib, the teeth broad, tapering to a sharp yellowish-brown point, the leaf sheath with a broadly expanded and rounded shoulder, this entire or with a few minute teeth, expanding into a broad sheath 5–6 mm long, 3–4 mm wide at the base; plants dioecious, the staminate flowers 3–4 mm long, with the single anther 4-celled, the pistillate flowers 3–4 mm long with 3 stigmas of slightly unequal lengths; fruits 4–5 mm long, ovoid, the seed smooth and shining but finely reticulated with numerous rows of minute areolae; $n = 6$.

Fresh water ponds, lakes or slow-moving streams; widespread Old World species of Eurasia and Africa introduced into many parts of the U.S. and now known from Calif., Nev., Utah, Ariz., N.M., N.D., Texas and the e. U.S.; rare and local in our area, occurring outside our boundary in ne. Calif. and s. Nev. (Ash Meadows, Nye Co.), and in scattered locations within, such as in Huntington Valley, Elko Co., Nev. and in n. Utah (Fish Springs, Tooele Co. and near Logan, Cache Co.). July–Aug.

2. **Najas caespitosa** (Maguire) Reveal

Najas flexilis subsp. *caespitosa* Maguire, Rhodora 44: 7. 1942. *N. caespitosa* Reveal, Great Basin Naturalist 34: 357. 1976. (*Maguire 19888,* Pelican Point, Fish Lake, Sevier Co., Utah, 3 Aug. 1940.)

Stems stout, densely branching throughout, 0.2–0.4 (0.5) dm long, unarmed; leaves narrowly linear, 0.3–1 cm long, 0.5–1.5 mm wide, entire, rarely with a few scattered minute spines, the leaf sheath with an obliquely sloping shoulder, the margin entire, expanding to a sheath 2–3 mm long and wide at the base; plants monoecious, the staminate flowers 2–2.5 mm long, with the single anther 1-celled, the pistillate flower 2–2.5 mm long, with (2) 3 stigmas of slightly unequal length; fruits 2–2.5 mm long, nar-

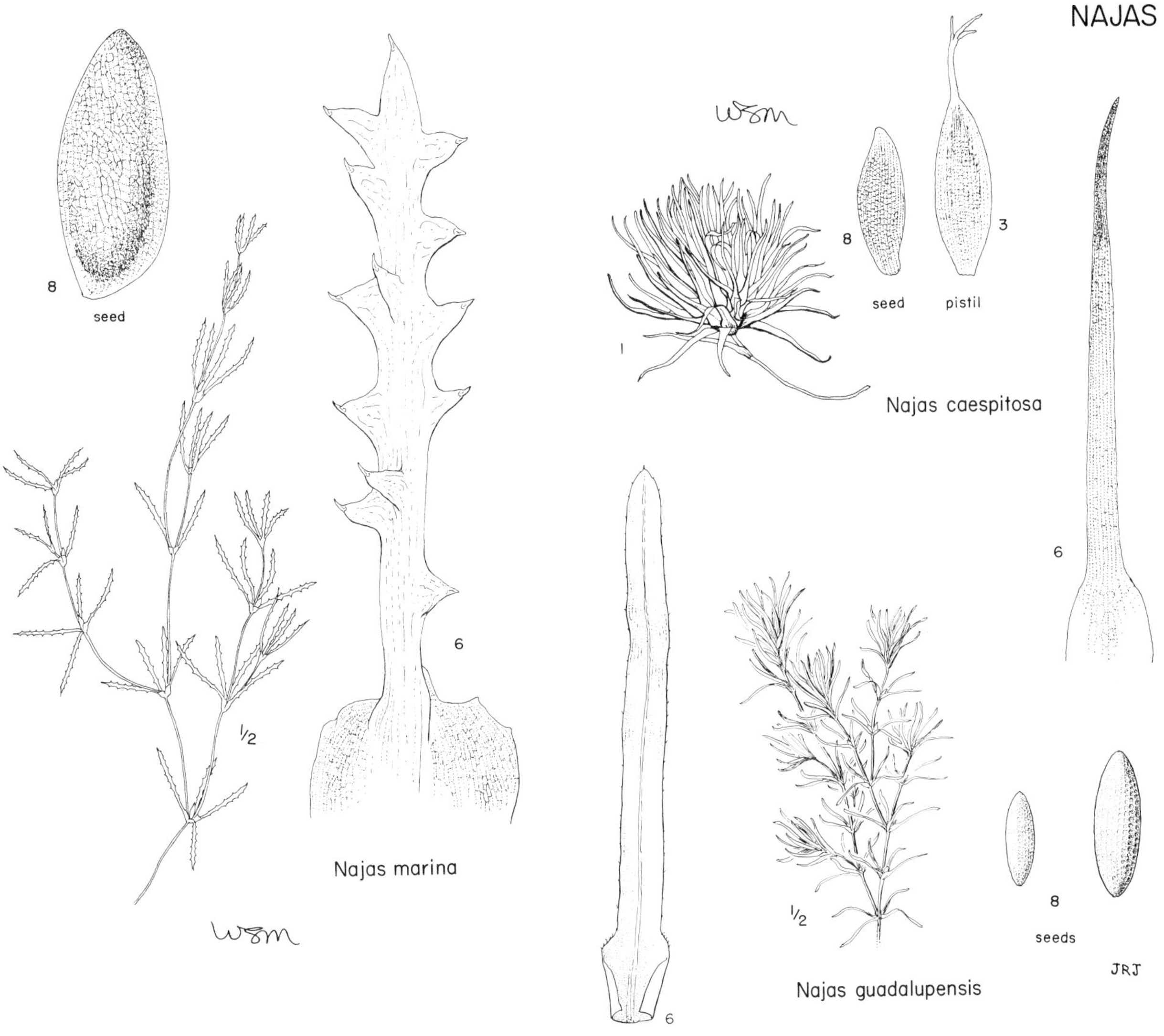

rowly elliptic, the seed smooth and shiny but finely reticulate with (40) 50–70 rows of minute areolae.

Fresh water lake shore of Fish Lake, Sevier Co., Utah, in water 1–3 dm deep. Endemic. July–Aug.

Najas caespitosa is a unique species in the *N. flexilis* Rostk. & Schmidt complex. Although *N. flexilis* occurs across North America and Eurasia, nowhere are the plants similar to those found in central Utah. Clausen noted in his treatment that *N. flexilis* is variable in overall size of the plants and the plumpness of the seeds. However, the extreme reduction seen in the Utah plants is rarely seen elsewhere, and the seed characteristics of *N. caespitosa* are unique. The range of *N. flexilis* in the American West is not as great as assumed by some authors. This species apparently does not occur south of Washington and northern Idaho, and even some of these stations are questionable. Most reports of *N. flexilis* in the west are based on specimens of *N. guadalupensis*.

3. **Najas guadalupensis** (Sprengel) Morong

Caulina guadalupensis Sprengel, Syst. Veg. 1: 20. 1826. *Najas flexilis* var. *guadalupensis* A. Braun, J. Bot. 2: 276. 1864. *N. microdon* var. *guadalupensis* A. Braun, Sitzungsber. Ges. Naturf. Freunde Berlin 1868: 17. 1868. *N. guadalupensis* Morong, Mem. Torrey Bot. Club 3(2): 60. 1893. (Guadalupe I.)

Stems slender, branching throughout, 3–6 (7) dm long, unarmed; leaves narrowly linear, 1–2.5 cm long, 0.5–1 mm wide, very minutely toothed, the teeth consisting of a single marginal cell, the leaf sheath with an oblique to rounded shoulder, the margin bearing only a few teeth, infrequently entire, expanding into a sheath 3–5 mm long, 2–3 mm wide at the base; plants monoecious, the staminate flowers 2–3 mm long, with the single anther 4-celled, the pistillate flowers 2–3 mm long, with 2–3 stigmas all of about equal length and usually with 1–2 additional sterile spiny processes; fruits 2.5–3 mm long, ellipsoid, the seed prominently pitted with 10–20 rows of conspicuous areolae, dull; $2n$ = 12, 36, 42, 48, 54, 60.

Fresh water pools, streams and lakes; N. and S. Amer., W.I.; Oregon, Idaho and Calif. s. to Baja Calif., and from Colo., the Dakotas and Texas e. through the e. U.S. and se. Can.; rare in n. Utah and apparently introduced in artificial ponds near Reno, Washoe Co., Nev. June–Aug.

Family **POTAMOGETONACEAE,** the Pondweed Family

Herbaceous aquatic perennials with numerous slender rhizomes bearing the terete to slightly flattened stems with alternate to somewhat subopposite or rarely opposite, 2-ranked leaves with prominent, occasionally sheathing, stipules, the blade either submersed or floating, rarely erect; flowers perfect, sessile on axillary pedunculate sheathed spikes in few to several more or less well defined whorls, 4-merous; perianth of 4 expanded, free, claw-like tepals, the 4 stamens sessile on these segments (often called "perianth segments" or "dilated sepaloid connectives"), with 2-celled extrorse anthers, the pistils 4 (rarely only 1), sessile, 1-celled and 1-ovuled, the stigma sessile or on a short, persistent, erect or curved style, the single ovule attached to the adaxial surface of the ovary; fruits ovoid to obovoid, more or less compressed achenes with a somewhat prominent keel on the dorsal or abaxial surface with two much less obvious to obscure lateral keels, the seed exalbuminous and usually strongly curved around a projection of the bony endocarp.

A monotypic family of fresh temp. waters throughout the world. The sometimes segregated, monotypic genus, *Groenlandia* Gay, from Eurasia and nw. Africa is based on the presence of opposite leaves.

1. **POTAMOGETON** L. Pondweed

Perhaps 90 species, widely distributed in fresh water in the temp. regions of the N. Hemisphere with some scattered populations in S. Amer. (An ancient Greek name, *potamogeiton,* from *potamos,* a river, and *geiton,* a neighbor, alluding to the habitat of another group of aquatic plants.)

The genus is extremely difficult for several reasons. The immature plants, or those which lack flowers or fruits, are almost impossible to reliably key out. Occasionally, even with these parts, the submersed leaves have disintegrated, or care was not taken to gather both submersed and floating leaves. Likewise, seasonal variation plays a role in altering the overall morphology of vegetative features so that leaf width and the degree of sharpness of the leaf apex will change during the growing season. New flushes of leaves are frequently produced in the summer which differ from those produced in the spring, and decaying leaves are often less leathery or rigid than otherwise expected as the plant matures. It is possible that salinity may play a role in altering the gross morphology of the plants, and it seems that many of the varieties recognized by some authors (notably St. John and Fernald) are little more than briefly expressed phenotypes that change with the season, salinity and maturity of the plants. Hybridization is reported to be relatively common among some species, and a few intermediate specimens have been seen from our area. In most cases of hybridization, there is a high percentage of sterile pollen and imperfectly formed fruits. This is not to be confused with some species, such as *P. robbinsii,* which rarely forms flowers, and no fruiting specimens of this species have been found in our area.

It is traditional for those presenting a floristic treatment of *Potamogeton* to comment on the unique perianth, or the lack of the same. I am in no position to judge the full merits of the numerous arguments and have simply chosen to call the four sterile segments associated with the flower "tepals." The fact that these segments are similar to, and perhaps evolutionarily the same as the "bracts" found in Juncaginaceae (*Lilaea*) is a fact I can not deny. However, to the average person using this Flora, the segments of *Lilaea* look like bracts, whereas those of *Potamogeton* appear more typical of tepals. A more technical review of this matter is presented in Cronquist's book, *The evolution and classification of flowering plants* (1968). In this connection, finally, it should be noted that if the tepals of *Potamogeton* are interpreted as "connectives," then the distinction between Potamogetonaceae and Ruppiaceae seems less significant and the two could be united into a single family as done by Melchior in *Engler's Syllabus der Pflanzenfamilien* (1964) and others.

References:

Correll, D. S., and H. B. Correll. Aquatic and wetland plants of southwestern United States. Environmental Protection Agency, Washington, D.C. 1972.

Fernald, M. L. The linear-leaved North American species of *Potamogeton,* section Axillares. Mem. Amer. Acad. Arts, II. 17: 1–183. 1932.

Hagström, J. O. Critical researches on the potamogetons. Kongl. Svenska Vetenskapsakad. Handl. 55(5): 1–281. 1916.

Haynes, R. R. A revision of North American *Potamogeton* subsection Pusilli (Potamogetonaceae). Rhodora 76: 564–649. 1975.

Mason, H. L. A flora of the marshes of California. University of California Press, Berkeley. 1957.

Ogden, E. C. The broad-leaved species of *Potamogeton* of North America north of Mexico. Rhodora 45: 57–105, 119–163, 171–214. 1943.

St. John, H. A revision of the North American species of *Potamogeton* of the section Coleophylli. Rhodora 18: 121–138. 1916.

1 Submersed leaves with stipules adnate to the base of the leaf and forming a sheath around the stem, the leaf blade or petiole not directly attached at the node of the stem.

2 Plants with submersed leaves only; fruiting spikes elongated and erect, the peduncle long and slender.

3 Leaves entire, up to 5 mm wide, mostly 1- or 3-nerved.

4 Leaves retuse, blunt or rarely short-apiculate; styles mostly wanting with the fruiting beak wart-like and central, or if styles present, then leaves 2–5 mm wide.

5 Main sheaths greenish, not inflated, usually subtending 1 or 2 branches, the free stipule tips projecting upwards into a distinct ligule; leaves filiform to linear, 0.2–5 mm wide; spikes with mostly unequally spaced whorls of flowers (at least in ours); common . . . . 1. *P. filiformis*

5 Main sheaths brownish, inflated at the base, usually subtending 2 or more branches, the free stipule tips not producing an obvious ligule; leaves filiform to narrowly linear, 1–2 mm wide; spikes with equally spaced whorls of flowers; rare in our area . . . . 2. *P. vaginatus*

4 Leaves sharply pointed; styles short with the stigma on the ventral margin of the fruit; leaf blade filiform, 0.2–0.8 (1) mm wide; spikes with unequally spaced flowers; common . . . . 3. *P. pectinatus*

3 Leaves minutely serrulate-spinulose toward the tip, 3–8 mm wide, linear to lanceolate, finely 20- to 60-nerved; plants usually sterile; infrequent .... 4. *P. robbinsii*

2 Plants with long-petioled floating leaves and sessile, linear, submersed ones; fruiting spikes short and somewhat capitate, usually found below the surface of the water, the peduncles, when present, often curving downwardly .... 11. *P. diversifolius*

1 Submersed leaves with stipules free from the rest of the leaf, the leaf blade or petiole directly attached to the node of the stem.

6 Submersed leaves crispate and finely serrulate, (3) 5–10 (12) mm wide; achenes with curved beak 2–3 mm long; introduced .... 5. *P. crispus*

6 Submersed leaves entire (or nearly so) and not serrulate; achenes with beak less than 2 mm long.

7 Submersed leaves linear or nearly so, less than 5 mm wide; floating leaves lacking.

8 Stems strongly flattened and winged; leaves (2) 3–5 mm wide with numerous nerves; fruits 4–5 (5.5) mm long .... 6. *P. zosteriformis*

8 Stems terete or nearly so, not obviously winged; leaves mostly less than 2.5 mm wide and less than 10-nerved; fruits less than 3 mm long.

9 Peduncles stout, 0.3–1.5 (2) cm long, bearing congested, subcapitate spikes 0.1–0.5 cm long; leaves glandless; fruiting beak 0.1–0.4 mm long .... 7. *P. foliosus*

9 Peduncles slender, (0.4) 1–9 cm long, bearing more open, elongated spikes 0.6–2 cm long; leaves often with 2 basal glands; fruiting beak 0.3–0.8 mm long.

10 Stipules obvious, 0.7–2 cm long, strongly fibrous; stems slender and simple or sparsely branched; rare in n. Utah and s. Idaho.

11 Leaves thin and translucent, (3) 5- to 7-nerved, 1.5–3.5 mm wide, with two yellowish glands at the base, the apex rounded; stipules 0.7–1.1 (2) cm long ... . 8. *P. friesii*

11 Leaves firm, often revolute, 3-nerved, 0.5–2 (2.5) mm wide, glandless, the apex gradually tapering to a slender tip; stipules 0.8–2 cm long .... 9. *P. strictifolius*

10 Stipules obscure, 0.3–1.4 cm long, not fibrous; stems very slender or capillary and much branched; leaves firm, 3-nerved, (0.3) 0.5–2 mm wide, with two small globose basal glands; common throughout the n. part of our area .... 10. *P. pusillus*

7 Submersed leaves more broadly linear to greatly elongated or ovate, or if linear and less than 2 mm wide, the plants with obvious, broad, leathery floating leaves.

12 Submersed leaves with stipules adnate to the base of the leaf (a feature often difficult to determine), linear, 0.5–1 (1.5) mm wide, the floating leaves elliptic to oblong-elliptic, (1) 1.5–3 (4) cm long, these sometimes not obvious on herbarium material or lacking due to fluctuating water level .... 11. *P. diversifolius*

12 Submersed leaves with stipules free from the rest of the leaf, the blade or petiole attached directly to the stem.

13 Submersed leaves petiolate or tapering to a sessile base, not strongly clasping the stem, the floating leaves often well developed and differentiated from the submersed ones.

14 Submersed leaves (1) 2–75 mm wide, mostly broadly linear to ovate.

15 Stems compressed; submersed leaves mostly (1) 2–10 mm wide, ribbon-like; fruits laterally compressed; stipules 1–3 (4) cm long .... 12. *P. epihydrus*

15 Stems terete; submersed leaves mostly wider, ovate to elliptic, or if narrower, then not ribbon-like; fruits with rounded sides.

16 Submersed leaves sessile, the floating leaves (when present) delicate and translucent, short petiolate, not markedly different from the submersed ones except wider; plants with a reddish tinge .... 13. *P. alpinus*

16 Submersed leaves sessile or petioled, obviously different from the thick, coriaceous floating leaves, or if not markedly different, then leaves strongly falcate.

17 Submersed leaves 25–75 mm wide, folded and strongly falcate, petiolate, the leaf margin entire; floating leaves mostly 25- to 50-nerved, ovate to ovate-elliptic .... 14. *P. amplifolius*

17 Submersed leaves mostly 3–35 mm wide, flat, sessile or petiolate, the leaf margin with minute fugacious, 1-celled translucent denticles; floating leaves with less than 20 nerves.

18 Submersed leaves on petioles 2–10 (13) cm long; floating leaves elliptic to oblong-elliptic, (3) 5–12 cm long, (1.5) 2–4 cm wide, on longer petioles; stipules (3) 4–8 cm long; fruits 3.5–4 mm long, reddish .... 15. *P. nodosus*

18 Submersed leaves sessile or on petioles up to 4 cm long; fruits 2–3.5 (4) mm long, greenish.

19 Stipules 0.5–3 cm long; submersed leaves (1) 3–10 (12) mm wide, sessile; floating leaves narrowly to broadly elliptic or oblong-elliptic, (1.5) 2–5 (7) cm long, 1–2 (2.5) cm wide, on shorter petioles; fruits 2–2.5 (2.8) mm long .... 16. *P. gramineus*

19 Stipules (2.5) 3–7 cm long; submersed leaves (10) 20–40 (50) mm wide, subsessile or tapering to a short petiole less than 4 cm long; floating leaves broadly elliptic to oblong-elliptic, (4) 6–12 cm long, 2–6 cm wide, on very short petioles; fruits 3–3.5 (4) mm long .... 17. *P. illinoensis*

14 Submersed leaves 1–2 mm wide, linear, sessile; floating leaves lance-ovate to ovate-elliptic, (3) 5–10 cm long, (1.5) 2.5–5 (6.5) cm wide, long petioled; stipules 4–10 cm long, strongly fibrous; fruits 3–5 mm long .... 18. *P. natans*

13 Submersed leaves sessile and clasping the stem, no specialized floating leaves present.

20 Stems usually zigzag; leaves oblong-lanceolate, (5) 10–25 (35) cm long, (10) 20–30 mm wide; stipules rigid, (3) 4–10 cm long, usually persistent; peduncles 10–30 (60) cm long; achenes 4–5 mm long .... 19. *P. praelongus*

20 Stems not zigzag; leaves lanceolate to ovate-lanceolate, 1.5–10 cm long, 5–20 mm wide; stipules coarse, 1–2 cm long, soon disintegrating into strong fibers; peduncles 1.5–25 cm long; achenes 2.5–3.5 mm long .... 20. *P. richardsonii*

1. Potamogeton filiformis Pers.

Potamogeton filiformis Pers. Syn. Pl. 1: 152. 1805. *Spirillus filiformis* Nieuwl. Amer. Midl. Naturalist 3: 18. 1913. *Ruppia filifolia* Skottsb. Kongl. Svenska Vetenskapsakad. Handl. 55(5): 171. 1916. (Denmark.)

P. borealis Raf. Med. Repos. II. 5: 354. 1808. *P. filiformis* var. *borealis* St. John, Rhodora 18: 134. 1916. (*Michaux,* St. Lawrence R., 1792.) = var. *alpinus.*

P. marinus var. *alpinus* Blytt, Norges Fl. 1: 370. 1861. *P. filiformis* var. *alpinus* Aschers. & Graebner, Syn. Mitteleur. Fl. 1: 353. 1897. *P. filiformis* f. *alpinus* (attrib. to Blytt by) Hagström, Kongl. Svenska Vetenskapsakad. Handl. 55(5): 17. 1916. (Norway.)

Potamogeton filiformis var. **latifolius** (J. W. Robbins) Reveal, hoc loc. *P. pectinatus* var. *latifolius* J. W. Robbins in King, Rep. Geol. Explor. 40th Parallel 5: 338. 1871. *P. latifolius* Morong, Mem. Torrey Bot. Club 3(2): 52. 1893. (*Bailey 1142,* in running brackish waters of the Humboldt R., below Humboldt Lake, Pershing Co., Nev., Aug. 1867.)

P. marinus var. *occidentalis* J. W. Robbins in King, Rep. Geol. Explor. 40th Parallel 5: 339. 1871. *P. filiformis* var. *occidentalis* Morong, Mem. Torrey Bot. Club 3(2): 51. 1893. *P. interior* Rydb. Agric. Exp. Sta. Agric. Coll. Colorado Bull. (Fl. Colorado) 13. 1906. (*Watson 1143,* Ruby Lake, Elko Co., Nev., Aug. 1868.)

P. marinus var. *macounii* Morong in Macoun, Cat. Canad. Pl. 2(4): 88. 1888. *P. filiformis* var. *macounii* Morong, Mem. Torrey Bot. Club 3(2): 50. 1893. (*Macoun 3008,* in salt marshes, Old Wives Lake, Sask., 23 July 1880, is the first collection cited.) = var. *occidentalis.*

Plants wholly submersed, the stem slender to stout and freely branching below, somewhat less so above, whitish to greenish, subterete, (1) 2–6 (9) dm long, arising from an extensively creeping slender rhizome, terminated by a slender white tuber 1–2 cm long; leaves all submersed, the blade filiform to linear, 1–10 (12) cm long, 0.2–4 (5) mm wide, the apex obtuse to acute, (1-) 3- to 5- (7)-nerved; stipules connate with the leaf and forming a closed, clasping sheath around the stem, the sheath 0.5–2 cm long, greenish to whitish, not swollen, slowly deciduous and bending back from the stem with the upper portion confluent across the top of the sheath to form a conspicuous hyaline ligule 1–7 mm long; peduncles filiform to slender, (2) 3–10 (15) cm long; spikes moniliform, 1–5 cm long, with 2–8 whorls of crowded to remote flowers, the lower ones up to 3 cm apart, the upper ones 0.3–1.2 cm apart, occasionally the whorls somewhat irregularly defined and the spike of 8–25 interrupted flowers with the lower ones well spaced and the upper ones very crowded; flowers small, the tepals brownish, 1–2 (2.5) mm long, the styles short to nearly lacking; fruits obovoid, 2–4 mm long, 1.5–3 mm wide, the dorsal keel low and rounded, the two lateral keels low and obsolete or lacking, the fruiting stigma capitate with a short, persistent, wart-like beak less than 0.5 mm long; $2n = 78$.

Shallow pools, lakes and slow-moving streams; Eurasia, Africa, N. Amer. and Australia; widespread and common from Alaska to Greenl., s. to central Calif., Ariz., Colo., N.M., Texas, Mich. and Pa.; se. Oregon, e. Calif. and most of Nev. and Utah. July–Sept.

Several interpretations of the variation in this species have been proposed. In Europe, it is divided into numerous varieties and forms whose merits are exceedingly dubious. In North America, St. John suggested that three varieties should be distinguished. He proposed the typical variety (and included within it var. *occidentalis* as a synonym), var. *borealis* and var. *macounii.* The latter two variants were differentiated by leaf width, with var. *borealis* having leaves less than 0.5 mm wide, and the var. *macounii* with leaves more than 0.5 mm wide. This type of variation is seen in Europe, with var. *filiformis* having leaves more than 0.5 mm wide, and var. *alpinus* having leaves less broad. A close examination of var. *alpinus* of Europe and var. *borealis* of North America has shown that they can not be separated, although at the extremes, var. *alpinus* of Norway, Finland and other northern countries has widely separated whorls of flowers, while var. *borealis,* in some locations at least, has more congested spikes. Across northern Canada, Alaska and most of northern Asia, even this slight difference disappears, and even some populations within the United States can not be distinguished from the plants of northern Europe. Consequently, the older European name, var. *alpinus,* is now applied to our plants.

The distinction between var. *occidentalis* and var. *macounii* is subtle at best. The type of var. *occidentalis* has relatively narrow leaves (0.6–0.9 mm wide), and is on the narrow side of the spectrum for the taxon which is defined as having leaves 0.5–2 mm wide. The type of var. *macounii* is on the broad side of the spectrum. St. John referred var. *occidentalis* to var. *filiformis* (and essentially made this collection the only known specimens of the typical phase in North America), an action which I disagree with, and I have taken up the older name, var. *occidentalis,* for what has passed under the name

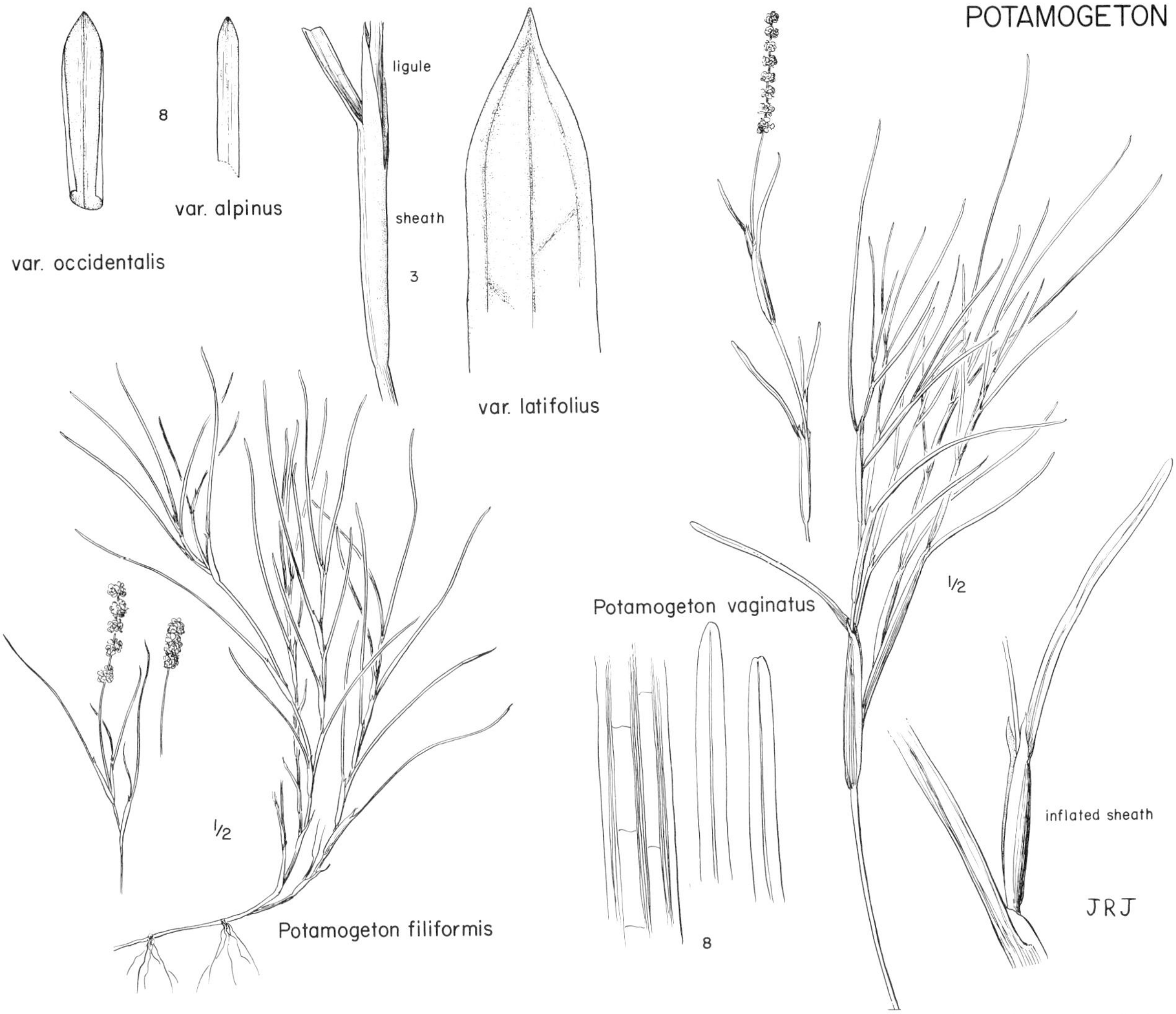

var. *macounii*. Our var. *occidentalis* is much coarser a plant than that of var. *filiformis* of Europe, and the two do not seem to be identical. Unlike the European phase, the leaves of our plants can be as much as 2 mm wide, are of a darker green, and usually have a rounded leaf apex.

The reduction of *P. latifolius* to a variety of *P. filiformis* is due primarily to the intergradation seen between various populations of var. *occidentalis* and *P. latifolius*. At their respective extremes, the two are quite distinct, but across the broad range of variation, the two forms gradually merge in nearly all features. The distinction between the two in northern Nevada and northwestern Utah is extremely difficult to make, and the same problem seems to exist in Texas, judging from herbarium material. Plants in Alaska may represent another taxon entirely or be a wide leaf form of var. *occidentalis*, or even a narrow leaf phase of var. *latifolius*—it is impossible to tell at this time. The type of var. *latifolius* is on the narrow side, with leaves 2–2.5 mm wide, although on some plants of the type there are leaves slightly less than 2 mm wide, and this represents the kind of difficulty one encounters in northern Nevada in distinguishing var. *occidentalis* and var. *latifolius*. Plants from northeastern California and southeastern Oregon are representative of var. *latifolius*, although, as reported by Mason, there are some narrow leaf plants in California that are transitional between the two variants.

In general, the three variants recognized in North America occupy distinct (but overlapping) geographical ranges and ecological niches. The var. *alpinus* is the northern element which occurs at higher elevations in the west-central United States. The var. *occidentalis* ranges from southern Canada to the southern United States, and usually occurs at middle elevations, while var. *latifolius* occurs more frequently to the south of var. *occidentalis* and at somewhat lower elevations. The phase of the species in central Alaska remains to be studied in more detail before it can be placed in the species.

In our area, the three variants can be distinguished as follows:

1 Leaves 0.2–2 mm wide; fruits 2–3 mm long.
  2 Leaves 0.5–2 mm wide; plants tall, the stems mostly 3–5 (9) dm long; common throughout our area . . . . var. *occidentalis* (J. W. Robbins) Morong
  2 Leaves 0.2–0.5 mm wide; plants short, the stems mostly 1–3 dm long; rare, high mts., Wasatch Plateau and Uinta Mts., becoming more common to the n. in Can., Alaska and n. Eurasia . . . . . var. *alpinus* (Blytt) Aschers. & Graebner
1 Leaves 2–4 (5) mm wide; fruits 3–4 mm long; plants 2–6 dm long; nw. Utah, n. Nev., ne. Calif. and se. Oregon; infrequent elsewhere to the s. and e. of us . . . . var. *latifolius* (J. W. Robbins) Reveal

2. Potamogeton vaginatus Turcz.

Potamogeton vaginatus Turcz. Fl. Baicalensi-Dahurica 2: 162. 1856. *P. pectinatus* var. *vaginatus* Aschers. & Graebner, Syn. Mitteleur. Fl. 1: 351. 1897. (c. Siberia.)

P. moniliformis St. John, Rhodora 18: 130. 1916. (*Drummond*, between Cumberland House and Hudson Bay, Sask., Aug. probably 1825.)

P. vaginatus var. *canadensis* Hagström, Kongl. Svenska Vetenskapsakad. Handl. 55(5): 36. 1916. (*Macoun*, near Banff, Alta., 1891, is one of three collections cited.)

Plants wholly submersed, the stem stout and rather freely branched throughout, greenish, terete, 2–5 (7) dm long, arising from a slender creeping rhizome terminated by a slender white tuber 3–4 cm long; leaves all submersed, the blade narrowly filiform to linear, 2–10 (15) cm long, 1–2 mm wide, the upper leaves narrower than the lower ones, 1- (rarely 3)-nerved, the apex rounded or obtuse to slightly retuse; stipules adnate to the base of the leaf and forming a closed, clasping sheath around the stem, the sheath 2–6 cm long, attached to the leaf for 1–5 cm, persistent but bending back from the leaf, the upper portion confluent across the top of the sheath to form a short obscure hyaline ligule, the sheath of the main stem usually brownish, swollen especially at the base, those of the secondary branches greenish-brown and less swollen with a slightly longer ligule; peduncles slender, 3–15 cm long; spikes moniliform, (1) 3–8 cm long, with 4–9 rather evenly spaced whorls of flowers, up to about 1 cm apart; flowers small, the tepals brownish, 1.5–2 mm long, the style lacking; fruits obovoid, 3–3.5 mm long, 2–2.5 mm wide, the dorsal keel low and inconspicuous, the two lateral keels essentially lacking, the fruiting stigma broad, sessile, persistent on a wart-like projection well to the side of the fruit; $2n = 78$, ca 88.

Shallow to deep ponds and streams or lakes; Eurasia and N. Amer.; Alaska to Newfl., s. to Oregon, Idaho and Utah, Wyo., Wisc. and N.J.; rare in our area in se. Oregon (Lake Sect.) and n. Utah. July–Sept.

Potamogeton vaginatus is closely related to *P. pectinatus,* and is often submerged under that older name (Mason, 1957). A re-examination of the material from Russia, Alaska and Canada has shown that *P. vaginatus* is distinctive throughout its range and does not tend to merge with *P. pectinatus*. However, during this study, it was found that much of the material called *P. vaginatus* is *P. filiformis* (either var. *occidentalis* or narrow leaf forms of var. *latifolius*), and that *P. vaginatus* more likely verges on *P. filiformis* than *P. pectinatus*. Plants in Alaska are particularly questionable, and the plants noted under the discussion of *P. filiformis* may represent some form of *P. filiformis* that has been influenced by *P. vaginatus*.

In our area, true *P. vaginatus* is rather rare, but occurs across the northern edge of the Intermountain Region.

3. Potamogeton pectinatus L.

Potamogeton pectinatus L. Sp. Pl. 127. 1753. *Buccaferrea pectinata* Bubani, Fl. Pyrén. 4: 17. 1901. (Europe.)

P. columbianus Suksd. Deutsche Bot. Monatsschr. 19: 92. 1901. (*Suksdorf 2062,* Columbia R., near Bingen, Klickitat Co., Wash., 15 Sept. and Oct. 1891.)

Plants wholly submersed, the stem filiform and freely branched throughout most of its length, whitish, terete, 3–8 (10) dm long, arising from a slender creeping rhizome terminated by a slender white tuber 1–1.5 cm long; leaves all submersed, the blade filiform, 2–12 (15) cm long, 0.2–0.8 (1) mm wide, 1- (rarely 3)-nerved, the apex sharply pointed; stipules adnate to the base of the leaf and forming a slender clasping sheath around the stem, the sheath (1) 2–3 cm long, greenish, persistent but occasionally bending back from the leaf, the upper portion confluent across the top of the sheath forming a short hyaline ligule; peduncles very slender, 1–15 (25) cm long; spikes moniliform, 1–3 cm long, with 2–6 usually unequally spaced whorls of flowers, the lower ones up to 1 cm apart, the upper ones more crowded; flowers small, the tepals brownish-green, 1–1.5 mm long, the style very short; fruits obliquely ovoid, 2.5–4.2 mm long, 2–3.5 mm wide, the dorsal keel low and rounded, the two lateral keels inconspicuous, the fruiting stigma broad, on a slender, slightly curved, persistent style near the ventral margin; $2n = 42, 78$.

Shallow ponds, streams, lakes and ditches; Eurasia, Africa, Australia and N. Amer.; Alaska to Newfl., s. to Calif., Nev., Ariz., Colo., Fla. and e. U.S.; common in our area. July–Sept.

Of the first three species mentioned here, this is the most common. It is equally common throughout much of the world and serves as a major food source for several species of wild birds. As with other widely distributed species of *Potamogeton, P. pectinatus* is frequently divided into numerous minor variants whose taxonomic usefulness is highly doubtful. Fortunately the several names applied to various elements of this species found in the eastern United States have not found their way into our botanical literature.

4. Potamogeton robbinsii Oakes

Potamogeton robbinsii Oakes, Mag. Hort. Bot. 7: 180. 1841. *Spirillus robbinsii* Nieuwl. Amer. Midl. Naturalist 3: 19. 1913. ("In Pondicherry Pond, Jefferson [Gloucester Co.], New Hampshire.")

Plants wholly submersed, the stem stout and usually much branched, greenish, terete, 1–5 (20) dm long, arising from a long slender creeping rhizome, not tuberous; leaves all submersed, the blade linear to lanceolate, 3–7 (12) cm long, 3–4 (8) mm wide, 20- to 60-nerved, the apex acute, auriculate at the base, the margin finely serrulate, stiffly 2-ranked and crowded on the stem; stipules adnate to the base of the leaf and forming a white clasping sheath around the stem, the sheath 0.5–1.5 (2) cm long, persistent and rarely bending back, the upper portion confluent across the top of the sheath forming a long hyaline ligule, this ultimately shredding into white fibers; peduncles slender, 3–5 (7) cm long; spikes stiff, interrupted, 0.7–2 cm long, few flowered, often in pairs at a single node; flowers small, the tepals reddish to greenish-yellow, 1–1.5 (1.8) mm long, the style short; fruits extremely rare and not known to occur in plants of our area, obliquely obovoid, 3.5–4 (5) mm long, 2–2.8 (3.3) mm wide, the dorsal keel sharp and prominent, the two lateral keels low and rounded, the fruiting stigma flattened on a centrally placed curved style 0.7–0.9 mm long; $n = 26$.

Slow-running deep watered streams, ponds and lakes; B.C. to Lab., s. to c. Calif., Utah, Mont., Wyo., n. Ind. and Pa.; infrequent in our area, approaching our area in s. Idaho and known from it only in the Wasatch Mountains Sect. of ne. Utah. Aug.–Sept.

5. Potamogeton crispus L.

Potamogeton crispus L. Sp. Pl. 126. 1753. *Buccaferrea crispata* Bubani, Fl. Pyrén. 4: 13. 1901. (Europe.)

Plants wholly submersed, the stem stout and freely branching above, often forming mats, greenish, somewhat flattened with channelled sides, 4–8 (15) dm long, arising from a slender creeping rhizome;

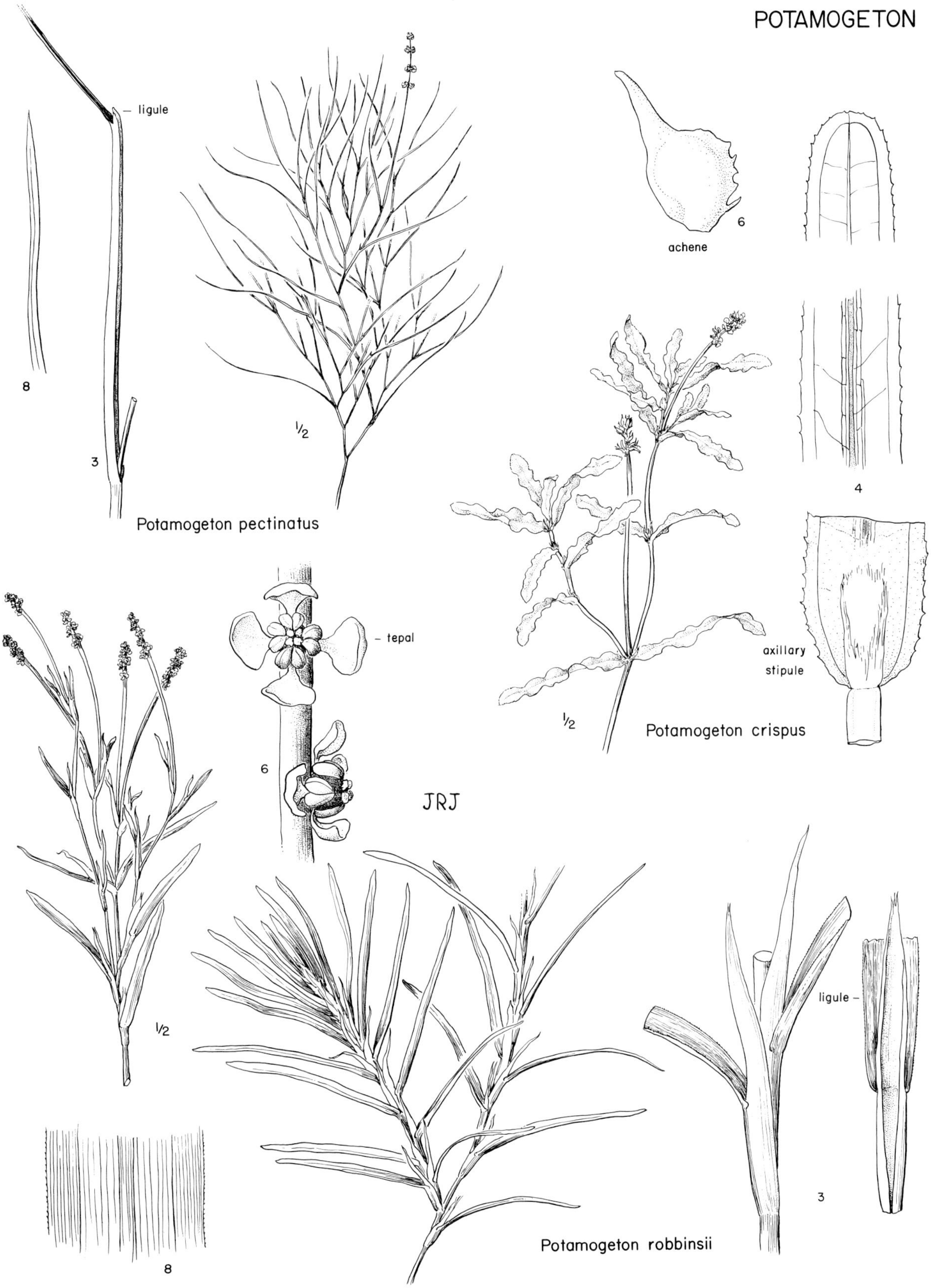
POTAMOGETON
ligule
8
3
1/2
Potamogeton pectinatus
achene
6
4
axillary
stipule
1/2
Potamogeton crispus
tepal
6
JRJ
1/2
8
ligule
3
Potamogeton robbinsii

leaves all submersed, the blade oblong, 3–8 cm long, (3) 5–10 (12) mm wide, 3- to 5- (7)-nerved, the apex rounded, the margin usually strongly crispate, finely serrulate especially near the apex, the base sessile or more or less clasping, attached directly to the stem; stipules axillary, delicate, becoming fibrillose early, 0.3–0.8 cm long, whitish, often united around the stem to form a sheath, becoming free and bending back in some, the sheath open on the adaxial surface, free of the blade; peduncles stout, 3–5 (7) cm long; spikes few flowered, 1–2 cm long; flowers large, the tepals yellowish, 1.8–2.5 (3) mm long, the style long; fruits ovoid, 5–6 mm long including the stout, erect or somewhat curved beak, this 2–3 mm long, the dorsal keel prominent, entire or slightly denticulate, the two lateral keels low and rounded to somewhat obscure; $2n$ = 36, 42, 50, 52, 72 and 78.

Ponds, streams and ditches; widely introduced throughout the world from Europe; in much of the e. U.S. and s. Can., infrequently found in Wash., Oregon, Calif., w. Nev. and n. Utah. June–Aug.

This European species has been introduced and naturalized in northern Utah and near Reno in western Nevada at gun clubs and game preserves. It is to be expected in other parts of our area in similar situations.

6. **Potamogeton zosteriformis** Fern.

Potamogeton zosterifolius var. *minor* Hook. Fl. Boreali-Amer. 2: 172. 1838. (*Richardson,* Portage la Loch, and *Drummond,* York Factory, Hudson Bay, Can., are both cited.)

P. zosterifolius var. *americanus* Benn. Trans. & Proc. Bot. Soc. Edinburgh 29: 46. 1924. (*Chase 1283,* Wolf Lake, Ind., 3 Sept. 1900.)

P. zosteriformis Fern. Mem. Amer. Acad. Arts, II. 17: 36. 1932. *P. zosterifolius* subsp. *zosteriformis* Hultén, Kongl. Svenska Vetenskapsakad. Handl. IV. 8(5): 168. 1962. (*C. A. Davis,* Alma, Mich., 12 July 1893.)

P. compressus sensu Amer. authors; not L. 1753.

P. zosterifolius sensu Amer. references; not Schumacher, 1801.

Plants wholly submersed, the stem slender and freely branching, greenish, flattened and winged, constricted at the nodes, 3–6 dm long, arising from a slender creeping rhizome; leaves all submersed, the blade linear to broadly linear, 5–20 cm long, (2) 3–5 mm wide, up to 35 fine nerves and 3 larger ones, the apex obtuse to acute or mucronate, the base sessile, attached directly to the stem, not at all clasping, the margin entire; stipules axillary, firm, fibrillose, 1–3 (4) cm long, greenish, clasping the stem at first but soon bending back, not forming a true sheath, open on the adaxial surface, free of the blade; peduncles slender to stoutish, flattened, 3–10 cm long; spikes compressed, 1.5–3 cm long, with several whorls of flowers; flowers small, the tepals greenish, 2–2.5 mm long, the style short; fruits obliquely oblong-elliptic, 4–5 (5.5) mm long, 3–3.5 mm wide, the dorsal keel sharp, undulate or denticulate, the two lateral keels obscure, the fruiting stigma flattened on a slightly curved persistent beak less than 1 mm long, positioned slightly to one side of center.

Ponds, slow-moving streams and lakes; B.C. to Que., s. to c. Calif., Idaho, Nebr., Ind., W.Va. and Va.; infrequent in sc. Oregon, ne. Calif. and e. Idaho, mostly, but not entirely restricted to the Lake Sect. July–Sept.

This species is part of a Northern Hemisphere complex that remains to be studied in detail. Our plants are only weakly separated from the European *P. zosterifolius* Schumacher (1801), which in turn seems to be merely a phase of *P. compressus* L. (1753), or even a synonym of the Linnaean species (i.e., vide Ohwi, Fl. Jap. 122. 1965). As the differences between these plants remain uncertain, it seems unwise to suggest any modification of the present nomenclature. It appears that our plants should not be specifically distinguished from the European plants, but until the European taxonomy is solved, what our plants should be called must remain in doubt.

7. **Potamogeton foliosus** Raf.

Potamogeton gramineus sensu Michx. [as *gramineum*?] Fl. Boreali-Amer. 1: 102. 1803; not L. 1753. *P. foliosus* Raf. Med. Repos. II. 5: 354. 1808. *P. pauciflorus* Pursh, Fl. Amer. Sept. 1: 121. 1814. *Spirillus foliosus* Lunell, Amer. Midl. Naturalist 4: 161. 1915. (*Michaux,* "in rivis affluente mari inundatis Carolinae inferioris.")

P. niagarensis Tuckerman, Amer. J. Sci. Arts, II. 7: 354. 1849. *P. pauciflorus* var. *niagarensis* J. W. Robbins in A. Gray, Manual Bot. ed. 2. 435. 1856. *P. foliosus* var. *niagarensis* Morong, Mem. Torrey Bot. Club 3(2): 39. 1893. *Spirillus foliosus* var. *niagarensis* Nieuwl. Amer. Midl. Naturalist 3: 18. 1913. *P. foliosus* f. *niagarensis* Hagström, Kongl. Svenska Vetenskapsakad. Handl. 55(5): 91. 1916. (*Tuckerman,* "Niagara falls, near the brink of the Hog-back," Niagara Co., N.Y., 1848.) = var. *foliosus.*

P. pauciflorus var. *californicus* Morong, Bot. Gaz. 10: 254. 1885. *P. foliosus* var. *californicus* Morong, Mem. Torrey Bot. Club 3(2): 40. 1893. *P. californicus* Piper, Contr. U.S. Natl. Herb. 11: 98. 1906. *P. foliosus* f. *californicus* Hagström, Kongl. Svenska Vetenskapsakad. Handl. 55(5): 91. 1916. (*Parish & Parish s.n.,* San Diego, Calif., May 1881, not *Parish & Parish 940,* San Bernardino, Calif., as indicated by Haynes.) = var. *foliosus.*

P. foliosus var. *macellus* Fern. Mem. Amer. Acad. Arts, II. 17: 46. 1932. (*Faxon & Faxon s.n.,* Cambridge, Mass., 23 July 1880.) = var. *foliosus.*

P. fibrillosus Fern. Mem. Amer. Acad. Arts, II. 17: 51. 1932. *P. foliosus* var. *fibrillosus* Haynes & Reveal, Rhodora 75: 76. 1973. (*Cusick 2598,* margin of Harney Valley, "P" Ranch, Harney Co., Oregon, 22 June 1901.)

Plants wholly submersed, the stem slender and freely branching, greenish, slightly compressed and obscurely winged, constricted at the nodes, 2–10 dm long, arising from a dense mat of slender rhizomes; leaves all submersed, the blade narrowly linear, 1–10 cm long, (0.3) 1–1.5 (2.5) mm wide, (1-) 3- to 5-nerved, the apex acute to cuspidate, tapering and sessile at the base, attached directly to the stem, often without basal glands; stipules connate into a tubular sheath, the sheath 0.5–1.5 cm long, this soon rupturing and deciduous, fibrillose, the fibers persistent in some, open on the adaxial surface near the apex, free of the blade; peduncles rather stout, 0.3–1.5 (2) cm long; spikes subglobose to subcapitate, congested, 0.1–0.5 cm long, the few flowers in 3–5 whorls of paired flowers at each node; flowers minute, the tepals greenish, 0.5–1 mm long, the style short; fruits obliquely obovoid, 1.5–2 (2.5) mm long, 1–2 mm wide, the dorsal keel prominent, undulate-dentate, the two lateral keels obscure, the fruiting stigma flattened on a short curved subterminal persistent style 0.1–0.4 mm long; $2n$ = 28.

Shallow, often warm water ponds, lakes, springs and slow-moving streams; B.C. to Que., s. to Calif., Fla., Mex. and C. Amer., W.I., and Hawaii; fairly common in se. Oregon, ne. Calif., n. Nev., s. Idaho and n. Utah. July–Sept.

For the record, I must note that it has become traditional in recent years to accept the place of publication for *P. foliosus* as Rafinesque's letter published in 1808 in spite of the fact that he

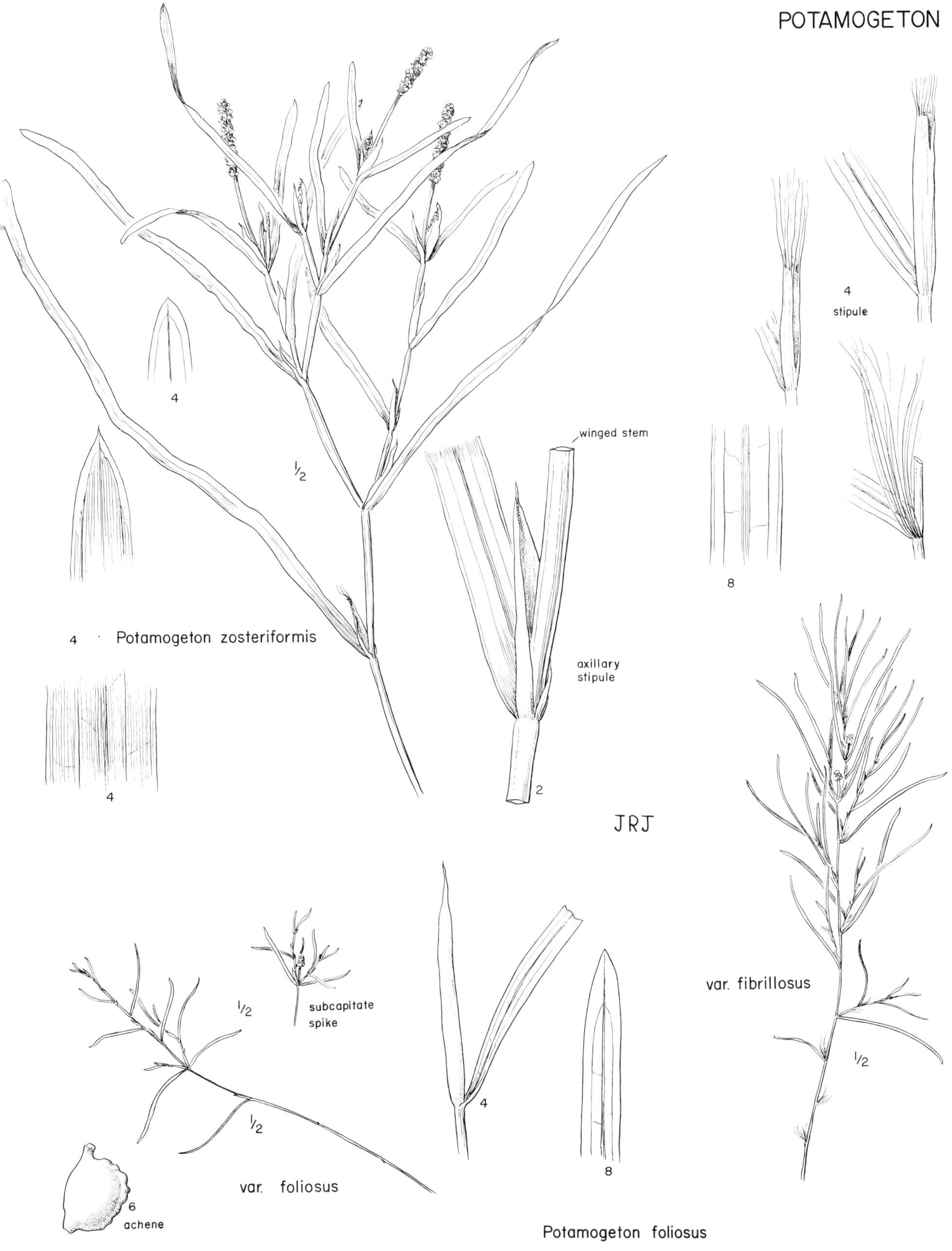
POTAMOGETON
4
½
4
4 · Potamogeton zosteriformis
4
winged stem
axillary stipule
2
4
stipule
8
JRJ
var. fibrillosus
½
½
subcapitate spike
½
var. foliosus
6
achene
4
8
Potamogeton foliosus

POTAMOGETON

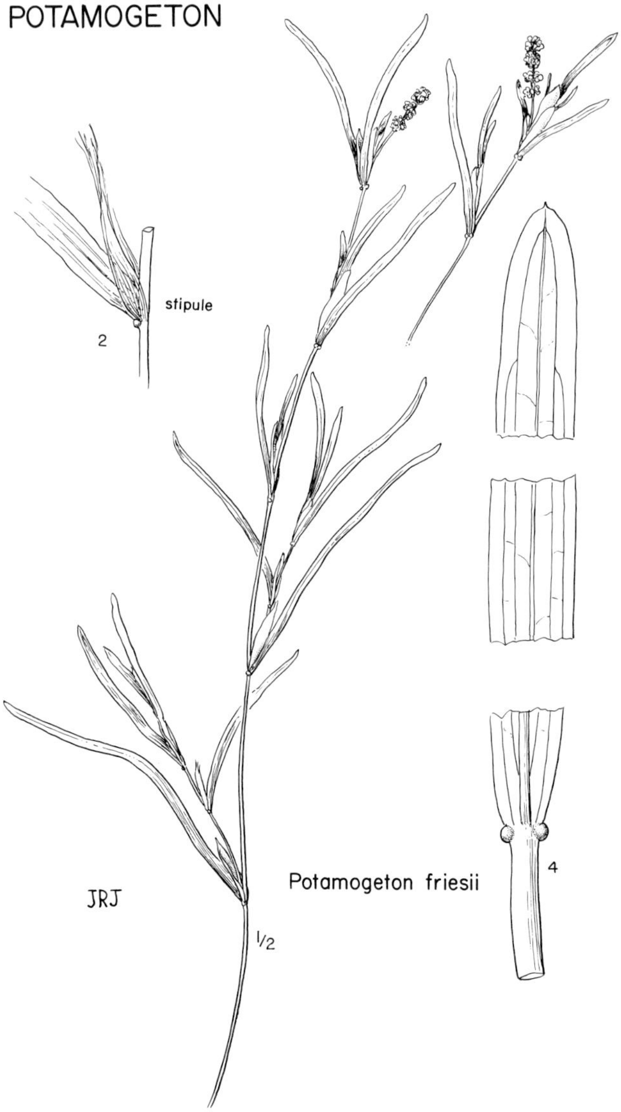

Potamogeton friesii

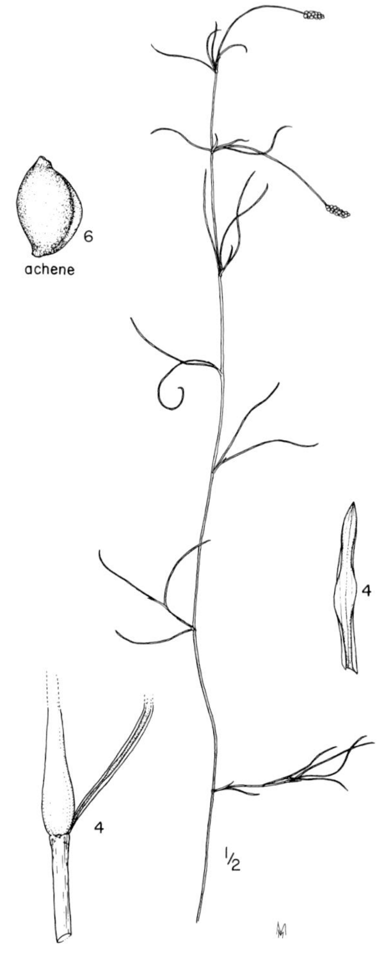

Potamogeton strictifolius

clearly stated that all of the names in the article are provisional, and thus, according to the Code, would not be validly published therein. This is clearly indicated by Rafinesque's later publication in 1811 when he did publish *P. foliosus* (as *P. foliorum*), and where he did indicate that he was publishing the names as new. For this Flora, the tradition is followed, but some significant nomenclatural changes may have to be made in the future, or at least the place of publication altered.

Fernald proposed two variants of *P. foliosus* which differed in a series of minor, overlapping features. In our area, and throughout the range of the species, this distinction is arbitrary and often artificial. The var. *macellus* would seem to be our common phase, with var. *foliosus* occurring mainly in southeastern Oregon and eastern California. Plants from northern Utah can be associated with var. *foliosus,* but those of the central part of the state can not be placed in either extreme. As this type of confusion exists all over, the var. *macellus* is reduced to synonymy under var. *foliosus.*

At the same time, Fernald proposed *P. fibrillosus,* which has since been shown by Haynes and Reveal (1973) to be a mere variant of *P. foliosus,* differing mainly in the nature of the sheath. The two variants can be separated as follows:

1 Stipules delicate, tending to completely disintegrate, the fibers not persistent; fruiting beaks 0.2–0.4 mm long; inflorescence not interrupted; glands not present at the base of the leaves; common across the n. half of the Region . . . . var. *foliosus*

1 Stipules firm, strongly fibrous, the fibers free at the tip of the stipule, becoming longer as the membranous part of the stipule disintegrates, the fibers soon becoming free, persistent; beaks shorter or lacking; inflorescence interrupted; glands usually present at the base of the leaves; nw. Calif., se. Oregon, sw. Wash., e. across s. Idaho and n. Utah to Yellowstone Natl. Park, Wyo. . . . . var. *fibrillosus* (Fern.) Haynes & Reveal

8. **Potamogeton friesii** Rupr.

Potamogeton friesii Rupr. Beitr. Pflanzenk. Russ. Reiches 4: 43. 1845. *P. pusillus* subsp. *friesii* Hook. f. Stud. Fl. Brit. Islands, ed. 3. 435. 1884. *Spirillus friesii* Nieuwl. Amer. Midl. Naturalist 3: 17. 1913. (St. Petersburg, Russia.)

Plants wholly submersed, the stem slender and simple or nearly so below, freely branching above, greenish, compressed, not constricted at the nodes, 3–10 dm long, arising from a short slender rhizome; leaves all submersed, the blade narrowly linear, 3–8 cm long, 1.5–2.5 (3.5) mm wide, (3-) 5- (7)-nerved,

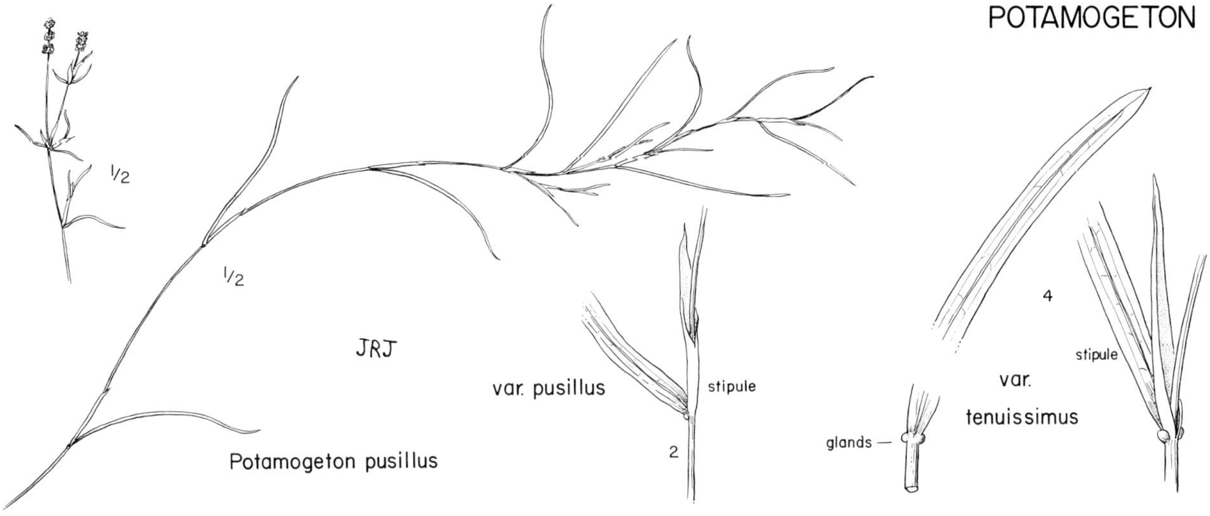

the apex rounded but with a short mucronate tip, tapering and sessile at the base, attached directly to the stem with two yellowish glands just below the point of attachment; stipules at first connate at the base but quickly rupturing, not forming a distinct sheath, 0.7–1.1 (2) cm long, white, strongly fibrillose and tardily shredding, often 2-cleft, free of the blade; peduncles slightly flattened and somewhat winged, 1.5–4 (5) cm long; spikes congested, (0.7) 1–2 cm long, the few flowers in 3–4 whorls of 2–3 flowers each; flowers small, the tepals greenish, 1.5–2.5 mm long, the style short; fruits obliquely ovoid to obovoid, (1.5) 2–2.5 mm long, 1–2 mm wide, the dorsal keel low and rounded, the two lateral keels obscure, the fruiting stigma flattened on a slightly curved persistent beak 0.3–0.6 (0.8) mm long.

In shallow pools and lakes; Eurasia and N. Amer.; Alaska to Newfl., s. to Wash., n. Utah, S.D., Ind. and Va.; infrequently found in our area, known only from Mud Lake, Jefferson Co., Idaho, in Summit Co. and in Fish Lake, Sevier Co., Utah. July–Aug.

9. Potamogeton strictifolius Benn.

Potamogeton strictifolius Benn. J. Bot. 40: 148. 1902. (*Hill 146,* Wolf Lake, Hammond, Lake Co., Ind., 3 Sept. 1900; lectotype by Fernald.)

P. strictifolius var. *rutiloides* Fern. Mem. Amer. Acad. Arts, II. 17: 57. 1932. *P. pusillus* var. *rutiloides* B. Boivin, Naturaliste Canad. 94: 527. 1967. (*Metcalf 1453,* Molly Lake, Crow Wing Co., Minn., 26 Aug. 1921.)

P. rutilus sensu Amer. authors; not Wolfg. 1827.

P. pusillus var. *vulgaris* sensu J. W. Robbins in King, Rep. Geol. Explor. 40th Parallel 5: 338. 1871; not Fries, 1828.

Plants wholly submersed, the stem slender and simple or sparsely branched below, slightly more so above, greenish, somewhat compressed, 1–10 dm long, arising from a short slender rhizome; leaves all submersed, the blade narrowly linear, 2–7 cm long, 0.5–2 (2.5) mm wide, firm, rigid in some, 3-nerved, the apex obtuse or abruptly mucronate, tapering and sessile at the base, attached directly to the stem, without basal glands; stipules connate into a tubular sheath, the sheath 0.8–2 cm long, this soon rupturing and deciduous, membranous, somewhat fibrillose, open on the adaxial side near the apex, free of the blade; peduncles filiform, 1–9 cm long; spikes interrupted, 0.8–1.5 cm long, with 3–4 remote whorls of few flowers; flowers small, the tepals greenish, 1.3–1.8 mm long, the style short; fruits obliquely obovoid, 2–3 mm long, 1.5–2 mm wide, the dorsal keel prominent, low and rounded, the two lateral keels obscure, the fruiting stigma flattened on a short, curved to slightly erect persistent style 0.3–0.7 mm long, positioned on the ventral margin.

Slow-moving streams and lakes; e. Can. and ne. U.S., s. to Pa., Ind. and Nebr. with an outlying population in n. Utah along the Bear R. July–Sept.

Our plant is dubiously called var. *rutiloides* Fern. which differs from the typical phase in the generally narrower leaves (0.5–1.2 mm wide) which are slightly more rigid. As these features are totally intergrading throughout the range of the species, it seems unwise to maintain the supposed distinction of our plants. To date, this species is known from a single collection in our area, taken by Watson along the Bear River in 1868.

10. Potamogeton pusillus L.

Potamogeton pusillus L. Sp. Pl. 127. 1753. *Buccaferrea pusilla* Bubani, Fl. Pyrén. 4: 16. 1901. *Spirillus pusillus* Nieuwl. Amer. Midl. Naturalist 3: 18. 1913. (Europe.)

P. pusillus var. *tenuissimus* Mert. & Koch in Röhling, Deutschl. Fl. ed. 3. 1: 857. 1823. *P. tenuissimus* Reichenb. Icon. Fl. German. & Helv. 7: 14. 1845. *P. pusillus* f. *tenuissimus* Hagström, Kongl. Svenska Vetenskapsakad. Handl. 55(5): 122. 1916. *P. berchtoldii* var. *tenuissimus* Fern. & Schubert, Rhodora 42: 246. 1940. *P. berchtoldii* subsp. *tenuissimus* Calder & Taylor, Canad. J. Bot. 43: 1388. 1965. (Germany.)

P. berchtoldii Fieber in Bercht. Oekon.-Techn. Fl. Boehm. 2: 277. 1838. *P. pusillus* var. *berchtoldii* Hagström, Kongl. Svenska Vetenskapsakad. Handl. 55(5): 124. 1916. (Czechoslovakia.) = var. *tenuissimus.*

P. berchtoldii var. *mucronatus* Fieber in Bercht. Oekon.-Techn. Fl. Boehm. 2: 277. 1838. *P. pusillus* var. *mucronatus* P. Graebner, Pflanzenr. IV. 11 (Heft 31): 115. 1907. (*Konrad,* Tepf [= Tepfenhand], Germany.) = var. *tenuissimus.*

P. panormitanus Biv. Nuove piante inedite 6. 1838. *P. pusillus* var. *panormitanus* Morong, Mem. Torrey Bot. Club 3(2): 46. 1893. *P. pusillus* proles *panormitanus* P. Graebner, Pflanzenr. IV. 11 (Heft 31): 116. 1907. ("Habitat Panormi *al fiume Oreto dietro il ponte della Grazia, al Gurgo di Rebuttone; Catanae.*") = var. *pusillus.*

P. panormitanus var. *minor* Biv. Nuove piante inedite 6. 1838. *P. pusillus* var. *cuspidatus* G. Fischer, Ber. Bayer. Bot. Ges. 11: 116. 1907, a new name for var. *minor,* and not a new taxon as proposed by Haynes. *P. pusillus* var. *minor* Fern.

& Schubert, Rhodora 50: 154. 1948. (No location given, but likely "Panormi.") = var. *pusillus.*

P. pusillus var. *vulgaris* subvar. *interruptus* J. W. Robbins in King, Rep. Geol. Explor. 40th Parallel 5: 338. 1871. (*Watson 1137,* Silver Cr., Parley's Peak area, Wasatch Mts., Summit Co., Utah, Aug. 1868.) = var. *pusillus.*

Plants wholly submersed, the stem very slender to capillary and usually much branched, more so above than below, greenish, terete or nearly so, 2–10 dm long, arising from a slender creeping rhizome; leaves all submersed, the blade narrowly linear, (0.8) 1–6 (7) cm long, (0.3) 0.5–2 mm wide, firm or nearly so, 3-nerved, the apex acute or obtuse, infrequently mucronate, tapering and sessile at the base, attached directly to the stem with two small globose glands just below the point of attachment; stipules connate at least half their length but soon rupturing and deciduous, or not connate at all, the sheath (when present as in our plants) 0.3–0.8 cm long with an additional free portion 0.4–0.8 cm long, somewhat fibrillose at the tip, open on the adaxial surface, free from the blade; peduncles very slender to filiform, axillary, (0.4) 1–5 (8) cm long; spikes somewhat congested, 0.6–1.2 (1.5) cm long with 1–3 (5) whorls of 2–3 flowers; flowers small, the tepals greenish, 1–2 mm long, the style very short; fruits obliquely obovoid, 2–2.5 (2.8) mm long, 1–1.9 mm wide, the dorsal keel obscure or lacking, the two lateral keels essentially lacking, the fruiting stigma flattened on a short curved persistent beak 0.3–0.5 mm long; $2n = 26$.

Shallow pools, lakes and slow-moving streams; Eurasia and N. Amer.; throughout most of s. Can., U.S. and n. Mex.; se. Oregon, s. Idaho, e. Calif., n. and w. Nev. and nearly all of Utah. July–Aug.

In Haynes's recent treatment of *Potamogeton pusillus,* he reduces the large number of species and varieties proposed mainly by Fernald in his 1932 monograph. During the present study of the genus for this Flora, the entire complex in North America was reviewed, and I can find no justification for separating *P. berchtoldii* from *P. pusillus*—a point also arrived at through the independent studies of Haynes (per. comm.; 1975). We have two phases of *P. pusillus* in our area, the common var. *pusillus,* and the rarely found var. *tenuissimus.* The two varieties may be distinguished by the following key.

1 Stipules connate on young plants or near the apex of older plants with newly formed leaves; peduncles 1.5–5 (8) cm long, 1–3 per plant; mature achenes with concave sides and a central placed beak; common throughout our area .... var. *pusillus*

1 Stipules not connate, mostly convolute; peduncles 0.4–3 cm long, usually more than 3 per plant; mature achenes with rounded sides and a forwardly placed beak; known currently only from ne. Nev. and e.-central Calif. .... var. *tenuissimus* Mert. & Koch

11. Potamogeton diversifolius Raf.

Potamogeton hybridum sensu Michx. Fl. Boreali-Amer. 1: 101. 1803; not Petagna, 1787. *P. diversifolius* Raf. Med. Repos. II. 5: 354. 1808. *P. diversifolius* var. *capitatus* Engelm. Amer. J. Sci. Arts 46: 102. 1844, proposed in the same sense as var. *diversifolius* would be today, not as a new variety. *Spirillus diversifolius* Nieuwl. Amer. Midl. Naturalist 3: 18. 1913. (*Michaux,* "Hab. in aquis stagnantibus Carolinae.")

P. diversifolius var. *spicatus* Engelm. Amer. J. Sci. Arts 46: 102. 1844. *P. hybridum* var. *spicatus* Engelm. in A. Gray, Manual Bot. ed. 2. 434. 1856. (*Geyer 117,* small ponds, St. Louis, Mo., May 1842.)

P. spirillus sensu western authors; not Tuckerman, 1848.

Plants partially submersed, the stem slender and freely branched, greenish, terete, 2–4 (5) dm long, arising from a creeping rhizome; leaves dimorphic, the submersed leaf-blade narrowly linear, 2–4 cm long, 0.5–1 (1.5) mm wide, thin and delicate, 1- (3)-nerved, the apex subobtuse to acute, sessile on the stipule, the floating leaf-blade elliptic to oblong-elliptic, (1) 1.5–3 (4) cm long, 0.3–1.7 (2) cm wide, firm, the apex rounded, short-petioled, sometimes not well differentiated from the submersed leaves; stipules of the submersed leaves connate with a tubular sheath around the stem, the sheath 0.2–0.6 cm long, adnate to the leaf for 0.1–0.3 (0.5) cm with an entire, erect, hyaline ligule 1–3 mm long, those of the floating leaves not connate into a sheath, (0.5) 1.5–2 (3) cm long, only slightly adnate or free, becoming fibrillose in age; spikes of two kinds, the lower submersed ones with short peduncles up to 1 cm long and congested with 1–6 flowers, the upper (and usually emersed) ones with longer peduncles (up to 1.5 cm long and more elongated) and 4–8 flowers; flowers minute, the tepals light brownish-green, 0.5–1 mm long, the style nearly lacking; fruits suborbicular to subreniform, 1–1.5 (1.8) mm long, 0.9–1.5 mm wide, the prominent dorsal keel sharp and undulate-dentate, the two lateral keels low and rounded, mostly entire, the fruiting stigma minute and capitate on a tiny persistent protuberant style 0.1–0.2 mm long.

Slow-moving streams, ponds and lakes; s. B.C., Mont. and S.D., s. to Mex. and e. to Pa., N.J. and W.Va.; se. Oregon (Steens Mts.), s. Idaho, ne. Calif. and n. Ariz. (Kaibab Plateau). July–Sept.

Potamogeton diversifolius is a clearly marked species in the western United States.

12. Potamogeton epihydrus Raf.

Potamogeton natans var. *β* Michx. Fl. Boreali-Amer. 1: 101. 1803. *P. epihydrus* Raf. Med. Repos. II. 5: 354. 1808. (*Michaux,* "in Canadae amnibus et in flumine *Ohio,*" the Ohio R. collection taken as the type by Rafinesque, Med. Repos. III. 2: 409. 1811.)

P. nuttalii [sic] Cham. & Schlecht. Linnaea 2: 226. 1827. *P. epihydrus* var. *nuttallii* Fern. Mem. Amer. Acad. Arts, II. 17: 115. 1932. *P. epihydrus* subsp. *nuttallii* Calder & Taylor, Canad. J. Bot. 43: 1388. 1965. (No collector or specimens cited, only "*P. natans*" of Nuttall is mentioned, the description is based mainly on the fruit.)

P. claytonii Tuckerman, Amer. J. Sci. Arts 45: 38. 1843. (*Tuckerman,* "Ponds and streams, Roxbury, Cambridge, Mass.")

P. nuttallii var. *ramosus* Peck, Annual Rep. New York State Mus. 47: 162. 1894. *P. epihydrus* var. *ramosus* H. D. House, New York State Mus. Bull. 254 (Ferns & Fl. Pls. New York State): 52. 1924. (*Peck,* Oswegatchie R. near Sternbergs, St. Lawrence Co., N.Y.)

Plants partially submersed, the stem slender and simple or sparsely branched, greenish, somewhat flattened but not compressed, 5–15 (20) dm long, arising from a long slender, somewhat matted rhizome; leaves dimorphic, the submersed leaf-blade linear, 5–25 cm long, (1) 3–5 (10) mm wide, thin and flaccid, 5- to 7- (13)-nerved, the apex narrowly acute, sessile on the stem, the floating leaf-blade elliptic to broadly oblong-elliptic, (2) 3–8 cm long, (0.5) 1–2 (3.5) cm wide, coriaceous, 15- to 25- (40)-nerved, the apex bluntly cuspidate or rounded, on a long flattened petiole arising from the stem well below the

POTAMOGETON

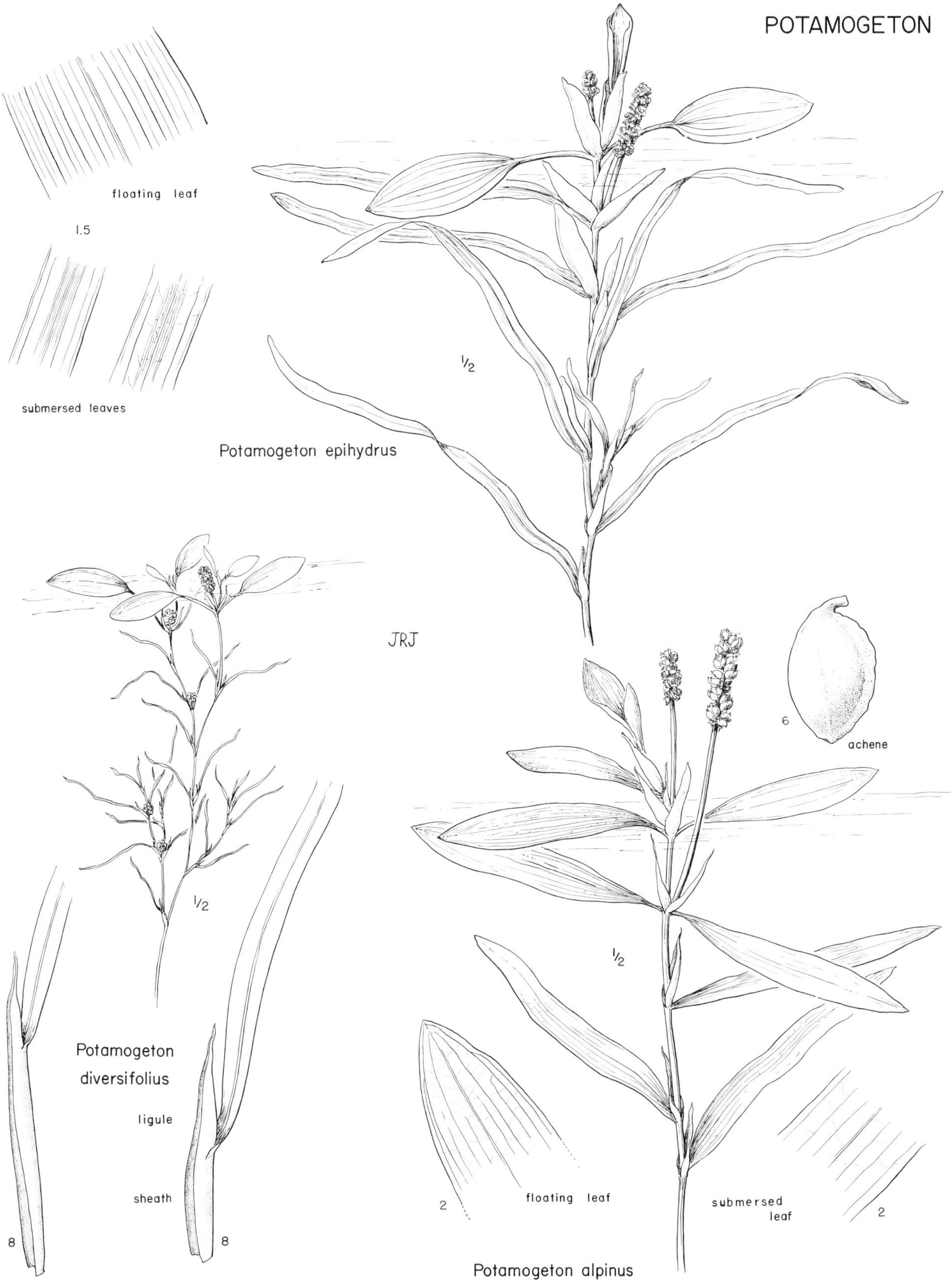

surface of the water; stipules of the submersed leaves free, not connate into a sheath, hyaline, 1–3 (4) cm long, obtuse, those of the floating leaves firmer and more attenuate; peduncles stout, (1) 3–8 (15) cm long; spikes rather congested, (1.5) 2–4 cm long, composed of numerous flowers; flowers large, the tepals greenish, 1.5–3 mm long, the style nearly lacking; fruits obliquely and broadly round-obovoid, 2–4 mm long, 2–3.5 mm wide, the surface of the achene depressed or pitted on the side, the dorsal keel prominent, sharply angled and mostly entire, the two lateral keels prominent but not sharply angled, the fruiting stigma sessile on a minute ventrally placed style; *n* = 13.

Shallow to deep water mainly in lakes and ponds; s. Alaska s. to n. Calif., e. across Idaho to Mont., s. to Colo. and from Minn. and Man. e. to Que. and Newfl., s. in the e. U.S. to Tenn. and s. Ga.; ne. Calif., w. Nev., se. Oregon and s. Idaho. June–July.

The distinction between those plants with submersed leaves 5–10 mm wide (as var. *epihydrus*) and those with narrower ones, (1) 2–8 mm wide (as var. *nuttallii*), seems hardly worth the effort, and the var. *nuttallii* is hereby reduced to synonymy.

13. Potamogeton alpinus Balb.

Potamogeton alpinus Balb. Misc. Bot. 13. 1804. (*Piotta,* Lake Chamollet, Italy.)

P. lucens sensu Michx. Fl. Boreali-Amer. 1: 101. 1803; not L. 1753. *P. tenuifolius* Raf. Med. Repos. II. 5: 354. 1808. *P. alpinus* subsp. *tenuifolius* Hultén, Fl. Aleutian Islands 65. 1937. *P. alpinus* var. *tenuifolius* Ogden, Rhodora 45: 90. 1943. (*Michaux,* Lake Mistassin, Que., Can., 1792.)

P. microstachys Wolfg. ex Schultes & Schultes, Mantissa 3: 360. 1827. *P. alpinus* proles *microstachys* P. Graebner, Pflanzenr. IV. 11 (Heft 31): 72. 1907. (*Eschscholtz,* Unalaska, 7–15 Sept. 1816.)

P. montanense Gandoger, Bull. Soc. Bot. France 66: 304. 1919. (*Rydberg & Bessey 3724,* Shoshone Lake, Yellowstone Natl. Park, Wyo., 10 Aug. 1897, originally cited as "*Rydberg n. 3724.*")

P. microstachys var. *subellipticus* Fern. Rhodora 32: 82. 1930. *P. tenuifolius* var. *subellipticus* Fern. Rhodora 33: 211. 1931. *P. alpinus* var. *subellipticus* Ogden, Rhodora 45: 94. 1943. (*Fernald, Long & St. John 6766,* Coffin I., Magdalen Islands, Can., 27 Aug. 1912.)

Plants partially submersed, the stem slender and simple or sparsely branched, usually reddish-brown or olivaceous, terete, 3–10 dm long, arising from a slender rhizome; leaves somewhat dimorphic, the submersed leaf-blade linear-lanceolate to linear-oblong or narrowly elliptic, (5) 7–15 (25) cm long, (5) 7–15 (20) mm wide, rather thin, 7- to 9-nerved, the apex tapering to rounded or broadly acutish, the base narrowing gradually and sessile on the stem, the floating leaf-blade transitional to and not greatly different from the submersed one, often lacking (or not markedly differentiated), oblanceolate to elliptic-oblanceolate or obovate, 4–6 (10) cm long, 1–2.5 (4) cm wide, 7- to 15-nerved, the apex mostly acute, narrowing gradually to a short petiole; stipules not connate or forming a sheath, (1) 1.5–2.5 (4) cm long, thin and hyaline, soon deteriorating and deciduous; peduncles slender, 7–15 cm long; spikes densely flowered and compact, (1) 1.5–3 (3.5) cm long with 5–9 whorls of flowers; flowers large, the tepals greenish-red to red, 1.5–2.5 mm long, the style short; fruits obliquely obovoid, plump, 3–3.5 mm long, 2–2.5 mm wide, the dorsal keel prominent and narrow, entire, the two lateral keels low, rounded and obscure, the fruiting stigma flattened on a subterminal curved style less than 0.6 mm long; *n* = 26.

Ponds and lakes; Eurasia and N. Amer.; Alaska to Newfl., s. to s. Calif., Utah, Colo., the Great Lakes area and the ne. U.S.; se. Oregon, e. Calif., s. Idaho, ne. Nev. and n. Utah. June–Aug.

The distinction between the populations of *P. alpinus* in Eurasia and North America has been a source of taxonomic difficulties. Both Fernald (Rhodora 32: 76–83. 1930 and 33: 209–211. 1931) and Ogden (1943) have noted the supposed differences and similarities between the two continental phases. Yet, both authors (and especially Ogden) repeatedly pointed out that the differences were minor, not consistent and could be found in some populations on both continents. How, with all their combined data, they were able to propose two variants of *P. alpinus* in North America is difficult to explain. In reviewing the North American plants, the differences between var. *tenuifolius* and var. *subellipticus* were found to be extremely dubious. Ogden reports, and his observations were confirmed during the present study, that the same type of leaf variation responsible for the two varieties in North America is found in Europe. In a small sample of plants from the Soviet Union, and in particular Siberia, similar variation was noted. However, the fine differences in fruit characters noted by Ogden seem to hold apart the extremes in Europe from the extremes in North America, but again, there is some overlap. The name, var. *subellipticus* (Fern.) Ogden, would be available for our plants if one wished to distinguish the New World *P. alpinus* from the Old World populations.

14. Potamogeton amplifolius Tuckerman

Potamogeton amplifolius Tuckerman, Amer. J. Sci. Arts, II. 6: 225. 1848. *Spirillus amplifolius* Nieuwl. Amer. Midl. Naturalist 3: 16. 1913. ("Cambridge and elsewhere, very common in New England, and I have seen it from New York," no specimens cited.)

Plants partially submersed, the stem stout and simple or branched near the top, often rufous, terete, 4–10 (12) dm long, arising from a rather stout spreading rhizome; leaves somewhat dimorphic, the submersed leaf-blade variable, the lower ones mostly lanceolate and short-lived, the upper ones broadly lanceolate to ovate, persistent, folded along the midvein and distinctly falcate in outline, the blade 8–20 cm long, 25–75 mm wide, rather thin and flaccid, the lower blades with some 20–25 nerves and the upper ones with about 25–35 nerves, the apex mostly acute, the base tapering into a distinct petiole, 1–6 cm long, this directly attached to the stem, the floating leaf-blade ovate to ovate-elliptic, 5–10 cm long, 2.5–5 cm wide, coriaceous, with 25–50 nerves, the apex rounded or blunt mucronate, the base mostly rounded, on a long stout petiole 8–20 cm long, attached to the stem well below the water surface; stipules free, not connate, 2-keeled, those of the submersed leaves subpersistent, 3.5–11 cm long, those of the floating leaves 8–20 cm long; peduncles stout, 4–15 (30) cm long; spikes congested, 2–6 (8) cm long, with numerous flowers; flowers large, the tepals greenish or with a reddish cast, 1.5–3 (3.5) mm long, the style short; fruits obliquely obovoid, 3.5–4.5 (5) mm long, 2.5–3.5 (3.7) mm wide, the dorsal keel sharply angled and prominent, the two lateral keels rounded and prominent, the fruiting stigma broadly flattened on a prominent beak 0.5–1 mm long; *n* = 26.

Slow-moving streams and lakes usually in deep water; B.C. s. to Calif. and e. to Mont., and from the Dakotas s. to Okla., Ark., Miss. and Va., common in the ne. U.S. and se. Can. to Newfl.; infrequently found in our area in e. Calif. and sw. Idaho. June–Aug.

POTAMOGETON

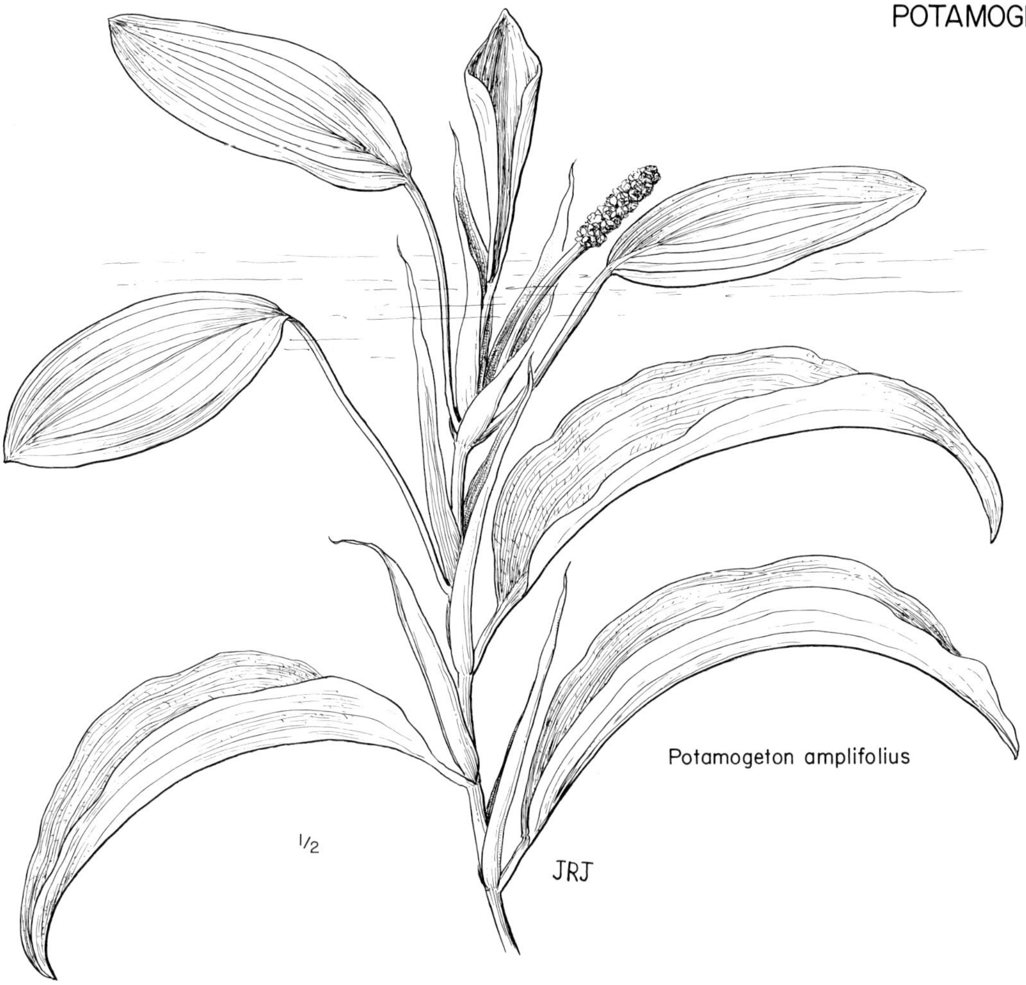

Potamogeton amplifolius

15. Potamogeton nodosus Poiret

Potamogeton fluitans Roth, Tent. Fl. Germany 1: 72. 1788. *P. natans* var. *fluitans* Cham. Adnot. 4. 1815. *P. oblongus* var. *fluitans* G. Meyer, Chloris Hanoverana 520. 1836. *P. lucens* var. *fluitans* A. Gray, Manual Bot. ed. 2. 435. 1856. ("Habitat in fossis profundis lente fleuntibus et in *Hunte* fluuio *Ducatus Oldenburgensis*.")—See comments below.

P. nodosus Poiret in Lam. Encycl. Méth. Bot. Suppl. 4: 535. 1816. (*Broussonet,* Canary Islands.)

P. occidentalis Sieber ex Cham. & Schlecht. Linnaea 2: 224. 1827. (*Sieber,* Martinique.)

P. americanus Cham. & Schlecht. Linnaea 2: 226. 1827. *P. fluitans* subsp. *americanus* P. Graebner, Pflanzenr. IV. 11 (Heft 31): 60. 1907. (No collection cited.)

P. lonchites sensu Tuckerman as emended in Amer. J. Sci. Arts, II. 7: 350. 1849 and subsequent authors, but not as originally described.

P. lonchites var. *novaeboracensis* Morong, Mem. Torrey Bot. Club 3(2): 20. 1893. *P. americanus* var. *novaeboracensis* Benn. J. Bot. 31: 297. 1893. *P. fluitans* proles *novaeboracensis* P. Graebner, Pflanzenr. IV. 11 (Heft 31): 62. 1907. (*Tuckerman,* Erie Canal near Seneca Lake, N.Y.)

P. rotundatus Hagström, Kongl. Svenska Vetenskapsakad. Handl. 55(5): 153. 1916. (*Fendler 837,* Santa Fe, N.M., 1847, is the first of four collections cited.)

Plants partially submersed, the stem stout and simple or branched near the top, greenish, subterete, 4–15 (20) dm long, arising from a rather stout spreading rhizome; leaves dimorphic, the submersed leaf-blade linear-lanceolate to lance-elliptic, 10–20 (30) cm long, 10–20 (35) mm wide, thin and flaccid, 7- to 15-nerved, the apex acute, the base tapering to a petiole 2–10 (13) cm long, the floating leaf-blade elliptic to oblong-elliptic, (3) 5–12 cm long, (1.5) 2–4 cm wide, coriaceous, mostly 10- to 20-nerved, the apex obtuse to mucronate, the base mostly cuneate or somewhat rounded in some, on a long stout petiole arising well below the surface of the water; stipules free, not connate, those of the submersed leaves delicate and soon deciduous, (3) 4–8 cm long, those of the floating leaves slightly broader; peduncles stout, (3) 5–15 cm long; spikes congested, becoming densely so in fruit, 3–6 cm long, composed of numerous flowers; flowers large, the tepals greenish or greenish-brown, 1.5–2.5 mm long, the style short; fruits obliquely obovoid, 3.5–4 mm long, 2.5–3 mm wide, the dorsal keel sharp and prominent, the two lateral keels rounded and less prominent, the fruiting stigma small and flattened on a short stout erect beak 0.3–0.5 mm long; $n = 26$.

In slow-moving streams and ditches, or in shallow to deep water of ponds and lakes; Eurasia, Africa, N. and S. Amer., W.I.; B.C. s. to Mex., and e. from Calif. to the e. U.S. and se. Can.; common in our area. June–Aug.

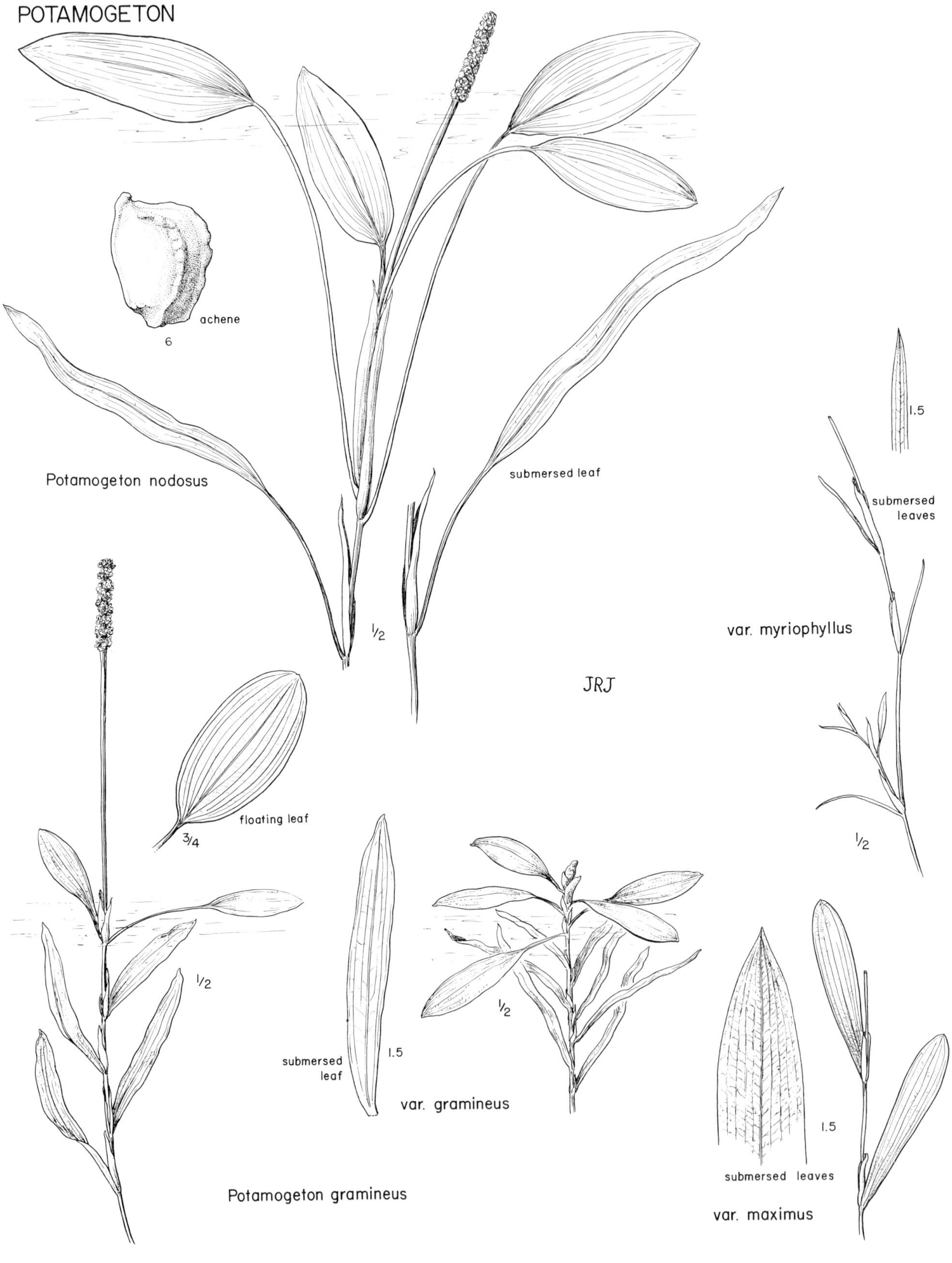
POTAMOGETON
achene
6
Potamogeton nodosus
submersed leaf
1/2
JRJ
1.5
submersed leaves
var. myriophyllus
1/2
floating leaf
3/4
1/2
submersed leaf
1.5
var. gramineus
1/2
1.5
submersed leaves
var. maximus
Potamogeton gramineus

This common and widely distributed species is in need of critical study, both to understand the nature of the variation seen in the species, and to determine the correct name for it. The precise application of the name *P. fluitans* remains to be determined. Ogden (1943) reviewed the literature on the problem, but his conclusion, and that of Hagström, that *P. fluitans* must be rejected as a nomen confusum can not stand, as authentic material does exist and it is possible to typify the name. Roth included both fertile and sterile plants in his original description; the fertile plants are what is here called *P. nodosus*, but the sterile plants represent a hybrid between *P. lucens* L. and *P. natans* L. Ogden and others have clearly shown that Roth meant to include both within his concept of *P. fluitans*, but what they failed to realize is that the Code calls for typification of such names on one of these two elements. I am reluctant to attempt such typification here in a flora, but feel certain nonetheless that *P. fluitans* should most likely be the proper name for this plant.

The exact identity of *P. nodosus* is taxonomically questioned by some authors, and rightly so, as the Canary Islands plants (which is the type locality of this species) do not conform well with the general trends in the species as a whole. Ogden (1943) noted this problem but felt that until the Canary Islands *P. nodosus* was shown *not* to be the same as the rest of the species, no change in the nomenclature should be attempted.

Past attempts to distinguish the North American plants from those in Europe have largely disappeared from our modern literature. Our material does not differ in any major way from the European plants, and the maintenance of any infraspecific taxon for our plants is unjustified.

16. Potamogeton gramineus L.

Potamogeton gramineus L. Sp. Pl. 127. 1753. (Europe.)

P. heterophyllus Schreber, Spic. Fl. Lipsicae 21. 1771. *P. proteus* var. *heterophyllus* Cham. & Schlecht. Linnaea 2: 202. 1827. *P. gramineus* var. *heterophyllus* Fries, Novit. Fl. Suec. 37. 1828. *Spirillus heterophyllus* Nieuwl. Amer. Midl. Naturalist 3: 17. 1913. ("In stagno ad Lindenthal," Germany.) = var. *gramineus*.

P. gramineus var. *graminifolius* Fries, Novit. Fl. Suec. 36. 1828. *P. heterophyllus* var. *graminifolius* Wats. & Coulter, Gray's Manual Bot. ed. 6. 561. 1890. *P. heterophyllus* f. *graminifolius* Morong, Mem. Torrey Bot. Club 3(2): 24. 1893. (Sweden.) = var. *gramineus*.

P. lonchites Tuckerman, Amer. J. Sci. Arts, II. 6: 226. 1848. *Spirillus lonchites* Nieuwl. Amer. Midl. Naturalist 3: 16. 1913. (*Tuckerman*, in the Charles R. at Newton and Natick, Mass.) = var. *maximus*.

P. gramineus var. *myriophyllus* J. W. Robbins in A. Gray, Manual Bot. ed. 5. 487. 1867. *P. heterophyllus* f. *myriophyllus* Morong, Mem. Torrey Bot. Club 3(2): 24. 1893. *P. gramineus* f. *myriophyllus* H. D. House, New York State Mus. Bull. 254 (Ferns & Fl. Pls. New York State): 54. 1924. (*Robbins*, Apponaug Pond, Apponaug, R.I., 14 Oct. 1865.)

P. gramineus var. *maximus* Morong ex Benn. J. Bot. 19: 241. 1881. *P. heterophyllus* f. *maximus* Morong, Mem. Torrey Bot. Club 3(2): 25. 1893. *P. gramineus* f. *maximus* H. D. House, New York State Mus. Bull. 254 (Ferns & Fl. Pls. New York State): 54. 1924. (*Macoun s.n.*, "Canada West.")

Plants partially submersed, the stem slender and much branched especially near the top, greenish, terete, (2) 3–8 (15) dm long, arising from a slender spreading rhizome; leaves dimorphic, the submersed leaf-blade linear to lanceolate or oblanceolate, (2) 3–9 (13) cm long, (1) 3–10 (12) mm wide, rather thin and flaccid, 3- to 9- (11)-nerved, entire but with minute fugacious, 1-celled translucent denticles, the apex acute and usually sharp-pointed, the base gradually tapering to the stem, sessile, the floating leaf-blade usually present, narrowly to broadly elliptic or oblong-elliptic, (1.5) 2–5 (7) cm long, 1–2 (2.5) cm wide, coriaceous, mostly 13- to 20-nerved, the apex obtuse or bluntly mucronate, the base cuneate to rounded, on a long stoutish petiole; stipules free and not connate, rather firm, persistent, 0.5–3 cm long; peduncles stout, 2–10 cm long, composed of numerous flowers; flowers large, the tepals greenish, 2–3 mm long, the style short; fruits obliquely obovate, 2–2.5 (2.8) mm long, 1.5–2 mm wide, the dorsal keel sharp and prominent, the two lateral keels rounded and obscure, the fruiting stigma small and flattened on a short slightly curved beak 0.3–0.5 mm long; $n = 26$.

In slow-moving streams and ditches, shallow ponds, lakes and swampy areas; Eurasia and N. Amer.; Alaska to Greenl., s. to Calif., Ariz. and Colo., and in the midw. and ne. U.S.; common in our area. June–Aug.

This species is extremely variable morphologically. It forms hybrids with just about any species of pondweed it comes in contact with, and in our area, easily blends into both *P. nodosus* and *P. illinoensis* to such an extent that clear differentiation is often quite impossible. Throughout the world-wide range of *P. gramineus*, considerable variation is found, and in North America, Ogden (1943) has proposed three varieties, all of which occur in our area. All of these variants grade into one another, and numerous intermediate populations are known. For what it is worth, the following key may be used when attempting to distinguish these dubious (but at their extremes quite distinct) variants.

1 Submersed leaves narrowly elliptic to oblanceolate, 3–9 (13) cm long, 3–12 mm wide, 5- to 11-nerved.
  2 Larger submersed leaves 3–6 cm long, 3–8 mm wide, 5- to 7-nerved; common phase in our area .... var. *gramineus*
  2 Larger submersed leaves 3–13 cm long, 6–12 mm wide, 7- to 11-nerved; known in our area only from c. Utah ... . var. *maximus* Morong ex Benn.
1 Submersed leaves linear, 2–3 (4) cm long, 1–3 mm wide, 3-nerved; a rather distinctive variant in the e. U.S., but scarcely so where it occurs in se. Oregon .... var. *myriophyllus* J. W. Robbins

17. Potamogeton illinoensis Morong

Potamogeton illinoensis Morong, Bot. Gaz. 5: 50. 1880. (*H. N. Patterson*, Mississippi R. bottoms, near Oquawka, Ill., 23 July 1873; lectotype selected here, deposited at NY.)

P. lucens sensu Amer. authors; not L. 1753.

P. zizii sensu Amer. authors; not Mert. & Koch, 1823.

Plants partially submersed, the stem slender to stoutish and simple to branched, greenish, terete, (3) 5–15 (20) dm long, arising from a spreading stoutish rhizome; leaves dimorphic, the submersed leaf-blade elliptic to oblong-elliptic or oblanceolate, (5) 10–20 cm long, (10) 20–40 (50) mm wide, rather firm, 9- to 17-nerved, entire but with minute fugacious, 1-celled translucent denticles, the apex acute and somewhat mucronate in most, the base tapering to a short petiole less than 4 cm long, the floating leaf-blade (when present) elliptic to ovate-elliptic or oblong-elliptic, (4) 6–12 (15) cm long, 2–6 cm wide, rather firm and somewhat coriaceous, mostly 15- to 25-nerved, the apex obtuse, the base cuneate or rounded, on a stoutish petiole 2–9 cm long, the uppermost leaves only gradually differentiated from the lower submersed ones; stipules free and not connate, persistent, those of the submersed leaves (2.5) 3–8 cm long, those of the floating leaves broader; peduncles stout, 4–12 (15) cm long, composed of numerous flowers; flowers large, the tepals greenish, 2.5–3.5 mm long, the style very short; fruits obliquely obovoid, 3–3.5 (4) mm long, 2–2.8 (3) mm wide, the dor-

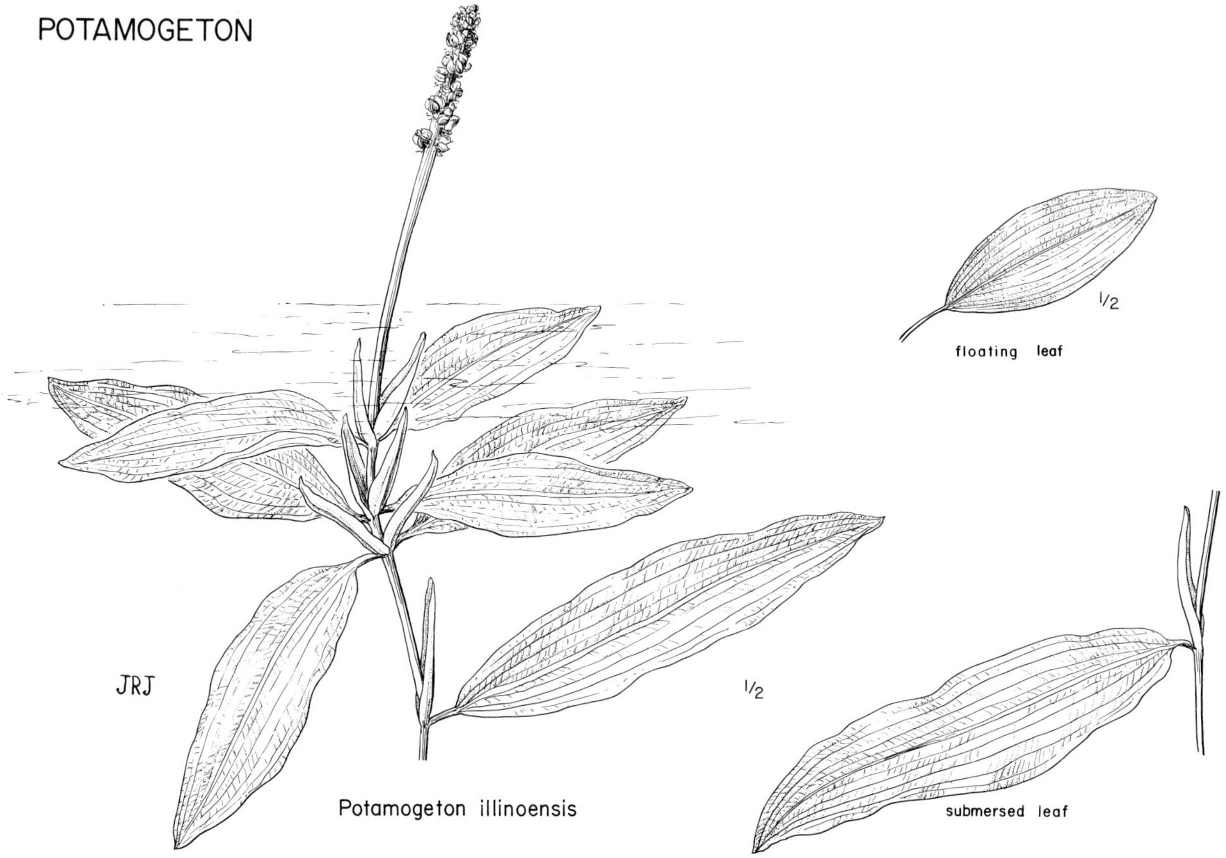

Potamogeton illinoensis

sal keel sharp and prominent, the two lateral keels rounded and obvious, the fruiting stigma rounded on the subapical beak less than 0.4 mm long; $2n = 52$.

In slow-moving streams and ditches, and in deep watered ponds and lakes; B.C. e. to Que., s. to s. Calif., Utah, N.M., Texas and most of the e. U.S.; se. Oregon, n. Nev., e.-central Calif. and most of Utah. June–Aug.

This is generally a well-marked species that occurs throughout much of our range, but it will be confused with *P. gramineus* in some locations.

18. Potamogeton natans L.

Potamogeton natans L. Sp. Pl. 126. 1753. *Buccaferrea natans* Bubani, Fl. Pyrén. 4: 11. 1901. *Spirillus natans* Nieuwl. Amer. Midl. Naturalist 3: 16. 1913. (Europe.)

Plants partially submersed, the stem slender and simple, rarely branched, greenish, terete or nearly so, (3) 5–15 (18) dm long, arising from a rather stout, spreading rhizome; leaves dimorphic, the submersed leaf-blade narrowly linear, subterete, 10–20 cm long, 1–2 mm wide, coriaceous, 3- to 5-nerved, entire, sessile on the stem, the floating leaf-blade lance-ovate to ovate-elliptic, (3) 4–10 cm long, (1.5) 2.5–5 (6.5) cm wide, coriaceous, mostly 20- to 35-nerved, the apex rounded, the base rounded or cordate, on a long, rather stout petiole; stipules free and not connate, persistent, strongly fibrous, those of the submersed leaves 4–10 cm long, those of the floating leaves nearly twice as broad and 6–11 cm long; peduncles stout, mostly 3–8 cm long; spikes densely congested, especially in fruit, 3–5 cm long, composed of numerous flowers; flowers large, the tepals greenish, 1.5–2.5 (3) mm long, the style short; fruits obliquely elliptic-obovoid, (3) 3.5–5 mm long, 2.5–3.5 mm wide, the keels lacking or the fruit merely rounded, occasionally with the dorsal keel somewhat prominent and rounded on dried specimens, the fruiting stigma rounded on a curved subterminal persistent beak 0.7–1 mm long; $n = 21$, $2n = 52$.

Shallow ponds and lakes, or in slow-moving streams where brackish; Eurasia and N. Amer.; Alaska to Newfl., s. to s. Calif., c. Ariz., n. N.M. and most of the midw. and ne. U.S.; widely scattered throughout our area. June–Aug.

19. Potamogeton praelongus Wulfen

Potamogeton praelongus Wulfen, Arch. Bot. (Leipzig) 3: 331. 1805. *Spirillus praelongus* Nieuwl. Amer. Midl. Naturalist 3: 17. 1913. (Europe.)

Plants mostly submersed, the stem stoutish and rather sparsely branched, whitish or olive-green, subterete, 10–30 dm long, the internodes usually zigzag, arising from a stoutish spreading rhizome; leaves all submersed, the leaf-blade oblong-lanceolate, (5) 10–25 (35) cm long, (10) 20–30 mm wide, firm, with 3–5 (7) prominent primary nerves and 10–20 less obvious ones, the apex rounded but cucullate so often becoming split upon pressing, the base sessile, cordate to rounded and clasping the stem; stipules free and not connate, (3) 4–10 cm long, whitish, tardily disintegrating; peduncles stout, 10–30 (60) cm long, usually extending well above the

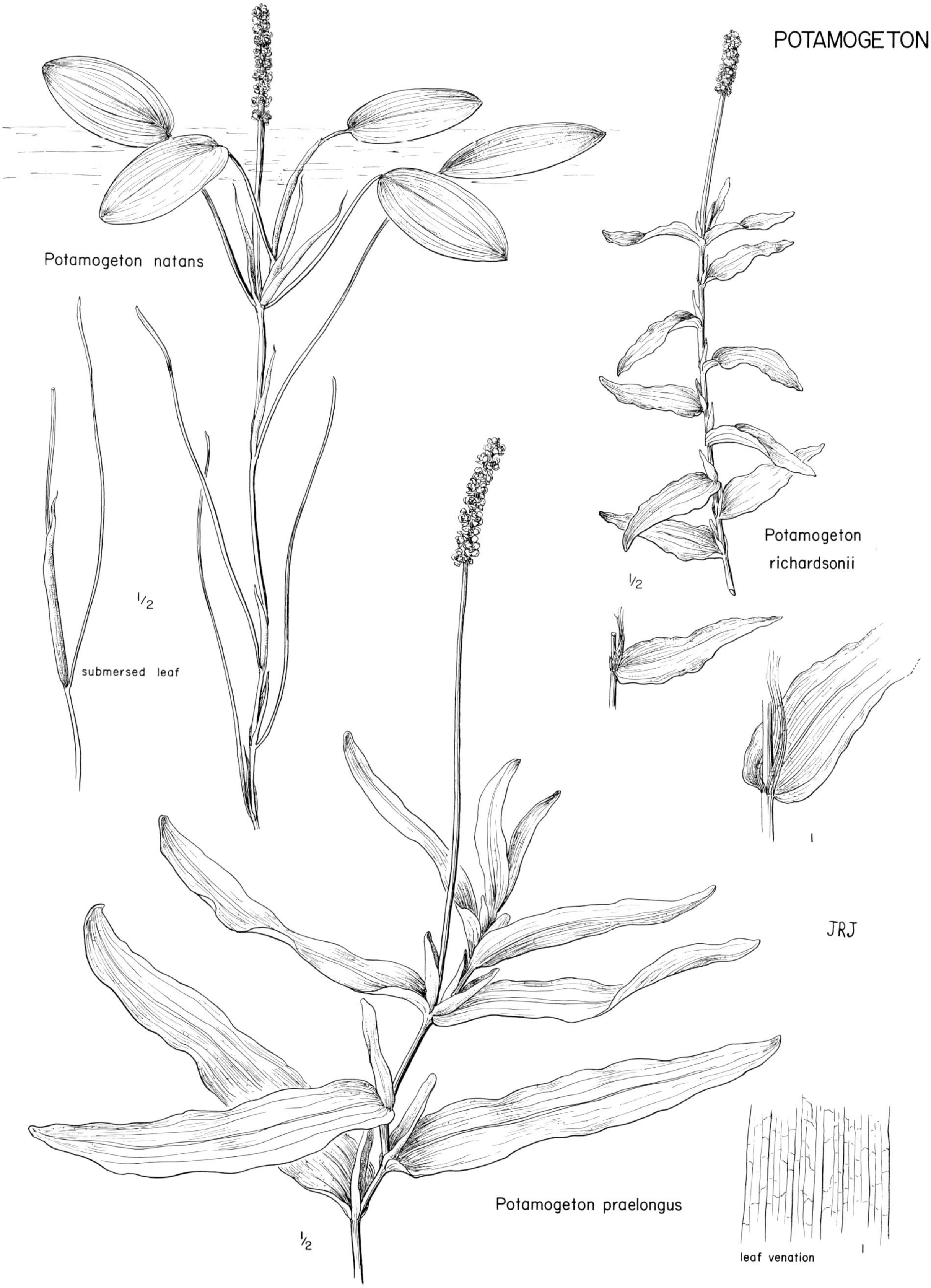
POTAMOGETON
Potamogeton natans
submersed leaf
1/2
Potamogeton
richardsonii
1/2
1
JRJ
Potamogeton praelongus
1/2
leaf venation
1

surface of the water; spikes congested, especially so in fruit, but not densely so, 3–5 cm long, composed of several flowers; flowers large, the tepals greenish, 2–4 mm long, the style short; fruits obovate, 4–5 mm long, 3–4 mm wide, the dorsal keel prominent and becoming sharp near the apex, the two lateral keels low, rounded and less obvious or lacking, the fruiting stigma round and flattened on an erect persistent beak 0.5–0.8 mm long; $2n = 52$.

Deep water in lakes and slow-moving streams; Eurasia and N. Amer.; Alaska to Newfl., s. to c. Calif., s. Utah, c. Colo., Nebr., n. Ind. and N.J.; rather common in scattered locations in Utah, s. Idaho and e.-central Calif. May–July.

20. Potamogeton richardsonii (Benn.) Rydb.

Potamogeton perfoliatus var. *lanceolatus* J. W. Robbins in A. Gray, Manual Bot. ed. 5. 488. 1867; not Blytt, 1861. *P. perfoliatus* var. *richardsonii* Benn. J. Bot. 27: 25. 1889. *P. richardsonii* Rydb. Bull. Torrey Bot. Club 32: 599. 1905. *Spirillus perfoliatus* var. *richardsonii* Nieuwl. Amer. Midl. Naturalist 3: 17. 1913. *P. perfoliatus* subsp. *richardsonii* Hultén, Fl. Alaska & Yukon 1: 102. 1941. (*Robbins*, fluvio Ontanagon, Lake Superior, 1863; lectotype selected here, deposited at NY.)

Plants mostly submersed, the stem stoutish, rather sparsely branched, greenish, terete, 3–6 (8) dm long, the internodes very rarely zigzag, arising from a slender spreading rhizome; leaves all submersed, the leaf-blade ovate-lanceolate to lanceolate, (1.5) 3–10 cm long, (5) 10–20 mm wide, firm, with slightly crisped margins, with 13–25 (30) fairly prominent nerves, the apex rounded to acute but not cucullate, the base sessile, cordate and clasping the stem; stipules free and not connate, 1–2 cm long, fibrous, soon shredding into numerous whitish fibers; peduncles stout, (1.5) 2–10 (25) cm long, extending just above the surface of the water; spikes congested, becoming slightly more so in fruit, 1.5–4 cm long, composed of several flowers; flowers large, the tepals greenish, 2.5–4 mm long, the style short; fruits obovate, 2.5–3.5 mm long, 2–3 mm wide, the dorsel keel low and rounded, the two lateral keels lacking, the fruiting stigma rounded and somewhat flattened on an erect persistent beak 0.5–0.8 mm long; $n = 26$.

In rather shallow water in slow-moving streams and lakes; Alaska to Lab., s. to c. Calif., Utah, c. Colo., Nebr., n. Ind., Pa. and N.Y.; in scattered locations in our area except for most of Nev. June–Aug.

Our species is clearly related to the coastal western European and eastern North American *P. perfoliatus* L. However, throughout the range of the two, no intermediate populations are known, and they are morphologically rather distinct.

Family **RUPPIACEAE**, the Ditch-grass Family

Submersed aquatic herbaceous perennials with filiform branching stems and largely alternate, delicate, linear leaves, these with sheathing, adnate stipules with free ligule-like tips; flowers perfect, sessile on a short spadix enclosed in sheathing leaf-bases, typically two per spike, naked, consisting of 2 stamens (occasionally with 1 stamen lacking) and 4 (rarely as many as 8) pistils, the stamens normally with one above and one below the pistil of each flower, on a short broad filament with the extrorse anther encircling about 3/4 the diam of the spadix, the pistil sessile at anthesis with a sessile or subsessile peltate-umbonate stigma, the ovary 1-celled and with a single, pendulous ovule; fruits small, drupe-like, indehiscent, ovoid and pointed, borne on a slender stipe that continually elongates as the fruit matures in most, the stipe often twisted, the peduncle also becoming much elongated after anthesis and often spirally coiled at maturity; seeds exalbuminous.

Only one genus.

1. RUPPIA L. Ditch-grass

A single complex and variable species that is often divided into several poorly defined and overlapping species, found in brackish and saline water throughout the temp. and subtrop. regions of the world. (Named for Heinrich Bernhard Ruppias [or Ruppius or Rupp], 1688–1719, a German botanist and the author of the Flora Jenensis.)

References:

Fernald, M. L., and K. M. Wiegand. The genus *Ruppia* in eastern North America. Rhodora 16: 119–127. 1914.
Setchell, W. A. The genus *Ruppia*. Proc. Calif. Acad. Sci. IV. 25: 469–478. 1946.

1. Ruppia maritima L.

Ruppia maritima L. Sp. Pl. 127. 1753. *Buccaferrea maritima* Lunell, Amer. Midl. Naturalist 4: 162. 1915. (Europe.)

R. maritima var. *rostrata* Agardh, Physiogr. Sällsk. Årsberätt. 6: 37. 1823. *R. rostellata* Koch ex Reichenb. Iconogr. Bot. 2: 66. 1824. *R. maritima* subsp. *rostellata* Aschers. & Graebner, Syn. Mitteleur. Fl. 1: 356. 1897. (No specimens cited, but presumably from Sweden.)

R. spiralis L. ex Dumort. Fl. Belg. 1: 164. 1827. *R. maritima* var. *spiralis* Moris, Stirpium Sardoarum Elenchus 1: 43. 1827. *R. maritima* subsp. *spiralis* Aschers. & Graebner, Syn. Mitteleur. Fl. 1: 356. 1897. (*Hope*, probably England.)

R. obliqua Schur ex Griseb. & Schenk, Arch. Naturgesch. 18: 355. 1852. *R. maritima* var. *obliqua* Aschers. & Graebner, Syn. Mitteleur. Fl. 1: 357. 1897 (Salzburg.)

R. intermedia Thed. Bot. Not. 1887: 83. 1887. *R. brachypus* subsp. *intermedia* Schlegel in Hartman's Handb. Skand. Fl. ed. 12. 57. 1889. *R. maritima* var. *intermedia* Aschers. & Graebner, Syn. Mitteleur. Fl. 1: 358. 1897. (*Sandén*, Södermanlands, Sweden.)

R. occidentalis S. Wats. Proc. Amer. Acad. Arts 25: 138. 1890. *R. maritima* var. *occidentalis* P. Graebner, Pflanzenr. IV. 11 (Heft 31): 145. 1907. (*Macoun s.n.*, near Kamloops, B.C., Can., June 1889.)

R. lacustris Macoun, Cat. Canad. Pl. 2(5): 372. 1890. (*Macoun s.n.*, near Kamloops, B.C., Can., June 1889, typonym of *R. occidentalis*.)

R. curvicarpa A. Nels. Bull. Torrey Bot. Club 26: 122. 1899. *R. maritima* proles *curvicarpa* P. Graebner, Pflanzenr. IV. 11 (Heft 31): 144. 1907. *R. maritima* var. *curvicarpa* Fern. & Wieg. Rhodora 16: 124. 1914. (*A. Nelson 2821*, Alkali Lakes, west of Laramie, Albany Co., Wyo., 24 Oct. 1896.)

R. pectinata Rydb. Mem. New York Bot. Gard. 1: 18. 1900. *R. maritima* f. *pectinata* Hagström, Bot. Not. 1911: 138. 1911. (*Tweedy 229*, Yellowstone Natl. Park, Wyo., in 1884.)

R. maritima var. *longipes* Hagström, Bot. Not. 1911: 138. 1911. *R. cirrhosa* subsp. *longipes* A. Löve, Taxon 19: 298. 1970. (*O. Poulsen*, Buchara.)

Plants wholly submersed, the stem capillary to

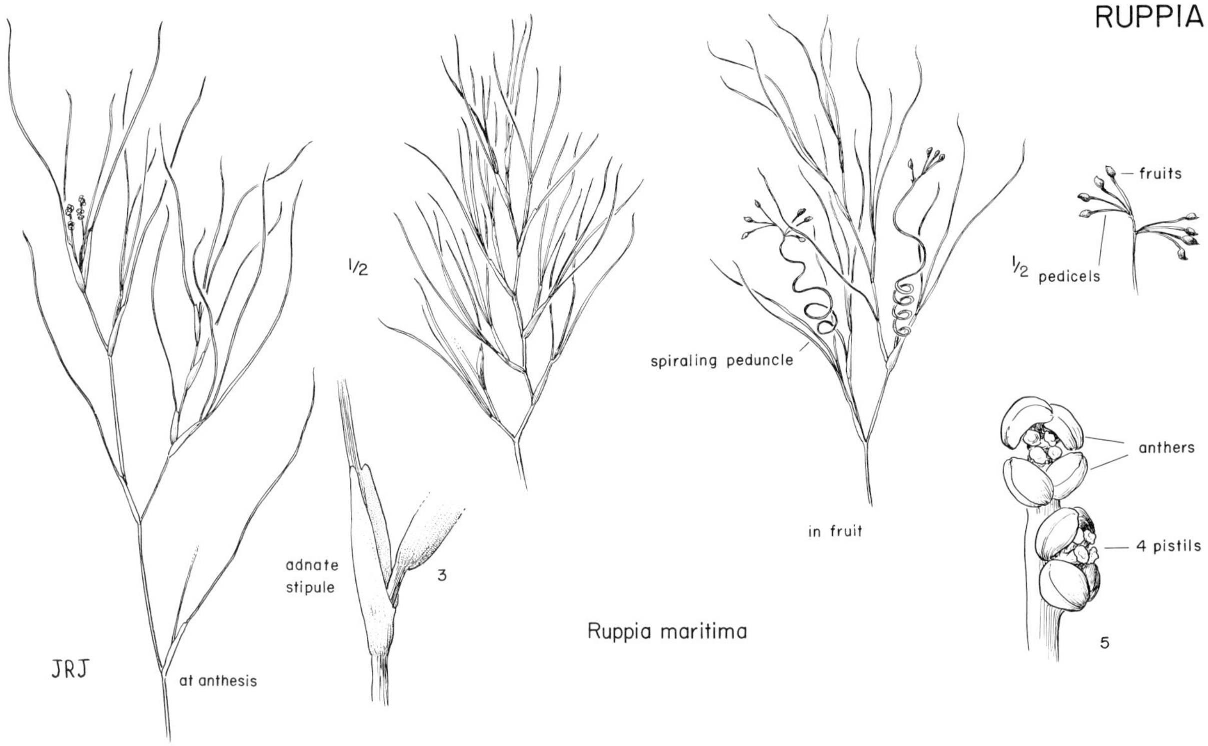

Ruppia maritima

slender and much branched, greenish to brownish, terete, (2) 4–8 (10) dm long, arising from an extensively creeping rhizome; leaves all submersed, scattered to tufted, the leaf-blade capillary to semicapillary, 1–10 cm long, 0.2–0.5 (0.9) mm wide, with an obtuse to rounded apex, bordered at the base with an adnate, more or less sheathing, membranous stipule, the sheath (0.2) 0.5–2 (7) cm long, with a ligule-like portion often lacking, but if present, then up to 3 cm long; peduncles axillary, slender, at first short, but (in some) becoming as much as 3 dm long in fruit, often spiraling or coiling in fruit, especially in the long-peduncled forms; flowers naked, 2 per peduncle, each composed of 2 large bicellular anthers and (usually) 4 pistils with sessile peltate stigmas, one flower above the other and on the opposite side of the peduncle, during early anthesis the peduncles and flowers sheathed by the uppermost leaves of the stem, this sheath ruptured by the elongating peduncle during anthesis as the flowers extend to a position slightly above the water surface; fruits ovoid to somewhat pyriform, symmetrical or asymmetrical, (1.5) 2–3 (4) mm long, with obtuse to pointed, erect or curved beak, the nut-like fruit on a slender elongated pedicel 4–15 (35) mm long; $n = 8$, $2n = 16, 20, 40$.

Alkaline, subsaline and brackish water along sea coasts or in ponds, ditches or inland lakes; cosmopolitan throughout the world but especially in Eurasia, Africa, and N. and S. Amer.; Alaska to Baja Calif., e. through Can. and the U.S.; found in our area in the n. half of the Intermountain Region. Apr.–July.

Ruppia is considered to be one of the more important aquatic plants because it provides excellent food and protection for fish, and all parts of the plant are eaten by waterfowl.

This is perhaps our most variable aquatic plant and it will require a world-wide study of the genus to fully understand the variation and determine the oldest available name for each entity that might deserve taxonomic recognition. Although our plants are covered by a multitude of names at all taxonomic ranks, it has been impossible to place our material satisfactorily. What names are used is determined by which sets of characters one selects. Those plants with short sheaths and short (0.2–2.5 cm) peduncles are referable to var. *maritima;* those with short sheaths and long (5–30 cm) peduncles can be called var. *spiralis* (L. ex Dumort.) Moris; while those with long sheaths (2–7 cm) can be called var. *occidentalis* (S. Wats.) P. Graebner. If one looks at fruiting characters, then a second set of names is used which may or may not circumscribe the same entities just noted above. Those plants with ovoid, essentially symmetrical fruits and long peduncles are called var. *obliqua* (Schur ex Griseb. & Schenk) Aschers. & Graebner, while those with short peduncles are var. *intermedia* (Thed.) Aschers. & Graebner. Those having asymmetrical fruits and long peduncles are var. *longipes* Hagström, but when the peduncles are short, these plants are commonly called var. *rostrata* Agardh. That all of this makes any real taxonomic sense is highly questionable, for all of these characteristics are based on dimensions of features which change as the plants mature and as the seasons change. That there are strongly expressed degrees of variation is not denied, but the best that I can do is to suggest that at this stage of development in the understanding of *R. maritima,* it is best not to attempt to make any infraspecific divisions of the species, but recognize that there is a possibility that some elements might exist which may be worthy of formal taxonomic recognition after a detailed, monographic study of *Ruppia* has been made.

Family **ZANNICHELLIACEAE,** the Horned Pondweed Family

Monoecious or dioecious, submersed herbaceous aquatic perennials with narrow, opposite or alternate leaves, with free or adnate, membranous, sheathing stipules often producing a ligule; flowers imperfect, axillary, solitary or cymose, naked or the staminate ones with 3 small transparent scales; stamens 1–3, with the (1) 2-celled, longitudinally dehiscent anthers on slender filaments; pistillate flowers of (1) 2–8 (9) distinct carpels, the stigma peltate or rarely 2- to 4-lobed on a short to long style, the ovary with a single pendulous ovule; fruit a 1-seeded, indehiscent, sessile or short-stipitate achene, the seed exalbuminous.

A family of 2–3 genera, widely distributed in fresh, brackish or salt waters throughout much of the world.

1. ZANNICHELLIA L. Horned Pondweed

Submersed aquatic perennials with opposite leaves; flowers imperfect, naked, axillary, the staminate and pistillate flowers usually in the same axil, enclosed in the bud by a hyaline, spathe-like sheath; staminate flowers solitary, composed of a single stamen on a slender filament which greatly exceeds the pistil; pistillate flowers mostly of (2) 3–5 (8) distinct carpels, usually sessile, subtended by a hyaline bract, each with a short style and peltate stigma; fruits obliquely oblong, compressed, the achene often with dentate margins.

Two species in fresh and brackish water; cosmopolitan. (Named for Gian Girolamo Zannichelli, 1662–1729, a Venetian botanist.)

1. Zannichellia palustris L.

Zannichellia palustris L. Sp. Pl. 969. 1753. *Algoides palustre* Lunell, Amer. Midl. Naturalist 4: 162. 1915. (Europe.)

Plants wholly submersed, the stem filiform to slender and usually much branched, light green (but often drying tan or light brown), 3–5 dm long, arising from an extensively creeping slender rhizome; leaves all submersed, the leaf-blade capillary, 2–8 (10) cm long, 0.2–0.5 mm wide, the apex acute, the base sessile, the free membranous sheath 1.5–4 mm long; flowers sessile, unisexual, the staminate flower of a single stamen, the pistillate flower of (2) 4 (6) distinct carpels; fruits obliquely oblong, flattened, (2) 3–4 mm long, short-pedicellate in some at maturity, often dentate on the back, the slender flask-shaped beak 0.8–1.5 mm long; n = 12, $2n$ = 12, 24, 28, 32, 36.

Fresh and brackish ponds, lakes and streams; cosmopolitan, throughout most of the world, including much of N. Amer.; in scattered locations in our area. Mar.–Nov.

Zannichellia is occasionally used by waterfowl and some fish (mostly trout) as food.

Subclass COMMELINIDAE

Monocotyledonous plants, generally with vessels in all vegetative organs; inflorescence of various sorts, but not a spadix; flowers sometimes with well differentiated sepals and petals (though in ours without nectar), but more commonly either with a chaffy or bristly perianth that is not differentiated into sepals and petals, or without perianth; all our families with a superior ovary that is (except in most Typhales) syncarpous; seeds mostly with starchy endosperm (except Typhaceae).

The subclass Commelinidae consists of 8 orders (four in our area), 25 families (six in our area), and nearly 19,000 species. The Cyperales, with about 12,000 species, are by far the largest order of the group.

Synoptical Arrangement of the Orders of Commelinidae
(as represented in our flora)

1 Flowers with more or less showy petals that are well differentiated from the sepals, perfect; pollination mostly by insects .... Order Commelinales, p. 44
1 Flowers perfect or often unisexual, without showy petals, the perianth sometimes in 2 series, but the 2 series much alike and not petaloid.
  2 Ovary either with more stigmas than locules, or with more than one ovule per locule, the ovules of various form and position; flowers perfect or unisexual; plants terrestrial to semi-aquatic.
    3 Ovary with 1–3 locules and 3–many ovules; fruit capsular; pollen in tetrads; flowers in open or compact inflorescences unlike those of the Cyperales .... Order Juncales, p. 47
    3 Ovary with a single locule and ovule; fruit indehiscent; pollen in monads; flowers in characteristic spikes or spikelets .... Order Cyperales, p. 67
  2 Ovary with a single locule, ovule and stigma, or with 2 locules, ovules and stigmas; flowers unisexual; plants emergent aquatics .... Order Typhales, p. 464

Order COMMELINALES

Flowers mostly perfect, rarely imperfect or unisexual, actinomorphic or zygomorphic, the perianth well differentiated into a series of 3 sepals and 3 petals, or with one series reduced to 2 sepals or petals, the sepals mostly herbaceous or chaffy, the petals thin and occasionally ephemeral (ours), mostly highly colored, without nectaries or nectar; stamens 6 or 3 or rarely 1, often reduced to staminodes, hypogynous or adnate to

ZANNICHELLIA
TRADESCANTIA

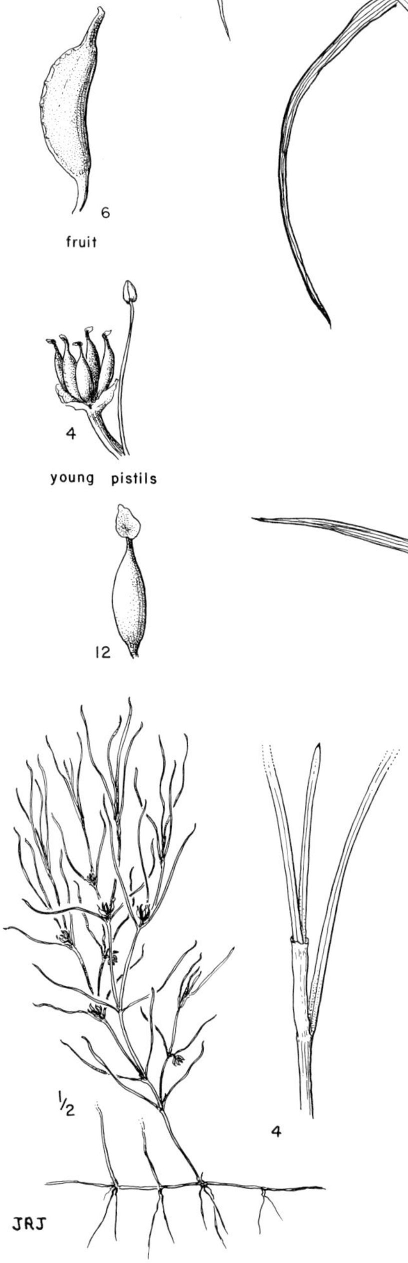

Zannichellia palustris

Tradescantia occidentalis

the corolla; ovary superior, 3-carpelled, unilocular to trilocular, the style usually one with 1–3 stigmatic lobes; pollen trinucleate; seeds mostly with a starchy endosperm; essentially herbaceous annual or perennial mesophytic to hydrophytic, xerophytic or infrequently epiphytic plants from fibrous to tuberous or somewhat bulbous roots or rootstocks, mostly adapted to insect pollination, with linear to slightly broader, spirally alternate or tufted basal leaves and alternate cauline leaves, parallel-veined, often well differentiated into an open or closed sheath and a blade.

An order of four families, all subtrop. or trop. with only one family found in our area.

Family **COMMELINACEAE,** the Spiderwort Family

Succulent perennials or annual herbs, acaulescent or with knotty and leafy stems from fibrous or sometimes much thickened and tuberous roots; leaves alternate, basal or cauline or both, mostly narrow, distinctly parallel-veined; flowers bisexual, usually actinomorphic but sometimes zygomorphic, mostly small but with often bright and showy parts, 3-merous or rarely 2-merous, strongly differentiated into sepals and petals; sepals 3, free or rarely slightly connate at the base, imbricate, herbaceous and mostly green; petals 3, free or united into a short tube, equal or with the third petal much reduced, alternate with the sepals and usually highly colored, mostly ephemeral and deliquescent; stamens 6 and in two series, or reduced to 3 functional ones and 3 abortive or staminoid, or rarely only one, hypogynous or adnate to the corolla tube, the filaments filiform and often flattened at the base, usually densely covered with long moniliform, variously colored hairs, the anthers mostly oblong to ovate, basifixed, mostly introrse or rarely dehiscing by terminal pores, often on slender to broad, straight to curved, mostly trapezoid connectives; ovary superior, sessile or rarely short-stipitate, 3- or rarely 2-celled with one to few or rarely several ovules in each locule with axile placentation, the single style with a single or 3-lobed stigma; fruits loculicidal capsules sometimes enclosed by fleshy sepals, or rarely fleshy and indehiscent; seeds one to few, with abundant endosperm.

A large family of some 40 genera and over 600 species found mainly in the trop. regions of the world. Many of the species are found in the horticultural trade and make excellent house plants.

We have only one genus in our area.

1. **TRADESCANTIA** L. Spiderwort

Tall, ascending to erect or infrequently trailing subsucculent perennial herbs, frequently producing stolons, arising from fleshy roots; leaves alternate, sessile, linear to oblong, produced in a perfoliated sheath; flowers white to blue, violet or magenta, on slender pedicels subtended by solitary or paired hyaline to slightly foliaceous bractlets, in umbelliform dichotomous helicoid cymes, terminal or lateral and subtended by 2 (3) subequal or unequal foliaceous or rarely scarious bracts, the sepals 3, persistent, separate, equal, more or less concave or navicular, foliaceous to petaloid or hyaline, the petals ephemeral, separate, equal; stamens 6, fertile, the filaments slender, pilose, the anthers mostly with reniform anther sacs at the tip of a broadly trapezoid connective, dehiscing laterally; ovary superior, 3-celled, with 3–6 ovules, with a long filiform style terminated by a capitate stigma; fruits dry loculicidal capsules; seeds few, oblong to spherical.

As now usually defined, a genus of perhaps 30 species mainly of the trop. regions of the New World. (Named in honor of John, the elder Tradescant, ?–1637, gardener to Charles I of England.)

Reference: Anderson, E., and R. E. Woodson. The species of *Tradescantia* indigenous to the United States. Contr. Arnold Arbor. 9: 1–132. 1935.

1. **Tradescantia occidentalis** (Britton) Smyth

Tradescantia virginiana var. *occidentalis* Britton in Britton & Brown, Ill. Fl. N. U. S., Canada & Brit. Possessions 1: 377. 1896. *T. occidentalis* Smyth, Trans. Kansas Acad. Sci. 16: 163. (June) 1899. ("Wisconsin to Missouri, Texas and New Mexico," no specimens cited.)

T. scopulorum Rose, Contr. U.S. Natl. Herb. 5: 205. (Oct.) 1899. *T. occidentalis* var. *scopulorum* Anders. & Woodson, Contr. Arnold Arbor. 9: 106. 1935. (*Pringle s.n.*. Santa Catalina Mts., Pima Co., Ariz., 16 May 1881.)

T. laramiensis Goodd. Bot. Gaz. 33: 68. 1902. (*A. Nelson 7455*, Halleck Canyon, Albany Co., Wyo., 4 July 1900.)

T. universitatis Cockerell, Nature 75: 7. 1906. (*Cockerell s.n.*, University of Colorado campus, Boulder, Boulder Co., Colo., 30 Sept. 1906.)

Tall, more or less erect perennial herbs 0.5–5 (6) dm tall with erect or ascending, straight, more or less branching, glabrous, glaucous and somewhat succulent stems from rather stout, fleshy roots accompanied by more slender, fibrous roots; leaves firm, membranaceous, linear-lanceolate, long-acuminate, 0.5–3 (5) dm long, 0.2–2 cm wide, stiff and widely divaricate in most, often falcate; sheaths thin and more or less transparent, rather turgid and inflated, 0.5–3 cm long, 0.5–3 (3.5) cm wide; inflorescences of umbelliform cymes, few to many-flowered, terminal, solitary and accompanied by one or more lateral pedunculate cymes in the upper nodes; bracts foliaceous, sharply reflexed or divaricate, 0.5–1.5 (2) dm long, 0.5–2 cm wide; pedicels erect to spreading or reflexed, 1–2 cm long, glandular-puberulent to glabrous; flowers mostly broadly campanulate, the sepals elliptic, acute to acuminate, 4–10 mm long, mostly greenish or suffused with rose or purple, glandular-puberulent to glabrous, the petals broadly ovate, 7–16 mm long, bright-blue to rose or magenta; stamens erect, the filament 3–5 mm long, densely pilose, the anther yellow; ovary ovoid, glandular-puberulent to glabrate or glabrous; capsules oblong, 4–7 mm long; seeds compressed, 2–4 mm long; $2n$ = 12, 24.

In sandy to rocky places in open grasslands to woodland areas from the Great Plains of N.D. s. to Texas, w. to e. Mont., c. Wyo. and Colo., and across N.M. into Ariz. and s. Utah; in our area mostly in moist places near streams, springs or meadows, infrequent on forest floors, across the s. tier of cos. in Utah, 1050–2000 m elev. May–Aug.

Most of our plants have glandular-puberulent pedicels and sepals, and represent the widespread var. *occidentalis*. Plants from Zion National Park have glabrous pedicels and sepals, and belong to the weakly defined var. *scopulorum* (Rose) Anders. & Woodson, which is otherwise confined to the south of our area.

Order **JUNCALES**

The order Juncales consists of two families, the Juncaceae and Thurniaceae, the latter with only three species. The Thurniaceae are distinguished from the Juncaceae by some vegetative anatomical peculiarities (vascular bundles of the leaf in vertical pairs, the phloem of the lower bundle uppermost and facing the phloem of the upper bundle; some cells of the leaf epidermis containing silica bodies) and by the densely capitate, leafy-bracteate inflorescence. Thus for the purposes of this flora the description provided for the family Juncaceae may also serve for the order Juncales.

Family **JUNCACEAE,** the Rush Family

Perennial (seldom annual) grasslike herbs; leaves with cylindrical, open or closed sheath and flat to channeled or terete blade, or the blade sometimes obsolete; flowers perfect (in all ours) or seldom unisexual, in diverse sorts of open or compact inflorescences, anemophilous or in some species secondarily entomophilous; perianth small and inconspicuous, the tepals ordinarily 6, in 2 more or less similar sets of 3, persistent, dry and more or less scarious or chartaceous or partly herbaceous or hyaline, commonly greenish to brownish or blackish, sometimes partly suffused with purple, but not petaloid; stamens 6 or less often 3, rarely only 1 or 2; pollen in tetrads; ovary superior, ordinarily tricarpellate, with one or 3 locules, sometimes with partial partitions (intruded placentas) that do not meet in the center; style short or elongate, with 3 elongate stigmatic branches; ovules anatropous, 3 and basal in a unilocular ovary, or numerous and parietal or axile in a partly or fully 3-locular ovary; fruit a loculicidal capsule; embryo embedded in the endosperm.

The family consists of 8 genera and about 350 species, widespread in temp. and cool regions of the earth; the two large genera, *Juncus* and *Luzula*, are best developed in the N. Hemisphere.

Reference: Buchenau, F. Juncaceae, in Engl. Pflanzenr. IV. 36 (Heft 25): 1–284. 1906.

1 Seeds numerous, on axile or intruded parietal placentas; leaf sheaths open; plants glabrous . . . . 1. *Juncus*, p. 47

1 Seeds 3, on a basal placenta in a unilocular ovary; leaf sheaths closed; leaf blades commonly with long hairs along the margins, at least when young . . . . 2. *Luzula*, p. 64

1. **JUNCUS** L. Rush

Annual or more often perennial, glabrous, grasslike herbs, cespitose or with more or less well developed rhizomes; stems rounded to sometimes flattened and even somewhat winged; leaves divided into a sheath and blade, the sheath open, often with hyaline or scarious margins, these margins sometimes projecting beyond the collar to form a pair of auricles, or the auricles confluent to form a ligule resembling that of grasses; blade in many species flat and grasslike or merely channeled on the upper side, but in others terete, hollow, and septate at intervals (said to be septate-nodulose), or equitant (i.e. folded along the midrib and with the margins connate from a little above the base, so that the edge formed by the connate margins faces the stem) and also more or less septate, or in some species the blade more or less strongly reduced or even obsolete; flowers perfect, (1–) several to many in an essentially cymose inflorescence that varies from open and paniculiform or corymbiform to congested and formed into 1–many cymose heads, generally subtended by 1–several more or less leafy or partly hyaline-scarious bracts that form something of an involucre, the lowest of these bracts sometimes elongate, firm, and terete, resembling a continuation of the stem, so that the inflorescence appears to be lateral rather than terminal; flowers trimerous (or seldom some of them casually dimerous), individually sessile or pedicellate in the axil of a small, usually hyaline-scarious bract, sometimes also closely subtended by a pair of hyaline-scarious bracts (prophylls), in the latter case said to be prophyllate; tepals ordinarily 6, in 2 more or less similar sets of 3, persistent, dry and more or less scarious or chartaceous or partly herbaceous or hyaline, commonly greenish to brownish or blackish, sometimes partly suffused with purple, but not at all petaloid; stamens (1–2) 3 or 6, when 3 mostly antesepalous, commonly persistent into fruit; ovary ordinarily tricarpellate, with a short to evident style and 3 more or less elongate stigmatic branches; fruit a small, loculicidal, often triquetrous capsule, trilocular or unilocular with slightly to strongly intruded placentae; seeds numerous, commonly ellipsoid or fusiform and minutely apiculate, sometimes with each end prolonged into a slender tail of softer texture that may be longer than the body.

A large, nearly cosmopolitan genus of more than 200 species, best developed in temp. and boreal regions with a moist climate. (The classical Latin name for these or similar plants.)

All measurements of floral parts in the key and descriptions are taken from dry herbarium material. The stamens are commonly persistent, so that they can be counted and measured even when the fruit is mature.

1 Plants annual.
  2 Stamens 6; well developed plants more than 5 cm tall and with the flowers lateral as well as terminal on the stems .... 21. *J. bufonius*
  2 Stamens (2) 3; plants 0.5–5 cm tall, with 1–3 flowers in a cluster at the top of the stem .... 22. *J. kelloggii*
1 Plants perennial.
  3 Flowers prophyllate (see generic description); leaves, when present, neither septate, nor equitant, nor terete and hollow.
    4 Inflorescence more or less evidently terminal, the lowest bract generally ascending or spreading and not appearing like a continuation of the stem.
      5 Seeds ca 0.5 mm long, merely apiculate, not tailed; flowers mostly more than 7 per stem; plants from the lowlands to middle altitudes in the mts.
        6 Capsule obtuse to truncate at the summit, only imperfectly 3-celled; inflorescence looser and often larger than in the next species, up to 5 cm long or more .... 1. *J. tenuis*
        6 Capsule distinctly retuse at the summit and fully 3-celled; inflorescence small, congested, up to ca 2 cm long .... 2. *J. confusus*
      5 Seeds ca 1 mm long or more, evidently tailed; flowers few, mostly 1–7 per stem; plants alpine or subalpine.
        7 Capsule acute; tepals acuminate, 5–7 mm long .... 5. *J. parryi*
        7 Capsule retuse; tepals acute, 4–5 (5.5) mm long .... 6. *J. hallii*
    4 Inflorescence seemingly lateral, the lowest (involucral) bract terete, erect, and appearing to be a continuation of the stem.
      8 Flowers more numerous than in the next group, commonly 5–50 or more; involucral bract often more than 5 cm long; seeds ca 0.4–0.8 mm long, not tailed; plants of lowlands to middle elev. in the mts., often more than 4 dm tall.
        9 Anthers large, commonly 1.5–2.5 mm long, much longer than the filaments, commonly 3–5 times as long; involucral bract moderately elongate, up to ca half as long as the proper stem; stem coarse, (1) 1.5–3 (4) mm thick .... 3. *J. balticus*
        9 Anthers small, ca 0.5–0.7 mm long, equaling or a little shorter than the filaments; involucral bract much elongate, half as long to longer than the proper stem; stem very slender, ca 1 mm thick or less .... 4. *J. filiformis*
      8 Flowers few, commonly 1–4 (7) per stem; involucral bract rarely as much as 5 cm long; seeds tailed, ca 1–2 mm long; plants alpine or subalpine, 1–4 dm tall.
        10 Uppermost basal sheaths with a well developed blade mostly 3–20 cm long.
          11 Capsule acute; tepals acuminate, 5–7 mm long .... 5. *J. parryi*
          11 Capsule retuse; tepals acute, 4–5 (5.5) mm long .... 6. *J. hallii*
        10 Uppermost basal sheaths bladeless or with a blade scarcely 1 cm long; capsule retuse or truncate .... 7. *J. drummondii*
  3 Flowers not prophyllate; leaves in most (not all) species septate and either equitant or terete and hollow.
    12 Leaves flat, neither septate nor equitant.
      13 Seeds merely apiculate, ca 0.5 mm long; heads seldom with more than 10 flowers.
        14 Tepals smooth and shining; sheath evidently auriculate at the summit and produced into a distinct ligule ca 1 mm long or more .... 8. *J. longistylis*
        14 Tepals minutely roughened and dull along the midstrip; sheath sometimes produced into small auricles at the summit, but not forming a distinct ligule, or this rudimentary and less than 0.5 mm long .... 9. *J. orthophyllus*
      13 Seeds evidently tailed, ca 1.5 mm long; at least the larger heads commonly with more than 10 flowers; tepals minutely roughened on the back as in *J. orthophyllus* .... 10. *J. regelii*
    12 Leaves either terete and hollow (at least distally) or flattened and equitant, in either case septate.
      15 Leaves flattened, equitant (see generic description) .... 11. *J. ensifolius*
      15 Leaves terete, hollow (only toward the tip in no. 12, marked by its very long-tailed seeds).

16 Stamens 3; seeds shortly tailed, 0.8–1.0 mm long .... 19. *J. tweedyi*
16 Stamens 6.
17 Seeds long-tailed, ca 1.5–4 mm long overall; plants alpine.
18 Plants rhizomatous, with scattered stems; leaves hollow and septate-nodose only near the tip, otherwise merely inrolled or folded, relatively coarse, commonly 1–2.5 mm wide (as rolled or folded) at midlength; seeds ca 2.5–4 mm long overall .... 12. *J. castaneus*
18 Plants densely tufted, without rhizomes; leaves hollow and septate-nodose for most of their length, very slender, commonly ca 0.5 mm thick at midlength; seeds ca 1.5–2 mm long overall .... 20. *J. triglumis*
17 Seeds merely apiculate, 0.4–0.6 mm long; habitat various.
19 Capsule subulate, tapering gradually into the stylar beak exserted from the perianth; flowers in the heads spreading in all directions.
20 Auricles free for 2–5 mm at the collar of the leaf; perianth mostly 4–5 mm long; plants robust, commonly (2) 4–10 dm tall .... 13. *J. torreyi*
20 Auricles minute, not over ca 1 mm long; perianth mostly 2.5–3.5 mm long; plants slender, commonly 1–4 dm tall .... 14. *J. nodosus*
19 Capsule more abruptly pointed, or rounded or subtruncate, longer or shorter than the perianth; flowers in the heads mostly spreading or ascending to erect, generally few if any of them definitely reflexed.
21 Flowers larger, the perianth mostly 3–5.5 mm long; capsule about equaling or shorter than the perianth; tepals (at least the outer) strongly acuminate; anthers longer or shorter than the filaments.
22 Heads mostly (2) 4–30; anthers mostly 1–2 mm long, longer than the filaments .... 15. *J. nevadensis*
22 Head solitary (2); anthers mostly 0.5–1.0 mm long, shorter than the filaments .... 16. *J. mertensianus*
21 Flowers smaller, the perianth mostly 2–3 mm long; capsule surpassing the perianth; tepals obtuse or rounded to acute or barely acuminate; anthers shorter than the filaments.
23 Tepals more or less acute, the inner equaling or longer than the outer .... 17. *J. articulatus*
23 Tepals mostly obtuse or broadly rounded, the inner shorter than the outer .... 18. *J. alpinus*

1. Juncus tenuis Willd.

Juncus tenuis Willd. Sp. Pl. 2: 214. 1799. (N. Amer.)

J. gracilis J. E. Smith, Engl. Bot. 31: t. 2174. July, 1810; not *J. gracilis* R. Br. Prodr. 259. April, 1810. *J. macer* S. F. Gray, Nat. Arr. Brit. Pl. 2: 164. 1821. (*G. Don,* mts. of Angusshire, England, in 1795 or 1796.)

J. tenuis var. *congestus* Engelm. Trans. Acad. Sci. St. Louis 2: 450. 1866. *J. tenuis* var. *occidentalis* Coville, Proc. Biol. Soc. Wash. 10: 129. 1896. *J. occidentalis* Wieg. Bull. Torrey Bot. Club 27: 521. 1900. (*Bolander,* San Francisco, is the first of 3 collections cited.)

J. interior Wieg. Bull. Torrey Bot. Club 27: 516. 1900. (*Vasey,* Richmond, Ill.)

J. dudleyi Wieg. Bull. Torrey Bot. Club 27: 524. 1900. *J. tenuis* var. *dudleyi* F. J. Hermann, J. Arnold Arbor. 25: 56. 1944. (*Wiegand,* Truxton, N.Y.)

J. arizonicus Wieg. Bull. Torrey Bot. Club 27: 517. 1900. *J. interior* var. *arizonicus* F. J. Hermann, J. Wash. Acad. Sci. 30: 217. 1940. (*Thurber,* Copper Mines, N.M.)

Path rush.

Habitally similar to *J. confusus,* the leaves sometimes a little wider, up to 1.5 mm; inflorescence generally larger and more open, 1–5 cm long or longer; perianth often greener; capsule obtuse to truncate, minutely apiculate by the persistent style-base, only imperfectly 3-celled, the partitions not meeting in the center; n = 15, 16.

Moist places, especially at lower alt.; widespread in N. Amer., and introduced in temp. regions elsewhere in the world; in the Intermountain Region found mainly toward the northeast, as in Fremont Co., Idaho, and in Rich, Cache, and Uintah cos., Utah, but also extending to San Juan and Washington cos., Utah.

Most of our plants have cartilaginous, somewhat yellowish auricles, and thus belong to the phase of the species that has been called var. *dudleyi* (Wieg.) F. J. Hermann, but the taxonomic significance of the distinction is doubtful.

Juncus vaseyi Engelm., a species of Canada and northern U.S., has been reported from Utah, but we have seen no specimens. It differs from *J. tenuis* most notably in having tailed seeds ca 1 mm long, and in having capsules equal to or longer than the perianth.

2. Juncus confusus Coville

Juncus confusus Coville, Proc. Biol. Soc. Wash. 10: 127. 1896. (*C. S. Crandall,* North Park, Colo., 6 Sept. 1890.)

Densely tufted perennial; stems numerous, terete, 1.5–4 dm tall; leaves all clustered at the base, the blade 5–15 cm X 0.4–0.8 mm, convex on the lower surface, channeled or nearly flat on the upper, the sheath prolonged into a pair of evident hyaline auricles; inflorescence small and compact but evidently branched, mostly 1–2 cm long, with 1–3 slender, leafy bracts at the base, the lowest one elongate, mostly 2–10 cm long, ascending or suberect, the other(s) much smaller and more spreading; flowers sessile or short-pedicellate, each one closely subtended by a pair of scarious or hyaline bracts ca 1–2 mm long; tepals 3–4.5 mm long, more or less strongly acute (but soft), with a green or stramineous mid-

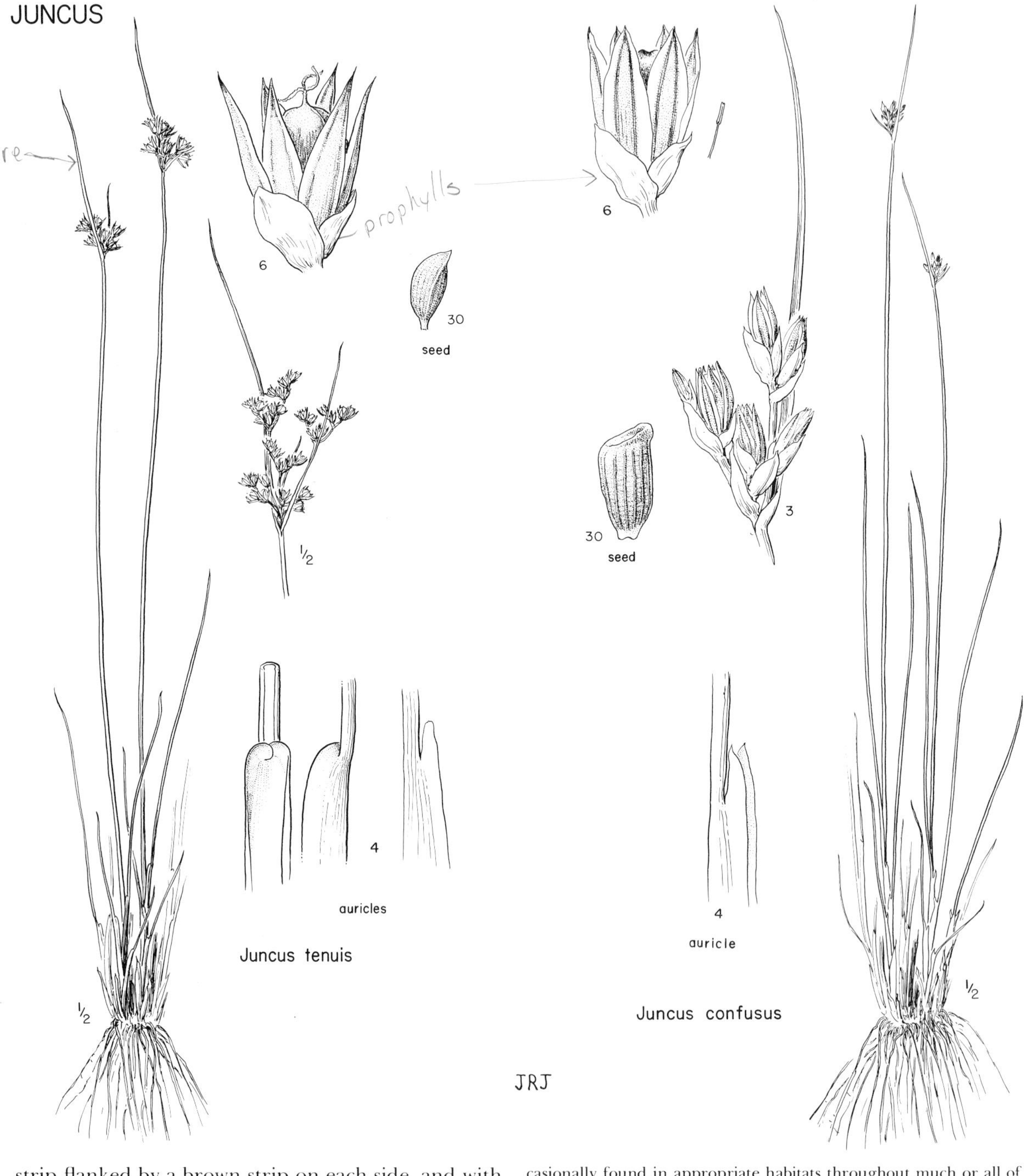

strip flanked by a brown strip on each side, and with pale, hyaline margins; stamens 6, the anthers 0.5–0.6 mm long, evidently shorter than the filaments; capsule equaling or a little shorter than the perianth, triquetrous above, shortly but distinctly retuse at the summit, scarcely or not at all apiculate, completely 3-locular; seeds ca 0.5 mm long, minutely apiculate.

Meadows, streambanks, parks, and moist woods at middle elev. in the mts.; cordilleran species, ranging from the Cypress Hills of Sask. to s. Alta., s. B.C., and e. Wash., s. to Calif. (Sierra Nevada), c. Nev. (Toiyabe Mts.) and n. Ariz. (Kaibab Plateau) and N.M., and occasionally found in appropriate habitats throughout much or all of our range. June–Aug.

It is possible that further study may show *J. confusus* to be a regional variety of the more widespread *J. tenuis,* which it largely replaces in our range.

3. **Juncus balticus** Willd.

Juncus balticus Willd. Ges. Naturf. Freunde Berlin Mag. Neuesten Entdeck. Gesammten Naturk. 3: 298. 1809. *J. arcticus* var. *balticus* Trautv. Acta Horti Petrop. 5: 119. 1877. *J. arcticus* subsp. *balticus* Hylander, Nord. Kärlväxtfl. 178. 1953. (Shores of the Baltic Sea, Warnemünde, Germany.)

J. mexicanus Willd. in Roemer & Schultes, Syst. Veg. 7: 178.

JUNCUS

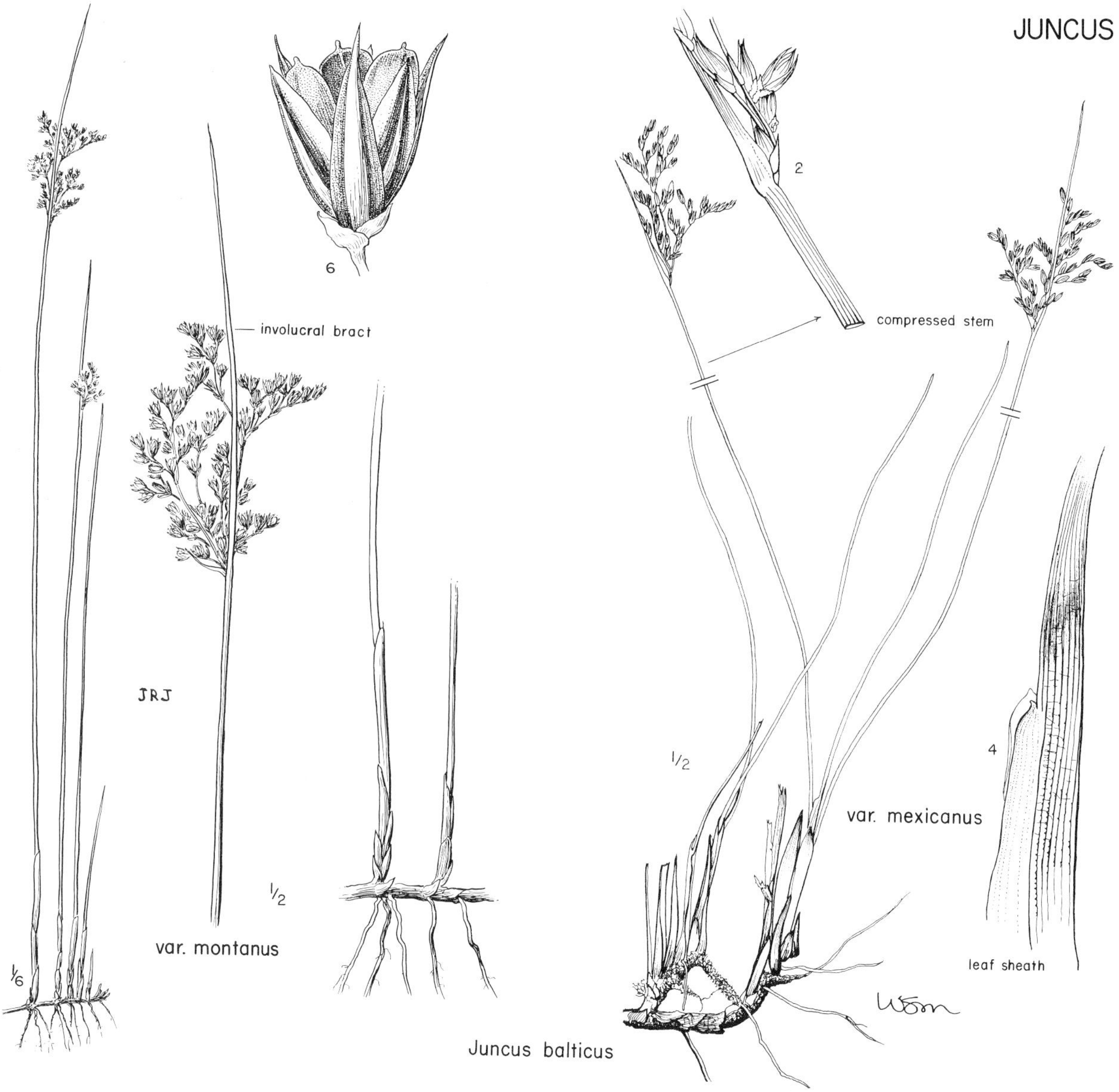

Juncus balticus

1829. *J. balticus* var. *mexicanus* Kuntze, Revisio Gen. Pl. 3(2): 320. 1898. *J. balticus* f. *mexicanus* Parish, Muhlenbergia 6: 119. 1910. (*Humboldt*, Mex.)

J. balticus var. *montanus* Engelm. Trans. Acad. Sci. St. Louis 2: 442. 1866; not *J. montanus* Lam. 1789. *J. ater* Rydb. Fl. Rocky Mts. 1060. 1917 [1918]. *J. arcticus* subsp. *ater* Hultén, Kongl. Svenska Vetenskapsakad. Handl. IV. 8(5): 24. 1962. (Western plains and mountains.)

J. balticus var. *vallicola* Rydb. Bull. Torrey Bot. Club 31: 399. 1904. *J. vallicola* Rydb. Fl. Rocky Mts. 1060. 1917 [1918]. (*Merrill & Wilcox 664*, Point of Rocks, Wyo.)

Stems firm, wiry, more or less terete (except var. *mexicanus*), (1) 3–9 dm tall, (1) 1.5–3 (4) mm thick, closely or loosely set on coarse, blackish, creeping, often sod-forming rhizomes; leaf sheaths clustered at the base, 2–15 cm long, light to dark brown or castaneous, bladeless or nearly so (except in var. *mexicanus*); lowest bract of the inflorescence terete (or somewhat flattened in var. *mexicanus*), 2–20 cm long, up to half as long as the proper stem, erect and appearing like a continuation of the stem, so that the inflorescence appears to be lateral near the summit or above the middle of the plant; flowers (5) 10–50 or more in a compact to loose and proliferating inflorescence up to ca 6 cm long and wide, some of them sessile or nearly so, others evidently pedicellate, each closely subtended by a pair of evident, hyaline-scarious bracts; tepals 3.5–5 mm long, variously pale or dark, lanceolate and acute to ovate, sometimes broadly hyaline-margined; stamens 6; anthers 1.5–2.5 mm long, (2) 3–5 times as long as the filaments; capsule from a little shorter to a little longer than the perianth; seeds ca 0.6–0.8 mm long, not tailed; $n = 20, 40$.

Low, wet, often alkaline places, from the valleys and plains up to middle elev. in the mts.; interruptedly circumboreal, and in appropriate habitats throughout our range; S. Amer. May–July.

Juncus balticus is a widespread, taxonomically complex species that should perhaps be considered a part of the otherwise strictly arctic, fully circumboreal species *J. arcticus* Willd. There has been no comprehensive treatment that characterizes the major variants of *J. balticus* or of *J. arcticus* in the expanded sense since that of Buchenau in 1906.

The characteristic phase in our range and throughout most of the western cordillera of the U.S. and Canada is generally referred to *J. balticus* var. *montanus* Englem., as principally described above. The name *J. arcticus* subsp. *ater* (Rydb.) Hultén is available for our plants, when they are referred to *J. arcticus* sensu lato, but no comparable varietal combination under *J. arcticus* has been proposed.

The chiefly more southern var. *mexicanus* (Willd.) Kuntze differs from var. *montanus* in its distinctly compressed (but not margined or thin-edged) stems and in often having a well developed, flattened leaf blade on the uppermost basal sheath, but the distinction is not always easily drawn. Var. *mexicanus* is common in the *Larrea* zone, where it extends north to the southern border of our range. Occasional specimens from as far north as the Craters of the Moon in Idaho have distinctly flattened stems as in var. *mexicanus*, but the only material I have seen from our range that has both the flattened stem and the well developed basal leaf blade of characteristic var. *mexicanus* comes from high altitudes in the White Mts. of Calif. There is a further problem in that although the name *J. mexicanus* is usually applied to plants with flattened stems, the original description specifies that the stems are terete. I have not examined the original Humboldt material of *J. mexicanus*. Resolution of the taxonomic and nomenclatural problems of our plants of *J. balticus* awaits monographic study of the group.

4. Juncus filiformis L.

Juncus filiformis L. Sp. Pl. 326. 1753. (Europe.)

Stems slender, ca 1 mm thick or less, (0.3) 1–4.5 dm tall, closely set on rather slender, often sod-forming rhizomes; leaf sheaths clustered at the base, 1–10 cm long, brown to stramineous or partly greenish, bladeless or nearly so; lowest bract of the inflorescence terete, much elongate, half as long to longer than the proper stem, erect and appearing like a continuation of the stem, so that the inflorescence appears to be lateral, from somewhat below to somewhat above the middle of the plant; flowers few, mostly 5–20 in a compact inflorescence up to ca 2 cm long and wide, some of them sessile or nearly so, others evidently pedicellate, each closely subtended by a pair of evident, more or less hyaline-scarious bracteoles; tepals 3–4.5 mm long, partly or wholly greenish or stramineous, the outer lance-acuminate, the inner commonly a little shorter and often blunter; stamens 6, the anthers short, mostly 0.5–0.7 mm long, equaling or commonly shorter than the filaments; seeds ca 0.4–0.5 mm long, not tailed; n = 20, 40.

Wet places, often in sandy soil; circumboreal, s. at middle elev. in the mts. to n. Oregon, Idaho, Utah (Uinta Mts.), w. Wyo., Mich., and Pa.

5. Juncus parryi Engelm.

Juncus parryi Engelm. Trans. Acad. Sci. St. Louis 2: 446. 1866. *J. drummondii* var. *parryi* M. E. Jones, Bull. Montana State Univ., Biol. Ser. 15: 22. 1910. (*C. C. Parry 360,* Colo., in 1861.)

Densely tufted perennial; stems numerous, terete, 1–3 dm tall; basal sheaths 1–4 cm long, the lower bladeless or with a bristle-like blade less than 1 cm long, the uppermost one (at least on many of the stems) with a slender, terete or channeled blade 3–10 cm long; lowest involucral bract terete, (0.5) 2–6 cm long, generally erect or nearly so, commonly appearing to be a continuation of the stem, the inflorescence thus seemingly lateral near the summit; flowers 1–4, fairly closely grouped, the first one sessile or nearly so, the others generally evidently short-pedicellate, each closely subtended by a pair of hyaline or subchartaceous bracts; tepals with greenish or brownish midstrip fading into hyaline margins, the outer lanceolate, acuminate, 5–7 mm long, the inner a little shorter and blunter; capsule distinctly acute, from a little shorter to a little longer than the perianth; seeds ca 2 mm long, the whitish membranous appendage (tail) at each end about equaling or a little longer than the body.

Meadows, wet low ground, and open rocky slopes at upper elev. in the mts., sometimes above timberline; B.C. and Alta. to Calif. (Sierra Nevada and White Mts.), Nev. (Elko Co.), Utah (s. to Iron Co.) and Colo. July–Aug.

6. Juncus hallii Engelm.

Juncus hallii Engelm. Trans. Acad. Sci. St. Louis 2: 446. 1866. (*Hall & Harbour 562,* near Lake Ranch, Colo.)

Densely tufted perennial; stems numerous, terete, 1–4 dm tall; basal sheaths bladeless or with a reduced, bristle-like blade, but one or two leaves a little above the base with an evident, slender, somewhat channeled blade 4–20 cm long; inflorescence either evidently terminal or seemingly lateral near the summit, the lowest involucral bract sometimes but not always more or less erect and like a continuation of the stem, up to ca 5 cm long; flowers mostly 2–7, fairly closely clustered, generally at least some of them evidently short-pedicellate, each closely subtended by two hyaline or subchartaceous bracts; tepals subequal or the outer a little the longer, 4–5 (5.5) mm long, lanceolate, acute or acuminate, mostly brownish or castaneous and hyaline-margined; stamens 6, the anthers scarcely 1 mm long; capsule castaneous, from a little shorter to a little longer than the perianth, distinctly retuse at the tip; seeds ca 1 mm long or a bit more, the whitish, membranous appendage at each end about half as long as the body.

Wet meadows and wooded, rocky slopes at upper elev. in the mts.; sw. Mont. to Colo. and the Uinta Mts. of Utah. June–Aug.

7. Juncus drummondii E. Meyer

Juncus arcticus var. *gracilis* Hook. Fl. Boreali-Amer. 2: 189. 1838; not *J. gracilis* R. Br., 1810. *J. drummondii* E. Meyer in Ledeb. Fl. Ross. 4: 235. 1853. (*Drummond,* at a great elevation in the Rocky Mts.)

Densely tufted perennial; stems numerous, terete, 1–4 dm tall, sometimes some of them sterile; basal sheaths 2–6 cm long, bladeless or with bristle-like blade up to 5 (10) cm long; lowest involucral bract terete, 1–4 cm long, erect or nearly so, appearing to be a continuation of the stem, the inflorescence thus seemingly lateral near the summit of the stem; flowers (1) 2–4, fairly closely aggregated, generally one or more of them sessile or nearly so and one or more of them evidently short-pedicellate, each closely subtended by two hyaline or subchartaceous bracts; tepals subequal, 5–7 mm long, lanceolate,

JUNCUS

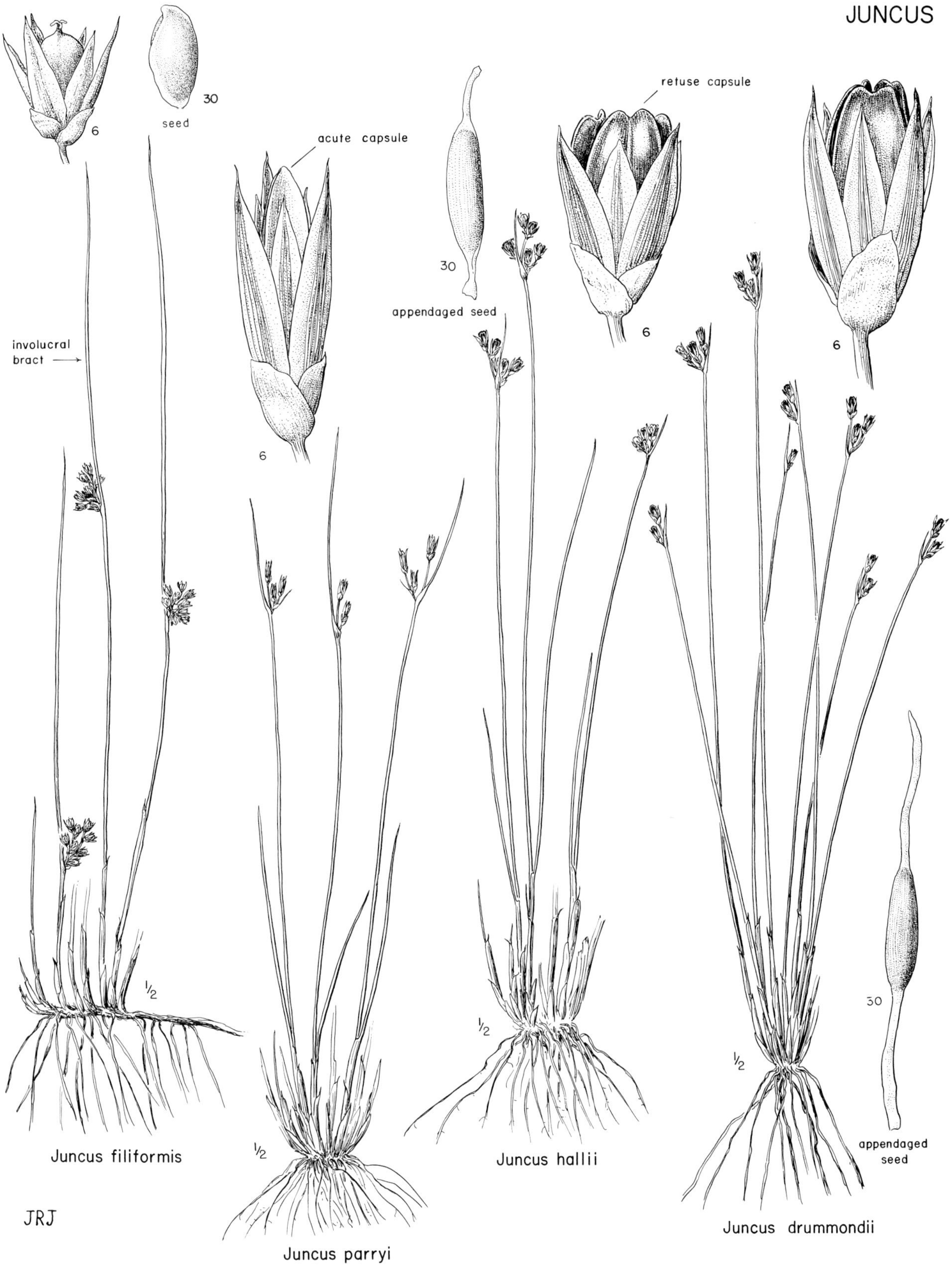
6
30
seed
acute capsule
30
appendaged seed
retuse capsule
6
6
involucral
bract
6
1/2
1/2
1/2
1/2
30
appendaged
seed
Juncus filiformis
Juncus hallii
JRJ
Juncus drummondii
Juncus parryi

acuminate, generally with greenish midstrip and thinner, castaneous (or marginally hyaline and colorless) borders; stamens 6, shorter than the perianth; capsule slender, from a little shorter to a little longer than the perianth, truncate or more often shortly retuse; seeds 1.5–2 mm long, the whitish, membranous appendage at each end about equaling or somewhat longer than the body.

Meadows, streambanks, and wooded slopes at upper elev. in the mts., sometimes above timberline; cordilleran species, from Alaska s. to the mts. of Colo., N.M., Utah (Bear River Mts., Deep Creek Mts., La Sal Mts., Uinta Mts., Wasatch Mts. and Plateau) and Nev. (Elko Co.). July–Aug.

8. **Juncus longistylis** Torr.

Juncus longistylis Torr. in Emory, Rep. U.S. & Mex. Boundary Surv. 223. 1859. (*Bigelow,* near the Copper Mines, N.M.)

Rhizomatous perennial with scattered or more or less clustered, terete stems (1) 2–6 dm tall; leaves mainly clustered toward the base, but some distinctly cauline (sometimes even above the middle of the stem), the sheath produced at the summit into a conspicuous, pale, often somewhat cartilaginous, rounded-auriculate ligule ca 1–2 mm long, the blade flat, neither septate nor equitant, mostly 4–15 cm X 1.5–3 (4) mm; inflorescence of (1) 2–8 heads on peduncles that are axillary to much reduced, bracteal leaves, each head provided with several broad, largely pale and hyaline-scarious bracts that individually subtend some or all of the flowers, but the flowers not prophyllate; heads commonly castaneous with a silvery sheen; flowers 3–10 (seldom more) in each head; tepals 5–7 mm long, green or anthocyanic with broad, hyaline, usually pale margins (the inner especially broad-margined), smooth and shining, finely cellular-striate under higher magnification, subequal or the outer a little the longer; stamens 6, the anthers 1.4–2.0 mm long, longer than the filaments; capsule broadly rounded to slightly retuse at the summit, shorter than (or seldom equaling) the perianth, incompletely 3-locular; seeds apiculate, ca 0.5 mm long.

Moist places, often along streams, from the valleys and foothills to middle elev. in the mts., s. B.C. to Sierran Calif., e. to s. Alta., N.D., S.D., Nev., and N.M., and locally across s. Can. to Que.; in appropriate habitats nearly throughout our range. June–Aug.

9. **Juncus orthophyllus** Coville

Juncus longistylis var. *latifolius* Engelm. Trans. Acad. Sci. St. Louis 2: 496. 1868. *J. latifolius* Buchenau, Bot. Jahrb. Syst. 12: 425. 1890; not Wulfen, 1789. *J. orthophyllus* Coville, Contr. U.S. Natl. Herb. 4: 207. 1893. (*Bolander 46,* Yosemite Valley, Calif., was the first of several collections cited.)

Rhizomatous perennial with scattered or more or less clustered, terete stems 1.5–6 dm tall; leaves often minutely scaberulous, mainly clustered toward the base, but more often than not some of them distinctly cauline (sometimes above the middle of the stem), the sheath with narrow (or no) hyaline margins that are sometimes prolonged into a pair of inconspicuous auricles up to ca 0.5 mm long, but not produced into a ligule, the blade flat, neither septate nor equitant, mostly 4–15 cm X 1.5–5.5 mm; inflorescence of (1) 2–9 heads on peduncles that are axillary to much reduced bracteal leaves, each head provided with several broad, largely pale and hyaline-scarious bracts that individually subtend some or all of the flowers, but the flowers not prophyllate; heads greenish to more often brown or castaneous, but mostly without the silvery sheen of *J. longistylis;* flowers 3–10 (seldom more) in each head; tepals 4.5–6 mm long, subequal or the inner a little the longer, with green, evidently (at 10X) cellular-roughened midstrip and brown or partly pale, hyaline-scarious margins, the inner sometimes more broadly hyaline-margined and more abruptly pointed, as in *J. regelii;* stamens 6, the anthers 1.6–3.2 mm long, much longer than the filaments; capsule broadly rounded to slightly retuse at the summit, shorter than the perianth, incompletely 3-locular; seeds apiculate, ca 0.5 mm long.

Meadows and streambanks, from the valleys and foothills to middle elev. in the mts.; Cascade-Sierran axis from s. Wash. to s. Calif., and inland to se. Wash., e. Oregon, sw. Idaho (Owyhee Co.), Nev. (Humboldt Co.), and the White Mts. of Calif. and Nev. June–Aug.

10. **Juncus regelii** Buchenau

Juncus regelii Buchenau, Bot. Jahrb. Syst. 12: 414. 1890. (*Suksdorf 678,* Wash.)

J. jonesii Rydb. Fl. Rocky Mts. 153, 1061. 1917 [1918]. (*M. E. Jones 1199,* Alta, Wasatch Mts., Utah, 6 Aug. 1879.)

Rhizomatous perennial with scattered or more or less clustered, terete stems 2–7 dm tall; leaves mainly clustered toward the base, but some distinctly cauline (sometimes above the middle of the stem), the sheath with narrow (or no) hyaline margins that are sometimes prolonged into a pair of inconspicuous auricles up to ca 0.5 mm long, but not produced into a ligule (or the ligule rudimentary and less than 0.5 mm long), the blade flat, neither septate nor equitant, mostly (5) 10–30 cm X 1.5–3 mm; inflorescence of 1–5 heads on peduncles that are axillary to much reduced, bracteal leaves, each head provided with several broad, largely pale and hyaline-scarious bracts that individually subtend some or all of the flowers, but the flowers not prophyllate; heads commonly brown or castaneous, at least the larger ones commonly with more than 10 flowers; tepals 4.5–6 mm long, subequal or the outer a little the longer, the outer narrow and more strongly acuminate than the broadly margined, more abruptly pointed inner, all with green (or less often brown) evidently (at 10X) cellular-roughened midstrip and brown or partly pale, hyaline-scarious margins; stamens 6, the anthers 1.2–1.8 mm long, longer than the filaments; capsule broadly rounded to slightly retuse at the summit, shorter than the perianth, incompletely 3-locular; seeds ca 1.5 mm long overall, conspicuously tailed at each end, each tail about as long as the body of the achene.

Streambanks and wet meadows at middle and upper elev. in the mts.; Cascade Mts. from s. Wash. to n. Calif., e. to w. Mont. and s. to Nev. (Elko Co.) and Utah (Wasatch and Uinta mts.). July–Aug.

JUNCUS

Juncus longistylis

Juncus orthophyllus

11. Juncus ensifolius Wikstr.

Juncus ensifolius Wikstr. Kongl. Vetensk. Acad. Handl. 2: 274. 1823. (Unalaska.)

J. xiphioides var. *montanus* Engelm. Trans. Acad. Sci. St. Louis 2: 481. 1868. *J. saximontanus* A. Nels. Bull. Torrey Bot. Club 29: 401. 1902. *J. ensifolius* var. *montanus* C. L. Hitchc. Univ. Wash. Publ. Biol. 17(1): 195. 1969. (*Spalding*, Oregon, is the first of many collections cited.)

J. xiphioides var. *triandrus* Engelm. Trans. Acad. Sci. St. Louis 2: 482. 1868. (*Eschscholtz*, Unalaska, is the first of several specimens cited.) = var. *ensifolius*.

Juncus ensifolius var. **brunnescens** (Rydb.) Cronq. hoc loc.

J. brunnescens Rydb. Bull. Torrey Bot. Club 31: 400. 1904. *J. saximontanus* f. *brunnescens* F. J. Hermann, J. Wash. Acad. Sci. 30: 218. 1940. (*C. F. Baker 245*, Pagosa Springs, Colo., July 1899.)

J. parous Rydb. Bull. Torrey Bot. Club 31: 401. 1904. (*C. L. Shear 3666*, Fort Garland, Colo.) = var. *montanus*, a form with shortly tailed seeds.

J. tracyi Rydb. Fl. Rocky Mts. 155, 1061. 1917 [1918]. (*S. M. Tracy 389*, Ogden, Utah, 2 Aug. 1887.) = var. *montanus*, a form with shortly tailed seeds.

J. utahensis R. F. Martin, Rhodora 40: 69. 1940. *J. tracyi* f. *utahensis* F. J. Hermann, J. Wash. Acad. Sci. 30: 218. 1940. (*Nord & Sargent 1*, Ashley Nat. Forest, Summit Co., Utah, 10 Sept. 1927.) = var. *brunnescens*, a form with shortly tailed seeds.

Stems scattered or loosely clustered on well developed creeping rhizomes, mostly 2–6 (7) dm tall, flattened and often narrowly margined; leaves mostly crowded on the lower half of the stem, the better developed ones mostly 10–40 cm long (including the sheath) and 1.5–6 mm wide, all equitant, folded along the midrib, the edges fully connate above the top of the narrowly prolonged and usually more or less auriculate and hyaline-scarious margins of the sheath, the edge of the leaf blade (representing the connate margins) facing the edge of the stem; leaf blade with scattered, irregular, often slanting, partial or complete partitions; heads (1) few and loosely clustered at the top of the stem, many-flowered, commonly ca 1 (1.5) cm thick, varying (in var. *brunnescens*) to not infrequently more numerous in an elongate, branching inflorescence and then smaller and with fewer flowers, commonly dark brown or blackish-purple, varying (especially in var. *brunnescens*) to paler brown or even greenish-stramineous; heads well separated or crowded and partly coalesced, the bracts subtending the heads or peduncles reduced and scarcely leafy, or the lowest one larger and leaflike; flowers axillary to evident, hyaline-scarious bracts but not prophyllate; tepals slightly unequal, the outer 2.5–4 mm long, evidently acuminate, the inner slightly (0.2–0.8 mm) shorter, tending to be a little wider and less strongly pointed; stamens 3 or 6, the antesepalous ones consistently well developed, the antepetalous ones often suppressed or sometimes imperfectly or irregularly developed; anthers 0.4–1.3 (1.5) mm long, from a little shorter to a little longer than the filaments; style 0.5–1.5 mm long, slender and fragile, the basal part persisting on the capsule as a short beak; capsule mostly a little shorter than the perianth, but sometimes equaling or even surpassing it, more or less abruptly contracted or occasionally more tapering to the stylar beak, incompletely 3-locular, the partial partitions not meeting in the center; seeds commonly apiculate at both ends and 0.4–0.6 mm long, but the apiculus occasionally prolonged into a short tail and the seed then up to 1.0 mm long overall; $n = 20$.

Wet places, often along streams, from the foothills to middle elev. in the mts., sometimes descending to lower elev., as in the hanging gardens along the Colorado River in Utah; widespread in the western cordillera, from Alaska to s. Calif. and n. Mex., e. to the Cypress Hills of Sask., the Black Hills of S.D., and the mts. of w. Texas; in appropriate habitats throughout our range.

Juncus ensifolius as here defined consists of three geographically significant but widely overlapping varieties. Var. *ensifolius* and var. *montanus* are very similar in appearance but differ sharply in the number of stamens. The staminal character might seem a priori to be of specific importance, but my observations support those of C. L. Hitchcock, who notes [Univ. Wash. Publ. Biol. 17(1): 195. 1969] that flowers with 3 and with 6 stamens sometimes occur in the same head. Inasmuch as the plants are otherwise much alike and have broadly overlapping ranges, it seems more useful to treat them as varieties of a single species, as Hitchcock has done. Var. *brunnescens*, on the other hand, is very distinctive in appearance but intergrades so much with var. *montanus*, and overlaps so much with it geographically, that the usefulness of the taxonomic separation might be questioned. Inasmuch as var. *brunnescens* does occupy a considerable area where var. *montanus* is rare or absent, however, I think the distinction is worth while.

Both var. *montanus* and var. *brunnescens* have a scattered form with more elongate, somewhat tailed seeds (0.7–1.0 mm long), elongate anthers (0.6–1.5 mm long), and elongate styles (ca 1–1.5 mm long), in contrast to the more ordinary members of the population, which have merely apiculate seeds 0.4–0.6 mm long, anthers 0.4–0.8 mm long, and styles 0.5–1 mm long. The three correlated differences all relate to length of the floral and fruiting structures and might well prove to reflect the action of the same gene or set of associated genes. Inasmuch as these long-seeded forms have neither a distinctive range, nor a distinctive habitat, nor a distinctive appearance, and inasmuch as there is no real discontinuity in measurements, they are here regarded as being without taxonomic significance. The record of *J. macrandrus* Coville from the White Mts. of California by Lloyd and Mitchell (Fl. White Mts. Calif. & Nevada) is based on two collections in flower that are here considered to represent the long-anthered, long-styled form of *J. ensifolius* var. *montanus*. No opinion is here expressed as to the taxonomic status of *J. macrandrus*.

The three varieties of *J. ensifolius* may be characterized as follows:

1 Stamens ordinarily 3, occasionally 6 in some flowers; inflorescence commonly as in var. *montanus*, but in southern forms occasionally approaching that of var. *brunnescens;* northern and western phase, occurring mostly beyond the borders of our range, but occasionally found as far south, in our range, as Harney Co., Oregon, Elko Co., Nev., and Salt Lake Co., Utah . . . . var. *ensifolius*

1 Stamens ordinarily 6, occasionally some of them imperfectly developed.

2 Heads relatively few (seldom more than 10) and many-flowered (flowers commonly 10–25), mostly 8–15 mm thick, usually dark in color, but sometimes paler as in var. *brunnescens;* the common and characteristic var. throughout our range and to the east and northeast . . . . var. *montanus* (Engelm.) C. L. Hitchc.

2 Heads relatively numerous, mostly (5) 10–50 and few flowered (flowers mostly 4–12), commonly 3–8 mm thick, usually rather light brown (varying to greenish-stramineous, or sometimes darker as in var. *montanus*); nearly throughout the range of var. *montanus*, but much commoner southward, the common phase in Ariz. and N.M., and the only phase in Texas . . . . var. *brunnescens* (Rydb.) Cronq.

The closely related but more robust, largely Californian species *J. xiphioides* E. Meyer (1822) extends into s. Nevada and has sometimes been reported from our range (as by F. J. Hermann, Contr. Toward Fl. Nevada 16: 17. 1940), but all the specimens so labeled that I have seen belong to *J. ensifolius* instead. *J. xiphioides* is commonly more than 6 dm tall, with leaves commonly more than 6 mm wide, and it has small, pale, usually more or less numerous heads of

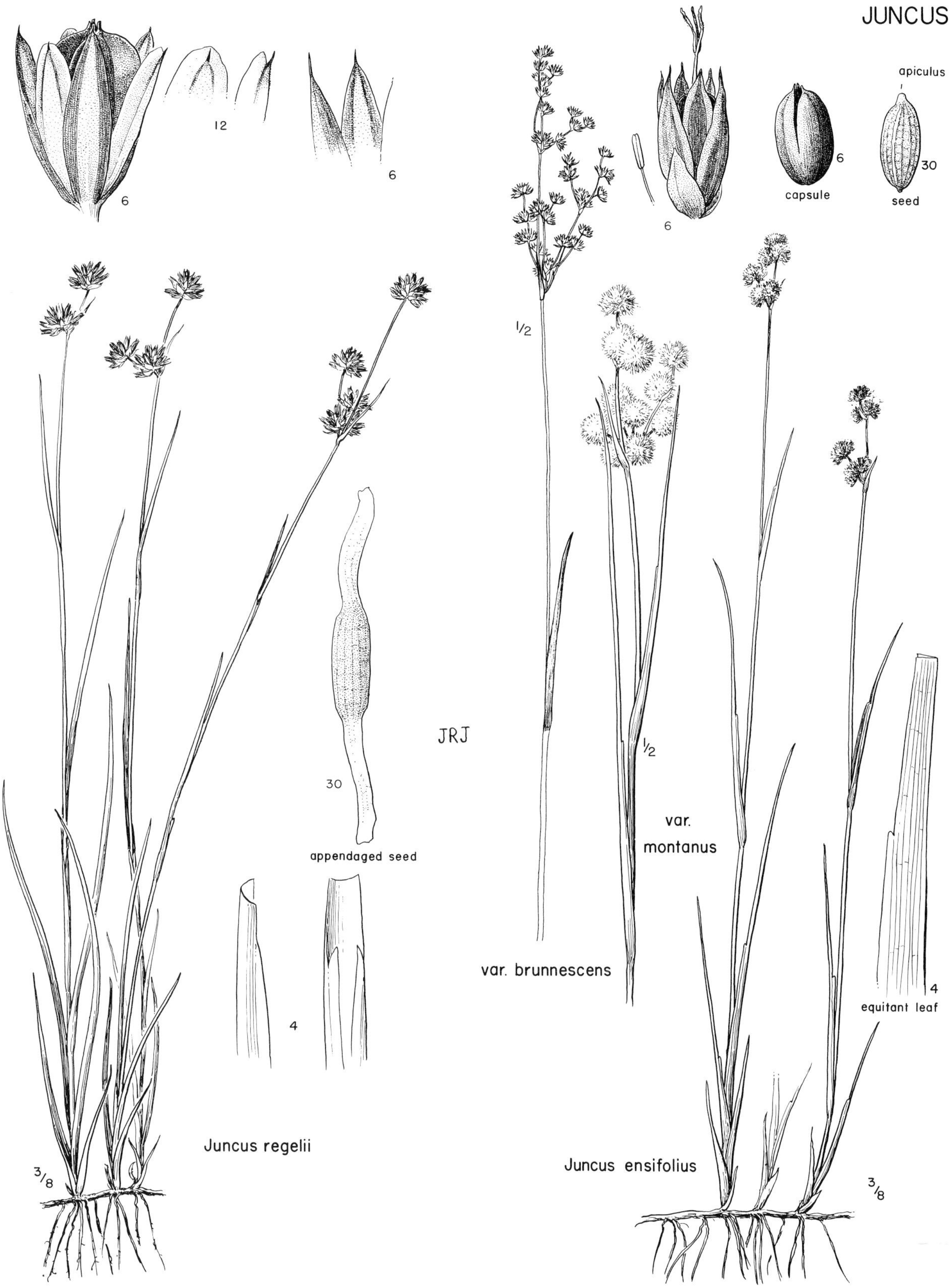
JUNCUS
12
6
6
apiculus
6
6
capsule
30
seed
6
1/2
1/2
JRJ
30
appendaged seed
var.
montanus
var. brunnescens
4
4
equitant leaf
Juncus regelii
Juncus ensifolius
3/8
3/8

flowers in which the inner tepals are often as long as the outer. It grows in moist places at lower elevations and/or in drier climates than *J. ensifolius*. It is possible that *J. ensifolius* sensu lato should be attached to *J. xiphioides* (the older name) as a series of varieties, but I am reluctant to do so without more compelling evidence.

12. **Juncus castaneus** J. E. Smith

Juncus castaneus J. E. Smith, Fl. Brit. 1: 383. 1800. (Scotland.)

Rhizomatous perennial with scattered, terete stems mostly 1–4.5 dm tall; leaves scattered along the stem, or sometimes confined to the lower portion, the sheath with narrow (or no) hyaline margins, not auriculate, often not sharply delimited from the blade, which is mostly 3–20 cm long and compressed but folded or strongly inrolled for most of its length so as to appear more or less terete, at midlength ca 1–2.5 mm wide as rolled or folded, and with membranous, cross-ribbed, largely hidden upper surface, only near the tip becoming terete, hollow, and septate; inflorescence of 1–4 compact heads that are shortly to strongly surpassed by one or more of the subtending, more or less leafy bracts; flowers mostly 3–15 in each head, not prophyllate; tepals castaneous or brownish-purple, mostly (4.5) 5–7.5 mm long, the outer a little longer and more pointed than the inner; stamens 6, the anthers mostly 0.8–1.5 mm long, much shorter than the elongate filaments; capsule rather elongate, well exserted from the perianth, unilocular but with somewhat intruded placentae, tapering to an evident stylar beak ca 1 mm long; seeds conspicuously long-tailed at each end, ca 2.5–4 mm long overall, each tail longer than the body of the achene; $n = 20, 30$.

Wet meadows and margins of streams and pools in arctic and alpine tundra; circumboreal at high latitudes, s. irregularly in the western cordillera to Mont., Utah (Uinta Mts.) and the s. Rocky Mts. of Colo. and s. Wyo. July–Aug.

13. **Juncus torreyi** Coville

Juncus nodosus var. *megacephalus* Torr. Fl. State New York 2: 326. 1843. *J. megacephalus* A. Wood, Cl.-Book Bot. new ed. of 1861. 724. 1861; not M. A. Curtis, 1835. *J. torreyi* Coville, Bull. Torrey Bot. Club 22: 303. 1895. (*A. Gray*, shores of Lake Ontario, Aug. 1834.)

Plants mostly (2) 4–10 dm tall, coarse, the stems terete, arising singly at the nodes of cord-like rhizomes that are evidently tuberous-thickened at intervals; leaves several, cauline, with a pair of prominent auricles mostly 2–5 mm long at the collar, the blade hollow, septate-nodulose, terete, or channeled and flattened on the upper side toward the base, mostly 10–30 cm long and 0.5–3.5 mm thick; heads mostly 2–10 in a congested terminal inflorescence, sometimes so closely aggregated as almost to form a secondary head, greenish-stramineous to medium brown, mostly 10–15 mm thick, many-flowered, the flowers axillary to hyaline-scarious, shortly awn-tipped bracts, but not prophyllate, spreading in all directions, the head seeming finely bristly; lowest bract commonly surpassing the inflorescence, or sometimes more or less suppressed; tepals firm and subspinulosely long-acuminate, the outer mostly 4–5 mm long, the inner somewhat shorter; stamens 6, the anthers mostly 0.5–0.8 mm long; capsule slender, strongly triquetrous, equaling or somewhat surpassing the perianth, tapering to the slender, only tardily dehiscent stylar beak, incompletely 3-celled, the partitions (placentae) only slightly intruded; seeds apiculate, 0.4–0.5 mm long; $n = 20$.

Banks of streams and ditches in the valleys and lowlands, sometimes in shallow standing water; widespread in the U.S. and s. Can., and extending into n. Mex.; common in appropriate habitats throughout our range.

Juncus torreyi is obviously related to the more slender *J. nodosus*, but the two are sharply distinct. I have seen no doubtful specimens.

14. **Juncus nodosus** L.

Juncus nodosus L. Sp. Pl. ed. 2. 466. 1762. (e. N. Amer.)

Plants mostly 1.5–4 dm tall, the slender, terete stems arising singly at the nodes of slender rhizomes that are evidently tuberous-thickened at intervals; leaves several, cauline, with a pair of small and inconspicuous auricles at the collar, the blade slender, hollow, septate-nodulose, terete, or channeled on the upper side toward the base, mostly 4–20 cm long and 0.7–1.5 mm thick; heads mostly 2–10 (15) in a loose or somewhat congested, proliferating inflorescence, greenish-stramineous to medium brown, mostly 6–9 mm thick, 5- to 25-flowered, the flowers spreading in all directions, individually axillary to hyaline-scarious bracts shorter than the perianth, but not prophyllate; lowest bract commonly equaling or surpassing the inflorescence; tepals acuminate, but broader than in *J. torreyi*, 2.5–3.5 mm long, the inner nearly or fully as long as the outer; stamens 6, the anthers 0.5–1.0 mm long; capsule slender, sharply triquetrous, equaling or commonly surpassing the perianth, tapering into a slender, only tardily dehiscent stylar beak, incompletely 3-locular, partitions (placentae) only slightly intruded; seeds apiculate, 0.4–0.5 mm long.

Streambanks, lake shores, and wet meadows, from the foothills and lowlands to middle elev. in the mts.; widespread in n. and c. U.S. and s. Can., s. in the cordilleran region to s. Calif., Ariz., N.M., and Texas, and found here and there throughout most or all of our range. July–Sept.

15. **Juncus nevadensis** S. Wats.

Juncus phaeocephalus var. *gracilis* Engelm. Trans. Acad. Sci. St. Louis 2: 484. 1868; not *J. gracilis* R. Br. 1810. *J. nevadensis* S. Wats. Proc. Amer. Acad. Arts 14: 303. 1879. *J. mertensianus* subsp. *gracilis* F. J. Hermann, Leafl. W. Bot. 10: 85. 1964. (*Hillebrand*, about the "Big Tree Grove," Calif., is the first collection cited.)

J. suksdorfii Rydb. Bull. Torrey Bot. Club 26: 541. 1899. *J. mertensianus* subsp. *gracilis* var. *suksdorfii* F. J. Hermann, Leafl. W. Bot. 10: 86. 1964. (*Suksdorf 217*, Falcon Valley, Klickitat Co., Wash., 30 June 1883.)

J. columbianus Coville, Proc. Biol. Soc. Wash. 14: 87. 1901. *J. nevadensis* var. *columbianus* St. John, Fl. S.E. Wash. & Adj. Idaho 82. 1937. *J. mertensianus* subsp. *gracilis* var. *columbianus* F. J. Hermann, Leafl. W. Bot. 10: 86. 1964. (*Elmer 235*, Pullman, Wash.)

J. badius Suksd. Deutsche Bot. Monatsschr. 19: 92. 1901. *J. mertensianus* subsp. *gracilis* var. *badius* F. J. Hermann, Leafl. W. Bot. 10: 86. 1964. *J. nevadensis* var. *badius* C. L. Hitchc. Univ. Wash. Publ. Biol. 17(1): 201. 1969. (*Suksdorf 2144*, Falcon Valley, Wash., 25 June 1892.)

Stems generally scattered on elongate rhizomes, or

JUNCUS

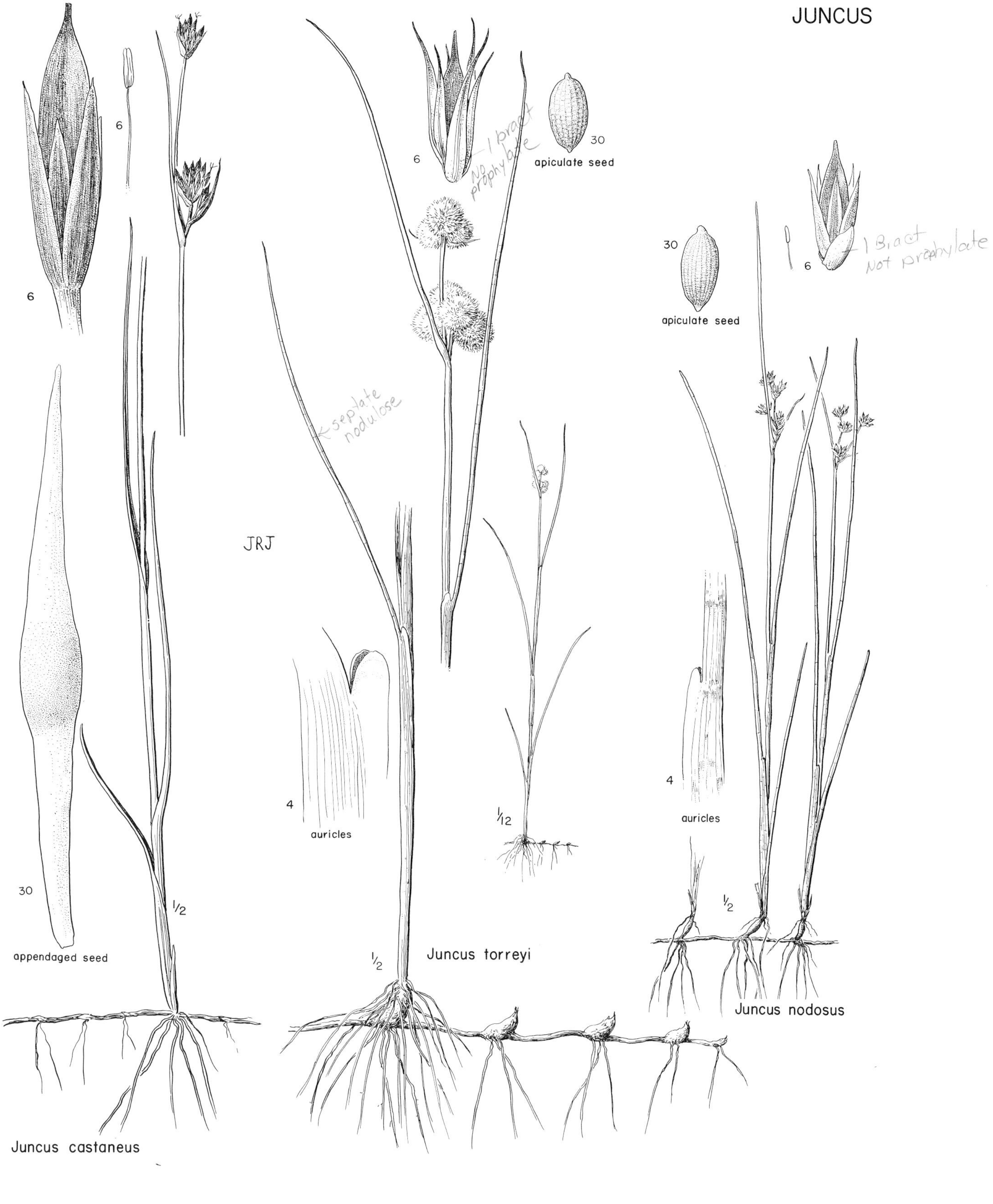
6
6
6
30
appendaged seed
1/2
Juncus castaneus
JRJ
6
—1 bract
No prophylate
30
apiculate seed
<— septate
nodulose
4
auricles
1/2
1/12
Juncus torreyi
30
apiculate seed
6
—1 Bract
Not prophylate
4
auricles
1/2
Juncus nodosus

sometimes more crowded on shorter rhizomes, terete, mostly 2–7 dm tall; leaves basal and cauline or all cauline (the lowermost sometimes reduced), the blade mostly 5–25 cm long and 0.5–2 mm thick, terete, hollow, and septate-nodulose at least distally, but often merely channeled on the upper side toward the base; hyaline margins of the sheath expanded upwards and prolonged into a pair of evident auricles 1–3 mm long; inflorescence of (2) 4–30 heads 5–10 (12) mm thick, usually surpassing the lowest bract; flowers several (seldom fewer than ca 7) or rather numerous in each head, axillary to hyaline-scarious bracts, but not prophyllate; tepals variously dark to light brown or purplish-brown, often evidently hyaline-margined, the outer gradually acuminate, (3) 3.5–5.5 mm long, the inner similar or a little broader, blunter, and often shorter; stamens 6; anthers mostly 1–2 mm long, longer than the filaments; capsule triquetrous, obtuse or broadly rounded to subtruncate or even slightly retuse beneath the stylar apiculus, slightly to evidently shorter than the perianth, 1-celled, the placentae only slightly intruded; seeds apiculate, ca 0.5 mm long.

Wet meadows, streambanks, and marshes, from the valleys to middle elev. in the mts.; s. B.C. to Alta., s. to s. Calif., Ariz., and N.M., and more or less throughout our range, but apparently not often collected in our area. June–Aug.

Although *Juncus nevadensis* is variable in general vigor and in length and color of the perianth members, I see no evidence that the species over most of its range (including our area) is divisible into geographically or ecologically significant populations with morphological markers. Therefore no infraspecific taxa are here recognized.

16. **Juncus mertensianus** Bong.

Juncus mertensianus Bong. Mem. Acad. Imp. Sci. St. Petersbourg, Ser. 6, Sci. Math. 2: 167. 1833. (*Mertens*, Sitka, Alaska.)

Stems crowded on a stout rhizome, terete, 1–2.5 dm tall; leaves all cauline, the lowermost ones reduced, the others with terete, hollow, septate-nodulose blade 4–15 cm long and 1–2 mm thick, the hyaline margins of the sheath expanded above and prolonged into a pair of evident auricles ca 1–1.5 mm long; inflorescence a single terminal head (seldom 2 heads) 8–12 mm thick, often shortly surpassed by the lowest bract; flowers numerous, axillary to hyaline-scarious bracts but not prophyllate; tepals shiny, dark purple-brown or blackish-purple, thin and soft, the outer ones gradually acuminate, (2.5) 3–3.5 (4) mm long, the inner similar or often a little shorter and less sharply pointed; stamens 6; anthers mostly 0.5–1.0 mm long, shorter than the filaments; capsule triquetrous, somewhat expanded upwards, obtuse to broadly rounded or subtruncate or even slightly retuse beneath the stylar apiculus, slightly to evidently shorter than the perianth, 1-celled, the placentae only slightly intruded; seeds apiculate, ca 0.5 mm long.

Moist or wet places, often along streams or in wet meadows or about seeps, in the mountains from middle (and sometimes lower) elev. to above timberline; widespread in the western cordillera from Alaska and Yukon to s. Calif., Nev., Utah, Colo., S.D., and reputedly Ariz. and N.M., and found in appropriate habitats nearly or quite throughout our range.

Juncus mertensianus is evidently allied to *J. nevadensis*, and there are a few doubtful or intermediate specimens, some of which may well be of hybrid origin. Nevertheless the two taxa are obviously different in a well correlated syndrome of characters, and they retain their identity over a wide area of range overlap.

17. **Juncus articulatus** L.

Juncus articulatus L. Sp. Pl. 327. 1753. (Europe.)

Stems rather closely set on a coarse rhizome, terete, 1–6 dm tall; leaves chiefly cauline, the blade mostly 2–15 cm long and 0.7–1.5 mm thick, terete, hollow, and septate-nodulose, the scarious margin of the sheath somewhat expanded distally and prolonged at the collar into a ligule or a pair of auricles (0.5) 1–2 mm long; heads green to brown or purplish, several to many in a proliferating inflorescence with more or less ascending to evidently divaricate branches, the lowest bract variously somewhat longer to often much shorter than the inflorescence; flowers mostly 3–10 per head, erect or ascending, axillary to evident, scarious bracts, but not prophyllate; tepals essentially equal, or the inner a little longer than the outer, (2) 2.5–3 mm long, more or less distinctly acute, varying to sometimes merely acutish or shortly subacuminate, sometimes with a minute awn-point; stamens 6; anthers 0.5–0.8 mm long, shorter than the filaments; capsule surpassing the perianth, triquetrous, more or less acutely tapered to the short stylar beak, unilocular, the placentae not much intruded; seeds ca 0.5 mm long, apiculate; $n = 40$.

Edges of ponds and streams; circumboreal and becoming more or less cosmopolitan by naturalization, the original limits of range now difficult to determine; in our area known from scattered stations at lower elev. in Idaho (Canyon Co.), Utah (Kane Co.), and Nev. (Elko Co.). June–Aug.

18. **Juncus alpinus** Vill.

Juncus alpinus Vill. Hist. Pl. Dauphiné 2: 233. 1787. *Phylloschoenus alpinus* Fourr. Ann. Soc. Linn. Lyon, II. 17: 172. 1869. (High mts. of France.)

J. alpinus var. *insignis* Fries ex Buchenau, Bot. Jahrb. Syst. 12: 374. 1890. (*Canby 336*, Rocky Mts., is the first of several specimens cited.)

J. alpinus var. *fuscescens* Fern. Rhodora 10: 48. 1908. (*Dudley 137*, Cayuga marshes, N.Y.)

Much like *J. articulatus*, averaging a little smaller, mostly 0.5–3 dm tall; branches of the inflorescence more or less closely ascending or erect; tepals mostly 2–2.5 mm long, the outer obtuse or rounded but sometimes minutely apiculate, the inner a little shorter, generally broadly rounded at the tip; capsule surpassing the perianth, obtuse or rounded to the stylar apiculation; $n = 20, 40$.

Marshes and shores of lakes and streams in the mts., but in our range not at high elev.; circumboreal species, extending s., in Amer. to Pa., Ind., Mo., Colo., Utah (Bear Lake, Navajo Lake) and Wash. July.

Several varieties of *J. alpinus* have been described, but I find it hard enough to distinguish *J. alpinus* from *J. articulatus*, without further dividing either species.

19. **Juncus tweedyi** Rydb.

Juncus tweedyi Rydb. Mem. New York Bot. Gard. 1: 90. 1900. (*Tweedy 223*, Yellowstone Natl. Park, in 1884.)

J. canadensis var. *kuntzei* Buchenau, Bot. Jahrb. Syst. 12: 272.

JUNCUS

30
seed
6
6
1/2
4
auricle
Juncus
nevadensis
1/2
JRJ
6
capsule
30
seed
4
auricle
1/2
Juncus mertensianus
6
30
seed
4
auricles
1/2
Juncus articulatus

JUNCUS

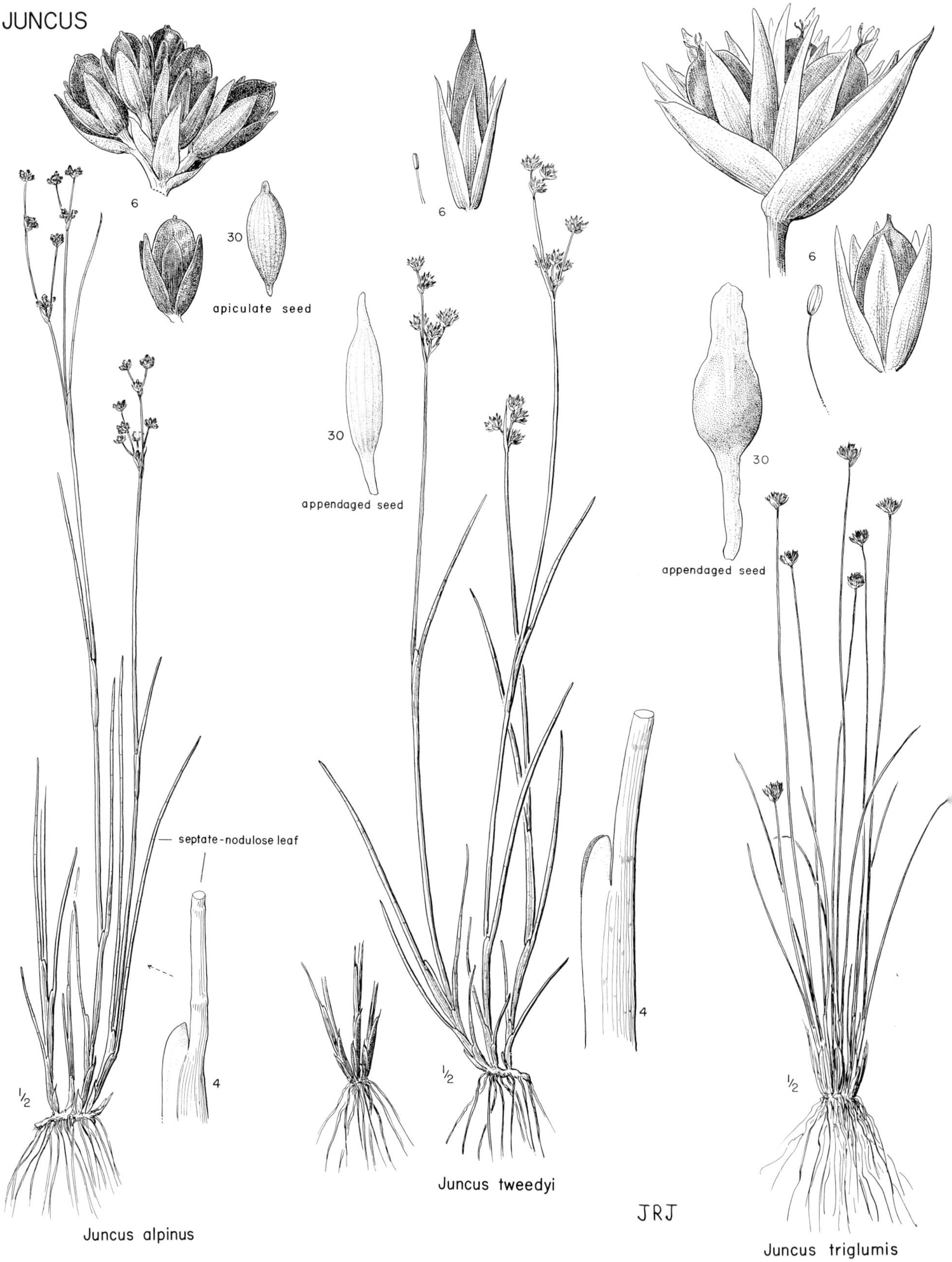
6
30
apiculate seed
septate-nodulose leaf
4
1/2
Juncus alpinus
6
30
appendaged seed
4
1/2
Juncus tweedyi
6
30
appendaged seed
1/2
Juncus triglumis
JRJ

1890. *J. kuntzei* Vierh. in Engl. & Prantl, Natürl. Pflanzenf. ed. 2. 15a: 218. 1930. (*Kuntze 3133,* near Corinne, Box Elder Co., Utah.)

Plants 2–4 dm tall, not rhizomatous, the stems clustered, terete; leaves basal and cauline, the blade mostly 5–12 cm long and 1–2.5 mm thick, hollow, septate-nodulose, terete or nearly so, the hyaline margins of the sheath prominent distally and terminating in a pair of evident auricles (0.5) 1–2 mm long at the collar; heads mostly 2–8 in a proliferating inflorescence, brown, mostly 3- to 8-flowered and 3–8 mm wide, the flowers all erect or ascending, axillary to evident scarious bracts, but not prophyllate; lowest bract shorter to somewhat longer than the inflorescence; tepals gradually acute, 3–4 mm long, the inner equaling or a little longer than the outer; stamens 3; anthers 0.5–0.7 mm long, shorter than the filaments; capsule slightly longer than the perianth, triquetrous, more or less acute (but not subulate) and tipped by a short stylar beak, imperfectly 3-celled, the partitions not meeting in the center; seeds with cylindric body and a short tail at each end, 0.8–1.0 mm long overall.

Wet places about hot springs; Yellowstone Natl. Park, Wyo., and perhaps extending into adjacent Mont.; Box Elder Co., Utah, the type station for *J. canadensis* var. *kuntzei.* I have seen Kuntze's type (NY) and am convinced that it is taxonomically identical with the plants from Yellowstone Park.

20. Juncus triglumis L.

Juncus triglumis L. Sp. Pl. 328. 1753. (Lappland.)

J. triglumis var. *albescens* Lange, Consp. Fl. Groenlandicae 123. 1880. *J. albescens* Fern. Rhodora 26: 202. 1924 [1925]. *J. triglumis* subsp. *albescens* Hultén, Kongl. Svenska Vetenskapsakad. Handl. IV. 8(5): 241. 1962. (Greenland.)

Stems slender, densely tufted, 5–20 cm tall; leaves all crowded toward the base; hyaline margins of the sheath expanded above and prolonged into a pair of broadly rounded ligular auricles ca 0.5 mm long; leaf blades mostly 1.5–7 cm long, very slender, commonly ca 0.5 mm thick, but hollow and septate-nodulose; head terminal and solitary, 2- to several-flowered; flowers axillary to more or less hyaline scarious (often largely whitish) bracts, but not prophyllate, the lowest bract often with a foliaceous awn-tip shortly surpassing the head, varying to merely acuminate and wholly scarious and barely surpassing the flowers, or even obtuse and shorter than the flowers; tepals mostly 3.5–5 mm long, often whitish, especially distally, but sometimes more castaneous, especially toward the base; stamens 6, the anthers 0.5–0.8 mm long, evidently shorter than the filaments; capsule mostly 2.5–4 mm long, blunt or subtruncate to slightly retuse, shorter to barely (0.5 mm) longer than the perianth; seeds conspicuously tailed at each end, ca 1.5–2 mm long overall, each tail nearly as long as the body; $2n$ = ca 130, ca 134 (Greenland).

Moist, often turfy or peaty places at high elev. in the mts., generally near or above timberline; circumboreal, s. to Utah (Uinta Mts.), N.M., and Que. July–Aug.

Our plants, as here described, belong to the largely North American var. *albescens* Lange. The Eurasian and high-boreal var. *triglumis* generally has more elongate capsules 4–6 mm long and exserted 0.5–2 mm beyond the tepals, with longer-tailed seeds mostly 2–3 mm long, and more often has the outermost bract blunt and shorter than the flowers, but none of these distinctions is wholly dependable.

21. Juncus bufonius L.

Juncus bufonius L. Sp. Pl. 328. 1753. (Europe.)

J. sphaerocarpus sensu Amer. authors, perhaps not of Nees ex Funk, Flora 1: 521. 1818.

Annual, (1) 5–30 cm tall, the stems terete, sometimes branched; leaves basal and cauline, or all basal in depauperate plants, capillary, 0.5–10 cm long and 0.2–1.1 mm wide, convex on the lower surface, flat or channeled on the upper; flowers scattered singly along the upper part (often more than half) of the stem, short-pedicellate or subsessile in the axils of hyaline-scarious or partly leafy bracts (seldom 2 or 3 together at some nodes), and each one also closely subtended by a pair of broad, hyaline-scarious bracts 1–2.5 mm long, the inflorescence sometimes reduced to a single terminal flower in depauperate extremes; tepals slender, acuminate, with greenish midstrip and hyaline margins, mostly (2.5) 3–6 (7) mm long, the outer a little longer than the inner; stamens 6; anthers 0.3–1.0 mm long, much shorter than to about as long as the filaments; style very short; capsule shorter than the tepals, imperfectly trilocular, the intruded placentae (partial partitions) not meeting in the center; seeds 0.3–0.6 mm long, apiculate; n = 15, 30, 60.

Moist or vernally wet or moist places, from the valleys and lowlands to middle elev. in the mts.; nearly cosmopolitan, and throughout our range.

The species exists at several ploidy-levels and is morphologically highly variable, but efforts to subdivide it have been unsuccessful. The depauperate extreme might be confused with *J. kelloggii* on casual inspection, but the difference in number of stamens appears to be constant. The smaller-flowered specimens have often been called *J. sphaerocarpus* Nees ex Funk, but that name may properly apply to a related, Mediterranean species that is not known in North America.

22. Juncus kelloggii Engelm.

Juncus kelloggii Engelm. Trans. Acad. Sci. St. Louis 2: 494. 1868. (*Kellogg,* San Francisco, Calif., 22 April 1866.)

J. triformis var. *brachystylus* Engelm. Trans. Acad. Sci. St. Louis 2: 492. 1868. *J. brachystylus* Piper, Contr. U.S. Natl. Herb. 11: 181. 1906. (*Bolander 31,* Ukiah, Mendocino Co., Calif.)

J. uncialis Greene, Pittonia 2: 105. 1890. (Low, moist fields near Suisun, Calif., May 1890.)

J. capillaris F. J. Hermann, Leafl. W. Bot. 5: 116. 1948. (*Parry,* Yosemite, Mariposa Co., Calif., in 1881.)

J. bryoides F. J. Hermann, Leafl. W. Bot. 5: 117. 1948. (*Parish 1859,* Bear Valley, San Bernardino Mts., Calif.)

J. hemiendytus F. J. Hermann, Leafl. W. Bot. 5: 118. 1948. (*Howell 13213,* Big Flat, Trinity Co., Calif.)

J. abjectus F. J. Hermann, Leafl. W. Bot. 5: 120. 1948. (*Peck 13893,* 7 mi. w. of Riley, Harney Co., Oregon.)

Diminutive annual, 0.5–5 cm tall; leaves clustered at the base, capillary, 0.4–4 cm long and 0.2–0.4 mm wide, rounded on the lower surface, flat or channeled on the upper; stems clustered, capillary;

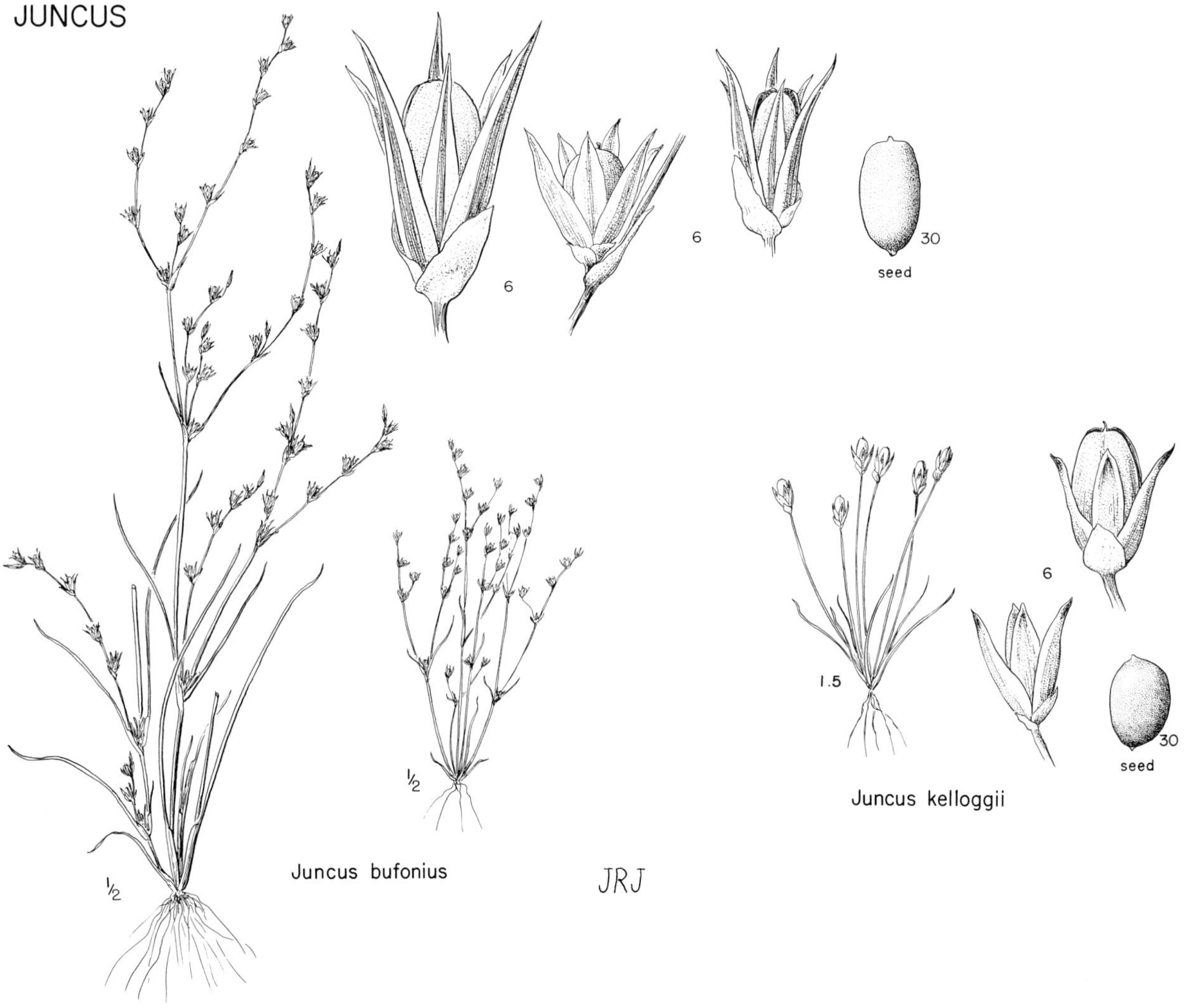

flowers 1–3 together at the top of the stem; when 2–3 flowers are present the inflorescence is collectively subtended by a pair of small, more or less hyaline bracts, but the flowers are not individually bracteolate; when only a single flower is present it is subtended by a pair of small, more or less hyaline bracts, but one or both of these bracts may be reduced or wanting; flowers trimerous, as in other species of the genus, or sometimes some of them casually dimerous; tepals 2–3 mm long, with greenish midstrip and hyaline margins; stamens (2) 3, antesepalous; anthers 0.2–0.6 mm long, shorter than the filaments; style short (to 0.25 mm long) or obsolete; capsule imperfectly trilocular, the intruded placentae (partial partitions) not meeting in the center; seeds 0.3–0.6 mm long, apiculate.

Moist or vernally wet places, often about seeps or around the margins or in the beds of vernal pools; s. B.C. to Baja Calif., e. to Oregon (Steens Mts.), c. and se. Idaho (Custer and Franklin cos.), Utah, and Colo. (Moffat Co.); evidently widespread in our range, but seldom collected; perhaps usually overlooked.

The several segregate species proposed by Hermann appear to be mere technical variants, often locally constant as in self-pollinated groups in other genera, but with widely overlapping ranges and similar habitat requirements.

2. LUZULA DC. Woodrush

Perennial, grasslike herbs, generally with some long, spreading hairs along the margins of the leaf blades, at least when young; leaves divided into sheath and blade, the sheath closed, often hyaline ventrally, the blade flat and grasslike; inflorescence varying from open, lax and subumbellately paniculiform to composed of several compact spikes or heads, these sometimes clustered to form a secondary spike, or head, flowers perfect, generally prophyllate as in species of *Juncus,* the flower or its pedicel axillary to a usually hyaline-scarious bract; tepals 6, in 2 more or less similar sets of 3, persistent, dry and more or less scarious or chartaceous to hyaline, commonly greenish or stramineous to brownish or blackish, not at all petaloid; stamens 6, commonly persistent into fruit; ovary tricarpellate, unilocular, with 3 ovules on a basal placenta;

capsule dehiscent at the angles (along the midlines of the carpels); seeds 3, relatively large. *Juncoides* Adans.; also spelled *Juncodes*.

About 75 species, mostly of temp. and cool regions, especially in the N. Hemisphere. (Name said to be from *Gramen Luzulae,* the classical Latin name for one of the species, which supposedly shines when wet with dew; *luzulae* is a diminutive from Latin *lux,* light.)

Reference: Hämet-Ahti, L. A synopsis of the species of *Luzula,* subgenus Anthelaea Griseb. (Juncaceae) indigenous in North America. Ann. Bot. Fenn. 8: 368–381. 1971.

1 Flowers borne singly or 2–3 together at the slender branch-tips of a lax, open, many-flowered inflorescence . . . . 1. *L. parviflora*
1 Flowers borne in 2–several compact (sometimes headlike) spikes, each spike generally with (5) 8–20 sessile or subsessile flowers.
 2 Seeds with a conspicuous caruncle mostly 0.3–0.6 mm long; plants of middle altitudes in the mts. (in our area), with the larger leaves mostly 3–6 mm wide . . . . 2. *L. campestris*
 2 Seeds without a caruncle, or the caruncle poorly developed, ill-defined, scarcely 0.2 mm long; alpine and subalpine plants with the leaves 1–3 mm wide . . . . 3. *L. spicata*

1. Luzula parviflora (Ehrh.) Desvaux

Juncus parviflorus Ehrh. Beitr. Naturk. 6: 139. 1791. *Luzula parviflora* Desvaux, J. Bot. (Desvaux) 1: 144. 1808. *L. spadicea* var. *parviflora* E. Meyer, Linnaea 22: 402. 1849. *Juncoides spadiceum* var. *parviflorum* Kuntze, Revisio Gen. Pl. 2: 724. 1891 (as *Juncodes*). *Juncoides parviflorum* Coville, Contr. U. S. Natl. Herb. 4: 209. 1893. (Europe.)

L. fastigiata E. Meyer, Syn. Luz. 9. 1823. *L. parviflora* var. *fastigiata* Buchenau, Krit. Verz. Juncaceen 83. 1880. *L. parviflora* subsp. *fastigiata* Hämet-Ahti, Ann. Bot. Fenn. 8: 374. 1971. (*Haenke,* Nutka.)

Stems more or less clustered on rather short rhizomes, 2.5–6 dm tall; leaves basal and cauline, the blade 4–20 cm long and 3–13 mm wide, glabrous at maturity; inflorescence subumbellately cymose-paniculate, loose and open, with slender, lax or nodding branches, 4–15 cm long and often nearly or fully as wide; flowers numerous, borne singly or in 2's or 3's at the branchlet-tips; tepals brownish, often partly hyaline-scarious, lanceolate or ovate, acute or acuminate, 1.5–2.5 mm long, the outer a little longer than the inner; prophylls evident, often half as long as the tepals, sometimes somewhat lacerate-ciliate; anthers 0.3–0.6 mm long, equaling or shorter than the filaments; capsule equaling or slightly surpassing the tepals; $n = 12$.

Along streams and in woodlands at middle and upper-middle alt. in the mts., in our area not extending above timberline; circumboreal, extending s. in Amer. to Calif., Ariz., N.M., Minn., and N.Y., and found in appropriate habitats nearly throughout our range. June–Aug.

Some of our plants of *L. parviflora* have often been incorrectly referred to *L. wahlenbergii* Rupr., a strictly northern, circumboreal species with much reduced cauline leaves and more strongly lacerate-fimbriate bracts and bracteoles, and to *L. piperi* (Coville) M. E. Jones, a cordilleran close relative of *L. wahlenbergii* that approaches our range from the north.

2. Luzula campestris (L.) DC.

Juncus campestris L. Sp. Pl. 329. 1753. *Luzula campestris* DC. in Lam. & DC. Fl. Fr. ed. 3. 3: 161. 1805. *Luciola campestris* J. E. Smith, Engl. Fl. 2: 181. 1824. *Gymnodes campestris* Fourr. Ann. Soc. Linn. Lyon, II. 17: 172. 1869. *Juncoides campestre* Kuntze, Revisio Gen. Pl. 2: 722. 1891 (as *Juncodes*). *Cyperella campestris* MacMillan, Metasp. Minnesota Valley 143. 1892. (Europe.)

Juncus multiflorus Ehrh. Calamariae Exsiccatae 127. 1791. *Luzula multiflora* Lej. Fl. Environs Spa 1: 169. 1811. *L. intermedia* var. *multiflora* Spenner, Fl. Friburg. 178. 1825. *Gymnodes multiflora* Fourr. Ann. Soc. Linn. Lyon, II. 17: 173. 1869. *L. campestris* var. *multiflora* Celak. Prodr. Fl. Böhm. 85. 1869. *Cyperella campestris* var. *multiflora* MacMillan, Metasp. Minnesota Valley 143. 1892. *Juncoides campestre* var. *multiflorum* Sheldon, Minnesota Bot. Stud. 1(9): 65. 1894 (as *Juncodes*). (Europe.)

Juncus intermedius Thuill. Fl. Environs Paris, ed. 2. 178. 1799. *Juncoides intermedium* Rydb. Bull. Torrey Bot. Club 32: 610. 1905. *L. intermedia* A. Nels. in Coulter & Nels. New Manual Bot. Centr. Rocky Mts. 109. 1909. (France.)

L. comosa E. Meyer, Syn. Luzularum 21. 1823. *Juncoides campestre* var. *comosum* Kuntze, Revisio Gen. Pl. 2: 724. 1891 (as *Juncodes*). *Juncoides comosum* Sheldon, Minnesota Bot. Stud. 1(9): 64. 1894 (as *Juncodes*). *L. campestris* var. *comosa* Fern. & Wieg. Rhodora 15: 41. 1913. *L. multiflora* subsp. *comosa* Hultén, Fl. Alaska & Yukon 3: 437. 1943. *L. multiflora* var. *comosa* St. John, Leafl. W. Bot. 9: 96. 1960. (*Haenke,* Nootka Sound.)

Stems tufted, 1–5 dm tall; leaves basal and cauline, the better developed ones mostly 4–12 cm long and 3–6 mm wide, some of the long marginal hairs generally persistent; inflorescence of 2–several compact, often short and headlike spikes, at least some of these generally slender-pedunculate, but sometimes all essentially sessile in a clump, although not forming the sort of slender, compound spike found in *L. spicata;* each spike with (5) 8–20 or more sessile or subsessile, prophyllate flowers axillary to pale, more or less hyaline-scarious, often fimbriate-lacerate bracts; tepals brownish or purplish-brown to greenish-stramineous, mostly 2.5–3 (3.5) mm long, acuminate but scarcely caudate, the outer a little longer than the inner, the prophylls evident, often half as long as the perianth, often more or less lacerate-ciliate; anthers mostly 0.5–1.1 mm long, the filaments about as long or a little shorter; capsule equaling or a little shorter than the tepals; seeds 1.4–1.7 mm long, with a conspicuous, whitish, basal caruncle mostly 0.3–0.6 mm long; $2n = 12, 13, 14, 16, 18, 24, 36, 48$.

Moist places, in our area mostly along streams at middle alt. in the mts., widespread in temp. and cool-temp. regions of the world, and known from scattered stations in Idaho, Utah, and Nev. June–Aug.

Luzula campestris as here defined is a highly variable polyploid complex. Various authors have proposed numerous segregate species, subspecies, and varieties, but there has been no comprehensive treatment on a world-wide basis since that of Buchenau in 1906. In the absence of such a treatment, the significance of the segregates and the proper application of the names for them are doubtful. Therefore in the synonymy given above I have avoided indicating the precise infraspecific disposition of *Juncus intermedius, Luzula comosa,* and their associated nomenclatural synonyms. Our material all

LUZULA

seed
15
15
seeds
basal caruncle
15
seed
8
8
8
1/2
1/2
1/2
4
Luzula parviflora
Luzula campestris
Luzula spicata
JRJ

appears to belong to the same ultimate taxon, here doubtfully called var. *multiflora* (Ehrh.) Celak. The name *Luzula campestris* var. *multiflora* is based on a European type, and it is possible that a new varietal name will have to be provided for the western American plants that are usually treated as *L. campestris* var. *multiflora* or *L. multiflora*.

3. Luzula spicata (L.) DC.

Juncus spicatus L. Sp. Pl. 330. 1753. *Luzula spicata* DC. in Lam. & DC. Fl. France, ed. 3. 3: 161. 1805. *Luciola spicata* J. E. Smith, Engl. Fl. 2: 182. 1824. *Gymnodes spicata* Fourr. Ann. Soc. Linn. Lyon, II. 17: 172. 1869. *Juncoides spicatum* Kuntze, Revisio Gen. Pl. 2: 725. 1891 (as *Juncodes*). (Lappland.)

Luzula cusickii Gandoger, Bull. Soc. Bot. France 66: 295. 1919. (*Cusick 2248*, Anthony's Creek, Blue Mts., Oregon.)

Stems tufted, 1–3 dm tall; leaves mainly basal or on the lower part of the stem, slender, mostly 3–10 cm long and 1–3 mm wide, some of the marginal hairs persistent; inflorescence of several sessile or subsessile heads or short spikes aggregated to form a continuous or basally interrupted compound spike 1–3 cm long, this often nodding; individual spikes several-flowered, the flowers sessile or nearly so in the axils of fimbriate-ciliate bracts, the prophylls similar to these bracts, but smaller and sometimes irregularly developed; tepals dark or partly pale and hyaline, caudately acuminate, 2–3 mm long, the outer a little longer than the inner; anthers mostly 0.3–0.5 mm long, half as long to about as long as the filaments; capsule a little shorter than the perianth; seeds 1.0–1.3 mm long, without a caruncle, or with a poorly developed and ill-defined caruncle scarcely 0.2 mm long; $n = 6, 7, 10, 11, 12$.

Open, rocky or grassy places at high alt. in the mts., commonly above timberline; circumboreal, and widespread in the western cordillera, from Alaska to Calif., Ariz., and Colo.; found in all the higher mts. of our range. June–Aug.

Order **CYPERALES**

Monocotyledonous plants with well developed vessels in all vegetative organs; leaves commonly with a basal, open or closed sheath and a terminal, elongate, narrow blade, often with the juncture of blade and sheath marked by a thickened collar on the outer side and a membranous appendage (ligule) on the inner, or the blade sometimes more or less reduced or wanting; flowers borne in characteristic, bracteate spikes or spikelets that are commonly arranged into a secondary inflorescence of various form and structure, perfect or unisexual; perianth wanting, or of 2 or 3 tiny scales, or of 1–many (often 6) short or elongate bristles; stamens mostly 3, seldom 1, 2, or 6; pollen in monads, which in the Cyperaceae are considered to be cryptotetrads with a single microspore of each initial tetrad surviving to become the pollen grain; ovary generally 2- or 3-carpellate, with as many styles or style branches or stigmas as carpels, but unilocular, with a single ovule; fruit indehiscent, generally dry, forming a caryopsis or achene; endosperm largely starchy.

The order Cyperales consists of two very large families, the Cyperaceae and Poaceae, both well represented in our range.

Synoptical Arrangement of the Families of Cyperales

1 Flowers spirally or less often distichously arranged on the axis of the spike or spikelet, usually each flower seemingly or actually subtended by a single bract, without an evident bract between the flower and the axis; seed coat generally free from the pericarp; leaf sheath usually closed; stem usually solid, often triangular; flowers often with a perianth of evident bristles; carpels 3 or less often 2; ovules anatropous; embryo embedded in the endosperm .... Cyperaceae, p. 67

1 Flowers distichously arranged on the axis of the spikelet (or only one per spikelet), each flower ordinarily subtended by a pair of bracts (lemma and palea), the palea inserted between the flower and the axis; seed coat generally adnate to the pericarp; leaf sheath usually open; stem usually hollow, never triangular; flowers without a perianth, unless the lodicules are so interpreted; carpels 2 (in all ours), rarely 3; ovules commonly orthotropous, varying to hemianatropous or even anatropous; embryo peripheral to the endosperm .... Poaceae, p. 175

Family **CYPERACEAE,** the Sedge Family

Fibrous-rooted, often rhizomatous herbs, frequently grasslike in aspect, the culm triangular to less often terete, solid or seldom hollow, the leaves mostly 3-ranked, with closed (rarely open) sheath and a parallel-veined, typically elongate and grasslike blade, or some or all the leaves with reduced or no blade; flowers much reduced, perfect or often unisexual (the plants then monoecious or seldom dioecious), sessile, in spikes or spikelets, axillary (or at least seemingly so) to small bracts called scales, spirally or less often distichously arranged on the axis of the spike or spikelet; perianth of 1–many (often 6) short or elongate bristles, or none; stamens 3, less commonly only 1 or 2, exserted at anthesis, the plants wind-pollinated; ovary superior, tricarpellate or less often bicarpellate, 1-celled with a single basal anatropous ovule, ripening into an achene; style trifid or bifid, according to the number of carpels; endosperm well developed; embryo embedded in the endosperm.

A large, cosmopolitan family with nearly 4000 species, commonest in moist or wet places.

Style: 3 > △   Style 2: ⬭

1 Achene not enclosed or enwrapped in a perigynium.
  2 Scales of the spikelet spirally arranged.
    3 Style not thickened, although the achene may have a stylar apiculus; spikelets several or many in most spp., solitary in a few.
      4 Achene subtended by 1–many bristles (or these rarely obsolete) in addition to the scale of the spikelet; stamens (2) 3; most species perennial; achenes lenticular or often trigonous.
        5 Perianth bristles 0–6, in most species not exceeding the scales (2 spp. of *Eleocharis,* with the basal stylar thickening not sharply set off from the body of the achene, might also be sought here) .... 1. *Scirpus,* p. 68
        5 Perianth bristles numerous (more than 10), conspicuous and much elongate .... 2. *Eriophorum,* p. 80
      4 Achene subtended by an inconspicuous, hyaline, adaxial scale (or this rarely obsolete) in addition to the ordinary scale of the spikelet; small annuals with 1 (2) stamens and lenticular achenes .... 3. *Hemicarpha,* p. 81
    3 Style thickened toward the base, the thickened part often persistent on the achene as a tubercle.
      6 Spikelet solitary; base of the style persistent on the achene as a tubercle; perianth bristles usually present .... 4. *Eleocharis,* p. 81
      6 Spikelets several or many; style wholly deciduous; perianth none .... 5. *Fimbristylis,* p. 88
  2 Scales of the spikelet distichous, arranged in 2 vertical ranks .... 6. *Cyperus,* p. 88
1 Achene enclosed or enwrapped in a perigynium, as well as subtended by a scale.
  7 Perigynium open, with unsealed margins, merely wrapped around the achene .... 7. *Kobresia,* p. 95
  7 Perigynium closed (except for the apical orifice), though often showing a dorsal suture distally .... 8. *Carex,* p. 95

1. SCIRPUS L. Bulrush

Herbs with triquetrous or terete, mostly solid stems, the leaves with closed sheath and elongate, grasslike blade, or the blade sometimes much reduced or obsolete, in some species the foliage represented only by scarcely sheathing basal scales; spikelets 1–many in capitate to umbelliform or paniculiform inflorescences of basically cymose nature, the inflorescence subtended by 1–several large and leafy to small and scale-like bracts, the spikelet sometimes solitary and with the subtending bract(s) not much different from the scales except for being empty; scales of the spikelet spirally arranged, scarious, with or without an excurrent awn-tip; flowers perfect, borne singly in the axils of the scales; perianth of 1–6 bristles, these shorter to much longer than the achene, sometimes conspicuously exserted from the spikelet, or sometimes more or less reduced or even obsolete; stamens 3, or sometimes fewer; style bifid or trifid, the achene accordingly lenticular or more or less trigonous; achenes commonly with a short, slender, stylar apiculus (this wanting in some species), but without an apical tubercle, the style deciduous.

A cosmopolitan genus of about 175 species. (The classical Latin name for some of the species.)

References:

Beetle, A. A. *Scirpus,* in N. Amer. Fl. 18(8): 481–504. 1947.
Koyama, T. Taxonomic study of the genus *Scirpus* Linné. J. Fac. Sci. Univ. Tokyo, Sect. 3, Bot. 7: 271–366. 1958.
———. The genus *Scirpus* Linn. Some North American aphylloid species. Canad. J. Bot. 40: 913–937. 1962.
———. The genus *Scirpus* Linn. Critical species of the section Pterolepis. Canad. J. Bot. 41: 1107–1131. 1963.
Schuyler, A. E. A taxonomic revision of North American leafy species of *Scirpus.* Proc. Acad. Nat. Sci. Philadelphia 119: 295–323. 1967.

1 Involucre very different from the scales, at least one of the bracts more or less elongate (at least 1 cm long) and either leaflike or resembling a prolongation of the culm.
  2 Well developed involucral bracts 2 or more, commonly very unequal, but spreading, green, and leafy-textured, the inflorescence thus evidently terminal.
    3 Spikelets small, mostly 3–6 mm long, very numerous (more than 100) in a branching inflorescence (sect. Taphrogeton).
      4 Pistil tricarpellate; sheaths not anthocyanic; midrib of the scales conspicuously exserted as a short, stiff awn .... 1. *S. pallidus*
      4 Pistil bicarpellate; sheaths mostly anthocyanic; midrib of the scales not exserted, or only very shortly and inconspicuously so .... 2. *S. microcarpus*
    3 Spikelets larger, mostly 12–25 mm long, not very numerous (commonly 3–40), either all sessile in a compact terminal cluster, or the principal cluster surpassed by one or more rays each bearing a subsidiary cluster or a single spikelet (sect. Bulboschoenus).
      5 Pistil tricarpellate .... 3. *S. fluviatilis*
      5 Pistil bicarpellate .... 4. *S. maritimus*

2 Well developed involucral bract solitary, green, leafy or more often erect and resembling a prolongation of the culm, so that the inflorescence appears to be lateral near the tip of the stem rather than truly terminal; smaller involucral bracts sometimes present, but scale-like and not chlorophyllous.
6 Rhizomatous perennials, often well over 3 dm tall; achenes smooth or cellular-reticulate (sect. Pterolepis, sensu lato).
7 Spikelets numerous (more than 15) in a branching inflorescence, often sessile in clusters at the ends of the branches; culms terete (sect. Pterolepis, sensu stricto).
8 Spikelets appearing dull gray-brown, the individual scales (at 10X) with prominent red-brown striolae on a pale, gray-white background; scales, at least the middle and lower ones, mostly (3) 3.5–4 mm long .... 5. *S. acutus*
8 Spikelets appearing more reddish-brown, the individual scales with a brownish or tawny ground-color, so that the red-brown striolae are usually not prominent; scales mostly (2) 2.5–3 (3.5) mm long .... 6. *S. validus*
7 Spikelets solitary or few (up to 10 or seldom 15), sessile in a sessile cluster (sect. Schoenoplectus).
9 Spikelets usually 2 or more, occasionally solitary on some culms; plants sometimes growing in shallow water, but the culms stiff and emergent for most of their length.
10 Achenes evidently apiculate, not cellular-reticulate even at 20X; culms more or less distinctly triquetrous.
11 Bract solitary; culms very sharply triquetrous, with notably concave sides; achenes 1.8–2.5 mm long and 1.4–1.7 mm wide .... 7. *S. americanus*
11 Bracts 2 or 3, the second and third ones resembling enlarged scales of the spikelet, but not subtending flowers; culms not very sharply triquetrous, the sides plane to slightly concave or slightly convex; achenes 2.2–3.3 mm long and 1.6–2.3 mm wide .... 8. *S. pungens*
10 Achenes not at all apiculate, evidently cellular-reticulate at 10X; culms subterete .... 9. *S. nevadensis*
9 Spikelet strictly solitary; plants usually aquatic and with flaccid, distally floating stems and leaves .... 10. *S. subterminalis*
6 Fibrous-rooted annuals, seldom over 3 dm tall; achenes cross-ridged (sect. Actaeogeton) .... 11. *S. supinus*
1 Involucre consisting only of 2 or 3 slightly modified empty scales less than 1 cm long at the base of the solitary spikelet, which appears to be strictly terminal; lowest scale with a broad, blunt-tipped awn 1–3 mm long that sometimes shortly surpasses the spikelet; culms 1–4 dm tall, densely tufted, without long rhizomes (sect. Baeothyron) .... 12. *S. cespitosus*

1. Scirpus pallidus (Britton) Fern.

Scirpus atrovirens var. *pallidus* Britton, Trans. New York Acad. Sci. 9: 14. 1889. *S. pallidus* Fern. Rhodora 8: 163. 1906. *S. atrovirens* subsp. *pallidus* D. Löve & Bernard, Svensk Bot. Tidskr. 53: 379. 1959. (*Palmer 358*, Indian Territory, the first specimen cited, is generally regarded as the type.)

Coarse, leafy-stemmed perennial from short, stout rhizomes, the stems tending to be loosely clustered, evidently triquetrous, (4) 6–15 (18) dm tall; leaves several, with flat, grasslike blade commonly 6–15 (18) mm wide and up to several dm long, the sheaths greenish or pale, not anthocyanic; spikelets mostly 3–4 mm long, very numerous, sessile in dense, often compound clusters that are usually at least 1 cm thick, the clusters arranged in an umbelliform, sometimes compound terminal cyme that is subtended by several conspicuous, leaflike, unequal sheathless bracts, the largest of these seldom less than 1 dm long, the inflorescence appearing less diffuse and with larger clusters than in *S. microcarpus;* scales with a distally prominent, thickened, commonly greenish or brownish midrib which is exserted as a short awn commonly about 0.5 mm long, the body of the scale otherwise greenish-black, hyaline, and cellular-reticulate; bristles mostly 6, slender, equaling or more often shorter than the achene, minutely and retrorsely barbellate above the middle; achene pale, unequally trigonous (the broad side ventral), commonly 1.0–1.2 mm long, including the minute stylar apiculus.

Wet, low ground, from the lowlands to moderate elev. in the mts., se. Wash., and ne. Oregon to Utah, N.M., and Texas, e. to Minn. and Mo. June–Aug.

Scirpus pallidus is closely related to the more eastern *S. atrovirens* Willd., of which it is sometimes regarded as a geographic variety. It is here retained as a species, following revisionary studies by A. E. Schuyler.

2. Scirpus microcarpus C. Presl

Scirpus microcarpus C. Presl, Reliquiae Haenkeanae 1: 195. 1828. *S. sylvaticus* var. *microcarpus* MacMillan, Metasp. Minnesota Valley 97. 1892. (*Haenke,* Nootka Sound.)

S. lenticularis Torr. Ann. Lyceum Nat. Hist. New York 3: 328. 1836. *S. sylvaticus* var. *digynus* Boeck. Linnaea 36: 727. 1870. *S. sylvaticus* subsp. *digynus* T. Koyama, Canad. J. Bot. 43: 1392. 1965. (*Scouler,* Observatory Inlet, Nw. coast of Amer.)

S. rubrotinctus Fern. Rhodora 2: 20. 1900. *S. microcarpus* var. *rubrotinctus* M. E. Jones, Bull. Montana State Univ., Biol. Ser. 15: 20. 1910. (*Fernald 301,* Foxcroft, Me.; lectotype by A. A. Beetle.)

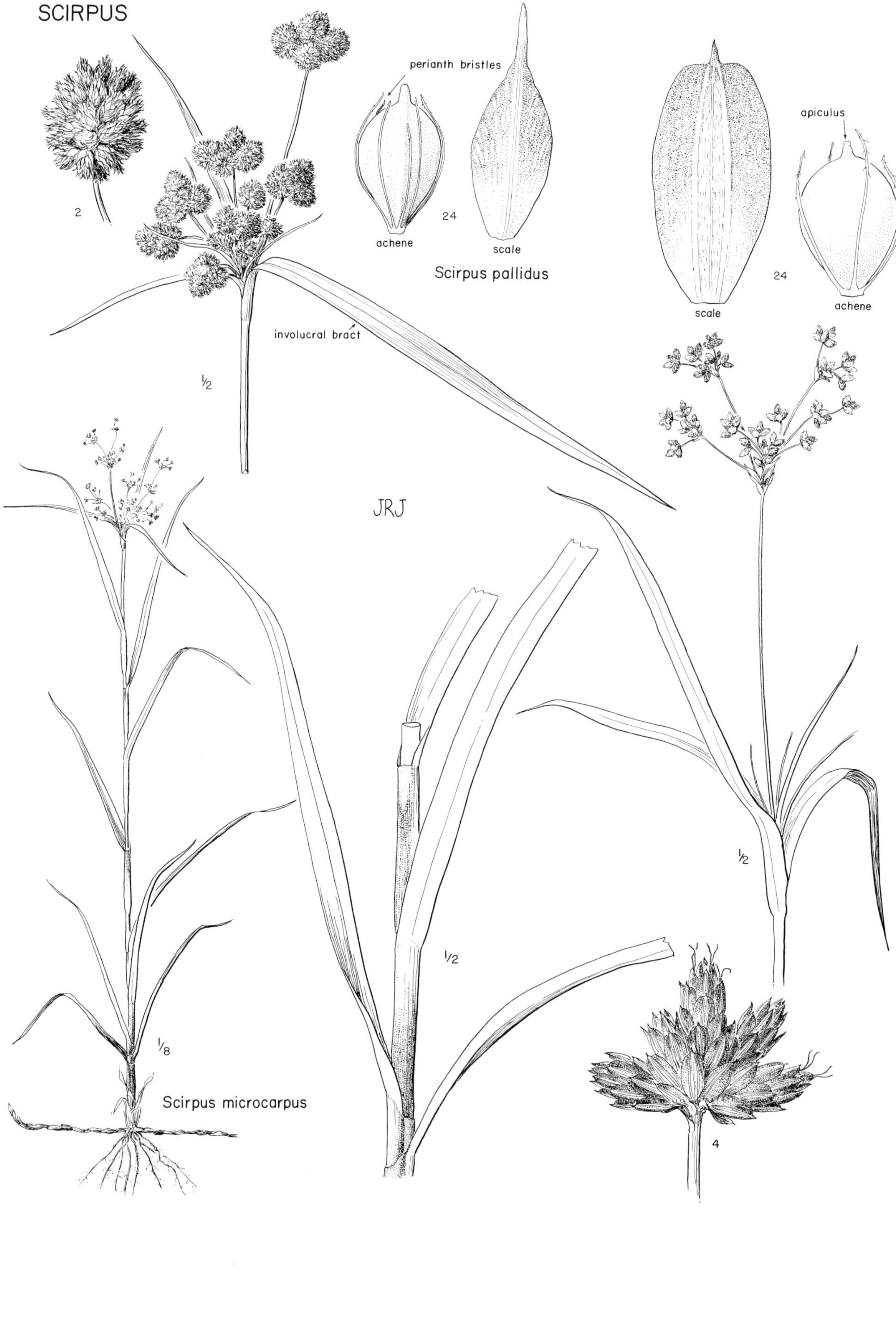
SCIRPUS
2
perianth bristles
achene
24
scale
Scirpus pallidus
apiculus
scale
24
achene
involucral bract
1/2
JRJ
1/2
1/2
1/8
Scirpus microcarpus
4

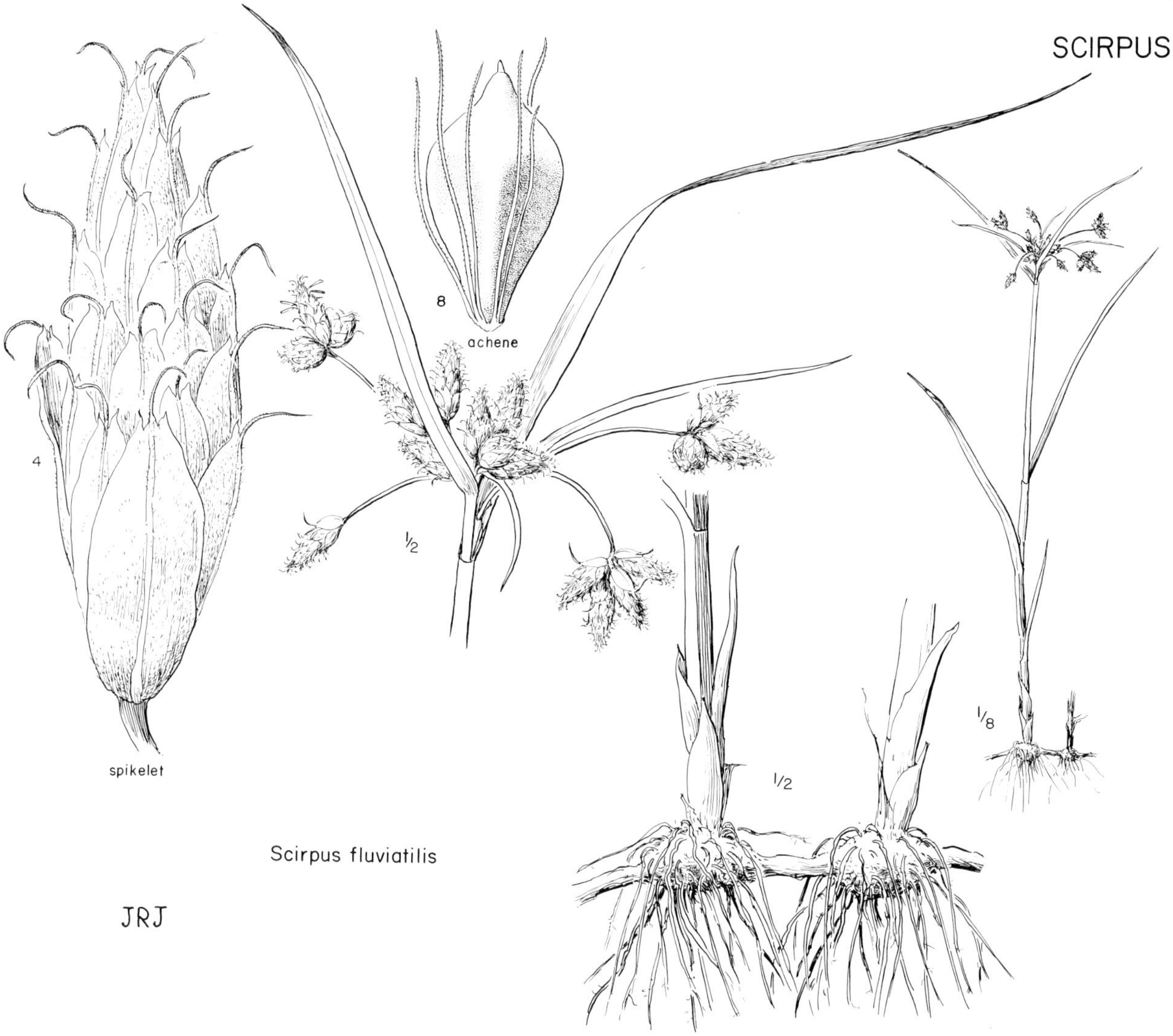

S. macounii T. H. Holm, Amer. J. Sci. 168: 21. 1904. (*Macoun*, Chilliwack Valley, B.C., in 1901.)

S. microcarpus var. *longispicatus* M. E. Peck, Proc. Biol. Soc. Wash. 47: 185. 1934. (*Peck 13332*, Rockaway, Clatsop Co., Oregon.) A form with unusually large spikelets, up to 8 mm long.

Coarse, leafy-stemmed perennials from well developed, stout, creeping rhizomes, the stems arising singly or few together, 6–15 dm tall, obscurely triquetrous; leaves several, with flat, grasslike blade commonly (6) 8–15 (20) mm wide and up to several dm long, the sheaths more or less tinted with reddish-purple; spikelets 4–6 (8) mm long, very numerous, sessile in small, pedunculate clusters in a compound, umbelliform, terminal cyme that is subtended by several conspicuous, leaflike, unequal, sheathless bracts, the largest of these seldom less than 1 dm long, sometimes as much as 3 dm; scales numerous, about 1.5 mm long, largely hyaline and cellular-reticulate, blackish or greenish-black in aspect, with greener, often ill-defined midstripe that may be very shortly excurrent as a minute point; bristles 4–6, slender, minutely and retrorsely barbellate, slightly surpassing the achene, but not exserted beyond the scale; achene lenticular, pale, commonly 1.0–1.2 mm long, including the minute stylar apiculus.

Wet, low ground, from the lowlands to moderate elev. in the mts.; in appropriate habitats throughout the w. cordilleran region, from s. Can. to Calif., Ariz.,-and N.M., and e. through s. Can. and n. U.S. to Newfl., New England, and W.Va. June–Aug.

3. **Scirpus fluviatilis** (Torr.) A. Gray

Scirpus maritimus var. ?*fluviatilis* Torr. Ann. Lyceum Nat. Hist. New York 3: 324. 1836. *S. fluviatilis* A. Gray, Manual Bot. 527. 1848. (w. N.Y., presumably collected by *Torrey*, possibly at Onondaga Lake.)

River bulrush.

Stout, rhizomatous perennial mostly 6–15 dm tall; culms sharply triquetrous, often with concave sides; leaves several, well developed, elongate, flat or nearly so, commonly 6–15 mm wide, well distributed along the stem; spikelets mostly 10–40, 1.2–2.5 cm long and 6–10 mm thick, commonly tapering to a point, sometimes all sessile in a compact terminal cluster, but more often the central cluster conspicuously surpassed by several slender, often flexuous rays up to about 7 cm long, each ray with a single terminal spikelet or a small terminal cluster of spikelets; involucral bracts several, very unequal, some of them elongate and leaflike; scales tan or light brown, thin and scarious, hirtellous on the back, short-awned from a tapering to more often lacerate or notched tip, the awn flexuous and more or less recurved, commonly 2–4 mm long; bristles 6, retrorsely barbellate, variable, often equaling the achene; style branches 3; achene trigonous, 3.5–5 mm long, apiculate, minutely cellular-reticulate; n = 55.

Marshes, shores and standing water (sometimes as much as 1 m deep), tolerant of alkali; widespread in the U.S., and s. Can., from coast to coast, but somewhat irregularly distributed westward; in our range known from Cache and Box Elder cos., Utah, and to be expected elsewhere. June–Aug.

4. Scirpus maritimus L.

Scirpus maritimus L. Sp. Pl. 51. 1753. *Schoenoplectus maritimus* K. Lye, Blyttia 29: 145. 1971. (Europe.)

Scirpus campestris Britton in Britton & Brown, Ill. Fl. N. U.S., Canada & Brit. Possessions 1: 267. 1896. *S. robustus* var. *campestris* Fern. Rhodora 2: 241. 1900. *S. brittonianus* Piper, Contr. U.S. Natl. Herb. 11: 157. 1906. (c. and w. U.S.; no type indicated.) Not *S. campestris* Roth, 1795.

S. paludosus A. Nels. Bull. Torrey Bot. Club 26: 5. 1899. *S. robustus* var. *paludosus* Fern. Rhodora 2: 241. 1900. *S. campestris* var. *paludosus* Fern. Rhodora 8: 162. 1906. *S. maritimus* var. *paludosus* Kükenthal, Repert. Spec. Nov. Regni Veg. 23: 200. 1926. *Bolboschoenus paludosus* Sóo, Acta Bot. Acad. Sci. Hung. 16: 368. 1970. (*Nelson 3874,* Granger, Wyo.)

"*S. robustus* Pursh," the name misapplied.

Alkali bulrush.

Stout, rhizomatous perennial 2–15 dm tall, the rhizomes commonly bearing firm tubers; culms sharply triquetrous, often with concave sides; leaves usually several, well developed, elongate, flat or nearly so, up to about 1 cm wide, well distributed along the stem, or all borne below the middle; spikes 3–20 or more, 1.2–2 cm long and 6–10 mm thick, usually rather blunt, all sessile in a compact terminal cluster, or the principal cluster somewhat surpassed by one or more short peduncles bearing each a subsidiary cluster; involucral bracts several, very unequal, 2–4 of them more or less elongate, green and leafy-textured (occasionally anthocyanic at the base), surpassing the inflorescence, the largest one generally much surpassing the others; scales tan or light brown or sometimes darker brown, thin and more or less hyaline-scarious, minutely hirtellous-puberulent on the back, short-awned from a notched (often somewhat irregularly lacerate) tip; bristles retrorsely barbellate, rather poorly developed, up to about half as long as the achene, or often obsolete; style branches 2; achene lenticular, 2.5–4 mm long, apiculate, minutely cellular-reticulate (obscurely so at 10X, somewhat more evidently so at 20X); n = 40, 43, 48, 53, 55, 57.

Marshes, wet meadows, and margins of ponds, especially in alkaline or saline places; widespread (but somewhat irregularly distributed) in the N. Hemisphere, especially along the seacoasts, and found essentially throughout our range. June–Sept.

Plants of w. and c. U.S., as here described, belong to the rather ill-defined var. *paludosus* (A. Nels.) Kükenthal. The Eurasian and e. Amer. var. *maritimus* tends to have more acute spikelets and often has some or all of the flowers tricarpellate, and the leafy involucral bracts (especially the inner) are more apt to be anthocyanic at the base.

5. Scirpus acutus Muhl. ex Bigel.

Scirpus acutus Muhl. ex Bigel. Fl. Bostoniensis 15. 1814. *Schoenoplectus acutus* A. Löve & D. Löve, Bull. Torrey Bot. Club 81: 33. 1954. (Fresh Pond, Boston, Mass.)

Scirpus lacustris var. *occidentalis* S. Wats. Bot. Calif. 2: 218. 1880. *S. occidentalis* Chase, Rhodora 6: 68. 1904. *S. acutus* var. *occidentalis* A. A. Beetle, Univ. Wyoming Publ. 13: 5. 1948. (w. U.S.; no specimens cited.)

S. malheurensis L. F. Henderson, Rhodora 32: 20. 1930. *S. acutus* var. *malheurensis* A. A. Beetle, Univ. Wyoming Publ. 13: 5. 1948. (*Henderson 8655,* Malheur Lake, Harney Co., Oregon.)

S. rubiginosus A. A. Beetle, Amer. J. Bot. 28: 697. 1941. (*Beetle & Constance 2686,* Stimson Beach, Marin Co., Calif.)

Hardstem bulrush, tule.

Stout, rhizomatous perennial mostly 1–3 m tall, reputedly sometimes to 5 m, forming large colonies; culms terete, stout, often over 1 cm (or even 2 cm) thick below the middle or toward the base, firm, not easily crushed between the fingers; leaves few, borne mainly or wholly toward the base of the culm, commonly with prominent, well developed sheath and short, poorly developed blade (or bladeless); principal involucral bract solitary, 2–10 cm long, erect or nearly so and resembling a prolongation of the culm, usually green; subsidiary bracts small and inconspicuous, very different from the principal bract, chaffy and brownish; spikelets dull gray-brown, mostly 8–15 mm long, more or less numerous in an open or compact, subumbellately branched inflorescence, all or nearly all of them sessile in small clusters at the ends of the mostly rather stiff and ascending or horizontal branches of the inflorescence; scales mostly (3) 3.5–4 mm long (or the upper a little smaller), thin and largely hyaline-scarious, with numerous short, linear, reddish-brown marks (these sometimes becoming raised and viscid) on a pale, gray-white background, the margins commonly somewhat lacerate or irregularly arachnoid-ciliate, the midrib firm, scabrous, commonly exserted (often from a broad notch) as a short awn-tip; bristles fragile, retrorsely barbellate, commonly equaling or slightly exceeding the achene; pistil bicarpellate or occasionally in some of the flowers tricarpellate, the style accordingly bifid or trifid and the achene planoconvex or unequally trigonous; achenes mostly 2.2–2.5 mm long (including stylar apiculus), more or less completely hidden by the scales even at maturity; n = 20, 21.

Marshes and muddy shores of lakes and streams at lower elev., sometimes in water as much as 1 m deep, tolerant of alkali; widespread in temp. N. Amer., commonest in w. U.S., and found in appropriate habitats throughout our range. June–Aug.

Scirpus acutus and *S. validus* are closely related and have widely overlapping ranges, without any very clear differentiation in habi-

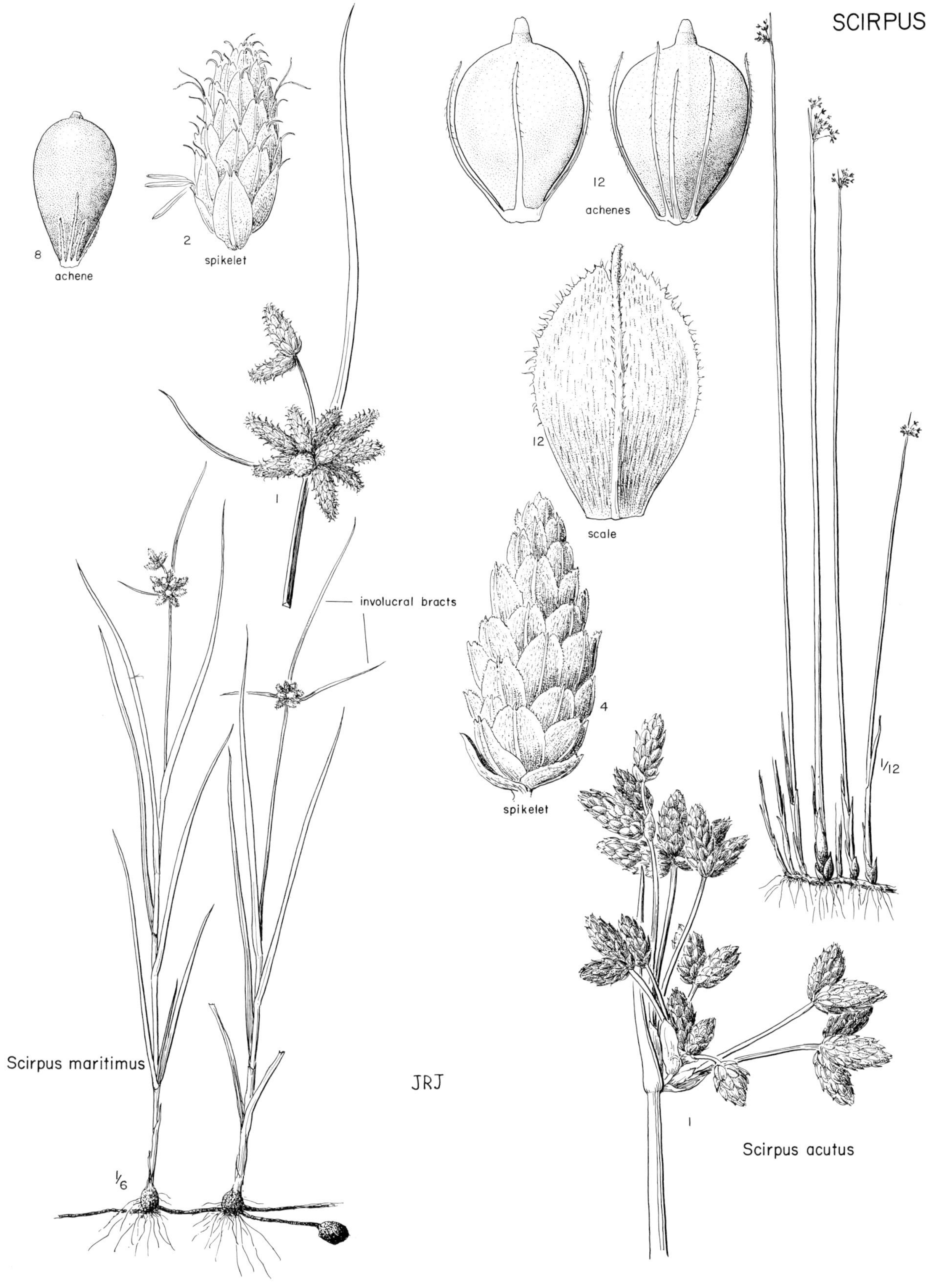
SCIRPUS
8
achene
2
spikelet
12
achenes
12
scale
4
spikelet
1
involucral bracts
1/12
Scirpus maritimus
JRJ
1/6
1
Scirpus acutus

tat. Reports as to the frequency of hybridization and the sharpness of the distinction between the populations in the field vary with different observers, but it is agreed that they maintain their separate identities even when growing together.

Our two taxa of this group belong to a cosmopolitan species-complex in which the oldest name is *Scirpus lacustris* L. Typical *S. lacustris* is a Eurasian plant that does not occur in North America. Another closely related Eurasian taxon of this complex, *S. tabernaemontani* Gmelin [= *S. lacustris* subsp. *glaucus* (J. E. Smith) Hartm.], is closely similar to North American *S. acutus*, and some authors consider these two to be taxonomically identical. Our taxa of this group clearly warrant separation at some level, and they are here for convenience retained as species under their accustomed names.

6. Scirpus validus Vahl

Scirpus validus Vahl, Enum. Pl. 2: 268. 1805. *S. lacustris* var. *validus* Kükenthal, Repert. Spec. Nov. Regni Veg. 23: 200. 1926. *Schoenoplectus validus* A. Löve & D. Löve, Bull. Torrey Bot. Club 81: 33. 1954. *Scirpus lacustris* subsp. *validus* T. Koyama, Canad. J. Bot. 40: 927. 1962. ("in Caribaeis.")

Scirpus validus var. *creber* Fern. Rhodora 45: 282. 1943. *S. validus* var. *condensatus* f. *creber* A. A. Beetle, Univ. Wyoming Publ. 13: 6. 1948. *Schoenoplectus validus* subsp. *creber* A. Löve & D. Löve, Bull. Torrey Bot. Club 81: 33. 1954. *Scirpus lacustris* subsp. *creber* T. Koyama, Phytologia 8: 457. 1963. (*St. John 2581,* Fisher's Island, N.Y.)

Scirpus validus var. *creber* f. *megastachyus* Fern. Rhodora 45: 283. 1943. (*Shear 239,* Joy Creek, Osborne Co., Kansas.)

Softstem bulrush, tule.

Similar to *S. acutus,* but averaging smaller and more slender, the stems soft, easily crushed between the fingers; inflorescence tending to be looser and more open, with longer, laxer, sometimes drooping rays, some of the spikelets often borne singly instead of all in sessile clusters; spikelets averaging smaller, seldom over 1 cm long, more shining and reddish-brown; scales mostly (2) 2.5–3 (3.5) mm long, with a brown or tawny ground-color, so that the fine, red-brown striolae are not usually prominent at 10X; margins of the scales entire to somewhat lacerate or shortly fringed-ciliolate, but less conspicuously so than in *S. acutus;* achenes not wholly concealed by the scales at maturity, mostly 1.8–2.3 mm long; $n = 21$.

Marshes and muddy shores of lakes and streams at lower elev., colonial but seldom forming large exclusive colonies in the fashion of *S. acutus;* tolerant of alkali; widespread in temp. N. Amer., and s. into trop. Amer., and found in appropriate habitats throughout our range, somewhat less commonly then *S. acutus.* June–Aug. See comment under *S. acutus.*

7. Scirpus americanus Pers.

Scirpus americanus Pers. Syn. Pl. 1: 68. 1805. *Heleophylax americanus* Schinz & Thell. Vierteljahrsschr. Naturf. Ges. Zürich 53: 587. 1908. ("in Carolina interiore.")

S. olneyi A. Gray, Boston J. Nat. Hist. 5: 238. 1845. *Schoenoplectus olneyi* Palla, Bot. Jahrb. Syst. 10: 299. 1888. (*Olney,* Providence, R.I.)

"*Scirpus chilensis* Nees & Meyen," the name probably misapplied.

Olney threesquare.

Rhizomatous perennial, the culms loosely clustered or somewhat scattered, mostly 5–15 dm tall, averaging stouter than in *S. pungens,* often 1 cm or more thick toward the base, very sharply triangular with conspicuously concave sides, easily flattened in pressing; leaves few, all borne on the lower part of the culm but not closely crowded, the sheaths normally developed, but the blades commonly short, seldom over 1 dm long, usually somewhat wider than in *S. pungens,* sometimes as much as 1 cm wide; spikelets mostly 2–15, essentially sessile in a compact cluster, mostly 6–15 mm long, subtended by a prominent green bract 1–3.5 (5.5) cm long which appears like a continuation of the stem, without additional bracts; scales yellowish-brown or reddish-brown to purplish-brown, thin, largely hyaline-scarious, the midrib firmer (at least distally) and commonly exserted as a short mucro; hypogynous bristles 4, sometimes 5 or 6, often unequal, retrorsely barbellate, not much if at all exceeding the achene; pistil 2-carpellate, the style bifid, the achene planoconvex, 1.8–2.5 mm long (including the evident apiculus of ca 0.3 mm) and 1.4–1.7 mm wide; $n = 39$.

Marshes, wet meadows, and other wet, low places, tolerant of alkali, but not requiring it; N.S. to Wash. (Grant Co.), s. to S. Amer.; less often collected, in our range, than *S. pungens.* June–Sept.

The name *S. americanus* was for many years consistently misapplied to the species here called *S. pungens.* The present application of the name follows Schuyler in Rhodora 76: 51–52. 1974.

Scirpus americanus is obviously related to *S. pungens,* but the two appear to be sharply distinct in nature. A rare putative hybrid has been called *S.* X *contortus* (Eames) T. Koyama, Canad. J. Bot. 41: 1123. 1963, based on *S. olneyi* var. *contortus* Eames, Rhodora 9: 220. 1907.

8. Scirpus pungens Vahl

Scirpus pungens Vahl, Enum. Pl. 2: 255. 1805. *Heleogiton pungens* Reichenb. Fl. German. Excursoria 78. 1830. *Schoenoplectus pungens* Palla, Verh. K. K. Zool.-Bot. Ges. Wien 38: 49. 1888. *Scirpus americanus* var. *pungens* Barros & Osten, Anales Mus. Nac. Montevideo, II. 3: 203. 1931. (*Commerson,* Borbon Isl., Europe.)

Scirpus pungens var. *polyphyllus* Boeck. Linnaea 36: 709. 1870. *S. americanus* var. *polyphyllus* A. A. Beetle, Amer. J. Bot. 30: 399. 1943. (No type indicated.) Varietal identity uncertain.

Scirpus pungens var. **longispicatus** (Britton) Cronq. hoc loc. *S. americanus* var. *longispicatus* Britton, Trans. New York Acad. Sci. 11: 78. 1892. *S. longispicatus* Smyth, Trans. Kansas Acad. Sci. 16: 163. 1899. (*Wright 1963,* N.M.)

Common threesquare.

Rhizomatous perennial, the culms more or less scattered, mostly 1.5–10 dm tall, triangular with flat to slightly concave or slightly convex sides; leaves usually several, all borne near the base, the blade elongate or more or less reduced (likely to be reduced when the plant grows in shallow water rather than merely in wet ground), firm, flat and mostly 2–4 mm wide, or more often deeply channeled above or folded and only 1–2 mm wide as folded; spikelets mostly 1–6, essentially sessile in a compact cluster, mostly 7–20 mm long, subtended by a prominent green bract (1) 2–15 cm long which appears like a continuation of the stem; one or two smaller bracts in addition to the main bract nearly always present, these relatively inconspicuous, resembling enlarged scales (and colored like the scales) but not subtending flowers; scales yellowish-brown to reddish-brown (or blackish-purple in some chiefly extralimital forms), thin, largely hyaline-scarious, the midrib firmer (especially distally) and exserted from a broad notch as a mucro or a short awn; bristles re-

SCIRPUS

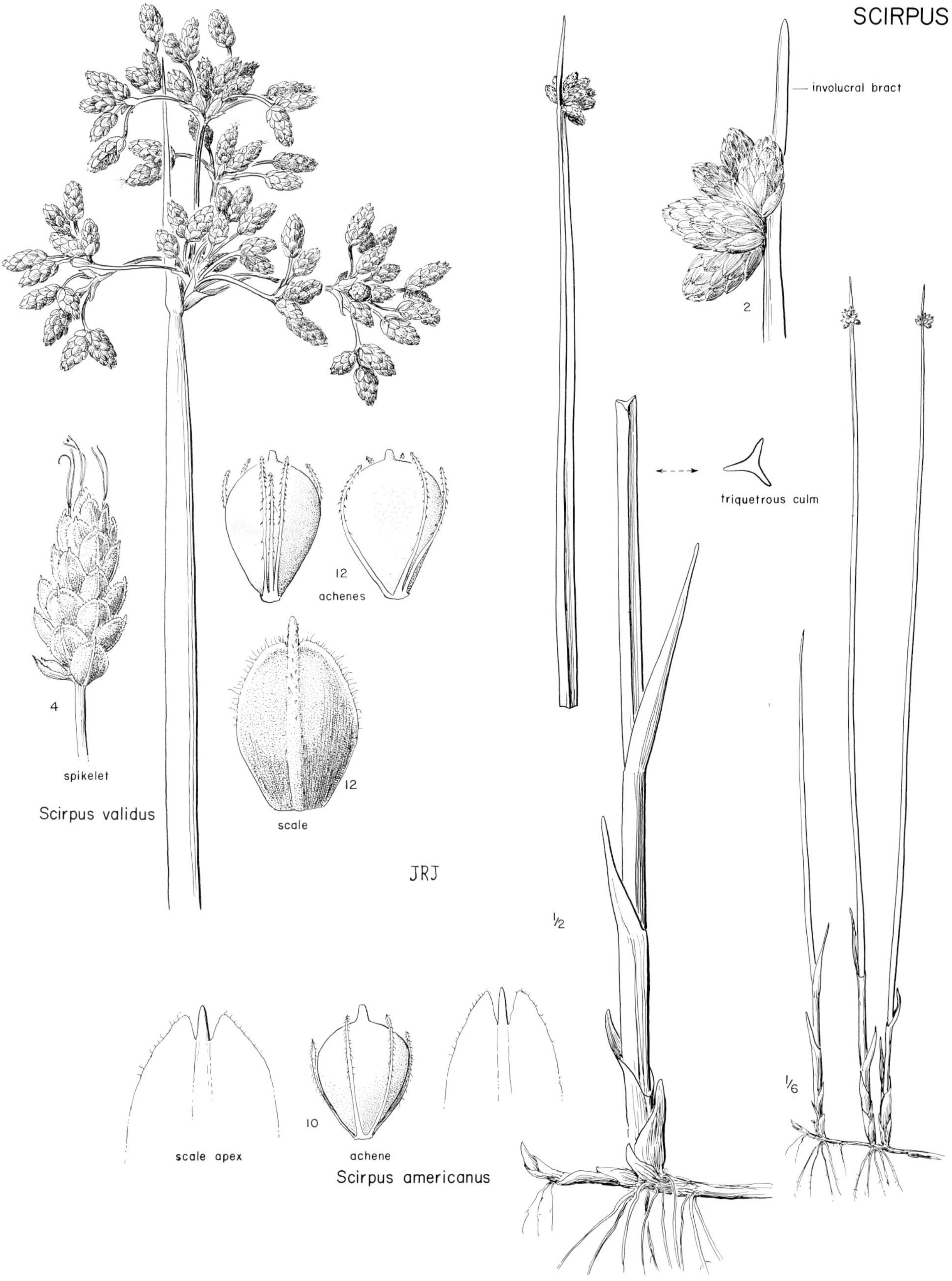
involucral bract
2
triquetrous culm
4
spikelet
Scirpus validus
12
achenes
12
scale
JRJ
1/2
1/6
scale apex
10
achene
Scirpus americanus

trorsely barbellate, 4–6, somewhat irregularly developed, often unequal (but still evident), not much if at all exceeding the achene; pistil 2- or 3-carpellate, the style accordingly bifid or trifid, and the achene lenticular or trigonous (commonly unequally trigonous, with the ventral face wider, or appearing more nearly planoconvex); achene 2.2–3.3 mm long (including the evident, slender stylar apiculus of about 0.5 mm) and 1.6–2.3 mm wide; n = 38, 39, 60+.

Marshes and wet low ground, tolerant of alkali, but not requiring it; widespread in the U.S. (and throughout our range) and s. Can., s. to S. Amer.; also in parts of Europe, New Zealand, and Australia. May–Aug.

Scirpus pungens is variable in a number of small technical details, and the variability is correlated to some extent with geography, but the infraspecific taxonomy is not yet fully clarified. Our plants, with reddish or brown scales that have a prominent mucro or short awn, and with many or all of the flowers tricarpellate, belong to var. *longispicatus* (Britton) Cronq. The var. *longispicatus* ranges e. to the Great Lakes region and the Ohio Valley. Var. *pungens*, differing from var. *longispicatus* in its strictly bicarpellate flowers, occurs in e. U.S. (chiefly from the Appalachian region eastward) and adj. Can., and in Europe, perhaps also in the Caribbean and in S. Amer.

9. **Scirpus nevadensis** S. Wats.

Scirpus nevadensis S. Wats. in King, Rep. Geol. Explor. 40th Parallel 5: 360. 1871. *Amphiscirpus nevadensis* Oteng-Yeboah, Notes Roy. Bot. Garden Edinburgh 33: 308. 1974. (*Watson 1213*, Soda Lake, Carson Desert, Nev.)

Plants vigorously rhizomatous, the culms scattered or in small tufts, tough and wiry, subterete, 1–4 dm tall; leaves several, clustered near the base, the lowest ones sometimes reduced to scales and the plant aphyllopodic in the manner of some spp. of *Carex;* old sheaths tending to be persistent; leaf blades firm, deeply channeled to thick and flat, mostly 1–2 mm wide, up to 2 dm long, often much shorter; spikelets (1) 2–10, sessile in a compact cluster, 1–2 cm long, subtended by a slender, stiff green bract 1.5–7 cm long and by 1 or more additional short anthocyanic bracts which resemble enlarged scales of the spikelet, the principal bract sometimes essentially erect and resembling a continuation of the culm, but more often diverging at an angle; scales shining brown, largely hyaline-scarious except for the paler, firm midrib which may be exserted as a minute mucro, but the scale not apically notched in the manner of *S. americanus* and *S. pungens;* hypogynous bristles 1–3 (4), retrorsely barbellate, slender and much reduced, seldom surpassing the middle of the achene, generally unequal; achene about 2 mm long, bicarpellate, planoconvex, pale, firm-walled, evidently cellular-reticulate at 10X, not at all apiculate, the style abruptly deciduous.

Moist or wet, alkaline sites in desert and semidesert regions; c. Wash. (e. of the Cascade Mts.) to Inyo Co., Calif., e. to Sask., N.D., Wyo., and Utah (Juab Co.), and reported also from sw. B.C. (mouth of the Fraser R.); also in Argentina. June–Aug.

Scirpus nevadensis looks superficially not unlike some of the more reduced forms of *S. pungens* but the achenes are wholly different, and other differences emerge on more careful comparison.

10. **Scirpus subterminalis** Torr.

Scirpus subterminalis Torr. Fl. N. & Middle Sect. U.S. 47. 1823. (*Cooley*, near Deerfield, Mass.)

S. subterminalis var. *terrestris* Paine, Annual Rep. New York State Mus. 18: 148. 1865. *S. subterminalis* f. *terrestris* Fern. Rhodora 45: 288. 1943. (*Paine*, Herkimer Co., N.Y.) The rare terrestrial form.

Water bulrush.

Rhizomatous, aquatic perennial with slender, subterete, flaccid stems 20–80 cm long, and elongate, slender, flaccid leaves from near the base, the leaves and stems usually floating distally rather than emergent, or the plant seldom more or less terrestrial and with erect or emergent stems and leaves; spikelet solitary, light brown, 7–12 mm long, subtended by a prominent bract (1) 1.5–6 cm long which appears like a continuation of the stem; scales 4–6 mm long, very thin and almost hyaline except for the somewhat firmer midrib, which may be minutely exserted; hypogynous bristles retrorsely barbellate, shorter than to occasionally equaling or slightly exceeding the achene; achene trigonous, 2.5–3.8 mm long, including the prominent (0.5 mm), slender beak-apiculus.

In quiet, shallow water commonly 2–8 dm deep; s. Alaska, to s. Oregon, chiefly w. of the Cascade summits, but also extending inland to n. Idaho and nw. Mont.; Newfl. to Ont., s. to S.C., Ga., and Mo.; an apparently isolated station in Utah. July–Aug.

11. **Scirpus supinus** L.

Scirpus supinus L. Sp. Pl. 49. 1753. *Isolepis supina* R. Br. Prodr. Fl. Novae Holl. & Insulae van-Diemen 77. 1810. (Paris, France.)

S. hallii A. Gray, Manual Bot. ed. 4. xcix. 1863. *S. supinus* var. *hallii* A. Gray, Manual Bot. ed. 5. 563. 1867. *S. uninodis* var. *hallii* A. A. Beetle, Amer. J. Bot. 29: 656. 1942. (*Hall*, Menard Co., Ill., in 1861.) The eastern American, chiefly bicarpellate var.

S. saximontanus Fern. Rhodora 3: 251. 1901. *S. supinus* var. *saximontanus* T. Koyama, Canad. J. Bot. 40: 920. 1962. (*Greene s.n.*, Greeley, Colo., is the first of several specimens cited.)

Sharpscale bulrush.

Fibrous-rooted annual; stems clustered, decumbent to erect, 0.3–3 dm tall, slender and wiry; leaves few, borne mostly below the middle of the stem, the sheath bladeless and merely mucronate-tipped, or the upper ones with a slender blade sometimes as much as 10 cm long; principal involucral bract solitary, erect or somewhat incurved, mostly 2–10 cm long, resembling a prolongation of the culm, the inflorescence appearing lateral; subsidiary involucral bract borne below the spikelets, inconspicuous, seldom as much as 3 cm long; spikelets 2–several, sessile or some of them short-pedunculate, 3–15 mm long; scales 2.5–3.5 mm long, hyaline-scarious, with firmer, greenish, shortly excurrent midrib; bristles present or absent; pistil in ours tricarpellate, the achene evidently trigonous, about 1.4 mm long, broadly obovoid, dark gray-brown to blackish at maturity, finely but conspicuously cross-ridged; n = 14.

Swales and shores, tolerant of alkali; interruptedly circumboreal and indeed almost cosmopolitan, but rare and irregularly distributed; in our range known only from a single old collection near the Great Salt Lake.

Koyama considers the plants of c. and w. U.S. to represent a distinct var., the var. *saximontanus* (Fern.) T. Koyama, in contrast to the Eurasian var. *supinus* and the e. Amer. var. *hallii* (A. Gray) A. Gray.

SCIRPUS

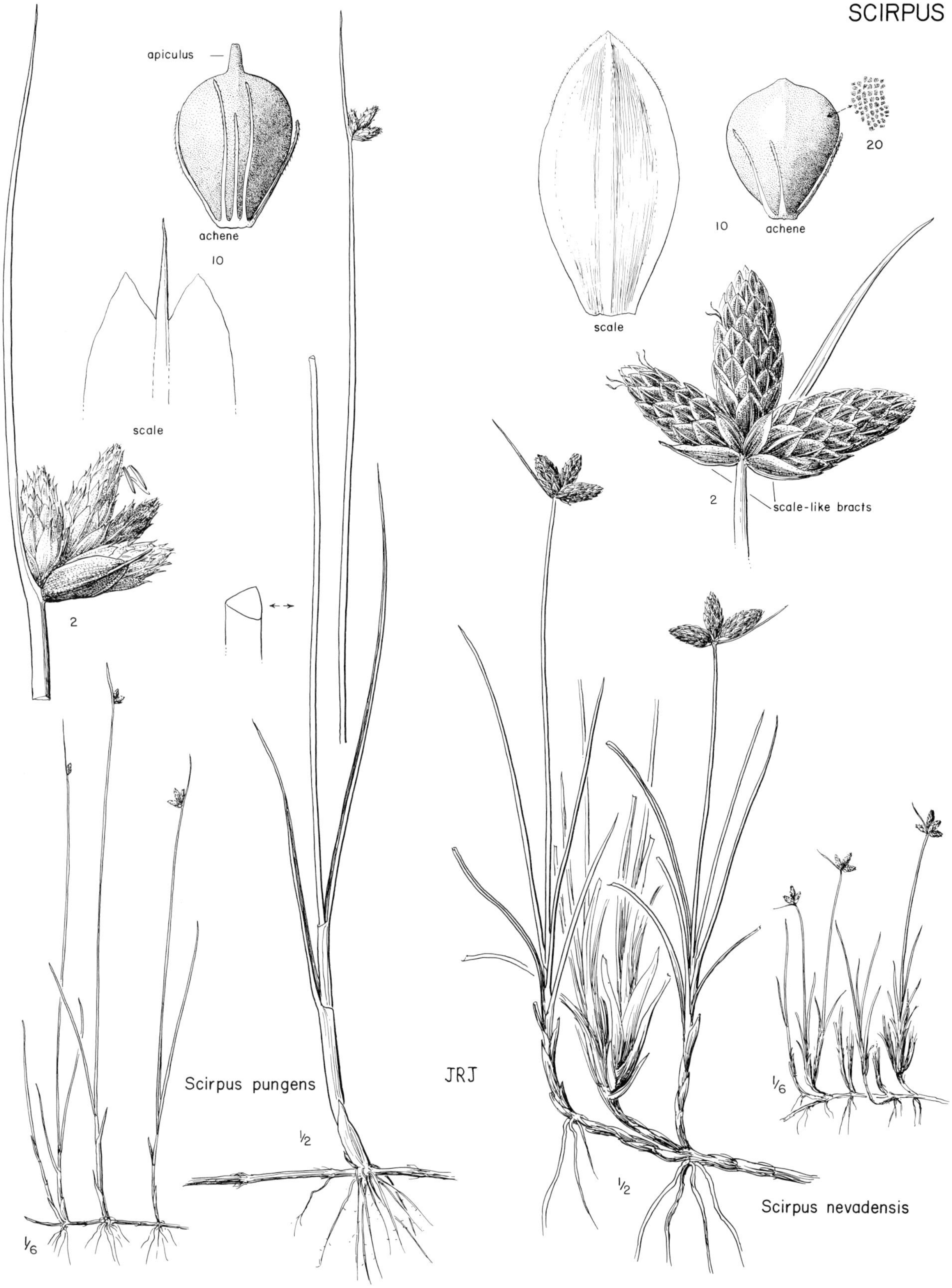

apiculus
achene
10
scale
2
Scirpus pungens
JRJ
1/2
1/6
scale
achene
10
20
scale-like bracts
2
1/6
1/2
Scirpus nevadensis

SCIRPUS

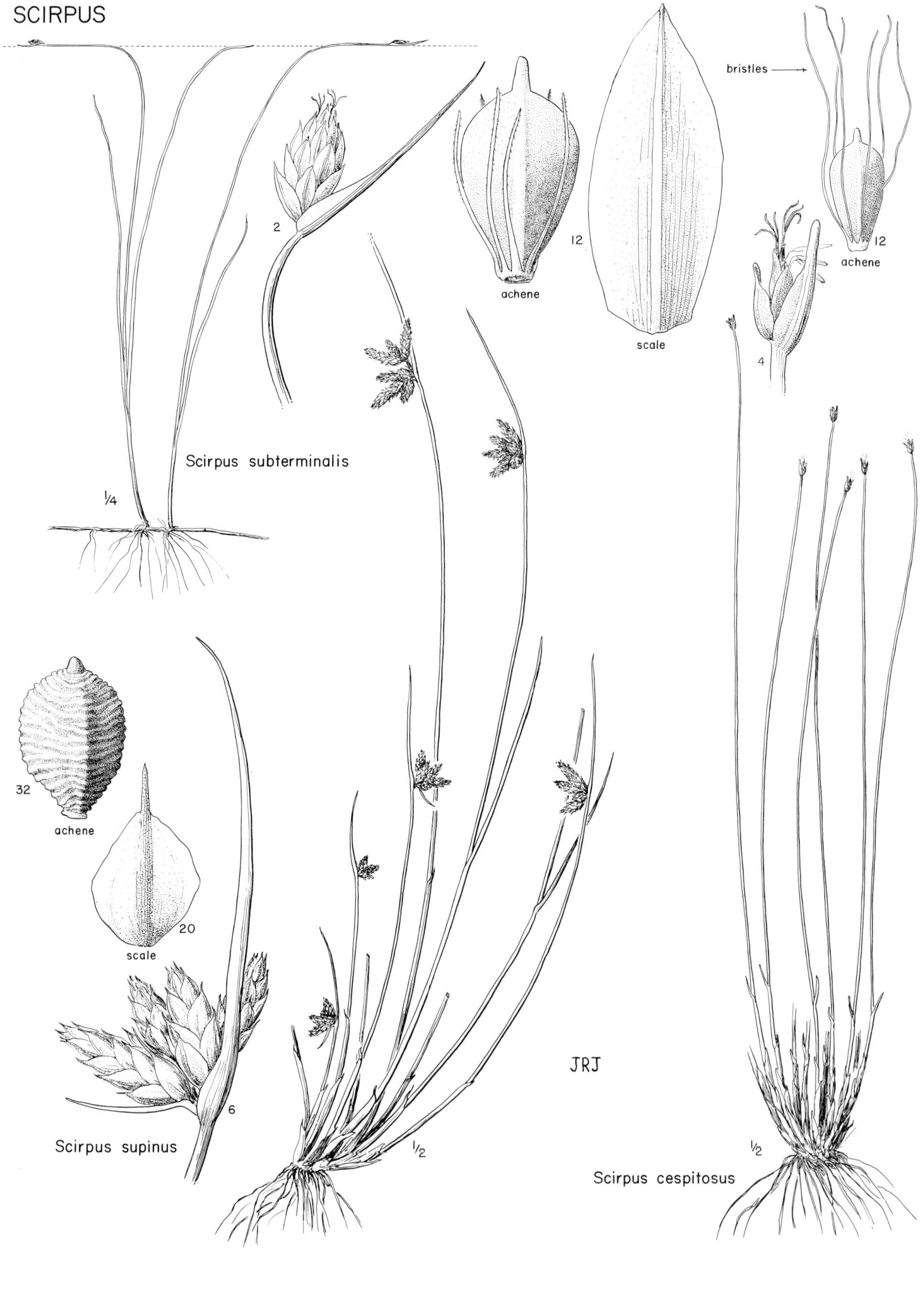
bristles
achene
12
scale
2
4
12
achene
Scirpus subterminalis
1/4
32
achene
20
scale
6
Scirpus supinus
1/2
JRJ
1/2
Scirpus cespitosus

ERIOPHORUM HEMICARPHA

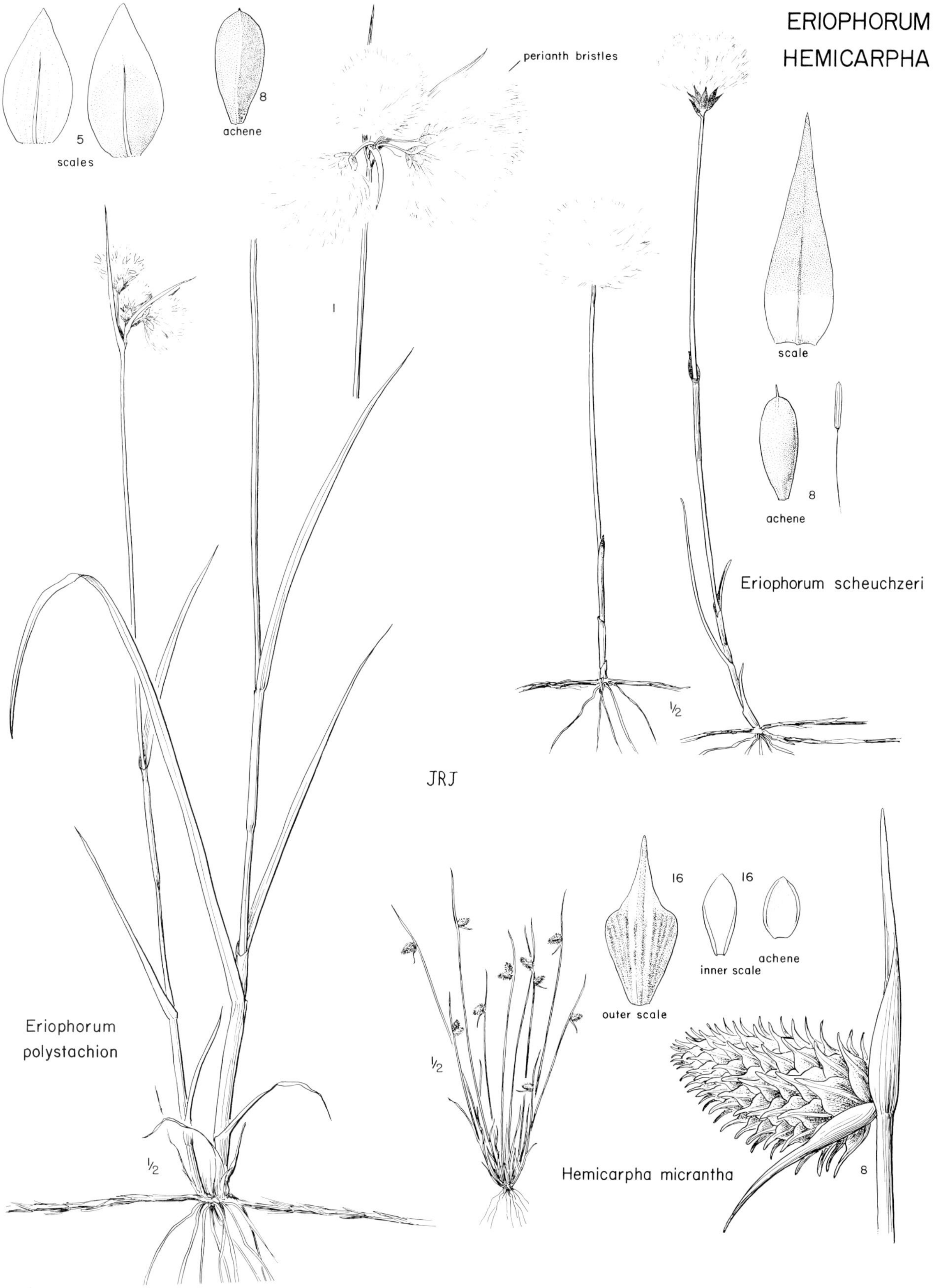

12. Scirpus cespitosus L.

Scirpus cespitosus L. Sp. Pl. 48. 1753. *Clavula cespitosa* Dumort. Fl. Belg. 143. 1827. *Limnochloa cespitosa* Reichenb. in Mössler, Handb. Gewächsk. ed. 2. 1808. 1829. *Baeothyron cespitosum* A. Dietr. Sp. Pl. 2: 89. 1833. *Eleocharis cespitosa* Nees, Linnaea 9: 294. 1834. *Trichophorum cespitosum* Schur, Verh. Mitth. Siebenbürg. Vereins Naturwiss. Hermannstadt 4: 78. 1853. (Europe.)

S. cespitosus var. *callosus* Bigel. Fl. Bostoniensis, ed. 2. 21. 1824. (*Bigelow,* White Hills of N.H.)

Trichophorum austriacum Palla, Ber. Deutsch. Bot. Ges. 15: 468. 1897. *S. cespitosus* var. *austriacus* Aschers. & Graebner, Syn. Mitteleur. Fl. 2(2): 300. 1904. (Europe.)

Culms very densely tufted, on a short, freely rooting rhizome, 1–4 dm tall, smooth, subterete, clothed with several conspicuous, light brown scale leaves at the base, and commonly with a single more normal leaf a little above the base, this leaf with typical sheath but the blade slender and short, only 4–6 mm long; spikelet solitary, terminal, medium to light brown, 4–6 mm long, several-flowered, the involucre represented only by 2 or 3 empty scales at the base of the spikelet (these scales often deciduous as the spikelet approaches full maturity), the lowest of these scales with a prominent, broad, blunt-tipped awn 1–3 mm long that sometimes shortly surpasses the spikelet; bristles 6, white or distally brownish, very fragile, usually exceeding the achene but scarcely if at all exceeding the scale; achene trigonous, about 1.5 mm long, minutely apiculate; $n = 52$.

Sphagnum bogs, marshes, and other very wet places; circumboreal, extending s., in Amer., to N.C., Tenn., ne. Utah (Uinta Mts.) and c. Oregon. July–Aug.

2. ERIOPHORUM L. Cottongrass

Herbs with triangular or terete, mostly solid stems, the leaves with closed sheath and elongate, grasslike blade, the upper sheaths often bladeless; spikelets 1–several in a terminal inflorescence of basically cymose nature, the inflorescence subtended by 1–several elongate and leafy to small and scale-like involucral bracts, these sometimes (especially in species with only a single spikelet) more or less resembling the scales of the spikelet and then referred to as sterile scales; scales of the spikelet spirally arranged, scarious, not awned; perianth of numerous (more than 10) conspicuous, persistent bristles, these white to rufous, much elongate at least in fruit, so that the fruiting spikelet forms a dense, cottony tuft commonly 2–4 cm long; stamens 3; style trifid; achenes unequally trigonous, often with a short, slender stylar apiculus.

About 20 species, occurring chiefly in the cooler parts of the N. Hemisphere, always in wet places. (Name from the Greek *erion,* wool or cotton, and *phoros,* bearing, referring to the cottony spikelets.)

Reference: Fernald, M. L. The North American species of *Eriophorum.* Rhodora 7: 81–92; 129–136. 1905.

1 Spikelets 2 or more; anthers mostly 2.5–4 mm long . . . . 1. *E. polystachion*
1 Spikelet solitary; anthers mostly 0.5–1 mm long . . . . 2. *E. scheuchzeri*

1. Eriophorum polystachion L.

Eriophorum polystachion L. Sp. Pl. 52. 1753. *Linagrostis polystachia* Scop. Fl. Carniolica, ed. 2. 1: 48. 1772. (Europe.)

E. angustifolium Honck. Vollständiges Syst. Verzeichnis Gewächse Teutschl. 1: 153. 1782. (non vidi). *E. polystachion* var. *angustifolium* A. Gray, Manual Bot. ed. 2. 502. 1856. *Scirpus angustifolius* T. Koyama, J. Fac. Sci. Univ. Tokyo, Sect. 3, Bot. 7: 356. 1958. (Presumably Germany.)

Extensively colonial from creeping rhizomes; culms subterete, 2–6 (9) dm tall; leaves basal and cauline, the blade well developed, more or less elongate (especially that of the lower leaves), 2–6 (8) mm wide, flat or nearly so for most of its length, but becoming narrow and triangular or channeled toward the tip; uppermost culm leaf with well developed blade usually equaling or exceeding the sheath; involucral bracts several, unequal, 2 or more of them evidently foliaceous at least above the broadened, more chartaceous base, the longest one generally surpassing (or at least equaling) the inflorescence; spikelets 2–8, most or all of them individually pedunculate, in a compact to open, umbelliform cyme, the peduncles more or less strongly compressed, smooth or sometimes minutely scabrous-hirtellous; scales tawny to brownish or blackish-green, very thin distally, the slender midrib attenuated distally and not reaching the tip; anthers mostly 2.5–4 mm long (dry); bristles numerous, white or nearly so; achenes blackish, 2–3 mm long, broadly oblanceolate to obovate, 2–3 times as long as wide; $n = 29, 30$.

Cold swamps and bogs at moderate to high elev. in the mts., commoner northward; circumboreal at high latitudes, extending s., in Amer., to N.Y., Mich., Iowa, Colo., n. N.M., ne. Utah (Uinta Mts.) and c. Oregon. July–Aug.

2. E[illegible]horum scheuchzeri Hoppe

[illegible]phorum scheuchzeri Hoppe, Bot. Taschenb. Anfänger Wiss. Apothekerkunst 1800: 104. 1800. (Germany.)

E. leucocephalum Boeck. Flora 41: 418. 1858. *Scirpus leucocephalus* T. Koyama, J. Fac. Sci. Univ. Tokyo, Sect. 3, Bot. 7: 360. 1958. (Labrador.)

Extensively colonial from creeping rhizomes, the culms short, only 1–4 dm tall; leaves few, borne mostly near the base, with well developed sheath and slender, often much reduced blade that is channeled or triangular and not much over 1 mm wide; uppermost sheath borne above or more often below the middle of the culm, dark-tipped and bladeless; spikelet solitary and terminal, the sterile scales rather inconspicuous, up to about 1 cm long, not markedly different from the fertile ones; fertile scales narrow, tapering gradually to a slender tip, the distal half or two-thirds of the scale wholly blackish-green, or sometimes very narrowly and inconspicuously pale-margined distally; anthers short, mostly 0.5–1 mm long (dry); bristles white or nearly so; achenes brown or blackish, about 2 mm long, oblanceolate, with a prominent stylar apiculus; $n = 29, 30$.

Cold, wet places; circumboreal, extending s., in Amer., to Newfl. and s. Alta., and at apparently isolated stations in the high mts. of Mont., Utah (Uinta Mts.) and Colo. July–Aug.

3. HEMICARPHA Nees

Delicate glabrous annuals with clustered stems up to 2 dm tall, the leaves few and slender, with closed sheaths; inflorescence of 1–several small spikes at the summit of the culm, subtended by 2 or 3 unequal involucral bracts, the longest of which resembles a continuation of the culm, so that the inflorescence appears to be lateral near the summit rather than evidently terminal; spikes compound, consisting of very numerous, spirally arranged, sessile spikelets, each spikelet with a single perfect flower subtended by a well developed outer scale and a smaller, very thin (often much reduced or obsolete) inner scale (between the flower and the axis of the spike) that has sometimes been interpreted as a modified perianth, the perianth otherwise wanting; stamen 1 (2); achene bicarpellate, dorsiventrally more or less compressed; style deeply bifid, with elongate stigmas, deciduous at maturity, leaving a small mucro at the tip of the achene.

A small genus of perhaps 5 species, occurring in both the New and the Old World, chiefly in trop. regions, only 2 reaching the U.S. (Name from the Greek *hemi,* half, and *karphos,* chaff, referring to the inner scale.)

Hemicarpha has sometimes been submerged in *Scirpus* as a section, but its relationships remain to be clarified.

Reference: Friedland, S. The American species of *Hemicarpha.* Amer. J. Bot. 28: 855–861. 1941.

1. Hemicarpha micrantha (Vahl) Britton

Scirpus micranthus Vahl, Enum. Pl. 2: 254. 1805. *Isolepis micrantha* Roemer & Schultes, Syst. Veg. 2: 110. 1817. *Hemicarpha micrantha* Britton, Bull. Torrey Bot. Club 15: 104. 1888. (*Richard,* thought to be from C. Amer.)

H. micrantha var. *aristulata* Coville, Bull. Torrey Bot. Club 21: 36. 1894. *H. aristulata* Smyth, Trans. Kansas Acad. Sci. 16: 163. 1899. (*Nealley,* w. Texas, in 1888.)

Delicate, tufted annuals 3–5 cm tall; leaves very narrow, scarcely 0.5 mm wide, resembling the slender stems; spikes mostly 1–3 per culm, sessile, ovoid, 2–6 mm long and 1.5–2.5 mm thick; scales firm, rather abruptly contracted to a blunt mucro or a short, stout awn-tip up to about 0.5 mm long, or the awn-tips of the lowermost scales sometimes slightly longer but still not reaching the middle of the spike; inner (adaxial) scales inconspicuous, hyaline, smaller than the outer (abaxial) scales, unvasculated or sometimes imperfectly vasculated, equaling or slightly exceeding the achenes, or more reduced or even obsolete; achenes minute, 0.5–0.7 mm long.

Wet bottomlands, sand bars, and beaches; trop. Amer., n. to s. Wash., s. Idaho, s. Wyo., Minn., and Me.; seldom collected in our range. June–Sept.

Most plants of c. and w. U.S. have the inner scale fairly well developed, about as long as or slightly longer than the achene. Such plants have been distinguished as var. *aristulata* Coville, in contrast to var. *micrantha,* chiefly of e. U.S. and tropical and subtropical Amer., with the inner scale more or less strongly reduced or obsolete. The morphological distinction is not sharp, however, and the geographical segregation is also imperfect. Some western plants have the inner scale much reduced and would therefore fall with var. *micrantha.* It has been suggested that the presence of one variety within the range of the other reflects casual introduction on the feet of migrating waterfowl, but this remains to be demonstrated. The taxonomic status of the purported varieties of *H. micrantha* merits reinvestigation.

4. ELEOCHARIS R. Br. Spikerush

Herbs with angular to terete or flattened stems, the leaves all basal or nearly so and reduced to mere sheaths or scarcely sheathing scales; spikelet solitary and terminal, without subtending bracts; scales spirally arranged (all ours) or sometimes distichous, largely scarious or hyaline, the lowest one(s) usually empty except in the *E. acicularis* group; flowers perfect, borne singly in the axils of the scales; perianth of 0–6 (9) bristles, these shorter to somewhat longer than the achenes, sometimes reduced or obsolete; stamens 3, sometimes fewer; style bifid or trifid, thickened toward the base, the thickened part persistent on the achene as a tubercle which is typically sharply differentiated from the body of the achene but which in a few species is confluent with the body of the achene and poorly differentiated from it; achene lenticular to planoconvex or more or less trigonous; chromosomes polycentric; $x = 4, 8, 9$.

A cosmopolitan genus with perhaps as many as 150 species. (Name from the Greek *helos,* a marsh, and *charis,* grace.) The name is often spelled *Heleocharis,* and indeed this form is etymologically preferable to the original spelling, which is here retained in accordance with majority practice. Names originally published under *Heleocharis* are here treated as if they had been published under *Eleocharis,* with no indication of the spelling used by the original author.

Eleocharis is closely related to *Scirpus,* and several species have been shifted back and forth between the two by authors who accept both genera. Three species with the habit of *Eleocharis* but with the tubercle poorly differentiated from the body of the achene are here treated under *Eleocharis,* following current opinion among cyperologists.

References:

Harms, L. J. Cytotaxonomic studies in *Eleocharis* subser. Palustres: Central United States taxa. Amer. J. Bot. 55: 966–974. 1968.

Strandhede, S.-O. *Eleocharis,* subser. Eleocharis in North America. Taxonomical comments and chromosome numbers. Bot. Not. 120: 355–368. 1967.

Svenson, H. K. *Eleocharis,* in N. Amer. Fl. 18(9): 509–540. 1957.

1 Stigmas 3; achenes more or less distinctly trigonous; flowers mostly 2–30 in each spikelet.
 2 Tubercle confluent with the achene, not forming a distinct apical cap.
  3 Achenes and scales small, the achenes 0.9–1.3 mm long (including the tubercle), the scales mostly 1.5–2 (2.5) mm long; very small plants, mostly 0.2–0.6 (1) dm tall; spikelets 2.5–4 (6) mm long, with 2–9 (20) flowers; lowest scale empty .... 1. *E. parvula*

3 Achenes and scales larger, the achenes 1.9–2.8 mm long, the scales mostly 2.5–5.5 mm long; larger plants, seldom less than 1 (0.5) dm tall.
4 Lowest scale, like the others, ordinarily subtending a flower; stems slender and short, (0.5) 1–3 (4) dm tall, seldom as much as 1 mm wide, not flattened, not proliferous; spikelets 4–8 mm long, with mostly 3–9 flowers .... 2. *E. pauciflora*
4 Lowest scale empty; stems coarser, mostly (2) 4–10 dm tall or more, more or less flattened at least distally and commonly 1–2 mm wide, some of them usually proliferous; spikelets (5) 8–13 mm long, with (5) 10–20 (25) flowers .... 3. *E. rostellata*
2 Tubercle forming a distinct apical cap well differentiated from the body of the achene.
5 Achene whitish or pale gray to ochroleucous, longitudinally many-ribbed, with numerous fine cross-ridges, forming ladderlike configurations; scales usually with green or stramineous midvein, not notably pale-tipped, the lowest one floriferous, not markedly broader or different from the others.
6 Perennial, rhizomatous; anthers 0.7–1.3 mm long (dry); scales (1.3) 1.5–2.2 mm long .... 4. *E. acicularis*
6 Annual, rarely rhizomatous; anthers 0.25–0.4 mm long; scales 1.0–1.3 mm long ...... 5. *E. bella*
5 Achene golden yellow (reputedly sometimes black), merely 3-ribbed, the surface finely cellular-roughened; scales (except the broader, often suborbicular, usually sterile lowest one) largely or wholly dark (often blackish purple) except for the conspicuously pale and hyaline tip.
7 Tubercle low and broad with a central apiculus, broader than high; rhizomes short and stout; spikelets mostly 2–3 times as long as wide, up to about 8 mm long; north of latitude 40 in our range .... 6. *E. bolanderi*
7 Tubercle narrow and high, about as high as wide; rhizomes slender (often lost when the plant is collected); spikelets mostly 3–7 times as long as wide, usually more than 8 mm long; chiefly south of latitude 40 in our range .... 7. *E. parishii*
1 Stigmas 2 (sometimes 3 in *E. ovata*, an annual with numerous flowers in each spikelet); achenes lenticular.
8 Rhizomatous perennial; anthers mostly 1.3–2.5 mm long; achenes mostly over 1.5 mm long (tubercle included); scales mostly 2–4.5 mm long .... 8. *E. palustris*
8 Annual; anthers mostly 0.25–0.8 mm long; achenes up to about 1.5 mm long; scales often less than 2 mm long.
9 Achene stramineous to dark brown, mostly 1.0–1.5 mm long (tubercle included) .... 9. *E. ovata*
9 Achene black, 0.7–1.0 mm long; widespread, chiefly trop. and subtrop. species to be sought along the s. border of our range .... *E. caribaea* (Rottb.) Blake

1. Eleocharis parvula (Roemer & Schultes) Link ex Bluff & Fingerh.

Scirpus parvulus Roemer & Schultes, Syst. Veg. 2: 124. 1817. *Eleogiton parvula* Link, Hort. Regius Bot. Berol. 1: 285. 1827. *Limnochloa parvula* Reichenb. in Mössler, Handb. Gewächsk. ed. 2. 1808. 1829. *Eleocharis parvula* Link ex Bluff & Fingerh. Compend. Fl. German. ed. 2. 1: 93. 1836. (Mannsfeld, Germany.)

E. pygmaea Torr. Ann. Lyceum Nat. Hist. New York 3: 313. 1836. *Chaetocyperus pygmaea* Walp. Ann. Bot. Syst. 3: 683. 1852. (N.Y.) = var. *parvula*.

E. pygmaea var. (?) *anachaeta* Torr. Ann. Lyceum Nat. Hist. New York 3: 441. 1836. *Scirpus nanus* var. *anachaetus* Britton, Trans. New York Acad. Sci. 11: 75. 1892. *E. parvula* var. *anachaeta* Svenson, Rhodora 36: 386. 1934. (*Drummond 409,* New Orleans, La.)

Isolepis leptos Steudel, Syn. Pl. Glum. 2: 91. 1855. *Scirpus leptos* C. Wright in Sauvolle, Anales Acad. Ci. Méd. Habana 8: 80. 1871. *E. leptos* Svenson, Rhodora 31: 176. 1929. (*Parry 130,* Mex.) = var. *anachaeta*. Not *E. lepta* C. B. Clarke, 1900.

Scirpus coloradoensis Britton, Torreya 4: 93. 1904. *E. leptos* var. *coloradoensis* Svenson, Rhodora 31: 177. 1929. *E. parvula* var. *coloradoensis* A. A. Beetle, Univ. Wyoming Publ. 5: 22. 1938. *E. coloradoensis* Gilly, Amer. Midl. Naturalist 26: 66. 1941. (*Cowen 2576,* Shore Lake, Larimer Co., Colo.) = var. *anachaeta*.

Diminutive, very slender perennial from inconspicuous slender rhizomes, forming dense mats; culms filiform, 2–6 (10) cm tall; spikelets 2.5–4 (6) mm long, with 2–9 (20) flowers; scales 1.5–2 (2.5) mm long, the lowest one empty; stigmas 3; achenes stramineous, trigonous (often unequally so), 0.9–1.3 mm long, including the short, inconspicuous tubercle which is 0.15–0.3 mm long and confluent with the body of the achene; $n = 4, 5$.

Wet, saline or alkaline sites; Newfl. to Vancouver I., s. to n. S. Amer.; also in w. Europe and along the Medit. Sea. June–Sept.

Our plants are var. *anachaeta* (Torr.) Svenson, with much reduced or obsolete perianth and often with cellular-roughened achenes, which occurs at scattered inland stations w. of the Mississippi R., s. to the Gulf Coast, Mex., and the Caribbean region; although seldom collected, it may be expected anywhere within our range. The var. *parvula,* with well developed perianth bristles (commonly equaling or surpassing the achene) and smooth achenes, occurs in salt marshes on both coasts of N. Amer. and at scattered inland stations chiefly e. of the Mississippi R., as well as in the Old World; it is not known in our range.

2. Eleocharis pauciflora (Lightf.) Link

Scirpus pauciflorus Lightf. Fl. Scotica 1078. 1777. *Eleocharis pauciflora* Link, Hort. Regius Bot. Berol. 1: 284. 1827. (*Stuart,* Malghyrdy, Breadalbane, Scotland.)

Perennial with clustered stems on rather short, stout rhizomes, and also producing more elongate, slender, stoloniform rhizomes with thickened terminal bud; culms (0.5) 1–3 (4) dm tall, slender, seldom as much as 1 mm thick distally, not flattened, not

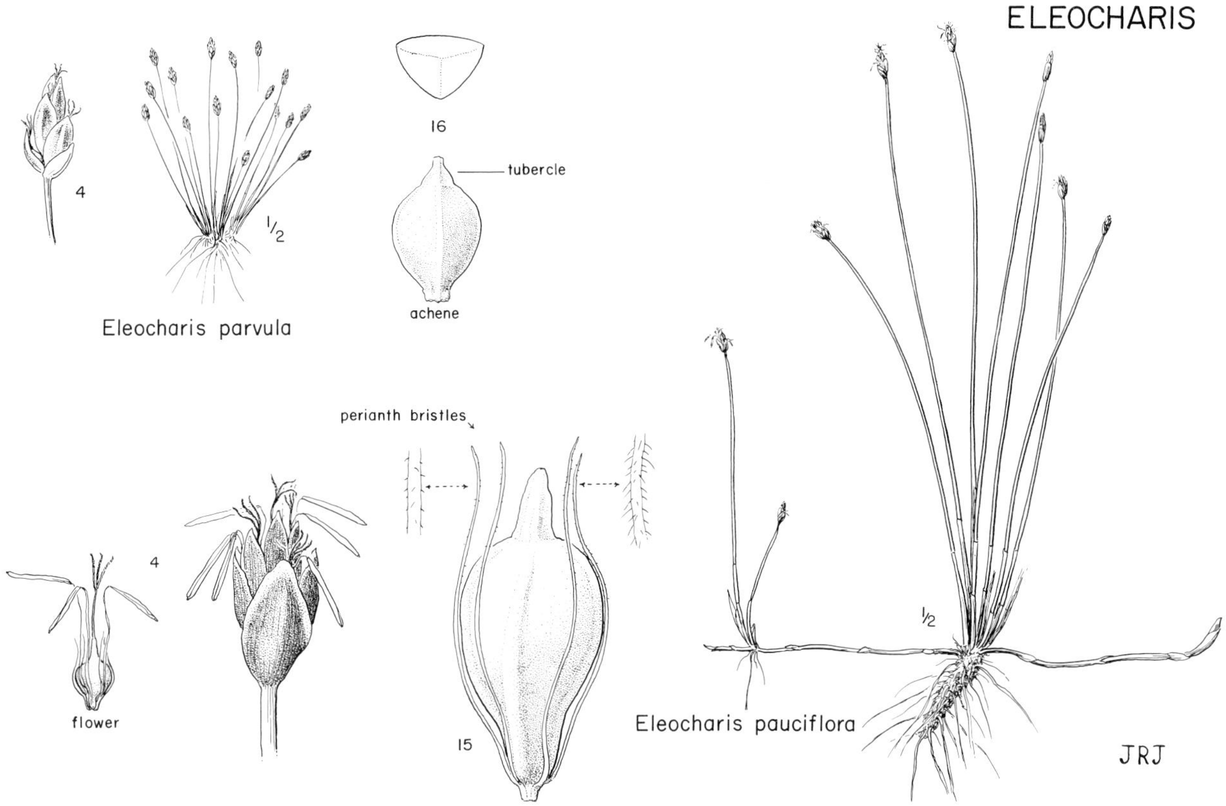

proliferous; spikelets 4–8 mm long, mostly 3- to 9-flowered; scales all floriferous (or the lowest one rarely empty), 2.5–5.5 mm long, the two lower ones larger than the others; perianth bristles equaling or exceeding the achene, or more or less reduced; stigmas 3; achene broadest above the middle, cellular-roughened, 1.9–2.6 mm long, including the short (to 0.6 mm), thickened stylar beak which is continuous with the body of the 3-ribbed, equally or unequally trigonous to merely planoconvex achene; $n = 10$.

Boggy or otherwise wet places, from the lowlands to high elev. in the mts., sometimes above timberline; tolerant of salt and alkali, circumboreal, s. in Amer. to Calif., N.M., Ill., and N.J. June–Aug.

3. **Eleocharis rostellata** (Torr.) Torr.

Scirpus rostellatus Torr. Ann. Lyceum Nat. Hist. New York 3: 318. 1836. *Eleocharis rostellata* Torr. Fl. State New York 2: 347. 1843. *Trichophyllum rostellatum* Farw. Rep. Michigan Acad. Sci. 15: 166. 1913 [1914]. (*Sartwell,* Penn-Yan, Yates Co., N.Y., is the first of two specimens cited.)

Perennial with clustered stems on short, stout, often ascending or nearly vertical rhizomes; culms (2) 4–10 dm tall or sometimes more, more or less flattened at least distally and commonly 1–2 mm wide, some of them commonly proliferous (rooting from an apical bulbil); spikelets (5) 8–13 mm long, (5) 10- to 20- (25)-flowered; scales equaling or surpassing the achene; stigmas 3; achene light greenish to medium brown, rounded-trigonous to planoconvex, smooth and shining or slightly cellular-roughened, 1.9–2.8 mm long, including the prominent, pale tubercle which is up to 0.75 mm long and is confluent with the body of the achene.

Salt marshes along the coast, and in alkaline or highly calcareous situations inland, often around hot springs; Vancouver I. to N.S., s. to n. Mex. and the Greater Antilles, and in the S. Amer. Andes. June–Aug.

4. **Eleocharis acicularis** (L.) Roemer & Schultes

Scirpus acicularis L. Sp. Pl. 48. 1753. *Mariscus acicularis* Moench, Meth. Pl. Horti Bot. & Agri Marburg 350. 1794. *Eleocharis acicularis* Roemer & Schultes, Syst. Veg. 2: 154. 1817. *Isolepis acicularis* Schlechtendal, Fl. Berol. 1: 36. 1823. *Clavula acicularis* Dumort. Fl. Belg. 143. 1827. *Scirpidium aciculare* Nees, Linnaea 9: 293. 1834. (Europe.)
E. acicularis var. *occidentalis* Svenson, Rhodora 31: 190. 1929. *E. acicularis* f. *occidentalis* A. A. Beetle, Univ. Wyoming Publ. 5: 23. 1938. (*Rothrock 80,* Santa Barbara, Calif.)

Diminutive, very slender perennial from slender, branching rhizomes, commonly forming dense tufts; culms filiform, 3–12 cm tall; basal sheaths pale or less often purplish; spikelets 2.5–7 (9) mm long, 3- to 15-flowered; scales (1.3) 1.5–2.2 mm long, with greenish midrib and more or less hyaline, paler or partly anthocyanic margins; bristles mostly 3 or 4, equaling or surpassing the achene, or more or less reduced or wanting; anthers 0.7–1.3 mm long (dry); stigmas 3; achenes white to pale gray or faintly yellowish, 0.7–1.1 mm long, including the well defined, more or less triangular-conic tubercle, obovoid, rounded-trigonous, longitudinally 8- to 18-ribbed (3

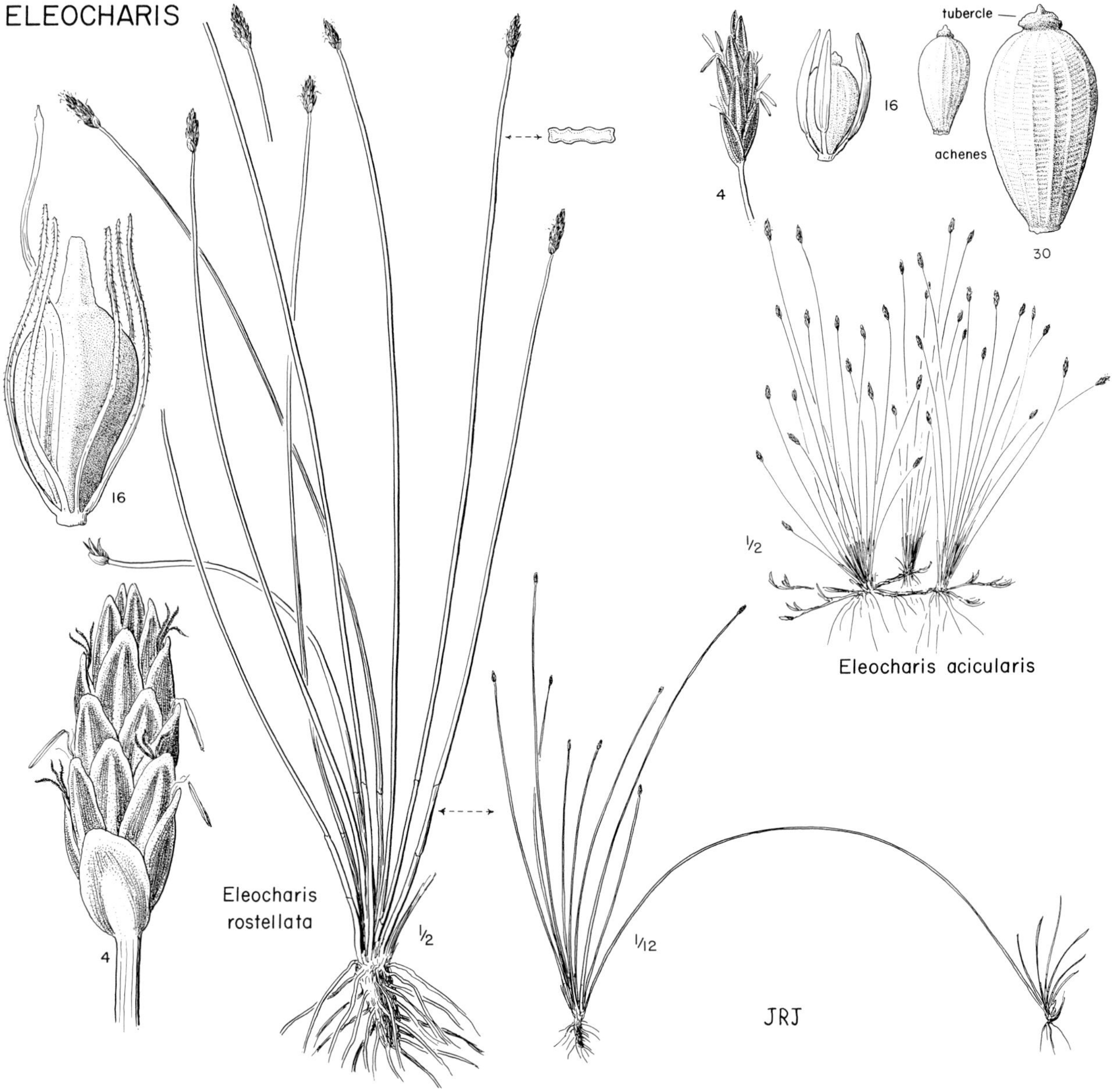

ribs stronger than the others), the ribs connected by very numerous fine, straight cross-ridges; $n = 10–29$.

Marshes, muddy shores, and other wet places, from the lowlands to rather high elev. in the mts.; circumboreal, s. to Fla. and Mex., and found throughout our range. June–Sept.

5. Eleocharis bella (Piper) Svenson

Eleocharis acicularis var. *minima* Torr. ex Britton, J. New York Microscop. Soc. 5: 104. 1889. (*Hall 566,* Oregon, in 1871.)

E. acicularis var. *bella* Piper in Piper & Beattie, Fl. Palouse Region 35. 1901. *E. bella* Svenson, Rhodora 31: 201. 1929. (*Piper 3055,* Pullman, Wash.)

Diminutive, very slender annual, forming dense tufts, seldom with slender rhizomes but probably still annual; culms filiform, 1–8 cm tall; spikelets 1.5–3 mm long, 3- to 15-flowered; scales 1.0–1.3 mm long, with greenish midrib and more or less hyaline, paler or partly anthocyanic margins; bristles none; anthers 0.25–0.4 mm long (dry); stigmas 3; achene white or slightly yellowish, 0.6–0.8 mm long, including the well defined, very short and broad tubercle, the body obovoid, rounded-trigonous, with rather numerous longitudinal ribs (3 of them stronger than the others), the ribs connected by numerous fine, straight cross-ridges, forming ladder-like patterns.

Borders of marshes and lakes, and other wet, muddy or springy places, from the lowlands to moderate elev. in the mts.; Wash. and n. Idaho to s. Calif. and N.M. June–Sept.

Eleocharis bella is obviously related to *E. acicularis*. The range of *E. bella* is wholly included within that of *E. acicularis,* and they occur in similar habitats. On the other hand, the critical distinctions between the two are well correlated, and I see no evidence of hybridization or populational intergradation. *E. bella* is therefore retained as a distinct species.

ELEOCHARIS

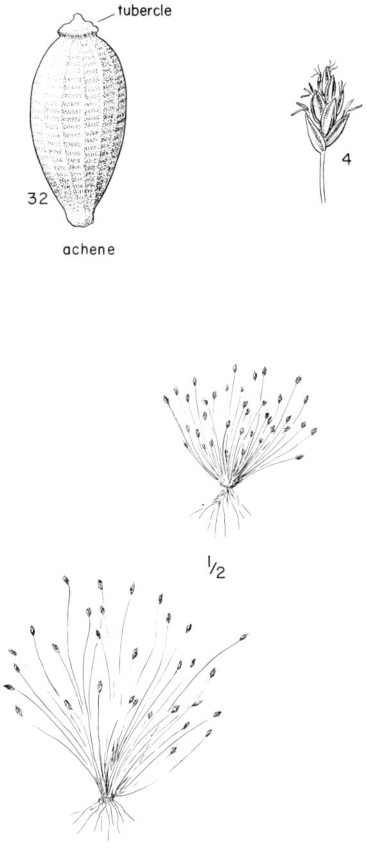

Eleocharis bella

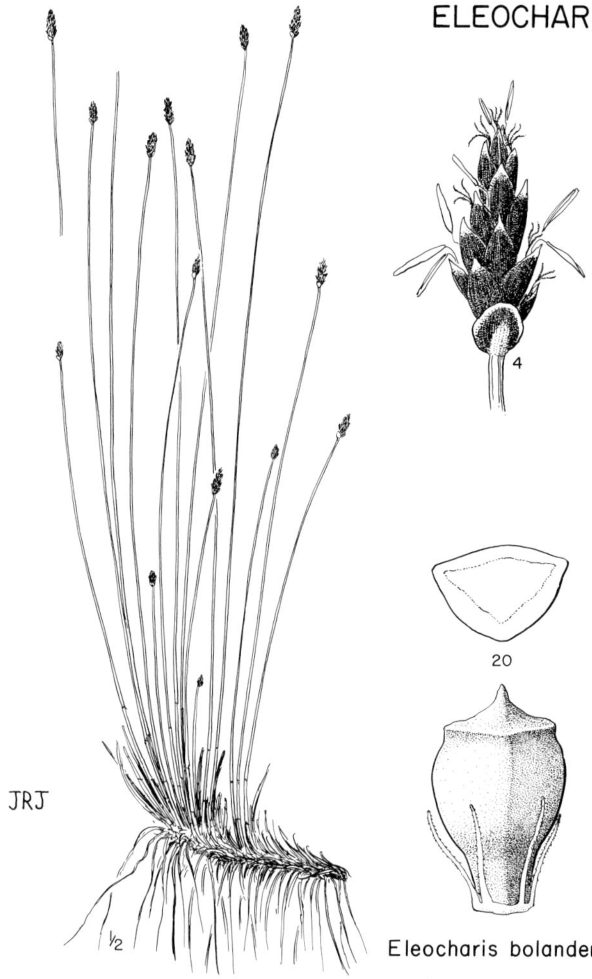

Eleocharis bolanderi

6. Eleocharis bolanderi A. Gray

Eleocharis bolanderi A. Gray, Proc. Amer. Acad. Arts 7: 392. 1868. *E. montevidensis* var. *bolanderi* V. Grant, Aliso 4: 87. 1958. (*Bolander 4869,* near Clark's, Mariposa Co., Calif.)

Culms numerous, slender, 0.8–3 dm tall, commonly about 6-angled and with as many vascular bundles, densely clustered on short, freely rooting rhizomes, the debris from previous years tending to be persistent around the base; spikelets mostly 3–8 mm long and 10- to 25-flowered; outermost scale relatively broad, more or less orbicular, usually with a firm, often greenish area toward the base medially, often sterile, the other scales narrower, mostly 1.8–2.7 mm long, largely or wholly dark (commonly atropurpureous) except for the pale, hyaline tip; bristles 3 or 4, reddish brown, retrorsely barbellate, distinctly shorter than the achene; stigmas 3; achene unequally trigonous, with broad inner face and blunt outer angle, obovoid from a broadly and shortly substipitate base, golden yellow (greenish yellow when submature) or reputedly sometimes black, minutely cellular-roughened, about 1.5 mm long, including the well defined, commonly broad and depressed tubercle which may appear almost as an apical scar with a central mucro.

Wet meadows and moist soil along streams, from the foothills to moderate elev. in the mts.; Sierra Nevada of Calif., n. to the Cascade Mts. of s. Oregon and the Wallowa Mts. of ne. Oregon, and e. through se. Oregon and n. Nev. to n. Utah (Cache Co.) and se. Idaho (Franklin Co.), in our range apparently wholly north of the 40th parallel. May–July.

7. Eleocharis parishii Britton

Eleocharis parishii Britton, J. New York Microscop. Soc. 5: 110. 1889. *E. montevidensis* var. *parishii* V. Grant, Aliso 4: 87. 1958. (*Parish 1569,* Agua Caliente, Colorado Desert, San Diego Co., Calif.)

"*E. montana* (H.B.K.) Roemer & Schultes" of authors, misapplied.

Culms numerous, slender, (0.5) 1–3 dm tall, mostly 6- to 8-ribbed, in small clusters on well developed slender rhizomes, these often lost in collecting; spikelets slender and elongate, mostly 7–15 mm long and 3–7 times as long as wide, mostly 20- to 40-flowered; scales mostly 2.4–3.0 mm long, largely purplish (often somewhat mottled) except for the paler, often greenish midvein and pale, hyaline tip, the lowest one tending to be shorter, broader, and firmer, and commonly sterile; bristles mostly 6 or 7, white or brownish, retrorsely barbellate, longer or shorter than the achene; stigmas 3; achene 3-nerved, planoconvex or unequally trigonous, yellow to brown, smooth at 10X, often visibly a little roughened at 20X, the body mostly 1.0–1.2 mm long, narrowed to the well developed tubercle (this commonly 0.3 mm long and as high as or higher than wide) that may be wholly sessile or basally constricted.

Wet places in desert regions, and up to middle elev. in the desert mts.; s. Calif. to N.M. and n. Chihuahua, n. to Utah and s. Nev. (Lincoln Co.), and in the Puget Trough and coast ranges apparently to sw. Oregon; chiefly s. of latitude 40° in our range, but known also from Cache Co., Utah. April–Sept.

8. **Eleocharis palustris** (L.) Roemer & Schultes

Scirpus palustris L. Sp. Pl. 47. 1753. *Bulbostylis palustris* Steven, Mém. Soc. Imp. Naturalistes Moscou 5: 355. 1814. *Eleocharis palustris* Roemer & Schultes, Syst. Veg. 2: 151. 1817. *Clavula palustris* Dumort. Fl. Belg. 143. 1827. *Trichophyllum palustre* Farw. Rep. Michigan Acad. Sci. 15: 166. 1913 [1914]. (Europe.)

Scirpus uniglumis Link, Jahrb. Gewächsk. 1(3): 77. 1820. *E. uniglumis* J. A. Schultes in Roemer & Schultes, Mantissa 2: 88. 1824. *Clavula uniglumis* Dumort. Fl. Belg. 143. 1827. *Scirpus palustris* var. *uniglumis* Junge, Jahrb. Hamburg. Wiss. Anst. 25 Beih. 3: 248. 1908. (Rostock, Germany.)

E. calva Torr. Fl. State New York 2: 346. 1843, nom. provis. Place and date of validation not ascertained. *E. palustris* var. *calva* A. Gray, Manual Bot. 522. 1848. *Trichophyllum palustre* var. *calvum* H. D. House, Amer. Midl. Naturalist 6: 205. 1920. *E. palustris* var. *glaucescens* subvar. *calva* Farw. Amer. Midl. Naturalist 12: 50. 1930. (w. N.Y.)

E. palustris var. *major* Sonder, Fl. Hamburg. 22. 1851. *E. smallii* var. *major* Seymour, Fl. New Engl. 113. 1969. (Hamburg, Germany.)

E. erythropoda Steudel, Syn. Pl. Glum. 2: 76. 1855. *Scirpus erythropodus* Kuntze, Revisio Gen. Pl. 2: 757. 1891. (*Frank,* Ohio.)

Scirpus mamillatus H. Lindb. Acta Soc. Fauna Fl. Fenn. 23(7): 7. 1902. *E. mamillata,* incorrectly attr. to H. Lindb. by authors: place and date of validation not ascertained. (n. Europe.)

E. macrostachya Britton ex Small, Fl. S.E. U. S. 184. 1903. (*Palmer 264a,* Indian Territory [Okla.].)

E. xyridiformis Fern. & Brackett, Rhodora 31: 76. 1929. (*Pringle 6817,* valley of Mex.)

Rhizomatous perennial; culms scattered or in small clusters on the rhizome, slender to very stout, 1–10 dm tall; spikelet 5–23 mm long, lanceolate or lance-ovate in outline, light to dark brown or chestnut; empty scales (glumes) at the base of the spikelet 1 or 2 (3), the lowest one often nearly encircling the culm; fertile scales mostly 2–4.5 mm long; perianth bristles retrorsely barbed, typically 4, less often 5 or 6, shortly surpassing the achene, or sometimes reduced or obsolete; anthers 1.3–2.5 mm long (dry); stigmas 2; achene lenticular, yellow to medium brown, finely cellular-roughened, mostly 1.5–2.5 mm long, including the well differentiated tubercle that is commonly 0.4–0.7 mm long and constricted at the base; n = 5–46.

Wet places from sea level to moderate elev. in the mts., tolerant of alkali; widespread in temp. and cold-temp. regions of the N. Hemisphere, and common throughout our range. May–Aug.

Eleocharis palustris is a highly variable polyploid complex, with chromosome numbers between as well as at the ploidy-levels. Numerous efforts have been made to subdivide it into varieties or segregate species, but these segregates overlap and intergrade morphologically, without clear ecogeographic correlations on anything beyond a local basis. The most obvious, and at first appearance potentially the most useful distinction within the complex is between specimens with 2 (or 3) glumes and those with only one. However, both types sometimes occur in the same collection. It has been alleged that in uniglumate forms the base of the glume encircles or nearly encircles the culm, in contrast to the first glume of biglumate types which extends only half way around the culm. The correlation is imperfect, however; some specimens with 2 glumes have the first glume encircling the culm. There is also a partial correlation of chromosome number with morphology, but identifications based on chromosome number do not always conform to identifications based on morphology, and even those who would divide the group (cf. Strandhede, 1967, and Harms, 1968) have found it necessary to include more than one cytotype within a single species. It is possible that careful biosystematic study on a world-wide basis may permit the recognition of several specific or infraspecific segregates from *E. palustris* sensu lato, but this remains to be demonstrated. At the present state of our knowledge it seems advisable to define the species broadly and not attempt any infraspecific segregation.

9. **Eleocharis ovata** (Roth) Roemer & Schultes

Scirpus ovatus Roth, Catalecta Bot. 1: 5. 1797. *Bulbostylis ovata* Steven, Mém. Soc. Imp. Naturalistes Moscou 5: 355. 1814. *Eleocharis ovata* Roemer & Schultes, Syst. Veg. 2: 152. 1817. *Clavula ovata* Dumort. Fl. Belg. 143. 1827. *Trichophyllum ovatum* H. D. House, Amer. Midl. Naturalist 6: 204. 1920. (Germany.)

Scirpus obtusus Willd. Enum. Pl. Hort. Regii Bot. Berol. 76. 1809. *E. obtusa* J. A. Schultes in Roemer & Schultes, Mantissa 2: 89. 1824. *Trichophyllum obtusum* H. D. House, Amer. Midl. Naturalist 6: 205. 1920. *E. ovata* var. *obtusa* Kükenthal ex Skottsb. Acta Horti Gothob. 2: 212. 1926. (N. Amer.; no specimen cited.)

E. engelmanni Steudel, Syn. Pl. Glum. 2: 79. 1855. *E. ovata* var. *engelmanni* Britton, J. New York Microscop. Soc. 5: 103. 1889. *Trichophyllum engelmanni* H. D. House, Amer. Midl. Naturalist 6: 204. 1920. *E. obtusa* var. *engelmanni* Gilly, Iowa State Coll. J. Sci. 21: 92. 1946. (*Engelmann,* St. Louis.)

E. monticola Fern. Proc. Amer. Acad. Arts 34: 496. 1899. *Trichophyllum monticolum* [sic] H. D. House, Amer. Midl. Naturalist 6: 205. 1920. *E. engelmanni* var. *monticola* Svenson, Rhodora 31: 209. 1929. (*Mrs. Pulsifer Ames,* Plumas Co., Calif. in 1876; lectotype by Svenson.)

E. monticola var. *leviseta* Fern. Proc. Amer. Acad. Arts 34: 496. 1899. *E. engelmanni* var. *monticola* f. *leviseta* Svenson, Rhodora 31: 210. 1929. (*Bush 585,* Montier, Mo., is the first of several specimens cited.)

E. monticola var. *pallida* St. John, Northw. Sci. 2: 81. 1928. (*A. D. Dunn s.n.,* Parker, Wash., 8 Aug. 1901.)

Tufted annual, mostly 0.5–5 dm tall; culms several- to many-ribbed, 0.5–2 mm thick; spikelets (2) 5–13 mm long, ovoid, many-flowered (flowers seldom less than 40); scales mostly 1.7–2.5 mm long, purplish or brownish, with greenish midstripe and paler, hyaline margins, the lowest one empty or floriferous; perianth bristles mostly 6 or 7, brownish, retrorsely barbellate, commonly equaling or exceeding the achene, or reduced or wanting; anthers 0.3–0.8 mm long; stigmas 2 or sometimes 3; achene lenticular (even when there are 3 stigmas), mostly 1.0–1.5 mm long, including the flattened, broad-based tubercle which is appressed to the summit of the stramineous to dark brown, smooth and shining body of the achene; n = 5.

Marshes and other wet places, from sea level to moderate elev. in the mts.; widespread in the N. Hemisphere, and found in appropriate habitats throughout our range. June–Sept.

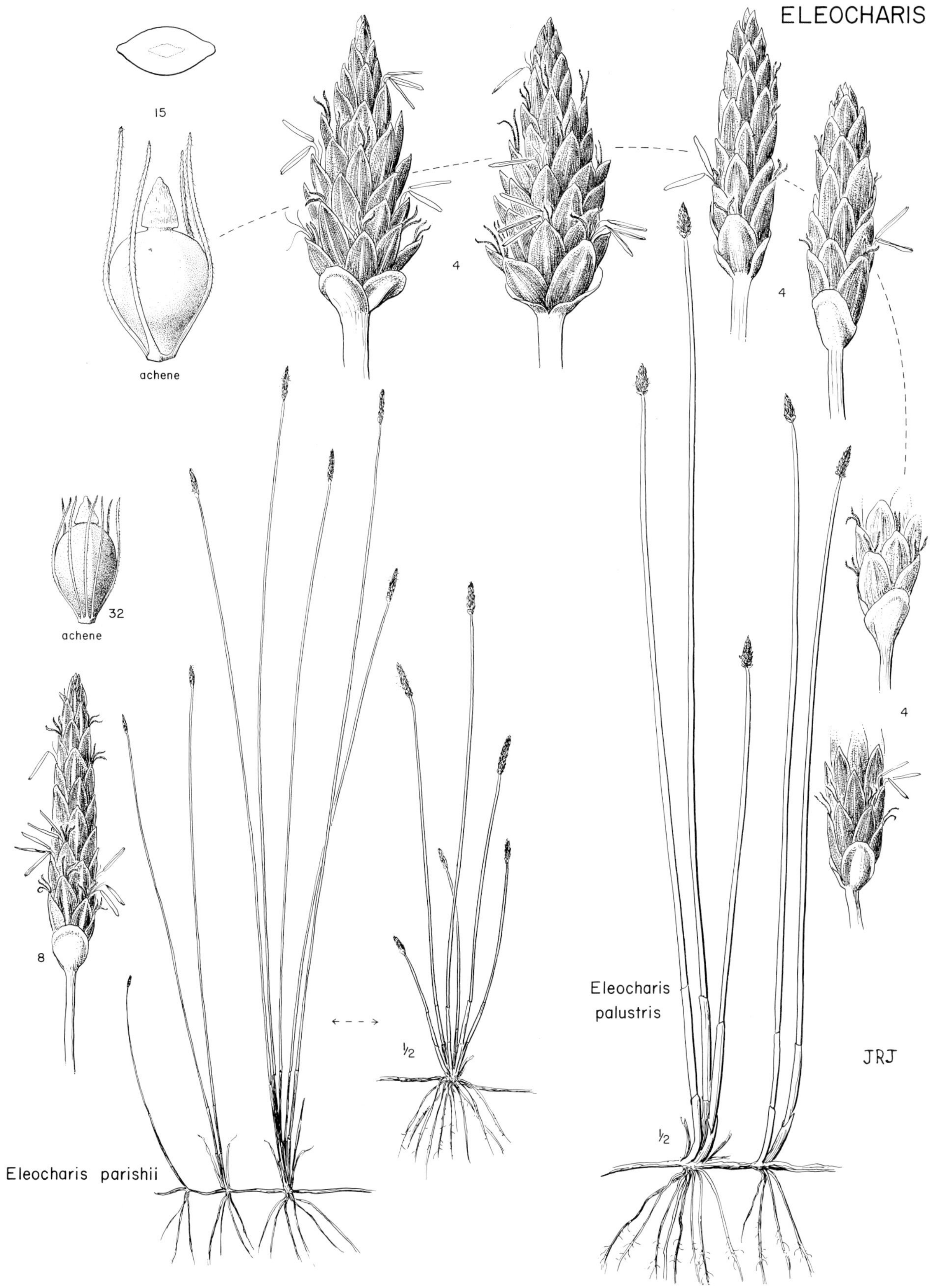
ELEOCHARIS
15
achene
4
4
4
32
achene
8
1/2
Eleocharis parishii
Eleocharis
palustris
1/2
JRJ

5. FIMBRISTYLIS Vahl

Annual or perennial herbs with triangular stems, the leaves grasslike, mostly clustered near the base; spikelets several to many in simple or compound, often umbelliform cymes in a terminal inflorescence that is subtended by a cluster of sheathless, leafy or scarious involucral bracts; scales spirally arranged; flowers perfect, borne singly in the axils of the scales; perianth wanting; stamens 1–3; style bifid (ours) or less often trifid, often hairy, enlarged toward the base, wholly deciduous.

More than 100 species, chiefly tropical. (Name from the Latin *fimbris*, fringe, and *stylus*, style.)

References:

Kral, R. A treatment of *Abildgardia, Bulbostylis*, and *Fimbristylis* (Cyperaceae) for North America. Sida 4: 57–227. 1971.
Svenson, H. K. *Fimbristylis*, in N. Amer. Fl. 18(9): 550–556. 1957.

1. Fimbristylis spadicea (L.) Vahl

Scirpus spadiceus L. Sp. Pl. 51. 1753. *Fimbristylis spadicea* Vahl, Enum. Pl. 2: 294. 1805. (Jamaica.)
Scirpus carolinianus Lam. Tableau Encycl. & Méth. 1: 142. 1791. *F. caroliniana* Fern. Rhodora 42: 246. 1940. (*Fraser*, Carolina.)
Scirpus puberulus Michx. Fl. Boreali-Amer. 1: 31. 1803. *F. puberula* Vahl, Enum. Pl. 2: 289. 1805. *F. spadicea* var. *puberula* Chapman, Fl. S. U. S. 522. 1860. *F. castanea* var. *puberula* Britton ex H. D. House, New York State Mus. Bull. 254 (Ferns & Fl. Pls. New York State): 141. 1924. (Carolina and Georgia.)
F. thermalis S. Wats. in King, Rep. Geol. Explor. 40th Parallel 5: 360. 1871. (*Watson 1216*, hot springs in Ruby Valley, Nev.) The type of this name was incorrectly given by Kral as *Parish 3667*, near San Bernardino, Calif.
F. interior Britton in Britton & Brown, Ill. Fl. N. U. S., Canada & Brit. Possessions, ed. 2. 1: 320. 1913. *F. puberula* var. *interior* Kral, Sida 4: 135. 1971. (*Osterhout s.n.*, Sterling, Logan Co., Colo.)

Culms 2–8 dm tall, borne in rather small tufts, the plant also producing slender rhizomes; leaves slender and elongate, up to 2 (4) mm wide, often folded, glabrous to short-hairy; inflorescence a simple or more often compound, umbelliform cyme, the sessile terminal spikelet subtended by a setaceous involucre with several slender rays up to about 6 cm long, each ray with a terminal spikelet which may be subtended by an involucel and one or more secondary rays with terminal spikelets; spikelets 6–20 mm long, 3–4 mm thick, the scales numerous, at least the outer ones usually short-hairy on the back, the midrib raised distally and excurrent as a short point; stamens 3; style bifid above the middle, evidently villosulous especially toward the bifurcation; achenes obovate, minutely apiculate, about 1.5 mm long, multiribbed and finely cross-rugulose.

Wet places in the valleys and lowlands, sometimes around hot springs; widespread in trop. Amer., n. to N.Y., Mich., Colo., Utah (Utah Lake), Nev. (Elko Co.) and Calif.; seldom collected in our range. July–Aug.

The name *F. spadicea* is here adopted for this broadly defined species, following Svenson in North American Flora. Under a narrower specific concept our plants are called *F. puberula* (Michx.) Vahl var. *interior* (Britton) Kral.

6. CYPERUS L.

Annual or more often perennial herbs with mostly triangular, solid stems, the leaves with closed sheath and usually with an elongate, grasslike blade; spikelets several to many, arranged in one or usually several capitate to spicate clusters, the terminal cluster commonly sessile or nearly so, the others borne on more or less elongate rays originating in the axils of the sheathless, leafy involucral bracts; scales of the spikelet distichous, arranged in 2 vertical rows, the lowermost scale empty and more or less modified; flowers perfect, borne singly in the axils of the scales; perianth wanting; stamens 3, less commonly 1 or 2; style bifid or more often trifid, the achene accordingly lenticular or trigonous; $x = 8, 9$.

About 600 species, of cosmopolitan distribution, commonest in trop. and warm-temp. regions. (From *Kypeiros*, the ancient Greek name.)

Certain characters, especially the length and number of the involucral bracts, the length and number of the rays, the number of scales in each spikelet, and the length of the spikelet are highly variable within species and thus are of little taxonomic significance. Our species are all sharply distinct among themselves. They all flower in mid- or late summer and fall.

References:

Corcoran, Sister Mary Lucy. A revision of the subgenus Pycreus in North and South America. Catholic Univ. Amer., Biol. Ser. 37. 68 pp. 1941.
Kükenthal, G. Cyperaceae-Scirpoideae-Cypereae, in Engl. Pflanzenr. IV. 20 (Heft 101). 671 pp. 1935–36.
McGivney, Sister M. Vincent De Paul. A revision of the subgenus Eucyperus found in the United States. Catholic Univ. Amer., Biol. Ser. 26. 74 pp. 1938.

1 Plants perennial (except in *C. erythrorhizos*); spikelets borne in more or less elongate and open, cylindric spikes (except in *C. schweinitzii* and sometimes *C. esculentus*); rachilla winged (or scarcely so in *C. schweinitzii*); stamens 3; pistil tricarpellate.

2 Perennial; scales relatively large, 2–6 mm long, not very closely set, the tip of each scale surpassing the one below it by 1–3 mm; achenes 1.3–2.5 mm long.

3 Wings of the rachilla clasping each achene toward the base; plants wholly without rhizomes; spikelets deciduous at maturity, the rachilla disarticulating just above the base, the individual scales not falling away from the rachilla .... 1. *C. strigosus*

3 Wings of the rachilla narrow, not clasping the achene, or the rachilla scarcely winged; plants rhizomatous; scales falling from the rachilla at maturity, the rachilla sometimes (*C. schweinitzii*) thereafter itself deciduous.

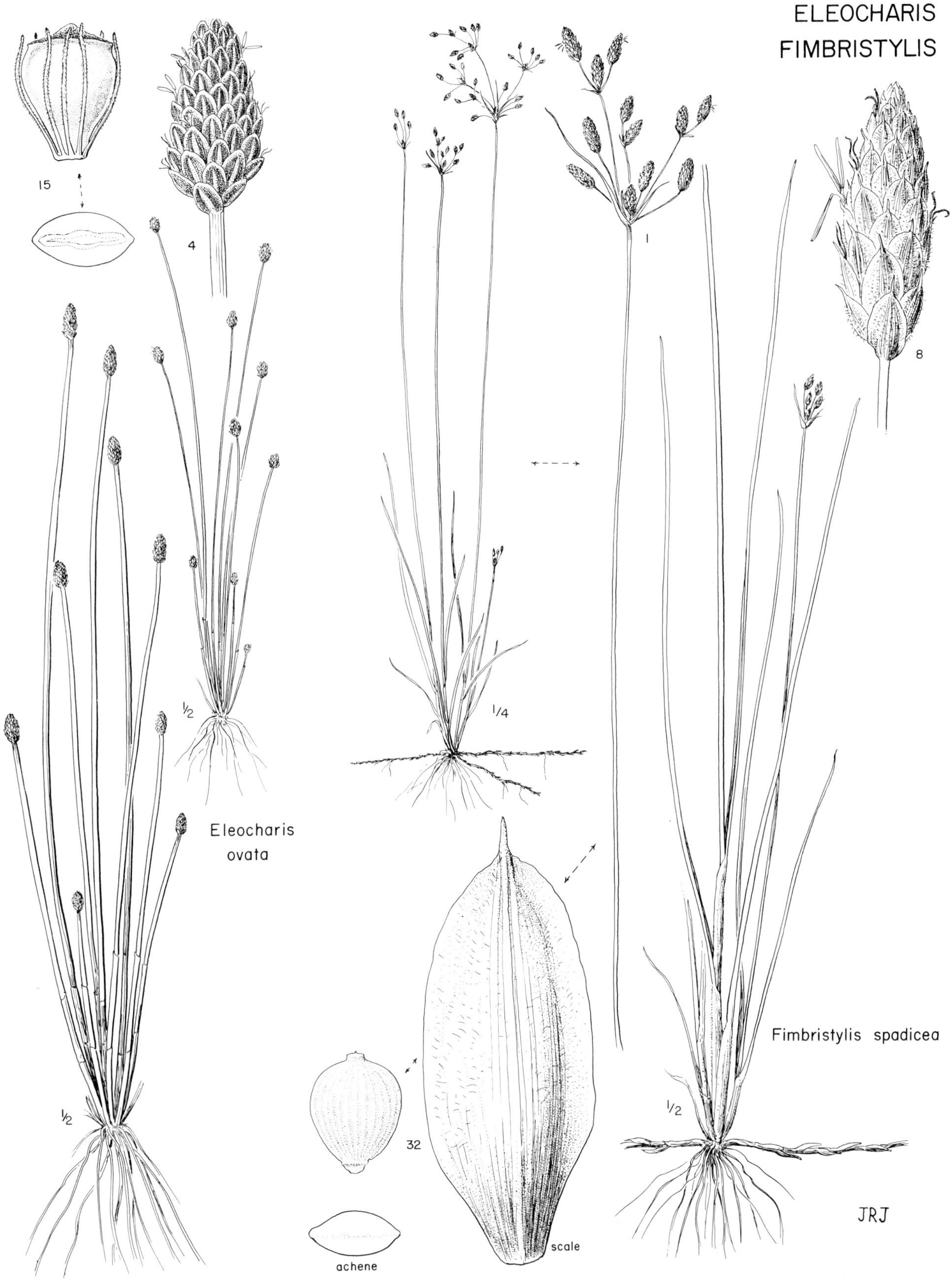
ELEOCHARIS
FIMBRISTYLIS
15
4
1/2
Eleocharis
ovata
1/2
1/4
1
8
32
achene
scale
Fimbristylis spadicea
1/2
JRJ

4 Rhizomes short, thickened at intervals, but scarcely tuber-bearing; clusters of spikelets relatively short and dense, subcapitate; rachilla narrowly or scarcely winged, the wing when present firm, not hyaline .... 2. *C. schweinitzii*

4 Rhizomes elongate, slender, with a terminal tuber; clusters of spikelets usually elongate, open, and more or less cylindric, seldom more condensed; rachilla evidently hyaline-winged .... 3. *C. esculentus*

2 Annual; scales small, mostly 1.2–1.5 mm long, closely set, the tip of each scale surpassing the one below it by 0.4–0.7 mm; achene 0.7–1.0 mm long .... 4. *C. erythrorhizos*

1 Plants annual; spikelets borne in very short, more or less capitate clusters with a very short rachis; rachilla wingless; stamens 1 or 2, rarely 3.

5 Pistil tricarpellate; stamen 1; scales sharp-pointed or awn-tipped, often outcurved distally, often less than 2 mm long.

6 Scales (5) 7- to 9-nerved, with a slender, squarrose-recurved, short but distinct awn-tip .... 5. *C. aristatus*

6 Scales 3-nerved, tending to be recurved-acuminate, but not awn-tipped .... 6. *C. acuminatus*

5 Pistil bicarpellate; stamens 2 (3); scales mostly 2–2.5 mm long, straight, blunt .... 7. *C. rivularis*

1. Cyperus strigosus L.

Cyperus strigosus L. Sp. Pl. 47. 1753. *Mariscus strigosus* C. B. Clarke, J. Linn. Soc., Bot. 30: 308. 1894. (Jamaica and Virginia.)

Short-lived perennial without rhizomes, the culms few or solitary, sharply triquetrous, mostly 1–6 (10) dm tall; leaves several, crowded toward the base, mostly 2–10 (15) mm wide; spikelets numerous, in short-cylindric spikes, spreading more or less at right angles from the more or less elongate rachis, 6–25 mm long and 1–2 mm wide, deciduous at maturity, the rachilla disarticulating just above the reduced lowermost pair of scales; scales mostly 3–6 mm long, several-nerved, tending to be narrowly hyaline-margined distally, persistent on the rachilla; rachilla rather broadly hyaline-winged, the wings clasping the lower part of each achene; stamens 3; style trifid; achene unequally trigonous, 1.5–2 mm long, 3–4 times as long as wide.

Wet places, as along the banks of large streams, in the valleys and lowlands; widespread in the U.S. and adj. Can., s. to trop. Amer.

2. Cyperus schweinitzii Torr.

Cyperus schweinitzii Torr. Ann. Lyceum Nat. Hist. New York 3: 276. 1836. *Chlorocyperus schweinitzii* Palla, Oesterr. Bot. Z. 58: 392. 1908. (*Bradley,* near Greece, Monroe Co., N.Y. is the first specimen cited.)

"Cyperus bushii Britton," the name sometimes misapplied to this species.

Perennial from short, irregularly thickened rhizomes; culms 1–7 dm tall, somewhat bulbous-based, triquetrous, often antrorsely scabrous toward the summit; leaves several, all crowded toward the base, with well developed, elongate blade mostly 1–4 mm wide; involucral bracts elongate, unequal, sometimes a little wider than the proper leaves; spikelets borne in 1–several fairly compact, subcapitate spikes with a rather short rachis, the terminal spike sessile, the others terminating slender rays up to 7 cm long; individual spikelets 0.5–2.5 cm long, compressed, mostly 2–4 mm wide; scales mostly 3–4 mm long (the lowest ones shorter), conspicuously striate-nerved, the midrib commonly excurrent as a short point, the individual scales well spaced, the tip of each scale surpassing the scale next below it by 1.5–3 mm; rachilla narrowly or obscurely winged, the wing sometimes scarious but scarcely hyaline, the scales falling from the rachilla at maturity, and the rachilla itself eventually deciduous; stamens 3; style trifid; achene dark, equably trigonous, 2–2.5 mm long.

Sandy places in the valleys and lowlands, sometimes along streambanks, sometimes in drier soil or even on dunes; w. N.Y. and s. Ont. to Texas, Mex., Utah (Kanab) and n. Idaho.

3. Cyperus esculentus L.

Cyperus esculentus L. Sp. Pl. 45. 1753. ("Habitat Monspelii, inque Italia, Oriente.")

Culms stout, sharply triquetrous, 1–7 dm tall, arising singly, producing many slender rhizomes which terminate in small tubers; herbage sweet-scented; leaves clustered at the base, with elongate blade mostly 3–8 mm wide; involucral bracts elongate, unequal, sometimes a little wider than the proper leaves; spikelets borne in open (seldom short and congested), cylindric spikes with more or less elongate rachis, the terminal spike or cluster of spikes sessile, the others borne singly or in small groups at the ends of rays up to 7 cm long; spikelets slender, 0.5–5 cm long, only 1–2 mm wide; scales mostly (2) 2.5–3 (4) mm long (the lowest ones shorter), several-nerved, broad and much overlapping laterally, but not closely set, the tip of each scale surpassing the one next below it by 1.1–1.7 mm; rachilla narrowly hyaline-winged, persistent, the scales falling from it at maturity; stamens 3; style trifid; achene unequally trigonous, 1.3–2.0 mm long; $n = 9$, ca. 54.

Moist, low ground along streams and ditches, sometimes also in drier ground away from streams, becoming weedy; widespread in trop. and temp. (but not cold-temp.) regions.

4. Cyperus erythrorhizos Muhl.

Cyperus erythrorhizos Muhl. Descript. Gram. 20. 1817. *Chlorocyperus erythrorhizos* Palla, Allg. Bot. Z. Syst. 9: 69. 1903. (Pa.)

Tufted annual; culms rather stout, mostly 1–7 dm tall; leaves crowded toward the base, with well developed, elongate blade mostly 2–9 mm wide; involucral bracts elongate, unequal, sometimes a little wider than the proper leaves; inflorescence of clustered, cylindric spikes, the terminal cluster sessile, the others

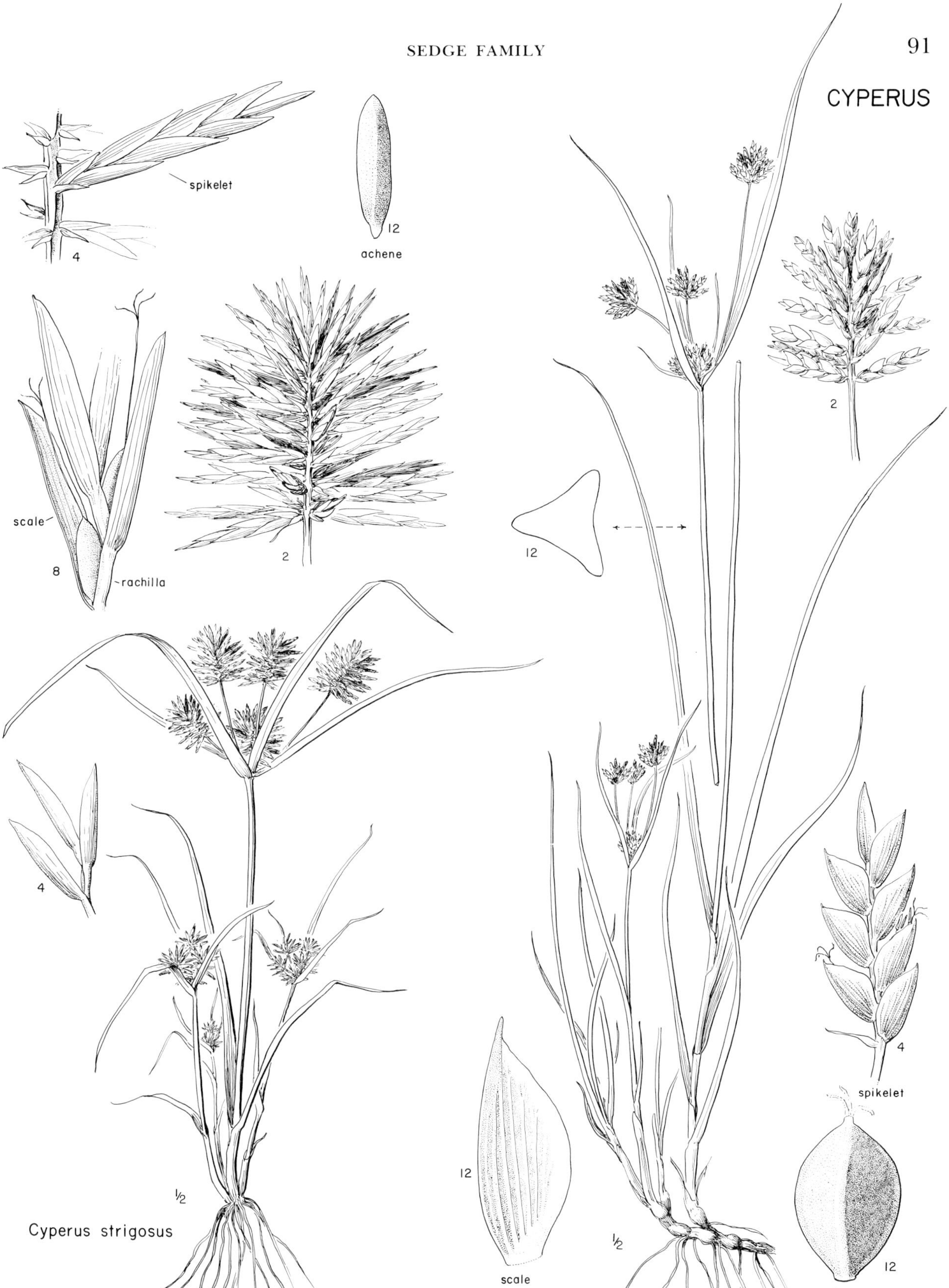
CYPERUS
spikelet
4
12
achene
8
scale
rachilla
2
2
12
4
4
spikelet
12
achene
12
scale
1/2
1/2
Cyperus strigosus
Cyperus schweinitzii
JRJ

CYPERUS

2
spikes
4
16
scale
achene
2
spike
8
4
spikelet
2
spikelet
12
achene
12
12
scales
Cyperus esculentus
terminal tuber
1/2
JRJ
1/2
Cyperus erythrorhizos

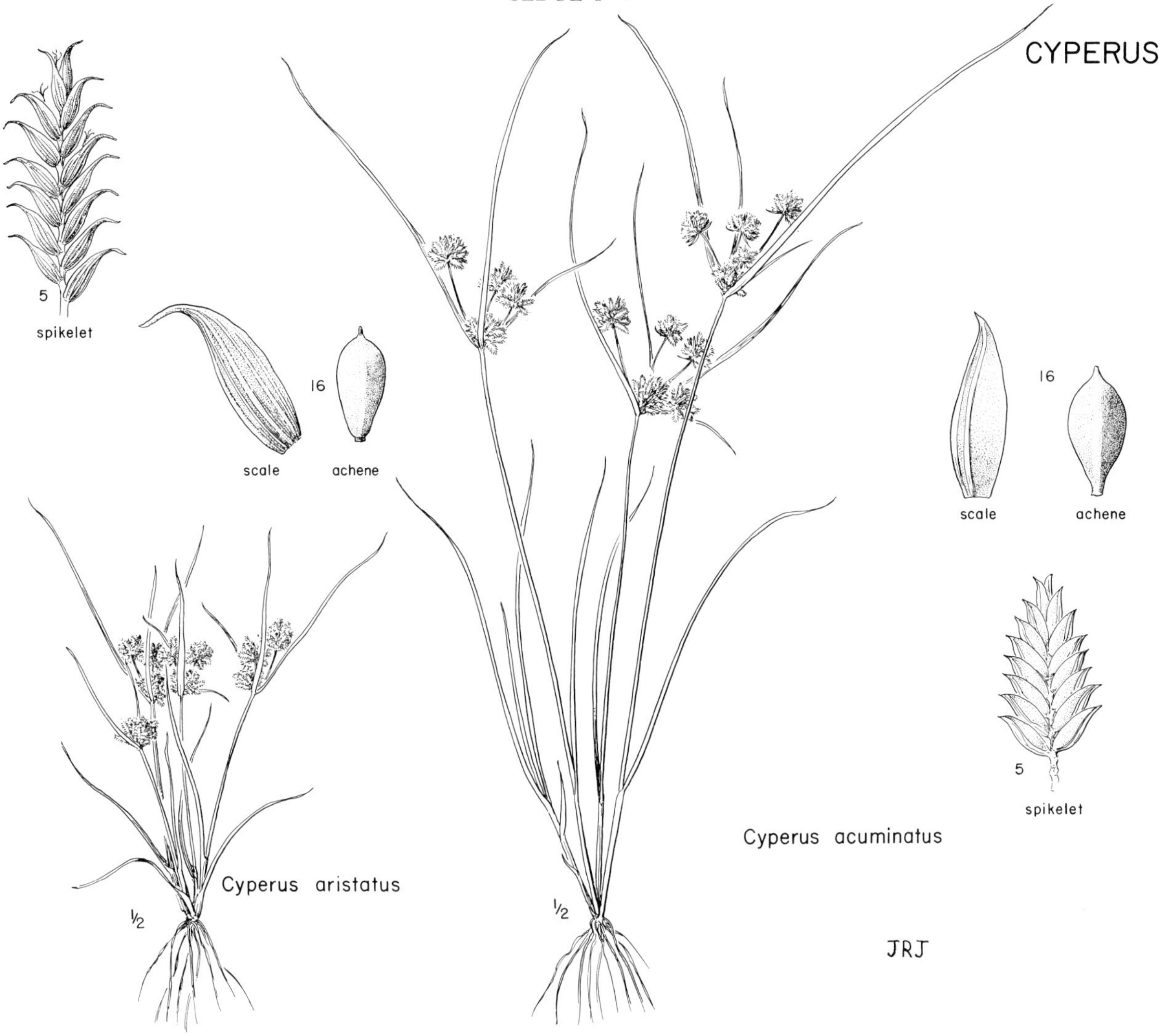

on rays up to 7 cm long (or some rays bearing only a single spike), each spike 1–4 cm long, with numerous spirally arranged spikelets on an elongate rachis; spikelets slender, 3–12 mm long, 1–1.5 mm wide; rachilla narrowly hyaline-winged, the wings readily deciduous in short segments at maturity; scales 1.2–1.5 mm long, with prominent, green, minutely excurrent keel and one or two pairs of much more slender and inconspicuous lateral nerves, closely set, the tip of each scale surpassing the one below it by 0.4–0.7 mm, the scales eventually deciduous from the persistent rachilla; stamens 3; style trifid; achene unequally trigonous, pale, shining, 0.7–1.0 mm long; $n = 48$.

Streambanks and other wet places in the valleys and lowlands, especially along the shores of major rivers; widespread in the U.S. and adj. Can., but seldom collected in our range.

5. **Cyperus aristatus** Rottb.

Cyperus aristatus Rottb. Descript. & Icon. Rariores 23. 1773. *Dichostylis aristata* Palla, Bot. Jahrb. Syst. 10: 296. 1888. (*Koenig,* India.)

C. inflexus Muhl. Descript. Gram. 16. 1817. *C. aristatus* f. *inflexus* C. B. Clarke, J. Linn. Soc., Bot. 21: 92. 1884. *Chlorocyperus inflexus* Palla, Allg. Bot. Z. Syst. 17: Beil 6. 1911. *Cyperus aristatus* var. *inflexus* Kükenthal in Engl. Pflanzenr. IV. 20 (Heft 101): 504. 1936. (Pa.)

Tufted annual; herbage sweet-scented; culms slender, 0.3–1.5 dm tall; leaves few, all borne near the base, mostly 0.5–2 (2.5) mm wide, up to as long as or a little longer than the culms; involucral bracts elongate, leafy, some or all of them much surpassing the inflorescence; spikelets borne in dense, capitate clusters, the terminal cluster sessile, the others (if present) on slender rays up to 3 cm long; spikelets 4–10 mm long, flattened; scales evidently (5) 7- to 9-nerved, 1.0–1.7 mm long exclusive of the recurved, slender awn-tip, this mostly 0.3–1 mm long; rachilla wingless, persistent, the scales falling from it at maturity; stamen solitary; style trifid; achene trigonous, 0.6–1.0 mm long.

Wet places in the valleys and lowlands; more or less cosmopolitan, except at high latitudes.

The American plants have sometimes been distinguished as a separate species, variety, or form, under the epithet *inflexus,* but they probably could not be sorted out if the labels were covered.

CYPERUS
KOBRESIA

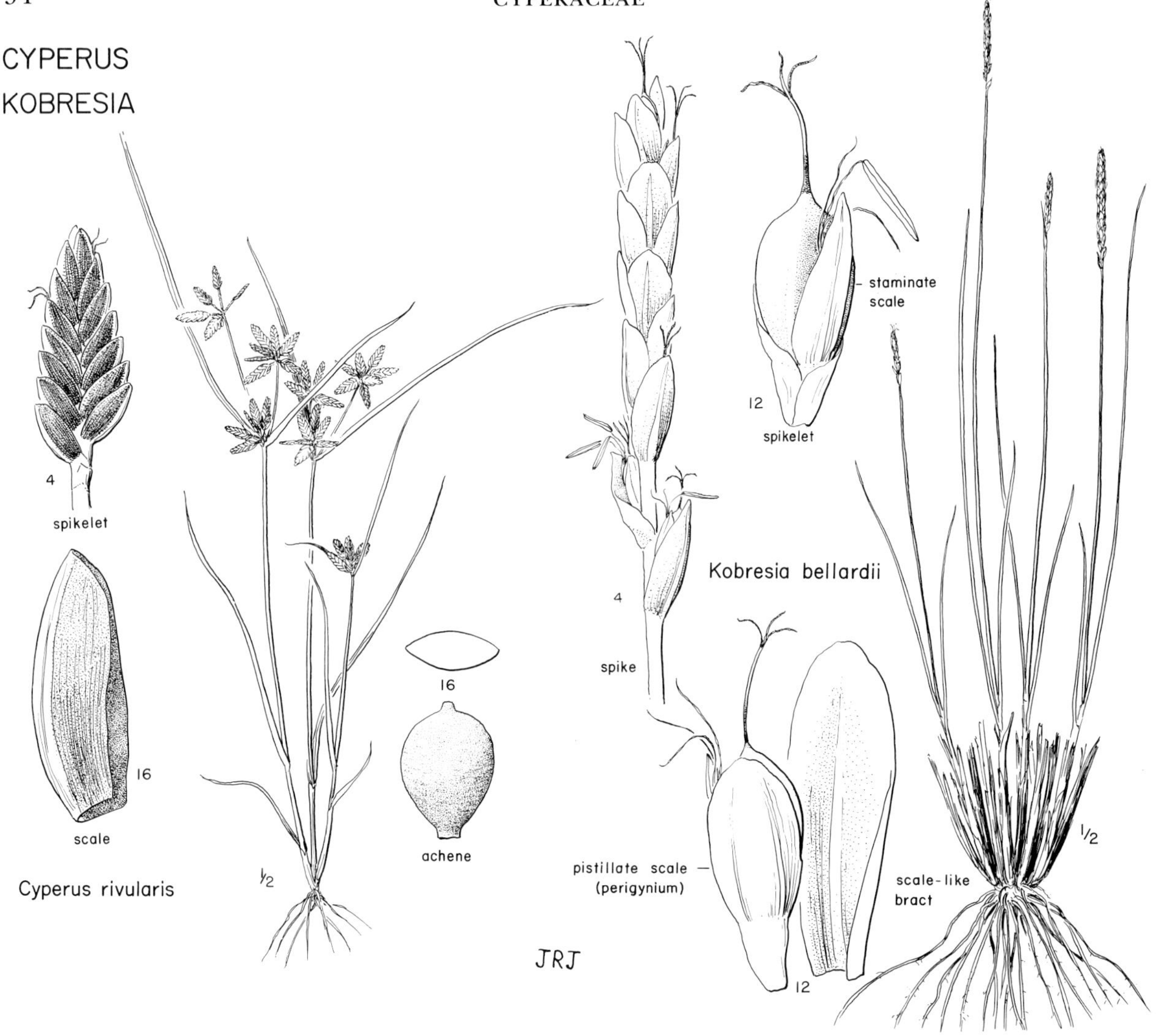

6. Cyperus acuminatus Torr. & Hook.

Cyperus acuminatus Torr. & Hook. Ann. Lyceum Nat. Hist. New York 3: 435. 1836. (*Drummond*, St. Louis, Mo.)

Tufted annual; culms slender, 0.5–4 dm tall, sometimes slightly thickened at the base; leaves few, all borne near the base, slender and more or less elongate, 1–2 (3) mm wide; involucral bracts unequal, elongate, some or all of them much surpassing the inflorescence; spikelets borne in very dense, globose clusters, the rachis of each cluster very short and concealed by the spikelets; spikelets 3–7 mm long, strongly flattened; scales 1.5–2 (2.5) mm long, very strongly 3-nerved, otherwise subhyaline and cellular-reticulate, tending to be shortly recurved-acuminate at the tip, closely set, the tip of each scale surpassing the one next below it by 0.4–0.7 mm; rachilla wingless, persistent, the scales falling from it at maturity; stamen solitary; style trifid; achene trigonous, 0.6–0.9 mm long.

Streambanks and other wet, low places in the valleys and lowlands, tolerant of alkali; N.D. to Ga., w. to the Pacific; seldom collected in our range.

7. Cyperus rivularis Kunth

Cyperus rivularis Kunth, Enum. Pl. 2: 6. 1837. *Chlorocyperus rivularis* Rikli, Jahrb. Wiss. Bot. 27: 564. 1895. *Pycreus rivularis* Palla, Allg. Bot. Z. Syst. 13: 48. 1907. *Cyperus niger* var. *rivularis* V. Grant, Aliso 4: 87. 1958. (Ga.)

Tufted annual; culms slender, 0.5–2 (3) dm tall; leaves few, all borne near the base, slender and more or less elongate, 0.5–2 mm wide; involucral bracts elongate, unequal, at least one of them much surpassing the inflorescence; spikelets in 1–several capitate clusters, the primary cluster sessile, the others (if present) on slender rays up to 3 (7) cm long; spikelets 3–15 mm long; scales mostly 2–2.5 mm long, blunt, with prominent, pale midrib, otherwise usually more or less strongly anthocyanic, deciduous at maturity from the persistent, wingless rachilla; stamens 2, rarely 3; style deeply bifid; achenes lenticular, 1.0–1.3 mm long, olivaceous to blackish, becoming minutely roughened or cross-ridged at maturity.

Streambanks and other wet, low places in the valleys and lowlands, tolerant of alkali; widespread in the U.S., and s. Can. (more commonly eastward), s. to S. Amer.; rarely collected in our range.

Plants from the U.S. and Can., as here described, are var. *rivularis*. Southward this gives way to another var. with mostly 3 stamens and broader achenes.

7. KOBRESIA Willd.

Densely cespitose, grass-like perennial herbs with narrow leaves, closed sheaths, and obtusely triangular, solid stems, much like *Carex,* but the inflorescence less reduced, its basic structure more evident; flowers unisexual, without perianth, solitary in the axils of small, scarious bracts called scales, arranged in small spikelets which are aggregated into a spike or spikes; each spikelet subtended by a scale on the axis of the spike (this scale not directly subtending a flower) and having a single proximal pistillate flower and one or more distal staminate flowers, or some of the spikelets unisexual, with only the single pistillate flower (and often a prolonged rachilla) or one or more staminate flowers; scale subtending the pistillate flower borne with its back to the axis of the spike, loosely wrapped about the trigonous achene, resembling and homologous with the perigynium of *Carex,* but open, with unsealed margins; stamens 3; stigmas 3; $x = 26$.

About 30 species, mostly of mountainous parts of Asia, a few circumboreal, or boreal American. (Named for Josef Paul von Cobres, 1747–1823, of Augsburg, naturalist and patron of botany, a contemporary of Willdenow.)

1. Kobresia bellardii (All.) Degland

Carex bellardi All. Fl. Pedemontana 2: 264. 1785. *Kobresia bellardi* Degland in Loisel. Fl. Gallica 626. 1807. *Scirpus bellardi* Wahlenb. Fl. Lappon. 15. 1812. *Elyna bellardi* Hartm. Kongl. Vetensk. Acad. Handl. 1814: 107. 1814. (*Bellardi,* near L'Auteret, in n. Italy.)

Carex myosuroides Vill. Hist. Pl. Dauphiné 2: 194. 1787. *Carex hermaphroditica* J. F. Gmelin, Syst. Nat. 2: 139. 1791. *K. myosuroides* Fiori & Paoli, Fl. Analitica Ital. 1: 125. 1896. (se. France.)

K. scirpina Willd. Sp. Pl. 4: 205. 1805. nom. illeg. *Elyna scirpina* Pax in Engl. & Prantl, Natürl. Pflanzenfam. 2(2): 122. 1887. *Carex scirpina* Missb. & Krause, Sturms Fl. Deutschl. 2: 63. 1900. (Mts. of c. Europe.)

Very densely tufted; old sheaths persistent and conspicuous, but generally lacking blades; culms (0.5) 1–3 (5) dm tall, longer or shorter than the leaves, these very slender and wiry, flat to channeled or inrolled, mostly 0.3–0.6 mm wide, sometimes scabrous along the margins, otherwise glabrous; spike solitary, terminal, 1–3 cm long, slender, seldom more than about 2 mm thick; spikelets 8–20, generally each (except the usually staminate terminal one) androgynous, with one staminate flower above the single pistillate flower; $n = 26$.

Open, dry or moderately moist places at high alt. in the mts., often above timberline; circumboreal, extending s. to Newfl., Que., and irregularly to Colo., Utah (Uinta Mts.), Oregon (Wallowa Mts.) and Calif. (s. Sierra Nevada). June–Aug.

Kobresia simpliciuscula (Wahlenb.) Mackenzie, another circumboreal species, has been collected near Driggs, Idaho, and may eventually prove to occur in our range. It is sharply distinguished from *K. bellardii* by having several spikes; furthermore, it occurs in wetter habitats, often at lower elevations, it averages somewhat more robust, and it usually has some of the dried leaves from the previous year persistent at flowering time; in typical plants many of the spikelets are uniflorous and pistillate.

8. CAREX L. Sedge

unisexual flowers

Grass-like perennial herbs with 3-ranked leaves, closed sheaths, and triangular or terete, solid stems; plants monoecious or seldom dioecious; flowers unisexual, without perianth, borne in spikes, each flower solitary in the axil of a usually small and scarious bract called a scale; spikes 1–many, each generally subtended by a large and leafy to much reduced and inconspicuous bract which may or may not have a sheathing base, or this bract sometimes wholly wanting, the individual spikes sessile or pedunculate, racemosely arranged in a loose to very compact terminal inflorescence which is rarely again branched, sometimes some of them well removed from the others and axillary to leaves near the base of the stem; spikes unisexual or bisexual, when bisexual the staminate flowers terminal (spikes androgynous) or basal (spikes gynaecandrous) or seldom more or less intermingled with the pistillate ones; stamens 3, seldom only 2; pistillate flowers individually enclosed by a sac-like scale called a perigynium, from the mouth of which the style or stigmas protrude, as well as being subtended by an open scale (the pistillate scale); stigmas 2 or 3, the achene accordingly lenticular or trigonous.

More than 1000 species, occurring in all parts of the world, most abundantly in moist regions of the N. Temp. Zone and in the arctic; tropical species are mostly montane. In 1935 Mackenzie recognized 533 species in North America alone. (The classical Latin name for the genus.)

The genus is usually divided into 4 well marked subgenera (3 of which occur in our range) and numerous sections which might perhaps better be treated as series. The key here presented is intended solely for the identification of our species, but, as an aid to understanding the organization of the genus, the subgeneric and sectional names are inserted in parentheses at appropriate places in the key. It has been thought useful to present a few species at more than one place in the key, and in one instance the subsidiary appearance is under a subgenus to which the species does not belong.

A few species of the subgenus Primocarex have a definite rachilla alongside the achene within the perigynium, showing that each pistillate flower represents a branch of an inflorescence that has been reduced to a single flower. The perigynium is a highly modified bract on the adaxial side of this short, uniflorous branch. This bract wraps around the pistillate flower, and its margins have become connate, so that the flower is enclosed in a sac with a minute apical opening. In two of the four subgenera (Primocarex and Vignea) the line of fusion of the bract margins can be seen as a suture or imperfection (the dorsal suture) toward the tip of the abaxial side of the perigynium. The side of the perigynium next to the pistillate scale (the abaxial side) is called the dorsal side; the side next to the axis of the spike (the adaxial side) is called the ventral side.

The lowermost leaves are frequently very much reduced, with a short, nongreen blade that is often firm, pointed, and more or less sheathing, the true closed sheath then often being short or absent. When the lowermost leaves are of this type, with the foliage leaves being borne farther up the stem, the plant is said to be aphyllopodic; when the lowest leaves are normally developed foliage leaves, the plant is said to be phyllopodic. The distinction is not absolutely sharp; many species are slightly aphyllopodic, with one or two reduced leaves at the bottom but with foliage leaves still arising from very near the base of the culm.

Many of the characteristically tristigmatic species of subgenus Carex occasionally have a few distigmatic flowers intermingled with the tristigmatic ones. Such specimens should be keyed as being tristigmatic. Outside of the subgenus Primocarex, and with the notable exception of *C. saxatilis,* our distigmatic species of the genus are more uniform in this regard and only rarely produce tristigmatic flowers.

The frequent variation in reported chromosome numbers of individual species of *Carex* doubtless reflects, at least in part, natural variability associated with the characteristic diffuse or polycentric centromere of the Cyperaceae. Misidentification of the material may also contribute to the reported diversity.

All students of *Carex* admit that the species must often be recognized by small technical differences, especially in the details of the structure of the perigynium. In some groups, notably the section Ovales, precise measurements of the mature perigynium and achene may be necessary for accurate identification. In some sections the species, though narrowly limited, are sharply defined. In others, again notably the Ovales, the boundaries are apt to be vague and there appears to be considerable intergradation between taxa which it seems necessary to hold at the specific level.

References:
Hermann, F. J. Manual of the Carices of the Rocky Mountains and Colorado Basin. U.S.D.A. Agric. Handb. No. 374. 1970.
Kükenthal, G. *Carex,* in Engl. Pflanzenr. IV. 20 (Heft 38): 67–821. 1909. A much more conservative treatment than that of Mackenzie.
Mackenzie, K. K. *Carex,* in N. Amer. Fl. 18: 9–478. 1931–35.
Murray, D. F. Taxonomy of *Carex* sect. Atratae (Cyperaceae) in the southern Rocky Mountains. Brittonia 21: 55–76. 1969.
———. *Carex podocarpa* and its allies in North America. Canad. J. Bot. 48: 313–324. 1970.

The treatment here presented approaches that of Mackenzie in some groups, and that of Kükenthal in others.

Key to the Species-Groups

1 Spike solitary, terminating the culm; perigynium often containing a vestigial rachilla as well as the achene (subg. Primocarex) .... Group I, p. 96
1 Spikes 2–many (in *C. backii,* of Group IV, one spike terminal and one, or more, basal or nearly so); perigynium never containing a rachilla.
  2 Stigmas 2 or 3 and the achene accordingly lenticular or trigonous; if stigmas 2, then the spikes elongate and cylindric (the larger ones seldom under 1.5 cm long), or at least some of them evidently pedunculate, or both; perigynium without a dorsal suture (subg. Carex).
    3 Style continuous with the achene and of the same bony texture, not withering, often becoming flexuous or contorted; stigmas 3 and achene trigonous, except in *C. saxatilis;* leaves often septate-nodulose .... Group II, p. 97
    3 Style deciduous; other characters various.
      4 Stigmas 3 and the achene trigonous. (A few lenticular, distigmatic achenes seldom intermingled with the others.)
        5 Perigynia more or less pubescent. (A few specimens of *C. scirpoidea,* a more or less dioecious species otherwise belonging to Group I, would be sought here.) .... Group III, p. 98
        5 Perigynia glabrous .... Group IV, p. 98
      4 Stigmas 2 and achenes lenticular .... Group V, p. 100
  2 Stigmas 2 and achenes lenticular; spikes sessile and relatively short (seldom over 1.5 cm long), not elongate and cylindric; perigynium usually with a more or less evident dorsal suture, at least distally (subg. Vignea).
    6 Spikes androgynous, or the plant more or less dioecious and most or all of the spikes unisexual .... Group VI, p. 100
    6 Spikes gynaecandrous, or some of the lateral ones wholly pistillate.
      7 Perigynia planoconvex, often with raised margins, but not thin-edged .... Group VII, p. 102
      7 Perigynia planoconvex or flattened, evidently thin-edged or wing-margined (sect. Ovales) .... Group VIII, p. 103

Key to the Species of Group I

1 Achene relatively very large, mostly 4–6 mm long; perigynia few, 5–8 mm long, beakless or nearly so (Firmiculmes).
  2 Plants distinctly rhizomatous, although the stems may be more or less clustered; lowest pistillate scale less than 1.5 (2) cm long; perigynium with 2 strong marginal nerves, otherwise nerveless or nearly so; widespread, but apparently absent from Nev. .... 1. *C. geyeri*
  2 Plants tufted, not at all rhizomatous; lowest pistillate scale elongate, (1.5) 2–15 cm long; perigynium commonly with some fine (or obscure) facial nerves in addition to the 2 strong marginal ones; s. Oregon to Calif., and in Elko Co., Nev. .... 2. *C. multicaulis*
1 Achene smaller, not over 3.5 mm long; perigynia various, often distinctly beaked.
  3 Rachilla well developed, at least half as long as the achene; leaves slender and wiry, not more than 1.5 mm wide.
    4 Plants more or less densely tufted, without rhizomes (rhizome present but short in *C. capitata,* a species with 2 stigmas).

5 Perigynium planoconvex, with sharply angled thin margins (especially distally), not wholly filled by the achene, obscurely serrulate on the margins distally, or wholly glabrous; scales not much if at all wider than the perigynia, not tending to wrap around a large part of the spike; stigmas 2 or 3; arctic and alpine species.

6 Plants aphyllopodic; perigynia relatively broad, mostly broadly ovate and more or less rounded toward the base; plants tufted but not very densely so, the stems crowded on very short rhizomes; stigmas 2 (Capitatae) .... 3. *C. capitata*

6 Plants phyllopodic, the old sheaths persistent and very conspicuous; perigynia narrower, elliptic or lanceolate to oblong-ovate, tapering to the base; plants very densely tufted, without rhizomes; stigmas 2 or 3 (Nardinae) .... 4. *C. nardina*

5 Perigynium more or less turgid, not sharp-margined, wholly filled by the mature achene, glabrous or short-hairy; scales evidently broader than the perigynia, tending to wrap around a large part of the slender spike; stigmas 3 (Filifoliae).

7 Perigynia glabrous or with only a very few inconspicuous short hairs near the beak; alpine species, from sw. Mont. to Colo., w. to Utah and Nev. .... 5. *C. elynoides*

7 Perigynia finely short-hairy, at least above the middle.

8 Perigynium with a definite short beak mostly 0.2–0.5 mm long; pistillate scales mostly about equaling the mature perigynia; dryland species, seldom at high elev. in the mts.; widespread, but not extending to Calif. .... 6. *C. filifolia*

8 Perigynium essentially beakless; pistillate scales mostly distinctly shorter than the mature perigynia; montane species of Calif., extending into adj. Oregon and Nev. .... 7. *C. exserta*

4 Plants not densely tufted, the stems arising singly or few together from long, creeping rhizomes; stigmas 3.

9 Perigynia thin-walled, not at all coriaceous, mostly rather numerous, (5) 10–40 or more; alpine species (Inflatae).

10 Perigynium relatively large, (4) 4.5–7.5 mm long, much larger than the achene; spike stout, commonly 6–12 mm wide and 1.2–2.5 times as long as wide .... 8. *C. breweri*

10 Perigynium smaller, 2.5–3.5 (4) mm long, somewhat (but not very conspicuously) larger than the achene; spike more slender, commonly 4–6 mm wide and 2.5–3 times as long as wide .... 9. *C. subnigricans*

9 Perigynia thick-walled and coriaceous, few, mostly 1–6; dryland species (Obtusatae) .... 10. *C. obtusata*

3 Rachilla obsolete; leaves various.

11 Pistillate scales deciduous as the perigynia approach maturity, the perigynia becoming spreading or reflexed (Callistachys).

12 Plants densely cespitose, without evident rhizomes; leaves slender, 0.5–1.5 mm wide .... 11. *C. pyrenaica*

12 Plants evidently rhizomatous, only loosely or not at all cespitose; leaves wider, mostly (1) 1.5–3 mm wide .... 12. *C. nigricans*

11 Pistillate scales persistent; perigynia widely spreading, or more appressed.

13 Perigynia rounded or emarginate at the tip, not at all beaked (Polytrichoideae) .... 13. *C. leptalea*

13 Perigynia more or less distinctly (though often only shortly) beaked.

14 Stigmas 2; rhizomes very slender; perigynia soon becoming widely spreading (Dioicae) .... 14. *C. dioica*

14 Stigmas normally 3, seldom only 2; rhizomes stouter; perigynia more or less appressed or ascending.

15 Spikes nearly or quite unisexual; perigynia mostly rather numerous, at least more than 15.

16 Perigynia evidently pubescent; mostly at high alt. in the mts. (Scirpinae) .... 15. *C. scirpoidea*

16 Perigynia glabrous or nearly so; lowlands to moderate elev. in the mts. (Atratae, pro min. parte) Occasional specimens of .... 35. *C. parryana*

15 Spikes evidently androgynous; perigynia mostly 5–15, glabrous; high altitudes in the mts. (Rupestres). (*C. pyrenaica* might sometimes be sought here, but has more numerous perigynia.) .... 16. *C. rupestris*

Key to the Species of Group II

1 Sheaths distinctly short-hairy, at least on the ventral surface; style straight; perigynia glabrous or hairy (Paludosae).

2 Perigynia short-hairy, 5–8 mm long, the teeth (0.5) 1–1.5 mm long .... 17. *C. sheldonii*
2 Perigynia glabrous, 7–10 mm long, the teeth 1.5–2.5 (3) mm long .... 18. *C. atherodes*
1 Sheaths and perigynia glabrous; style generally becoming flexuous or contorted as the achene matures.
3 Stigmas usually 2 and achenes lenticular, seldom 3 and achenes trigonous; perigynia with 2 marginal nerves only, or with a few faint additional nerves on the dorsal surface, scarcely inflated, the achene filling the lower $^2/_3$ or $^3/_4$ of the body (Vesicariae, pars) .... 19. *C. saxatilis*
3 Stigmas 3 and achenes trigonous; perigynia evidently several- to many-nerved, scarcely to strongly inflated, the achene loose in the lower half of the body.
4 Pistillate spikes (at least the lower ones) more or less nodding on slender peduncles; pistillate scales with a short, largely scarious or hyaline body only 1–2 mm long and a firm awn 2–6 mm long; perigynia lanceolate or lance-ovate, not much inflated, strongly (12-) 15- to 20-nerved, 5–7 mm long (Pseudocypereae) .... 20. *C. hystricina*
4 Pistillate spikes erect or stiffly ascending, subsessile or (especially the lowest) slender-pedunculate but still more or less erect; pistillate scales and perigynia otherwise, the perigynia distinctly inflated (Vesicariae, pro max. parte).
5 Perigynia ascending, lanceolate to lance-ovate, tapering gradually to the often poorly defined beak; inflorescence loose as in *C. rostrata;* stems more or less clustered on rather short rhizomes .... 21. *C. vesicaria*
5 Perigynia more or less strongly spreading at maturity (or the lowest ones reflexed), the body ellipsoid or ovoid to subglobose, more or less abruptly contracted to the conspicuous beak.
6 Leafy bract subtending the lowest spike 1–2 times as long as the whole inflorescence, the inflorescence more or less elongate and the spikes remote; perigynia 4–7 mm long; plants sod-forming, the stems arising singly or few together from long rhizomes .... 22. *C. rostrata*
6 Leafy bract subtending the lowest spike 2–several times as long as the whole inflorescence, the spikes crowded, or only the lowest one remote; perigynia 7–10 mm long; plants densely tufted from a very short rhizome .... 23. *C. retrorsa*

Key to the Species of Group III

1 Achene filling the body of the perigynium; spikes all short, the pistillate ones up to about 1 cm and the staminate ones to about 1.5 cm long; some of the pistillate spikes basal or nearly so (Montanae).
2 Soft-leaved, lax plants, (0.5) 1–3 (4) dm tall, of the coniferous forest zones; bract subtending the lowest nonbasal pistillate spike surpassing the staminate spike; widespread cordilleran sp. .... 24. *C. rossii*
2 Firm-leaved, low plants of the pinyon-juniper and chaparral zones, not over 1 dm tall; bract subtending the lowest nonbasal pistillate spike seldom surpassing the staminate spike; c. Utah and sw. Colo. to Ariz. and N.M. .... 25. *C. geophila*
1 Achene loose in the perigynium, the upper part of the perigynium empty; spikes mostly longer, commonly 1–5 cm long, none of them near-basal.
3 Perigynium densely velvety or velvety-sericeous, thick-walled and more or less turgid, not strongly compressed; bract subtending the lowest pistillate spike with short (to 1 cm) or no sheath and with an elongate blade commonly more than 8 cm long and very often surpassing the inflorescence; rhizomatous, strongly aphyllopodic plants, not tufted (Hirtae) .... 34. *C. lanuginosa*
3 Perigynium only sparsely and inconspicuously hirtellous, except for the serrulate beak and distal margins, thin-walled, more or less strongly compressed, not turgid; bract subtending the lowest pistillate spike with a well developed (1–4.5 cm) sheath and a short (1–8 cm) blade which is much surpassed by the inflorescence; plants tufted, not strongly aphyllopodic (Ferrugineae, pars) .... 28. *C. luzulina*

Key to the Species of Group IV

1 Lower pistillate scales much enlarged and leaflike, commonly 2–7 cm long; achene about 3 mm long, without a stylar apiculus (Phyllostachyae) .... 27. *C. backii*
1 Lower pistillate scales much smaller and not at all leaflike, seldom as much as 1 cm long; achene usually smaller and with a stylar apiculus.
2 Bracts subtending the spikes (or at least the bract subtending the lowest spike) with a well developed sheath, this usually at least 5 mm long.

3 Perigynia distinctly compressed, narrow, lanceolate or lance-ovate, usually ciliolate-serrulate on the margins distally (Ferrugineae).
4 Terminal spike staminate or androgynous; leaves 3–9 mm wide; montane but scarcely alpine . . . . 28. *C. luzulina*
4 Terminal spike gynaecandrous, or sometimes wholly pistillate; leaves 1–3 (4) mm wide; arctic-alpine . . . . 29. *C. misandra*
3 Perigynia scarcely or not at all compressed, variously shaped, not ciliolate-serrulate.
5 None of the bracts conspicuously surpassing the inflorescence; all of the pistillate spikes, or all but the uppermost one, subtended by conspicuously sheathing bracts; perigynia short-beaked or beakless.
6 Culms densely tufted; lateral spikes on lax, capillary, nodding or loosely spreading peduncles; perigynium essentially nerveless except for the 2 marginal nerves (Capillares) . . . . 30. *C. capillaris*
6 Culms arising singly or few together from creeping rhizomes; spikes all erect, short-pedunculate, or the lowest ones long-peduncled; perigynium more or less evidently 15- to 25-nerved (Granulares) . . . . 31. *C. crawei*
5 One or more of the bracts elongate and conspicuously surpassing the inflorescence; sheathing base present only on those bracts which subtend the 1 or 2 pistillate spikes (if any) that are remote from the terminal cluster of spikes; perigynia crowded, widely spreading or reflexed, prominently beaked (Extensae) . . . . 32. *C. oederi*
2 Bracts subtending the spikes all sheathless or nearly so.
7 Perigynium with a conspicuous beak 0.8–1.2 mm long.
8 Stems 1–4 dm tall, clustered, without rhizomes; leaves narrow, mostly 1–3 mm wide (Extensae) . . . . 32. *C. oederi*
8 Stems 5–10 dm tall, from creeping rhizomes; leaves very broad, mostly 8–20 mm wide (Anomalae) . . . . 33. *C. amplifolia*
7 Perigynium beakless, or with a short beak up to about 0.6 mm long.
9 Roots covered with a yellowish-brown felt; lateral spikes all nodding on slender peduncles; perigynia densely and conspicuously papillate; spp. of *Sphagnum* bogs (Limosae).
10 Plants strongly aphyllopodic; leaves tending to be canaliculate; pistillate spikes 1–2.5 cm long, sometimes with a few staminate flowers at the tip, never at the base . . . . 45. *C. limosa*
10 Plants phyllopodic and with the old leaves more or less persistent at the base; leaves flat; pistillate spikes 0.7–1.5 cm long, often with a few staminate flowers at the base, never at the tip . . . . 46. *C. paupercula*
9 Roots glabrous or inconspicuously hairy, not felty-tomentose; lateral spikes sessile or pedunculate, erect to nodding; perigynia in most species (with the notable exception of *C. buxbaumii*) not strongly papillate, although the cellular reticulum is sometimes readily visible; plants not of *Sphagnum* bogs (Atratae).
11 Terminal spike staminate or androgynous; bisexual spikes, if any, always androgynous.
12 Perigynia small, only 1.9–3.0 mm long; achene 1.4–1.8 mm long, only slightly smaller than the perigynial cavity; spikes all erect or closely ascending . . . . 35. *C. parryana*
12 Perigynia larger, mostly 3.0–5.5 mm long; achenes either at least 2.0 mm long, or notably smaller than the perigynial cavity, or both; lateral spikes erect or nodding.
13 Perigynium somewhat inflated, its marginal nerves not displaced . . . . 36. *C. raynoldsii*
13 Perigynium strongly flattened, though often distended around the achene, its "marginal" nerves displaced so as to be distinctly inframarginal on the abaxial side . . . . 37. *C. paysonis*
11 Terminal spike gynaecandrous, or in *C. parryana* often wholly pistillate or with more or less intermingled male and female flowers; bisexual spikes never androgynous.
14 Plants producing elongate rhizomes; stems of the season strongly aphyllopodic and arising singly or few together, not surrounded by the dried sheaths of previous years; perigynia densely and conspicuously papillate; boreal sp. of wet places . . . . 44. *C. buxbaumii*
14 Plants without long rhizomes; stems of the season clustered and surrounded by the dried sheaths of previous years; perigynia moderately to not at all papillate.

15 Perigynia small, mostly only 1.9–3.0 mm long; spikes more or less approximate (but not capitate-crowded), erect or closely ascending.
16 Spikes relatively short, the terminal one 6–14 mm long; pistillate scales very dark, blackish purple or brownish black; montane .... 38. *C. norvegica*
16 Spikes longer, the terminal one (10) 15–30 mm long; pistillate scales brown or stramineous, not at all blackish; valleys and foothills .... 35. *C. parryana*
15 Perigynia larger, mostly 3.0–5.5 mm long (or as little as 2.7 mm in spp. with capitate inflorescence or nodding spikes).
17 Spikes more or less closely aggregated into a dense head; plants strictly alpine and subalpine.
18 Spikes all sessile or very nearly so and very closely crowded, the lateral ones rather widely ascending or spreading; achenes tending to be stipitate; pistillate scales about equaling the perigynia, only very narrowly and obscurely hyaline-margined, or marginless; plants (1) 2–6 dm tall .... 39. *C. nova*
18 Spikes a little less densely crowded, more closely ascending, the lowest one often short-pedunculate; achenes sessile or nearly so; plants 1–3 dm tall.
19 Pistillate scales narrow and elongate, evidently surpassing the perigynia, but leaving the margins exposed, the scales not evidently hyaline-margined; Calif. and adj. Nev. .... 40. *C. helleri*
19 Pistillate scales shorter and broader, scarcely if at all exceeding the perigynia and not obviously narrower than the perigynia, usually conspicuously hyaline-margined; more eastern sp. .... 41. *C. albonigra*
17 Spikes more or less approximate, but not closely aggregated into a dense head; plants 1.5–8 dm tall, montane or submontane to alpine.
20 Spikes tending to be erect or stiffly ascending (or the whole inflorescence sometimes nodding), not individually nodding; achene relatively small, 1.2–1.8 mm long, reaching only to within 0.6–1.5 mm of the tip of the perigynial cavity; widespread in the mts. .... 42. *C. atrata*
20 Spikes (at least the lower) individually more or less nodding; achene larger, 1.8–2.4 mm long, reaching to within 0.2–0.7 mm of the tip of the perigynial cavity; southern sp., extending n. to c. Utah and c. Colo. .... 43. *C. bella*

Key to the Species of Group V

1 Bract subtending the lowest spike with a well developed sheath at the base; perigynia hardly at all compressed; plants small, not over 4 dm tall (Bicolores) .... 26. *C. aurea*
1 Bract subtending the lowest spike sheathless or nearly so; other characters various (Acutae).
2 Perigynia evidently nerved on both faces as well as on the margins; lowest bract usually equaling or surpassing the terminal spike, seldom somewhat shorter.
3 Plants more or less densely tufted, without long rhizomes; perigynium membranous, its beak entire; leaves relatively narrow, seldom over 4 mm wide .... 47. *C. lenticularis*
3 Plants scarcely tufted, vigorously rhizomatous; perigynium coriaceous, its beak tending to be bidentate; leaves broader, the larger ones mostly (3) 4–10 mm wide .... 48. *C. nebrascensis*
2 Perigynia nerveless or nearly so on both faces, or occasionally some of them with one or two irregular nerves.
4 Lowest bract short, up to about 7 cm long, surpassed by the terminal spike; plants relatively low, mostly 1–4 (6) dm tall .... 49. *C. scopulorum*
4 Lowest bract elongate, mostly 7–50 cm long, usually equaling or surpassing the terminal spike; plants taller, mostly 3–10 (15) dm tall .... 50. *C. aquatilis*

Key to the Species of Group VI

1 Plants rhizomatous, the culms arising singly or only loosely clustered.
2 Inflorescence interrupted-linear, the individual spikes mostly well separated from each other,

very small and few-flowered, about 5 mm long or less, with mostly only 1–3 perigynia and about as many staminate flowers; soft, slender plants of swamps and streambanks (Heleonastes, pro parte) .... 67. *C. disperma*

2 Inflorescence more or less headlike, though sometimes elongate, the individual spikes, or most of them, more or less closely aggregated, often more than 5 mm long and with more flowers than in *C. disperma;* various habitats, but seldom in swamps or along streambanks.

3 Spikes all very closely aggregated into a compact head, not individually distinguishable to the naked eye; dwarf, alpine and subalpine plants up to about 2.5 dm tall (Foetidae).

4 Perigynium relatively broad, mostly 1.9–2.6 mm wide and 1.5–2 times as long as wide, nearly or fully twice as wide as the achene, not stipitate at the base; leaves narrow, commonly equaling or surpassing the culms, seldom at all shorter; above timberline in the s. Rocky Mts. .... 51. *C. perglobosa*

4 Perigynium narrower, mostly 1.0–1.6 mm wide and (2) 2.5–4.5 times as long as wide, not more than half again as wide as the achene, more or less strongly stipitate at the base; leaves wider, mostly 2–4 mm wide, shorter than the culms, irregularly distributed at high elev., not necessarily above timberline, from Colo. to Calif. and Wash. ..... 52. *C. foetida*

3 Spikes not so closely aggregated, the lower ones, at least, readily distinguishable to the naked eye; mostly taller plants of more moderate elev.

5 Plants phyllopodic to slightly aphyllopodic, the leaves all clustered at or near the base; perigynia 1.7–4.3 mm long, essentially nerveless ventrally, often sharp-edged but not thin-margined (Divisae).

6 Perigynia very small, 1.7–2.4 (2.6) mm long, the beak 0.2–0.4 (0.6) mm long, the body very firm and thick-walled, especially proximally .... 53. *C. simulata*

6 Perigynia larger, 2.6–4.3 mm long, the beak 0.5–1.5 mm long, the perigynial wall sometimes fairly firm but not especially thick.

7 Spikes at least ordinarily all unisexual and the plants dioecious; spikes mostly 10–15 mm long, the head mostly 1.5–4.5 cm long; perigynia 3.5–4.6 mm long, with a prominent beak mostly 1.0–1.5 mm long .... 54. *C. douglasii*

7 Spikes (or most of them) ordinarily androgynous, rarely all unisexual and the plants dioecious, the spikes, heads, and perigynia then smaller than in *C. douglasii.*

8 Rhizome slender, brownish; plants slender, 0.5–2 dm tall, with very narrow leaves only 0.3–1.5 mm wide; perigynia 2.6–3.3 (3.5) mm long .... 55. *C. stenophylla*

8 Rhizome coarse, black or brownish black; plants (1) 3–7 dm tall, with leaves 1–3 mm wide; perigynia 2.9–3.8 mm long .... 56. *C. praegracilis*

5 Plants strongly aphyllopodic, the leaves not forming a basal cluster; perigynia 4.5–6.2 mm long, more or less evidently nerved ventrally, the body tending to be thin-margined distally (Arenariae) .... 57. *C. siccata*

1 Plants more or less densely tufted, with short or no creeping rhizomes.

9 Perigynia variously shaped (but not lance-triangular), generally broadest distinctly above the base, the body more or less abruptly tapering or contracted into the beak, which may also be of different texture from the body.

10 Spikes few, mostly 10 or less; sheaths not red-dotted (Bracteosae).

11 Body of the perigynium strongly planoconvex, broadly rounded on the convex side (the marginal nerves tending to appear ventromarginal), very abruptly contracted to the beak and wholly filled by the mature achene, not evidently serrulate-margined (except often on the beak) at 10X magnification; spikes loosely aggregated into an irregular, oblong-cylindric head .... 58. *C. vallicola*

11 Body of the perigynium moderately planoconvex (the marginal nerves only occasionally appearing ventromarginal), less abruptly contracted to the beak, the upper part not wholly filled by the achene, more or less evidently serrulate-margined above the middle.

12 Spikes loosely aggregated into an irregular or interrupted, oblong-cylindric to nearly linear head, at least the lower ones readily distinguishable to the naked eye; perigynia usually not so markedly bicolored as in *C. hoodii* .... 59. *C. occidentalis*

12 Spikes closely aggregated into a dense, oblong-cylindric to ovoid head, not readily distinguishable to the naked eye; perigynia tending to be coppery-brown except for the green or greenish margins .... 60. *C. hoodii*

10 Spikes more or less numerous (at least more than 10) in a usually compound inflorescence; sheaths somewhat red-dotted ventrally.

13 Midrib of the pistillate scale not excurrent, or excurrent only as a minute point not more than 0.5 mm long; sheaths not cross-rugose (Paniculatae).

14 Perigynium very firm and thick-walled, the dorsal side with a median thin (often depressed and pale) strip that extends all the way to the base; inflorescence short, mostly 2–3.5 (5) cm long; circumboreal sp., rare with us .... 61. *C. diandra*

14 Perigynium rather thin-walled, lacking a median thinner strip; inflorescence more or less elongate, mostly 3–10 cm long; southwestern species, just entering the s. part of our range in Nev. .... 62. *C. alma*

13 Midrib of the pistillate scale excurrent as a definite awn mostly 1–5 mm long; sheaths generally cross-rugose ventrally; low elev. (Multiflorae) .... 63. *C. vulpinoidea*

9 Perigynium broadest at or near the base, commonly lance-triangular, tapering very gradually and evenly into the often ill-defined beak, or only slightly more abruptly contracted distally (Vulpinae).

15 Perigynia relatively large, (3.6) 4.0–5.2 mm long; leaves broad, the larger ones mostly 5–10 mm wide; inflorescence relatively large, (2) 3–10 cm long; sheaths generally cross-corrugated .... 64. *C. stipata*

15 Perigynia smaller, mostly 2.9–3.8 mm long; leaves narrower, mostly 1–5 mm wide; inflorescence smaller, 0.8–2.5 (3) cm long; sheaths cross-corrugated or not.

16 Leaves clustered near the base; ventral surface of the sheath only seldom cross-corrugated, usually truncate or concave at the mouth and not cartilaginous-thickened, often hidden in the leaf-tuft .... 65. *C. jonesii*

16 Leaves not closely clustered; ventral surface of at least some of the sheaths readily visible, usually cross-corrugated, tending to be convex or prolonged (and often cartilaginous) at the mouth .... 66. *C. neurophora*

Key to the Species of Group VII

1 Perigynia ascending or ascending-spreading at maturity. (*C. leporinella,* of the section Ovales, might sometimes be sought here.)

2 Perigynium 1.7–3.4 mm long, short-beaked or nearly beakless, the beak often less than 0.5 mm long, less often up to 1.0 mm, only inconspicuously or not at all serrulate-margined; leaves mostly 1–2.5 mm wide (to 3.5 mm in *C. canescens*) (Heleonastes, pro max. parte).

3 Spikes 2–4, all approximate; pistillate scales fully equaling the body of the perigynium, though sometimes minutely surpassed by the beak; perigynia mostly 2.4–3.4 mm long, with a prominent dorsal suture 0.7–1.5 mm; alpine plants, only 1–3 dm tall, rare in our range .... 68. *C. bipartita*

3 Spikes 4–8 (10), the lower ones (or all of them) often more or less remote; pistillate scales tending to be shorter than the body of the perigynium; perigynia mostly 1.7–2.5 mm long.

4 Spikes relatively small and few-flowered, with only 5–10 (15) perigynia; wall of the distal part of the perigynium-body very thin and easily ruptured; ventral surface of the perigynium nerveless or obscurely nerved; dorsal suture of the perigynium prominent, extending the length of the beak-apiculation and encroaching onto the distal part of the body .... 69. *C. brunnescens*

4 Spikes larger and more numerously flowered, with mostly (10) 15–30 perigynia; wall of the distal part of the perigynium not especially thin and fragile; ventral surface of the perigynium evidently nerved.

5 Spikes 4–6, brownish in aspect, all closely aggregated; dorsal suture usually extending the length of the beak-apiculation, but scarcely encroaching onto the body of the perigynium; plants 1–3 dm tall, occurring at high elev. in the mts. ..... 70. *C. praeceptorum*

5 Spikes 4–8, silvery green to pale grayish or stramineous in aspect, the lower ones, at least, usually somewhat remote, so that the inflorescence is interrupted and not headlike; dorsal suture of the perigynium short and poorly developed, commonly restricted to the distal part of the beak-apiculation; plants of various alt., often over 3 dm tall .... 71. *C. canescens*

2 Perigynium (3.1) 3.5–4.8 mm long, relatively long beaked, the beak mostly (0.8) 1.1–1.9 mm long and serrulate-margined; leaves mostly 2–5 mm wide (Deweyanae) .... 72. *C. deweyana*

1 Perigynia widely spreading at maturity, or the lower ones reflexed.

6 Perigynium more or less evidently serrulate on the beak (and often also on the distal part of the body) at 10X magnification (Stellulatae).

7 Beak of the perigynium short and stout, one fourth to one half as long as the body, up to about 1 mm long, only very shortly and inconspicuously bidentate; perigynia mostly 2.2–3.2 mm long .... 73. *C. interior*

7 Beak of the perigynium longer and relatively more slender, one half as long to almost as long as the body, mostly 1–2 mm long; perigynia mostly 2.8–3.5 (4.0) mm long ... 74. *C. muricata*

6 Perigynium strictly entire-margined, not serrulate on either the body or the beak, 2.5–3.2 mm long (Ovales, pro min. parte) .... 75. *C. illota*

Key to the Species of Group VIII

1 Perigynia very small, only 2.5–3.2 mm long.
  2 Margins of the perigynium sharp but neither winged nor serrulate; perigynia widely spreading (unique among our spp. of the section in these regards) .... 75. *C. illota*
  2 Margins of the perigynium more or less winged or serrulate, at least toward the distal part of the body; perigynia more nearly appressed, at least not widely spreading.
    3 Inflorescence small and compact, mostly 7–15 mm long, the spikes scarcely distinguishable to the naked eye; nw. montane sp., in our range known only from Elko Co., Nev. .... 77. *C. limnophila*
    3 Inflorescence larger and looser, mostly 15–30 mm long, the spikes readily distinguishable to the naked eye; w. and s. sp., in our range known only from s. Utah .... 78. *C. subfusca*
1 Perigynia larger, 3.2–7.9 mm long.
  4 Perigynium evidently slender-beaked, the beak rather narrowly margined and serrulate in the proximal portion only, the distal 0.5 mm or more generally subterete, marginless, and entire or nearly so; inflorescence pale to often dark or bicolored.
    5 Lowest bract of the inflorescence elongate, (1) 2–8 (10) cm long, equaling or usually much surpassing the inflorescence .... 76. *C. athrostachya*
    5 Bracts all short and inconspicuous, or the lowest one seldom setaceous-prolonged and up to about as long as the inflorescence.
      6 Pistillate scales distinctly shorter and narrower than the perigynia, largely exposing at least the distal margins as well as the beak of the perigynium.
        7 Perigynium more or less distinctly plano-convex, the perigynial cavity nearly filled by the plump achene; w. and s. spp. (occasional specimens of the inland sp. *C. petasata* might be sought here, but have mostly larger perigynia and less crowded inflorescence than these spp.).
          8 Perigynia 3.2–5.1 mm long, few- to several-nerved (nerves fewer than 10) dorsally, the ventral side similarly nerved or fewer-nerved or nerveless.
            9 Perigynia up to 3.6 mm long, usually evidently nerved ventrally as well as dorsally; perigynia and scales commonly paler than in the next sp.; occasional specimens of .... 78. *C. subfusca*
            9 Perigynia mostly 3.6–5.1 mm long, the ventral side usually nerveless or obscurely few-nerved .... 79. *C. pachystachya*
          8 Perigynia mostly 4.5–6.2 mm long, obscurely to conspicuously multi-nerved (nerves 10 or more) dorsally and usually also ventrally, or the ventral nerves sometimes fewer or obsolete .... 88. *C. multicostata*
        7 Perigynium more or less strongly flattened except where distended by the relatively small achene; inland spp.
          10 Perigynia middle-sized, mostly 3.2–5.0 mm long and 2–3 (4) times as long as wide, generally paler than the pistillate scales, and the spike thus conspicuously bicolored; plants mostly (1.5) 2–6 (8) dm tall, occurring from the foothills to near timberline in the mts. .... 80. *C. microptera*
          10 Perigynia larger, mostly (4.5) 5.0–7.2 mm long, often nearly or quite as dark as the scales; plants mostly 1–3 (4) dm tall, occurring near or above timberline in the mts.
            11 Perigynium mostly 1.8–3.3 times as long as wide, tending to have a pair of prominent inframarginal nerves delimiting the wing-margins; stipelike base of the achene (below the joint) inconspicuous, seldom over 0.5 mm long, or obsolete; widespread, but more common northward .... 81. *C. haydeniana*
            11 Perigynium mostly 3.4–5.0 times as long as wide, without prominent inframarginal nerves delimiting a wing-margin; achene jointed to a prominent, stipelike base commonly about (0.5) 1 mm long; s. Rocky Mts., to Utah and Ariz. .... 82. *C. ebenea*
      6 Pistillate scales nearly or quite as long and wide as the perigynia and more or less completely concealing them in dorsal view, the tip of the beak sometimes exposed.
        12 Perigynia large, mostly 5.8–7.9 mm long, planoconvex, nerved on both sides, 2.6–4.0 times as long as wide; moderate and lower elev., chiefly in drylands .... 83. *C. petasata*
        12 Perigynia smaller, 3.3–5.5 (6.0) mm long, variously shaped and nerved.

13 Perigynium small, only 3.2–4.0 mm long, 0.8–1.2 mm wide, and 3.3–4.0 times as long as wide, sharp-edged but scarcely wing-margined, the beak rather ill-defined; inflorescence compact, the first internode mostly 2–6 (9) mm long, the second one 1–5 (6) mm long .... 93. *C. leporinella*

13 Perigynium larger, mostly (3.6) 4.0–5.5 (6.0) mm long, 1.3–2.0 mm wide, and 2.2–4.0 times as long as wide, more or less distinctly wing-margined at least toward the distal part of the body, the beak often but not always well defined.

14 Plants of lowlands to moderate elev. in the mts., often more than 4 dm tall; inflorescence relatively loose and flexuous, the two lowest internodes each 5–10 (15) mm long; leaves (1) 2–4 mm wide .... 84. *C. praticola*

14 Plants of high elev. in the mts., near or above timberline, only 1–3 (4) dm tall; inflorescence more compact, the second internode mostly 2–5 (6) mm long, the first of similar length or sometimes longer and up to 10 mm; leaves 0.5–2 (2.5) mm wide .... 85. *C. phaeocephala*

4 Perigynium with a flattened, broadly margined and serrulate, often ill-defined beak, this sometimes margined and serrulate throughout, or sometimes with a minute, subterete, marginless and entire tip less than 0.5 mm long; inflorescence (except in *C. arapahoensis*) relatively pale, greenish or stramineous to light brown in aspect.

15 Pistillate scales distinctly shorter and narrower than the perigynia, largely exposing at least the distal margins as well as the beak of the perigynium.

16 Dorsal surface of the perigynium evidently few-nerved, the nerves fewer than 10.

17 Perigynium 1–1.5 mm wide, 2.0–2.7 (3.0) times as long as wide, planoconvex, tapering to a rather ill-defined beak; achenes 0.6–0.8 mm wide .... 86. *C. bebbii*

17 Perigynium 2.2–3.4 mm wide, (1.2) 1.3–1.7 (1.8) times as long as wide, strongly flattened, the body suborbicular and more or less strongly contracted to the well defined beak; achenes 1.3–1.9 mm wide .... 87. *C. brevior*

16 Dorsal surface of the perigynium obscurely to conspicuously multinerved, the nerves 10 or more.

18 Perigynia more or less strongly planoconvex; chiefly Cascade-Sierran sp., occurring also in the mts. of e. Oregon and c. Idaho .... 88. *C. multicostata*

18 Perigynia more or less strongly flattened.

19 Spikes borne in a compact head, the first two internodes collectively only 4–7 (9) mm long; high elev. in the mts.

20 Perigynia middle-sized, mostly 4.1–5.0 mm long and 1.5–2.3 mm wide; Cascade-Sierran sp., occurring also in the Wasatch Mts. of Utah .... 89. *C. straminiformis*

20 Perigynia large, mostly 5.5–7.4 mm long and 2.1–3.5 mm wide; sp. of the s. Rocky Mts. and Colo. Plateau region .... 90. *C. egglestonii*

19 Spikes borne in a more elongate, spiciform inflorescence, the first two internodes collectively (8) 10–18 mm long; mostly at middle elev. in the mts., seldom above timberline .... 91. *C. xerantica*

15 Pistillate scales nearly or quite as long and wide as the perigynia and more or less completely concealing them in dorsal view, the tip of the beak sometimes exposed.

21 Perigynia larger, (4.2) 4.5–7.0 mm long and 1.9–2.8 mm wide, evidently wing-margined; spp. in our range confined to the Colo. Plateau region.

22 Spikes borne in a more or less elongate, spiciform inflorescence, the first two internodes collectively (8) 10–18 mm long; plants mostly 3–7 dm tall, commonly occurring at middle elev. in the mts., seldom above timberline .... 91. *C. xerantica*

22 Spikes aggregated into a compact head, the first two internodes collectively only 1.5–6 mm long; plants mostly 1.5–3 (4) dm tall, generally growing above timberline .... 92. *C. arapahoensis*

21 Perigynia smaller, 3.2–4.0 mm long and 0.8–1.2 mm wide, sharp-edged but scarcely wing-margined; chiefly Cascade-Sierran sp., irregularly e. to the Wasatch Mts. of Utah .... 93. *C. leporinella*

1. Carex geyeri F. Boott

Carex geyeri F. Boott, Trans. Linn. Soc. London 20: 118. 1846. (*Geyer 332*, Rocky Mts., in n. Idaho, in 1844.)

Elk sedge.

Culms loosely to rather densely clustered on a system of branching rhizomes, mostly 1.5–5 dm tall, aphyllopodic; leaves elongate, not much if at all shorter than the culms, flat, mostly 1.5–3 mm wide; spike solitary, bractless, androgynous, with 1–3 perigynia at the base, the terminal staminate part compact, slender, mostly 1–2.5 cm long, light brown; pistillate scales brownish, often with paler midrib, hyaline-margined, the lower short-awned and sur-

CAREX

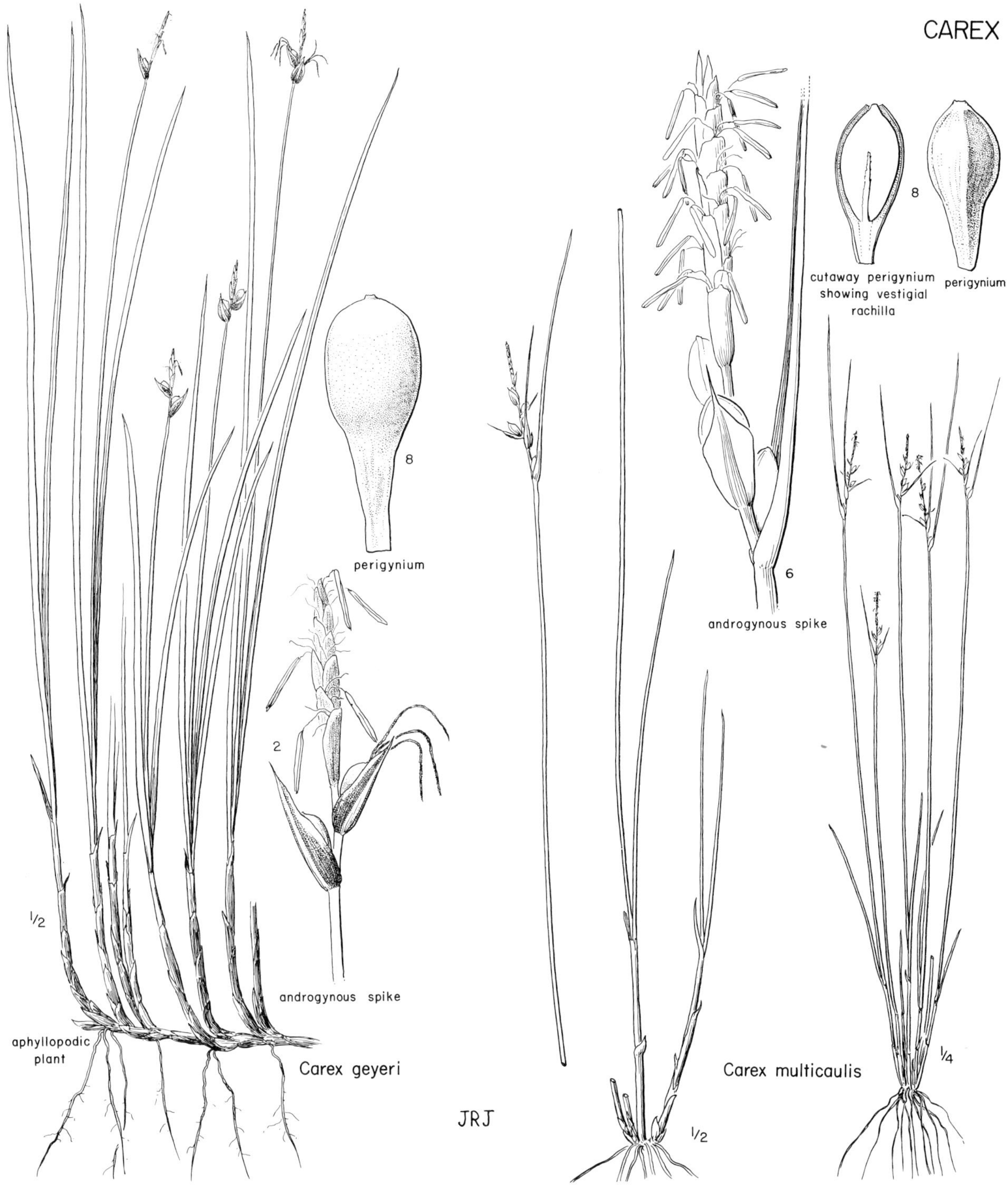

passing the perigynium, the upper shorter and blunter; perigynia tending to be somewhat remote, and often separated from the staminate flowers by a segment of bare rachis, greenish or somewhat brownish, ellipsoid or obovoid, 2-ribbed, otherwise nerveless, 5–6 mm long, tapering to a somewhat spongy base, rather abruptly contracted above to the scarcely beaked tip; stigmas 3; achene trigonous, obpyramidal, filling the perigynium, 4–5 mm long; rachilla up to half as long as the achene, or obsolete.

Woodlands, open slopes, and dry meadows, in the foothills and at moderate or sometimes high elev. in the mts.; Alta. and B.C. to Wash., Oregon, n. Calif., Utah, and Colo., and disjunct (presumably introduced) in Pa.

A common, readily recognized species, and an important winter food for elk.

2. Carex multicaulis L. H. Bailey

Carex multicaulis L. H. Bailey, Bot. Gaz. 9: 118. 1884. (*Torrey 544,* Yosemite Valley, Calif., is the first collection cited and was designated by Mackenzie as the type.)

Plants in dense but rather small tufts, without creeping rhizomes, the numerous stems mostly 1.5–6 (10) dm tall, aphyllopodic; leaves narrow, mostly 1–2 mm wide, sometimes all relatively short and not over 10 or 15 cm long, sometimes some of them elongate and surpassing the stems; spike solitary, androgynous, bractless, but with elongate, bract-like lower pistillate scales, the lowest of these 1.5–15 cm long; terminal, staminate part of the spike compact, slender, 1–2.5 cm long, light brown, the pistillate part loose and often more or less interrupted, with 1–6 rather remote perigynia; pistillate scales progressively reduced upwards, the uppermost barely equaling or a little shorter than the perigynia, with green center and broad hyaline margins; perigynia pale greenish or somewhat brownish, ellipsoid or obovoid, 2-ribbed and with rather numerous fine and often obscure nerves on each face (or these partly obsolete), 5–8 mm long, tapering to a stipe-like spongy base, rather abruptly contracted above to the scarcely beaked tip; stigmas 3; achene trigonous, obpyramidal, 4–6 mm long, filling the perigynium and commonly slightly exserted at the tip; rachilla well developed, up to about half as long as the achene.

Dry, open woods; Cascade Mts. and Klamath regions of s. Oregon, s. through the Sierra Nevada and into the mts. of s. Calif.; an isolated station in the mts. of Elko Co., Nev.

3. Carex capitata L.

Carex capitata L. Syst. Nat. ed. 10. 1261. 1759. *Vignea capitata* Reichenb. in Mössler, Handb. Gewächsk. ed. 2. 1606. 1829. *Diemisia capitata* Raf. Good Book 27. 1840. (*Solander,* n. Europe.)

C. arctogena H. Smith, Acta Phytogeogr. Suec. 13: 193. 1940. *C. capitata* f. *arctogena* Raymond, Contr. Inst. Bot. Univ. Montréal 64: 38. 1949. *C. capitata* subsp. *arctogena* Boecher, Meddel. Grønland 147(9): 54. 1952. *C. capitata* var. *arctogena* Hultén, Kongl. Svensk. Vetenskapsakad. Handl. IV. 7(1): 38. 1958. (*H. Smith,* Swedish Lappland.) A compact phase with short, dark pistillate scales, reputed to occur with typical *C. capitata* but in drier sites.

Plants tufted, the stems clustered on very short rhizomes, mostly (0.5) 1–3 (4) dm tall, aphyllopodic, the lower sheaths bladeless or nearly so; leaves elongate and slender, but a little coarser than in *C. nardina,* up to about 1 mm wide, glabrous or with somewhat scaberulous margins, shorter than the culms in tall plants, often longer than the culms in dwarf ones; spike solitary, androgynous, ovoid to lanceolate, 5–15 mm long, bractless, the terminal staminate portion short or elongate; pistillate scales brown or tan, with more or less hyaline margins, from about as long and wide as the perigynia to often somewhat shorter and narrower; perigynia 6–25, ovate to suborbicular, usually broadly rounded below, ascending, not filled by the achene, 2–3.5 mm long, flattened, with sharply angled, thin margins, essentially nerveless, glabrous, tipped with a terete, obliquely cleft beak 0.3–0.7 mm long; stigmas 2; achene lenticular; rachilla well developed, more than half as long as the achene; $n = 25$.

Open, dry or wet places at high alt. in the mts., generally above timberline; circumboreal at high latitudes, extending s. in the western cordillera at increasing elev. to s. B.C. and Alta., and irregularly southward to the Sierra Nevada of Calif. and adj. Nev., the Beartooth Mts. of sw. Mont. and adj. Wyo., the Uinta Mts. of Utah, and reputedly to the mts. of Colo.; Chihuahua; reputedly in s. S. Amer.

4. Carex nardina Fries

Carex nardina Fries, Novit. Fl. Suec. Mantissa 2: 55. 1839. (*J. Angstrom,* Lappland.)

C. hepburnii F. Boott in Hook. Fl. Boreali-Amer. 2: 209. 1839. *C. nardina* var. *hepburnii* Kükenthal in Engl. Pflanzenr. IV. 20 (Heft 38): 70. 1909. *C. nardina* subsp. *hepburnii* Löve, Löve, & Kapoor, Arctic & Alpine Res. 3: 145. 1971. (*Drummond,* Rocky Mts. of Can.)

C. stantonensis M. E. Jones, Bull. Montana State Univ., Biol. Ser. 15: 20. 1910. nom. provis. (*R. S. Williams s.n.,* mts. of Stanton Lake, Mont., 11 Aug. 1894.)

Plants densely tufted, without creeping rhizomes, dwarf, the very slender, wiry culms up to 1 or sometimes 1.5 dm tall; old leaf bases persistent; leaves mostly basal, very slender but stiff and wiry, commonly about 0.5 mm wide, resembling the culms, glabrous or with somewhat scaberulous margins, from a little shorter to more often a little longer than the culms; spike solitary, androgynous, oblong or elliptic-oblong to ovoid, up to about 1.5 cm long, bractless, the terminal staminate portion usually well developed; pistillate scales dark with a pale midvein, ovate or obovate to orbicular, not much if at all wider than the perigynia, acute to obtuse or the lowermost ones shortly mucronate-awned; perigynia 5–15, elliptic or lanceolate to oblong-obovate, tapering to the base, ascending or nearly erect, not wholly filled by the achene, 3.5–4.5 mm long, planoconvex, with sharply angled thin margins (especially distally), otherwise nerveless or obscurely many-nerved at full maturity, obscurely serrulate on the margins distally, or wholly glabrous, rather abruptly short-beaked, the beak terete, up to 0.8 mm long, obliquely cleft at the tip; stigmas 2 or 3, the achene accordingly lenticular or trigonous; rachilla well developed, nearly or quite as long as the achene; $n = 34, 35$.

Cliff crevices, talus, and open, dry slopes at high alt. in the mts., commonly above timberline; circumboreal in the arctic, extending s. in the western cordillera to s. Wash., Nev. (Elko Co.), s. Utah (Tushar Mts., LaSal Mts.), and Colo.

Cordilleran plants frequently have 3 stigmas, in contrast to the more uniformly distigmatic arctic ones, and the arctic plants tend to have slightly broader spikes than the cordilleran ones, but many of the specimens could not be correctly placed in the absence of geographic data. The cordilleran plants have often been segregated as *C. hepburnii* F. Boott, but Boott himself recognized, in a letter to John Torrey written in 1848, that *C. hepburnii* properly falls into the synonymy of *C. nardina* Fries, published in the same year as *C. hepburnii.*

See comment under *C. elynoides.*

5. Carex elynoides T. H. Holm

Carex elynoides T. H. Holm, Amer. J. Sci. 159: 356. 1900. (*Holm,* Mt. Kelso and Mt. Elber, middle Colo.)

C. filifolia var. *miser* L. H. Bailey, Proc. Amer. Acad. Arts 22: 122. 1887. (*Watson 1220,* Clover Mts., Nev., is the first of several collections cited.)

Plants densely tufted, without creeping rhizomes, dwarf, the very slender, wiry culms up to about 1.5 dm tall; old leaf-bases persistent; leaves mostly basal, very slender but stiff and wiry, commonly about 0.5

CAREX

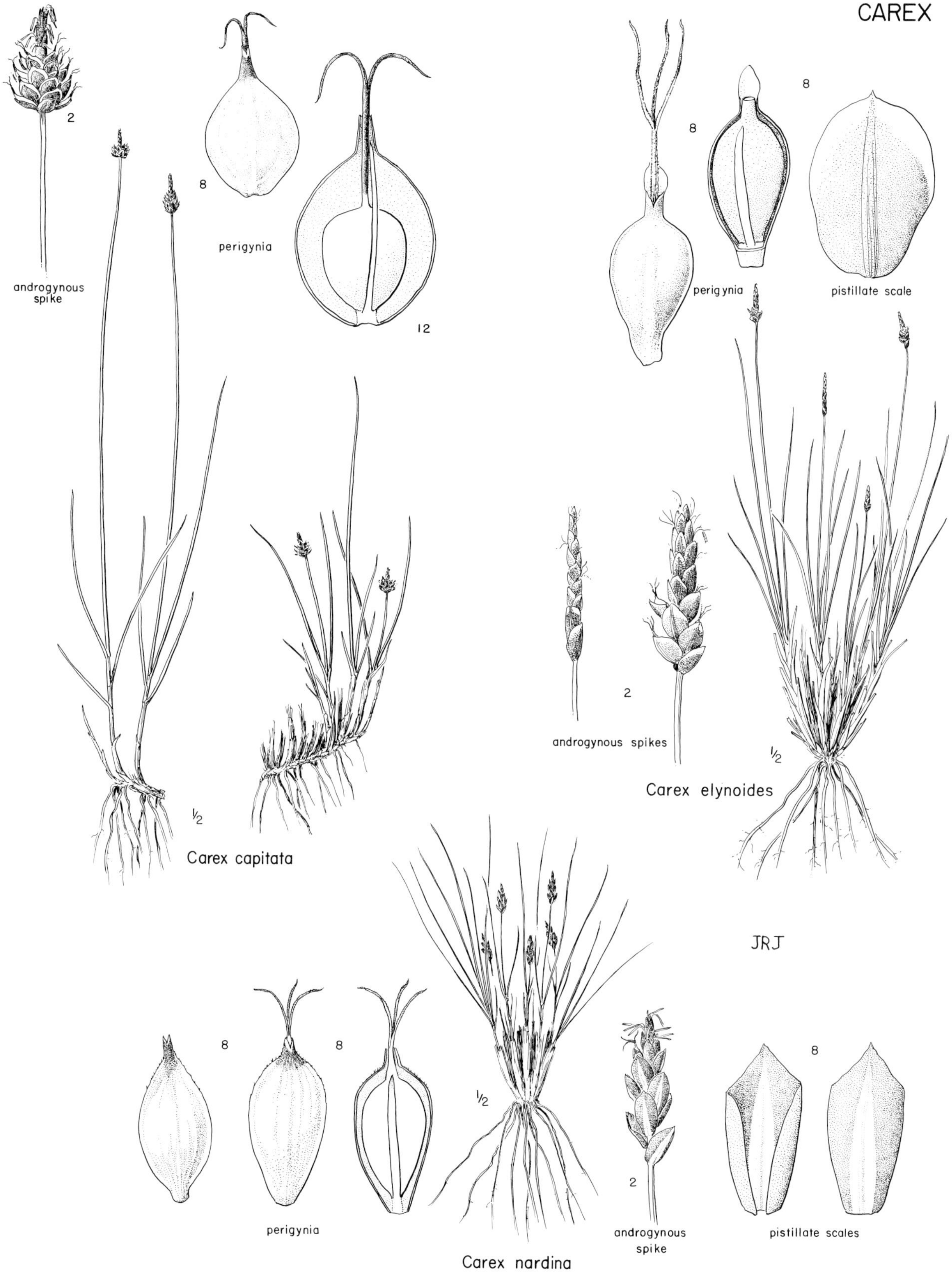
androgynous spike
2
8
perigynia
12
1/2
Carex capitata
8
perigynia
8
pistillate scale
androgynous spikes
2
1/2
Carex elynoides
JRJ
8
8
perigynia
1/2
2
androgynous spike
8
pistillate scales
Carex nardina

mm wide, resembling the culms, glabrous or with somewhat scaberulous margins, from a little shorter to a little longer than the culms; spike solitary, androgynous, narrowly and compactly cylindrical, or at maturity a little broader below and narrowly conic, mostly 1–1.5 cm long, bractless, the terminal staminate portion well developed; pistillate scales very broad, tending to wrap around the spike (especially before the perigynia are ripe), acutish to broadly rounded above (or the lowermost ones mucronate-awned), light to dark brown, with broad, pale, hyaline margins, and sometimes with a paler midvein; perigynia mostly 4–12, erect or closely ascending, wholly filled by the achene at maturity, thus somewhat turgid and not much compressed, obscurely 2-ribbed, obpyramidal or ellipsoid, 2.5–4.5 mm long, rather abruptly contracted into a short, cylindric, obliquely cut beak 0.4–1.0 mm long, glabrous or with a few very short and inconspicuous hairs near the beak; stigmas 3; achenes trigonous; rachilla well developed, from half as long to fully as long as the achene.

Open, dry slopes at high alt. in the mts., commonly above timberline; Colo. to sw. Mont., and w. across Utah to the mts. of Elko and White Pine cos., Nev.

One of the commonest alpine species of *Carex* in Utah, and extensively grazed by sheep.

Carex elynoides is similar in many respects to *C. nardina,* and the two have often been confused. The several differences between them may require some skill in interpretation, but the populational distinction in nature is probably sharp enough. *C. elynoides* is actually more closely related to *C. filifolia,* but here the pubescence of the perigynium provides a clear separation if adequate magnification (10X or preferably 20X) is available.

6. Carex filifolia Nutt.

Carex filifolia Nutt. Gen. N. Amer. Pls. 2: 204. 1818. *Olotrema filifolia* Raf. Good Book 25. 1840. *Uncinia filifolia* Nees in Neuwied, Reise Inn. Nord-Amer. 2: 450. 1841. (*Nuttall,* on the dry plains and gravelly hills of the Missouri, in 1813.)

Plants densely tufted, without creeping rhizomes, the very slender, wiry culms (0.5) 1–3 dm tall; old leaf-bases persistent; leaves mostly basal, very slender but stiff and wiry, commonly about 0.5 mm wide, resembling the culms, glabrous or with somewhat scaberulous margins, from a little shorter to a little longer than the culms; spike solitary, androgynous, narrowly and compactly cylindric at first, at maturity somewhat broader, narrowly conic or narrowly elliptic, mostly 1–2 cm long, up to about 6 mm wide, bractless, the terminal staminate portion well developed; pistillate scales mostly about equaling the mature perigynia, very broad, much broader than the perigynia, mostly broadly rounded or erose distally (or the lowest ones more pointed), with very broad and conspicuous, pale, hyaline margins, the central portion brown but sometimes with paler midvein; perigynia mostly 5–15, erect or ascending, wholly filled by the achene at maturity, thus somewhat turgid and not much compressed, rather obscurely 2-ribbed, obovoid or spherical-obovoid to ellipsoid or obpyramidal, mostly 3–4.5 mm long, rather abruptly contracted into a short, cylindric, obliquely cut beak 0.2–0.5 mm long, copiously but very minutely hirtellous-puberulent at least above the middle, the hairs visible at 20X and usually at 10X magnification; stigmas 3; achenes trigonous; rachilla well developed, from half as long to fully as long as the achene.

Dry, open places, often with sagebrush or in grassy prairies or dry meadows, from the plains and valleys to moderate or occasionally high elev. in the mts.; Yukon and Mack. to Wash. (e. of the Cascade Mts.), Oregon (Harney and Malheur cos.), Calif. (White Mts.), Nev. (Elko Co.), Utah (Emery Co.), and N.M., and e. through the Rocky Mts. and onto the plains of Man., Nebr., and Texas.

7. Carex exserta Mackenzie

Carex filifolia var. *erostrata* Kükenthal in Engl. Pflanzenr. IV. 20 (Heft 38): 86. 1909. *C. exserta* Mackenzie, Bull. Torrey Bot. Club 42: 620. 1915. (*Brainerd,* Echo Lake, El Dorado Co., Calif., 11 July 1897; lectotype by Mackenzie.) Not *C. erostrata* F. Boott ex C. B. Clarke, 1894.

Very much like *C. filifolia,* but the perigynia only 2–3 mm long, obovoid or spherical-obovoid, essentially beakless, and the pistillate scales relatively short, mostly distinctly shorter than the mature perigynia; plants averaging smaller than *C. filifolia,* rarely over 2 dm tall, the spike seldom more than 1.5 cm long.

Dry to moist (not wet) soil and talus, from middle elev. (2300 m [7500 ft.]) in the mts. to a little above timberline; Sierra Nevada of Calif., extending also to the Warner Mts. of Oregon and the White Mts. of Nev., the range approaching or merging with that of *C. filifolia* in Oregon.

Specimens with beaked perigynia rarely occur with the typical form of *C. exserta,* and are hardly to be distinguished from *C. filifolia.* A specimen from Crane Mt., near Lakeview, Lake Co., Oregon (*Thompson 13219*) has the perigynia of *C. filifolia* and the scales of *C. exserta. C. exserta* might with good reason be treated as a variety of *C. filifolia,* but it should be noted that, in addition to being segregated geographically, *C. exserta* is distinctly more montane than *C. filifolia.* Varieties of one species often occupy different habitats within the same range, or similar habitats in different ranges, but they seldom differ in both of these respects at once. *C. scirpoidea* with its var. *curatorum* does provide an example of the latter situation in *Carex,* however. The question of the proper taxonomic status of *C. exserta* is here commended to a future monographer.

8. Carex breweri F. Boott

Carex breweri F. Boott, Ill. Gen. Carex 142, pl. 455. 1867. (*Brewer 1422,* Mt. Shasta, Calif.)

C. engelmannii L. H. Bailey, Proc. Amer. Acad. Arts 22: 132. 1887. (*Engelmann,* Clear Creek, Colo.) = var. *paddoensis.*

C. paddoensis Suksd. Allg. Bot. Z. Syst. 12: 43. 1906. *C. engelmannii* var. *paddoensis* Kneucker, Allg. Bot. Z. Syst. 17: 90. 1911. *C. breweri* var. *paddoensis* Cronq. Univ. Wash. Publ. Biol. 17(1): 249. 1969. (*Suksdorf 1296,* Mt. Paddo [Adams], Wash.)

Stems arising singly or few together from creeping rhizomes, 0.5–3 dm tall; leaves crowded toward the base, slender, stiff, and wiry, tending to be involute or deeply channeled, up to about 15 (20) cm long and 1 mm wide, glabrous or with slightly scaberulous margins; spike solitary, lustrous, androgynous, bractless, stout, typically ellipsoid or ovoid, commonly 1–2 cm long, 6–12 mm wide, and 1.2–2.5 times as long as wide, the terminal staminate portion short and inconspicuous; pistillate scales thin, about as long as the perigynia, or somewhat shorter, light to dark brown, acute to more often acuminate, often notably pale at the tip, 1-nerved; perigynia mostly rather

CAREX

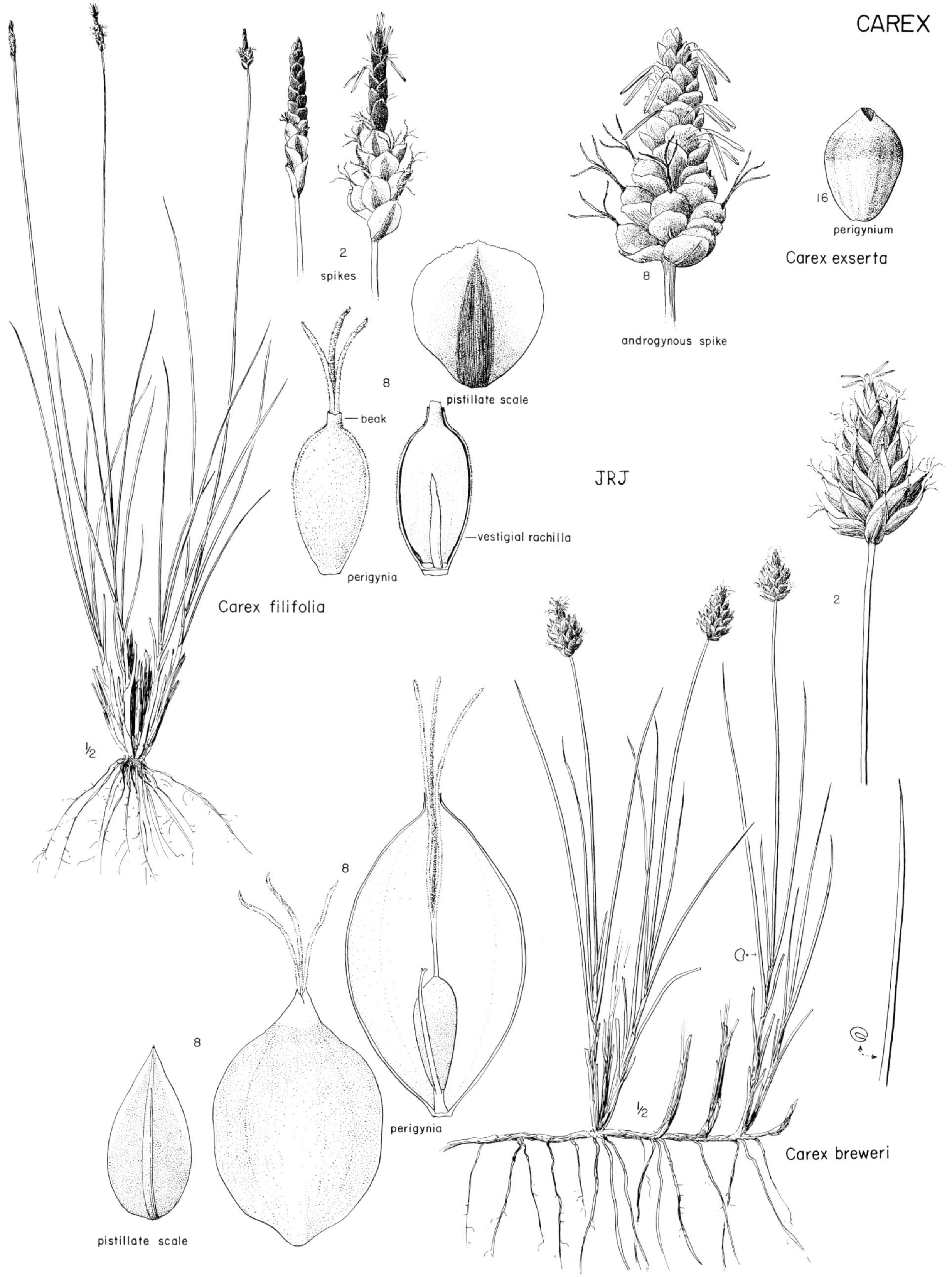
2
spikes
8
pistillate scale
beak
perigynia
vestigial rachilla
Carex filifolia
1/2
8
androgynous spike
16
perigynium
Carex exserta
JRJ
2
8
8
pistillate scale
perigynia
1/2
Carex breweri

numerous, glabrous, ascending, thin-walled, nerveless or obscurely nerved, more or less ovate or elliptic, often broadly so, 4–7.5 mm long and 1.8–4.5 mm wide, abruptly contracted or more gradually narrowed into the short, obliquely cleft beak; stigmas 3; achene trigonous, about 2 mm long or less, much smaller than the perigynium; rachilla well developed, more than half as long as the achene, often surpassing it.

In wet or dry soil or talus at high alt. in the mts., near or above timberline; Cascade-Sierran axis from s. B.C. to Calif. and adj. Nev., and rarely also in w. Wyo., Colo., and Utah (Bear R. Mts., Wasatch Mts., Tushar Mts.).

Our plants, as here described, belong to the var. *paddoensis* (Suksd.) Cronq. Typical *C. breweri,* with 3-nerved pistillate scales, is restricted to the Cascade-Sierran region from Mt. Adams (Wash.) southward.

9. Carex subnigricans Stacey

Carex subnigricans Stacey, Leafl. W. Bot. 2: 167. 1939. (*J. T. Howell 14519,* Mt. Dana, Tuolumne Co., Calif.)

C. rachillis Maguire, Brittonia 5: 199. 1944. (*Maguire et al. 14668,* Gilbert Peak, Uinta Mts., Summit Co., Utah.)

Plants sod-forming, the stems arising singly or few together from long, creeping rhizomes, 0.5–2 dm tall; leaves mostly basal, slender but stiff and wiry, up to about 12 cm long and 1 mm wide, glabrous or with slightly scaberulous margins; spike solitary, androgynous, bractless, relatively narrow, mostly cylindric to narrowly elliptic or conic, commonly 1–2 cm long and 4–6 mm thick, 2.5–3 times as long as wide, the terminal staminate portion well developed; pistillate scales thin and fragile, tan to dark brown, tending to be paler and somewhat hyaline at the margins and tip, 1-nerved or obscurely 3-nerved to nearly nerveless, acute to obtuse, about equaling the perigynia; perigynia rather numerous, (5) 10–40, glabrous, ascending, thin-walled, nerveless or nearly so, elliptic to ovate, mostly 2.5–3.5 (4) mm long and up to 1.5 mm wide, with a short, obliquely cut beak; stigmas 3; achene trigonous, not wholly filling the perigynium; rachilla well developed, more than half as long as the achene.

Moist meadows at high alt. in the mts., near or above timberline; Sierra Nevada and the Sweetwater and White mts. of Calif.; also at more or less isolated stations in the Steens and Wallowa mts. of e. Oregon, the mts. of c. Idaho, the Wind River Mts. of Wyo., the Uinta and Tushar mts. of Utah, and the Toquima Mts. of Nev.

In the original description this species was compared to *C. nigricans* and *C. pyrenaica,* but it is actually much more closely allied to *C. breweri.*

10. Carex obtusata Liljeblad

Carex obtusata Liljeblad, Kongl. Vetensk. Acad. Nya Handl. 14: 69. 1793. *Genersichia obtusata* Heuffel, Flora 27: 528. 1844. (Sweden.)

Culms scattered on elongate, dark, scaly, tough but mostly rather slender rhizomes, (0.5) 1–1.5 (2) dm tall; leaves crowded near the ground level, slender, flat or nearly so, (0.5) 1–1.5 mm wide; spike solitary, androgynous, mostly 8–15 mm long, bractless, the terminal staminate portion well developed, mostly longer than the pistillate portion; pistillate scales from a little shorter to a little longer than the perigynia, sharply acute or cuspidate, brownish, with hyaline margins and often paler midvein; perigynia few, mostly 1–6, ascending, plump, ellipsoid, thick-walled and coriaceous, obscurely to conspicuously ribbed and sulcate, usually becoming rich, shining brown, 3–4 mm long, including the prominent, hyaline, obliquely cleft to somewhat bidentate beak; stigmas 3; achene trigonous, filling the perigynium; rachilla well developed, more than half as long as the achene, often with a flattened terminal appendage.

Open, fairly dry, often grassy places, from the plains and foothills to fairly high elev. in the mts.; Yukon and Mack. to B.C., Mont., Wyo., Utah (Aquarius Plateau), Colo., and N.M., and e. across the plains to Minn. and Nebr.; also widespread in Eurasia.

11. Carex pyrenaica Wahlenb.

Carex pyrenaica Wahlenb. Kongl. Vetensk. Acad. Nya Handl. 24: 139. 1803. *Callistachys pyrenaica* Heuffel, Flora 27: 528. 1844. *Psyllophora pyrenaica* Schur, Enum. Pl. Transsilv. 697. 1866. (*La Peyrouse,* Pyrenees Mts.)

Plants densely tufted, without rhizomes, the slender stems 0.5–3 dm tall, somewhat aphyllopodic, the lowest sheaths bladeless; leaves usually 2–4 per culm, sometimes more numerous, crowded near the base, wiry, flat or channeled, mostly 2–10 cm long and 0.5–1.5 mm wide, glabrous or with somewhat scaberulous margins; spike solitary, androgynous, bractless, more or less oblong or elliptic, mostly 7–20 mm long and 4–7 mm thick, densely many-flowered, commonly appearing a little more slender than in *C. nigricans,* and often narrower-based; staminate flowers usually few; pistillate scales, perigynia, and achenes much as in *C. nigricans,* except that the perigynia are usually less widely spreading; pistillate scales sometimes only tardily deciduous.

Cliff crevices, ledges, open slopes, and meadows, at high elev. in the mts., often above timberline; cordilleran region of w. N. Amer., s. to Wash. (Cascade Mts.), Oregon (Wallowa Mts.), Utah (Uinta Mts.) and Colo., and interruptedly across Asia to Europe. Not yet recorded from Nev.

Carex pyrenaica is obviously closely related to *C. nigricans,* but the two appear to be sharply distinct.

12. Carex nigricans C. A. Meyer

Carex nigricans C. A. Meyer, Mém. Acad. Imp. Sci. St. Pétersbourg, Sér. 6, Sci. Math. 1: 211, pl. 7. 1831. (Unalaschka.)

C. nigricans f. *feminea* Kneucker, Allg. Bot. Z. Syst. 17: 90. 1911. (*Suksdorf,* Mt. Adams.)

Plants rhizomatous, only loosely or not at all cespitose, the stems crowded to fairly well spaced, (0.5) 1–3 dm tall, somewhat aphyllopodic, the lowest sheaths bladeless; leaves rather numerous, usually 4–9 per culm, crowded near the base, firm, flat or nearly so, mostly 4–13 cm long and (1) 1.5–3 mm wide, glabrous or with somewhat scaberulous margins; spikes solitary, androgynous (rarely wholly or almost wholly staminate or pistillate), bractless, broadly oblong to ellipsoid or oblong-ovoid, mostly (0.7) 1–2 cm long and 6–10 mm thick, densely many-flowered; staminate flowers with persistent scales; pistillate scales equaling or shorter than the perigynia, dark brown or blackish, thin-margined, sometimes with pale midrib, soon becoming spreading, and deciduous as the fruit approaches maturity,

CAREX

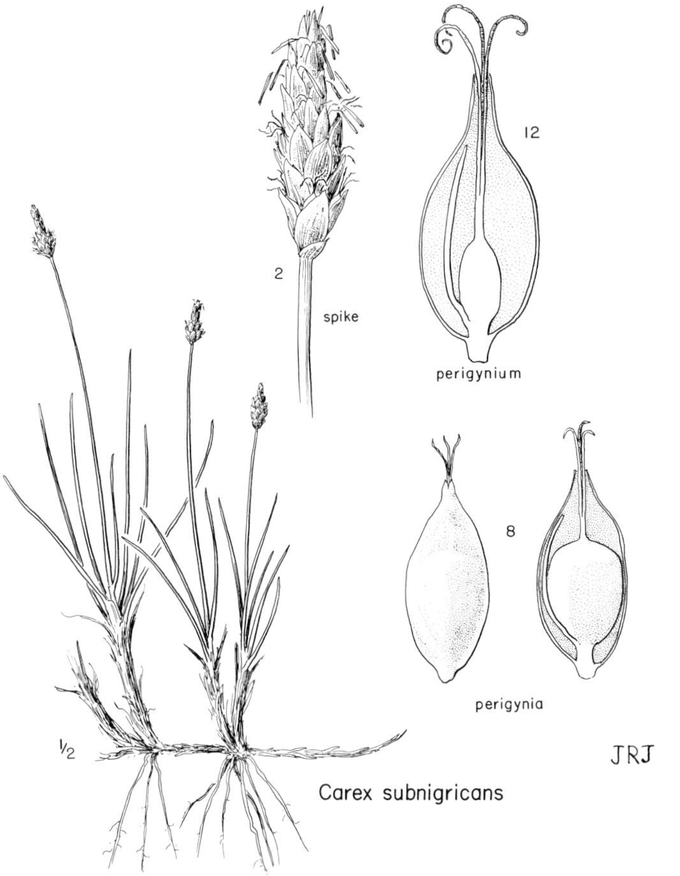

Carex subnigricans

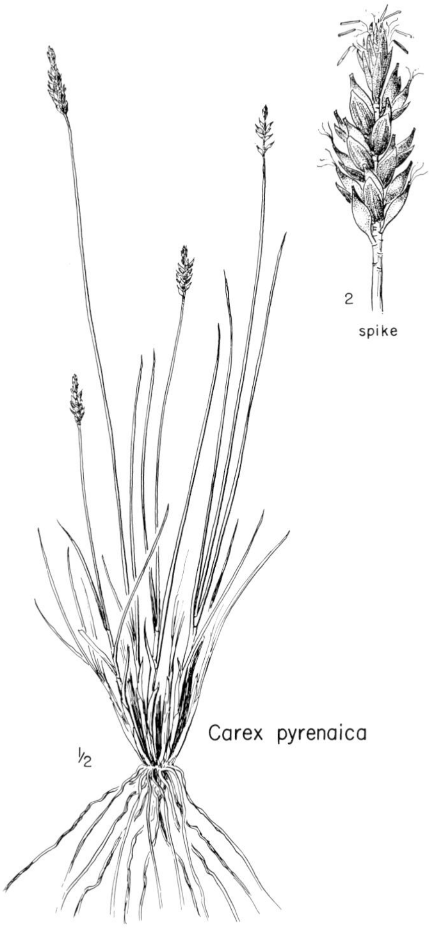

Carex pyrenaica

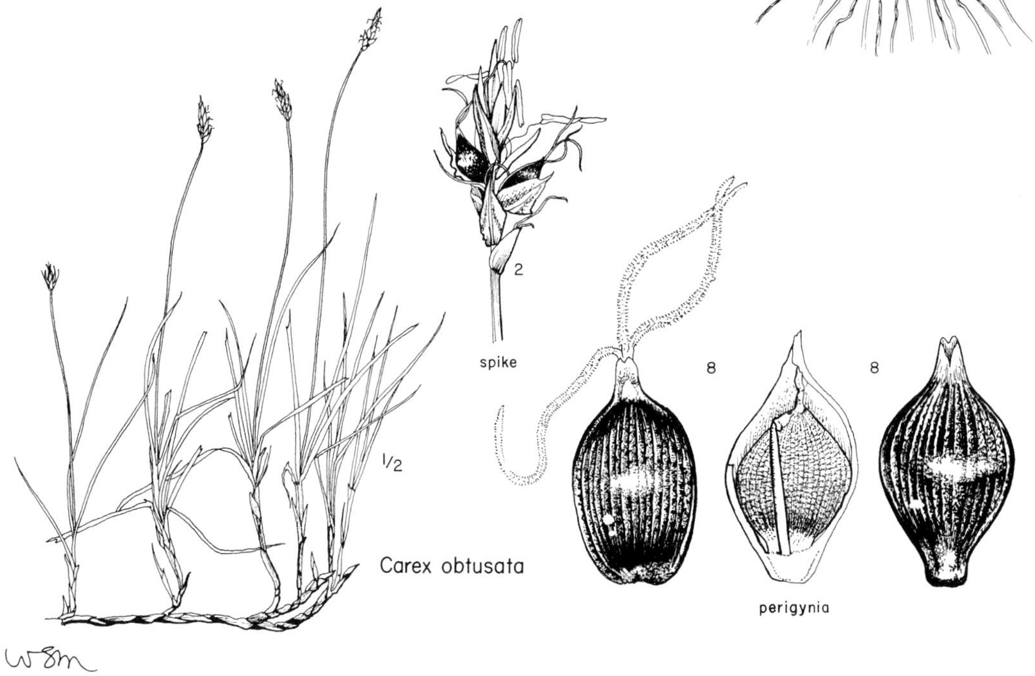

Carex obtusata

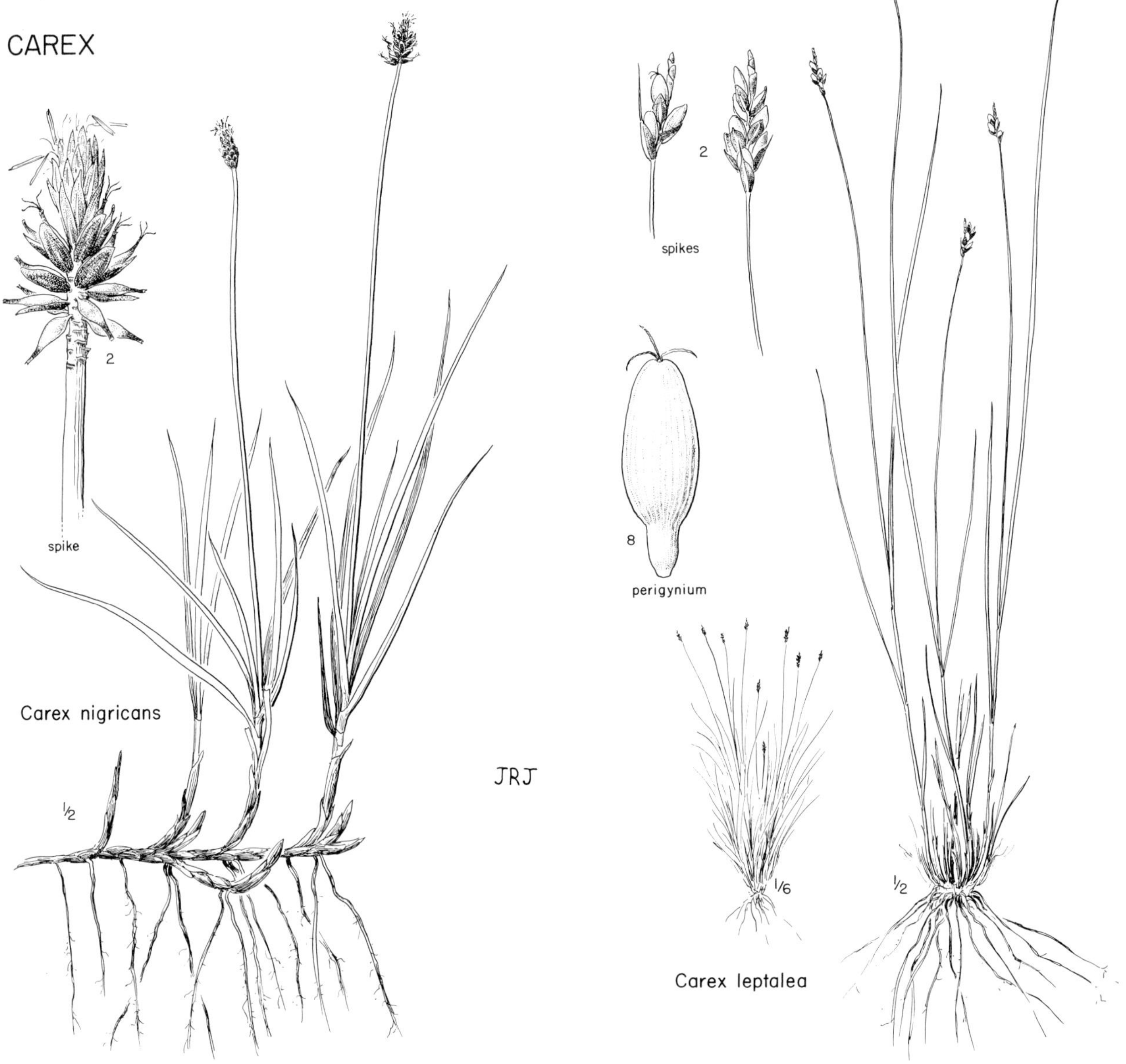

each leaving a conspicuous ridge on the axis of the spike; perigynia mostly rather numerous, typically 20–50, at first appressed-ascending, but becoming spreading, and deciduous at maturity, dark brown or blackish distally, paler and more stramineous or greenish toward the base, 3–4.5 mm long, including the slender, usually more or less stipitate base, the body lanceolate to elliptic-ovate, smooth and shining, with a pair of marginal nerves only, short-beaked or merely tapering to the narrow, obliquely cleft tip; stigmas 3 (or sometimes only 2); achene trigonous, not filling the perigynium, jointed to the style; rachilla obsolete.

Meadows, streambanks, and other moist or wet places at high alt. in the mts., often above timberline; western cordilleran region, from Alaska and the Commander Isl. to Calif., Utah (Uinta and reputedly Wasatch mts.) and Colo.; in the mts. of e. Oregon, as well as in the Cascades, but not yet recorded from Nev.

13. Carex leptalea Wahlenb.

Carex leptalea Wahlenb. Kongl. Vetensk. Acad. Nya Handl. 24: 139. 1803. ("in America borealis, secundum herbarium *Cl. Torneri*.")

Stems densely clustered on slender, freely branched rhizomes, very slender, 1.5–6 dm tall, loosely erect; leaves shorter than the culms, flat or nearly so but slender, only 0.7–1.2 mm wide; spike solitary, 0.5–1.5 cm long, bractless (but the lowest pistillate scale sometimes elongate, to 5 cm, and resembling a setaceous bract), androgynous, the staminate part often short and inconspicuous; pistillate scales greenish to brownish, obtuse to acute or short-awned, distinctly shorter than the perigynia, or the lowest ones larger and more prominently awned, the scales persistent or sometimes deciduous as the perigynia approach maturity; perigynia 1–10, ap-

CAREX

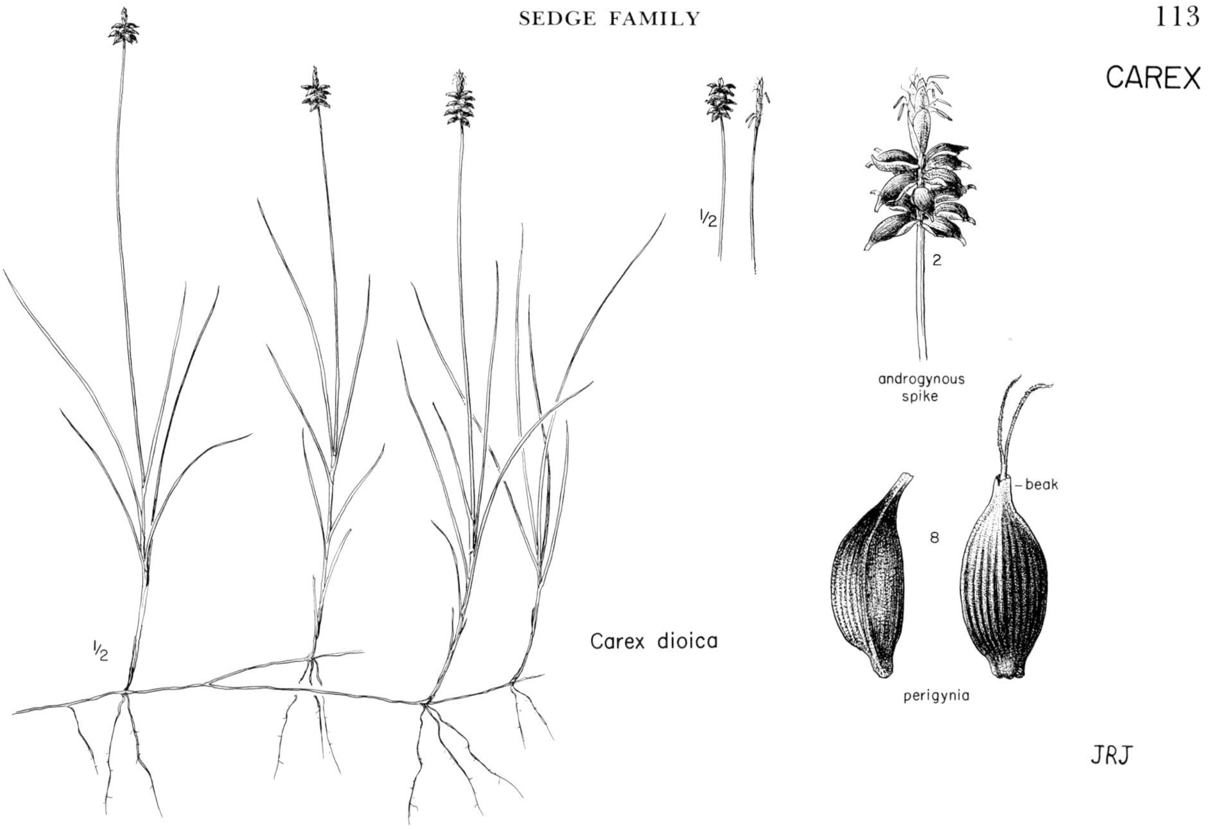

pressed-ascending (not becoming spreading), often rather remote, mostly pale greenish or stramineous, 2.5–4.5 mm long, more or less elliptic or lance-elliptic, with a narrow, often substipitate, spongy base 0.5–1 mm long, rounded or emarginate at the tip and not at all beaked, with 2 marginal nerves, and many finer nerves on each face; stigmas 3; achene trigonous, 1.3–1.8 mm long, loosely filling the lower half to four-fifths of the cavity of the perigynium, the style deciduous; rachilla wanting; $n = 26$.

Sphagnum bogs, swamps, lake shores, and wet low ground; Alaska to Lab., s. to n. Calif., Idaho, ne. Utah (Uinta Mts.), Colo., Texas, and Fla. May–Aug.

14. Carex dioica L.

Carex dioica L. Sp. Pl. 972. 1753. (Europe.)

C. gynocrates Wormsk. ex Drejer, Naturhist. Tidsskr. 3: 434. 1841. *C. dioica* var. *gynocrates* Ostenf. Fl. Arctica 60. 1902. *C. dioica* subsp. *gynocrates* Hultén, Kongl. Svenska Vetenskapsakad. Handl. IV. 8(5): 170. 1962. (*Wormskjold,* Greenland.)

Stems 0.5–2 dm tall, arising singly from long, very slender rhizomes that are only about 0.5 mm thick; leaves crowded toward the base, long and very slender, mostly 3–15 cm long and 0.4–0.9 mm wide; spike solitary, bractless, androgynous, varying to sometimes almost wholly staminate or almost wholly pistillate, 1–1.5 cm long, the staminate portion very slender, only 1–2 mm thick; pistillate scales persistent, light brownish, somewhat scarious or almost hyaline, mostly shorter but broader than the perigynia; perigynia crowded, plump, 3–3.5 mm long, inconspicuously to very evidently multistriate, abruptly contracted to a short, obliquely cleft beak, glabrous, commonly chestnut brown at maturity, soon becoming widely spreading and often with the terminal part further recurved, the wall (especially ventrally) thick and spongy; stigmas 2; achene lenticular, filling the cavity of the perigynium; rachilla obsolete; $n = 26$.

Sphagnum bogs and wet, marshy places, in our range rare and restricted to middle and upper elev. in the mts.; circumboreal at high latitudes, extending s. in N. Amer. to Pa., Mich., Sask., Colo., Utah (Iron Co.), Nev. (Elko Co.), and ne. Oregon (Wallowa Mts.); apparently missing from Wash., but known from Marble Mts. of s. B.C.

Our plants, as here described, represent the var. *gynocrates* (Wormsk.) Ostenf., of North America and Siberia. Var. *dioica,* of Europe, is more nearly dioecious, tends to have somewhat darker pistillate scales, and tends to have the perigynia less plump on the ventral side. No sharp distinction can be drawn on these bases, however.

15. Carex scirpoidea Michx.

Carex scirpoidea Michx. Fl. Boreali-Amer. 2: 171. 1803. *C. michauxii* Schwein. Ann. Lyceum Nat. Hist. New York 1: 64. 1824. *C. scirpina* Tuckerman, Enum. Meth. Caricum Quarundum 8. 1843. (Hudson Bay.)

C. pseudoscirpoidea Rydb. Mem. New York Bot. Gard. 1: 78. 1900. *C. scirpoidea* var. *pseudoscirpoidea* Cronq. Univ. Wash. Publ. Biol. 17(1): 325. 1969. (*Rydberg 3064,* Spanish Basin, Mont.; lectotype by Mackenzie.)

C. scirpiformis Mackenzie, Bull. Torrey Bot. Club 35: 270. 1908. *C. scirpoidea* var. *scirpiformis* O'Neill & Duman, Rhodora 43: 417. 1941. (*McCalla 2348,* Banff, Alta.) = var. *scirpoidea,* but the name sometimes extended to cover some of our plants of var. *pseudoscirpoidea.*

Carex scirpoidea var. **curatorum** (Stacey) Cronq. hoc loc. *C. curatorium* Stacey, Leafl. W. Bot. 2: 13. 1937. (*Eastwood & Howell 1100,* Kaibab Trail to Roaring Springs, Grand Canyon Natl. Park, Ariz.)

Culms clustered or fairly well spaced, on a branching system of stout, mostly rather short rhizomes, 1–4 (8) dm tall, phyllopodic to aphyllopodic; leaves clustered near the base (or seldom well above the base), firm, mostly flat, often elongate, 1.5–3 (4) mm wide; spikes bractless or with an inconspicuous basal bract, solitary or occasionally with one or two shorter sessile spikes crowded at the base of the main spike, unisexual or nearly so, the plants more or less completely dioecious; staminate spikes similar to the pistillate ones in size, shape, and color; pistillate spike densely flowered, cylindric, 1.5–4 cm (to 5 cm in var. *curatorum*) long and 3–6 mm thick, the scales medium brown to blackish, with thin, often pale and more or less hyaline margin that may also be fringed-ciliolate, sometimes short-hairy on the back as well; perigynia crowded, appressed-ascending, evidently short-hirsute or hispidulous, especially distally, 2–3.5 mm long and 1.5–2.5 times as long as wide, obscurely several-nerved, broadest near or above the middle and somewhat tapered to the base, rather abruptly contracted above into an evident, short, obliquely cleft beak; stigmas 3; achene trigonous, loosely to sometimes rather tightly filling the perigynium; rachilla obsolete; n = 31, 32, 34.

Meadows, streambanks, and open, rocky slopes at high alt. in the mts., often above timberline (var. *curatorum* at lower elev.); Greenl. to Alaska, s. to Me., N.Y., Mich., Sask., Colo., s. Utah (Iron Co.), Ariz. (Kaibab Plateau), Nev. (Elko and White Pine cos.), and Calif.; irregularly across Siberia to n. Europe.

Most of our plants belong to var. *pseudoscirpoidea* (Rydb.) Cronq., the characteristic Southern Rocky Mt. and Great Basin form of the species. On the Kaibab Plateau in Arizona the species is represented by a distinctive local ecotype, the var. *curatorum*. Aside from its relatively lowland habitat, however, the var. *curatorum* overlaps in all respects with var. *pseudoscirpoidea*, and it seems best treated at the varietal level. The two varieties that occur in our range may be distinguished as follows:

1 Pistillate scales tending to be narrower and shorter than the perigynia, so that much of the perigynium is exposed; plants relatively robust, phyllopodic to strongly aphyllopodic, 2.5–8 dm tall, with spikes mostly 2.5–5 cm long, one or two short spikes in addition to the terminal one frequently present; Kaibab Plateau n. to hanging gardens at the junction of the San Juan and Colorado rivers .... var. *curatorum* (Stacey) Cronq.

1 Pistillate scales usually about as long and wide as the perigynia, seldom distinctly narrower, or shorter, or both; plants phyllopodic, mostly less robust, 1–4 (5) dm tall, only occasionally with more than one spike; widespread at high alt. (and in an unusual, highly alkaline habitat at 1830 m [6000 ft.] at Monte Neva Hot Springs, White Pine Co., Nev.), but not known from Ariz. .... var. *pseudoscirpoidea* (Rydb.) Cronq.

16. **Carex rupestris** All.

Carex rupestris All. Fl. Pedemontana 2: 264. 1785. *Edritria rupestris* Raf. Good Book 26. 1840. *Caricinella rupestris* St.-Lager in Cariot, Étude Fleurs Bot. Element. Descript. & Usuelle, ed. 8. 2: 882. 1897. (Italian Alps.)

Carex drummondiana Dewey, Amer. J. Sci. 29: 251. 1836. *C. rupestris* var. *drummondiana* L. H. Bailey, Cat. N. Amer. Car. 4. 1884, as a nomen nudum; place and date of validation not ascertained. (*Richardson*, Fort Norman and the Rocky Mts.)

Curly sedge.

Stems 4–15 dm tall, arising singly or in small tufts from long, creeping rhizomes; leaves crowded at the base, flat or nearly so, but narrow, 1–3 mm wide, shorter than the culms, tapering to a long, slender, circinate or flexuous tip; spike solitary, bractless, androgynous, 1–2 cm long; pistillate scales broad, a little shorter to a little longer than the perigynia and largely concealing them, acute to obtuse or broadly rounded, usually brownish and more or less scarious or hyaline; perigynia mostly 5–15, 3–4 mm long, glabrous, elliptic, with a narrow or substipitate base and an abrupt, short truncate beak only about 0.2 mm long, thin-walled, several-nerved; stigmas 3; achene trigonous, filling the perigynium; rachilla wanting; n = 25, 26.

Open slopes at high alt. in the mts., near or above timberline; circumboreal at high latitudes, extending s. in Amer. to Que. and Alta., and irregularly to Wyo., Colo., and ne. Utah (Uinta Mts.), seemingly skipping Mont.

17. **Carex sheldonii** Mackenzie

Carex sheldonii Mackenzie, Bull. Torrey Bot. Club 42: 618. 1915. (*Sheldon 8854*, Clark's Creek, Oregon.)

Culms 4–10 dm tall, arising singly or in small tufts from deep-seated creeping rhizomes; leaf-sheaths villous-hirsute, at least on the ventral side, especially toward the summit; leaf-blades elongate, firm, flat, 3–5 mm wide, glabrous or often villous-hirsute like the ventral side of the sheath, inconspicuously or not at all septate-nodulose; leaves subtending the pistillate spikes sheathless or nearly so, but with well developed, elongate blade, those subtending the staminate spikes much reduced or more often wanting; spikes several, sessile or nearly so, remote, the lower pistillate and the upper staminate, or one of them androgynous; pistillate spikes cylindric, 2–6 cm long, about 1 cm wide or a little less, the scales lanceolate or lance-ovate, narrowly acute, or shortly-awned, greenish or brownish and often largely scarious, narrower and mostly shorter than the perigynia; perigynia crowded, ascending, conspicuously 12- to 20-ribbed and copiously hirsutulous, lanceolate or lance-ovate, turgid below, more flattened distally and more or less strongly beaked, 5–8 mm long, including the usually prominent and unequal teeth, these (0.5) 1–1.5 mm long; stigmas 3; achenes plump, trigonous, 2–2.5 mm long, rather loosely filling the lower part of the perigynium, continuous with the straight, bony style; staminate spikes 1.5–4.5 cm long, stramineous or light brownish.

Streambanks and moist meadows in the foothills and at moderate elev. in the mts.; Blue Mt. region of ne. Oregon, s. through se. Oregon to the Warner Mts. of Calif., and e. to Blaine and Twin Falls cos., Idaho (but apparently avoiding the main body of the Snake River Plains).

18. **Carex atherodes** Sprengel

Carex aristata R. Br. in Franklin, Narr. Journey Shores Polar Sea, Bot. Appendix 751. 1823. *C. atherodes* Sprengel, Syst. Veg. 3: 828. 1826. *C. trichocarpa* var. *aristata* L. H. Bailey, Bot. Gaz. 10: 294. 1885. *C. aristata* var. *browniana* Aschers. Ber. Deutsch. Bot. Ges. 6: 290. 1888. (Type not given in the original description, but doubtless based on *Richardson*, Cumberland House, nw. Can.) Not *C. aristata* Honck. 1792.

Culms 4–15 dm tall, arising singly or in small tufts

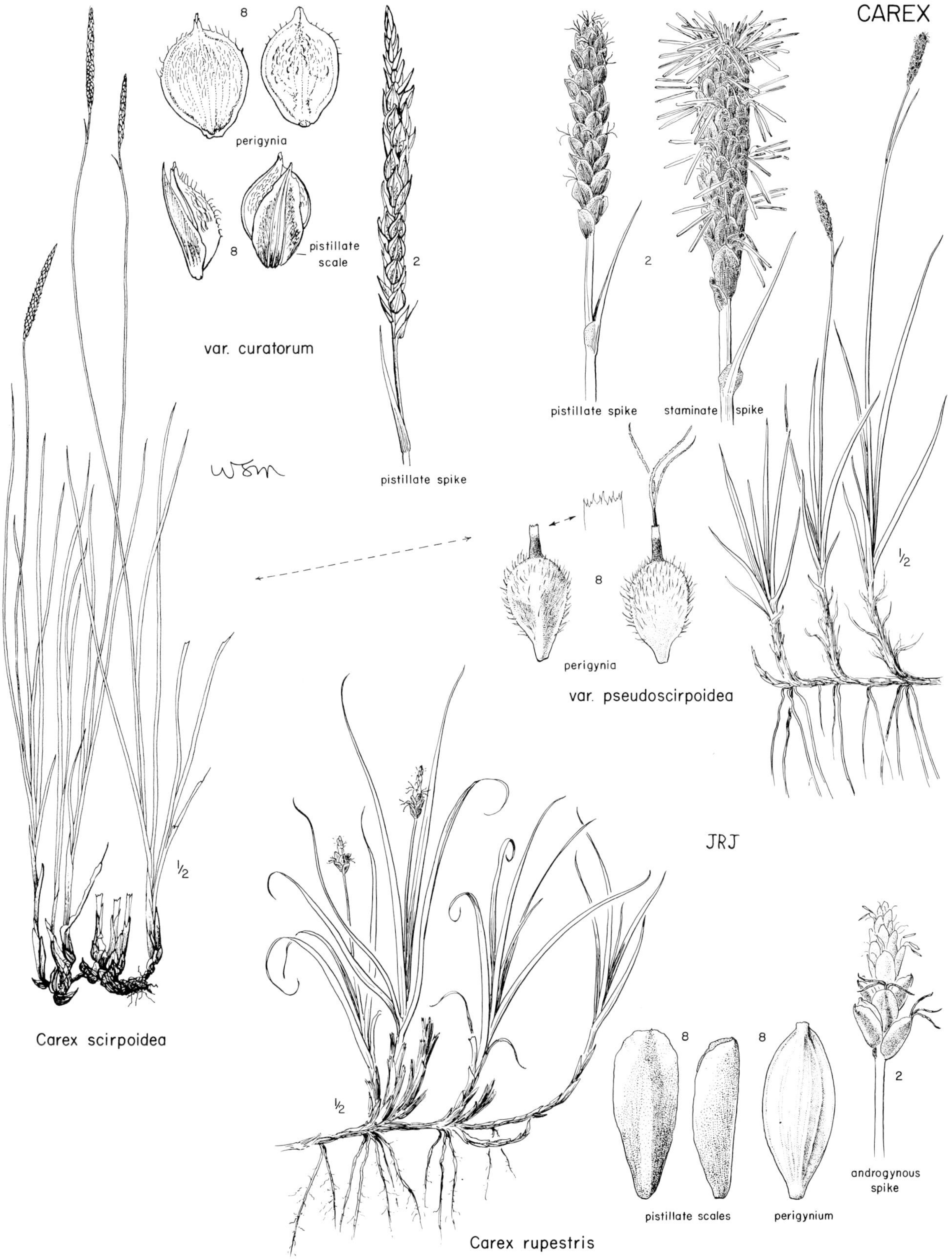
CAREX
8
perigynia
8
pistillate scale
2
var. curatorum
pistillate spike
2
pistillate spike
staminate spike
8
perigynia
var. pseudoscirpoidea
1/2
1/2
Carex scirpoidea
JRJ
1/2
8
8
2
pistillate scales
perigynium
androgynous spike
Carex rupestris

CAREX

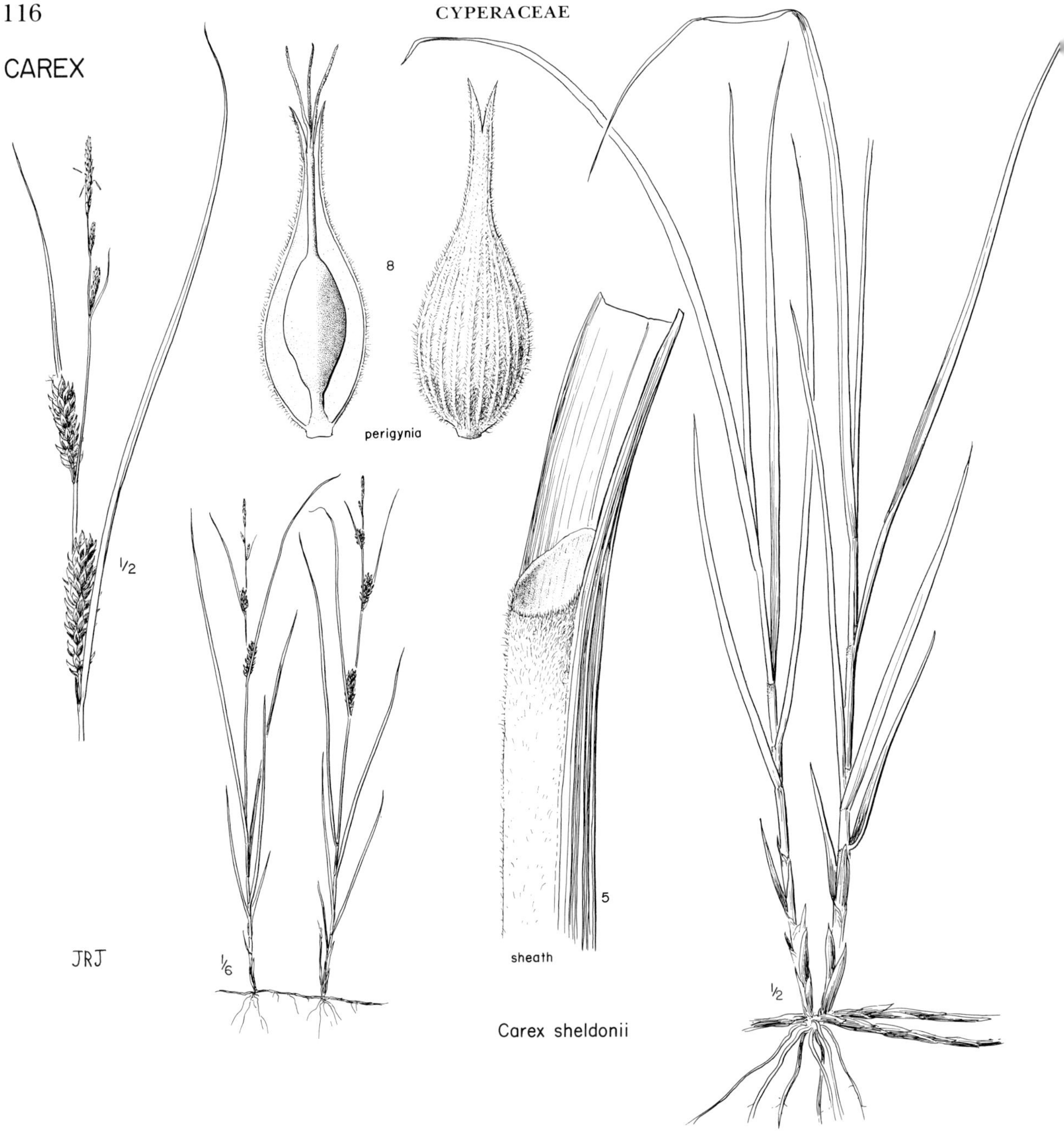

from coarse, deep-seated, creeping rhizomes; leaf-sheaths villous-hirsute all around, or sometimes only toward the ventral summit, the lower sheaths breaking ventrally and becoming filamentose; leaf-blades elongate, firm, flat, 4–10 mm wide, glabrous or often hirsute on the lower side toward the base, tending to be septate-nodulose; leaves subtending the pistillate spikes sheathless or nearly so, but with well developed, elongate blade, those subtending the staminate spikes much reduced or wanting; spikes several, sessile or inconspicuously short-pedunculate, remote, the lower pistillate and the upper staminate, or the middle one(s) androgynous; staminate spikes 2–6 cm long, pale, commonly stramineous, the scales awned; pistillate spikes cylindric, 2–10 cm long, about 1 cm wide, the scales stramineous or pale greenish and often largely scarious, narrow, lanceolate to lance-ovate, tipped by a well developed rough awn 1–5 (9) mm long; perigynia crowded (except often at the base of the spike), ascending, pale greenish to stramineous, conspicuously 12- to 20-ribbed, glabrous, firm, 7–10 mm long, lanceolate or lance-ovate, turgid-inflated below, more flattened at the well developed beak, this with prominent, often divergent teeth 1.5–2.5 (3) mm long; stigmas 3; achenes trigonous, 2–2.5 mm long, loosely filling the lower part of the perigynium, continuous with the straight, bony style.

CAREX

JRJ
sheaths
1/2
pistillate spike
8
pistillate scale
2
teeth
8
perigynium
1/10
Carex atherodes
1/2
8
perigynia
Carex saxatilis

In wet low ground or in standing water up to nearly 1 m deep, from the lowlands to fairly high elev. in the mts.; circumboreal, extending s. in Amer. to N.Y., Mo., Colo., Utah (Sevier Co.), Nev. (Elko Co.), and s. Oregon.

19. Carex saxatilis L.

Carex saxatilis L. Sp. Pl. 976. 1753. *C. vesicaria* subsp. *saxatilis* Kükenthal in Engl. Pflanzenr. IV. 20 (Heft 38): 727. 1909. (Europe.)

C. physocarpa C. Presl, Reliquiae Haenkeanae 1: 205. 1828. *C. vesicaria* subsp. *saxatilis* var. *physocarpa* Kükenthal in Engl. Pflanzenr. IV. 20 (Heft 38): 728. 1909. (*Haenke,* Nootka Sound or Port Mulgrave.) = var. *major.*

C. saxatilis var. *major* Olney in King, Rep. Geol. Explor. 40th Parallel 5: 370. 1871. (*Watson 1248,* Uinta Mts., Utah.)

Plants turf-forming, the stems arising singly or few together from well developed creeping rhizomes, 2–8 dm tall, slender; leaves glabrous, often septate-nodulose (especially the sheaths), the blade 2–4 mm wide; spikes 2–4, the lower pistillate, the upper staminate, the lowest one generally on a slender peduncle 0.5–3 cm long and more or less nodding; leaf subtending the lowest spike reduced to a sheathless or nearly sheathless bract 3–15 cm long, the other spikes with smaller bracts or bractless; staminate spikes 1 or 2, 1–4 cm long, 3–5 mm thick, dark, the scales usually castaneous or fuscous; pistillate spikes 1 or 2, 1–3 cm long and 5–10 mm thick, usually castaneous or fuscous, sometimes pale; pistillate scales slightly to distinctly shorter and narrower than the perigynia, usually castaneous or fuscous, with pale, hyaline, often erose tip, rarely stramineous throughout; perigynia crowded, castaneous or fuscous (especially distally) or occasionally greenish-stramineous, 3.5–5 mm long, with a pair of marginal nerves, the ventral surface nerveless, the dorsal surface obscurely few-nerved or essentially nerveless, the body elliptic or elliptic-ovate, biconvex, abruptly contracted to a short beak about 0.5 mm long, this obliquely cleft at the tip and scarcely bidentate; stigmas 2; achene lenticular, about 2 mm long or a little less, filling the lower part of the perigynium, the upper quarter or third (half) of the perigynium empty; seldom stigmas 3 and achenes trigonous; style continuous with the achene, firm, becoming flexuous or contorted as the achene matures; $n = 30$, 40.

Wet meadows and margins of streams and ponds, from middle alt. in the mts. up to about timberline; circumboreal, extending s. in Amer. to Lab., Hudson's Bay, Colo., Utah (Kane Co.), Nev. (Elko Co.), and Wash. (Clallam Co.).

The cordilleran plants of *C. saxatilis,* as here described, tend to differ in several minor respects (more robust; lowest spike more peduncled and more or less nodding; perigynia a little larger, often with better developed nerves) from more characteristic *C. saxatilis* as represented in Greenland and e. Can. (and Eurasia). No firm distinction can be drawn on these bases, however, and the geographical origin of many specimens could not be correctly determined from the morphology alone. By stretching a point, the cordilleran plants may be distinguished as var. *major* Olney.

20. Carex hystricina Muhl. ex Willd.

Carex hystricina Muhl. ex Willd. Sp. Pl. 4: 282. 1805. (Pa.)

Stems clustered on a short, stout rhizome, mostly 3–6 (10) dm tall; leaves glabrous, tending to be septate-nodulose (especially the sheaths), the blade elongate, flat, 3–9 mm wide; leaf subtending the lowest pistillate spike sheathless or with only a short sheath, the blade well developed, elongate, somewhat surpassing the inflorescence, the leaves subtending the other spikes much reduced or wanting; pistillate spikes several, 1.5–4 cm long and 1–1.5 cm thick, borne fairly close together or the lowest one remote, the lower ones loosely nodding on slender, flexuous peduncles, the uppermost one(s) often subsessile and loosely ascending-spreading; staminate spike terminal, solitary, 2–4 cm long; pistillate scales with a short, largely scarious or hyaline body only 1–2 mm long and a prominent, rough awn-tip 2–6 mm long; perigynia very numerous, densely crowded, spreading or ascending-spreading (or the lowermost ones retrorse), usually pale greenish, 5–7 mm long, prominently (12) 15- to 20-nerved, lanceolate or lance-ovate, not stipitate, thin-textured, slightly inflated, the beak slender, conspicuous, obliquely cleft, with short straight teeth 0.3–0.9 mm long; stigmas 3; achene trigonous, 1.2–1.6 mm long, loose in the lower half of the perigynium, continuous with the persistent, bony style, which becomes flexuous or contorted as the achene matures; $n = 29$.

Wet ground near streams at lower elev.; N.B. to Wash., s. to Va., Mo., Texas, N.M., and Calif. Seldom collected in our range.

21. Carex vesicaria L.

Carex vesicaria L. Sp. Pl. 979. 1753. *Trasus vesicarius* S. F. Gray, Nat. Arr. Brit. Pl. 2: 59. 1821. (Europe.)

C. vesicaria var. *major* F. Boott in Hook. Fl. Boreali-Amer. 2: 221. 1839. *C. exsiccata* L. H. Bailey, Mem. Torrey Bot. Club 1: 6. 1889. ("Columbia River. *Douglas. Scouler.*")

Stems loosely to densely clustered on a system of rather short, stout, branching rhizomes, 3–10 dm tall; leaf blades and sheaths glabrous, tending to be septate-nodulose, the blade elongate, flat, 3–8 mm wide; leaves subtending the pistillate spikes sheathless or nearly so, but with well developed, elongate blade, those subtending the staminate spikes much reduced or wanting; spikes several, sessile or inconspicuously short-pedunculate, erect or nearly so, remote, the lower pistillate, the upper staminate, or one of them androgynous; staminate spikes 2–7 cm long, slender, only 2–3 (4) mm thick; pistillate spikes mostly 2–7 cm long and 1–1.5 (2) cm thick at maturity; pistillate scales shorter and narrower than the perigynia, thin and often more or less scarious, stramineous to castaneous, variously acutish to long-acuminate but soft-tipped and scarcely awned; perigynia crowded, ascending, pale greenish or stramineous to castaneous, strongly 10- to 20-ribbed, glabrous, lanceolate or lance-ovate in outline, turgid-inflated below, gradually tapering to the flatter, often poorly defined beak, 5–8 mm long, including the short (0.3–1.2 mm) teeth; stigmas 3; achenes trigonous, 1.7–2.4 mm long, loose in the lower part of the perigynium, continuous with the bony style, which becomes flexuous to strongly contorted as the achene matures; $n = 37$, 41.

In wet soil or shallow water in bogs or swamps and at the margins of ponds or streams, from the lowlands to moderate elev.

CAREX

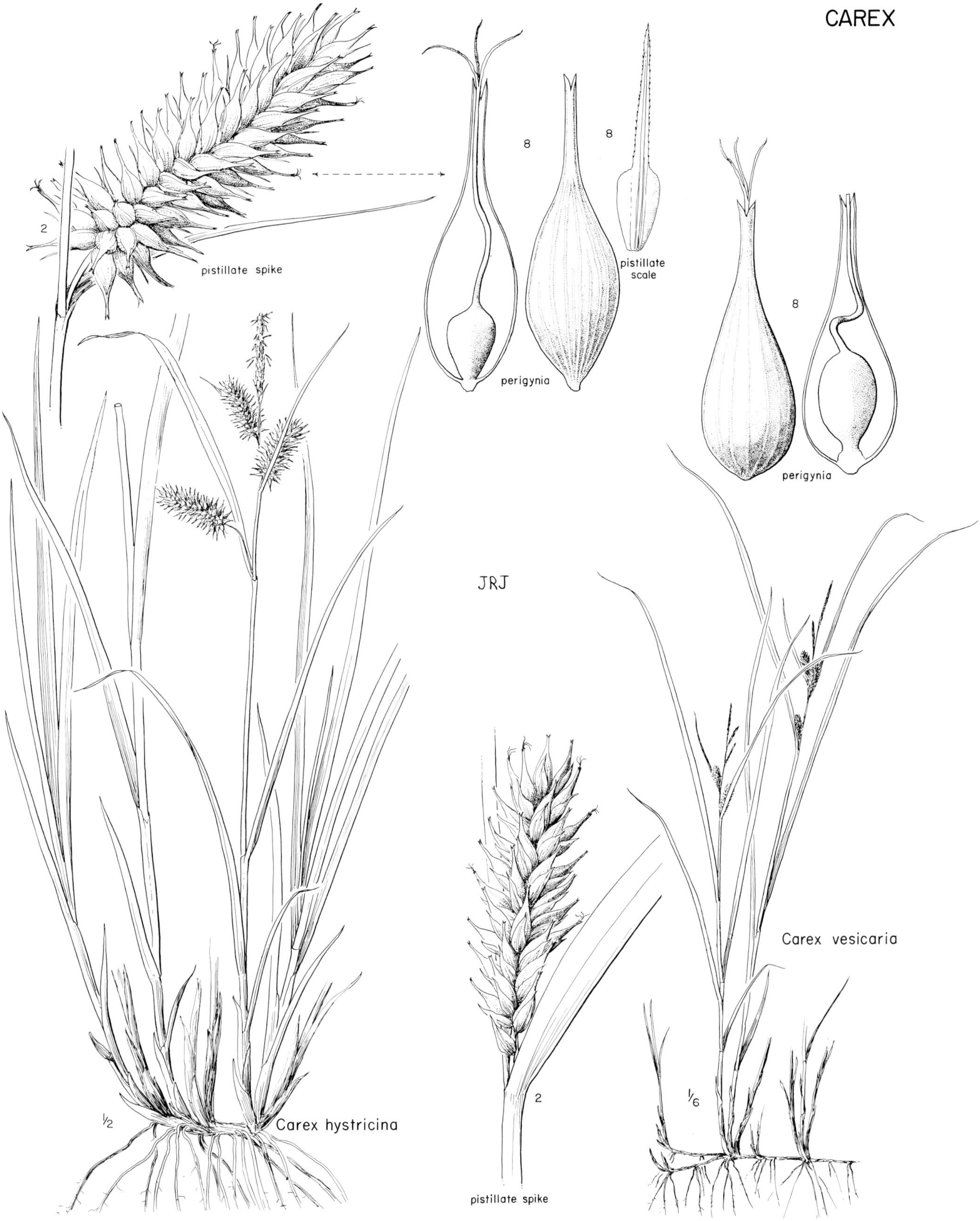
2
pistillate spike
8
8
pistillate scale
perigynia
8
perigynia
JRJ
1/2
Carex hystricina
2
pistillate spike
Carex vesicaria
1/6

in the mts.; circumboreal, in Amer. extending s. to Del., Ind., Mo., N.M., Ariz., and Calif.

The common phase over most of the U.S. and Can., including our region, appears to be essentially identical with typical European *C. vesicaria* and may be referred to var. *vesicaria*. The var. *major* F. Boott, with larger perigynia, occurs mainly near the Pacific Coast and is not found in our region.

22. **Carex rostrata** Stokes ex With.

Carex rostrata Stokes ex With. Bot. Arr. Brit. Pl. ed. 2. 2: 1059. 1787. (England.)

C. inflata Huds. Fl. Anglica 354. 1762, misapplied. (See Fernald in Rhodora 44: 324. 1942, and Koyama in J. Fac. Sci. Univ. Tokyo, Sect. 3, Bot. 8: 241. 1962.)

C. utriculata F. Boott in Hook. Fl. Boreali-Amer. 2: 221. 1839. *C. vesicaria* var. *utriculata* Dewey in A. Wood, Cl.-Book Bot. 428. 1845. *C. ampullacea* var. *utriculata* Carey in A. Gray, Manual Bot. 566. 1848. *C. rostrata* var. *utriculata* L. H. Bailey, Proc. Amer. Acad. Arts 22: 67. 1887. *C. inflata* var. *utriculata* Druce, Bot. Soc. Exch. Club Brit. Isles 9: 141. 1930. ("British America. *Dr. Richardson. Drummond.*")

C. utriculata var. *minor* F. Boott in Hook. Fl. Boreali-Amer. 2: 221. 1839. *C. rostrata* var. *utriculata* f. *minor* Kükenthal in Engl. Pflanzenr. IV. 20 (Heft 38): 722. 1909. ("British America. *Dr. Richardson.* Rocky Mountains. *Drummond.*")

Stems arising singly or few together from stout, deep-seated, elongate creeping rhizomes, sometimes forming a dense sod, mostly (3) 5–12 dm tall; leaves glabrous, septate-nodulose (especially the sheaths), the blades elongate, flat, mostly (2) 4–12 (15) mm wide; leaves subtending the pistillate spikes sheathless or with a very short sheath, but the lowest one, at least, with an elongate blade 1–2 times as long as the inflorescence; leaves subtending the staminate spikes much reduced or wanting; spikes several, remote, forming an elongate inflorescence, erect or nearly so, the lower ones pistillate, the upper staminate, or one or two of the middle ones androgynous, the lowest one on a slender peduncle 1–8 cm long, or subsessile, the others sessile or nearly so; staminate spikes mostly 2–7 cm long and 2–4 mm thick, stramineous to sometimes castaneous; pistillate spikes mostly 2–10 cm long and about 1 (1.5) cm thick at maturity; pistillate scales stramineous to sometimes castaneous, narrower and usually shorter than the perigynia, acutish to more often acuminate or short-awned; perigynia densely crowded, at first ascending, later more or less strongly spreading (often at right angles), glabrous, shining, usually pale greenish-stramineous, sometimes more or less castaneous, especially distally, more or less strongly 8- to 16-nerved, 4–7 mm long, the body inflated, broadly ellipsoid to ovoid or subglobose, rather abruptly contracted to an evident beak, this 1–2 mm long, including the short (0.2–0.8 mm) teeth; stigmas 3; achene trigonous, 1.3–2.0 mm long, loose in the lower half of the body of the perigynium; style persistent, bony, becoming strongly flexuous or contorted as the achene matures; $n = 36, 37, 38, 41$.

In wet soil or shallow water at the margins of streams, lakes, and ponds, sometimes in swamps or wet meadows, from the lowlands to moderate elev. in the mts., sometimes up to near timberline; circumboreal, extending s. in Amer. to Del., Ind., Nebr., N.M., s. Utah, Nev. (Lander Co.) and Calif.

One of the commonest species of *Carex* in our range, but not a prime source of fodder.

23. **Carex retrorsa** Schwein.

Carex retrorsa Schwein. Ann. Lyceum Nat. Hist. New York 1: 71. 1824. (*Dewey,* Mass.)

Culms densely clustered on a very short rhizome, 3–10 dm tall; leaves glabrous, septate-nodulose (especially the sheaths), the blades elongate, flat, mostly 4–10 mm wide; leaf subtending the lowest pistillate spike sheathless or shortly sheathing, with a well developed, elongate blade usually several times as long as the whole inflorescence; spikes several, crowded, sessile or nearly so, or the lowest one slender-pedunculate and remote from the others, the lower spikes pistillate, the upper staminate or androgynous; pistillate spikes 1.5–5 cm long, about 1.5–2 cm thick; pistillate scales inconspicuous, shorter and narrower than the perigynia; perigynia numerous, densely crowded, widely spreading or the lowest ones retrorse, glabrous, shining, evidently 8- to 12-nerved, 7–10 mm long, the body inflated, ellipsoid to subglobose and often somewhat oblique, narrowed to a prominent slender beak, this 2–3 (4) mm long, including the short (0.3–0.9 mm) teeth; stigmas 3; achene trigonous, 1.9–2.2 mm long, loose in the lower half of the body of the perigynium, continuous with the persistent bony style, which becomes contorted as the achene matures; $n = 35$.

Swamps, bogs, wet meadows, and edges of streams and lakes, in the foothills and lowlands; Que. to N.J., w. across s. Can. and n. U.S. to s. B.C., Wash., and ne. Oregon, entering our range on the northern part of the Snake River Plains, as at Idaho Falls, and reported from Weber Co., Utah; also known from Colo.

24. **Carex rossii** F. Boott

Carex rossii F. Boott in Hook. Fl. Boreali-Amer. 2: 222. 1839. *C. novae-angliae* var. *rossii* L. H. Bailey, Bot. Gaz. 10: 207. 1885. *C. deflexa* var. *rossii* L. H. Bailey, Mem. Torrey Bot. Club 1: 43. 1889. ("N.W. Coast. *Douglas.* Rocky Mountains. *Drummond.*")

C. brevipes W. Boott in S. Wats. Bot. Calif. 2: 246. 1880. *C. deflexa* var. *boottii* L. H. Bailey, Mem. Torrey Bot. Club 1: 43. 1889. *C. rossii* var. *brevipes* Kükenthal in Engl. Pflanzenr. IV. 20 (Heft 38): 452. 1909. (*Kellogg,* Lake Tahoe to Bear Valley, in the Sierra Nevada of Calif.) A chiefly Californian form, seldom found in our range, with small, short-beaked perigynia.

C. deflexa var. *media* L. H. Bailey, Mem. Torrey Bot. Club 1: 43. 1889. *C. deflexa* (var.) *farwellii* Britton in Britton & Brown, Ill. Fl. N. U.S., Canada & Brit. Possessions 1: 334. 1896. *C. farwellii* Mackenzie, Bull. Torrey Bot. Club 37: 244. 1910. ("Montana and Colorado to Oregon.")

Stems slender, (0.5) 1–3 (4) dm tall, densely clustered, somewhat aphyllopodic, the basal parts often clothed with old sheaths and appearing somewhat rhizome-like, but vertical or nearly so, the plants without creeping rhizomes; leaves elongate and rather lax, often surpassing the stems, glabrous or somewhat scabrous, 1–4 mm wide; staminate spike terminal, short and slender, 5–12 (15) mm long; 1–4 short, few-flowered, sessile or short-pedunculate pistillate spikes borne near the staminate spike, the lowest of these subtended by a leaf-like (but sheathless or nearly so) bract which generally surpasses the staminate spike; one or more small pistillate spikes also borne on slender peduncles arising from near the base of the culm, well removed from the termi-

CAREX

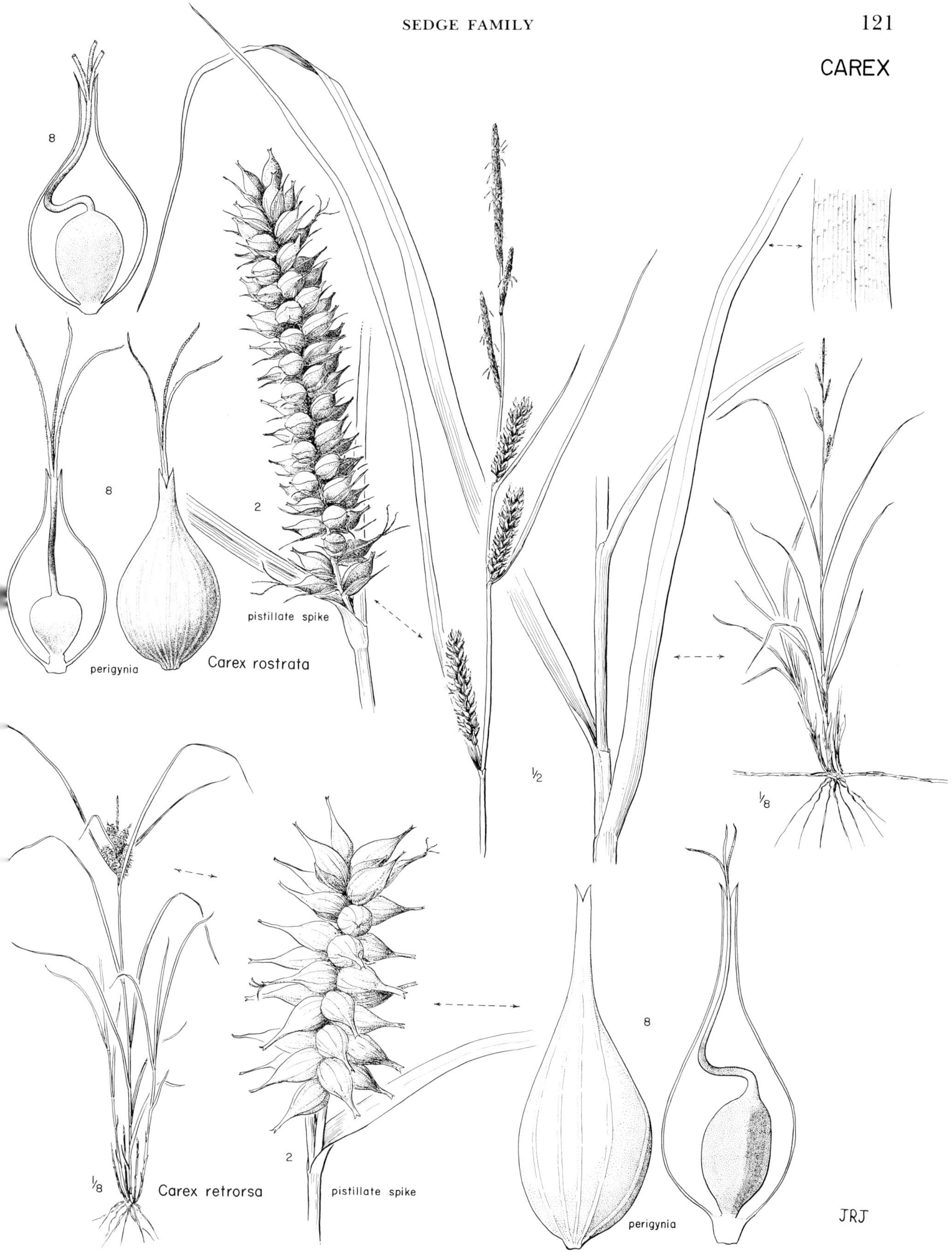
8
8
2
pistillate spike
perigynia
Carex rostrata
1/2
1/8
1/8
Carex retrorsa
2
pistillate spike
8
perigynia
JRJ

nal cluster of spikes; pistillate scales a little wider but usually shorter than the mature perigynia; perigynia finely puberulent, (2.7) 3.0–4.5 mm long, including the 0.7–1.8 mm beak, the body subglobose and with a short, stipe-like base, or obovoid and narrowed to the base, with 2 prominent marginal nerves, otherwise nerveless or with a few obscure facial nerves, abruptly contracted to the beak, this obliquely cleft and often also bidentate, the teeth up to about 0.6 mm long; stigmas 3; achene trigonous, 1.5–2.0 mm long, plump, the sides convex above, filling the body of the perigynium.

In dry to moderately moist sites in areas which will support coniferous forest (not including pinyon-juniper woodland), and in adj. meadows and prairies, from near sea level in w. Oregon and Wash. to about timberline in the mts.; widespread in the western cordilleran region, from Yukon to Calif., Ariz. (Kaibab Plateau) and Colo., and irregularly e. to n. Mich.

25. Carex geophila Mackenzie

Carex geophila Mackenzie, Bull. Torrey Bot. Club 40: 546. 1913. (*Eggleston 6584,* Tierra Amarilla, Rio Arriba Co., N.M.)

C. pityophila Mackenzie, Bull. Torrey Bot. Club 40: 545. 1913. (*Eggleston 6605,* Tierra Amarilla, Rio Arriba Co., N.M.)

Plants densely tufted, forming large clumps, without creeping rhizomes; stems slender, 2–10 cm tall, often surpassed by the leaves; leaves numerous, firm, 3–15 (20) cm long and 1–2.5 mm wide; staminate spike solitary, terminal, 5–12 mm long; one or two short, few-flowered, sessile or short-pedunculate pistillate spikes usually borne near the staminate one, the bract subtending the lowest of these spikes sheathless or only shortly sheathing, usually relatively short and surpassed by the staminate spike, seldom longer and surpassing it; one or more small, few-flowered pistillate spikes borne on short peduncles at or near the base of the stem, well removed from the terminal cluster of spikes; pistillate scales equaling or shorter than the perigynia; perigynia puberulent, at least distally, 2.7–4.0 mm long, including the 0.4–1.2 mm beak, the body subglobose to ellipsoid or obovoid, often tapering to a narrow or substipitate base, with 2 prominent lateral nerves, otherwise nerveless or nearly so; stigmas 3; achene trigonous, with convex sides, 1.6–2.8 mm long, rounded and apiculate to retuse at the tip, filling the perigynium.

Dry, open places in the pinyon-juniper zone; Ariz. and N.M., n. to sw. Colo. and s. Utah (San Juan and possibly Sevier cos.).

This species is still known from relatively few collections, and the limits between it and *C. rossii* are not yet clear, but it does appear to constitute a distinctive population which deserves some sort of taxonomic recognition.

26. Carex aurea Nutt.

Carex aurea Nutt. Gen. N. Amer. Pls. 2: 205. 1818. *Neskiza aurea* Raf. Good Book 27. 1840. (*Nuttall,* Shores of Lake Michigan.)

C. aurea var. *androgyna* Olney in King, Rep. Geol. Explor. 40th Parallel 5: 371. 1871. *C. garberi* Fern. Rhodora 37: 253. 1935. (*Garber s.n.,* Presque Isle, Erie Co., Pa., 9 June 1869.)

C. aurea var. *celsa* L. H. Bailey, Mem. Torrey Bot. Club 1: 75. 1889. *C. celsa* Piper in Piper & Beattie, Fl. N.W. Coast 79. 1915. (*Vasey s.n.,* San Bernardino Mts., Calif.) The robust extreme.

C. hassei L. H. Bailey, Bot. Gaz. 21: 5. 1896. (*Hasse s.n.,* San Antonio Cañon, San Bernardino Mts., Calif., June 1894.)

C. garberi var. *bifaria* Fern. Rhodora 37: 255. 1935. *C. garberi* subsp. *bifaria* Hultén, Ark. Bot. II, 7(1): 27. 1968. (*Collins & Fernald s.n.,* River Ste. Anne des Monts, Que., Aug. 1905.)

Stems slender, 0.3–4 dm tall, often somewhat aphyllopodic, arising singly or in small clumps from well developed creeping rhizomes; leaves narrow, flat or nearly so, mostly 1–4 mm wide, often surpassing the stems, especially in shorter plants; spikes 2–several, approximate or more or less remote (sometimes some of them borne near the base of the stem), erect or loose on slender, sometimes elongate peduncles, the lowest non-basal spike subtended by a prominent bract with a well developed sheath, the sheath generally 3–12 mm long, the blade usually more or less elongate and often surpassing the inflorescence, or in small plants sometimes all the bracts except those subtending the basal spikes or clusters of spikes reduced and sheathless or nearly so; spikes 5–20 mm long, the terminal one staminate, or less often gynaecandrous, rarely androgynous, the lateral ones pistillate; pistillate scales half to fully as long as the perigynium, stramineous to brown, often with greener, more herbaceous midrib and white-hyaline margins, rounded to acute or apiculate; perigynia ellipsoid to more often obovoid or obovoid-globose, sometimes broadly substipitate at the base, more or less rounded and beakless distally, only slightly or not at all compressed, 1.7–3.0 mm long, obscurely to more often prominently 12- to 20-ribbed, or sometimes appearing essentially nerveless, strongly whitish-papillate-glaucous on a light green background when young, tending to turn golden or yellow-brown with increasing maturity, becoming somewhat fleshy (especially toward the base) and only obscurely papillate, these changes beginning proximally and progressing distally, often leaving the distal half of the perigynium relatively unaffected, or the perigynium sometimes remaining whitish-papillate throughout and not becoming at all fleshy; stigmas 2 (seldom 3 on some flowers); achene lenticular (trigonous), often thickly so, 1.3–2.0 mm long, often nearly or quite as wide, filling the perigynial cavity, except often at the tip; $n = 26$.

In various sorts of moist or wet places from the lowlands to near timberline in the mts.; Newfl. to Alaska, s. to Calif., s. Nev., Utah, n. N.M., Nebr., n. Ind., and Pa.

It has been customary in recent years to attempt to distinguish *C. hassei* and/or *C. garberi* from *C. aurea* on the following characters: Plants more robust, with more crowded spikes and more numerous and more crowded flowers in each spike, the terminal spike androgynous (actually gynaecandrous) instead of wholly staminate, the pistillate scales darker in color and nearly as long as the perigynia, the perigynia narrower, obscurely instead of conspicuously ribbed, remaining white-papillate and dry at maturity instead of turning golden yellow and becoming somewhat fleshy. Each of these characters represents, in my opinion, one end of the normal variation in the population of *C. aurea,* and they are not well correlated among themselves. Neither is there any obvious geographic or ecologic correlation; *C. aurea* and *C. garberi-hassei* (as indicated by authoritative determinations) have essentially the same geographic range and they occur in the same set of habitats. The changes in

CAREX

8
perigynia
2
2
1/2
8
perigynium
Carex rossii
6
6
1/2
Carex aurea
1/2
12
perigynia
Carex geophila
JRJ

CAREX

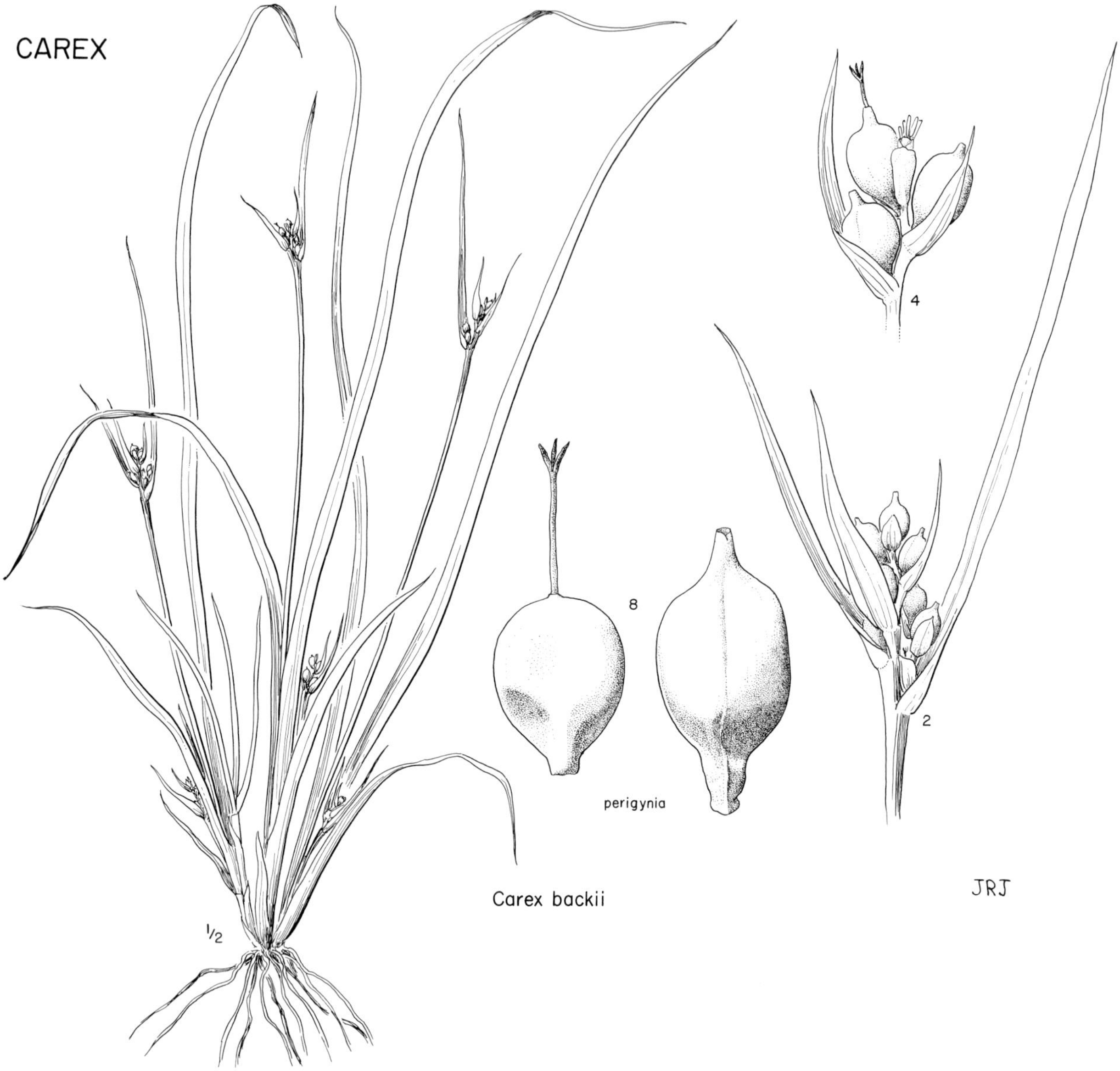

Carex backii

color and texture of the perigynium are, as indicated in the foregoing description, partly but not wholly related to maturation.

The supposed distinctions between *C. hassei* and *C. garberi,* and between typical *C. garberi* and the var. *bifaria* are recondite as well as inconstant, and they appear to have little or no populational significance.

27. Carex backii F. Boott

Carex backii F. Boott in Hook. Fl. Boreali-Amer. 2: 210. 1839. *C. durifolia* L. H. Bailey, Bull. Torrey Bot. Club 20: 428. 1893. (*Richardson,* Carlton House, Lake Winnipeg; lectotype by Mackenzie.) Not *C. backana* Dewey, 1836.

C. saximontana Mackenzie, Bull. Torrey Bot. Club 33: 439. 1906. *C. backii* var. *saximontana* B. Boivin, Naturaliste Canad. 94: 522. 1967. (*Baker,* Fort Collins, Colo.)

Plants tufted, without creeping rhizomes, somewhat aphyllopodic, but the leaves still clustered and arising from near the ground level; leaves flat, mostly 2–6 mm wide, elongate (the longer ones 10–40 cm long), surpassing the slender, strongly triangular culms, these mostly 0.5–3 dm tall; spikes terminal and basal or near-basal, the latter on slender but rather short peduncles from the axils of foliage leaves, or terminating very short stems, the former solitary and bractless, although the lower scales are bractlike in size and appearance; stem becoming concavely winged-triquetrous beneath the solitary terminal spike, the lower part of the rachis likewise green and concavely winged-triquetrous; spikes androgynous, the pistillate flowers few, commonly 2–5, the lower ones subtended by broad, elongate, leaflike scales mostly (1.5) 2–7 cm long, the upper progressively reduced but still generally surpassing the perigynia; staminate flowers few, their scales very small, forming an inconspicuous terminal cylinder only about 2 mm long and surpassed by the uppermost perigynium; perigynia turgid, rounded-triquetrous, green or greenish, inconspicuously nerved, 4.0–5.4 mm long, rhombic-ovoid and obscurely beaked to subglobose or obovoid and conspicuously

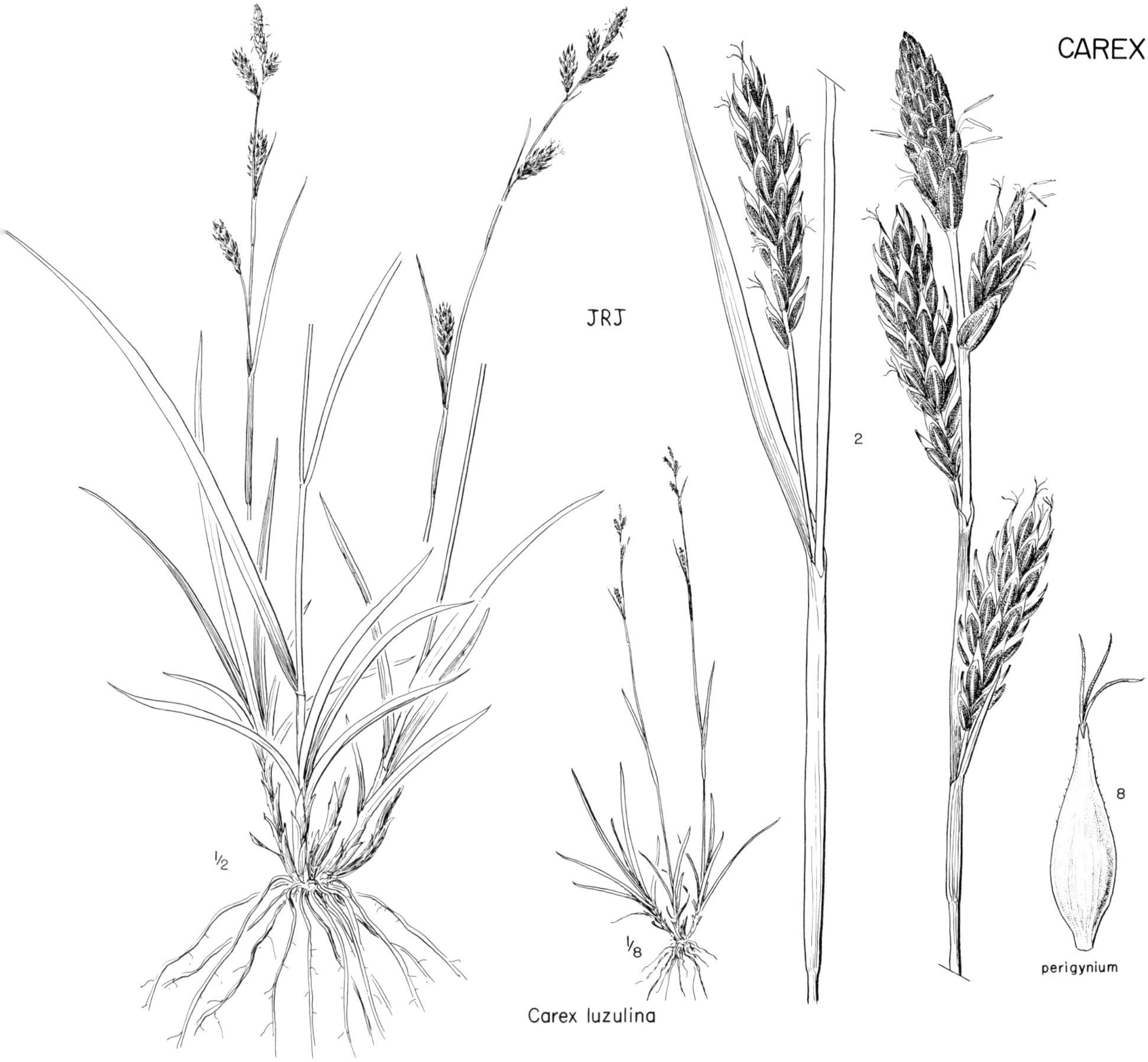

Carex luzulina

beaked, the beak when well defined mostly 1–2 mm long, obliquely cleft at the tip; stigmas 3; achene rounded-trigonous, more or less obovoid and broadly short-stipitate, about 3 mm long, broadly rounded or subtruncate distally, without a stylar apiculus.

Moist, often shady places, commonly in woods or thickets, from the lowlands to moderate elev. in the mts., infrequent in our range; Que. to N.Y., w. to Alta., s. B.C., Wash., Oregon (Wallowa Mts.), and Utah (Cache, Salt Lake and Utah cos.).

It has been customary to separate the cordilleran specimens as a distinct species, *C. saximontana* Mackenzie, but the purported distinguishing characters are fictitious.

28. Carex luzulina Olney

Carex luzulina Olney, Proc. Amer. Acad. Arts 7: 395. 1868. (*Bolander 4740,* Mendocino City, Calif.)

C. ablata L. H. Bailey, Bot. Gaz. 13: 82. 1888. *C. luzulaefolia* var. *ablata* Kükenthal in Engl. Pflanzenr. IV. 20 (Heft 38): 558. 1909. *C. luzulina* var. *ablata* F. J. Hermann, Rhodora 70: 420. 1968. (*Macoun 13401,* Mt. Mark, Vancouver I., B.C.; lectotype by Mackenzie.)

C. fissuricola Mackenzie, Muhlenbergia 5: 53. 1909. (*Heller 9429,* head of south fork of Humboldt R., Elko Co., Nev.) The form with slightly hirtellous perigynia.

C. owyheensis A. Nels. Bot. Gaz. 53: 219. 1912. (*Macbride 422,* Owyhee Mts., near Silver City, Idaho.)

Plants in tight but often rather small tufts, the stems 1.5–6 dm tall; leaves generally crowded near the base, flat and relatively broad, 3–9 mm wide, 6–15 (25) cm long, commonly much shorter than the culms; terminal staminate spike mostly 1–2 cm long, sometimes with a few perigynia at the base; pistillate spikes (2) 3–5 (6), 1–3 cm long, the upper ones sessile or nearly so, crowded close to each other and to the staminate spike, sometimes staminate toward the tip, the lower one or two generally more remote (but not basal) and long-pedunculate, but still erect; bract subtending the lowest pistillate spike with a well developed (1–4.5 cm) sheath and a short (1–8 cm) blade which is much surpassed by the inflorescence;

pistillate scales brown with pale (often green) midrib, mostly shorter and sometimes also narrower than the perigynia, glabrous or obscurely scabrous-hirtellous especially along the midrib; perigynia lanceolate or lance-ovate, mostly 3.0–5.0 (5.5) mm long and 1.0–1.5 mm wide, distinctly compressed and with 2 prominent marginal nerves, sometimes evidently several-nerved on the faces as well, the margins usually ciliolate-serrulate distally, the body often also with some very short, stiff hairs at least when young, but never densely or conspicuously hairy, tapering to an often ill-defined beak, this generally 1 mm long or more, scarcely or shortly (to 0.3 mm) bifid; stigmas 3; achene trigonous, 1.4–1.7 mm long, only loosely enveloped by the perigynium.

Bogs, wet meadows, and other wet places, from the foothills to near timberline in the mts.; s. B.C. to Calif., e. to Idaho, sw. Mont., w. Wyo., the mts. of Elko Co., Nev., and the Wasatch and Tushar mts. of Utah.

29. Carex misandra R. Br.

Carex misandra R. Br. Chloris Melvilliana 25. 1823. *C. fuliginosa* var. *misandra* O. F. Lang, Linnaea 24: 597. 1851. (*Ross,* Melville I.)

Culms 0.5–2 (3) dm tall, densely tufted, not at all rhizomatous, strongly phyllopodic, the leaves tufted at the base, thick and firm, flat or slightly canaliculate, mostly 4–10 cm long and 1–3 (4) mm wide; spikes 2–4, mostly 1–1.5 cm long, the terminal one gynaecandrous or on some culms sometimes wholly pistillate, the lateral ones pistillate, loose or nodding on slender peduncles; bract subtending the lowest spike with a well developed sheath 0.7–2 cm long, its blade from distinctly shorter to slightly longer than the sheath, the upper bracts smaller and often nearly bladeless; pistillate scales shorter but commonly wider than the perigynia, medium to dark brown or dark purplish, with paler, hyaline margins, and sometimes with paler midrib as well; perigynia lanceolate, 3.3–5.0 mm long, tapering to an often more or less stipitate base and to a slender but poorly defined, obliquely cleft, scarcely bidentate beak, strongly flattened, especially distally, ciliolate-serrulate on the margins above the middle, otherwise glabrous, nerveless or nearly so except for the well developed marginal nerves, dark distally, pale toward the base; stigmas 3; achene trigonous, 1.4–1.9 mm long, loosely enclosed by the perigynium; $n = 20$.

Meadows and open slopes above timberline; circumboreal, s. in Amer. to Que. and Alta., disjunct in Colo. and Utah (Uinta and reputedly Wasatch mts.).

30. Carex capillaris L.

Carex capillaris L. Sp. Pl. 977. 1753. *Trasus capillaris* S. F. Gray, Nat. Arr. Brit. Pl. 2: 66. 1821. (Sweden.)

Culms slender and lax, 1–6 dm tall, densely tufted, strongly phyllopodic, the leaves mainly in a basal cluster, 1–3 mm wide, up to about 15 cm long, generally much shorter than the culms; terminal spike staminate or rarely gynaecandrous, 4–10 mm long; lateral spikes 1–4, usually more or less remote, on lax, capillary, nodding or loosely spreading peduncles, 5- to 25-flowered, 0.5–1.5 cm long, sometimes compound; pistillate spikes all subtended by sheathing bracts, the sheath 0.5–2.5 cm long, or the uppermost one more reduced, the lowermost of these bracts generally with a definite blade 1–6 cm long, the others with the blade often more or less reduced or obsolete; pistillate scales shorter but often wider than the perigynia, white-hyaline distally and usually also marginally, otherwise light brown or greenish, eventually deciduous; perigynia more or less elliptic or lance-ovate, 2.4–3.3 mm long, tapering to a short, poorly defined, hyaline-tipped and obliquely cleft beak, obliquely 2-nerved, not much if at all compressed, loosely enclosing the achene and empty distally, glabrous, or faintly scabrous-serrulate toward the beak, shining brown to olive green; stigmas 3; achene trigonous, 1.2–1.5 mm long, jointed to the style; $n = 18, 27, 28$.

Streambanks, wet meadows and wet ledges, from moderate to high elev. in the mts., up to about timberline, seldom higher; circumboreal, extending s. in Amer. to s. B.C. (Selkirk Mts.), ne. Oregon (Wallowa Mts.), ne. Nev. (Elko Co.), s. Utah (Iron Co.), n. N.M., Mich., and N.Y.; apparently absent from Wash.

31. Carex crawei Dewey

Carex crawei Dewey, Amer. J. Sci. 52: 246. 1846. (*Crawe s.n.,* Jefferson Co., N.Y.)

Stems arising singly or few together from well developed creeping rhizomes, 1–3 (4) dm tall, phyllopodic or only slightly aphyllopodic; leaves flat or nearly so, mostly 1.5–3 (4) mm wide; spikes cylindric, 1–2.5 cm long, the terminal one staminate, the lateral ones 2–4 (6), pistillate or the upper androgynous or seldom staminate, compactly (5) 15- to 50-flowered, generally more or less remote, the lowest one often nearly basal, all erect and short-pedunculate, the peduncle often scarcely surpassing the subtending sheath, or the lowest ones on long peduncles; lower spikes subtended by bract-leaves with well developed sheath and blade, the sheath mostly 5–15 mm long, the blade elongate but surpassed by the terminal spike; upper bracts with more reduced blade but still with a well developed sheath; pistillate scales shorter and often narrower than the perigynia, with firm, pale or greenish, often shortly excurrent midrib and brownish, hyaline-scarious body; perigynia elliptic or ovate-elliptic, greenish-stramineous to tan or light brownish, often finely reddish-brown-speckled, 2.3–3.8 mm long, including the very short, pale, truncate or slightly bidentate beak that is up to 0.4 mm long or essentially obsolete, evidently to rather obscurely but coarsely 15- to 25-nerved, sometimes flattened on the adaxial side, otherwise subinflated and not at all compressed, only the angles of the achene in contact with the perigynial wall, the distal part of the perigynium empty; stigmas 3; achene trigonous, 1.3–1.7 mm long, the style deciduous, leaving a straight or bent recurved apiculus on the achene.

Wet places in the foothills, valleys and plains, often associated with limestone or with marl-bogs; Que. to se. B.C. (Kootenay R.), Mont. (Beaverhead Co.) and reputedly Wash., s. to N.J., Ala., Mo., Wyo., and Utah (Kane Co.), but seldom collected in our range.

CAREX
8
gynaecandrous
terminal spike
perigynium
2
Carex misandra
1/2
2
scale
8
8
perigynia
2
Carex capillaris
1/2
8
perigynia
1/2
Carex crawei
JRJ

32. Carex oederi Retz.

Carex oederi Retz. Fl. Scand. Prodr. 179. 1779. *Onkerma oederi* Raf. Good Book 27. 1840. (Denmark.)

C. viridula Michx. Fl. Boreali-Amer. 2: 170. 1803. *Edritria viridula* Raf. Good Book 26. 1840. *C. flava* var. *viridula* L. H. Bailey, Mem. Torrey Bot. Club 1: 31. 1889. *C. oederi* var. *viridula* Kükenthal in Engl. Pflanzenr. IV. 20 (Heft 38): 674. 1909. *C. oederi* subsp. *viridula* Hultén, Ark. Bot. II. 7(1): 30. 1968. (Can.)

C. flava var. *recterostrata* L. H. Bailey, Bot. Gaz. 13: 84. 1888. *C. oederi* var. *viridula* f. *recterostrata* Kükenthal in Engl. Pflanzenr. IV. 20 (Heft 38): 674. 1909. (*Macoun*, Vancouver I.) Not *C. flava* var. *rectirostra* Gaudich. 1830.

Stems 1–4 dm tall, clustered, not at all rhizomatous, phyllopodic; leaves basal and cauline, or essentially all basal, narrow and elongate, often equaling or surpassing the culms, mostly 1–3 mm wide, tending to be canaliculate, but sometimes distinctly flat; basal sheaths pale at the base; terminal spike slender, staminate, 7–21 mm long; lateral spikes pistillate (or the upper androgynous), 2–4, short and stout, 5–14 mm long, sometimes compound, all sessile or short-pedunculate and crowded close to each other and to the staminate spike, or 1 or 2 of the lower ones more or less remote and more evidently pedunculate; one or more bracts with conspicuous, elongate, erect to spreading blade much surpassing the inflorescence, the bracts subtending the spikes in the terminal cluster sheathless or nearly so, those subtending the more remote spikes (when these are present) with a well developed sheath as well as a long blade; pistillate scales shorter than the perigynia, though sometimes longer than the body, largely scarious and medium-brown to stramineous except for the firmer, often light green midrib and narrow, ill-defined, more or less white-hyaline margins, or wholly pale and hyaline-scarious except for the midrib; perigynia 2.2–3.3 mm long, straight or nearly so, mostly spreading or the lower reflexed, the body pale greenish to stramineous or somewhat yellowish, obovoid, distended by the achene but distally empty and slightly inflated, irregularly several-nerved, or with 2 more or less evidently marginal nerves and 4–6 often less prominent nerves on each face, the abaxial face often less bulged and less prominently nerved than the adaxial one, the beak 0.8–1.2 mm long, slender, very shortly bidentate, not oblique at the tip; stigmas 3; achene trigonous, 1.1–1.5 mm long, jointed to the style; $n = 33$–36.

Wet, low ground, in a wide variety of soils, sometimes on sandy beaches or seashores, or in *Sphagnum* bogs, or in saline sites, from sea level to moderate elev. in the mts. (to 2750 m [9000 ft.] in s. Utah); circumboreal, extending s. in Amer. to N.J., Ind., n. N.M., s. Utah (Kane Co.), s. Nev., and nw. Calif.

The American and east-Asian plants, as described above, are weakly differentiated from typical *C. oederi* of n. Europe and may be distinguished as var. *viridula* (Michx.) Kükenthal. The var. *viridula* tends to have the ventral side of the bract-sheaths concave at the mouth, in contrast to typical material from Europe, in which at least the lower bracts tend to have the sheath truncate or upwardly prolonged on the ventral side at the mouth. Many of the European plants also have a slightly longer beak on the perigynium than does var. *viridula*. These differences are inconstant, however, and no specific distinction can properly be founded on them, especially in view of the fact that the American plants are so precisely like those from Europe in many other characters.

33. Carex amplifolia F. Boott

Carex amplifolia F. Boott in Hook. Fl. Boreali-Amer. 2: 228. 1839. (*Douglas*, Columbia River.)

Stems arising singly or few together from long, stout, creeping rhizomes, sharply triangular, coarse and stout, 5–10 dm tall, somewhat aphyllopodic; leaves flat, glabrous except for the scabrous margins, notably large, 8–20 mm wide and up to as much as 5 dm long, well distributed along the stem; spikes several, elongate, mostly (2) 4–10 cm long, not crowded, the terminal one staminate, the others pistillate, erect, subsessile or the lower on erect slender peduncles, the lower ones subtended by progressively larger leafy bracts which often surpass the staminate spike but are nearly or quite sheathless; pistillate scales usually with pale or greenish midstripe and darker, scarious margins, narrower and either shorter or longer than the perigynia, often shortly awn-tipped; perigynia glabrous, 2.6–3.3 mm long, including the 0.8–1.2 mm beak, this obliquely cleft or hyaline-erose, scarcely bidentate, the body strongly 2-nerved, thin-walled, slightly inflated but tending to collapse around the achene and become somewhat trigonous; stigmas 3; achene trigonous, 1.4–1.8 mm long, not wholly filling the perigynium.

Swamps, bogs, and other wet places, from the lowlands to moderate elev. in the mts.; s. B.C. to Calif., e. to Idaho (n. of the Snake River Plains) and the Steens Mts. of Oregon.

A distinctive, readily recognized species, which apparently has no synonyms.

34. Carex lanuginosa Michx.

Carex lanuginosa Michx. Fl. Boreali-Amer. 2: 175. 1803. *C. filiformis* var. *latifolia* Boeck. Linnaea 41: 309. 1877. *C. filiformis* var. *lanuginosa* Britton, Sterns, & Poggenb. Prelim. Cat. Antho. & Pterid. within 100 Mi. New York City 63. 1888. *C. lasiocarpa* var. *lanuginosa* Kükenthal in Engl. Pflanzenr. IV. 20 (Heft 38): 748. 1909. *C. lasiocarpa* subsp. *lanuginosa* Clausen & Wahl, Rhodora 41: 31. 1939. *C. lasiocarpa* var. *latifolia* Gilly, Iowa State Coll. J. Sci. 21: 125. 1946. (*Michaux*, Lake Mistassini, Que.)

Stems arising singly or few together from well developed creeping rhizomes, 3–10 dm tall, strongly aphyllopodic; leaves with well developed blades few, borne well above the base of the stem, elongate, flat or nearly so, 2–5 mm wide; terminal spike staminate, 2–5 cm long, often closely subtended by one or two shorter, sessile staminate spikes, these 1 cm long or more; pistillate spikes 2 or 3, 1–4 cm long, rather remote, sessile or the lowest one on an erect slender peduncle, leafy-bracteate, the lowest bract sheathless or with a sheath up to 1 cm long, the blade well developed, elongate, very often surpassing the staminate spikes; pistillate scales with a mostly 3-nerved, usually pale midstripe and brownish or occasionally more purplish hyaline-scarious margins, or seldom uniformly anthocyanic, usually narrower than the perigynia and acute or shortly awn-tipped; perigynia 3.3–5.0 mm long, densely velutinous or velutinous-sericeous, multi-ribbed, but the ribs often somewhat obscured by the pubescence, the body ovoid to broadly ellipsoid or subglobose, turgid, not much compressed, firm and thick-walled, abruptly con-

CAREX

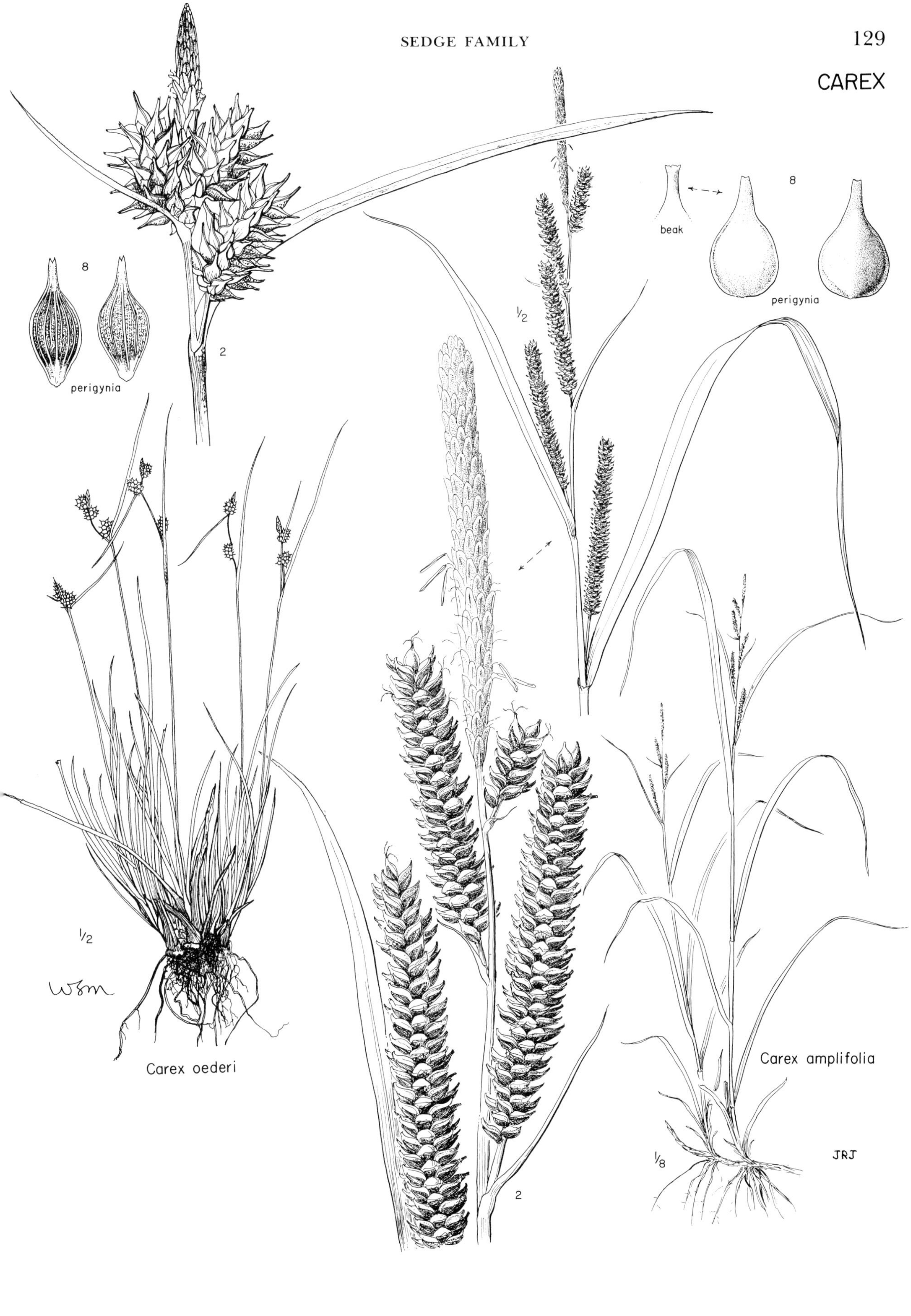
8
perigynia
2
1/2
wsm
Carex oederi
beak
8
perigynia
1/2
2
1/8
JRJ
Carex amplifolia

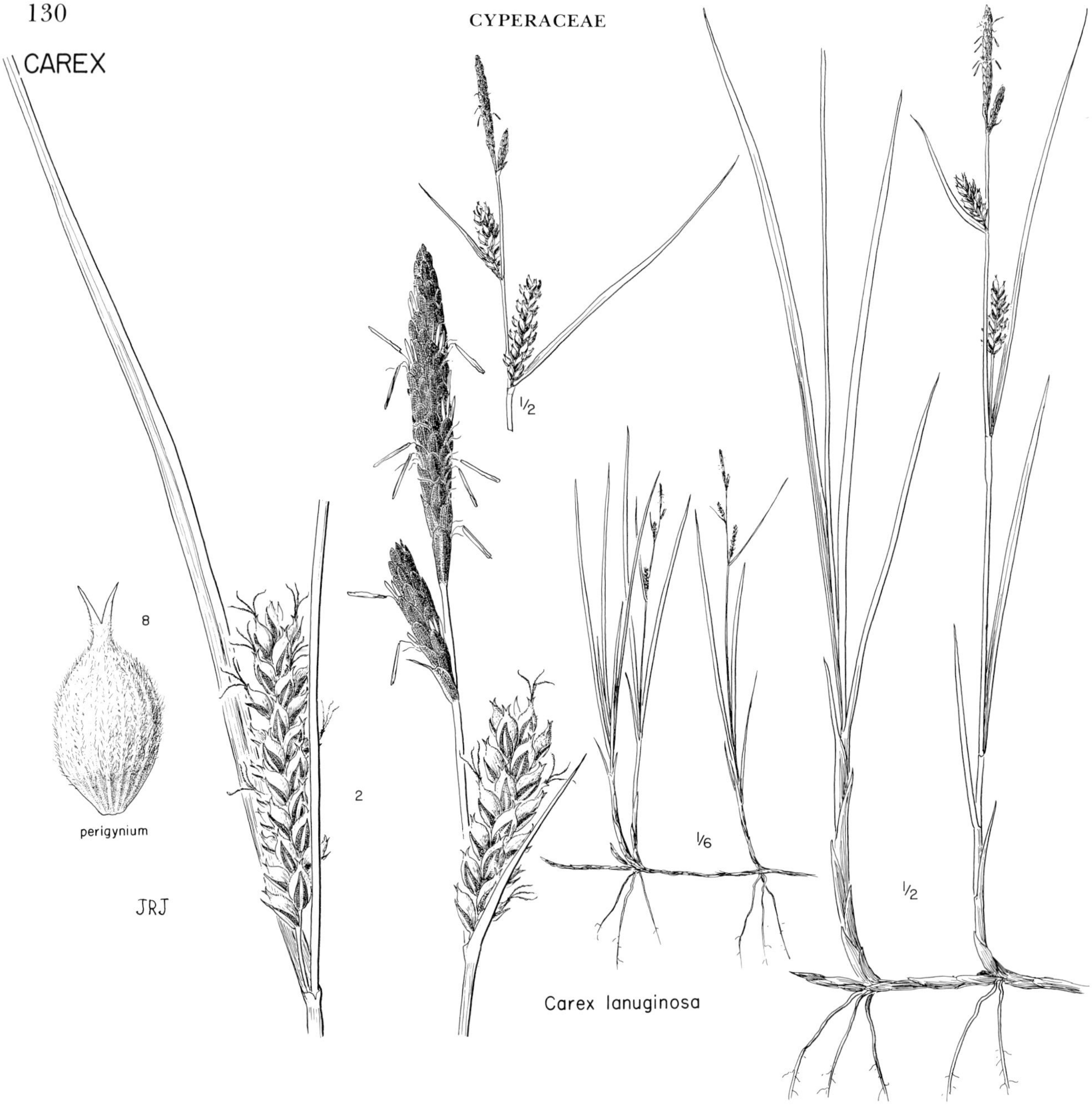

tracted to the beak, this 0.8–1.5 mm long, including the 0.3–0.9 mm teeth; stigmas 3; achene trigonous, 1.5–2.0 mm long, only loosely filling the perigynium; $n = 39$.

Wet meadows and other wet places, sometimes in shallow water, from the lowlands to moderate elev. in the mts.; N.B., and Que. to B.C., s. to Tenn., Ark., Texas, and s. Calif., occurring in appropriate habitats essentially throughout our range.

The species is palatable to livestock.

35. **Carex parryana** Dewey

Carex parryana Dewey, Amer. J. Sci. 27: 239. 1835. (*Richardson*, Hudson's Bay; in the present province of Manitoba.)
C. hallii Olney in F. V. Hayden, Rep. U.S. Geol. Surv. Territories 5: 496. 1872. *C. parryana* var. *unica* L. H. Bailey, Mem. Torrey Bot. Club 1: 54. 1889. *C. parryana* var. *hallii* Kükenthal in Engl. Pflanzenr. IV. 20 (Heft 38): 388. 1909. *C. parryana* subsp. *hallii* D. F. Murray, Brittonia 21: 71. 1969. (*Hayden s.n.*, Pleasant Valley, in e. Idaho; lectotype by Mackenzie.)
C. idahoa L. H. Bailey, Bot. Gaz. 21: 5. 1896. *C. parryana* subsp. *idahoa* D. F. Murray, Brittonia 21: 73. 1969. (*Rydberg 2339*, Beaver Canyon, Clark Co., Idaho.) A form with relatively long, narrow, and acute pistillate scales.
C. elrodi M. E. Jones, Bull. Montana State Univ., Biol. Ser. 15: 70. 1910. (*Jones s.n.*, Monida, Mont., 8 July 1909.)
C. parryana var. *statoni* M. E. Jones, op. cit. p. 72. (*Jones s.n.*, Deer Lodge Valley, Mont.)
C. parryana var. *brevisquama* F. J. Hermann, Rhodora 70: 420. 1968. (Typification recondite.)

Plants loosely tufted and with short, scaly, creeping rhizomes, (1) 2–6 dm tall, phyllopodic, the principal leaves crowded near the base, firm, flat to sometimes channeled and revolute-margined, up to about 4 mm wide, much shorter than the culms, sel-

CAREX
2
24
8
perigynium
24
1/2
JRJ
Carex parryana
2
8
perigynia
1/2
Carex raynoldsii

dom over 15 (25) cm long; spikes 1–5, cylindric, the terminal one (1) 1.5–3 cm long, often longer than the lateral ones, which are sessile or stiffly short-pedunculate and erect or nearly so and neither closely crowded nor notably remote; spikes often all wholly pistillate, or the terminal one frequently staminate, or gynaecandrous, or with intermingled ♂ and ♀ flowers; bract subtending the lowest spike relatively short, inconspicuous, and scarcely sheathing at the base, rarely as long as the inflorescence; pistillate scales stramineous to more often brown with pale, hyaline margins, broadly obtuse and shorter than the perigynia to narrower, acute, and surpassing the perigynia; perigynia more or less distinctly obovate, often broadly so, 1.9–3.0 mm long, including the short (0.2–0.6 mm) or obsolete beak, wholly glabrous, or often shortly strigose-hirtellous or inconspicuously scabrous distally, especially about the base of the beak; marginal nerves of the perigynium prominent, the faces nerveless or inconspicuously several-nerved; stigmas 3; achene trigonous, 1.4–1.8 mm long, rather loosely filling the perigynium.

Meadows, swales, and moist low ground around streams and lakes, on the prairies and high plains, extending upward to moderate elev. in the mts.; ne. Man. to s. Mack. and ne. B.C., s. to Colo., chiefly e. of the continental divide, but extending onto the Pacific slope in c. and e. Idaho and n. Utah (Cache Co.) and apparently isolated in c. Nev. (Monitor Valley, Nye Co.); reported also from sw. Yukon and s. Alaska.

36. Carex raynoldsii Dewey

Carex raynoldsii Dewey, Amer. J. Sci. 82: 39. 1861. (*Hayden s.n.,* Pierre's Hole, valley of the Snake River, Fremont Co., Idaho, 20 June 1860; lectotype by D. F. Murray.)
C. lyallii F. Boott, Ill. Gen. Carex 150, pl. 483. 1867. (*Lyall,* e. side of the Cascade Mts. at 49 degrees North Latitude.)

Stems rather loosely tufted on short, stout rhizomes, mostly 2–7 dm tall, phyllopodic or only slightly aphyllopodic, and the old leaf-bases from previous years tending to be persistent; leaves flat, 2–7 mm wide, more or less elongate and sometimes nearly equaling the stem, the plant appearing much leafier than *C. parryana;* spikes more or less approximate, sometimes almost capitate-crowded, the terminal one staminate, mostly 1–1.5 cm long, the lateral ones 2–5, pistillate, mostly 1–2 cm long or the uppermost ones shorter, all sessile or nearly so and erect or closely ascending, or the lower ones often slender-pedunculate (the peduncle up to as much as 3 cm long) but still erect, seldom one spike well removed from the others and long-pedunculate from the axil of a leaf with a well developed sheath borne below the middle of the stem; bract subtending the lowest spike of the terminal cluster sheathless or nearly so, shorter than or about equaling the inflorescence; pistillate scales conspicuously dark, black to purplish-black or very deep reddish-brown, acute, varying to obtuse or acuminate, tending to be shorter and (at least distally) narrower than the perigynia, which are usually pale greenish, so that the spike is bicolored; perigynia elliptic or elliptic-obovate, somewhat inflated, especially distally, rather firm-walled, glabrous, (3.0) 3.3–4.4 mm long, including the abrupt, short beak which is about 0.5 mm long or less and only shortly or scarcely bidentate, the body coarsely several-nerved on each face, as well as with 2 marginal nerves, or the facial nerves sometimes obscure; stigmas 3; style deciduous; achene trigonous, 1.8–2.4 mm long, the upper part of the perigynium empty; $n = 29$.

Moist or dry meadows and open or wooded slopes, from the foothills to fairly high elev. in the mts., but seldom above timberline; Alta. and B.C. to Colo., Utah, Nev., and the Sierra Nevada of Calif.

The species is palatable to livestock, but sensitive to grazing.

37. Carex paysonis Clokey

Carex paysonis Clokey, Amer. J. Sci. 203: 89. 1922. (*Payson 2224,* Grand Teton, Wyo.)

Culms stiffly erect, 1–5 dm tall, borne singly or few together on rather coarse, creeping rhizomes, somewhat aphyllopodic, but the leaves still clustered near the base, firm, flat or nearly so, 2–8 mm wide, the old sheaths more or less persistent and becoming fibrous; spikes mostly 3–5, stout, (1) 1.5–2.5 cm long, approximate or the lowest one somewhat remote, all sessile or subsessile and more or less erect, or the lower ones evidently pedunculate and sometimes nodding, the terminal one staminate or androgynous, the lateral ones pistillate or some of the upper ones androgynous; bract subtending the lowest spike sheathless or nearly so, shorter than or about equaling the inflorescence; pistillate scales dark reddish brown to purplish black, about equaling or somewhat shorter (and commonly also narrower) than the perigynia; perigynia pale or more often partly or wholly purplish, broadly elliptic to elliptic-ovate or subrotund, mostly 3.3–4.3 mm long and well over half as long as wide, abruptly contracted to the short (0.2–0.5 mm), shallowly or scarcely bidentate beak, glabrous (though tending to be finely papillate), flattened, the "marginal" nerves distinctly inframarginal on the abaxial side, the actual margins nerveless, the surfaces otherwise nerveless or lightly few-nerved; stigmas 3; achene trigonous, much narrower and shorter than the perigynium, mostly 1.2–1.7 mm long, sessile or nearly so.

Moist meadows and open often rocky slopes at high alt. in the mts., often above timberline; sw. Alta. to w. Wyo. and ne. Utah (Uinta Mts.), w. across c. Idaho to ne. Oregon (Wallowa and Elkhorn mts.), and in the Snake Range of e. Nev.

This species has generally passed as *C. podocarpa* R. Br., or *C. tolmiei* F. Boott. The type of *C. tolmiei* is clearly referable to *C. spectabilis* Dewey. The true identity of *C. podocarpa* is more doubtful, and the name has been applied at various times by different authors to several other species as well as to this one. The holotype of *C. podocarpa,* which I have seen, clearly does not belong with *C. paysonis;* it may prove to belong instead with *C. spectabilis* Dewey, a northwestern species that does not reach our range.

38. Carex norvegica Retz.

Carex norvegica Retz. Fl. Scand. Prodr. 179. 1779. (Norway.)
C. halleri Gunnerus, Fl. Norveg. 2: 106. 1772, in part only but not as to the type.
C. alpina Liljeblad, Utkast Svensk Fl. ed. 2. 26. 1798. (Lappland.) Not *C. alpina* Schrank, 1787.
C. vahlii Schkuhr, Beschreibung & Abbild. Arten Riedgräsern 87. 1801. (Norway.)
C. alpina var. *inferalpina* Wahlenb. Fl. Lappon. 241. 1812. *C.*

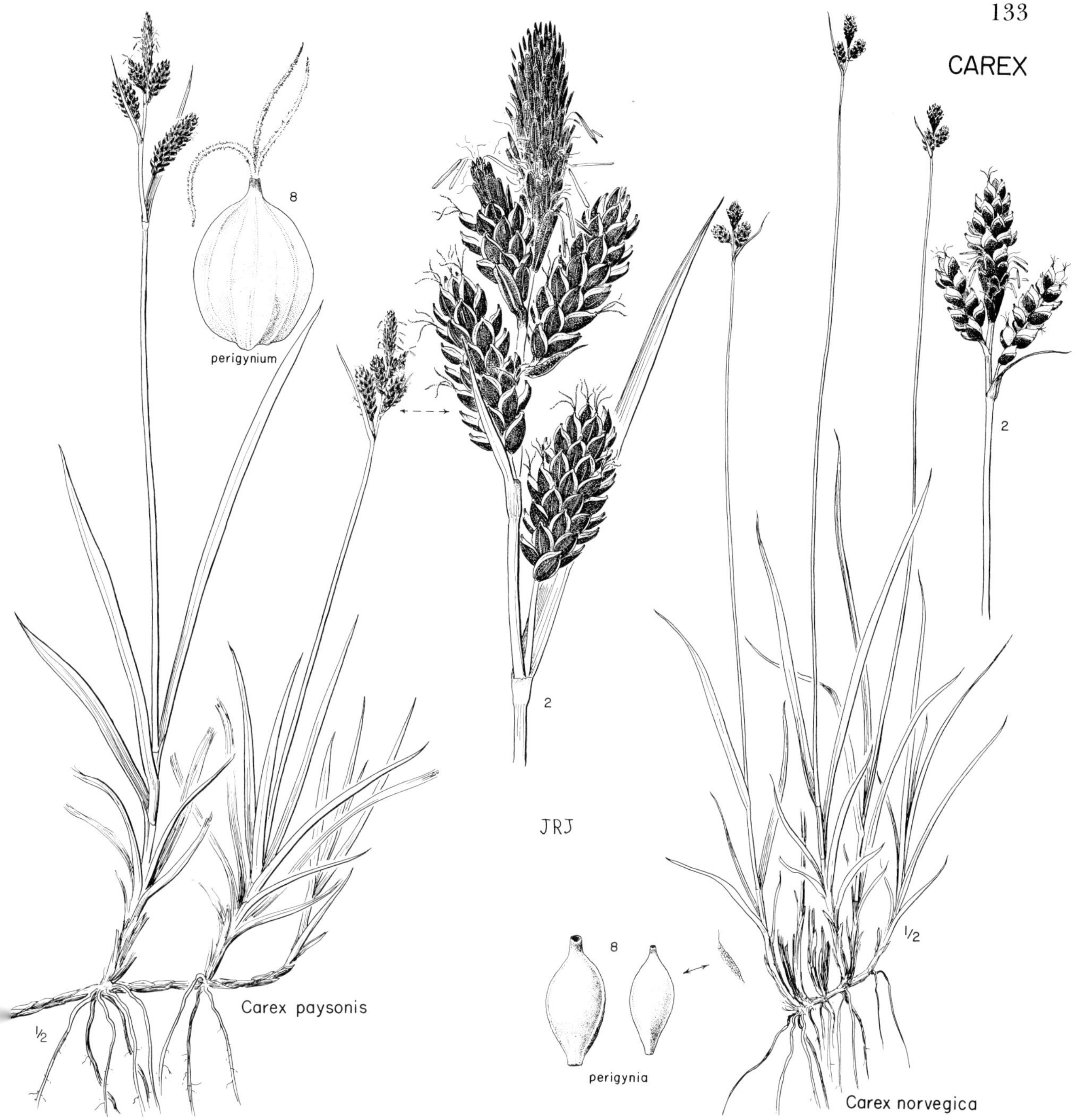

alpina f. *inferalpina* Kükenthal in Engl. Pflanzenr. IV. 20 (Heft 38): 386. 1909. *C. vahlii* var. *inferalpina* Fern. Rhodora 35: 398. 1933. *C. norvegica* subsp. *inferalpina* Hultén, Fl. Alaska & Yukon 2: 348. 1942. *C. norvegica* var. *inferalpina* B. Boivin, Naturaliste Canad. 94: 523. 1967. (Lappland.)

C. media R. Br. in Franklin, Narr. Journey Shores Polar Sea, Bot. Appendix 750. 1823. (*Richardson,* between 54 and 64 degrees North Latitude in nw. Can.)

C. angarae Steudel, Syn. Pl. Glum. 2: 190. 1855. (Angara and Lena rivers in Siberia.)

C. alpina var. *stevenii* T. H. Holm, Amer. J. Sci. 166: 27. 1903. *C. vahlii* var. *stevenii* Fern. Rhodora 35: 398. 1933. *C. angarae* var. *stevenii* A. E. Porsild, Rhodora 41: 204. 1939. *C. media* var. *stevenii* Fern. Rhodora 44: 303. 1942. *C. stevenii* Kalela, Ann. Bot. Soc. Zool.-Bot. Fenn. "Vanamo" 19(3): 11. 1944. *C. norvegica* subsp. *stevenii* D. F. Murray, Brittonia 21: 74. 1969. (*Holm s.n.,* Lamb's Ranch on Long's Peak, 13 Aug. 1899; lectotype by D. F. Murray.)

Culms loosely to densely tufted on a compact system of branched, rather slender rhizomes, somewhat aphyllopodic, slender and rather lax, (1) 2–7 dm tall; leaves glabrous, flat or nearly so, soft, mostly 1.5–3 mm wide, some of them generally distinctly cauline; spikes mostly 2–5, approximate, erect or closely ascending, sessile or short-pedunculate, relatively small, the terminal one gynaecandrous, 6–14 mm long, the lateral ones pistillate, shorter, seldom over 10 mm long; bract subtending the lowest spike sheathless or nearly so, from distinctly shorter to slightly longer than the inflorescence; pistillate scales purplish black or brownish black, prominently white-hyaline-margined, equaling or distinctly shorter than the perigynia, usually fully as wide as the perigynia, seldom somewhat narrower; perigynia elliptic, commonly rather narrowly so, to more or less obovate, 2.1–3.0 mm long, including the short (0.3–0.4 mm) generally blackish beak, distended and nearly filled by the achene, empty only just beneath the beak, coppery yellowish or light green to sometimes dark purple, with well developed marginal nerves, otherwise nerveless or inconspicuously few-nerved, glabrous but minutely rough-textured, the cellular reticulum more or less evident at 10 diam. magnification; stigmas 3; achene trigonous, sessile, 1.3–1.7 mm long; n = 27, 28.

Streambanks, seepage areas, and moist meadows, at moderate to high elev. in the mts., sometimes above timberline; circumboreal, extending s. in Amer. to Que., Wisc., Minn., N.M., Idaho, and Utah (Garfield Co.) and B.C., but missing from Nev. and the Pacific states.

39. Carex nova L. H. Bailey

Carex nova L. H. Bailey, J. Bot. 26: 322. 1888. (*Patterson 22*, mts. about the headquarters of Clear Creek, Colo.; lectotype by D. F. Murray.)

C. nelsonii Mackenzie in Rydb. Fl. Rocky Mts. 137, 1060. 1917 [1918]. (*A. Nelson & E. Nelson 5264*, La Plata Mines, Albany Co., Wyo.) An extreme, perhaps distinctive phase, with narrow perigynia not more than half as wide as long (beak included) and with more elongate beak up to 0.8 mm (reputedly 1.0 mm) long. The type and some similar plants come from the mts. of Wyo., but Murray (1969) considers that *C. nelsonii* (which he treats as a distinct sp.) occurs also in Colo., the Uinta Mts. of Utah, and perhaps elsewhere. The Utah specimen (*Maguire et al. 14467*) cited by Murray to validate the stated range is less extreme than the type of *C. nelsonii* and in my opinion should in any case remain within *C. nova* as here understood.

C. pelocarpa F. J. Hermann, Rhodora 39: 492. 1937. (*Hermann 5983*, Lamotte Peak, Summit Co., Utah.) The type specimen is of ordinary appearance and falls well within the range of variability of the population of *C. nova*.

Culms (1) 2–6 dm tall, more or less densely tufted on a system of short, branched, sometimes scarcely evident rhizomes, only slightly or scarcely aphyllopodic, the basal sheaths tending to be anthocyanic, those from previous years tending to be persistent; leaves short or elongate, flat, mostly 2–5 mm wide, tending to be clustered at the base and often much shorter than the culms; inflorescence a densely capitate terminal cluster of 3–5 (6) short, broad spikes which are so closely crowded that the lateral ones are rather widely spreading and difficult to differentiate, toward the base, from each other and from the terminal spike; bract subtending the lowest spike inconspicuous, slender and hardly leaflike, sheathless or nearly so, mostly shorter than the inflorescence; terminal spike gynaecandrous, with relatively few staminate flowers, commonly about 1 cm long or a little more; lateral spikes pistillate, shorter, rarely over 1 cm long; pistillate scales dark brown to purplish-black, not markedly pale-margined, equaling or shorter than the perigynia, about as wide as the perigynia, or sometimes distinctly narrower distally; perigynia elliptic (commonly rather broadly so) to elliptic-obovate or even subrotund, seldom less than half as wide as long, more or less strongly flattened, or seldom subinflated distally, commonly with a few scabrous points on the distal margins near the beak, but these not rarely obscure or wanting, otherwise glabrous (though sometimes minutely papillate), nerveless except for the 2 marginal nerves, mostly 2.8–4.5 mm long, including the black or blackish, shallowly or obscurely bidentate beak, this 0.3–0.6 (0.8) mm long, the body of the perigynium sometimes wholly stramineous or pale greenish, but more often dark purplish or brownish except for the pale margins; stigmas 3; achene trigonous, 1.5–1.8 mm long, much smaller than the perigynium, tending to be evidently stipitate, seldom sessile or nearly so.

Streambanks, meadows and moist open slopes at high alt. in the mts., often above timberline; Colo. and n. N.M., w. to Utah and Elko Co., Nev., and n. to Wyo., w. Mont., and c. Idaho.

The species is reported to be a good forage plant for livestock.

Carex nova does not stand out very sharply from *C. albonigra* and *C. atrata* on the basis of technical characters, but the distinction in nature appears to be sharp enough.

40. Carex helleri Mackenzie

Carex helleri Mackenzie, Erythea 8: 70. 1922. (*Heller 9975*, Mt. Rose, Washoe Co., Nev.)

Similar to *C. albonigra;* leaves averaging narrower, mostly 1.5–3 mm wide, and often longer, frequently 15 cm long or more; pistillate scales long and narrow, evidently surpassing the perigynia, but leaving the margins exposed, acuminate or tapering to a short awn-tip, purplish-black or blackish-brown, usually with paler (commonly greenish) midrib, without white-hyaline margins; perigynia strongly flattened, sometimes lightly nerved on the faces, usually with pale (often greenish) margins and darker center, and with a smoother surface texture than *C. albonigra*.

Open, often rocky places at high alt. in the mts., commonly above timberline; Sierra Nevada of Calif. and adj. Nev., Sweetwater Mts. of Calif., and the White Mts. along the Calif.-Nev. border; also in the Toquima Mts. of c. Nev., and to be expected in some other Nev. ranges.

Carex helleri is scarcely more different from *C. albonigra* than the varieties of *C. atrata* are from each other, but broad-scale populational intergradation between these two dwarf alpine types remains to be demonstrated, whereas the varieties of *C. atrata* are wholly confluent. The specimens of *C. helleri* from c. Nevada have the pistillate scales less conspicuously elongate than typical materials from the Sierra Nevada, and look very much like a mere alpine ecotype of *C. atrata* var. *erecta*. The taxonomy of this complex remains to be clarified.

41. Carex albonigra Mackenzie

Carex albonigra Mackenzie in Rydb. Fl. Rocky Mts. 137, 1060. 1917 [1918]. (*Cary 613*, Needle Mt., Wyo.)

Culms 1–3 dm tall, rather loosely tufted on short rhizomes, only slightly aphyllopodic, the remains of old leaves more or less persistent at the base; sheaths tending to be anthocyanic; leaves clustered at or near the base, firm and rather short, seldom much over 10 cm long, flat, mostly 2–7 mm wide; bract subtending the lowest spike sheathless or nearly so,

CAREX

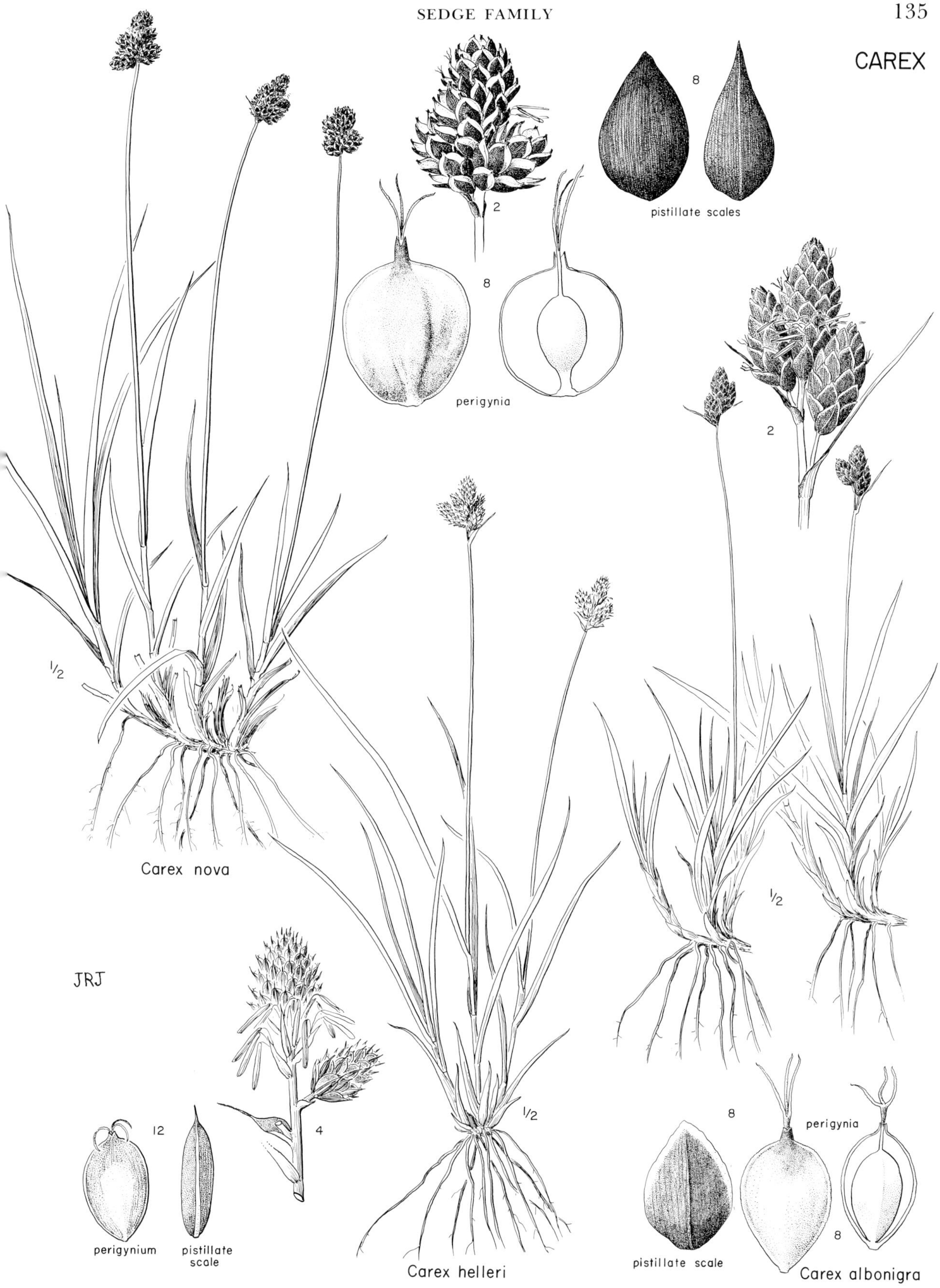
2
8
pistillate scales
8
perigynia
2
1/2
Carex nova
JRJ
12
4
perigynium
pistillate scale
1/2
Carex helleri
1/2
8
perigynia
pistillate scale
8
Carex albonigra

shorter than or about equaling the inflorescence; spikes mostly 2–4, close aggregated into a headlike inflorescence, but still individually distinguishable and rather closely ascending instead of spreading, the terminal one gynaecandrous, 1–1.5 (2) cm long, the lateral ones pistillate, smaller, commonly about 1 cm long, sessile or inconspicuously short-pedunculate; pistillate scales about equaling or somewhat shorter (but not notably narrower) than the perigynia, dark reddish-brown or blackish-purple, with more or less prominent, pale, hyaline margins; perigynia usually dark like the scales, flattened, elliptic-obovate, often rather broadly so, 2.7–3.6 mm long, including the minute (0.2–0.4 mm) only obscurely bidentate beak, nerveless except for the 2 marginal nerves, minutely papillate-roughened, sometimes only obscurely so, otherwise glabrous; stigmas 3; achene trigonous, sessile, 1.5–1.9 mm long, distinctly shorter and also narrower than the perigynium; *n* = 26.

Open, often dry or rocky slopes and summits at high alt. in the mts., commonly above timberline; Alaska, Yukon and Mack., s. to Wash. (Cascade and Olympic mts.), Idaho, Colo., Utah, Nev. (Elko and White Pine cos.), and Ariz. (San Francisco Mt.); rarely also in the Sierra Nevada and White Mts. of Calif.

Carex albonigra is obviously related to *C. atrata,* of which it might with some reason be considered an alpine ecotype. However, it spans the range of three different varieties of *C. atrata,* without itself being divisible. Clearly it is not just the dwarf alpine phase of each of these three geographic varieties. It therefore seems advisable to retain *C. albonigra* in specific status, despite the meagerness of the distinction from *C. atrata.*

42. Carex atrata L.

Carex atrata L. Sp. Pl. 976. 1753. *Trasus atratus* S. F. Gray, Nat. Arr. Brit. Pl. 2: 56. 1821. ("in Alpibus Europae.")

C. atrata var. *erecta* W. Boott in S. Wats. Bot. Calif. 2: 239. 1880. (Calif.)

C. heteroneura W. Boott, loc. cit. (*Kellogg,* Lake Tahoe to Bear Valley, Sierra Nevada of Calif.) = var. *erecta.*

C. chalciolepis T. H. Holm, Amer. J. Sci. 166: 28. 1903. *C. atrata* var. *chalciolepis* Kükenthal in Engl. Pflanzenr. IV. 20 (Heft 38): 399. 1909. *C. heteroneura* var. *chalciolepis* F. J. Hermann, Rhodora 70: 421. 1968. (*Holm s.n.,* Mt. Kelso, Colo., 7 Sept. 1899; lectotype by D. F. Murray.)

C. epapillosa Mackenzie in Rydb. Fl. Rocky Mts. 138, 1060. 1917 [1918]. *C. heteroneura* var. *epapillosa* F. J. Hermann, Rhodora 70: 421. 1968. (*M. E. Jones 5345,* Marysvale, Utah.) = var. *erecta.*

C. heteroneura var. *brevisquama* F. J. Hermann, Rhodora 70: 421. 1968. Without Latin, and untypified. Name intended to apply to American plants previously referred to *C. atrata* sensu stricto.

Plants in dense, small to large tufts, without creeping rhizomes, 1.5–8 dm tall, more or less distinctly aphyllopodic, the basal sheaths or scales often somewhat anthocyanic; leaves glabrous, flat or nearly so, mostly 2–7 mm wide; bract subtending the lowest pistillate spike sheathless or nearly so, about equaling or often somewhat shorter than the inflorescence; spikes mostly 2–5, 1–2.5 (3) cm long, approximate or somewhat crowded, but not densely capitate-clustered, all erect or closely ascending, sessile or subsessile or the lowest one often on a stiff, erect peduncle that may be longer than the spike; terminal spike gynaecandrous, sometimes only a few of the terminal flowers pistillate; lateral spikes pistillate, or gynaecandrous with a few staminate flowers at the base; pistillate scales dark reddish-brown to almost black, longer or shorter than the perigynia, generally rather narrow, the distal margins of the perigynia commonly exposed; perigynia more or less strongly flattened, often with pale green margins and dark center, glabrous but often more or less papillate-roughened, and with a short, slender, scarcely cleft beak; stigmas 3; style deciduous; achene trigonous, tending to be shortly stipitate, 1.2–1.8 mm long, reaching only to within about 0.6–1.5 mm of the tip of the perigynial cavity, often also much narrower than the perigynium; *n* = 24–28.

Wet meadows to open, rather dry slopes, at moderate to high elev. in the mts., sometimes above timberline; circumboreal; extending s. in Amer. to Vt., Mich., Colo., n. Ariz., and Calif.

The species is reported to be a good forage plant for livestock.

Carex atrata consists of a number of geographical varieties, at least four of which occur in North America, only two in our range. The var. *erecta* W. Boott, occurring from the Sierra Nevada of California, e. to the Wasatch and Uinta regions of Utah, and n. to nw. Montana and s. Washington, has the perigynia 3.0–4.2 mm long, half to seven-eighths as long as wide, broadly rounded distally and abruptly beaked, the scales not obviously longer than the perigynia. The var. *chalciolepis* (T. H. Holm) Kükenthal, of the southern Rocky Mt. region from Colorado to s. Wyoming, se. Utah (LaSal and Abajo mts.) and n. Arizona, has relatively long, narrow pistillate scales evidently surpassing the perigynia, and the perigynia average a little larger (to 5 mm) and are often more tapered distally. The whole inflorescence (but not the individual spikes) is often nodding in this var. Many of the plants from the Wasatch region of Utah, and southward in the adjacent high plateaus, are more or less intermediate between var. *erecta* and var. *chalciolepis,* but are here arbitrarily referred to var. *erecta.* Some of the plants from the Tushar Mts. (Beaver Co., Utah) are perhaps best referred to var. *chalciolepis.*

43. Carex bella L. H. Bailey

Carex bella L. H. Bailey, Bot. Gaz. 17: 152. 1892. *C. atrata* var. *discolor* L. H. Bailey, J. Bot. 26: 321. 1888. (*Greene 4,* San Francisco Mts., Ariz.; lectotype by D. F. Murray.)

Resembling *C. atrata,* but apparently sharply distinct; lateral spikes more often gynaecandrous; basal sheaths seldom evidently anthocyanic; inflorescence looser, the spikes more or less approximate but not crowded, averaging longer (often to 3 cm) and more slender, at least the lowest ones nodding on slender peduncles which often exceed the spikes; pistillate scales dark reddish brown to blackish brown, usually hyaline-margined and often with pale or greenish midrib, equaling or somewhat shorter than the pale, greenish-stramineous perigynia, at least the distal margins of which are evidently exposed; perigynia obovate, 2.8–3.6 mm long, including the short (0.2–0.5 mm), scarcely bifid beak, from a little less to a little more than half as wide as long, nerveless except for the two marginal nerves, distended by the achene, which is trigonous, 1.8–2.4 mm long, and reaches to within 0.2–0.7 mm of the tip of the perigynial cavity, the achene thus both relatively and absolutely larger than that of *C. atrata,* and more nearly filling the perigynium; *n* = 20.

Streambanks, meadows, and moist woods or open slopes at high alt. in the mts., often above timberline; high mts. of Ariz. and N.M., n. to c. Utah and n. Colo., and reported to be disjunct in the Uinta Mts. of Utah, the Black Hills of S.D., and in Nuevo Leon, Mex.

CAREX

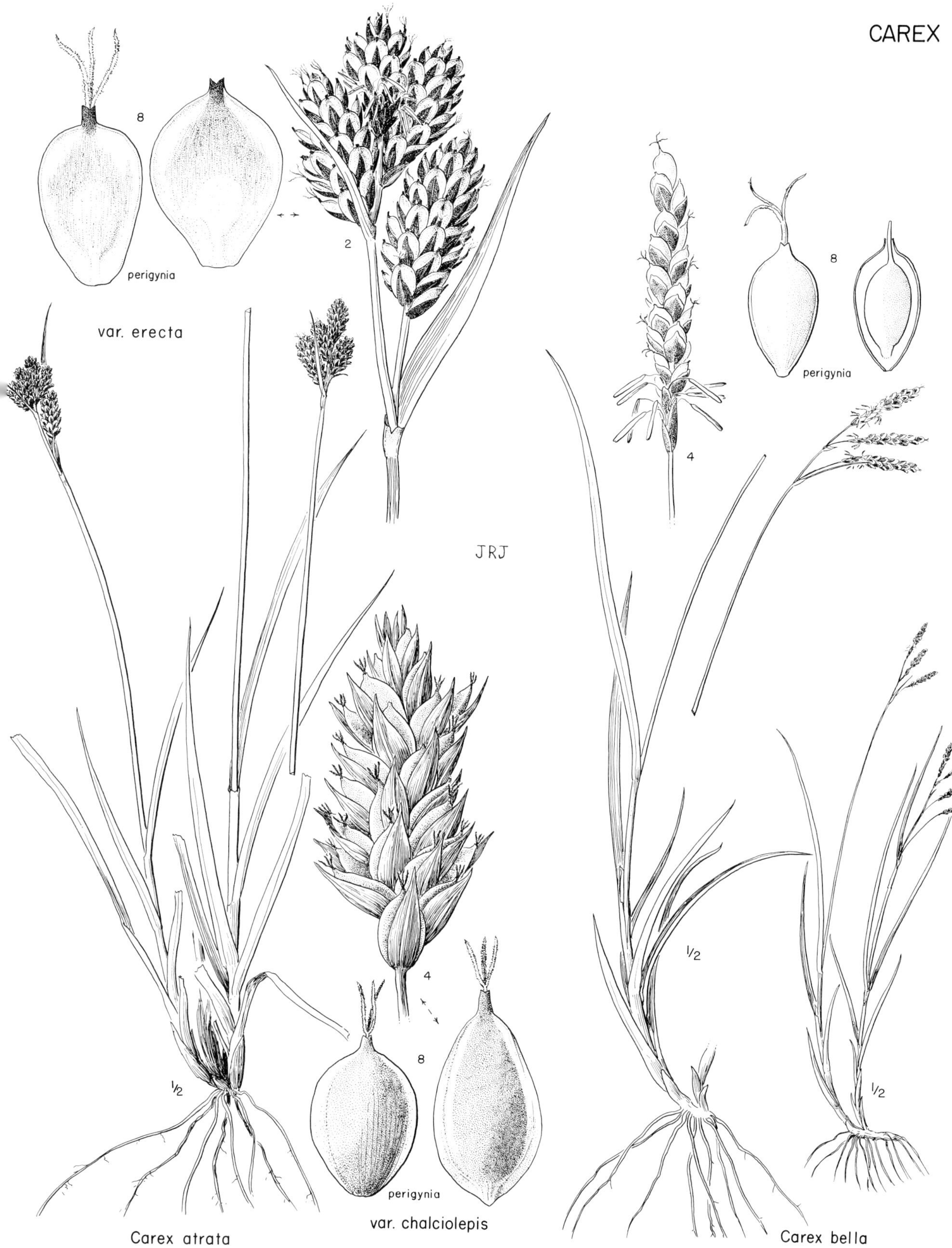
8
perigynia
2
var. erecta
4
8
perigynia
JRJ
4
8
perigynia
1/2
1/2
1/2
Carex atrata
var. chalciolepis
Carex bella

44. Carex buxbaumii Wahlenb.

Carex buxbaumii Wahlenb. Kongl. Vetensk. Acad. Nya Handl. 24: 163. 1803. *Physiglochis buxbaumii* Raf. Good Book 27. 1840. (Sweden and Lappland.)

C. holmiana Mackenzie, Bull. Torrey Bot. Club 36: 481. 1909. (*Vreeland 1121,* John's Lake, n. Mont.)

Culms arising singly or few together from well developed creeping rhizomes, mostly 3–10 dm tall, strongly aphyllopodic, not surrounded by old sheaths from previous years (though these are often persistent separately from the new stems); leaves glabrous, elongate, mostly 2–4 mm wide; spikes mostly 2–5, approximate or somewhat remote, erect or closely ascending, sessile or (especially the lower) with more or less well developed peduncle, the terminal spike gynaecandrous, 1–3 cm long, the lateral ones pistillate, about the same length or somewhat shorter; bract subtending the lowest spike sheathless or nearly so, from distinctly shorter to somewhat longer than the inflorescence; pistillate scales lanceolate to lance-ovate, brown to purplish black with a usually paler midrib, surpassing the perigynia, tapering to an awn-tip 0.5–3 mm long; perigynia 2.7–4.3 mm long, beakless or very shortly beaked, rather narrowly elliptic to sometimes elliptic-obovate or elliptic-ovate, up to barely over half as wide as long, firm-walled, not strongly flattened, light gray-green, densely papillate all over, with prominent marginal nerves and 6–8 inconspicuous or obscure nerves on each face; stigmas 3; achene trigonous, 1.4–1.9 mm long, somewhat narrower and much shorter than the perigynial cavity; n = 37, ca 50.

Peat bogs, marshes, wet meadows, and other wet places, from near sea level on the northwest coast to moderate elev. in the mts.; widespread, but relatively uncommon and infrequently collected; circumboreal, s. in Amer. to N.C., Ky., Ark., Colo., Utah (Uinta Mts.), and c. Calif.; not recorded from Nev.

Carex buxbaumii is a well marked and wholly distinct species. It apparently requires a moist atmosphere as well as wet feet, but even where these conditions are met its occurrence is irregular.

45. Carex limosa L.

Carex limosa L. Sp. Pl. 977. 1753. *Trasus limosus* S. F. Gray, Nat. Arr. Brit. Pl. 2: 67. 1821. *Facolos limosa* Raf. Good Book 26. 1840. (n. Europe.)

Stems arising singly or few together from long, creeping rhizomes, mostly 2–6 dm tall, strongly aphyllopodic; roots covered with a yellowish brown, felty tomentum; leaves few, 1–2 mm wide, tending to be canaliculate rather than truly flat; staminate spike solitary, terminal, 1–3 cm long; pistillate spikes 1–3, not crowded, nodding on slender peduncles, occasionally with a few staminate flowers at the tip, 1–2.5 cm long, the lowest usually subtended by a leafy bract 2–10 cm long, or this occasionally shorter and setaceous, in any case sheathless or with only a short sheath up to about 4 mm long; bracts of the other pistillate spikes more or less reduced; pistillate scales light to dark brown (not black), variable in size and shape, but most commonly about as long and wide as the perigynia, obtuse to narrowly acute or acuminate, sometimes short-cuspidate; perigynia pale, commonly greenish or stramineous, glabrous but densely papillate, elliptic to ovate, somewhat compressed, with prominent marginal nerves and 4–7 more or less evident nerves on each face, 2.3–4.2 mm long, beakless or with a very short beak up to about 0.2 mm long; stigmas 3; achene trigonous, 1.5–2.2 mm long, rather loosely enveloped by the perigynium, the upper one-eighth to one-third of which is empty; n = 28, 31, 32.

Sphagnum bogs; circumboreal, s. at increasing elev. to Mont., Idaho, Oregon, and the Sierra Nevada of Calif.; isolated stations in the mts. of Elko Co., Nev., and the Uinta Mts. of Utah.

46. Carex paupercula Michx.

Carex paupercula Michx. Fl. Boreali-Amer. 2: 172. 1803. (*Michaux,* Lake Mistassini, Que., in 1792.)

Stems 1.5–7 dm tall, loosely clustered in small tufts on short or long rhizomes, phyllopodic, and the remains of old leaves commonly persistent around the base; roots covered with a yellowish brown, felty tomentum; leaves flat, 1–3 mm wide, glabrous; staminate spike solitary, terminal, 0.7–1.5 cm long; pistillate spikes 1–4, not crowded, nodding on slender peduncles, often with a few staminate flowers at the base, 7–15 mm long, the lowest one subtended by a leafy bract 2–10 cm long, this sheathless or with only a short sheath up to about 4 mm long; bracts of the other pistillate spikes more or less reduced; pistillate scales light to dark brown, often with green midstripe, generally longer and narrower than the perigynia and tapering to a long, narrow point, sometimes with a short (to 1 mm) awn; perigynia pale, commonly greenish or stramineous, glabrous but densely papillate, elliptic to ovate, somewhat compressed, with prominent marginal nerves and 3–6 evident to obscure nerves on each face, 2.2–3.1 (3.8) mm long, beakless or with a very short beak only 0.1 mm long; stigmas 3; achene trigonous, 1.4–1.9 mm long, rather loosely enveloped in the perigynium, the upper one-eighth to one-fifth of which is empty; n = 29, ca 30.

Sphagnum bogs; circumboreal, s. at increasing elev. to Colo., Utah (Uinta Mts.), n. Idaho, and reputedly Wash.

Carex paupercula is evidently allied to *C. limosa.* The two maintain their separate identities, however, over a large area in which they occur in similar habitats, and the limits of their ranges are not the same. In spite of some problems in the herbarium, it appears that the two are sharply distinct in nature. *Carex magellanica* Lam. (1791), of s. South America, is even more closely allied to *C. paupercula* (1803), and it is possible that further study will require the union of these two under Lamarck's name.

47. Carex lenticularis Michx.

Carex lenticularis Michx. Fl. Boreali-Amer. 2: 172. 1803. ("per tractus montium, a sinu Hudsonis ad Canadam, praesertim ad lacum Cycnorum dictum.")

C. kelloggii W. Boott in S. Wats. Bot. Calif. 2: 240. 1880. (*Kellogg,* Sierra Nevada of California, is the first specimen cited.) = var. *lenticularis.*

C. interrupta var. *impressa* L. H. Bailey, Mem. Torrey Bot. Club 1: 18. 1889. *C. paucicostata* Mackenzie, Erythea 8: 74. 1922. (*Kellogg,* Summit Camp, Calif., is the first specimen cited.) = var. *lenticularis.*

C. lenticularis var. *paullifructus* Kukenthal in Engl. Pflanzenr. IV. 20 (Heft 38): 308. 1909. (*Elmer 881,* Palouse Creek, Wash.) = var. *lenticularis.*

C. hindsii var. *brevigluma* Kükenthal in Engl. Pflanzenr. IV. 20 (Heft 38): 307. 1909. (*Sandberg 933,* Lake Pend d'Oreille, Idaho.) = var. *lenticularis.*

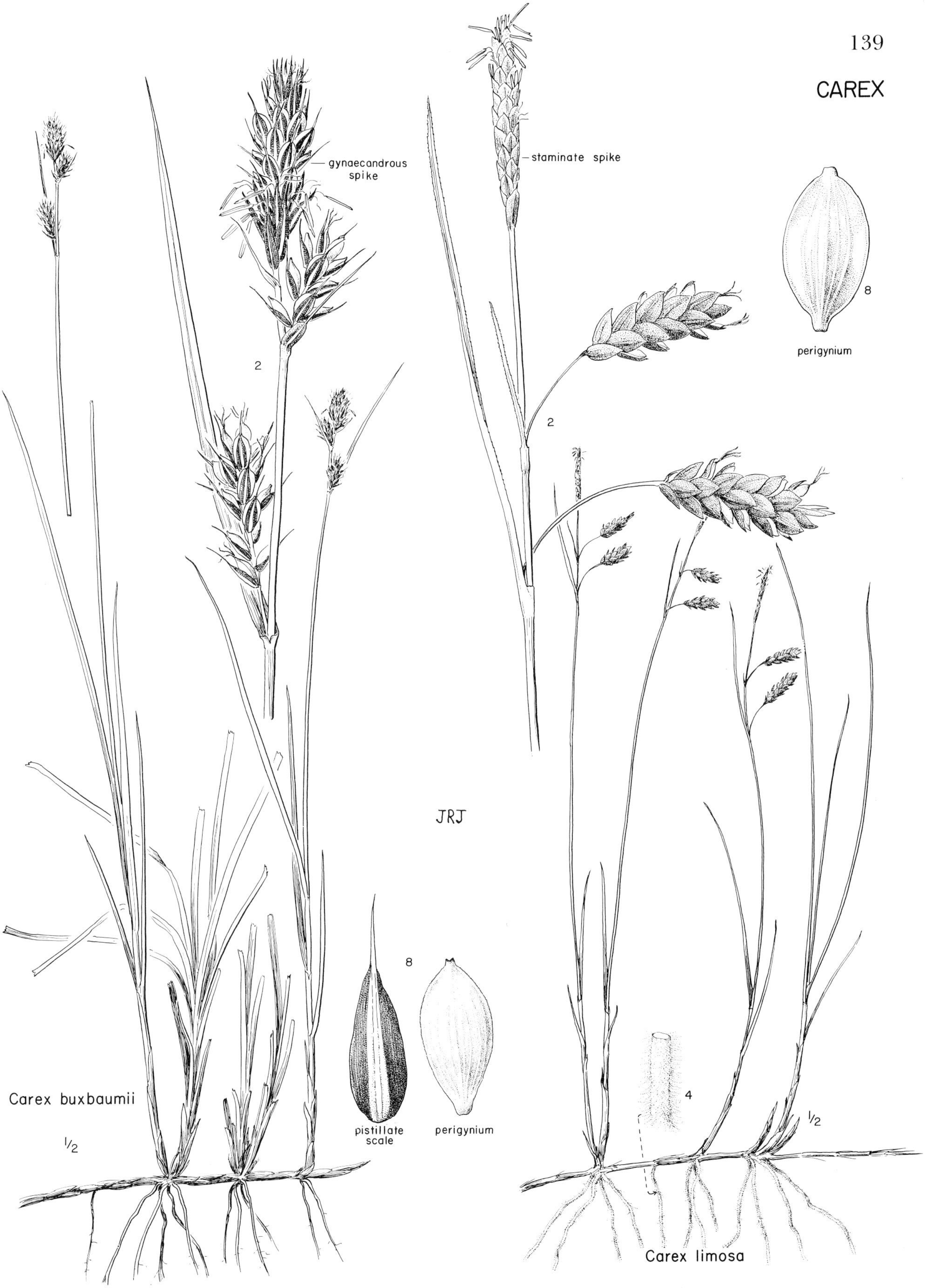
gynaecandrous spike
2
staminate spike
perigynium
8
2
JRJ
8
pistillate scale
perigynium
Carex buxbaumii
1/2
4
1/2
Carex limosa

CAREX

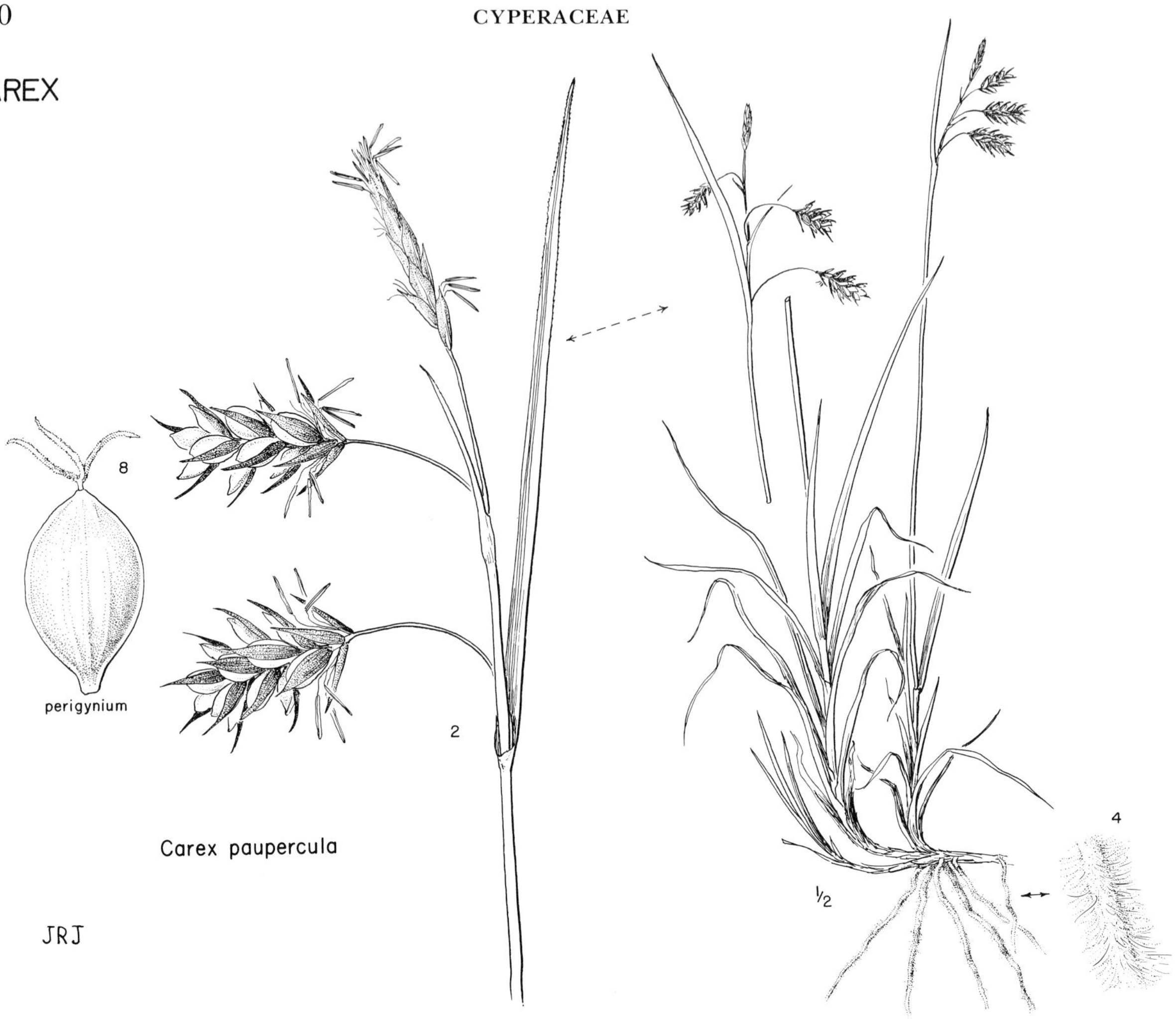

Plants densely tufted, not obviously rhizomatous, or with very short rhizomes only, (1) 2–8 dm tall, tending to be somewhat aphyllopodic; leaves narrow and elongate, flat or nearly so, mostly 2–4 mm wide; spikes 3–6, approximate or more or less remote, erect, sessile or the lower ones more or less evidently pedunculate, all slender and more or less elongate, mostly (1) 1.5–5 cm long and seldom appreciably over 5 mm thick, the terminal one commonly staminate, less often gynaecandrous or with a few basal perigynia, rarely with the flowers more irregularly disposed, the other spikes generally pistillate; bract subtending the lowest spike elongate and leafy, equaling or commonly surpassing the inflorescence, sheathless or nearly so, the other bracts more reduced; pistillate scales commonly scarious and blackish or dark brown, with paler, more herbaceous, commonly greenish midstripe, and often with narrow, white-hyaline margins, or the scarious part seldom (in our range) paler and more reddish-brown, the scales commonly distinctly shorter than and largely exposing the perigynia; perigynia mostly 1.9–3.0 mm long, somewhat compressed, lanceolate to ovate or broadly elliptic, most commonly rather narrowly ovate, 3- to 7-nerved on each face, often 3-nerved adaxially even when 4- to 7-nerved abaxially, the stipe well defined, 0.15–0.4 mm long, the beak slender, terete, 0.1–0.35 mm long, entire, often black tipped, the body of the perigynium commonly greenish; stigmas 2; achene lenticular, 1.0–1.5 mm long, distinctly shorter than the perigynial cavity; $n = 34$.

Wet meadows, streambanks, lake shores, and other wet places, from sea level to fairly well up in the mts., but not reaching above timberline; Lab. to Alaska, s. to Mass., Mich., Minn., Sask., and in the mts. to Colo., Utah (Wasatch and Uinta mts. and reputedly Aquarius Plateau), Nev. (mts. of Elko Co.), and Calif. (Sierra Nevada); apparently also in e. Asia, where passing under other names such as *C. thunbergii* Steudel.

The foregoing description applies mainly to the cordilleran phase of var. *lenticularis*. Typical var. *lenticularis*, of ne. U.S. and se. Can., tends to have paler scales and perigynia, the scarious part of the scales often reddish-brown. The differences, however, are merely statistical, and many of the eastern plants have dark scales and green perigynia like the cordilleran ones. Conversely, some of the cordilleran plants are colored like the eastern ones. No clear line can be drawn on these bases, and other alleged differences appear to be fictitious. It is conceivable that critical study might warrant varietal distinction for the cordilleran plants, but no such treatment is here proposed.

Seashore plants, from the Aleutian Islands to northern California, constitute a fairly well marked ecotype which has been distinguished as *C. lenticularis* var. *limnophila* (T. H. Holm) Cronq. or *C. hindsii* C. B. Clarke.

CAREX

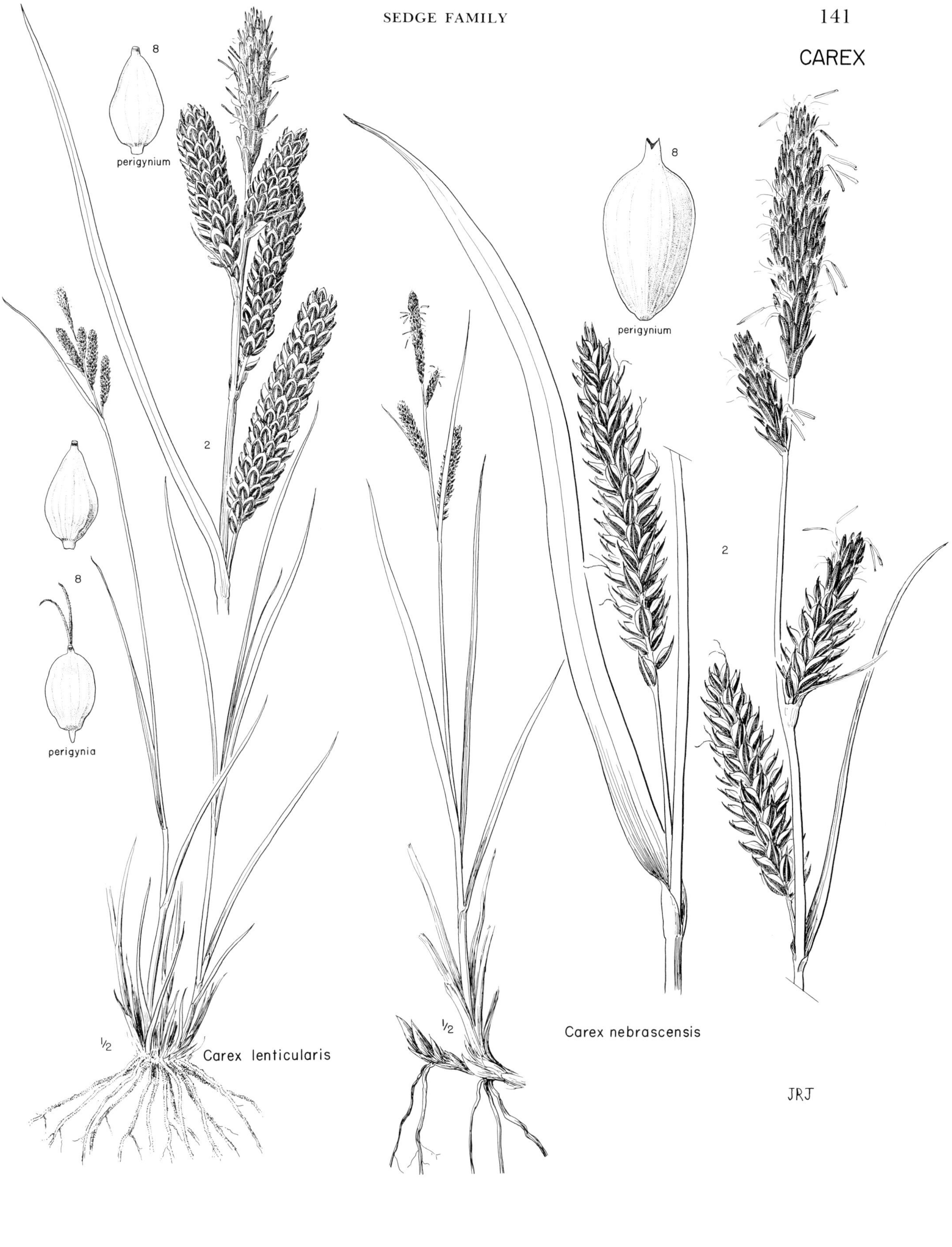
8
perigynium
2
8
perigynia
1/2
Carex lenticularis
8
perigynium
2
1/2
Carex nebrascensis
JRJ

48. **Carex nebrascensis** Dewey

Carex nebrascensis Dewey, Amer. J. Sci. 68: 102. 1854. *C. jamesii* var. *nebraskensis* L. H. Bailey, Cat. N. Amer. Carices Suppl. 1. 1884. (No specimens or locality cited; presumably *Hayden s.n.*, Nebraska, in 1853.)

C. jamesii Torr. & Gray, Ann. Lyceum Nat. Hist. New York 3: 398. 1836. *C. nebraskensis* var. *praevia* L. H. Bailey, Mem. Torrey Bot. Club 1: 49. 1889. (*James*, Rocky Mts.) Not *C. jamesii* Schwein. 1824.

Plants coarse, the stout, sharply triangular stems 2–10 dm tall, arising singly or few together from long, stout, scaly rhizomes; leaves firm, blue-glaucous, the larger ones (3) 4–10 mm wide; spikes 3–6, approximate or more or less remote, erect, sessile or the lower ones evidently pedunculate, all cylindric and more or less elongate, mostly 1.5–7 cm long and 5–9 mm thick at maturity, the terminal one or two spikes staminate, the others pistillate or the upper with some distal staminate flowers; bract subtending the lowest spike sheathless or nearly so, usually elongate and leafy, commonly about equaling or surpassing the inflorescence, seldom somewhat shorter, the other bracts more reduced; pistillate scales with green or pale midrib and light brown to often nearly black scarious margins, typically slender and elongate, equaling or surpassing the perigynia and tapering to a narrowly acute point, seldom shorter or blunter; perigynia elliptic to obovate, coriaceous, 5- to 10-nerved on each face and with prominent marginal nerves, often slightly inflated, but still compressed, stramineous or yellowish brown to light brown, sessile or very nearly so, mostly (2.6) 3.0–4.1 mm long, including the (0.2) 0.4–0.6 mm beak, the beak more or less distinctly bidentate; stigmas 2; achene lenticular, 1.3–2.2 mm long, the distal part of the perigynial cavity empty.

Wet meadows and other wet places, often in alkaline situations, from the lowlands to moderate elev. in the mts.; Kansas, Nebr., and S.D. to Mont., Wash., Oregon, Calif., Ariz., and N.M., and found throughout our range.

Unlike some other moisture-loving species of *Carex, C. nebrascensis* readily tolerates dry air and intense insolation, so long as its feet are wet. It is regarded as an important forage species, although it is not so palatable as some others in the genus.

49. **Carex scopulorum** T. H. Holm

Carex scopulorum T. H. Holm, Amer. J. Sci. 164: 422. 1902. (*Holm*, region of Clear Creek Canyon and also near Leadville, Colo.)

C. stylosa var. *virens* L. H. Bailey, Proc. Amer. Acad. Arts 22: 79. 1887. *C. spreta* L. H. Bailey, Mem. Torrey Bot. Club 1: 6. 1889. *C. accedens* T. H. Holm, Amer. J. Sci. 166: 457. 1903. (All as to the *Howell* specimen from Mt. Adams, Wash., selected by Mackenzie in 1935 as the type, but not as to the *Howell* specimen from Sauvie's Island in the Columbia River at the mouth of the Willamette, selected by inference as the type by Bailey when he published *C. spreta*.) Not *C. spreta* Steudel, 1855.

C. vulgaris var. *bracteosa* L. H. Bailey, Proc. Amer. Acad. Arts 22: 81. 1887. *C. rigida* var. *bracteosa* L. H. Bailey, J. Bot. 28: 173. 1890. *C. rigida* (subsp.) *hesperia* Piper, Contr. U.S. Natl. Herb. 11: 173. 1906. *C. scopulorum* var. *bracteosa* F. J. Hermann, Leafl. W. Bot. 9: 16. 1959. (*Brewer 2015*, Ebbett's Pass, Calif.)

C. gymnoclada T. H. Holm, Amer. J. Sci. 164: 424. 1902. (*Cusick 2487*, Hurricane Creek, Wallowa Mts., Oregon.)

C. chimaphila T. H. Holm, Amer. J. Sci. 166: 33. 1903. *C. scopulorum* var. *chimaphila* Kükenthal in Engl. Pflanzenr. IV. 20 (Heft 38): 303. 1909. (No specimens or locality given by Holm; *Holm*, Long's Peak, Colo., fide Mackenzie.)

C. campylocarpa T. H. Holm, Amer. J. Sci. 170: 304. 1905. (*Coville 1457*, Cathedral Spring, Crater Lake Nat. Park, Oregon.)

C. miserabilis Mackenzie, N. Amer. Fl. 18: 385. 1935. (*Eggleston 13567*, Chiwaukum Lake, Chelan Co., Wash.)

C. campylocarpa subsp. *affinis* Maguire & A. Holmgren, Leafl. W. Bot. 4: 262. 1946. (*Maguire & Holmgren 21947*, Deep Creek Mts., Utah.)

Plants 1–4 (6) dm tall, sod-forming, with freely branched rhizomes and scattered or loosely clustered stems, phyllopodic or somewhat aphyllopodic; leaves firm, flat or nearly so, mostly 2–6 mm wide, not greatly elongate, the longer ones mostly (0.5) 1–3 (4) dm long; bracts sheathless or nearly so, small, the lowest ones 1–7 cm long, foliaceous or setaceous, shorter than the inflorescence; spikes mostly 3–6, the terminal one staminate or sometimes androgynous or even gynaecandrous or with the perigynia in the middle, the others pistillate or the upper androgynous, all more or less crowded and approximate (but not forming a head) or the lowest one a little removed from the others, all cylindric, erect, mostly 1–2.5 (3) cm long, the lateral ones sessile or the lower ones short-pedunculate; pistillate scales mostly narrower than and surpassed by or surpassing the perigynia, black or blackish-purple to dark reddish-brown, generally scarious throughout; perigynia elliptic to obovate, often broadly so, pale greenish to more often coppery-tan or more or less suffused with dark purple, seldom purple-flecked on a pale background as in *C. aquatilis*, compressed but tending to be somewhat inflated distally and only loosely enclosing the achene, nerveless except for the well developed marginal nerves, mostly 1.8–3.3 mm long, including the short (to 0.4 mm) or nearly obsolete beak; stigmas 2; achene lenticular, 1.4–1.6 mm long; a few trigonous, tristigmatic achenes rarely intermingled with the others.

Wet meadows, streambanks, lakeshores, and moist, open slopes, at middle and upper elev. in the mts., often above timberline; s. B.C. to Calif. (Sierra Nevada), Nev. (Washoe, Nye, Elko, and White Pine cos.), Utah (Deep Creek Mts.), and Colo., but apparently missing from large parts of Utah and Nev. where the habitat appears suitable for it.

See comment under *C. aquatilis*, with which *C. scopulorum* appears to intergrade.

50. **Carex aquatilis** Wahlenb.

Carex aquatilis Wahlenb. Kongl. Vetensk. Acad. Nya Handl. 24: 165. 1803. *Neskiza aquatilis* Raf. Good Book 27. 1840. (Lappland.)

C. variabilis var. *elatior* L. H. Bailey, Mem. Torrey Bot. Club 1: 19. 1889. *C. variabilis* var. *altior* Rydb. Mem. New York Bot. Gard. 1: 76. 1900. *C. aquatilis* var. *altior* Fern. Rhodora 44: 295. 1942. *C. aquatilis* subsp. *altior* Hultén, Kongl. Svenska Vetenskapsakad. Handl. IV. 8(5): 74. 1962. (*Brandegee*, Cañon City, Colo.; lectotype by Mackenzie.) Not *C. aquatilis* var. *elatior* Bab. 1843. Name variously applied to the more robust, southern phase of the species, or only to those plants with obovate perigynia.

C. aquatilis var. *substricta* Kükenthal in Engl. Pflanzenr. IV. 20 (Heft 38): 309. 1909. *C. substricta* Mackenzie in Rydb. Fl. Rocky Mts. 1060. 1917 [1918]. (*Sartwell 56*, Junius, Seneca Co., N.Y.; lectotype by Mackenzie.)

CAREX

lateral spike
terminal spike
2
2
8
perigynia
8
perigynium
1/2
1/6
1/2
Carex scopulorum
JRJ
Carex aquatilis

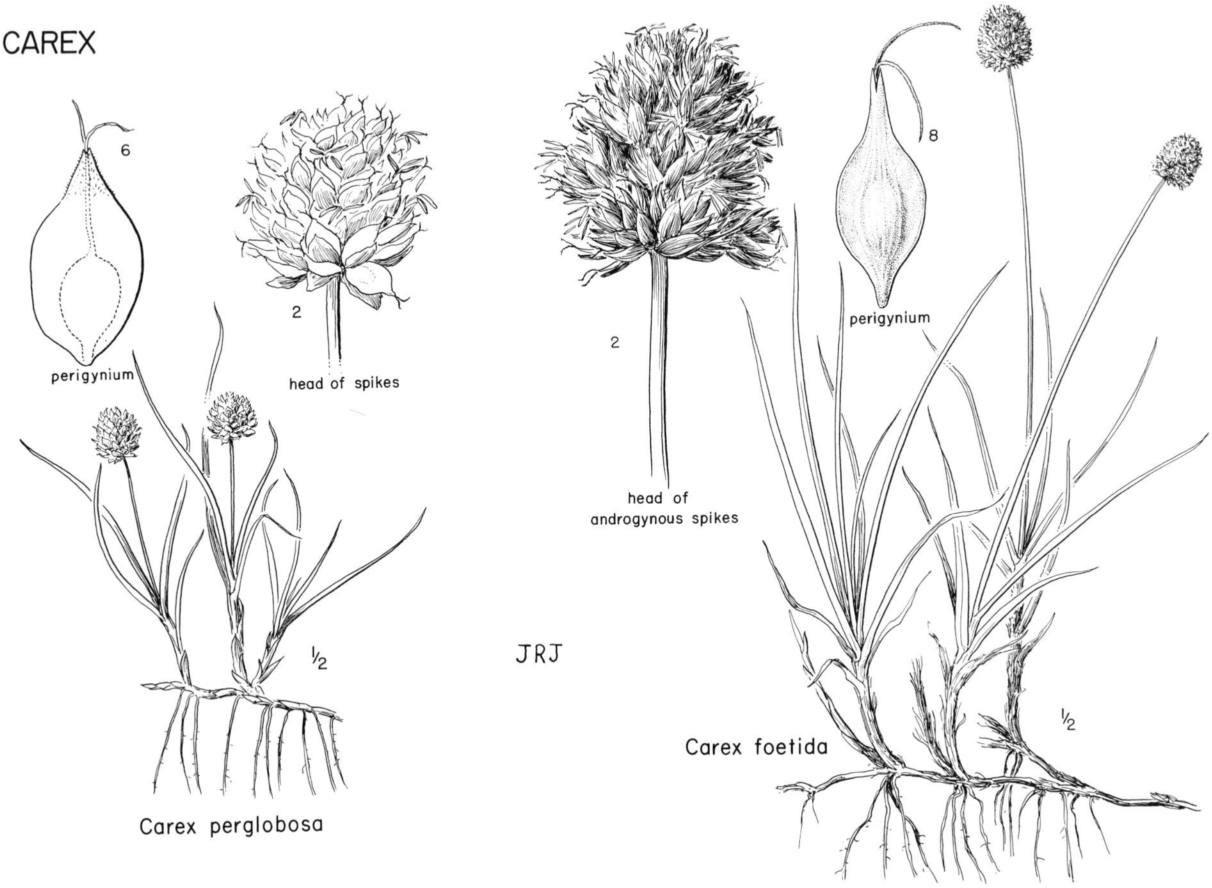

C. interimus Maguire, Brittonia 5: 200. 1944. (*Maguire 16098,* Tony Grove Lake, Bear River Range, Cache Co., Utah.) A diseased form.

Stems mostly 3–10 (15) dm tall, borne singly or few together on well developed, stout, deep-seated creeping rhizomes (these seldom collected), often forming large clumps or a turf; leaves elongate, flat or nearly so, mostly 2–7 mm wide; spikes mostly 3–7, cylindric, approximate to more or less remote, but not closely crowded, mostly (1) 1.5–5 cm long, the terminal one staminate (sometimes with a few basal perigynia), sometimes with a smaller staminate spike at its base, the others pistillate or (especially the upper) sometimes androgynous, erect, appearing somewhat stouter than in *C. lenticularis,* all sessile or nearly so, or the lower ones often evidently pedunculate, closely flowered throughout, or sometimes loosely flowered toward the base; bracts subtending the spikes sheathless or nearly so, the lowest one more or less elongate and leafy, mostly 7–25 cm long, usually equaling or longer than the inflorescence, occasionally somewhat shorter; pistillate scales reddish brown to often purplish black, usually with a paler midrib, generally narrower than the perigynia, and shorter than or about equaling to sometimes longer than the perigynia; perigynia elliptic to obovate, more or less strongly flattened, nerveless except for the well developed marginal nerves, commonly speckled with reddish brown on a stramineous or pale coppery-stramineous to pale greenish background, or sometimes more or less suffused with reddish brown on a paler background, but generally not green as in *C. lenticularis,* sessile or very shortly stipitate, 2.0–3.3 mm long, including the beak, this 0.1–0.3 mm long, entire; stigmas 2; achene lenticular, 1.2–1.5 mm long, distinctly shorter and often also narrower than the perigynial cavity; n = 13, 37, 38, 42.

In shallow water or wet soil, often in swampy places or at the margins of ponds and lakes, from the foothills to near timberline in the mts.; circumboreal at high latitudes, extending s. in Amer. to N.J., Ind., Nebr., N.M., Ariz., and Calif., and found in appropriate habitats throughout our range.

The species is regarded as a fairly good forage plant, and is often a significant constituent of meadow hay.

It is difficult to draw precise lines in the herbarium among the several species of the section Acutae. In our range, *C. aquatilis* and *C. scopulorum* do not always appear to be sharply distinct, and in other areas *C. aquatilis* becomes difficult to distinguish from certain other species. It remains to be determined how much of the difficulty reflects real intergradation among the several taxa (based on hybridization or lack of complete initial differentiation) and how much merely reflects inadequacy of the herbarium specimens or over-emphasis of convenient key characters.

51. **Carex perglobosa** Mackenzie

Carex perglobosa Mackenzie, Bull. Torrey Bot. Club 34: 606. 1908. (*Mackenzie 167,* Mt. Baldy, Summit Co., Colo.)

C. incurva var. *chartacea* Kükenthal in Engl. Pflanzenr. IV. 20 (Heft 38): 114. 1909. (*Parry 385,* Middle Park, Colo.)

Stems borne singly or loosely aggregated on a system of slender, branching rhizomes, 3–15 cm tall, phyllopodic; leaves thick and firm, narrow but flat or nearly so, mostly 1–2 mm wide, tapering to a long, slender tip, commonly equaling or surpassing the culms, seldom all somewhat shorter than the culms; spikes numerous, small, sessile, very closely aggregated into a compact, globose or ovoid head 10–18 mm long, the head sometimes subtended by an inconspicuous short bract, the spikes otherwise bractless and nearly indistinguishable; pistillate scales shining brown, wholly hyaline-scarious except for the slender midrib, not much if at all shorter than the perigynia; perigynia broadly elliptic or elliptic-ovate, 3.6–4.7 mm long and 1.9–2.6 mm wide, 1.5–2 times as long as wide, not stipitate, tapering distally to a thin, obliquely cleft, often poorly defined beak, shining brown or partly greenish-stramineous, very thin-walled, strongly flattened and with well developed marginal nerves, finely about 10-nerved toward the base on each side, or the facial nerves more or less obsolete; stigmas 2; achene lenticular, 1.4–1.7 mm long and about 1.1 or 1.2 mm wide, much narrower than the perigynium and not extending beyond about the middle of the perigynial cavity.

Dry, open slopes at high alt. in the mts., commonly above timberline; Colo. and se. Utah (LaSal Mts.).

Some of the material from the LaSal Mts. (*Maguire et al. 16475, 16476*) is habitally like *C. perglobosa,* but has the perigynia of *C. foetida* var. *vernacula. C. perglobosa* is well represented in the LaSals, but unequivocal *C. foetida* var. *vernacula* is not known there, though it might be expected. The proper taxonomic disposition of those "intermediate" specimens remains to be determined.

52. Carex foetida All.

Carex foetida All. Fl. Pedemontana 2: 265. 1785. *Vignea foetida* Reichenb. Fl. German. Excursoria 56. 1830. (Italian Alps.)

C. vernacula L. H. Bailey, Bull. Torrey Bot. Club 20: 417. 1893. *C. foetida* var. *vernacula* Kükenthal in Engl. Pflanzenr. IV. 20 (Heft 38): 115. 1909. ("Mountains from Colorado and Wyoming westward"; no type specified.)

Culms loosely aggregated on rather short rhizomes, or borne singly on somewhat more elongate rhizomes, triangular, 0.5–2.5 dm tall, phyllopodic; leaves all borne at or near the base, thick and firm but flat or nearly so, mostly (1.5) 2–4 mm wide, gradually tapering to a long, slender tip, distinctly shorter than the culms; spikes numerous, small, sessile, very closely aggregated into a compact, globose or ovoid head 8–16 mm long, the head sometimes subtended by an inconspicuous short bract, the spikes otherwise bractless and nearly indistinguishable; spikes few-flowered, androgynous; scales shining brown, hyaline-scarious throughout except for the slender, often paler midrib, not much if at all shorter than the perigynia; perigynia lanceolate or rather narrowly elliptic to sometimes more or less obovate, 3.3–4.7 mm long (including the slender, stipitate base and the prominent beak) and 1.0–1.6 mm wide, (2) 2.5–4.5 times as long as wide and not more than about half again as wide as the achene, commonly brownish, often with paler, greenish margins, thin-walled, flattened and with well developed marginal nerves, the faces nerveless or 5- to 10-nerved especially toward the base; beak of the perigynium mostly 0.8–1.7 mm long, obliquely cleft and often also bidentate; stigmas 2; achene lenticular, 1.1–1.4 mm long and 0.7–1.1 mm wide; $n = 29$.

Moist or wet places at high elev. in the mts., near or above timberline; irregularly distributed in the higher mts. of w. U.S. from s. Wash. to the Sierra Nevada of Calif., and in the adj. White Mts. of Calif., also in the Wallowa Mts. of ne. Oregon, in some of the mts. of Utah, and in the southern Rocky Mts. of Colo. and adj. Wyo.; also in the mts. of Europe.

Carex foetida consists of two geographically disjunct but otherwise only very weakly characterized phases, one in the mountains of Europe, the other in the mountains of the western U.S. The European plants tend to have longer, often slightly narrower leaves, not much if at all shorter than the culm, in contrast to the American plants as here described. Other alleged differences appear to be largely fictitious. The American plants have usually been treated as a distinct species under the name *C. vernacula* L. H. Bailey, but they are better treated as a variety, *C. foetida* var. *vernacula* (L. H. Bailey) Kükenthal. See also comment under *C. perglobosa.*

53. Carex simulata Mackenzie

Carex simulata Mackenzie, Bull. Torrey Bot. Club 34: 604. 1908. (*Nelson 7316,* Chug Creek, Albany Co., Wyo.)

Culms arising singly or few together from well developed, elongate, rather stout, brownish rhizomes, somewhat aphyllopodic, (1) 2–9 dm tall; leaves flat or nearly so, 1–2.5 mm wide, often much elongate; spikes 8–25 or more, sessile, short, less than 1 cm long, aggregated into a terminal, ovoid to subcylindric head (1) 1.5–3 cm long and up to 1.5 cm wide, the spikes more or less distinguishable to the naked eye, each subtended by an inconspicuous bract not much different from the scales; plants subdioecious, some individuals producing only spikes that are largely or wholly staminate, others producing only spikes that are largely or wholly pistillate, the spikes, when bisexual, always androgynous; pistillate scales largely brown or brownish and hyaline-scarious, with a firmer, pale midrib that is often excurrent as a short awn, the scale distinctly longer than the subtended perigynium; perigynia relatively small, only 1.7–2.4 (2.6) mm long, ovate or elliptic-ovate, often broadly so, with a very short, obliquely cleft beak only 0.2–0.4 (0.6) mm long, plano-convex, tending to be slightly serrulate along the margins distally, shining brown, the dorsal surface commonly with a few nerves well removed from the margins and often not reaching the tip, the ventral surface nerveless or often with several very short but obvious nerves at the base only, or both surfaces wholly nerveless; perigynial wall very thick and firm, especially proximally, closely enveloping the achene; stigmas 2; achene lenticular, rhombic-elliptic to obovate, 1.2–1.5 mm long.

Wet meadows, from the foothills to moderate elev. in the mts., not reaching timberline; Mont. and s. Alta. to Wash. (e. of the Cascade summits), s. to N.M., Utah, and Calif.; widespread but not especially common.

The small, hard, short-beaked, shining brown perigynia of this species are very distinctive.

54. Carex douglasii F. Boott

Carex douglasii F. Boott in Hook. Fl. Boreali-Amer. 2: 213. 1839. (*Douglas,* Nw. Coast.)

Plants vigorously but rather slenderly rhizomatous, producing more or less scattered single stems, these (0.5) 1–3 (4) dm tall, phyllopodic; rhizome and often also the persistent basal sheaths tending to be dirty brown or blackish; leaves mostly basal or nearly so, firm, narrow, flat or involute, mostly 1–2.5 mm wide, tapering to a long, slender point, from distinctly shorter than to sometimes equaling or even surpassing the culms; plants dioecious, reputedly sometimes incompletely so and some of the spikes on the staminate plant androgynous; spikes several or rather numerous, sessile, more or less ovoid-fusiform, mostly 1–1.5 cm long, rather closely aggregated into a rhomboidal or ovoid to subcylindric head 1.5–4.5 cm long, the individual spikes clearly visible, each subtended by a bract that is somewhat larger than but otherwise not much different from the scales; scales elongate and rather narrow, acute or acuminate and sometimes shortly awn-tipped, commonly with green midrib flanked on each side by a light brown, hyaline-scarious strip that fades into the broad or narrow white-hyaline border, the effect greenish-stramineous to light brown; stamens conspicuous; perigynia mostly shorter than the scales and more or less hidden by them, stramineous to pale brownish, rather narrowly elliptic to elliptic-ovate, plano-convex, becoming sharp-edged and finely serrulate distally, firm-walled, obscurely many-striate and with well developed marginal nerves, 3.5–4.6 mm long, including the prominent beak, this mostly 1–1.5 mm long, obliquely cleft and sometimes also bidentate; stigmas 2, elongate; achene lenticular, 1.6–2.0 mm long, closely enveloped by the perigynium, except around the distal end.

Dry to fairly wet places, often with sagebrush but not in the drier parts of the desert, from the valleys and plains to moderate elev. in the mts., tolerant of alkali; Man. to s. B.C. and Wash., s. to Iowa, Nebr., N.M., and Calif., and found in appropriate habitats essentially throughout our range.

A very distinctive species which is rarely confused with anything else.

Carex douglasii is not highly palatable to livestock, but because of its abundance and its resistance to grazing pressure it is regarded as an important forage plant.

55. Carex stenophylla Wahlenb.

Carex stenophylla Wahlenb. Kongl. Vetensk. Acad. Nya Handl. 24: 142. 1803. *Vignea stenophylla* Reichenb. Fl. German. Excursoria 56. 1830. (No type given; presumably from c. Europe.)

C. eleocharis L. H. Bailey, Mem. Torrey Bot. Club 1: 6. 1889. *C. stenophylla* subsp. *eleocharis* Hultén, Fl. Alaska & Yukon 1: 314. 1941. *C. stenophylla* var. *eleocharis* Breit. Canad. Field-Naturalist 71: 48. 1957. (*Macoun 1665,* Sask. plains.)

Culms slender, erect, 0.5–2 dm tall, arising singly or few together from well developed, slender, brownish rhizomes, phyllopodic or somewhat aphyllopodic; leaves all borne at or near the base, firm, slender, flat or more often involute or canaliculate, 0.3–1.5 mm wide, tapering to a long, slender tip, usually surpassed by the culms; spikes several, sessile, small, 4–9 mm long, closely aggregated into an ovoid or oblong-cylindric head 8–17 mm long, the spikes individually distinguishable to the naked eye (though sometimes with difficulty) and each subtended by an inconspicuous bract not much different from the scales, androgynous, or rarely all unisexual and the plants dioecious; pistillate scales thin, stramineous to brownish, with firm, often shortly exserted midrib, about equaling or slightly surpassing the perigynia; perigynia brownish, rather narrowly fusiform-elliptic to broadly elliptic-ovate, 2.6–3.3 (3.5) mm long, often broadly short-stipitate, plano-convex, evidently to very obscurely serrulate along the prominent marginal nerves distally, essentially nerveless on both faces, or often more or less evidently several nerved on the back (the nerves commonly about 10 when present), less often finely few-nerved ventrally as well; beak of the perigynium short, mostly 0.5–1.0 (1.2) mm long, obliquely cleft and sometimes also bidentate; stigmas 2; achene lenticular, 1.5–2.1 mm long, closely enveloped by the perigynium; $n = 31$.

Open, dry to moderately moist, often grassy places, from the plains to fairly high elev. in the mts., but mostly below timberline; typically a species of the n. Great Plains, from Iowa and Nebr. w. to the e. base of the Rocky Mts., and thence at scattered stations to se. B.C., Idaho (Clark Co.), Oregon (Crook Co.), Calif. (White Mts.), Nev. (Lander Co.), and Ariz. (Apache Co.), n. irregularly to s. Yukon, and widespread in similar habitats in Eurasia.

It has been customary to distinguish the American plants as a separate species, *C. eleocharis,* but I cannot sort the specimens without reference to geography.

56. Carex praegracilis W. Boott

Carex praegracilis W. Boott, Bot. Gaz. 9: 87. 1884. (*Scott,* San Diego, Calif.)

C. marcida F. Boott in Hook. Fl. Boreali-Amer. 2: 212. 1839. *C. camporum* Mackenzie, Bull. Torrey Bot. Club 37: 244. 1910. (*Scouler,* Columbia River.) Not *C. marcida* J. F. Gmelin, 1791.

C. douglasii var. *brunnea* Olney in King, Rep. Geol. Explor. 40th Parallel 5: 363. 1871. *C. usta* L. H. Bailey, Mem. Torrey Bot. Club 1: 20. 1889. *C. douglasii* f. *brunnea* (attr. to Olney by) Kükenthal in Engl. Pflanzenr. IV. 20 (Heft 38): 122. 1909. (*Coulter 805,* Calif., is the first collection cited.) Not *C. brunnea* Thunb. 1784.

C. marcida var. *debilis* L. H. Bailey, Proc. Amer. Acad. Arts 22: 136. 1887. *C. marcida* f. *debilis* (attr. to L. H. Bailey by) Kükenthal in Engl. Pflanzenr. IV. 20 (Heft 38): 124. 1909. (*Howell,* Harney Valley, Oregon.)

Stems (1) 3–7 dm tall, arising singly or few together from elongate, stout, black or brownish-black creeping rhizomes, somewhat aphyllopodic, the old sheaths or rudiments dark like the rhizomes; leaves firm, more or less elongate, all borne at or near the base, flat or nearly so, mostly 1–3 mm wide; spikes mostly 6–25, androgynous or some of them wholly pistillate (sometimes some plants largely staminate and others largely pistillate, but not truly dioecious), sessile, less than 1 cm long, aggregated into a thick-cylindric to more or less elliptic or ovoid head 1.5–3.5 (5) cm long and about 1 (1.5) cm thick or less, the spikes individually fairly readily distinguishable to the naked eye (the lowest ones sometimes somewhat removed from the main cluster), inconspicuously bracteate; scales stramineous to light

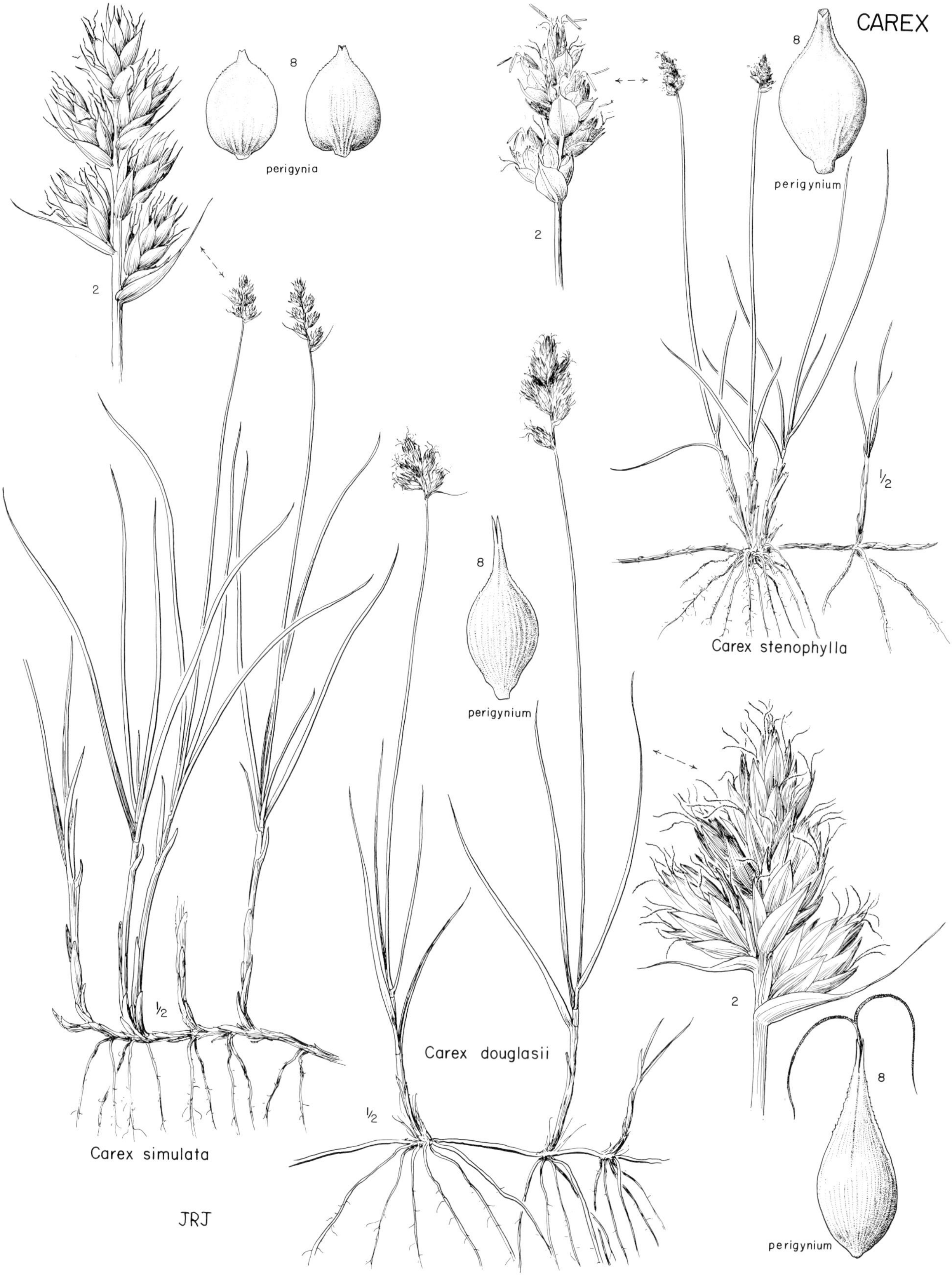
CAREX
perigynia
perigynium
Carex stenophylla
perigynium
Carex douglasii
Carex simulata
perigynium
JRJ

brown or seldom dark brown, often with paler or green midrib, and tending to be hyaline-margined, nearly equaling to often surpassing and generally concealing the perigynia; perigynium light to dark brown, commonly with paler, often greenish margins, 2.9–3.8 mm long, more or less ovate or lance-ovate to elliptic, usually distinctly short-stipitate, tending to be spongy-based but otherwise thin-walled, plano-convex, sharp-edged and more or less serrulate distally, nerveless ventrally, inconspicuously several-nerved dorsally, the beak prominent, 0.6–1.3 mm long, obliquely cleft and sometimes also bidentate, tending to be hyaline at the tip; stigmas 2; achene lenticular, 1.2–1.7 mm long, nearly or quite filling the perigynium.

Open, moist or wet, often alkaline places, frequently in subalkaline meadows, but not confined to alkaline sites, from the plains and lowlands to moderate elev. in the mts., not reaching the spruce-fir zone; common and widespread in interior N. Amer., from n. Mich., Minn. and Man. to s. B.C., s. to Iowa, Okla., Texas, Calif., and Mex., and found in appropriate habitats throughout our range.

The species is a valuable forage plant.

57. Carex siccata Dewey

Carex siccata Dewey, Amer. J. Sci. 10: 278. 1826. *Vignea siccata* Raf. Good Book 27. 1840. (*E. Davis,* Westfield, Mass.)

C. foenea sensu Svenson, Rhodora 40: 325. 1938; perhaps not Willd. 1809.

C. foenea var. *tuberculata* F. J. Hermann, Rhodora 70: 419. 1968. (*Garrison et al. 80–56,* Mission Peak headwaters, Chelan Co., Wash.)

Culms slender, mostly 1.5–4 (6) dm tall, arising singly from well developed, light to medium brown, creeping rhizomes, distinctly aphyllopodic; foliage leaves several, shorter than the culms, generally all borne on the lower one-fourth of the culm, but not forming a basal or near-basal cluster, flat or nearly so, mostly (1) 1.5–3 mm wide; spikes relatively few, mostly (3) 6–12, sessile, about 1 cm long or less, inconspicuously bracteate, aggregated into a compact, usually more or less cylindric or clavate-cylindric inflorescence mostly 1.5–3.5 cm long and about 1 cm thick or less, individually fairly readily distinguishable to the naked eye, the lowest ones sometimes a little removed from the main cluster; spikes, or some of them, androgynous, the lowest ones often largely or wholly staminate, the uppermost one sometimes wholly pistillate and often closely subtended by one or more very short staminate spikes and thus seemingly but falsely gynaecandrous; pistillate scales largely brownish, with firmer, often pale or greenish midvein and tending to be hyaline-margined; perigynia 4.5–6.2 mm long, including the prominent long beak, this well over half as long (sometimes fully as long) as the body, generally conspicuously surpassing the scales, obliquely cleft and more or less bidentate, flattened, serrulate nearly throughout, and somewhat wing-margined below; body of the perigynium lance-ovate to more or less elliptic, plano-convex, sharp-edged proximally, becoming thin-margined distally, serrulate over most of its length, evidently several-nerved (the nerves commonly ca 10) on the dorsal surface, evidently to obscurely few-nerved ventrally, sometimes somewhat granular-tuberculate ventrally; stigmas 2; achene lenticular, 1.7–2.2 mm long, filling the perigynium; $n = 35$.

Dry to moderately moist, open or thinly wooded, often grassy slopes, from the base of the mts. up nearly to timberline; Me. to N.J., w. to s. Mack., Alta., Wash., Wyo., Utah (LaSal and Abajo mts., and reputedly more generally distributed), and Ariz.

58. Carex vallicola Dewey

Carex vallicola Dewey, Amer. J. Sci. 82: 40. 1861. (*Hayden,* Jackson's Hole, Wyo.)

C. vagans T. H. Holm, Amer. J. Sci. 167: 301. 1904. (*Leiberg 2558,* Steens Mts., Oregon.)

C. phaeolepis T. H. Holm, Amer. J. Sci. 167: 302. 1904. (*Leiberg 335,* Bear Butte, Crook Co., Oregon.)

C. rusbyi Mackenzie, Smithsonian Misc. Collect. 65(7): 2. 1915. *C. vallicola* var. *rusbyi* F. J. Hermann, Rhodora 70: 419. 1968. (*Rusby 859,* Yavapai Co., Ariz., in 1883.)

Stems 1.5–4 dm tall, slender, slightly aphyllopodic, clustered on a short, stout, freely rooting rhizome, or the rhizome not evident; leaves all clustered near the base, generally shorter than the culms, slender, flat or the narrower ones canaliculate, 0.5–2 mm wide; spikes several (less than 10), small, sessile, loosely aggregated into an irregular, narrowly oblong-cylindric head 1–3 cm long and 4–8 mm wide; bracts subtending the spikes short, inconspicuous, and sheathless, or the lowest ones sometimes filiform-elongate and up to 3 cm long; spikes few-flowered, androgynous, the loose or spreading perigynia equaling or often evidently longer than the pale, largely hyaline-scarious scales, with the result that the mature inflorescence tends to look ragged or uneven, with the individual perigynia evident but the spikes not sharply set off from each other to the naked eye; perigynia generally pale greenish or greenish-stramineous, or partly but usually weakly coppery-tinged, plano-convex, broadly rounded on the convex side, the strong marginal nerves tending to be displaced somewhat onto the ventral surface and appearing ventro-marginal, otherwise nerveless or nearly so, the body abruptly contracted into a short (0.6–1.0 mm), obliquely cleft beak that is only very slightly or not at all bidentate, the margins of the beak often very finely serrulate, but the serrulations scarcely or not at all extending onto the body of the perigynium, at least as seen at 10X magnification; stigmas 2; achene lenticular, 1.9–2.4 mm long, filling the perigynium, jointed to the style.

Moist or moderately dry slopes from the foothills to moderate elev. in the mts., often with sagebrush or under aspen, e. Oregon (southward from Crook and Baker cos.) to c. Idaho (Custer Co.) and w. S.D., s. to Calif. (Mono Co.), Nev. (Clark Co.), and Utah (Garfield and San Juan cos.), and to the mts. of Ariz., N.M., and Hidalgo, Mex.

Carex vallicola is common and fairly palatable, but sensitive to grazing.

59. Carex occidentalis L. H. Bailey

Carex occidentalis L. H. Bailey, Mem. Torrey Bot. Club 1: 14. 1889. *C. hookerana* var. *occidentalis* Kükenthal in Engl. Pflanzenr. IV. 20 (Heft 38): 162. 1909. ("Montana to Arizona"; typification recondite.)

C. muricata var. *americana* L. H. Bailey, Proc. Amer. Acad. Arts 22: 140. 1887. ("Colorado to New Mexico and Arizona"; typification recondite.)

CAREX
8
perigynia
2
2
2
8
perigynium
8
1/2
perigynia
8
1/2
1/2
JRJ
Carex praegracilis
Carex siccata
Carex vallicola

Stems 1.5–8 dm tall, aphyllopodic, densely clustered, the rhizome sometimes evident but short and freely rooting; leaves more or less elongate but shorter than the culms, flat or nearly so, 1–2.5 mm wide, all borne on the lower part of the stem, but generally not closely clustered; spikes several, small, few-flowered, androgynous, sessile, rather loosely aggregated into a slender, usually oblong-cylindric head 1.5–3.5 cm long that may be irregular or interrupted especially toward the base, at least the lower spikes readily distinguishable to the naked eye; bracts sheathless, poorly developed, seldom any of them over 1 (2.5) cm long, only the lowest one, or none of them, surpassing the subtended spike; pistillate scales greenish-stramineous to coppery-brownish, largely hyaline-scarious, about equaling or somewhat surpassing the perigynia and tending to be acuminate or shortly awn-tipped; perigynia greenish-stramineous to coppery-brown, often with greener margins, broadly lance-elliptic to elliptic or elliptic-ovate, 3.0–4.1 mm long, a third to half as wide, plano-convex, the marginal nerves only seldom appearing to be ventromarginal instead of truly marginal; body of the perigynium more or less evidently serrulate along the marginal nerves above the middle, tending to be inconspicuously several-nerved on each side, or the surfaces nearly nerveless, fairly abruptly contracted to a rather ill-defined beak, this shortly bidentate, the teeth mostly only 0.2–0.3 mm long; stigmas 2; achene lenticular, jointed to the style, 1.7–2.4 mm long, not wholly filling the perigynium, the upper part of the body of the perigynium (where it tapers to the beak) empty.

Open or lightly wooded slopes, less often in meadows, at middle and upper alt. in the mts., sometimes above timberline; southern Rocky Mt. region, from s. Wyo. to N.M., and w. to Utah (Duchesne Co. to Washington Co.), Nevada (Elko Co. and Nye Co.), and Ariz., and reported from s. Calif. The Nevada record in the literature has been doubtful, but is confirmed by a collection from the Ruby Mts. in Elko Co. (*Maguire & Holmgren 22648*) and by several collections from south-central Nye Co. by Janice Beatley.

Carex occidentalis is very closely allied to *C. hookerana* Dewey, of the northern Great Plains, and *C. tumulicola* Mackenzie, of the Pacific states, and it might be questioned whether these three should be held as distinct species. The ranges, however, are wholly disjunct, and no taxonomic or nomenclatural change is proposed at this time.

60. **Carex hoodii** F. Boott

Carex hoodii F. Boott in Hook. Fl. Boreali-Amer. 2: 211. 1839. ("Columbia River. *Douglas. Scouler.*")

Stems 3–8 dm tall, strongly aphyllopodic, densely clustered; leaves more or less elongate but shorter than the culms, flat or nearly so, mostly 1.5–3.5 mm wide, all borne on the lower part of the stem, but not closely clustered; spikes several, small, few-flowered, androgynous, sessile, closely aggregated into a dense, oblong-cylindric to often ovoid head 1–2 cm long, the individual spikes hardly distinguishable to the naked eye, the perigynia at least in the lower part of the head commonly more or less spreading away from the main axis; bracts sheathless, inconspicuous, the lowest ones sometimes as much as 1.5 cm long, the others more reduced; pistillate scales equaling or somewhat shorter than the perigynia, largely brown and scarious, but the midrib firmer and often greenish; perigynia typically coppery-brown with broad green margins, narrowly elliptic to rather broadly elliptic-ovate, 3.4–5.0 mm long, a third to half as wide, plano-convex, the marginal nerves sometimes appearing somewhat ventromarginal rather than truly marginal; body of the perigynium evidently serrulate along the marginal nerves down to the middle or below, tending to be inconspicuously several-nerved on both sides, or the surfaces nearly nerveless, the body fairly abruptly contracted to a prominent beak, this mostly 0.7–1.5 mm long and evidently bidentate, the teeth mostly 0.3–0.6 mm long; stigmas 2; achene lenticular, jointed to the style, mostly 1.7–2.1 mm long, not wholly filling the body of the perigynium, which is empty in the upper part where it tapers to the beak; $n = 29, 30$.

Common and widespread in the mts. from the foothills to near timberline, on open or thinly wooded slopes and in dry or moderately moist meadows and forest openings; s. Sask. to s. B.C., s. to S.D. (Black Hills), Colo., Utah (San Juan and Beaver cos.), Nev. (Washoe and Elko cos.), and Calif.

The characters separating *C. hoodii* from *C. vallicola* and *C. occidentalis* do not seem very strong, but the species (*C. hoodii*) nevertheless appears to be sharply limited, and there is only rarely any difficulty in distinguishing it from its relatives.

Carex hoodii furnishes considerable forage for livestock but it is rather sensitive to grazing. Its presence in quantity is considered to indicate that the range is in good condition.

61. **Carex diandra** Schrank

Carex diandra Schrank, Centurie Bot. Anmerk. 57. 1781. (s. Bavaria.)

C. teretiuscula Gooden. Trans. Linn. Soc. London 2: 163. 1794. *Vignea teretiuscula* Reichenb. in Mössler, Handb. Gewächsk. ed. 2. 1621. 1829. *Physiglochis teretiuscula* Raf. Good Book 27. 1840. (Norwich, England.)

Stems densely clustered, 3–10 dm tall, strongly aphyllopodic; foliage leaves more or less elongate, sometimes equaling or even surpassing the culms, flat or nearly so, mostly 1–2.5 (3) mm wide; sheaths somewhat red-dotted ventrally; spikes individually small, few-flowered, androgynous and sessile, more or less numerous in a branched but compact inflorescence mostly 2–3.5 (5) cm long and seldom over 1 cm thick; bracts subtending the spikes inconspicuous, small, sheathless, often hardly different from the pistillate scales, or the lowest ones sometimes more elongate and setaceous; pistillate scales largely hyaline-scarious and stramineous or brownish, with firmer, sometimes shortly excurrent midrib; perigynia more or less ovate or lance-ovate, 2.4–3.0 mm long and about half as wide, spreading or ascending-spreading, the ventral surface slightly convex and often exposed in the spike, the body very firm and thickwalled, commonly dark brown (sometimes pale), the dorsal surface strongly convex with a median thin strip (an extension of the dorsal suture) that is bordered by a pair of veins and extends all the way to the base, this thin strip often depressed and pale; beak of the perigynium more or less abruptly differentiated from the body, coarse, serru-

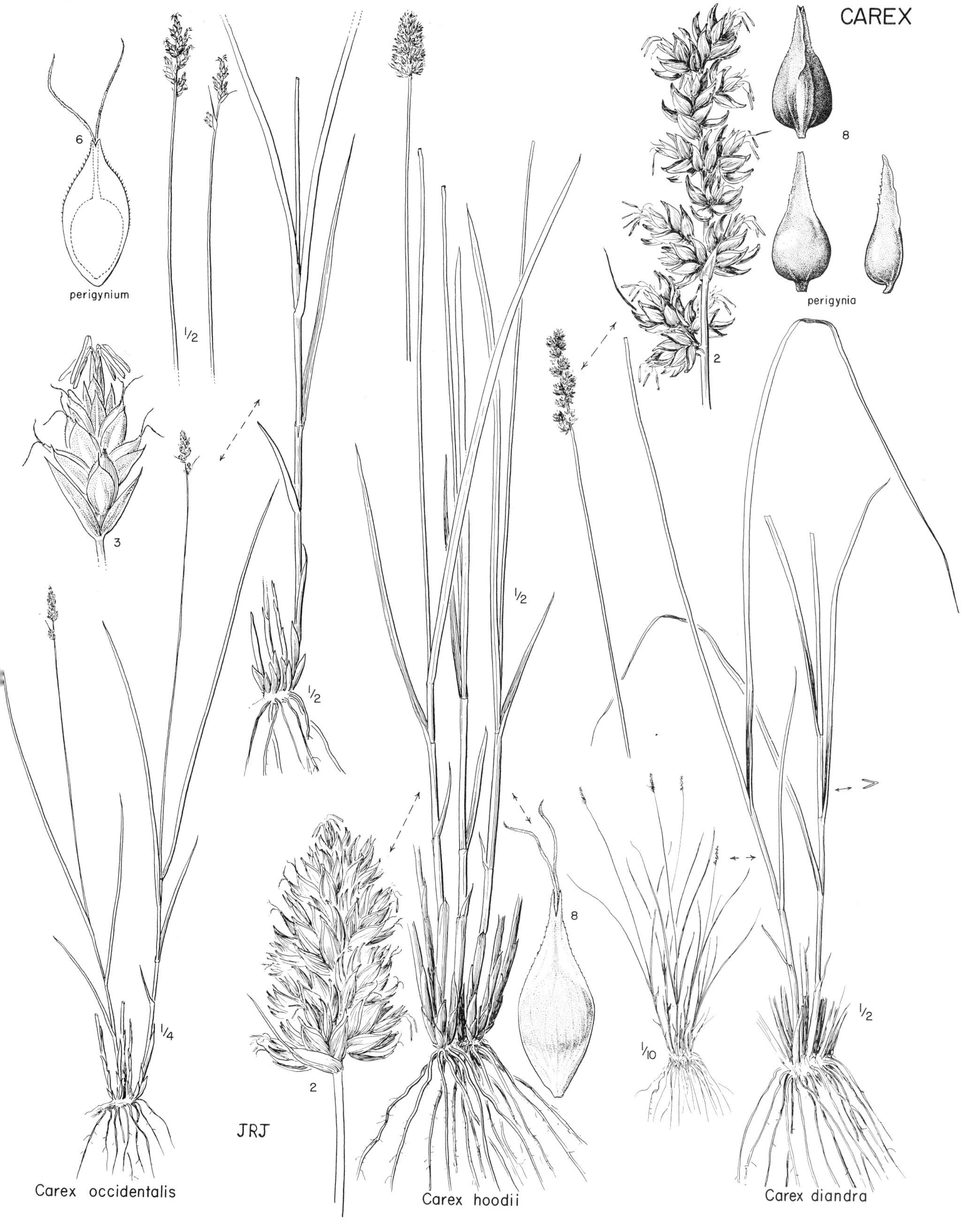
CAREX
6
perigynium
1/2
3
1/4
8
perigynia
2
1/2
1/2
2
8
1/10
1/2
JRJ
Carex occidentalis
Carex hoodii
Carex diandra

late-margined, thinner-walled than the body and often pale or greenish; stigmas 2; achene lenticular, jointed to the style, filling the perigynial cavity; $n = 30$.

Swamps, wet meadows, and *Sphagnum* bogs; circumboreal, but not at the highest latitudes, extending s. in Amer. to Pa., Iowa, Nev., Colo., Utah, and Calif.; in the cordilleran region only seldom collected s. of the Canadian border, the single Utah station (Aquarius Plateau) well isolated.

The name *C. diandra* is a misnomer, inasmuch as the species has 3 stamens like other Carices.

Carex cusickii Mackenzie in Piper & Beattie, Fl. N.W. Coast 72. 1915, is not definitely known in our range. It is closely related to *C. diandra,* but more robust, with wider leaves (the larger ones 3–5 mm wide), larger inflorescences (3–8 cm long and up to 2 cm thick) and sometimes larger perigynia (up to 3.5 mm long). The sheaths are usually coppery-tinged toward the mouth ventrally, as well as red-dotted. It occurs from s. B.C. to nw. Mont., s. to nw. Wyo., c. Idaho, and c. Calif., being commonest in and w. of the Cascade Mts.

62. Carex alma L. H. Bailey

Carex alma L. H. Bailey, Mem. Torrey Bot. Club 1: 50. 1889. (*Parry & Lemmon 396,* San Bernardino Co., Calif.)

Stems densely clustered, mostly 4–12 dm tall, aphyllopodic; foliage leaves more or less elongate, but seldom surpassing the culms, flat or nearly so, 2–6 mm wide; sheaths somewhat red-dotted ventrally, not cross-rugulose; spikes individually small, few-flowered, androgynous and sessile, more or less numerous in a branched but compact (or sometimes more elongate and interrupted) inflorescence mostly 3–10 cm long and 1–2 cm thick; bracts subtending the spikes inconspicuous, small, sheathless, often hardly differentiated from the pistillate scales, or the lowest ones sometimes more elongate and setaceous-tipped; pistillate scales about equaling or a little longer than the perigynia, with a median greenish or partly castaneous strip and broad white-hyaline margins, sometimes with an inconspicuous exserted point less than 0.5 mm long; perigynia more or less ovate, strongly plano-convex, mostly 3.0–4.2 mm long, ascending or somewhat spreading, brown or partly greenish, nerveless (or nearly so) ventrally, several-nerved dorsally, rather thin-walled, and without a thinner median strip; beak of the perigynium about 1 mm long, somewhat flattened, serrulate, obliquely and shortly bidentate; stigmas 2; achene lenticular, jointed to the style, filling the perigynial cavity.

About springs and in other moist places in the desert mts. of se. Calif., w. Ariz., and s. Nev., barely reaching our range in s. Nye Co., Nev.

63. Carex vulpinoidea Michx.

Carex vulpinoidea Michx. Fl. Boreali-Amer. 2: 169. 1803. (Can. and New England.)

Stems rather stout, densely clustered, somewhat aphyllopodic, 3–10 dm tall, often surpassed by the leaves; leaves scattered along the lower half or two-thirds of the stem, flat or nearly so, elongate, mostly 2–5 mm wide; ventral side of the sheaths sparsely red-dotted and usually conspicuously cross-rugose ventrally; spikes small, sessile, few-flowered, androgynous, numerous, more or less densely aggregated into an often irregular or interrupted narrow head 5–10 cm long and up to 1.5 cm wide, the upper spikes, at least hardly distinguishable to the naked eye, and at least the lower part of the head generally compound; bracts subtending the spikes small, setaceous, and sheathless, only the lowest ones sometimes elongate and as much as 5 cm long, the others often shorter than their subtended spikes and not much different from the pistillate scales; pistillate scales slender, with brownish or stramineous, hyaline-scarious margins and firmer midrib that is excurrent as a setaceous, often greenish awn 1–5 mm long; perigynia more or less stramineous or light brown, or greenish distally and toward the margins, flattened or somewhat plano-convex, 2.0–3.5 mm long, 1.5–2.5 times as long as wide, the body narrowly ovate to rotund-ovate or sometimes even obovate, nerveless on both sides, or often with a few usually inconspicuous nerves especially on the dorsal side, more or less abruptly contracted to a conspicuous, flattened, shortly bidentate beak half to fully as long as the body, the beak and the distal part of the body tending to be serrulate on the margins; stigmas 2; achene lenticular, jointed to the style, 1.3–1.6 mm long, usually somewhat narrower as well as a little shorter than the perigynial cavity; $n = 26, 27$.

In marshes or standing water and other wet low places in the foothills and lowlands; Newfl. to Fla., w. to sw. B.C., Wash. and Oregon (only e. of the Cascade summits in Wash. and Oregon), Colo., and Ariz.; apparently missing from Utah, Nev., and Calif., and entering our range only along the n. margin of the Snake River Plains in Idaho and adj. Oregon.

64. Carex stipata Muhl. ex Willd.

Carex stipata Muhl. ex Willd. Sp. Pl. 4: 233. 1805. *Loncoperis stipata* Raf. Good Book 27. 1840. (Pa.)

Stems stout, triangular, densely clustered, somewhat aphyllopodic, (2) 3–10 dm tall; leaves coarse and often elongate, flat, the larger ones mostly 5–11 mm wide; ventral side of the sheaths thin and easily ruptured, usually more or less conspicuously cross-corrugated; spikes individually very small, few-flowered, androgynous, and sessile, numerous or very numerous in a very compact but obviously compound inflorescence (2) 3–10 cm long and 1–3 cm thick, the individual spikes scarcely distinguishable; bracts sheathless, short and inconspicuous, or some of them setaceous and surpassing the spikes; pistillate scales mostly shorter than the perigynia, largely pale or brownish and hyaline-scarious except for the often greenish midvein, which may be excurrent as a short awn-tip; perigynia greenish or greenish-stramineous, widely spreading (the inflorescence appearing prickly because of the prominent perigynial beaks), lance-triangular or lance-ovate, broadest near the base, tapering gradually to the tip, or a little more abruptly narrowed above the middle, tending to be serrulate-margined distally, (3.6) 4.0–5.2 mm long, plano-convex, somewhat spongy-thickened at the base, prominently nerved-striate on both sides, the dorsal nerves more numerous than the ventral ones, or the ventral side seldom ap-

CAREX

Carex alma

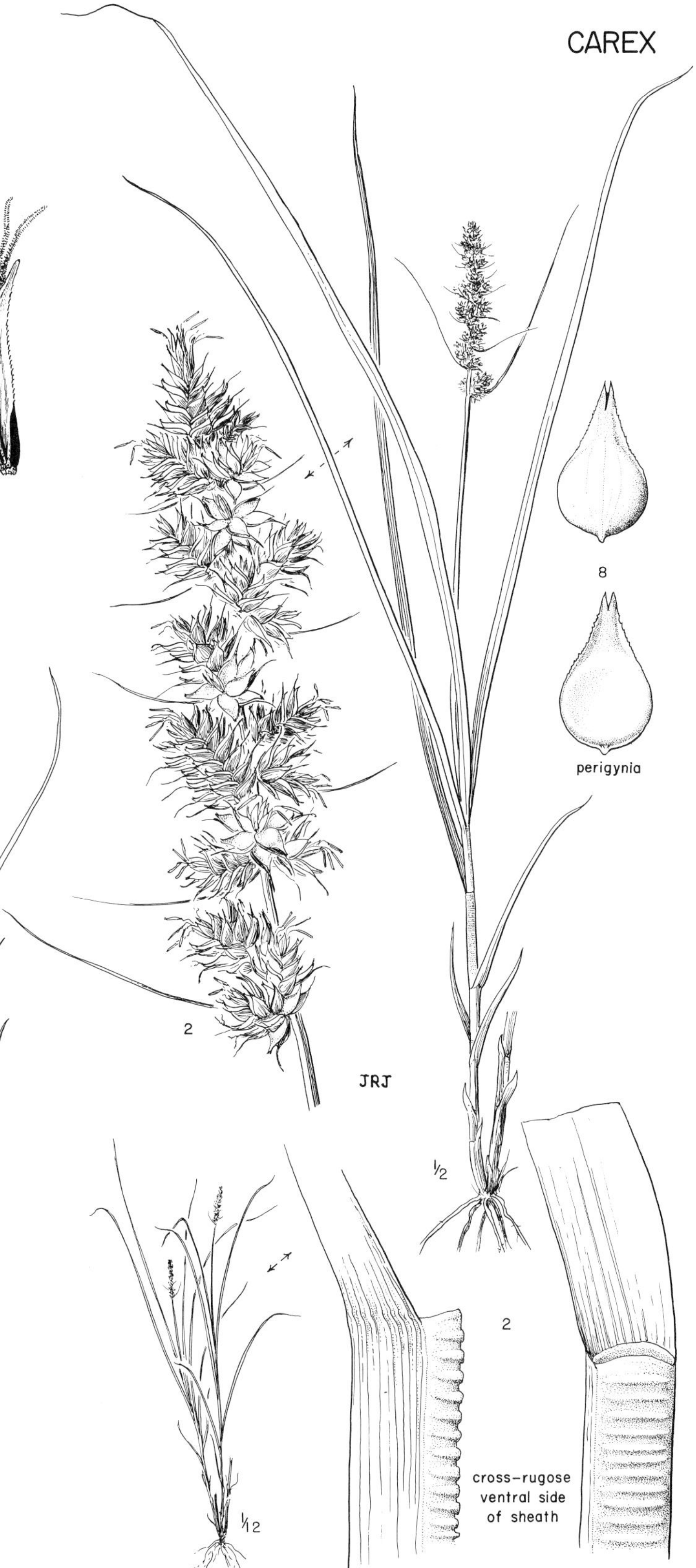

Carex vulpinoidea

CAREX

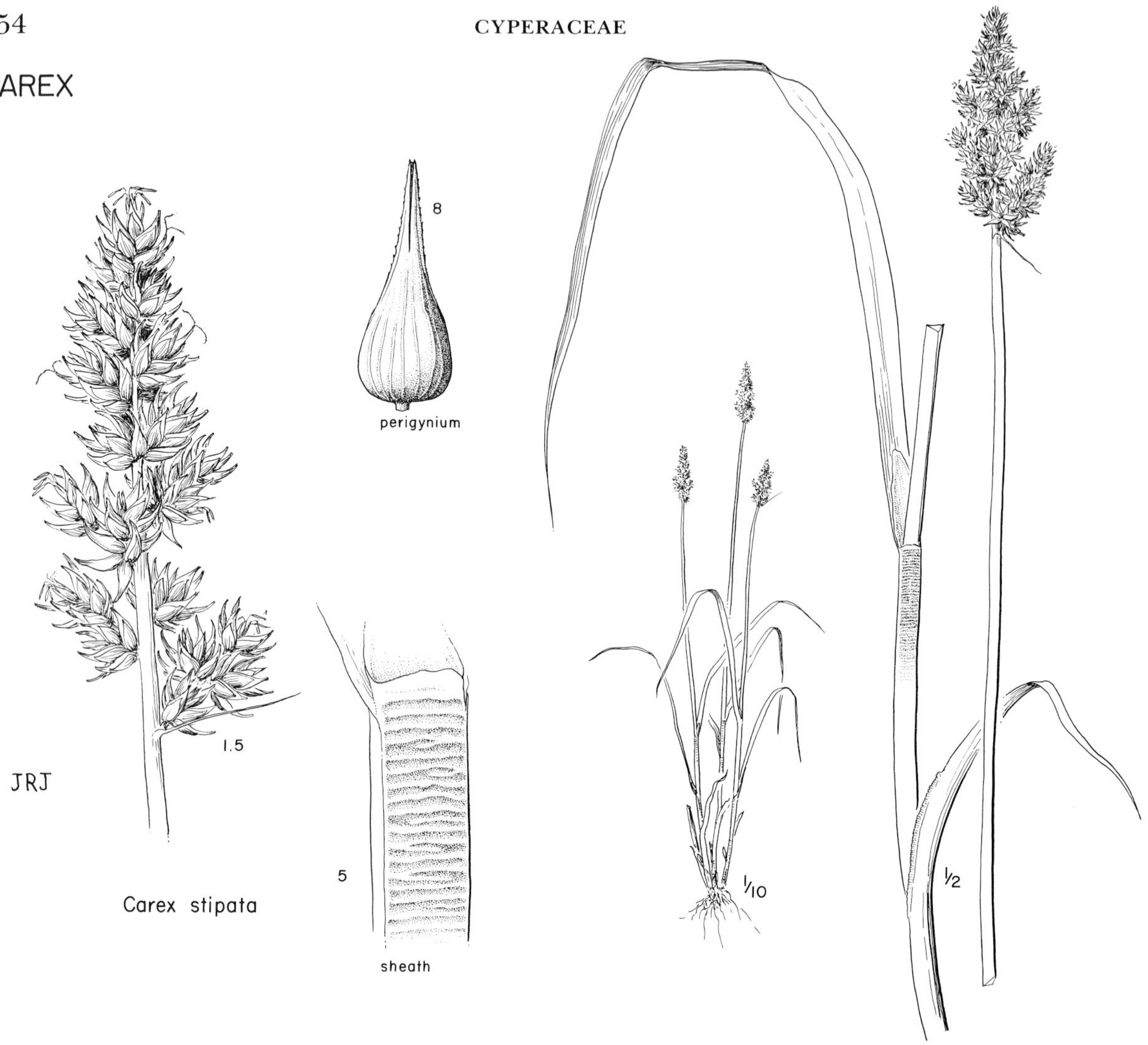

Carex stipata

parently nerveless; stigmas 2; achene lenticular, 1.3–1.7 mm long, broadly ovate, jointed to the style; $n = 24, 26$.

Wet low ground, from the lowlands to moderate elev. in the mts.; Newfl. to s. Alaska, s. to Fla., N.M., and Calif., but missing from most of our range; with us known mainly along our northern border, and from several collections in n. Utah; also in Japan.

Our plants, as here described, belong to the widespread var. *stipata*. The var. *uberior* C. Mohr, of the coastal plain of se. U.S., is more robust, with broader leaves and longer perigynia with notably longer beak.

65. **Carex jonesii** L. H. Bailey

Carex jonesii L. H. Bailey, Mem. Torrey Bot. Club 1: 16. 1889. *C. nervina* var. *jonesii* Kükenthal in Engl. Pflanzenr. IV. 20 (Heft 38): 167. 1909. (*M. E. Jones*, Soda Springs, Nevada Co., Calif., 7000 ft., 1881.)

Stems clustered on a short, compactly branched rhizome, 1.5–3 (4) dm tall, slender, somewhat aphyllopodic; leaves 1–3 mm wide, flat or nearly so, all clustered near the base; ventral side of the leaf sheaths pale, thin, truncate or concave at the mouth, plane or seldom cross-corrugated; spikes individually small, few-flowered, androgynous, and sessile, more or less numerous in a usually compound, thick-cylindric to often broadly ovoid head that is 8–20 mm long and 6–12 mm thick, the individual spikes scarcely distinguishable; bracts small, sheathless, and inconspicuous, often scarcely larger than the pistillate scales, these largely hyaline-scarious and brownish or stramineous, equaling or somewhat smaller than the perigynia; perigynia shining brownish-stramineous or sometimes partly greenish, lance-triangular to lance-ovate or broadly lanceolate, mostly 2.9–3.8 mm long and 0.9–1.4 mm wide, tapering gradually to the ill-defined beak, or a little more abruptly narrowed above the middle, the beak obliquely cleft and shortly or scarcely bidentate; body of the perigynium somewhat spongy at the base, several-nerved dorsally, few-nerved ventrally, or the ventral nerves seldom obscure; stigmas 2; achene lenticular, elliptic, 1.2–1.5 mm long, jointed to the style.

Moist or wet meadows, streambanks, and bogs, at moderate to rather high elev. in the mts., extending up to about timberline; Cascade-Sierran region from s. Wash. to s. Calif., s. through the mts. of e. Oregon to c. Idaho, and at scattered stations in the mts. of Nev. (Elko Co.), Utah (Salt Lake Co.), s. Wyo., and n. Colo.

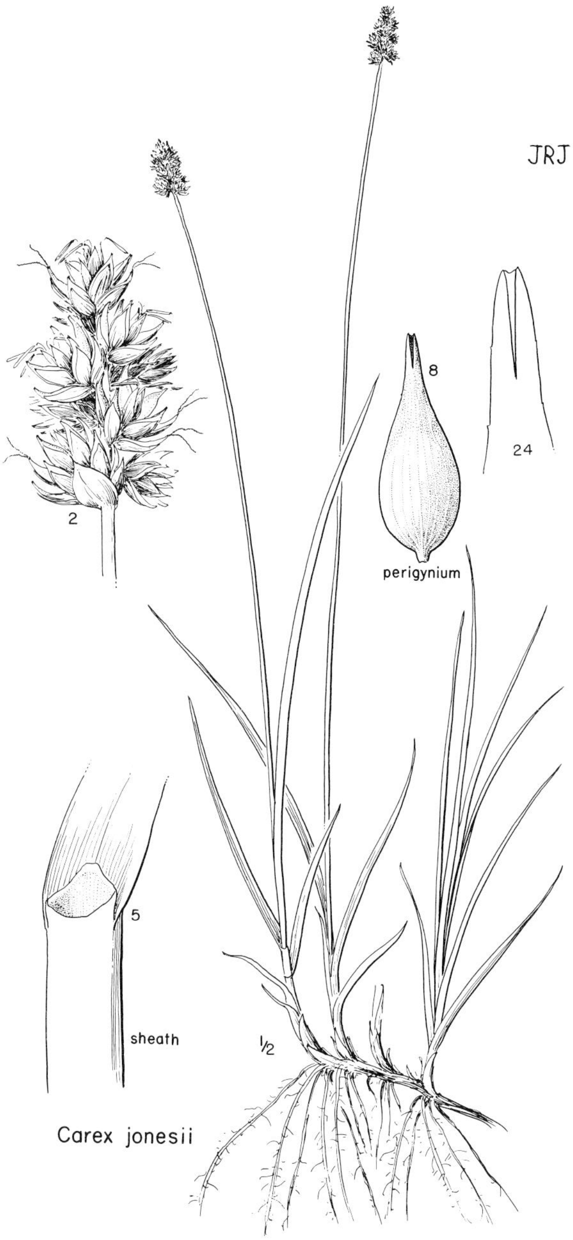

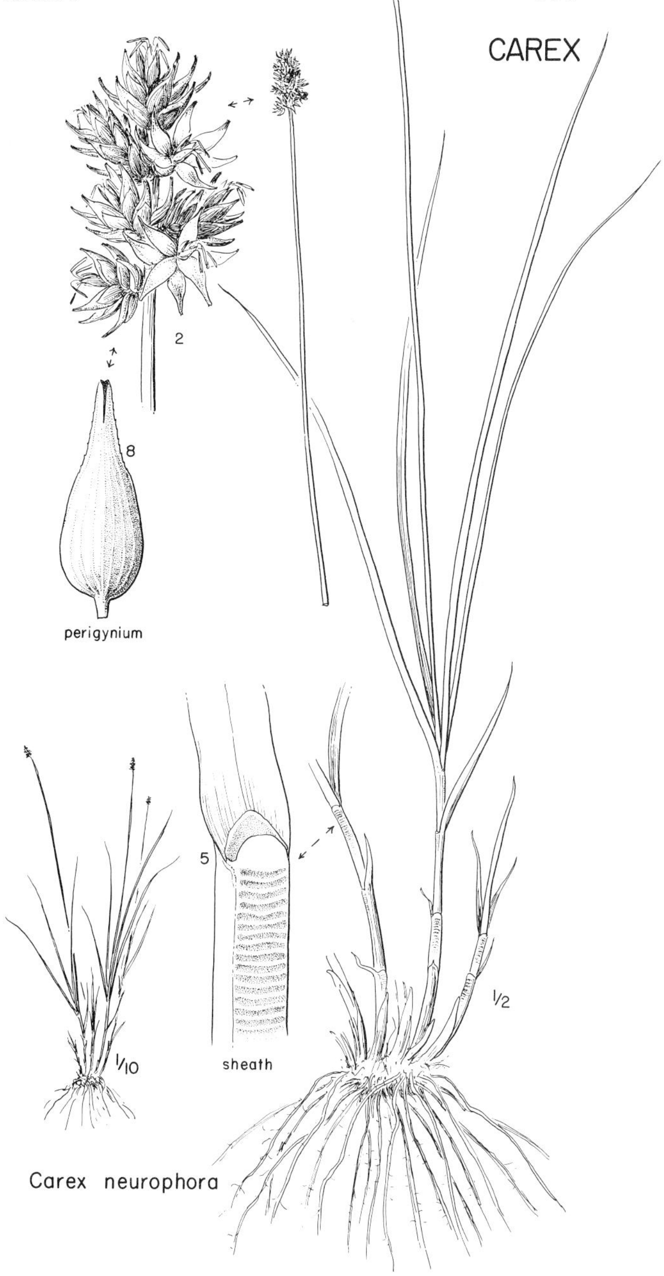

66. Carex neurophora Mackenzie

Carex neurophora Mackenzie in Abrams, Ill. Fl. Pacific States 1: 298. 1923. (*Sandberg & Leiberg 773,* Stevens Pass, Wash.)

C. vernacula var. *hobsonii* Maguire, Brittonia 5: 199. 1944. (*Maguire et al. 14013,* White Pine Lake, Cache Co., Utah.)

Stems clustered, 2–8 dm tall, strongly aphyllopodic, not notably slender; leaves flat or nearly so, 1.5–3 mm wide, all borne on the lower part of the stem, but not forming a close cluster; ventral side of the leaf sheaths pale (often with darker veins), usually rather conspicuously cross-corrugate especially distally, and commonly convex or prolonged (and often cartilaginous) at the mouth, occasionally some or even all of the sheaths plane instead of corrugated; spikes individually small, few-flowered, androgynous, and sessile, more or less numerous in a usually compound, thick-cylindric to broadly ovoid head that is 10–25 mm long and 6–15 mm thick, the individual spikes scarcely distinguishable; bracts small, sheathless, and inconspicuous, often scarcely larger than the pistillate scales, these largely hyaline-scarious and brownish or brownish-stramineous, equaling or somewhat smaller than the perigynia; perigynia commonly shining brownish-stramineous, lance-triangular, often rather broadly so, mostly 2.9–3.8 mm long and barely if at all over 1 mm wide, tapering evenly from base to apex, or a little more abruptly narrowed above the middle, the beak obliquely cleft and shortly or scarcely bidentate, often ill-defined; body of the perigynium somewhat spongy at the base, several-nerved dorsally, few-nerved ventrally, or the ventral nerves seldom obscure; stigmas 2; achene lenticular, 1.1–1.3 mm long, jointed to the style.

Swamps, moist or wet meadows, and streambanks at moderate elev. in the mts.; Cascade region from Wash. to s. Oregon, e. across n. and c. Idaho, to sw. Mont., and s. through the mts. of Wyo. to n. Colo.; also in n. Utah (Cache and Salt Lake cos.), and reported from Elko Co., Nev.

Carex neurophora, C. jonesii, and *C. nervina* L. H. Bailey (of Calif. and s. Oregon) form a closely related trio with overlapping but distinctive ranges. Although the characters separating these three are subject to some failure, the taxa are real enough, and at the present state of our knowledge it seems appropriate to continue to recognize them as species.

67. Carex disperma Dewey

Carex disperma Dewey, Amer. J. Sci. 8: 266. 1824. *Vignea disperma* Raf. Good Book 27. 1840. (*Dewey,* Mass.)

Soft, slender plants, the stems arising singly or few together from a branching system of slender rhizomes, somewhat aphyllopodic, 1–4 dm tall; leaves flat, mostly 1–2 mm wide, all borne on the lower part of the stem, but not densely tufted-clustered; spikes several, sessile, in a slender, interrupted-linear inflorescence, only the uppermost ones more or less contiguous; bracts subtending the spikes sheathless, inconspicuous and scarcely differing from the pistillate scales, or the lower ones filiform-foliaceous and up to about 2 cm long; spikes androgynous, very small and few-flowered, up to about 5 mm long, with mostly 1–3 perigynia and about as many staminate flowers; pistillate scales stramineous or paler, largely scarious or hyaline-scarious, equaling or often distinctly shorter than the perigynia; perigynia 2–3 mm long, elliptic, plano-convex, obscurely to evidently many-nerved, rather abruptly contracted to a short, obliquely cleft beak; stigmas 2; achene lenticular, filling the perigynium; style (below the branches) short (scarcely 0.5 mm), semi-persistent as a slender apiculus on the achene; $n = 35$.

Swamps, bogs, wet meadows, and shady woods in the mts.; circumboreal, but not at the highest latitudes, in Amer. from Newfl. and Lab. to Yukon and s. Alaska., s. to N.J., Ind., N.M., Utah (San Juan and Beaver cos.), Nev. (White Pine Co. and White Mts.), and Oregon (Klamath Co.).

68. Carex bipartita All.

Carex bipartita All. Fl. Pedemontana 2: 265. 1785. *Kobresia bipartita* Britton, Mem. Torrey Bot. Club 5: 101. 1894. (Piedmont of Italy.)

C. lachenalii Schkuhr, Beschreibung & Abbild. Arten Riedgräsern 51. 1801. (c. Europe.)

C. lagopina Wahlenb. Kongl. Vetensk. Acad. Nya Handl. 24: 145. 1803. (Lappland.)

C. bipartita var. *austromontana* F. J. Hermann, Leafl. W. Bot. 10: 16. 1963. (*Hermann 17059,* 8 mi. nw. of Nederland, Boulder Co., Colo.) The long-beaked extreme, represented by a collection from the Uinta Mts. of Utah (*Maguire 14497*) in addition to the type. More typical material of *C. bipartita* also occurs in Colo.

Culms 0.5–3 dm tall, clustered, only slightly aphyllopodic; leaves clustered at the base, flat or nearly so, equaling or often shorter than the culms, 1–2.5 mm wide; bracts inconspicuous, short, sheathless, and largely hyaline-scarious, sometimes shortly awn-tipped; spikes 2–4, all approximate but easily distinguishable to the naked eye, sessile, gynaecandrous, 6–15 mm long; pistillate scales largely brownish and hyaline-scarious, often with firmer midrib and sometimes with pale, hyaline margins, about equaling (or even surpassing) and largely concealing the perigynia, or minutely exceeded by the perigynial beaks; perigynia mostly (10) 15–30, rather closely ascending, often somewhat paler than the scales, mostly 2.4–3.4 mm long, plano-convex or almost flat, but not especially thin-margined, finely and sometimes obscurely several-nerved on each side, minutely rough-textured, elliptic to obovate, with a short beak or beak-apiculation up to 0.5 mm long, or seldom with a more prominent, slender beak up to 1.0 mm long; dorsal suture of the perigynium evident, 0.7–1.5 mm long, extending well onto the distal part of the body; stigmas 2; achene lenticular, 1.2–1.5 mm long, often not filling the distal part of the perigynium; $n = 29, 30, 31, 32$.

Wet places at high alt. in the mts.; circumboreal at high latitudes, extending s. at increasing elev. to Que., Alta., B.C. and n. Mont.; disjunct in the Uinta Mts. of Utah and the Rocky Mts. of Colo., apparently skipping Wyo.

69. Carex brunnescens (Pers.) Poiret

Carex curta var. *brunnescens* Pers. Syn. Pl. 2: 539. 1807. *C. brunnescens* Poiret in Lam. Encycl. Méth. Bot. Suppl. 3: 286. 1813. *C. canescens* var. *brunnescens* Koch, Syn. Fl. German. & Helv. 754. 1837. *Facolos brunnescens* Raf. Good Book 26. 1840. *C. canescens* subsp. *C. brunnescens* Aschers. & Graebner, Syn. Mitteleur. Fl. 2(2): 61. 1902. (Switzerland.)

C. vitilis Fries, Novit. Fl. Suec. Mantissa 3: 137. 1842. *C. canescens* var. *vitilis* Carey in A. Gray, Manual Bot. ed. 2. 514. 1856. (Norway & Sweden.)

Culms mostly (1) 2–6 dm tall, more or less densely clustered, without long rhizomes, somewhat aphyllopodic; leaves clustered near the base, equaling or often shorter than the culms, flat or nearly so, mostly 1–2.5 mm wide; bracts inconspicuous, short, and sheathless, or the lowest ones setaceous-tipped and surpassing the spikes; spikes mostly 4–9, sessile, gynaecandrous, 4–8 mm long, the lower ones generally somewhat remote, so that the inflorescence is interrupted, the uppermost ones often approximate, or seldom all spikes approximate in a narrow but continuous inflorescence, sometimes pale as in *C. canescens,* sometimes brownish as in *C. praeceptorum;* pistillate scales largely hyaline-scarious except for the mostly firmer and often greenish midvein, commonly shorter than the perigynia; perigynia mostly 5–10 (15), tending to be a little more spreading than in *C. canescens,* so that the beak-apiculations interrupt the outline of the spike; perigynia mostly 1.7–2.5 mm long, plano-convex, not thin-edged, elliptic or elliptic-ovate, with a short beak-apiculation up to about 0.5 mm long, evidently but rather finely several-nerved on the back, only obscurely nerved or practically nerveless ventrally, smoother-textured than in *C. canescens,* the wall very thin distally and easily ruptured; dorsal suture of the perigynium rel-

CAREX

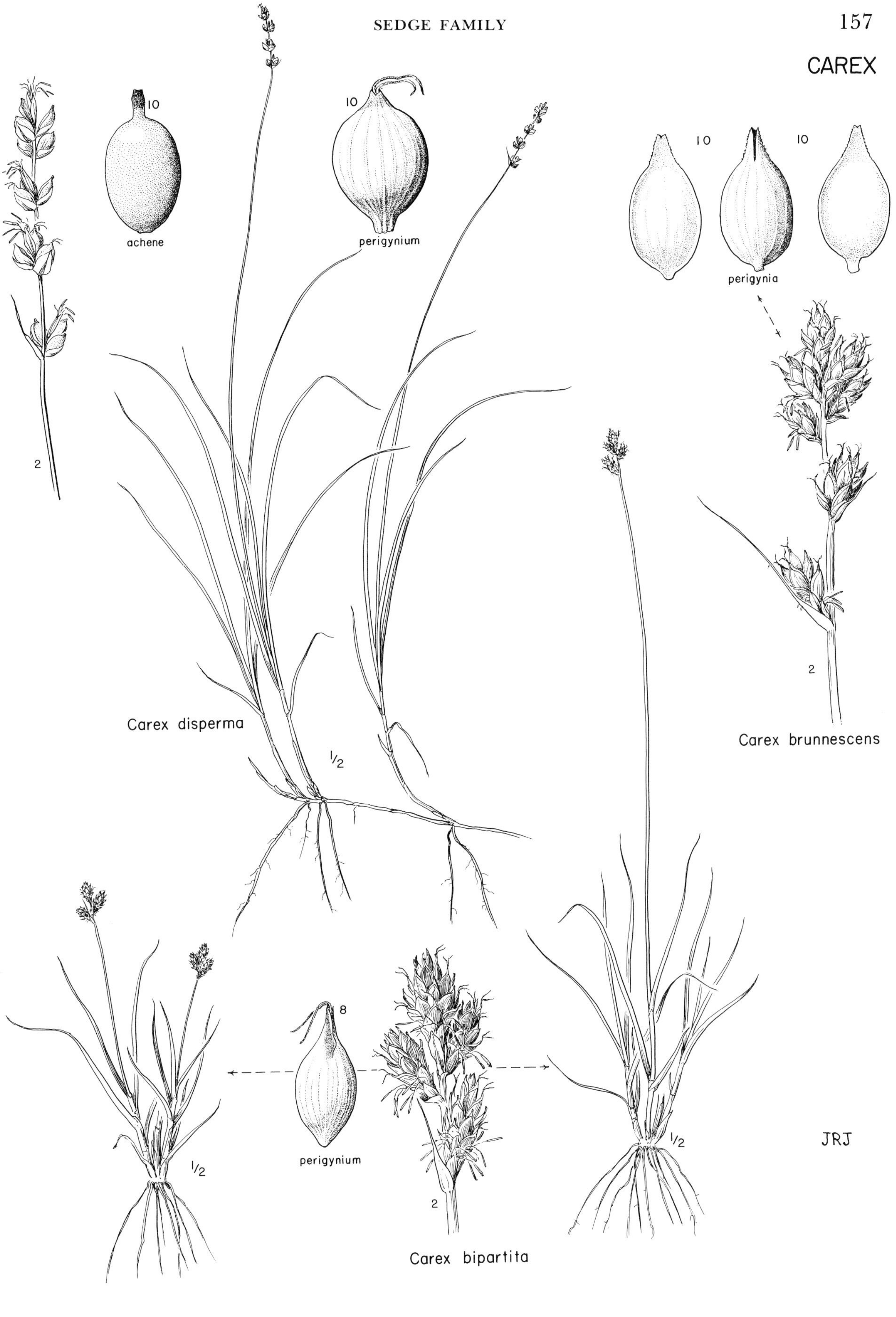
10
achene
10
perigynium
2
10
10
perigynia
2
Carex brunnescens
Carex disperma
1/2
8
perigynium
1/2
2
1/2
Carex bipartita
JRJ

atively well developed and prominent, commonly 0.4–0.8 mm long and extending onto the distal end of the body, often with a narrow, white-hyaline, overlapping, longitudinal flap; stigmas 2; achene lenticular, 1.2–1.4 mm long; *n* = 28.

Wet places in the mts. (with us only at high elev.); circumboreal at high latitudes, extending s. in Amer. to Va., Tenn., Wyo., Utah (Uinta Mts. and reputedly Tushar Mts.), and through the Cascade Mts. of Wash. and Oregon to the Siskiyou Mts. of sw. Oregon. Seldom collected in our range.

Carex brunnescens is evidently allied to *C. canescens,* and at first consideration the two scarcely appear to be specifically distinct. They differ, however, in several well correlated small characters. None of these provides a sharp enough distinction to serve as a wholly reliable key character, but collectively they mark an apparently real hiatus between the two taxa. *C. canescens* and *C. brunnescens* maintain their separate identities throughout their broadly overlapping (but not identical) ranges. In our area the separation between them is complicated by the existence of *C. praeceptorum* (q.v.), which cannot well be included in either one.

70. Carex praeceptorum Mackenzie

Carex praeceptorium Mackenzie, N. Amer. Fl. 18: 95. 1931. (*Peck 13,* Simcoe Mts., 15 mi. n. of Goldendale, Wash.)

Stems loosely to densely tufted on rather short to very short rhizomes, 1–3 dm tall, slightly to evidently aphyllopodic; leaves all borne near the base, equaling or shorter than the culms, mostly 1.5–2.5 mm wide; bracts inconspicuous, short, sheathless, and largely hyaline-scarious, sometimes shortly awn-tipped; spikes mostly 4–6, all approximate but readily distinguishable to the naked eye, sessile, gynaecandrous (or sometimes some of them apparently wholly pistillate), 5–12 mm long; pistillate scales brownish, largely hyaline-scarious, with a firmer, often greenish midrib, mostly somewhat shorter than the perigynia; perigynia mostly (10) 15–25, ascending, not notably loose-tipped, commonly somewhat paler than the scales but the exposed parts often brownish, mostly 1.7–2.5 mm long, plano-convex, neither much thickened nor evidently thin-margined, evidently several-nerved on each side, the nerves slightly raised and often somewhat darker than the rest of the body, which is very minutely rough-textured, more or less elliptic, with a short beak-apiculation that is less than 0.5 mm long; dorsal suture of the perigynium short but evident, 0.35–0.6 mm long, often darker brownish than the surrounding tissues, commonly extending the length of the beak-apiculation but only slightly if at all encroaching onto the body of the perigynium, which is not especially thin-walled; stigmas 2; achene lenticular, 1.1–1.4 mm long, filling the body of the perigynium.

Wet meadows and other wet places, mostly at rather high elev. in the mts., commonly a little below timberline; s. Wash. (Simcoe Mts.), e. across c. Idaho to nw. Wyo., s. to the Sierra Nevada of Calif., the mts. of Nev. (Elko Co.), Utah (Piute Co.), and Colo. Rather seldom collected.

Carex praeceptorum is obviously related to *C. canescens, C. brunnescens,* and *C. bipartita,* but it cannot be included in any of these species without seriously compromising the distinctions among them. The spelling *praeceptorium,* used by Mackenzie in the original publication of the species, is here treated as an unintentional orthographic error, to be corrected under Article 73 of the International Code of Botanical Nomenclature.

71. Carex canescens L.

Carex canescens L. Sp. Pl. 974. 1753. *Vignea canescens* Reichenb. in Mössler, Handb. Gewächsk. ed. 2. 1612. 1829. (n. Europe.)

C. canescens var. *dubia* L. H. Bailey, Bot. Gaz. 9: 119. 1884. (*Watson 1231,* Bear River Canyon, Utah.)

Culms mostly (1) 2–6 dm tall, loosely to densely clustered on rather short to very short rhizomes, somewhat aphyllopodic; leaves clustered near the base, equaling or often shorter than the culms, flat or nearly so, 1.5–3.5 mm wide; bracts inconspicuous, short, sheathless, and largely hyaline-scarious, or the lowest ones setaceous-tipped and surpassing the spikes; spikes 4–8, sessile, gynaecandrous, 5–10 mm long, varying from more or less approximate to somewhat remote, often the upper ones more crowded than the lower, silvery greenish to pale grayish or pale stramineous in aspect; pistillate scales pale and largely hyaline-scarious except for the mostly firmer and often greenish midvein, commonly shorter than the perigynia; perigynia mostly (10) 15–30, ascending (the short beaks not conspicuously interrupting the outline of the spike), light greenish to dull silvery or pale stramineous, mostly 1.8–2.5 mm long, plano-convex, not thin-edged, minutely rough-textured, somewhat corky-thickened below, not especially thin-walled distally, elliptic or elliptic-ovate, with a short beak-apiculation up to 0.5 mm long, evidently but rather finely nerved on both faces, the ventral nerves fewer than the dorsal; dorsal suture of the perigynium short and inconspicuous, up to about 0.4 (0.7) mm long, or practically obsolete; stigmas 2; achene lenticular, 1.2–1.4 mm long, filling the body of the perigynium; *n* = 27, 28.

Wet meadows and other wet places at moderate to fairly high elev. in the mts., descending to the lowlands in more humid regions beyond our range; circumboreal, extending s. to Va., Ariz., and Calif.; only occasional in our range, where known from the Uinta, LaSal, and Wasatch ranges of Utah and the mts. of Elko Co., Nev.

72. Carex deweyana Schwein.

Carex deweyana Schwein. Ann. Lyceum Nat. Hist. New York 1: 65. 1824. (s. New England; no specimens cited.)

C. bolanderi Olney, Proc. Amer. Acad. Arts 7: 393. 1868. *C. deweyana* var. *bolanderi* W. Boott in S. Wats. Bot. Calif. 2: 236. 1880. (*Bolander 6209,* Yosemite, Calif.; lectotype by Mackenzie.)

C. leptopoda Mackenzie in Rydb. Fl. Rocky Mts. 124, 1060. 1917 [1918]. *C. deweyana* subsp. *leptopoda* Calder & Taylor, Canad. J. Bot. 43: 1389. 1965. *C. deweyana* var. *leptopoda* B. Boivin, Naturaliste Canad. 94: 522. 1967. (*Heller 10052,* near Oswego, Clackamas Co., Oregon.)

Stems mostly 2–10 dm tall, slightly aphyllopodic, in small to large, loose to dense clumps, without long, creeping rhizomes; leaves borne on the lower part of the stem, but not closely tufted, flat or nearly so, mostly 2–5 mm wide, the plant appearing more leafy than in related species; ligule longer than wide, often acute; spikes 4–10, gynaecandrous, mostly 7–20 mm long, sessile (but the inflorescence sometimes branched), approximate and loosely aggregated, or the lowest ones somewhat remote, forming a loose to moderately compact inflorescence 2–8 cm long, light green or greenish tan in aspect; bracts subtending the spikes reduced and inconspicuous, or

CAREX

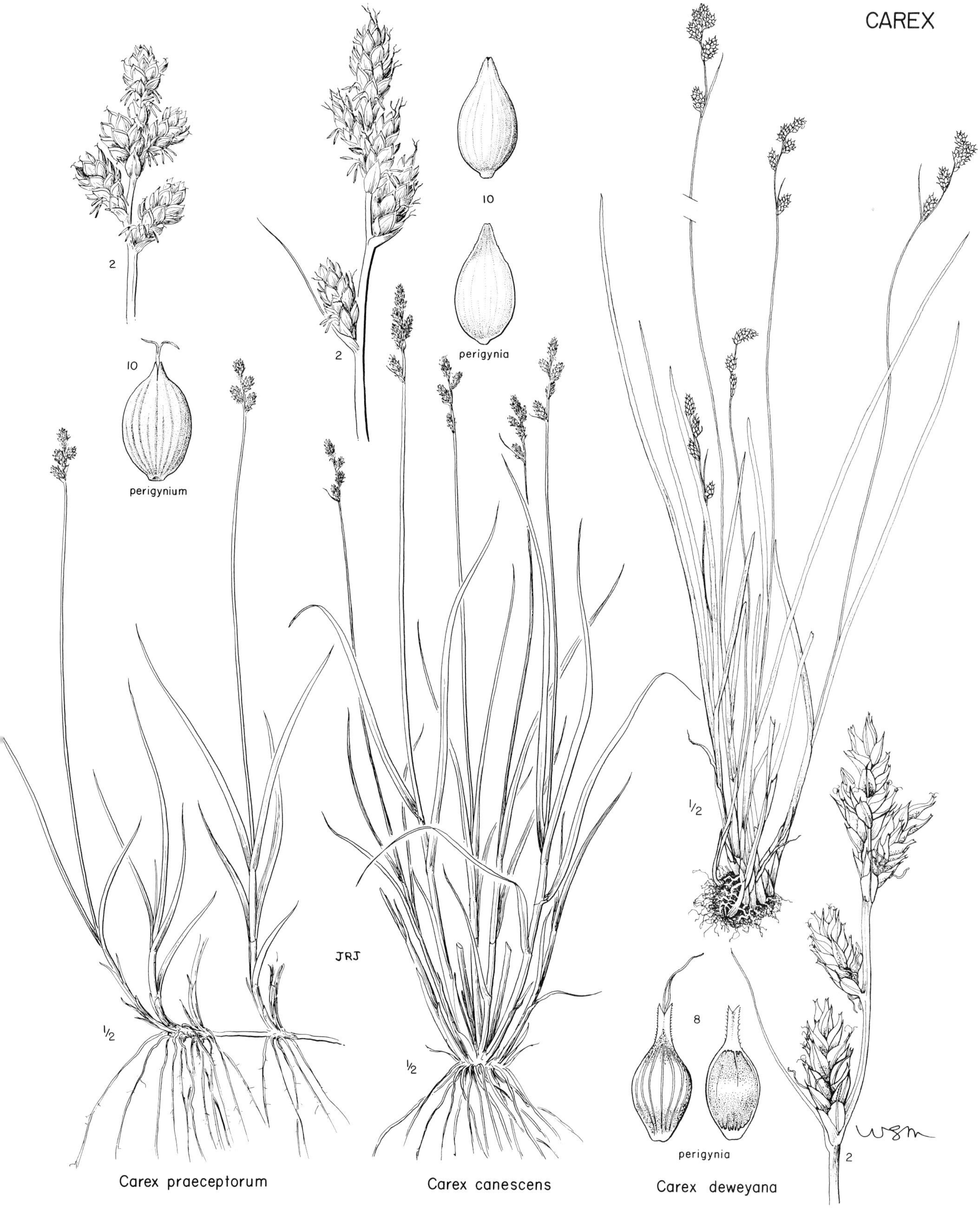

Carex praeceptorum

Carex canescens

Carex deweyana

the lower ones setaceous-prolonged; pistillate scales largely hyaline or scarious and whitish to light brown, with firmer, commonly greenish midrib which may be excurrent as a short awn-tip, surpassed by the beak of the perigynium, but often wholly covering the body; perigynia mostly (5) 10–25 per spike, loosely ascending, lance-elliptic to rather narrowly elliptic, (3.1) 3.5–4.8 mm long, the body commonly tan or stramineous, becoming more greenish distally and marginally, or greenish throughout, plano-convex, few- to several-nerved dorsally, 0- to few-nerved ventrally, spongy-thickened at the base, otherwise rather thin-walled, with prominent, raised marginal veins, but not thin-edged, the body tapering into a prominent, slender, conspicuously bidentate beak (0.8) 1.1–1.9 mm long, the beak (and often also the distal part of the body) serrulate-margined; stigmas 2; achenes lenticular, 1.5–2.0 mm long; $n = 27$.

Streambanks and moist woodlands or forest openings, from near sea level (in moist regions) to near timberline in the mts.; Lab. and Newfl. to Mack. and B.C., s. to Pa., Iowa, N.M., Ariz., and Calif., occurring at scattered stations in the mts. of our range; also in e. Asia.

The cordilleran phase of the species, as here described, may be distinguished, with some difficulty, as var. *bolanderi* (Olney) W. Boott, in contrast to the more eastern var. *deweyana,* with mostly larger (to 2.5 mm) achenes and short, broadly rounded ligule. For those who prefer the subspecific category, our plants become *C. deweyana* subsp. *leptopoda* (Mackenzie) Calder & Taylor.

73. Carex interior L. H. Bailey

Carex interior L. H. Bailey, Bull. Torrey Bot. Club 20: 426. 1893. (*Sartwell 36,* Penn Yan, N.Y.)

Stems slender, densely tufted, somewhat aphyllopodic, 2–5 dm tall; leaves all borne on the lower part of the stem but not closely bunched, slender and elongate, sometimes equaling the culms, flat or nearly so, mostly 1–2 mm wide; spikes mostly 3–6, sessile, small and few-flowered, contiguous or more or less remote, gynaecandrous, or some of the lateral ones wholly pistillate, but the terminal one with a conspicuous, slender, staminate base; bracts small and inconspicuous, scarcely different from the pistillate scales, these shorter than the perigynia and largely hyaline or scarious except for the firmer, often greenish midvein; perigynia mostly (1) 5–15 per spike, crowded, widely spreading at maturity, or the lower ones reflexed, green or tan, broadest near the base, ovate or triangular-ovate to broadly lance-triangular, coriaceous and shining, spongy-thickened at the base, 4- to 12-nerved dorsally, 0- to 7-nerved ventrally, plano-convex, with prominent, raised marginal nerves, serrulate-margined distally and on the beak, 2.2–3.2 mm long, with a short, broad, very shallowly bidentate beak 0.5–1.0 mm long, a fourth to half as long as the body; stigmas 2; achene lenticular, about 1.5 mm long, commonly rather broadly trapezoid-ovate, jointed to the style.

Swamps, bogs, and other wet places, from the lowlands (in moist regions) to moderate elev. in the mts.; Lab. and Newfl. to s. B.C., s. to Pa., Ind., Kansas, and in the cordilleran region to n. Mex.; seldom collected in our area, but to be expected in appropriate habitats throughout our range.

74. Carex muricata L.

Carex muricata L. Sp. Pl. 974. 1753. (Europe.)

C. stellulata Gooden. Trans. Linn. Soc. London 2: 144. 1794. *Vignea stellulata* Reichenb. in Mössler, Handb. Gewächsk. ed. 2. 1610. 1829. *Caricina stellulata* St.-Lager in Cariot, Étude Fleurs Bot. Element. Descript. & Usuelle, ed. 8. 2: 878. 1897. (England.)

C. stellulata var. *angustata* Carey in A. Gray, Manual Bot. 544. 1848. *C. echinata* var. *angustata* L. H. Bailey, Mem. Torrey Bot. Club 1: 59. 1889. *C. sterilis* var. *angustata* L. H. Bailey, Bull. Torrey Bot. Club 20: 425. 1893. *C. angustior* Mackenzie in Rydb. Fl. Rocky Mts. 124, 1060. 1917 [1918]. (Fairfield, N.Y., fide Mackenzie.)

Stems slender, densely tufted, somewhat aphyllopodic, 1–6 dm tall; leaves all borne on the lower part of the stem, but not closely bunched, slender and commonly more or less elongate, sometimes equaling the culms, or nearly so, mostly 1–2 mm wide; spikes mostly 3–6 (7), sessile, small, few-flowered, contiguous or more or less remote, gynaecandrous or some of lateral ones wholly pistillate, but the terminal one with a conspicuous, slender, staminate base; bracts small and inconspicuous, scarcely different from the pistillate scales, or a little larger and shortly awn-tipped; pistillate scales shorter than the perigynia and largely hyaline or scarious (commonly brownish-tinged) except for the firmer, often greenish midvein; perigynia mostly 5–15 per spike, crowded, widely spreading at maturity, or the lower ones reflexed, green or tan (the beak commonly greener than the body), narrowly lance-triangular to more commonly lance-ovate or even ovate, slightly corky-thickened at the base, otherwise rather thin-walled, lightly several- to many-nerved on the back, few- to several-nerved or practically nerveless ventrally, plano-convex, with prominent marginal nerves (but not thin-edged), serrulate-margined on the beak and often also on the distal part of the body, 2.8–3.5 (4.0) mm long, with a prominent, slender, evidently bidentate beak 1.1–1.6 mm long, this half as long to almost as long as the body; stigmas 2; achene lenticular, 1.1–1.6 mm long, varying from much longer than wide to about as wide as long; $n = 28, 29$.

Swamps, bogs, and other wet places, from near sea level to moderate elev. in the mts.; circumboreal, but not at the highest latitudes, in Amer. extending s. to N.C., Wisc., Colo., Utah, and c. Oregon, and s. in the Cascade-Sierran axis to s. Calif., mainly in relatively humid regions; seldom collected in our range.

Carex muricata as here defined is no more variable than many other species of *Carex,* but it has been customary to maintain several specific or varietal segregates based on technical details of the pistillate scales, perigynium, and achenes. These segregates are not sharply delimited, and they have widely overlapping to almost coextensive ranges. Only *C. angustior* is admitted by Mackenzie to occur in our range, but the distinction between this and typical European *C. muricata* is obscure. At least until a thorough monographic study is made, under modern taxonomic concepts, it seems best to maintain *C. muricata* as a single broadly defined species without infraspecific taxa.

75. Carex illota L. H. Bailey

Carex bonplandii var. *minor* F. Boott, Proc. Acad. Nat. Sci. Philadelphia 1863: 77. 1864. *C. illota* L. H. Bailey, Mem. Torrey Bot. Club I: 15. 1889. (*Hall & Harbour 591,* Colo.)

Plants tufted, without creeping rhizomes; culms slightly aphyllopodic, 1–3 (4) dm tall; leaves all clus-

CAREX

Carex interior

Carex muricata

Carex illota

tered towards the base, shorter than the culms, flat or nearly so, 1–3 mm wide; spikes several, sessile, gynaecandrous, 4–7 mm long, inconspicuously bracteate, closely crowded into an ovoid or ellipsoid head 8–13 mm long, the individual spikes often scarcely distinguishable to the naked eye, but the lowest ones often more spreading than the others so that the inflorescence has a lobate outline, the head in any case tending to appear ragged or serrate because of the separate tips of the spreading perigynia; pistillate scales shorter and often somewhat narrower than the perigynia, largely or wholly hyaline-scarious, shining, rather dark brown to greenish-black; perigynia lanceolate to narrowly ovate, greenish-fuscous to more or less brownish, or sometimes brownish-stramineous, 2.5–3.2 mm long, plano-convex, sharp-edged but not wing-margined, tending to be lightly several-nerved dorsally, nerveless or obscurely few-nerved ventrally, tapering to an ill-defined, obliquely cleft, dark beak, entire-margined throughout, not serrulate; stigmas 2; achene lenticular, 1.3–1.5 mm long, including the substipitate base; $n = 32$.

Bogs, wet meadows, and other wet places at high alt. in the mts., often above timberline; Wash. and adj. B.C. to nw. Mont., s. to Calif., Nev. (mts. of Elko Co.), Utah (Uinta and occasionally Wasatch mts.), and Colo.

Carex illota is a distinctive and well-marked species which might about as well be referred to the Stellulatae as to the Ovales.

76. **Carex athrostachya** Olney

Carex athrostachya Olney, Proc. Amer. Acad. Arts 7: 393. 1868. (*Brewer 1650,* Yosemite Valley, Calif., is the first specimen cited.)

C. macloviana var. *pachystachya* f. *involucrata* Kükenthal in Engl. Pflanzenr. IV. 20 (Heft 38): 197. 1909. (*Rydberg 3041,* Mont., is the first specimen cited.)

Plants densely tufted, without creeping rhizomes; culms aphyllopodic, 1.5–10 dm tall; leaves flat or nearly so, 1.5–4 mm wide, mostly shorter than the culms, or borne on the lower half of the culm but not closely crowded; spikes several, sessile, gynaecandrous, 6–10 mm long, closely crowded in an ovoid or globose-ovoid head 1–2 cm long, often scarcely distinguishable to the naked eye, the inflorescence greenish-stramineous to medium brown in aspect, obviously terminal, but one or more of the lowest bracts setaceous-prolonged, (1) 2–8 (10) cm long and equaling or generally surpassing the inflorescence, the other bracts reduced and inconspicuous; pistillate scales largely brownish and hyaline-scarious except for the firmer midrib, often shortly awn-pointed, slightly to conspicuously shorter and narrower than the perigynia, or the lowest ones in the spike somewhat longer than their perigynia; perigynia pale green to tan, often with a brown dorsal midstripe distally, flattened or somewhat plano-convex, lanceolate or rather narrowly ovate, lightly several-nerved dorsally, usually fewer-nerved or nerveless ventrally, mostly (2.8) 3.2–4.6 (4.8) mm long, 0.9–1.5 (1.8) mm wide, and 2.5–4.5 times as long as wide, wing-margined and serrulate, tapering to an obliquely cleft beak, the beak narrowly margined and serrulate below, marginless, subterete, and usually entire in the distal 0.5–0.9 mm; stigmas 2; achene lenticular, 1.1–1.6 mm long and 0.7–0.9 mm wide.

Moist or wet places, from the lowlands to moderate elev. in the mts., not reaching timberline; s. Alaska to s. Calif., e. to Sask., N.D., and Colo., and found in appropriate habitats throughout our range.

The species furnishes fair to good forage for cattle and horses.

77. **Carex limnophila** F. J. Hermann

Carex limnophila F. J. Hermann, Leafl. W. Bot. 8: 28. 1956. (*Hermann 12252,* Half Moon Lake, Wind R. Mts., Sublette Co., Wyo.)

Plants tufted, without creeping rhizomes; culms slightly aphyllopodic, 1.5–4.5 dm tall; leaves all borne on the lower part of the culm, shorter than the culms, flat or nearly so, 1–3 mm wide; spikes several, sessile, gynaecandrous, 4–7 mm long, inconspicuously bracteate, closely crowded into an elliptic to globose-ovoid head 7–15 mm long, the individual spikes scarcely distinguishable to the naked eye; pistillate scales largely or wholly hyaline-scarious, medium or dark brown to greenish-black, from somewhat shorter and narrower than the perigynia to fully as wide and nearly as long as the perigynia, with only the beak-tip exserted; perigynia lanceolate or rather narrowly ovate, appressed-ascending, brownish to greenish, 2.5–3.2 mm long, plano-convex, distinctly wing-margined and serrulate, especially toward the distal part of the body, lightly several-nerved or nerveless dorsally, usually nerveless ventrally and often with a transverse ventral fold near or below the middle, tapering or more abruptly contracted into a slender, obliquely cleft beak, of which at least the distal 0.5 mm is subterete, marginless, and entire; stigmas 2; achene lenticular, 1.0–1.4 mm long.

Moist or wet places at moderate elev. in the mts., seldom collected; Alta. to Wash., s. to Colo., ne. Oregon, and Elko Co., Nev.

78. **Carex subfusca** W. Boott

Carex subfusca W. Boott in S. Wats. Bot. Calif. 2: 234. 1880.

C. macloviana var. *subfusca* Kükenthal in Engl. Pflanzenr. IV. 20 (Heft 38): 197. 1909. (*Kellogg,* Lake Tahoe, is the first of two specimens cited.)

C. festiva var. *stricta* L. H. Bailey, Mem. Torrey Bot. Club 1: 51. 1889. *C. festiva* (subsp.) *horneri* Piper, Contr. U.S. Natl. Herb. 11: 164. 1906. *C. macloviana* var. *stricta* (attr. to L. H. Bailey by) Kükenthal in Engl. Pflanzenr. IV. 20 (Heft 38): 196. 1909. (*Palmer 389,* Calif.; lectotype by Mackenzie.)

C. teneraeformis Mackenzie, Bull. Torrey Bot. Club 43: 609. 1917. (*Hall 9781,* Jonesville, Butte Co., Calif.) A form with the perigynia nearly nerveless ventrally.

Similar in most respects to *C. pachystachya;* perigynia and scales commonly paler, sometimes largely greenish-stramineous; perigynia smaller, mostly 2.7–3.2 (3.6) mm long and 1.0–1.5 mm wide, usually evidently nerved ventrally as well as dorsally.

Moist or dry soil at moderate elev. in the mts.; w. Oregon (in and w. of the Cascade Mts.) s. to the mts. of s. Calif., and thence e. to the mts. of Ariz., s. Utah (Wayne & Washington cos.), and probably N.M.

CARER

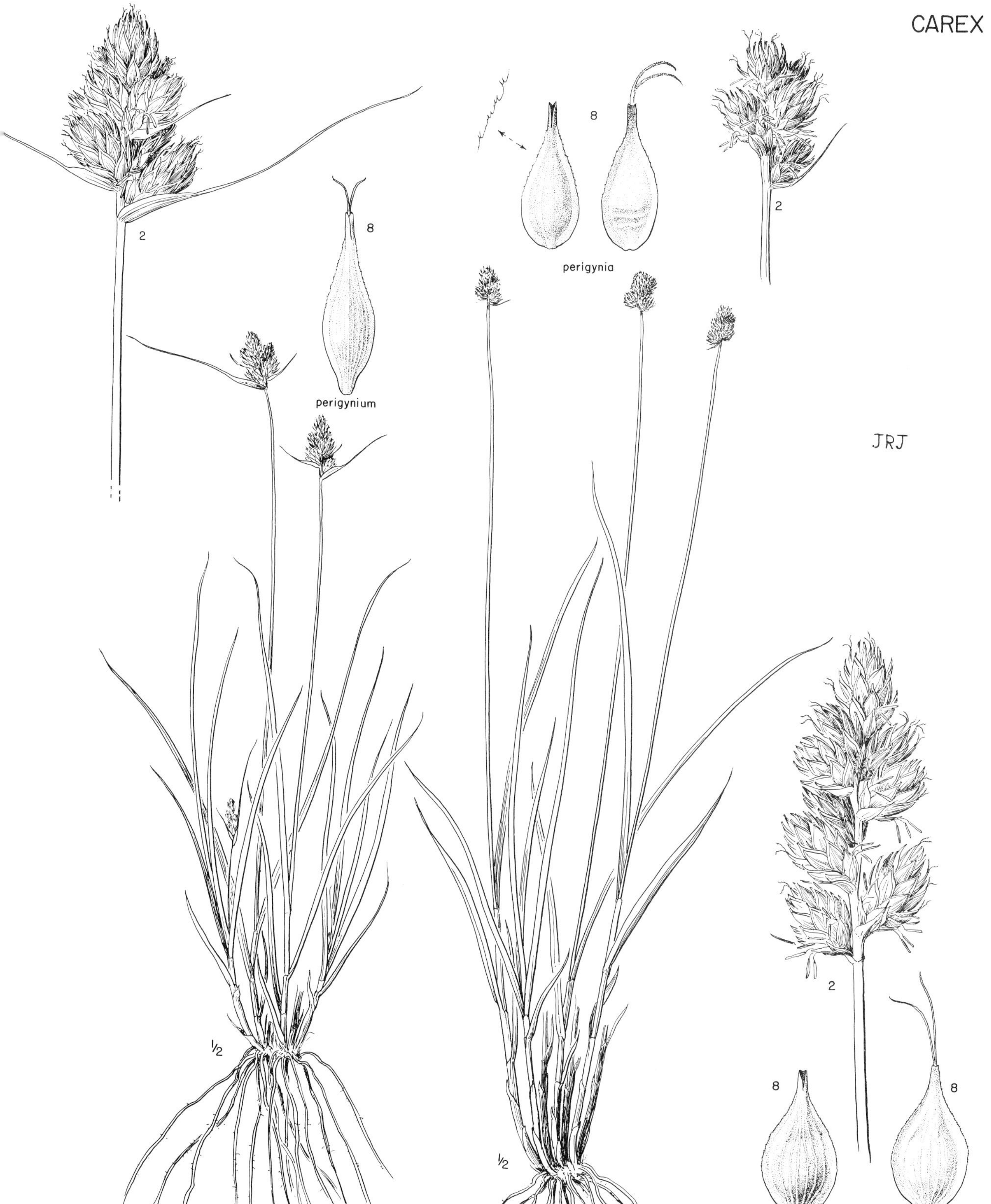

Carex athrostachya

Carex limnophila

Carex subfusca

79. Carex pachystachya Cham. ex Steudel

Carex pachystachya Cham. ex Steudel, Syn. Pl. Glum. 2: 197. 1855. *C. festiva* var. *pachystachya* L. H. Bailey, Mem. Torrey Bot. Club 1: 51. 1889. *C. macloviana* var. *pachystachya* Kükenthal in Engl. Pflanzenr. IV. 20 (Heft 38): 197. 1909. *C. macloviana* subsp. *pachystachya* Hultén, Fl. Alaska & Yukon 2: 318. 1942. (*Chamisso*, Unalaschka.)

C. preslii Steudel, Syn. Pl. Glum. 2: 242. 1855. (*Haenke*, Nootka Sound.)

C. festiva var. *gracilis* Olney ex W. Boott in S. Wats. Bot. Calif. 2: 234. 1880. *C. multimoda* L. H. Bailey, Bot. Gaz. 21: 5. 1896. *C. macloviana* var. *gracilis* Kükenthal in Engl. Pflanzenr. IV. 20 (Heft 38): 197. 1909. *C. pachystachya* var. *gracilis* Mackenzie, N. Amer. Fl. 18: 136. 1931. (Typification recondite; *Hall 586*, Oregon, fide Mackenzie.)

C. olympica Mackenzie, Bull. Torrey Bot. Club 43: 610. 1917. (Presumably *Elmer 2700*, Olympic Mts., Wash.)

C. platylepis Mackenzie, N. Amer. Fl. 18: 142. 1931. (*Williams 2951*, Ten Sleep Lakes, Big Horn Mts., Wyo.)

C. pachystachya var. *monds-coulteri* Kelso, Biol. Leafl. 64: 2. 1953. *C. pachystachya* f. *monds-coulteri* F. J. Hermann, Leafl. W. Bot. 9: 16. 1959. (*Kelso 6662*, Aspen, Pitkin Co., Colo.)

Plants tufted, without creeping rhizomes; culms somewhat aphyllopodic, 1.5–7 dm tall; leaves all borne on the lower part of the stem but generally not appearing tufted at the base, flat or nearly so, shorter than or often nearly equaling the culms, mostly (1.5) 2–5 mm wide; spikes 3–10, sessile, gynaecandrous, 7–12 mm long, inconspicuously bracteate, more or less closely crowded into a typically rather elongate-ovoid inflorescence 1–2 (3) cm long, the inflorescence varying to globose-ovoid or narrower and compactly short-spiciform; pistillate scales largely or wholly hyaline-scarious, medium brown, often with green midrib, shorter and often somewhat narrower than the perigynia, exposing their beaks and commonly also their distal margins; perigynia with tan center and pale green or greenish-stramineous margins, varying to largely or wholly golden brown, the beak in any case dark-tipped, the whole perigynium 3.6–5.1 mm long and 1.5–3 times as long as wide, narrowly to broadly ovate, with winged, serrulate margins, thin-walled but more or less distinctly plano-convex, the ventral side flat and usually nerveless or obscurely few-nerved, seldom evidently nerved, the dorsal side distended by the achene, finely several-nerved, and tending to have a pair of prominent inframarginal nerves which delimit the wing from the perigynial cavity; beak of the perigynium well defined, up to about 1.5 mm long, shortly or scarcely bidentate, the lower part often flattened and serrulate like the body, but the distal part (0.5 mm or more) subterete and scarcely serrulate; stigmas 2; achene lenticular, 1.4–1.9 mm long, 0.9–1.5 mm wide, flat ventrally, more convex dorsally, commonly not much smaller than the perigynial cavity; $n = 39$.

Open, often rather dry slopes, or less often in wet places, from near sea level to timberline in the mts.; moist regions, from coastal Alaska and the Aleutian Isl., s. to Wash. and Oregon (especially in and w. of the Cascade Mts.) and occasionally to the Sierra Nevada of Calif. and the Mt. Rose corner of Nev., e. to sw. Alta., w. Mont. and occasionally n. Wyo., and in the Wallowa Mts. of Oregon and the Seven Devils & Sawtooth mts. of Idaho, reaching our range only in Owyhee Co., Idaho.

Carex pachystachya is closely allied to *C. microptera*, from which it differs in often having a somewhat more elongate inflorescence with less closely crowded spikes (the individual spikes more readily distinguishable to the naked eye) and more especially in its plumper, more or less obviously plano-convex perigynia which are more nearly filled by the achene, the achene itself being distinctly larger than that of *C. microptera*. *C. pachystachya* is a plant of moister climates than *C. microptera*, showing the characteristic range pattern of many Cascadian and coastal species that also extend e. across n. Idaho to nw. Mont. The ranges of the two would appear more nearly discrete on a dot map than they do in a mere generalized statement. Where the ranges do overlap, it is not always easy to draw the distinction. It is not improbable that monographic study will force the subordination of both *C. pachystachya* and *C. microptera* to the high-boreal and antarctic species *C. macloviana* Urv., but no such treatment is here proposed.

Plants from the Sierra Nevada that are much like *C. pachystachya*, but with the perigynia more evidently nerved ventrally, have been distinguished as *C. mariposana* L. H. Bailey (if the lower spikes are more or less separate from the closely crowded upper ones) or *C. abrupta* Mackenzie (if the inflorescence is more densely capitate). Other alleged differences are not confirmed by examination of the specimens. Specimens with the perigynia evidently nerved ventrally also occur occasionally within the range of typical *C. pachystachya*, as for example in central Idaho, and *C. abrupta* has been reported from the White Mts. of California. It seems likely that *C. abrupta* and *C. mariposana* collectively are no more than varietally distinct from *C. pachystachya*, but in view of the possibility that *C. pachystachya* itself should be subordinated to a more widespread and broadly defined *C. macloviana* Urv., no new combinations are here proposed.

80. Carex microptera Mackenzie

Carex festiva var. *viridis* L. H. Bailey, Mem. Torrey Bot. Club 1: 51. 1889. *C. macloviana* var. *stricta* f. *viridis* (erroneously attr. to L. H. Bailey by) Kükenthal in Engl. Pflanzenr. IV. 20 (Heft 38): 197. 1909. (*Scribner 315*, 16 Mile Creek, Park Co., Mont., is the first of two specimens cited.)

C. microptera Mackenzie, Muhlenbergia 5: 56. 1909. *C. macloviana* var. *microptera* B. Boivin, Naturaliste Canad. 94: 523. 1967. (*Heller 9067*, near Deeth, Elko Co., Nev.)

C. festivella Mackenzie, Bull. Torrey Bot. Club 42: 609. 1915. (*A. Nelson 3275*, Albany Co., Wyo.)

Plants densely tufted, without creeping rhizomes; culms numerous, somewhat aphyllopodic, mostly (1.5) 2–6 (8) dm tall; leaves all borne on the lower part of the stem, but not densely tufted, flat or nearly so, mostly 2–4 (6) mm wide, generally shorter than the culms; spikes several, gynaecandrous, sessile, individually not over about 1 cm long, closely crowded into a compact, typically globose-ovoid head (1) 1.5–2 (2.5) cm long; lowest bract sheathless, setaceous above an expanded, scarious-margined base, shorter to slightly longer than the head, or reduced and not much different from the pistillate scales; bracts subtending the other spikes reduced and inconspicuous, often resembling the pistillate scales, these light to dark brown, largely scarious or hyaline-scarious, mostly shorter and narrower than the perigynia, exposing their tips and distal margins; perigynia typically pale greenish-stramineous (varying to more greenish or more stramineous) except for the brown-tipped beak, but often tan or brownish medially and sometimes also along the very edge, lanceolate to ovate, 3.2–5.0 mm long and 2–3 (4) times as long as wide, rather lightly or obscurely several-nerved on each side, strongly flattened except where distended by the achene, thin-edged and serrulate-margined, but the distal 0.5–1.0 mm of the beak more or less terete and not markedly serrulate, the beak-tip obliquely cleft and only obscurely (or

Carex pachystachya

Carex microptera

not at all) bidentate; stigmas 2; achene lenticular, 1.1–1.5 mm long, much smaller than perigynium; $n = 41$.

Moist or wet places, from the foothills to moderately high elev. in the mts., seldom approaching timberline; B.C. to Sask. and Man., s. to Calif., Ariz., and N.M., and in the Black Hills of S.D.

This is one of the commonest sedges at moderate elev. in the mts. of our range. It is only moderately palatable to livestock, and is sensitive to grazing.

The characters by which *C. festivella* is purportedly to be distinguished from *C. microptera* are not well correlated among themselves and do not individually display any obviously bimodal distribution. I see no indication that more than one taxon is involved.

Carex microptera, C. haydeniana, and *C. ebenea* form a trio with a complex pattern of relationship. *C. haydeniana* is not sharply distinguished from *C. microptera,* of which it might with some justification be treated as merely an alpine ecotype. However, toward the southern part of the range of *C. microptera* the position of *C. haydeniana* as its alpine correlative is largely taken over by *C. ebenea,* which is fairly sharply distinguished from *C. microptera* but intergrades to some extent with *C. haydeniana.* Occasional specimens from far north of the range of *C. ebenea,* which apparently represent merely the extreme of variation in *C. haydeniana,* would probably pass as *C. ebenea* if they had been collected in Colorado.

It is possible that *C. microptera* (1909), *C. ebenea* (1901) and *C. haydeniana* (1871) should be regarded as varieties of a single species, which would then inescapably have to be merged with *C. macloviana* Urv. (1826), a high-boreal and antarctic species. *C. pachystachya* (1855) would then also have to be included, and this in turn would bring in *C. subfusca* (1880). With a few more unavoidable additions from Calif., *C. macloviana* would be expanded essentially to the limit set by Kükenthal in 1909 and would include 20 Mackenzie species—all of his subsection Festivae except *C. illota.* A return to such an expanded concept of *C. macloviana* may ultimately prove to be necessary or desirable, but it should probably not be undertaken in a floristic work such as this. The alternative is to hold the several recognizable taxa of the group as distinct species, with the understanding that they are not sharply distinguished from each other and that a more conservative treatment may eventually be preferable.

81. Carex haydeniana Olney

Carex haydeniana Olney in King, Rep. Geol. Explor. 40th Parallel 5: 366. 1871. *C. festiva* var. *haydeniana* W. Boott in S. Wats. Bot. Calif. 2: 234. 1880. *C. macloviana* var. *haydeniana* T. H. Holm, Amer. J. Sci. 160: 266. 1900. (*Hayden s.n.*, Uinta Mts. of Utah, in 1870, is one of two collections cited.)

C. festiva var. *decumbens* T. H. Holm, Amer. J. Sci. 166: 20, 26. 1903. *C. macloviana* var. *stricta* f. *decumbens* (erroneously attr. to T. H. Holm by) Kükenthal in Engl. Pflanzenr. IV. 20 (Heft 38): 197. 1909. *C. nubicola* Mackenzie, Bull. Torrey Bot. Club 36: 480. 1909. (*Baker 232*, Pagosa Peak, Colo.; lectotype by Mackenzie.)

Plants more or less densely tufted, without creeping rhizomes; culms somewhat aphyllopodic, mostly 1–3 (4) dm tall; leaves all borne on the lower part of the stem, tending to be tufted near the base, flat or nearly so, mostly 2–3.5 mm wide, generally shorter than the culms; spikes several, gynaecandrous, sessile, mostly 7–10 mm long, closely crowded into a compact, typically globose-ovoid head 1–2 cm long; bracts as in *C. microptera;* pistillate scales medium to dark brown, largely scarious or hyaline-scarious, mostly shorter and narrower than the perigynia, exposing their tips and at least the distal margins; perigynia brownish at least in the center, the margins commonly paler, sometimes merely tan or light brown, sometimes greenish-stramineous as in *C. microptera,* the beak in any case dark-tipped; perigynia lance-ovate to ovate, elliptic, or subrotund, prominently beaked, (4.5) 5.0–6.2 mm long and 1.8–3.3 times as long as wide, tending to have a pair of more or less prominent inframarginal nerves (visible on both sides) delimiting the wing from the perigynial cavity, otherwise lightly rather many-nerved dorsally and usually nerveless or nearly so ventrally, but sometimes lightly few-nerved ventrally as well as dorsally, strongly flattened except where distended by the achene, thin-edged and serrulate-margined, but the distal 0.5–1.0 mm of the beak more or less terete and not markedly serrulate, the beak-tip obliquely cleft and only obscurely (or not at all) bidentate; stigmas 2; achene lenticular, 1.4–1.7 mm long, much smaller than the perigynium, attached essentially to the base of the perigynium or jointed to a very short and inconspicuous stipe seldom over 0.5 mm long that traverses a loosely spongy area at the very base of the perigynium.

Wet to moderately dry places at high alt. in the mts., near or above timberline; s. B.C. (Marble Mts.) to sw. Alta. (Banff), s. to Colo., Utah (LaSal & Tushar mts.), c. Nev. (Toiyabe Mts.), and Calif.; chiefly e. of the Cascade-Sierran axis, and apparently skipping Wash. entirely, but known from the Three Sisters in Oregon and from scattered stations in the White Mts. and Sierra Nevada of Calif.

See comment under *C. microptera*.

82. Carex ebenea Rydb.

Carex ebenea Rydb. Bull. Torrey Bot. Club 28: 266. 1901. *C. festiva* var. *ebenea* J. K. Henry, Fl. S. Brit. Columb. & Vancouver I. 58. 1915. *C. haydeniana* var. *ebenea* Kelso, Biol. Leafl. 66: 23. 1953. (*Clements s.n.*, Pike's Peak, Colo., in 1900.)

Plants tufted, without creeping rhizomes; culms slightly aphyllopodic, stout, 1–4 dm tall; leaves all borne on the lower part of the stem and often somewhat tufted at the base, flat or nearly so, mostly 2–4 mm wide, generally shorter than the culms; spikes several, gynaecandrous, sessile, 7–15 mm long, closely crowded into a compact, typically globose-ovoid or broadly ellipsoid head 1.2–3 cm long; lowest bract sheathless, shorter than the head, setaceous above an expanded, scarious-margined base, or reduced and not much different from the pistillate scales; bracts subtending the other spikes reduced and inconspicuous, often resembling the pistillate scales, these rather light brown to more often dark brown, largely or wholly hyaline-scarious, shorter and distally often narrower than the perigynia, exposing their beaks and distal margins; perigynia coppery-brown with green margins, or golden brown throughout (the head often less markedly bicolored than in *C. microptera*), lanceolate to narrowly elliptic or even oblanceolate, 5.2–7.2 mm long, 3.4–5 times as long as wide, thin and flat except as distended by the achene, lightly several- to many-nerved on each side, thin-edged and serrulate-margined, with a prominent beak, this mostly 1.5–2.5 mm long, the distal part dark, terete or nearly so, and not markedly serrulate, the beak-tip obliquely cleft and only shallowly (or not at all) bidentate; achene lenticular, 1.5–1.8 mm long, jointed to a prominent stipe about (0.5) 1 mm long which traverses a loosely spongy area at the base of the perigynium; $n = 42$.

Wet to moderately dry places at high alt. in the mts., near or above timberline; s. Rocky Mt. region, from w. Wyo., to n. N.M., and extending w. into Utah (Uinta, LaSal and Abajo mts., and occasionally as far w. as the Tushar Mts.) and n. Ariz. (San Francisco Mts.).

See comment under *C. microptera*.

83. Carex petasata Dewey

Carex petasata Dewey, Amer. J. Sci. 29: 246. 1836. (*Drummond,* Rocky Mts.)

C. liddoni F. Boott in Hook. Fl. Boreali-Amer. 2: 214. 1839. (*Scouler,* Columbia River.)

Plants densely tufted, without creeping rhizomes; culms aphyllopodic, 3–9 dm tall; leaves flat or nearly so, mostly (1) 2–4 (5) mm wide, much shorter than the culms, more or less clustered towards the base; spikes mostly 3–6, gynaecandrous, sessile or very nearly so, 9–18 mm long (the terminal one often longer than the others), the staminate part well developed and often slender, all aggregated into a rather compact spike 2–5 cm long, the first two internodes each mostly (2) 3–7 mm long, or seldom longer and up to 15 mm; bracts subtending the spikes sheathless, short and inconspicuous, largely brownish-scarious, not much different from the scales, or the lowest one sometimes setaceous tipped; pistillate scales stramineous to medium brown, largely scarious, somewhat paler and more hyaline towards the margins, nearly or fully as long and

CAREX
2
8
8
perigynia
2
6
6
perigynia
2
8
perigynia
1/2
Carex haydeniana
JRJ
1/2
Carex ebenea
1/4
1/2
Carex petasata

wide as the mature perigynia, and largely concealing them in dorsal view, the tip of the beak sometimes exserted; perigynia loosely ascending-appressed, stramineous or pale greenish to brownish, relatively large, mostly 5.8–7.9 mm long, nerved on both sides, 2.6–4.0 times as long as wide, plano-convex, wing-margined and serrulate, tapering to the beak, the distal 0.5–1.5 mm of which is slender, marginless, and subentire, the beak-tip obliquely cleft; stigmas 2; achene lenticular, 2.2–2.7 mm long.

Open places, commonly in grassland or sagebrush-grass mixture, sometimes in dry or even wet meadows, from the plains to fairly high elev. in the mts., sometimes reaching timberline; B.C., Alta., and Sask. (Cypress Hills), s. to Oregon, Nev., Utah, and Colo., and found in appropriate habitats essentially throughout our range.

The species is palatable to livestock, but rather sensitive to grazing.

Carex petasata is a distinctive taxon, but it approaches or passes into both *C. praticola* and *C. phaeocephala.*

84. **Carex praticola** Rydb.

Carex pratensis Drejer, Naturhist. Tidsskr. 3: 442. 1841. *C. praticola* Rydb. Mem. New York Bot. Gard. 1: 84. 1900. (*Vahl,* Greenl.) Not *C. pratensis* Hose, 1797.

C. pratensis var. *furva* L. H. Bailey in Macoun, Cat. Canad. Pl. 5: 377. 1890. *C. pratericola* var. *furva* A. A. Heller, Cat. N. Amer. Pl. ed. 2. 3. 1900. *C. furva* Piper, Contr. U. S. Natl. Herb. 11: 166. 1906. *C. pratensis* f. *furva* (attr. to L. H. Bailey by) Kükenthal in Engl. Pflanzenr. IV. 20 (Heft 38): 198. 1909. *C. piperi* Mackenzie ex Piper & Beattie, Fl. N.W. Coast 75. 1915. (*Macoun,* s. Vancouver I.) Not *C. furva* Webb, 1838.

Plants densely tufted, without creeping rhizomes; culms aphyllopodic, (2) 3–8 dm tall; leaves flat or nearly so, (1) 2–4 mm wide, shorter than the culms, often softer than in *C. petasata,* all borne on the lower part of the stem, but mostly not closely clustered; spikes (3) 4–7, gynaecandrous, sessile, 7–18 mm long, often slender-based, borne in a slender, rather loose and flexuous spike, the two lowest internodes each (4) 5–10 (15) mm long, the inflorescence tending to be somewhat nodding in life; bracts subtending the spikes sheathless, short and inconspicuous, largely brownish-scarious or the lowest one greenish and setaceous-prolonged but still shorter than the spike; pistillate scales brownish or sometimes merely stramineous, largely scarious, paler and more hyaline towards the margins, nearly or fully as long and wide as the mature perigynia, and largely concealing them in dorsal view, but the tip of the beak often exserted; perigynia loosely ascending-appressed, pale green to stramineous or brownish, (4.0) 4.3–5.7 (6.0) mm long and (1.2) 1.3–2.0 (2.1) mm wide, 2.6–4.0 times as long as wide, plano-convex, wing-margined and serrulate, evidently nerved dorsally, commonly nerveless or nearly so ventrally, but sometimes evidently nerved, the beak commonly brownish, the distal 0.5 mm or more slender and entire or nearly so, obliquely cleft; stigmas 2; achene lenticular, 1.7–2.2 mm long.

Moist or wet meadows, swamps, streambanks, and moist woods, from the lowlands to moderate elev. in the mts., seldom approaching timberline; s. Greenl. to Alaska, s. to nw. Calif., ne. Oregon, c. and se. Idaho, ne. Utah, Colo., and N.D.

Carex praticola approaches or passes into 5 other taxa, *C. petasata, C. phaeocephala, C. leporina* L., *C. aenea* Fern. and *C. leporinella.*

85. **Carex phaeocephala** Piper

Carex leporina var. *americana* Olney ex L. H. Bailey, Proc. Amer. Acad. Arts 22: 152. 1887. *C. phaeocephala* Piper, Contr. U. S. Natl. Herb. 11: 172. 1906. (*Hall 583,* Oregon.)

C. eastwoodiana Stacey, Leafl. W. Bot. 2: 121. 1938. *C. phaeocephala* f. *eastwoodiana* F. J. Hermann, Leafl. W. Bot. 9: 15. 1959. (*Henderson 5583,* Dixie Mt., Grant Co., Oregon.) A form with the perigynia evidently nerved on both sides.

Plants densely tufted, without creeping rhizomes, 0.5–3 (4) dm tall, somewhat aphyllopodic, but still with the leaves clustered at the base; leaves shorter than or often equaling the culms, relatively narrow and firm, 0.5–2 (2.5) mm wide, flat or folded proximally, becoming trigonous distally, often channeled for some distance below the trigonous part; spikes 3–7, gynaecandrous, sessile or nearly so, 7–14 mm long, borne in a rather stiff and compact spike or narrow head mostly (1) 1.5–3 (3.5) cm long, the second internode mostly 2–5 (6) mm long, the first one of similar length or sometimes up to 10 mm; bracts subtending the spikes sheathless, short and inconspicuous, largely brownish and scarious, often scarcely different from the pistillate scales, or the lowest one firmer and awn-tipped but still no longer than its spike; pistillate scales light to dark brownish, largely scarious, paler and more hyaline towards the margins, nearly or fully as long and wide as the mature perigynia, but the tip of the beak often exposed; perigynia loosely ascending-appressed, commonly tan or brown, often with greener margins, (3.4) 4.0–5.5 (5.8) mm long and (1.2) 1.3–2.0 mm wide, 2.2–3.5 times as long as wide, slightly to strongly plano-convex, wing-margined (especially distally) and serrulate, evidently nerved dorsally, commonly nerveless or nearly so ventrally, but sometimes evidently nerved, the beak commonly dark brown, the distal 0.4 mm or more, slender and often entire, obliquely cleft; stigmas 2; achene lenticular, 1.5–2.1 mm long; n = ca 42.

Open, often rocky places at high alt. in the mts., near or commonly above timberline; Alta. and B.C. to Calif., and Colo., and occurring in appropriate habitats throughout our range.

One of our commoner alpine species of *Carex,* but not highly palatable to livestock.

86. **Carex bebbii** Olney ex Fern.

Carex tribuloides var. *bebbii* L. H. Bailey, Mem. Torrey Bot. Club. 1: 55. 1889. *C. bebbii* Olney ex Fern. Proc. Amer. Acad. Arts 37: 478. 1902. (*Bebb,* Ill.) The name *C. bebbii* is here attributed to Olney ex Fern., rather than to (L. H. Bailey) Olney ex Fern., because Fernald was validating Olney's nom. nudum of 1871 (Carices Boreali-Amer. 3).

Plants densely tufted, without creeping rhizomes; culms 2–9 dm tall, distinctly aphyllopodic; leaves elongate, well scattered on the lower half or third of the culm, flat or nearly so, mostly 2–4 mm wide; spikes mostly 4–12, gynaecandrous, sessile, 5–9 mm long, borne in a compact, crowded spike or head (1) 1.5–3 cm long and up to 1.5 cm wide, the internodes all short, the inflorescence pale greenish to stramineous or light brown in aspect; bracts sheathless, short and inconspicuous, or the lowest one somewhat setaceous-prolonged but still shorter than the inflores-

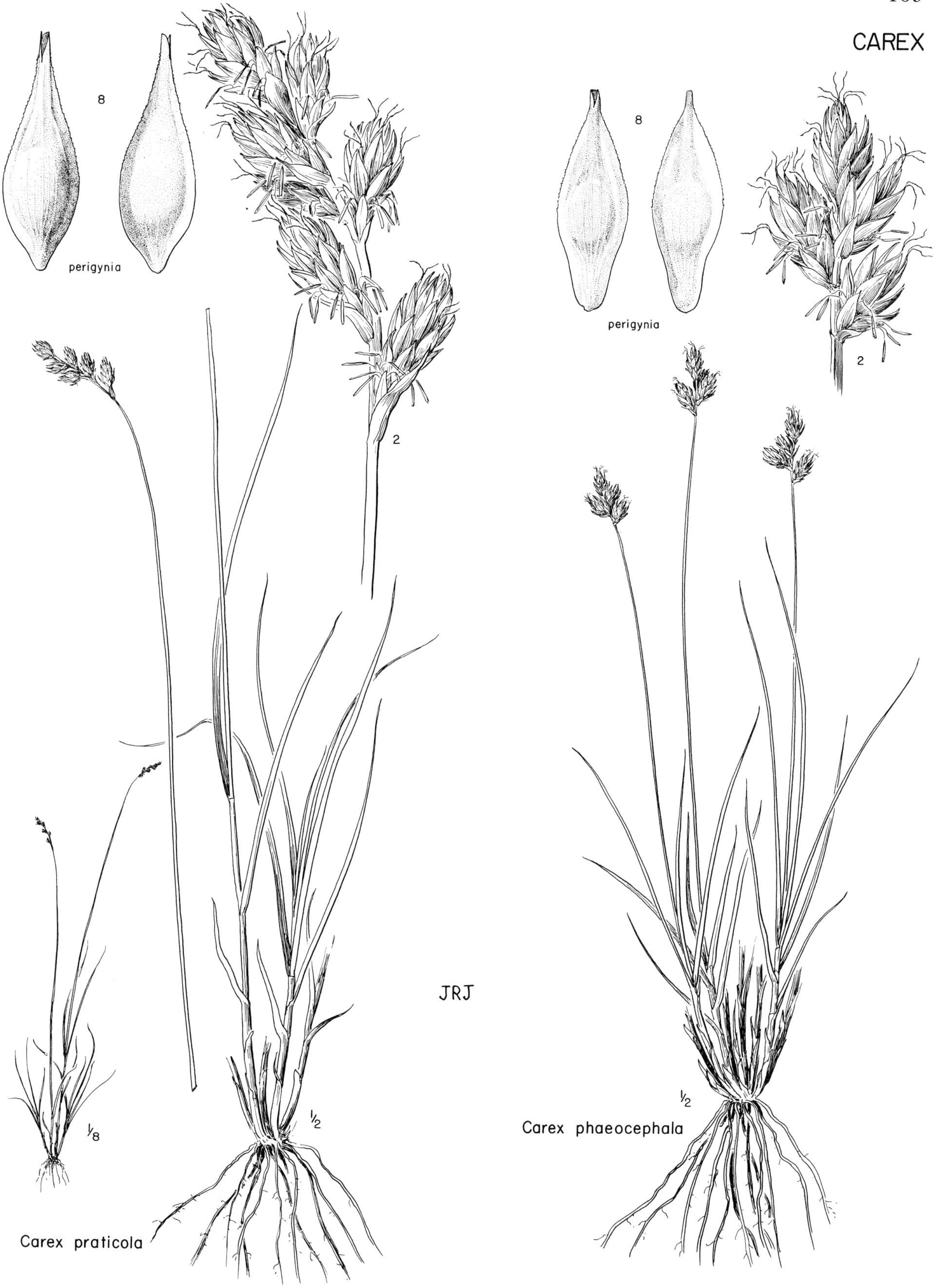
CAREX
8
perigynia
2
1/8
1/2
JRJ
Carex praticola
8
perigynia
2
1/2
Carex phaeocephala

cence; pistillate scales stramineous to light brown, largely hyaline-scarious except for the firmer, often greenish midrib, somewhat shorter and narrower than the perigynia; perigynia crowded, stiffly ascending (the beaks often standing out from the body of the spike), pale green to stramineous or tan, 2.7–3.7 mm long and 1.1–1.5 mm wide, 2.0–2.7 (3.0) times as long as wide, plano-convex, evidently nerved on both sides or nerveless ventrally, wing-margined and serrulate, tapering to a commonly rather ill-defined, obliquely cleft or shortly bidentate beak that is flattened and serrulate to the tip; stigmas 2; achene lenticular, 1.1–1.5 mm long and 0.6–0.8 mm wide; $n = 34$.

Wet meadows and other wet places, from the valleys and lowlands to moderate elev. in the mts., mostly in rather moist regions; Newfl. to B.C., s. to N.J., Ill., S.D., Mont., and n. Oregon, and at isolated stations in Utah (Uinta Mts.) and Colo.

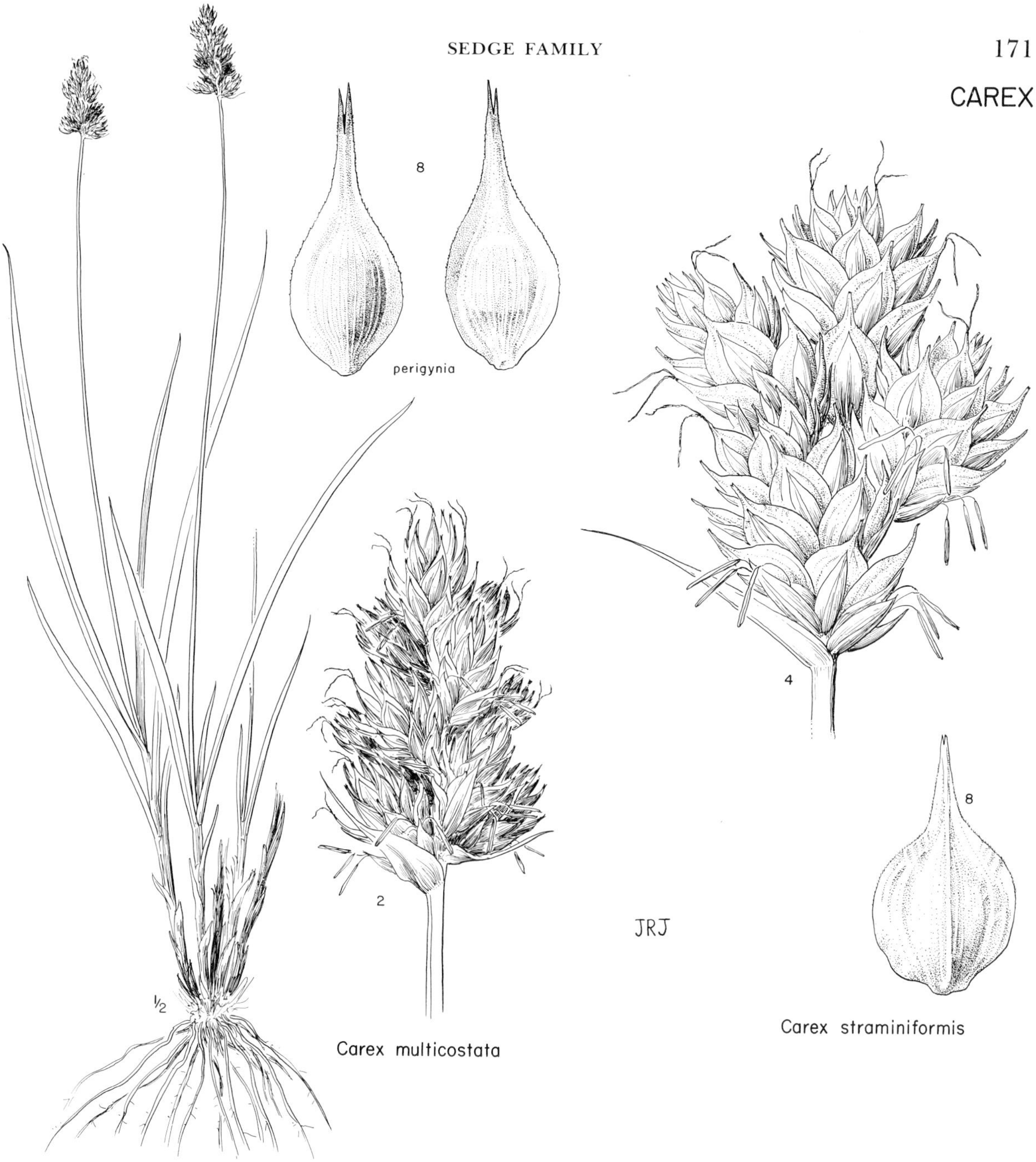

87. Carex brevior (Dewey) Mackenzie ex Lunell

Carex straminea var. *brevior* Dewey, Amer. J. Sci. 11: 158. 1826. *C. festucacea* var. *brevior* Fern. Proc. Amer. Acad. Arts 37: 477. 1902. *C. brevior* Mackenzie ex Lunell, Amer. Midl. Naturalist 4: 235. 1915. (w. Mass.)

Plants densely tufted, without creeping rhizomes; culms 2–10 dm tall, aphyllopodic; leaves coarse and firm, flat or sometimes canaliculate, (1) 2–4 mm wide, much shorter than the culms, sometimes all crowded near the base, sometimes more distributed over the lower third of the culm; spikes mostly 3–6, gynaecandrous, sessile, stout, 6–10 (15) mm long, borne in a moniliform or interrupted to occasionally more congested spike 1.5–5 cm long, the inflorescence greenish-stramineous to tan or light brown in aspect; bracts sheathless, short and inconspicuous, or the lowest one somewhat setaceous-prolonged but still shorter than the inflorescence; pistillate scales largely hyaline-scarious, stramineous or light brown to nearly colorless except for the firmer, often greenish midrib, distinctly narrower and tending to be shorter than the perigynia; perigynia crowded, stiffly ascending (the beaks tending to stand out from the body of the spike), pale green to stramineous or tan, 3.2–4.8 mm long and 2.2–3.4 mm wide, (1.2) 1.3–1.7 (1.8) times as long as wide, the body more or less strongly flattened, suborbicular, nerved on both sides (the ventral nerves fewer and fainter than the dorsal ones) or more commonly nearly nerveless ventrally, wing-margined all around and serrulate distally, more or less abruptly contracted to the beak, this 0.8–1.5 mm long, shortly bidentate, flattened throughout, often cuneate in outline, margined and serrulate to the tip or to within 0.5 mm of the tip; stigmas 2; achene lenticular, 1.6–2.2 mm long and 1.3–1.9 (most commonly 1.5) mm wide.

Streambanks, moist meadows, and swales, from the valleys and plains to moderate elev. in the mts.; Que. to D.C., w. to s. B.C., Wash., and N.M., chiefly e. of the continental divide, but crossing over in n. Mont. and found along the major water-courses as far w. as se. B.C. and e. Wash., also along the Snake R. at St. Anthony, Fremont Co., Idaho (the only station in our range).

88. **Carex multicostata** Mackenzie

Carex multicostata Mackenzie, Bull. Torrey Bot. Club 43: 604. 1917. (*Parish 3609,* Bear Valley Dam, San Bernardino Co., Calif.)

Plants loosely to densely tufted, without creeping rhizomes; culms aphyllopodic, 1.5–6 dm tall; leaves flat or nearly so, distinctly shorter than the culms, mostly 2–3.5 mm wide, more or less crowded toward the base of the culms, but not forming a dense tuft; spikes 3–7, sessile, gynaecandrous, 6–10 mm long, greenish to dark brown, crowded in a narrow to globose-ovoid head 1–3.5 cm long and up to 1.5 cm thick, the individual spikes more readily distinguishable to the naked eye than in *C. microptera,* less so than in *C. straminiformis;* bracts sheathless, short and inconspicuous; pistillate scales largely scarious, with firmer midrib, light to rather dark brown, with paler, more hyaline margins, somewhat narrower and shorter than the perigynia; perigynia green to stramineous or brownish, 4.5–6.2 mm long and 1.6–2.4 mm wide, 1.8–3.1 times as long as wide, distinctly plano-convex, obscurely to conspicuously multinerved (nerves 10 or more) dorsally and usually also ventrally, or the ventral nerves sometimes fewer or obsolete, rather thick-walled and tending to become longitudinally sulcate-wrinkled especially dorsally, wing-margined and serrulate, more or less ovate, tapering to the prominent, slender but proximally flattened, obliquely cleft to bidentate beak, of which the distal 0.5–0.6 (1) mm is marginless, entire, and often subterete; stigmas 2; achene lenticular, 1.8–2.4 mm long and 1.3–1.6 mm wide.

Moist or dry meadows, streambanks, and open, moist slopes at moderate elev. in the mts., occasionally ascending to near timberline; Cascade-Sierran region from c. Wash. s. to the mts. of s. Calif., and in the mts. of e. Oregon and c. Idaho.

Carex multicostata is much like *C. microptera* in aspect, but it is probably more closely related to *C. egglestonii* and *C. straminiformis.*

89. **Carex straminiformis** L. H. Bailey

Carex straminea var. *congesta* F. Boott ex Olney, Proc. Amer. Acad. Arts 7: 393. 1868. *C. straminiformis* L. H. Bailey, Mem. Torrey Bot. Club 1: 24. 1889. (*Brewer 1375, 1397, 1398,* and *1399,* Mt. Shasta, Calif.)

Similar to *C. egglestonii;* plants up to 4 dm tall; spikes and perigynia smaller, the spikes about 1 cm long or less, often paler in aspect than in *C. egglestonii,* the perigynia often merely green, mostly 4.1–5.0 mm long and 1.5–2.3 mm wide, 1.5–2.3 times as long as wide; achenes 1.3–2.0 mm long and 1.0–1.4 mm wide.

Open slopes, often near persistent snowbanks, at high alt. in the mts., mostly near or above timberline; Sierra Nevada and White Mts. of Calif. and adj. Nev., n. irregularly to s. Wash.; disjunct in the Wasatch Mts. of Utah (Salt Lake and Utah cos.), and in c. Idaho.

90. **Carex egglestonii** Mackenzie

Carex egglestonii Mackenzie, Bull. Torrey Bot. Club 42: 614. 1915. (*Eggleston 6181,* Kebler Pass, Mt. Carbon, Gunnison Co., Colo.)

Plants densely tufted, without creeping rhizomes; culms somewhat aphyllopodic, 1.5–6 dm tall; leaves crowded near the base, generally much shorter than the culms, rather firm, flat or nearly so and mostly 2–4 mm wide; spikes 3–6, gynaecandrous, sessile, stout, 9–15 (18) mm long, closely crowded into a compact head (but still readily distinguishable), the first two internodes collectively only 4–7 (9) mm long; bracts sheathless, short and inconspicuous; pistillate scales distinctly shorter and narrower than the perigynia, largely hyaline-scarious except for the firmer midrib, mainly medium brown to bright chestnut brown, with paler, hyaline margins; perigynia crowded, thin and flat, usually somewhat coppery-tinged at least toward the margins, otherwise greenish, 5.5–7.4 mm long and 2.1–3.5 mm wide, 1.8–2.5 (2.7) times as long as wide, finely and rather obscurely many-nerved on both sides or on the dorsal side only, or nearly nerveless except for the pair of prominent, inframarginal nerves delimiting the wing-margin, tapering to the prominent, obliquely cleft or minutely bidentate beak, which is flattened, margined, and serrulate essentially to the tip, or the distal 0.5 mm sometimes more slender and scarcely margined, though generally still serrulate; stigmas 2; achene lenticular, 1.9–2.3 mm long and 1.4–2.0 mm wide.

Dry meadows and open slopes at rather high elev. in the mts., from the spruce-fir zone to above timberline; s. Rocky Mts. and Colo. Plateau region, from s. Wyo. and Colo. w. to Utah, and reported from Nuevo Leon, Mex. In Utah known from the Uinta, LaSal, and Abajo mts., and from the chain of high plateaus stretching s. from the Wasatch Mts.; reported also from the Bear River Mts.

This palatable species is abundant enough on the high plateaus of Utah to furnish considerable forage for cattle.

91. **Carex xerantica** L. H. Bailey

Carex xerantica L. H. Bailey, Bot. Gaz. 17: 151. 1892. *C. foenea* var. *xerantica* Kükenthal in Engl. Pflanzenr. IV. 20 (Heft 38): 205. 1909. (*Macoun s.n.,* File Hills, s. Sask., 4 July 1879.)

Plants densely tufted, without creeping rhizomes; culms aphyllopodic, 3–7 dm tall; leaves tending to be somewhat crowded toward the base, but not densely tufted, generally much shorter than the culms, rather firm, flat or nearly so and mostly 2–4 mm wide; spikes (2) 3–6, gynaecandrous, sessile, 8–17 mm long, borne in a stiff spike mostly 2–4 cm long, the first two internodes collectively mostly (8) 10–18 mm long, the inflorescence pale greenish or stramin-

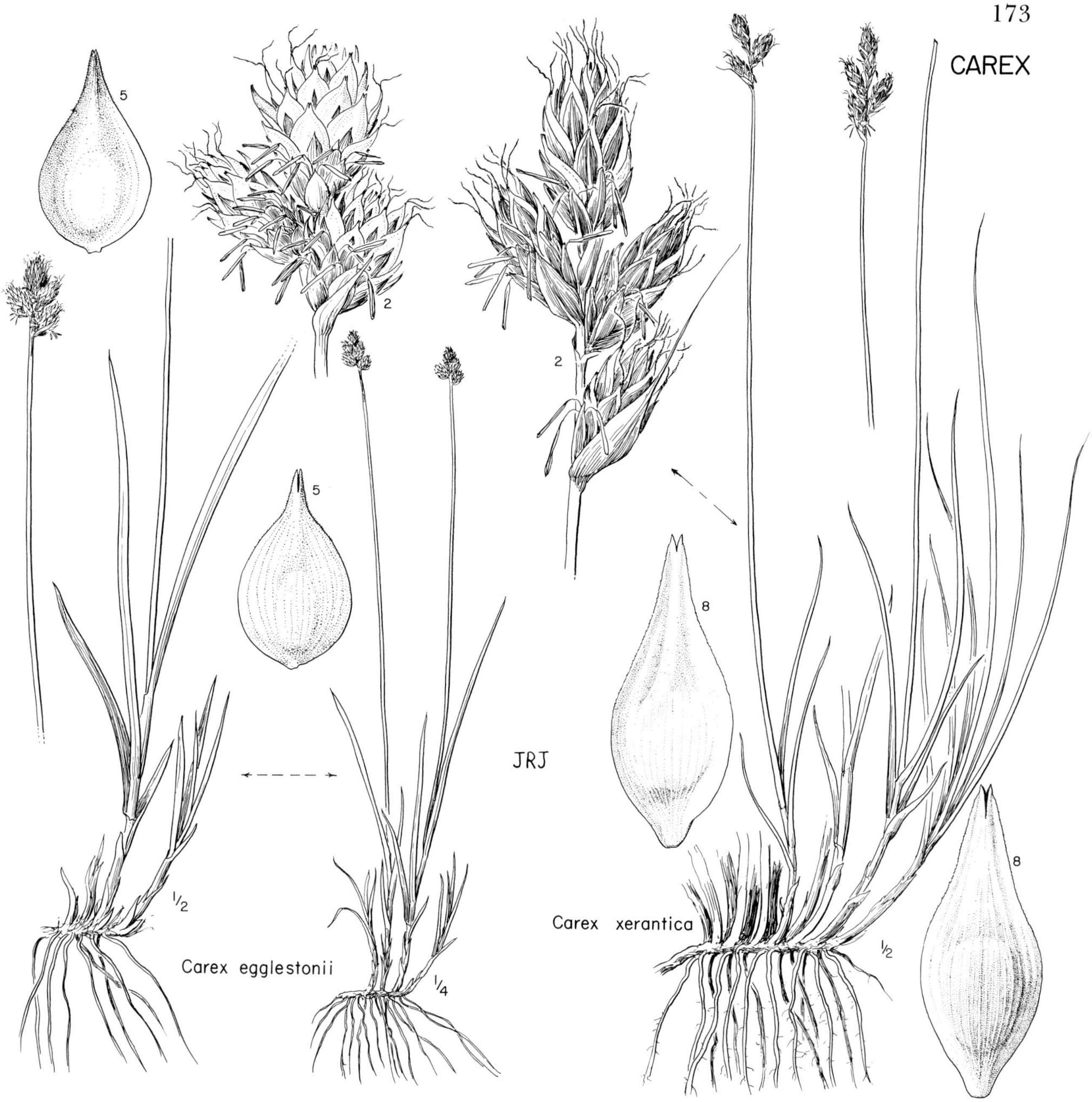

eous to medium brown in aspect; bracts sheathless, short and inconspicuous, or the lowest one somewhat setaceous-prolonged but still shorter than the inflorescence; pistillate scales varying from fully as long and wide as to distinctly narrower and sometimes also shorter than the perigynia, largely hyaline-scarious except for the firmer midrib, light to medium brown with paler, hyaline margins, or largely white-hyaline; perigynia more or less strongly appressed, elliptic or ovate, slightly to strongly plano-convex, (4.2) 4.5–7.0 (7.4) mm long and 1.9–2.8 mm wide, 2.1–2.9 times as long as wide, lightly to evidently multinerved (nerves mostly more than 10) on both sides, or the ventral side with fewer nerves or nearly nerveless, the body wing-margined and distally serrulate, carrying its width well upwards, tapering to the mostly ill-defined, flattened, strongly margined and serrulate, obliquely cleft or minutely bidentate beak, the distal 0.2 mm of which is often abruptly narrower, entire, and hyaline; stigmas 2; achene lenticular, 2.0–3.0 mm long and 1.2–1.6 mm wide.

Grasslands, open slopes, and mountain parks, from the high plains to the spruce-fir zone in the mts.; Minn. and Man. to Alta., s. to Colo. and reputedly N.M., crossing the continental divide in Colo. and extending w. to the high plateaus of c. and s. Utah and the Kaibab Plateau of Ariz.; reported from Durango, Mex. Rather infrequently collected.

The critical distinction between *C. xerantica* and *C. egglestonii* is here considered to reside in the shape of the inflorescence rather than in the size of the pistillate scales with relation to the perigynia.

CAREX

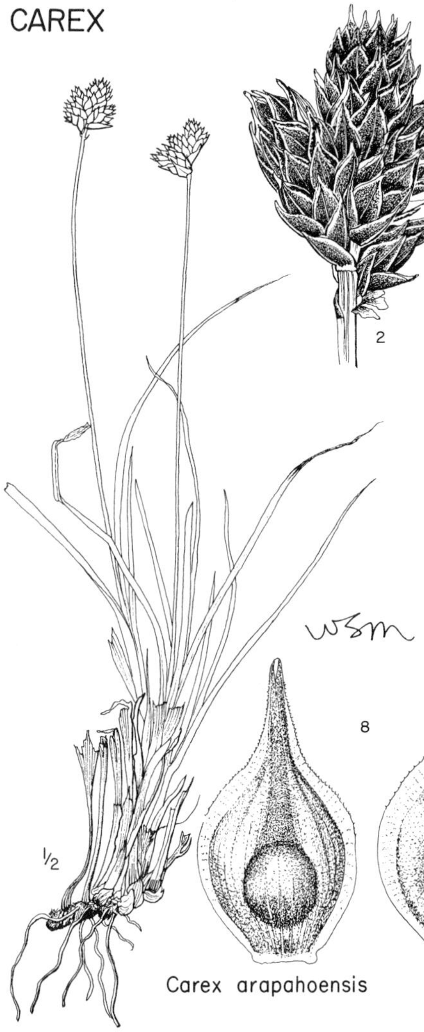

Carex arapahoensis

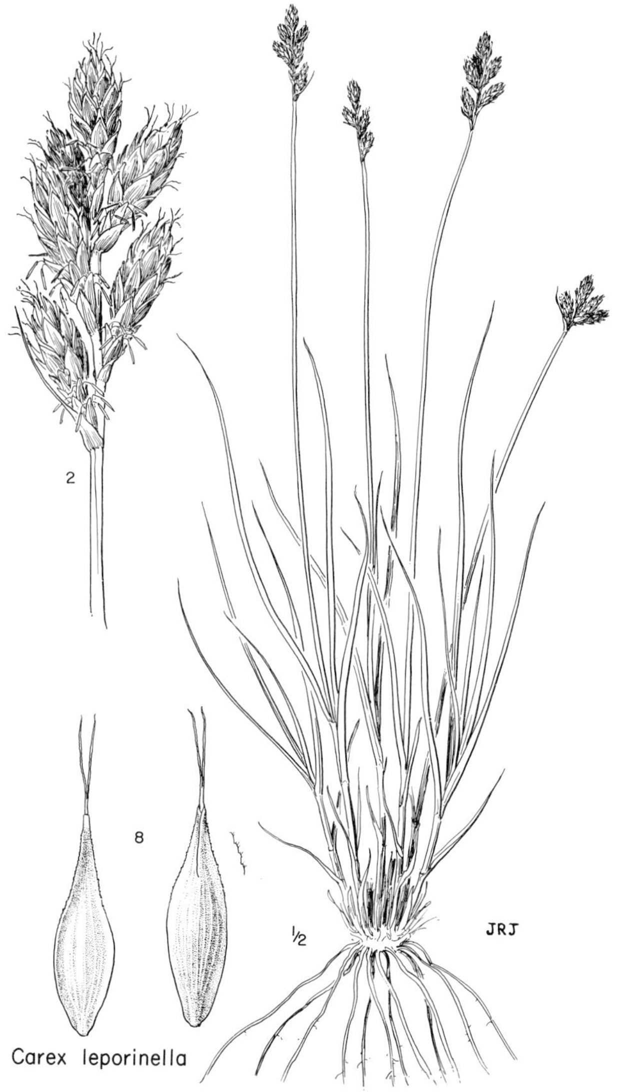

Carex leporinella

92. Carex arapahoensis Clokey
Carex arapahoensis Clokey, Rhodora 21: 83. 1919. (*Clokey 3227,* Mt. Arapahoe, Boulder Co., Colo.)

Plants densely tufted, without creeping rhizomes; culms aphyllopodic, mostly 1.5–3 (4) dm tall; leaves tending to be somewhat crowded toward the base, not densely tufted, shorter than or sometimes about equaling the culms, firm, flat or nearly so and mostly 1.5–4 mm wide, commonly slenderly long-tapering distally; spikes mostly 3–6, gynaecandrous, sessile, 8–12 mm long, aggregated into a more or less definite, often ovoid-globose head, the first two internodes collectively only 1.5–6 mm long, the inflorescence brown in aspect; bracts sheathless, short and not very conspicuous; pistillate scales brown with paler, hyaline margins, ovate, obtuse, largely covering the perigynia, the beak commonly shortly exserted; perigynia more or less appressed, brown to brownish-stramineous or partly greenish, more or less elliptic to ovate to elliptic-obovate, plano-convex, evidently wing-margined, mostly 4.5–5.5 mm long and 2–2.5 mm wide, finely multinerved on both sides, tapering to the evident, obliquely cleft or minutely bidentate beak, which is flattened, margined, and serrulate throughout or very shortly terete and marginless at the tip; stigmas 2; achenes lenticular, 1.7–2 mm long and 1.2–1.5 mm wide.

Open, often rocky slopes above timberline in the mts.; southern Rocky Mt. species, occurring in Colo., s. Wyo., and the LaSal Mts. of Utah (Mt. Peale).

93. Carex leporinella Mackenzie
Carex leporinella Mackenzie, Bull. Torrey Bot. Club 43: 605. 1917. (*Hall & Chandler 4716,* Pyramid Peak, El Dorado Co., Calif.)

Plants densely tufted, without creeping rhizomes; culms aphyllopodic, 1–3 dm tall; leaves slender and elongate, often some of them equaling the culms, flat to deeply channeled, 1–2 mm wide, all borne on the lower part of the stem; spikes 3–6, sessile, gynaecandrous, 6–12 mm long, aggregated into a compact spike or loose, ellipsoid to ovoid head 12–25 (35) mm long, the internodes mostly 1–5 (10) mm long;

lowest bract sheathless, green or brown, setaceous-tipped, up to about as long as the inflorescence; pistillate scales largely brown and scarious, with paler, hyaline margins, sometimes with green midrib, very nearly or fully as long and wide as the perigynia and largely concealing them in dorsal view, the tip of the beak sometimes exserted; perigynia more or less closely appressed, greenish to stramineous or brownish, relatively small, only 3.2–4.0 mm long, 0.8–1.2 mm wide, and 3.3–4.0 times as long as wide, sharp-edged but scarcely wing-margined, serrulate, evidently nerved dorsally, lightly nerved to essentially nerveless ventrally, tapering to a short, ill-defined beak, of which only the distal 0.3–0.5 mm is slender and entire or nearly so; stigmas 2; achene lenticular, 1.4–1.6 mm long and 0.7–0.9 mm wide.

Moist meadows, well up in the mts., sometimes above timberline; s. Wash. to the Sierra Nevada and White Mts. of Calif., e. to the Wallowa-Blue Mt. region of ne. Oregon, and to c. Idaho; an apparently isolated station in Wasatch Co., Utah; only doubtfully in Mineral Co., Nev. (Wassuk Range).

Family **POACEAE**, the Grass Family

Annual or perennial herbs, often rhizomatous or sometimes stoloniferous; culms (stems) usually terete, infrequently flattened or grooved, the branches occurring mostly at or near the base, rarely branched at the upper nodes, the internodes hollow to solid; leaves alternate, 2-ranked, often crowded at the base of the culms, differentiated into a sheath, ligule, and blade; sheath encircling the culm, the margins usually free, but usually overlapping or sometimes connate, glabrous or pubescent, the upper margins at the throat often more hairy than below, sometimes the upper margins prolonged into sheath auricles; ligules (appendages produced on the ventral [adaxial] surface of the summit of the sheaths) varying from membranous to membranous at the base and ciliate fringed, or entirely of hairs, rarely absent; blade usually narrow and elongate, flat, folded or rolled (involute), the dorsal (abaxial) area of the junction of sheath and blade, called the collar, usually marked in some way as less heavy venation, or by difference in color or pubescence; auricles occasionally present as small, rounded to acuminate projections on one or both sides of the base of the blades; inflorescence a panicle, or variously modified into compound or simple racemes, spikes or spicate-racemes; spikelets, the basic units of the inflorescence, consisting of 1 or more flowers and associated bracts, i.e., glumes, lemmas and paleas, the florets, when more than 1, borne alternately on opposite sides of a rachilla forming a spike (the spikelet as a whole is treated as a flower in inflorescence terminology), the rachilla sometimes prolonged behind the palea of the uppermost floret; glumes 2 subopposite bracts at the base of the spikelet, the lower glume often shorter than the upper, infrequently only 1 present, or rarely both wanting, awned or awnless; florets consisting of 2 bracts, the lemma and palea, and the enclosed flower parts, the basal floret or terminal florets of a spikelet sometimes variously reduced, sometimes represented by no more than an empty lemma; lemma oriented abaxial to the rachilla, rounded or keeled on back, glabrous or variously pubescent, 1- to many-nerved, awned or awnless, the basal portion often elongated and forming a hardened callus; palea adaxial to the rachilla, usually smaller than and clasped by the lemma, usually 2-keeled, sometimes reduced or wanting; perianth represented by 2, rarely 3, much reduced and modified sepals called lodicules (wanting in some species), the lodicules situated between the ovary and the lemma, often becoming distended at anthesis forcing the lemma and palea apart, exposing the anthers and stigmas; stamens usually 3, sometimes 1, 2, or 6, with 2-locular anthers; pistil with a superior, tricarpellate, 1-celled ovary with 1 anatropous ovule and usually 2 distinct, plumose stigmas, the stigmas may be sessile, or elevated on a single or on separate styles; fruit a 1-seeded caryopsis or grain with the seed fused to the ovary wall (pericarp), or rarely an achene with the seed separate from the pericarp; embryo consisting of a scutellum of haustorial tissue that serves as a grass cotyledon, opposite this in some genera is a flange of tissue called an epiblast, the coleorhiza (sheath) encloses the primary root, while the coleoptile protects the embryonic shoot.

The Poaceae family is one of the largest of the families of vascular plants with an estimated 600 to 620 genera and 7,500 to 10,000 species. The family ranks third in number of genera behind the Asteraceae and Orchidaceae and fourth in number of species behind Orchidaceae, Asteraceae, and Fabaceae. In terms of number of individual plants it is likely the largest and most widely distributed family, and in terms of economic importance it is, by all odds, the most important family.

Grasses are used for food by man and domestic animals. Man cultivates oats, barley, wheat, rye, millet, sugar cane, sorghum, corn and rice for food, and livestock eat grain, corn, hay, and graze pasture and range grasses.

Throughout most of the world grass is used in lawns for aesthetic and sporting pleasures. In some parts of the world grasses are used for building materials (bamboos). Several rhizomatous species are planted to arrest erosion on roadbanks, ditchbanks and ocean beaches. Also the many species listed as noxious weeds and hay fever plants add to the family's importance.

Several species of grasses are becoming popular garden subjects in our area. *Arundo donax* L. (giant reed) from warm regions of the Old World is the tallest of our ornamental grasses, often attaining heights to 6 m. This handsome grass rarely produces inflorescences in our area, but is grown from rhizomes or more often from culm cuttings. *Cortaderia selloana* (J. A. Schultes) Aschers. & Graebner (pampas grass) from Argentina is a beautiful ornamental that does well in the St. George area of Utah. *Erianthus ravennae* (L.) Beauv. (Ravenna grass or plume grass) is a much larger grass from southern Europe that surprisingly survives the cold winters of northern Utah. These are the tall grasses used effectively for special purposes such as framing large entrances or in garden corners. *Pennisetum setaceum* (Forssk.) Chiov., often listed as *P. ruppelii* Steudel in garden catalogues, is the fountain grass of our formal gardens. This popular perennial lives as an annual in our area. Grasses often grown for dry bouquets are *Briza maxima* L. (quaking grass), *Coix lacryma-jobi* L. (job's tears), *Holcus lanatus* L. (velvetgrass), *Lagurus ovatus* L. (hares-tail grass), and *Rhynchelytrum repens* (Willd.) C. E. Hubbard (ruby grass). *Phalaris arundinacea* f. *picta* (L.) Aschers. & Graebner (ribbon grass) is a common sight in wet places of old-fashioned gardens. *Avena sterilis* L. (animated oats) is often grown as a curiosity as the prominent awns exhibit hygroscopic action. For diagnostic characteristics of the preceding grasses see the index for page numbers.

From 1935 to recent years botanists in the United States have followed the grass classification of A. S. Hitchcock. His system is based almost exclusively on characteristics of the inflorescence and spikelets. His arrangement is adequate for the purpose of identification, but is now known to be artificial in respect to phylogenetic relationships of many genera. Much has been learned about the family since 1935. Important characteristics indicating relationships have been found in the position and shape of root hair cells, structure of the leaf and epidermis, nature of starch grains, structure of the embryo, chromosome number and size, persistence of nucleoli during nuclear division, reactions of seedlings to certain chemicals, geographical distribution, and other features not visible under the hand lens. The use of most of these new taxonomic tools in the Poaceae is reviewed by Stebbins (1956).

The following arrangement of subfamilies, tribes and genera is that of Gould (1968) with a few modifications. Gould's system reflects the renaissance that has occurred in grass classification. His arrangement of tribes and genera is a modification of the system proposed by Stebbins and Crampton (1961), with the sequence of subfamilies following Reeder's (1957) arrangement of embryo types. Unfortunately a system based mostly on anatomical, cytological and physiological characters is not the most convenient one for purposes of identification. The following artificial key to the genera is given to make identification as easy as possible with only a hand lens and metric ruler. A technical key to the subfamilies is also provided (p. 183) with keys to tribes and the genera of tribes within the text. The tribal keys to genera are often based on different characters than those found in the following artificial key and serve as additional aids in identification.

References:

Gould, F. W. Grass systematics. McGraw-Hill Book Co. 382 pp. 1968.

Hitchcock, A. S. Manual of the grasses of the United States. U.S.D.A. Misc. Publ. 200: 1–1040. 1935.

———. Manual of the grasses of the United States. ed. 2, revision by A. Chase. U.S.D.A. Misc. Publ. 200: 1–1051. 1951.

Hubbard, C. E. The genera of British grasses. *In:* J. Hutchinson. British flowering plants, pp. 284–348. Richard Clay & Co., Ltd. 1948.

———. Gramineae. *In:* J. Hutchinson. The families of flowering plants. Monocotyledons. 2: 710–741. Oxford, Clarendon Press, 1959.

Pilger, R. Das System der Gramineae. Bot. Jahrb. Syst. 76: 281–384. 1954.

Reeder, J. R. The embryo in grass systematics. Amer. J. Bot. 44: 756–768. 1957.

Stebbins, G. L., Jr. Cytogenetics and the evolution of the grass family. Amer. J. Bot. 43: 890–905. 1956.

———, and B. Crampton. A suggested revision of the grass genera of temperate North America. Recent Advances Bot. 1: 133–145. 1961.

Tateoka, T. Miscellaneous papers on the phylogeny of Poaceae (10). Proposition of a new phylogenetic system of Poaceae. J. Jap. Bot. 32: 275–287. 1957.

Artificial Key to the Genera of Poaceae

The following key is intended entirely for identification purposes. It reflects natural groupings of only some of the genera.

1 Spikelets dorsally compressed, the glumes and fertile lemma more or less flattened or merely arching, the spikelets sometimes terete, but then the caryopsis somewhat plano-convex; perfect floret 1, terminal in the spikelet (except 2 in the monoecious genus *Zea*), usually subtended by a sterile or staminate floret; spikelets disarticulating below the glumes, the spikelets falling entire, either singly or in pairs with segments of the rachis (Subfamily Panicoideae) .... Group I, p. 177

1 Spikelet usually more or less laterally compressed, sometimes terete; perfect florets 1 to many, the reduced florets, when present, above the perfect ones, except in *Arrhenatherum, Hierochloë, Phalaris,* and *Blepharidachne,* the latter 3 having 2 subtending rudimentary or staminate florets; spikelets disarticulating either above or below the glumes.

 2 Glumes lacking; spikelets very flat, 1-flowered; palea subequal to the lemma and 3-nerved .... 81. *Leersia,* p. 457

 2 Glumes present (only 1 in *Lolium*); spikelets various; palea, if present, 2-nerved.

  3 Spikelets usually with 2 staminate or rudimentary florets below the terminal fertile floret.

   4 Lower florets staminate; spikelets brown and shiny .... 24. *Hierochloë,* p. 282

   4 Lower florets empty and much reduced; spikelets green or yellowish .... 25. *Phalaris,* p. 284

  3 Spikelets without reduced floret below the perfect floret (except in *Blepharidachne,* which has a rudimentary floret above the perfect one as well).

   5 Spikelets sessile or subsessile (when more than one at a node some of the group usually pedicelled); inflorescence a single spike, or 2 or more digitately or racemosely arranged spicate branches (the spicate branches of *Beckmannia* sometimes arranged in panicles) .... Group II, p. 178

   5 Spikelets pedicellate; inflorescence an open or contracted, sometimes spike-like panicle, rarely a raceme, usually reduced to a single spikelet in *Danthonia unispicata.*

    6 Spikelets 2- to many-flowered.

     7 Glumes shorter than the first floret; lemmas awnless or awned from the tip or from a bifid apex, the awn rarely slightly dorsal .... Group III, p. 179

     7 Glumes, at least the second glume as long or longer than the lowest floret, often as long as the spikelet (sometimes slightly shorter than the first floret in *Koeleria* and *Sphenopholis*); lemmas with dorsal awns or awnless (awned from a bifid apex in *Danthonia, Schismus,* and *Koeleria*) .... Group IV, p. 180

    6 Spikelets 1-flowered, rarely with a reduced lemma on the extended rachilla of *Calamagrostis* .... Group V, p. 181

Group I

Spikelet dorsally compressed, the glumes and fertile lemma more or less flattened or merely arching, the spikelet sometimes terete, but then the caryopsis plano-convex; disarticulation below the glumes, the spikelet falling entire, either singly or in pairs with segments of the rachis; perfect floret 1, terminal in the spikelet (except in the monoecious genus *Zea*), usually subtended by a sterile or staminate floret. (Includes all of Subfamily Panicoideae.)

1 Glumes membranous; sterile lemma resembling the second glume and appearing like a third glume; fertile lemma and palea indurate or at least firmer than the glumes (Tribe 6. Paniceae).
  2 Spikelets subtended by bristles or enclosed in a prickly bur.
    3 Spikelets subtended by 1 to many bristles, the spikelets deciduous and the bristles persistent . . . . 49. *Setaria,* p. 378
    3 Spikelets enclosed in a prickly bur (involucre), the burs dropping intact from the rachis at maturity . . . . 50. *Cenchrus,* p. 382
  2 Spikelets not subtended by bristles or enclosed in a prickly bur.
    4 Glumes and sterile lemma awnless.
      5 Spikelets in 1-sided spikes or racemes.
        6 First glume much reduced to absent, the second glume from 1/3 to as long as the spikelet.
          7 Plants annual; lemma of the fertile floret firm but not hard, the margins hyaline; racemes more or less digitate . . . . 44. *Digitaria,* p. 364
          7 Plants perennial; lemma of the fertile floret relatively thick and rigid; racemes single or paired, not digitate . . . . 45. *Paspalum,* p. 366
        6 Glumes subequal and as long as the spikelet . . . . 46. *Panicum obtusum,* p. 369
      5 Spikelets in open or compact panicles.
        8 Second glume and sterile lemma glabrous, or sometimes scaberulous only on the nerves towards the apex; basal leaves similar to the culm leaves, not forming a winter rosette; spikelets all fertile . . . . 46. *Panicum,* p. 368
        8 Second glume and sterile lemma short pilose all over (sometimes glabrous in *D. oligosanthes*); basal leaves distinctly different from the culm leaves, forming a winter rosette; plants passing from one growth phase to another, the so-called vernal phase characterized by simple stems and terminal panicles becoming transformed to the autumnal phase with branched culms and axillary, often cleistogamous panicles . . . . 47. *Dichanthelium,* p. 374
    4 Glumes and sterile lemma usually awned; spikelets in irregular clusters along one side of the panicle branches, the spikelets stiffly hispid . . . . 48. *Echinochloa,* p. 376
1 Glumes indurate, except in *Imperata* and *Zea;* sterile lemma, fertile lemma, and palea hyaline or membranous, the 2 lemmas usually alike in texture (Tribe 7. Andropogoneae).
  9 Plants monoecious; staminate and pistillate spikelets in separate inflorescences; cultivated corn . . . . *Zea mays* L., p. 394
  9 Plants not monoecious; spikelets in pairs, one sessile, perfect, the other pedicellate, usually staminate or neuter, the pedicellate one sometimes obsolete, rarely both spikelets pedicellate.
    10 Spikelets all alike and fertile.
      11 Pedicelled spikelets present, similar to the sessile spikelets; panicle silky-pubescent with hairs 8–11 mm long . . . . 51. *Imperata,* p. 384
      11 Pedicelled spikelets completely absent, represented only by a pedicel; panicle branchlets and pedicels yellowish-villous with hairs 0.5–1.2 mm long . . . . 53. *Sorghastrum,* p. 386
    10 Spikelets not all alike, the pedicelled ones staminate or sterile.
      12 Inflorescence an open or contracted panicle with spikelets alternately arranged in 1- to few-jointed rachises . . . . 52. *Sorghum,* p. 385
      12 Inflorescence of 1 to several racemes, the racemes of several joints (internodes).
        13 Racemes solitary on each peduncle; rachis joints oblique and hollow at the summit . . . . 55. *Schizachyrium,* p. 390
        13 Racemes 2 to several on each peduncle.
          14 Racemes 2–6 on each peduncle, digitate; rachis joints slender, sometimes with a shallow groove on 1 side . . . . 54. *Andropogon,* p. 388
          14 Racemes numerous, clustered in a leafless panicle, usually on a relatively long axis; pedicels and rachis joints flat, each with a central, translucent groove . . . . 56. *Bothriochloa,* p. 393

Group II

Inflorescence a single spike or 2 or more digitately or racemosely arranged spicate branches (the subspicate branches of *Beckmannia* sometimes arranged in panicles); spikelets sessile or subsessile (when more than 1 at a node some of the group usually pedicelled), more or less laterally compressed or terete (in *Aegilops*), 1- to several-flowered.

1 Spikelets or spikelet clusters alternate on opposite sides of the continuous or discontinuous rachis; spike terminal, solitary; leaves often auriculate.
2 First glume lacking, except in the terminal spikelet; spikelets placed edgewise to the rachis . . . . . 4. *Lolium,* p. 215
2 First glume present, sometimes much reduced; spikelets placed flatwise to the rachis, except in *Hilaria.*
3 Plants annual; introduced weeds and cultivated plants.
4 Spikelets cylindric, fitting into concavities of the rachis . . . . . 35. *Aegilops,* p. 332
4 Spikelets laterally compressed, the rachis not concave above each joint.
5 Spikelets 2 or more at each node of the rachis.
6 Spikelets 2 per node; rachis not disarticulating; glumes connate at the base . . . . . 31. *Taeniatherum,* p. 310
6 Spikelets 3 per node; rachis disarticulating (except in *H. vulgare*); glumes not connate at the base . . . . . 32. *Hordeum,* p. 310
5 Spikelets solitary at each node of the rachis.
7 Spikes small, 0.8–2 cm long; glumes with rather thickened, indurate margins, somewhat connate at the base . . . . . 34. *Eremopyrum,* p. 332
7 Spikes larger, 3–17 cm long; glumes with scarious margins, not at all connate at the base.
8 Glumes ovate, 3- or more-nerved; spikelets (2-) 3- to 5-flowered . . . . . 36. *Triticum,* p. 334
8 Glumes subulate, 1- (3)-nerved; spikelets 2-flowered . . . . . 37. *Secale,* p. 336
3 Plants perennial; mostly native species.
9 Rachis continuous, disarticulation occurring above the glumes (below the glumes in *Hilaria*).
10 Spikelets in groups of 3, the spikelet clusters falling entire from a zig-zag rachis; central spikelet fertile, 1-flowered, the 2 lateral spikelets staminate, mostly 2-flowered; glumes broad, obtuse to truncate, irregularly lacerate or awned . . . . . 77. *Hilaria,* p. 448
10 Spikelets 1 to several at a node, the rachis not zig-zag; spikelets all similar, usually 2- to many-flowered; glumes entire, usually acute or acuminate.
11 Glumes subulate to narrowly lanceolate; lemmas usually arranged 90° from normal position due to twisting of the rachilla, thus the florets not strictly in 2 ranks; spikelets 1 or usually 2 or more at each node of the rachis . . . . . 29. *Elymus,* p. 298
11 Glumes narrowly lanceolate to broader; lemmas arranged in normal position above the glumes so the florets are in 2 ranks; spikelets mostly solitary at each node of the rachis or rarely 2 in robust plants . . . . . 33. *Agropyron,* p. 318
9 Rachis disarticulating at maturity.
12 Spikelets more than 1 at each node of the rachis.
13 Spikelets 3 to a node, 1-flowered, the lateral pair of spikelets usually pedicellate and reduced, the central one sessile and fertile (all 3 sessile and usually fertile in the annual *H. vulgare*) . . . . . 32. *Hordeum,* p. 310
13 Spikelets mostly 2 to a node, if 1 or 3 (rare) not as above, usually alike, (1-) 2- to 6-flowered.
14 Glumes 3–7 mm long; awn of the lemma 0.5–3 mm long . . . . . 29. *Elymus junceus,* p. 299
14 Glumes 12–85 mm long; awn of the lemma 5–120 mm long.
15 Spikes 8–10 mm broad; body of the lemma 8–10 mm long.
16 Plants 1–4.5 (6.5) dm tall; glumes subulate, 25–85 mm long . . . . . 30. *Sitanion,* p. 306
16 Plants (5) 6–10 dm tall, those less than 6 dm tall either with glumes linear-lanceolate and 2- to 4-nerved or the glumes less than 24 mm long; sterile hybrids . . . . . X *Elysitanion,* p. 295
15 Spikes about 5 mm broad; body of the lemma 6–8 mm long; sterile hybrids . . . . . X *Agrohordeum,* p. 295
12 Spikelets solitary at each node of the rachis.
17 Culms erect; sterile hybrids of low to midmontane elevs. . . . . . X *Agrositanion,* p. 295

17 Culms prostrate-spreading, or at least decumbent at the base; fertile subalpine or alpine species . . . . . 33. *Agropyron scribneri,* p. 331
1 Spikelets in 2 rows, along 2 sides of a 3-angled, continuous rachis, forming 1-sided spikes or racemes, these spicate branches usually 2 to several (1 in *Sclerochloa*), digitately, racemosely or paniculately arranged on a central axis; leaves non-auriculate.
18 Spikelets with 3–12 perfect florets ([1-] 2- to 3-flowered in *Leptochloa filiformis*).
19 Spike solitary, terminal . . . . . 7. *Sclerochloa,* p. 218
19 Spikes 2 to several, digitate to racemosely arranged on a central axis.
20 Spikes 2–4, digitate, sometimes with 1 or 2 spikes racemosely arranged below . . . . . 70. *Eleusine,* p. 433
20 Spikes numerous, racemosely arranged on an elongated axis . . . . . 71. *Leptochloa,* p. 434
18 Spikelets with only 1 perfect floret, often with additional imperfect florets, these often only rudimentary.
21 Spikelets with 1 or more modified florets above the perfect one.
22 Spikes digitate or nearly so . . . . . 74. *Chloris,* p. 436
22 Spikes racemose along a main axis, or reduced to 1 . . . . . 75. *Bouteloua,* p. 438
21 Spikelets without additional modified florets, the rachilla sometimes prolonged.
23 Spikelets disarticulating above the glumes.
24 Spikes racemose along a wiry axis; lemma 3–4.7 mm long; rachilla not prolonged . . . . . 72. *Schedonnardus,* p. 435
24 Spikes digitate; lemma 1.8–2.5 mm long; rachilla prolonged behind the palea . . . . . 73. *Cynodon,* p. 436
23 Spikelets disarticulating below the glumes and falling entire.
25 Spikes (2) 4–15 cm long; glumes narrow-lanceolate to subulate, 3.5–20 mm long; lemma 6.2–9.5 mm long . . . . . 76. *Spartina,* p. 446
25 Spikes less than 1.8 cm long; glumes broad, rounded-triangular, 2–3 mm long; lemma 1.8–3.5 mm long . . . . . 28. *Beckmannia,* p. 292

Group III

Inflorescence an open or contracted, sometimes spike-like panicle; spikelets 2- to many-flowered, more or less laterally compressed; glumes shorter than the first floret; lemmas awnless or awned from the tip or from a bifid apex, the awn rarely slightly dorsal.

1 Plants dioecious or monoecious; lemmas glabrous to scabrous, not villous.
2 Plants densely tufted, erect from short rhizomes; lemmas scabrous; dry mountain slopes . . . . . 5. *Leucopoa,* p. 216
2 Plants not densely tufted, spreading by long, creeping rhizomes; lemmas glabrous; saline or alkaline places . . . . . 78. *Distichlis,* p. 450
1 Plants not dioecious, except in some species of *Poa* with villous lemmas.
3 Plants tall, stout reeds, 2–3 m tall, with large, plume-like panicles; rachilla with long, spreading hairs . . . . . 82. *Phragmites,* p. 458
3 Plants not reed-like, rarely up to 2 m tall; rachilla without long, spreading hairs.
4 Lemmas mostly 5- or more-nerved, if 3-nerved the nerves usually obscure.
5 Callus of the florets bearded.
6 Lemmas awnless, entire . . . . . 9. *Poa,* p. 224
6 Lemmas awned from a bifid apex, the awn 8–15 mm long . . . . . 41. *Schizachne,* p. 344
5 Callus of the florets not bearded.
7 Lemmas keeled on the back.
8 Spikelets strongly compressed, crowded in dense, 1-sided panicles . . . . . 10. *Dactylis,* p. 252
8 Spikelets not strongly compressed (except some *Bromus*), not crowded in dense 1-sided panicles.
9 Spikelets large, mostly more than 15 mm long; lemmas awned, often from a minutely bifid apex, or nearly awnless . . . . . 1. *Bromus,* p. 185
9 Spikelets small, mostly less than 10 mm long.
10 Lemmas with an awn 2.5–3 mm long . . . . . 3. *Festuca dasyclada,* p. 210
10 Lemmas awnless . . . . . 9. *Poa,* p. 224
7 Lemmas rounded on the back.
11 Glumes papery; spikelet disarticulating above or below the glumes; terminal floret similar to the fertile florets, or often reduced to an obovoid, blunt rudiment . . . . . 38. *Melica,* p. 336
11 Glumes not papery; spikelets disarticulating above the glumes; reduced florets not unlike the others.

12 Nerves of the lemma parallel, not converging at the summit, or but only slightly so.
13 Sheaths of the leaves usually open; nerves of the lemma (3) 5 (7), mostly obscure; often of saline or alkaline soils .... 8. *Puccinellia,* p. 220
13 Sheaths of the leaves usually closed; nerves of the lemma usually 7, prominent; fresh water marshes .... 39. *Glyceria,* p. 338
12 Nerves of the lemma converging toward the summit.
14 Lemmas tapered to a point, usually awned or sometimes awn-tipped (this appearance often exaggerated by involution of the lemma); sheaths of the leaves open.
15 Plants annual .... 2. *Vulpia,* p. 201
15 Plants perennial .... 3. *Festuca,* p. 205
14 Lemmas broader, or if somewhat tapered then awned from a bifid apex; sheaths of the leaves closed, at least in the lower half.
16 Spikelets large, mostly more than 15 mm long; lemmas often awned from a bifid apex, or nearly awnless .... 1. *Bromus,* p. 185
16 Spikelets small, mostly less than 10 mm long; lemmas awnless ..... 9. *Poa,* p. 224
4 Lemmas 3-nerved, the nerves usually prominent.
17 Spikelets 2- or 3-flowered, rarely 5- or 6-flowered in *Redfieldia* and *Munroa.*
18 Inflorescence small, reduced, hidden among sharp-pointed leaves ..... 60. *Munroa,* p. 403
18 Inflorescence not concealed, or only partially so.
19 Lemmas broad, truncate, the nerves parallel; plants decumbent and rooting at the nodes .... 40. *Catabrosa,* p. 344
19 Lemmas lanceolate, acute or acuminate, the nerves converging towards the apex; plants erect.
20 Callus of the floret densely silky-pilose .... 61. *Redfieldia,* p. 404
20 Callus of the floret glabrous .... 66. *Muhlenbergia,* p. 408
17 Spikelets (5-) 6- to many-flowered, as few as 5 in some *Eragrostis curvula* and *Tridens muticus.*
21 Lemmas with glabrous or scaberulous lateral nerves, awnless; glumes 0.5–2.5 (2.8) mm long .... 57. *Eragrostis,* p. 395
21 Lemmas conspicuously long-hairy along the nerves, at least below, awned; glumes 3–8.5 mm long .... 58. ~~*Tridens*~~, p. 400

Continue couplet # 4 of Group 4

Group IV

Inflorescence an open or contracted, sometimes spike-like, panicle or raceme, sometimes reduced to a single spikelet in *Danthonia unispicata;* spikelets 2- to many-flowered, usually more or less compressed; glumes, at least the second as long or longer than the lowest floret, often as long as the spikelet (sometimes slightly shorter than the first floret in *Koeleria* and *Sphenopholis*); lemmas awned or awnless, the awn dorsal or apical.

1 Plants dioecious or monoecious, the sexes very dissimilar, the pistillate inflorescence superficially resembling *Sitanion* with its long awned lemmas and the staminate inflorescence with awnless lemmas .... 62. *Scleropogon,* p. 404
1 Plants not dioecious or monoecious.
2 Lemmas 3-nerved.
3 Spikelets 4-flowered, the lower 2 sterile, the third fertile and the fourth reduced to a rudiment; lemmas deeply 3-lobed .... 63. *Blepharidachne,* p. 406
3 Spikelets 5- to 18-flowered, all but the terminal floret fertile; lemmas not deeply 3-lobed, except in *Erioneuron pulchellum.*
4 Panicle elongate and somewhat open, (4) 6–15 cm long; leaves not crowded at the base .... 58. *Tridens,* p. 400
4 Panicle short, compact, ovoid or oblong, 1.5–4 (8) cm long; leaves crowded at the base .... 59. *Erioneuron,* p. 402
2 Lemmas 5- to 11-nerved (sometimes 3-nerved in some *Helictotrichon* and *Holcus*).
5 Lemma divided into 9 plumose awns .... 79. *Enneapogon,* p. 452
5 Lemmas with a single awn or awnless.
6 Awn of the lemma from the apex, between 2 lobes or absent.
7 Lemmas awned.
8 Awn of the lemma arising between 2 lobes.
9 Sheath margins united to near the throat; panicles (8) 10–15 (20) cm long; awns 10–24 mm long .... 1. *Bromus trinii,* p. 201

9 Sheath margins free to the base; panicles 1–6 (9) cm long; awns 3–10 (12.5) mm long.
10 Awns conspicuous, flat, bent; lemmas 6–15 mm long .... 83. *Danthonia,* p. 460
10 Awns minute, nearly obsolete, only visible with magnification; lemmas small, 1.5–2 mm long .... 84. *Schismus,* p. 462
8 Awn of the lemma arising nearly imperceptibly from a narrow, acuminate, unlobed apex .... 2. *Vulpia,* p. 201
7 Lemmas awnless or at most terminating in a mucro no more than 1 mm long.
11 Spikelets 18–30 mm long .... 15. *Avena sativa,* p. 264
11 Spikelets 1.5–11 mm long.
12 Spikelets mostly more than 7 mm long, or if smaller, plants annual (*Sclerochloa*).
13 Plants stout, rhizomatous perennials, 10–15 dm tall .... 6. *Scolochloa,* p. 218
13 Plants low, tufted annuals, 0.3–2 dm tall .... 7. *Sclerochloa,* p. 218
12 Spikelets mostly less than 7 mm long, or if up to 8 mm long plants densely tufted perennials.
14 Lemmas 1.5–2 mm long; glumes 4–5 (6) mm long .... 84. *Schismus,* p. 462
14 Lemmas 2–6 mm long, or if smaller the glumes 3 mm or less long.
15 First glume 1.5–4.5 (5) mm long; lemmas 2–5 (5.5) mm long.
16 Plants small, 0.3–2 (2.5) dm tall.
17 Glumes dissimilar, the first 1.5 mm long, the second 2.5–3 mm long; lemmas about 2 mm long; annuals of moist alkaline habitats in valley bottoms .... 8. *Puccinellia simplex,* p. 224
17 Glumes subequal, 3–5 mm long; lemmas 3–5 mm long; perennials of alpine habitats .... 9. *Poa suksdorfii,* p. 245
16 Plants (2) 2.5–6.5 dm tall.
18 Lemmas 3.2–5.2 mm long; glumes similar in shape, the second 3.2–5.5 mm long; spikelets disarticulating above the glumes .. .. 11. *Koeleria,* p. 254
18 Lemmas 2–3 (3.2) mm long; glumes dissimilar, the first linear, the second oblanceolate to broadly obovate, 2–2.8 (3.2) mm long; spikelets disarticulating below the glumes .... 12. *Sphenopholis,* p. 255
15 First glume and the lemmas 5–7 mm long .... 13. *Trisetum wolfii,* p. 256
6 Awn of the lemma dorsally attached.
19 Spikelets with 2 or more perfect florets, the reduced florets, when present, terminal in the spikelet.
20 Spikelets not large, the glumes 2–7.5 (8) mm long.
21 Awn of the lemma attached above the middle; lemmas (3.5) 4–9 mm long ... . 13. *Trisetum,* p. 256
21 Awn of the lemma attached below the middle; lemmas 1.7–4 (4.5) mm long .... 14. *Deschampsia,* p. 260
20 Spikelets large, the glumes 8–30 mm long.
22 Plants large annuals, (2.5) 4–19 dm tall; glumes 17–30 mm long; panicles open, drooping, 10–38 cm long .... 15. *Avena,* p. 262
22 Plants small perennials, 0.8–1.8 (2) dm tall; glumes 8–11.5 mm long; panicles narrow, erect, 3–5 (7.5) cm long .... 16. *Helictotrichon,* p. 264
19 Spikelets with 1 perfect floret and 1 staminate floret.
23 Lower floret staminate and awned, the awn twisted, geniculate, exserted; upper floret perfect, awnless or with a small, hair-like, nearly terminal awn .... 17. *Arrhenatherum,* p. 264
23 Lower floret perfect, awnless; upper floret staminate, with a hooked awn .... 18. *Holcus,* p. 266

Group V

Inflorescence an open or contracted, sometimes spike-like panicle; spikelets 1-flowered, rarely with a reduced lemma on the extended rachilla of *Calamagrostis,* more or less laterally compressed to terete.

1 Lemma indurate, awned, permanently enclosing the palea and caryopsis, the callus well developed.
2 Awn of the lemma 3-branched .... 80. *Aristida,* p. 452
2 Awn of the lemma unbranched.

3 Awn deciduous, mostly less than 10 mm long, not twisted, or only very slightly, sometimes bent, rarely more than 3 or 4 times as long as the plump lemma; callus short, usually obtuse.

4 Awns 7–18 mm long with villous lemmas 2.7–5.5 mm long; sterile hybrids .... X *Stiporyzopsis,* p. 346

4 Awns 2–7 mm long, up to 9.5 mm in *Oryzopsis micrantha* which has glabrous lemmas; fertile species.

5 Glumes 2.5–7.5 (8) mm long .... 42. *Oryzopsis,* p. 348

5 Glumes (7) 7.5–10 mm long .... 43. *Stipa webberi,* p. 353

3 Awn usually persistent, (10) 14–200 mm long, usually twisted, several to many times longer than the narrow lemma; callus sharp-pointed .... 43. *Stipa,* p. 352

1 Lemma not evidently indurate, usually not permanently enclosing the palea and caryopsis.

6 Glumes, at least the second, equalling or usually longer than the floret.

7 Spikelets disarticulating above the glumes.

8 Glumes abruptly short-awned from an obtuse apex, ciliate on the prominent keels .... 27. *Phleum,* p. 288

8 Glumes not abruptly awned, if awned at all.

9 Lemmas 1-nerved, awnless .... 67. *Sporobolus,* p. 422

9 Lemmas 3- to 5-nerved, awned or sometimes awnless.

10 Lemma awned from an entire or minutely cleft apex, or awnless the body of the lemma 3-nerved .... 66. *Muhlenbergia,* p. 408

10 Lemma with a dorsal awn, or if awnless, 5-nerved; glumes exceeding the lemma.

11 Awn of the lemma less than 4.5 mm long or absent, or if longer, arising at or below the midlength of the lemma (in *C. purpurascens*); lemma membranous.

12 Callus of the floret evidently bearded, the hairs more than 0.5 mm long; palea usually well developed; rachilla prolonged behind the palea .... 19. *Calamagrostis,* p. 266

12 Callus of the floret glabrous to minutely bearded; palea often absent or at least poorly developed; rachilla usually not prolonged behind the palea .... 21. *Agrostis,* p. 273

11 Awn of the lemma evident, more than 5.5 mm long, arising from just below the apex of the lemma; lemma firm to subindurate .... 20. *Apera,* p. 270

7 Spikelets disarticulating below the glumes, the spikelet falling entire.

13 Glumes awned.

14 Lemmas 0.7–1.5 mm long; anthers small, 0.4–0.7 mm long .... 22. *Polypogon,* p. 280

14 Lemmas (1.2) 1.7–2.5 mm long; anthers 1.1–2.3 mm long .... 27. *Phleum,* p. 288

13 Glumes awnless.

15 Lemma with a short awn from the notched apex; panicle usually somewhat open .... 23. *Cinna,* p. 282

15 Lemma awned from the middle or below; panicle dense, cylindrical, spike-like .... 26. *Alopecurus,* p. 286

6 Glumes shorter than the floret.

16 Glumes or lemma awned.

17 Spikelets disarticulating below the glumes, the spikelets falling in pairs, the lower spikelet sterile; first glume usually 2- (3)-awned .... 64. *Lycurus,* p. 406

17 Spikelets disarticulating above the glumes, the spikelets all alike; glumes each with a single awn or awnless .... 66. *Muhlenbergia,* p. 408

16 Glumes and lemma awnless.

18 Spikelets disarticulating below the glumes.

19 Panicles open to congested, but never spike-like, (2.5) 5–35 cm long; ligules membranous, 1.5–8.5 mm long.

20 Glumes dissimilar, the first linear, the second broadly oblanceolate to obovate; ligules 1.5–3.5 mm long .... 12. *Sphenopholis,* p. 255

20 Glumes subequal, semicircular; ligules 5–8.5 mm long .... 28. *Beckmannia,* p. 292

19 Panicles dense and spike-like, short, 0.7–4.5 (6.5) cm long; ligules of hairs, 0.2–1 mm long .... 69. *Crypsis,* p. 430

18 Spikelets disarticulating above the glumes.

21 Callus of the floret bearded; spikelets 6.5 mm or more long .... 65. *Calamovilfa,* p. 408

21 Callus of the floret glabrous to scabrous; spikelets 1–4.5 (5.5) mm long.

22 Lemma 1-nerved; fruit not a true caryopsis as the ovary wall and seed coat are not fused .... 67. *Sporobolus,* p. 422

22 Lemma 3-nerved; fruit a true caryopsis.
  23 Nerves of the lemma glabrous or scabrous.
    24 Sheath margins united to near the throat; nerves of the lemma parallel, not converging towards the apex .... 40. *Catabrosa,* p. 344
    24 Sheath margins free to the base; nerves of the lemma converging towards the apex .... 66. *Muhlenbergia,* p. 408
  23 Nerves of the lemma densely pubescent .... 68. *Blepharoneuron,* p. 430

Technical Key to Subfamilies
(as represented in our flora)

1 Spikelet dorsally compressed (most evident in the caryopsis), the glumes and fertile lemma more or less flattened or merely arching; perfect floret 1, terminal in the spikelet (except 2 in the monoecious genus *Zea*), a second sterile, imperfect or rudimentary one borne below it; spikelets disarticulating below the glumes, falling entire .... Subfamily II. Panicoideae, p. 362
1 Spikelet usually more or less laterally compressed to terete; perfect florets 1 to many, the reduced florets, when present, above the perfect ones (note exceptions in *Arrhenatherum, Hierochloë, Phalaris* and *Blepharidachne*); spikelets usually disarticulating above the glumes and leaving the glumes as persistent husks, if disarticulating below the glumes the spikelets definitely laterally compressed.
  2 Glumes lacking (in ours); spikelets falling entire, very flat, 1-flowered; lemma and palea about equal, the palea in ours 3-nerved, both 1-keeled; embryonic leaf margins overlapping; plants of moist or marshy areas .... Subfamily IV. Oryzoideae, p. 456
  2 Glumes usually both present, sometimes very short or rarely 1 lacking as in *Lolium;* palea 2-nerved, 2-keeled; embryonic leaf margins not overlapping; plants of various habitats.
    3 Epidermal cells in the region of root hairs alternately long and short with only the short ones giving rise to root hairs; embryo with vascular traces to the scutellum and coleoptile diverging at approximately the same point and the lower part of the scutellum fused to the coleorhiza .... Subfamily I. Pooideae, p. 183
    3 Epidermal cells in the region of root hairs all alike and all capable of giving rise to root hairs; embryo with vascular traces to the scutellum and coleoptile separated by an internode and the lower part of the scutellum and coleorhiza separated by a cleft.
      4 Epiblast present, except in tribe Aristideae which has 1-flowered spikelets and 3-branched awns .... Subfamily III. Eragrostoideae, p. 394
      4 Epiblast lacking (in ours) .... Subfamily V. Arundinoideae, p. 458

Subfamily I. **POOIDEAE**

(Festucoideae)

Epidermal cells in the region of root hairs alternately long and short with only the short ones giving rise to root hairs; culm internodes typically hollow, but some solid; sheath margins free or connate; ligule usually membranous, sometimes a fringe of hairs or a low crown; blades with auricles in some; inflorescence a panicle, raceme (*Sclerochloa*) or bilateral spike; spikelet usually laterally compressed, some terete, 1- to several-flowered, the reduced florets, when present, usually terminal (except below in *Arrhenatherum, Hierochloë* and *Phalaris*); disarticulation usually above the glumes, sometimes below; lemmas (3-) 5- to several-nerved; lodicules usually elongate, pointed or less often truncate, thick at the base and membranous above, with little or no vasculation; stamens usually 3, sometimes 2 or 1; embryo relatively small, $^1/_6$ to $^1/_4$ the length of the caryopsis, with the typical pooid morphology with vascular traces to the scutellum and coleoptile diverging at approximately the same point, the epiblast usually present (except absent in Tribe Stipeae), the lower part of the scutellum fused to the coleorhiza, and the embryonic leaf margins not overlapping; $x = 7$ in most genera, 6, 8, 10, 11 in some, the chromosomes typically large.

The Pooideae subfamily has its greatest development in temperate regions of both hemispheres. The species that extend into the tropics occur mostly in the mountains. The few species growing in the lowlands do their growing only during the cooler season. The other subfamilies center in the tropics and extend north and south in the lowlands. It is not surprising then that the Pooideae subfamily contains the majority of the grass species for our area.

Such characters as overall pattern of the inflorescence, number of florets per spikelet, ratio between glumes and lemma length and position of the awns of the lemma, although de-emphasized in modern grass systematics, are important here in the separation of tribes.

Key to the Tribes of Pooideae

1 Inflorescence a bilateral spike or spicate raceme with both sessile and pedicelled spikelets at the rachis nodes; spikelets placed flatwise to the rachis .... Tribe 3. Triticeae, p. 292

1 Inflorescence a panicle, often compressed and appearing spike-like, or rarely reduced to a raceme, if a spike then the spikelets placed edgewise to the rachis (*Lolium*).
  2 Spikelets 1-flowered; lemma indurate and awned from the tip; embryo lacking an epiblast .... Tribe 5. STIPEAE, p. 346
  2 Spikelets usually more than 1-flowered, or if 1-flowered then the lemma membranous and awned from the back or awnless; embryo with an epiblast (except in *Bromus*).
    3 Glumes, at least the second, usually longer than the lowermost lemma; awn of the lemma, when present, dorsally placed .... Tribe 2. AVENEAE, p. 253
    3 Glumes mostly shorter than the lowermost lemma; awn of the lemma, when present, terminally placed or arising from a bifid apex.
      4 Lodicules mostly pointed; silica cells present in the leaf epidermis; $x = 7$ .... Tribe 1. POEAE, p. 184
      4 Lodicules truncate; silica cells absent or scarce in the leaf epidermis; $x = 8, 9, 10$ .... Tribe 4. MELICEAE, p. 336

Tribe 1. **POEAE**

(Festuceae)

Annuals or perennials; culms hollow; sheath margins free or connate (in *Bromus*, *Sclerochloa*, *Poa* and *Dactylis*); ligule membranous; blades with or without auricles (present in *Lolium*, some *Festuca* and some *Bromus*); inflorescence usually a panicle, infrequently a raceme (*Sclerochloa*) or spike (*Lolium*); spikelets mostly 2- to several-flowered, rarely 1; reduced florets, when present, above the fertile florets; disarticulation above the glumes (except in *Sclerochloa*); glumes usually unequal and usually shorter than the lowermost lemma; lemmas (3-) 5- to several-nerved, the nerves often obscure, awnless or awned from a bifid or entire apex, sometimes subdorsal in some *Bromus;* lodicules 2, mostly pointed; starch grains of the endosperm compound, except in *Bromus* which has simple rounded grains resembling those of tribe Triticeae; embryo pooid; $x = 7$.

The Poeae tribe is a homogeneous group except for *Bromus* and the closely related *Brachypodium* (not in our area). These two genera have simple rather than compound starch grains, pubescent terminal appendages on the ovary, and styles attached laterally to this appendage, all characters of the Triticeae tribe. *Bromus* also contains a levorotatory glucoside which is related to triticin found in *Agropyron* and which is not found in other Poeae. The inflorescence of *Brachypodium* is reduced to a strict raceme with the spikelets nearly sessile. These genera may be as closely related to the Triticeae as to the remainder of tribe Poeae. They are sometimes placed in a separate tribe Brachypodieae.

A. S. Hitchcock's (Manual Grasses U.S., 1951) Festuceae tribe, held together on the basis of similarities in spikelet morphology, included genera which are now placed in the widely separated tribes Meliceae, Eragrosteae (sensu stricto), Aeluropodeae, Pappophoreae and Arundineae. *Lolium* was included in the Triticeae by A. S. Hitchcock because of its spicate inflorescence. This spike is very different from that of other members of that tribe, however, with the sessile spikelets placed edgewise to the rachis and bearing only the second glume. *Lolium* has the compound starch grains common in the Poeae, and species of *Lolium* have been successfully crossed with several species of *Festuca*.

Briza maxima L., big quaking grass, is sometimes planted as an ornamental in our area. It can be recognized by its broad spikelets with horizontally spreading glumes and lemmas, and the spikelets drooping on slender pedicels.

1 Inflorescence a bilateral spike; first glume lacking, except in the terminal spikelet .... 4. *Lolium*
1 Inflorescence a panicle or raceme; both glumes present.
  2 Plants dioecious; leaves strongly striate .... 5. *Leucopoa*
  2 Plants usually not dioecious, or if so, the leaves with 2 median lines above (in *Poa*).
    3 Callus of the floret bearded.
      4 Callus hairs straight, erect .... 6. *Scolochloa*
      4 Callus hairs cobwebby, crimped .... 9. *Poa*
    3 Callus of the floret not bearded.
      5 Lemmas keeled on back.
        6 Panicle 1-sided, the branches and branchlets arising along 2 sides of 3-angled rachises, the spikelets in congested fascicles terminating each branchlet .... 10. *Dactylis*
        6 Panicle not 1-sided, nor compressed and branched as described above.
          7 Spikelets large, (11) 14–43 mm long; lemmas awned from a bifid apex or awnless .... 1. *Bromus*
          7 Spikelets small, 2.5–8 (11) mm long; lemmas awnless .... 9. *Poa*
      5 Lemmas rounded on back (somewhat keeled in *Festuca dasyclada*).
        8 Spikelets in racemes; disarticulation below the glumes; low tufted annuals, less than 2 dm tall .... 7. *Sclerochloa*
        8 Spikelets in panicles, or if racemes then the lemmas awned; disarticulation above the glumes; habit of plants various.
          9 Nerves of the lemma parallel, not converging at the summit, or but slightly so; lemmas mostly obtuse, awnless .... 8. *Puccinellia*
          9 Nerves of the lemma converging toward the summit; lemmas awned or pointed.

10 Lemmas long-tapered to a point, usually awned or awn-tipped (this appearance often exaggerated by involution of the lemma); sheaths open.
11 Plants annual .... 2. *Vulpia*
11 Plants perennial .... 3. *Festuca*
10 Lemmas broader, or if somewhat tapered then awned from a minute bifid apex; sheaths closed, at least below.
12 Spikelets large, (11) 14–43 mm long; lemmas awned, often from a bifid apex, or nearly awnless .... 1. *Bromus*
12 Spikelets small, 2.5–8 (11) mm long; lemmas awnless .... 9. *Poa*

1. BROMUS L. Bromegrass

Low to tall annuals, biennials, or perennials, rarely producing rhizomes; culms hollow; sheaths usually closed nearly to the top, pubescent or glabrous (this character often inconsistent within a species); ligules membranous, short, entire to lacerate; blades flat to somewhat involute, often with soft spreading hairs; auricles present or absent; inflorescence an open to much contracted panicle, usually lax and drooping (some depauperate plants reduced to a raceme); spikelets large, many-flowered, (12) 14–30 (45) mm long, oval to narrowly oblong, strongly compressed to nearly terete, disarticulating above the glumes and usually between the florets; glumes acute to blunt, rarely awn-tipped, unequal, each shorter than the first lemma, the first 1- or 3-nerved, the second 3- or 5- (7)-nerved; lemmas strongly keeled to rounded on back, glabrous, scabrous to pubescent with soft spreading hairs, (3-) 5- to 9-nerved, the nerves faint to evident and converging, often bifid at the apex, or entire and appearing bifid as a result of tearing along both sides of the mid-nerve, the awn subdorsal, sometimes lacking; palea shorter than the lemma to about as long, strongly ciliate on the 2 keels, adhering to the caryopsis; lodicules 2, small, entire; stamens usually 3, rarely 2, the anthers sometimes included (cleistogamous) and much reduced, or exserted (chasmogamous) and much larger; ovary with a hairy, terminal appendage with the styles inserted laterally below it; caryopsis with a narrow hilum extending the full length; $x = 7$.

A genus of about 100 species found in temp. regions of the world. (Greek *bromus*, ancient name for the oat, denoting food.)

Bromegrasses are important mostly as forage in mountain regions. *Bromus inermis*, an introduction from Europe, is widely used as a pasture and meadow grass, and *B. willdenowii*, an introduction from South America, is cultivated in the southern states for winter forage. *Bromus brizaeformis* is occasionally used as an ornamental and often becomes a part of dry bouquets. The annual bromes are weedy species, mostly of European origin. Ripgut grass, *B. rigidus*, may pierce the intestines of animals. Other similar species may cause inflammation of the eyes, nostrils, and mouths of livestock.

Bromegrasses comprise several groups of closely related species that are sometimes treated on a generic level. Taxonomically the genus is difficult. Some species of section Bromus are very difficult to distinguish. The treatment presented here is a conservative one that has come about through many years of field work. This genus is sometimes placed in tribe Brachypodieae on the basis of the conspicuous, hairy terminal appendage on the ovary. The styles arise laterally and below this appendage. *Bromus* contains a levorotatory glucoside which differs chemically and physically from that found in many other grasses, including *Festuca*. Also, *Bromus* starch grains are all simple while *Festuca* has compound starch grains.

References:
Harlan, J. R. Cleistogamy and chasmogamy in *Bromus carinatus* Hook. & Arn. Amer. J. Bot. 32: 66–72. 1945.
Shear, C. L. A revision of the North American species of *Bromus* occurring north of Mexico. U.S.D.A. Div. Agrostol. Bull. 23: 1–66. 1900.
Stebbins, G. L., Jr. The origin of the complex of *Bromus carinatus* and its phytogeographic implications. Contr. Gray Herb. 165: 42–55. 1947.
Wagnon, H. K. Natural breeding structure in the *Bromus carinatus* complex as determined by population analyses. Amer. J. Bot. 32: 142–148. 1945.
———. A revision of the genus *Bromus*, section Bromopsis of North America. Brittonia 7: 415–480. 1952.

1 Plants perennial, or if annual (forms of *B. willdenowii*) the second glume lanceolate to broadly lanceolate and 9–13 mm long.
2 Spikelets strongly flattened; lemmas usually carinate-keeled (Ceratochloa).
3 Lemmas awnless or with awns up to 2.5 mm long; an introduced species entering the region at the s. end and in the Lahontan Basin .... 1. *B. willdenowii*
3 Lemmas with awns exceeding 3 mm in length; a widespread native species .... 2. *B. carinatus*
2 Spikelets not strongly flattened, terete, at least before anthesis; lemmas always rounded on the back (Bromopsis).
4 Rhizomes present; lemmas usually awnless or sometimes with awns up to 2 mm long; panicles mostly erect .... 3. *B. inermis*
4 Rhizomes wanting; lemmas with awns mostly more than 2 mm long; panicles mostly nodding.
5 Awns (5) 6–10 mm long; ligules (1.2) 2–4.5 mm long; known in our area only from the c. Wasatch Range .... 4. *B. vulgaris*
5 Awns 1–4 (5.5) mm long; ligules 0.3–1 (1.5) mm long; widespread.

6 Lemmas with hairs concentrated on the marginal zones only, occasionally across the back or at the base; glumes glabrous or with scabrous nerves, the first glume usually only 1-nerved; leaves broad, (3) 4–13 mm wide . . . . 5. *B. ciliatus*

6 Lemmas with hairs on the margins and across the back, rarely only on the margins; glumes often pubescent, the first glume mainly 3-nerved; leaves narrower, 2–5 mm wide . . . . 6. *B. anomalus*

1 Plants annual; second glume either broadly lanceolate to ovate and 6–8.5 mm long or linear to narrowly lanceolate and (8) 9–28 mm long.

7 Lemmas short, 6–9 (10) mm long, and broad, (1.2) 1.7–4 mm wide in side view, scarcely bifid at the apex with teeth mostly less than 2 mm long; first glume usually 3- or more-nerved; awns, if present, 0.5–12 mm long (Bromium).

8 Lemmas mostly with long awns; spikelets 4- to 8- (12)-flowered; florets not spreading; lemmas 1.2–2.5 mm broad in side view.

9 Lemmas rounded, the margins usually rolled around the caryopsis; florets distant with openings at the base; awns mostly less than 5 mm long, often curved; palea about as long as the lemma . . . . 11. *B. secalinus*

9 Lemmas somewhat flattened, the margins not rolled around the caryopsis; florets more imbricated; awns usually more than 5 mm long, straight or more or less recurved; palea mostly shorter than the lemma (equal in *B. mollis*).

10 Panicles open, (4) 7–20 cm long; pedicels usually longer than the spikelets.

11 Branches of the panicles rather stiffly spreading, not flexuous; awns straight or only slightly recurved . . . . 7. *B. commutatus*

11 Branches of the panicles more or less flexuous, rather slender; awns usually distinctly divergent . . . . 8. *B. japonicus*

10 Panicles compact, 3–8 (10) cm long; pedicels mostly shorter than the spikelets . . . . 9. *B. mollis*

8 Lemmas awnless or only minutely awn-tipped; spikelets (7-) 9- to 15-flowered; florets widely spreading; lemmas exceptionally broad, 2.8–4 mm broad in side view . . . . 10. *B. brizaeformis*

7 Lemmas long, (9) 10–30 mm long, and narrow, 0.5–1.5 mm wide in side view, acute to acuminate, the narrow apical teeth mostly more than 2 mm long; first glume usually 1-nerved; awns longer, (10) 12–60 mm long (short in some *B. tectorum* and *B. trinii*).

12 Awns straight or only slightly curved or twisted; lemma teeth membranous; second glume shorter than the first lemma (Bromus).

13 Panicle contracted, the branches stout, ascending . . . . 12. *B. rubens*

13 Panicle open, the branches spreading or drooping (erect in *B. rigidus*).

14 Lemmas elongate, mostly 13–30 mm long; spikelets (23) 25–43 mm long; awns 18–60 mm long.

15 Lemmas 22–30 mm long; first glume 13–18 (22) mm long; awns (30) 35–60 mm long; in our area in w. Utah and w. Nev. . . . . 13. *B. rigidus*

15 Lemmas 13–18 (20) mm long; first glume 7–11 mm long; awns 18–26 (31) mm long; widespread . . . . 14. *B. sterilis*

14 Lemmas relatively short, 10–12.5 (17.5) mm long; spikelets only 12–20 (24) mm long; awns (7) 10–17 mm long . . . . 15. *B. tectorum*

12 Awns geniculate and strongly twisted near the base; lemma teeth aristate; second glume often as long or longer than the first lemma (Neobromus) . . . . 16. *B. trinii*

1. Bromus willdenowii Kunth

Bromus catharticus Vahl, Symb. Bot. 2: 22. 1791, nomen confusum. *Ceratochloa cathartica* Herter, Revista Sudamer. Bot. 6: 144. 1940. (Lima, Peru.)

Festuca unioloides Willd. Hort. Berol. 1: 3, pl. 3. 1803. *Ceratochloa unioloides* Beauv. Essai Nouv. Agrostogr. 75. 1812. *B. unioloides* Rasp. Ann. Sci. Nat. (Paris) 5: 439. 1825; not H.B.K. 1816. *B. willdenowii* Kunth, Révision Gram. 1: 134. 1829. (Described from a plant grown at Berlin from seed from "Carolina," where it must have been cultivated.)

Rescue grass.

Stout annuals or short-lived perennials; culms (2.5) 4–7 (11) dm tall; sheaths hirsute to pilose, rarely glabrous, often ciliate at the throat; ligules long, 2.5–5 mm, erose, glabrous to pilose; auricles, if any at all, very small; blades flat, relatively broad, 3–6 mm wide; panicle long, (8) 12–22 (30) cm, loosely contracted; spikelets relatively long, (16) 20–30 (33) mm, 4- to 7- (9)-flowered, strongly flattened; glumes many-nerved, strongly keeled, the first glume (7) 8–11.5 mm long, the second 9–13 mm long; lemmas 9–14 (17) mm long, lanceolate, carinate-keeled, many-nerved, minutely scaberulous to glabrous, the apex minutely bifid; awns, when present, up to only 2.5 mm long; anthers about 3 mm long; $2n = 28, 42, 56$.

Dry or moderately moist waste places; introduced from S. Amer. and escaping from cultivation in n. Mex. and across s. U.S. from the s. half of Calif., s. and w. Nev., s. Utah and N.M., e. to D.C. and Fla. Late March–early June.

Bromus willdenowii is cultivated in the southern states as a winter forage grass. It will live for 2 or 3 years if mowed to prevent its producing seed.

BROMUS

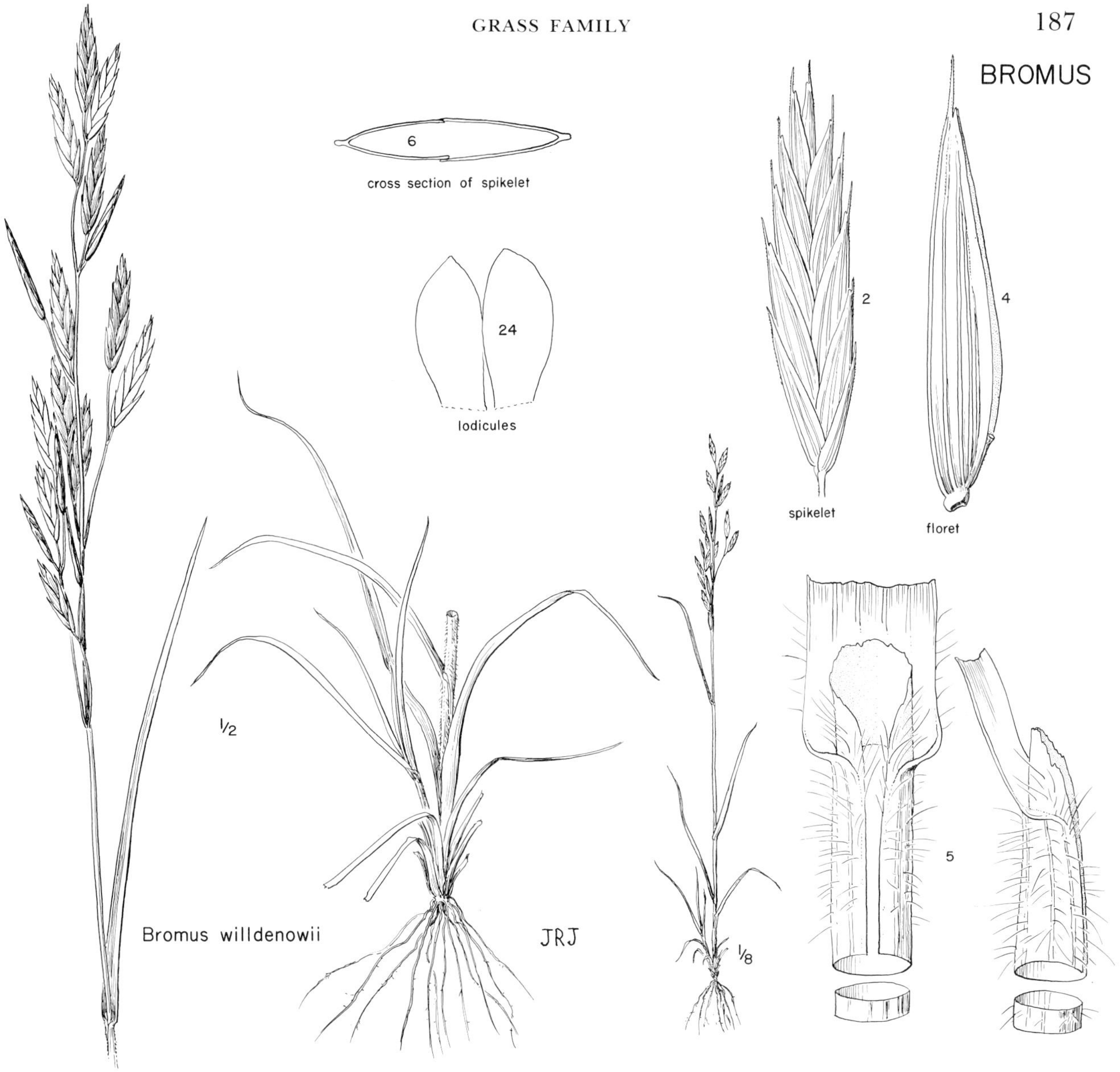

This taxon has been known in the past as *B. catharticus* Vahl, a nomen confusum, and *B. unioloides* H.B.K., which turns out to be a different species according to P. Raven (Brittonia 12: 219–221. 1960). Gould (The Grasses of Texas. 85. 1975) refers to it as *B. unioloides* without explanation.

2. Bromus carinatus Hook. & Arn.

Bromus carinatus Hook. & Arn. Bot. Beechey's Voyage 403. 1840. *Ceratochloa carinata* Tutin in Clapham, Tutin & Warb. Fl. Brit. Isles 1458. 1952. (*Douglas*, Nova California, in 1833.)

Ceratochloa brevi-aristata Hook. Fl. Boreali-Amer. 2: 253, pl. 234. 1840. *B. breviaristatus* Thurber in King, Rep. Geol. Explor. 40th Parallel 5: 389. 1871; not Buckley, 1863. *Forasaccus brebiaristatus* [sic] Lunell, Amer. Midl. Naturalist 4: 225. 1915. (*Douglas*, "Lewis' and Clarke's [Snake] River and near the sources of the Columbia," 24 July 1826.) = *marginatus* phase with exceptionally short awns.

Ceratochloa grandiflora Hook. Fl. Boreali-Amer. 2: 253, pl. 235. 1840; not *B. grandiflorus* Weigel, 1772. *B. hookerianus* Thurber in Wilkes, U. S. Explor. Exped. Bot. 17(2): 493. 1874. *B. carinatus* var. *hookerianus* Shear, U.S.D.A. Div. Agrostol. Bull. 23: 60, fig. 38. 1900. (*Scouler, Douglas,* "Common on the plains of the Columbia," Oregon.) = *carinatus* phase.

B. marginatus Nees ex Steudel, Syn. Pl. Glum. 1: 322. 1854. *B. hookeri* var. *marginatus* Fourn. Mex. Pl. 2: 127. 1886. *Ceratochloa marginata* Nees ex Steudel ex B. D. Jackson, Index Kew. 1: 487. 1893. *Forasaccus marginatus* Lunell, Amer. Midl. Naturalist 4: 225. 1915. *B. sitchensis* var. *marginatus* B. Boivin, Naturaliste Canad. 94: 521. 1967. (*Douglas*, Columbia River.)

B. breviaristatus Buckley, Proc. Acad. Nat. Sci. Philadelphia 1862: 98. 1863. *B. parviflorus* Nutt. ex A. Gray, ibid. 336. 1863, in synonymy; not Desf. 1829. *B. pauciflorus* [sic] Nutt. ex Shear, U.S.D.A. Div. Agrostol. Bull. 23: 53. 1900; not Hackel, 1899. (*Nuttall*, Rocky Mts., 1834.) = *carinatus* phase.

B. multiflorus Scribn. in A. Nels. U.S.D.A. Div. Agrostol. Bull. 13: 46. 1898; not Weigel, 1772. *B. polyanthus* Scribn. ex Shear, U.S.D.A. Div. Agrostol. Bull. 23: 56, fig. 34. 1900. (*Nelson 4021*, Battle Lake, Carbon Co., Wyo., 16 Aug. 1897.)

B. subvelutinus Shear, U.S.D.A. Div. Agrostol. Bull. 23: 52, fig. 32. 1900. (*Tracy 249*, Reno, Nev., in 1887.) = *marginatus* phase.

BROMUS

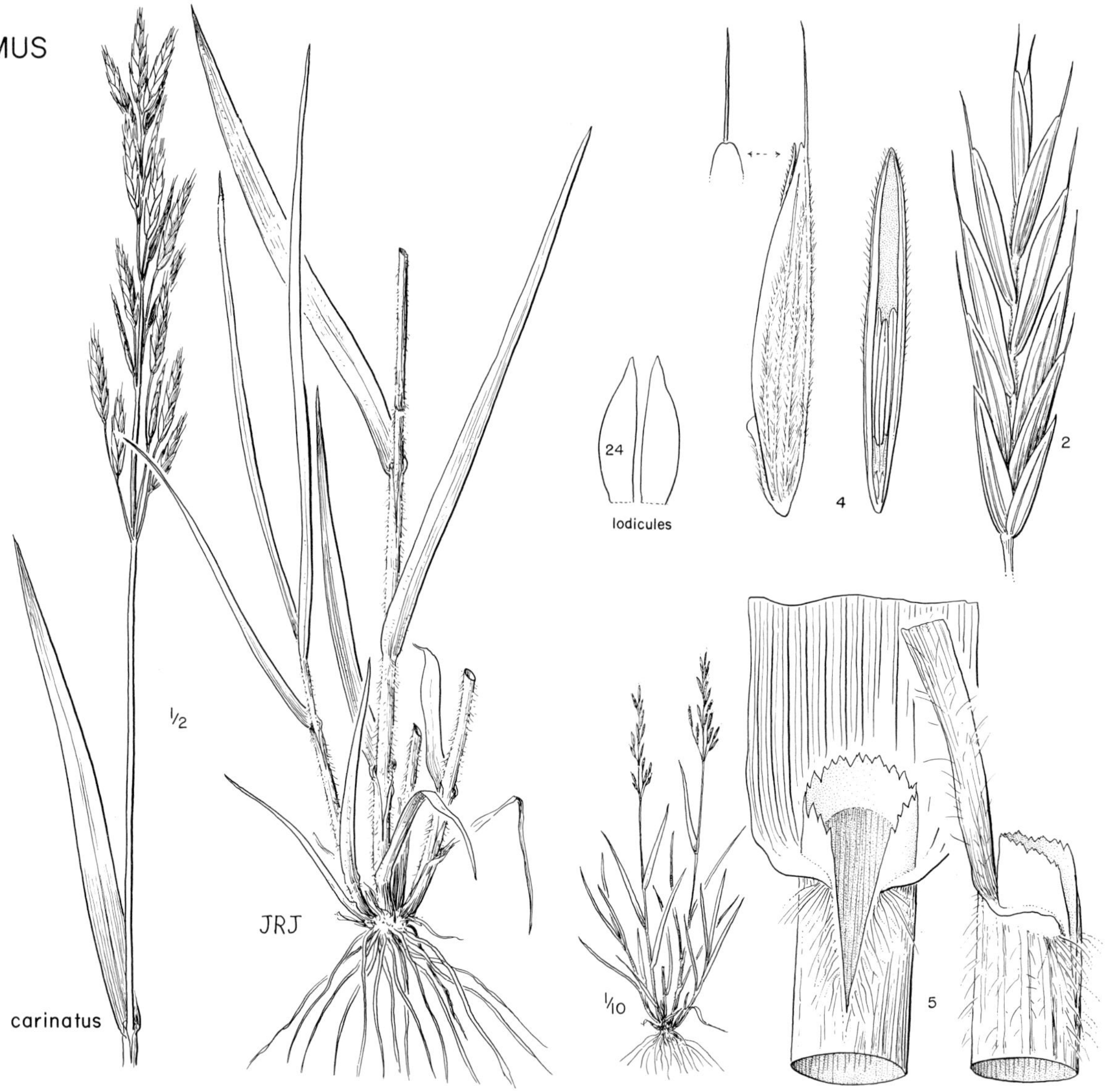

Bromus carinatus

B. marginatus var. *seminudus* Shear, U.S.D.A. Div. Agrostol. Bull. 23: 55. 1900. (*Shear 1811,* open mt. side, "5 mi. above Wallowa Lake, Oregon," 23 Aug. 1899.) = *polyanthus* phase.

B. marginatus var. *latior* Shear, U.S.D.A. Div. Agrostol. Bull. 23: 55. 1900. *B. latior* Rydb. Fl. Rocky Mts. 89, 1060. 1917 [1918]. (*Shear 1615,* "Edge of wheat field," Walla Walla, Wash., 29 June 1899.) = *marginatus* phase.

B. polyanthus var. *paniculatus* Shear, U.S.D.A. Div. Agrostol. Bull. 23: 57, fig. 35. 1900. *B. paniculatus* Rydb. Fl. Rocky Mts. 90, 1060. 1917 [1918]. (*Baker, Earle & Tracy 333,* "West Mancos Canon," Colo., 4 July 1898.) = *polyanthus* phase.

B. aleutensis Rydb. Mem. New York Bot. Gard. 1: 61. 1900; not Trin. 1853. *B. flodmanii* Rydb. Bull. Torrey Bot. Club 36: 538. 1909. (*Flodman 187,* Sheep Cr., Mont., 8 Aug. 1896.) = *polyanthus* phase.

California brome.

Tall perennials, often flowering the first season, without rhizomes; culms (3) 6–9 (14) dm tall, glabrous to pubescent; sheaths pilose, at least at the throat, rarely all glabrous; ligules relatively long, 1–3 (4) mm, erose, glabrous to pilose; auricles, if any at all, very small; blades flat and broad, (2) 3.5–9 (12) mm wide, rarely involute, scabrous to pilose or glabrous; panicles usually relatively long, (5) 10–20 (30) cm, loosely contracted and nodding; spikelets relatively long, 20–30 (40) mm, 4- to 7- (10)-flowered, strongly compressed; glumes lanceolate, strongly keeled, the first glume 7–11 mm long, 3- to 5-nerved, the second 9–13 mm long, 5- to 7-nerved; lemmas 11–15 (17) mm long, carinate-keeled, glabrous to scabrous or hirsute, the apex slightly bifid to entire; awns (3) 4–6 (8) mm long; anthers 1.5–3.5 mm long, sometimes small and included (cleistogamous) within the lemma; $2n$ = 28, 42, 56, 70.

Common in moderately moist habitats from sagebrush valleys to subalpine forests and meadows; widespread in w. N. Amer. from Alaska, B.C., and Alta., s. to Calif., w. Texas and throughout most of Mex.; introduced in Ill., Iowa and Kansas. Late May–early Aug.

Bromus carinatus is a highly polymorphic taxon that is recognized as 3 separate species by some. Customarily these 3 have been separated by the characters in the following key, characters which are often found in the same population.

BROMUS

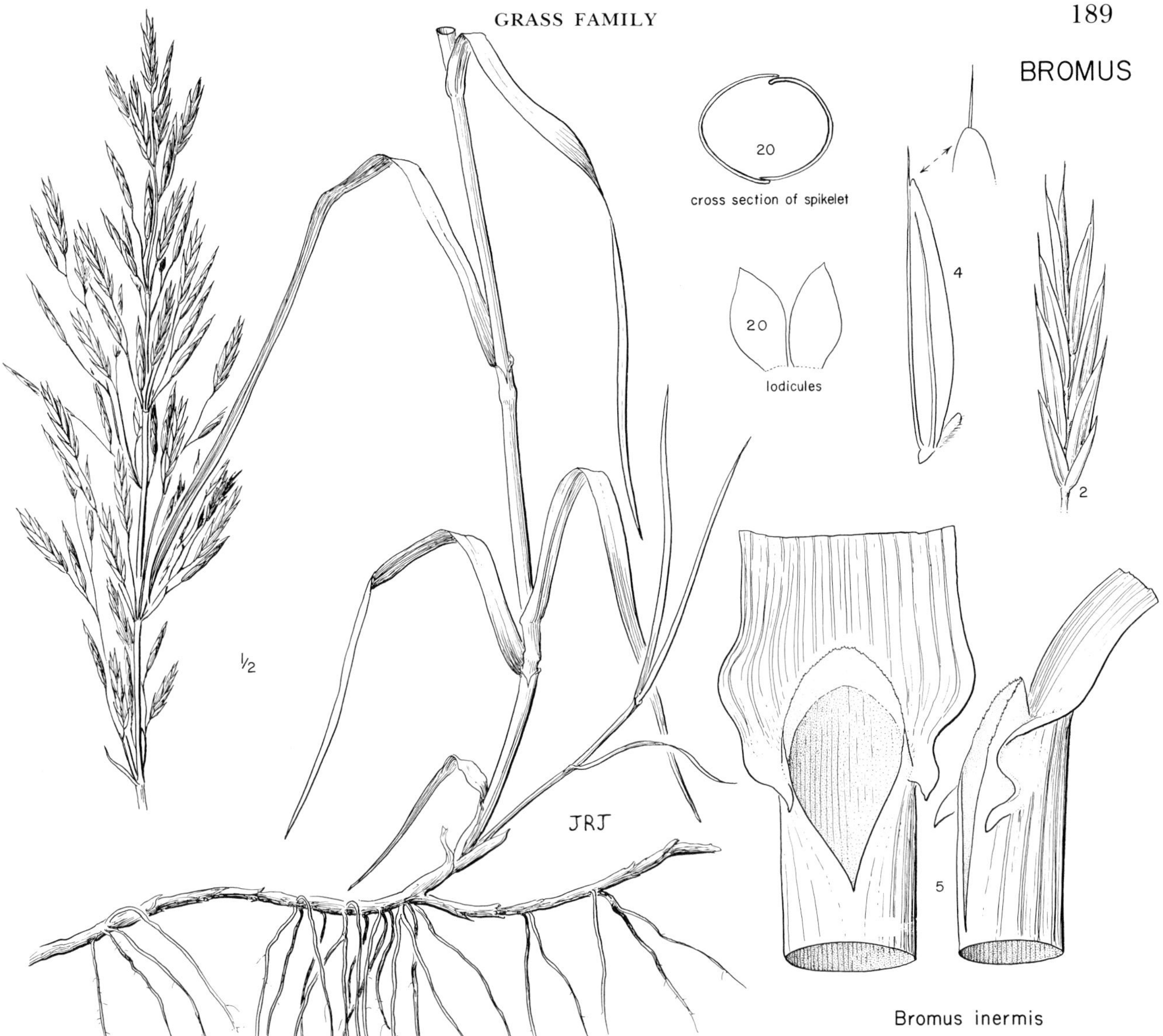

Bromus inermis

1 Plants annual or biennial, 3–10 dm tall; awns usually more than 7 mm long . . . . *B. carinatus*
1 Plants perennial, 8–12 dm tall; awns usually less than 7 mm long.
  2 Plants mostly pubescent throughout . . . . . . *B. marginatus* Nees
  2 Plants glabrous, at least in the spikelets . . *B. polyanthus* Scribn.

3. Bromus inermis Leysser

Bromus inermis Leysser, Fl. Halensis 16. 1761. *Festuca inermis* Lam. & DC. Fl. Franç. ed. 3. 3: 49. 1805. *Schedonorus inermis* Beauv. Essai Nouv. Agrostogr. 99, 177. 1812. *Forasaccus inermis* Lunell, Amer. Midl. Naturalist 4: 225. 1915. *Zerna inermis* Lindman, Svensk Fanerogamfl. 101. 1918. *Bromopsis inermis* Holub, Folia Geobot. & Phytotax. 8: 167. 1973. (Europe.) = var. *inermis*.

Bromus purgans var. *purpurascens* Hook. Fl. Boreali-Amer. 2: 252. 1840. *B. inermis* var. *purpurascens* Wagnon, Rhodora 52: 211. 1950. (*Richardson 130,* Bear Lake to Arctic seacoast.)

B. purgans var. *longispicata* Hook. Fl. Boreali-Amer. 2: 252. 1840. (*Drummond,* Rocky Mts., Can.) = var. *purpurascens*.

B. pumpellianus Scribn. Bull. Torrey Bot. Club 15: 9, pl. 76D, figs. 1–7. 1888. *Forasaccus pumpellianus* Lunell, Amer. Midl. Naturalist 4: 225. 1915. *B. inermis* subsp. *pumpellianus* Wagnon, Rhodora 52: 211. 1950. *Zerna pumpelliana* Tzvelev, Fl. Arct. U.R.S.S. 2: 225. 1964. *Bromopsis pumpelliana* Holub, Folia Geobot. & Phytotax. 8: 168. 1973. (*Scribner 418,* Belt Mts., Meagher Co., Mont., 9 Aug. 1883.) = var. *purpurascens*.

Bromus ciliatus var. *coloradensis* Vasey in Wheeler, Cat. Pl. Geogr. & Geol. Explor. & Surv. W. 100th Meridian 56. 1874, nomen nudum; Vasey ex Scribn. Bull. Torrey Bot. Club 15: 10. 1888, nomen nudum; Vasey ex Beal, Grasses N. Amer. 2: 619. 1896. (*Wolf 1158,* near Gray's Peak, Colo.; lectotype by A. S. Hitchc.) = var. *purpurascens*.

B. pumpellianus var. *tweedyi* Scribn. in Beal, Grasses N. Amer. 2: 622. 1896. *B. inermis* subsp. *pumpellianus* var. *tweedyi* C. L. Hitchc. Univ. Wash. Publ. Biol. 17(1): 507. 1969. (*Tweedy 587,* Slough Creek, Yellowstone Park, Mont., in 1885.) = var. *purpurascens*.

B. pumpellianus var. *melicoides* Shear, U.S.D.A. Div. Agrostol. Bull. 23: 50. 1900. (*Pammel 45,* Beaver Creek Camp, Colo., 8 July 1896.) = var. *purpurascens*.

Smooth, common or Hungarian brome.

Strongly rhizomatous perennials; culms few, tall, 4.5–10 (15) dm tall; sheaths glabrous or sometimes pilose; ligules small, 0.5–1 (1.5) mm long, erose to ciliolate; auricles sometimes present but small, up to 0.5 mm long; blades flat and broad, (4) 5–10 (13) mm wide, rarely involute, glabrous, rarely pilose on the upper side, scabrous on the margins and nerves; panicles relatively long, (7) 9–20 (23) cm, moderately open, oblong, erect, the branches ascending, sometimes the upper appressed; spikelets (16) 20–24 (32) mm long, (4-) 5- to 6- (10)-flowered, very narrow, terete when young, becoming only somewhat compressed; glumes glabrous, the first glume (4.5) 5–7 (8) mm long, narrow-lanceolate, 1- (3)-nerved, the second (6) 7–9 (10) mm long, lanceolate, 3- (5)-nerved; lemmas 9–12 mm long, rounded on the back, glabrous, rarely puberulent, greenish to purplish tinged, the apex shallowly bidentate to nearly obtuse; awns, when present, up to only 2 mm long; anthers 2.5–5 (6) mm long, often exserted; $2n$ = 28, 42, 49, 54–58, 70.

Common in fields, meadows and moderately moist waste places such as banks of irrigation ditches and roadsides, from the valleys to the subalpine zone; introduced from Europe and escaping cultivation from B.C. to Newfl., s. to Calif., Nev., Utah and Colo., and across the n. half of the c. and e. states. June–Aug. (early Sept.).

Bromus inermis var. *inermis* was introduced in the 1890s from Eurasia and planted as hay and pasture grass and used for reseeding western mountain ranges. This species, which escaped and is now widespread in our region, has good drought resistance which makes it especially good in semi-arid regions. The Rocky Mountain var. *purpurascens* (subsp. *pumpellianus*), which skirts our area to the north and east, often merges with var. *inermis* when the two grow together. Variety *purpurascens* has more pubescence on the herbage, longer ligules (1.2–5 mm long as opposed to 0.5–1 mm long), and somewhat broader glumes than does var. *inermis*. Its distribution is in the Rocky Mountains from Alaska to Colorado.

4. Bromus vulgaris (Hook.) Shear

Bromus purgans var. *vulgaris* Hook. Fl. Boreali-Amer. 2: 252. 1840. *B. vulgaris* Shear, U.S.D.A. Div. Agrostol. Bull. 23: 43, fig. 24. 1900. *B. eximius* subsp. *umbraticus* Piper, Contr. U. S. Natl. Herb. 11: 144. 1906. *Zerna vulgaris* Henrard, Blumea 4: 498. 1941. *Bromopsis vulgaris* Holub, Folia Geobot. & Phytotax. 8: 169. 1973. (*Scouler,* Columbia [River]; lectotype by Wagnon.)

Bromus vulgaris var. *eximius* Shear, U.S.D.A. Div. Agrostol. Bull. 23: 44. 1900. *B. eximius* Piper, Contr. U. S. Natl. Herb. 11: 143. 1906. (*Shear 1791,* 4 mi. above Wallowa Lake, Oregon, 23 Aug. 1899.)

Tall perennials without rhizomes; culms 6–10 (12) dm tall, pilose, at least at the nodes; sheaths pilose; ligules (1.2) 2–4.5 mm long, erose-ciliate; auricles lacking; blades flat, lax, 4–9 (12) mm broad, pilose or sometimes glabrous; panicles relatively long, 10–18 (20) cm, open and usually nodding; spikelets (16) 20–30 mm long, (3-) 5- to 7- (10)-flowered, not or only slightly compressed; glumes linear to narrow lanceolate, glabrous to pilose, the first glume (5) 6.5–8.5 (9.2) mm long, 1-nerved, the second (7) 8–11 mm long, 3-nerved; lemmas (9.5) 10.5–14 mm long, rounded on back, the hirsute pubescence across the back or concentrated towards the margins, the full length or only at the base, the apex bifid; awns (5) 6–9 (10) mm long; anthers 1.8–2.8 mm long; $2n$ = 14.

Moist hillsides in woods or meadows; B.C. and Alta., s. to Calif. and nw. Wyo., and from Red Butte Canyon e. of Salt Lake City, Utah (*Arnow 2007*), our first record. July–Aug.

5. Bromus ciliatus L.

Bromus ciliatus L. Sp. Pl. 76. 1753. *B. inermis* var. *ciliata* Trautv. Trudy Imp. S.-Petersburgsk. Bot. Sada 5: 135. 1877. *B. hookeri* var. *ciliatus* Fourn. Mex. Pl. 2: 128. 1886. *B. purgans* var. *ciliatus* Kuntze, Revisio Gen. Pl. 2: 763. 1891. *Forasaccus ciliatus* Lunell, Amer. Midl. Naturalist 4: 225. 1915. *Zerna ciliata* Henrard, Blumea 4: 498. 1941. *Bromopsis ciliata* Holub, Folia Geobot. & Phytotax. 8: 167. 1973. (Grown at Uppsala from seed collected by *Kalm* in Can.)

Bromus richardsonii Link, Hortus Regius Bot. Berol. 2: 281. 1833. *Zerna richardsoni* Nevski, Trudy Sredne-Aziatsk. Gosud. Univ., Ser. 8b, Bot. 17: 17. 1934. *B. ciliatus* var. *richardsonii* B. Boivin, Naturaliste Canad. 94: 521. 1967. *Bromopsis richardsonii* Holub, Folia Geobot. & Phytotax. 8: 168. 1973. (Grown at Berlin from seed sent by *Richardson* from nw. Can.)

Fringed brome.

Tall perennials, without rhizomes; culms 5–10 (16) dm tall, usually hairy at the nodes; sheaths pilose, rarely glabrous; ligules small, 0.3–1 (1.5) mm long, minutely erose-ciliate; auricles lacking to barely evident; blades flat, lax, (3) 4–10 (13) mm broad, glabrous or sometimes pilose; panicles relatively long, (6) 9–18 (24) cm long, loose and usually nodding; spikelets (15) 18–24 (30) mm long, 4- to 7- (9)-flowered, only slightly compressed, the rachilla often visible; glumes narrow-lanceolate to lanceolate, acute to obtuse, the first glume 5–8 (10) mm long, 1-nerved, sometimes with a faint lateral pair, the second 7–10 (12) mm long, 3- (5)-nerved; lemmas 10–13 mm long, rounded on the back, hirsute pubescence concentrated on the margins and basally, the dorsal surface sometimes glabrous, the apex bifid to entire; awns 2.5–5.5 mm long; anthers 1.2–2.5 mm long; $2n$ = 14, 28, 56.

In the mts. from 1830 to 3350 m (6000 to 11,000 ft.) elev., usually in moist wooded habitats; from Alaska e. to Newfl., s. through the w. states to n. Mex., and e. across the n. half of the U.S. to Me. and N.C. Mid-July–early Sept.

6. Bromus anomalus Rupr. ex Fourn.

Bromus anomalus Rupr. Bull. Acad. Roy. Sci. Bruxelles 9(2): 236. 1842, nomen nudum; Rupr. ex Fourn. Mex. Pl. 2: 126. 1886. *Zerna anomala* Henrard, Blumea 4: 499. 1941. *Bromopsis anomala* Holub, Folia Geobot. & Phytotax. 8: 167. 1973. (*Galeotti 5757,* Cerro San Felipe; *5815,* Real del Monte, Mex.)

Bromus kalmii var. *porteri* J. M. Coulter, Manual Bot. Rocky Mt. Region 425. 1885. *B. ciliatus* var. *porteri* Rydb. Contr. U. S. Natl. Herb. 3: 192. 1895. *B. porteri* Nash, Bull. Torrey Bot. Club 22: 512. 1895. *Bromopsis porteri* Holub, Folia Geobot. & Phytotax. 8: 168. 1973. (*T. C. Porter,* Twin Lakes, Colo.; *Coulter,* Buffalo Peaks, Colo., 23 July 1873.)

Bromus ciliatus var. *montanus* Vasey in Wheeler, Rep. U. S. Geogr. Surv. W. 100th Meridian 6: 292. 1878, nomen nudum; Vasey ex Beal, Grasses N. Amer. 2: 619. 1896. (*Patterson 264,* Empire Pass, Colo., in Aug. 1892.)

B. scabratus Scribn. in A. Nels. U.S.D.A. Div. Agrostol. Bull. 13: 46. 1898; not Link, 1843. (*A. Nelson 3800,* "Head of Vermilion Creek," Sweetwater Co., Wyo., 20 July 1897.)

Nodding brome.

Tall perennials without rhizomes; culms (3) 5–9 (12) dm tall, glabrous or the nodes commonly puberulent; sheaths glabrous or sometimes sparsely pilose; ligules small, 0.5–1 mm long, slightly erose; auricles lacking; blades flat, (2) 3–5 mm broad, sometimes slightly involute, glabrous or sometimes pilose;

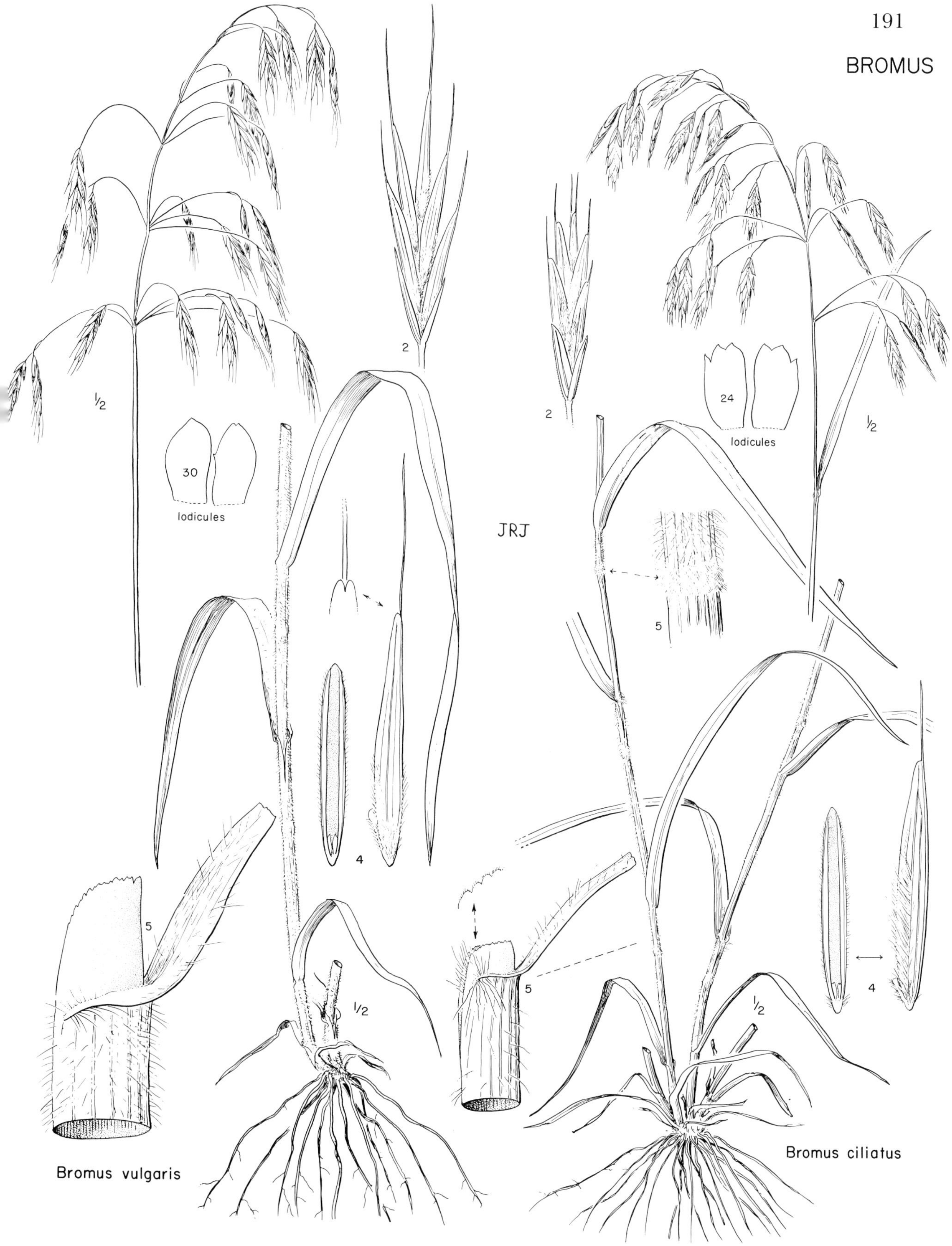
2
½
30
lodicules
2
24
lodicules
½
JRJ
5
4
5
½
5
½
4
Bromus vulgaris
Bromus ciliatus

panicles (5) 7–14 (18) cm long, loose and usually nodding; spikelets 17–25 (30) mm long, 5- to 10-flowered, slightly compressed; glumes mostly puberulent, sometimes glabrous, the first glume 5–7 (8) mm long, 3-nerved (rarely only 1), the second 6.5–8 (10) mm long, 3- to 5-nerved; lemmas 9–12 (13) mm long, 5- to 7-nerved, rounded on the back, relatively evenly pubescent all over, the apex rounded to emarginate; awns relatively short, 1.5–3 mm long; anthers 2–2.7 mm long, exserted; $2n = 14, 28$.

Dry to moderately moist sagebrush and grass meadows and open woods; in our area from 2000 to 3350 m (6500 to over 11,000 ft.) elev.; from se. B.C. to Sask., and s. through Idaho, Mont. and w. S.D., to Calif., Ariz., N.M. and w. Texas, and across most of Mex. Late June–Aug.

Bromus frondosus (Shear) Wooton & Standley is reported to be in southern Utah by Kearney & Peebles (Ariz. Fl., 1951). The record is probably based on a misidentified, robust specimen of *B. anomalus*.

7. Bromus commutatus Schrader

Bromus pratensis Ehrh. Beitr. Naturk. 6: 84. 1791, nomen nudum; Hoffmann, Deutschl. Fl. ed. 2. 1: 53. 1800; not Lam. 1785. *B. mutabilis* var. *pratensis* F. W. Schultz, Flora 32: 234. 1849. *B. hordeaceus* var. *pratensis* Fiori, Nuova Fl. Analitica Ital. 1: 149. 1923. (Europe.)

B. commutatus Schrader, Fl. German. 353. 1806. *Brachypodium commutatum* Beauv. Essai Nouv. Agrostogr. 101, 155. 1812. *Serrafalcus commutatus* Bab. Manual Brit. Bot. 374. 1843. *Bromus mutabilis* var. *commutatus* F. W. Schultz, Flora 32: 234. 1849. *B. racemosus* var. *commutatus* Coss. & Durieu, Explor. Sci. Alger. 2: 165. 1855. *B. mollis* var. *commutatus* Sanio, Verh. Bot. Vereins Prov. Brandenburg 23: Abt. (3). 31. 1882. *Serrafalcus racemosus* var. *commutatus* Husnot, Gram. France, Belg., Iles Brit., Suisse 72. 1899. *Forasaccus commutatus* Bubani, Fl. Pyrén. 4: 387. 1901. *B. hordeaceus* var. *commutatus* Fiori, Nuova Fl. Analitica Ital. 1: 149. 1923. (Germany.)

B. commutatus var. *apricorum* Simonkai, Enum. Fl. Transsilv. 583. 1886. *B. pratensis* var. *apricorum* Druce, Fl. Oxfordshire, ed. 2. 496. 1927. (Europe.) A form with pubescent lemmas.

Hairy chess.

Annuals; culms 2–8 (10) dm tall, puberulent, at least at the nodes; sheaths soft pilose with retrorse hairs; ligules small, 0.5–1.5 mm long, erose-ciliolate, often pubescent; auricles lacking; blades flat, 1–4 (6) mm broad, pilose; panicles usually relatively short, 4–10 (20) cm long, open, oval, the branches rather stiffly spreading, not flexuous or drooping, the pedicels mostly longer than the spikelets; spikelets usually small, 14–18 (30) mm long, 4- to 8- (12)-flowered, slightly compressed, oblong-lanceolate; glumes scabrous to occasionally puberulent, at least on the nerves, relatively short, the first glume 5–6 (7) mm long, lanceolate, 3- to 5-nerved, the second 6–7 (8.5) mm long, oblong-lanceolate, 5- to 9-nerved; lemmas relatively short, (7) 8–9 (10) mm long, 1.7–2.5 mm broad in side view, the nerves relatively faint, the apex bifid; awns 5–12 mm long, straight to somewhat geniculate; palea slightly shorter than the lemma; anthers 0.7–1.7 mm long; $2n = 14, 28, 56$.

Dry or moderately moist, disturbed sites in the valleys and foothills; native of Europe; in w. N. Amer. from s. Alaska, B.C. and Alta., s. to Calif., Nev., Utah and Texas, and e. occurring sporadically to the Atlantic seaboard. Late May–early July.

Bromus commutatus often intergrades with *B. japonicus*.

8. Bromus japonicus Thunb.

Bromus japonicus Thunb. Fl. Jap. 52, tab. 11. 1784. (Japan.)

B. japonicus var. *porrectus* Hackel, Magyar Bot. Lapok 2: 58. 1903. (Eurasia.)

Japanese chess.

Annuals; culms 2–7 (10) dm tall; sheaths retrorsely pilose; ligules small, 0.5–1.5 mm long, erose-ciliate, often pilose; auricles lacking; blades flat, 1.5–4 (6.5) mm broad, usually pilulose; panicles (5) 8–16 (20) cm long, open, pyramidal, the branches slender, flexuous, spreading to drooping, the pedicels usually longer than the spikelets; spikelets (13) 15–24 mm long, 5- to 8- (11)-flowered, glabrous or rarely scaberulose, greenish; glumes relatively short, the first glume 4.5–5.5 mm long, lanceolate, 3-nerved, the second 5.5–7 (8) mm long, 5-nerved, oblong-lanceolate; lemmas relatively short, (6) 7.5–9 mm long, 1.2–2.2 mm broad in side view, broadly oblong-lanceolate, somewhat flattened, rounded on back, faintly 7- to 9-nerved, the apex obtuse to bifid; awns 7–12 (14) mm long, somewhat twisted and slightly geniculate at maturity; palea shorter than the lemma; anthers 0.5–1.2 mm long; $2n = 14, 28$.

Moderately moist or dry weedy areas in the valleys and foothills; native to the Old World; B.C., s. to Calif., e. to Ont., Vt. and Fla. June–early Aug.

Bromus arvensis L. was reported from Nevada by A. S. Hitchcock (Manual Grasses U. S., 1935). It would key to *B. japonicus*, from which it is distinguished by its narrower, less turgid lemmas and its longer palea which equals the lemma in length. *Bromus arenarius* Labill., reported in Nevada by Swallen (Contr. Toward Fl. Nevada 1: 18. 1940), would also key here. It can be distinguished from *B. japonicus* by its more conspicuously flexuous panicle branches and pedicels and its pubescent lemmas. We were unable to locate Nevada specimens of either species at the U. S. National Herbarium.

9. Bromus mollis L.

Bromus mollis L. Sp. Pl. ed. 2. 1: 112. 1762. *Serrafalcus mollis* Parl. Rar. Pl. Cogn. Sicil. 2: 11. 1840. *Forasaccus mollis* Bubani, Fl. Pyrén. 4: 386. 1901. *B. hordeaceus* var. *mollis* Fiori, Nuova Fl. Analitica Ital. 1: 149. 1923. *B. hordeaceus* subsp. *mollis* Maire in Jahandiez & Maire, Cat. Pl. Maroc 4: 943. 1941. (Europe.)

B. racemosus L. Sp. Pl. ed. 2. 1: 114. 1762. *Serrafalcus racemosus* Parl. Rar. Pl. Cogn. Sicil. 2: 14. 1840. *B. arvensis* var. *racemosus* Neilr. Fl. Nieder-Oesterr. 81. 1859. *B. squarrosus* var. *racemosus* Regel, Trudy Imp. S.-Peterburgsk. Bot. Sada 7: 602. 1881. *B. mollis* var. *racemosus* Fiori, Fl. Analitica Ital. 1: 100. 1896. *Forasaccus racemosus* Bubani, Fl. Pyrén. 4: 387. 1901. *B. hordeaceus* var. *racemosus* Fiori, Nuova Fl. Analitica Ital. 1: 149. 1923. (Europe.)

B. hordeaceus sensu Amer. authors; not L. 1753.

Soft chess.

Annuals; culms 1.5–6.5 (10) dm tall, glabrous or pubescent, sometimes only on the nodes; sheaths soft-pubescent; ligules 0.5–2 mm long, finely erose-ciliate, glabrous or usually hairy; auricles lacking; blades flat, relatively narrow, 1.5–4 mm broad, sometimes involute; panicle crowded, erect, 3–8 (10) cm long, the pedicels usually shorter than the spikelets; spikelets usually small, (11) 14–20 (23) mm long, 5- to 10- (12)-flowered, soft-pubescent to glabrous; glumes relatively short, the first glume 5–7.5 mm long, lanceolate, 3- to 5-nerved, the second 6–8.5 mm long, ovate-lanceolate, 5- to 7-nerved; lemmas

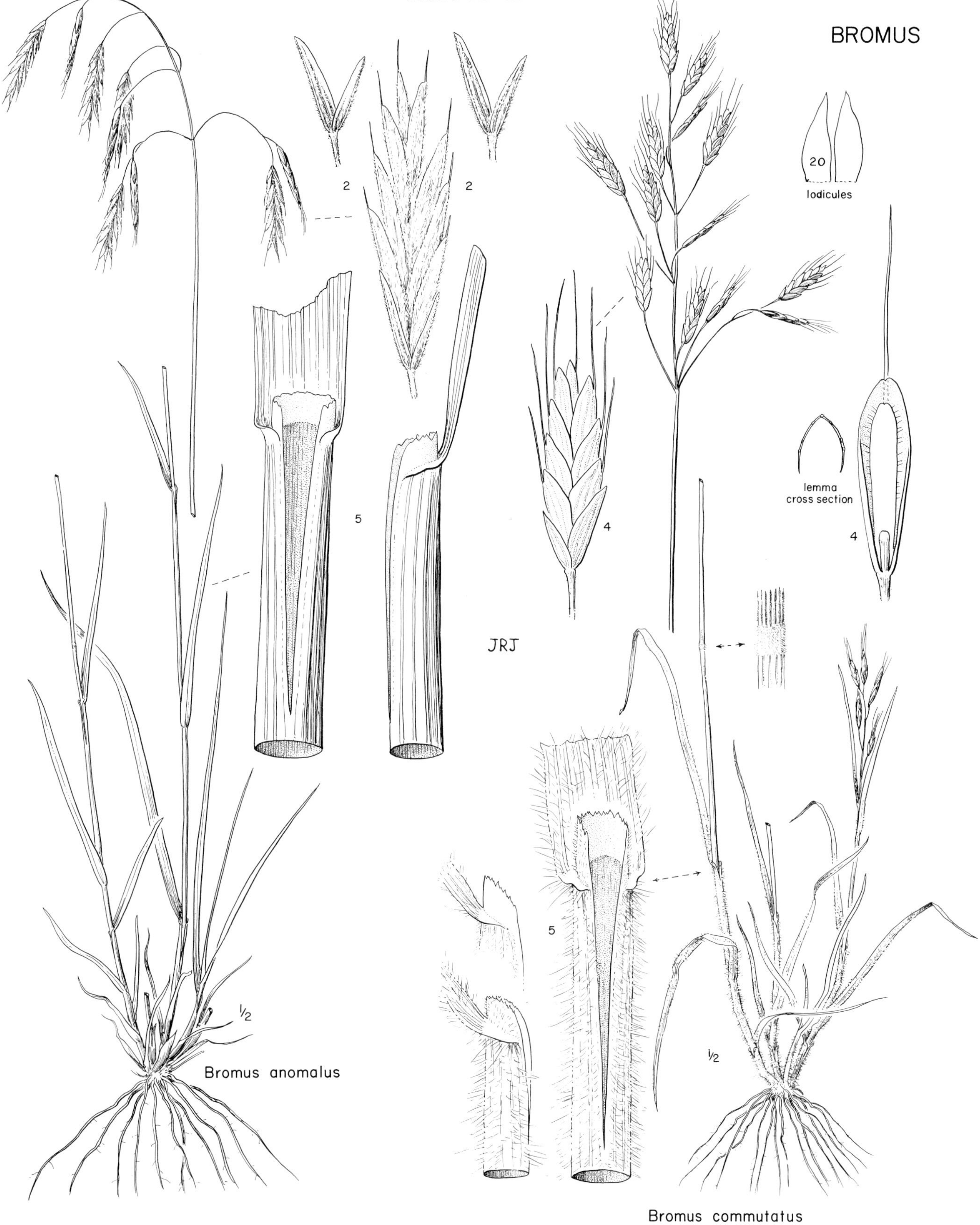
BROMUS
20
lodicules
2
2
5
4
lemma
cross section
4
JRJ
5
1/2
Bromus anomalus
1/2
Bromus commutatus

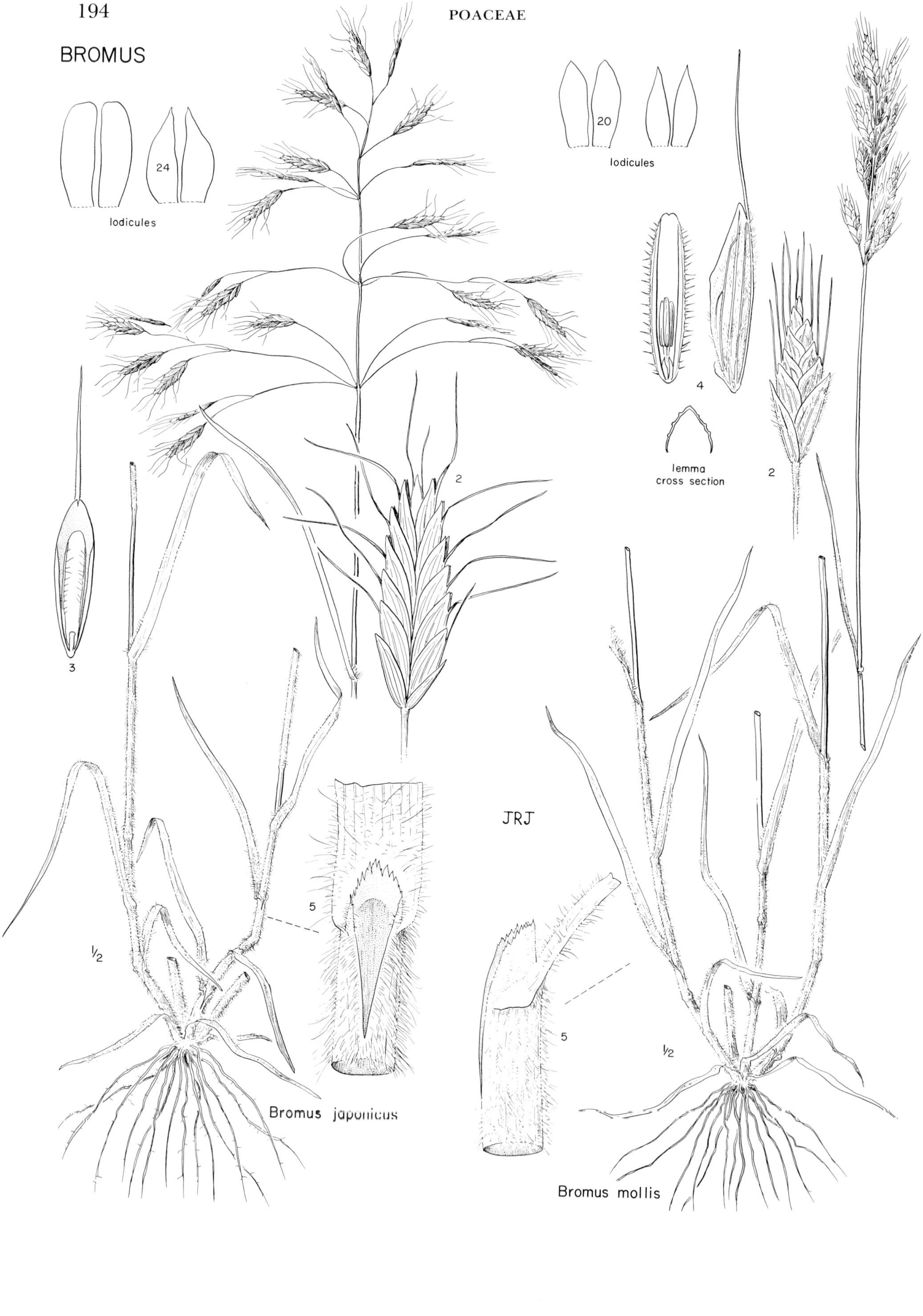
BROMUS
lodicules
24
20
lodicules
4
lemma
cross section
2
2
3
5
JRJ
1/2
5
1/2
Bromus japonicus
Bromus mollis

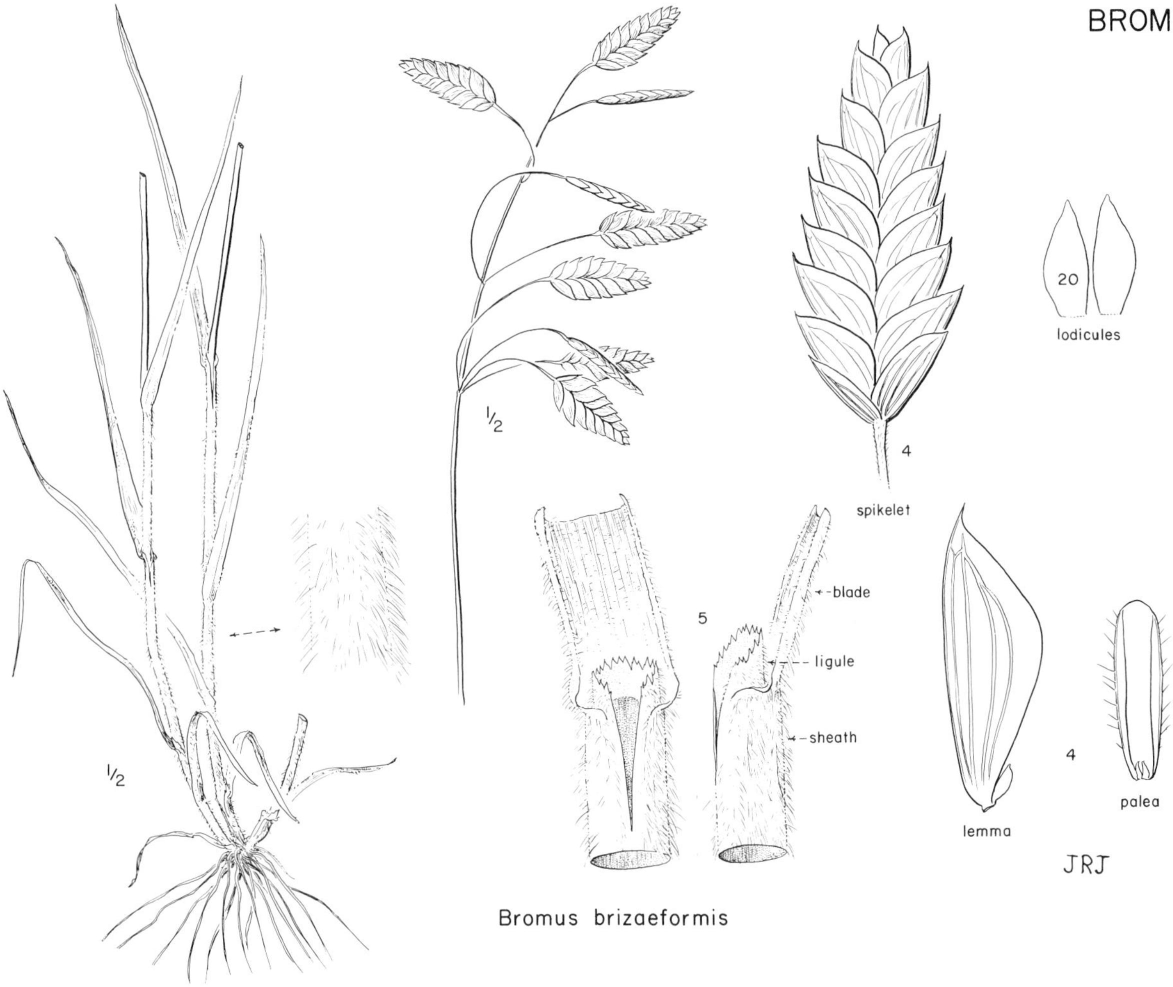

Bromus brizaeformis

relatively short, 7–9 (10) mm long, 1.5–2.5 mm broad in side view, somewhat flattened, the nerves relatively prominent, the apex scarcely bifid; awns 4–9 mm long, usually straight; palea about as long as the lemma; anthers 0.5–1 (2) mm long; $2n = 14, 28$.

Dry fields and waste places in the sagebrush zone with *B. tectorum,* low valleys and foothills up to about 2000 m (6500 ft.) elev.; native of the Old World; sporadically distributed from Alaska to N.S., and s. throughout most of the U.S. May–mid-July.

Bromus mollis is a good hay grass in the South.

10. Bromus brizaeformis Fischer & Meyer

Bromus brizaeformis Fischer & Meyer in Fischer, Meyer & Trautv. Index Seminum Hort. Bot. Imp. Petrop. 3: 1. 1837. (Europe.)

Rattlesnake chess.

Small annuals; culms 1.5–5.5 (8.5) dm tall; sheaths usually densely retrorsely hispid; ligules 0.5–2 mm long, usually 1 mm, erose, pubescent; auricles lacking; blades flat, (1) 2–4.5 (6) mm broad, sometimes involute, pilose; panicles usually relatively short, 3–10 (20) cm long, nodding, and somewhat secund, usually very few spikelets; spikelets (15) 20–26 (30) mm long, strongly compressed, oblong-ovate, 8–12 mm broad, many-flowered, (7) 9–15, the florets widely spreading, glabrous, rarely scabrous; glumes oblong-lanceolate, the nerves faint, relatively short, the first glume 4.5–6 mm long, 3- to 5-nerved, the second 6.5–7.5 mm long, (5-) 7- to 9-nerved; lemmas relatively short, (7) 8–9.5 mm long and exceptionally broad, 2.8–4 mm in side view, rhombic-ovate, strongly inflated, the nerves several, very faint, the apex scarcely bifid; awns, when present, up to only 1 mm long; palea slender; anthers usually enclosed (cleistogamous), about 0.8 mm long; $2n = 14$.

Usually in disturbed dry sites in sagebrush valleys and foothills in sandy or gravelly soil; native of Europe; occasional throughout the w. states and rare in the midw. and e. states, not represented in the se. Late May–early July.

Bromus brizaeformis is sometimes cultivated for ornament.

11. Bromus secalinus L.

Bromus secalinus L. Sp. Pl. 76. 1753. *B. mollis* var. *secalinus* Huds. Fl. Anglica, ed. 2. 49. 1778. *Avena secalina* Salisb. Prodr. Stirpium 22. 1796. *Serrafalcus secalinus* Bab. Manual Brit. Bot. 374. 1843. *Forasaccus secalinus* Bubani, Fl. Pyrén. 4: 388. 1901. (Europe.)

Chess or cheat.

Annuals; culms 3–10 dm tall, glabrous or pubescent at the nodes only; sheaths usually pilose to hirsute, sometimes glabrous; ligules relatively long in most, 0.5–3 mm, erose, glabrous or sparsely pubescent; auricles lacking; blades flat, 2–5 (9) mm broad, glabrous to pilose; panicles 5–17 (20) cm long, loose, usually nodding and somewhat secund; spikelets 13–20 (23) mm long, 4- to 8- (10)-flowered, glabrous or sometimes puberulent, strongly compressed, the florets spreading and exposing the rachilla joints; glumes relatively short, the nerves often faint, the first glume lanceolate, 4–5.5 mm long, 3- to 5-nerved, the second oblong to ovate-lanceolate, 5–7 mm long, 5- to 7-nerved; lemmas relatively short, 6.5–9 mm long, 1.7–2.5 mm broad in side view, strongly rounded on back, the margins curved around the caryopsis, the apex shallowly bifid; awns 0.5–8 mm long, fragile, often curved; palea usually about equalling or exceeding the lemma; anthers 0.8–1.6 mm long; $2n = 14, 28$.

Dry waste areas at middle elevs.; native of w. Europe; from Alaska and B.C., e. to Ont. and more or less throughout the U.S., less frequent in the s. states. June–July.

Bromus secalinus is usually a troublesome weed in grain fields, but it is occasionally grown for hay in the Pacific Northwest. Farmers at one time thought that cheat was a degenerate wheat. In all probability the more drought-resistant cheat competed with wheat in dry years or in poor soils.

12. **Bromus rubens** L.

Bromus rubens L. Centuria Pl. 1: 5. 1755. *Festuca rubens* Pers. Syn. Pl. 1: 94. 1805. *B. scoparius* var. *rubens* Saint-Amans, Fl. Agénaise 45. 1821. *B. sterilis* var. *rubens* Kuntze, Trudy Imp. S.-Peterburgsk. Bot. Sada 10: 249, 251. 1887. *B. madritensis* subsp. *rubens* Husnot, Gram. France, Belg., Iles Brit., Suisse 71. 1899. *Anisantha rubens* Nevski, Trudy Sredne-Aziatsk. Gosud. Univ., Ser. 8b, Bot. 17: 19. 1934. *Zerna rubens* Grossh. Trudy Bot. Inst. Akad. Nauk SSSR, Ser. 1, Fl. Sist. Vysš. Rast. 8: 306. 1939. (*Loefling*, Spain.)

Foxtail chess.

Small annuals with fine retrorse pubescence; culms (1) 1.5–4.5 (8) dm tall; ligules 1–2.5 (3) mm long, erose-lacerate; auricles lacking; blades flat, relatively narrow, 1.5–4 (5) mm broad, sometimes involute; panicles contracted, ovoid, short, (3) 4–8 (10) cm long (measurement including the awns), often purplish, the short branches and pedicels stout, ascending; spikelets 18–25 (30) mm long, 3- to 8- (10)-flowered, scaberulous to finely hirsute; glumes linear to narrow-lanceolate, membranous margined, the first glume (5) 6.5–10 mm long, 1-nerved, the second 9–13 mm long, 3-nerved; lemmas slightly carinate, relatively long, 12–17 mm, 0.5–1.2 mm broad in side view, 3- to 5-nerved, the apical teeth slender, 4–5 mm long; awns 12–20 (24) mm long, straight or only slightly twisted; anthers 0.4–0.7 (1) mm long, included (cleistogamous) in the lemma; $2n = 28$.

Deserts and shadscale valleys, usually in disturbed sites; native to the Medit. region; from Wash. to s. Calif., e. to sw. Idaho, w. Utah and Ariz.; reported in the e. from Mass. and N.Y. March–June.

Bromus madritensis L., a close relative of *B. rubens,* may have escaped from cultivation near Reno, Nevada. It differs from *B. rubens* in its glabrous culms and less densely congested panicle.

13. **Bromus rigidus** Roth

Bromus villosus Forssk. Fl. Aegypt.-Arabica 23. 1775; not Scop. 1772. (Egypt.)

B. rigidus Roth, Bot. Mag. (Römer & Usteri) 10: 21. 1790. *B. rubens* var. *rigidus* Mutel, Fl. Franç. 4: 133. 1837. *B. madritensis* var. *rigidus* Bab. ex Syme, Engl. Bot. ed. 3. 11: 161. 1873. *B. sterilis* var. *rigidus* Kuntze, Trudy Imp. S.-Peterburgsk. Bot. Sada 10: 249, 251. 1887. *B. gussonei* var. *rigidus* H. Lindberg, Acta Soc. Sci. Fenn., Ser. B., Opera Biol. 1(2): 15. 1932. *B. villosus* var. *rigidus* Aschers. & Graebner ex A. S. Hitchc. Manual Grasses U.S. 816. 1935, in synonymy (incorrectly attributed to Aschers. & Graebner, Syn. Mitteleur. Fl. 2: 596. 1901). *Anisantha rigida* Hylander, Uppsala Univ. Årsskr. 7: 32. 1945. (Europe.)

B. gussoni Parl. Rar. Pl. Cogn. Sicil. 2: 8. 1840. *B. maximus* var. *gussonii* Parl. Fl. Ital. 1: 407. 1848. *B. rigidus* var. *gussonii* Coss. & Durieu, Explor. Sci. Algér. 2: 159. 1855. *B. villosus* var. *maximus* subvar. *gussonii* Aschers. & Graebner, Syn. Mitteleur. Fl. 2: 595. 1901. *Zerna gussonii* Grossh. Trudy Azerbajdžansk. Fil. 8: 305. 1939. *B. rigidus* subsp. *gussonei* Maire, Cuénod. Fl. Tunisie 144. 1954. (Europe.)

Ripgut grass.

Small annuals; culms 2–4.5 (9) dm tall; ligules long, 2–4 (5.5) mm, erose-lacerate; auricles lacking; blades flat, 2–6 (8) mm broad, sometimes involute; panicles usually relatively long, (7) 10–20 cm, open, the branches elongate, erect or spreading, sometimes the lowermost reflexed; spikelets large, (28) 33–40 (43) mm long, (3-) 4- to 6- (8)-flowered, all floret parts exceptionally long; glumes glabrous, linear to narrow lanceolate, the first glume 13–18 (22) mm long, 1- (3)-nerved, the second 20–25 (28) mm long, 3-nerved; lemmas (22) 25–30 mm long, 1–1.5 mm broad in side view, scabrous to puberulent, the apical teeth 3–6 mm long; awns (30) 35–60 mm long; anthers (0.7) 1–1.5 mm long; $2n = 28, 42, 56, 70$.

Dry disturbed areas at low elevs.; native of the Medit. region; Baja Calif., n. to sw. B.C. and e. to Idaho, Colo., and Okla. and sporadically across the e. states; extending into our region from the s. into the Lahontan and Bonneville basins. April–June.

14. **Bromus sterilis** L.

Bromus sterilis L. Sp. Pl. 77. 1753. *Schedonorus sterilis* Fries, Bot. Not. 131. 1843. *Zerna sterilis* Panzer ex B. D. Jackson, Index Kew. 2: 1249. 1895, in synonymy. *Anisantha sterilis* Nevski, Trudy Sredne-Aziatsk. Gosud. Univ., Ser. 8b, Bot. 17: 20. 1934. (Europe.)

Barren brome.

Annuals with soft pubescence throughout, sometimes glabrous, except the sheaths; culms (3) 3.5–8 (10) dm tall; ligules relatively long, 1.5–2.5 (3.5) mm long, erose-ciliate or lacerate; auricles lacking; blades flat, relatively narrow, 2–4 (5) mm broad, sometimes involute; panicles relatively long, 10–18 (24) cm, open, drooping, the branches often bearing only 1 or 2 spikelets, the lower branches usually somewhat reflexed, the upper spreading; spikelets often large, (23) 25–31 (36) mm long, 4- to 8-flowered, glabrous to puberulent-strigulose; glumes linear to narrow-lanceolate, the first glume 7–10 (11) mm long, 1-nerved, the second 9–14 (17) mm long, 3-nerved; lemmas relatively long, 13–18 (20) mm, slightly carinate, 0.8–1.2 mm broad in side view, the slender apical teeth 2–2.5 mm long; awns 18–26 (31) mm long; anthers 0.9–1.6 mm long; $2n = 14, 28$.

Waste places; native of s. Europe; sporadically distributed from B.C. to Calif., e. to Mass. and N.C.; in our region near the mouth of Red Butte Canyon, e. of Salt Lake City, Utah. May–July.

BROMUS

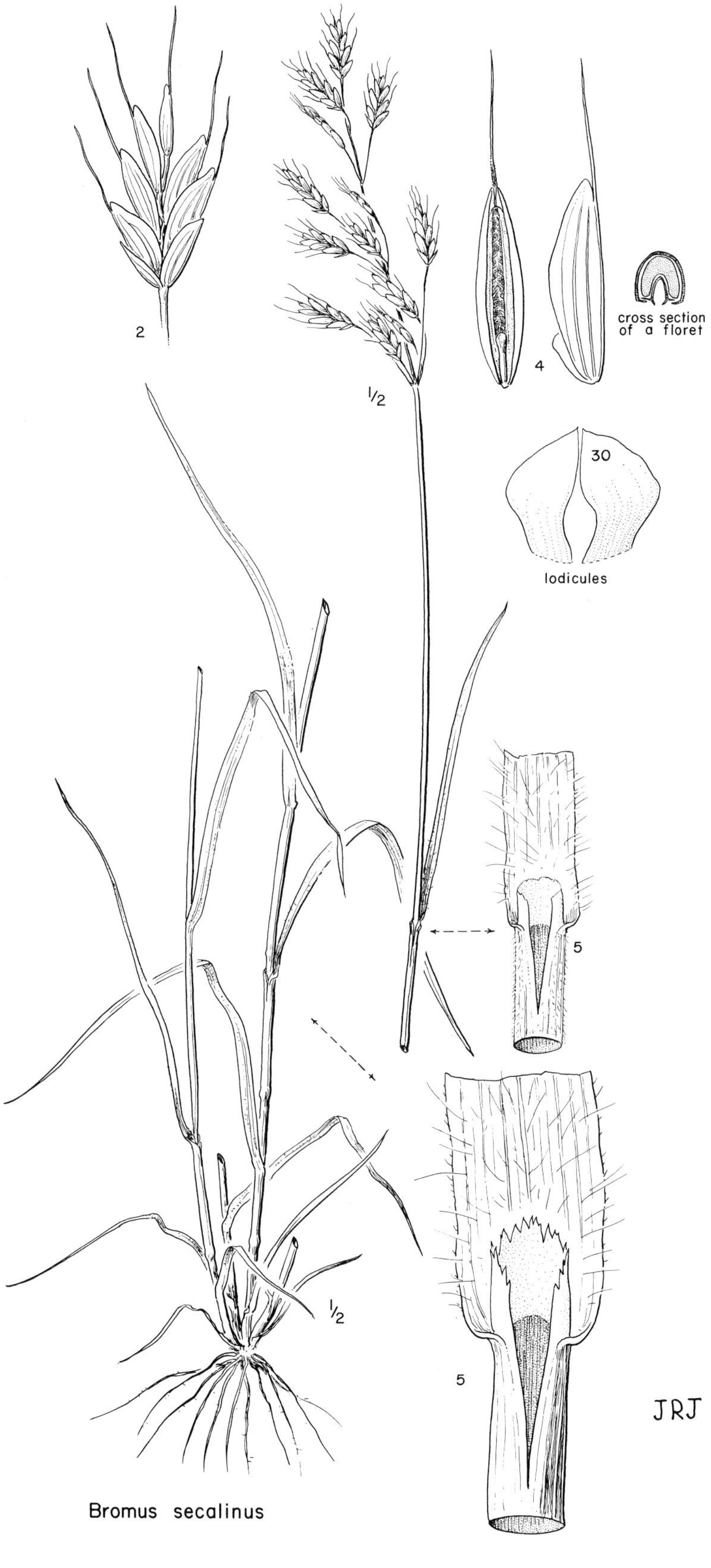

Bromus secalinus

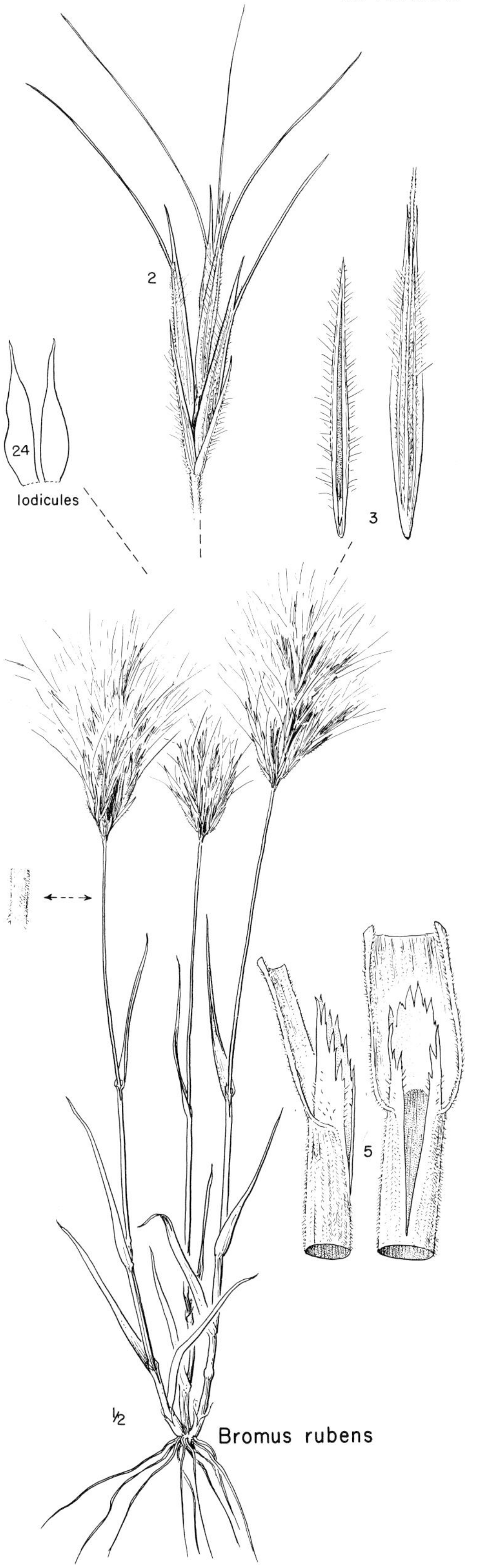

Bromus rubens

BROMUS

Bromus rigidus

BROMUS

Bromus sterilis

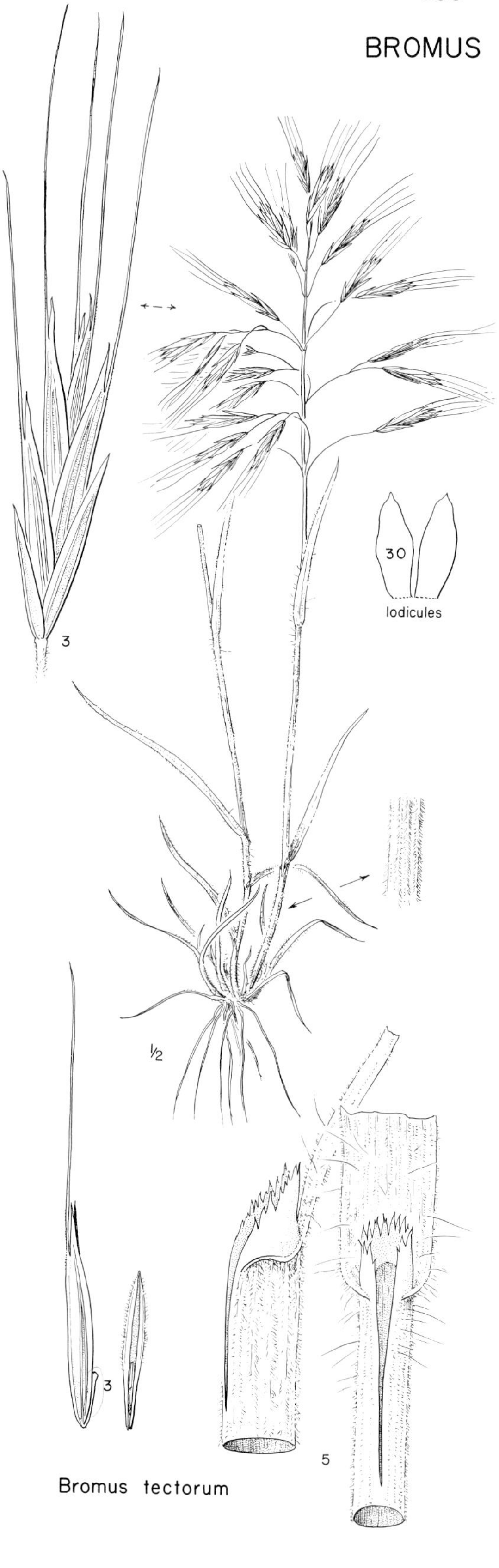

Bromus tectorum

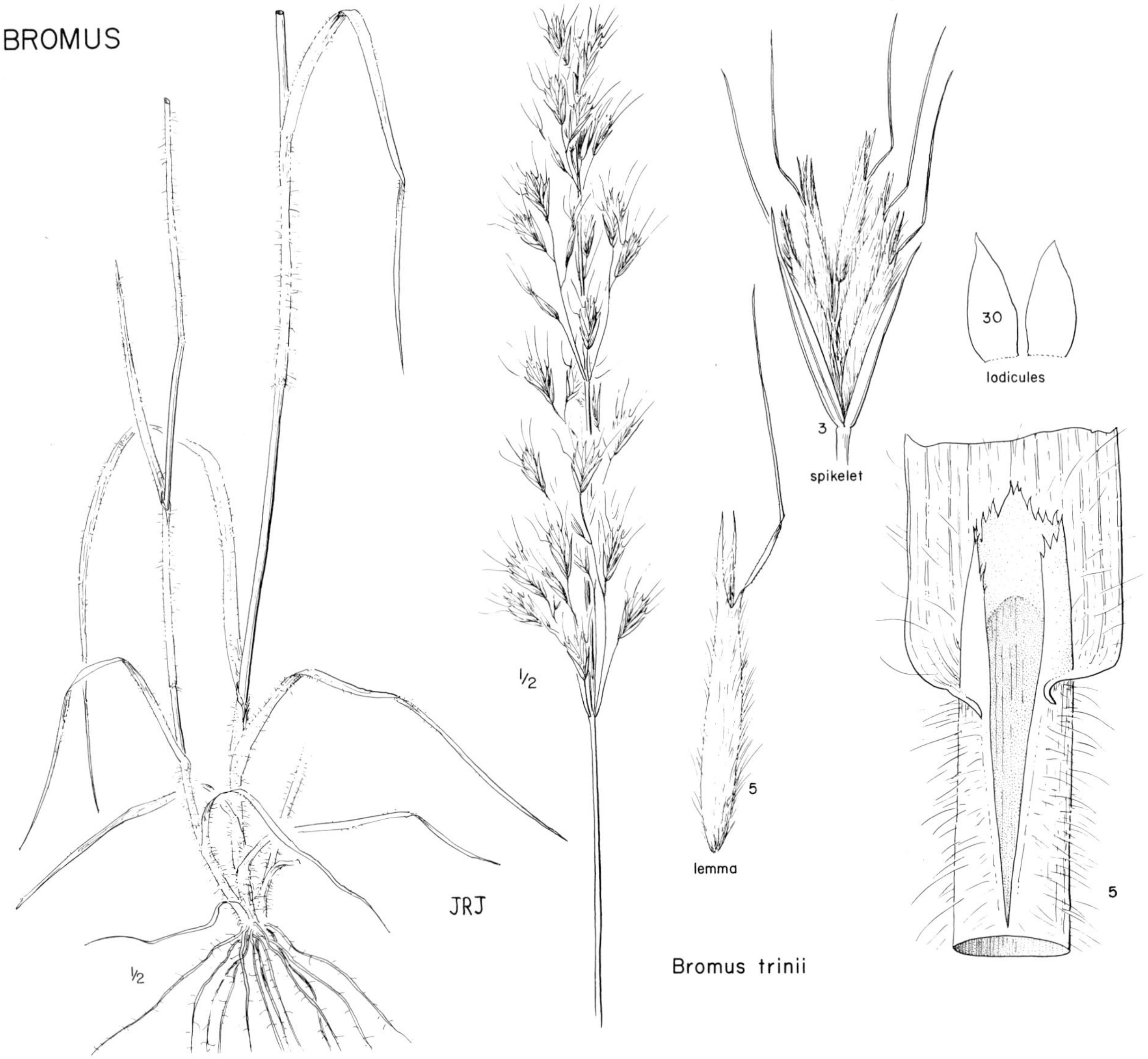

15. Bromus tectorum L.

Bromus tectorum L. Sp. Pl. 77. 1753. *Schedonorus tectorum* Fries, Bot. Not. 131. 1843. *B. sterilis* var. *tectorum* Kuntze, Trudy Imp. S.-Peterburgsk. Bot. Sada 10: 249, 250. 1887. *Zerna tectorum* Panzer ex B. D. Jackson, Index Kew. 2: 1249. 1895, in synonymy. *Anisantha tectorum* Nevski, Trudy Sredne-Aziatsk. Gosud. Univ., Ser. 8b, Bot. 17: 20, 22. 1934. (Europe.)

B. tectorum var. *glabratus* Spenner, Fl. Friburg. 1: 152. 1825. *B. tectorum* f. *glabratus* St. John, Fl. S.E. Wash. & Adj. Idaho, ed. 2. 536. 1956. (Germany.)

B. tectorum var. *nudus* Klett & Richter, Fl. Leipzig 109. 1830. *B. nudus* Anon. Proc. Biol. Soc. Wash. 18: 257. 1905, in index. *B. tectorum* f. *nudus* Hiit. Suomen Kasvis 220. 1933. (Germany.)

B. setaceus Buckley, Proc. Acad. Nat. Sci. Philadelphia 1862: 98. 1863. (*Buckley,* n. Texas.)

Downy chess, junegrass, cheatgrass or broncograss.

Small annuals with soft hispid pubescence throughout; culms 1–5 (7) dm tall; ligules 1–2.5 (4) mm long, lacerate-erose; auricles lacking; blades flat, relatively narrow, 2–3.5 (7.5) mm broad; panicle (4) 6–12 (15) cm long, nodding, often densely branched, the lower one spreading or reflexed; spikelets small, (12) 15–20 (24) mm long, few flowered, 3–6, the nerves usually prominent; glumes narrow-lanceolate to lanceolate, the first glume 5.5–7 mm long, 1-nerved, the second 8–11 mm long, 3-nerved; lemmas 10–12.5 (17.5) mm long, rounded to slightly carinate on back, 1–1.5 mm broad, glabrous to villose, the narrow apical teeth 1–3 mm long; awns (7) 10–17 mm long, straight or very slightly geniculate; anthers 0.5–0.7 mm long, usually included in the lemma (cleistogamous); $2n = 14$.

Abundant in the sagebrush zone in dry disturbed places, sometimes turning entire landscapes purplish-brown and eventually straw-brown; less common in alkaline soils; native of s. Europe; common in w. N. Amer. from Alaska, B.C. and Alta., s. to Calif., and sporadically e. to Vt. and Ga. May–June (sometimes –July or –Aug. during wet summers).

A form with glabrous spikelets is frequently referred to as var. *glabratus* Spenner.

16. Bromus trinii Desvaux

Trisetum hirtum Trin. Linnaea 10: 300. 1836; not *B. hirtus* Licht. 1817. *Bromus trinii* Desvaux in C. Gay, Fl. Chilena 6: 441. 1853. *Trisetum trinii* Louis-Marie, Rhodora 30: 243. 1928. *Trisetobromus hirtus* Nevski, Trudy Sredne-Aziatsk. Gosud. Univ., Ser. 8b, Bot. 17: 15. 1934. (*Poeppig*, "Andes de Chile austral.")

B. trinii var. *excelsus* Shear, U.S.D.A. Div. Agrostol. Bull. 23: 25. 1900. (*Coville & Funston 522*, Panamint Mts., near Pete's Garden, Johnson Canyon, Inyo Co., Calif., in 1891.)

Chilean chess.

Annual; culms (2) 3.5–7 (12) dm tall, glabrous except for the nodes; leaves sparsely to densely pilose; ligules 1–2.5 (4) mm long, erose-ciliate; auricles absent to often present and up to 0.8 mm long; blades flat, 1.5–5 (10) mm broad; panicles (8) 10–15 (20) cm long, the branches spreading; spikelets mostly small, 14–20 (24) mm long, few flowered, 3–5 (6), densely soft pubescent; glumes linear, the first glume 7.5–12 mm long, 1-nerved, the second (8) 10.5–15 mm long, often longer than the first floret, 3- to 5-nerved, strongly carinate; lemmas 9.5–13 (16) mm long, slightly carinate, 3- to 5-nerved, the apical teeth 2–5 mm long, linear-aristate; awns 10–20 (24) mm long, geniculate and strongly twisted near the base, arising from between the lemma teeth, not subdorsal; anthers 2–2.5 mm long; $2n = 42$.

Dry hillsides; native of Chile; Baja Calif., n. to Oregon and e. to Colo. and Ariz. April–May.

Bromus trinii is sometimes mistaken for *Trisetum* of Tribe Aveneae because of its glumes sometimes enclosing the florets and its geniculate, twisted awns which appear to be subdorsally attached.

2. VULPIA C. C. Gmelin Annual Fescue

Small spring annuals; culms hollow; sheaths open; ligules membranous, narrow and collar-like; blades more or less involute, filiform, non-auriculate; inflorescence a narrow panicle, sometimes contracted into a spike-like raceme; spikelets 1- to 15-flowered, most of the florets cleistogamous, slightly laterally compressed, disarticulating above the glumes and between the florets; glumes narrow-lanceolate to lanceolate, usually markedly unequal, the first glume 1-nerved, sometimes greatly reduced, the second usually 3-nerved and often shortly awned; lemma rounded on back, narrow-lanceolate, attenuate to long-awned, glabrous to scabrous, inconspicuously 5-nerved; palea from slightly shorter to longer than the lemma, bidentate, ciliate on the 2 keels; lodicules 2, entire, narrow lanceolate to ovate; stamens often solitary (sometimes 2 or rarely 3), often included in the lemma; caryopsis linear-cylindric, attenuate at both ends; $x = 7$.

A genus of about 30 species of weedy annuals, widely distributed in Europe, n. Africa, sw. Asia and temp. N. and S. Amer. (Name from *vulpes*, fox, for the resemblance to a fox tail from the many long awns of the panicle.)

The annual fescues (*Vulpia*) are weakly distinguished from the perennial fescues (*Festuca*). Most American authors in the past have treated the group as *Festuca* subg. *Vulpia* (C. C. Gmelin) Hackel. Recently, however, American authors have been following European workers in giving the annual fescues generic status.

References:

Henrard, J. T. A study in the genus *Vulpia*. Blumea 2: 299–326. 1937.
Lonard, R. I., and F. W. Gould. The North American species of *Vulpia* (Gramineae). Madroño 22: 217–230. 1974.
Piper, C. V. North American species of *Festuca*. Contr. U. S. Natl. Herb. 10: 1–48. 1906.
Saint-Yves, A. *Festuca* de l'Amérique du Nord et du Mexique. Candollea 2: 229–316. 1925.

1 Spikelets 6- to 15-flowered; awns 1.5–4 (5) mm long .... 1. *V. octoflora*
1 Spikelets 1- to 5- (8)-flowered; awns (4) 7–17 mm long.
  2 First glume short, 0.3–1.5 mm long, less than half as long as the second glume .... 2. *V. myuros*
  2 First glume 3–7 mm long, more than half as long as the second glume.
    3 Panicles narrow, the branches ascending or appressed-ascending, the branches and pedicels all lacking axillary calluses .... 3. *V. bromoides*
    3 Panicles broader, the branches (at least the lower ones) and sometimes the spikelets spreading or reflexed, the branches with a callus in the axil .... 4. *V. microstachys*

1. Vulpia octoflora (Walter) Rydb.

Festuca octoflora Walter, Fl. Carol. 81. 1788. *Vulpia octoflora* Rydb. Bull. Torrey Bot. Club 36: 538. 1909. *Gnomonia octoflora* Lunell, Amer. Midl. Naturalist 4: 224. 1915. (S.C.)

Festuca tenella Schreb. ex Muhl. Trans. Amer. Philos. Soc. 3: 161. 1793, nomen nudum; Willd. Sp. Pl. 1: 419. 1797. *Schedonorus tenellus* Beauv. Essai Nouv. Agrostogr. 99, 163, 177. 1812. *Brachypodium festucoides* Link, Enum. Pl. Hort. Regii Bot. Berol. 1: 95. 1821; not Beauv. 1812. *V. tenella* Heynh. Nomencl. Bot. Hort. 1: 854. 1840. *Festuca octoflora* var. *tenella* Fern. Rhodora 34: 209. 1932. *V. octoflora* var. *tenella* Fern. Rhodora 47: 107. 1945. (N. Amer. [Pa.].) = var. *glauca*.

Festuca tenella var. *glauca* Nutt. Trans. Amer. Philos. Soc. II. 5: 147. 1834. *Festuca octoflora* var. *glauca* Fern. Rhodora 34: 209. 1932. *V. octoflora* var. *glauca* Fern. Rhodora 47: 107. 1945. (*Nuttall*, Fort Smith, Ark.)

Festuca pusilla Buckley, Proc. Acad. Nat. Sci. Philadelphia 1862: 98. 1863. (*Nuttall*, upper Calif., in 1835.) = var. *hirtella*.

Festuca octoflora subsp. *hirtella* Piper, Contr. U. S. Natl. Herb. 10: 12. 1906. *V. octoflora* f. *hirtella* Blom, Acta Horti Gothob. 5: 89. 1930. *Festuca octoflora* var. *hirtella* (erroneously attributed to Piper by) A. S. Hitchc. Manual Grasses U. S. 60. 1935. *V. octoflora* var. *hirtella* Henrard, Blumea 2: 320. 1937. (*Shear 1962*, Santa Catalina Mts., Ariz., 10 April 1901.)

Six-weeks fescue, eight-flowered fescue.

Tufted annuals; culms 0.5–2 (4) dm tall, often geniculate at the lower nodes, otherwise erect, glabrous or pubescent; sheaths often pubescent; ligules erose-ciliolate, 0.2–0.7 mm long, usually longer on the sides; blades narrow, involute; panicles (1) 2–5 (7) cm long, narrow, erect, sometimes racemose

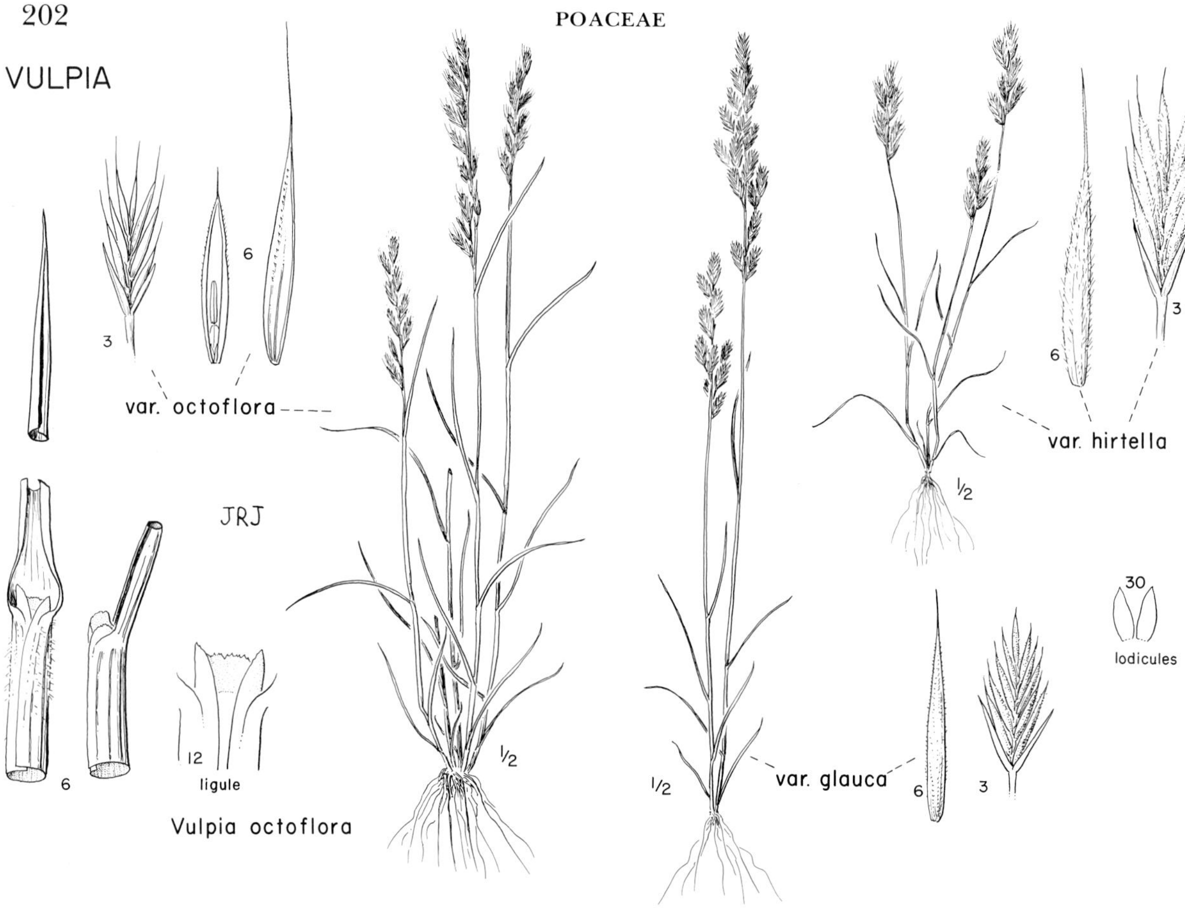

above; spikelets laterally compressed, densely 6- to 12- (15)-flowered; glumes linear-lanceolate to subulate, the small first glume 2–4.5 mm long, 1- or 3-nerved, the second 3.5–5.5 (6.5) mm long, 3-nerved; lemmas 3–5 (6.5) mm long, lanceolate, firm, rounded on back and with inrolled margins, glabrous, scabrous to puberulent, gradually tapering into relatively short awns, the awns 1.5–4 (5) mm long; anther 1, rarely 3; caryopsis 1.7–3.3 mm long; $2n = 14$.

Locally common from dry sagebrush deserts to pinyon-juniper woodlands, often in sandy soil in disturbed habitats; B.C. to Ont., s. throughout the U.S. (April) May–June.

Three races which show some geographical tendencies are separable by the following key characters:

1 Lemmas usually less than 4 mm long; awns usually no more than 2 mm long; florets usually more than 8 per spikelet; mostly in e. and c. U.S., and only occasionally in the s. half of w. U.S. .... var. *glauca* (Nutt.) Fern.
1 Lemmas usually more than 4 mm long; awns mostly 2–5 mm long; florets often less than 9 per spikelet.
  2 Plants usually less than 2 dm tall; lemmas pubescent or puberulent-scabrous; mostly in sw. U.S. and Baja Calif., but occasionally n. to Wash. and Mont. and e. to Texas .... var. *hirtella* (Piper) Henrard
  2 Plants often more than 2 dm tall; lemmas mostly glabrous or scaberulous; s. Can. and throughout the U.S. .... var. *octoflora*

2. **Vulpia myuros** (L.) C. C. Gmelin

Festuca myuros L. Sp. Pl. 74. 1753. *Avena muralis* Salisb. Prodr. Stirpium 22. 1796. *Vulpia myuros* C. C. Gmelin, Fl. Badensis 1: 8. 1805. *Distomomischus myuros* Dulac, Fl. Dépt. Hautes-Pyrén. 91. 1867. *Zerna myuros* Panzer ex B. D. Jackson, Index Kew. 2: 1249. 1895. (England, Italy.)
Festuca megalura Nutt. Proc. Acad. Nat. Sci. Philadelphia 4: 24. 1848. *V. megalura* Rydb. Bull. Torrey Bot. Club 36: 538. 1909. (*Gambel,* Santa Barbara, Calif., in 1842.)
V. myuros var. *hirsuta* Hackel, Cat. Gram. Portugal 24. 1880. *Festuca myuros* var. *hirsuta* Aschers. & Graebner, Syn. Mittel-eur. Fl. 2: 558. 1901. ("Coimbra," Portugal.)

Foxtail fescue, rattail fescue.

Single or loosely tufted annuals; culms 1.5–5 (7) dm tall, decumbent basally and geniculate at the lower nodes, otherwise erect or ascending, glabrous; sheaths glabrous; ligules erose-ciliolate, 0.2–0.6 mm long, usually longer on the sides; blades filiform or linear, involute, glabrous below and scabrous-puberulent above; panicles elongate, (4) 5–18 (20) cm long, narrow with ascending or appressed-ascending branches, often secund, the lower branches often remote and enclosed in the upper sheath; spikelets 3- to 5- (8)-flowered; glumes thin, glabrous, extremely unequal, the first sometimes scale-like, very small, 0.3–1.5 mm long, 1-nerved, the second 2.5–5.5 mm long, 1- to 3-nerved, lanceolate; lemmas 4.5–7.5 mm long, linear-lanceolate, scabrous to puberulent, often long ciliate towards the apex, produced into awns 8–15 (17) mm long; anthers usually 1, rarely 2 or 3, 0.5–1.5 mm long, included to partially protruding; $2n = 14, 42, 63$.

Dry disturbed areas; native of Europe; Alaska to Baja Calif., e. to Nev. and Ariz.; occasional in the c. states and common in the e. states; widely distributed throughout the world. April–July.

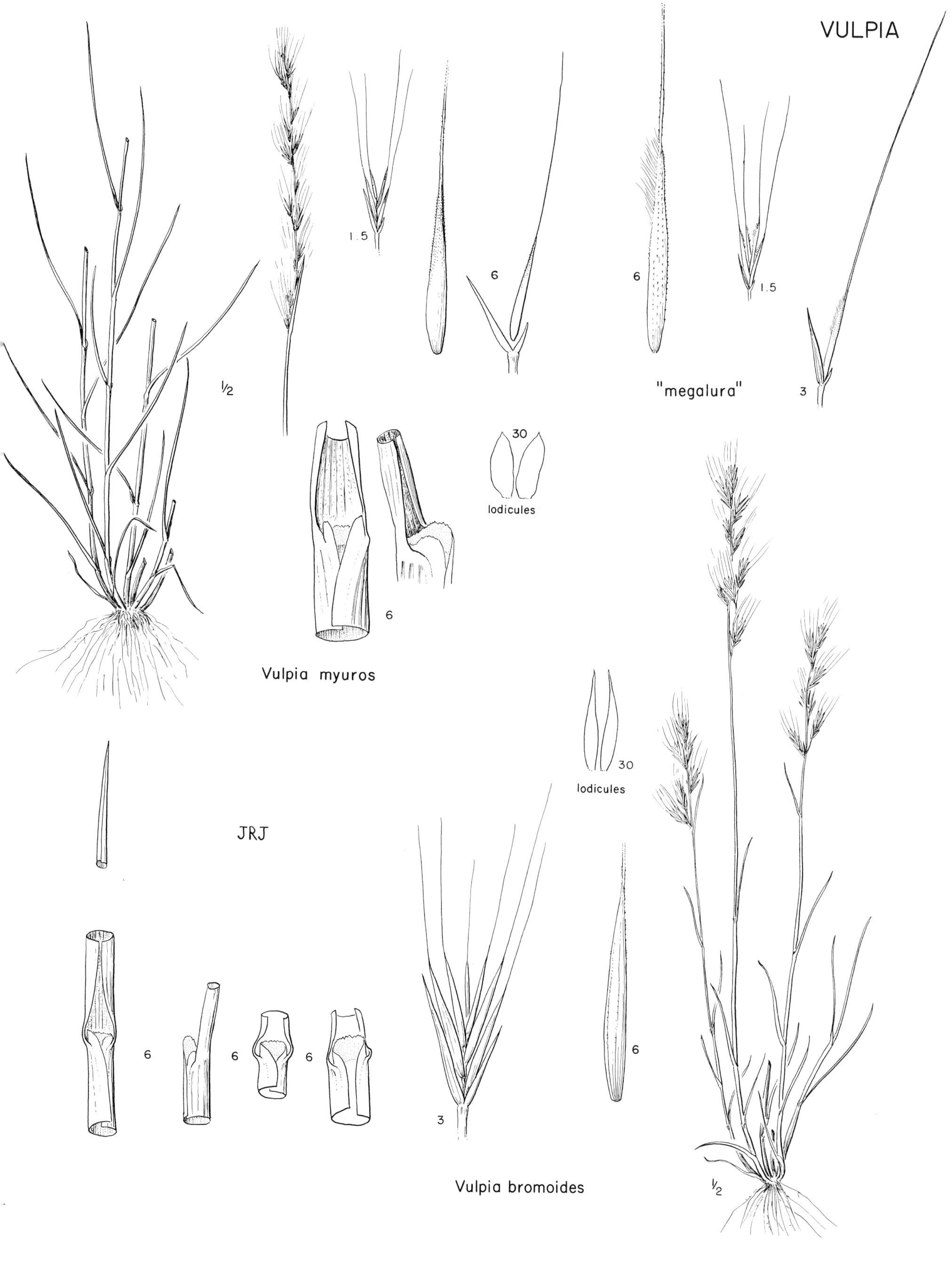
VULPIA
1.5
6
6
1.5
"megalura"
3
1/2
30
lodicules
6
Vulpia myuros
30
lodicules
JRJ
6
6
6
3
6
Vulpia bromoides
1/2

VULPIA

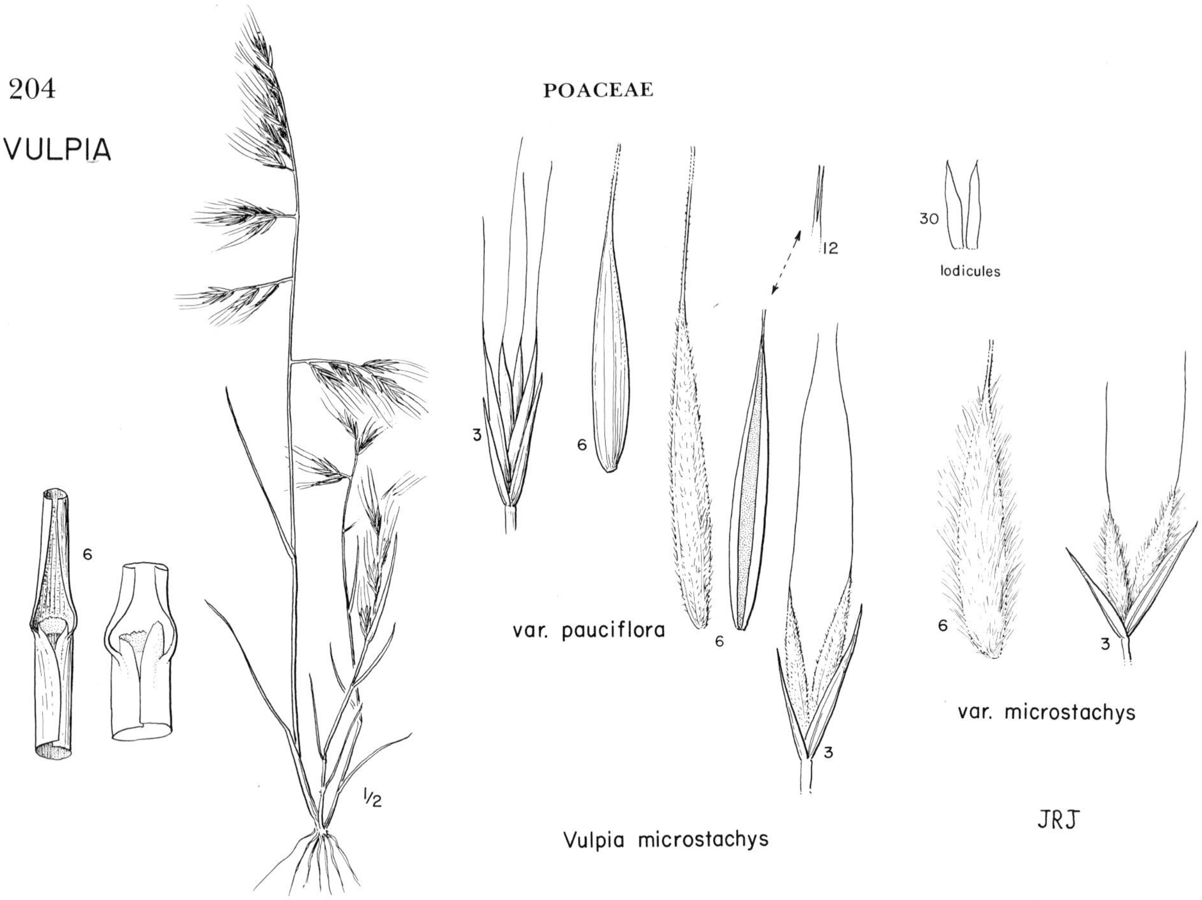

Vulpia megalura (Nutt.) Rydb. has long been considered a native to western North America and closely related to what was considered a European introduction, *V. myuros*. The distinguishing character has been the ciliate lemma of *V. megalura* versus the non-ciliate lemma of *V. myuros*. Ciliate lemmas occur, however, on a few European specimens under the name *V. myuros* var. *hirsuta* Hackel.

3. Vulpia bromoides (L.) S. F. Gray

Festuca bromoides L. Sp. Pl. 75. 1753. *Vulpia bromoides* S. F. Gray, Nat. Arr. Brit. Pl. 2: 124. 1821. *Festuca myuros* var. *bromoides* Wimmer & Grabowski, Fl. Siles. 1: 83. 1827. *V. myuros* var. *bromoides* Parl. Pl. Novae Minus Notae 46. 1842, description only; Parl. Fl. Palermo 1: 186. 1845, description only; Parl. Fl. Ital. 1: 419. 1848. *Zerna bromoides* Panzer ex B. D. Jackson, Index Kew. 2: 1249. 1895. (Europe.)

Bromus dertonensis All. Fl. Pedemontana 2: 249. 1785. *Festuca dertonensis* Aschers. & Graebner, Syn. Mitteleur. Fl. 2: 558. 1901. *V. dertonensis* Gola, Malpighia 18: 364. 1904. *V. myuros* var. *dertonensis* Fiori, Nuova Fl. Analitica Ital. 1: 142. 1923. (Italy.)

Loosely tufted annuals, mostly glabrous, except for the ventral leaf-surfaces; culms 1–3 (5.5) dm tall, shortly decumbent basally and geniculate at the lower nodes; ligules erose-ciliolate, 0.2–0.6 mm long, usually longest in the center; blades usually involute, filiform; panicles 2–8 (12) cm long, narrow, with short, ascending branches and pedicels, sometimes the lower branches diverging or secund; spikelets 3- to 5- (6)-flowered; glumes unequal, the first 3.5–5 (7) mm long, 1-nerved, linear-filiform, the second 4.5–8 mm long, with 3 relatively prominent nerves, lanceolate, sometimes slightly awn-tipped; lemmas (4) 5.5–7.5 mm long, minutely scaberulous or glabrous, not ciliate, the awns (4) 6–11 (15) mm long; palea long and narrow, bidentate; anthers usually only 1, 0.3–0.5 (2) mm long; caryopsis 3.5–4 mm long; $2n$ = 14, 42.

Dry disturbed places; native of Europe; s. B.C. to Calif., e. to Idaho and Utah and occasional in the East. April–mid-July.

Vulpia bromoides appears to be closely related to *V. myuros*, but the latter is readily distinguished by its greatly reduced first glume.

4. Vulpia microstachys (Nutt.) Benth.

Festuca microstachys Nutt. Proc. Acad. Nat. Sci. Philadelphia 4: 24. 1848. *Vulpia microstachya* Munro ex Benth. Pl. Hartwegianus 342. 1857. (*Gambel*, Los Angeles, Calif., in 1841 or 1842.)

Festuca reflexa Buckley, Proc. Acad. Nat. Sci. Philadelphia 1862: 98. 1863. *V. reflexa* Rydb. Bull. Torrey Bot. Club 36: 538. 1909. (*Nuttall*, "Upper California," in 1836.) = var. *pauciflora*.

Festuca microstachys var. *pauciflora* Scribn. ex Vasey, Descript. Cat. Grasses U. S. 90. 1885, nomen nudum; Scribn. ex Beal, Grasses N. Amer. 2: 586. 1896. *V. microstachys* var. *pauciflora* Lonard & Gould, Madroño 22: 226. 1974. (*Howell*, Portland, Oregon, May 1886.)

Festuca arida Elmer, Bot. Gaz. 36: 52. 1903. *V. arida* Henrard, Blumea 2: 323. 1937. (*Henderson 2196*, North Yakima, Wash., May 1892.) = var. *microstachys*.

Festuca pacifica Piper, Contr. U. S. Natl. Herb. 10: 12. 1906. *V. pacifica* Rydb. Bull. Torrey Bot. Club 36: 538. 1909. (*Elmer 262*, "on rich bottom lands," Pullman, Whitman Co., Wash., 20 June 1896.) = var. *pauciflora*.

Festuca dives Suksd. Werdenda 1(2): 3. 1923; not F. Muell. 1863. (*Suksdorf 6153*, "sandy shore of the Columbia River at Bingen," Klickitat Co., Wash.) = var. *pauciflora*.

Annuals; culms (0.5) 1.5–3.5 (4) dm tall, erect or geniculate at the base, glabrous to sometimes pubescent at the base; ligules erose-ciliolate, 0.2–0.5 mm long, usually longest in the center; blades glabrous, narrowly-linear to filiform, loosely involute; panicles (3) 5–11 (13) cm long, broader than the preceding species, sometimes racemose above with spreading to reflexed branches below, more or less secund; spikelets (1-) 3- to 5- (7)-flowered; glumes glabrous, the first 3–6 mm long, 1-nerved, subulate to linear, the second 5–8 mm long, strongly 3-nerved, linear-lanceolate; lemmas 5.5–7.5 mm long, lanceolate, mostly hirsute to scabrous, sometimes glabrous, not ciliate, the awns (7) 8–13 mm long; palea acute or attenuate and strongly bilobed; anthers usually 1, sometimes 3, about 1.5 mm long; caryopsis 3.5–5.5 mm long; $2n = 28$.

Dry to moderately moist disturbed areas; s. B.C. and Mont., s. to Baja Calif. and N.M. May–June.

The *V. microstachys* complex is probably best treated as a single polymorphic species as did Lonard and Gould (1974). Under this interpretation we have two varieties as follows:

1 Spikelets pubescent; Wash. to s. Calif., extending e. to w. Nev. and sw. Idaho .... var. *microstachys*

1 Spikelets glabrous or rarely the lemmas scaberulous; B.C. and w. Mont., s. to Baja Calif. and Ariz. .... var. *pauciflora* (Scribn. ex Beal) Lonard & Gould

Kannenberg and Allard (Evolution 22: 517–528. 1968), on the other hand, considered the complex to consist of eight autogamous taxa that appear to carry on no apparent gene exchange, even when growing side by side. The infrequent appearance of chasmogamy among the members of the complex did not seem to result in any F_1 progeny. The names *V. pacifica, V. reflexa* and *V. arida* have been applied to plants in our region. They would separate by the following key.

1 Lemmas glabrous to scabrous .... *V. pacifica* (Piper) Rydb.
1 Lemmas hirsute.
  2 Panicle branches spreading to reflexed, the pedicels usually ascending; spikelets 3- to 7-flowered; near Hagerman, Gooding Co., Idaho (*Piemeisel*, 13 May 1930) .... *V. arida* (Elmer) Henrard
  2 Panicle branches and pedicels divaricate to reflexed; spikelets only 1- to 2- (5)-flowered; Smoke Cr., Washoe Co., Nev. (*Griffiths & Hunter 514*) ..... *V. reflexa* (Buckley) Rydb.

3. FESTUCA L. Fescue

Tufted perennials, even when short-rhizomatous; culms hollow; sheaths open; ligules membranous, short, often truncate; blades thin, flat or narrow and involute; auricles present or absent; inflorescence an open or contracted panicle, sometimes drooping; spikelets (4-) 5- to 15- (20)-flowered, disarticulating above the glumes and between the florets; glumes unequal, narrow, acute or acuminate, 1- to 3-nerved; lemmas (3.5) 4–10 (12) mm long, rounded on back (except somewhat keeled in *F. dasyclada*), (3-) 5- to 7-nerved, awned from a narrow, entire or rarely minutely bifid apex or awnless; palea about as long as the lemma, the 2 keels ciliate or scabrous, free from the caryopsis; lodicules 2, entire, serrate or lobed, acuminate to truncate; stamens 3; $x = 7$.

A genus of about 100 species in temp. and cool regions of the world. (The ancient Latin name for a grass.)

A few European species of *Festuca* have been introduced into the United States and planted in pastures. Most of the native species are extremely valuable for forage in more arid habitats.

References:
Piper, C. V. North American species of *Festuca*. Contr. U. S. Natl. Herb. 10: 1–48. 1906.
Saint-Yves, A. *Festuca* de l'Amérique du Nord et du Mexique. Candollea 2: 229–316. 1925.

1 Blades mostly flat, lax and soft, mostly more than 3 mm broad, mostly cauline.
  2 Awn of the lemma none or very short; plants short rhizomatous.
    3 Spikelets (4-) 5- to 9-flowered, 10–17 mm long; auricles usually present.
      4 Lemmas 7–10 mm long; first glume (3) 4–6 mm long, the second (4) 5–7 mm long .... 1. *F. arundinacea*
      4 Lemmas 4–7 mm long; first glume 2.5–4 mm long, the second 3.5–5 mm long .... 2. *F. pratensis*
    3 Spikelets 2- to 4- (5)-flowered, 8–11 mm long; auricles lacking .... 3. *F. sororia*
  2 Awn of the lemma 5–20 mm long; plants non-rhizomatous .... 4. *F. subulata*
1 Blades involute, or if flat then mostly less than 3 mm broad and rather stiff.
  5 Ligules 2.5–9 mm long .... 6. *F. thurberi*
  5 Ligules less than 2 mm long.
    6 Lemmas 7.5–12 mm long .... 5. *F. scabrella*
    6 Lemmas 3.5–7 (8) mm long.
      7 Panicle branches divaricate branched and densely ciliate on the angles; lemmas somewhat keeled above; c. Utah .... 7. *F. dasyclada*
      7 Panicle branches ascending or appressed, not densely ciliate on the angles; lemmas rounded on back.
        8 Awn of the lemma usually less than 1 mm long, rarely up to 2 (3) mm; plants glaucous; barely entering s. Utah on Navajo Mt. .... 9. *F. arizonica*
        8 Awn of the lemma usually (1) 2 mm or more long.
          9 Culms decumbent at the usually red, fibrillose base; awn of the lemma shorter than the body; widespread .... 8. *F. rubra*
          9 Culms erect.
            10 Plants (3) 4–10 dm tall; panicles open, (5) 7–12 (15) cm long; spikelets (4-) 5- to 7-flowered; mostly in the n. half of the region, extending s. in the LaSal Mts. of Utah and the Ruby and Toquima mts. of Nev. .... 10. *F. idahoensis*
            10 Plants small, 0.8–3.5 (5) dm tall; panicles narrow and short, 1.5–7 (10) cm long; spikelets 2- to 4- (5)-flowered; widespread .... 11. *F. ovina*

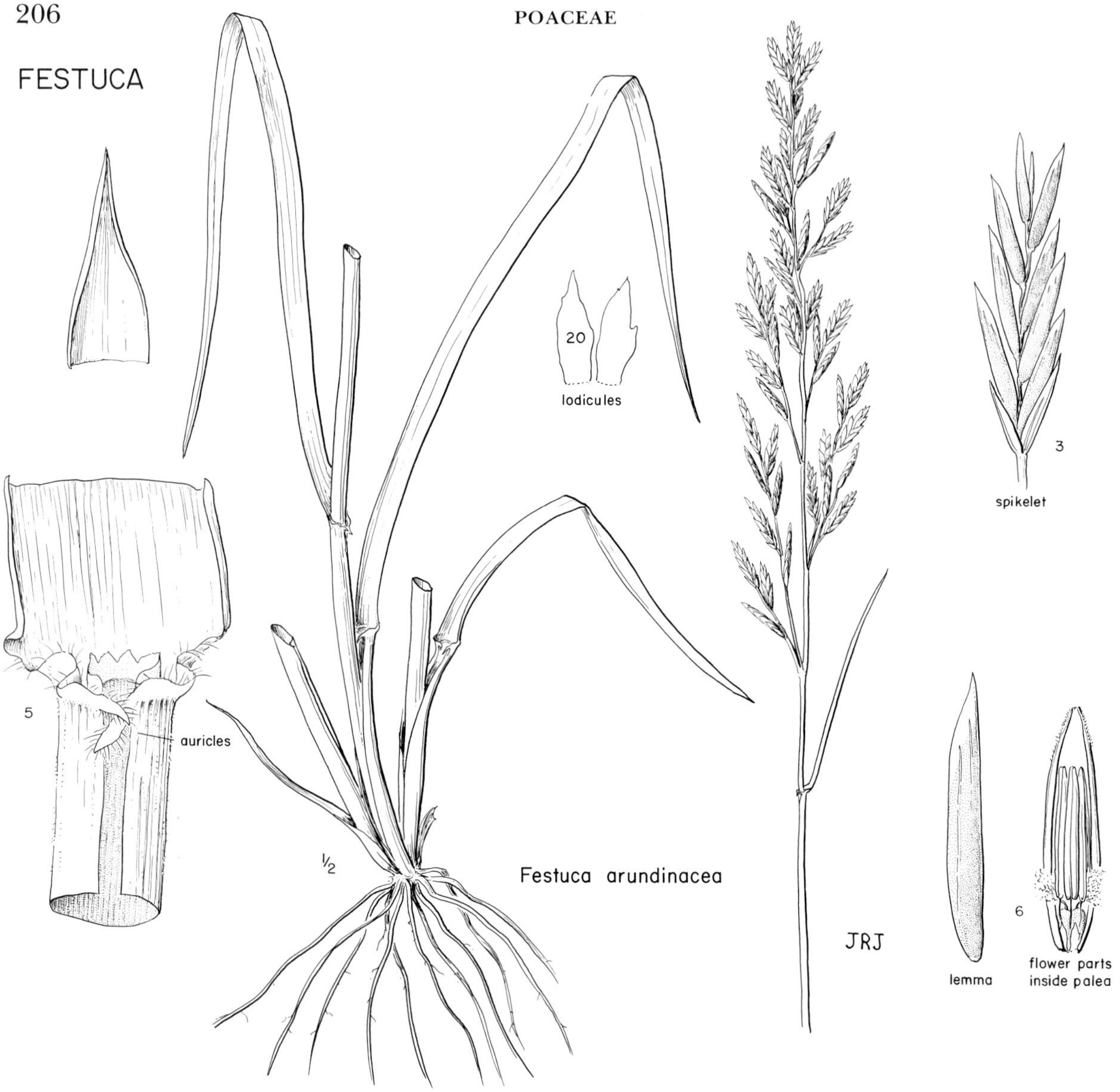

1. Festuca arundinacea Schreber

Festuca arundinacea Schreber, Spic. Fl. Lipsicae 57. 1771. *Bromus arundinaceus* Roth, Tent. Fl. German. 2: 141. 1789. *F. elatior* var. *arundinacea* Wimmer, Fl. Schlesien, ed. 3. 59. 1857. *F. elatior* subsp. *arundinacea* Hackel, Bot. Centralbl. 8: 407. 1881. (Germany.)

Reed fescue, alta fescue, tall fescue.

Stout perennials with short rhizomes, but often strongly tufted; culms ascending to erect, over 1 dm tall; sheaths smooth or scabrous; ligules about 0.2 mm long; auricles prominent, usually ciliate; blades stiff, flat or somewhat involute, 3–12 mm broad, scabrous above; panicles 15–30 (35) cm long, narrow, often somewhat nodding; spikelets 10–15 mm long, about 5- (9)-flowered; glumes lanceolate, the first (3) 4–6 mm long, 1- to 3-nerved, the second (4) 5–7 mm long, 3- to 5-nerved, with membranous margins; lemmas 7–10 mm long, 5–7 more or less faint nerves, the lip hyaline and acute, the awns absent or up to 2 mm long; palea about as broad as the lemma; anthers (1) 3, 2.5–4 mm long; $2n = 28, 42, 56, 63, 70$.

Along irrigation ditches and in fallow fields and roadsides in moderately moist situations; native of Europe; occasionally introduced in the U.S., known in our region from Owyhee Co., Idaho on the Snake River Plains, and likely to be an occasional constituent of pasture and seed mixtures. Late May–July.

Festuca arundinacea is used for hay and pastures in Europe and sometimes in the United States.

2. Festuca pratensis Huds.

Festuca elatior L. Sp. Pl. 75. 1753, nomen ambiguum. *Avena secunda* Salisb. Prodr. Stirpium 22. 1796. *Bromus elatior* Koeler, Descript. Gram. Gallia & German. 214. 1802. *Schedonorus elatior* Beauv. Essai Nouv. Agrostogr. 99, 156, 177. 1812. *Bucetum elatius* Parnell, Grasses Scotl. 107, pl. 46. 1842. *Tragus elatior* Panzer ex B. D. Jackson, Index Kew. 2: 1098. 1895, in synonymy. *Gnomonia elatior* Lunell, Amer. Midl. Naturalist 4: 224. 1915. (Europe.)

F. pratensis Huds. Fl. Anglica 37. 1762. *F. fluitans* var. *pratensis* Huds. Fl. Anglica, ed. 2. 47. 1778. *Schedonorus pratensis* Beauv. Essai Nouv. Agrostogr. 99, 163, 177. 1812. *Bromus pratensis* Huds. ex Sprengel, Syst. Veg. 1: 359. 1825; not Lam. 1785. *Bucetum pratense* Parnell, Grasses Scotl. 105, pl.

FESTUCA

6
3
20
lodicules
JRJ
1/2
3
6
20
lodicules
20
1/2
5
Festuca pratensis
6
Festuca sororia
1/2

46. 1842. *F. elatior* var. *pratensis* A. Gray, Manual Bot. ed. 5. 634. 1867. *F. elatior* subsp. *pratensis* Hackel, Bot. Centralbl. 8: 407. 1881. (England.)

Meadow fescue, English bluegrass.

Stout perennials with short rhizomes, but tufted; culms mostly 6–12 dm tall, glabrous; ligules about 0.2 mm long; auricles well developed, not ciliate; blades flat or somewhat involute, 3–5 (7.5) mm broad, usually scabrous; panicles 10–20 (24) cm long, narrow; spikelets 10–15 (17) mm long, 4- to 9-flowered (mostly 5 or 6); glumes lanceolate, the first 2.5–4 mm long, 1-nerved, the second 3.5–5 mm long, 3- to 5-nerved, with hyaline margins; lemmas (4) 5.5–7 mm long, usually glabrous, 5-nerved, the tip hyaline, acute, rarely awned, if so up to 2 mm long; anthers 3, 1.5–3.5 mm long; $2n$ = 14, 28, 42, 70.

Moist or moderately moist meadows, roadsides, ditchbanks and waste places; native of Europe; B.C. to N.S. and throughout the U.S. June–July (–Aug).

This species is often referred to *F. elatior* L., a name that has been confused with the closely related *F. arundinacea* Schreber. The Linnaean name has been abandoned as a nomen ambiguum in favor of the next oldest name, *F. pratensis*. This species is often planted in meadows and pastures.

3. **Festuca sororia** Piper

Festuca sororia Piper, Contr. U. S. Natl. Herb. 16: 197. 1913. *F. subulata* var. *sororia* St.-Yves, Candollea 2: 285. 1925. (*Nealley 177,* in ravines, Rincon Mts., Ariz., Sept. 1891.)

Stout perennials with short rhizomes, but tufted; culms 7.5–13 dm tall, glabrous; ligules 0.5–1 (1.5) mm long; auricles lacking; blades flat, 2.5–6 mm broad, glabrous, except for the scabrous margins and nerves; panicles 10–20 cm long, lax, somewhat nodding, the branches capillary and often curved, the lower branches often reflexed; spikelets 8–11 mm long, 2–4 (5)-flowered; glumes unequal, the first 3–4 mm long, 1-nerved, narrow-lanceolate, the second 4.5–5.5 mm long, 3-nerved, lanceolate; lemmas 6.5–8 (8.5) mm long, scabrous, 5-nerved, narrowly acuminate, the awns, when present, up to 1 mm long; anthers 1.5–2 mm long.

Open woods and meadows from middle elevs. to the subalpine zone; s. Utah (Abajo Mts., *Rydberg & Garrett 9852*), s. Colo., N.M. and Ariz. Aug.–early Sept.

4. **Festuca subulata** Trin.

Festuca subulata Trin. in Bong. Mém. Acad. Imp. Sci. St.-Pétersbourg, Ser. 6, Sci. Math. 2: 173. 1832. *Bromelica subulata* Farw. Rhodora 21: 78. 1919, as to basionym, not as to concept. (*Mertens,* Sitka, Alaska.)

F. jonesii Vasey, Grasses U. S. 43. 1883, nomen nudum; Contr. U. S. Natl. Herb. 1: 278. 1893. *F. subulata* var. *jonesii* St.-Yves, Candollea 2: 284. 1925. (*M. E. Jones s.n.,* City Creek Canyon, above Salt Lake City, Utah, 17 July 1880.)

Bearded fescue.

Tufted perennials, caulescent, glabrous to more or less scaberulous throughout, non-rhizomatous; culms 7–10 dm tall, decumbent at the base; ligules 0.3–0.7 mm long, erose-ciliolate; auricles lacking; blades usually flat and lax, 5–8 (10) mm broad, strongly nerved, scaberulous; panicles (10) 15–35 (40) cm long, lax, the branches slender and more or less drooping; spikelets (6) 7–8 (10) mm long, loosely 2- to 4- (6)-flowered; glumes linear to linear-lanceolate, the first glume (1.5) 2–3.5 mm long, 1-nerved, the second 3–5.5 mm long, 1- to 3-nerved; lemmas 5–7 mm long, minutely scaberulous, with 3 faint nerves, the slender awns (5) 6–10 (13) mm long; palea slightly longer than the lemma, very narrow; anthers 3, 1.5–2 (3) mm long; $2n$ = 28.

Moist, shady areas at middle elevs.; Alaska, s. in the mts. through B.C. and Alta. to the Sierra Nevada of Calif., n. Nev. (Carson Range and mts. of Elko Co.), n. Utah (Wasatch Range) and Wyo. June–Aug.

5. **Festuca scabrella** Torr. ex Hook.

Festuca scabrella Torr. ex Hook. Fl. Boreali-Amer. 2: 252, pl. 233. 1840. *F. altaica* subsp. *scabrella* Hultén, Fl. Alaska & Yukon 1: 241. 1942. *F. altaica* var. *scabrella* Breit. Amer. Midl. Naturalist 58: 12. 1957. (*Drummond,* alpine districts of the Rocky Mts., Can.)

Melica hallii Vasey, Bot. Gaz. 6: 296. 1881. *F. hallii* Piper, Contr. U. S. Natl. Herb. 10: 31. 1906. *Daluca hallii* Lunell, Amer. Midl. Naturalist 4: 221. 1915. *F. altaica* subsp. *eualtaica* var. *genuina* subvar. *hallii* St.-Yves, Candollea 2: 271. 1925. *F. scabrella* subsp. *hallii* W. A. Weber, Univ. Colorado Stud. Ser. Biol. 7: 8. 1961. (*Hall 621,* Rocky Mts., n. half of Colo., in 1862.)

F. scabrella var. *major* Vasey, Contr. U. S. Natl. Herb. 1: 278. 1893. *F. campestris* Rydb. Mem. New York Bot. Gard. 1: 57. 1900. *Daluca campestris* Lunell, Amer. Midl. Naturalist 5: 234. 1918. *F. altaica* var. *major* Gleason, Phytologia 4: 21. 1952. (*Suksdorf 118,* "Common on prairies," Spokane Co., Wash., 18 June 1884.)

F. confinis subsp. *rabiosa* Piper, Contr. U. S. Natl. Herb. 10: 41. 1906. *F. kingii* var. *rabiosa* A. S. Hitchc. Amer. J. Bot. 21: 128. 1934. *Hesperochloa kingii* var. *rabiosa* Swallen, Proc. Biol. Soc. Wash. 54: 45. 1941. (*Williams & Griffiths 25,* Branch Crazy Woman's Cr., Wyo., 30 July 1898.)

Rough fescue.

Cespitose perennials producing thick mats of persistent sheaths and culm bases; culms (4) 6–10 (15) dm tall; leaves mostly basal, the tuft (20) 30–40 (60) cm high; ligules short, 0.5–1 mm long, somewhat erose-lacerate and often ciliate, longer on the sides than in the center; blades stiff, loosely involute, sometimes becoming flat, 2.5–4 (4.5) mm broad and without auricles, scabrous on the ventral surface; panicles (7) 10–20 cm long, somewhat compressed, the lower branches often spreading; spikelets 9–15 (20) mm long, 3- to 5- (7)-flowered; glumes lanceolate, 1-nerved, the first glume (4) 5–7.5 mm long, the second 5.5–7 (10) mm long, sometimes with a pair of faint lateral nerves; lemmas 7.5–10 (12) mm long, scaberulous, 5-nerved, acute, awnless or sometimes minutely awn-tipped; anthers 4–6 mm long; $2n$ = 28, 56.

Mountain meadows and open woodlands in w. U.S. and prairies in midw. U.S.; in the Rocky Mt. cordillera from Alaska to e. Wash., e. Oregon and Colo.; entering our region in the Steens Mts., Harney Co., Oregon. June–Aug.

The name *F. scabrella* var. *major* Vasey has often been applied to the robust forms.

6. **Festuca thurberi** Vasey

Festuca thurberi Vasey in Rothr. Cat. Pl. Geogr. & Geol. Explor. & Surv. W. 100th Meridian 56. 1874. *F. tolucensis* subsp. *thurberi* St.-Yves, Candollea 2: 304. 1925. (*Wolfe 1154,* South Park, Colo., in 1873.)

Poa festucoides M. E. Jones, Proc. Calif. Acad. Sci. II. 5: 723.

FESTUCA

Festuca subulata

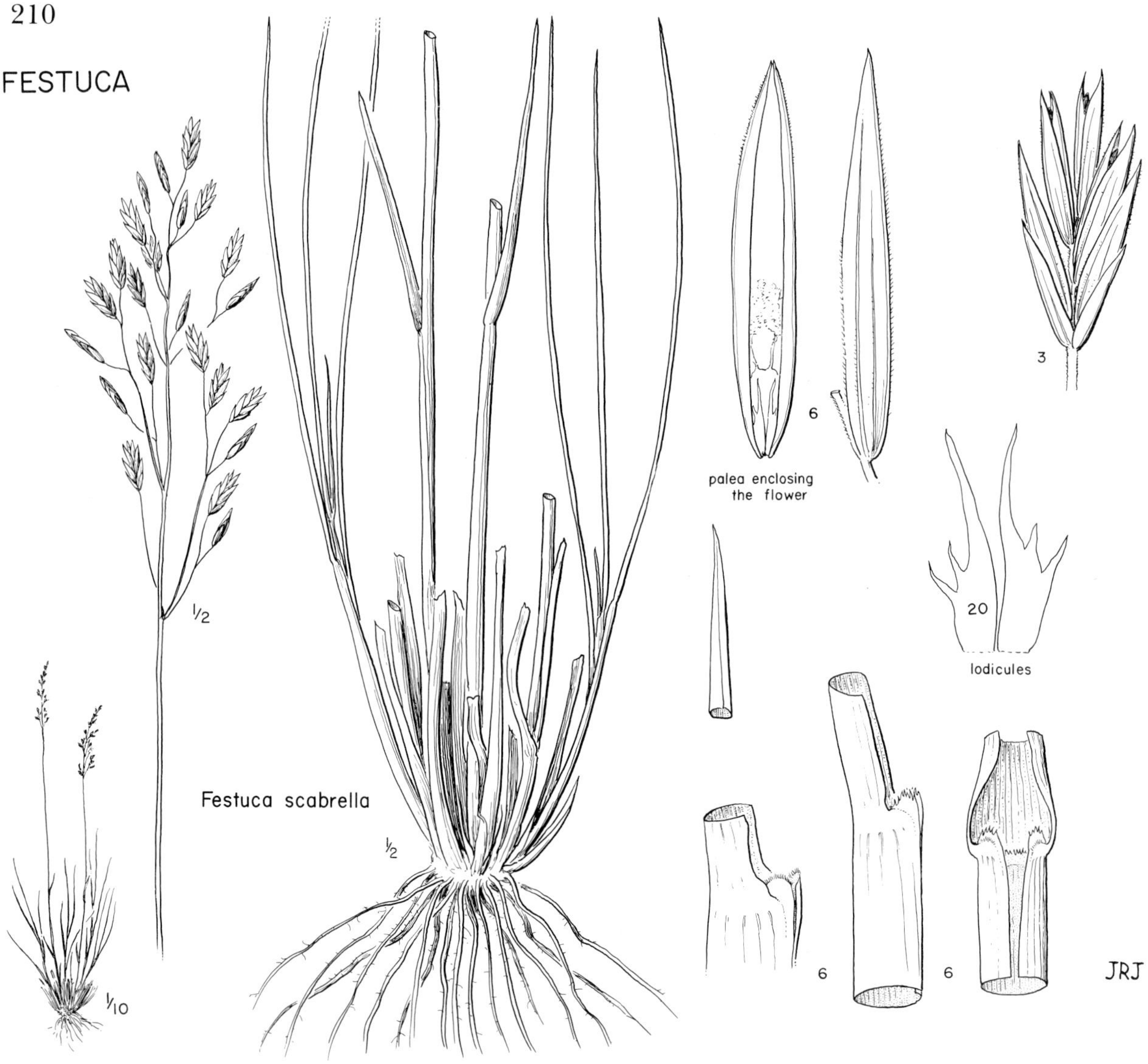

1895; not Lam. 1791. *Poa kaibensis* M. E. Jones, Erythea 4: 36. 1896. (*Jones 5671,* Mt. Ellen, Henry Mts., Utah, 25 July 1894.)

Thurber fescue.

Densely tufted perennials with persistent dried sheaths and culms at the base; culms erect, 4–9.5 (12) dm tall; leaves mostly basal, the tuft 30–45 cm high, usually scabrous; ligules conspicuous, 2.5–5 (9) mm long, decurrent, scarious, often lacerate; blades involute and without auricles; panicles (8) 10–15 (17) cm long, loose, slightly drooping; spikelets (8) 10–14 mm long, 3- to 5- (6)-flowered; glumes smooth to scaberulous, usually scabrous on the nerves, the first glume (3) 4–5 mm long, 1-nerved, the second 4.5–5.5 (7) mm long, 3-nerved; lemmas (6) 7–8 (10) mm long, elliptic-lanceolate, faintly 5-nerved, glabrous to scaberulous, sometimes minutely awn tipped; anthers 1.8–4.5 mm long; $2n$ = 28, 42.

From oakbrush woodlands in the foothills to subalpine meadows; s. Wyo., Colo., n. N.M. and se. Utah (Colorado Plateau). July–Aug. *Festuca thurberi* is very closely related to *F. scabrella* Torr. ex Hook. and resembles it in all respects except for the prominent ligules and statistically shorter glumes and lemmas.

7. Festuca dasyclada Hackel ex Beal

Festuca dasyclada Hackel ex Beal, Grasses N. Amer. 2: 602. 1896. (*Parry 93,* c. Utah, in 1875.)

Perennials; culms erect or somewhat geniculate, 2–4 dm tall, nearly smooth; sheaths glabrous; ligules short; auricles lacking; blades narrow, soft, folded, glabrous, acute, 5–15 cm long; panicles open, erect, the branches in 2s, 3s or 4s, ciliate on the prominent angles; spikelets 6–7 mm long, usually 2-flowered, pale green; glumes scabrous to puberulent, the first glume 3–4.5 mm long, narrow-lanceolate, 1-nerved, the second 5.5–6 mm long, lanceolate, 3-nerved; lemmas 5.5–6 mm long, lanceolate, green, membranous, 5-nerved, somewhat keeled, scabrous-puberulent all over the back, the awns 2.5–3 mm long; palea about as long as the lemma; anthers about 2 mm long.

Rocky slopes; endemic to c. Utah. A very distinct species that has been recollected only once in the Wasatch Plateau (*J. A. Willey s.n.,* s. slope Wagon Road Ridge, Emory Co., Utah).

FESTUCA

6
12
40
lodicules
JRJ
1/2
12
1/2
Festuca thurberi
6
12
24
24
12
1/2
Festuca dasyclada

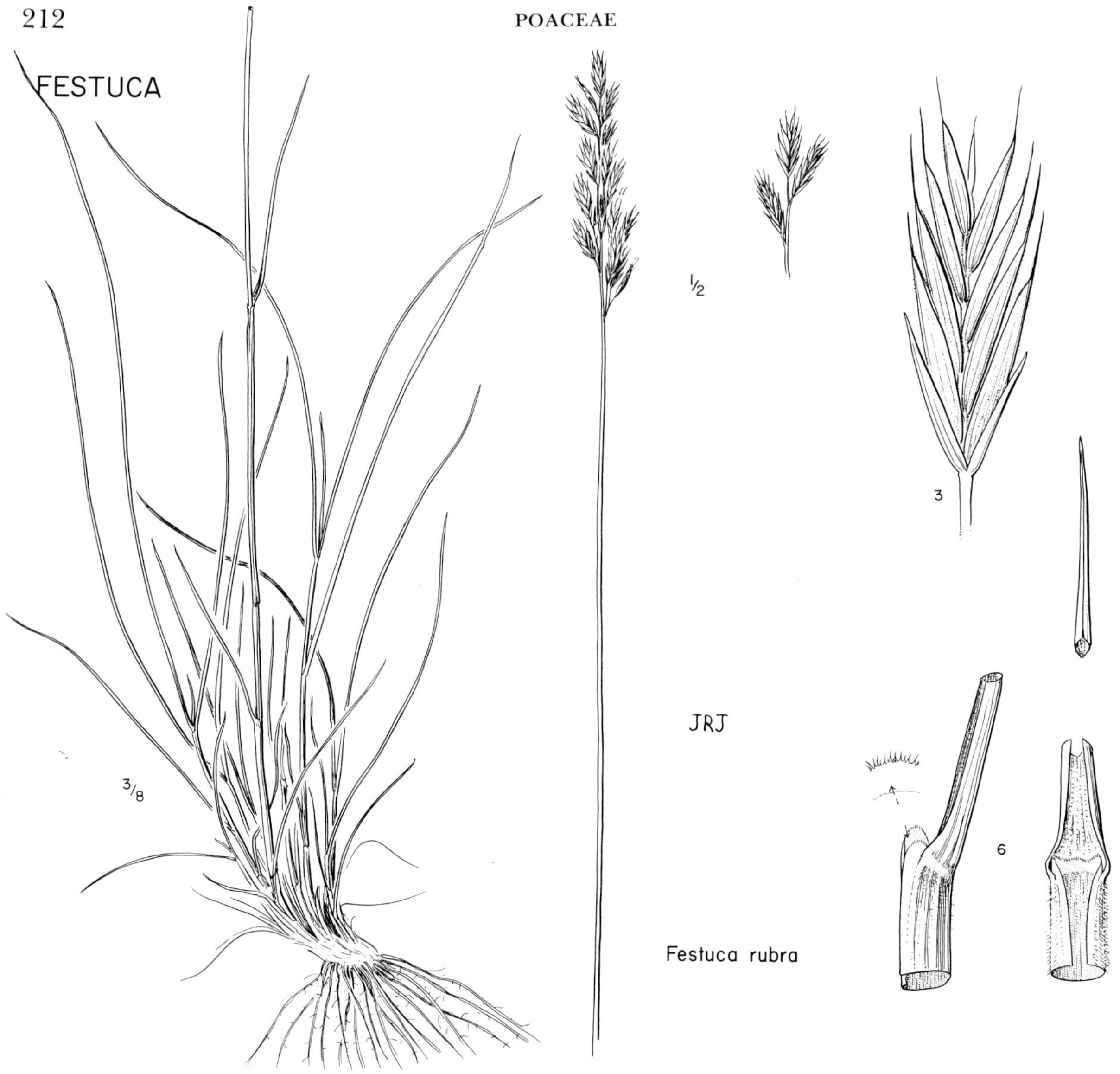

8. Festuca rubra L.

Festuca rubra L. Sp. Pl. 74. 1753. *F. ovina* var. *rubra* J. E. Smith, Engl. Fl. 1: 139. 1824. *F. duriuscula* var. *rubra* A. Wood, Amer. Bot. and Florist 2: 399. 1871; not Parnell 1842. *F. ovina* subsp. *rubra* J. D. Hook. Stud. Fl. Brit. Islands, ed. 3. 497. 1884. (Europe.)

F. rubra var. *lanuginosa* Mert. & Koch in Röhling, Deutschl. Fl. ed. 3. 1: 654. 1823. (Prussia.)

F. rubra var. *commutata* Gaudin, Fl. Helv. 1: 287. 1828. *F. rubra* subsp. *eurubra* var. *commutata* subvar. *eucommutata* St.-Yves, Annuaire Conserv. Jard. Bot. Genève 17: 129. 1913. *F. rubra* subsp. *eurubra* var. *caespitosa* subvar. *eucommutata* Litard. Candollea 10: 109. 1945. (Switzerland.)

F. oregona Vasey, Bot. Gaz. 2: 126. 1877. *Poa tenuifolia* var. *oregona* Vasey, Descript. Cat. Grasses U. S. 85. 1885. (*Oregon Worlds Fair Commission*, Oregon, in 1892.)

Red fescue.

Perennials from loose clumps, usually with rhizomes; culms (1.5) 3–9 (12) dm tall, decumbent at the base; sheaths glabrous to soft pubescent, the lower ones thin and papery but with fibrous, large veins, at first reddish-purple, aging brown and disintegrating into curled fibers; ligules usually less than 0.5 mm long, higher on the sides than in the center, ciliolate; auricles lacking; blades usually folded-involute, rarely flat, 1–1.5 (3) mm broad, the basal leaves 12–30 cm long; panicles 5–13 (17) cm long, usually compressed; spikelets 7.5–10 (13) mm long, 4- to 6- (9)-flowered, often reddish-purple or glaucous green; glumes lanceolate, the first 2.5–3.5 (4.5) mm long, 1-nerved, the second 3.5–4.5 (5.5) mm long, 1- to 3-nerved; lemmas 4.5–7 (8) mm long, glabrous to scaberulous-puberulent, the awns 1–5 (8) mm long; anthers 3, 2–3 (4) mm long; $2n$ = 14, 21, 28, 42, 46, 49, 53, 56, 64, 70.

Moist meadows and open woods, often along streams, from the sagebrush zone to the subalpine zone; Eurasia; Arctic Amer., s. through the mts. to s. Calif., w. Texas and n. Mex., e. across the n. half of the c. states and s. in the e. states to S.C. July–Aug. (–early Sept.).

Festuca rubra is included with lawngrass mixes and is sometimes used in pasture plantings. These biotypes produce short rhizomes. *Festuca rubra* var. *commutata* Gaudin, Chewings fescue, a form with more erect culms, is sometimes planted in our region as a pasture grass. The name *F. rubra* var. *lanuginosa* Mert. & Koch has been used for a form with pubescent lemmas, although the fragment of

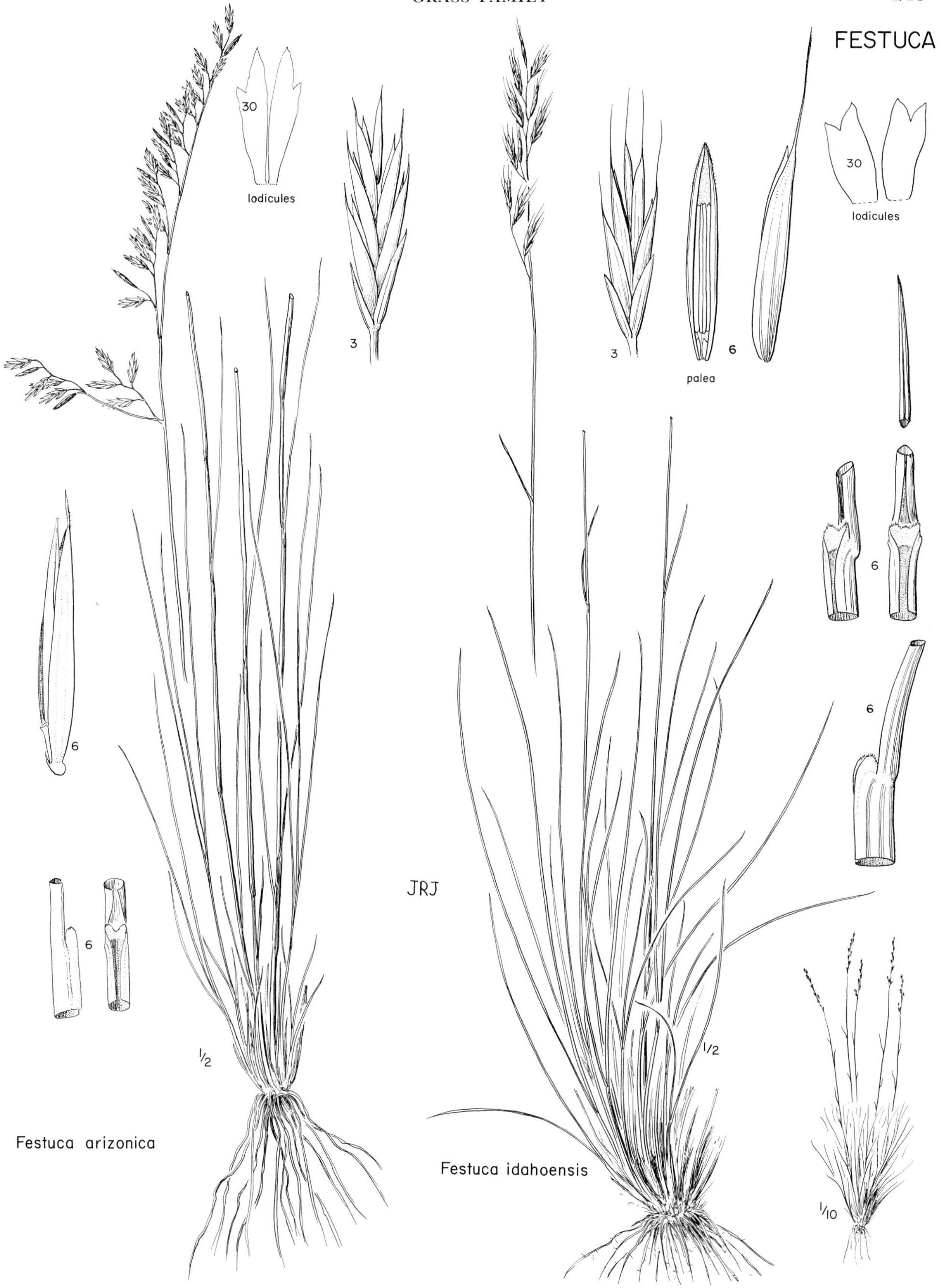
FESTUCA
30
lodicules
3
3
6
palea
30
lodicules
6
6
6
6
JRJ
1/2
1/2
1/10
Festuca arizonica
Festuca idahoensis

the Linnaean type of *F. rubra* at U.S. National Herbarium has very pubescent lemmas.

9. Festuca arizonica Vasey

Festuca arizonica Vasey, Contr. U. S. Natl. Herb. 1: 277. 1893. *F. ovina* var. *arizonica* Hackel ex Beal, Grasses N. Amer. 2: 598. 1896. *F. altaica* subsp. *arizonica* St.-Yves, Candollea 2: 267. 1925. (*Tracy 118,* near Flagstaff, Ariz., in 1887.)
F. vaseyana Hackel ex Beal, Grasses N. Amer. 2: 601. 1896. (*Vasey,* Veta Pass, Colo., in 1884.)
F. amethystina var. *asperrima* Hackel ex Beal, Grasses N. Amer. 2: 601. 1896; not Link, 1882. *F. ovina* subsp. *calligera* Piper, Contr. U. S. Natl. Herb. 10: 27. 1906. *F. calligera* Rydb. Bull. Torrey Bot. Club 36: 537. 1909. *F. ovina* subsp. *frigida* var. *calligera* St.-Yves, Candollea 2: 250. 1925. (*Rusby 901,* Ariz., in 1883.) A form with awns similar to *F. idahoensis.*

Arizona fescue, pinegrass.

Cespitose, glaucous perennials without rhizomes; culms 3.5–7.5 (9) dm tall; leaves mostly basal, the tuft 20–40 cm high, usually more than half the length of the culms, scabrous, the sheaths remaining firm and entire; ligules up to 2 mm long; blades filiform, involute, scabrous; panicles (4) 6–12 (20) cm long, loosely compressed, sometimes somewhat secund; spikelets 9–12 (15) mm long, (3-) 4- to 6-flowered; glumes linear to lanceolate, the first glume 3–4.5 mm long, 1-nerved, sometimes with a faint pair of lateral nerves, the second 4.5–6 (7) mm long, 3-nerved; lemmas 5.5–7 (8) mm long, obscurely 5-nerved, scabrous to smooth, the awns up to 2 mm long (rarely 3); anthers 3–3.5 mm long; $2n = 42$.

Dry meadows and openings of aspen–Douglas fir forests in gravelly rocky soil from 2135 to 3200 m (7000 to 10,500 ft.) elev.; s. Utah (Navajo Mt.) and Ariz., e. to Colo., N.M. and w. Texas. July–Aug.

Festuca arizonica is closely related to *F. idahoensis* Elmer, and some long-awned forms in Arizona may link the two in intergrading fashion that would warrant making *F. idahoensis* a variety of *F. arizonica.* However, no new combination is proposed here.

10. Festuca idahoensis Elmer

Festuca ovina var. *ingrata* Hackel ex Beal, Grasses N. Amer. 2: 598. 1896. *F. ingrata* Rydb. Bull. Torrey Bot. Club 32: 608. 1905. *F. occidentalis* var. *ingrata* B. Boivin, Naturaliste Canad. 94: 524. 1967. (*Howell 23,* Oregon, in 1880.)
F. ovina var. *oregona* Hackel ex Beal, Grasses N. Amer. 2: 599. 1896. *F. idahoensis* var. *oregona* C. L. Hitchc. Univ. Wash. Publ. Biol. 17(1): 577. 1969. (*Cusick 753,* "common over the hills," Oregon, in 1884.)
F. idahoensis Elmer, Bot. Gaz. 36: 53. 1903. *F. amethystina* var. *asperrima* subvar. *idahoensis* St.-Yves, Candollea 2: 260. 1925. (*Abrams 688,* Smiths Meadow, Shoshone Co., Idaho, July 1900.)
F. ingrata var. *nudata* Rydb. Agric. Exp. Sta. Agric. Coll. Colorado Bull. (Fl. Colorado) 100: 50. 1906. *F. ovina nudata* Vasey ex Rydb. ibid., as a synonym. (*Beardslee,* Middle Park, Colo., in 1892.)

Idaho fescue, bluebunch fescue.

Cespitose perennials, glabrous to scaberulous (sometimes more or less villosulose below), nonrhizomatous; culms (3) 4–6.6 (10) dm tall; leaves mostly basal, the tuft 15–25 (35) cm high, usually more than half the length of the culms, the sheaths remaining firm and entire; ligules mostly 0.3–0.6 mm long, ciliolate, higher on the sides than in the center; blades firm, filiform, folded-involute, less than 1 mm broad and without auricles; panicles (5) 7–12 (15) cm long, loosely compressed and sometimes somewhat secund; spikelets loosely 8–12 (13) mm long, (4-) 5- to 6- (7)-flowered, the rachilla joints usually visible; glumes linear-lanceolate to lanceolate, acute to obtuse, the first glume (2.5) 3–4 (4.5) mm long, the second 4–5.5 mm long; lemmas 5–7 mm long, rounded, glabrous to scaberulous (especially distally), the awns 2–6 mm long; palea often pubescent; anthers 2.5–4 mm long; $2n = 28$.

Dry sagebrush plains to the dry subalpine sagebrush meadows; a component of the sagebrush-grass community with *Artemisia tridentata, Poa sandbergii* and *Agropyron spicatum;* Yukon, B.C. and Alta., s. to the n. half of Calif., n. Nev., Utah and Colo. Late May–early Aug.

Festuca idahoensis is very closely related to *F. ovina* L. Some might regard it as the native low elevation phase of *F. ovina;* if so the name *F. ovina* var. *ingrata* Hackel ex Beal would be used.

11. Festuca ovina L.

Festuca ovina L. Sp. Pl. 73. 1753. *Bromus ovinus* Scop. Fl. Carniolica 1: 77. 1772. *Avena ovina* Salisb. Prodr. Stirpium 22. 1796. *Gnomonia ovina* Lunell, Amer. Midl. Naturalist 4: 224. 1915. (Europe.)
F. brevifolia R. Br. Chloris Melvilliana 31. 1823; not Muhl. 1817. *F. brachyphylla* J. A. Schultes, Mantissa 3 (Addit. 1): 646. 1827. *F. ovina* var. *brevifolia* S. Wats. in King, Rep. Geol. Explor. 40th Parallel 5: 389. 1871. *F. ovina* subsp. *brevifolia* Hackel, Bot. Centralbl. 8: 406. 1881. *F. ovina* subsp. *brachyphylla* Piper, Contr. U. S. Natl. Herb. 10: 27. 1906. *F. ovina* var. *brachyphylla* (incorrectly attributed to Piper by) A. S. Hitchc. Manual Grasses U. S. 860. 1935. (Melville Isl., Arctic Amer.)
F. minutiflora Rydb. Bull. Torrey Bot. Club 32: 608. 1905. *F. ovina* var. *minutiflora* J. T. Howell, Leafl. W. Bot. 6: 151. 1951. (*Baker,* Cameron Pass, alt. 10,000 ft., Colo., 13 July 1896.) = var. *brevifolia.*
F. saximontana Rydb. Bull. Torrey Bot. Club 36: 536. 1909. *F. ovina* subsp. *saximontana* St.-Yves, Candollea 2: 245. 1925. *F. ovina* subsp. *saximontana* var. *rydbergii* St.-Yves, ibid. *F. ovina* f. *saximontana* Carpenter in Dole, Fl. Vermont, ed. 3. 41. 1937. *F. brachyphylla* subsp. *saximontana* Hultén, Fl. Alaska & Yukon 1: 242. 1942. *F. ovina* var. *saximontana* Gleason, Phytologia 4: 21. 1952. (*MacCalla 2331,* vicinity of Banff, alt. 4600 ft., Alta., 28 July 1899.) Awns very short.
F. ovina subsp. *saximontana* var. *purpusiana* St.-Yves, Candollea 2: 247. 1925. (*Purpus 3076 & 5117,* Farewell Gap, 11,000 ft. elev., Calif., in 1894.) = var. *brevifolia.*
F. brevifolia var. *endotera* St.-Yves, Candollea 2: 254. 1925. *F. brachyphylla* var. *endotera* Litard. Candollea 10: 108. 1945. (Numerous collections cited from w. U.S. & Can.) = var. *brevifolia.*
F. brevifolia var. *utahensis* St.-Yves, Candollea 2: 257, fig. 17. 1925. *F. brachyphylla* var. *utahensis* Litard. Candollea 10: 108. 1945. (*M. E. D.* [possibly *M. E. Jones*], Alta, Wasatch Mts., alt. 5000 ft., Utah, 12 Aug. 1895; *Baker 175,* Pagosa Pk., Colo.) = var. *brevifolia.*

Sheep fescue, alpine fescue.

Short cespitose perennials with dense tufts of basal leaves; culms 0.8–3.5 (5) dm tall; leaves forming dense basal tufts, 5–15 cm high, generally less than half the length of the culms, glabrous to scabrous, the sheaths remaining firm and entire; ligules scarcely 0.4 mm long, ciliolate, higher on the sides than in the center; blades stiff, filiform-involute and without auricles; panicles 1.5–7 (10) cm long, mostly strongly compressed (somewhat open in var. *ovina*); spikelets (4) 5–7 (9) mm long, 2- to 4- (5)-flowered; glumes narrowly to broadly lanceolate, the first glume 2–3 (3.5) mm long, 1-nerved, the second

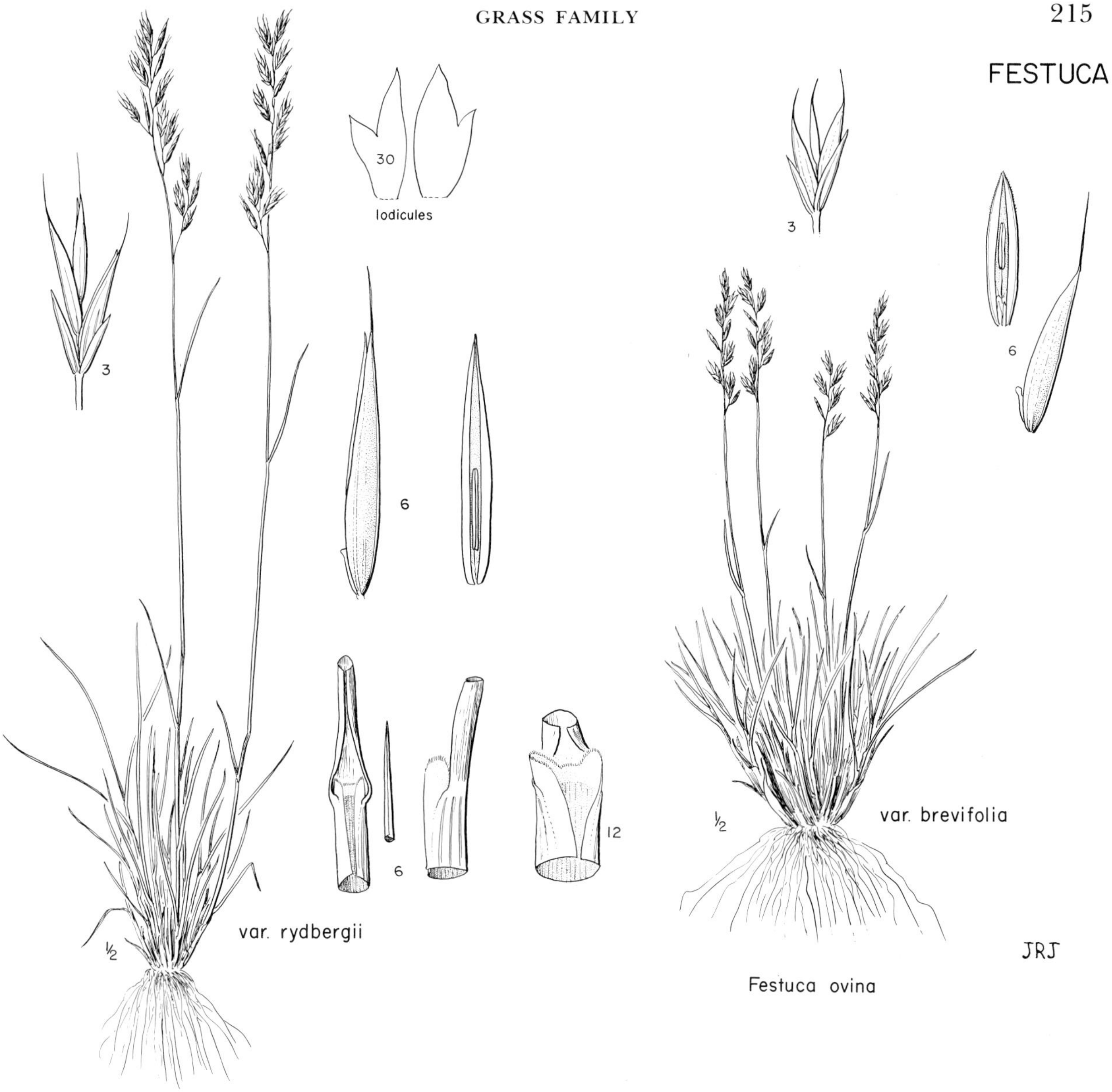

3–4.5 mm long, 3-nerved; lemmas 3.5–5.5 (6) mm long, glabrous to scaberulous, especially distally, the awns 1–3 (5) mm long; anthers 0.8–2 mm long; $2n$ = 14, 21, 28, 35, 36, 42, 44, 49, 56, 70.

Moderately moist ground from the sagebrush zone to the alpine tundra zone, usually in rocky open areas; arctic N. Amer., s. in the mts. to Calif., n. N.M. and in the c. and e. states s. to Kansas and N.C. July–Aug.

Festuca ovina is a highly polymorphic species in which 3 varieties can be distinguished. Variety *ovina* is native to Europe and is often found in lawn plantings. It rarely escapes in our area, however. The var. *brevifolia* looks as if it is phenotypically dwarfed by its subalpine environment, but it remains distinct when grown in a garden at lower elevations. The following key will separate these varieties:

1 Panicles strongly congested; plants native.
  2 Culms mostly over 2.5 dm tall, 2 to 3 times the height of the basal leaves; anthers usually more than 1 mm long; foothills and subalpine slopes .... var. *rydbergii* St.-Yves
  2 Culms short, 0.5–2.5 dm tall, usually less than twice the height of the basal leaves; anthers less than 1 mm long; alpine and subalpine habitats .. var. *brevifolia* (R. Br.) S. Wats.
1 Panicle more open and spreading; introduced from Europe, rare .... var. *ovina*

4. LOLIUM L. Ryegrass

Annuals or short-lived perennials; culms hollow; sheaths open; ligules short membranous; blades slender, flat or folded, with well developed auricles; inflorescence an elongate spike, the spikelets sessile and solitary at each node, appressed edgewise with the backs of the lemmas to the rachis; spikelets many-flowered, laterally compressed, disarticulating above the glumes and between the florets; glumes solitary, the first glume absent (except in the terminal spikelet), present only on the outer side, large, equaling or exceeding the first floret, stiff, 5- to 9-nerved, awnless; lemmas 5-nerved, rounded on back, hyaline above, obtuse to acute, awnless or awned; palea large, firm, usually as long as the lemma, the 2 keels ciliate; lodicules 2, mostly 1–1.5 mm long, glabrous, entire or unevenly bilobed; stamens 3; stigmas plumose, sessile; caryopsis fused to the palea, oblong, subconvex on the inner side, glabrous; $x = 7$.

A genus of about 10 species, native to temp. regions of Europe and Asia. (An old Latin name for darnel.)

Lolium perenne and *L. multiflorum* are among the most important forage grasses in Europe. They have been introduced and cultivated in the U.S. to a limited extent only. They are planted most often in the South where used for winter forage.

Reference: Terrell, E. E. Taxonomic implications of genetics in ryegrasses (*Lolium*). Bot. Rev. 32: 138–164. 1966.

1 Lemmas awnless; leaf blades relatively narrow, 1–3 (4) mm broad; perennials with numerous basal innovations .... 1. *L. perenne*

1 Lemmas awned, the awns 1–8 mm long; leaf blades (3) 3.5–10 mm broad; annuals or short-lived perennials, usually without innovations at the base .... 2. *L. multiflorum*

1. Lolium perenne L.

Lolium perenne L. Sp. Pl. 83. 1753. (Europe.)

Perennial ryegrass, English ryegrass.

Perennials with numerous basal innovations; culms ascending or erect, (3) 4–9 dm tall, perfectly glabrous and smooth; lower leaves forming tufts up to 1.5–3 dm high; sheaths smooth; ligules entire, 0.5–1.5 mm long; auricles usually well developed on most leaves, up to 1.5 (2) mm long; blades narrowly-linear, 1–3 (4) mm broad, glabrous and smooth, sometimes scaberulous on the inside; spikes erect, (8) 10–20 (25) cm long, narrow; spikelets (8) 9.5–15 (20) mm long, (3-) 5- to 7- (11)-flowered; glume solitary, (4) 6–10 (12) mm long, nearly as long as the spikelet, linear-lanceolate, ovate to rounded, 7- to 9-nerved, 3 or 5 of which are more prominent; lemmas (4) 5–7 (8) mm long, acute, awnless; palea about as long as the lemma; anthers 3–4 mm long; $2n = 14, 28$.

Moderately moist roadsides and waste places; introduced from Europe; widely distributed across s. Can. and throughout the U.S. June–Aug.

Perennial ryegrass has the distinction of probably being the first species to be cultivated as a pasture grass in Europe (Gould, Grass Systematics, p. 151. 1968).

2. Lolium multiflorum Lam.

Lolium multiflorum Lam. Fl. Franç. 3: 621. 1778. *L. perenne* var. *multiflorum* Parnell, Grasses Brit. 302, pl. 140. 1845. *L. temulentum* var. *multiflorum* Kuntze, Revisio Gen. Pl. 2: 779. 1891. *L. perenne* subsp. *multiflorum* Husnot, Gram. France, Belg., Iles Brit., Suisse 4: 85. 1899. (France.)

L. italicum A. Braun, Flora 17: 243. 1834. *L. perenne* var. *italicum* Parnell, Grasses Scotl. 1(1): 142, pl. 65. 1842; Scribn. Mem. Torrey Bot. Club 5: 57. 1894. *L. multiflorum* var. *italicum* Beck, Wiss. Mitt. Bosner Herzegowina 9: 459. 1904. *L. multiflorum* subsp. *italicum* Schinz & Keller, Fl. Schweiz, ed. 2. 1: 65. 1905. (Europe.)

Italian ryegrass.

Annuals or short-lived perennials, usually without basal innovations; culms ascending or erect, 5–10 dm tall, smooth to scabrous below; ligules 0.5–1.5 mm long; auricles usually well developed on most of the leaves, up to 1.5 (2) mm long; blades flat or slightly involute, (3) 3.5–7 (10) mm broad, scabrous on the upper surface; spikes erect, (10) 14–27 (30) cm long; spikelets 9–15 (20) mm long, 6- to 10- (15)-flowered; glume solitary, 4–9 (12) mm long, usually about half as long as the spikelet, narrow-lanceolate, obtuse to rounded; lemmas 5–6.5 (7.5) mm long, lanceolate, 5-nerved, at least upper florets with awns, the awns 1–5 (8) mm long, from a minutely bifid apex; palea about as long as the lemma; anthers 3–4 mm long; $2n = 14$.

Moderately moist waste places; introduced from Europe; B.C. to Me., s. throughout the U.S. June–July.

Italian ryegrass was probably bred from *Lolium perenne* stock by early Europeans as an agricultural strain. Some botanists consider it as a variety of *L. perenne*. Since *L. multiflorum* is easily distinguished from *L. perenne* by its awned lemmas and broader leaf blades, we treat it at the species level.

5. LEUCOPOA Griseb.

Incompletely dioecious, rhizomatous perennials; culms hollow; sheaths open; ligules membranous; blades flat, coarsely striate, non-auriculate; inflorescence an open to congested panicle; spikelets 3- to 5-flowered, laterally compressed, disarticulating above the glumes and between the florets; staminate florets with a reduced pistil and pistillate florets with rudiments of the stamens; glumes broadly lanceolate to ovate, somewhat keeled, semi-transparent, membranous, light-colored, shorter than the first lemma, the first glume 1-nerved, the second 3-nerved, the lateral nerves sometimes obscure; lemmas membranous, rounded-keeled, 5 inconspicuous converging nerves, entire or irregularly serrate at the acute or attenuate apex, awnless; palea about equalling the lemma, the 2 keels scabrous-ciliate; lodicules 2, 1–1.5 mm long, ovate, often with serrate margins; staminate florets with 3 exserted stamens; pistillate florets with long, plumose stigmas; caryopsis bidentate; $x = 7$.

A genus of 5 or 6 Asiatic species with a single w. Amer. representative. (From the Greek *leucos*, white, and *poa*, bluegrass.)

Our species has been variously treated as a member of *Festuca* or as a monotypic genus, *Hesperochloa kingii* (S. Wats.) Rydb. Weber has shown that it belongs to the small Asiatic genus *Leucopoa*, a group of fescue-like dioecious grasses. Our species most closely resembles *L. caucasica* (Hackel) Krecz. & Bobrov and *L. karatavica* (Bunge) Krecz. & Bobrov of Eurasia.

Reference: Weber, W. A. Additions to the flora of Colorado. IV. Univ. Colorado Stud. Ser. Biol. 23: 2. 1966.

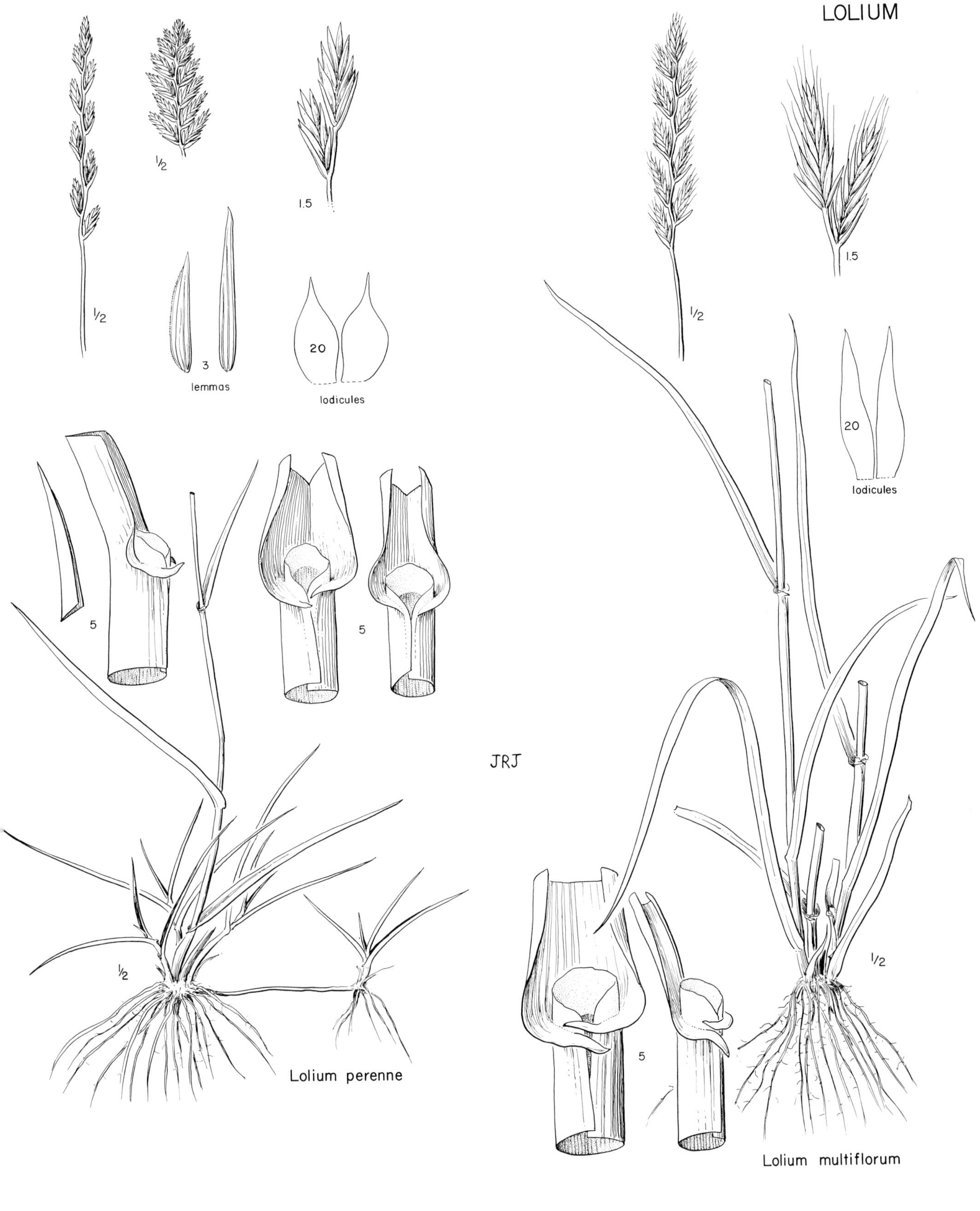
LOLIUM
1/2
1/2
1.5
3
lemmas
20
lodicules
5
5
JRJ
1/2
Lolium perenne
1/2
1.5
1/2
20
lodicules
1/2
5
Lolium multiflorum

1. Leucopoa kingii (S. Wats.) W. A. Weber

Poa kingii S. Wats. in King, Rep. Geol. Explor. 40th Parallel 5: 387. 1871. *Festuca kingii* Cassidy, State Agric. Coll. Colorado Bull. 12: 36. 1890; Scribn. U.S.D.A. Div. Agrostol. Bull. 5: 36. 1897. *Festuca watsoni* Nash in Britton, Manual Fl. N. States & Canad. 148. 1901. *Hesperochloa kingii* Rydb. Bull. Torrey Bot. Club 39: 106. 1912. *Wasatchia kingii* M. E. Jones, Contr. W. Bot. 14: 16. 1912. *Leucopoa kingii* W. A. Weber, Univ. Colorado Stud. Ser. Biol. 23: 2. 1966. (*Watson 1317,* East Humboldt [Ruby] Mts., Elko Co., Nev., July 1865.)

Festuca confinis Vasey, Bull. Torrey Bot. Club 11: 126. 1884. (*Vasey,* Pen Gulch, Colo., in 1884.)

Spikegrass, spike fescue.

Strongly rhizomatous, coarse perennials, often growing in ring-like tufts up to 2 m in diam; culms stout, 3–10 dm tall, arising from dense clumps of straw-colored remnants of old sheaths and culms; sheaths smooth, striate; ligules scarious, 1–3.5 (7) mm long, erose-ciliolate; blades firm, flat, 2–3.5 (9) mm broad and 20–40 (50) cm long, erect, glaucous, coarsely striate; panicles congested but occasionally open, 3–15 (20) cm long, spikelet bearing nearly to the base; spikelets 6–10 (12) mm long, 3- (5)-flowered, the staminate spikelets somewhat longer than the pistillate; glumes broadly lanceolate, subscarious to smooth and shining, the first 3.5–5.5 mm long, 1-nerved, the second 4–6.5 mm long, 3-nerved; lemmas 4.5–8 mm long, ovate, acute or acuminate, sometimes slightly awned, uniformly scabrous, faintly 5-nerved; palea about as long as the lemma, ciliate on the keels; anthers 2.5–4.5 mm long; stigmas long and bushy appearing; caryopsis hispidulous at the apex; $2n = 56$.

Rocky or gravelly mt. sides and open ridges from the foothills to broad mt. tops up to 3353 m (11,000 ft.) elev.; Mont. and Idaho, s. through Utah, se. Oregon (Steens Mts.), Nev. and the Basin Ranges of adjacent Calif.; also in the San Bernardino Mts. of s. Calif. and in the Black Hills of S.D. June–Aug.

Leucopoa kingii is an important and dominant grass species on the high, flat-topped mountains such as the Monitor Range in central Nevada. Rhizomes are rarely present on herbarium specimens, but they are nearly always present in large ring-like growth patterns so common in loose soils.

6. SCOLOCHLOA Link, nom. conserv.

Stout rhizomatous perennials; culms erect, hollow; sheaths open; ligules membranous; blades flat and broad to slightly involute, scabrous, non-auriculate; inflorescence an open panicle; spikelets closely 3- to 4-flowered, disarticulating above the glumes and between the florets; glumes lanceolate, rounded on the back, membranous-scarious and often lacerate at the tip, sometimes slightly awned, large, unequal, the first glume 1- to 3-nerved, the second slightly longer than the first lemma, 3- to 5-nerved; lemmas firm, rounded on back, with a tuft of hairs on the callus, the hairs 1–1.5 mm long, with 5 fairly evident nerves extending into a scarious-lacerate apex, unawned; palea narrow, about as long as the lemma; lodicules 2, lobed or entire; stamens 3; caryopsis oblong with a short, linear hilum, villous on the tip; $x = 7$.

A genus of 2 species, 1 in e. Siberia and the other in n. Eurasia and n. N. Amer. (From the Greek *scolops,* a prickle, and *chloa,* grass.)

1. Scolochloa festucacea (Willd.) Link

Arundo festucacea Willd. Enum. Pl. Hort. Regii Bot. Berol. 1: 126. 1809. *Scolochloa festucacea* Link, Hort. Regius Bot. Berol. 1: 137. 1827. *Triodia festucacea* Roth, Enum. Pl. Phaenog. German. 1(1): 382. 1827. *Graphephorum festucaceum* A. Gray, Proc. Amer. Acad. Arts 5: 191. 1861. *Fluminea festucacea* A. S. Hitchc. U.S.D.A. Bull. (1915–23) 772: 38, fig. 11. 1920. (Germany.)

Fluminea.

Stout perennials with thick rhizomes; culms 10–15 dm tall; ligules 3.5–7 (9) mm long, lacerate; blades flat, 5–10 mm broad and 2–4.5 cm long, firm, elongate and narrowed gradually to a slender tip; panicles 15–30 cm long, the branches appressed; spikelets 7.5–10 mm long; glumes lanceolate, proportionately large, the first glume 5–7.5 mm long, the second 6.5–9 mm long; lemmas (4.5) 5–8 mm long, more or less evidently (5-) 9-nerved; anthers 2–4 mm long; $2n = 28, 42$.

Marshes and shallow water of lake shores; e. Oregon, se. Idaho (Dingle Swamp at mouth of Bloomington Cr., Bear Lake Co.) and the n. Great Plains from Nebr. and Iowa, n. into Can. June–Aug.

7. SCLEROCHLOA Beauv. Hardgrass

Low tufted annuals, less than 2 dm tall; culms hollow; sheaths compressed, partially closed; ligules membranous; blades more or less boat-shaped at the tip, non-auriculate; inflorescence a 1-sided spicate-raceme, the spikelets subsessile, sometimes a few branches below, often partially enclosed in the upper leaf sheath; spikelets 3- (5)-flowered, disarticulating below the glumes, falling entire; glumes broad and firm, blunt-tipped, the first 1- to 3-nerved, the second 7-nerved, hyaline margined; lemmas broad, rounded on back, hyaline margined above with an obtuse apex, prominently 5-nerved, awnless; palea smaller and shorter than the lemma, hyaline, the 2 keels ciliate; lodicules 2, entire; stamens 3; ovary glabrous, the style terminal; $x = 7$.

A monotypic genus represented by *Sclerochloa dura* (L.) Beauv. which is native to s. Europe and adventive in a few scattered localities in the U.S. (Name from the Greek *skleros,* hard, and *chloa,* grass, in reference to the firm glumes.)

1. Sclerochloa dura (L.) Beauv.

Cynosurus durus L. Sp. Pl. 72. 1753. *Poa dura* Scop. Fl. Carniolica, ed. 2. 1: 70. 1772. *Eleusine dura* Lam. Tableau Encycl. & Méth. 1: 203. 1791. *Sclerochloa dura* Beauv. Essai Nouv. Agrostogr. 98, 174, 177. 1812. (s. Europe.)

Crassipes annuus Swallen, Amer. J. Bot. 18: 685, figs. 1–4. 1931. (*Fallas,* between Salt Lake City and Ogden, foot of Wasatch Mts., Utah, 26 April 1928.) Described under the misapprehension that it was a native American species.

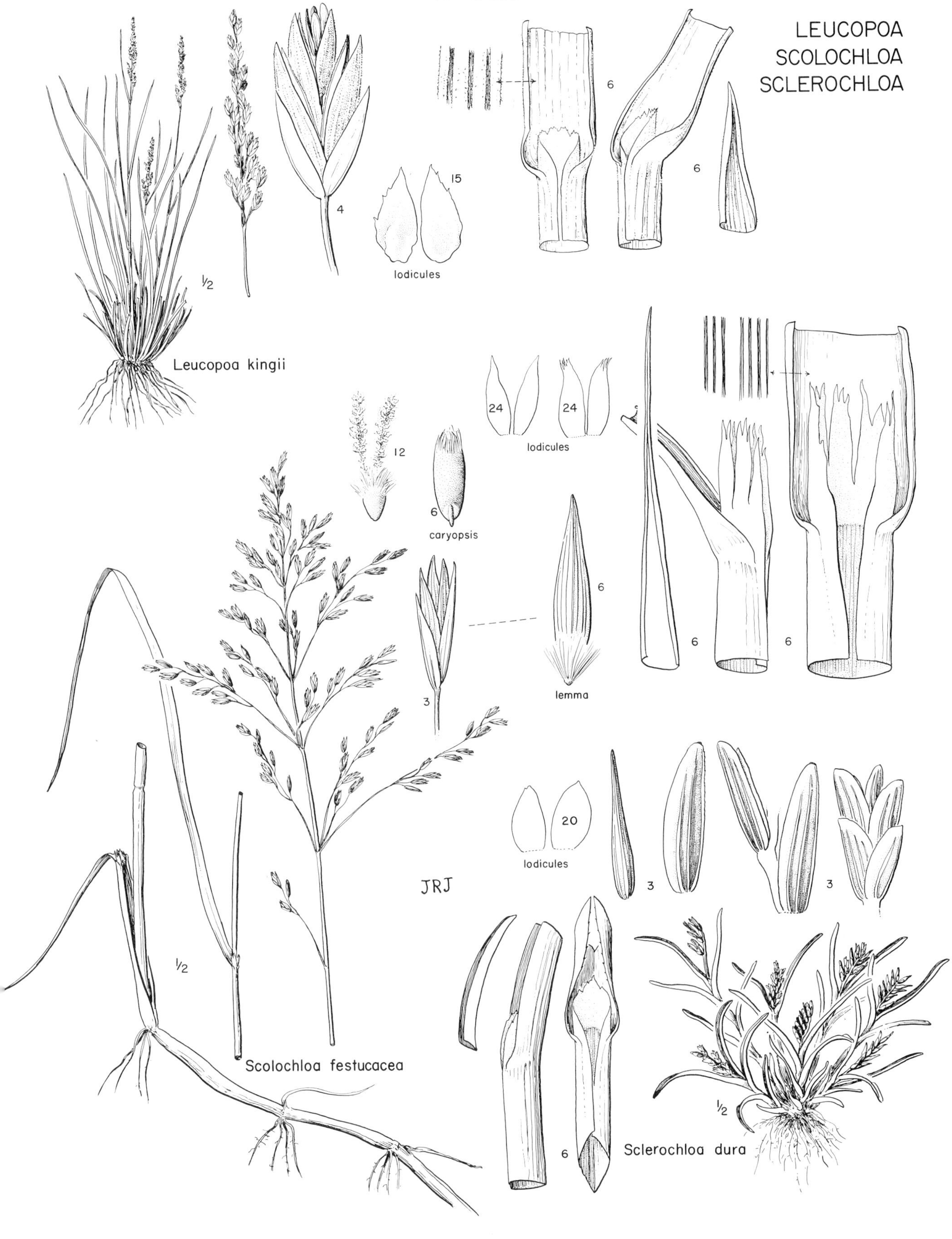
LEUCOPOA
SCOLOCHLOA
SCLEROCHLOA
Leucopoa kingii
1/2
4
15
lodicules
6
6
lodicules
24
24
12
6
caryopsis
lemma
3
6
6
6
Scolochloa festucacea
1/2
JRJ
20
lodicules
3
3
6
Sclerochloa dura
1/2

Hardgrass.

Small, glabrous annuals; culms 0.3–1.2 (2) dm tall, geniculate-ascending; ligules 0.5–1.5 mm long, lacerate; blades small, 2–6 cm long, linear-lanceolate; racemes 1.8–2.5 (5) cm long, dense, usually exceeded by the upper leaves; spikelets (6) 7–9 (11) mm long, 3- (5)-flowered; glumes ovate, hyaline margined, the first glume 2–3.5 mm long, the second 4–7.5 mm long; lemmas 3–7 mm long, coriaceous, ovate, awnless; anthers 1–1.5 mm long; $2n$ = 14.

Moist weedy habitats such as ditch banks and roadsides; native of s. Europe; Wash., n. Calif., Idaho, Utah and Colo.; also in Texas and N.Y. May–early July.

8. PUCCINELLIA Parl. nom. conserv. Alkali-grass

Annuals or perennials, sometimes rhizomatous; culms decumbent, hollow; sheaths free, but overlapping; ligules membranous; blades short, involute; inflorescence a narrow to open panicle; spikelets small, 3- to 9-flowered, the florets imbricate, 2-ranked, disarticulating above the glumes and between the florets; glumes firm, rounded on back, usually shorter than the lemmas, unequal, the first glume shorter, strongly 1-nerved, the second obscurely 3-nerved; lemmas firm, rounded on back, obscurely or evidently (3-) 5- (7)-nerved, the nerves not converging at the apex, glabrous or puberulent towards the base and on the margins, truncate, obtuse to acute, mostly hyaline on the margins, the apex often finely erose-ciliolate, awnless; palea about as long as the lemma, emarginate, the 2 keels ciliate; lodicules 2, free, usually about 0.5 mm long or less, hyaline; stamens 3; stigmas sessile, plumose; caryopsis ovoid to oblong, not furrowed, falling together with the lemma and palea; x = 7.

A genus of about 30 species, mostly of wetland habitats in temp. and arctic regions of N. Amer., Europe and Asia. There are about 11 species in the U.S. (Commemorating Benedetto Puccinelli, 1808–1850, an Italian botanist.)

Species of *Puccinellia,* especially *P. nuttalliana,* are excellent forage grasses in wet marshy areas.

Five of our six species are of alkaline sites and the sixth, *P. pauciflora* (C. Presl) Munz, of fresh water. *Puccinellia pauciflora* and its closely related species (from outside our region) were once considered members of the genus *Glyceria.* Church proposed a new genus, *Torreyochloa,* to accommodate them. More recently they have been transferred to *Puccinellia,* with which they agree in chromosome number, n = 7, similar unjoined sheaths and similar-shaped lodicules. *Glyceria* has a basic chromosome number of n = 10, fused sheaths and much different lodicules.

References:

Church, G. L. A cytotaxonomic study of *Glyceria* and *Puccinellia.* Amer. J. Bot. 36: 155–165. 1949.
Clausen, R. T. Suggestion for the assignment of *Torreyochloa* to *Puccinellia.* Rhodora 54: 42–45. 1952.
Fernald, M. L., and C. A. Weatherby. The genus *Puccinellia* in eastern North America. Rhodora 18: 1–23. 1916.

1 Plants diminutive annuals .... 6. *P. simplex*
1 Plants perennials.
 2 Lemmas with conspicuous nerves; plants with creeping rhizomes; leaves flat and broad, (3) 4–15 mm wide; fresh water habitats (*Torreyochloa*) .... 1. *P. pauciflora*
 2 Lemmas with obscure or indistinct nerves; plants tufted, not having creeping rhizomes; leaves usually involute, if flat then narrow and 0.3–3 (5) mm wide; usually alkaline or saline habitats (*Puccinellia*).
  3 Blades usually over 1 mm broad (at least more than 0.5 mm), more or less distributed along the stem.
   4 Panicles open, pyramidal or elongate, the lowest branches all or chiefly naked below the middle.
    5 Lower panicle branches erect or spreading, the panicles 10–25 (30) cm long; lemmas (1.8) 2–3 mm long, narrowed into an obtuse apex; anthers 0.7–1.2 mm long; native species .... 2. *P. nuttalliana*
    5 Lower panicle branches usually reflexed, the panicles 5.5–12 (15) cm long; lemmas 1.4–2.2 mm long, broadly rounded to truncate, erose and hyaline at the apex; anthers 0.5–0.8 mm long; introduced species .... 3. *P. distans*
   4 Panicles contracted, ellipsoid or ovoid, the lowest branches bearing spikelets nearly to the base; introduced species .... 4. *P. fasciculata*
  3 Blades filiform (scarcely 0.5 mm broad in their involuted condition), mostly in a short basal tuft .... 5. *P. lemmonii*

1. Puccinellia pauciflora (C. Presl) Munz

Glyceria pauciflora C. Presl, Reliquiae Haenkeanae 1: 257. 1830. *Panicularia pauciflora* Kuntze, Revisio Gen. Pl. 2: 783. 1891. *Torreyochloa pauciflora* Church, Amer. J. Bot. 36: 163. 1949. *Puccinellia pauciflora* Munz, Aliso 4: 87. 1958. (*Haenke,* Nootka Sound, Vancouver I., B.C.)

Glyceria microtheca Buckley, Proc. Acad. Nat. Sci. Philadelphia 1862: 96. 1863. *Puccinellia pauciflora* var. *microtheca* C. L. Hitchc. Univ. Wash. Publ. Biol. 17(1): 689. 1969. (*Nuttall,* Columbia River, Oregon.)

Panicularia holmii Beal, Torreya 1: 43. 1901. *Puccinellia pauciflora* var. *holmii* C. L. Hitchc. Univ. Wash. Publ. Biol. 17(1): 689. 1969. (*Holm 249,* near Lamb's Ranch, Long's Peak, Colo., 8 July 1899.)

Panicularia multifolia Elmer, Bot. Gaz. 36: 54. 1903. (*Elmer 1939,* Olympic Mts., Clallam Co., Wash., Aug. 1900.)

Weak mannagrass.

Strongly rhizomatous perennials; herbage glabrous to slightly scabridulous; culms often decumbent, 2.5–14 dm tall; sheaths cross-septate; ligules

PUCCINELLIA
15
6
1/2
5
JRJ
1/2
Puccinellia pauciflora
6
15
lodicules
6
6
1/2
6
Puccinellia nuttalliana

3–9 mm long, acute, erose-lacerate; blades flat and broad, (3) 4–10 (15) mm wide, sometimes scabridulous only on the upper surface; panicles 8–20 (28) cm long, loose, the branches mostly 2 or 3 per node, ascending, spreading or drooping; spikelets 2.5–5.5 mm long, 3- to 7-flowered; glumes broadly ovate, membranous, erose-ciliolate, 1–2 mm long, the first glume only slightly smaller than the second; lemmas 1.5–4 mm long, strongly 5- (7)-nerved, scabrid-puberulent, especially on the nerves, usually with a purplish band below the membranous, erose-ciliolate apex; anthers 0.5–0.7 mm long; $2n = 14$.

Marshes and streambanks, usually in fresh water; canyons to subalpine meadows and bogs; w. cordillera from Alaska, B.C. and Alta., s. to Calif. and n. N.M. June–Aug.

2. Puccinellia nuttalliana (J. A. Schultes) A. S. Hitchc.

Poa airoides Nutt. Gen. N. Amer. Pls. 1: 68. 1818; not Koeler, 1802. *Poa nuttalliana* J. A. Schultes, Mantissa 2: 303. 1824. *Festuca nuttalliana* Kunth, Révision Gram. 1: 129. 1829. *Glyceria airoides* Fries, Novit. Fl. Suec. Mantissa 3: Add. 176. 1832; not Reichenb. 1829. *Puccinellia airoides* Wats. & Coulter, Gray's Manual Bot. ed. 6. 668. 1889. *Panicularia distans* (var.) *airoides* Scribn. Mem. Torrey Bot. Club 5: 54. 1894. *Atropis airoides* T. H. Holm, Bot. Gaz. 46: 427. 1908. *Puccinellia nuttalliana* A. S. Hitchc. in Jepson, Fl. Calif. 1: 162. 1912. *Atropis nuttalliana* Pilger, Notizbl. Bot. Gart. Berlin-Dahlem 9: 291. 1925. (*Nuttall,* Mandan, N.D.)

Glyceria montana Buckley, Proc. Acad. Nat. Sci. Philadelphia 1862: 96. 1863. (*Nuttall,* Rocky Mts.)

Nuttall's alkali-grass.

Cespitose perennials, without rhizomes; culms erect, (2) 3.5–7 dm tall; ligules obtuse, 1–3 mm long; blades (0.6) 1–3 mm broad, often involuted; panicles 10–25 (30) cm long, open, pyramidal or elongate at maturity, the branches widely divergent, rarely reflexed, the lowest bearing spikelets chiefly above the middle; spikelets slender, 4–8 mm long, (2-) 3- to 7-flowered; glumes narrowly ovate, the first glume 0.6–1.5 mm long, the second 1–2.4 mm long; lemmas (1.8) 2–3 mm long, broadly oblong, narrowed into an obtuse apex, erose-ciliate, obscurely nerved; palea about as long as the lemma; anthers 0.7–1.2 mm long; $2n = 42, 56$.

Moist, usually alkaline soils, sometimes in standing water; Alaska to Greenl., s. to Calif., n. half of Nev. and Utah, e. across Colo. and Kansas to Mich. and adventive to New England. Late May–early Aug.

3. Puccinellia distans (L.) Parl.

Poa distans L. Mantissa Pl. 1: 32. 1767. *Aira aquatica* var. *distans* Huds. Fl. Anglica, ed. 2. 34. 1778. *Hydrochloa distans* Hartm. Gen. Gram. Scand. Indigenorum Recognita 8. 1819. *Glyceria distans* Wahlenb. Fl. Upsal. 36. 1820. *Festuca distans* Kunth, Révision Gram. 1: 129. 1829. *Sclerochloa distans* Bab. Manual Brit. Bot. 370. 1843. *Catabrosa distans* Link ex Heynh. Nomencl. Bot. Hort. 2: 126. 1846. *Puccinellia distans* Parl. Fl. Ital. 1: 367. 1848. *Atropis distans* Griseb. in Ledeb. Fl. Ross. 4: 388. 1853. *Sclerochloa multiculmis* subsp. *distans* Syme, Engl. Bot. ed. 3. 11: 104. 1873. *Panicularia distans* Kuntze, Revisio Gen. Pl. 2: 782. 1891. *Phippsia distans* A. Löve & D. Löve, Folia Geobot. & Phytotax. 10: 274. 1975. (Europe.)

Glyceria distans var. *tenuis* Uechtr. in Crépin, Notes Pl. Rares Crit. Belg. 229. 1865. *Atropis distans* var. *tenuis* Rouy, Fl. France 14: 195. 1913. *Puccinellia distans* var. *tenuis* Fern. & Weatherby, Rhodora 18: 12. 1916. (Germany.) An involute leaved, small panicled form.

Puccinellia suksdorfii St. John, Northw. Sci. 2: 80. 1928. (*Suksdorf 5089,* border of a pond near Rockland, Klickitat Co., Wash., 2 June 1905.)

European alkali-grass.

Cespitose perennials, without rhizomes; herbage glabrous to scaberulose; culms 1.5–3.5 (5) dm tall, decumbent at the base; ligules about 1 mm long, entire and obtuse; blades flat, becoming involuted, (0.5) 1–3 (4) mm broad; panicles 5.5–12 (15) cm long, open, ovoid, loosely branched, the lower branches often reflexed and bearing spikelets only above the middle; spikelets (2.5) 3–5 mm long, 2- to 5- (6)-flowered; glumes ovate, erose at the apex, the first glume 0.5–1.2 mm long, the second 1.1–1.8 mm long; lemmas 1.4–2.2 mm long, broadly ovate, faintly nerved, finely erose-ciliate, green or purplish below, the apex broadly rounded to truncate, hyaline; palea about as long or longer than the lemma; anthers 0.5–0.8 mm long; $2n = 14, 28, 42$.

Moist, more or less alkaline habitats; adventive and becoming naturalized in n. N. Amer.; Alaska to Newfl., s. in the w. to Calif., Nev., n. Ariz., N.M. and in the c. states to Nebr. and Wisc. and in the e. to N.J. June–Aug.

4. Puccinellia fasciculata (Torr.) E. P. Bicknell

Poa fasciculata Torr. Fl. N. & Middle Sect. U. S. 1: 107. 1823. *Sclerochloa arenaria* var. *fasciculata* A. Gray, Manual Bot. 594. 1848. *Puccinellia fasciculata* E. P. Bicknell, Bull. Torrey Bot. Club 35: 197. 1908. *Phippsia fasciculata* A. Löve & D. Löve, Folia Geobot. & Phytotax. 10: 274. 1975. (*Torrey,* salt marshes near N.Y.)

Festuca borreri Bab. Trans. Linn. Soc. London 17: 565. 1837. *Glyceria borreri* Bab. in J. E. Smith, Suppl. Engl. Bot. 3: pl. 2797. 1843. *Sclerochloa borreri* Bab. Manual Brit. Bot. 370. 1843. *Poa borreri* Parnell, Grasses Brit. 220, pl. 98. 1845. *Sclerochloa multiculmis* subsp. *borreri* Syme, Engl. Bot. ed. 3. 11: 105. 1873. *Atropis borreri* K. Richter, Pl. Eur. 1: 92. 1890. *Puccinellia borreri* A. S. Hitchc. in B. L. Robinson, Rhodora 10: 65. 1908. (England.)

Cespitose perennials, without rhizomes; culms 2–6 dm tall, often geniculate below; blades 2–5 mm broad (0.5–1 when involuted); panicles 4–15 cm long, contracted, narrowly ellipsoid or ovoid, the lowest branches bearing spikelets nearly to the base; spikelets 3.5–6 mm long, 2- to 5-flowered; glumes ovate to broadly ovate, the first glume 0.8–1.5 mm long, 1-nerved, the second 1.2–2.2 mm long, 3-nerved; lemmas 1.9–2.5 (3) mm long, firm, ovate-oblong, obscurely 3- to 5-nerved; anthers 0.7–0.8 mm long; $2n = 28$.

Alkaline marshes in our area and salt marshes and sandy shores on the e. coast; native of Europe; adventive along the Atlantic seaboard from N.S. to Del. and in the w. in the Bonneville Basin along the Wasatch front and in the Lahontan Basin of w. Nev. May–July.

5. Puccinellia lemmonii (Vasey) Scribn.

Poa lemmoni Vasey, Bot. Gaz. 3: 13. 1878. *Glyceria lemmoni* Vasey, Descript. Cat. Grasses U. S. 88. 1885, nomen; Vasey, Bull. Torrey Bot. Club 13: 119. 1886. *Atropis lemmoni* Vasey, U.S.D.A. Div. Bot. Bull. 13(2): pl. 90. 1893. *Puccinellia lemmoni* Scribn. U.S.D.A. Div. Agrostol. Bull. 17: 276, fig. 572. 1899. (*Lemmon,* Sierra Co., Calif.)

Puccinellia rubida Elmer, Bot. Gaz. 36: 56. 1903. (*Cusick 2621,* moist alkaline meadow at Cold Spring, on the Burns-Prineville road, Crook Co., Oregon, 29 June 1901.)

PUCCINELLIA

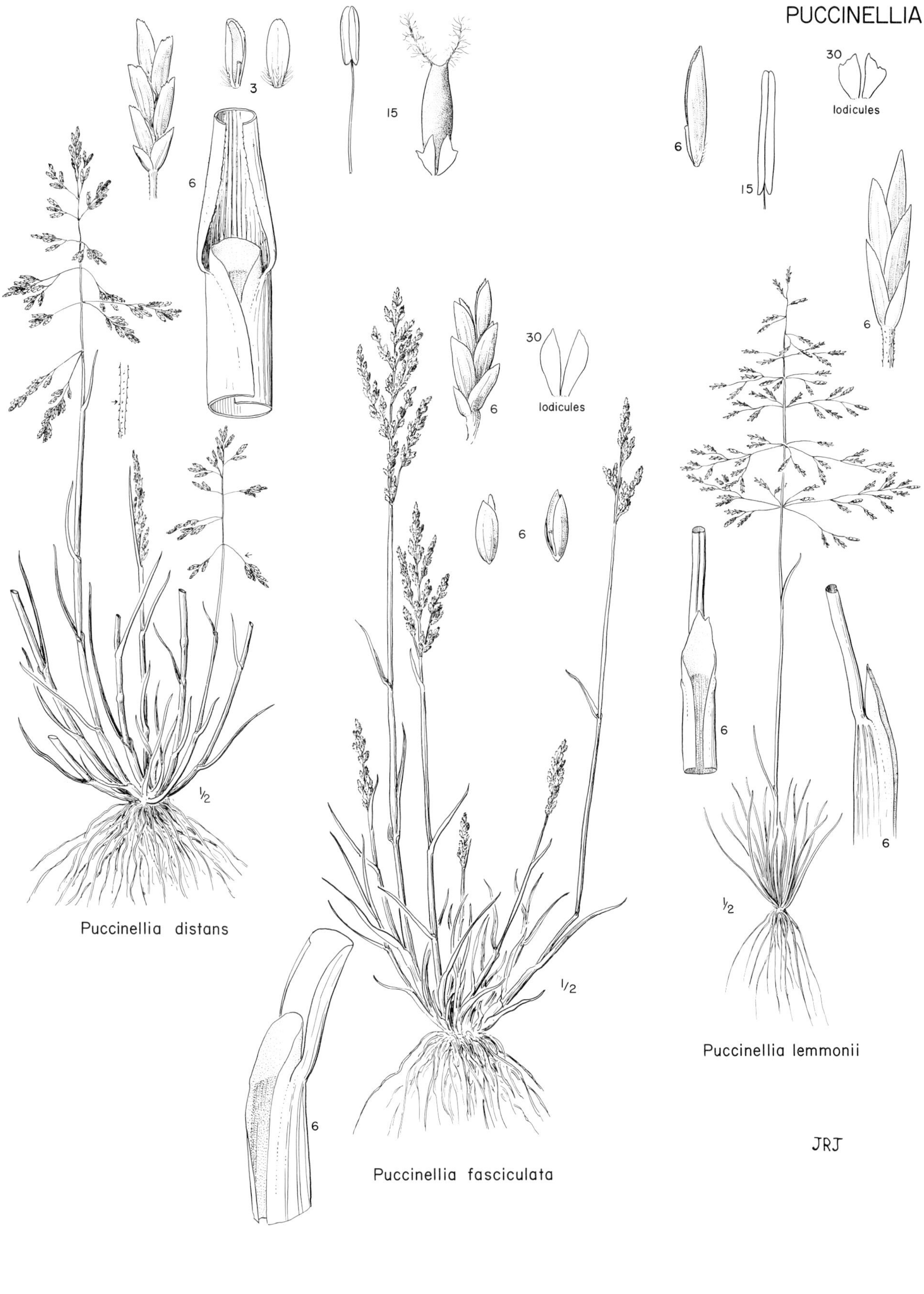

Puccinellia distans

Puccinellia fasciculata

Puccinellia lemmonii

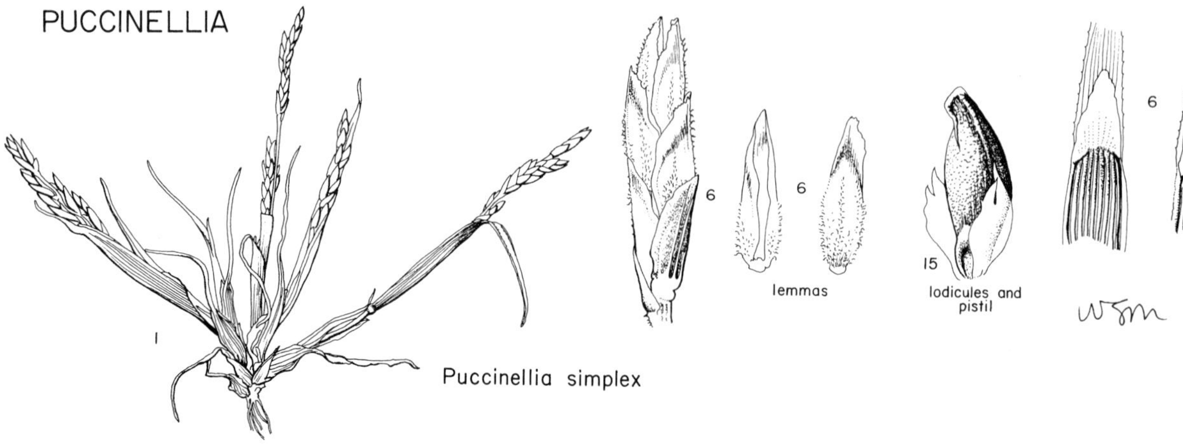

Puccinellia simplex

Lemmon's alkali-grass.

Cespitose perennials, without rhizomes; culms 1.5–4 dm tall, slender; leaves mostly basal, 4–10 cm long; ligules usually 1–2 mm long, erose-lacerate; blades filiform, involuted, mostly less than 0.5 mm broad in their involuted condition, more or less glaucous; panicles 4–10 (14) cm long, pyramidal to rather narrow, the branches slender, flexuous, fascicled, the lower naked in the lower half; spikelets (3) 5–7 mm long, 2- to 4- (5)-flowered; glumes narrow-ovate, the first glume 0.8–1.5 (2) mm long, the second 1.5–2.5 (3) mm long; lemmas 2.2–3.3 mm long, glabrous, acute, rarely obtuse, faintly 3- (5)-nerved; palea about as long as the lemma; anthers 1.2–2 mm long; $2n = 14$.

Valley bottoms in alkaline meadows and flats; mostly in the n. Great Basin; s. Idaho, ne. Nev., se. Oregon and ne. Calif., and n. to se. Wash. June–July.

6. **Puccinellia simplex** Scribn.

Puccinellia simplex Scribn. U.S.D.A. Div. Agrostol. Circ. 16: 1, fig. 1. 1899. (*Blankinship,* Woodland, Yolo Co., Calif.)

Dwarf annuals, 0.3–0.6 (1) dm tall; sheaths broad, loosely investing the culm; ligules acute, about 0.5 mm long, decurrent to the hyaline margins of the sheath; blades flat to subinvolute, less than 1 mm broad, 1–1.5 cm long; panicles narrow, 1–4 cm long, often scarcely extended beyond the broad sheath; spikelets 3–5 mm long, 3- to 6-flowered; glumes scarious on the margins and acute apex, the first glume about 1.5 mm long, 1-nerved, the second 2.5–3 mm long, with 3 prominent nerves; lemmas about 2 mm long, obtuse to truncate, scarious and more or less erose at the summit, pubescent on the lateral nerves nearly to the apex and on the intermediate nerves on the lower third; rachilla usually prolonged behind the palea of the terminal floret; caryopsis about 1.5 mm long; $2n = 56$.

Alkaline seeps and salt marshes; Central Valley and interior valleys of the Coast Ranges, Calif. and salt flats 1 mi. e. of Little Mt., Weber Co., Utah. May–June.

This rare little grass has been collected only once in our area (*Arnow 3986*).

9. **POA** L. Bluegrass

Low or moderately tall annuals or more often cespitose to stoloniferous or rhizomatous perennials; culms hollow; sheaths partially closed; ligules membranous or hyaline; blades flat to folded or sometimes involute, the upper surface with 2 median lines and the tips usually curved and boat-shaped, non-auriculate; inflorescence an open to contracted panicle, rarely reduced to a raceme; spikelets relatively small, (2.5) 3–8 (11) mm long, (1-) 2- to 8-flowered, the florets usually perfect, but often completely to incompletely dioecious plants are found with the pistillate flowers having vestigial stamens, these often about normal size, disarticulating above the glumes and between the florets; glumes relatively broad, acute, keeled or more or less rounded, usually shorter than the first lemma, somewhat unequal, the first usually shorter, 1- (3)-nerved, (1.2) 2–4.5 (6) mm long, the second often 3-nerved, (1.7) 2.2–5 (6.2) mm long; lemmas (1.5) 2.5–6 (7.5) mm long, keeled to more or less rounded, acute or acutish (rarely obtuse), membranous, usually somewhat scarious at the summit and anthocyanous in a band near the middle, 5-nerved, the nerves converging, usually obscure, often with a tuft of cobwebby hairs at the base, the surface glabrous but more often scabrous to pubescent or pilose on the keel and marginal nerves; palea usually slightly shorter than the lemma, the 2 keels often somewhat ciliolate; lodicules 2, 0.4–1 mm long, entire to lobed, acute to truncate; stamens 3; $x = 7$.

A genus of 150 to 200 species, widespread, but mostly of temp. and arctic regions. (*Poa,* an ancient Greek name for grass or fodder.) Bluegrasses are economically important as several species furnish considerable amounts of forage on the range and as hay and pasture grasses. Kentucky bluegrass, *P. pratensis,* is a desirable lawn grass throughout our area. Such species as *P. pratensis, P. fendleriana, P. sandbergii* and *P. nevadensis* make up a major portion of the important livestock forage at low and middle elevations in our region. At higher elevations *P. nervosa, P. leptocoma, P. alpina* and *P. rupicola* are important.

Several species of *Poa* present difficult taxonomic problems. Apomixis, a process whereby seed production takes place without the fusion of gametes, is largely responsible for many of the difficulties, especially in the Scabrellae and Nevadenses groups. *Poa pratensis* is often represented by several chromosome numbers in one small population. Hybridization, which frequently obscures species limits, is known to occur. *Poa fendleriana* and *P. ampla* are among several species that occasionally produce short rhizomes, and careful observations in the area where they are growing will often suggest introgression of a rhizomatous *Poa* into that species. Others, such as *P. palustris* and *P. leptocoma*, may appear to be rhizomatous or stoloniferous when they are merely decumbent at the base and rooting at the nodes. Still other problems are undoubtedly due to the simple fact that botanists have attempted to recognize too many taxa.

The section names that appear in the key below are given more for the sake of convenience in constructing a key than in assembling groups of close genetic affinity. *Poa ampla* of the Nevadenses group often produces apparently fertile hybrids with members of the Pratenses group.

References:
Marsh, V. L. A taxonomic revision of the genus *Poa* of United States and southern Canada. Ph.D. thesis. Univ. of Washington. 1950.
———. A taxonomic revision of the genus *Poa* of the United States and southern Canada. Amer. Midl. Naturalist 47: 202–250. 1952.

1 Plants annuals (some winter annuals), the remains of old culms never present (Annuae).
  2 Lemmas with a tuft of cobwebby hairs at the base; second glume tapered from the base upwards; panicles 4–16 (22) cm long.
    3 Panicles usually open; lemmas glabrous, except for the cobwebby tuft at the base; spikelets (1-) 2- to 3-flowered; w. Nev. and Idaho and the Pacific states .... 1. *P. bolanderi*
    3 Panicles narrow, contracted; lemmas villous on the nerves below, as well as cobwebby at the base; spikelets (3-) 4- to 6-flowered; desert plants of the Southwest, e. to Okla. ..... 2. *P. bigelovii*
  2 Lemmas not webbed at the base, pubescent on the nerves; second glume broadened above the middle; panicles short, 1.5–3.5 (8) cm long .... 3. *P. annua*
1 Plants perennials, often flowering the first year, usually bearing the remains of old culms.
  4 Creeping rhizomes present, usually extensively developed (Poa, Pratenses).
    5 Culms strongly flattened, 2-edged .... 4. *P. compressa*
    5 Culms mostly terete, not 2-edged.
      6 Lemmas glabrous to pubescent, not webbed at the base.
        7 Panicles open, the branches mostly spreading; lemmas from glabrous throughout to finely scabrid-puberulent over the back and puberulent on the keel and marginal nerves; ligules 1–2 mm long.
          8 Panicle branches becoming reflexed; blades broad and short; rhizomes short; florets mostly perfect .... 5. *P. curta*
          8 Panicle branches ascending or spreading; blades elongate; rhizomes long; florets pistillate.. 6. *P. nervosa*
        7 Panicles contracted, with ascending or appressed branches; lemmas villous on the keel and marginal nerves, at least towards the base; ligules 1.5–4.5 mm long.
          9 Culms mostly 2–4 dm tall; blades mostly involute-folded; anthers 1.2–2 mm long .... 7. *P. arida*
          9 Culms (4) 5–8 dm tall; blades mostly flat; anthers 2–2.5 mm long .... 8. *P. glaucifolia*
      6 Lemmas pubescent on the keel and marginal nerves and webbed at the base, the webbing sometimes scant and not readily detected.
        10 Lemmas with copious cobwebby hairs at the base; lower branches of the panicles usually in 4s or 5s (2–9), ascending or sometimes spreading; first glume 1.8–2.5 (3) mm long; plants usually more than 3 dm tall .... 9. *P. pratensis*
        10 Lemmas usually with scant amount of cobwebby hairs at the base; lower branches of the panicles usually in 2s (1–4), spreading to somewhat reflexed; first glume (2) 2.5–4.5 mm long; plants usually less than 3 dm tall .... 10. *P. arctica*
  4 Creeping rhizomes wanting; plants often becoming decumbent and stoloniferous in some.
    11 Florets mostly or all converted into bulblets with dark purple bases; culms becoming bulb-like at the base (Palustres, in part) .... 11. *P. bulbosa*
    11 Florets producing caryopses; culms usually not bulb-like at the base.
      12 Lemmas webbed at the base (Palustres).
        13 Lemmas glabrous, except for the tuft of cobwebby hairs and the slightly pubescent keel near the base .... 12. *P. trivialis*
        13 Lemmas pubescent on the keel and marginal nerves.
          14 Lower panicle branches usually more than 2 at a node, not capillary.
            15 Culms loosely tufted, decumbent from a purplish base; panicles usually more than 10 cm long; ligules (0.5) 1–5 mm long .... 13. *P. palustris*
            15 Culms tufted, erect from a greenish base; panicles mostly less than 10 cm long; ligules 0.5–1.2 (2) mm long .... 14. *P. interior*
          14 Lower panicle branches single or in pairs, capillary.
            16 Lemmas 2.2–3 (3.5) mm long, not acuminate; spikelets 3–4 (5) mm long; lower panicle branches reflexed at maturity; glumes subequal ...... 15. *P. reflexa*

16 Lemmas (3) 3.5–5 mm long, acuminate; spikelets 4–7.5 mm long; lower panicle branches usually not reflexed; glumes distinctly unequal . . 16. *P. leptocoma*

12 Lemmas not webbed at the base, rarely with a scant tuft in some *P. pattersonii.*

17 Spikelets strongly compressed; glumes and lemmas mostly keeled to the base.

18 Lemmas pubescent on the keels or marginal nerves or both (Alpinae).

19 Plants (2) 2.5–7.5 dm tall; panicles (3) 5–15 cm long; spikelets mostly pistillate; lemmas 4–7 mm long .... 17. *P. fendleriana*

19 Plants 0.5–2.5 (5) dm tall; panicles 1–5 (9) cm long; spikelets usually with perfect flowers; lemmas 2.5–4 (5.5) mm long.

20 Spikelets (3-) 4- to 6-flowered .... 18. *P. alpina*

20 Spikelets 2- to 3- (4)-flowered.

21 Leaf blades 5–10 cm long; culms scarcely exceeding the basal leaves; second glume mostly more than 3 mm long ([2.5] 3–5), nearly equalling the first lemma .... 19. *P. pattersonii*

21 Leaf blades 1–5 cm long; culms much longer than the basal leaves; second glume mostly less than 3 mm long (2–3 [3.5]), evidently shorter than the first lemma .... 20. *P. rupicola*

18 Lemmas glabrous to scaberulous (Epiles).

22 Lemmas (3) 4–7.5 mm long (3–4 mm in *P. suksdorfii*); spikelets 5–11 mm long; plants (0.5) 1–5.5 dm tall.

23 Plants mostly more than 2.5 dm tall.

24 Blades filiform, less than 1 mm wide, scabrous; plants of middle elevs. .... 21. *P. cusickii*

24 Blades wider, 1–3 mm broad, glabrous, the blades on dried specimens sometimes becoming filiform; plants of high elevs. ..... 22. *P. epilis*

23 Plants mostly less than 2.5 dm tall.

25 Glumes shorter than the first floret; lemmas 4–6 mm long; anthers 2–3 mm long; plants of sagebrush and juniper woodland habitats .... 24. *P. leibergii*

25 Glumes often as long as the first floret; lemmas 3–4 (5) mm long; anthers 0.8–1.3 mm long; plants of alpine habitats .... 25. *P. suksdorfii*

22 Lemmas 2–3 (3.5) mm long; spikelets 2.5–3.7 (4.5) mm long; plants 0.3–0.8 (1.2) dm tall .... 23. *P. lettermanii*

17 Spikelets little compressed, narrow; keels of lemmas obscure.

26 Lemmas crisp puberulent to nearly glabrous, the pubescence sometimes confined to the very base (Scabrellae).

27 Panicles open, the lower branches nearly at right angles to the axis; culms often decumbent at the base; plants summer flowering .... 26. *P. gracillima*

27 Panicles contracted or somewhat open; culms usually erect.

28 Plants relatively small, mostly less than 3 dm tall; basal leaves forming a short dense tuft, 3–10 (14) cm high; panicles 2–7 (10) cm long; mostly spring flowering .... 28. *P. sandbergii*

28 Plants larger, mostly more than 3 dm tall; basal leaves looser, the tuft (3) 5–30 cm high; panicles (4) 6–16 (20) cm long; mostly summer flowering.

29 Basal tuft of leaves 3–15 (25) cm high; panicles 4–12 (16) cm long .... 27. *P. scabrella*

29 Basal tuft of leaves (10) 15–30 cm high; panicles (7) 9–16 (20) cm long .... 29. *P. canbyi*

26 Lemmas usually glabrous, sometimes minutely scaberulous (Nevadenses).

30 Ligules decurrent, long, 1.5–6.5 mm long, acuminate or sharply acute .... 30. *P. nevadensis*

30 Ligules not obviously decurrent, short, 1–2 mm long, rounded or obtuse to truncate.

31 Blades involute, mostly less than 1.5 mm broad, greenish; plants 2–7 (8.5) dm tall; usually growing in alkaline soils .... 31. *P. juncifolia*

31 Blades mostly flat, 1.5–3.5 mm broad, often glaucous; plants robust, (4) 6–18 dm tall; growing in non-alkaline soils .... 32. *P. ampla*

1. Poa bolanderi Vasey

Poa bolanderi Vasey, Bot. Gaz. 7: 32. 1882. (*Bolander 6115,* Yosemite Natl. Park, Calif., July 1866.)

P. howellii var. *chandleri* Davy in H. M. Hall, Univ. Calif. Publ. Bot. 1: 60. 1902. *P. bolanderi* subsp. *chandleri* Piper, Contr. U.S. Natl. Herb. 11: 132. 1906. (*Chandler 1703,* Shackle-

POA

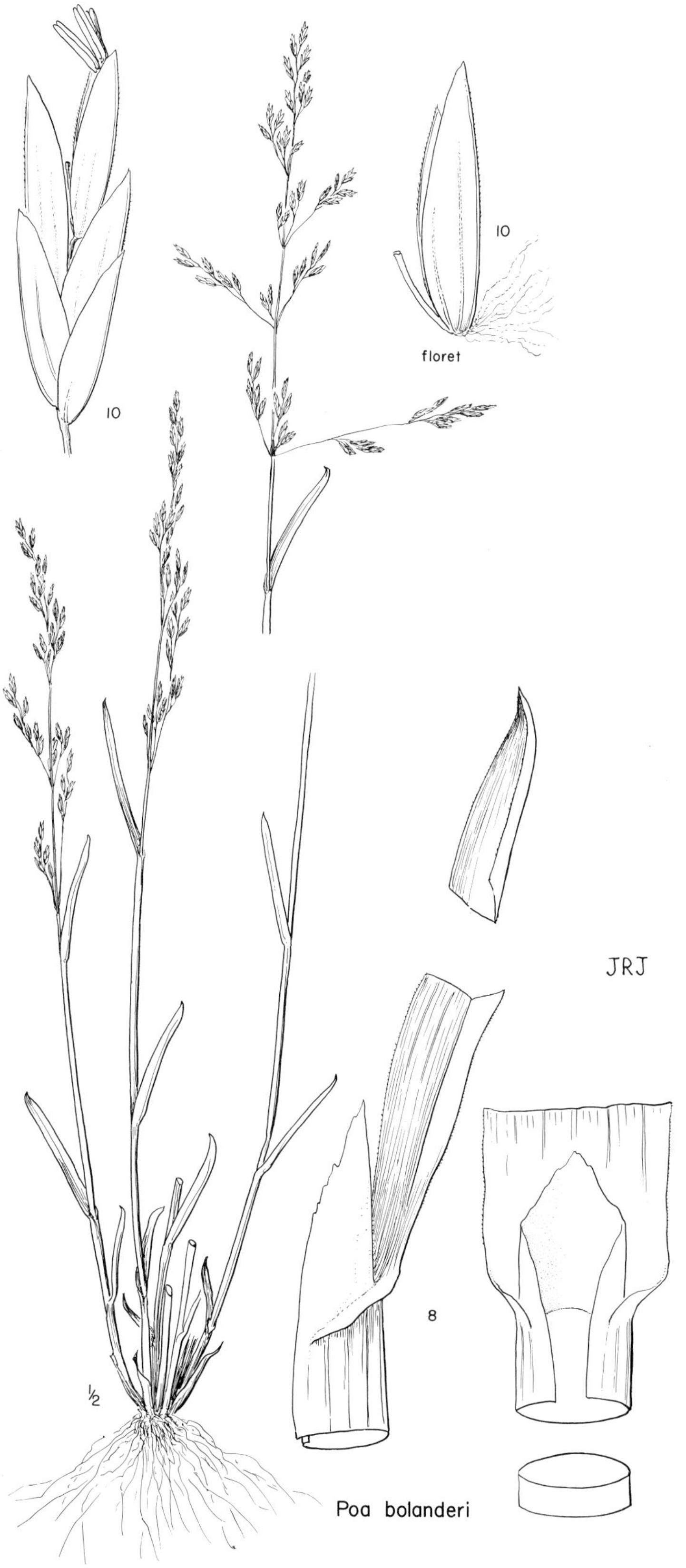

Poa bolanderi

Poa bigelovii

ford Canyon, near Marble Mt., Siskiyou Co., Calif., June 1901.)

P. horneri St. John, Fl. S.E. Wash. & Adj. Idaho 54. 1937. (*Darlington,* "Indian Corral Spring, T.8 N., R.41 E.," Columbia Co., Wash., 23 July 1913.)

Bolander bluegrass.

Slender glabrous annuals, often winter annuals; culms erect, 2–4.5 (8) dm tall; sheaths closed about half their length; ligules of lower leaves membranous, rounded, mostly entire, 1.2–2 mm long, those of the culm leaves becoming acuminate, 3–3.5 (4) mm long; blades usually flat, 2–4 mm broad, relatively short, about 2–5 cm long, scabrous, the tip strongly boat-shaped; panicles 8–16 (20) cm long, with few stiff branches, appressed at first but soon opening at nearly right angles with few branches at a whorl, the whorls somewhat distant, the lower branches occasionally becoming reflexed, the spikelets mostly at the tips of branches; spikelets 3.5–5.5 (6.5) mm long, (1-) 2- to 3-flowered; glumes relatively broad, green with scarious margins, the first glume 1.5–2.8 mm long, 1- to sometimes 3-nerved, lanceolate, the second 2–3.5 mm long, 3-nerved, broadly lanceolate to ovate; lemmas 1.5–4.5 mm long, obscurely 5-nerved, glabrous except for the tuft of cobwebby hairs at the base, somewhat scabridulous on the keel; anthers 0.7–1.5 mm long; lodicules about 0.7 mm long; $2n = 28$.

Dry to moderately moist places in the mts. from about 1524 to 3048 m (5000 to 10,000 ft.) elev.; c. Idaho, s. to w. Nev. and s. Calif. July–Aug.

No specimens collected within the Intermountain Region have been seen, but *P. bolanderi* is commonly encountered near Reno and in areas in Oregon and Idaho very near our boundary lines and likely will be found within the Region.

2. Poa bigelovii Vasey & Scribn.

Poa annua var. *stricta* Vasey, Bull. Torrey Bot. Club 10: 31. 1883. (*Pringle,* "Banks of the Rillita" River, Pima Co., Ariz., 9 April 1881.)

P. bigelovii Vasey & Scribn. in Vasey, Descript. Cat. Grasses U. S. 81. 1885, nomen nudum; Vasey & Scribn. in Vasey, Contr. U. S. Natl. Herb. 1: 270. 1893. (*Fendler 931,* N.M., in 1847.)

Bigelow bluegrass.

Light green winter annuals; culms erect, leafy, 1–5 (7) dm tall (tall in favorable sites); ligules thin, membranous, 1–3 (4) mm long; blades soft, flat, 1–4 mm broad, relatively short, 5–10 cm long; panicles long and narrow, often somewhat interrupted, 4–15 (22) cm long, with short, appressed branches, the lower in 2s or 3s, spikelet bearing mostly to the base; spikelets (4) 4.5–6 mm long, (3-) 4- to 6-flowered, broadly oblong, strongly compressed; glumes acuminate, glabrous, somewhat gibbous at the base, becoming hyaline or white margined, the keel scabrous, the first glume 2–3 (4) mm long, (1-) 3-nerved, the second 2.5–4 mm long, 3-nerved, the lateral nerves sometimes obscure; lemmas 2.7–3.5 (4.5) mm long, 5-nerved, broadly oblong, with a tuft of cobwebby hairs at the base, the keel and marginal nerves villous to the middle, sometimes hairy on the internerves; anthers 0.4–0.8 mm long, possibly cleistogamous.

Open areas at medium elevs. in wash bottoms and other moist habitats; arid regions of s. Calif., w. and s. Nev., s. Utah, e. to Okla. and s. to Texas and n. Mex. March–May.

Poa bigelovii is very closely related to *P. annua.* Bigelow bluegrass is well adapted to the arid regions. Its seeds germinate abundantly after summer rains and its life cycle is completed before the habitat dries again. It is an important winter forage grass in the Southwest.

3. Poa annua L.

Poa annua L. Sp. Pl. 68. 1753. (Europe.)

Annual bluegrass, wintergrass.

Bright green annuals, often winter annuals, erect or more often spreading, rooting at the nodes and forming mats, with numerous leaves; culms 0.3–2 (3.5) dm tall, strongly flattened; sheaths open about half their length, with overlapping margins; ligules 1–3 mm long, obtuse or truncate, often slightly erose; blades lax, folded or flat, 1–2.5 (3) mm broad, scaberulous on the margins but otherwise glabrous, 1–10 cm long; panicles 1.5–3.5 (8) cm long, pyramidal, open, the spreading branches usually rather rigid; spikelets (3) 4–5.5 (7) mm long, 3- to 6-flowered; florets rather distant making the rachilla often visible; glumes glabrous, the first glume 1.5–2.2 mm long, 1-nerved, narrow-lanceolate, acute, the second only slightly longer, 1.8–2.7 mm long, 1- to 3-nerved, oblanceolate to obovate, broadest above the middle, obtusish; lemmas 2.5–3.5 mm long, scarious margined above, 5-nerved, the nerves usually relatively prominent, pilose below, not webbed at the base; anthers about 0.7–1.1 mm long; $2n = 14$, 24–26, 28.

A common weed of garden and lawns and generally in areas receiving water in excess of natural precipitation, i.e. irrigated areas throughout our region; introduced from Europe; Alaska to Newfl., and s. throughout most of the U.S. April–Sept.

Annual bluegrass is a common lawn weed in our area. Its objectionable qualities are over-exaggerated, because its bright green color makes it a better lawngrass than no grass at all. However, it does turn brown in late summer if not watered.

4. Poa compressa L.

Poa compressa L. Sp. Pl. 69. 1753. *Paneion compressum* Lunell, Amer. Midl. Naturalist 4: 222. 1915. (Europe and N. Amer.)

Canada bluegrass.

Rhizomatous perennials; culms 2–5 (8) dm tall, scattered, strongly flattened and 2-edged, decumbent at the base and often geniculate at the nodes, wiry, bluish-green throughout; sheaths open to near the base, strongly compressed with the flat enclosed culms so as to appear winged; ligules 0.5–1.5 (2) mm long, fairly stiff, mostly entire, rounded to nearly truncate; blades flat to folded, 1.5–3.5 mm broad, relatively short, the tips boat-shaped; panicles 3–7 (9) cm long, usually compact, dense and narrow but occasionally becoming somewhat open; spikelets 3–6 (7.5) mm long, crowded on short branches, often appearing nearly sessile, mostly 3- to 7-flowered; glumes subequal, 1.8–2.7 (3.2) mm long, the first glume 1- to 3-nerved, the second broader, 3-nerved; lemmas (2) 2.5–3 mm long, firm, sometimes nearly coriaceous, strongly keeled, obtusish, usually pur-

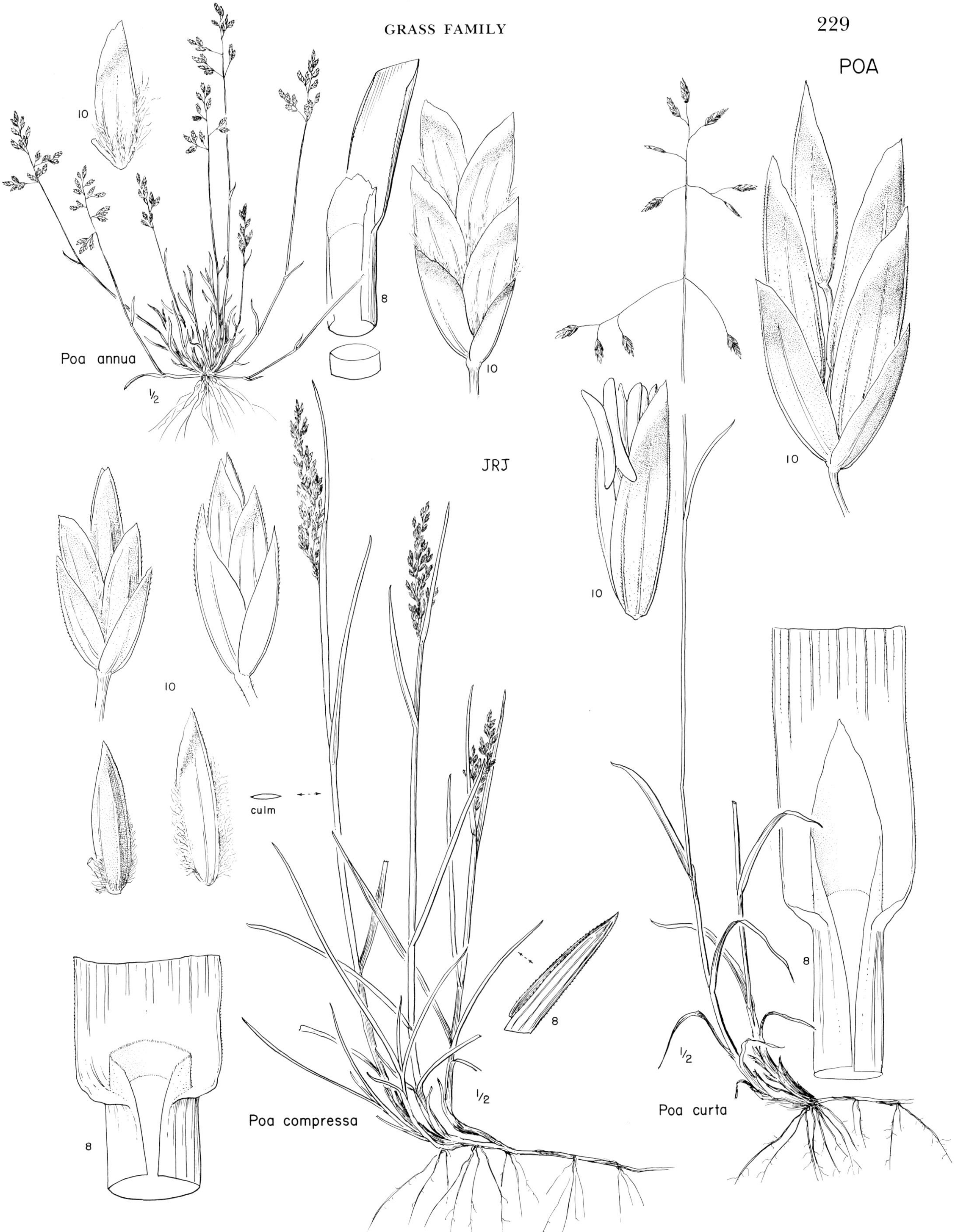
POA
10
8
10
Poa annua
1/2
JRJ
10
10
10
culm
8
1/2
Poa compressa
8
1/2
Poa curta
8

plish near the tip, the keel and marginal nerves with sericeous hairs, especially toward the base, the web lacking or if present then very scant; anthers 0.9–1.7 mm long; lodicules barely exceeding 0.5 mm; $2n$ = 14, 35, 39, 42 (the usual number), 45, 49, 50, 56.

Often in poor soils along roadsides, gardens and meadows, usually where there is an ample amount of moisture; said to have been introduced from Europe; Alaska to Newfl. and s. throughout most of the U.S. June–Aug.

Most of the Canada bluegrass in our area seems to have been introduced, but some populations may well represent native stands. This is an especially valuable species for planting along roadsides and in areas where the soil is too poor for *P. pratensis*. Some specimens of *P. compressa* have spikelets that suggest those of an awnless *Festuca*, but a careful look will prevent a mistaken identity.

5. **Poa curta** Rydb.

Poa curta Rydb. Bull. Torrey Bot. Club 36: 534. 1909. (*Tweedy 13*, Spread Creek [Jackson Hole], Teton Co., Wyo., July 1897.)

Shortleaf bluegrass.

Rhizomatous perennials; culms 4–8 dm tall, few in a loose tuft; sheaths closed most of their length, glabrous or minutely roughened; ligules 0.5–2 mm long, truncate to obtuse and becoming acutish on cauline leaves, erose; blades flat, lax, 2–4 (5) mm broad, those of the culms short; panicles 5–15 cm long, loose and open after nodding, the few branches spreading or more often reflexed, the lower branches 1–3 per node, the spikelets confined mostly to the tips of the branches; spikelets 4.5–7.5 (9.5) mm long, 2- to 4- (6)-flowered; florets mostly perfect; glumes unequal, the first glume (2) 2.5–3.5 (4) mm long, 1-nerved, the second (2.5) 3–4 (5) mm long, 3-nerved; lemmas (3.5) 4–5 (6) mm long, lanceolate, strongly keeled, prominently nerved, often slightly scaberulose at the base with somewhat longer hairs on the keel, without cobwebby hairs at the base; anthers 2–2.8 mm long; lodicules about 0.8 mm long.

Moist to wet places on mossy rocks or in dense shade on shallow soils and often in mt. meadows at middle elevs.; s. Idaho, nw. Wyo., and n. Utah (Wasatch Mts.). Late May–July.

Poa curta may represent the sexual phase of *P. nervosa*. The softer texture of the somewhat longer ligules, fewer culms and much shorter cauline leaves are among other reasons for recognizing this plant as a distinct species. There are some populations, however, where these distinctions break down.

6. **Poa nervosa** (Hook.) Vasey

Festuca nervosa Hook. Fl. Boreali-Amer. 2: 251, pl. 232. 1840. *Poa nervosa* Vasey, U.S.D.A. Div. Bot. Bull. 13(2): pl. 81. 1893. (*Scouler*, Nootka Sound, Vancouver I., B.C.)

P. wheeleri Vasey in Rothr. Cat. Pl. Geogr. & Geol. Explor. & Surv. W. 100th Meridian 55. 1874. *P. cuspidata* Vasey ex Scribn. U.S.D.A. Div. Agrostol. Circ. 9: 6. 1899; not Nutt. 1818. *P. nervosa* var. *wheeleri* C. L. Hitchc. Univ. Wash. Publ. Biol. 17(1): 671. 1969. (*Wolf 1131a*, "In timber," South Park, Colo., in 1873.)

P. vaseyana Scribn. ex Beal, Grasses N. Amer. 2: 532. 1896. *P. wheeleri vaseyana* (attributed to T. A. Williams by) Pammel, Proc. Iowa Acad. Sci. 20: 144. 1913, this combination not valid under the rules without any reference to rank or basionym. (*Patterson 8*, woods near timberline, Georgetown, Colo., 2 Sept. 1885.)

P. olneyae Piper, Erythea 7: 101. 1899. (*Piper 2820*, base of basalt cliffs in pine woods, Spokane, Wash., 28 May 1898.)

P. subreflexa Rydb. Bull. Torrey Bot. Club 36: 535. 1909. (*State Agricultural College 3731*, "mountain north of Steamboat Springs," Colo., 26 July 1891.)

Wheeler bluegrass.

Rhizomatous perennials, often producing large tufts, usually with numerous basal leaves; culms erect, 3–6 (8) dm tall, usually with 3 leaves; sheaths glabrous or those of the basal leaves with retrorsely pubescent hairs, closed to above the middle, commonly reddish-purple below, the culm sheaths glabrous or merely scabrous; ligules of basal leaves very short, truncate, mostly less than 1 mm, those of the culm leaves longer, about 1–2 mm long, acute to obtusish, the thickened ligules of the basal and culm leaves very distinctive; blades 1.2–3 (4) mm broad, flat to folded, somewhat scabrous above, the tip strongly boat-shaped, those of the innovations smaller and narrower; panicles 4–10 (12) cm long, usually open and pyramidal with a nodding apex but sometimes narrow, with (2) 3 or 4 branches below, the branches filiform and often flexuous, ascending or spreading; spikelets 5.5–10 mm long, ovate, strongly compressed, relatively large, loosely flowered and exposing the rachilla and reduced terminal floret, 3- to 5- (8)-flowered; florets in our plants seemingly always pistillate; glumes broad with narrow hyaline margins, the first glume 2.5–4 mm long, 1- to 3-nerved, lanceolate, the second 3–4.5 mm long, 3-nerved, oblong-oblanceolate; lemmas 4–5.5 mm long, mostly strongly nerved, glabrous or pubescent on the lower part of the lemma, not webbed at the base, with scarious margins and apex and often purplish in the middle; lodicules about 1 mm long; $2n$ = 28, 29, 56, 61–91.

Grassy slopes, ridges, open woods and meadows from middle to high elevs. (3048 m), sometimes in dry places but more often where there is ample moisture; B.C. to Alta., s. to s. Calif., Nev., Utah and N.M. April–Aug.

Herbarium specimens of this species are often found without rhizomes, but carefully collected specimens will nearly always show well-developed rhizomes. The thickened, short ligule of this *Poa* makes field identification nearly positive with only vegetative material. There are definite clinal variations within the species. Plants with strongly nerved lemmas, glabrous sheaths and drooping flexuous panicles are found in the Northwest, while plants from our region and the Rocky Mountains have less strongly nerved lemmas with pubescent nerves, retrorsely pubescent sheaths and more or less rigid panicles. The Rocky Mountain phase is referable to var. *wheeleri* (Vasey) C. L. Hitchc.

7. **Poa arida** Vasey

Poa arida Vasey, Contr. U. S. Natl. Herb. 1: 270. 1893. *P. pratericola* Rydb. & Nash in Rydb. Mem. New York Bot. Gard. 1: 51. 1900. *P. fendleriana* var. *arida* M. E. Jones, Contr. W. Bot. 14: 14. 1912. *Paneion aridum* Lunell, Amer. Midl. Naturalist 4: 222. 1915. *Paneion pratericolum* Lunell, ibid., p. 223. (*Vasey*, Socorro, N.M., in 1881.)

Poa sheldoni Vasey, Contr. U. S. Natl. Herb. 1: 276. 1893. (*Sheldon 615*, "Along Cottonwood Creek, Buena Vista," Chaffee Co., Colo., 4 July 1892.)

P. pseudopratensis Scribn. & Rydb. in Rydb. Contr. U. S. Natl. Herb. 3: 531, pl. 20. 1896; not Beyer, 1819. *P. pratensis* var. *pseudopratensis* M. E. Jones, Contr. W. Bot. 14: 15. 1912. *P. pratensiformis* Rydb. Fl. Rocky Mts. 79, 1060. 1917 [1918]. *Paneion pratensiforme* Lunell, Amer. Midl. Naturalist 5: 234. 1919. (*Rydberg 1151*, Hot Springs, Black Hills, Fall River Co., S.D., 13 June 1892.)

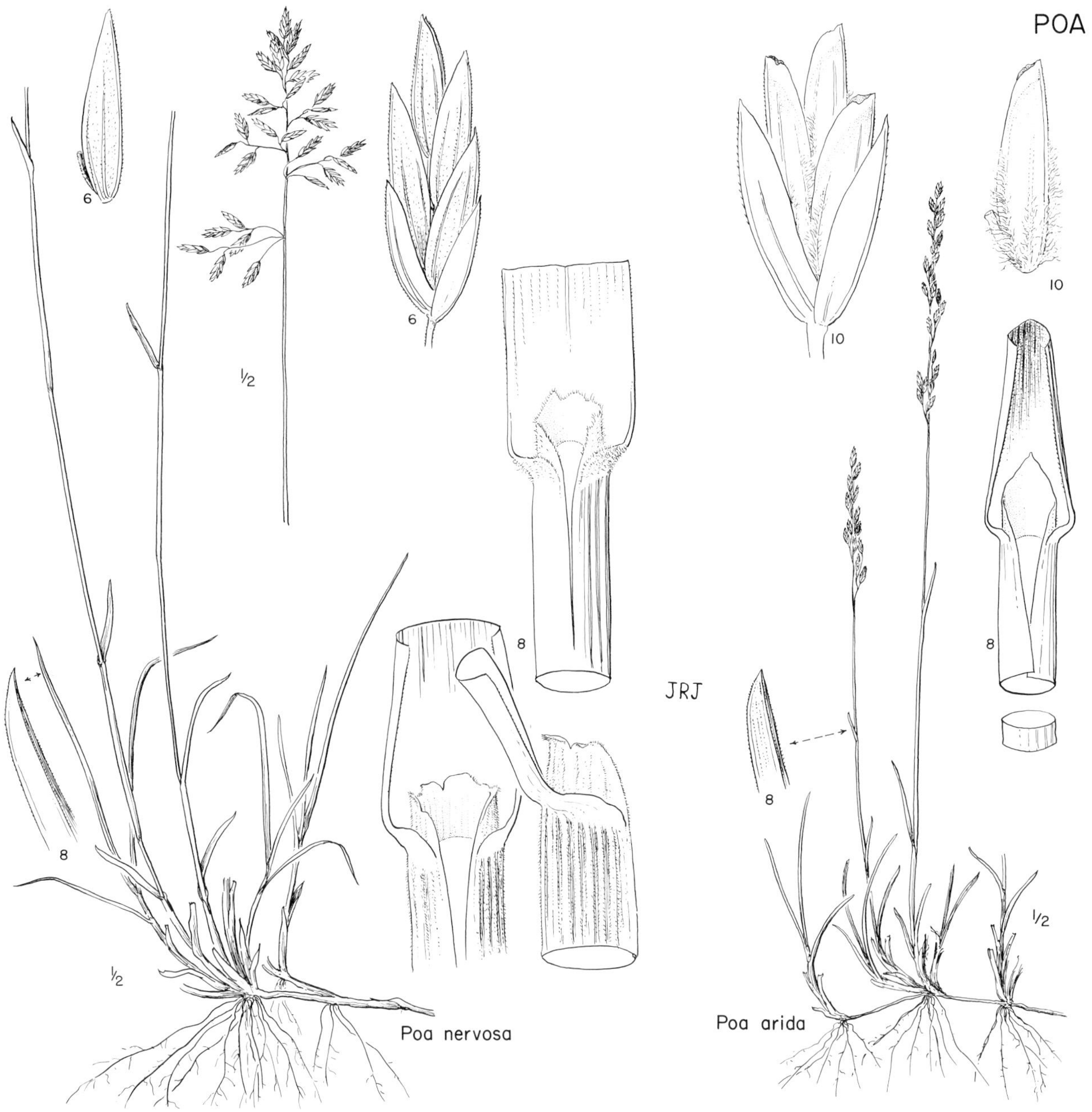

Plains bluegrass.

Tufted perennials from short rhizomes; culms 1–4.5 (6) dm tall, rarely taller; leaves mostly basal; sheaths smooth, open most of their length; ligules 1.5–3.5 (4) mm long, thick, conspicuous, acute to acuminate, sometimes decurrent; blades 1–3 mm broad, sometimes involute, appearing rather rigid, sharp pointed, little if at all boat-shaped; panicles 3–12 cm long, narrow, somewhat contracted, with short and stiffly erect branches; spikelets (4) 5.5–7.5 mm long on short pedicels, rather crowded, ovate, 2- to 6-flowered; glumes glabrous, green, 1- (3)-nerved, the first glume 2.5–4.2 mm long, the second 3–5 mm long; lemmas 3.2–5 (5.5) mm long, usually keeled but sometimes rounded, obscurely 5-nerved, scarious at the apex, densely villous on the keel and marginal nerves, also often slightly villous on the internerves near the base, without cobwebby hairs at the base; palea nearly as long as the lemma, scabrous on the nerves; anthers 1.2–2 mm long; lodicules about 0.7 mm long; $2n = 63–103$.

Dry to moist places from sagebrush deserts to alpine tundra above 3350 m (11,000 ft.) in the Uinta Mts.; Man. to Alta., s. through Mont. and w. Iowa to Texas, N.M., and Utah. June–Aug.

Plants collected east of the Intermountain Region are considerably taller than ours, up to 7 dm, and often have lemmas that are little if at all compressed. Our plants seem to produce rhizomes rather consistently while those from the Plains often seem to be without them. Some sort of infraspecific segregation may eventually prove to be useful.

8. Poa glaucifolia Scribn. & Williams

Poa planifolia Scribn. & Williams in Scribn. U.S.D.A. Div. Agrostol. Circ. 9: 3. 1899; not Kuntze, 1898. *P. glaucifolia* Scribn. & Williams, ibid. 10: 6. 1899. (*Williams 2814,* moist banks, Spring Creek, Big Horn Basin, Washakie Co., Wyo., 4 Aug. 1897.)

Glaucous perennials, producing culms in loose tufts, but usually with short rhizomes; culms 4–8 dm tall, smooth except at the nodes; foliage smooth to scabrous; sheaths open most of the way, closed only near the very base; ligules 1.5–3.5 (4.5) mm long, obtuse to acute, those of the upper culm leaves up to 5 mm long and often acuminate; blades more often flat or folded, 2–4 mm broad, glabrous but definitely scabridulous on the margins; panicles (8) 10–17 cm long, usually tight and narrow but sometimes opening with branches in 3s or 4s; spikelets (4.5) 5–7.5 (9) mm long, 2- to 4-flowered, lanceolate to ovate; glumes often scabrous keeled, the first glume 2.5–4 mm long, 1- to 3-nerved, the second 3–4.5 mm long, 3- to 5-nerved; lemmas 3–5 (5.5) mm long, usually keeled but sometimes rounded, the 3 to 5 nerves usually rather prominent, the keel and marginal nerves often villose on the lower half, the internerves usually glabrous but sometimes pubescent near the base, scarious at the apex with a purplish band near the middle, not at all webbed at the base; anthers 2–2.5 mm long; lodicules about 0.7 mm long; $2n$ = 50, 56.

Moist places along ditches and in meadows, and often on rocky slopes from middle to high elevs. (up to 3350 m); se. B.C. and s. Alta., s. to Nev., Ariz., N.M. and e. to Nebr. and Minn. July–Aug.

Poa glaucifolia is a poorly defined species which is not as well represented in herbaria as the wide distribution would suggest. Glaucous plants with short rhizomes, lemmas with pubescent keels and marginal nerves, and lacking cobwebby hairs tend to find their way into *P. glaucifolia* folders. Specimens with more or less rounded lemmas and very short rhizomes may represent *P. sandbergii* or *P. scabrella* with introgression of genes from a member of the Pratenses group.

9. Poa pratensis L.

Poa pratensis L. Sp. Pl. 67. 1753. *P. angustifolia* var. *pratensis* Simonkai, Enum. Fl. Transsilv. 580. 1886. *Paneion pratense* Lunell, Amer. Midl. Naturalist 4: 222. 1915. (Europe.)

Poa peckii Chase, J. Wash. Acad. Sci. 28: 54, fig. 2. 1938. (*Peck 19804,* "Dry woods, along Metolius River near Camp Sherman," Jefferson Co., Oregon, 11 July 1937.)

P. agassizensis Boivin & D. Löve in D. Löve & Bernard, Svensk Bot. Tidskr. 53: 371. 1959, nomen nudum; Boivin & D. Löve, Naturaliste Canad. 87: 176, fig. 1. 1960. (*Boivin et al. 9167,* 3 mi. nw. of Brunkild, Man., 6 June 1952.)

Kentucky bluegrass.

Perennials with extensive creeping rhizomes, forming dense sods; culms (2) 3–6.5 (10) dm tall, slightly compressed, erect, with numerous innovations; sheaths smooth but occasionally somewhat scabrous, closed about half their length; ligules shorter than wide, 0.5–1.7 mm long, truncate, mostly entire, those on culm leaves may be as long as 3 mm; blades numerous, soft, usually green but occasionally glaucescent, flat or usually folded, about 1–3 (4.5) mm broad, usually with scabrous margins, sometimes slightly pubescent and purplish near the collar, the tips strongly boat-shaped; panicles open, often pyramidal, (2) 4–11 (16) cm long, with spreading or ascending branches usually in whorls of 4 or 5 (2–9); spikelets (3) 4–5 (6) mm long, ovate, crowded, 2- to 4-flowered, green or purplish, strongly compressed; glumes scabrous on the keel, the first glume 1.8–2.5 (3) mm long, 1-nerved, the second 2.2–3.3 mm long, 3-nerved; lemmas 2.5–4 mm long, 3- to 5-nerved, the lateral nerves weak, obtuse or acute, with copious cobwebby hairs at the base, the keel and marginal nerves with sericeous hairs, glabrous between the nerves; palea nearly equalling the lemma; anthers 0.8–1.9 mm long; lodicules about 0.7 mm long; $2n$ = 21–147 (it is not at all unusual to find several chromosome numbers represented in a single meadow).

Moderately moist to dry roadsides, meadows and open woods; said to have been introduced from Eurasia; widely distributed throughout temp. N. Amer. May–Aug. (–Sept.).

Kentucky bluegrass is clearly the most important cultivated species of *Poa*. It is the most satisfactory lawn grass in our area, especially where there is natural lime in the soil. Its vigorous rhizomes carry it from lawns onto pavements and into flower borders where it often becomes one of our most noxious weeds. It is only a fair pasture grass, however, as many other species are considered to be more desirable. The name Kentucky bluegrass has been used in North America because of the extensive bluegrass pastures in that state.

Poa pratensis is considered to be entirely a European introduction by many botanists, but its occurrence on remote meadows in areas like the Uinta Mountains and numerous places throughout the region must be by native populations. Weber (Rocky Mt. Fl. 362. 1967) treats *P. agassizensis* Boivin & D. Löve as the native counterpart of *P. pratensis*. He separates *P. agassizensis* by the following characters:

1 Basal leaves bright green, 2–3 mm broad, flat or folded, soon withering in a few years; spikelets mostly 3-flowered; lemmas very cobwebby at the base .... *P. pratensis*
1 Basal leaves glaucous, 0.8–2 mm broad, folded to involute, remaining intact for many years; spikelets mostly 2-flowered; lemmas only slightly cobwebby ...... *P agassizensis*

Our apparent native stands cannot, however, be separated from the introduced material with any degree of satisfaction.

10. Poa arctica R. Br.

Poa arctica R. Br. Chloris Melvilliana 30. 1823. *P. cenisia* var. *arctica* K. Richter, Pl. Eur. 1: 83. 1890. (*Parry,* Melville Isl., Arctic Amer., in 1820.)

P. grayana Vasey, Contr. U. S. Natl. Herb. 1: 272. 1893. *P. arctica* subsp. *grayana* A. Löve, D. Löve & Kapoor, Arctic & Alpine Res. 3: 143. 1971. (*Patterson 14,* "High mountains about Gray's Peak, near timberline," Colo., 2 Sept. 1885.)

P. laxa var. *occidentalis* Vasey ex Rydb. & Shear, U.S.D.A. Div. Agrostol. Bull. 5: 32. 1897, nomen nudum. (*Shear 690* and *Rydberg 2440,* Grays Peak, Colo.)

P. longipila Nash in Rydb. Mem. New York Bot. Gard. 1: 46. 1900. (*Rydberg & Bessey 3614,* Electric Peak, Yellowstone Natl. Park, Mont., 18 Aug. 1897.)

P. aperta Scribn. & Merr. U.S.D.A. Div. Agrostol. Circ. 35: 4. 1901. (*Shear & Shear 98,* open mt. side, 9500ft. elev., Telluride, Colo., 1 Sept. 1900.)

P. callichroa Rydb. Bull. Torrey Bot. Club 32: 603. 1905. (*Clements & Clements 457,* "Dead Lake, near Pike's Peak," Colo., 14 Aug. 1901.)

P. phoenicea Rydb. Bull. Torrey Bot. Club 32: 605. 1905. (*Clements & Clements 466,* "Pike's Peak Valley," Colo., 21 Aug. 1901.)

P. tricholepis Rydb. Bull. Torrey Bot. Club 32: 606. 1905. (*Baker 210,* "Near Pagosa Peak," Colo., Aug. 1899.)

P. cenisia of Amer. authors; not All. 1789.

Arctic bluegrass.

POA

Decumbent perennials with culms in small tufts, usually producing short rhizomes, the innovations apparently both intravaginal and extravaginal; culms 1–4.5 (6.5) dm tall; sheaths mostly open to below midlength, those of the basal leaves loose, rather short; ligules (1) 2–3.5 (5) mm long, thin, acuminate, obtuse, or even truncate; blades 1.5–3.5 mm broad, flat or usually folded, the tips strongly boat-shaped; panicles loose, sometimes pyramidal, 3.5–9 (15) cm long, the lower branches usually in 2s (1–4), spreading to somewhat reflexed, bearing only a few spikelets towards the tips of the branches; spikelets (3.5) 5–7 (8.5) mm long, ovate, 2- to 4- (5)-flowered, mostly purplish; florets usually perfect, but sometimes apparently functionally only pistillate; glumes lanceolate, acuminate, purplish towards the apex, the first glume (2) 2.5–4.5 mm long, the second (2.5) 3–5 mm long; lemmas 3.5–5 (6) mm long, lanceolate, strongly keeled, purplish in the middle and scarious above, sericeous on the nerves and somewhat pubescent on the internerves on the lower half, usually with a scant tuft of cobwebby hairs at the base, these rarely lacking entirely; palea nearly as long as the lemma; anthers 1.4–2.5 mm long; lodicules about 0.7 mm long; $2n = 38–92$.

Moist meadows, streambanks and open ridges, mostly at or above timberline; circumboreal regions; in N. Amer., s. in the Rocky Mts. to n. N.M., in our area in the Uinta Mts., the Utah Plateaus, and Abajo Mts., Utah and in the Ruby Mts., Nev., reported from the Cascades of Wash. and Oregon, and Inyo Co., Calif. July–Aug.

Arctic bluegrass may well be the most variable bluegrass in our flora. The Rocky Mountain plants are taller than the arctic plants and have broader, often flat leaves. Some botanists recognize the Rocky Mountain plants as a distinct taxon, *P. grayana* Vasey or *P. arctica* subsp. *grayana* (Vasey) A. Löve, D. Löve & Kapoor. Plants from the Uinta and Ruby mountains more closely resemble the arctic phase, whereas those from the Abajo Mountains and the Utah Plateaus are referable to the *P. grayana* phase. Some tall plants with reflexed panicle branches may be influenced by genes of *P. reflexa*.

11. Poa bulbosa L.

Poa bulbosa L. Sp. Pl. 70. 1753. (France.)

P. bulbosa var. *vivipara* Koeler, Descript. Gram. Gallia & German. 189. 1802. *Paneion bulbosum* var. *viviparum* Lunell, Amer. Midl. Naturalist 4: 222. 1915. (Europe.)

Bulbous bluegrass.

Densely tufted perennials from somewhat bulbous bases, purplish below, the rhizomes absent; culms wiry, (2) 3–5 (6) dm tall; innovations numerous; sheaths open nearly to the base, smooth; ligules 1.5–4 (6.5) mm long, obtusish to acuminate, entire or somewhat erose; blades flat to usually folded or loosely involute on drying, 1.5–4 mm broad, somewhat scabrous; panicles (3) 4–7 (12) cm long, usually compact and crowded, sometimes ovoid, the branches ascending and scabrous; spikelets nearly always proliferous with the 4 to 6 florets produced into bulblets with broad, shiny, dark purple bases and protruding foliaceous linear tips, the bulbils as long as 2 mm; glumes narrow, the first 2–3 (3.5) mm long, the second 2.2–3.5 (4) mm long; lemmas (2.5) 3–5 (6.5) mm long; normal lemmas rarely present, but if present 4–6 per spikelet, pubescent on the nerves and cobwebby at the base, the modified spikelets usually producing a single bulbil that is capable of rooting immediately and asexually producing a new plant; $2n = 14, 21, 28, 39, 40–58$.

Pastures and other disturbed areas at low to middle elevs.; introduced from Europe; rare in the e. states, locally common from N.D. and Okla. westward. April–July.

The great number of basal leaves and wiry culms of this species soon dry, making this a rather unimportant range plant. Bulbous bluegrass has been planted in some areas, however, where it tends to replace even less desirable annual grasses. It is palatable to sheep early in the season.

12. Poa trivialis L.

Poa trivialis L. Sp. Pl. 67. 1753. (Europe.)

Rough bluegrass, rough-stalked meadow grass.

Decumbent, often strongly geniculate, slender, sometimes stoloniferous perennials, often rooting at several nodes and appearing rhizomatous though not actually so; culms 4–10 (11.5) dm tall, usually smooth but sometimes scabrous; sheaths often overlapping, closed about 1/4 their length, usually more or less retrorsely roughened, especially toward the summit; ligules (1.5) 3–5 (10) mm long, usually obtuse but sometimes acutish, entire; blades flat, lax, bright green, 1.7–5 (7.5) mm broad and up to 20 cm long, usually noticeably scabrous on the upper surface, the tips scarcely boat-shaped; panicles oblong, loose, (4) 6–14 (22) cm long, the lower branches often in 5s, the distant branches usually ascending, not at all reflexed; spikelets 3–4.5 (5) mm long, strongly compressed, usually 2- or 3-flowered, ovate or elliptic, apparently with perfect flowers; glumes lanceolate, acute, curved inward, the first glume 1.7–2.5 (3) mm long, 1-nerved, especially inward curved, the second 2–3 (3.8) mm long, 1- to 2-nerved, scabridulous on the keel; lemmas 2.5–3.5 (4.5) mm long, thin, strongly keeled and nerved (5), with sparse sericeous hairs at the base of the keel, otherwise glabrous on the body, prominently webbed at the base; palea nearly as long as the lemma; anthers 1.4–2 mm long; lodicules slightly more than 0.5 mm long; $2n = 14, 15, 28, 42$.

Often becoming a dominant grass around springs or moist places in woods or meadows; introduced from Europe; Alaska to Newfl., s. to n. Calif., Utah and Colo. and in the e. to La. and N.C.; seldom collected in our region. May–July.

This species may be confused with the very similar *P. palustris*. Our plants of *P. trivialis* nearly always root at the nodes on the decumbent base, have more strongly nerved and glabrous lemmas and are more consistently 2-flowered. The slightly longer ligules of this species are also helpful in making the separation.

13. Poa palustris L.

Poa palustris L. Syst. Nat. ed. 10. 874. 1759. *P. trivialis* var. *palustris* Huds. Fl. Anglica, ed. 2. 40. 1778. *Paneion palustre* Lunell, Amer. Midl. Naturalist 5: 94. 1917. (Europe.)

Poa triflora Gilib. Excercita Phytol. 2: 531. 1792. *Paneion triflorum* Lunell, Amer. Midl. Naturalist 4: 223. 1915. (Europe.)

Poa crocata Michx. Fl. Boreali-Amer. 1: 68. 1803. *P. glauca* var. *crocata* M. E. Jones, Contr. W. Bot. 14: 15. 1912. (*Michaux*, Lake Mistassini, Que., in 1792.)

Fowl bluegrass, fowl meadow grass.

Decumbent perennials with loosely tufted culms, often rooting at the sparingly branched, usually pur-

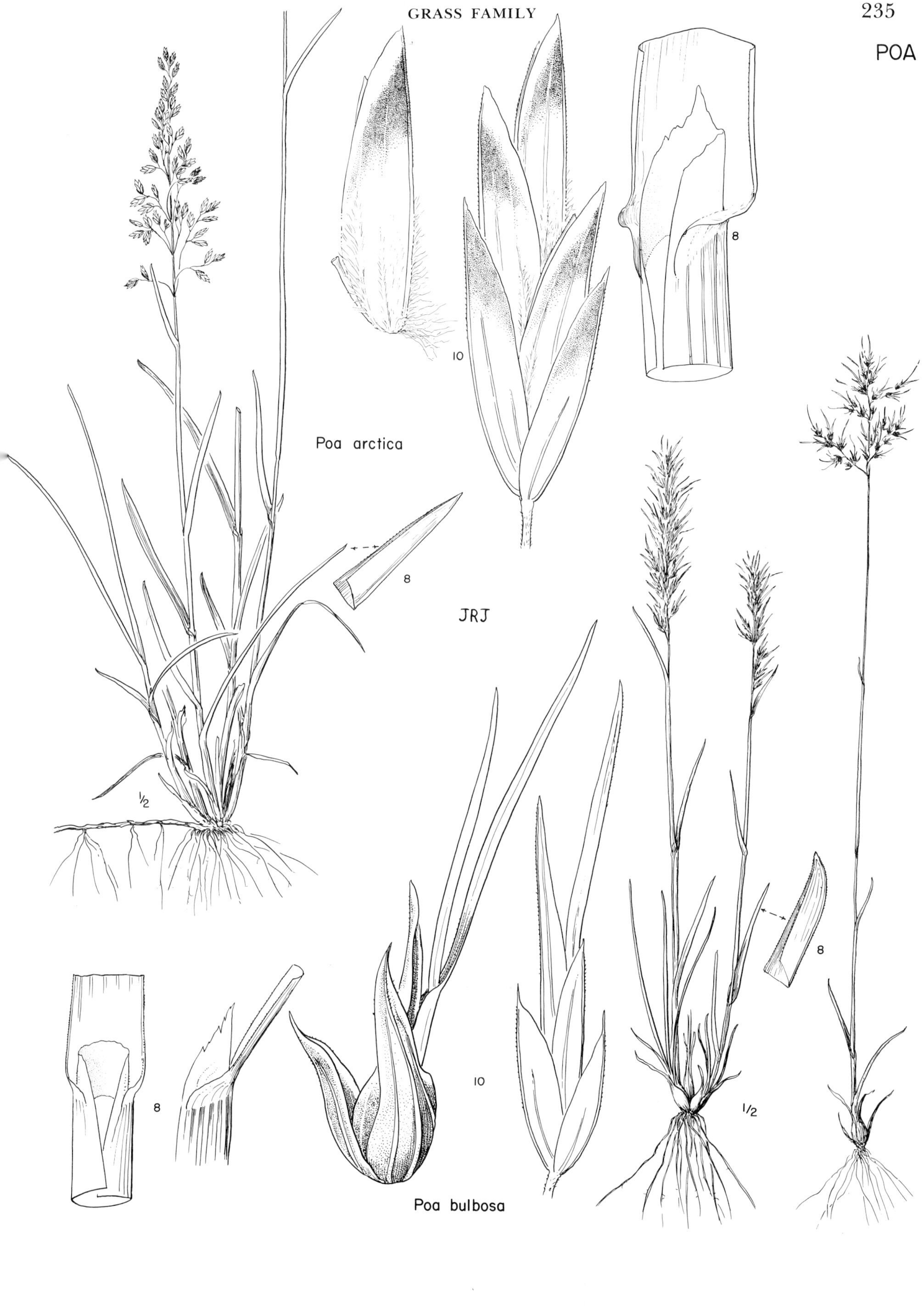
POA
10
8
Poa arctica
8
JRJ
1/2
8
8
10
1/2
Poa bulbosa

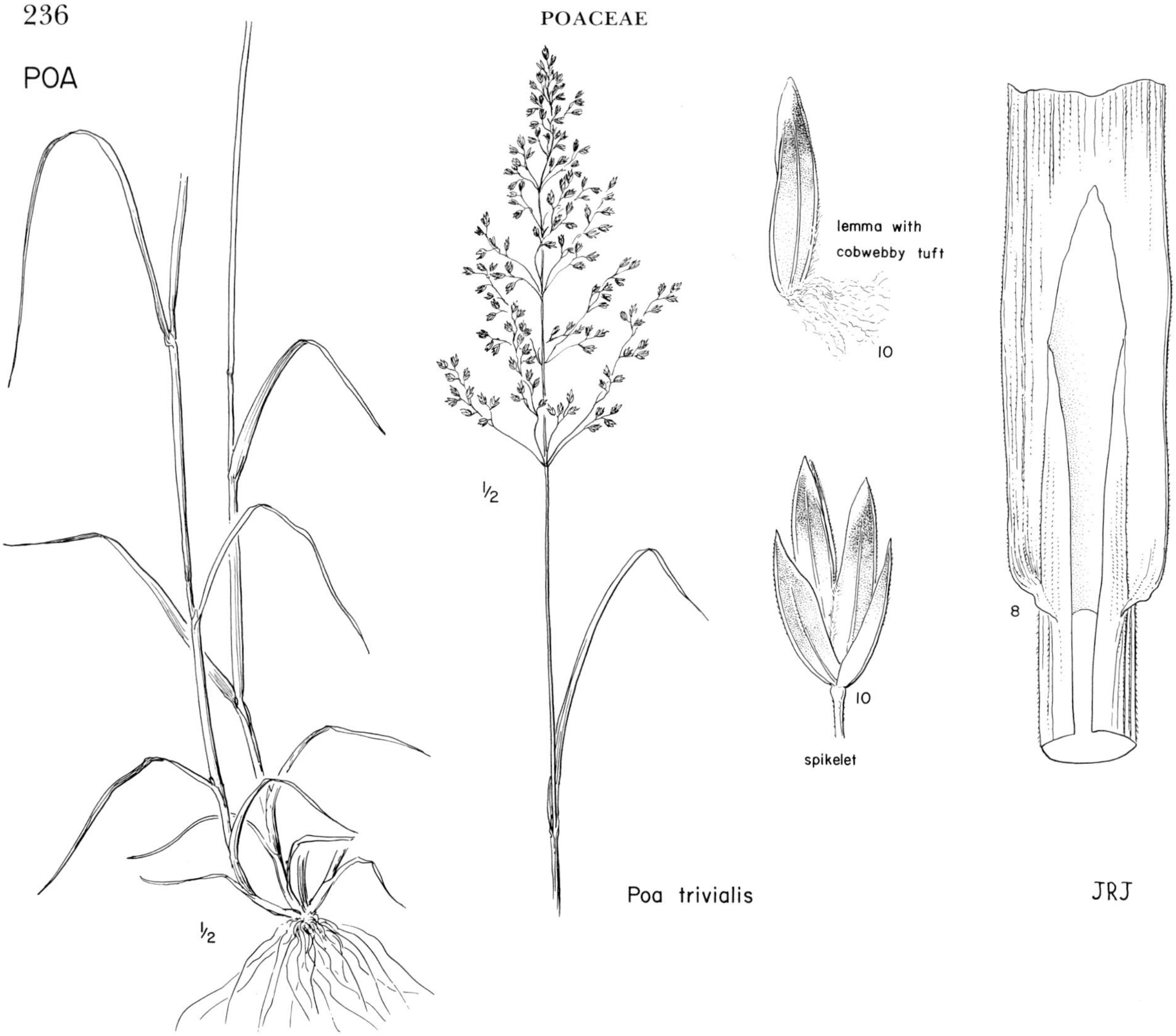

plish base, thereby simulating a rhizomatous plant; culms 2.5–9.5 (13) dm tall, smooth; sheaths keeled, glabrous or sometimes scaberulous, closed about 1/4 the length; ligules (0.5) 1–5 mm long, acutish to obtuse; blades 1–3.5 mm broad, soft, flat or folded, up to 20 cm long, often scabrous on the upper surface, with a slender boat-shaped tip; panicles pyramidal or oblong, open and loose, often nodding, elongate, 7–25 (30) cm long, yellowish-green or purplish, in rather distant whorls, the lower panicle branches (2) 3–5 (6) at a node; spikelets 3.2–5 (6) mm long, strongly keeled, 2- to 4-flowered; glumes acute, the first 1.7–2.8 mm long, 1-nerved, the second 2–3 mm long, 1- to 3-nerved; lemmas (2) 2.5–3.5 mm long, obscurely 5-nerved, bronze-tipped, villous on the keel and marginal nerves, with a cobwebby tuft of hairs at the base; palea nearly as long as the lemma, finely scabrous on the keels; anthers 0.8–1.2 mm long; lodicules about 0.6 mm long; $2n = 28, 42$.

Moist places in meadows, along ditches and open ground from low to middle elevs.; introduced from Europe and Eurasia; Alaska to Newfl. and s. in the w. to the Sierra Nevada of Calif., Nev., Utah and N.M. and in the e. to Mo. and Va. May–Aug.

Small plants of this species are difficult to separate from *P. interior.* The stoloniferous habit is sometimes mistaken for a rhizomatous one. An unconfirmed report of *P. occidentalis* Vasey (Contr. U. S. Natl. Herb. 1: 274. 1893) from Utah belongs, in all likelihood, to *P. palustris. Poa occidentalis,* which is a close relative of *P. palustris,* can be distinguished by its broader leaves and larger lemmas.

14. Poa interior Rydb.

Poa interior Rydb. Bull. Torrey Bot. Club 32: 604. 1905. *Paneion interius* Lunell, Amer. Midl. Naturalist 4: 222. 1915. *Poa nemoralis* var. *interior* Butters & Abbe, Rhodora 49: 6. 1947. (*Tweedy 3706,* "Eastern slope of the Big Horn Mountains. Headwaters of Clear Creek and Crazy Woman River," Johnson Co., Wyo., 20 July–15 Aug. 1900.)

P. subtrivialis Rydb. Bull. Torrey Bot. Club 36: 535. 1909. (*Tweedy 2141,* Big Horn Mts., Sheridan Co., Wyo., July 1899.)

Inland bluegrass.

Stiffly erect, tufted perennials, often scabrous throughout; culms 2–4.5 (7.5) dm tall, greenish at the base; sheaths somewhat keeled or terete, open almost to the base; ligules truncate, 0.5–1.2 (2) mm long, obtuse; blades rather lax, crowded at the base, 1–2 mm broad, usually flat but becoming involute in dried specimens; panicles 4–12 (17) cm long, narrow-pyramidal, with spreading to ascending

POA

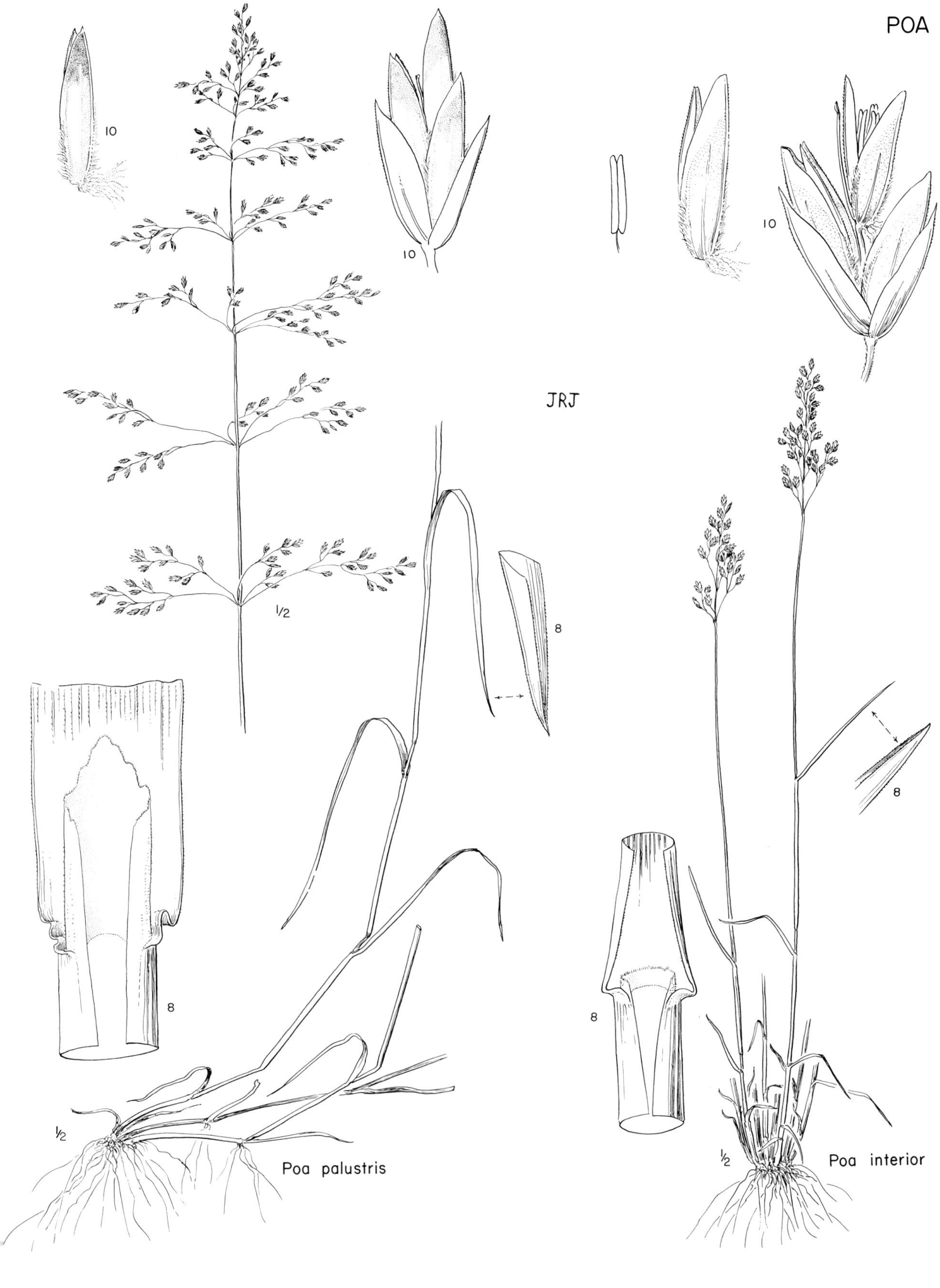

branches, the lower branches 2–5 to a node; spikelets 3–5 mm long, 2- to 3- (4)-flowered, somewhat crowded; glumes acute to acuminate, the nerves prominent, the first glume 1.8–2.8 mm long, 1- (3)-nerved, narrow-lanceolate, the second 2–3 mm long, 3-nerved, lanceolate; lemmas 2.5–3 (3.5) mm long, strongly keeled, villous on the keel and marginal nerves, with a cobwebby tuft at the base, the web sometimes so scant as to be nearly undetected; anthers 1–1.5 mm long; lodicules about 0.4 mm long; $2n$ = 28, 42, 43, 56.

Moist places on open slopes and woods at middle elevs., sometimes up to timberline; Alaska to Que., s. to e. Wash., Idaho, Ariz., N.M., w. Texas, the Dakotas and the Great Lakes region. June–Aug.

Plants of smaller stature with lemmas having very scant cobwebby hairs may be mistaken for timberline bluegrass, *P. rupicola*. The timberline bluegrass, however, consistently lacks the cobwebby hairs. Some forms of *P. pattersonii* with scant cobwebby hairs on the lemma may key here. These plants may be separated from *P. interior* on the basis of the papery, persisting sheaths and the second lemma being nearly as long as the first lemma.

15. **Poa reflexa** Vasey & Scribn.

Poa reflexa Vasey & Scribn. in Vasey, Descript. Cat. Grasses U. S. 83. 1885, nomen nudum; Vasey & Scribn. Contr. U. S. Natl. Herb. 1: 276. 1893. *P. leptocoma* var. *reflexa* M. E. Jones, Contr. W. Bot. 14: 15. 1912. (*Letterman,* on Kelso Mt., near Torrey Peak, Colo., 14 Aug. 1885.)

P. acuminata Scribn. ex Beal, Grasses N. Amer. 2: 538. 1896. (*Tweedy 1027,* subalpine bogs, Mt. Blackmore, Gallatin Co., Mont., July 1886.)

P. pudica Rydb. Bull. Torrey Bot. Club 32: 603. 1905. (*Rydberg 2443,* "Stephen's Mine, below Gray's Peak," Colo., 23 Aug. 1895.)

Nodding bluegrass.

Perennials; culms in small tufts, often solitary, 1.5–5 (6.5) dm tall, erect or sometimes decumbent at the base and rooting at the nodes; basal leaves well developed; sheaths closed about half their length, smooth or sometimes scaberulous; ligules truncate and no longer than 1 mm on basal leaves, becoming as long as 3 mm on the culm leaves, more or less acutish; blades flat, 1–3.5 mm broad, relatively short, boat-shaped at the tip, bright green; panicles 4–12 (15) cm long, open, usually in remote pairs, the lower branches single or in pairs and becoming strongly reflexed, with an abrupt geniculate, somewhat swollen base; spikelets 3–4 (5) mm long, (1-) 2- to 3- (4)-flowered, strongly compressed, the rachilla not showing; glumes subequal, 1.8–3 mm long, the first glume 1-nerved, narrowly lanceolate, the second 3-nerved, lanceolate; lemmas 2.2–3 (3.5) mm long, oblong-elliptic, more or less blunt at the apex, 5-nerved, scarious above and purplish in the middle, villous on the keel and marginal nerves and sometimes on internerves, copiously cobwebby at the base; anthers 0.4–0.8 (1.1) mm long; lodicules about 0.6 mm long.

Wet places on streambanks, open slopes and meadows, subalpine to alpine elevs.; in the Rocky Mts. from B.C. and Alta. s. to N.M., also in e. Oregon, Nev. and Ariz. July–Aug.

Poa reflexa and *P. leptocoma* are so similar that recognizing them as varieties rather than species might be more realistic. If they are so treated, the combination *P. leptocoma* var. *reflexa* (Vasey & Scribn.) M. E. Jones is available. It is not at all uncommon to find specimens of *P. leptocoma* with lemmas slightly less than 3.5 mm long and hardly acuminate, and with lower panicle branches slightly reflexed. In *P. reflexa* the glumes are nearly equal in length and shape, while in *P. leptocoma* the glumes are very unequal.

16. **Poa leptocoma** Trin.

Poa leptocoma Trin. Mém. Acad. Imp. Sci. St.-Pétersbourg, Ser. 6, Sci. Math. 1: 374. 1830. *P. stenantha* var. *leptocoma* Griseb. in Ledeb. Fl. Ross. 4: 373. 1853. (*Mertens,* Sitka, Alaska, in 1829.)

Bog bluegrass.

Perennials; culms few in a tuft or solitary, sometimes appearing rhizomatous because of decumbent bases with roots at the nodes, 1.5–5.5 (7.5) dm tall; sheaths usually scabrous, but sometimes smooth, closed about half their length; ligules truncate to obtuse, sometimes apiculate, 1–3.5 (4) mm long; blades flat, lax, relatively short, bright green, 1–4 mm broad, boat-shaped at the tip, more or less scabrous; panicles (3) 4.5–11 (18) cm long, the capillary branches mostly in pairs, ascending, spreading or rarely somewhat reflexed on the lower pair; spikelets 4–6.5 (7.5) mm long, 2- to 4- (6)-flowered, mostly purplish, strongly compressed; glumes unequal, acute, the first glume 2–3 (3.7) mm long, 1- (3)-nerved, the second 2.5–4 mm long, 3-nerved; lemmas (3) 3.5–4.5 (5) mm long, usually acuminate and pointed, sericeous on the keel and marginal nerves, with cobwebby hairs at the base; anthers 0.5–0.7 (1.7) mm long; lodicules about 0.6 mm long; $2n$ = 42.

Bogs, streambanks and meadows from subalpine to alpine elevs.; Alaska and Yukon, s. to the Sierra Nevada of Calif., Nev., Utah, Ariz. and n. N.M. July–Aug.

See statement regarding similarities between *P. leptocoma* and the preceding species under the discussion for *P. reflexa*.

17. **Poa fendleriana** (Steudel) Vasey

Eragrostis fendleriana Steudel, Syn. Pl. Glum. 1: 278. 1854. *Panicularia fendleriana* Kuntze, Revisio Gen. Pl. 2: 782. 1891. *Poa fendleriana* Vasey, U.S.D.A. Div. Bot. Bull. 13(2): pl. 74. 1893. *Atropis fendleriana* Beal, Grasses N. Amer. 2: 576. 1896. *Puccinellia fendleriana* Ponert, Feddes Repert. 84: 739. 1974. (*Fendler 932,* N.M., in 1847.)

Poa eatoni S. Wats. in King, Rep. Geol. Explor. 40th Parallel 5: 386. 1871. (*D. C. Eaton,* rocky gulch of Cottonwood Canyon, Wasatch Mts., Utah, in 1869.)

P. montana Vasey, Monthly Rep. Dept. Agric. 155. 1874; not All. 1785. *P. alpina* var. (?) S. Wats. in King, Rep. Geol. Explor. 40th Parallel 5: 386. 1871. (*Watson 1312, 1313,* "West Humboldt, Clover, and East Humboldt Mountains, Nevada.")

P. andina var. *spicata* Vasey in Wheeler, Rep. U. S. Geogr. Surv. W. 100th Meridian 6: 290. 1879. *P. fendleriana* var. *spicata* Scribn. in Rydb. & Shear, U.S.D.A. Div. Agrostol. Bull. 5: 31. 1897. (*Wolf 1135,* Colo., in 1873.)

P. longepedunculata Scribn. U.S.D.A. Div. Agrostol. Bull. 11: 54, pl. 11. 1898. (*Nelson 3292,* "summit of Sheep Mt.," Albany Co., Wyo., 3 July 1897.)

P. brevipaniculata Scribn. & Williams in Scribn. U.S.D.A. Div. Agrostol. Circ. 9: 2. 1899. (*Breninger 554,* Table Rock, Colo., 25 May 1891.)

P. longiligula Scribn. & Williams in Scribn. U.S.D.A. Div. Agrostol. Circ. 9: 3. 1899. *Paneion longiligulum* Lunell, Amer. Midl. Naturalist 4: 222. 1915. *Poa fendleriana* var. *longiligula* Gould, Madroño 10: 94. 1949. (*Jones 5149,* Silver Reef, Washington Co., Utah, 3 May 1894.)

P. longiligula var. *wyomingensis* T. A. Williams, U.S.D.A. Div.

POA
10
10
10
JRJ
8
8
8
1/2
Poa reflexa
1/2
Poa leptocoma

Agrostol. Circ. 10: 3. 1899. (*Nelson 4799a,* Tipton, Sweetwater Co., Wyo., 17 June 1898.)

P. scabriuscula T. A. Williams, U.S.D.A. Div. Agrostol. Circ. 10: 4. 1899. (*Ward 136,* "mountains south of Glenwood, Utah," 29 May 1895.)

Mutton grass.

Perennials forming large old tufts, (2) 2.5–5.5 (7.5) dm tall; basal leaves erect, long, many, usually strongly glaucous, firm and stiff; sheaths at the base papery and persisting for many years, somewhat scabrous, open nearly their full length; ligules highly variable, often truncate and as short as 0.5–1 mm long, so as to be scarcely visible when viewed from the side, to much elongate, acuminate and sometimes even exceeding 10 mm in length, mostly 3–8 mm long; blades stiff, usually scabrous, sometimes flat but more often folded or involute on the margins, 1.5–3 (4) mm broad, the tip pointed, but boat-shaped; culm blades becoming increasingly reduced and often lacking so the ligule tops the bladeless sheath; plants usually dioecious and mostly pistillate with an occasional plant seemingly with functionally perfect flowers; panicles (3) 5–10 (15) cm long and 1–2 cm wide, oblong, contracted, usually carried well above the basal leaves, pale to deep purple or sometimes tawny; spikelets usually closely flowered, relatively large, (5) 6–8 (10.5) mm long, 3- to 6-flowered, strongly compressed and keeled throughout, often appearing papery; first glume 2.8–4.5 mm long, the second glume broader, 3–5 (5.5) mm long; lemmas 4–6 (7) mm long, blunt on the usually erose apex, with sericeous hairs on the prominently nerved keel and marginal nerves, the intermediate nerves obscure, the internerves usually glabrous, not at all webbed at the base; anthers on the minute filaments usually less than 1 mm long; lodicules about 0.7 mm long.

Relatively dry sagebrush valleys and foothills to wooded areas at middle elevs.; B.C. to Man., s. to Calif., Nev., Ariz., N.M., w. Texas and Nebr. May–Aug.

Mutton grass is a widely distributed and often abundant grass in pinyon-juniper woodlands where it is an important forage species.

Occasional plants with rhizomes are found, and careful field observations will usually suggest that a species of the Pratenses group has contributed them. No attempt is made here to recognize *P. longiligula* Scribn. & Williams as a species or even as a variety as the continuous variation in length of the ligule has no other associated characters, including geographical distribution, to justify any kind of separation. There is some doubt that our plants undergo actual fertilization. All specimens appear to be pistillate. Some plants may reproduce sexually from pollen contributed by a species of the Pratenses group.

18. **Poa alpina** L.

Poa alpina L. Sp. Pl. 67. 1753. *P. pratensis* var. *alpina* Huds. Fl. Anglica, ed. 2. 39. 1778. (Lapland.)

P. alpina var. *minor* Scribn. ex Beal, Grasses N. Amer. 2: 543. 1896; not Koch, 1837. (*Scribner 388,* Little Belt Mts., Mont., 12 Aug. 1883.)

Alpine bluegrass.

Densely tufted perennials with thick mats of basal leaves; culms 1–2 (3.5) dm tall; sheaths persisting for more than one year and often densely clothing each bunch, open nearly to the base, smooth; ligules 1–4 mm long, truncate to obtuse; blades short, usually flat, 1.5–3 (5) mm broad, the basal leaves 3–5 cm long, the culm leaves 1 or rarely 2 on lower half leaving most of the culm naked; panicles 2–4.5 (6) cm long and as wide as long, usually pyramidal, sometimes ovoid, rather compact, the short branches capillary and usually spreading or sometimes slightly reflexed; spikelets 4.5–6 (7) mm long, strongly compressed, purplish, (3-) 4- to 5- (6)-flowered, broad in full anthesis, often subcordate; glumes ovate, abruptly acute, more than half as wide as long, both faintly 3-nerved, the first glume 2.2–3 (3.5) mm long, the second 2.5–3.5 mm long; lemmas 3–4 mm long, faintly nerved, strongly scarious-margined, abruptly acuminate, turned inward, villous on the keel and marginal nerves, pubescent on the internerves near the base, not at all webbed at the base; anthers 1.2–2 mm long; lodicules about 0.6 mm long; $2n = 14, 21–74$.

Subalpine and alpine meadows and slopes; Alaska to Que. and s. to e. Oregon, Idaho, n. Utah and Colo., and n. Mich.; also Mex. and Eurasia. July–Aug.

Alpine bluegrass is only known in our flora from the Uinta Mountains where it is fairly uniform with dense tufts of short basal leaves and old sheaths and broadly pyramidal panicles.

19. **Poa pattersonii** Vasey

Poa pattersoni Vasey, Contr. U. S. Natl. Herb. 1: 275. 1893. *P. abbreviata* subsp. *pattersonii* A. Löve, D. Löve & Kapoor, Arctic & Alpine Res. 3: 142. 1971. (*Patterson 154,* "Top of Mt. McClellan, near Gray's Peak," Colo., 19 Aug. 1885.)

Densely tufted alpine perennials with numerous basal leaves; culms 0.5–2.5 (5) dm tall; sheaths elongate and papery, persisting for a long time, open mostly the full length, glabrous to scabrous; ligules 1–2.5 (4) mm long, round or obtuse to acute; blades usually folded, rather lax and gracefully curving, mostly less than 10 cm long, about 1–2 (2.5) mm broad; panicles dense but soft and lax, narrow, greenish or purplish, the branches mostly in 2s, (1.5) 2.5–6 (9) cm long; spikelets often nearly sessile, (3.2) 4–6.5 (7.5) mm long, strongly compressed, 2- to 3- (4)-flowered; first glume 2–4 (4.5) mm long, 1-nerved, the second glume (2.5) 3–5 mm long, 1- to 3-nerved, often equalling the first lemma; lemmas 2.5–5 (5.5) mm long, acute, strongly pubescent on the keel and marginal nerves and with shorter hairs on the internerves, the body pubescent at the base, usually with a purplish band below the scarious apex, and sometimes with a scant tuft of cobwebby hairs at the base; palea about 3/4 as long as the lemma; anthers 0.8–1.8 mm long; lodicules about 0.6 mm long.

Subalpine and alpine slopes and meadows, mostly near or above timberline; s. Alta. to Idaho, Mont. and s. to Nev., n. Ariz. and Colo. July–Aug.

This species is sometimes represented by plants having lemmas with cobwebby hairs at the base that are so scant as to be easily missed. The type specimen has the cobwebby hairs, but the type description states that the lemma is not webbed. Specimens with cobwebby hairs may be confused with *P. interior;* see discussion under that species above. *Poa pattersonii* may be separated from the closely related *P. rupicola* on the basis of the papery, persisting sheaths and the second lemma being nearly as long as the first lemma.

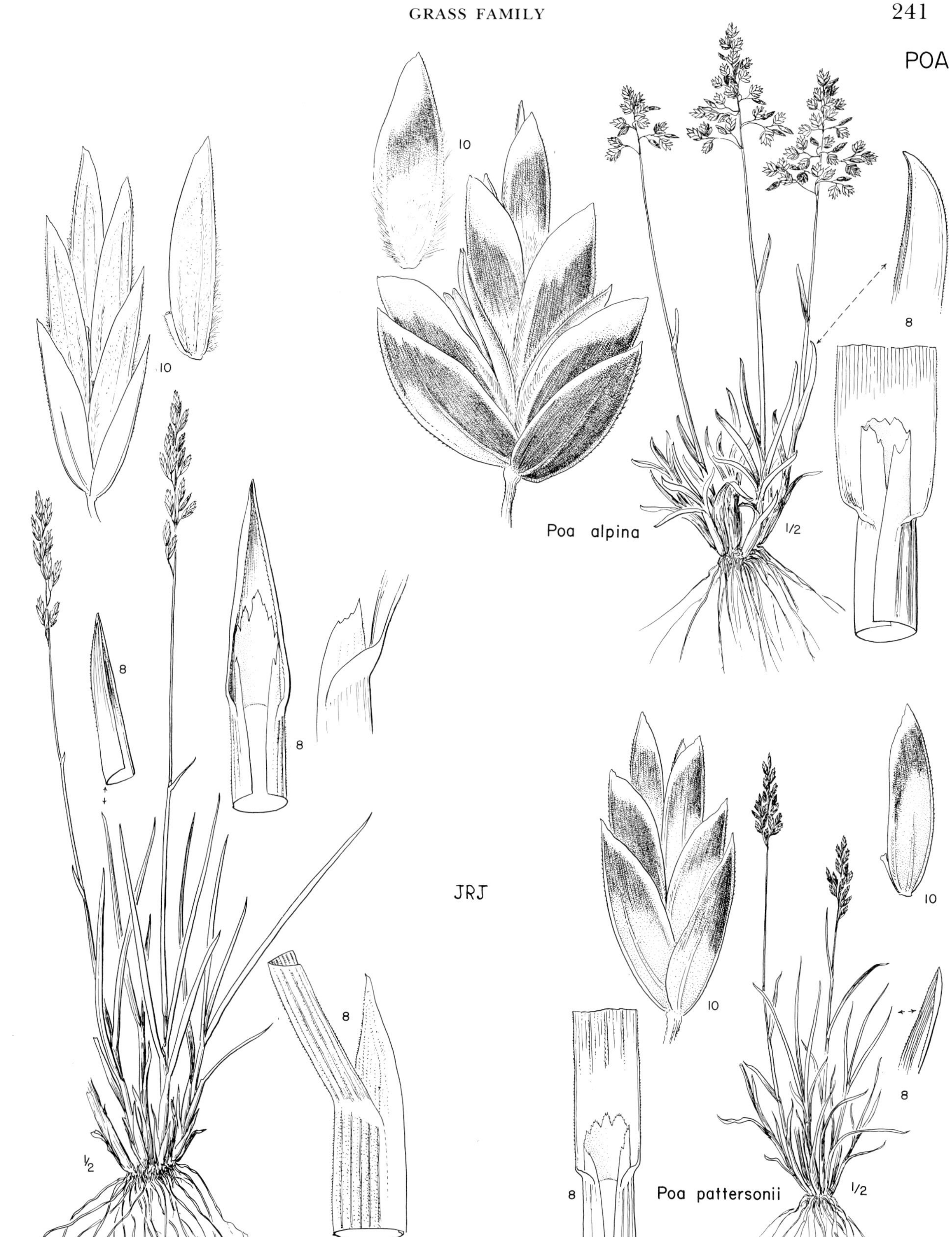
POA
10
10
8
1/2
Poa alpina
8
8
JRJ
10
10
8
8
1/2
Poa fendleriana
8
Poa pattersonii
1/2

20. Poa rupicola Nash

Poa rupestris Vasey, Bull. Torrey Bot. Club 14: 94. 1887; not With. 1796. *P. rupicola* Nash in Rydb. Mem. New York Bot. Gard. 1: 49. 1900. *P. glauca* var. *rupicola* B. Boivin, Naturaliste Canad. 94: 527. 1967. (*Wolf 341,* South Park, Colo., in 1873; lectotype by A. S. Hitchcock.)

Timberline bluegrass.

Rather rigid, densely tufted perennials; culms 0.5–2.5 (3.5) dm tall; basal leaves numerous, greatly exceeded by the culms; sheaths smooth, closed at the base only; ligules truncate to rounded, 0.5–2.2 mm long; blades involute, short and stiff, 1–2.2 mm broad and 1–5 cm long; panicles 1–5 (7) cm long, narrow and compact, usually purplish, with short ascending or appressed branches; spikelets 3.5–6 (7) mm long, 2- to 4-flowered, strongly compressed; glumes broad, often prominently 3-nerved, the first glume 2–3 (3.5) mm long, the second 2–3.5 mm long; lemmas 2.5–4 (5) mm long, strongly keeled, villous below on the keel and marginal nerves, and often pubescent on the internerves of lower half, not at all cobwebby at the base; anthers 1–1.5 mm long; lodicules about 0.4 mm long.

A common bluegrass on rocky ridges, slopes and meadows at or above timberline; B.C. and Mont., s. to the White Mts. and Sierra Nevada of Calif., n. Ariz. and N.M. July–Aug.

Tall specimens of this species often resemble *P. interior,* a species with cobwebby hairs at the base of the lemma. Timberline bluegrass is usually found at higher elevations than *P. interior.*

21. Poa cusickii Vasey

Poa subaristata Scribn. ex Macoun, Cat. Canad. Pl. 4: 227. 1888, nomen nudum; Scribn. ex Beal, Grasses N. Amer. 2: 533. 1896; not Phil. 1896 (slightly earlier). (*Tweedy 633,* East Fork, Yellowstone Park, Wyo., Aug. 1885.)

P. cusickii Vasey, Contr. U. S. Natl. Herb. 1: 271. 1893. (*Cusick 1219,* "Moist hill sides," Oregon, May 1885.)

P. filifolia Vasey, Contr. U. S. Natl. Herb. 1: 271. 1893; not Schur, 1866. *P. idahoensis* Beal, Grasses N. Amer. 2: 539. 1896. *P. scabrifolia* A. A. Heller, Bull. Torrey Bot. Club 24: 310. 1897. (*Sandberg 138,* Hatwai Cr., Nez Perce Co., Idaho, in 1892.)

P. spillmani Piper, Erythea 7: 102. 1899. (*Spillman,* between Coulee City and Waterville, Douglas Co., Wash., 27 May 1896.)

P. nematophylla Rydb. Bull. Torrey Bot. Club 32: 606. 1905. (*Osterhout 2601,* Meeker, Rio Blanco Co., Colo., 8 June 1902.)

P. scaberrima Rydb. Bull. Torrey Bot. Club 36: 534. 1909. (*Rydberg 2055,* Beaver Canyon, Idaho, 27 June 1895.)

Cusick bluegrass.

Strongly tufted, dioecious perennials, usually with abundant basal leaves from numerous innovations, often producing large bunches; culms (1) 2–5 dm tall, numerous, usually producing two short leaves near the base; sheaths closed about half their length, often papery and persistent for several years; ligules usually acute, 0.5–3 (6) mm long, those on the innovations usually very short and nearly truncate; blades numerous, filiform, less than 1 mm broad, usually elongate, up to 25 cm long, more or less scabrous but sometimes obscurely so, the culm leaves with long sheaths and short blades; panicles 3–7 (11) cm long, usually ovoid, compact, or in some forms becoming rather open, usually pale and shining but sometimes with a pale purplish tinge; spikelets (5) 6–8.5 (10.5) mm long, strongly compressed, (2) 3- to 4- (5)-flowered, usually on filiform pedicels; florets usually pistillate, with some plants producing vestiges of stamens or some plants even seemingly perfect, the perfect florets usually associated with rather open panicles; glumes broad, broadly scarious-margined, shorter than the first floret, the first glume 2.5–5 (6) mm long, 1-nerved, the second 3–6 mm long, 3-nerved; lemmas (4) 4.5–7 mm long, usually prominently 5-nerved, often sharply acute, scabrous or occasionally smooth, sometimes with a few hairs on the keel, rather thin, scarious on the margins and apex, often with a pale purplish tinge at about the middle; palea nearly as long as the lemmas and scabrous on the keels; anthers 2.2–3 mm long; lodicules about 0.8 mm long; $2n = 28, 42, 56, 59$.

Sagebrush slopes and open wooded areas in dry places, usually at middle elevs. but sometimes ranging up to timberline; B. C. to Sask., s. to Calif. (Sierra Nevada), c. Nev., n. Utah, n. Colo. and N.D. May–July.

Parts of large old clumps of *P. cusickii* are sometimes interconnected by short rhizomes, especially in loose, friable soils. Cusick bluegrass undoubtedly is apomictic, since all specimens examined are pistillate. *Poa cusickii* may be mistaken for *P. fendleriana,* another species that usually produces only pistillate flowers, as both occupy similar ecological niches and do superficially resemble one another. The lemmas of *P. cusickii* rarely have any pubescence, but if present the hairs are confined to the keel. The sharply acute lemmas are also in contrast to the more blunt lemmas of *P. fendleriana.* In some areas the basal leaves are very similar, and care must be exercised in having a close look at the lemmas. The two species are certainly more closely related than their placement in so-called different sections (*P. cusickii* in the Epiles and *P. fendleriana* in the Alpinae) would suggest. Plants intermediate with *P. sandbergii* of the *P. scabrellae* group are sometimes encountered in northern Nevada and southeastern Oregon. In these cases our specimens have been named *P. cusickii* if the flowers are pistillate and *P. sandbergii* if the flowers are perfect. The possibility of *P. sandbergii* furnishing pollen for *P. cusickii* may give an answer to these problems. *Poa cusickii* begins growth very early in the season, and it is not at all unusual to find plants that have produced two crops of inflorescences in a season, the first one having been nipped by a late spring frost.

22. Poa epilis Scribn.

Poa purpurascens Vasey, Bot. Gaz. 6: 297. 1881; not Sprengel, 1819. *P. alpina* var. *purpurascens* Vasey, Descript. Cat. Grasses U.S. 79. 1885. *P. paddensis* T. A. Williams in Scribn. U.S.D.A. Div. Agrostol. Bull. 17(ed. 2): 261, fig. 557. 1901. *P. subpurpurea* Rydb. Bull. Torrey Bot. Club 32: 606. 1905. *P. cusickii* var. *purpurascens* C. L. Hitchc. Univ. Wash. Publ. Biol. 17(1): 659. 1969. (*Howell,* "Base of Mount Hood," Oregon, in 1881.)

P. epilis Scribn. U.S.D.A. Div. Agrostol. Circ. 9: 5. 1899. *P. purpurascens* var. *epilis* M. E. Jones, Contr. W. Bot. 14: 14. 1912. *P. cusickii* var. *epilis* C. L. Hitchc. Univ. Wash. Publ. Biol. 17(1): 659. 1969. (*Shear & Bessey 1457,* "Open places in thin moist timberland," Buffalo Pass, Routt Co., Colo., 13 Aug. 1898.)

Skyline bluegrass.

Loosely tufted perennials; culms erect, not especially dense, 1.5–4.5 (5.5) dm tall; sheaths closed about half their length, smooth and often shiny at the base, becoming fibrous or papery and often persisting; ligules 1.5–5 mm long, sometimes long attenuate; blades usually bright green, usually folded or sometimes involute in drying, 1–3 mm broad and 6–20 cm long, the culm blades 2 or 3, flat or folded, rarely exceeding 6 cm long, smooth; panicles 2.5–7 (10.5) cm long, usually condensed, ovoid or oblong,

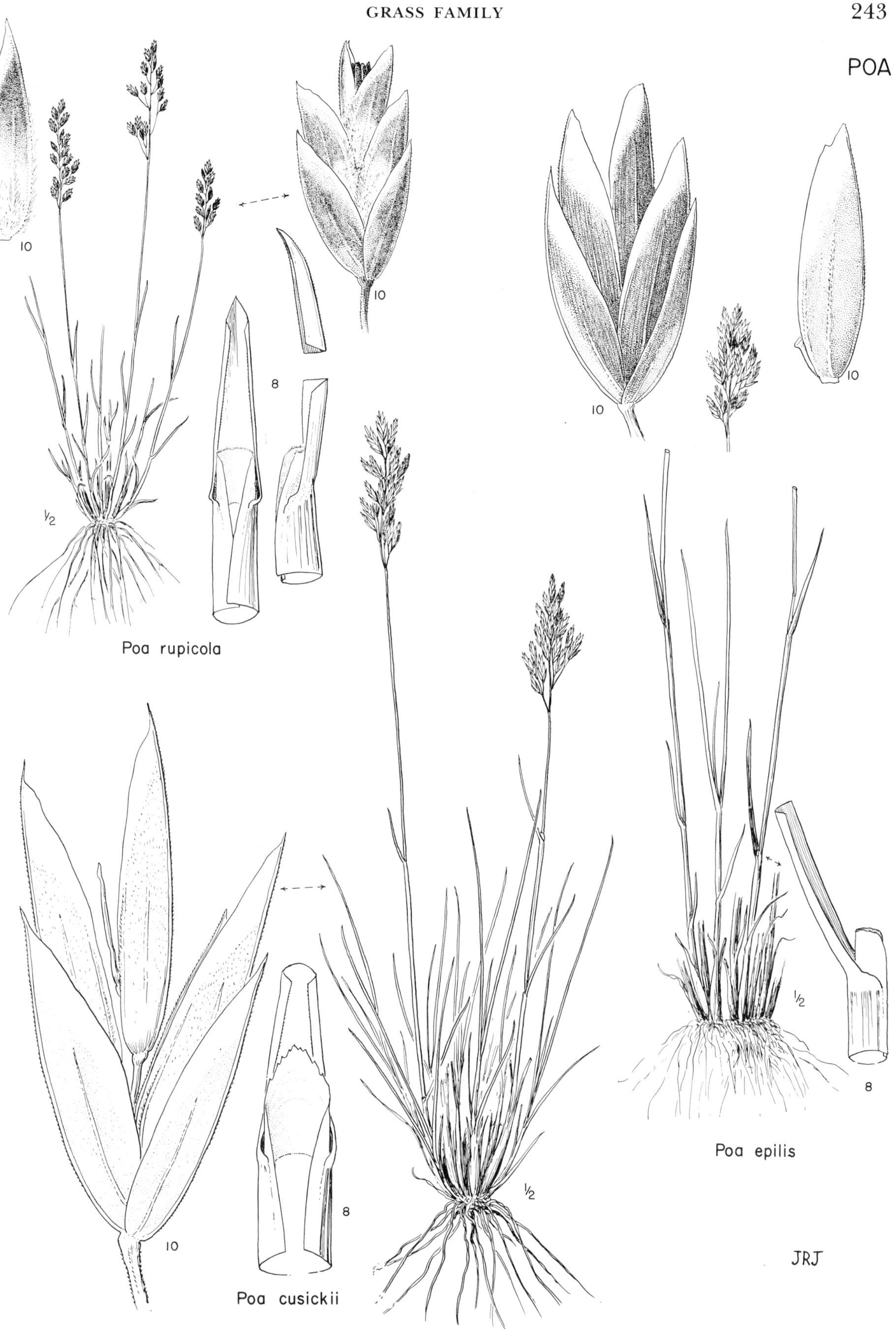
POA
Poa rupicola
Poa epilis
Poa cusickii
JRJ

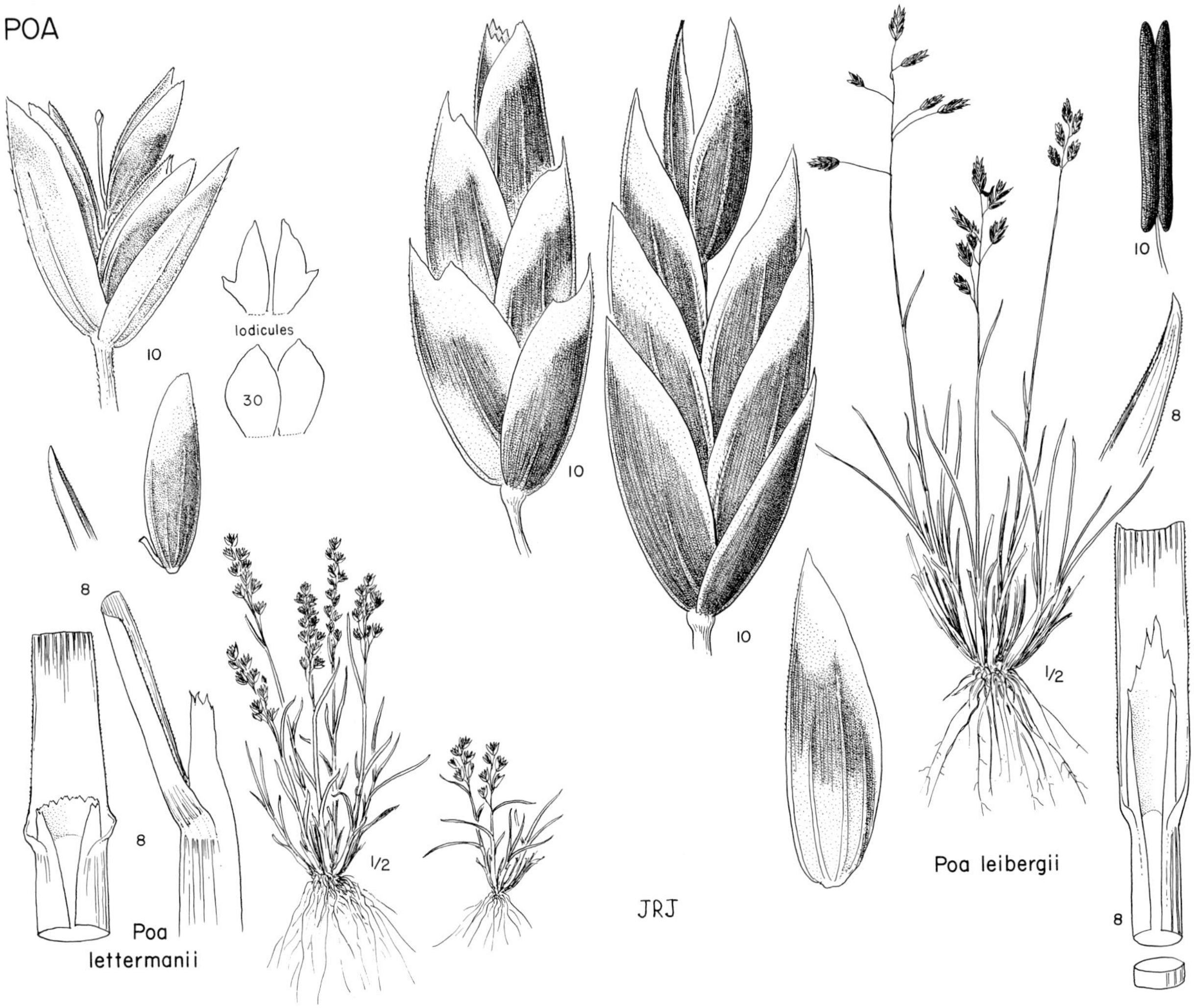

the branches usually appressed but sometimes spreading, usually deep purple but sometimes green; spikelets 5–9 mm long, strongly compressed, 2- to 5-flowered; florets usually pistillate; glumes shorter than the florets, the first 2.2–4.5 (5) mm long, 1-nerved, the second 2.8–5 mm long, 1-nerved or obscurely 3-nerved; lemmas 4–6 (6.5) mm long, often minutely scabrous, acute, 5-nerved, the internerves often obscure, with scarious margins and apex, not cobwebby at the base; palea about $^{2}/_{3}$ the length of the lemma; anthers about 2.5 mm long; lodicules about 0.8 mm long; $2n$ = 56, 84.

Subalpine and alpine meadows or rocky ridges and slopes; B.C. and Alta., s. to Calif. (Sierra Nevada), Nev., Utah and Colo. Late June–Aug.

Poa cusickii and *P. epilis* are considered here as distinct species and can be separated with more satisfaction than several other species in our area. The species are ecologically and cytologically distinctive.

23. **Poa lettermanii** Vasey

Poa lettermani Vasey, Contr. U.S. Natl. Herb. 1: 273. 1893. *Atropis lettermani* Beal, Grasses N. Amer. 2: 579. 1896. *Puccinellia lettermanii* Ponert, Feddes Repert. 84: 740. 1974. (*Letterman 7,* Gray's Peak, Colo., July 1885.)

Poa brandegei Scribn. ex Beal, Grasses N. Amer. 2: 544. 1896. (*Jones 714,* Gray's Peak, Colo., 29 Aug. 1878.)

Letterman bluegrass.

Dwarf, densely tufted alpine perennials; culms 0.3–0.8 (1.2) dm tall; basal leaves often exceeding the inflorescence, or the culms not much longer than the basal rosette; sheaths open their full length or nearly so; ligules usually truncate, erose, 0.5 mm on basal leaves to about 2 (3) mm on the culm leaves, acute to truncate; blades lax, folded, or if flat 0.5–2 mm broad; panicles 0.7–2 cm long, dense, narrow; spikelets 2.5–4 (4.5) mm long, strongly compressed, usually purplish, (1-) 2- to 3-flowered; glumes nearly as long as the spikelet, subequal, 2–3.2 mm long, acuminate, glabrous, the first glume 1-nerved, the second 1- to 3-nerved; lemmas glabrous, (2) 2.2–3 (3.5) mm long, hyaline at the apex, blunt, more or less erose, not cobwebby at the base; anthers about 0.5 mm long; lodicules 0.4 mm long.

Rocky ridges and ledges mostly above timberline in the Uinta (Utah) and Ruby (Nev.) mts.; B.C., Wash. and Calif. (Sierra Nevada), and e. to Wyo. and Colo. Aug.

This species is rarely encountered in our region. It is the smallest

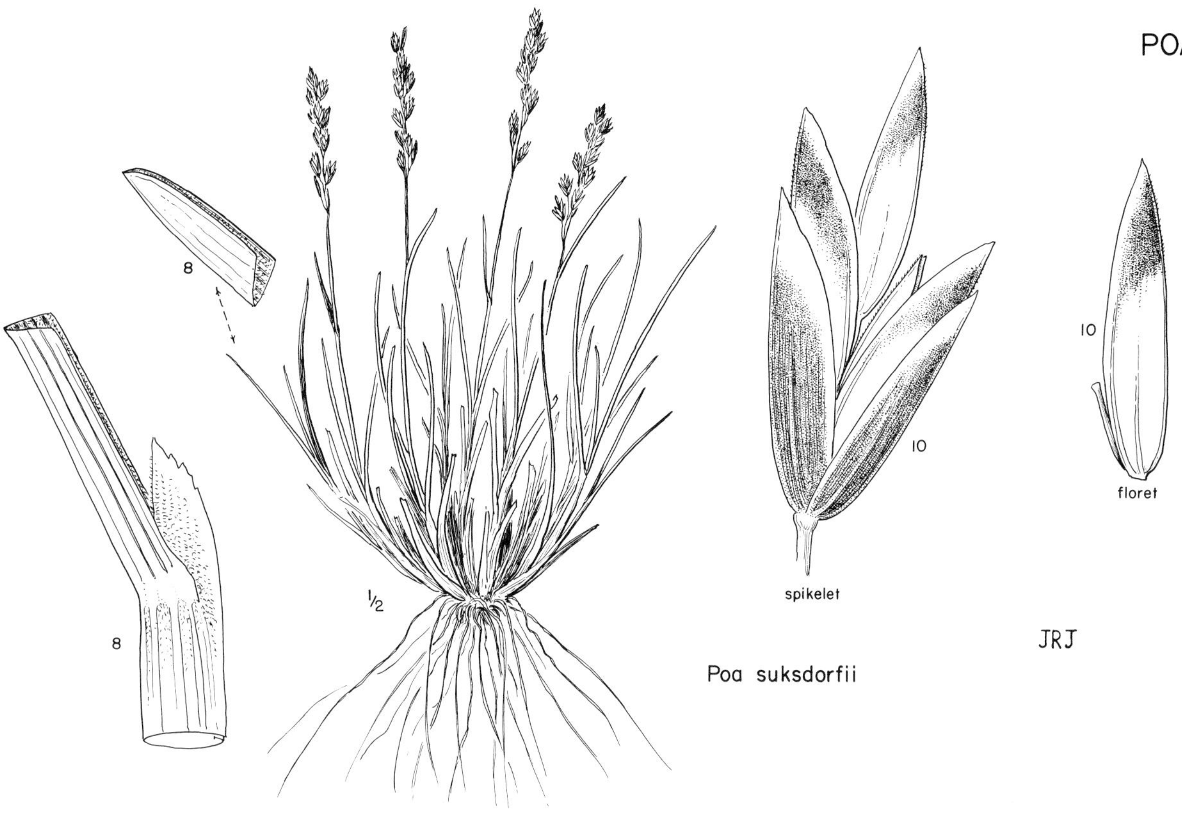

bluegrass in our flora. Plants of this species may go through their very short growing season without producing flowering culms. Tiny rosettes may be found around flowering specimens. These go undetected unless inflorescences are produced.

24. Poa leibergii Scribn.

Poa leibergii Scribn. U.S.D.A. Div. Agrostol. Bull. 8: 6, pl. 2. 1897. (*Leiberg 2171,* Barren Valley, Owyhee-Malheur Divide, Malheur Co., Oregon, 31 May 1896.)

Leiberg bluegrass.

Small densely tufted perennials; culms 0.5–2.5 (4) dm tall; leaves nearly all basal; sheaths glabrous, closed about half their length; ligules on the basal leaves about 1 mm long, rounded and entire, to about 3 (5) mm long and acute on the culm leaves; blades soft, folded, or if flat 1–2 mm broad, becoming more or less involute upon drying, the tips decidedly boat-shaped; panicles narrow, 1–5 (7) cm long; spikelets (4.5) 6–9 (11) mm long, 3- to 7-flowered, strongly compressed, usually purplish; florets perfect or seemingly dioecious; first glume 2.2–4.2 mm long, 1-nerved, the second glume (2.5) 3–4.5 mm long, 3-nerved; lemmas 4–6 mm long, smooth or scaberulous, not cobwebby at the base; anthers 2–3 mm long, apparently functional; lodicules about 1 mm long.

Moist meadows and rocky slopes from the sagebrush zone to the western juniper zone; s. Wash., s. to Malheur Co., Oregon and possibly adjacent Idaho. Late March–May.

The Columbia River Gorge element may turn out to be a distinctive species under the name *Poa vaseyochloa* Scribn. *Poa vaseyochloa* may be distinguished from *P. leibergii* by the open inflorescence, sometimes even with reflexed lower branches, broader spikelets with loosely arranged florets and exposed rachilla.

25. Poa suksdorfii (Beal) Vasey ex Piper

Atropis suksdorfii Beal, Grasses N. Amer. 2: 574. 1896. *Poa suksdorfii* Vasey ex Beal, ibid., in synonymy; Vasey ex Piper, Contr. U. S. Natl. Herb. 11: 135. 1906. (*Suksdorf 1116,* "Gravelly places near glaciers," Mt. Adams, Wash., 3 Sept. 1891.)

Small, densely tufted perennials; culms 0.5–2 (2.5) dm tall; leaves nearly all basal; sheaths glabrous to sometimes slightly pubescent near the collar, closed half their length; ligules about 1 mm long, rounded to obtuse, entire; blades involute, often filiform, glabrous to scabrous ventrally; panicles narrow, 2–5 (6) cm long, scarcely exceeding the basal tuft of leaves; spikelets 4.2–7 (8) mm long, 2- to 3- (4)-flowered, somewhat compressed, usually purplish; glumes 3–4.5 (5) mm long, usually equalling or exceeding the first lemma, glabrous to scaberulous, the first glume 1-nerved, the second 3-nerved, slightly longer; lemmas 3–4 (5) mm long, scaberulous, not cobwebby at the base; anthers 0.8–1.3 mm long; lodicules about 0.8 mm long.

Alpine tundra on talus slopes, meadows and summits; high mts. of Wash. and Oregon, s. to the c. Sierra Nevada and White Mts. of Calif. Mid-July–Aug.

The very similar and closely related *P. hansenii* Scribn. was reported from the White Mountains, Inyo Co., California (Lloyd & Mitchell, Fl. White Mts., Calif. & Nevada 183. 1973), but the specimens cited turn out to be a high elevation form of *P. sandbergii*. *Poa hansenii* is a Sierra Nevada endemic found at subalpine rather than alpine elevations. Morphologically it differs from *P. suksdorfii* in having culms greatly exceeding the basal leaf tufts.

26. Poa gracillima Vasey

Poa gracillima Vasey, Contr. U. S. Natl. Herb. 1: 272. 1893. (*Suksdorf 33,* Mt. Paddo [Adams], Wash., 10 Aug. 1882.)

Atropis tenuifolia var. *stenophylla* Vasey ex Beal, Grasses N. Amer. 2: 580. 1896. *P. buckleyana* var. *stenophylla* M. E. Jones, Contr. W. Bot. 14: 14. 1912. (*Howell,* Roseburg, Oregon, 4 May 1887.)

P. saxatilis Scribn. & Williams in Scribn. U.S.D.A. Div. Agrostol. Circ. 9: 1. 1899. *P. gracillima* var. *saxatilis* Hackel, Allg. Bot. Z. Syst. 21: 79. 1915. (*Piper 1964,* "dry rocky places," Mount Rainier, Wash., Aug. 1895.)

Pacific bluegrass, slender bluegrass.

Loosely tufted perennials from erect or decumbent bases; culms slender, often numerous, 2–5 (6) dm tall; leaves mostly all basal; sheaths tawny or purplish, closed less than half their length, glabrous, often appearing papery and persisting; ligules acuminate, often laciniate, about 1.5 mm long on the basal leaves and 3–5.5 mm on the culm leaves; blades pale green or glaucescent, flat, folded, involute, or sometimes filiform, 0.7–2 mm broad and 5–15 cm long, the tips more or less boat-shaped on broader blades; panicles 4–9 (12) cm long, open and usually pyramidal, sometimes ovate, the branches usually spreading, often at right angles and then the 4 or 5 whorls very evident, or rarely with appressed branches; spikelets 5–8 (9) mm long, 2- to 4-flowered, more or less terete; glumes acute, with a prominent hyaline margin, the first glume 2.5–4 mm long, faintly 1-nerved, the second 3–5 mm long, faintly 3-nerved; lemmas (3) 4–5 (6) mm long, somewhat keeled, only rounded at the base, usually narrow and acute to slightly obtusish, usually crisp-pubescent to nearly villous on the keel and marginal nerves and with shorter hairs on the body at the base or less often glabrous, without a web; anthers 1.2–2.2 mm long; lodicules about 0.7 mm long; $2n = 81, 84, 86$.

Meadows, streambanks and rocky slopes from middle elevs. to timberline; B.C. to Alta., s. to the Sierra Nevada of Calif., n. Nev., n. Utah and Colo. May–Sept.

Poa gracillima is closely related to *P. sandbergii,* and the identity of some plants has been confused. *Poa gracillima* is summer active and usually has a decumbent base, flat, lax blades that soon wither, an open, pyramidal panicle and lemmas with decidedly longer hairs on the keels and marginal nerves. The two are maintained here as distinct because the multiple characters that must be used to separate them usually result in satisfactory identifications. Both are apparently sexually reproducing species in most areas but are effectively isolated by the later flowering and higher altitude of *P. gracillima.*

27. Poa scabrella (Thurber) Benth. ex Vasey

Poa tenuifolia Buckley, Proc. Acad. Nat. Sci. Philadelphia 1862: 96. 1863; not A. Rich. 1851. *Atropis tenuifolia* Thurber in S. Wats. Bot. Calif. 2: 310. 1880. *Panicularia nuttaliana* Kuntze, Revisio Gen. Pl. 2: 783. 1891. *Poa buckleyana* Nash, Bull. Torrey Bot. Club 22: 465. 1895. (*Nuttall,* Columbia River, July–Aug. 1835.)

Atropis scabrella Thurber in S. Wats. Bot. Calif. 2: 310. 1880. *P. scabrella* Benth. ex Vasey, Grasses U. S. 42. 1883. *Panicularia scabrella* Kuntze, Revisio Gen. Pl. 2: 783. 1891. *Puccinellia scabrella* Ponert, Feddes Repert. 84: 740. 1974. (*Bolander,* dry meadows, Oakland, Calif., in 1875.)

Pine bluegrass, Malpais bluegrass.

Medium sized tufted perennials; culms (2) 3–6 (9) dm tall, smooth or scabrid; leaves numerous, mostly basal and often gracefully curved; sheaths closed only at the base, usually scabrous but sometimes smooth, often purplish, fibrous or papery and densely clothing the base of plants; ligules 2–6.5 mm long, acuminate, becoming lacerate, sometimes noticeably decurrent; blades flat or folded, lax, 1–3 mm broad, more or less scabrous, sometimes on the margins only, but more often very scabrous on both surfaces, the basal tuft (3) 8–15 (25) cm tall; panicles (4) 6–12 (16) cm long, usually rather narrow and contracted, occasionally opening, green with tinges of purple but more often brown; spikelets (5) 6–8 (9.5) mm long, 2- to 5- (6)-flowered, very little compressed and nearly terete; glumes obscurely 3-nerved, glabrous or more often slightly scabrous, acute or acuminate, scarious-margined, the first glume 2.7–4.5 mm long, the second 3–5 mm long; lemmas (3.5) 4–5.5 mm long, erose at the subacute apex, rounded on the back, becoming keeled above, becoming increasingly puberulent towards the base, scarious on the margins and especially towards the apex, not at all webbed at the base; palea nearly as long as the lemma, ciliate-scabrous on the keels; anthers 0.6–2.3 mm long; lodicules about 0.6 mm long; $2n = 44, 62–64, 81, 82, 84, 86, 91, 104$.

Relatively dry places on sagebrush hills and woodlands at low and middle elevs.; B.C. to Alta., s. to Baja Calif., Nev., Utah and Colo. and e. to Minn. April–July.

This is a highly variable species that often dominates some plant communities. The species is spring flowering and summer dormant, which more or less isolates it from the closely related *P. canbyi* which is actively growing during the summer. *Poa scabrella* is distinct from the norm of *P. sandbergii* in cismontane California, but throughout the Intermountain Region, where the two come together, they intergrade so freely that it is difficult to find a clear-cut identity for *P. scabrella.*

28. Poa sandbergii Vasey

Poa andina Nutt. ex S. Wats. in King, Rep. Geol. Explor. 40th Parallel 388. 1871; not Trin. 1836. *P. californica* Munro ex J. M. Coulter, Manual Bot. Rocky Mt. Region 420. 1885; not Steudel, 1854. (*Watson 1319,* "Colorado. East and West Humboldt Mountains and in Clover Mountains, Nevada; also in the Trinity Mountains.")

P. sandbergii Vasey, Contr. U. S. Natl. Herb. 1: 276. 1893. *P. buckleyana* var. *sandbergii* M. E. Jones, Contr. W. Bot. 14: 14. 1912. *Paneion sandbergii* Lunell, Amer. Midl. Naturalist 4: 223. 1915. (*Sandberg 164,* "near Lewiston, Idaho," 14 May 1892.)

Poa incurva Scribn. & Williams in Scribn. U.S.D.A Div. Agrostol. Circ. 9: 6. 1899. (*Piper 1989,* "Morain of Duckaloose Glacier," Olympic Mts., Wash., Aug. 1895.)

P. secunda of Amer. authors; not Presl, 1830.

Sandberg bluegrass.

Small, densely tufted perennials, often producing extensive tufts of basal leaves that wither and disappear early; plants often more or less completely purple-tinged; culms 2–3.5 (4.5) dm tall, wiry; culm leaves usually 1 or sometimes 2; sheaths glabrous to scabrous, closed only at the base, persisting and often in thick rings; ligules acute, usually rather prominent, those on the basal leaves mostly less than 1.5 mm long and those of the culm leaves usually more than 2.5 mm; blades soft, folded or involute, rarely flat, 0.5–1.5 mm broad and 3–5 cm long, often

POA
10
10
10
10
10
JRJ
8
10
8
8
8
1/2
1/2
1/2
Poa sandbergii
Poa gracillima
Poa scabrella

becoming curled, the tip usually definitely boat-shaped, the basal cluster of leaves 3–10 (14) cm high; panicles (2) 3–7 (10) cm long, narrow to rather open, but usually not dense, the branches short, usually ascending; spikelets (4) 5.5–8.5 (9.5) mm long, 2- to 4-(6)-flowered, little compressed and rather terete, narrow and pointed, usually strongly tinged with purple; glumes acute, hyaline-margined, scabrous, the first glume 2.2–4 (5) mm long, 1-nerved, the second 3–5 mm long, 3-nerved; lemmas (3) 4–5 (5.5) mm long, rounded below and only slightly keeled above, becoming increasingly crisp-puberulent below, often with slightly longer hairs on the central and marginal nerves, scarious on the margins and the acutish apex, not at all webbed at the base; anthers 1.5–2 mm long; lodicules about 0.8 mm long; $2n$ = 74, 81, 82, 84, 86, 87.

Relatively dry habitats in sagebrush valleys and foothills and into wooded areas at middle elevs., rarely into alpine places; Alaska to B.C. and Sask., and s. to Calif. (Sierra Nevada), Nev., n. Ariz., N.M. and Nebr. April–June.

Sandberg bluegrass is by far the most common bluegrass of our area. It is an important forage species on spring-fall ranges, and the dense tufts of basal leaves and penetrating fibrous roots make this species an effective soil binder.

Poa sandbergii is the common early season desert bunchgrass in our area. It differs from *P. scabrella* in its shorter height, shorter basal leaves and shorter panicles. *Poa gracillima, P. scabrella* and the remaining species of *Poa* (sections Scabrellae and Nevadenses) are very closely related and all seem to revolve about *P. sandbergii.* All are typically bunch grasses with only slightly keeled, non-webbed lemmas. They are apparently mostly self-pollinated, as pollination has usually already occurred when the florets open. Some poas with round lemmas occasionally bear rhizomes, which can often be traced to the presence of members of the Pratenses group and suspected hybridization. Facultative apomixis occurs within several species of this closely knit group. Perhaps it is unnecessary to recognize both the Scabrellae and Nevadenses sections when the species are so closely related, but it does lend to a useful scheme in separating the species.

29. Poa canbyi (Scribn.) Howell

Aira brevifolia Pursh, Fl. Amer. Sept. 1: 76. 1814; not *Poa brevifolia* DC. 1806. *Airopsis brevifolia* Roemer & Schultes, Syst. Veg. 2: 578. 1817. (*Lewis,* "In the Plains of the Missouri," probably collected on the Clearwater–Lolo Creek Canyon divide, ne. of Kamiah, Idaho Co., Idaho, 10 June 1806.)

Poa tenuifolia Nutt. ex S. Wats. in King, Rep. Geol. Explor. 40th Parallel 5: 387. 1871; not A. Rich. 1851. (*Watson 1318,* Diamond Mts., Nev., July 1868.)

P. tenuifolia var. *rigida* Vasey in Wheeler, Rep. U. S. Geogr. Surv. W. 100th Meridian 6: 290. 1879, nomen nudum. (*Rothrock 1138,* Nev., in 1872; *Wolf 1140,* Twin Lakes, Colo., in 1873.)

P. tenuifolia var. *elongata* Vasey in Wheeler, Rep. U. S. Geogr. Surv. W. 100th Meridian 6: 290. 1879. *P. buckleyana* var. *elongata* M. E. Jones, Contr. W. Bot. 14: 14. 1912. (*Rothrock 1141,* Nev., in 1872; *Wolf,* Colo. [Twin Lakes], in 1873.)

Glyceria canbyi Scribn. Bull. Torrey Bot. Club 10: 77, figs. 1–4. 1883. *Atropis canbyi* Beal, Grasses N. Amer. 2: 580. 1896. *P. canbyi* Howell, Fl. N. W. Amer. 764. 1903. *Puccinellia canbyi* Ponert, Feddes Repert. 84: 739. 1974. (*Tweedy & Brandegee,* Cascade Mts., Wash., Aug. 1882.)

Poa laevis Vasey, Contr. U. S. Natl. Herb. 1: 273. 1893; not R. Br. 1810. *Atropis laevis* Beal, Grasses N. Amer. 2: 577. 1896. *P. laevigata* Scribn. in Rydb. & Shear, U.S.D.A. Div. Agrostol. Bull. 5: 31. 1897. *P. nevadensis* var. *laevigata* M. E. Jones, Contr. W. Bot. 14: 14. 1912. *Puccinellia laevis* Ponert, Feddes Repert. 84: 739. 1974. (*Scribner,* North Fork Smith River, Mont., 19 July 1883.)

Atropis laevis var. *rigida* Beal, Grasses N. Amer. 2: 578. 1896. (*Jones 1021,* Lake Point, Tooele Co., Utah, 19 July 1879.)

Poa wyomingensis Scribn. in Pammel, Proc. Davenport Acad. Nat. Sci. 7: 242. 1899. (*Pammel 192,* "In clay soil above Big Horn," Sheridan Co., Wyo., July 1897.)

P. leckenbyi Scribn. U.S.D.A. Div. Agrostol. Circ. 9: 2. 1899. *P. nevadensis* var. *leckenbyi* M. E. Jones, Contr. W. Bot. 14: 14. 1912. (*Leckenby,* Scott, Klickitat Co., Wash., 5 June 1898.)

P. helleri Rydb. Bull. Torrey Bot. Club 36: 534. 1909. (*Heller & Heller 3274,* about Lake Waha, Nez Perce Co., Idaho, 20 June 1896.)

Canby bluegrass.

Green or glaucous, densely tufted perennials; culms (3) 4.5–8 (10) dm tall, often tinged with purple below; plants usually naked in upper half, the leaves forming large basal rosettes; sheaths usually smooth, closed about 1/4 their length, persisting and becoming papery, usually shorter than the internodes; blades flat or folded, 1–2.5 (3) mm broad, the basal cluster of leaves (10) 15–30 cm high, the few culm leaf blades 5–7 cm long; panicles (7) 9–16 (20) cm long, from loose to rather compact, with short, erect branches; spikelets (5) 6–8.5 (9.5) mm long, 2- to 5-flowered, little compressed, narrow and pointed, tawny and shiny, brownish, or sometimes tinged with purple; glumes lanceolate, acute, thin, scarious-margined, the first glume 2.5–4 (4.5) mm long, 1-nerved, the second 3–5 mm long, 3-nerved; lemmas mostly lanceolate, 4–5.5 mm long, more or less rounded, and if keeled only towards the apex, the margins and the acute or obtusish tip scarious, crisp-puberulent on the lower half, sometimes more strongly so on the center and marginal nerves, the intermediate nerves obscure, without a web at the base; palea slightly shorter than the lemma, scabrous on the keels; anthers 1.5–2.5 mm long; lodicules about 0.6 mm long; $2n$ = 72, 82–85, 90, 93, 94, 105, 106.

Relatively dry to moist places on open grassy or sagebrush slopes at middle elevs.; Alaska to Que. and s. to n. Calif., Nev., Utah, Colo., Nebr., Minn. and n. Mich. June–Aug.

Poa canbyi is an important range species that is actively growing during the summer months. This is in contrast to *P. sandbergii,* which flowers much earlier. Scabrous plants of *P. canbyi* have longer leaves than *P. scabrella.*

30. Poa nevadensis Vasey ex Scribn.

Atropis pauciflora Thurber in S. Wats. Bot. Calif. 2: 310. 1880. *Poa pauciflora* Benth. ex Vasey, Grasses U. S. 42. 1883; not Roemer & Schultes, 1817. *Panicularia thurberiana* Kuntze, Revisio Gen. Pl. 2: 783. 1891. *Poa thurberiana* Vasey, U.S.D.A. Div. Bot. Bull. 13(2): pl. 84. 1893. (*Lemmon,* "Wet meadows, Sierra Valley," Calif., in 1871.)

Poa nevadensis Vasey ex Scribn. Bull. Torrey Bot. Club 10: 66. 1883. *P. tenuifolia* var. *scabra* Vasey ex Scribn., ibid., in synonymy. *Atropis nevadensis* Beal, Grasses N. Amer. 2: 577. 1896. *Puccinellia nevadensis* Ponert, Feddes Repert. 84: 740. 1974. (*Palmer 474,* Nev. [or Ariz.].)

Poa limosa Scribn. & Williams in Scribn. U.S.D.A. Div. Agrostol. Circ. 9: 5. 1899. (*Bolander 6114,* Mono Lake, Mono Co., Calif.) Type bearing a rhizome.

Nevada bluegrass.

Densely tufted perennials; culms many, erect, 3–10 dm tall; plants leafy throughout but mostly in a

POA

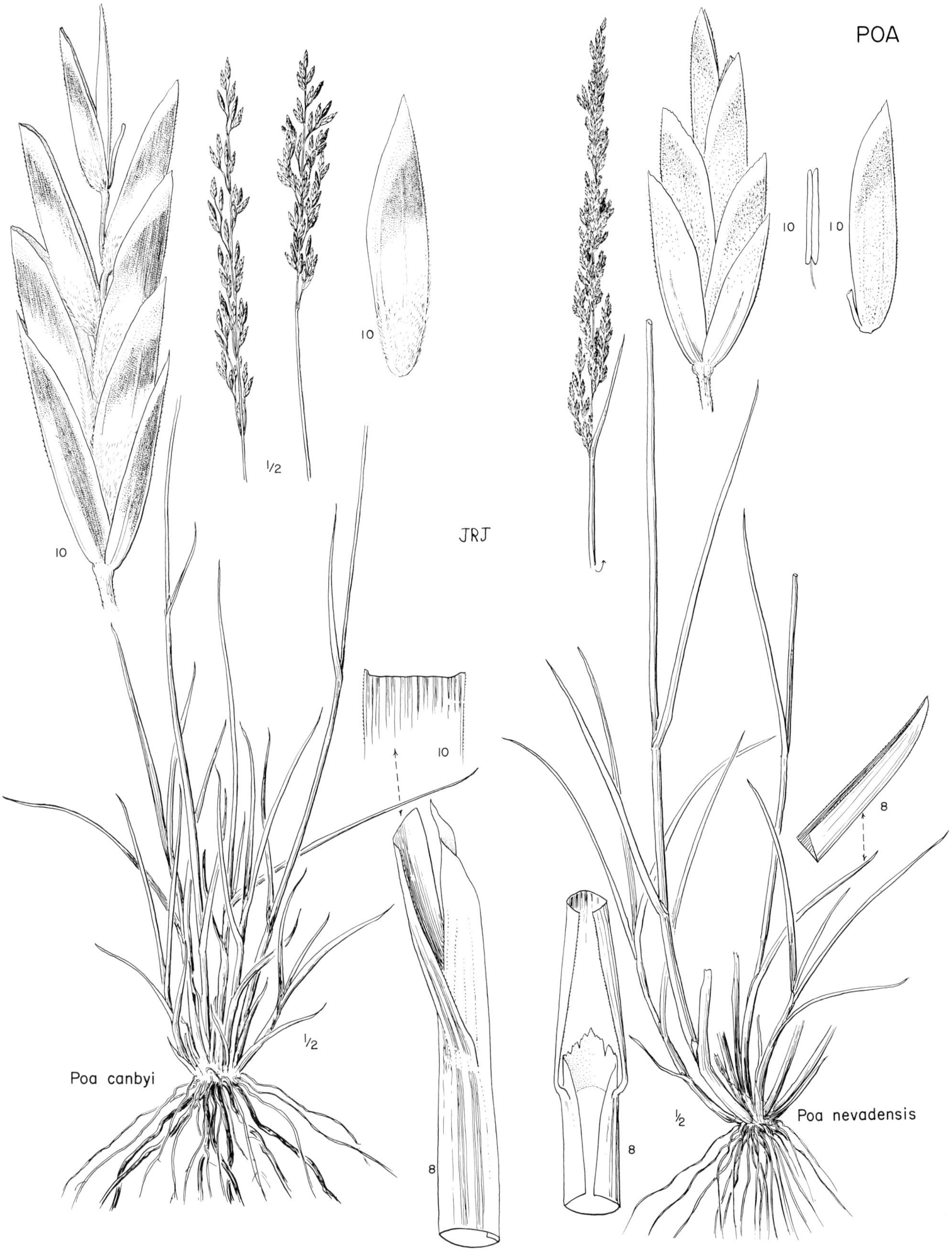

basal cluster, 13–25 (50) cm high; sheaths glabrous to strongly scabrous, closed little more than 1/4 their length, becoming broad and papery and persistent; ligules acute to acuminate, 1.5–4.5 (6.5) mm long, those of the culms up to 6 mm long, much shorter on the innovations, decidedly decurrent and wider than the blade; blades flat or folded or sometimes becoming involute on drying plants, 1–3 mm broad, bright green or pale, usually elongate, up to 30 cm long in robust specimens; panicles (8) 10–18 (24) cm long, narrow, elongate, usually yellowish-green but sometimes tinged with purple, the short branches appressed; spikelets 5–8 (11) mm long, relatively narrow, pointed, terete, 2- to 6-flowered; glumes scabrous, the first 2.7–4.2 (4.7) mm long, 1- to 3-nerved, the second 3–5 (5.7) mm long, about equal to the first floret; lemmas 3.7–5.2 (5.7) mm long, glabrous or sometimes scaberulous throughout or sometimes even slightly crisp-puberulent at the base, sometimes slightly purplish below the obtusish, scarious apex, decidedly rounded on the back below and becoming slightly keeled above or rounded throughout; palea ciliate on the keels and scabrous otherwise; anthers 1.6–2.2 (3) mm long; lodicules about 0.6 mm long; $2n$ = 62–70.

Relatively moist places in sagebrush valleys and edges of meadows at middle elevs.; Alaska and Yukon, s. to s. Calif., Nev., Ariz. and Colo. June–July.

Nevada bluegrass is an important range grass that may be separated from its very close relatives by its long, pointed ligules that are nearly always decurrent, running into the margins of the sheaths. This and the other members of the Nevadenses group are actively growing during the summer months.

31. **Poa juncifolia** Scribn.

Poa juncifolia Scribn. U.S.D.A. Div. Agrostol. Bull. 11: 52, pl. 8. 1898. *P. fendleriana* var. *juncifolia* M. E. Jones, Contr. W. Bot. 14: 14. 1912. (*A. Nelson 3721,* Black Rock Springs, Point of Rocks, Sweetwater Co., Wyo., 13 July 1897.)

P. brachyglossa Piper, Proc. Biol. Soc. Wash. 18: 145. 1905. (*Sandberg & Leiberg 267,* Douglas Co., Wash., 22 June 1893 [according to label data on one of the isotypes it was collected at junction of Crab and Wilson creeks in present-day Grant Co.].)

P. juncifolia subsp. *porteri* Keck in C. L. Porter, Wyoming Agric. Exp. Sta. Bull. 418: 24. 1964. (*C. L. Porter 3249,* Pole Mt. region, Albany Co., Wyo., 6 July 1943.)

Alkali bluegrass.

Light green or glaucescent, densely tufted perennials; culms slender, (2) 3–7 (8.5) dm tall; sheaths smooth, glabrous, sometimes papery, usually shorter than the internodes, open about 1/4 their length; ligules small, 1–2 mm long, truncate or more often rounded or obtuse, those of the innovations hardly visible when viewed from the side; blades tightly involute, rather stiff, greenish, up to 20 cm long and usually less than 2 mm broad, smooth or scabrous, sharp-pointed; panicles narrow, elongate, (5) 7–15 (17.5) cm long, the branches short, strictly ascending, pale green or sometimes tinged with purple; spikelets 5–7.5 (9) mm long, 2- to 5-flowered, relatively narrow, terete; glumes smooth, thin, with scarious margins, acute, the first glume 2.8–4.5 mm long, lanceolate, the second 3.2–5 mm long, broader; lemmas 3.5–5.2 mm long, obtuse at the apex, the obscure nerves disappearing into the scarious margins, usually rounded on back towards the base and often slightly keeled above, smooth, minutely scabrous, crisp-puberulent, firm in texture, without a basal web; palea nearly as long as the lemma; anthers 1.8–2.5 mm long; lodicules about 0.7 mm long; $2n$ = 62–64, 78, 84.

Moist or dry alkaline meadows from the sagebrush zone to the montane zone; B.C. to Mont. and S.D., s. to Calif. (to the San Bernardino Mts.), Nev., Utah and N.M. May–July.

This species is very closely related to *P. ampla,* but the two are kept apart here as our material can be separated with more certainty than is often the case in poas. *Poa juncifolia* generally has many more basal leaves than *P. ampla.* Furthermore, the leaves are more revolute, thereby being more drought resistant and not withering as early as those of *P. ampla.* Alkali bluegrass usually has smaller spikelets with fewer florets, and the lemmas are not quite as blunt. However, C. L. Hitchcock [Univ. Wash. Publ. Biol. 17(1): 666. 1969] indicates that the two kinds tend to lose their distinctions when grown under uniform garden conditions at the grass nursery in Pullman, Washington. Careful observations have been made in the field after reading the above statement, and it still seems desirable to treat them as species instead of varieties of a single taxon.

32. **Poa ampla** Merr.

Poa ampla Merr. Rhodora 4: 145. 1902. (*Vasey 3009,* Steptoe, Wash., 3 July 1901.)

P. laeviculmis T. A. Williams ex Elmer, Bot. Gaz. 36: 55. 1903. (*Vasey 3026,* Steptoe, Wash., 25 June 1900.)

P. confusa Rydb. Bull. Torrey Bot. Club 32: 607. 1905. (*Nelson 7787,* Nash's Fork, Medicine Bow Mts., Albany Co., Wyo., 28 July 1900.)

P. truncata Rydb. Bull. Torrey Bot. Club 32: 607. 1905. (*Clements 373,* Dillon, Summit Co., Colo.)

Big bluegrass.

Robust, densely tufted, glabrous to puberulent, green or glaucous perennials, with intravaginal innovations; culms (4) 6–13 (18) dm tall; leaves usually many, largely basal, often withering early, the basal cluster 20–40 cm high; sheaths smooth to scaberulous, usually shorter than the internodes, open less than 1/4 their length, those of the culms often with conspicuous hard auricles at the mouth; ligules short and thick, 1–2 mm long, truncate or more often rounded; blades 1.5–3.5 mm broad, usually flat but becoming involute on drying plants, usually scabrous only on the margins, glaucous; panicles narrow, elongate, (8) 10–18 (22) cm long, rather densely flowered, the branches short ascending, or on very robust plants becoming more open with branches in interrupted verticils, usually pale green but sometimes tinged with purple; spikelets (5.5) 6–9.5 (11) mm long, 3- to 5-flowered, relatively narrow, pointed, terete; glumes acuminate, the first 3–4.5 mm long, the second 3.5–5 (5.5) mm long, about as long as the first floret; lemmas 4.2–5.5 (6) mm long, obtuse or rounded on apex, usually rounded on the back below and becoming somewhat keeled above, yellowish-green, with a scarious margin, often brownish, usually glabrous or scaberulous, but sometimes finely crisp-puberulent below, the nerves sometimes more noticeably scabrous than the body of the lemma, without a basal web; palea nearly as long as the lemma; anthers usually slightly less than 3 mm

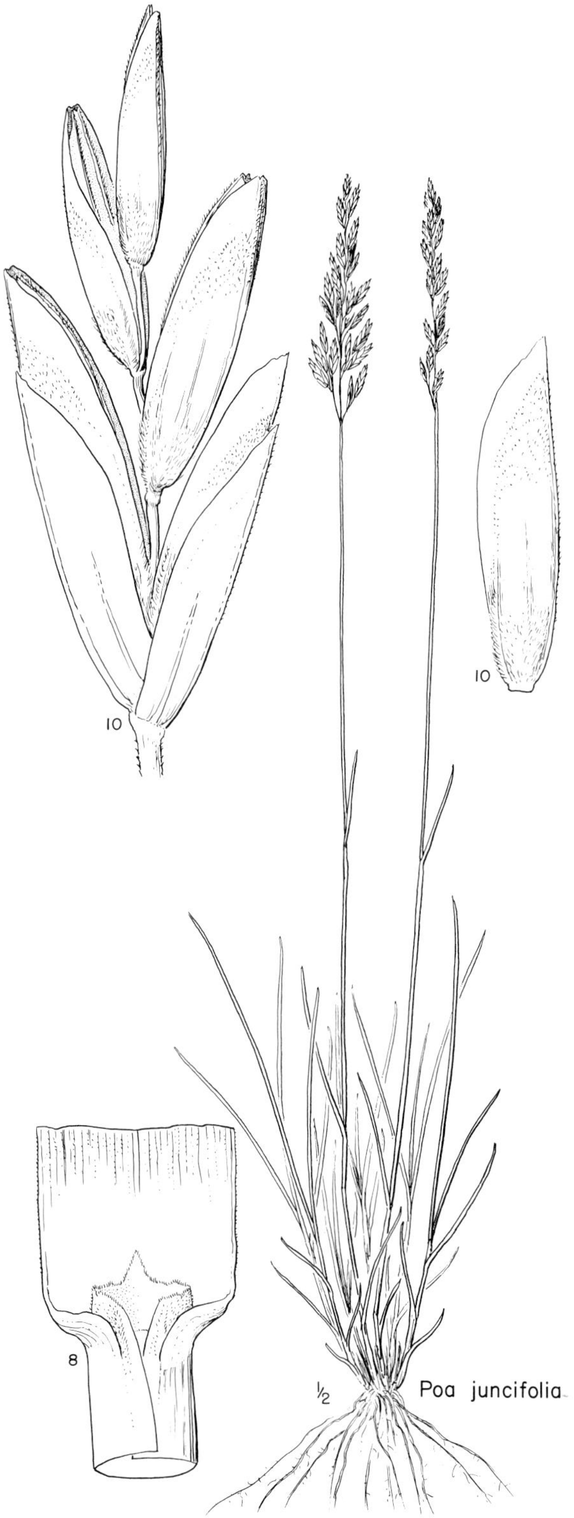

long; lodicules about 0.7 mm long; $2n$ = 62–65, 68, 70, 71, 97, ca 100.

Moderately moist to dry sagebrush slopes, open places in aspen and often in meadows at middle elevs.; Alaska to N.D. and s. to ne. Calif., Nev., Ariz. and N.M. May–July.

Poa ampla will hybridize rather readily with rhizome-producing species and this may be the reason for plants often being found with short rhizomes. This is an important forage species, deserving the attention of the grass breeder.

DACTYLIS

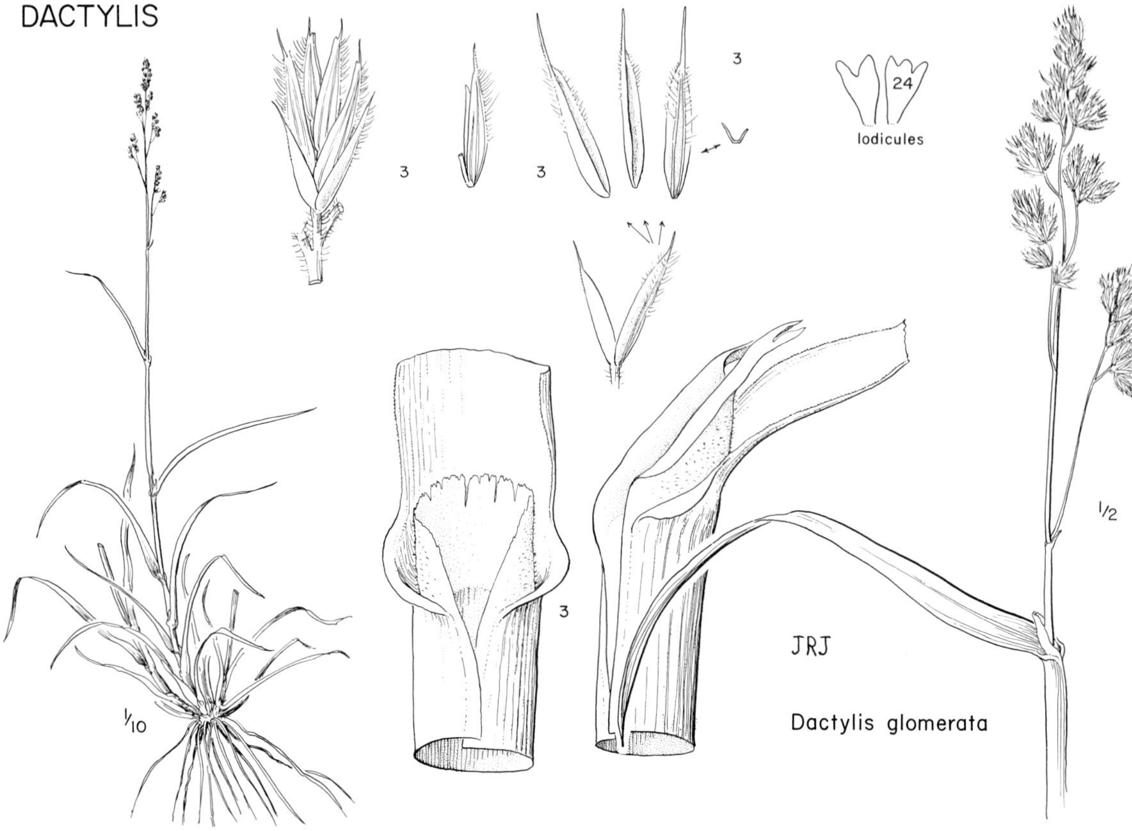

10. DACTYLIS L. Orchard Grass

Erect, tufted perennials; culms hollow; sheaths keeled, open; ligules membranous; blades broad; inflorescence a 1-sided panicle, the branches and branchlets arising along 2 sides of 3-angled rachises, the spikelets in compact fascicles terminating each branchlet; spikelets (2-) 3- (5)-flowered, subsessile or pedicelled, laterally compressed, disarticulating above the glumes and between the florets; glumes acute, carinate, unequal to subequal, shorter than the first lemma, sometimes with 2 lateral nerves, terminating in a short awn; lemmas lanceolate, acuminate, with 5 hispid-ciliate nerves, terminating in a short awn; palea well developed, bidentate at the apex, the 2 keels ciliate; lodicules 2, bi- or tridentate; stamens 3; ovary with an elongated style and plumose stigmas; caryopsis oblong, furrowed on the inside; $x = 7$.

A genus with 1 to 3 species of temp. and cold regions of Europe, n. Africa and Asia. (A name used by Pliny for a grass with digitate spikes, from the Greek *dactulos,* a finger.)

Dactylis glomerata, a native of the Old World, has been cultivated in the United States and has become extensively naturalized. Stebbins and Zohary consider *Dactylis* to be a monotypic genus with diploid and tetraploid subspecies.

Reference: Stebbins, G. L., Jr., and D. Zohary. Cytogenetic and evolutionary studies in the genus *Dactylis.* Univ. Calif. Publ. Bot. 31: 1–40. 1959.

1. Dactylis glomerata L.

Dactylis glomerata L. Sp. Pl. 71. 1753. *Bromus glomeratus* Scop. Fl. Carniolica, ed. 2. 1: 76. 1772. *Festuca glomerata* All. Fl. Pedemontana 2: 252. 1785. *Limnetis glomerata* Eat. Manual Bot. N. States 14. 1817. *Trachypoa vulgaris* Bubani, Fl. Pyrén. 4: 359. 1901. (Europe.)

Orchard grass.

Perennials with short rhizomes; culms 3–12 dm tall; sheaths mostly shorter than the internodes; ligules 4–8 mm long, oblong-ovate, fimbriate-lacerate; blades of the culm leaves 4–6 (10) mm broad, lax, scabrous; panicles 4.5–14 (20) cm long, laterally compressed, the lowest internode long, the branches of the first order long (especially the lower ones) and the second order very short; spikelets 5–8 mm long, (2-) 3- (5)-flowered, oblong-obovoid, aggregated in compact fascicles; glumes 3–6 mm long, 1-nerved, the first glume often glabrous, the second with a pubescent body and ciliate keels or glabrous, awn tipped; lemmas (3) 4–6.5 (7) mm long, 5-nerved, oblong-lanceolate, the keel short ciliate, the margins hyaline, produced into a short awn at the apex; palea lanceolate; anthers 2–3.5 mm long; $2n$ = 14, 21, 27–31, 42.

Fields, canyons and mt. sides; native of Europe, widely distributed in N. Amer. June–Aug.

Orchard grass is recommended for and thrives in moist shaded situations.

Tribe 2. AVENEAE

Annuals or mostly perennials; culms hollow; sheath margins free; ligule membranous; blades usually non-auriculate, sometimes short-auricled in *Phalaris* and *Phleum;* inflorescence a panicle, sometimes congested and spike-like; spikelets 1- to several-flowered, the reduced florets, when present, above the fertile ones (except below in *Arrhenatherum, Hierochloë* and *Phalaris*); rachilla sometimes prolonged behind the upper floret as a slender bristle; disarticulation above or below the glumes; glumes, at least the second, longer than the lowermost lemma (except in some *Koeleria, Sphenopholis, Arrhenatherum, Alopecurus* and *Beckmannia*); lemmas 3- to several-nerved, awnless or dorsally awned from near the base to near the apex (in *Koeleria*); lodicules 2, usually pointed; endosperm liquid in some, in the solid ones the starch grains compound; embryo pooid; $x = 7$.

The Aveneae tribe is closely related to the Poeae, as is shown by the difficulty in placing *Koeleria nitida* in one tribe or the other. In modern grass systematics, the Aveneae tribe has been expanded to include the Agrostideae and Phalarideae tribes of A. S. Hitchcock (Manual Grasses U.S., 1951). The Agrostideae included the species with only one floret per spikelet, a condition now interpreted to have occurred by reduction more than once in the expanded Aveneae. *Hierochloë* and *Phalaris,* which Hitchcock put in tribe Phalarideae, differ from other Aveneae only in the reduction of the lower two florets. *Arrhenatherum* and *Holcus* of the old Aveneae also have reduced florets, one perfect and one staminate floret. These genera are interpreted as intermediates in a reduction series. *Beckmannia,* considered a member of the tribe Chlorideae by Hitchcock, contains pooid characters throughout, with affinities in the Aveneae (Reeder, Bull. Torrey Bot. Club 80: 187–196. 1953).

Lagurus ovatus L., hares-tail, a native of the Mediterranean region, is sometimes grown as an ornamental in our region. It can be recognized by its densely flowered, ovoid or oblong, woolly inflorescence.

1 Spikelets sessile or subsessile, arranged along 2 sides of a 3-angled, continuous rachis, forming 1-sided spicate branches, the spicate branches racemosely arranged on a central axis, the spikelets disarticulating below the glumes . . . . . 28. *Beckmannia*
1 Spikelets pedicellate in an open or contracted panicle, the spikelets disarticulating above or below the glumes.
  2 Spikelets with 2 sterile or staminate florets below the fertile one, no florets above (Phalarideae, sensu stricto).
    3 Lower florets staminate; spikelets bronze-colored . . . . . 24. *Hierochloë*
    3 Lower florets reduced to small, scale-like lemmas; spikelets green or yellowish . . . . . 25. *Phalaris*
  2 Spikelets without sterile florets below the perfect floret.
    4 Spikelets 2- to many-flowered (sometimes 1 in some *Sphenopholis* and *Deschampsia*) (Aveneae, sensu stricto).
      5 Lemmas awnless, or with a very short awn up to 2 mm long.
        6 Spikelets disarticulating above the glumes; glumes similar in shape, the second glume 3.2–7 mm long, lanceolate.
          7 First glume 2.8–4.5 (5) mm long; rachilla pilose . . . . . 11. *Koeleria*
          7 First glume 5–7.2 mm long; rachilla sparsely hirsute . . . . . 13. *Trisetum wolfii*
        6 Spikelets disarticulating below the glumes; glumes distinctly different in shape, the second glume 2–2.8 (3.2) mm long, obovate . . . . . 12. *Sphenopholis*
      5 Lemmas awned (at least on 1 of the florets in the spikelet).
        8 Perfect florets 2 or more.
          9 Spikelets small, the glumes less than 8 (8.5) mm long.
            10 Awns attached above the midlength of the lemma; lemmas keeled, (3.5) 4–9 mm long . . . . . 13. *Trisetum*
            10 Awns attached below the midlength of the lemma; lemmas rounded and small, 2–4 (4.5) mm long . . . . . 14. *Deschampsia*
          9 Spikelets larger, the glumes 8 mm or more long.
            11 Plants annual; glumes 17 mm or longer . . . . . 15. *Avena*
            11 Plants perennial; glumes less than 17 mm long . . . . . 16. *Helictotrichon*
        8 Perfect florets 1 (1 perfect and 1 staminate).
          12 Lower floret staminate with an exserted, geniculate awn, the upper floret perfect, rounded, awnless or with a short awn . . . . . 17. *Arrhenatherum*
          12 Lower floret perfect, unawned, the upper floret staminate with a short, hooked awn . . . . . 18. *Holcus*
    4 Spikelets 1-flowered, rarely 2-flowered in *Calamagrostis* (Agrostideae, sensu stricto).
      13 Callus evidently bearded . . . . . 19. *Calamagrostis*
      13 Callus not bearded or only minutely so.
        14 Spikelets disarticulating above the glumes.
          15 Glumes not compressed-carinate, awnless; panicles open to compressed, but not spike-like.
            16 Lemma evidently awned, the awn 5.5–16 mm long, attached near the apex, the lemma firm to subindurate; palea well developed, nearly as long as the lemma . . . . . 20. *Apera*

16 Lemma awnless or with an awn less than 5 mm long, attached near the midlength, the lemma membranous; palea often absent or no more than 2/3 the length of the lemma .... 21. *Agrostis*
15 Glumes compressed-carinate, abruptly short-awned from an obtuse apex; panicles dense, spike-like .... 27. *Phleum*
14 Spikelets disarticulating below the glumes, the spikelet falling entire.
17 Glumes awned.
18 Lemmas 0.7–1.5 mm long; anthers small, 0.4–0.7 mm long .... 22. *Polypogon*
18 Lemmas (1.2) 1.7–2.5 mm long; anthers 1.1–2.3 mm long .... 27. *Phleum*
17 Glumes awnless, sometimes aristate in *Cinna*.
19 Panicles open, drooping; lemmas short-awned from a notched apex; rachilla prolonged behind the palea .... 23. *Cinna*
19 Panicles dense, spike-like; lemmas awned from the middle or below; rachilla not prolonged behind the palea .... 26. *Alopecurus*

11. KOELERIA Pers. Junegrass

Tufted annuals and perennials, mostly cespitose; culms hollow; sheaths open to the base; ligules membranous; blades narrow, flat to involute, non-auriculate; inflorescence usually contracted into a spike-like panicle; spikelets usually 2- to sometimes 4-flowered, flattened, the florets perfect, the rachilla joints short, less than 1 mm long, usually prolonged beyond the upper floret as a slender bristle or bearing a vestigial floret, disarticulating above the glumes and between the florets; glumes relatively large, keeled, thin, acute, unequal, the first glume 1-nerved, the second broader, often slightly longer, obscurely 3- (5)-nerved; lemmas thin, 5-nerved, shiny, glabrous to somewhat scabrous, the first lemma as long to slightly longer than the second glume, awnless or short-awned from a minutely bifid apex; palea 2-keeled, nearly as long as the lemma, scarious and colorless; lodicules entire to bifid; stamens 3; ovary glabrous, the styles terminal; $x = 7$.

About 20 species of temp. and arctic regions of the N. Hemisphere; two species in the U.S., *K. nitida,* a native perennial, and *K. phleoides* (Vill.) Pers., an introduced annual. (Named in honor of the German agrostologist from Mainz, Georg Ludwig Koeler, 1765–1807.)

Koeleria shows close affinities with the Poeae tribe but is a member of the Aveneae with closer affinities to *Sphenopholis* and *Trisetum*.

1. Koeleria nitida Nutt.

Aira cristata L. Sp. Pl. 63. 1753. *Poa cristata* L. Syst. Nat. ed. 12. 2: 94. 1767. *Festuca cristata* Vill. Hist. Pl. Dauphiné 1: 250. 1786; not L. 1753. *Airochloa cristata* Link, Hort. Regius Bot. Berol. 1: 127. 1827. *Rostraria cristata* Tzvelev. Novit. Syst. Pl. Vasc. 7: 47. 1970 [1971]. ("Habitat in Angliae, Galliae, Helvetiae.")

Koeleria cristata Pers. Syn. Pl. 1: 97. 1805, based on "*Poa cristata* auctorum"; not *Poa cristata* (L.) L. *Brachystylus cristatus* Dulac, Fl. Dépt. Hautes-Pyrén. 85. 1867. (Typification recondite.) *K. cristata* is an illegitimate name.

K. gracilis Pers. Syn. Pl. 1: 97. 1805. *Aira gracilis* Trin. Fund. Agrostogr. 144. 1820. *Airochloa gracilis* Link, Hort. Regius Bot. Berol. 2: 276. 1833. *K. cristata* var. *gracilis* A. Gray, Manual Bot. 591. 1848. (Europe.) *K. gracilis* is an illegitimate name also since Persoon included *Poa nitida* Lam. in the synonymy.

K. nitida Nutt. Gen. N. Amer. Pls. 1: 74. 1818. *K. gracilis* subsp. *nitida* Domin, Biblioth. Bot. 65: 229. 1907. (*Nuttall,* "On the plains of Missouri," in 1811.)

K. cristata var. *nuttalii* A. Wood, Cl.-Book Bot. ed. of 1847. 613. 1847. (Mich.)

K. gracilis subsp. *idahensis* Domin, Biblioth. Bot. 65: 175, 237. 1907. (*Heller 309* [error for *3091*], Lewiston, Nez Perce Co., Idaho, 20 May 1896.)

Junegrass.

Cespitose perennials; culms (2) 2.5–6.5 dm tall; leaves mostly basal, the culm leaves few and small; sheaths hispid, sometimes densely so, rarely glabrous; ligules 0.5–1.5 (2) mm long, erose to subentire, ciliolate, sometimes puberulent; blades usually folded or involute, if flat 1–2 (2.5) mm broad, hispid to glabrous, with prow-shaped tips; panicle (2.5) 4–11 (15) cm long, contracted, spike-like, branches short, appressed, the rachis densely puberulent; spikelets 4–5 mm long, 2- (3- to 4)-flowered, the rachilla pilose; glumes subequal, scabrous on the keel and sometimes scaberulous all over, the first glume 2.8–4.5 (5) mm long, narrow-lanceolate to lanceolate, 1-nerved, the second 3.2–5 (5.5) mm long, lanceolate, 1- to 3-nerved, the lateral nerves obscure; lemmas about as long as the second glume, 3.2–5 (5.2) mm long, lanceolate, 5-nerved, the lateral nerves obscure, scaberulous, sometimes with a short awn near the tip; palea often as long as the lemma, membranous; lodicules about 0.8 mm long; anthers 1.2–2.5 mm long; $2n$ = 14, 15, 16, 28.

Sagebrush foothills to dry open forests and ridges at subalpine elevs.; B.C., N.W.T., and Ont., s. throughout the w. states and n. Mex. and e. through Texas, Mo. and Ill. to Del. Late June–Aug.

Koeleria nitida bears close resemblance to *Poa fendleriana* of tribe Poeae from which it can be distinguished by its fewer florets per spikelet, densely puberulent rachis and shining white hyaline palea.

Shinners (Rhodora 58: 93–96. 1956) pointed out the illegitimacy of the name *K. cristata* (L.) Pers. *Koeleria cristata* is to be regarded as a new name by Persoon, based on "*Poa cristata* auctorum" (typification recondite), rather than a transfer from *Poa cristata* (L.) L. which he referred to *K. phleoides.* Shinners believed the oldest valid name for the taxon to be *K. macrantha* (Ledeb.) J. A. Schultes, based on the Asian *Aira macrantha* Ledeb. (1812). Johnston (in Correll & Johnston, Manual of the Vascular Plants of Texas. 139. 1970) and Gould (Grasses of Texas. 126. 1975) used the name *K. pyramidata* (Lam.) Beauv., based on the European *Poa pyramidata* Lam. (1791). To avoid confusion and to encourage a careful comparison with the European material, we have chosen the oldest New World name, *K. nitida* Nutt.

Junegrass is a good native forage grass but is seldom in enough abundance to be important.

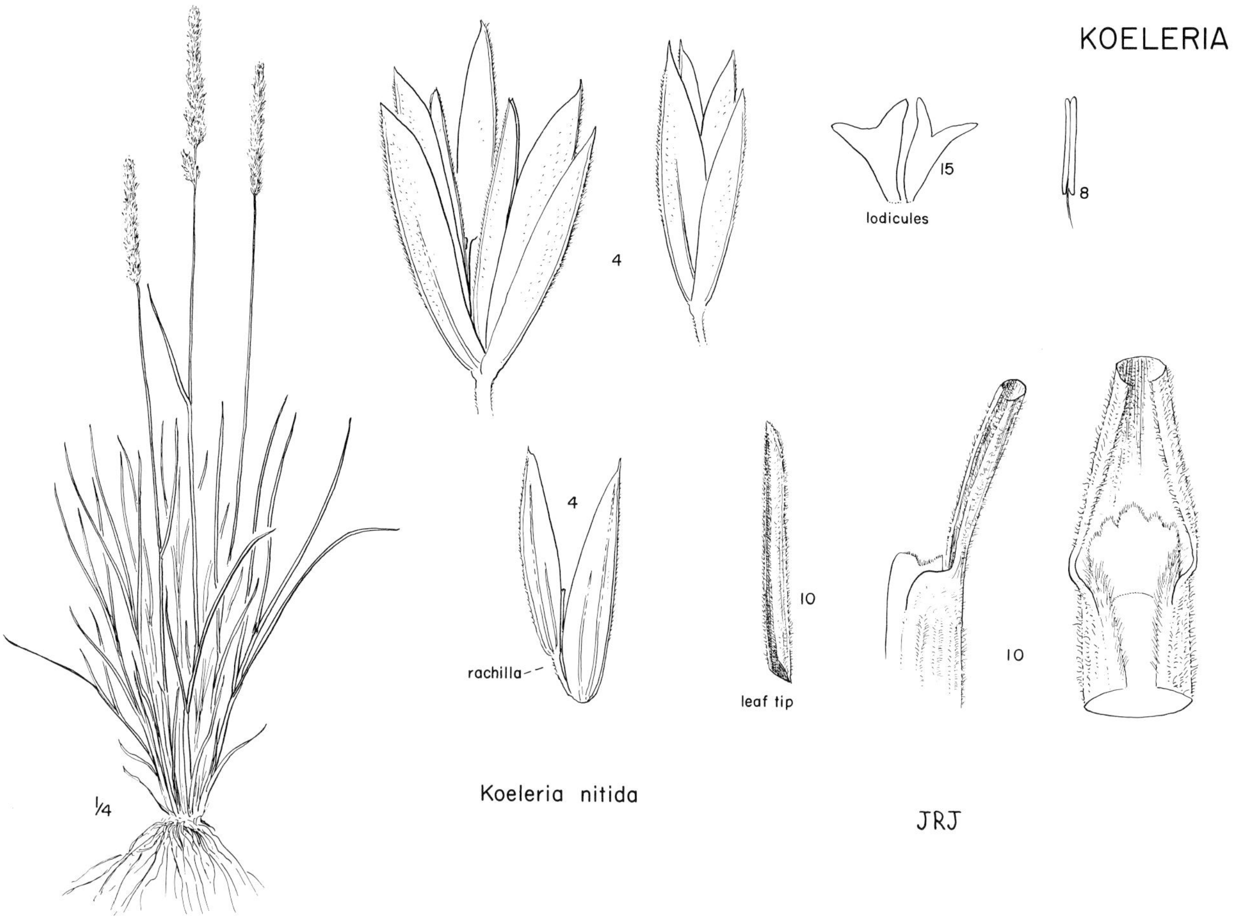

12. SPHENOPHOLIS Scribn. Wedgegrass

Tufted annuals or short-lived perennials; culms hollow; sheaths open to the base, glabrous to pubescent; ligules membranous; blades flat, soft, non-auriculate, the collar often oblique; inflorescence a dense or loose, erect or nodding panicle; spikelets (1-) 2- (3)-flowered, the florets perfect, the rachilla produced beyond the upper floret, often with a vestigial lemma or as a slender bristle, disarticulating below the glumes and falling as a unit; glumes slightly keeled and compressed, very unlike in size and shape, the first glume narrow-lanceolate, acute, 1-nerved, the second broadly obovate to oblanceolate, 3- (5)-nerved, obtuse or broadly acute at the apex, scarious margined, usually slightly shorter than the lowermost lemma; lemmas firm, obscurely, if at all, 5-nerved, rounded on the back, smooth to scabrous, awnless; palea shorter than the lemma, thin, membranous, colorless, 2-nerved; lodicules membranous, obovate, often lobed; stamens 3; ovary glabrous, the styles terminal; $x = 7$.

A genus of 4 species, 3 in the c. and e. U.S. and *S. obtusata,* widespread nearly throughout N. Amer. and in the Caribbean. (From the Greek *sphen,* a wedge, and *pholis,* scale, in reference to the broadly obovate second glume.)

The fact that *S. obtusata* and its synonyms have been variously placed in *Koeleria, Trisetum, Poa, Festuca* and *Agrostis* points up the fact that it shares with *Koeleria* a somewhat transitional position between tribes Poeae and Aveneae.

Reference: Erdman, K. S. Taxonomy of the genus *Sphenopholis.* Iowa State Coll. J. Sci. 39: 289–336. 1965.

1. Sphenopholis obtusata (Michx.) Scribn.

Aira obtusata Michx. Fl. Boreali-Amer. 1: 62. 1803. *Airopsis obtusata* Desvaux, J. Bot. (Desvaux) 1: 200. 1808. *Festuca obtusata* Michx. ex Beauv. Essai Nouv. Agrostogr. 163. 1812. *Poa obtusata* Link, Hort. Regius Bot. Berol. 1: 176. 1827. *Agrostis obtusata* Steudel, Nomencl. Bot. Hort. ed. 2. 1: 41. 1840. *Koeleria obtusata* Trin. ex Steudel, ibid. 849, in synonymy. *Reboulea obtusata* A. Gray, Manual Bot. 591. 1848. *Eatonia obtusata* A. Gray, Manual Bot. ed. 2. 558. 1856. *Sphenopholis obtusata* Scribn. Rhodora 8: 144. 1906. (*Michaux,* "Hab. in sabulosis Carolinae, Georgiae, Floridae.")

Koeleria truncata var. *major* Torr. Fl. N. & Middle Sect. U. S. 1: 117. 1823. *Koeleria pennsylvanica* var. *major* Torr. Fl. State New York 2: 469. 1843. *Reboulea pennsylvanica* var. *major* A. Gray, Manual Bot. 591. 1848. *Eatonia pennsylvanica* var. *major* A. Gray, Manual Bot. ed. 2. 558. 1856. *S. pallens* subsp. *major* Scribn. Rhodora 8: 145. 1906. *S. pallens* var. *major* Scribn. ex B. L. Robinson, Rhodora 10: 65. 1908. *Reboulea pallens* var. *major* Farw. Rep. Michigan Acad. Sci. 17: 182. 1916. *S. obtusata* var. *major* K. S. Erdm. Iowa State Coll. J. Sci. 39: 310. 1965. (*Cooley,* "Near Deerfield, Mass.")

Trisetum lobatum Trin. Mem. Acad. Imp. Sci. St.-Pétersbourg,

Sér. 6, Sci. Math. 1: 66. 1830. *S. obtusata* subsp. *lobata* Scribn. Rhodora 8: 144. 1906. *S. obtusata* var. *lobata* Scribn. ex B. L. Robinson, Rhodora 10: 65. 1908. *Reboulea obtusata* var. *lobata* Farw. Rep. Michigan Acad. Sci. 17: 182. 1916. (N. Amer.) = var. *obtusata.*

Eatonia intermedia Rydb. Bull. Torrey Bot. Club 32: 602. 1905. *S. intermedia* Rydb. Bull. Torrey Bot. Club 36: 533. 1909. (*Rydberg 3174,* East Gallatin Swamps, Mont., 24 July 1896.) = var. *major.*

S. pallens sensu Scribn. Rhodora 8: 145. 1906. = var. *major.*

Prairie wedgescale.

Annuals or short-lived perennials; culms (2) 3.5–8.5 (15) dm tall; sheaths glabrous to scabrous; ligules 1.5–3.5 mm long, finely erose-ciliate, glabrous to scaberulous, often departing from an oblique collar; blades flat, 1.5–5 (8) mm broad, glabrous or scabrous; panicle (2.5) 5–15 (23) cm long, dense and spike-like to loose and open; spikelets 1.5–5 mm long, 1- to 2- (3)-flowered; glumes very different in shape, the first glume 1.7–2.5 (3) mm long, linear, the keel scabrous, the second 2–2.8 (3.2) mm long, oblanceolate to broadly obovate, rounded to subacute, the nerves often prominent and scabrous; lemmas 2–3 (3.2) mm long, lanceolate, awnless; palea shorter than the lemma; anthers very small, 0.3–0.7 mm long; $2n = 14$.

Moist meadows, streambanks and shores of ponds and lakes; Alaska to Newfl., s. throughout most of the U.S. and across most of Mex. and in the Caribbean, less common in w. U.S. July–early Sept.

Sphenopholis obtusata is a highly variable species which has been divided into several species and varieties. A. S. Hitchcock (Manual Grasses U. S. 283. 1951) recognized 2 species in the western states, *S. obtusata* and *S. intermedia,* while Erdman (1965) sees them as varieties, *S. obtusata* var. *obtusata* and var. *major* (Torr.) K. S. Erdm. These taxa occupy the same geographical range and appear more like extremes in variation with intermediates being very common. Most collections may fit into one or the other category, however, based on the following characters.

1 Second glume rounded and broadly obovate, 1/3 to 1/2 as wide as broad; panicle dense and spike-like .... var. *obtusata*
1 Second glume blunt to acute, oblanceolate, 1/6 to 1/3 as wide as broad; panicle loose .... var. *major* (Torr.) K. S. Erdm.

13. **TRISETUM** Pers. Trisetum

Tufted perennials (ours) or annuals, mostly cespitose; culms slender, hollow; sheaths open to the base; ligules membranous, erose; blades flat to involute, non-auriculate; inflorescence an open to usually narrow, spike-like panicle; spikelets 2- (3- to 4)-flowered, the florets perfect, the rachilla villose, the joints evident, 1.5 mm or more long, usually prolonged above the uppermost floret, often with a vestigial lemma, disarticulating above the glumes and between the florets (ours) or sometimes below the glumes; glumes keeled, acute, awnless, subequal to very unequal, the first glume 1-nerved, the second 3-nerved; lemmas usually shorter than the second glume, short bearded at the base, obscurely 5-nerved, keeled, bifid at the apex, usually awned or awnless (in *T. wolfii*); awn straight or bent from just below the base of the notch, attached above the midlength of the lemma; palea shorter than the lemma, the 2 nerves extending into bristle tips; lodicules 2, oblong, shallowly and usually slenderly lobed to fimbriate; stamens 3; ovary glabrous or pubescent, the styles terminal; $x = 7$.

A genus of from 50 to 65 species in the temp. and arctic regions of the Old and New World (also montane-tropical); 8 species native to the U.S. and 1 introduced from Europe. (From the Latin *tres,* three, and *seta,* a bristle, in reference to the awn and two teeth of the lemma of the type species.)

Trisetum shows close affinities with *Sphenopholis* and *Koeleria,* especially through the awnless or short-awned *T. wolfii.*

Reference:. Father Louis-Marie O.C. The genus *Trisetum* in America. Rhodora 30: 209–228, 231–245. 1928.

1 Lemmas awnless or with short awns less than 2 mm long .... 1. *T. wolfii*
1 Lemmas with awns 4–13 mm long.
  2 Lemmas 5.8–9 mm long; panicle open .... 2. *T. canescens*
  2 Lemmas 3.5–5.5 (6.5) mm long.
    3 Panicle dense and spike-like; leaves usually basal, the culms 0.5–5 (7.5) dm tall ...... 3. *T. spicatum*
    3 Panicle loose and open, neither spreading nor spike-like; culms usually leafy and taller, (3) 4–8 dm tall .... 4. *T. montanum*

1. **Trisetum wolfii** Vasey

Trisetum wolfii Vasey, Monthly Rep. Dept. Agric., Feb. Mar. 156. 1874. *Graphephorum wolfii* Vasey ex J. M. Coulter, Manual Bot. Rocky Mt. Region 423. 1885. (*Wolf,* Twin Lakes, Colo., in 1873.)

T. subspicatum var. *muticum* Boland. ex Thurber in S. Wats. Bot. Calif. 2: 296. 1880. *T. muticum* Scribn. U.S.D.A. Div. Agrostol. Bull. 11: 50, fig. 10. 1898. *Graphephorum muticum* A. A. Heller, Cat. N. Amer. Pl. ed. 2. 31. 1900, invalid; Greene, Pl. Bakerianae 2: 40. 1901. *T. wolfii* subsp. *muticum* Scribn. Rhodora 8: 88. 1906. *T. wolfii* var. *brandegei* f. *muticum* Louis-Marie, Rhodora 30: 241. 1928. (*Bolander 5019,* "On the Upper Tuolumne [River]," Calif.)

T. brandegei Scribn. Bull. Torrey Bot. Club 10: 64, figs. 4–7. 1883. *Graphephorum brandegei* Rydb. Fl. Rocky Mts. 61, 1060. 1917 [1918]. *T. wolfii* var. *brandegei* Louis-Marie, Rhodora 30: 241. 1928. (*Brandegee & Tweedy,* Cascade Mts., Aug. 1882.)

Wolf's trisetum.

Cespitose, sometimes short rhizomatous perennials; culms (2) 3–8 (10) dm tall; sheaths usually scaberulous to sometimes pilose; ligules 1.2–4.5 (6) mm long, truncate and erose-ciliate, sometimes departing from an oblique collar; blades flat, 2–5 (7) mm broad, scabrid to pilose; panicle (4) 6.5–15 (18) cm long, narrow, the branches erect or ascending; spikelets 2- to 3-flowered, the rachilla sparsely hirsute, often purplish; glumes acute, 1-nerved, nearly equal, the first glume 5–7 mm long, the second usually slightly shorter, 5–6.5 (7) mm long; lemmas 4.7–6 mm long, awnless or short-awned, the callus sparsely hispid; awns, if present, 0.5–2 mm long, attached

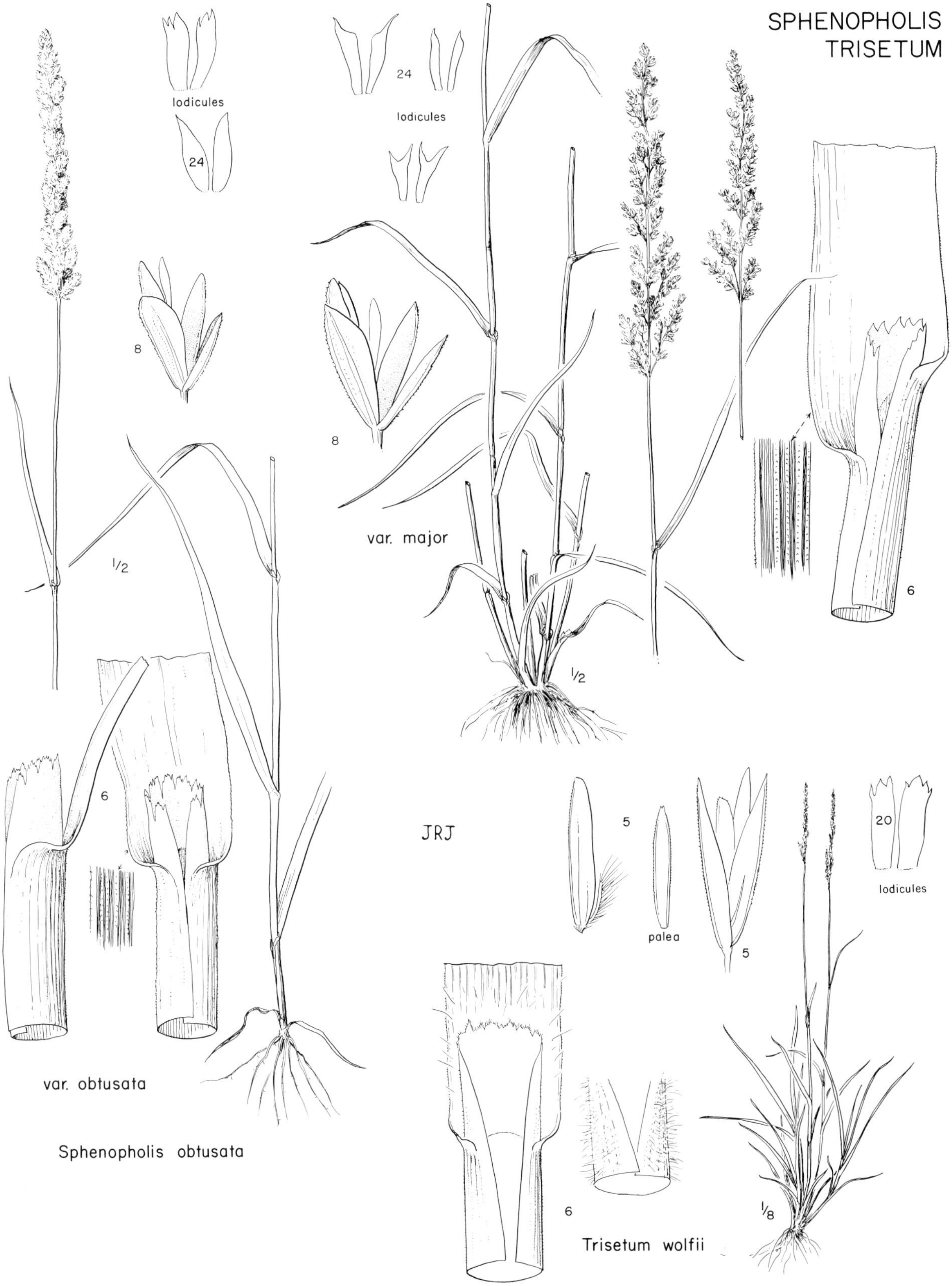
SPHENOPHOLIS
TRISETUM
lodicules
24
24
lodicules
8
8
1/2
var. major
1/2
6
6
JRJ
var. obtusata
Sphenopholis obtusata
5
palea
5
20
lodicules
6
1/8
Trisetum wolfii

near the apex; palea shorter than the lemma; lodicules about 1 mm long, oblong, slenderly lobed to fimbriate; anthers 0.6–1.2 mm long; ovary pubescent.

Moist meadows and streambanks in the mts. from middle to subalpine elevs.; s. Alta., s. through the Pacific and Rocky Mt. cordillera to Calif., Nev., Utah and N.M.; Uinta Mts., Utah and mts. of ne. Nev. July–early Sept.

2. Trisetum canescens Buckley

Trisetum canescens Buckley, Proc. Acad. Nat. Sci. Philadelphia 1862: 100. 1863. *T. elatum* Nutt. ex A. Gray, ibid., 337. 1863, as a synonym. *T. cernuum* var. *canescens* Beal, Grasses N. Amer. 2: 380. 1896. *T. cernuum* subsp. *canescens* Calder & Taylor, Canad. J. Bot. 43: 1389. 1965. (*Nuttall*, Columbia Plains, Oregon, July–early Aug. 1835.)

Tall trisetum.

Cespitose perennials; culms (3) 5–11 (14) dm tall; sheaths usually pilose, sometimes only scabrous; ligules (1) 1.5–4 (5.5) mm long, erose; blades flat, (1) 2–6 (10) mm broad, scaberulous to pilose; panicle (6) 8–17 (22) cm long, narrow and loose; spikelets mostly 2-flowered, sometimes 3, the rachilla with long hairs; glumes unequal, the first (3.2) 4–5 mm long, narrow-lanceolate to linear, the second (4.5) 5–7 (7.5) mm long, lanceolate, shorter than the first floret; lemmas 5.8–8 (9) mm long, the callus beard of shorter hairs than on the rachilla; awns 6–13 mm long, geniculate; palea nearly as long as the lemma; lodicules oblong, truncate, 1–1.5 mm long, usually fimbriate; anthers 1.2–2.2 mm long; ovary often pubescent above; $2n = 28, 42$.

Moist to moderately moist, shady hillsides and open meadows at low elevs. in the mts.; s. Alaska and B.C., s. to Calif., c. Idaho and Mont. with a disjunct station in Red Butte Canyon of the Wasatch Range, Utah. Late May–July.

3. Trisetum spicatum (L.) K. Richter

Aira spicata L. Sp. Pl. 64. 1753. *Aira subspicata* L. Syst. Nat. ed. 10. 2: 873. 1759. *Avena airoides* Koeler, Descript. Gram. 298. 1802. *Avena subspicata* Clairv. Manuel Herborisation Suisse & Valais 17. 1811. *Trisetum subspicatum* Beauv. Essai Nouv. Agrostogr. 88, 149. 1812. *Trisetaria airoides* Baumg. Enum. Stirp. Magno Transsilv. Princ. 3: 265. 1816. *Trisetum airoides* Beauv. ex Roemer & Schultes, Syst. Veg. 2: 666. 1817. *Koeleria subspicata* Reichenb. Fl. German. Excursoria 49. 1830. *Koeleria spicata* Reichenb. ex Willk. & Lange, Prodr. Fl. Hispan. 1: 72. 1861, as a synonym. *Trisetum spicatum* K. Richter, Pl. Eur. 1: 59. 1890. *Avena spicata* Fedtsch. Trudy Imp. S.-Petersburgsk. Bot. Sada 28: 76. 1908; not L. 1753. *Panicum spicatum* Farw. Rhodora 32: 262. 1930; not Roxb. 1820. *Trisetaria spicata* Paunero, Anales Jard. Bot. Madrid 9: 516. 1950. ("Habitat in Lapponiae.")

Avena mollis Michx. Fl. Boreali-Amer. 1: 72. 1803; not Salisb. 1796. *Trisetum molle* Kunth, Révision Gram. 1: 101. 1829. *Koeleria canescens* Torr. ex Trin. Mém. Acad. Imp. Sci. Saint-Pétersbourg, Sér. 6, Sci. Math, Seconde Pt. Sci. Nat. 2(1): 13. 1836, as a synonym. *T. subspicatum* var. *molle* A. Gray, Manual Bot. ed. 2. 572. 1856. *Rupestrina pubescens* Prov. Fl. Canad. 689. 1862. *T. spicatum* var. *molle* Beal, Grasses N. Amer. 2: 377. 1896. *T. spicatum* subsp. *molle* Piper, Contr. U. S. Natl. Herb. 11: 125. 1906. *T. spicatum* var. *michauxii* St. John, Fl. S.E. Wash. and Adj. Idaho 62. 1937. (Canada.)

T. congdoni Scribn. & Merr. Bull. Torrey Bot. Club 29: 470. 1902. *T. spicatum* var. *congdoni* A. S. Hitchc. Proc. Biol. Soc. Wash. 41: 160. 1928. *T. spicatum* subsp. *congdonii* Hultén, Svensk Bot. Tidskr. 53: 218. 1959. (*Congdon*, Shadow Lake Trail, Mariposa Co., Calif., in 1899.)

T. americanum Gandoger, Bull. Soc. Bot. France 49: 182. 1902. (*Baker, Crandall, Osterhout & Tracy*, Colo.; *Rydberg & Heller*, Idaho.)

T. majus Rydb. Agric. Exp. Sta. Agric. Coll. Colorado Bull. (Fl. Colorado) 100: 34. 1906. *T. subspicatum major* Vasey ex Rydb., ibid., as a synonym. *T. spicatum* var. *majus* Farw. Rep. Michigan Acad. Sci. 21: 352. 1920. *T. spicatum* subsp. *majus* Hultén, Svensk Bot. Tidskr. 53: 218. 1959. (*Vasey*, Pen Gulch, Colo., in 1884.)

Spike trisetum.

Cespitose perennials, the pubescence of the herbage extremely variable, from puberulent to densely pilose or villous; culms (0.5) 1–5 (7.5) dm tall; ligules 0.5–2.5 (3.5) mm long, erose, finely ciliate, truncate; blades flat or folded, 1.5–4 (5.5) mm broad; panicle 2–12 (17) cm long, narrow and spike-like; spikelets 2- (3)-flowered, purplish, tawny or silvery; glumes lanceolate, acute, unequal, the first glume 3–5 (5.8) mm long, 1-nerved, the second 4.5–6 (6.8) mm long, 3-nerved, usually about as long as the upper floret; lemmas (3.5) 4–5.5 mm long, 5-nerved, scaberulous, bifid at the apex with narrow teeth, the callus hairs short; awns 4–6 (7) mm long, geniculate, arising just below the tip of the lemma; palea bifid, nearly as long as the lemma; lodicules 0.7–1 mm long, acutely lobed; anthers 0.7–1.4 mm long; ovary glabrous; $2n = 14, 28, 42$.

Moist to moderately moist subalpine and alpine meadows; throughout Arctic Amer. from Alaska to Lab., s. in the mts. of w. U.S. to Calif., Nev., n. Ariz. and n. N.M. and e. across Minn., Mich., Penn. and N.Y.; also Mex., antarctic S. Amer. and arctic alpine Eurasia. Late July–Sept.

Trisetum spicatum is an important forage grass in the high mountain country. The highly variable *T. spicatum* is a circumboreal species that has been subdivided into several subspecies and varieties by Hultén (Svensk Bot. Tidskr. 53: 203–228. 1959). Of the 14 subspecies, 3 were indicated as entering our area. Those with shorter and broader glumes were referred to subsp. *congdonii* (Scribn. & Merr.) Hultén, and those with longer and narrower glumes were referred to subsp. *molle* (Kunth) Hultén and subsp. *majus* (Rydb.) Hultén. Subspecies *molle* differs from *majus* in being more pubescent. Böcher (Bot. Tidsskr. 55: 23–29. 1959) referred most of the North American material to var. *molle* (Kunth) Beal by separating it on the basis of chromosome number from var. *spicatum* of the Arctic and Eurasia.

4. Trisetum montanum Vasey

Trisetum montanum Vasey, Bull. Torrey Bot. Club 13: 118. 1886. *T. canescens* var. *montanum* A. S. Hitchc. Proc. Biol. Soc. Wash. 41: 160. 1928. (*Vasey*, mts. w. of Las Vegas, N.M., in 1881.)

T. argenteum Scribn. U.S.D.A. Div. Agrostol. Bull. 11: 49, fig. 8. 1898; not Roemer & Schultes, 1817. *T. shearii* Scribn. U.S.D.A. Div. Agrostol. Circ. 30: 8. 1901. *Graphephorum shearii* Rydb. Bull. Torrey Bot. Club 32: 602. 1905. *T. montanum* var. *shearii* Louis-Marie, Rhodora 30: 213. 1928. (*Shear 1214*, "Among rocks, Las Animas Canyon below Silverton, Colo.," 4 Aug. 1897.) A form with short, straight, nearly terminal awns.

T. montanum var. *pilosum* Louis-Marie, Rhodora 30: 212. 1928. (*Standley 4536*, "near Caroles [Cowles?], N. Mex.," 26 July 1908.)

Rocky Mountain trisetum.

Relatively stout perennials; culms (3) 4–8 dm tall; sheaths scabrous near the collar, glabrous below; ligules (0.7) 2–3 (4.3) mm long, erose-ciliate, lacerate, truncate; blades flat, 1.5–5 mm broad; panicle (7)

TRISETUM

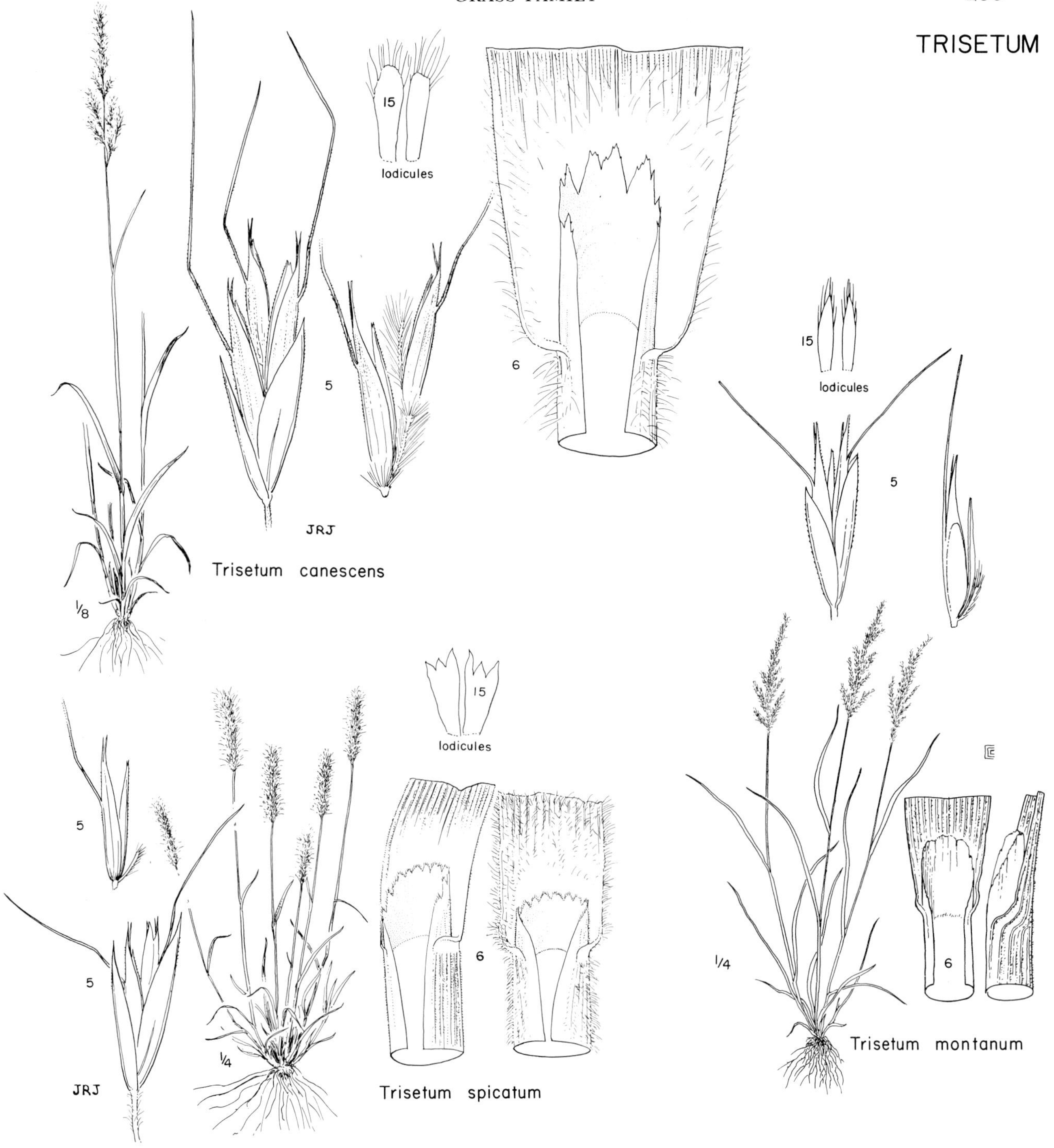

9–15 (20) cm long, loose and open but not spreading; spikelets 2-flowered; glumes lanceolate, glabrous, except the scabrous keel, unequal, the first glume 3–5.5 mm long, 1-nerved, the second 4–6 (7) mm long, 3-nerved; lemmas 4.5–5.5 (6.5) mm long, 5-nerved, the lateral nerves sometimes obscure, scaberulous, lanceolate, bifid at the apex with narrow teeth, the callus hairs often 1 mm long; awns 4–6 (8) mm long, geniculate, arising just below the tip of the lemma; palea bifid, nearly as long as the lemma; anthers 0.8–1.2 mm long.

Mountain meadows and moderately moist wooded slopes from the chaparral zone to the subalpine zone; Rocky Mts. of Colo. and N.M. and the Uinta and Abajo mts. of Utah. Late July–early Sept.

Trisetum montanum of middle elevations is closely related to the subalpine and alpine meadow species, *T. spicatum.* The two seem to agree in all characters except those in the key. Because of the complexity in *T. spicatum,* both morphologically and nomenclaturally, we hesitate to make a varietal combination.

14. DESCHAMPSIA Beauv. Hairgrass

Tufted perennials and some annuals; culms hollow, slender; sheaths open to the base; ligules narrow and long (ours), puberulent and decurrent; blades narrow, flat to involute, non-auriculate; inflorescence an open or contracted panicle, commonly with slender, wiry branches and pedicels; spikelets (1-) 2- (3)-flowered, the florets perfect, the rachilla usually hairy, prolonged beyond the terminal floret, sometimes bearing a vestigial floret, disarticulating above the glumes and between the florets; glumes 1- to 3-nerved, lanceolate, acute, subequal to sometimes unequal, exceeding the first floret and sometimes the uppermost as well; lemmas shiny, rounded on back, mostly 5- to 7-nerved, but the nerves often obscure, truncate and 2- to 4-toothed at the apex, bearded on the callus; awn slender, usually twisted and geniculate, attached below the middle of the lemma; palea nearly as long as the lemma, 2-keeled, the keels often scabrous; lodicules lanceolate to ovate-lanceolate, usually entire; stamens 3; ovary glabrous, the styles terminal; $x = 7$.

A genus of about 30 species in temp. and arctic regions of both hemispheres, mostly at high elevs. in the warmer latitudes; 7 species in the U.S. (Named in honor of Jean Louiseleur-Deslongchamps, 1774–1849, a French botanist.)

Deschampsia is closely related to the European genus *Aira* and has often been included in that genus in the past.

1 Plants perennial; lemmas larger, 2.5–4 (4.5) mm long; awns 1.5–5 (6.5) mm long.
  2 Panicle open, if narrow not more than 1/4 the length of the culm; blades flat or folded, 1–3 (5) mm broad; anthers 1.2–2.2 mm long .... 1. *D. cespitosa*
  2 Panicle narrow; blades usually filiform, the basal leaves about 1 mm broad; anthers small, 0.3–0.7 mm long .... 2. *D. elongata*
1 Plants annual; lemmas very small, 2–2.8 (3) mm long; awns (4) 5–7.5 mm long ...... 3. *D. danthonioides*

1. Deschampsia cespitosa (L.) Beauv.

Aira cespitosa L. Sp. Pl. 64. 1753. *Agrostis caespitosa* Salisb. Prodr. Stirpium 25. 1796. *Deschampsia caespitosa* Beauv. Essai Nouv. Agrostogr. 91, 149, 160. 1812. *Campella caespitosa* Link, Hort. Regius Bot. Berol. 1: 122. 1827. *Aira caespitosa* var. *genuina* Reichenb. Icon. Fl. German. & Helv. 1: pl. 96, fig. 1682. 1834. *Podionapus caespitosus* Dulac, Fl. Dépt. Hautes-Pyrén. 82. 1867. *Avena caespitosa* Kuntze, Taschen-fl. Leipzig 45. 1867. *Aira major* subsp. *caespitosa* Syme, Engl. Bot. ed. 3. 11: 64. 1873. *D. caespitosa* var *genuina* Volk in Rübel, Bot. Jahrb. Syst. 47: 312. 1912. *D. caespitosa* subsp. *genuina* W. E. Lawrence, Amer. J. Bot. 32: 302. 1945. ("Habitat in Europae.")

Aira caespitosa var. *montana* Vasey in Wheeler, Rep. U.S. Geogr. Surv. W. 100th Meridian 6: 294. 1879; not Reichenb. 1850. (Utah, 1871 and 1872; Colo., South Park, 1873, *"1175"*; Ariz., Willow Spring, 1874, *"230."*)

D. caespitosa var. *alpina* Vasey ex Beal, Grasses N. Amer. 2: 368. 1896; not Gaudin in Ducomm. 1869, nor *D. alpina* (L.) Roemer & Schultes, 1817. *D. alpicola* Rydb. Bull. Torrey Bot. Club 32: 601. 1905. *Aira alpicola* Rydb. Fl. Rocky Mts. ed. 2. 1112. 1923. *D. caespitosa* var. *alpicola* A. Löve, D. Löve & Kapoor, Arctic & Alpine Res. 3: 143. 1971. (*Elliott*, Alaska; *Letterman*, Gray's Peak, Colo., 15 Aug. 1885.)

D. caespitosa var. *confinis* Vasey ex Beal, Grasses N. Amer. 2: 369. 1896. *D. confinis* Rydb. Bull. Torrey Bot. Club 36: 533. 1909. (*Palmer 231*, s. Calif., in 1888.)

D. pungens Rydb. Bull. Torrey Bot. Club 39: 103. 1912. *Aira pungens* Rydb. Fl. Rocky Mts. ed. 2. 1112. 1923. *D. caespitosa* var. *pungens* B. Boivin, Naturaliste Canad. 75: 84. 1948. (*McCalla 2309*, "Along stream below warm sulphur springs," vicinity of Banff, 10 July 1899.)

Tufted hairgrass.

Strongly cespitose, tufted perennials; culms 2–8 (13) dm tall; sheaths glabrous to scabrous; ligules 3–7.5 (9.5) mm long, narrow, acuminate, often lacerate, decurrent with the sheath; blades firm, flat or folded, 1–3 (5) mm wide, often scabrous ventrally; panicle (4) 5–20 (30) cm long, loose, often nodding, open to narrow, the branches capillary, finely villose; spikelets 2- (3)-flowered, usually shiny, purplish; glumes lanceolate, acute, glabrous or scaberulous, the first glume (2) 3–5 mm long, 1-nerved, the second (2.5) 3.2–5.2 mm long, 1- to 3-nerved; lemmas 2.5–4 (4.5) mm long, often purplish basally, 5-nerved, the lateral 4 leading into 4 apical teeth, the callus villose; awns 2–5 (6.5) mm long, attached below the midlength of the lemma; palea more or less as long as the lemma; lodicules ovate-lanceolate, 0.7–1 mm long; anthers (1.2) 1.5–2.2 mm long; $2n = 24–28, 52$.

Streambanks, lake shores and wet mt. meadows at subalpine and alpine elevs.; Alaska to Newfl., s. throughout the w. states and e. from S.D. through the Great Lakes states to N.J. and N.C.; also in Mex. and in the Old World. Late July–Sept.

This extremely variable and widespread species has caught the fancy of several botanists who have described at least 30 races, variants or even species. *Deschampsia cespitosa* was the subject of an ecotypic study by Lawrence (Amer. J. Bot. 32: 298–314. 1945) and a cytogeographic study by Kawano (Canad. J. Bot. 41: 719–742. 1963). Lawrence distinguished 5 climatic ecotypes of which his *D. cespitosa* subsp. *genuina* (based on typical *D. cespitosa*) is the only one in our region.

2. Deschampsia elongata (Hook.) Munro ex Benth.

Aira elongata Hook. Fl. Boreali-Amer. 2: 243, pl. 228. 1840. *Deschampsia elongata* Munro ex Benth. Pl. Hartwegianus 342. 1857. (*Douglas*, "on the sandy islands of the river Columbia.")

D. elongata var. *ciliata* Vasey ex Beal, Grasses N. Amer. 2: 371. 1896. *D. ciliata* Rydb. Fl. Rocky Mts. 60, 1060. 1917 [1918]. *Aira vaseyana* Rydb. ibid. ed. 2. 1112. 1923. (*Anderson*, Santa Cruz, Calif., in 1887; *Howell*, Oregon.)

Slender hairgrass.

Cespitose perennials with numerous tufted basal leaves and culms, the herbage usually glabrous below the inflorescence; culms 1–6 (10) dm tall; ligules 2.5–7 (8) mm long, narrow-lanceolate, acute to acuminate, decurrent with the sheath; blades mostly filiform, flat or folded, the basal leaves about 1 mm broad, the cauline leaves sometimes up to 5 mm; panicle (3) 5–30 (35) mm long, the capillary branches ascending; spikelets 2- (3)-flowered, pale greenish to purplish, the hairs of the rachilla about 1 mm long; glumes narrow lanceolate, 3-nerved, scaberulose, the hairs sometimes confined to the nerves, usually ex-

DESCHAMPSIA

ceeding the upper floret in length, the first glume (3) 3.7–5.5 (6.7) mm long, the second slightly shorter, (3) 3.5–5 (6) mm long; lemmas 1.7–2.5 (3) mm long, faintly 5-nerved, shiny, deeply bifid, the lobes blunt, weakly toothed or erose, the callus bearded with hairs about 1 mm in length; awns 1.5–5 (6) mm long, nearly straight, attached just below the midlength of the lemma; palea nearly as long as the lemma; lodicules lanceolate, entire, 0.1–0.2 mm long; anthers very small, 0.3–0.5 (0.7) mm long; ovary glabrous; $2n = 26$.

Streambanks, lake shores, wet meadows and moist shady mt. slopes; w. N. Amer. from Alaska and Yukon, s. to Calif., n. Nev., n. Utah and Wyo., sporadically distributed in Ariz.; also in Mex. and Chile. Late June–Aug.

3. Deschampsia danthonioides (Trin.) Munro ex Benth.

Aira danthonioides Trin. Mém. Acad. Imp. Sci. St.-Pétersbourg, Sér. 6, Sci. Math. 1: 57. 1830. *Deschampsia danthonioides* Munro ex Benth. Pl. Hartwegianus 342. 1857. (w. N. Amer.)

D. calycina C. Presl, Reliquiae Haenkeanae 1: 251. 1830. *Aira calycina* Steudel, Synopsis Pl. Glum. 1: 220. 1854. (*Haenke*, "Monte Rey," Calif., 13–23 Sept. 1791 [on the type but Presl gave "Peru" as the locality].)

Annual hairgrass.

Delicate, simple to somewhat tufted annuals, the herbage glabrous; culms 1–4 (7) dm tall; leaves mostly basal; ligules (0.5) 2–3 (4) mm long, acute to acuminate, entire, decurrent with the sheath; blades filiform, involute or flat, 1 mm broad; panicle (3) 5–15 (20) cm long, the branches erect to spreading; spikelets (1-) 2-flowered, the rachilla hairy; glumes acuminate, linear, 3-nerved, the lateral nerves close to the midrib, glabrous to scaberulous, the first glume (4) 5–7.5 (8.5) mm long, the second usually slightly shorter, (3.5) 5–7.5 (8) mm long; lemmas very small, 2–2.8 (3) mm long, shiny, pale green or purplish, the apex blunt, erose and ciliate, the callus with hairs 0.7–1.2 mm long; awns (4) 5–7.5 mm long, geniculate in the middle, brown and thick below, pale and thin above, attached near the base of the lemma; palea slightly shorter than the lemma; lodicules 0.2–0.3 mm long, oblong-lanceolate; anthers 0.3–0.5 mm long; ovary glabrous; $2n = 26$.

Wet or drying meadows, streambanks, drying streambeds and vernal pools; w. N. Amer. from Alaska and Yukon, s. through Wash., Idaho and w. Mont. to Baja Calif., Nev. and n. Utah and disjunct stations in Coconino and Cochise cos., Ariz.; also in Chile and Argentina. May–July (–Aug.)

There appears to be a Wasatch phase with smaller glumes and lemmas.

15. AVENA L. Oats

Stout annuals; culms tall, hollow; sheaths open to the base; ligules short and membranous; blades flat, non-auriculate; inflorescence an open, drooping panicle; spikelets usually 2- to 3-flowered, large, pendulous, the florets perfect, the rachilla hirsute to glabrous, prolonged behind the upper floret, disarticulating above the glumes and between the florets; glumes subequal, usually more than 2 cm long, exceeding the first floret and often surpassing the second; lemmas 1.5 cm or more long, 5- to 7-nerved, glabrous or pilose on the lower half, tough and indurate; awns stout, twisted and geniculate, sometimes small and straight or absent, attached about at the midlength; palea shorter than the lemma, 2-keeled; lodicules 2, linear-lanceolate; stamens 3; ovary hairy; styles subterminal; $x = 7$.

A genus of about 10 to 15 species, native to Europe and Asia and widely cultivated. (The classical Latin name.)

Avena sterilis L., animated oats, is sometimes grown in our region for its large spikelets (35–45 mm long). The prominent awns (50–70 mm long) are used to demonstrate hygroscopic action. A fourth introduced species in w. U.S., *A. barbata* Brot., remains to the s. and w. of our region.

1 Awns present on the first and second florets, stout and strongly geniculate; lemmas with brown hairs . . . . 1. *A. fatua*

1 Awns lacking or present on the first floret only, short, straight or curved, not geniculate; lemmas glabrous to sometimes scabrous . . . . 2. *A. sativa*

1. Avena fatua L.

Avena fatua L. Sp. Pl. 80. 1753. *A. sativa* var. *fatua* Fiori, Nuova Fl. Analitica Ital. 1: 109. 1923. ("Habitat in Europae.")

A. fatua var. *glabrata* Peterm. Fl. Bienitz 13. 1841. (Europe.)

Wild oats.

Stout annuals; culms (2.5) 4.5–9 (19) dm tall; ligules 2–5.5 (7) mm long, obtuse, erose-ciliolate, often pubescent; blades flat, 3–10 (15) mm broad, scabrous to somewhat pilose; panicle 10–25 (38) cm long, open; spikelets 2- to 3- (4)-flowered, the rachilla densely hairy; glumes prominently (7-) 9- to 11-nerved, the first glume 19–26 (29) mm long, narrow-lanceolate, acuminate, the second slightly longer, 20–28 (30) mm long, lanceolate, acuminate; lemmas 14–20 (23) mm long, 5- to 7-nerved, the nerves often obscure on the shiny indurated lower half and prominent on the greenish upper half, glabrous, sparsely pilose to pilose, shortly bilobed (for less than 1 mm), the callus usually densely pilose; awns (18) 28–45 (52) mm long, stout, twisted below the knee, thin and tapering above, attached above the midlength of the lemma, usually only on the first 2 florets; palea firm, the 2 keels ciliate; lodicules about 2 mm long; anthers 1.6–4.2 (6) mm long; $2n = 42, 48$.

A weed along roadsides and in fallow fields; introduced from Europe, Alaska to Newfl., s. sporadically throughout the U.S. and Latin Amer., especially common in cismontane Calif. May–Aug.

Avena fatua var. *glabrata* Peterm., with an extra floret, subglabrous lemmas and more twisted, geniculate awns, may be a hybrid between the wild oats and cultivated oats.

DESCHAMPSIA
AVENA
24
lodicules
6
JRJ
5
1/2
Deschampsia danthonioides
20
lodicules
2
2
4
1/2
Avena fatua

2. Avena sativa L.

Avena sativa L. Sp. Pl. 79. 1753. *A. fatua* var. *sativa* Hausskn. Mitt. Geogr. Ges. (Thüringen) Jena 3: 238. 1885. *A. fatua* subsp. *sativa* Thell. Vierteljahrsschr. Naturf. Ges. Zürich 56: 325. 1911. (Europe.)

Cultivated oats.

Similar to *A. fatua;* stout annuals, the herbage glabrous to scabrous; culms (2.5) 4–8 (12.5) dm tall; ligules 2–5.5 mm long, obtuse, puberulent, finely erose-ciliolate; blades (4) 5–10 (17) mm broad; panicle 12–30 cm long, open; spikelets 2- (3)-flowered, the rachilla glabrous or only sparsely hirsute, not readily disarticulating; glumes usually exceeding the florets, 9- to 11- (13)-nerved, the first glume (17) 18–26 mm long, the second (18) 20–28 mm long; lemmas (12) 14–20 mm long, glabrous to sometimes scabrous, 3- to 7-nerved, the nerves obscure on the indurate, tawny lower part and prominent on the herbaceous upper third, entire or very shallowly bilobed at the apex, the callus naked or only sparsely bearded; awns, when present, (15) 22–35 mm long, then only on the first floret, not geniculate; palea firm, the 2 keels ciliate; lodicules linear, about 1.8 mm long; anthers 1.7–4.3 mm long; $2n = 20$, 42, 48, 63.

Commonly cultivated and only occasionally escaping short distances from areas of cultivation; introduced from Europe; Alaska to Newfl., s. throughout most of U.S. and Latin Amer. June–Aug.

Avena sativa is cultivated throughout central and western United States and occasionally in the east. It sometimes escapes but will not persist for long out of cultivation. The cultivated oat has its origin from *A. fatua* through selection by man. The two are morphologically very similar and their distinction often breaks down. Some taxonomists recognize the combination *A. fatua* var. *sativa* (L.) Hausskn., but we see enough differences that appear to withstand sympatry to consider them as separate species.

16. HELICTOTRICHON Besser Perennial Oatgrass

Tufted cespitose perennials; culms short (ours) to moderately tall, hollow; sheaths open to the base; ligules short membranous; blades flat to involute, non-auriculate; inflorescence a contracted panicle; spikelets 3- to 7-flowered, large, the florets perfect, the rachilla villous, disarticulating above the glumes and between the florets; glumes large, thin and membranous at the acute apex, subequal, the first glume somewhat smaller, 1-nerved, the second 3-nerved, longer than the first floret; lemmas rounded on the back, few- to several-nerved, firm below, membranous above and entire or usually toothed at the apex; awns stout, twisted and geniculate, attached about at the midlength of the lemma; palea shorter than the lemma, thin, 2-keeled; lodicules 2, linear; stamens 3, the anthers large; ovary hairy at the tip, the styles terminal; $x = 7$.

A genus of 30 to 40 species, mostly Eurasian and African; 2 native species in the U.S., our alpine S. Rocky Mt. species, *H. mortonianum,* and the low elev. *H. hookeri* (Scribn.) Henrard of the prairies and foothills from Colo. n. through Mont., Alta. to N.W.T. and e. across the n. plains to Minn. (From the Greek *helicos,* twisted, and *trichon,* hair, referring to the twisted awn.)

Helictotrichon is so closely related to *Avena* that it had long been under sect. Avenastrum Koch of that genus.

1. Helictotrichon mortonianum (Scribn.) Henrard

Avena mortoniana Scribn. Bot. Gaz. 21: 133, pl. 11. 1896. *Helictotrichon mortonianum* Henrard, Blumea 3: 429. 1940. *Arrhenatherum mortonianum* Potztal, Willdenowia 4: 400. 1968. (*Shear 697,* Silver Plume, Colo., alt. 13,000 ft., 24 Aug. 1895; lectotype by A. S. Hitchcock.)

Alpine oat.

Densely tufted, cespitose perennials; culms 0.8–1.8 (2) dm tall; leaves becoming pale straw-colored, puberulent; ligules 0.5–1 mm long, membranous, ciliolate, puberulent; blades filiform, involute, 1.5–2 mm broad when unrolled; panicle 3–5 (7.5) cm long, narrow, the short branches erect; spikelets pale or purplish, mostly 2-flowered, the rachilla densely villous, extending beyond the upper floret; glumes hyaline, at least on the margins, acuminate, scabrous on the keel and nerves, nearly equal, the first glume 8–10.5 mm long, 1-nerved, the second 8.5–11.5 mm long, 3-nerved; lemmas 7–9.5 mm long, firm, indistinctly 3-nerved, cleft into 4 bristled teeth at the apex, the callus bearded with hairs 1–2 mm long; awns 10–16 mm long, attached near the midlength of the lemma, geniculate, twisted below the knee, dark brown; palea shorter than the lemma; anthers about 1.5 mm long.

Alpine meadows; S. Rocky Mts. of Colo. and n. N.M. and in the Uinta Mts. of Utah. July–Aug.

Named for J. Sterling Morton, the Secretary of Agriculture in the late 1880s.

17. ARRHENATHERUM Beauv. Tall Oatgrass

Tall perennials; culms hollow; sheaths open to the base; ligules membranous; blades flat, non-auriculate; inflorescence a narrow panicle; spikelets usually 2-flowered, the upper floret perfect, the lower staminate, the rachilla prolonged beyond the upper floret as a bristle, disarticulating above the glumes, the florets falling together; glumes broad, thin, keeled, acute and unequal, the first glume 1-nerved, the second larger, 3-nerved, usually equaling the second floret; lemmas 5- to 7-nerved, the callus bearded; awn of the first floret attached about at the midlength, stout, geniculate, that of the second floret absent or hair-like, attached from just below the tip (in ours); palea 2-keeled, the keels scabrous or hairy; lodicules 2, linear; stamens 3; ovary pubescent, the styles terminal; caryopsis narrowly-oblong; $x = 7$.

A group of 6 species, native to temp. regions of Europe and Asia. (From the Greek *arrhen,* masculine, and *ather,* awn, in reference to the awned staminate floret.)

AVENA
HELICTOTRICHON

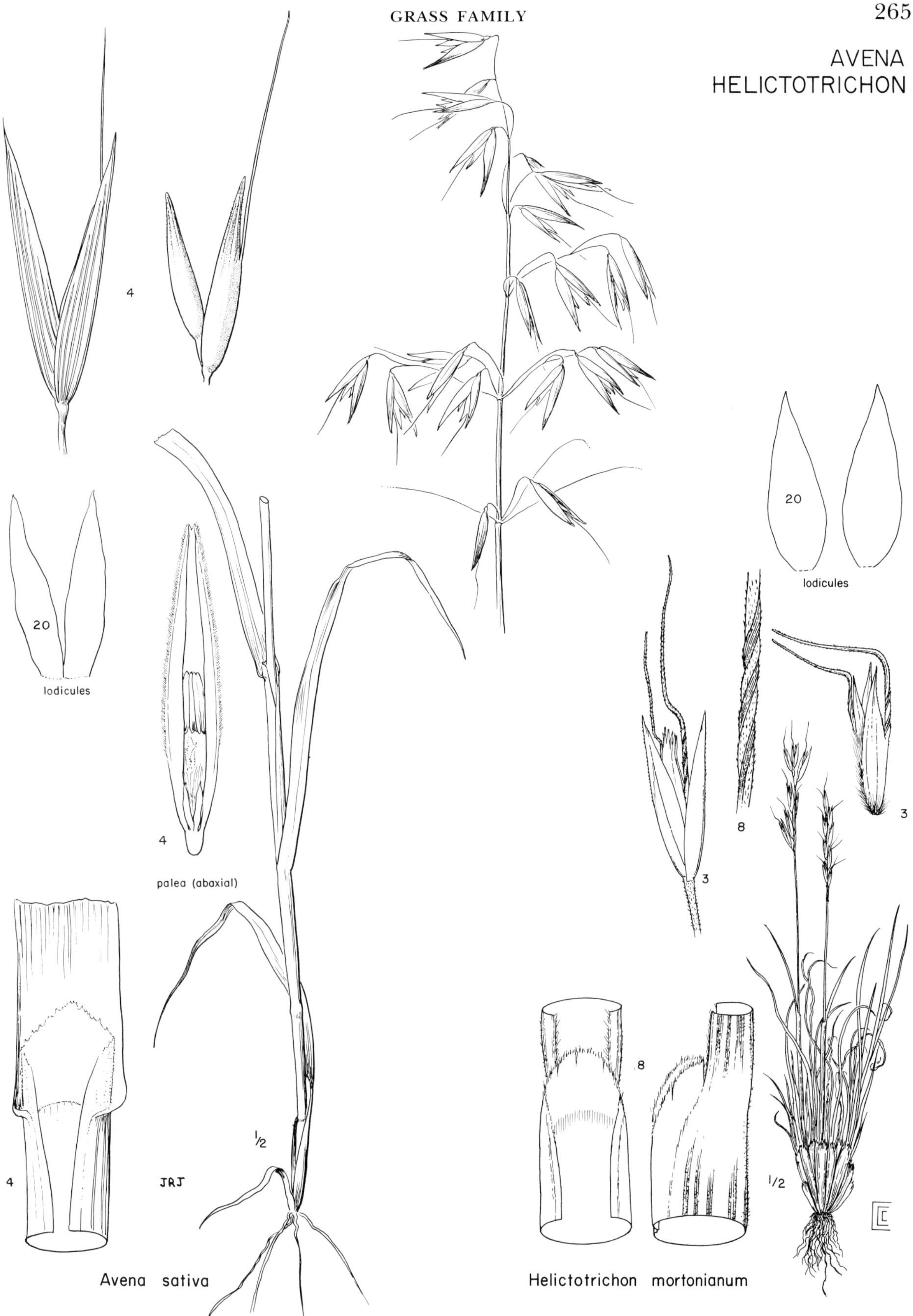

Avena sativa

Helictotrichon mortonianum

1. Arrhenatherum elatius (L.) J. & C. Presl

Avena elatior L. Sp. Pl. 79. 1753. *Holcus avenaceus* Scop. Fl. Carniolica, ed. 2. 2: 276. 1772. *Avena elata* Salisb. Prodr. Stirpium 23. 1796; not Forssk. 1775. *Arrhenatherum americanum* Beauv. Essai Nouv. Agrostogr. 56, 152, 165. 1812, name only. *Arrhenatherum avenaceum* Beauv. ibid. 55, 152, 153, 164. 1812. *Hordeum avenaceum* Wigg. ex Beauv. ibid. 165. 1812, name only. *Arrhenatherum elatius* J. & C. Presl, Fl. Čech. 17. 1819. (Europe.)

Avena bulbosa Willd. Ges. Naturf. Freunde Berlin Neue Schriften 2: 116. 1799. *Holcus bulbosus* Schrader, Fl. German. 1: 248. 1806. *Holcus avenaceus* var. *bulbosus* Gaudin, Agrostol. Helv. 1: 136. 1811. *Arrhenatherum bulbosum* C. Presl, Cyper. & Gram. Siculae 29. 1820. *Arrhenatherum avenaceum* var. *bulbosum* Hartm. Handb. Skand. Fl. Stockholm 51. 1820. *Avena elatior* var. *bulbosa* Saint-Amans, Fl. Agénaise 47. 1821. *Arrhenatherum elatius* var. *bulbosum* Spenner, Fl. Friburg. 1: 113. 1825. *Arrhenatherum elatius* subsp. *bulbosum* Hylander, Nord. Karlvaxtfl. 1: 290. 1953. *Avena elatior* subsp. *bulbosa* Litard. Candollea 15: 13. 1955. (Switzerland.)

Tall oatgrass.

Short lived perennials; culms (7.5) 10–17 dm tall, glabrous or sometimes pubescent at the nodes, the base sometimes enlarged into bulbs; sheaths smooth; ligules 1–3 mm long, obtuse, erose-ciliolate, finely puberulent; blades flat, 3–6 (10) mm broad, glabrous to scaberulous; panicle (7) 14–25 (30) cm long, narrow, the branches floriferous to the base; spikelets shiny, 2-flowered, the first floret staminate and the second perfect, the florets disarticulating as a unit; glumes membranous, scabrous, unequal, the first glume (3.5) 5–7 mm long, 1-nerved, the second (7) 7.5–9.5 (10) mm long, 3-nerved; lemmas 7–9.5 mm long, 7-nerved, scabrous; awn of the first lemma 10–14 (20) mm long, attached about at the midlength, geniculate, twisted below the knee, the awn of the second lemma small, straight, attached just below the apex of the lemma, rarely absent; palea hyaline, acuminate, slightly shorter than the lemmas; lodicules 2.5–3.5 mm long; anthers about 4 mm long; $2n$ = 14, 28, 42.

Along roadsides, fence rows, and moderately moist fallow fields; native of Europe; widespread across s. Can. and s. in the U.S. to the arid Southwest deserts. June–July.

Tall oatgrass has long been grown as a pasture grass in sandy soils in many parts of the United States and has frequently escaped cultivation.

18. HOLCUS L. nom. conserv. Velvetgrass

Low to moderately tall perennials with velvety pubescent herbage; culms hollow; sheaths open to the base; ligules membranous, short, truncate, puberulent; blades flat, broad, non-auriculate; inflorescence a contracted panicle; spikelets 2-flowered, the lower floret perfect, the upper staminate, disarticulating below the glumes; glumes large, thin, keeled, subequal and enclosing the florets, the first glume 1-nerved, the second 3-nerved, much broader than the first; lemmas obscurely 3- to 5-nerved, shiny and somewhat coriaceous, the lower lemma awnless, blunt at the apex, the upper with a short, hooked awn just below the bifid tip; palea thin, nearly as long as the lemma; lodicules 2, lanceolate; stamens 3; ovary glabrous; x = 7.

About 8 species native to Europe and Africa. Two species, *H. lanatus* and *H. mollis*, have been introduced into the U.S., the former entering our area. (The Latin name used by Pliny for some grass of uncertain identity.)

Most old American literature referred our species to the genus *Notholcus* Nash.

1. Holcus lanatus L.

Holcus lanatus L. Sp. Pl. 1048. 1753. *Aira holcus-lanata* Vill. Hist. Pl. Dauphiné 2: 87. 1787. *Avena pallida* Salisb. Prodr. Stirpium 24. 1796; not Thunb. 1794. *Avena lanata* Koeler, Descript. Gram. Gallia & German. 303. 1802. *Ginannia pubescens* Bubani, Fl. Pyrén. 4: 321. 1901. *Notholcus lanatus* Nash ex A. S. Hitchc. in Jepson, Fl. Calif. 1: 126. 1912. *Nothoholcus lanatus* Nash in Britton & Brown, Ill. Fl. N. U.S., Canada & Brit. Possessions, ed. 2. 1: 214. 1913. *Ginannia lanata* Hubbard, Rhodora 18: 234. 1916. ("Habitat in Europae pascuis arenosis.")

Common velvetgrass.

Cespitose, tufted perennials with grayish, soft pubescence; culms (3) 4–10 (20) dm tall, sometimes decumbent at the base; ligules 1.5–3 (4.2) mm long, truncate, erose-ciliate, puberulent; blades flat, 3.5–8 (12) mm broad; panicle 5–12 (20) cm long, contracted, branches and pedicels short and ascending, hirsute; spikelets 2-flowered, the first floret perfect, the second staminate; glumes hirsute, the nerves ciliate, grayish, sometimes with a purplish tinge, the first glume 3.5–5 mm long, lanceolate, 1-nerved, the second 3.8–5.5 (6) mm long, much broader, ovate, prominently 3-nerved, sometimes glabrous between the midnerves and margins, sometimes tipped with a small awn; lemmas small, 1.7–2.5 mm long, enclosed within the glumes, shiny, smooth, the upper one with a short hooked dorsal awn attached below a shallow cleft; lodicules broadly lanceolate-acuminate, up to 0.5 mm long; anthers 1.4–2.2 (2.5) mm long; $2n$ = 14.

Moist waste places such as ditch banks and roadsides, mostly at low elevs.; introduced from Europe; sporadically from B.C. to N.S., s. throughout most of the U.S. May–June (–Sept.).

Velvetgrass is occasionally cultivated as a meadow grass.

19. CALAMAGROSTIS Adans. Reedgrass

Moderately large, mostly rhizomatous perennials; culms hollow; sheaths open to the base; ligules relatively large, often puberulent; blades flat to involute, non-auriculate; inflorescence open to contracted and spike-like panicles, spikelets 1-flowered (rarely 2), the rachilla prolonged behind the palea as a slender, hairy bristle (in ours), disarticulating above the glumes; glumes subequal, longer than the floret, acute to acuminate, somewhat keeled, the first glume 1-nerved, the second usually 3-nerved; lemmas (ours 2.2–6 mm long) 5-nerved, the lateral nerves often extending into 4 small teeth, the callus bearded, the hairs sometimes as

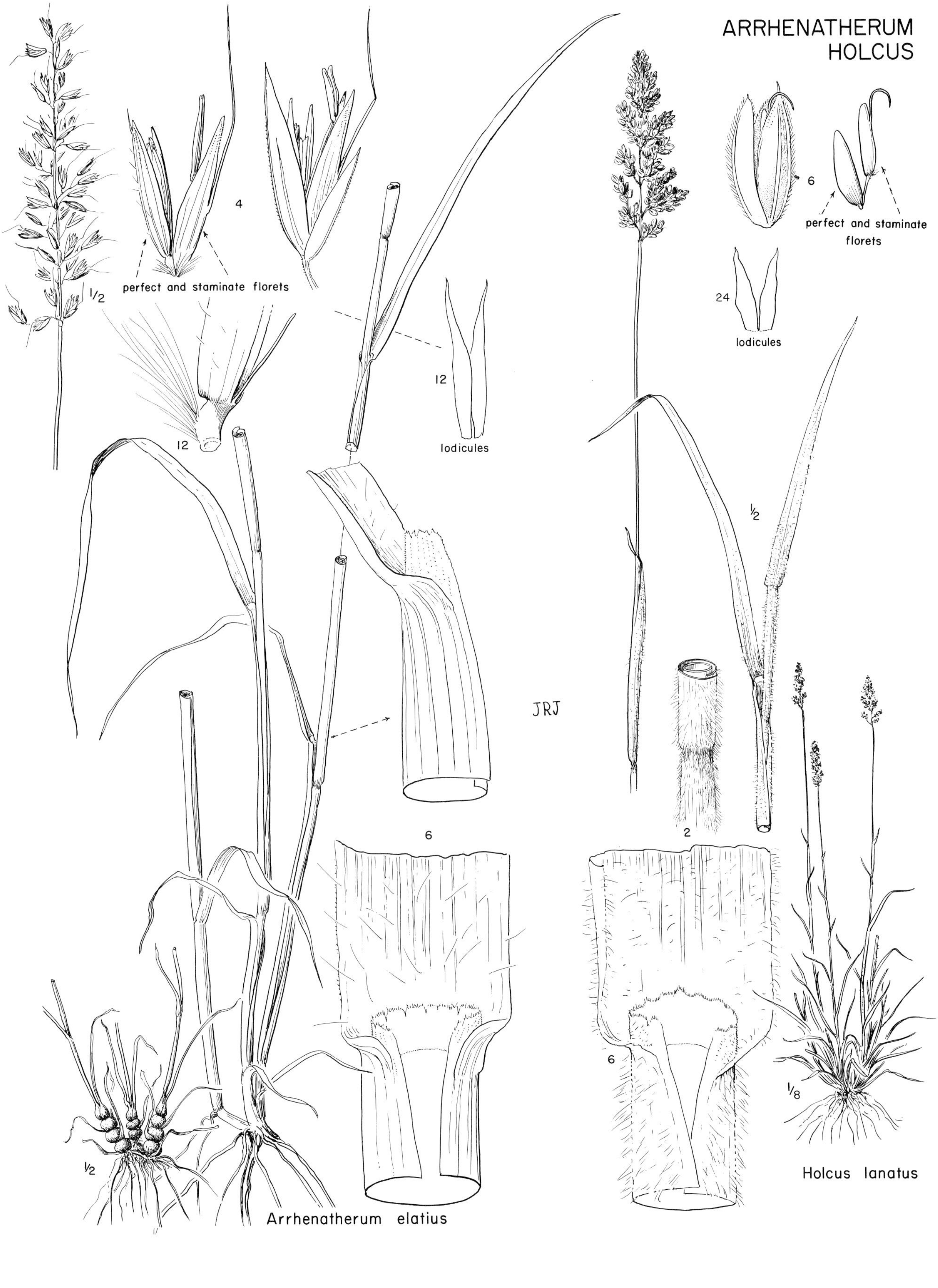
ARRHENATHERUM
HOLCUS
1/2
4
perfect and staminate florets
12
12
lodicules
6
perfect and staminate florets
24
lodicules
1/2
JRJ
2
6
6
1/8
1/2
Holcus lanatus
Arrhenatherum elatius

long as the lemmas; awns slender or geniculate, arising from the dorsal side of the lemma, usually below the middle; palea usually slightly shorter than the lemma; lodicules 2, membranous, ovate-lanceolate to narrowly oblong; stamens 3; ovary glabrous; $x = 7$.

More than 100 species in cool and temp. regions of both hemispheres, about 25 native to the U.S. Several of the U.S. species are relatively narrow endemics. (Name from the Greek *calamos,* a reed, and *agrostis,* a grass.)

Apomixis is known to occur in *C. canadensis, C. neglecta* and *C. purpurascens,* which has added to their complexity.

References:

Kawano, S. *Calamagrostis purpurascens* R. Br. and its identity. Acta Phytotax. Geobot. 21: 73–89. 1965.

Kearney, T. H., Jr. A revision of the North American species of *Calamagrostis.* U.S.D.A. Div. Agrostol. Bull. 11: 7–42. 1898.

Nygren, A. Investigations on North American *Calamagrostis.* I. Hereditas 40: 377–397. 1954.

———. Investigations on North American *Calamagrostis.* II. Lantbrukshögskolans Ann. 24: 363–368. 1958.

Stebbins, G. L., Jr. A revision of some North American species of *Calamagrostis.* Rhodora 32: 35–57. 1930.

1 Awn straight, 1–2.5 (3) mm long (often absent in *C. scopulorum*); callus hairs 1.5–3.5 mm long, half to nearly as long as the lemma.
  2 Panicle rather loose and open; leaves soft .... 1. *C. canadensis*
  2 Panicle more or less contracted; leaves ridged.
    3 Glumes 2.5–4 mm long; lemmas 2.2–3.2 mm long; widespread .... 2. *C. neglecta*
    3 Glumes 4.2–6 mm long; lemmas 3.5–5 mm long; s. Wyo., Colo., N.M., Utah and n. Ariz. .... 3. *C. scopulorum*
1 Awn geniculate, (2.2) 2.5–8 mm long; callus hairs 0.5–1.5 mm long, less than half the length of the lemma.
  4 Awn 4.5–8 mm long, longer than the glumes; glumes 5.5–7.5 mm long .... 4. *C. purpurascens*
  4 Awn 2.2–3.5 (4.5) mm long, included or scarcely extending beyond the glumes; glumes shorter, 3–5 mm long .... 5. *C. rubescens*

1. Calamagrostis canadensis (Michx.) Beauv.

Arundo canadensis Michx. Fl. Boreali-Amer. 1: 73. 1803. *Calamagrostis canadensis* Beauv. Essai Nouv. Agrostogr. 15, 152, 157. 1812. *C. michauxii* Trin. ex Steudel, Nomencl. Bot. ed. 2. 1: 249, 250. 1840. *Arundo fissa* Willd. ex Steudel, Nomencl. Bot. ed. 2. 1: 144. 1840, as a synonym of *C. michauxii. Deyeuxia canadensis* Munro ex Hook. f. Trans. Linn. Soc. London 23: 345. 1861. (*Michaux,* Can., in 1792.)

C. canadensis var. *acuminata* Vasey ex Shear & Rydb. U.S.D.A. Div. Agrostol. Bull. 5: 26. 1897. *C. langsdorfii* var. *acuminata* Litw. Schaedae Herb. Fl. Ross. 8: 170. 1918. (*Shear 615,* Georgetown, Colo., 17 Aug. 1895.)

Bluejoint, Canada reedgrass.

Strongly rhizomatous perennials; culms stout, 6–12 (15) dm tall, smooth; sheaths smooth; ligules 3–7.5 mm long, membranous, usually lacerate and erose-ciliate; blades usually flat, 2–6 (7.5) mm broad, scabrous; panicle 8–17 (24) cm long, relatively open; spikelets usually with only 1 floret; glumes lanceolate, acuminate, scabrous, at least on the keel, usually purplish or sometimes greenish or straw-colored, 1-nerved or with faint lateral pairs, the first glume aristate, 3–4.5 (5.2) mm long, the second slightly shorter, 2.7–4.2 mm long; lemmas 2.5–4 mm long, obscurely nerved; callus hairs abundant, long, nearly as long as the lemma; awn 1.2–2 (3) mm long, slender, straight, attached below or at about the middle of the lemma; palea usually 3/4 to 4/5 as long as the lemma; rachilla vestige 0.1–1.3 mm long; lodicules 0.6–0.8 mm long, slender; anthers 1.2–2 mm long; $2n = 28, 42–66$.

Wet places in meadows and streambanks from low to high elevs. in the mts.; Arctic Amer. extending s. in the w. to n. N.M., n. Ariz., Nev., Sierra Nevada and n. Coast Ranges of Calif., and e. across the U.S. through Kansas, Mo., Ky. and N.C. July–Aug.

Bluejoint furnishes much forage for cattle and horses but tends to be tough and unpalatable when the herbage is mature.

Throughout most of its range *C. canadensis* shows considerable complexity with several intergrading geographical races. Plants from the Intermountain Region can be sorted more or less into two varieties but with no real satisfaction. Variety *acuminata* Vasey has acuminate glumes that average about 4 mm in length and the var. *canadensis* has acute glumes that average 3.5 mm in length. There appears to be no geographical or ecological correlation, however.

2. Calamagrostis neglecta (Ehrh.) Gaertn., Meyer & Scherb.

Arundo neglecta Ehrh. Beitr. Naturk. 6: 137. 1791. *Calamagrostis neglecta* Gaertn., Meyer & Scherb. Oekon.-Techn. Fl. Wetterau 1: 94. 1799. *Deyeuxia neglecta* Kunth, Révision Gram. 1: 76. 1829. ("Germania, Suecia.")

C. inexpansa A. Gray, N. Amer. Gram. & Cyper. 1: No. 20. 1834. *C. neglecta* var. *inexpansa* M. E. Jones, Contr. W. Bot. 14: 9. 1912. *C. expansa* Rickett & Gilly, Bull. Torrey Bot. Club 69: 464. 1942, orthographic error. (*Sartwell,* Penn Yan, N.Y.)

C. stricta var. *robusta* Vasey in Wheeler, Rep. U.S. Geogr. Surv. W. 100th Meridian 6: 285. 1879. *C. robusta* Vasey, Contr. U.S. Natl. Herb. 3: 82. 1892; not C. Muell. 1861. *C. inexpansa* var. *robusta* Stebbins, Rhodora 32: 48. 1930. (*Wolf 1099,* Twin Lakes, Colo.)

C. stricta var. *brevior* Vasey in Wheeler, Rep. U. S. Geogr. Surv. W. 100th Meridian 6: 285. 1879. *C. inexpansa* var. *brevior* Stebbins, Rhodora 32: 50. 1930. (*Wolf 1098,* "Mosquito, South Park, Colo., 1873.")

C. stricta sensu Amer. authors; not Trin.

Rhizomatous perennials; culms 3–9 (13) dm tall, smooth to scabrous; sheaths smooth; ligules 1–3.5 (10) mm long, truncate, erose-ciliate, nearly lacerate, puberulent; blades involute, stiff, folded, or if flat up to 4 (5.5) mm broad, scabrous to pilose above and often glabrous below; panicle 6–12 (18) cm long, contracted, 1–1.5 (2.5) cm broad when pressed; glumes subequal, (2.5) 3–4 mm long, glabrous to scaberulous, pale-green, sometimes purple-tinged, the first glume usually slightly longer, lanceolate, 1-nerved, the second broadly lanceolate, 3-nerved; lemmas 2.2–3.2 mm long, smooth to scaberulous, 5-nerved, often obscurely so, the lateral nerves projecting into 4 minute apical teeth; callus hairs 1.5–2.5

CALAMAGROSTIS

Calamagrostis canadensis

Calamagrostis neglecta

mm long, $^2/_3$ to $^3/_4$ the length of the lemma; awn (1) 1.5–2.5 mm long, straight, attached below the middle of the lemma; palea shorter than the lemma, membranous; rachilla vestige bearded; lodicules about 0.7 mm long, acute; anthers 1.3–2 mm long; $2n$ = 28, 56, 58, 70, 84–105.

Moist and wet places, streambanks, meadows, marshes; Newfl. to Alaska, s. through the Rocky Mts. through Mont., Idaho, Wyo. and Colo., and e. across n. U.S. through Nebr., Iowa, Mich., Ohio, Pa. to N.Y. and scattered sites in Utah, Nev., Oregon and Calif. July–Aug.

Calamagrostis inexpansa is considered here to be inseparable from *C. neglecta.* Traditionally *C. inexpansa* has been distinguished as having somewhat longer ligules, broader blades and more pubescent lemmas than *C. neglecta.*

3. Calamagrostis scopulorum M. E. Jones

Calamagrostis scopulorum M. E. Jones, Proc. Calif. Acad. Sci. II. 5: 722. 1895. (*Jones 6075,* "at base of sandstone cliffs along the Virgin river," Springdale, Utah, 25 Sept. 1894; POM 115105.)

C. scopulorum var. *lucidula* Kearney, U.S.D.A. Div. Agrostol. Bull. 11: 33. 1898. (*Jones 1145,* "Altah, Wahsatch Mountains, Utah," in 1879.)

C. scopulorum var. *bakeri* Stebbins, Rhodora 32: 47. 1930. (*Baker 162,* "Near Pagosa Peak" at 9000 ft., Colo., Aug. 1899.)

Perennials with short rhizomes, sometimes from loose tufts; culms 4–9 dm tall, smooth; sheaths smooth to scaberulous; ligules 2.5–5 mm long, obtuse, erose-ciliate, sometimes lacerate; blades flat, 2–4.5 mm broad, firm, erect, scaberulous, mostly only on the upper surface; panicle 7–15 (18) cm long, compressed, spike-like; glumes acuminate, glabrous, the first 4.5–6 mm long, 1-nerved, the second slightly shorter, 4.2–5.5 (6) mm long, 3-nerved; lemmas relatively large, 3.5–5 mm long, glabrous to scaberulous, 5-nerved, the lateral nerves terminating in 4 very small apical teeth; callus hairs 2–3 mm long; awn sometimes absent or up to 2 mm long, straight, arising from above the middle of the lemma; palea nearly as long as the lemma; rachilla vestige bearded; lodicules about 0.8 mm long, ovate-acuminate; anthers 2.2–3 mm long; $2n$ = 28.

Moist places on exposed rocky slopes and at the bases of cliffs, subalpine and alpine zones; Rocky Mts. from s. Wyo. and Colo. to N.M. and in the mts. and plateaus of Utah and n. Ariz. July–Aug.

Jones's type material of *C. scopulorum* is atypical for the species. The New York isotype could pass for a large spikelet form of *C. canadensis,* but the holotype from Rancho Santa Ana (POM) is morphologically closer to biologically typical *C. scopulorum.*

4. Calamagrostis purpurascens R. Br.

Calamagrostis purpurascens R. Br. in Franklin, Narr. Journey Shores Polar Sea, Bot. Appendix 731. 1823. *Arundo purpurascens* J. A. Schultes, Mantissa 3 (Addit. 1): 603. 1827. *Deyeuxia purpurascens* Kunth, Révision Gram. 1: 77. 1829. *C. arundinacea* f. *purpurascens* Gelert in Ostenf. Fl. Arctica 1: 103. 1902. *C. arundinacea* var. *purpurascens* Porsild, Meddel. Grønland 47: 261. 1910. (N. British Amer.)

C. sylvatica var. *americana* Vasey, Contr. U.S. Natl. Herb. 3: 83. 1892. (*Vasey,* Pen Gulch, Colo., in 1884.)

C. sylvatica sensu Amer. authors in part; not DC.

Purple reedgrass.

Strongly cespitose perennials with short thick rhizomes; culms 2–5.5 (7) dm tall; sheaths smooth to scabrous, usually scabrous around the collar; ligules 2–5 (6.5) mm long, truncate to obtuse, lacerate and erose-ciliolate, puberulent; blades involute, or if flat 2.5–4 (5) mm broad, scabrous or hirsutulous, at least on the upper surface; panicle relatively short, 4–10 cm long, compressed; spikelets 1- or rarely 2-flowered; glumes subequal, relatively long, 5.5–7.5 mm, scabrid, at least on the keel, purplish, sometimes straw-colored, the first glume 1-nerved, the second 3-nerved, the lateral nerves usually obscure; lemmas large, 4.5–5.5 (6) mm long, scaberulous or glabrous, 4-toothed at the apex; callus hairs relatively short, 1–1.5 mm long; awn 4.5–8 mm long, geniculate, twisted below, attached at the middle or slightly below the middle of the lemma; rachilla vestige densely villous with hairs about 1 mm long; lodicules about 0.6 mm long, oblong; anthers 2–3 mm long; $2n$ = 28, 40–58, 84.

Dry to moderately moist rocky alpine slopes; Arctic Amer., extending s. in the w. to Colo., Utah, n. $^2/_3$ of Nev. and the Sierra Nevada of Calif., and e. to S.D., Minn., Sask. and Newfl. July–Aug.

5. Calamagrostis rubescens Buckley

Calamagrostis rubescens Buckley, Proc. Acad. Nat. Sci. Philadelphia 1862: 92. 1863. *Deyeuxia rubescens* Vasey, Grasses U. S. 28. 1883. (*Nuttall,* Oregon, in 1835.)

Deyeuxia cusickii Vasey, Bot. Gaz. 10: 224. 1885. *C. cusickii* Vasey, Contr. U. S. Natl. Herb. 3: 81. 1892. (*Cusick 1159,* Eagle Creek Mts., e. Oregon.)

C. sylvatica sensu Amer. authors in part; not DC.

Pinegrass.

Rhizomatous perennials; culms (4) 7–11 dm tall, smooth; sheaths glabrous to pilose, the collar puberulent to hirsute, at least toward the margins; ligules 1–4 (5) mm long, obtuse or truncate, lacerate and erose-ciliolate; blades usually flat, 1.5–5 (5.5) mm broad, scabrous to pilose, at least above; panicle 6–12 (20) cm long, somewhat compressed; glumes subequal, (3) 4–5 mm long, lanceolate, acuminate, glabrous to minutely scaberulous, pale greenish to sometimes purplish, the first glume 1-nerved, the second 3-nerved, the lateral nerves often obscure; lemmas 2.5–3.2 mm long, smooth to finely scaberulous, 5-nerved, the nerves sometimes faint, terminating as 4 short teeth; callus hairs sparse and relatively short, 0.7–1.2 mm long; awn 2.2–3.5 (4.5) mm long, twisted at the base, geniculate from below the middle of the lemma; palea shorter than the lemma, membranous; rachilla vestige bearded; lodicules about 0.6 mm long, oblong; anthers 1.8–2.8 mm long; $2n$ = 28, 42, 56.

Dry to moderately moist sagebrush or timbered slopes at lower montane elevs.; w. N. Amer. from Man. to B.C., s. to c. Calif., Oregon and n. Colo. July–Aug.

Calamagrostis montanensis (Scribn.) Scribn. has been reported from Utah. We have not seen any specimens of this more northern species, but if it does appear it would key to *C. rubescens,* from which it differs in having narrow, involute leaf blades and shorter culms, mostly less than 4 dm tall [*C. rubescens* (4) 7–11 dm tall].

20. APERA Adans.

Tufted annuals; culms hollow; sheaths open; ligules membranous, relatively large; blades flat to involute, non-auriculate; inflorescence an open or contracted, densely flowered panicle; spikelets small, 1-

CALAMAGROSTIS

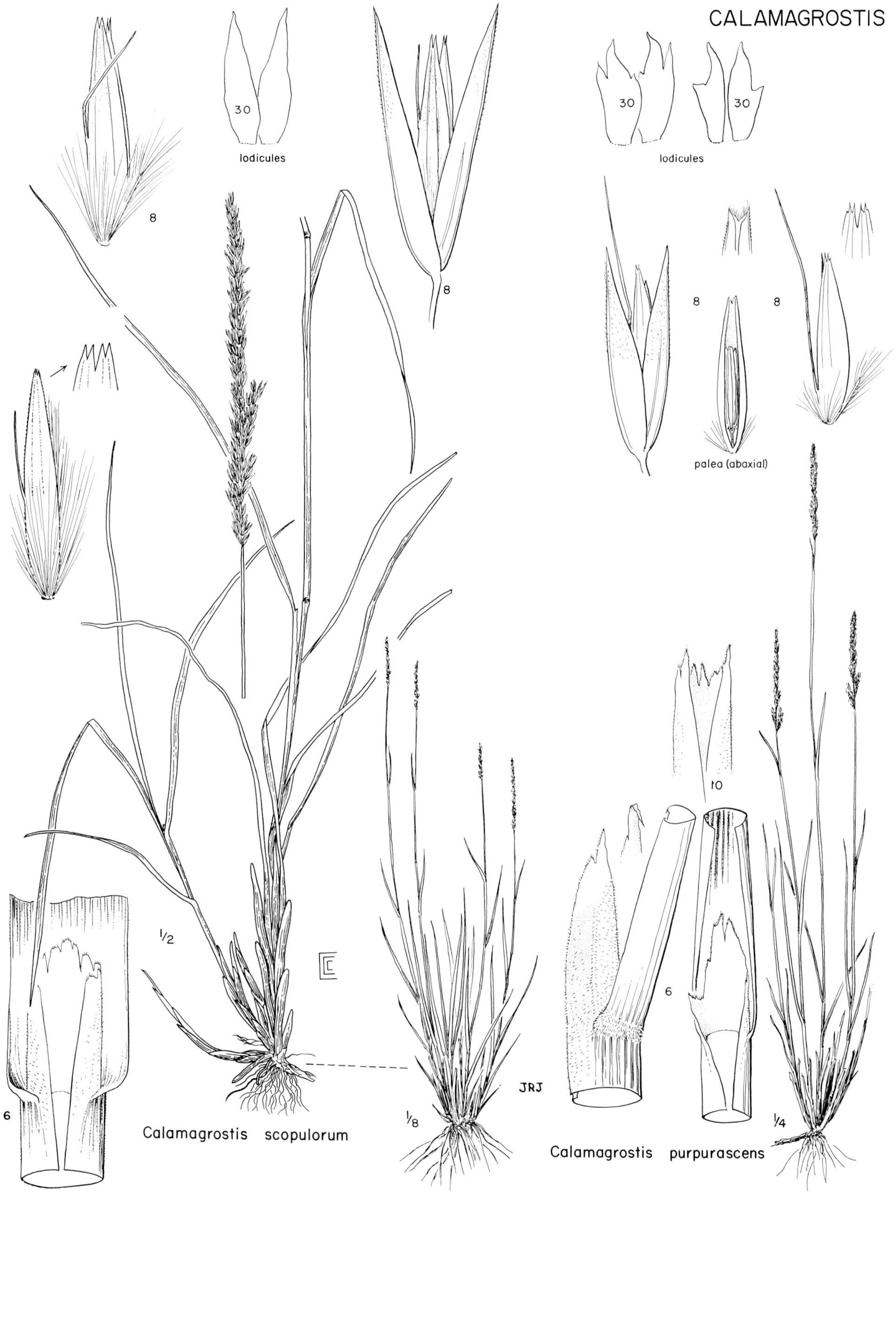

CALAMAGROSTIS
APERA

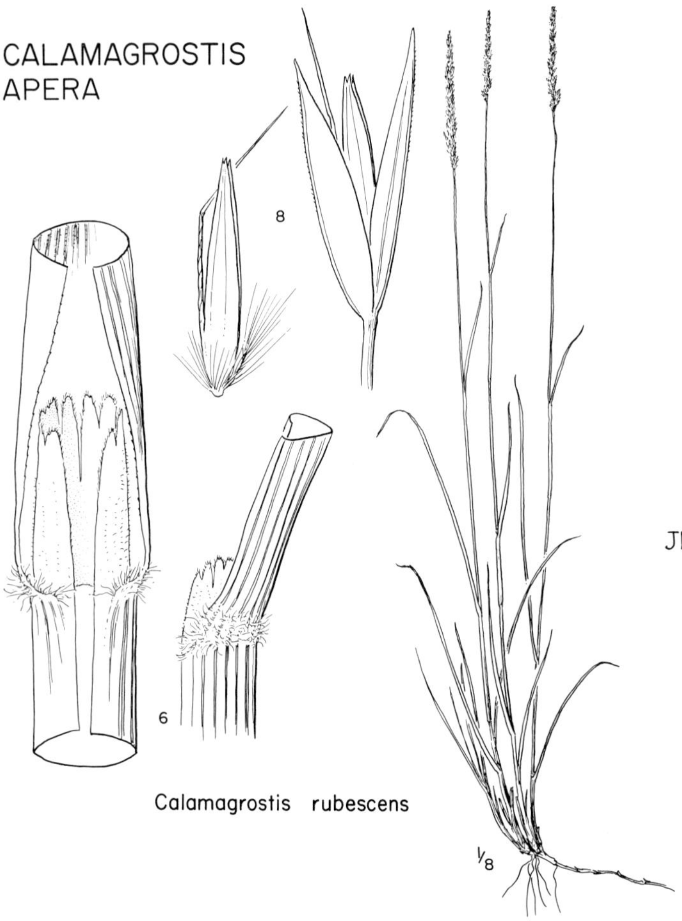

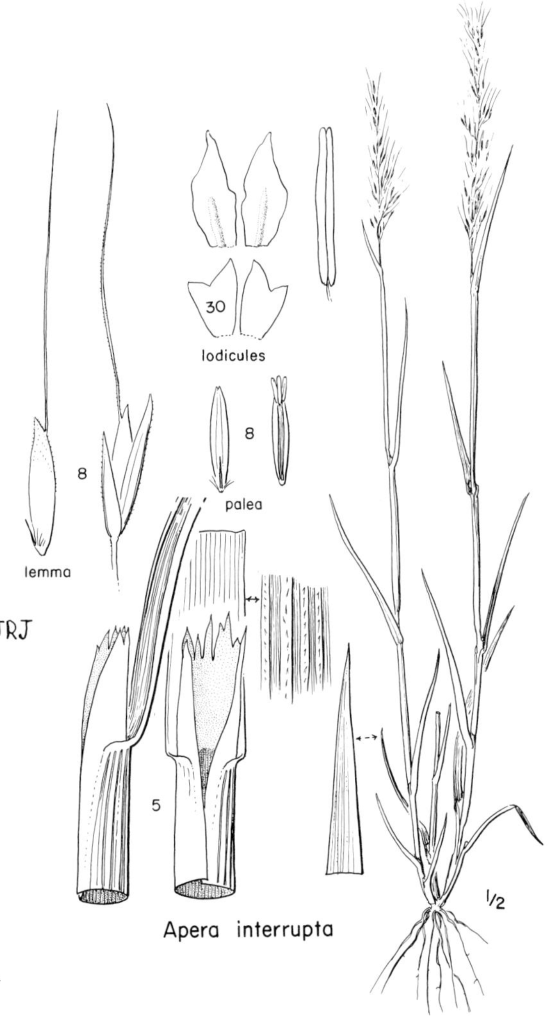

flowered, the rachilla prolonged behind the palea as a slender bristle, disarticulating above the glumes; glumes longer than the floret, membranous, lanceolate, acuminate (ours) to aristate, sometimes somewhat keeled, the first glume 1-nerved, the second 1- to 3-nerved and longer than the first; lemma [ours 1.6–2 (2.2) mm long] becoming subindurate at maturity, rounded in back, faintly 5-nerved, scaberulous near the apex, the callus short bearded; awn straight or flexuous, arising from the dorsal side of the lemma just below the apex; palea well developed, 2-nerved, as long as the lemma or slightly shorter; lodicules 2; stamens 3; caryopsis narrowly oblong, tightly enclosed between the lemma and palea; $x = 7$.

A small genus of 3 species, native to the cooler parts of Europe and Asia. (Name from the Greek *a,* not, and *peros,* maimed, possibly in reference to the well-developed awn or the rachilla elongation, which sometimes bears a rudiment of a second floret.)

The species of *Apera* have been included in *Agrostis* by various American botanists. The firm lemmas, the long, subterminal awn and the well-developed palea are the distinguishing characters of *Apera.*

1. Apera interrupta (L.) Beauv.

Agrostis interrupta L. Syst. Nat. ed. 10. 2: 872. 1759. *Apera interrupta* Beauv. Essai Nouv. Agrostogr. 31, 151. 1812. *Anemagrostis interrupta* Trin. Fund. Agrostogr. 129. 1820. *Muhlenbergia interrupta* Steudel, Syn. Pl. Glum. 1: 177. 1854. *Agrostis spica-venti* subsp. *interrupta* Hook. f. Stud. Fl. Brit. Islands 432. 1870. *Agrostis anemagrostis* subsp. *interrupta* Syme in Sowerby, Engl. Bot. ed. 3. 11: 44. 1873. *Apera spica-venti* var. *interrupta* Beal, Grasses N. Amer. 2: 357. 1896. *Agrestis interrupta* Bubani, Fl. Pyrén. 4: 289. 1901. (Europe.)

Tufted annuals; culms 1–5.5 (7) dm tall, erect, sometimes decumbent at the base, glabrous; sheaths smooth, somewhat keeled; ligules 1.5–3.5 (4.5) mm long, acute, lacerate-erose, the margins decurrent with the sheath; blades flat, folded or usually involute, 1–2 (4) mm broad, scaberulous above and glabrous below; panicles (5) 8–15 (20) cm long, narrow, densely flowered, interrupted below, the branches ascending, often spikelet bearing to the base; spikelets 1-flowered, greenish; glumes lanceolate, acuminate, scabrous on the midnerves, the first glume 1.7–2.2 mm long, 1-nerved, the second 2–2.5 (2.8) mm long, (1-) 3-nerved; lemma 1.6–2 (2.2) mm long, lanceolate, acute, finely 5-nerved; callus minutely pilose; palea about $^3/_4$ to as long as the lemma, 2-nerved; rachilla prolonged as a bristle 0.2–0.6 mm long; awn 5.5–10 (16) mm long, attached near the tip of the lemma, straight or flexuous; lodicules about 0.5 mm long; anthers very small, 0.3–0.4 mm long, often purplish-brown; $2n = 14, 28$.

Moist or moderately moist, often disturbed sites; relatively recently introduced to the U.S. from Europe (Constance, L. & L. A. Dillon, Madroño 3: 171. 1935), widespread in the Old World; s. B.C., Wash., n. Idaho, Oregon, and ne. Calif.; collected in our region n. of the Salt Lake City Airport (*Arnow 3108*). Late May–July.

21. AGROSTIS L. Bentgrass

Low to moderately tall annuals or perennials, tufted to rhizomatous or stoloniferous; culms slender, hollow; sheaths open; ligules membranous, more or less puberulent; blades flat, folded or involute, nonauriculate; inflorescence an open or contracted and spike-like panicle; spikelets small, 1-flowered, the rachilla usually not hairy, sometimes prolonged behind the palea as a slender, hairy bristle, disarticulating above the glumes; glumes thin, lanceolate, acute to acuminate, sometimes the first aristate or short-awned, subequal or equal, both usually 1-nerved, the nerves usually scabrous, usually exceeding the lemma in length; lemma [ours 0.8–2.2 (2.5) mm long] thin, (3-) 5-nerved, the lateral nerves often projecting into minute teeth at the obtuse or truncate apex, awnless or awned from the back, the awn usually geniculate when long, the callus glabrous or minutely bearded (ours); palea lacking or vestigial to nearly as long as the lemma, nerveless or 2-nerved; lodicules 2, usually oblong and acute, sometimes shallowly bilobed; stamens 3; caryopsis narrowly-oblong, loosely enclosed by the lemma, glabrous; $x = 7$.

A genus of about 100 species of temp. and subarctic regions of the world and extending into the tropics at high elevs.; about 40 species in the U.S. (From the Greek word *agros*, a field, the place of growth of some grass; the word agrostology comes from the same root.)

Several species are important forage grasses and some are important as turfgrass. Strains of *A. stolonifera* var. *stolonifera* (red top) and var. *palustris* (creeping bentgrass) are used extensively as lawn and turfgrass in cooler regions of North America. The species of *Agrostis* are difficult to distinguish. Identification often requires careful dissection of the florets and an exploration of both leads of the key for character combinations when measurements fall in the overlap.

References:

Beetle, A. A. A new section Microphyllae in *Agrostis*. Bull. Torrey Bot. Club 72: 541–549. 1945.

Carlbom, C. G. A biosystematic study of some North American species of *Agrostis* L. and *Podagrostis* (Griseb.) Scribn. & Merr. Ph.D. thesis, Oregon State University. 233 pp. 1967.

Hitchcock, A. S. North American species of *Agrostis*. U.S.D.A. Bur. Pl. Industr. Bull. 68: 1–68. 1905.

1 Palea well developed, 0.5–2 mm long, 1/2 to 2/3 as long as the lemma.
- 2 Anthers (0.6) 0.8–1.4 mm long.
  - 3 Panicles usually closing after anthesis, the branches floriferous from the base; blades 2–6 mm broad . . . . . . 1. *A. stolonifera*
  - 3 Panicles remaining open after anthesis, the branches naked at the base; blades 1–2.5 (3) mm broad . . . . . . 2. *A. tenuis*
- 2 Anthers 0.3–0.7 mm long.
  - 4 Lemma 0.8–1.2 mm long; glumes scabrous over the back; panicles compact, the branches spikelet-bearing to the base . . . . . . 3. *A. semiverticillata*
  - 4 Lemma 1.5–2.2 mm long; glumes glabrous or scabrous on the nerve only; panicles often open, the branches rarely spikelet-bearing to the base.
    - 5 Plants usually over 1.5 dm tall; panicles (3.5) 5–10 cm long, more or less open at anthesis; rachilla often prolonged behind the palea as a tiny bristle, up to 0.5 mm long.
      - 6 Anthers 0.6–1.5 mm long; rachilla prolongation lacking; palea 0.7–1.2 mm long; plants of lower elevs. . . . . . . 2. *A. tenuis*
      - 6 Anthers 0.5–0.6 mm long; rachilla prolonged behind the palea as a tiny bristle, up to 0.5 mm long; palea 1.2–2 mm long; plants of subalpine and alpine elevs. . . . . . . 4. *A. thurberiana*
    - 5 Plants usually less than 1.5 dm tall; panicles 1.5–5 cm long, often compressed at anthesis; rachilla prolongation lacking or only up to 0.15 mm long . . . . . . 5. *A. humilis*

1 Palea obsolete or a small nerveless scale, no longer than 0.3 mm and never as much as 1/2 the length of the lemma.
- 7 Anthers 1.2–1.8 mm long; plants strongly rhizomatous . . . . . . 6. *A. diegoensis*
- 7 Anthers merely 0.3–0.7 mm long; plants not rhizomatous (sometimes shortly rhizomatous in *A. exarata*).
  - 8 Panicles narrow, contracted, several times as long as broad, at least some branches spikelet-bearing from the base.
    - 9 Plants usually less than 2 dm tall; glumes rarely scaberulous over the back, acute, neither aristate nor awn-tipped; leaves mostly less than 2 mm broad . . . . . . 7. *A. variabilis*
    - 9 Plants usually more than 2 dm tall; glumes scaberulous over the back, at least basally, often aristate or awn-tipped; leaves rarely less than 2 mm broad . . . . . . 8. *A. exarata*
  - 8 Panicles open, often not more than 2 to 3 times as long as broad, the branches naked at the base.
    - 10 Plants (2) 3–9 dm tall; spikelets (2.2) 2.5–3.2 mm long; panicles diffuse, the branches forking above the middle . . . . . . 9. *A. scabra*
    - 10 Plants smaller, 0.5–3 (4) dm tall; spikelets 1.3–2.3 (2.6) mm long; panicles not diffuse, the branches usually forking below the middle.
      - 11 Lemma awnless . . . . . . 10. *A. idahoensis*
      - 11 Lemma awned; reported from Utah but not seen . . . . . . *A. borealis* Hartm.

1. Agrostis stolonifera L.

Agrostis stolonifera L. Sp. Pl. 62. 1753. *A polymorpha* var. *stolonifera* Huds. Fl. Anglica, ed. 2. 31. 1778. *Decandolia stolonifera* Bast. Essai Fl. Dept. Maine & Loire 29. 1809. *Vilfa stolonifera* Beauv. Essai Nouv. Agrostogr. 16, 148, 182. 1812. *A. alba* var. *stolonifera* J. E. Smith, Engl. Fl. 1: 93. 1824. *A. vulgaris* var. *stolonifera* Koch, Syn. Fl. German. & Helv. 782. 1837. *A. palustris* var. *stolonifera* Druce, Fl. Oxfordshire, ed. 2. 473. 1927. ("Habitat in Europa.")

A. alba of authors; not L. which is based on a *Poa* (see discussion below) = var. *stolonifera*.

A. palustris Huds. Fl. Anglica 27. 1762. *A. polymorpha* var. *palustris* Huds. Fl. Anglica, ed. 2. 32. 1778. *A. alba* var. *palustris* Pers. Syn. Pl. 1: 76. 1805. *Apera palustris* S. F. Gray, Nat. Arr. Brit. Pl. 2: 148. 1821. *Agrostis stolonifera* var. *palustris* Farw. Annual Rep. Michigan Acad. Sci. 21: 351. 1920. (England.)

A. alba var. *major* Gaudin, Fl. Helv. 1: 189. 1828. *A. stolonifera* var. *major* Farw. Annual Rep. Michigan Acad. Sci. 21: 351. 1920. (Switzerland.) = var. *stolonifera*.

A. exarata var. *stolonifera* Vasey, Bull. Torrey Bot. Club 13: 54. 1886. *A. reptans* Rydb. Fl. Rocky Mts. 54, 1060. 1917 [1918]. (*Suksdorf 40,* "Bottom lands of the Columbia River," Klickitat Co., Wash., Aug. 1883.) = var. *palustris*.

A. depressa Vasey, Bull. Torrey Bot. Club 13: 54. 1886. (*Patterson,* "Clear Creek Canyon, near Georgetown," Colo. in 1885.) = var. *palustris*.

Redtop bentgrass, creeping bentgrass.

Rhizomatous or stoloniferous perennials; culms relatively tall, (3) 4–10 (13) dm, decumbent; ligules relatively long, 1.5–5 (8.5) mm, truncate to obtuse, erose-ciliolate, often lacerate; blades flat, folded or involute, relatively broad, 2–5 (6) mm (opened); panicles (7) 10–18 (25) cm long, open, sometimes closing after anthesis, the branches verticillate, ascending, floriferous from the base; glumes lanceolate, acute, rounded on the back, the nerves scabrous to ciliate, purplish, the first glume 2–2.5 mm long, the second 1.8–2.2 (2.5) mm long; lemma 1.4–1.8 (2) mm long, 3- to 5-nerved, membranous, awnless; callus minutely bearded; palea 0.7–1.5 mm long; lodicules about 0.4 mm long; anthers (0.6) 0.8–1.4 mm long; $2n$ = 28–46, 56.

Streambanks and meadows and moderately moist sagebrush slopes from low to middle elevs.; native to Europe; escaped from cultivation from Alaska to Newfl. and throughout the U.S. (var. *palustris* rare in the sw. states and not seen in the se. states). Mid June–early Sept.

This species is cultivated for meadows, pastures and lawns, especially golf courses.

According to Niles (Contr. U.S. Natl. Herb. 24: 155. 1925) and Phillipson (J. Linn. Soc. Bot. 51: 91. 1937), Linnaeus based his *A. alba* on a *Poa* (apparently *P. nemoralis*) so we take the name *A. stolonifera* for the species. Two varieties can be recognized.

1 Panicles open, usually over 1.5 cm broad when pressed, the branches more spreading, not appressed ...... var. *stolonifera*
1 Panicles compressed, usually less than 1.5 cm broad when pressed, the branches short, appressed to ascending .... var. *palustris* (Huds.) Farw.

2. Agrostis tenuis Sibth.

Agrostis capillaris Huds. Fl. Anglica 27. 1762; not L. 1753. *A. tenuis* Sibth. Fl. Oxoniensis 36. 1794. (England.)

A. vulgaris With. Bot. Arr. Veg. Gr. Brit. ed. 3. 2: 132. 1796. *Vilfa vulgaris* Beauv. Essai Nouv. Agrostogr. 16. 1812. *A. alba* var. *vulgaris* Coss. & Durieu, Explor. Sci. Algér. 2: 63. 1855. *A. stolonifera* var *vulgaris* Čelak. Prodr. Fl. Böhm. 710. 1881; not Heuffel, 1858. (Europe.)

A. vulgaris var. *aristata* Parnell, Grasses Scotl. 1: 34, pl. 13. 1842. *A. tenuis* var. *aristata* Druce, List. Brit. Pl. 79. 1908. *A. tenuis* f. *aristata* Wieg. Rhodora 26: 2. 1924. *A. capillaris* var. *aristata* Druce, Fl. Oxfordshire, ed. 2. 474. 1927. (Scotland.) Awned phase.

Colonial bentgrass, common bentgrass.

Tufted perennials, sometimes with stolons or short rhizomes; culms (2) 3–6 (8) dm tall; sheaths smooth; ligules 0.5–2.5 (3.5) mm long, truncate to obtuse, erose-ciliolate; blades flat or folded, 1–2.5 (3) mm broad, scaberulous; panicles 4–15 (20) cm long, ellipsoid or ovoid, remaining open after anthesis, the branches naked at the base; glumes subequal, 1.5–2.5 (3) mm long, lanceolate, acute, 1- (3)-nerved, glabrous or scabrous on the nerve, often purplish, the first glume slightly longer than the second; lemma 1.5–2 mm long, membranous, 3- to 5-nerved, the nerves faint, awnless or with a short awn from the back; callus minutely bearded; palea 0.7–1.2 mm long; rachilla vestige lacking; lodicules about 0.4 mm long; anthers 0.6–1.5 mm long; $2n$ = 28–35, 41.

Wet to moderately moist meadows, streambanks and shores of ponds and lakes; in the west on the Pacific slope from Alaska to Calif., and scattered sites e. to n. Idaho and w. Mont., and our only record near Salt Lake City (*Arnow 467*), Utah; in the east from Que. to Newfl., s. to the Great Lakes region and Atlantic states to N.C.; Eurasia. July–Aug.

Agrostis tenuis is commonly used in lawns and as a pasture grass in northeastern United States.

3. Agrostis semiverticillata (Forssk.) Christensen

Phalaris semiverticillata Forssk. Fl. Aegypt.-Arabica 17. 1775 *Agrostis semiverticillata* Christensen, Dansk Bot. Ark. 4(3): 12. 1922. *Nowodworskya semiverticillata* Nevski, Trudy Bot. Inst. Akad. Nauk SSSR, Ser. 1, Fl. Sist. Vysš. Rast. 4: 339. 1937. *Polypogon semiverticillatus* Hylander, Uppsala Univ. Årsskr. 7: 74. 1945. (Egypt.)

A. verticillata Vill. Prospectus Hist. Pl. Dauphiné 16. 1779. *A. alba* var. *verticillata* Pers. Syn. Pl. 1: 76. 1805. *A. villarsii* Poiret in Lam. Encycl. Méth. Bot. Suppl. 1: 251. 1810. *Vilfa verticillata* Beauv. Essai Nouv. Agrostogr. 16, 148, 182. 1812. *A. stolonifera* var. *verticillata* Saint-Amans, Fl. Agénaise 28. 1821. *A. condensata* Willd. ex Steudel, Nomencl. Bot. ed. 2. 1: 40, 43. 1840, in synonymy. *Agrestis verticillata* Bubani, Fl. Pyrén. 4: 282. 1901. *Agrostis alba* subsp. *verticillata* Douin in Bonnier, Fl. Compl. France, Suisse & Belgique 11: 142. 1932. *Nowodworskya verticillata* Nevski, Trudy Bot. Inst. Akad. Nauk SSSR, Ser. 1, Fl. Sist. Vysš. Rast. 3: 143. 1936. (France.)

Water bentgrass.

Perennials; culms (1.5) 3–6 (8) dm long, strongly decumbent at the base and rooting at the nodes, geniculate; sheaths smooth, broad; ligules 1.5–4 (6) mm long, truncate, erose-ciliolate and somewhat lacerate; blades flat, relatively broad, (1.5) 2.5–4 (7) mm, smooth or scabrous on the upper surface; panicles (2) 4–10 (16) cm long, compact, interrupted, the branches ascending to erect, spikelet-bearing to the base; spikelets disarticulating below the glumes and the panicle branches becoming naked in age; glumes small, subequal, (1.2) 1.5–1.8 (2) mm long, 1-nerved, strongly scabrous, the first glume ovate and the second lanceolate; lemma small, 0.8–1.2 mm long, smooth, membranous, awnless, minutely apiculate by the excurrent nerves; callus glabrous and blunt; palea 0.5–1 mm long; rachilla not prolonged; lodicules about 0.3 mm long; anthers 0.3–0.5 mm long; $2n$ = 14, 28.

AGROSTIS

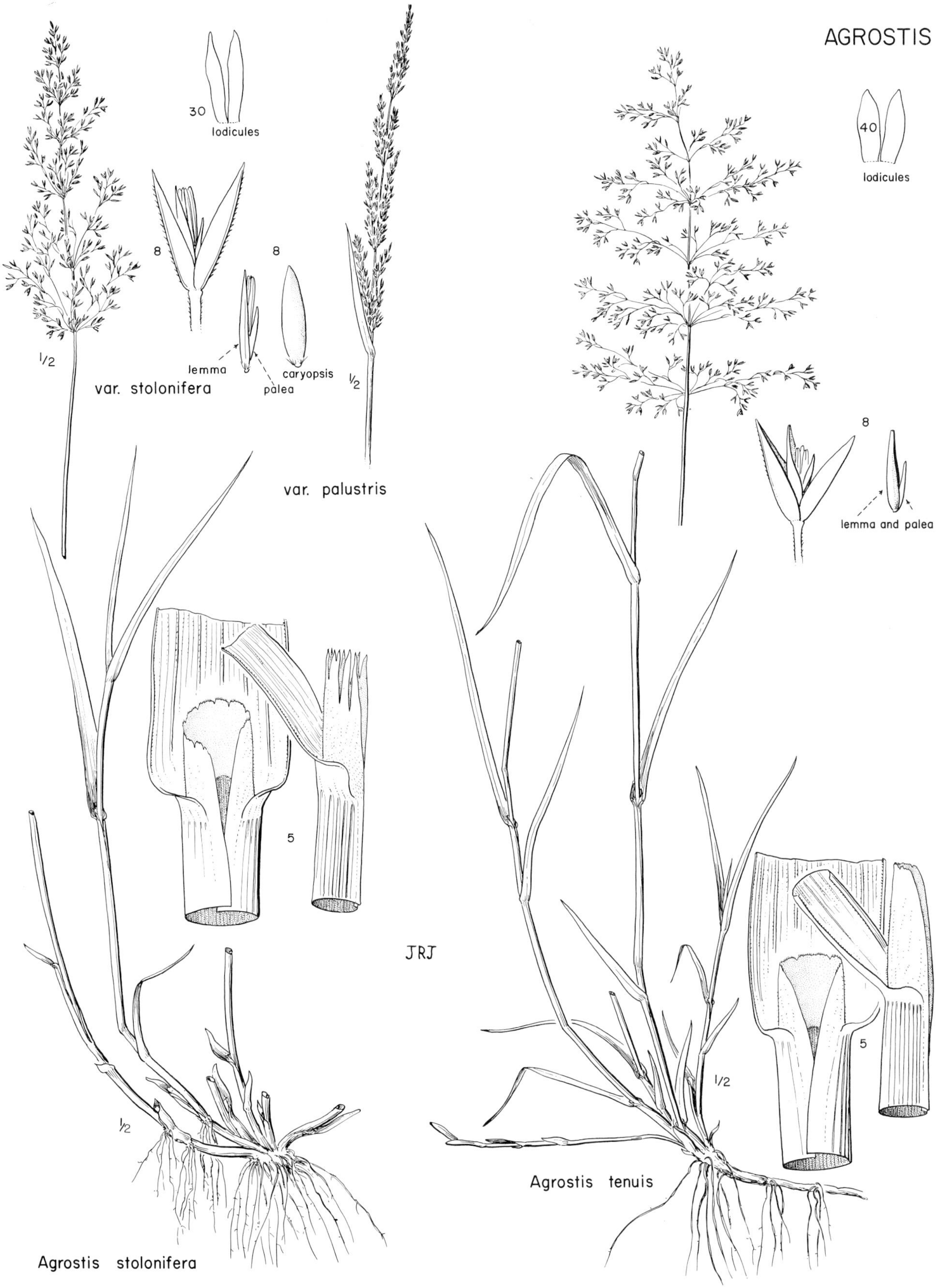
30
lodicules
8
8
1/2
lemma
caryopsis
palea
1/2
var. stolonifera
var. palustris
40
lodicules
8
lemma and palea
5
JRJ
5
1/2
1/2
Agrostis tenuis
Agrostis stolonifera

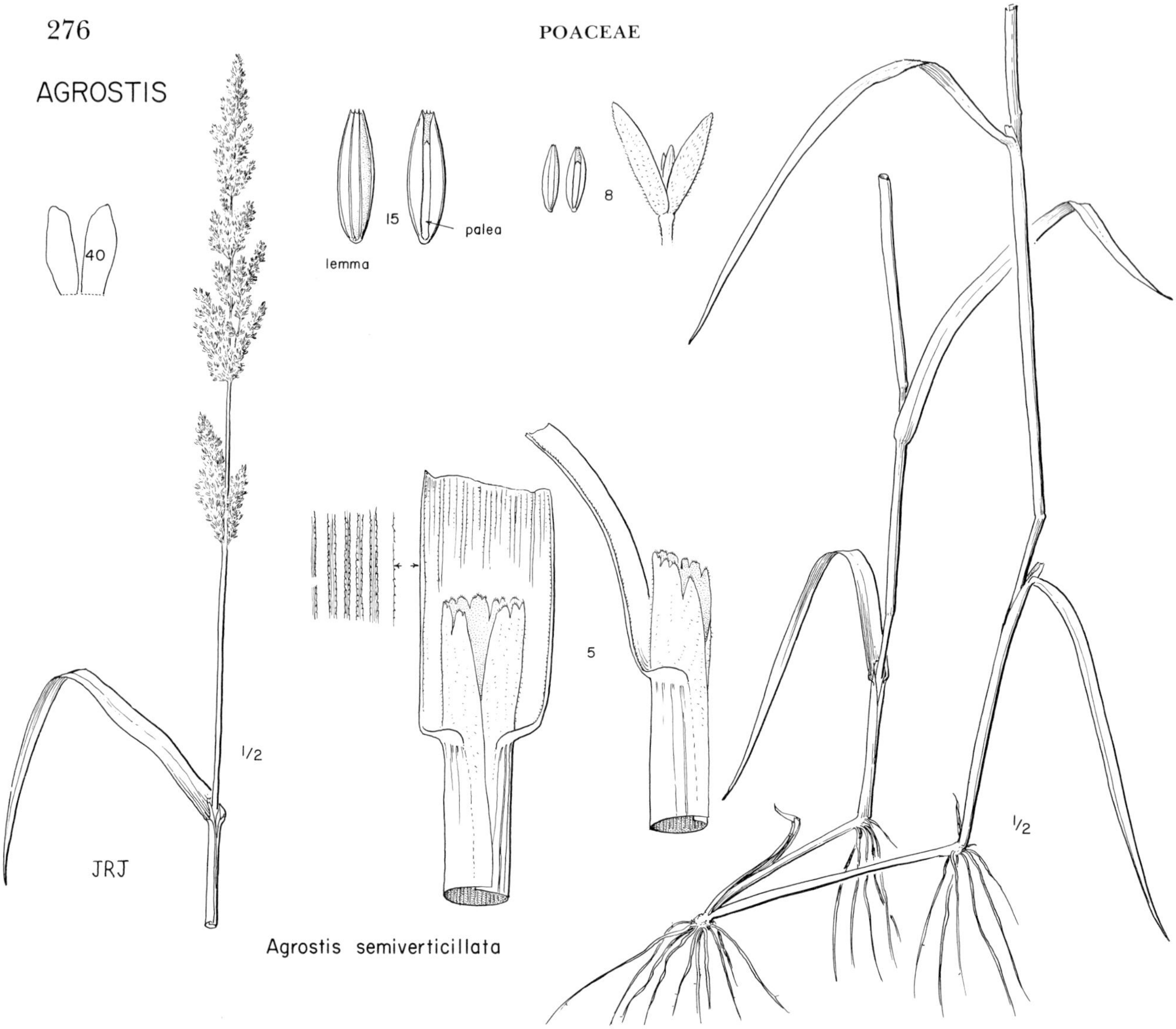

Agrostis semiverticillata

Moist places along the streams and rivers at low elevs.; native of the Old World; common in Mex. and the sw. U.S., n. to Texas, s. Colo., s. Utah (also in the c. Wasatch Range), s. Nev., s. Calif. and along the Pacific slope to s. Wash. May–Aug.

4. **Agrostis thurberiana** A. S. Hitchc.

Agrostis thurberiana A. S. Hitchc. U.S.D.A. Bur. Pl. Industr. Bull. 68: 23, pl. 1, fig. 1. 1905. *Podagrostis thurberiana* Hultén, Fl. Aleutian Islands 75. 1937. (*Suksdorf 1021,* "Wet places on mountains," Skamania Co., Wash., 28 Aug. 1890.)

Thurber redtop.

Tufted perennials, often with short rhizomes; culms (1.5) 2–4 (6) dm tall; sheaths smooth, the collar sometimes oblique; ligules short, 0.5–3 mm long, obtuse to truncate, sometimes lacerate; blades flat, 0.5–2.5 (3) mm broad; panicles (3.5) 5–10 cm long, somewhat open at anthesis, sometimes closing up later, the capillary branches ascending; spikelets 1-flowered; glumes narrow-lanceolate to lanceolate, acute, 1-nerved, often purplish, subequal, 1.7–2.5 mm long; lemma 1.6–2 (2.2) mm long, awnless, blunt, sometimes minutely toothed, 3- to 5-nerved; callus minutely bearded; palea 1.2–1.5 (2) mm long; rachilla usually prolonged behind the palea to 0.5 mm long; lodicules 0.4–0.6 mm long; anthers 0.5–0.6 mm long; $2n = 14$.

Streambanks, meadows, bogs and around springs, from the subalpine to the alpine zones in our areas; Alaska, B.C. and Alta., s. in the mts. to Calif., Utah (Wasatch, Uintas) and Colo. July–Aug.

Agrostis thurberiana is closely related to *A. aequivalvis* (Trin.) Trin. of coastal habitats from the Olympic Peninsula to Alaska.

5. **Agrostis humilis** Vasey

Agrostis humilis Vasey, Bull. Torrey Bot. Club 10: 21. 1883. *Podagrostis humilis* Björkman, Symb. Bot. Upsal. 17(1): 15. 1960. (*Suksdorf 25,* Mt. Paddo [Adams], Wash.; *Howell 85,* Mt. Adams, Wash., Aug. 1882.)

Small tufted perennials; culms low, only 0.3–1.8 (2.4) dm tall; sheaths smooth; ligules short, 0.5–1.5 (2.5) mm long, obtuse to truncate; blades flat to folded or filiform, 0.5–1.2 mm broad, mostly basal; panicles short, 1.5–4 (5) cm long, loosely contracted; glumes subequal, 1.5–2.2 mm long, narrow-lan-

AGROSTIS

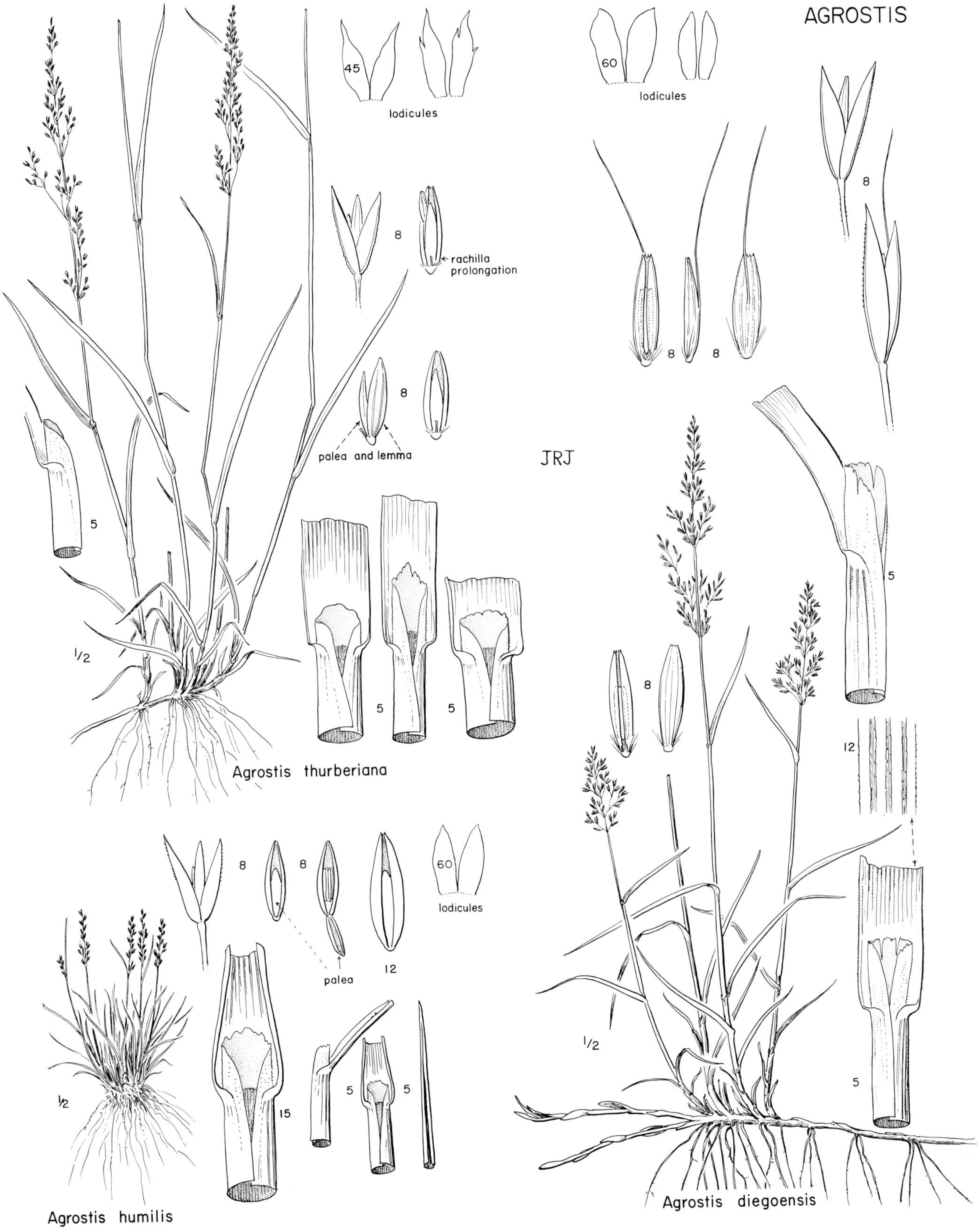
45
lodicules
60
lodicules
8
rachilla prolongation
8
palea and lemma
JRJ
5
1/2
5
5
Agrostis thurberiana
8
8
lodicules
60
palea
12
1/2
15
5
5
Agrostis humilis
12
5
1/2
Agrostis diegoensis

ceolate to lanceolate, acute, purple; lemma 1.5–1.8 mm long, awnless; callus subglabrous; palea 1–1.5 mm long; rachilla vestige lacking or very short, only up to 0.15 mm long; lodicules about 0.3 mm long; anthers 0.5–0.7 mm long; $2n = 14$.

Alpine and subalpine meadows and bogs; Cascade Range of s. B.C., Wash. (and Olympics), to Oregon, across n. Nev. and n. Utah (Uinta Mts.) to the Rocky Mts. from Mont. to N.M. July–Aug.

6. Agrostis diegoensis Vasey

Agrostis diegoensis Vasey, Bull. Torrey Bot. Club 13: 55. 1886. (*Orcutt,* Smiths Mt., San Diego Co., Calif., Aug. 1884.)

Tufted, rhizomatous perennials; culms (1.5) 2.5–6 (10) dm tall, erect or decumbent, leafy; sheaths smooth; ligules 1–3 (4.5) mm long, obtuse, erose, often lacerate, minutely scaberulous; blades flat, 1.5–3 (4.5) mm broad, glabrous below and scaberulous above; panicles 5–15 (22) cm long, relatively open, oblong; glumes subequal, 2.5–3.5 mm long, lanceolate, acute, rounded on the back, 1-nerved, glabrous with scabrous nerves, usually purplish, the margins membranous; lemma 1.8–2.7 mm long, membranous, 5-nerved, the lateral nerves terminating in 4 minute teeth, awnless or with a delicate awn from about the midlength or below; callus somewhat bearded; palea wanting; rachilla not prolonged; lodicules scarcely 0.3 mm long; anthers large, 1.2–1.8 mm long; $2n = 42$.

Moist to moderately dry meadows and woods; B.C., Wash., n. Idaho, nw. Mont., Oregon, Calif. and adjacent Nev. and in the Intermountain Region in White Pine and Elko cos., Nev. May–July.

Sometimes a specimen of *A. stolonifera* may accidentally be keyed to *A. diegoensis* when the presence of the palea is elusive. There is a general similarity between the two species. *A. diegoensis* can be separated on the basis of its usually larger glumes which measure 2.5–3.5 mm in length as compared to 1.8–2.5 mm for *A. stolonifera.*

7. Agrostis variabilis Rydb.

Agrostis varians Trin. Mém. Acad. Imp. Sci. Saint-Pétersbourg, Sér. 6, Sci. Math., Seconde Pt. Sci. Nat. 4(1): 314. 1841; not Thuill. 1799. *A. variabilis* Rydb. Mem. New York Bot. Gard. 1: 32. 1900. (*Hooker 217,* "Rocky Mountains," probably collected by *Drummond.*)

A. rossiae of most authors; not Vasey, 1892.

Densely tufted perennials; culms low, 0.8–2 (3) dm tall; leaves mostly basal; ligules short, 0.5–2.5 (3.5) mm long, truncate to obtuse, finely erose-ciliolate and lacerate; blades involute, folded, or if flat up to 2 mm broad, short, smooth or scabrous; panicles short, 2–4.5 (8) cm long, strongly compressed; glumes lanceolate, acute, 1-nerved, scabrous on the keel, rarely scaberulous on the back, usually purple, the first glume 2–2.5 mm long, the second 1.8–2.3 mm long; lemma 1.5–2 mm long, 5-nerved, membranous, mostly awnless, rarely with an awn from about the mid-length; callus very slightly bearded; palea minute; rachilla not prolonged; lodicules 0.3–0.5 mm long; anthers 0.4–0.7 mm long; $2n = 28$.

Moist to moderately moist meadows and open ridges at subalpine and alpine elevs., B.C. and Alta., s. to Calif., n. Nev., n. Utah and Colo. July–Aug.

Agrostis variabilis has been confused with the annual, *A. rossiae,* an endemic species confined to warm soil surrounding hot springs in Yellowstone National Park (Swallen, J. R. Leafl. W. Bot. 5: 123–124. 1948). The perennial *A. variabilis* is one of the most widespread native species of *Agrostis* in western North America.

8. Agrostis exarata Trin.

Agrostis exarata Trin. Gram. Unifloris & Sesquifloris 207. 1824. (*"Eschholz,* Unalaschka," in 1816.)

A. exarata var. *minor* Hook. Fl. Boreali-Amer. 2: 239. 1839. *A. drummondi* Torr. ex Hook., ibid., given as an herbarium name. *A. exarata* subsp. *minor* C. L. Hitchc. Univ. Wash. Publ. Biol. 17(1): 467. 1969. (*Drummond, Douglas,* "Valleys of the Rocky Mountains.")

A. grandis Trin. Mém. Acad. Imp. Sci. Saint-Pétersbourg, Sér. 6, Sci. Math., Seconde Pt. Sci. Nat. 4(1): 316. 1841. (*Hooker,* Columbia [probably collected by *Douglas*].) = var. *exarata.*

A. asperifolia Trin. Mém. Acad. Imp. Sci. Saint-Pétersbourg, Sér. 6, Sci. Math., Seconde Pt. Sci. Nat. 4(1): 317. 1841. *A. exarata* f. *asperifolia* Vasey, U.S.D.A. Div. Bot. Bull. 13(1): pl. 31. 1892. (*Hooker,* Rocky Mts. [probably collected by *Drummond*].) = var. *exarata.*

Polypogon monspeliensis var. *monolepis* Torr. in Williamson, Explor. & Surv. Railroad Route from Mississippi River to Pacific Ocean 5(2): 366. 1858. *A. exarata* var. *monolepis* A. S. Hitchc. Amer. J. Bot. 21: 136. 1934. *A. ampla* f. *monolepis* A. A. Beetle, Bull. Torrey Bot. Club 72: 544. 1945. (*Blake,* "Posé Creek, Walker's Pass," Calif., Aug. 1853.)

A microphylla var. *major* Vasey, Contr. U.S. Natl. Herb. 3: 58, 72. 1892. *A. exarata* var. *microphylla* S. Wats. ex Vasey, ibid., as a synonym. (*Watson 1284,* Truckee Valley, Nev., July 1867.) = var. *monolepis.*

A. ampla A. S. Hitchc. U.S.D.A. Bur. Pl. Industr. Bull. 68: 38, pl. 20. 1905. *A. exarata* var. *ampla* A. S. Hitchc. Amer. J. Bot. 2: 303. 1915. (*Suksdorf 135,* "Wet rocks near Rooster Rock," Multnomah Co., Oregon, 16 July 1885.) = var. *monolepis.*

Spike redtop.

Tufted perennials, sometimes developing short rhizomes; culms 2–7 (12) dm tall, erect or sometimes decumbent at the base and rooting at the nodes; sheaths smooth to scaberulous, the collar sometimes oblique; ligules 2–5 (8.5) mm long, truncate to obtuse, erose-ciliate and lacerate, puberulent; blades flat, relatively broad, 1.5–6 (10) mm broad, scabrous; panicles (3.5) 5–14 (25) cm long, compressed, often spike-like, sometimes interrupted; glumes with scabrid nerves, the internerves scaberulous, at least basally, green to purplish, relatively long, the first glume (1.8) 2–4 (5) mm long, the tip often narrowed into an awn 0.5 mm long, the second 1.8–3.5 (3.8) mm long, rarely aristate; lemma 1.2–2.2 (2.5) mm long, 5-nerved, membranous, sometimes awned from about the middle or above, the awn short and straight or 3–5 mm long and geniculate; callus with a beard; palea minute; rachilla not prolonged; lodicules 0.2–0.5 mm long; anthers 0.3–0.6 mm long; $2n = 28, 42, 56$.

Moist to moderately moist habitats, mostly in the subalpine zone; Alaska to Sask. and s. in the mts. to Baja Calif., n. Mex. and w. Texas. July–Aug.

Spike redtop is one of the commonest species of *Agrostis* in the West and is an important range grass. It is our most complex species with a large number of ecotypes over its distributional range. There are intergradations with the annual *A. microphylla* Steudel along the Pacific coast through var. *monolepis.* Carlbom (1967) has observed sterile hybrids between spike redtop and *Agrostis stolonifera.* The following varieties can be distinguished in our region.

1 Lemma unawned or bearing a short straight awn.
  2 Glumes (2.5) 2.8–4 mm long, aristate; in the Pacific cordillera from Alaska to the n. half of Calif. and e. intergrading with var. *minor* to Sask., Mont., Wyo. and Colo., entering our area in n. Nev. and se. Oregon .... var. *exarata*
  2 Glumes 1.8–2.7 mm long, acute, sometimes sharply so, but not aristate; s. B.C. and s. Alta., s. to Baja Calif., Ariz., N.M. and w. Texas .... var. *minor* Hook.

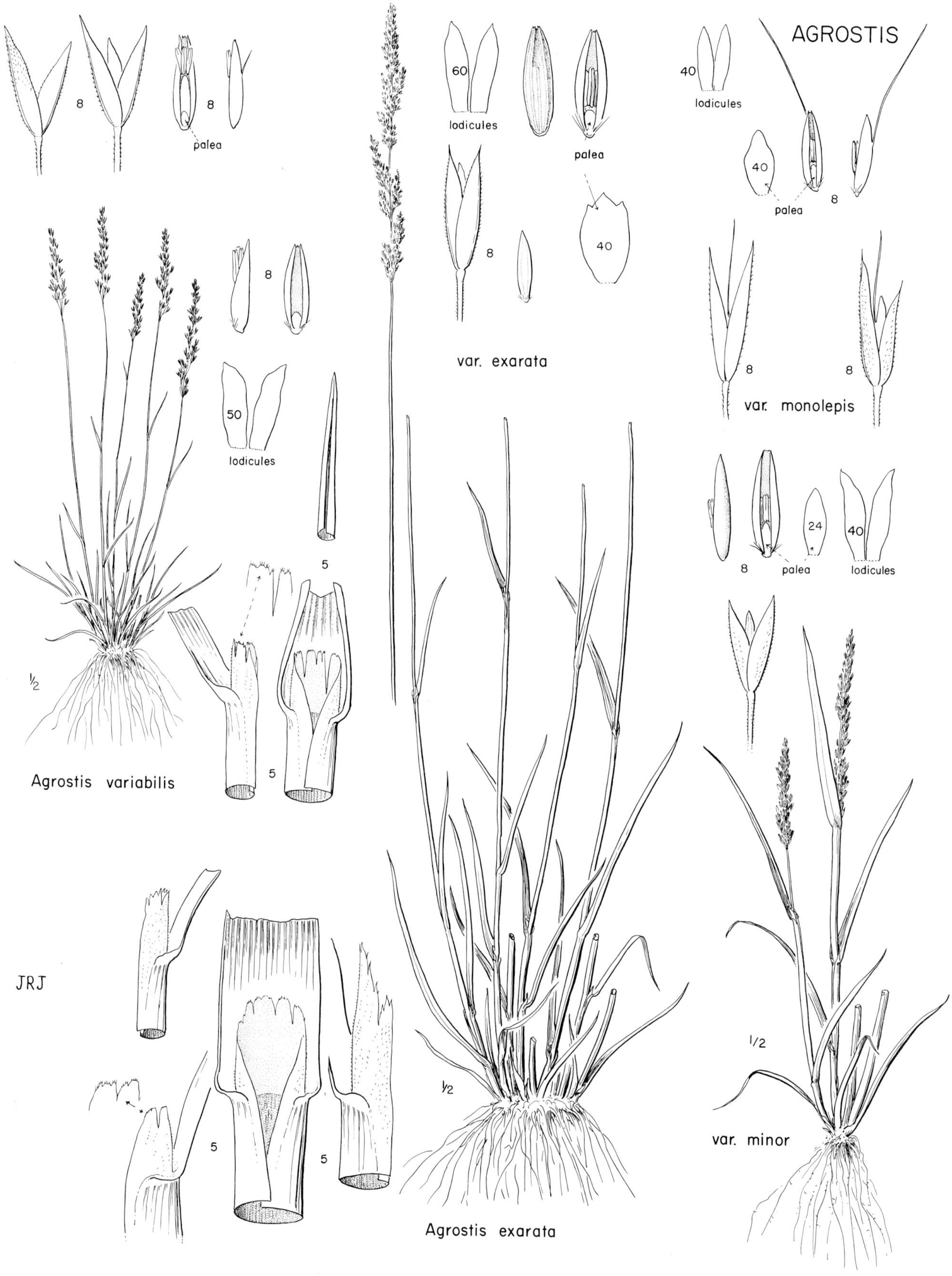
AGROSTIS
8
8
palea
60
lodicules
palea
40
lodicules
40
palea
8
8
40
8
50
lodicules
var. exarata
8
8
var. monolepis
8
palea
24
40
lodicules
5
1/2
5
Agrostis variabilis
JRJ
5
5
1/2
1/2
var. minor
Agrostis exarata

1 Lemma prominently awned, the awn 3–5 mm long, geniculate; Vancouver I. to s. Calif., w. of the Cascade-Sierran crest, except in sw. Nev. and Elmore Co., Idaho .... var. *monolepis* (Torr.) A. S. Hitchc.

9. Agrostis scabra Willd.

Agrostis scabra Willd. Sp. Pl. 1: 370. 1797. *Vilfa scabra* Beauv. Essai Nouv. Agrostogr. 16, 148, 182. 1812. *Trichodium scabrum* Muhl. Cat. Pl. N. Amer. 10. 1813. *A. laxiflora* var. *scabra* Torr. Fl. State New York 2: 442. 1843. *A. hiemalis* var. *scabra* Blomq. Grasses N. Carol. 82. 1948. ("Habitat in America-boreali.")

A. geminata Trin. Gram. Unifloris & Sesquifloris 207. 1824. *A. hyemalis* var. *geminata* A. S. Hitchc. U.S.D.A. Bur. Pl. Industr. Bull. 68: 44, pl. 28. 1905. *A. scabra* var. *geminata* Swallen, Proc. Biol. Soc. Wash. 54: 45. 1941. (*"Eschholz*, Unalaschka.")

A. hiemalis of authors; not (Walter) Britton, Sterns, & Poggenb. 1888.

Ticklegrass.

Tufted perennials; culms (2) 3–5 (9) dm tall; sheaths scabrous; ligules short, 1–2.5 (3.5) mm long, truncate to obtuse, erose-ciliolate, sometimes lacerate; blades usually flat, narrow, 0.8–1.5 (2) mm broad, scabrous, stiff; panicles long, (9) 14–25 (30) cm long, diffuse, the branches spreading to erect, with spikelets borne mostly near the tips; spikelets purplish; glumes lanceolate, 1-nerved, the nerve scabrous, occasionally scaberulous across the back, the first glume 2.5–3.2 mm long, sometimes aristate, the second 2.2–3 mm long; lemma (1) 1.5–1.8 mm long, 5-nerved, membranous, awnless or sometimes with a short awn from above the midlength, the awn up to 2 mm long; callus bearded; palea lacking or to 0.3 mm long; rachilla not prolonged; lodicules about 0.4 mm long; anthers 0.3–0.6 mm long; $2n$ = 28, 42, 43.

Wet meadows, streambanks and shores of ponds and lakes; Alaska to Newfl., s. throughout the U.S. (except sparingly in the se.) to n. Mex. July–Aug.

Agrostis oregonensis Vasey, from the north and west of us, has been reported from the region, but we have not seen any specimens that would fit the description. This species falls between *A. scabra* and *A. idahoensis*. It has the height and spikelet size of *A. scabra* but its panicle is less diffuse with branches forking below its middle as in *A. idahoensis*.

10. Agrostis idahoensis Nash

Agrostis idahoensis Nash, Bull. Torrey Bot. Club 24: 42. 1897. (*Heller & Heller 3431*, "collected about Forest," Nez Perce Co., Idaho, 16 July 1896.)

A. filiculmis M. E. Jones, Contr. W. Bot. 14: 13. 1912. (*Jones 6056bb*, "Little De Motte Park on the Kaibab," Coconino Co., Ariz., 19 Nov. 1894.)

Small tufted perennials; culms relatively low, 0.5–3 (4) dm tall; ligules short, (0.5) 1–2 (3) mm long, obtuse, erose-ciliolate, sometimes lacerate; blades flat or involute, 0.5–1.5 (2) mm broad; panicles ovate, 1.5–10 (12) cm long, the capillary branches ascending, forking below the midlength and bearing rather few spikelets; spikelets green or purple; glumes lanceolate, 1-nerved, the nerves scabrous, the first glume 1.5–2.3 (2.6) mm long, slightly aristate, the second 1.3–2 mm long, acute; lemma 1.2–1.8 mm long, 5-nerved, membranous, awnless; callus glabrous or with a slight beard; palea lacking or only up to 0.2 mm long; lodicules 0.2–0.3 mm long; anthers 0.3–0.5 mm long; $2n$ = 28.

Meadows and bogs and moderately moist slopes (rock crevices); s. B.C. to Mont. and s. to Calif., Nev., n. Ariz. and n. N.M. July–Aug.

Some specimens identified as *A. idahoensis* turn out to have a well-developed palea and belong to *A. humilis*.

22. POLYPOGON Desf. Beardgrass

Low to moderately tall annuals and perennials; culms hollow, weak, erect to decumbent, often rooting at the lower nodes; sheaths open; ligules prominent, membranous; blades thin and flat, non-auriculate; inflorescence a dense, contracted, spike-like panicle; spikelets 1-flowered, disarticulating below the glumes, bearing a stipe and falling entire; glumes subequal, keeled and flattened, 1-nerved, awned from an entire or bilobed apex, the awn terminal or from between the lobes; lemma much shorter than the glumes, broad, smooth and shiny, mostly 5-nerved, sometimes awned from the broad, often minutely toothed apex, the awn mostly straight, short and delicate; palea slightly shorter than the lemma, 2-nerved; rachilla not prolonged behind the palea; lodicules 2, lanceolate; stamens 1–3; ovary glabrous; x = 7.

About 10 species throughout the temp. regions of the world, especially in Europe and Asia. Both our species have been introduced from Europe. (From the Greek *poly*, much, and *pogon*, bearded, alluding to the bristly appearance of the inflorescence from the awns of the glumes.)

Polypogon australis Brongn. was reported from the Nevada Test Site, just outside our range (Raven, P. Leafl. W. Bot. 10: 117. 1964). The specimens annotated by Raven belong to *P. interruptus*.

1 Glumes shortly bilobed, the awns (4) 4.5–10 mm long; annuals .... 1. *P. monspeliensis*
1 Glumes acute or rounded, rarely bilobed, the awns 1–3 (5) mm long; perennials .... 2. *P. interruptus*

1. Polypogon monspeliensis (L.) Desf.

Alopecurus monspeliensis L. Sp. Pl. 61. 1753. *Phleum crinitum* Schreber, Beschreibung Gräser 1: 151. 1769. *Alopecurus aristatus* var. *monspeliensis* Huds. Fl. Anglica, ed. 2. 28. 1778. *Agrostis alopecuroides* Lam. Tableau Encycl. & Méth. 1: 160. 1791. *Polypogon monspeliensis* Desf. Fl. Atlantica 1: 67. 1798. *Phleum monspeliense* Koeler, Descript. Gram. Gallia & German. 57. 1802. *Polypogon crinitus* Nutt. Gen. N. Amer. Pls. 1: 50. 1818. *Santia monspeliensis* Parl. Fl. Palermo 1: 73. 1845. ("Habitat Monspelii" [France].)

Rabbitfoot grass, beardgrass.

Tufted annuals; culms 0.3–4 (8) dm tall, often decumbent at the base and rooting at the nodes; sheaths smooth to scabridulous; ligules 2–6 (12) mm long, puberulent, acute, lacerate-erose at the tip; blades usually flat, 1.5–5 (12) mm broad, glabrous to scaberulous; panicles 1–6.5 (10) cm long, densely compact, the branches short, appressed-ascending; spikelets 1-flowered, spreading upwards; glumes subequal, 1.5–2.5 mm long, narrowly elliptic, scabrous to hispidulous with longer hairs on the keel

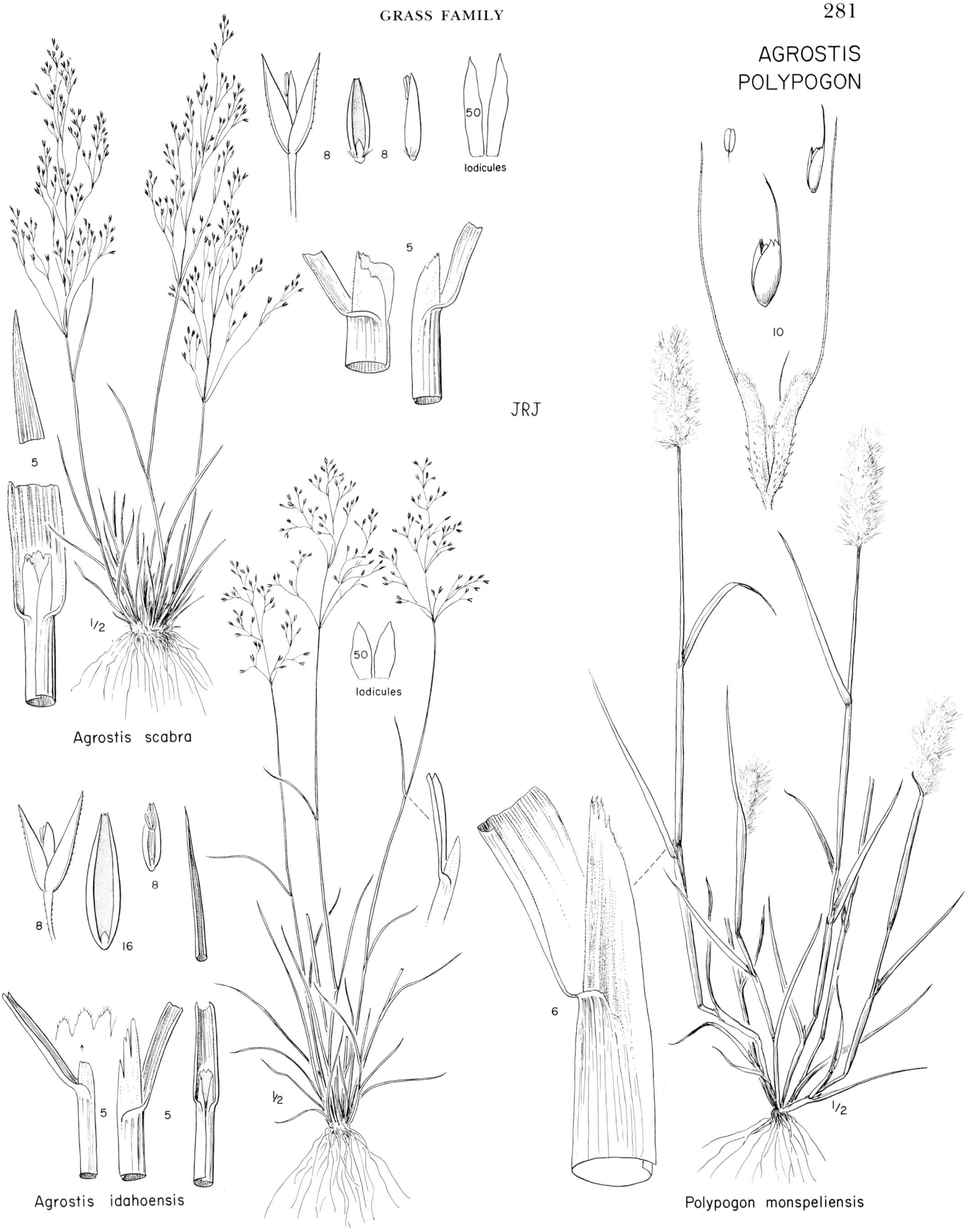
AGROSTIS
POLYPOGON
8
8
50
lodicules
5
JRJ
10
5
1/2
50
lodicules
Agrostis scabra
8
16
8
6
5
5
1/2
1/2
Agrostis idahoensis
Polypogon monspeliensis

and margins, especially towards the base, awned from between rounded lobes of a bifid apex, the awns (4) 4.5–9.5 (10) mm long; lemma about half the length of the glumes, 0.7–1.2 mm long, ovate, smooth and shiny, minutely toothed at the broad apex; awn absent or very short and straight; palea nearly as long as the lemma; lodicules oblong-lanceolate, about 0.2 mm long; anthers 0.4–0.5 mm long; $2n$ = 14, 28.

Moist soil or in water along streams, in meadows and marshes, often in saline, alkaline or highly mineralized hot springs, usually in the valleys; native to Europe; widespread in N. Amer. from Yukon and N.H., southward. June–Aug.

Polypogon monspeliensis is a palatable forage grass which only occasionally is of sufficient abundance to be important.

2. Polypogon interruptus H.B.K.

Polypogon interruptus H.B.K. Nova Gen. & Sp. Pl. 1: 134, tab. 44. 1815. *Alopecurus interruptus* Poiret in Lam. Encycl. Méth. Bot. Suppl. 5: 495. 1817. *P. litoralis* var. *interruptus* Kuntze, Revisio Gen. Pl. 3(2): 367. 1898. (*Humboldt & Bonpland,* Venezuela.)

P. lutosus of Amer. authors; not (Poiret) A. S. Hitchc. 1920.

Ditch polypogon.

Tufted perennials; culms 0.5–6 (9) dm tall, sometimes decumbent at the base and rooting at the nodes; sheaths smooth to scaberulous; ligules (1.5) 3–7 (9.5) mm long, stout and nerved, narrowed to an obtuse to truncate apex, entire to lacerate-erose, puberulent; blades 2–5 (9) mm broad, flat, scabrous; panicles 2–9 (15) cm long, less compact, lobed or interrupted, the branches erect, longer; spikelets 1-flowered, lanceolate; glumes subequal, 1.7–2.5 mm long, keeled, hispidulous, narrowed abruptly into terminal awns, sometimes minutely bilobed, the awns (1) 1.5–3 (5) mm long; lemma (0.7) 1–1.5 mm long, narrowly-lanceolate, smooth and shiny, minutely toothed at the apex, awned from near the tip, the awn about 3 mm long; palea nearly as long as the lemma; lodicules lanceolate, about 0.3 mm long; anthers 0.4–0.7 mm long; $2n$ = 28, 42.

Moist soil or in water along streams and in marshes and meadows in the valleys; introduced from S. Amer.; in N. Amer. from B.C. to Baja Calif., sporadically distributed e. to Nebr., Texas and La. May–Aug.

23. CINNA L. Woodreed

Tall perennials; culms hollow and slender; sheaths open; ligules membranous; blades thin, flat and lax, non-auriculate; inflorescence an open or contracted panicle; spikelets 1-flowered, flattened, disarticulating below the glumes; glumes subequal, the first usually somewhat shorter, lanceolate, strongly keeled, 1- to 3-nerved, acute at the apex, awnless, hispidulous on the keel; lemma similar to the glumes, flattened, acute or narrowly rounded and notched at the tip, 3- to 5-nerved, the mid-nerve sometimes prolonged into a short, straight awn from just below the apex; palea slightly shorter than the lemma, 2- or appearing 1-nerved with the nerves close together; rachilla projecting beyond the palea as a minute bristle; callus not bearded; lodicules short, broader at the apex and erose-toothed; stamens 1–3; ovary glabrous; x = 7.

A genus of 3 or 4 species of N. Amer. and Eurasia; mostly of wet places. (From the word *kinni,* a name used by Dioscorides for some grass.)

1. Cinna latifolia (Trev. ex Goeppert) Griseb.

Agrostis latifolia Trev. ex Goeppert, Beschreibung Bot. Gart. Königl. Univ. Brelau 82. 1830. *Cinna latifolia* Griseb. in Ledeb. Fl. Ross. 4: 435. 1853. (Europe.)

Drooping woodreed.

Rhizomatous perennials; culms 6–20 dm tall, bearing most of the leaves; sheaths glabrous to scaberulous; ligules (2) 3.5–6 (8) mm long, erose and sometimes lacerate; blades flat, 5.5–12 (16.5) mm broad, scaberulous; panicles 12–30 (38) cm long, loose, the branches spreading to drooping; spikelets strongly compressed, lanceolate, 1-flowered; glumes narrow lanceolate, keeled, scaberulous over the back and scabrous on the keel, the first glume 2.5–3.3 mm long, the second 2.8–3.7 mm long, sometimes aristate; lemma (2) 2.5–3.2 mm long, lanceolate, scaberulous over the back, scabrid on the keel, 3-nerved; awn absent or very short from near the tip of the lemma and up to 1.2 mm long, straight; palea about 2.5 mm long; lodicules about 0.3 mm long, cuneate-obovate, dentate; anthers 0.6–1 mm long; $2n$ = 28.

Moist streams, meadows and springs, often in shaded places from lower montane to subalpine elevs.; Alaska to Newfl., s. in the w. mts. to Calif. (n. Coast Range and Sierra Nevada), Nev., Utah and Colo., in the middle states to the Great Lakes region and in the e. in the mts. to N.C. July–Aug.

Drooping woodreed is palatable but too scarce to be of any importance as forage.

24. HIEROCHLOË R. Br. nom. conserv. Sweetgrass

Low to moderately tall rhizomatous perennials; herbage fragrant due to the presence of coumarin; culms hollow; sheaths open; ligules membranous, fringed; blades narrow to broad and flat to involute, non-auriculate; inflorescence an open or contracted, mostly few-flowered panicle; spikelets with 3 well-developed florets, the lower 2 staminate, the uppermost perfect, disarticulating above the glumes and the 3 florets falling together; glumes subequal, as long as the florets or slightly shorter, broad, thin, membranous, smooth and shiny, 1- to 3-nerved, awnless; lemmas all 3 about equal in length, rounded on the back and entire to deeply bifid, the lemmas of the staminate florets broadly elliptic, 5-nerved, awned or awnless or with a short awn from a bifid apex, the lemma of the perfect floret 3- to 5-nerved, awnless, short-hairy at the apex, becoming indurate; paleas of the staminate florets 2-nerved, enclosing 3 stamens, the palea of the perfect

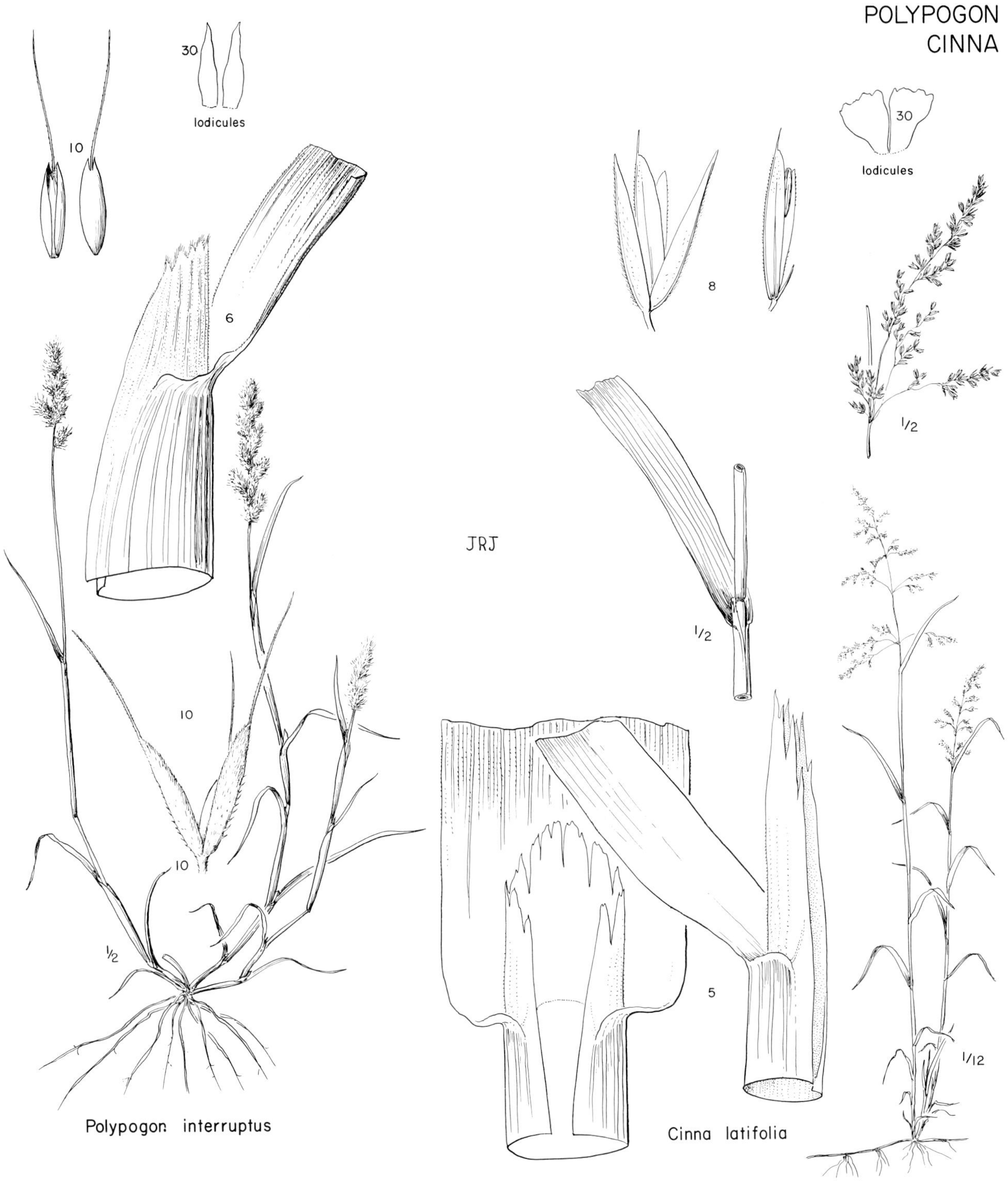

floret apparently 1- or 3-nerved, rounded on the back and enclosing 2 stamens; lodicules 2, ovate-lanceolate and somewhat toothed, thin and membranous; $x = 7$.

A genus of 10 to 20 species, these in the temp. and cool regions of the world, mostly in the N. Hemisphere. Three species are reported from the U.S. (From the Greek *hieros,* sacred, and *chloë,* grass; "holy-grass," *H. borealis* and *H. odorata,* were strewn in front of churches in northern Europe on Saints' Days.)

Hierochloë, until recently, was placed in the tribe Phalaridae (see under *Phalaris* below).

1. Hierochloë odorata (L.) Beauv.

Holcus odoratus L. Sp. Pl. 1048. 1753. *Avena odorata* Koeler, Descript. Gram. Gallia & German. 299. 1802. *Hierochloë odorata* Beauv. Essai Nouv. Agrostogr. 62, 164. 1812. *Savastana odorata* Scribn. Mem. Torrey Bot. Club 5: 34. 1894. *Torresia odorata* A. S. Hitchc. Amer. J. Bot. 2: 301. 1915. ("Habitat in Europae.")

Sweetgrass, holygrass, vanilla grass.

Rhizomatous perennials; culms 1.5–5 (7.5) dm tall; sheaths glabrous to puberulent on the innovations, especially at the collar; ligules 2–4 (5) mm long, obtuse to acute, slightly erose-lacerate and ciliolate; blades flat, (2) 3–6 mm broad; panicles (2) 4–7 (12) cm long, open-pyramidal; spikelets broadly ovate, 3-flowered, the first 2 florets staminate and the third perfect; glumes subequal, (3) 3.5–5.5 (6.5) mm long, membranous, shiny, faintly nerved, the first glume ovate-lanceolate, 1- to 3-nerved, the second broadly ovate, 3-nerved; lemmas of 2 types, the 2 staminate lemmas subequal, 3–5 mm long, ovate, appressed-pubescent, especially on the margins, shiny, 5- to 7-nerved, the terminal perfect lemma 2.2–3.7 (4) mm long, lanceolate, shiny, pilose above, glabrous below, keeled, all 3 lemmas unawned; paleas of all 3 florets more or less as long as their lemmas; lodicules oblong-lanceolate, nearly entire, about 1.5 mm long; anthers 1.4–2 (2.5) mm long; $2n$ = 28, 42, 56.

Moist slopes, meadows and streambanks from the foothills to subalpine elevs.; Alaska to Newfl. (across Yukon and N.W.T.), s. in the w. to Oregon, Nev., n. Ariz. and n. N.M., in the c. states to Nebr. and the Great Lakes Region, and in the e. states to N.C. June–Aug.

Sweetgrass was used by the Blackfoot Indians as incense in some of their ceremonies, and an extract of it was used in a hair tonic. Some Indians wove fragrant baskets from its culms.

25. PHALARIS L. Canary Grass

Annuals and perennials, the perennials rather coarse, densely cespitose or in small clusters from rhizomatous bases; culms hollow; sheaths open; ligules membranous; blades flat and mostly glabrous, often with short auricles at the base; inflorescence a contracted, usually spike-like panicle; spikelets with 1 terminal perfect floret and 1 or 2 scale-like vestiges of florets below, strongly compressed, disarticulating above the glumes; glumes subequal, large, strongly compressed and dorsally keeled, the keel often with a thin membranous wing, 3- (7)-nerved, the marginal nerves often close together and appearing as one, awnless; sterile lemmas greatly reduced and sometimes lacking (present in ours); fertile lemma coriaceous and shiny, shorter and firmer than the glumes, often more or less hairy, permanently enclosing the palea and caryopsis, indistinctly 5-nerved, awnless; paleas absent from the sterile florets, that of the fertile floret about as long as the lemma, faintly 2-nerved; lodicules 2, membranous, oblong-ovate, sometimes lobed or toothed; stamens 3; x = 6, 7.

A genus of about 15 species in temp. regions of Europe, Asia, Africa and the Americas. Nine species are reported for the U.S., 5 native and 4 introduced. (From the Greek name *Phalaris*, for some grass.)

Hitchcock (Manual Grasses U.S., 1951) treated *Phalaris* along with *Hierochloë* and *Anthoxanthum* as the tribe Phalaridae with their "2 (rarely 1) staminate, neuter, or rudimentary lemmas . . . , below the fertile lemma." In modern grass systematics (see Stebbins, G. L., and B. Crampton, Recent Advances Bot. 1: 133–145. 1961), the tribe Phalaridae has been abandoned, and its genera are placed in the Aveneae tribe.

Reference: Anderson, D. E. Taxonomy and distribution of the genus *Phalaris*. Iowa State Coll. J. Sci. 36: 1–96. 1961.

1 Plants perennial with rhizomes . . . . 1. *P. arundinacea*
1 Plants annual.
  2 Glumes winged toward the summit, wingless below; sterile florets subulate, usually less than half as long as the fertile lemma; a native of the se. states known as close to our region as along the Virgin River, Nev. . . . . *P. caroliniana* Walt.
  2 Glumes broadly winged; sterile florets broad and chaffy, more than half as long as the fertile lemma . . . . 2. *P. canariensis*

1. Phalaris arundinacea L.

Phalaris arundinacea L. Sp. Pl. 55. 1753. *Arundo colorata* Aiton, Hort. Kew. 1: 116. 1789. *Typhoides arundinacea* Moench, Meth. Pl. Horti Bot. & Agri Marburg. 202. 1794. *Arundo riparia* Salisb. Prodr. Stirpium 24. 1796. *Calamagrostis variegata* With. Arr. Brit. Pl. ed. 3. 2: 124. 1796. *Baldingera colorata* Gaertn., Meyer & Scherb. Oekon.-Techn. Fl. Wetterau 1: 96. 1799. *Digraphis arundinacea* Trin. Fund. Agrostogr. 127. 1820. *Baldingera arundinacea* Dumort. Observ. Gram. Fl. Belg. 130, tab. 10, fig. 40. 1823. *Digraphis americana* Ell. ex Loudon, Hort. Brit. 27. 1830. *Phalaridantha arundinacea* St.-Lager in Cariot, Étude Fleurs Bot. Element. Descript. & Usuelle, ed. 8. 2: 900. 1897. *Endallex arundinacea* Raf. ex B. D. Jackson, Index Kew. 1: 839. 1893. *Phalaroides arundinacea* Rausch, Feddes Repert. 79: 409. 1969 ("Habitat in Europae.")

Phalaris arundinacea var. *picta* L. Sp. Pl. 55. 1753. *P. americana* var. *picta* Eat. & Wright, N. Amer. Bot. ed. 8. 352. 1840. *Digraphis arundinacea* var. *picta* Pacher, Jahrb. Naturhist. Landesmus. Kärnten 14: 119. 1880. *P. arundinacea* f. *picta* ——— Aschers. & Graebner, Syn. Mitteleur. Fl. 2: 24. 1898. (Europe.)

Reed canary grass.

Strongly rhizomatous perennials; culms 5–15 (20) dm tall; ligules 3–8 (10) mm long, obtuse, usually lacerate, sometimes pubescent; blades flat, 6–15 (18) mm broad; panicles 7–15 (20) cm long, compact, sometimes lobed at the base; spikelets ovate, 3-flowered, the first 2 florets sterile and the third perfect; glumes subequal, (4) 4.5–6 (7.5) mm long, strongly compressed, especially distally, keeled, 3-nerved, the lateral nerves prominent, the keel scabrous, acute, usually wingless; fertile lemma (3) 3.5–4 mm long, ovate, brownish at maturity, shiny, appressed-hairy, faintly 3- to 5-nerved, the 2 subtending sterile lemmas subequal, 1–2 mm long, su-

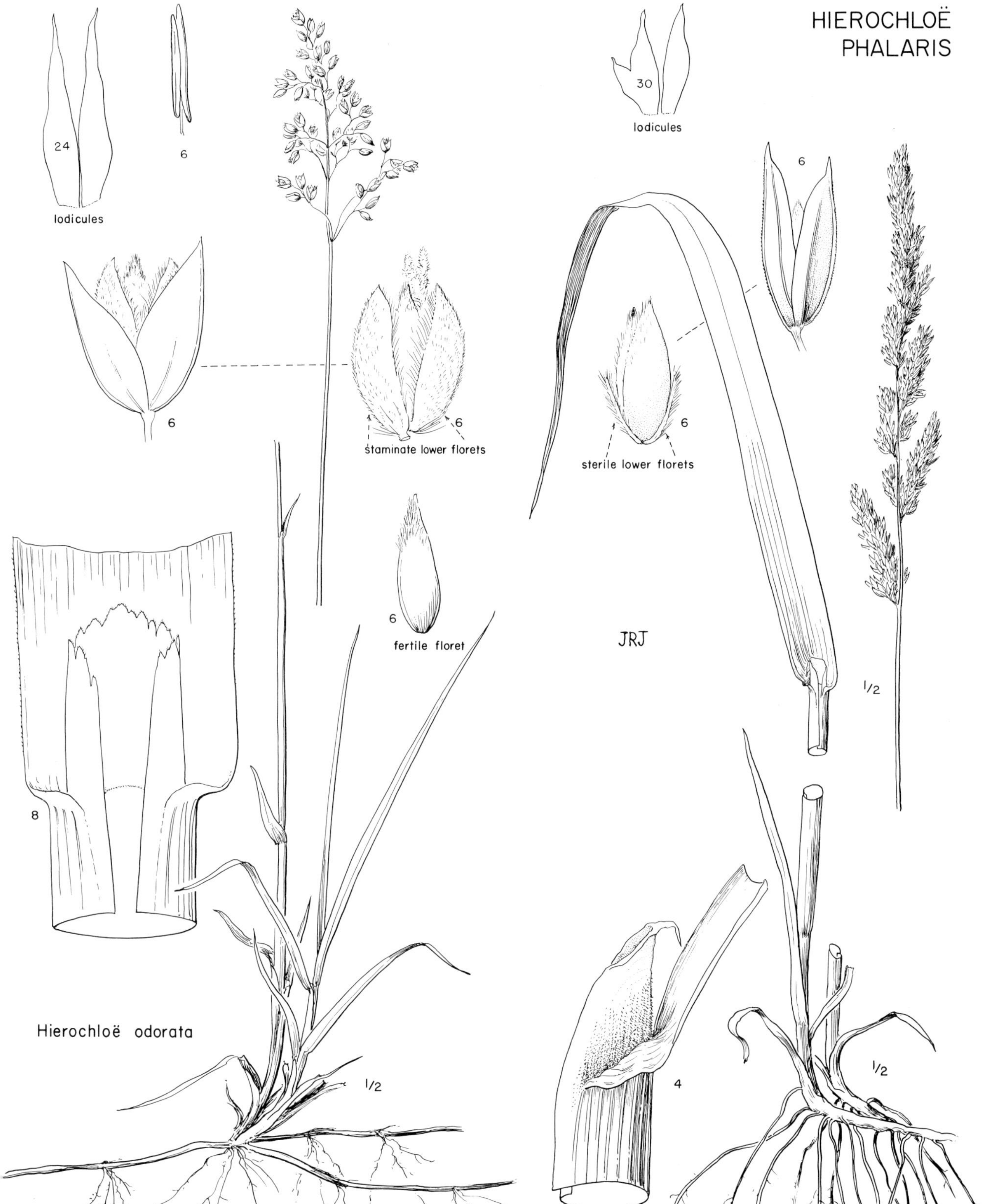
HIEROCHLOË
PHALARIS
24
lodicules
6
6
6
staminate lower florets
6
fertile floret
8
Hierochloë odorata
1/2
30
lodicules
6
6
sterile lower florets
JRJ
1/2
4
1/2
Phalaris arundinacea

bulate, hairy; palea well developed, 1- to 2-nerved, shiny, with same type of hairs as on the lemma; lodicules lanceolate, about 0.5 mm long; anthers 2.5–3 mm long; $2n$ = 14, 27–42, 56.

Wet places along ditch and stream banks, and in marshes and meadows; Alaska and N.W.T. to Newfl., s. in the w. states to Calif., Nev., n. Ariz., n. N.M. and Okla., in the c. states around the Great Lakes and in the e. states to N.C.; also in Mex. and S. Amer. Late June–Aug.

Phalaris arundinacea is an important hay grass of moist meadows. A form, f. *picta* (L.) Aschers. & Graebner, with white striped leaves is sometimes grown as an ornament in wet gardens.

2. Phalaris canariensis L.

Phalaris canariensis L. Sp. Pl. 54. 1753. *P. avicularis* Salisb. Prodr. Stirpium 17. 1796. ("Habitat in Europa australi, Canariis.")

Canary grass.

Robust annuals; culms (2) 3–6 (10) dm tall; ligules 4–8 mm long, truncate to rounded, usually deeply lacerate, puberulent externally; blades flat, (3) 4–10 mm broad, with short auricles sometimes present; panicles 1.5–3.2 (4) cm long, greatly contracted, ovoid to cylindric-ovoid; spikelets obovate, imbricate, with 3 florets, the first 2 sterile and the third perfect; glumes subequal, (6) 7–8 (9.5) mm long, abruptly short-acute, 2- (5)-nerved, the keels strongly winged, the wings about 1 mm broad, broadest above the middle, tawny except for the green lateral nerves, loosely pilose distally on the wingless part; fertile lemma (4.5) 5–5.5 (6.2) mm long, lanceolate, densely appressed-pubescent, faintly 5-nerved, the 2 subtending sterile lemmas linear, chaffy, sparsely pubescent, unequal, the first 2.5–3.2 (4.5) mm long, the second 3–4 (5) mm long; palea 4–5 mm long; lodicules lacking; anthers 2.5–3 (3.5) mm long; caryopsis often 4 mm long; $2n$ = 12, 28.

Wet streambanks and moist wasteplaces; introduced from the Medit. region; Alaska across s. Can. to N.B. and s., sporadically occurring throughout the U.S., Mex. and S. Amer. July–Aug.

Seed of this species, often mixed with seed of *Panicum miliaceum* L., is used commercially for canary seed.

26. ALOPECURUS L. Foxtail

Tufted annuals and perennials, some rhizomatous; culms hollow; sheaths open; ligules membranous, truncate to acute, erose-ciliate to entire or lacerate and somewhat puberulent; blades flat, non-auriculate; inflorescence a compact, cylindrical, spike-like panicle; spikelets 1-flowered, strongly flattened, disarticulating below the glumes and falling entire; glumes subequal, opposite and often connate below, ciliate on the keels, pubescent to villous on the lateral nerves and glabrous to silky or lanate-villous between the nerves, acute to obtuse or rounded at the apex, awnless; lemma more or less as long as the glumes, strongly flattened, firm, 3- or 5-nerved, awned from the middle or below, the margins more or less connate at the base; awn straight to strongly twisted and geniculate, from short to more than twice the length of the glumes; palea and prolongated rachilla lacking; lodicules absent; stamens 3; ovary glabrous; x = 7.

A genus of about 25 species, distributed in temp. regions of the N. Hemisphere, 6 native to the U.S. and 4 introduced or adventive. (Name from the Greek *alopex,* fox, and *oura,* tail, referring to the appearance of the cylindrical panicle.)

All of our species of *Alopecurus* are palatable forage grasses, but they are more scarce than are their already scarce wetland habitats in the Intermountain Region.

1 Plants annual; lemma 1.8–2.2 mm long; anthers 0.3–0.5 mm long .... 5. *A. carolinianus*
1 Plants perennial; lemma 2–6 mm long, as short as 1.8 mm in *A. aequalis;* anthers 0.5–3.5 mm long.
  2 Spikelets densely woolly; ligules 1.5–3.5 (5) mm long; lemmas 2.5–6 mm long; anthers 1.8–3.5 mm long.
    3 Inflorescence several times longer than broad, 3.5–8.5 (10) cm long and 5–7.5 (10) mm broad .... 1. *A. pratensis*
    3 Inflorescence only about twice as long as broad, 1–3.5 (4) cm long and 6–11 mm broad .... 2. *A. alpinus*
  2 Spikelets not woolly; ligules (1.8) 3–8 mm long; lemmas 1.8–3 mm long; anthers 0.5–2 mm long.
    4 Awn distinctly exserted beyond the glumes, the awns (3) 5–8 mm long .... 3. *A. geniculatus*
    4 Awn scarcely, if at all, exceeding the glumes, the awns 0.7–2.5 mm long .... 4. *A. aequalis*

1. Alopecurus pratensis L.

Alopecurus pratensis L. Sp. Pl. 60. 1753. ("Habitat in Europae.")

Meadow foxtail.

Stout perennials; culms (2.5) 4–8 (10) dm tall, sometimes rooting at the lower nodes; sheaths glabrous; ligules (1.5) 2–3.5 (5) mm long, truncate to obtuse, erose-ciliolate to lacerate, sometimes entire, often puberulent; blades flat, 3–6 (10) mm broad, scaberulous on both surfaces; panicles 3.5–8.5 (10) cm long and 5–7.5 (10) mm broad; spikelets 1-flowered, ovate; glumes subequal, relatively large, 3–5.2 (6) mm long, connate for 1–1.5 mm, 3-nerved, the nerves often green, long, villous-ciliate on the keel and appressed-villous on the lateral nerves; lemma 3–5 (6) mm long, the margins connate up to about 1 mm, lance-ovate, glabrous, sometimes pubescent on the keel above, 5-nerved; awn 5–8 (10.5) mm long, geniculate, attached near the base of the lemma; palea absent; anthers relatively large, 2.2–3.5 mm long; $2n$ = 14, 28, 42.

Moist meadows, ditches and streams; introduced from Europe; Alaska to Newfl., s. to Calif., Utah, Mont. and e. across n. U.S. June–July.

PHALARIS
ALOPECURUS
6
6
sterile lower florets
1/2
4
Phalaris canariensis
JRJ
6
4
6
1/2
1/2
5
5
Alopecurus pratensis

Alopecurus pratensis was introduced from Europe as a pasture-grass and has become well established in fields and other disturbed sites throughout the northern and mountain states.

2. Alopecurus alpinus J. E. Smith

Alopecurus alpinus J. E. Smith, Engl. Bot. pl. 1126. 1803. (Scotland.)

A. occidentalis Scribn. & Tweedy, Bot. Gaz. 11: 170. 1886. (*Tweedy* [*591*], Mirror Lake Plateau, Yellowstone National Park, Wyo.)

Alpine foxtail.

Short rhizomatous or stoloniferous perennials; culms (1) 2.5–8 (10) dm tall; sheaths glabrous; ligules 1.5–4 mm long, truncate, finely erose, sometimes becoming lacerate; blades flat, (2) 3–6 mm broad, scaberulous on both surfaces, relatively short; panicles 1–3.5 (4) cm long and 6–11 mm broad, short and ovoid; spikelets 1-flowered, ovate; glumes subequal, 2.8–4 mm long, connate for 1–1.2 mm, densely villous all over, 3-nerved; lemma 2.5–4 mm long, ovate, the margins connate for about 1.5 mm, puberulent above, glabrous below; awn 3.5–6 (7.5) mm long, slightly geniculate, attached just below the midlength of the lemma; palea absent; anthers 1.8–2.2 mm long; stigmas long and plumose; $2n$ = 98, ca 100, 105, 110, 112–130.

Along stream courses and wet meadows at subalpine and alpine elevs.; Alaska and across n. Can. to Greenl., s. in the Rocky Mts. to Colo. and in the Uinta Mts. of Utah; also in Arctic and n. Eurasia. Late June–Aug.

3. Alopecurus geniculatus L.

Alopecurus geniculatus L. Sp. Pl. 60. 1753. *A. palustris* subsp. *geniculatus* Syme, Engl. Bot. ed. 3. 11: 25. 1873. *Tozzettia geniculata* Bubani, Fl. Pyrén. 4: 275. 1901. ("Habitat in Europae.")

A. pallescens Piper & Beattie, Fl. Palouse Region 18. 1901. (*Piper 1743*, Pullman, Wash., 20 June 1893.)

Water foxtail, marsh foxtail.

Perennials; culms (1) 2–6 dm tall, decumbent and rooting at the nodes; sheaths glabrous; ligules (1.8) 3–6 (8) mm long, membranous, narrowed, erose-toothed to lacerate at the narrow tip; blades flat, 2–4 (7) mm broad, scaberulous above, less scaberulous to glabrous below; panicles 2–7 (8) cm long, compact, spike-like with short pedicelled spikelets; spikelets 1-flowered, ovate; glumes subequal, 2–3.5 (4) mm long, connate for about 0.5 mm, ciliate-villous on the keel, silky-appressed pubescent on the sides; lemma 2–3 mm long, ovate, the margins connate to nearly the midlength, glabrous; awn (3) 5–7 (8) mm long, geniculate, attached near the base of the lemma; palea absent; anthers 1.2–2 mm long; $2n$ = 14, 28.

Wet meadows and marshy areas at edges of ponds and lakes; Alaska to Newfl., s. to Calif., Ariz. and N.M. in the w., and e. across n. U.S. to Va. June–Aug.

4. Alopecurus aequalis Sobol.

Alopecurus aequalis Sobol. Fl. Petrop. 16. 1799. *A. geniculatus* var. *aequalis* Fiori, Nuova Fl. Analitica Ital. 1: 92. 1923. (Greece.)

A. aristulatus Michx. Fl. Boreali-Amer. 1: 43. 1803. *A. geniculatus* var. *aristulatus* Torr. Fl. N. & Middle Sect. U.S. 1: 97. 1823. (*Michaux*, "in paludosis Canadae," 1792.)

A. caespitosus Trin. Sp. Gram. Icon. Descript. Ill. 3: pl. 241. 1836. (*Douglas*, Nw. Amer.)

Short-awn foxtail.

Perennial or sometimes flowering as a winter annual; culms 0.8–4 (6.5) dm tall, erect or decumbent and rooting at the nodes; sheaths glabrous; ligules relatively long, 3.5–6 (8) mm and attenuate, lacerate or entire; blades flat, 1.5–4 (5) mm broad, glabrous or scaberulous on the upper surface; panicles narrow-cylindric, 2–6 (8) cm long and 4–5 mm broad; spikelets 1-flowered, ovate; glumes subequal, relatively small, 1.7–2.2 (2.5) mm long, connate up to 0.5 mm, 3-nerved, ciliate on the keels and appressed-pilose on the lateral nerves; lemma small, 1.8–2.2 mm long, lance-ovate, the margins connate for slightly less than half the length, glabrous; awn very short, 0.7–2.5 mm long, straight or only slightly bent, often included in the glumes, attached 0.7–1 mm from the base of the lemma; palea absent; anthers small, 0.5–0.9 mm long; $2n$ = 14.

Marshes, wet meadows, edges of lakes and ponds and along water courses in the higher valleys and up to subalpine elevs.; Alaska to Newfl., s. to Calif., n. Ariz., N.M. and e. across Kansas, Mo., Ill., Iowa, Mich. and Pa. June–Aug.

Alopecurus aequalis is a good forage grass but is not plentiful.

5. Alopecurus carolinianus Walter

Alopecurus carolinianus Walter, Fl. Carol. 74. 1788. (South Carolina.)

Carolina foxtail.

Small, tufted annuals; culms 1–4 dm tall, decumbent at the base, geniculate; sheaths glabrous; ligules 2–3.5 (4) mm long, attenuate to a narrow, erose tip, decurrent; blades flat, scaberulous above, glabrous below; panicles 2–5 (6) cm long and 2.5–4 mm broad; spikelets 1-flowered, ovate; glumes subequal, relatively small, 2–2.5 mm long, connate only at the very base, obtuse, hirsute-ciliate on the keel and sparsely appressed-pubescent on the sides; lemma small, 1.8–2.2 mm long, ovate, the margins connate to mid-length, glabrous; awn 3.5–5 mm long, geniculate; palea absent; anthers small, 0.3–0.5 mm long.

Moist areas in meadows and near streams and ponds; B.C. and nearly throughout the U.S., except the New England states. June–July.

27. PHLEUM L. Timothy

Tufted annuals and perennials (ours); culms hollow; sheaths open; ligules membranous; blades flat, often with small rounded auricles; inflorescence a contracted, dense, spike-like panicle; spikelets 1-flowered, laterally flattened, disarticulating above the glumes, but sometimes below the glumes later on; glumes subequal, strongly keeled, often ciliate on the keel, abruptly short-awned from a truncate apex; lemma much shorter than the glumes, broad and blunt, membranous, 3- to 5-nerved, awnless or sometimes minutely

ALOPECURUS

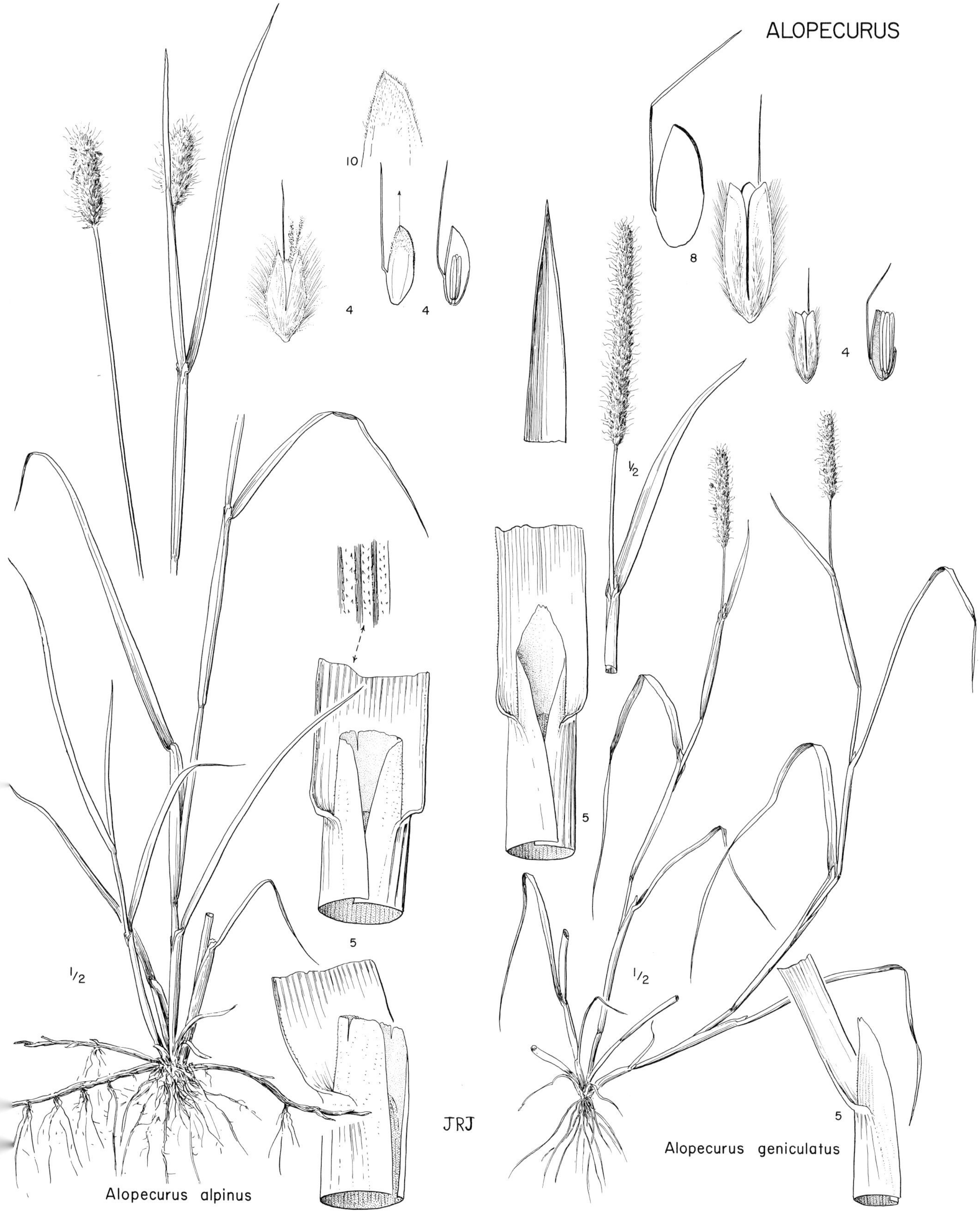
10
4
4
8
4
1/2
5
5
1/2
1/2
5
JRJ
5
Alopecurus alpinus
Alopecurus geniculatus

ALOPECURUS

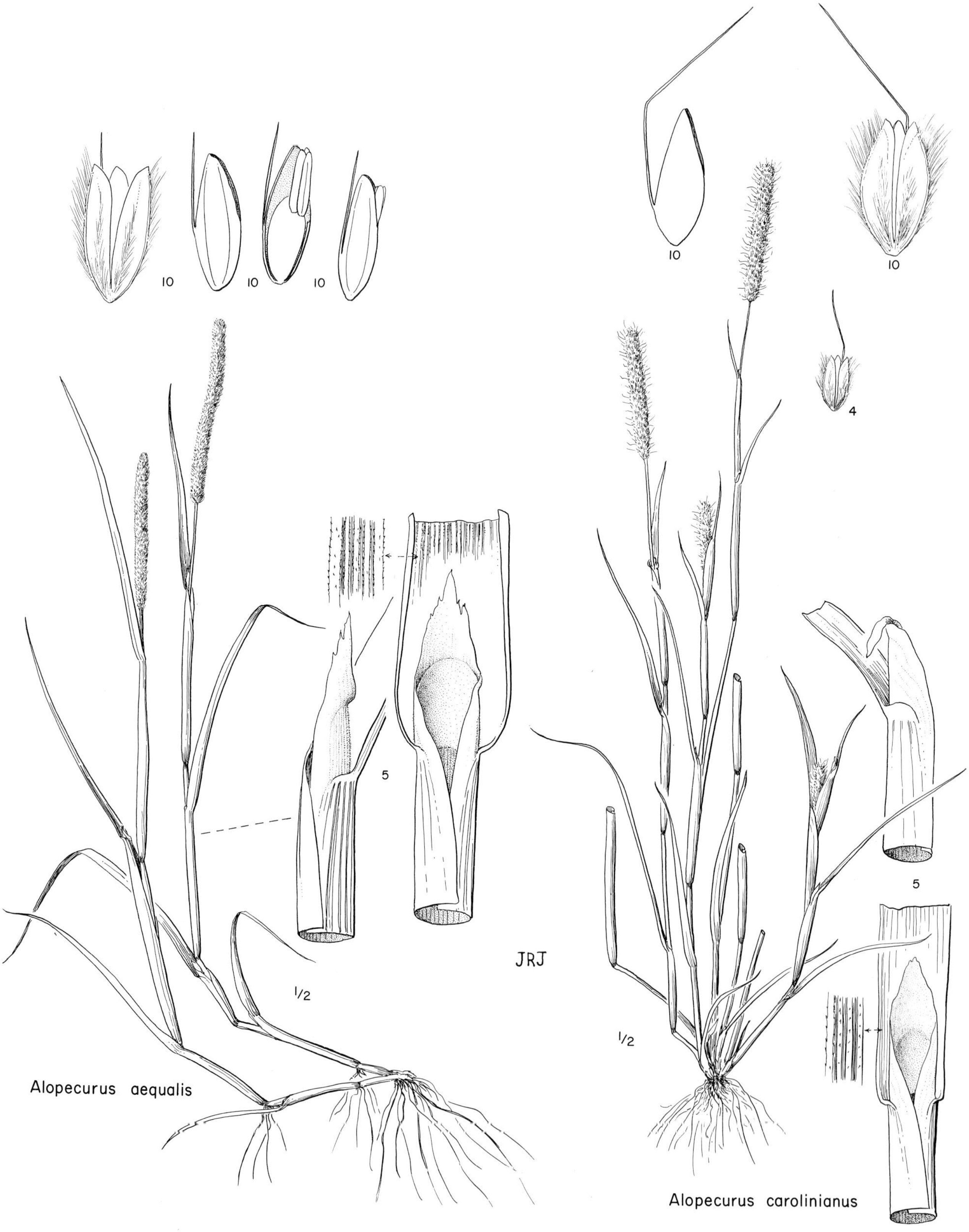
10
10
10
10
10
4
5
5
1/2
1/2
JRJ
Alopecurus aequalis
Alopecurus carolinianus

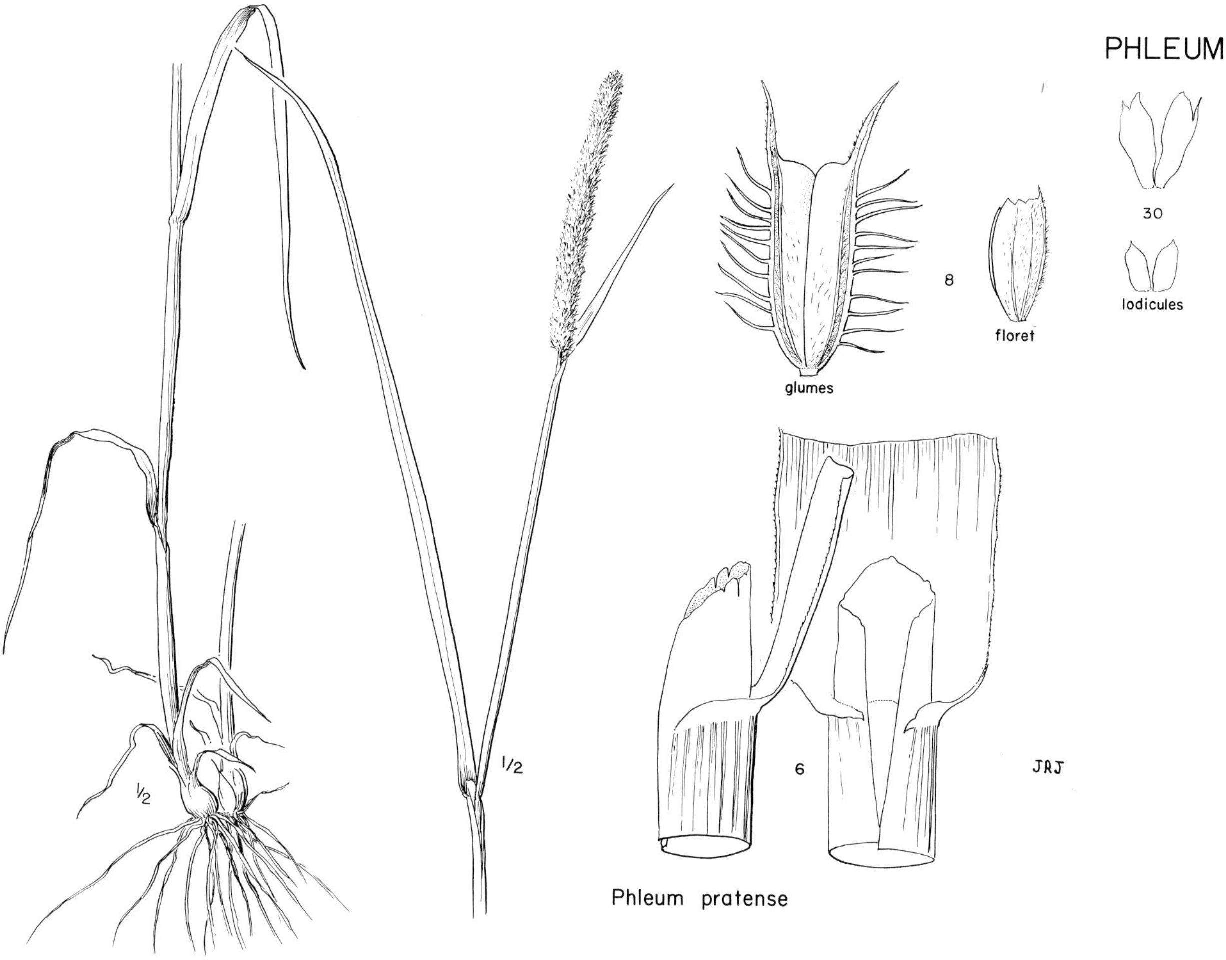

Phleum pratense

awned; palea membranous, nearly as long as the lemma; lodicules 2, oblong, sometimes lobed or toothed; stamens 3; ovary glabrous; $x = 7$.

About 10 species in the temp. regions of both hemispheres, 4 species occur in the U.S., but only *Phleum alpinum* is native. *Phleum pratense,* introduced from Europe, is one of the most important pasture grasses cultivated in temp. N. Amer. (Name from the Greek word *phleos,* an old name for a marsh reed.)

1 Panicles several times longer than wide, (3) 4–16 cm long and 5–7.5 (10) mm wide; culms usually more than 5 dm tall, erect from a swollen, bulb-like base . . . . 1. *P. pratense*
1 Panicles only 2 or 3 times longer than wide, 1–5 (6) cm long and (7) 8–12 mm wide; culms usually less than 5 dm tall with a somewhat decumbent, creeping, non bulb-like base . . . . 2. *P. alpinum*

1. Phleum pratense L.

Phleum pratense L. Sp. Pl. 59. 1753. *P. nodosum* var. *pratense* Saint-Amans, Fl. Agénaise 23. 1821. *Plantinia pratensis* Bubani, Fl. Pyrén. 4: 270. 1901. *Stelephuras pratensis* Lunell, Amer. Midl. Naturalist 4: 216. 1915. ("Habitat in Europae.")

Timothy.

Tufted perennials; culms (2) 5–10 (15) dm tall, usually bulbous at the base; sheaths glabrous; ligules (1.5) 2–3 (5) mm long, subentire, obtuse, sometimes lacerate; blades flat, 3–6.5 (8) mm broad, scabrous-margined, sometimes with small auricles; panicles (3) 4–11 (16) cm long and 5–7.5 (10) mm wide when pressed, compressed-cylindrical; spikelets 1-flowered, elliptic; glumes subequal, 2–3.2 mm long, abruptly truncate, with an awn, the keel strongly pectinate-ciliate and often scabrous on the sides, 3-nerved, the lateral nerves close to the keel, the first glume slightly narrower and sometimes villous margined, the awns (0.6) 1–1.2 (1.6) mm long; lemma (1.2) 1.7–2 mm long, ovate, truncate and erose, faintly 5-nerved, puberulent, sometimes with a minute awn; palea nearly as long as the lemma; lodicules about 0.3 mm long, oblong; anthers 1.2–1.7 (2.3) mm long; $2n = 14, 21, 28–84$.

Moist streambanks, meadows, woodlands and forests from the foothills to subalpine elevs.; native of Europe, cultivated and well established from Alaska to Newfl. and s. throughout the rest of N. Amer. Late June–Sept.

Phleum pratense is a very important meadow and hay grass.

2. Phleum alpinum L.

Phleum alpinum L. Sp. Pl. 59. 1753. *P. pratense* var. *alpinum* Celak. Prodr. Fl. Böhm. 38. 1867. *Plantinia alpina* Bubani, Fl. Pyrén. 4: 272. 1901. ("Habitat in Alpibus," Europe.)

Mountain timothy.

Tufted perennials; culms 2–4.5 (6) dm tall, often decumbent; sheaths glabrous, the collar oblique; ligules 0.5–3 (4.5) mm long, truncate, subentire; blades flat, 2.5–6 (7) mm broad, scabrous on the margins, sometimes on the surfaces; auricles sometimes present, small and blunt to rounded; panicles small, 1–5 (6) cm long and (7) 8–12 mm broad when pressed, ovoid to cylindric; spikelets 1-flowered, elliptic; glumes subequal, 2.5–3.5 mm long, abruptly truncate-attenuate to the stout awns, the keels pectinate-ciliate, the sides scabrous, the first glume sometimes with a ciliate margin, the awns (1.2) 1.5–2.2 (3.2) mm long; lemma 1.7–2.5 mm long, lance-ovate, truncate, glabrous, shiny to sometimes puberulent, erose-toothed by the 5 excurrent nerves; palea slightly shorter than the lemma; lodicules about 1 mm long; anthers 1.1–1.5 (2) mm long; $2n = 14, 28$.

Meadows and moist woods at subalpine and alpine elevs.; Alaska to Greenl., s. to Calif., Ariz. and N.M. in the w. and Nebr. and N.H. in the c. and e. states; also Mex., Eurasia and arctic and alpine regions of the S. Hemisphere. Late June–Aug.

Phleum alpinum is a good forage grass in the mountains.

28. BECKMANNIA Host Sloughgrass

Stout annuals; culms hollow and moderately thick; sheaths open; ligules membranous; blades flat and broad, non-auriculate; inflorescence a narrow, loosely contracted panicle with the spikelets closely imbricate in 2 rows on short spicate branches; spikelets 1- (2 in the European species)-flowered, suborbicular, laterally compressed, subsessile, disarticulating below the glumes and falling entire; glumes subequal, strongly compressed, broad, obovate, firm, 3-nerved, keeled, abruptly apiculate at the apex; lemma about as long as the glumes, lanceolate, faintly 5-nerved, tapering to a slender, pointed tip; palea narrow, slightly shorter than the lemma; lodicules 2, linear; stamens 3; $x = 7$.

A genus of 2 species, *B. erucaeformis* (L.) Host of n. Europe and Asia and *B. syzigachne* (Steudel) Fern., widespread in cooler parts of Eurasia and N. Amer. (Named for Johann Beckmann, 1739–1811, a German botanist from Goettingen.)

Beckmannia was included in the tribe Chlorideae by A. S. Hitchcock (Manual Grasses U.S., 1951) but differs from that group in many fundamental characteristics as has been noted by Reeder (1953).

References:

Fernald, M. L. The American and eastern Asiatic *Beckmannia*. Rhodora 30: 24–27. 1928.
Reeder, J. R. Affinities of the grass genus *Beckmannia* Host. Bull. Torrey Bot. Club 80: 187–196. 1953.

1. Beckmannia syzigachne (Steudel) Fern.

Panicum syzigachne Steudel, Flora 29: 19. 1846. *Beckmannia syzigachne* Fern. Rhodora 30: 27. 1928. *B. erucaeformis* subsp. *syzigachne* Breit. Amer. Midl. Naturalist 58: 10. 1957. (Japan.)
B. erucaeformis of Amer. authors; not (L.) Host.

American sloughgrass.

Robust annuals, often stoloniferous; culms 4–12 dm tall, erect; sheaths glabrous; ligules 5–8.5 mm long, entire or lacerate, pubescent; blades flat, 5–10 (12.5) mm broad; panicles (7) 10–27 (35) cm long, narrow, congested; spikelets 1-flowered, the rachilla not prolonged, broad-pyriform; glumes subequal, 2–3 mm long, rounded-triangular, broadest above the middle, laterally compressed, keeled, with a weak pair of lateral nerves, glabrous or sometimes scaberulous on back, the margins hyaline; lemma (1.8) 2.5–3.5 mm long, lanceolate, acuminate, usually barely projected beyond the glumes; palea nearly as long as the lemma; lodicules almost 1 mm long, lobed or dentate; anthers 0.6–1.2 mm long; $2n = 14$.

Marshes, ditches, streams and edges of ponds and lakes; Alaska, Yukon and N.W.T. to Ont. and s. in the w. states to Calif., Nev., n. Ariz. and n. N.M., and in the c. and e. states to Iowa, Ill., Ohio and N.Y. July–Aug.

Beckmannia syzigachne is a palatable forage grass but is not plentiful enough in the Intermountain Region to be of any importance.

Tribe 3. TRITICEAE

(Hordeae)

Annuals or perennials; culms hollow; sheath margins free; ligules membranous; blades usually auriculate; inflorescence a bilateral spike or occasionally a spicate-raceme with both sessile and pedicelled spikelets at the rachis nodes (in *Hordeum* and some *Elymus*); spikelets 1- to several-flowered (typically 1-flowered in *Hordeum*), disarticulating above the glumes, or at the rachis nodes, with the spikelet or spikelet cluster falling with the rachis segment; glumes, when there are 2 or more spikelets per node, borne parallel in front of the spikelets, simulating an involucre; lemmas (3-) 5- to 7-nerved, awnless or awned from the tip; lodicules 2, variously shaped, often large, pubescent to glabrous, but mostly conspicuously ciliate-margined; endosperm with simple, rounded, separate starch grains; embryo pooid; $x = 7$.

The Triticeae is a small but economically important tribe, sometimes referred to as the cereal or wheat-grass tribe.

We know many of the shortcomings in recognizing some of the groups in the tribe as genera. Hybrids among these "genera" have been suspected for a long time, and the number of controlled hybrids produced under carefully planned conditions show a need for reconsidering generic limits in the tribe. It has been suggested more than once that all the genera of the tribe be combined in a single genus. This would be a large and unruly genus. There are some morphologically distinct genera that represent natural groupings and should not be merged. However, it has been evident for many years that the traditional separation of *Elymus, Sitanion* and *Agropyron,* on the basis of glume

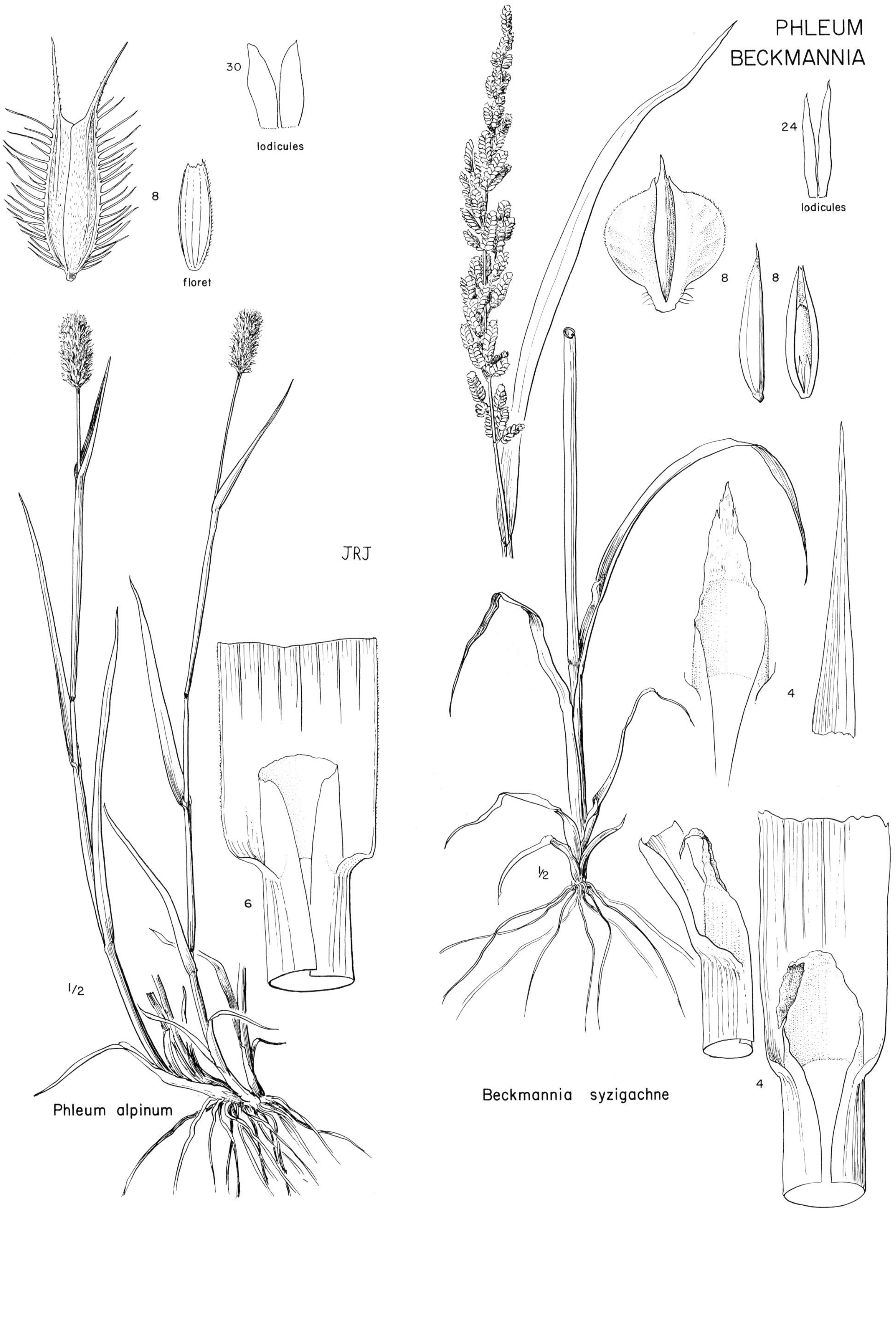
PHLEUM
BECKMANNIA
30
lodicules
8
floret
24
lodicules
8
8
JRJ
4
4
6
1/2
1/2
4
Phleum alpinum
Beckmannia syzigachne

breadth, number of spikelets per node and continuous versus disarticulating rachis is artificial. The proposal to lump these three genera into one large genus (Runemark & Heneen, 1968) ignores the phylogenetic differences that have been emerging through genome analyses of interspecific and intergeneric hybrids. Dewey (1974) suggests a realignment of the species of these genera rather than a mere consolidation. He has found that large segments of *Elymus* and *Agropyron* and all of *Sitanion* contain related genomes that are not shared by the other species of *Elymus* and *Agropyron.* This genomic group consists of self-fertilizing, small-anthered, awned, cespitose taxa such as *E. sibiricus* L. (the type species of the genus *Elymus,* not in our region), *E. canadensis, E. virginicus, E. glaucus, Agropyron trachycaulum,* and *A. scribneri,* as well as the species of *Sitanion* and *Hystrix.* Another unrelated genomic group consists of cross-pollinating, long-anthered, frequently rhizomatous and awnless taxa. This group, which includes *E. flavescens, E. triticoides, E. simplex, E. salina, E. ambiguus* and *E. cinereus,* could be transferred to the genus *Leymus* Hochst. Such genera as *Eremopyrum* and *Taeniatherum* should remain separated from *Agropyron* and *Elymus.* They are distinctive morphologically and both are diploids with their own unique chromosome morphology. The diploid chromosome number is rare among species of *Agropyron* and *Elymus,* and nonexistent in *Sitanion.* The splitting of *Agropyron* into *Roegneria, Agropyron* and *Elytrigia,* as is done by most Asian botanists, has some merit. An adequate overall revision must await clarification of other species relationships. Until this has been accomplished we will treat the genera in the classical sense.

Five sterile intergeneric hybrids are occasionally encountered in our area. Because of their occasional recurrence the key is made to accommodate them, and they are treated after the following key.

References:

Bowden, W. M. Taxonomy of intergeneric hybrids of the tribe Triticeae from North America. Canad. J. Bot. 45: 711–724. 1967.

Dewey, D. R. Cytogenetics of *Elymus sibiricus* and its hybrids with *Agropyron tauri, Elymus canadensis* and *Agropyron caninum.* Bot. Gaz. 135: 80–87. 1974.

Runemark, H., and W. K. Heneen. *Elymus* and *Agropyron,* a problem of generic delimitation. Bot. Not. 121: 51–79. 1968.

Stebbins, G. L., Jr., J. I. Valencia, and R. M. Valencia. Artificial and natural hybrids in the Gramineae, tribe Hordeae. I. *Elymus, Sitanion* and *Agropyron.* Amer. J. Bot. 33: 338–351. 1946.

1 Spikelets 2 or more at each node of the rachis.
- 2 Spikelets 3 at each node of the rachis, the lateral pair pedicelled and sterile (except sessile and fertile in *H. vulgare*), each spikelet with a single floret .... 32. *Hordeum*
- 2 Spikelets mostly 2 at each node of the rachis, up to 6 in some *Elymus* spp., if 3 not as above, each spikelet with 2 or more florets, rarely 1 in some *Taeniatherum* or *Sitanion.*
  - 3 Plants perennial; spikes (3) 4–25 cm long; spikelets usually with 2 to several fertile florets.
    - 4 Rachis usually continuous, the disarticulation occurring above the glumes .... 29. *Elymus*
    - 4 Rachis disarticulating at maturity.
      - 5 Glumes 3–7 mm long; awn of the lemma 0.5–3 mm long; introduced species .... 29. *Elymus junceus*
      - 5 Glumes 12–100 mm long; awn of the lemma 5–120 mm long (discussion of intergeneric hybrids following this key).
        - 6 Glumes 12–24 mm long; sterile hybrids.
          - 7 Spikes about 5 mm broad; body of the lemma 6–8 mm long .... X *Agrohordeum macounii*
          - 7 Spikes 8–10 mm broad; body of the lemma 8–10 mm long .... X *Elysitanion aristatum*
        - 6 Glumes 25–100 mm long.
          - 8 Glumes linear-lanceolate, 2- to 4-nerved; plants 6–10 dm tall .... X *Elysitanion hansenii*
          - 8 Glumes subulate, 1- or obscurely 2-nerved; plants 1–4.5 (6.5) dm tall .... 30. *Sitanion*
  - 3 Plants annual; spikes small, 1.5–4 cm long; spikelets with 1 fertile and 1 rudimentary floret .... 31. *Taeniatherum*

1 Spikelets solitary at each node of the rachis (occasionally 2 in *Agropyron,* but then only at a few nodes).
- 9 Spikelets borne edgewise to the rachis; inner glume absent, except in the terminal spikelet (belonging to tribe Poeae) .... 4. *Lolium*
- 9 Spikelets borne flatwise to the rachis; both glumes present on all spikelets.
  - 10 Plants perennial; mostly native.
    - 11 Rachis continuous.
      - 12 Glumes subulate to narrowly lanceolate; lemmas usually arranged 90° from the normal position due to twisting of the rachilla .... 29. *Elymus*
      - 12 Glumes narrowly lanceolate to broader; lemmas arranged in normal position above the glumes, i.e., the florets 2-ranked .... 33. *Agropyron*
    - 11 Rachis tardily disarticulating.
      - 13 Culms prostrate-spreading, or at least decumbent at the base; fertile subalpine or alpine species .... 33. *Agropyron scribneri*
      - 13 Culms èrect; sterile hybrids of low to midmontane elevs. (discussion of intergeneric hybrids following this key).
        - 14 Awns of the lemma long and spreading, often recurved, mostly (14) 18–37 mm long; internodes of the rachis mostly 7–10 mm long .... X *Agrositanion saxicola*
        - 14 Awns of the lemma usually shorter, 4–17 mm long, up to 35 mm long in the Rocky Mts. of Colo., usually straight and ascending; internodes of the rachis 2.5–6 (7) mm long .... X *Agrositanion saundersii*
  - 10 Plants annual; introduced weeds and cultivated plants.
    - 15 Spikes very short, 0.8–2 cm long; lemmas 5–7.5 mm long .... 34. *Eremopyrum*

15 Spikes larger, (3) 6–17 cm long; lemmas larger, (7) 8–16 mm long.
16 Spikelets cylindrical, fitting into concavities of the rachis .... 35. *Aegilops*
16 Spikelets laterally compressed, the rachis not evidently concave above each joint.
17 Glumes ovate, indurate, (3-) 5- to 7-nerved; spikelets 3- to 5-flowered, rarely 2 .... 36. *Triticum*
17 Glumes narrowly-lanceolate to subulate, not indurate, 1- (3)-nerved; spikelets 2-flowered .... 37. *Secale*

X AGROHORDEUM Camus

Reference: Boyle, W. S., and A. H. Holmgren. A cytogenetic study of natural and controlled hybrids between *Agropyron trachycaulum* and *Hordeum jubatum*. Genetics 40: 539–545. 1955.

X Agrohordeum macounii (Vasey) Lepage
Elymus macounii Vasey, Grasses U.S. 46. 1883, nomen nudum; Bull. Torrey Bot. Club 13: 119. 1886. *Terrellia macounii* Lunell, Amer. Midl. Naturalist 4: 228. 1915. X *Agrohordeum macounii* Lepage, Naturaliste Canad. 79: 242. 1952. (*Macoun,* Great Plains of B.C.)

Macoun wildrye.

X *Agrohordeum macounii* refers to plants which are known to be sterile hybrids of *Agropyron trachycaulum* X *Hordeum jubatum.* It has been collected from Alaska to Minn., s. to Calif., Utah, N.M. and Kansas. A controlled hybrid has been produced using the above parents, and the resulting F_1 is similar to the type specimen and plants found in the field (Boyle & Holmgren, 1955). Macoun wildrye shows intermediate characteristics of the parents with its tardily disarticulating rachis and paired spikelets.

X AGROSITANION Bowden

References:
Bowden, W. M. Taxonomy of intergeneric hybrids of the tribe Triticeae from North America. Canad. J. Bot. 45: 711–724. 1967.
Dewey, D. R. Natural and synthetic hybrids of *Agropyron spicatum* X *Sitanion hystrix.* Bull. Torrey Bot. Club 91: 396–405. 1964.
Stebbins, G. L., Jr., J. I. Valencia, and R. M. Valencia. Artificial and natural hybrids in the Gramineae, tribe Hordeae. I. *Elymus, Sitanion,* and *Agropyron.* Amer. J. Bot. 33: 338–351. 1946.

1 Awns of the lemma long and spreading, often recurved, mostly (14) 18–37 mm long; internodes of the rachis mostly 7–10 mm long .... X *A. saxicola*
1 Awns of the lemma usually shorter, 4–17 mm long, up to 35 mm long in the Rocky Mts. of Colo., straight; internodes of the rachis 2.5–6 (7) mm long .... X *A. saundersii*

X Agrositanion saxicola (Scribn. & Smith) Bowden
Elymus saxicolus Scribn. & Smith in Scribn. U.S.D.A. Div. Agrostol. Bull. 11: 56, pl. 15. 1898. *Agropyron saxicola* Piper, Contr. U. S. Natl. Herb. 11: 148. 1906. X *Agrositanion saxicola* Bowden, Canad. J. Bot. 45: 721. 1967. (*Elmer 554,* "On summit of Mount Chapaca," Okanogan Co., Wash., Aug. 1897.)
Sitanion insulare J. G. Smith, U.S.D.A. Div. Agrostol. Bull. 18: 14. 1899. *Elymus insularis* M. E. Jones, Contr. W. Bot. 14: 20. 1912. (*Watson 1338,* Carrington Island, Great Salt Lake, Utah, June 1869.)
Sitanion lanceolatum J. G. Smith, U.S.D.A. Div. Agrostol. Bull. 18: 20. 1899. (*Rydberg 3381,* Barker, Mont., 17 Aug. 1896.)
Sitanion flexuosum Piper, Erythea 7: 99. 1899. *Agropyron flexuosum* Piper, Proc. Biol. Soc. Wash. 18: 149. 1905. (*Piper 3004,* Wawawai, Wash., 17 June 1899.)

The sterile F_1 hybrids, referable to X *Agrositanion saxicola,* have been collected from Wash. to w. S.D., s. to Sierran Calif. and n. Ariz. The narrow glumes, long divergent awns, and tardily disarticulating rachis suggest a species of *Sitanion* as one of the parents. Dewey (1964) crossed the tetraploid *S. hystrix* with the diploid *Agropyron spicatum* and came up with a sterile triploid hybrid that matched the X *Agrositanion saxicola* type. The types listed in the synonymy are probably hybrids of different parentage. The isotypes of *Sitanion flexuosum* and *S. insulare,* besides having the above characters, show bifid and sometimes trifid glumes and an occasional paired spikelet, suggesting a *Sitanion jubatum* X *Agropyron spicatum* parentage [C. L. Hitchcock, Univ. Wash. Publ. Biol. 17(1): 457. 1969].

X Agrositanion saundersii (Vasey) Bowden
Elymus saundersii Vasey, Bull. Torrey Bot. Club 11: 126. 1884. *Agropyron saundersii* A. S. Hitchc. Proc. Biol. Soc. Wash. 41: 159. 1928. X *Agrositanion saundersii* Bowden, Canad. J. Bot. 45: 720. 1967. (*Vasey,* Veta Pass, Colo., in 1884.)
Elymus saundersii var. *californicus* Hoover, Leafl. W. Bot. 3: 255. 1943. (*Hoover 5568,* "Near head of Cottonwood Creek," Sweetwater Mts., Mono Co., Calif., 29 July 1941.) The common short-awn form of our area.

This sterile hybrid has been collected from Idaho to Wyo., s. to Calif., n. Ariz. and Colo. Stebbins, et al. (1946) attributed the parentage of X *Agrositanion saundersii* to *Agropyron trachycaulum* X *Sitanion jubatum,* and Bowden (1967) suggests the *Sitanion* is *S. hystrix.* The lowermost floret is sometimes reduced to a glume-like structure as in some *S. hystrix.* The type specimen has much longer awns than most of the specimens identified as X *Agrositanion saundersii.*

X ELYSITANION Bowden

References:
Dewey, D. R., and A. H. Holmgren. Natural hybrids of *Elymus cinereus* X *Sitanion hystrix.* Bull. Torrey Bot. Club 89: 217–228. 1962.
Stebbins, G. L., Jr., J. I. Valencia, and R. M. Valencia. Artificial and natural hybrids in the Gramineae, tribe Hordeae. I. *Elymus, Sitanion* and *Agropyron.* Amer. J. Bot. 33: 338–351. 1946.

1 Glumes 12–24 mm long .... X *E. aristatum*
1 Glumes 25–85 mm long .... X *E. hansenii*

X AGROHORDEUM
X AGROSITANION

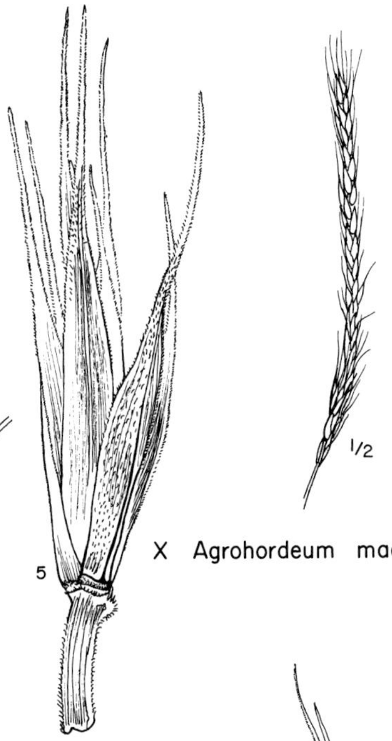

X Agrohordeum macounii

X Agrosilanion saxicola

X Agrositanion saundersii

X ELYSITANION

X Elysitanion aristatum

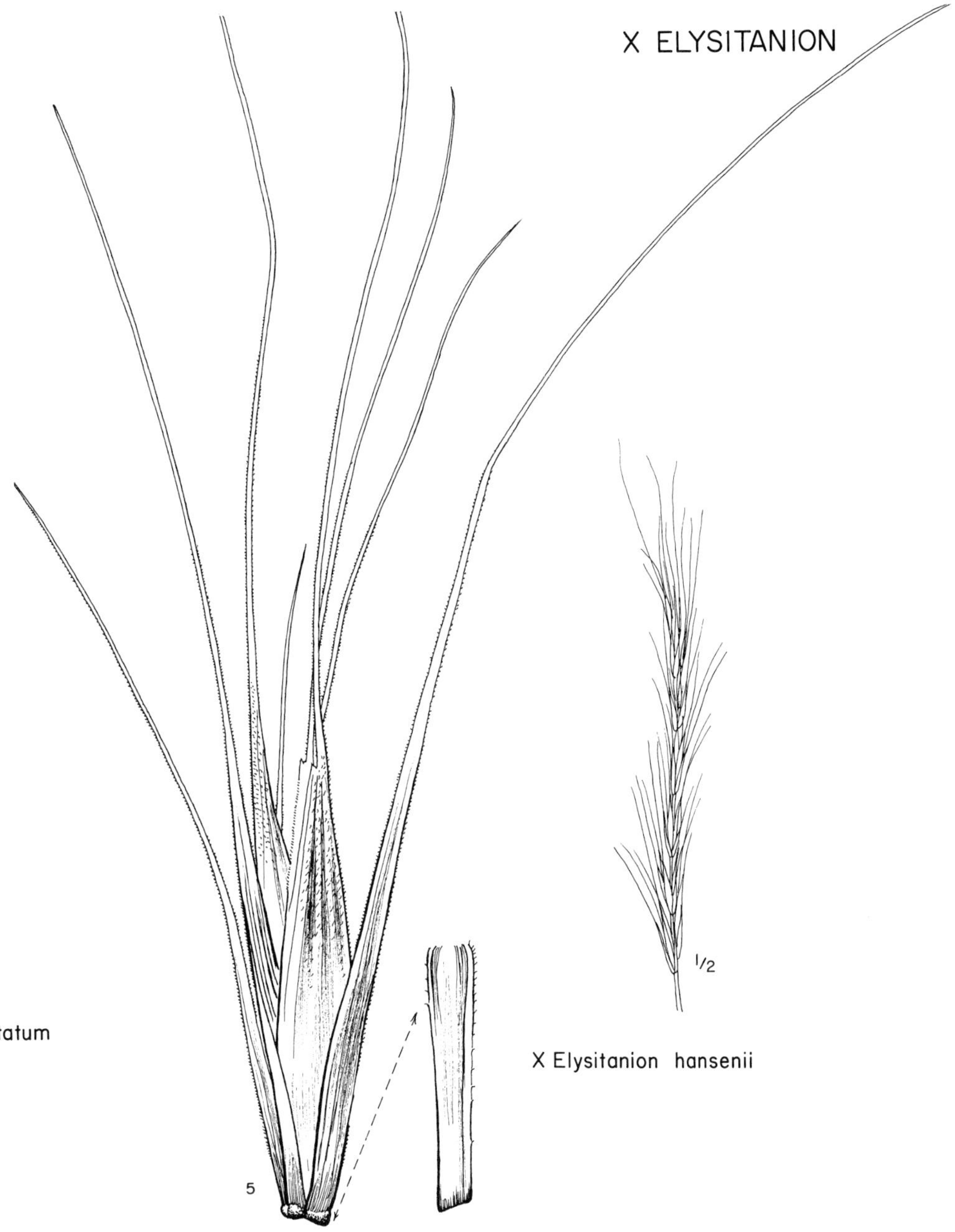

X Elysitanion hansenii

X Elysitanion aristatum (Merr.) Bowden

Elymus aristatus Merr. Rhodora 4: 147. 1902. *Elymus glaucus* var. *aristatus* A. S. Hitchc. in Abrams, Ill. Fl. Pacific States 1: 252. 1923. X *Elysitanion aristatum* Bowden, Canad. J. Bot. 45: 722. 1967. (*Cusick 2712,* "in large clumps, Silver Creek, Harney Co., Oregon," 31 July 1901.)

Sterile plants referable to X *Elysitanion aristatum* have been collected from Wash. to Wyo., s. to Calif. and Nev. Stebbins, et al. (1946) suggested that the sterile *Elymus aristatus* may be the result of *Elymus glaucus* X *Hordeum brachyantherum.* As such it would belong to Bowden's X *Elyhordeum stebbinsianum.* The New York isotype of *Elymus aristatus* has 3 spikelets at each rachis node, with subsessile lateral spikelets, at least in the lower nodes, and the spikelets have as few as 2 to 3 (4) florets which suggest a *Hordeum* parent. Dewey & Holmgren (1962) studied a hybrid involving *Elymus cinereus* X *Sitanion hystrix* in which the sterile F_1 was referable to *Elymus aristatus.* They suggested that the Cusick type could have been an intermediate between the rhizomatous *Elymus triticoides* and *S. hystrix,* since the isotype at Washington State University shows short rhizomes and subulate glumes, neither of which would suggest *Elymus glaucus.*

X Elysitanion hansenii (Scribn.) Bowden

Elymus hanseni Scribn. U.S.D.A. Div. Agrostol. Bull. 11: 56. 1898. *Sitanion hanseni* Scribn. ibid., 18: 20. 1899. X *Elysitanion hansenii* Bowden, Canad. J. Bot. 45: 721. 1967. (*Hansen 1742,* "very scarce in dry open ground," Agric. Station, Amador Co., Calif., 17 June 1896.)

Hansen squirreltail.

X *Elysitanion hansenii* is a common hybrid in nature resulting from a cross between *Elymus glaucus* Buckley and a *Sitanion.* It has been collected from e. Wash. to Wyo., s. to Calif., Nev. and Utah. These hybrids are extremely variable due to introgressive hybridization, but for the most part, will key to this category. The glumes of the type have the breadth of *E. glaucus* and some of them have the lateral cleft of *Sitanion.*

29. ELYMUS L. Wildrye

Cespitose or rhizomatous perennials, the herbage green or glaucous; culms erect and often tall, often rather coarse, hollow; sheaths open; ligules membranous, from rather short to prominent; blades flat to involute, often strongly striate and stiff, often with well developed auricles; inflorescence typically a spike with spikelets in 2s or 3s at the nodes, sometimes the spikelets solitary or some with 4 or 6 or more, the rachis continuous, or in some disarticulating, the spikelets sessile or sometimes pedicellate; spikelets (1-) 2- to 6- (10)-flowered, disarticulating above the glumes and between the florets, the rachilla distorted at the base bringing the florets more or less dorsiventral to the rachis, thus the spikelets do not appear to be 2-ranked; glumes subequal, subulate, narrowly lanceolate or rarely broader, acute to slenderly awned, usually rigid and sometimes indurate and cartilaginous below, sometimes soft, 1- to 5-nerved, the glumes of the contiguous spikelets often parallel and borne in front of the rest of the spikelets simulating an involucre; lemmas about equalling the glumes or longer, rounded on back, faintly 5-nerved, glabrous to pubescent or even soft villous, tapering to a short or long awn, the awns straight or divergent, less frequently awnless; palea well developed, often as long as the body of the lemma; lodicules 2, often very prominent and hairy at the tip; stamens 3, the anthers mostly long; caryopsis linear-oblong, furrowed in back, hairy at the summit, adhering to the palea; $x = 7$.

A genus of about 60 species of arctic and temp. regions of the N. Hemisphere and about 10 species in Chile and Argentina; there are about 20 species in the U.S. (Name from the Greek *elumos,* the ancient Greek name for some kind of millet.)

Many of the species of *Elymus* are fair forage species, some often occurring in dense stands. Some of the rhizomatous species are important in stabilizing shifting sand and are used in binding road embankments.

The genus *Elymus,* as it is traditionally treated, has been found to consist of two phylogenetically distinct elements (Dewey, 1974). The broader-glumed, small-anthered, self-fertilizing, non-rhizomatous species, including *E. canadensis, E. virginicus,* and *E. glaucus,* show close genetic affinity to the self-fertilizing species of *Agropyron* and both species of *Sitanion.* The normally cross-pollinating, large-anthered, subulate-glumed species, represented by *E. flavescens, E. triticoides,* and *E. cinereus* are sometimes referred to the genus *Leymus* Hochst. *Elymus simplex, E. salina,* and *E. ambiguus* also belong with these species. They contain a different pair of genomes from the former group. Further genome analysis will be necessary before a revision is made, so we follow the classical treatment here.

Two sterile hybrids, treated as species of *Elymus* in earlier manuals, are sometimes encountered in our area. These hybrids are of intergeneric origin and are treated as X *Elysitanion aristatum* (Merr.) Bowden and X *Elysitanion hansenii* (Scribn.) Bowden. They are discussed above following the tribal key to the genera (p. 295).

References:

Bowden, W. M. Cytotaxonomy of the species and interspecific hybrids of the genus *Elymus* in Canada and neighboring areas. Canad. J. Bot. 42: 547–601. 1964.

Dewey, D. R. Cytogenetics of *Elymus sibiricus* and its hybrids with *Agropyron tauri, Elymus canadensis* and *Agropyron caninum.* Bot. Gaz. 135: 80–87. 1974.

1 Rachis disarticulating at maturity, sometimes tardily so . . . . 1. *E. junceus*
1 Rachis continuous, disarticulation occurring above the glumes.
  2 Rhizomes present, slender, creeping.
    3 Lemmas hirsute to densely villous with long, yellowish or brownish hairs; plants of loose, sandy soils; Snake River Plains to the Columbia Basin . . . . 2. *E. flavescens*
    3 Lemmas glabrous, scabrous, or rarely sparsely strigose.
      4 Lemmas awnless or awn-tipped, the awn rarely approaching 2.5 mm in length.
        5 Plants with long creeping rhizomes; glumes (4) 7–11 mm long; spikelets usually paired at a rachis node, sometimes solitary; plants principally of valley bottoms in moist meadows or saline soils; widespread throughout the west . . . . 3. *E. triticoides*
        5 Plants mostly with short rhizomes; glumes 1.2–8 (10) mm long or sometimes nearly obsolete; spikelets usually solitary at a rachis node; plants of mt. sides, often in heavy clay soils; sw. Wyo., e. Utah and adjacent Colo., s. to s. Calif., s. Nev. and n. Ariz. . . . . 5. *E. salina*
      4 Lemmas with awns 2.5–8 mm long; spikelets mostly solitary at a rachis node; plants of loose sandy soils; entering our area from Wyo. along the Green River in Daggett Co., Utah . . . . 4. *E. simplex*
  2 Rhizomes wanting, or if present (in some *E. salina* and *E. cinereus*) short and the plants forming clumps and appearing as bunchgrasses.
    6 Glumes subulate, not broadened above the base, the nerves obscure; anthers (3) 4–7.5 mm long.
      7 Spikelets mostly solitary at a rachis node, distant; leaves mostly involute . . . . 5. *E. salina*
      7 Spikelets 2 or more per node (often solitary towards the base and apex in *E. ambiguus*); leaves mostly flat.
        8 Spikes usually slender, 2–5 (6) mm broad; spikelets two per rachis node or solitary towards the base and apex; culms mostly less than 7 dm tall . . . . 6. *E. ambiguus*
        8 Spikes thick, (4) 5–12 mm broad; spikelets in 2s or more per rachis node.
          9 Culms usually less than 7 dm tall; blades 1.5–4 mm broad; glumes 3–6.5 (7) mm long; introduced species . . . . 1. *E. junceus*

ELYMUS

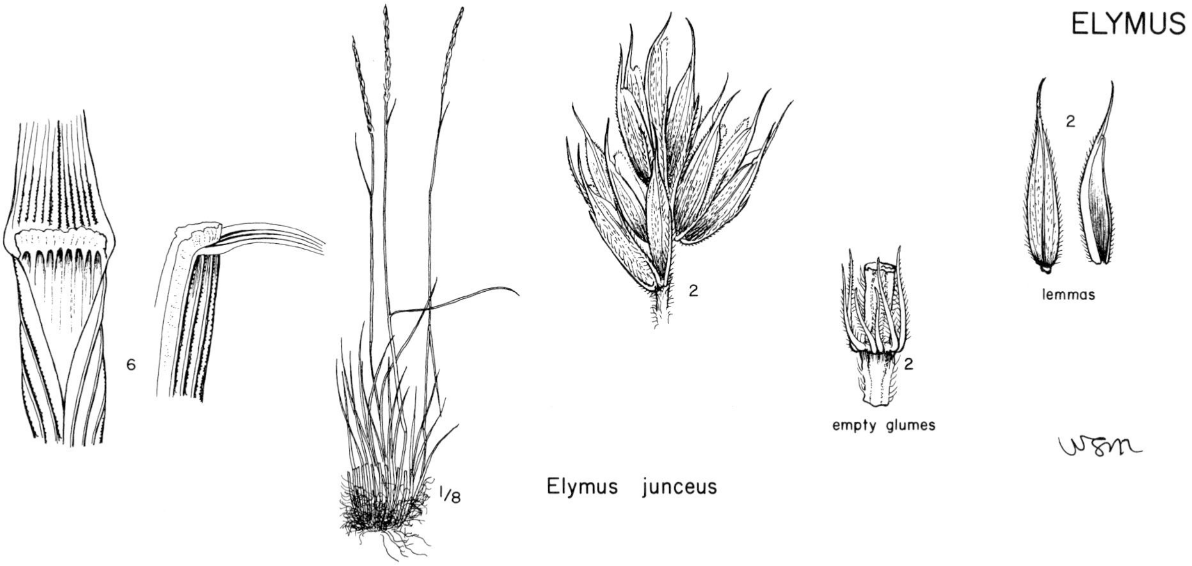

Elymus junceus

9 Culms usually more than 7 dm tall; blades 4.5–15 mm broad; glumes 7–16 mm long; a common native species . . . . . 7. *E. cinereus*

6 Glumes broader, broadened above the base, the nerves evident (except in some *E. virginicus*); anthers 1.5–3 mm long.

10 Awns of the lemmas curving outward at maturity; lemmas scabrous-pubescent . . 8. *E. canadensis*

10 Awns of the lemmas straight or nearly so; lemmas glabrous to crisp-puberulent.

11 Glumes firm, indurate at the base, the bases bowed out, yellowish; lemmas 6–9 mm long . . . . . 9. *E. virginicus*

11 Glumes thin, flat, not indurate at the base, parallel; lemmas 8.5–14 mm long . . . . . 10. *E. glaucus*

1. Elymus junceus Fischer

Elymus junceus Fischer, Zap. Obšč. Isp. Prir. Imp. Moskovsk. Univ. 1: 45, pl. 4. 1806. *Psathyrostachys juncea* Nevski in Komarov, Fl. USSR 2: 714. 1934. *E. cretaceus* Zing ex Nevski in Komarov, ibid., as a synonym. (Lower Volga Steppes, U.S.S.R.)

Russian wildrye.

Densely tufted perennials, clothed at the base with remnants of old sheaths that eventually separate into fibers; culms 4–7 (11) dm tall, glabrous or scabrous below the spike; leaves mostly basal; sheaths glabrous; ligules very short, 0.2–1 mm long, truncate, ciliate; blades flat or involute, 1.5–4 mm broad, glaucous, scabrous, the auricles developed; spikes (4) 7–11 (13) cm long and 4–9 (12) mm broad, erect, the rachis disarticulating between the spikelets; spikelets usually 3 or sometimes 2 per node, (0.7) 8–10 mm long, only 2- (3)-flowered; glumes subequal, short, (3) 4–6.5 (7) mm long, subulate, scabrous or short-hairy, often glabrous and shiny at the base; lemmas small, 6.7–8.5 (10) mm long, scabrous to hirsute, 5- to 7-nerved, tapering into an awn 0.5–2 (3) mm long; anthers 4.2–5.2 mm long; $2n = 14$.

Dry sagebrush slopes; recently introduced from n. Asia and becoming naturalized; known from a few localities in Box Elder Co., Utah, Wyo., Sask. and Man. June–July.

2. Elymus flavescens Scribn. & Smith

Elymus flavescens Scribn. & Smith in Scribn. U.S.D.A. Div. Agrostol. Bull. 8: 8, fig. 1. 1897. *Leymus flavescens* Pilger, Bot. Jahrb. Syst. 74: 6. 1947. (*Suksdorf 916,* "Dry sandy grounds near Columbus," Klickitat Co., Wash., 11 June 1886.)

E. arenicolus Scribn. & Smith in Scribn. U.S.D.A. Div. Agrostol. Circ. 9: 7. 1899. *Leymus arenicola* Pilger, Bot. Jahrb. Syst. 74: 6. 1947. (*Leckenby,* "in sand dunes along the Columbia River," Suferts, Wash., 25 Nov. 1898.)

Yellow wildrye.

Strongly rhizomatous, usually glaucous perennials arising from vertical rhizomes attached to deep, slender, brown, scaly, horizontal rhizomes, the horizontal rhizomes often extending for considerable distances; plants often forming large clumps, thus appearing as a bunchgrass; culms (4) 7–10 (12) dm tall, densely pubescent below the spike and below each node, the nodes dark; sheaths glabrous; ligules about 0.5–1 mm long, erose-ciliolate; blades involute, 4–6 mm broad, firm, glabrous beneath and usually scabrous on the upper surface, the auricles mostly lacking, the collar and auricle areas dark brown or purple; spikes relatively long, 10–20 (25) cm long, erect or somewhat nodding, compound; spikelets mostly in pairs at each node with one of the pair often pedicellate, overlapping or the lower ones more distant, 10–20 mm long, 3- to 4- (6)-flowered; glumes subequal, 8–11 (15) mm long, narrowly-lanceolate to subulate, nerveless or faintly 3-nerved, usually shorter than the first lemma; lemmas large, 10–15 mm long, strongly 5- to 7-nerved, copiously

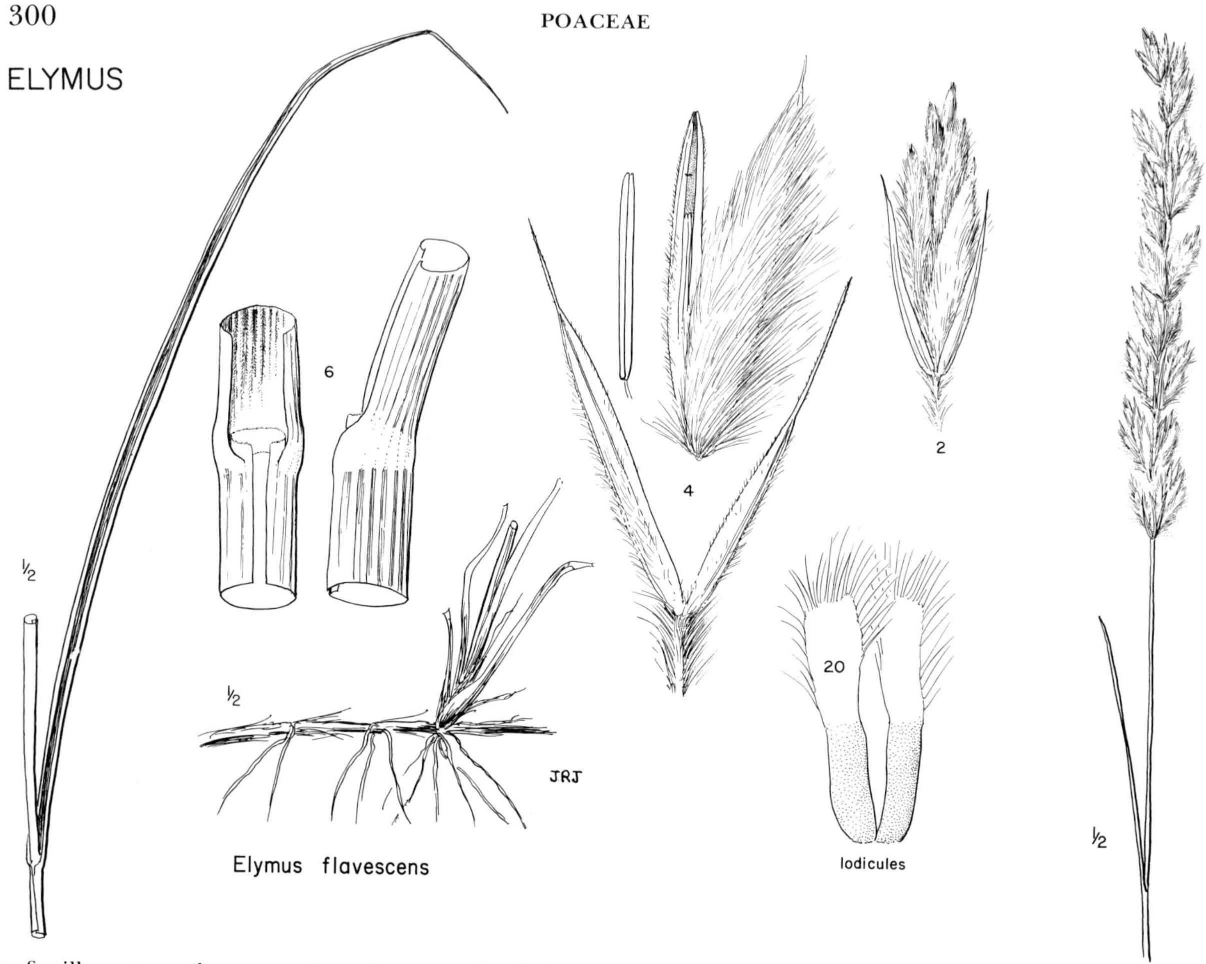

Elymus flavescens

soft-villous to rather sparsely pubescent, the hairs often yellowish or brownish, awnless and sharp-pointed or sometimes with a short awn 0.5–1 (4) mm long; anthers 4–6 mm long; $2n = 28$.

Sand dunes and open sandy flats where often the dominant grass; c. Wash., e. Oregon and the Snake River Plains of Idaho with a disjunct in the Black Hills of S.D. June–Aug.

Elymus flavescens is effective in stabilizing loose, sandy soils.

3. Elymus triticoides Buckley

Elymus triticoides Buckley, Proc. Acad. Nat. Sci. Philadelphia 1862: 99. 1863. *E. condensatus* var. *triticoides* Thurber in S. Wats. Bot. Calif. 2: 326. 1880. *Leymus triticoides* Pilger, Bot. Jahrb. Syst. 74: 6. 1947. (*Nuttall,* "Rocky Mountains.")

E. triticoides var. *pubescens* A. S. Hitchc. in Jepson, Fl. Calif. 1: 186. 1912. (*Elmer 3748,* Griffin, Ventura Co., Calif., July 1902.)

Beardless wildrye, creeping wildrye.

Glaucous or sometimes green perennials from extensively creeping rhizomes; culms 3–7 (12) dm tall; sheaths glabrous, scabrous or puberulent with spreading hairs; ligules very short, 0.2–0.7 mm long, truncate, erose-ciliolate; blades flat or usually involute, 2.5–5 (7) mm broad, rather stiff, scaberulous, sometimes glabrous below, the auricles well developed and often clasping the culm; spikes relatively short, 3.5–7 (9) cm long, erect, loose and open to rather dense, slender, sometimes compound; spikelets paired or sometimes solitary at each node, 10–16 (18) mm long, 3- to 5- (8)-flowered, greenish, brownish to purple; glumes subequal, (4) 7–10 (11) mm long, narrow to subulate, firm, 1- to 3-nerved, mostly scabrous; lemmas small, 6–9 (10) mm long, often shiny and smooth, glabrous or puberulent, sometimes puberulent apically only, faintly to prominently 5- to 7-nerved, rounded on back or keeled toward the tip, awnless or with a short awn 0.5–2.5 mm long; anthers 3.5–5 mm long, purplish; $2n = 28$.

Dry to moist places in valley bottoms, often in saline meadows; Wash. to Mont., s. to Baja Calif., Ariz., N.M. and Trans-Pecos Texas. Late May–Aug.

Beardless wildrye often fails to set seed. Perhaps this is due to the need for the species to be cross pollinated. Large stands may represent plants of a single clone. *Elymus triticoides* is sometimes confused with *Agropyron smithii* when the spikelets occur one at a node. This pitfall will be eliminated if care is taken in observing that *E. triticoides* has exceedingly narrow glumes, and the florets are out of their normal position due to the distorted rachilla that twists about 90°. Two variants seem worthy of recognition and may be separated as follows:

1 Plants glabrous or nearly so throughout; range of the species . . . . var. *triticoides*

1 Plants pubescent throughout, usually with spreading hairs; Wash., Oregon and Nev. . . . . var. *pubescens* A. S. Hitchc.

4. Elymus simplex Scribn. & Williams

Elymus simplex Scribn. & Williams in Scribn. U.S.D.A. Div. Agrostol. Bull. 11: 57, pl. 17. 1898. *E. triticoides* var. *simplex* A. S. Hitchc. Amer. J. Bot. 21: 132. 1934. (*T. A. Williams*

ELYMUS

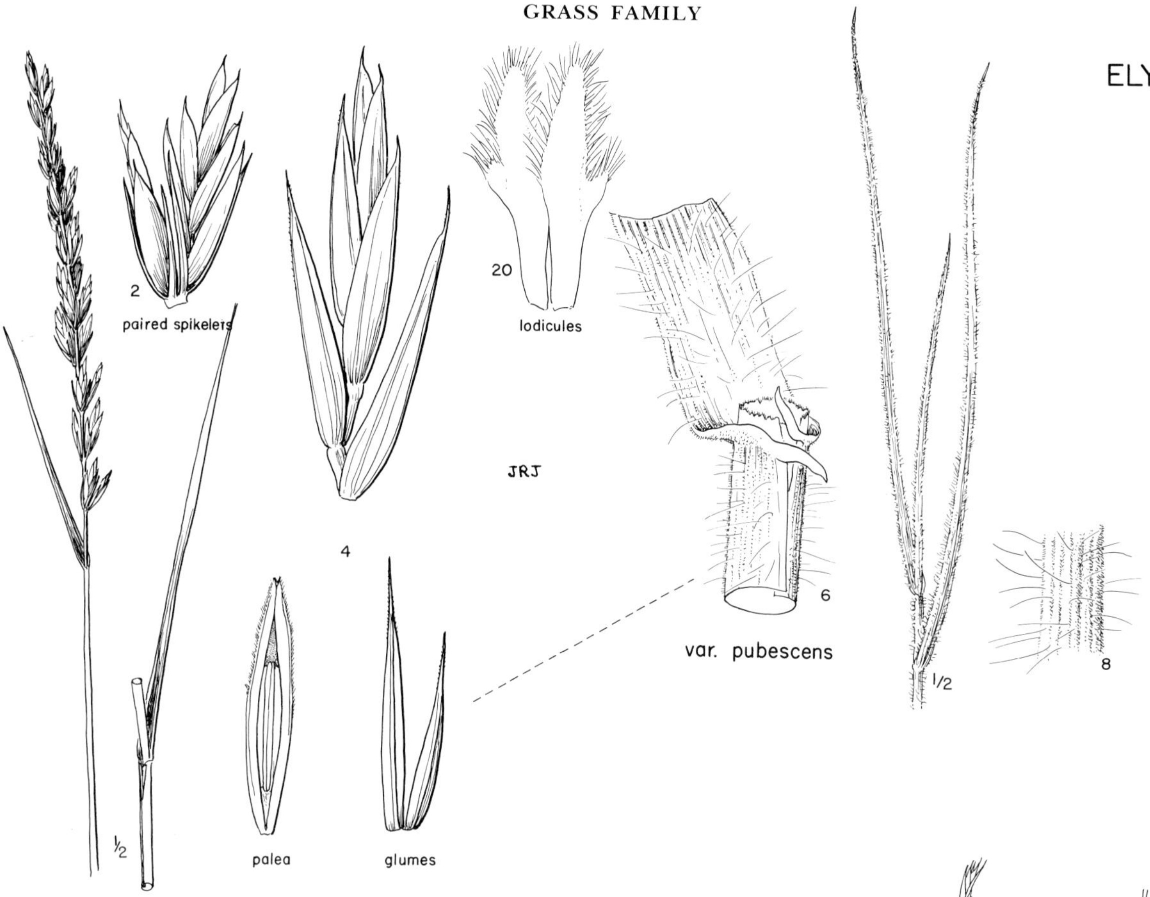

Elymus triticoides

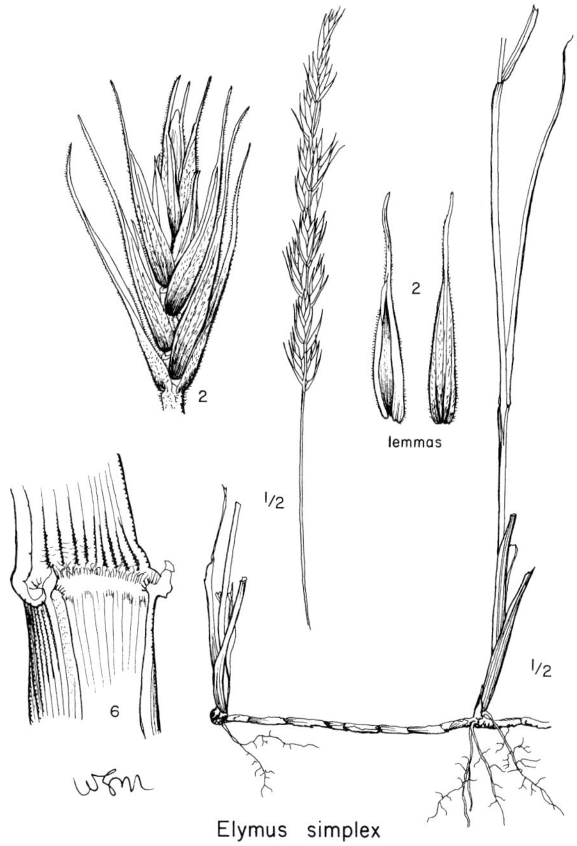

Elymus simplex

ELYMUS

Elymus salina

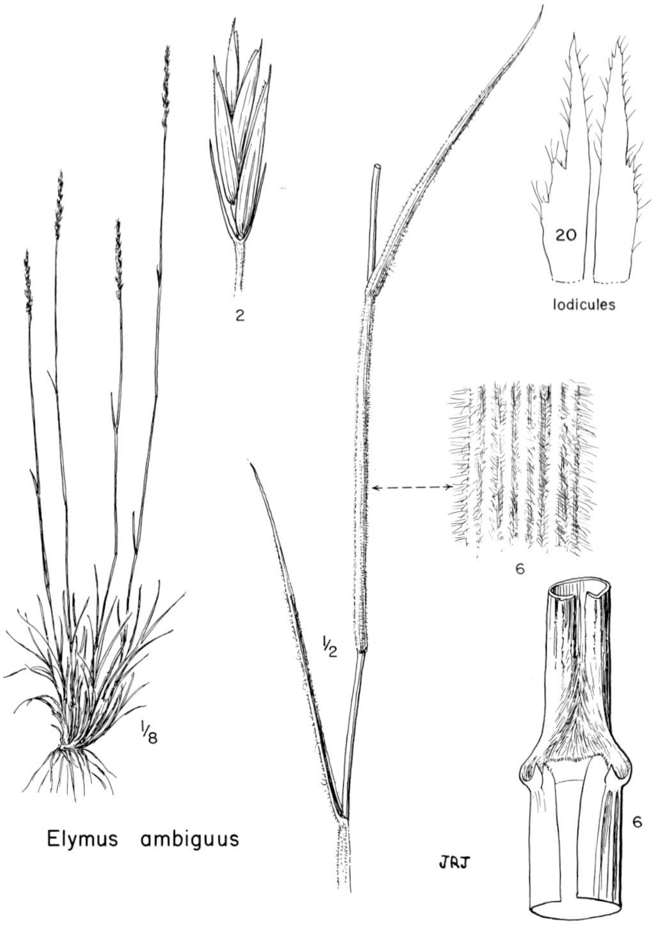

Elymus ambiguus

2334, "banks of Green River," Wyo., 8 July 1897.)
E. simplex var. *luxurians* Scribn. & Williams in Scribn. U.S.D.A. Div. Agrostol. Bull. 11: 58. 1898. (*T. A. Williams 2338,* Green River, Wyo., 8 July 1897.)

Glaucous perennials from extensive creeping rhizomes, the rhizomes up to 5 m long; culms 4–6 (7) dm tall; leaves mostly basal with only a few on the culm; sheaths glabrous and often anthocyanous at the base; ligules very short, about 0.5 mm long, truncate, ciliolate; blades usually involute or more or less flat, 1–3 mm broad, rather rigid, scabrous or sparsely pilose on the upper surface, the auricles often poorly developed or sometimes to 1 mm long, usually anthocyanous; spikes (6) 8–12 (16) cm long, erect, usually rather slender; spikelets usually solitary or sometimes 2 per node, then 1 sessile and 1 on a short pedicel, especially in the lower part of the spike, sometimes remote and not at all overlapping, but usually slightly overlapping, 11–17 (20) mm long, (3-) 4- to 6- (10)-flowered, the rachilla often elongated and quite visible between the florets; glumes subequal, (5.5) 9–14 (16) mm long, subulate to slightly broader at the base, glabrous to scabrous, sometimes with an extra glume at a one spikelet node, nearly nerveless or 1- to 3-nerved; lemmas 7.5–9.5 mm long, usually glabrous and shiny-green to minutely puberulent, rarely with a purplish tinge, nerveless below, becoming faintly 5- (7)-nerved above, tapering into an awn 2.5–8 mm long; anthers 3.6–5 mm long; $2n = 28$.

Moist sandy soils along river banks and sometimes in drifting sands; Utah, along the Green River in Daggett Co., to Wyo. and Colo. June–Aug.

Elymus simplex, as interpreted here, has a rather restricted distribution. Barely entering Utah, it is not otherwise found west of Colorado and Wyoming. Small plants of *E. triticoides* with solitary spikelets are often confused with this species. The type specimens cited above and our collections have spikelets solitary (rarely paired) and awns on the lemmas mostly about 5 mm long. *Elymus simplex* is also confused in herbaria with *E. salina* when it is discovered that *E. salina* usually produces rhizomes. *Elymus simplex* is a grass of sandy bottom lands whereas *E. salina* usually occurs on mountainsides and often in heavy clay soils. Because of the clay soils, *E. salina* is usually collected without rhizomes.

5. Elymus salina M. E. Jones
Elymus salina M. E. Jones, Proc. Calif. Acad. Sci. II. 5: 725. 1895. *E. ambiguus* var. *salina* C. L. Hitchc. Univ. Wash. Publ. Biol. 17(1): 558. 1969. (*Jones 5447,* "Top of Salina Pass, Utah," 16 June 1894.)

Salina wildrye.

Densely tufted perennials, usually bearing short rhizomes; culms 3.5–7 (10) dm tall, glabrous; leaves glabrous, scabrous to pubescent; ligules usually very short, 0.2–0.7 (1.5) mm long, truncate, ciliate; blades flat or usually involute, 2–3 (4) mm broad, often ciliate near the throat, the auricles well developed and often clasping the culm; spikes 5–10 (15) cm long,

2.5–4 (8) mm broad, erect, slender; spikelets usually solitary at each node, somewhat remote, usually slightly overlapping, (9) 12–17 (20) mm long, 2- to 3-(6)-flowered; glumes subequal, (1.2) 4–8 (10) mm long, subulate, glabrous to scabrous, sometimes broader at the base and 3-nerved, sometimes with an extra glume at a one-spikelet node; lemmas (7) 8–10.5 mm long, 5-nerved, at least above, glabrous to scaberulous, the midnerve and awn scabrous, tapering into a short awn 0.2–2 mm long; anthers long, 4–7.5 mm long; $2n = 28, 56$.

Dry slopes, usually in clay or loamy soils, often a dominant grass in the pinyon-juniper and gambel oak communities; sw. Wyo., e. Utah, adjacent Colo. and n. Ariz. (Colorado Plateau), also s. Nev. and s. Calif. June–July.

The name is spelled "salina" in Jones' text and "salinus" in his index. The epithet refers to Salina Pass and the original spelling is maintained here and not considered as a slip of the pen. The type specimen has conspicuous rhizomes as do nearly all of the plants in the topotype collection. Plants collected carefully usually show the presence of rhizomes although in most areas the plants form large tufts and give the appearance of being non-rhizomatous, especially in heavy clay soils. Reports of *E. salina* from Nevada and Idaho are probably based on specimens of *E. ambiguus* var. *salmonis*.

6. Elymus ambiguus Vasey & Scribn.

Elymus ambiguus Vasey & Scribn. Contr. U.S. Natl. Herb. 1: 280. 1893. (*Vasey*, Pen Gulch, Colo., in 1884.)

E. ambiguus var. *salmonis* C. L. Hitchc. Univ. Wash. Publ. Biol. 17(1): 558. 1969. (*Hitchcock & Muhlick 22305*, shale cliff, 9 mi. s. of Challis, along e. side of Salmon River, Custer Co., Idaho, 31 May 1962.)

Tufted, soft-pubescent perennials; culms 3–7 dm tall, soft-pubescent towards the base; leaves mostly basal, densely spreading soft pilose; ligules very short, 0.2–0.5 mm long; blades involute or rarely flat, 2–5 mm broad, the auricles usually absent; spikes 5.5–9 cm long, 2- 5 (6) mm broad, erect, slender; spikelets paired or solitary at each node, 10–18 mm long, 3- to 5- (10)-flowered; glumes subequal, short, 0.8–6.5 (10) mm long, subulate, scabrous, 1-(3)-nerved; lemmas 6–9 (12) mm long, scabrous or puberulent, at least apically, nerveless below and becoming 5-nerved above, tapering into a short awn 0.5–1.5 mm long; anthers 4–5 mm long; $2n = 28$.

Dry sagebrush slopes at low elevs.; (var. *salmonis*) s. Idaho, n. Nev. and w. Utah. Late May–July.

Our plants belong to var. *salmonis* C. L. Hitchc. which is distinct enough to be raised to species rank (E. Jensen, Ph.D. thesis, Utah State University, 1972). The binomial is avoided here in order not to anticipate Jensen's formal publication.

7. Elymus cinereus Scribn. & Merr.

Elymus condensatus (var.) *pubens* Piper, Erythea 7: 101. 1899. *E. condensatus* f. *pubens* St. John, Fl. S.E. Wash. & Adj. Idaho 42. 1937. *E. piperi* Bowden, Canad. J. Bot. 42: 592. 1964. *E. cinereus* var. *pubens* C. L. Hitchc. Univ. Wash. Publ. Biol. 17(1): 561. 1969. (*Piper 2591*, "in strong alkali soil near Yakima City, Wash.," 10 July 1897.)

E. cinereus Scribn. & Merr. Bull. Torrey Bot. Club 29: 467. 1902. (*Purpus 6050*, "dry plains, Pahrump Valley," Nye Co., Nev., in 1898.)

Basin wildrye, Great Basin wildrye.

Robust, densely tufted perennials, often forming clumps up to 1 m across, typically without rhizomes, but rarely producing short, thick rhizomes; culms usually more than 7 dm tall, often as tall as 20 dm, glabrous or more often harshly puberulent, especially so at the nodes; sheaths glabrous to soft hairy with appressed or spreading hairs; ligules relatively long, (1) 2–7 mm, membranous; blades flat or nearly so, 4.5–15 mm broad, firm, strongly nerved, the auricles usually well developed to nearly lacking on some of the leaves; spikes relatively long, (7) 11–20 (23) cm long, 7–12 mm broad, stiff, erect, sometimes becoming compound and branching with up to 6 spikelet bearing nodes on a branch, often interrupted below; spikelets usually 3–6 per node, occasionally only paired, (9) 11–20 mm long, (2-) 3- to 5-flowered; glumes subequal, 7–13 (16) mm long, narrow, nearly subulate, tapered, often as long as the spikelet, scabrous on the angles, sometimes slightly bowed apart at the base; lemmas (7) 8–10 (12) mm long, hirsutulous, rarely glabrous, usually nerveless below, becoming 5- to 7-nerved above, awnless or more often with a short awn up to 5 mm long; anthers (3) 4.5–6.2 mm long; $2n = 28, 56$.

Dry places in valley bottoms, along roadsides, streams and in gulleys, among sagebrush to the edges of wooded areas; B.C. to Sask., s. to Calif., n. Ariz. and Colo., reported as far e. as Minn. June–July (–Aug.).

The seeds of Basin wildrye were sometimes used for food by Indians. *Elymus cinereus* is undoubtedly the grass that the old-time cowboys referred to as being stirrup high. Basin wildrye has often suggested the local place name, Ryegrass Flat, as the name is often encountered in our area. It is a coarse grass but nevertheless utilized to some extent by all classes of livestock. It is also important in stabilizing soil and holding embankments. Black sclerotia of ergot (*Claviceps purpurea*) commonly protrude from the spikelets of this species. If occurring in abundance, this fungus should be considered as an important poisonous species that may cause abortion in animals or even death to adult animals.

Elymus cinereus was lumped with *E. condensatus* Presl in some earlier manuals. *Elymus condensatus* differs in having thicker culms, broader leaves, thicker spikes with more spikelets per node and by being restricted to the coastal area of California. Two varieties of *E. cinereus* have been recognized on the following characters:

1 Sheaths glabrous to finely puberulent; blades mostly glabrous to scabrous; the wide ranging plant of our area .... var. *cinereus*

1 Sheaths with appressed to spreading hairs; blades pubescent; occasionally encountered across the n. part of our area ... . var. *pubens* (Piper) C. L. Hitchc.

8. Elymus canadensis L.

Elymus canadensis L. Sp. Pl. 83. 1753. *Hordeum patulum* Moench, Meth. Pl. Horti Bot. & Agri Marburg. 199. 1794. *Hordeum canadense* Aschers. & Graebner, Syn. Mitteleur. Fl. 2: 745. 1902. *Terrellia canadensis* Lunell, Amer. Midl. Naturalist 4: 228. 1915. *Clinelymus canadensis* Nevski, Izv. Bot. Sada Akad. Nauk SSSR 30: 650. 1932. (*Kalm*, "Habitat in Canada.")

Sitanion brodiei Piper, Erythea 7: 100. 1899. (*Brodie*, Kellys Orchard, Bishop's Bar, Snake River, Whitman Co., Wash., 7 July 1898.)

Canada wildrye.

Erect, open, tufted perennials; culms 8–15 dm tall; leaves scattered along the culm; sheaths glabrous or sometimes hirsute to pilose; ligules 0.2–1.2 (2) mm long, truncate, ciliolate; blades flat, 3–10 (15) mm broad, scabrous above, the auricles well developed, usually about 1–2 mm long and often clasping the culm; spikes 7–17 (25) cm long, nodding to droop-

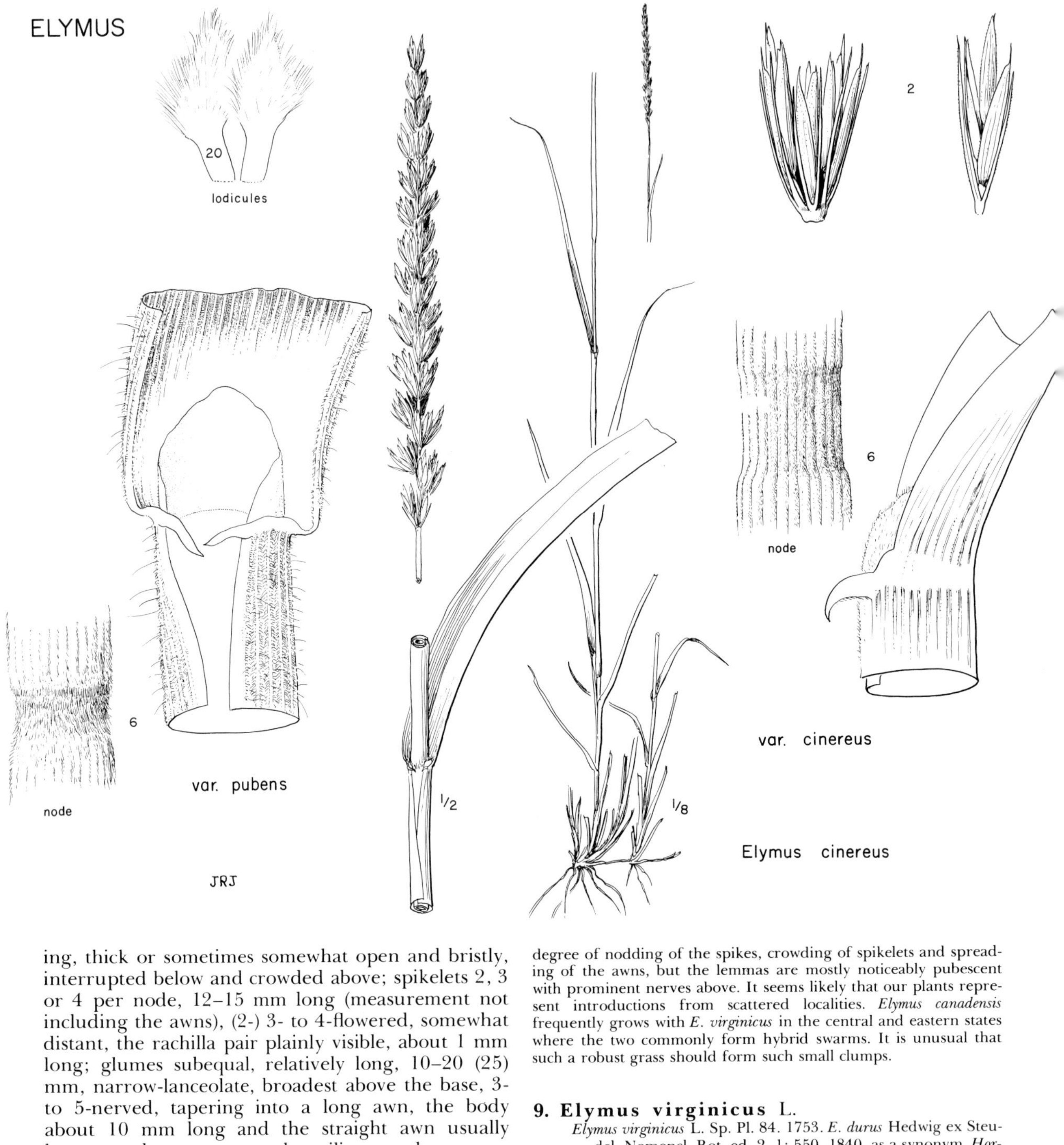

ing, thick or sometimes somewhat open and bristly, interrupted below and crowded above; spikelets 2, 3 or 4 per node, 12–15 mm long (measurement not including the awns), (2-) 3- to 4-flowered, somewhat distant, the rachilla pair plainly visible, about 1 mm long; glumes subequal, relatively long, 10–20 (25) mm, narrow-lanceolate, broadest above the base, 3- to 5-nerved, tapering into a long awn, the body about 10 mm long and the straight awn usually longer, scabrous to somewhat ciliate on the nerves; lemmas 8.5–14 mm long, strongly scabrous-pubescent, 5- to 7-nerved above, gradually tapering into a long, spreading, flexuous, scabrous awn 16–30 (35) mm long; anthers relatively small, 2.5–3 mm long; $2n = 28, 42$.

Streambanks and dry to moist fields and meadows, a common weed along irrigation ditches and roadsides; s. Alaska to N.B., s. to n. Calif., Ariz., N.M., Texas, Mo. and S.C. (June–) July–Aug.

Our plants are extremely variable as to size, width of blades, degree of nodding of the spikes, crowding of spikelets and spreading of the awns, but the lemmas are mostly noticeably pubescent with prominent nerves above. It seems likely that our plants represent introductions from scattered localities. *Elymus canadensis* frequently grows with *E. virginicus* in the central and eastern states where the two commonly form hybrid swarms. It is unusual that such a robust grass should form such small clumps.

9. **Elymus virginicus** L.

Elymus virginicus L. Sp. Pl. 84. 1753. *E. durus* Hedwig ex Steudel, Nomencl. Bot. ed. 2. 1: 550. 1840, as a synonym. *Hordeum virginicum* M. Schenck. Bot. Jahrb. Syst. 40: 109. 1907. *Terrellia virginica* Lunell, Amer. Midl. Naturalist 4: 228. 1915. *Terrella virginica* Nevski, Izv. Bot. Sada Akad. Nauk SSSR 30: 639. 1932. ("Habitat in Virginia.")

E. virginicus var. *submuticus* Hook. Fl. Boreali-Amer. 2: 255. 1840. *E. submuticus* B. B. Smyth & L. C. R. Smyth, Trans. Kansas Acad. Sci 25: 99. 1913. *Terrellia virginica* var. *submutica* Lunell, Amer. Midl. Naturalist 4: 228. 1915. *E. virginicus* f. *submuticus* R. W. Pohl, Amer. Midl. Naturalist 38: 549. 1947. (*Drummond,* Cumberland House Fort, Sask., in 1829.)

ELYMUS

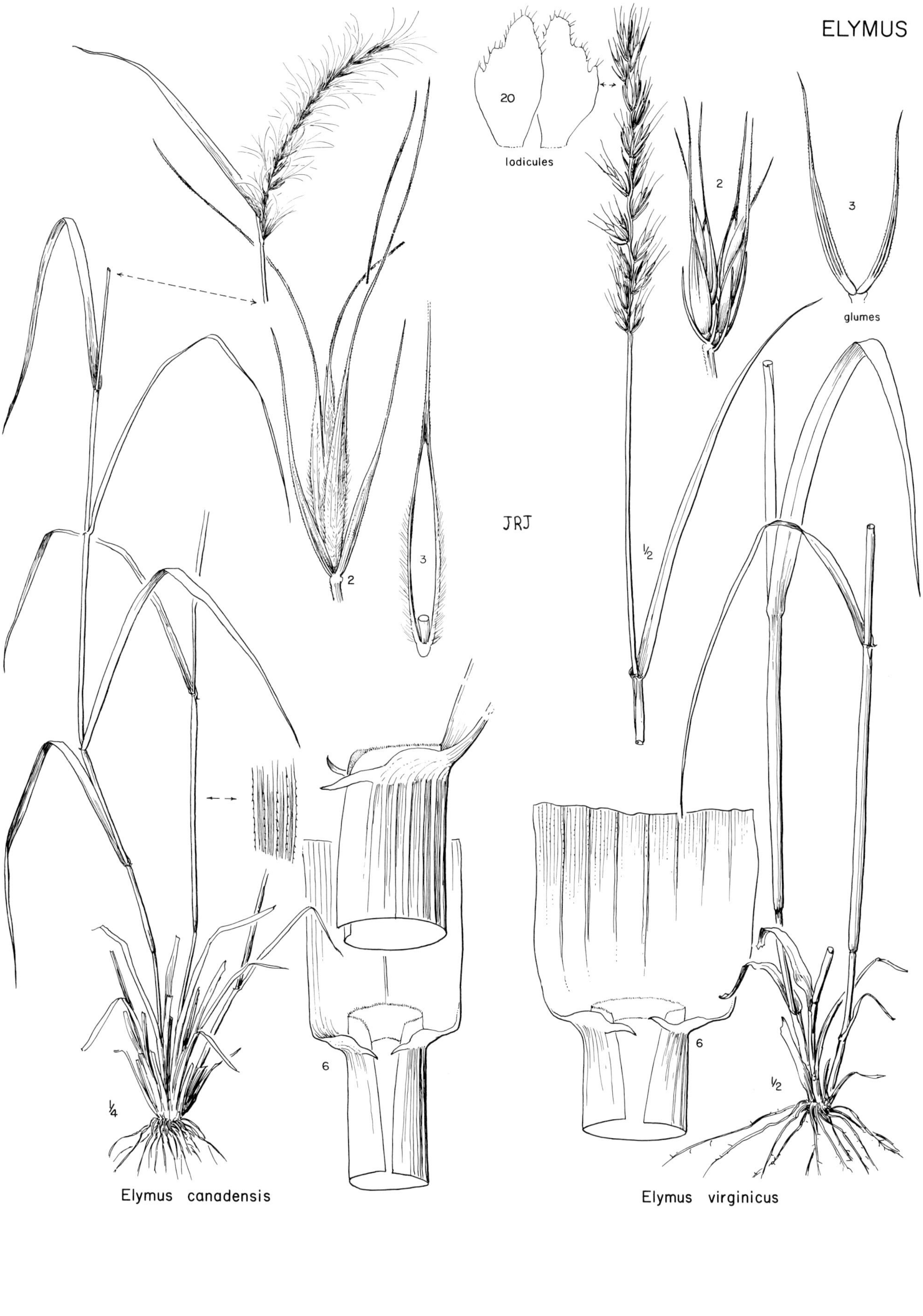

Elymus canadensis

Elymus virginicus

Virginia wildrye.

Tufted perennials; culms 6–12 dm tall; sheaths glabrous; ligules very short, 0.2–0.7 mm long, ciliolate; blades flat, (2) 4–10 mm broad, scabrous on both sides, the auricles well developed, up to 1 mm long or sometimes lacking; spikes 4–12 (16) cm long, erect, rigid, exserted or sometimes slightly included in the upper sheath; spikelets mostly 2 per node, mostly 11–12 mm long, (1-) 4- (5)-flowered; glumes subequal, relatively long, 7–15 (22) mm, firm, narrow-lanceolate or sometimes very narrow, broadest above the base, 2- to 5-nerved, bowed out at the indurate, yellow base, exposing the base of the enclosed florets, ciliolate on the nerves, tapering to an awn tip; lemmas small, 6–9 mm long, glabrous to crisp puberulent, nerveless below, scabrous and becoming 5-nerved above the middle, long-awned or merely awn-tipped in ours with awns only 0.5–2.5 mm long; anthers relatively small, 1.5–3 mm long, very fragile; $2n = 28$.

Mostly along irrigation ditches in our area where it is found in n. Utah; Alta. to Newfl., s. throughout the contiguous states, except in Calif. and Nev. July–Sept.

Our plants have glumes and lemmas nearly awnless to short-awned and are referable to the var. *submuticus* Hook.

10. Elymus glaucus Buckley

Elymus glaucus Buckley, Proc. Acad. Nat. Sci. Philadelphia 1862: 99. 1863. *E. sibiricus* var. *glaucus* Ramaley, Minnesota Bot. Stud. 9: 112. 1894. *Terrellia glauca* Lunell, Amer. Midl. Naturalist 4: 228. 1915. *Clinelymus glaucus* Nevski, Izv. Bot. Sada Akad. Nauk SSSR 30: 648. 1932. (*Nuttall,* Columbia River, Oregon, 1835.)

E. americanus Vasey & Scribn. ex Macoun, Cat. Canad. Pl. 4: 245. 1888, nomen nudum; Vasey & Scribn. ex Cassidy, State Agric. Coll. Agric. Exp. Sta. Bull. 12: 57. 1890. (Arapahoe Pass, Colo.)

E. glaucus var. *jepsoni* Davy in Jepson, Fl. W. Middle Calif. 79. 1901. *E. glaucus* f. *jepsoni* St. John, Fl. S.E. Wash. & Adj. Idaho 42. 1937. *E. glaucus* subsp. *jepsonii* Gould, Madroño 9: 126. 1947. (*Jepson,* Napa Valley, Calif.)

Blue wildrye.

Green to glaucous, loose to densely tufted perennials, sometimes short stoloniferous; culms (6) 8–14 dm tall; leaves scattered along the culm; sheaths glabrous, retrorsely puberulent or pilose, the collar often purple; ligules short, 0.3–1 mm long, truncate, erose-ciliolate or sometimes entire; blades flat to sometimes slightly involute, 4–12 mm broad, mostly scabrous or sparsely pilose, sometimes glabrous below, the auricles mostly well developed, about 2 mm long and clasping the culm; spikes 6–16 cm long (measurement not including the awns), erect, somewhat loose or interrupted below and overlapping above or dense throughout, long-exserted; spikelets mostly 2 per node, 10–16 mm long, 2- to 3- (4)-flowered; glumes subequal, (6.5) 9–14 (19) mm long, narrow-lanceolate, broadest above the base, 3- to 5-nerved, almost parallel and concealing the base of the enclosed florets, glabrous to scabrous, tapering into a short awn, sometimes slightly indurate at the base but membranous above; lemmas 8.5–12 (14) mm long, 5-nerved above, glabrous to scabrous on the nerves, tapering into a long, slender and mostly straight, rarely spreading awn (10) 14–22 (30) mm long; anthers relatively small, 1.7–3 mm long; $2n = 28$.

Open woods and moist meadows to dry hillsides at middle elevs.; common from Alaska to Sask., s. to Calif., n. Ariz. and N.M., less common from Ont. to w. N.Y. and s. to Ark. July–Aug.

Plants of this taxon are often misidentified in herbaria as *Agropyron trachycaulum* var. *unilaterale*. *Elymus glaucus* has the *Elymus* characters of 2 spikelets per node, the narrow glumes and the distorted rachilla. The leaf blades are also much broader in blue wildrye. However, there is a bewildering array of variation in this species. Some biotypes seem to cross readily with *Sitanion hystrix* while others do not. It seems possible that plants referred to *Elymus glaucus* have had different origins and that we are here dealing with plants of a polyphyletic origin, i.e., some of these look-alikes may have come from different parent combinations than others. The name X *Elysitanion hansenii* (Scribn.) Bowden has been used for *Elymus glaucus* X *Sitanion hystrix* hybrids. The following variants are recognizable:

1 Sheaths and blades hirsute or pilose with spreading to somewhat retrorse hairs; B.C. to Mont., s. to Calif., Nev. and Idaho .... var. *jepsonii* Davy
1 Sheaths and blades glabrous to scabrous; range of the species .... var. *glaucus*

30. SITANION Raf. Squirreltail

Cespitose perennials; culms mostly erect, hollow; sheaths glabrous to densely villous, open; ligules membranous, short, truncate, entire or lacerate; blades flat and up to 6 mm broad or folded to involute, glabrous or glaucous to pubescent, sometimes densely villous, usually with well developed auricles; inflorescence a spike with (1) 2 (4) sessile spikelets per node, the number often variable throughout the spike, at maturity the rachis readily disarticulating above each node and the rachis segment falling with the spikelet pair, the spreading awns mechanically separating the inflorescence into 1-node units; spikelets borne flatwise to the rachis, 1- to 6-flowered, some of the florets sometimes sterile and the lemma reduced to glume-like structures; glumes subulate to narrowly lanceolate, cartilaginous at the base, 1- to 3-nerved, entire, usually bifid or deeply cleft into 1 to several, scabrous, elongate, straight or flexuous awns, the glume of contiguous spikelets often parallel and borne in front of the rest of the spikelets simulating an involucre; lemmas firm, lanceolate, rounded on back, usually faintly 3- to 5-nerved, the mid-nerves extending into a scabrous, straight or curved awn, the lateral nerves sometimes extended into awn-like bristles, glabrous or scaberulous-pubescent; palea firm, well developed, subequal to the body of the lemma, the 2 nerves sometimes extending into bristles up to 5 mm long; lodicules 2, oblanceolate, often ciliate; stamens 3; $x = 7$.

A genus of 2 or 3 species of w. temp. N. Amer. (Name from the Greek *sitos,* meaning a grain.)

It has been known for some time now that *Sitanion* has very close affinities with *Elymus* and *Agropyron*. The species of *Sitanion* are known to hybridize freely with certain members of these genera, and the sterile hybrids have been named. The putative *Elymus* and

ELYMUS

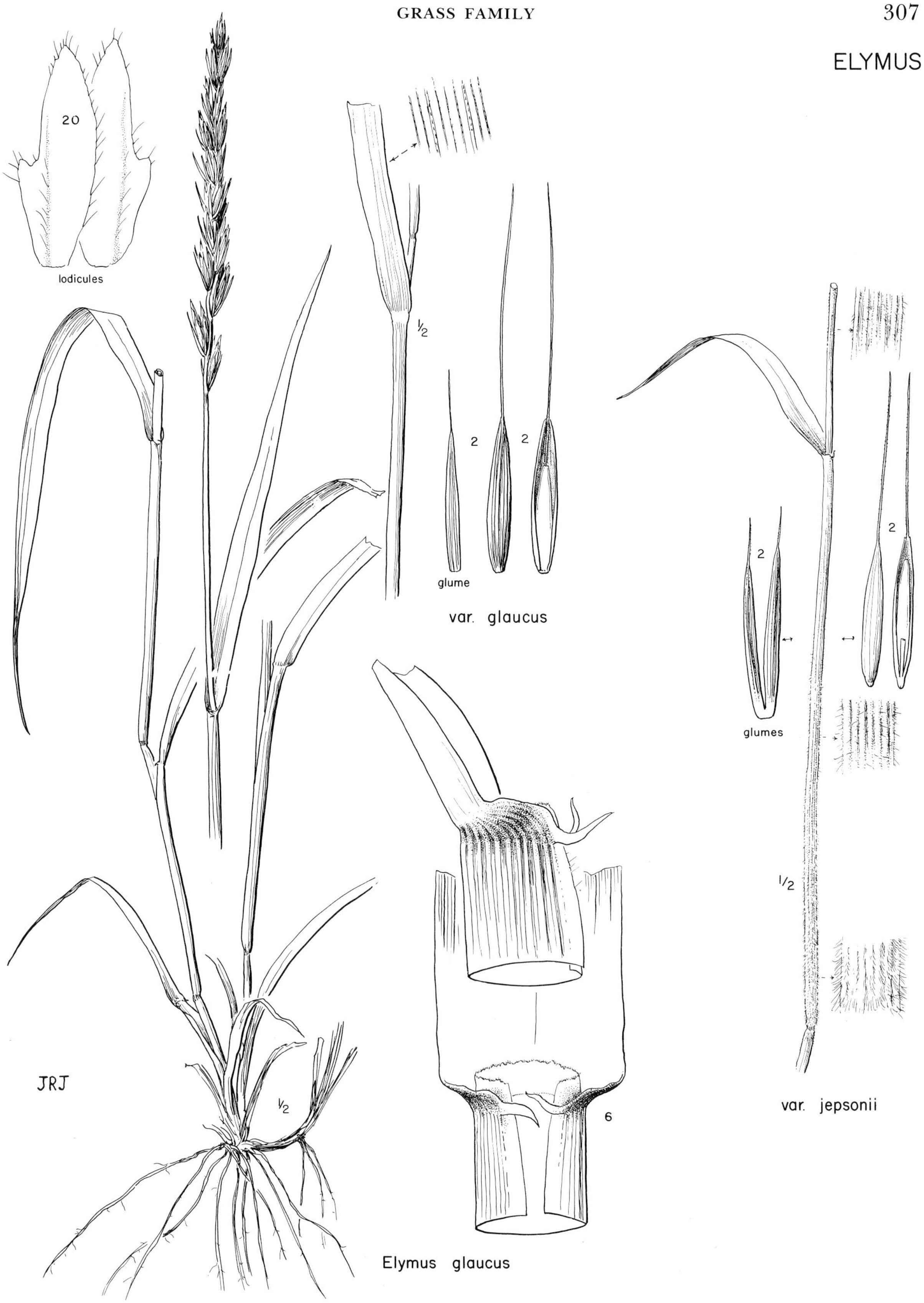

Elymus glaucus

Agropyron parents and intergeneric names of the hybrids involving *Sitanion* spp. are as follows: *E. glaucus* Buckley → X *Elysitanion hansenii* (Scribn.) Bowden; *E. cinereus* Scribn. & Merr. → X *Elysitanion aristatum* (Merr.) Bowden; *A. trachycaulum* (Link) Malte → X *Agrositanion saundersii* (Vasey) Bowden; and *A. spicatum* (Pursh) Scribn. & Smith → X *Agrositanion saxicola* (Scribn. & Smith) Bowden. Since these sterile hybrids are sometimes encountered in our area, they are keyed and discussed in the tribal treatment above (p. 294). Studies of these hybrids have shown that *Sitanion* shares similar genomes with both *Elymus* and *Agropyron*. Wilson (1963) suggests the possibility that *Sitanion* may have originated either as a direct amphiploid between a diploid *Elymus* and a diploid *Agropyron*, or as a secondary product of hybridization between two original tetraploids of the two genera.

The treatment as presented here recognizes 2 species rather than 5 as recognized by Wilson (1963) or 23 as recognized by Smith (1899). It would even seem reasonable to reduce the whole genus to a single species with 3 or 4 varieties.

References:
Smith, J. G. A synopsis of the genus *Sitanion*. U.S.D.A. Div. Agrostol. Bull. 18: 5–21. 1899.
Wilson, F. D. Revision of *Sitanion* (Triticeae, Gramineae). Brittonia 15: 303–323. 1963.

1 Glumes entire or 2-cleft, rarely 3-cleft with 2 short lateral awns; auricles mostly less than 1 mm long and often so inconspicuous as to appear wanting .... 1. *S. hystrix*
1 Glumes 3- to many-cleft; auricles mostly apparent, some approaching 1 mm in length .... 2. *S. jubatum*

1. Sitanion hystrix (Nutt.) J. G. Smith

Aegilops hystrix Nutt. Gen. N. Amer. Pls. 1: 86. 1818. *Polyanthrix hystrix* Nees, Ann. Nat. Hist. 1: 284. 1838 (misapplied to *S. jubatum*). *Sitanion hystrix* J. G. Smith, U.S.D.A. Div. Agrostol. Bull. 18: 15. 1899. *Elymus difformis* Nutt. ex J. G. Smith, ibid., as a synonym of *S. hystrix*. *Chretomeris trichoides* Nutt. ex. J. G. Smith, ibid., as a synonym of *S. hystrix*. *Elymus hystrix* M. E. Jones, Contr. W. Bot. 14: 20. 1912; not L. 1753. (*Nuttall*, "On the arid plains of the Missouri," in 1811.)

S. elymoides Raf. J. Phys. Chim. Hist. Nat. Arts 89: 103. 1819. *Elymus sitanion* J. A. Schultes, Mantissa 2: 426. 1824. *Elymus elymoides* Swezey, Nebraska Loc. Fl. Pl. Herb. Doane Coll. 15. 1891. *Hordeum elymoides* M. Schenck, Bot. Jahrb. Syst. 40: 109. 1907. (*Bradbury*, Missouri River, in 1811.) = var. *hystrix*.

S. californicum J. G. Smith, U.S.D.A. Div. Agrostol. Bull. 18: 13. 1899. *S. minus* var. *californicum* I. M. Johnston, Pl. World 22: 81. 1919. *S. rigidum* var. *californicum* Smiley, Univ. Calif. Publ. Bot. 9: 99. 1921. *S. hystrix* var. *californicum* F. D. Wilson, Brittonia 15: 316. 1963. (*Parish 3295*, Bear Valley, San Bernardino Mts., Calif., 23 June 1894.)

S. rigidum J. G. Smith, U.S.D.A. Div. Agrostol. Bull. 18: 13. 1899. (*Allen 178*, Goat Rocks, Cascade Mts., Yakima Co., Wash., 5 Aug. 1896.) = var. *californicum*.

S. glabrum J. G. Smith, U.S.D.A. Div. Agrostol. Bull. 18: 14. 1899. *Elymus glaber* Davy in H. M. Hall, Univ. Calif. Publ. Bot. 1: 57. 1902. (*Coville & Funston 914*, near Crystal Spring, Coso Mts., Inyo Co., Calif., 12 June 1891.) = var. *hystrix*.

S. cinereum J. G. Smith, U.S.D.A. Div. Agrostol. Bull. 18: 14. 1899. (*Tracy 222*, Reno, Washoe Co., Nev., in 1887.) = var. *hystrix*.

S. montanum J. G. Smith, U.S.D.A. Div. Agrostol. Bull. 18: 16. 1899. (*Rydberg 3091*, Spanish Creek, Mont., 15 July 1896.) = var. *californicum*.

S. brevifolium J. G. Smith, U.S.D.A. Div. Agrostol. Bull. 18: 17, pl. 3. 1899. *Elymus brevifolius* T. Holm, Bot. Gaz. 46: 427. 1908. *S. hystrix* var. *brevifolium* C. L. Hitchc. Univ. Wash. Publ. Biol. 17(1): 701. 1969. (*Toumey 797*, Tucson, Pima Co., Ariz., in 1892.)

S. longifolium J. G. Smith, U.S.D.A. Div. Agrostol. Bull. 18: 18. 1899. *Elymus longifolius* Gould, Brittonia 26: 60. 1974. (*Shear 1213*, "among rocks on open sides of canyon near Silverton," San Juan Co., Colo., 4 Aug. 1897.) = var. *brevifolium*.

S. hordeoides Suksd. Werdenda 1(2): 4. 1923. *S. hystrix* var. *hordeoides* C. L. Hitchc. Univ. Wash. Publ. Biol. 17(1): 701. 1969. (*Suksdorf 8705*, "on dry prairie north of Spangle," Spokane Co., Wash., 29 June 1916.)

Squirreltail.

Loosely to densely tufted perennials, subglabrous or puberulent to white villous throughout; culms erect to spreading, 1–4.5 (6.5) dm tall; sheaths glabrous to pilose; ligules short, mostly less than 0.5 mm long, truncate; blades flat to folded or involute, 1–4 (6) mm broad, hirsute to villous, sometimes glabrous below, the auricles inconspicuous, sometimes up to 0.5 (1) mm long, often purplish; spikes 3–12 (15) cm long (measurement excluding the awns), from very compact to loose and open, often included in the upper sheaths or in robust specimens entirely exserted, often at maturity with only the lower part remaining, the rachis disarticulating and pulled apart by the spreading awns; spikelets mostly in pairs, sometimes in 3s, or often solitary in the upper part of the spike, mostly 1- to 6-flowered, the placement of reduced florets varying, sometimes all fertile, but sometimes the lower one sterile and glume-like, the awns of the glumes and lemmas 20–70 (120) mm long, varying from green to red or often purplish; glumes long, 35–85 mm long, subulate or sometimes broader, extending into slender, spreading, scabrous awns, entire or sometimes divided into 2 (3) short, unequal segments; lemmas 8–10 mm long, glabrous to strongly puberulent, faintly 3- to 5-nerved with the main nerve extending into a slender to stout, divergent or straight to flexuous awn, with 2 of the lateral nerves also extended into bristles nearly 10 mm long; palea subequal to the body of the lemma, 2-nerved, the nerves often extending as bristles; anthers about 2 mm long; $2n = 28$.

From the dry sagebrush deserts to the alpine tundra in many different habitats; B.C. to S.D., s. throughout the mt. and plains states and into Mex. as far as Hidalgo. Late May–Aug.

Squirreltail is often the dominant grass in pinyon-juniper areas where overgrazing has eliminated the more favorable range grasses. It is palatable early in the season, but later in the year the disarticulating spikes with pointed rachis joints and long, antrorsely scabrous awns cause mouth, eye and ear injury to grazing animals.

The varieties of *S. hystrix* have been accorded specific rank or recognized as one to several species in the past. Some morphological races of var. *brevifolium* are among the most robust plants of *Sitanion* and show close affinities with *Elymus*. Variety *hordeoides* shows characters in its triad of spikelets, such as sterility in the lateral pairs, that suggest a *Sitanion* X *Hordeum* ancestry. Except for the long awns and very short rachis internodes, var. *hordeoides* resembles X *Elysitanion aristatum*. The following key will more or less distinguish the varieties that have been recognized in our area.

1 Spikelets usually 2 at each node of the rachis, if more then some florets of the lateral spikelets fertile.
2 Lowermost floret of one or both spikelets at each node sterile and reduced to a glume-like structure.
3 Glumes tending to be 2 cleft, at least one of the glumes at each node; awns of the glumes longer than those of the lemmas; low lands from s. B.C. and sw. Mont., s. through e. Oregon, Idaho and sw. Wyo. to s. Calif., nw. Ariz., Utah and w. Colo. .... var. *hystrix*

SITANION

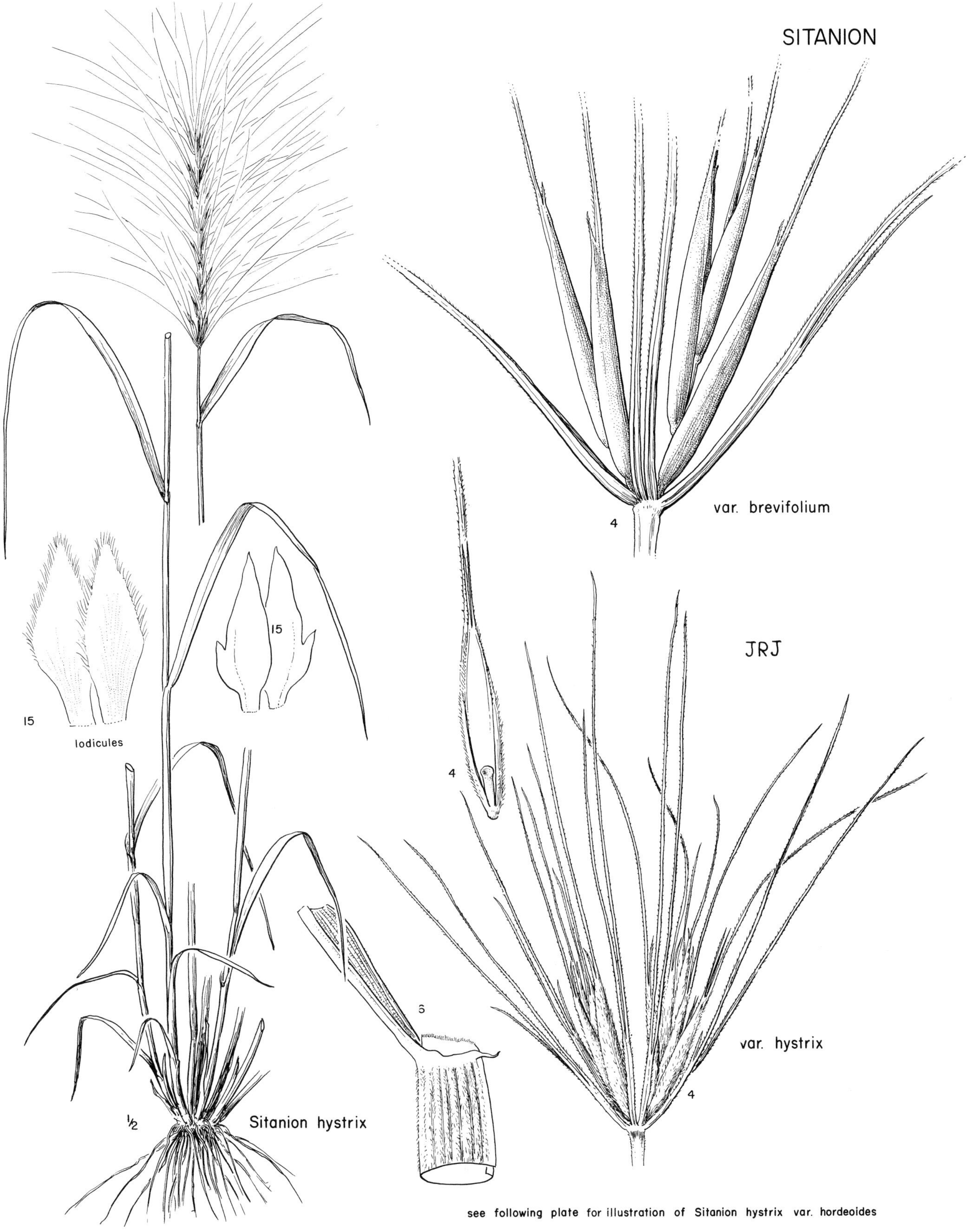

see following plate for illustration of Sitanion hystrix var. hordeoides

3 Glumes entire; awns of the lemmas longer than those of the glumes; mid-elevs. to arctic-alpine places from B.C. to Mont., s. to Calif., Nev. and Utah .... var. *californicum* (J. G. Smith) F. D. Wilson

2 Lowermost floret fertile, not at all reduced; an ecologically diverse var. ranging from the deserts to the highest mts.; Oregon, s. Mont. and S.D., s. to Calif., Ariz., N.M. and Texas........... var. *brevifolium* (J. G. Smith) C. L. Hitchc.

1 Spikelets 3 at each node of the rachis, the florets of the central spikelet fertile, those of the lateral spikelets rudimentary; low lands of e. Wash., e. Oregon and adj. Idaho, s. to n. Calif. and n. Nev. .... var. *hordeoides* (Suksd.) C. L. Hitchc.

2. Sitanion jubatum J. G. Smith

Sitanion jubatum J. G. Smith, U.S.D.A. Div. Agrostol. Bull. 18: 10. 1899. *Elymus sitanion jubatum* J. G. Smith, ibid., as a synonym. (*Horner 573,* Waitsburg, Walla Walla Co., Wash., 27 May 1897.)

S. breviaristatum J. G. Smith, U.S.D.A. Div. Agrostol. Bull. 18: 12. 1899. (*Coville & Funston 833,* "Willow Creek Cañon, Panamint Mts.," Calif., 22 May 1891.)

Big squirreltail.

Densely cespitose perennials with many flowering culms; culms 2–4 (6) dm tall, puberulent; sheaths glabrous to pilose; ligules short, 0.2–0.7 (1) mm long, mostly entire; blades flat to involute, 1.5–3.5 (5) mm broad, scabrous to pilose, the auricles generally well developed, 0.5–1.5 mm long, usually ascending; spikes (3) 5–10 (17) cm long (measurement excluding the awns), dense, with a bristly appearance, exserted to slightly included in the uppermost leaf sheaths; spikelets mostly in 2s, rarely in 3s or 4s, few- to several-flowered, the lowermost florets of one or sometimes both spikelets sterile, reduced to a structure similar to the glumes, the awns of the glumes and lemmas 25–70 (90) mm long, mostly red or purplish, rarely green; glumes long, 25–100 mm long, 3- to 9-cleft, the divisions extending into scabrous, divergent awns of unequal lengths; lemmas 8–10 mm long, faintly 3- to 5-nerved, tapering into a straight, scabrous awn up to 100 mm long, the 2 lateral nerves often extending into bristles of awns up to 20 mm long; palea subequal to the lemma, the 2 nerves extending as bristles up to 2 mm long; anthers 1–2 mm long; $2n = 28$.

At middle elevs. in generally more mesic habitats than *S. hystrix;* e. Wash. and Idaho, s. through Nev. and Utah to s. Calif. and nw. Ariz. Late May–June (July).

Sitanion jubatum is very similar to *S. hystrix* and probably should be treated as a variety of it. However, no new combination is proposed here.

31. TAENIATHERUM Nevski

Annuals; culms ascending to erect, hollow; sheaths glabrous, open; ligules membranous, short; blades flat to involute, the auricles usually small; inflorescence a short, dense, bristly spike with a continuous rachis; spikelets arranged in pairs, with 1 perfect floret, the rachilla prolonged behind and often bearing a vestigial floret, disarticulating above the glumes; glumes subulate, indurate, united at the base and tapering into slender, stiff awns; lemmas lanceolate, scabrous with a long, divergent, flattened awn; palea subequal to the lemma; lodicules ciliate, often lobed; caryopsis elongate, deeply furrowed in back, hairy at the summit; $x = 7$.

A genus of 2 or 3 weedy species of temp. regions of Europe and Asia. (Name from the Greek *taenia,* ribbon, and *ather,* awn.)

1. Taeniatherum caput-medusae (L.) Nevski

Elymus caput medus L. Sp. Pl. 84. 1753. *Hordeum caput-medusae* Coss. & Durieu, Explor. Sci. Algér. 2: 198. 1855. *Taeniatherum caput-medusae* Nevski, Trudy Sredne-Aziatsk. Gosud. Univ., Ser. 8b, Bot. 17: 38. 1934. ("Habitat in Lusitaniae, Hispaniae maritimis.")

Medusahead rye.

Annuals; culms ascending from a decumbent base, (1.5) 2.5–5 (6) dm tall, slender; sheaths slightly inflated, glabrous; ligules very short, 0.2–0.5 mm long; blades more or less involute, narrow, 1–2.5 mm broad, short, glabrous to puberulent, the margins sometimes ciliate, the auricles very short and inconspicuous; spikes small, 1.5–4 cm long (measurement excluding the awns), dense; spikelets in 2s, with 1 perfect floret, the second floret much reduced, sometimes obsolete; glumes long, 15–35 mm long, usually much shorter than the awns of the lemma, stiff, subulate and awn-like throughout, connate at the base, usually glabrous and shiny below and scabrous apically; lemmas (5) 6–8 mm long, narrow-lanceolate, very scabrous throughout, prolonged into a long, flattened and divergent awn (20) 30–70 mm long; lodicules about 0.8 mm long, oblanceolate, ciliate; anthers 0.6–1 mm long; $2n = 14$.

A weed of fields, dry roadsides and abused sagebrush slopes; introduced from Asia; e. Wash. and c. Idaho, s. through Oregon to Calif., entering our region along the n. edge of the Snake River Plains and in Washoe Co., Nev. May–June.

Taeniatherum caput-medusae is often referred to the genus *Elymus,* but it differs in so many ways from that genus of perennial grasses in both morphology and cytology that it seems desirable to resort to the above treatment. It has been suggested that our species is *T. asperum* (Simonkai) Nevski rather than *T. caput-medusae,* a problem we cannot evaluate. Medusahead rye is a worthless range weed that has received much attention. We have every reason to believe it has not yet invaded all of its potential sites.

32. HORDEUM L. Barley, Foxtail

Cespitose annuals and perennials without rhizomes; culms hollow; sheaths open; ligules membranous, mostly short, truncate; blades flat, mostly broad and lax, strongly auriculate to non-auriculate; inflorescence a dense spike or spicate raceme, usually with both sessile and short pedicelled spikelets, the spikelets forming a triad at each node with a central, fertile, usually sessile spikelet flanked by a pair of lateral, usually pedicelled and staminate or sterile (fertile in *H. vulgare*) spikelets, the rachis usually fragmentary at maturity and disarticulating above each node, the rachis segment remaining with the spikelet triad; spikelets 1-flowered,

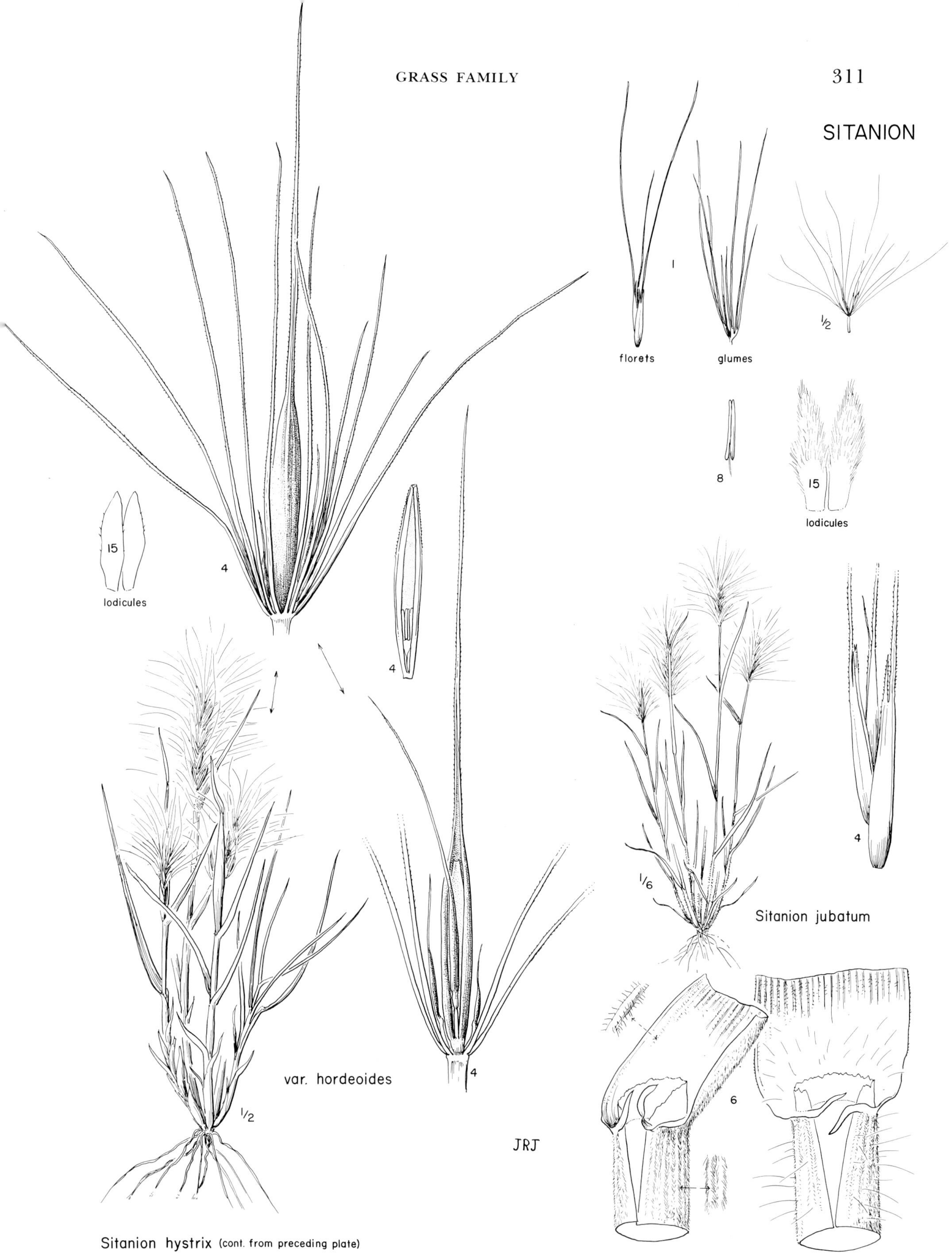
SITANION
1
florets
glumes
1/2
8
15
lodicules
15
lodicules
4
4
1/6
Sitanion jubatum
4
6
var. hordeoides
1/2
4
JRJ
Sitanion hystrix (cont. from preceding plate)

the rachilla of the central spikelet prolonged behind the palea as a bristle which rarely bears a vestige of a second floret, the lateral spikelets sometimes represented by glumes only; glumes narrow, usually subulate or awn-like, rigid, scabrous or ciliate, the inner glumes of the lateral spikelets shorter than and sometimes broader than the outer ones; lemma firm, rounded on back, faintly 5-nerved, tapering into a long awn; palea slightly shorter than the lemma; lodicules 2; stamens 3; caryopsis oblong, furrowed in back, hairy at the summit, adhering to the palea; $x = 7$.

A genus of about 35 species, 5 native to N. Amer., 10 to S. Amer. and 20 to Eurasia. (The classical Latin name for barley.)

The genus *Hordeum* includes the widely cultivated barley, but with few exceptions the species of our area are undesirable grasses that are usually considered as noxious weeds. *Hordeum jubatum* often increases in poorly managed pastures and brings about deterioration until reseeding becomes necessary. All the species are very undesirable in hay, and their presence will often cause sore eyes and abscessed jaws in animals in feed lots.

The perennial species of *Hordeum* are interfertile with species of *Elymus* and *Agropyron*. The name X *Elyhordeum* Mansf. was proposed for hybrids with species of *Elymus* and X *Agrohordeum* Camus with *Agropyron*. See comments following the tribe description (p. 295).

References:

Bowden, W. M. Natural and artificial X *Elymordeum* hybrids. Canad. J. Bot. 36: 101–123. 1958.

———. Cytotaxonomy of the native and adventive species of *Hordeum, Eremopyrum, Secale, Sitanion,* and *Triticum* in Canada. Canad. J. Bot. 40: 1675–1711. 1962.

———. Taxonomy of intergeneric hybrids of the tribe Triticeae from North America. Canad. J. Bot. 45: 711–724. 1967.

Boyle, W. S., and A. H. Holmgren. A cytogenetic study of natural and controlled hybrids between *Agropyron trachycaulum* and *Hordeum jubatum*. Genetics 40: 539–545. 1955.

Covas, G. Taxonomic observations on the North American species of *Hordeum*. Madroño 10: 1–21. 1949.

1 Plants perennials; awns slender.
  2 Lemmas 8–10 mm long . . . . *Sitanion hystrix* var. *hordeoides*
  2 Lemmas 5.5–8 mm long.
    3 Glumes 25–150 mm long; awns of the lemmas 10–60 mm long; auricles sometimes present but very short, about 0.5 mm long, or lacking . . . . 1. *H. jubatum*
    3 Glumes 7–20 mm long; awns of the lemma 5–10 (20) mm long; auricles lacking . . . . 2. *H. brachyantherum*
1 Plants annuals; awns stout.
  4 Rachis not disarticulating; blades 5–16 mm broad; awns of the lemma 60–160 mm long; cultivated plants . . . . 9. *H. vulgare*
  4 Rachis disarticulating; blades 1–5 (9) mm broad; awns of the lemma (4) 5–45 mm long; weedy plants mostly of waste places.
    5 Glumes of the central spikelet at most scabrous margined, not ciliate; auricles usually lacking or when present less than 1 mm long.
      6 Glumes unlike, those of the central spikelet and inner glumes of the lateral spikelets broadened above the base, while the outermost glumes are awn-like from the base . . . . 3. *H. pusillum*
      6 Glumes all awn-like.
        7 Lateral spikelets with an awned floret; glumes more or less spreading; auricles lacking or tiny (less than 0.5 mm long) . . . . 4. *H. geniculatum*
        7 Lateral spikelets with a vestigial, awnless floret; glumes more or less erect; auricles lacking . . . . 5. *H. depressum*
    5 Glumes of the central spikelet with ciliate margins; auricles well-developed on most of the leaves, mostly more than 1 mm long.
      8 Floret of the central spikelet sessile or subsessile, the rachilla joint not nearly as long as the pedicels of the lateral spikelets . . . . 6. *H. murinum*
      8 Floret of the central spikelet with a rachilla joint as long or longer than the pedicels of the lateral spikelets, or the spikelet with a pedicel.
        9 Joints of the rachis mostly less than 2 mm long . . . . 7. *H. glaucum*
        9 Joints of the rachis (1.5) 2 mm long or more . . . . 8. *H. leporinum*

1. Hordeum jubatum L.

Hordeum jubatum L. Sp. Pl. 85. 1753. *Elymus jubatus* Link, Hort. Regius Bot. Berol. 1: 19. 1827. *Critesion jubatum* Nevski in Komarov, Fl. USSR 2: 721. 1934. ("Habitat in Canada. *Kalm.*")

H. caespitosum Scribn. in Pammel, Proc. Davenport Acad. Nat. Sci. 7: 245. 1899. *H. jubatum* var. *caespitosum* A. S. Hitchc. Proc. Biol. Soc. Wash. 41: 160. 1928. (*Pammel 157*, Geranium Park, Wyo., July 1897.) Reported to be a hybrid between *H. jubatum* and *H. brachyantherum;* see discussion below.

Foxtail barley.

Tufted perennials, sometimes flowering the first year, from glabrous to densely soft pubescent throughout; culms erect or decumbent, (2) 3–7 dm tall; sheaths glabrous to pilose, sometimes velvety; ligules 0.5–1 mm long, truncate, erose or entire, ciliolate; blades flat to involute, 1.5–4 (5) mm broad, scabrous to hirsutulous, the auricles usually lacking, or when present less than 0.5 mm long; spikes 4–10 cm long (measurement excluding the awns), nodding, often purplish at maturity, the rachis disarticulating and pulled apart by the spreading awns, the central spikelet sessile, the lateral spikelets borne on curved pedicels 0.7–1.2 mm long, their florets

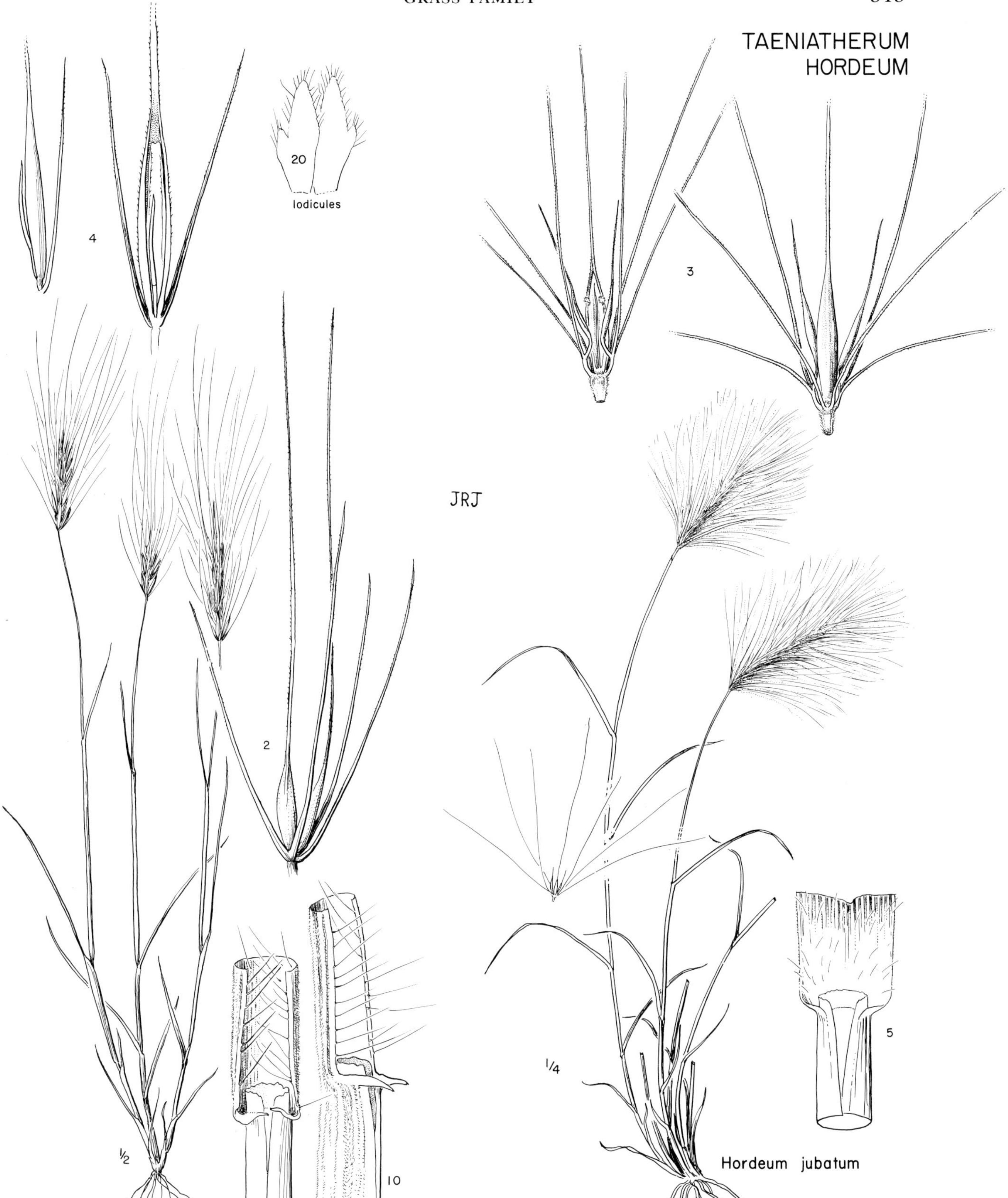
TAENIATHERUM
HORDEUM
4
20
lodicules
3
JRJ
2
1/2
10
1/4
5
Hordeum jubatum
Taeniatherum caput-medusae

reduced, borne on rachilla joints about 0.7 mm long; glumes long and awn-like, 25–60 (150) mm long; lemma of the fertile floret 5.5–8 mm long, faintly 5-nerved, tapering into a long awn 10–60 mm long; palea subequal to the lemma; anthers 1–1.5 mm long; $2n$ = 14, 28, 42.

A weed along roadsides and often becoming abundant in moist to wet meadows; Alaska to Newfl., s. in the mt. and prairie states to Calif., Ariz., N.M. and Texas and s. in the e. to Mo., Ill., Ind., Pa. and Del.; introduced into the Old World. June–Aug.

Mitchell and Wilton (Madroño 17: 269–280. 1964) have presented rather conclusive evidence that in Alaska *H. caespitosum* Scribn. is a short-lived and only partially fertile hybrid of *H. brachyantherum* and *H. jubatum*. It has been suggested by Bowden (1962) and Rajhathy (Madroño 18: 243–244. 1966) that *H. brachyantherum* and *H. jubatum* be considered conspecific because of their interfertility. This view, if carried out in the grass family and especially in this tribe, would result in drastic changes that few agrostologists could accept. The hybrids are only partially fertile and the two species are easily separated in a key and readily recognized in the field. Some intergeneric hybrids involving *H. jubatum* that may occasionally occur in our region have been named. The cross with *Elymus triticoides* Buckley has been named X *Elyhordeum piperi* (Bowden) Bowden, with *E. canadensis* L. is X *Elyhordeum dakotense* (Bowden) Bowden, with *E. virginicus* L. is X *Elyhordeum montanense* (Scribn.) Bowden, and with *Agropyron trachycaulum* (Link) Malte is X *Agrohordeum macounii* (Vasey) Lepage.

2. Hordeum brachyantherum Nevski

Hordeum boreale Scribn. & Smith, U.S.D.A. Div. Agrostol. Bull. 4: 24. 1897; not Gandoger, 1881. *H. nodosum* var. *boreale* A. S. Hitchc. Amer. J. Bot. 21: 134. 1934. *H. brachyantherum* Nevski, Trudy Bot. Inst. Akad. Nauk SSSR, Ser. 1, Fl. Sist. Vysš. Rast. 2: 61. 1936. *H. jubatum* subsp. *breviaristatum* Bowden, Canad. J. Bot. 40: 1690. 1962. *H. jubatum* var. *boreale* B. Boivin, Naturaliste Canad. 94: 525. 1967. (*L. M. Turner 1193,* Atka Island, Aleutian Islands, May–July 1879.)

H. nodosum of Amer. authors; not L. 1762.

Meadow barley.

Tufted perennials; culms erect or sometimes spreading, 3–7 dm tall; sheaths glabrous, strigillose or with spreading pubescence; ligules short, 0.2–0.7 mm long, truncate, ciliolate; blades flat, 2–5 (7) mm broad, scabrous to pilose, at least above, the auricles lacking; spikes 2–5 (8) cm long (measurement excluding the awns), erect, green or brownish-purple in matured plants, the rachis disarticulating, the central spikelet sessile, the lateral spikelets borne on curved pedicels 0.7–1 mm long, their florets much reduced, rarely staminate; glumes awn-like, 7–20 mm long; lemma of the fertile floret 6.5–8 mm long, faintly 5-nerved, glabrous to scaberulous above, tapering into a short awn 5–10 (20) mm long; anthers 1–1.8 mm long; $2n$ = 28.

Moist mt. meadows and along streambanks, less often among sagebrush vegetation or on rocky slopes; Alaska to Mont., s. to Calif., Ariz. and N.M.; introduced into scattered localities in the east. June–Aug.

In the past our taxon was called *H. nodosum* L. which, according to Covas (1949), is a closely related European species that does not occur in our area. The European *H. nodosum* has well developed auricles, longer awns and anthers about twice the length of those in our species. Meadow barley is sometimes a good forage plant of higher ranges. See discussion at the end of *H. jubatum* on the separation of our perennial species of *Hordeum*.

A couple of named intergeneric hybrids involving *H. brachyantherum* are X *Agrohordeum macounii* nm. *valencianum* Bowden with *Agropyron trachycaulum* (Link) Malte as the other parent, and X *Elyhordeum stebbinsianum* (Bowden) Bowden with *Elymus glaucus* Buckley as the other parent.

3. Hordeum pusillum Nutt.

Hordeum pusillum Nutt. Gen. N. Amer. Pls. 1: 87. 1818. (*Nuttall,* "On the arid and saline plains of the Missouri," in 1811.)

H. pusillum var. *pubens* A. S. Hitchc. J. Wash. Acad. Sci. 23: 453. 1933. (*Jones 5196W,* La Verkin, Washington Co., Utah, 8 May 1894.)

Little barley.

Annuals; culms erect, (1) 2.5–4 (6) dm tall, rarely taller than 3.5 dm in our area; sheaths glabrous to pilose; ligules short, 0.3–0.5 mm long, somewhat erose; blades flat, 2–4 mm broad, erect, scabrous to hispidulous, at least above, the auricles lacking; spikes 2–7 cm long (measurement excluding the awns), erect, the rachis disarticulating, the central spikelet sessile and the lateral spikelets borne on curved pedicels 0.3–0.7 (1) mm long, their florets reduced, awn-tipped and borne on curved rachilla joints about 0.5 mm long; glumes 7–12 (15) mm long, those of the central spikelet and inner glumes of the lateral spikelets slightly to conspicuously broadened above the base before narrowing into slender awns while the outermost glumes awn-like from the base and somewhat shorter, scabrous to puberulent; lemma of the fertile floret relatively small, 5–7 mm long, scabrous, faintly 5-nerved, tapering into a short awn (4) 5–8 mm long; anthers small, 0.5–0.8 mm long; $2n$ = 14, 28.

A weed of pasturelands, roadsides and moderately moist waste places, often in saline soil; Wash. to Del. (introduced to Me. and Pa.), s. to s. Calif., Fla. and n. Mex.; also in Argentina. May–Aug.

Little barley is our most widespread native species of *Hordeum*. It is frequently associated with the weedy annual *Vulpia octoflora* (Walter) Rydb. *Hordeum pusillum* var. *pubens* A. S. Hitchc. may be an acceptable variety with pubescent glumes and lemmas and with a distribution in the Southwest, entering our area in southern Utah.

4. Hordeum geniculatum All.

Hordeum geniculatum All. Fl. Pedemontana 2: 259, and 3: pl. 91, fig. 3. 1785. ("Ballardi legit ad litora maris Nicaeensis prope Portum di Limpia" [Turkey].)

H. hystrix Roth, Catalecta Bot. 1: 23. 1797. ("Habitat in Hispania.")

H. gussoneanum Parl. Fl. Palermo 1: 246. 1845. *H. maritimum* var. *gussonianum* K. Richter, Pl. Eur. 1: 131. 1890. *H. maritimum* subsp. *gussoneanum* Aschers. & Graebner, Syn. Mitteleur. Fl. 2(1): 737. 1902. *H. marinum* subsp. *gussoneanum* Thell. Vierteljahrsschr. Naturf. Ges. Zürich 52: 441. 1908. (Italy.)

Mediterranean barley.

Small, grayish-green annuals; culms usually geniculate, 1–2.5 (4.5) dm tall; sheaths glabrous to soft pubescent, the upper sheaths somewhat inflated; ligules short, 0.2–0.5 mm long, truncate, erose, ciliolate; blades flat, narrow, 1–2.5 (4) mm broad, short, scabrous to hirsutulose, the auricles usually lacking, but an occasional one sometimes present and then less than 0.5 mm long; spikes small, 1.5–4 cm long (measurement excluding the awns), erect, usually well exserted from the sheath in mature plants, the rachis disarticulating, the central spikelet sessile, the lateral spikelets borne on straight or slightly curved

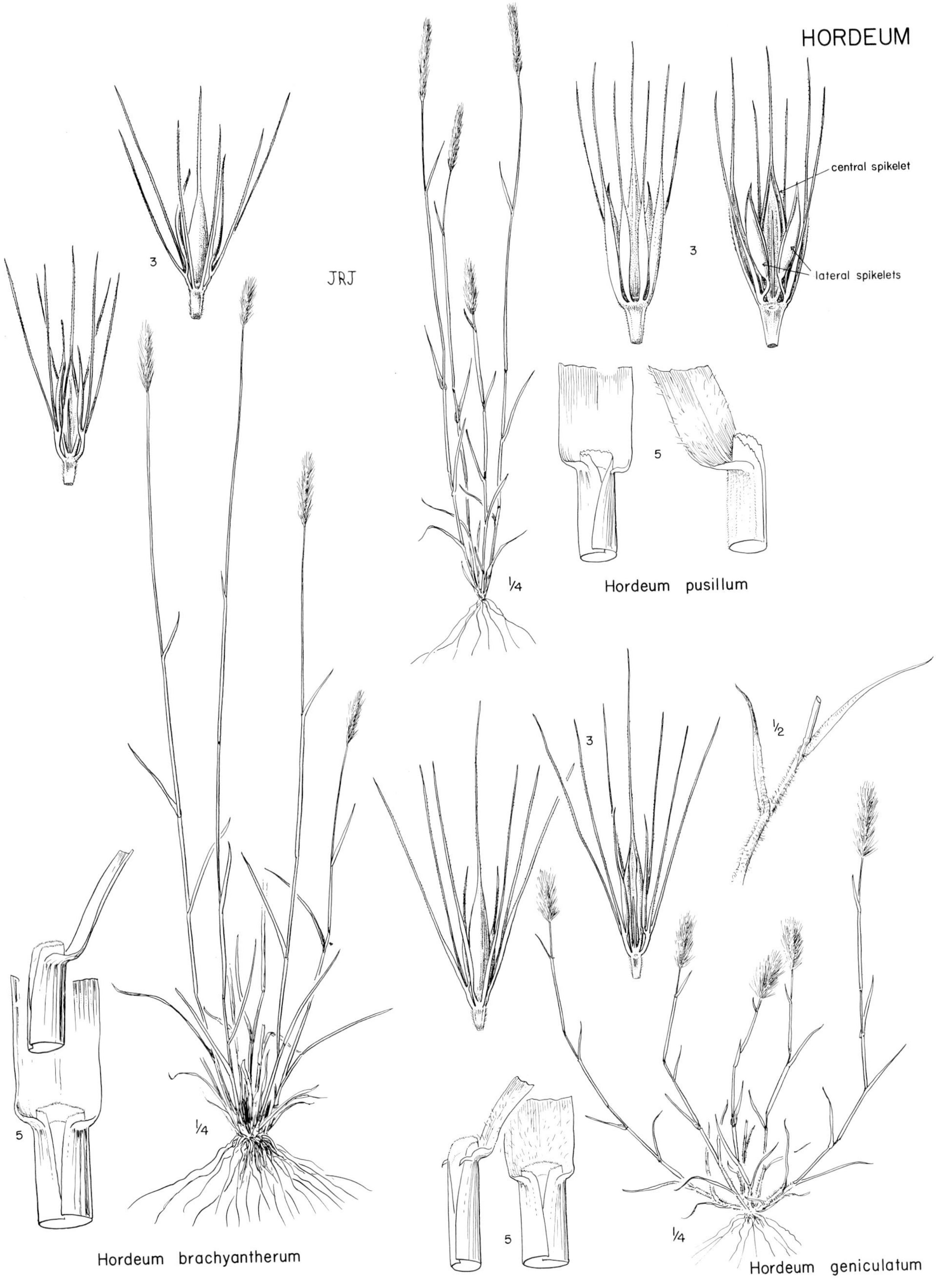
HORDEUM
central spikelet
lateral spikelets
3
JRJ
1/4
5
Hordeum pusillum
1/2
Hordeum brachyantherum
Hordeum geniculatum

pedicels 0.7–1 mm long, their florets much reduced, short-awned and borne on rachilla joints 0.1–0.5 mm long; glumes (6) 9–16 mm long, all awn-like, those of the lateral spikelets subequal; lemma of the fertile floret relatively small, 5.5–6.5 mm long, glabrous, faintly 5-nerved, tapering into a short awn 8–16 mm long, exceeding and stouter than the awns of the glumes; anthers 0.6–1 mm long; $2n$ = 14, 28.

Moderately moist, often alkaline waste places; introduced from Europe; s. B.C. through Wash. and Idaho to Calif. and Ariz. and sporadic in the east. May–June.

Hordeum geniculatum is perhaps the most common of the annual species of *Hordeum* in our area. Plants growing in dense stands tend to be erect and tall, while isolated individuals will be strongly geniculate and much branched from the base.

5. Hordeum depressum (Scribn. & Smith) Rydb.

Hordeum nodosum var. *depressum* Scribn. & Smith, U.S.D.A. Div. Agrostol. Bull. 4: 24. 1897. *H. depressum* Rydb. Bull. Torrey Bot. Club 36: 539. 1909. (*Leiberg 39,* near Lexington, Morrow Co., Oregon, 8 May 1894.)

Small annuals; culms often geniculate at the base, sometimes erect, 0.5–4.5 dm tall; sheaths glabrous to hispid, the upper ones usually somewhat inflated; ligules short, 0.1–0.5 mm long; blades flat or sometimes folded to involute, narrow, 1–3 (4) mm broad, the auricles lacking; spikes small, 1.5–3.5 (6) cm long (measurement excluding the awns), erect, slightly enclosed by the upper sheath or more often exserted, the rachis disarticulating, the central spikelet sessile, the lateral spikelets borne on short, straight pedicels 0.5–1 mm long, their florets rudimentary and awnless, borne on curved rachilla joints about 1 mm long; glumes all awn-like, 8.5–15 mm long, hardly at all broadened below; lemma of the fertile floret 5–8 (10) mm long, glabrous or sometimes scabrous above, faintly 5-nerved, tapering into a short awn 5–10 (20) mm long; palea often slightly longer than the body of the lemma; anthers 0.6–1 mm long; $2n$ = 28.

Moist waste places, often in saline soil; s. B.C., s. to Calif., may enter our region in sw. Idaho. May–June.

Hordeum depressum is very similar to *H. geniculatum.* About the only constant difference between the two is the awnless lemma of the rudimentary florets of the lateral spikelets in *H. depressum* versus the consistently awned reduced lemmas in *H. geniculatum. Hordeum depressum* is apparently a native species, while *H. geniculatum* has come to us by way of Europe.

6. Hordeum murinum L.

Hordeum murinum L. Sp. Pl. 85. 1753. *Triticum murale* Salisb. Prodr. Stirpium 27. 1796. *Zeocriton murinum* Beauv. Essai Nouv. Agrostogr. 115, 182. 1812. *H. leporinum* var. *murinum* Lojac. Fl. Sicula 3: 364. 1909. ("Habitat in Europae locis ruderatis.")

Small annuals; culms geniculate at the base, 1–4 (7) dm tall, glabrous; sheaths glabrous; ligules 0.5–1 (1.5) mm long, truncate, erose or entire, ciliolate; blades flat to folded, 2–4.5 mm broad, scabrous to soft pubescent, sometimes glabrous on back, the auricles well developed, 1–2 (3) mm long; spikes 3–8 cm long (measurement excluding the awns), erect, often partially enclosed by the inflated upper sheaths, the rachis disarticulating, all 3 spikelets of the triad well developed, the central spikelet sessile but with the floret borne on a shorter rachilla joint, the lateral spikelets short-pedicellate with subsessile florets; glumes 13–25 (40) mm long, those of the central spikelet and inner glumes of the lateral spikelets somewhat broadened below and ciliate, the outer glumes of the lateral spikelets awn-like and scabrous, longer than the inner ones; lemma of the central spikelet 6.5–9 (12) mm long, fertile, faintly 5-nerved, the lemmas of the lateral spikelets larger, 10–18 mm long, but sterile, all 3 lemmas tapering into stout awns 15–25 (45) mm long; anthers about 1 mm long; $2n$ = 14, 28, 42.

Roadsides, ditch banks and waste areas, usually where there is good moisture; introduced from Europe; common in the w. from B.C., Wash. and Idaho s. to Baja Calif., Ariz. and N.M., sporadic to rare in the c. and e. states. May–July.

Hordeum murinum and the following two species, *H. glaucum* and *H. leporinum,* are very closely related and sometimes difficult to distinguish.

7. Hordeum glaucum Steudel

Hordeum glaucum Steudel, Syn. Pl. Glum. 1: 352. 1854. (*Schimper 383,* Sinai Peninsula, Egypt.)
H. stebbinsii Covas, Madroño 10: 17. 1949. (*Stebbins & Covas 3927,* 5 mi. se. of Middletown, Lake Co., Calif., 9 May 1948.)

Small annuals; culms often geniculate at the base, 1.5–4.5 (6) dm tall, glabrous; sheaths glabrous; ligules short, 0.2–0.7 (1) mm long, truncate, erose or entire, ciliolate; blades flat, 1.5–4 (9) mm broad, scabrous to pilose, usually sparsely so, the auricles well developed, 1–2.5 (4) mm long; spikes 3.5–6.5 (10) cm long (measurement excluding the awns), erect, usually partially enclosed in the inflated upper sheath, the rachis disarticulating, the segment 1–1.8 mm long, all 3 spikelets of the triad well developed, the central spikelet sessile but with the floret borne on a rachilla joint that is as long or longer than the pedicels of the lateral spikelets, these pedicels 1–2 mm long; glumes 11–22 (30) mm long, those of the central spikelet and the inner glumes of the lateral spikelets broadened at the base and ciliate, often with 3 scabrous nerves, the outer glumes of the lateral spikelets awn-like, scabrous and longer than the inner ones; lemma of the central spikelet 6–10 mm long, fertile, glabrous, 5-nerved, the nerves faint below, more conspicuous above, the lemmas of the lateral spikelets larger, 8–12 mm long, sterile, all 3 lemmas tapering into awns 10–20 (35) mm long; paleas of all 3 florets pilose; anthers small, about 0.5 mm long; $2n$ = 14.

Disturbed areas; introduced from the Medit. region; c. Wash., Idaho and Oregon, s. to S. Amer., entering our region along the s. border. May–June.

8. Hordeum leporinum Link

Hordeum leporinum Link, Linnaea 9: 133. 1834. *H. murinum* var. *leporinum* Arcang. Compend. Fl. Ital. 805. 1882. *H. murinum* subsp. *leporinum* Aschers. & Graebner, Syn. Mittel eur. Fl. 2(1): 739. 1902. (Greece.)

Light green annuals; culms geniculate to erect, (1) 2–4 (6) dm tall; sheaths mostly glabrous, sometimes pilose; ligules 0.5–1 mm long, truncate erose, cilio-

HORDEUM

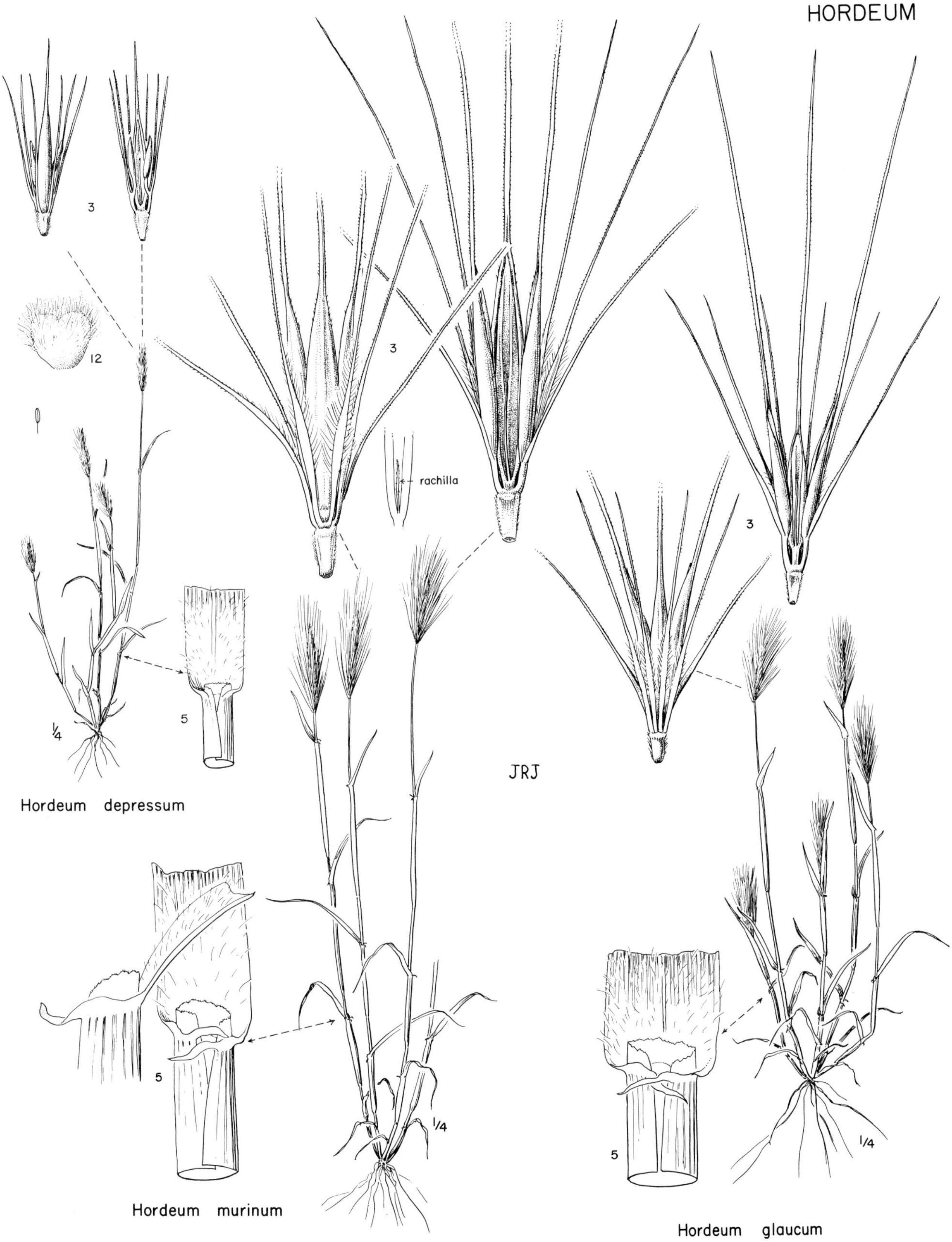
3
12
rachilla
3
3
5
1/4
Hordeum depressum
JRJ
5
1/4
Hordeum murinum
5
1/4
Hordeum glaucum

late; blades flat, 2–5 mm broad, glabrous to pilose, the auricles well developed, up to 3.5 mm long; spikes 4–7 (10) cm long (measurement excluding the awns), erect, partially enclosed by the inflated upper sheath or sometimes well exserted, the rachis readily disarticulating, the segments (1.5) 2–3 mm long, all 3 spikelets of the triad well developed, the central spikelet sessile, the floret borne on an elongate rachilla joint 1–2 mm long, equaling or even exceeding the pedicels of the lateral spikelets, the lateral spikelets usually staminate; glumes 15–22 (28) mm long, those of the central spikelet and the inner glumes of the lateral spikelets narrow lanceolate, 3-nerved and ciliate on the margins, the nerves scabrous, the outer glumes of the lateral spikelets awn-like and somewhat longer, scabrous; lemma of the central spikelet (6) 7–9 mm long, fertile, glabrous, 5-nerved, the nerves faint below, the lemmas of the lateral spikelets somewhat larger, 7–10 mm long, sterile, all 3 lemmas tapering into awns 18–30 mm long; anthers 0.8–1.2 mm long; $2n$ = 14, 28, 42.

Moderately moist waste places; introduced from s. Europe; B.C. to Me., s. to n. Mex., widespread in w. N. Amer. and absent from most of the midw. May–July.

9. Hordeum vulgare L.

Hordeum vulgare L. Sp. Pl. 84. 1753. *H. sativum* Pers. Syn. Pl. 1: 108. 1805, in synonymy. *H. polystichum* var. *vulgare* Doell, Rheinische Fl. 67. 1843. *H. sativum* subsp. *vulgare* Hackel in Engl. & Prantl, Natürl. Pflanzenfam. 7: 86. 1887. *H. distichum* var. *vulgare* Cout. Bol. Soc. Brot. II. 10: 62. 1935. ("Habitat. . . .")

H. coeleste var. *trifurcatum* Schlechtendal, Linnaea 11: 543. 1837. *H. trifurcatum* Wenderoth, Flora 26: 233. 1843. *H. vulgare* var. *trifurcatum* Alef. Landwirthschaftl. Fl. 341. 1866. *H. polystichum* subsp. *trifurcatum* Aschers. & Graebner, Syn. Mitteleur. Fl. 2(1): 731. 1902. (Cultivated plant grown at Halle.)

Barley.

Large cultivated annuals; culms erect, 6–12 (14) dm tall, glabrous; sheaths smooth; ligules 0.5–1.2 mm long, erose-lacerate, ciliolate; blades flat, 5–12 (16) mm broad, scabrous to glabrous, the auricles well developed, up to 6 mm long; spikes stout, 6–9 (12) cm long (measurement excluding the awns), the rachis continuous, all three spikelets of the triad sessile and fertile; glumes subequal, 6.5–20 mm long, broadened below, linear, 3-nerved, ascending-pilose, tapering into scabrous awns; lemmas of the 3 spikelets fertile and subequal, 6.5–12 mm long, 5-nerved, often faintly so, glabrous, sometimes scabrous above, tapering into long, stout, flattened, scabrous margined awns 60–160 mm long, awnless and curiously 3-lobed at the tip in var. *trifurcatum;* anthers 2–2.5 mm long; $2n$ = 14, 28.

Widely cultivated and sometimes occurring for short periods as roadside weeds; throughout the U.S. May–June.

Barley is widely cultivated. It is the only important species of *Hordeum,* as the rest mostly occur as weeds. Beardless or pearl barley, var. *trifurcatum* (Schlechtendal) Alef., is sometimes planted in our area. *Hordeum distichon* L., in which only the central spikelet of the triad is fertile (two-rowed barley), may be found near dwellings.

33. AGROPYRON Gaertn. Wheatgrass

Cespitose to strongly rhizomatous perennials; culms hollow; sheaths open; ligules membranous, short, ciliolate; blades involute or flat, usually pubescent on the upper surface, the auricles usually well developed; inflorescence a bilateral spike with solitary spikelets (rarely 2 at a node) and alternating on 2 sides of a continuous or sometimes disarticulating rachis, the spikelets borne flatwise to the rachis; spikelets (2-) 3- to 8- (12)-flowered, laterally compressed, disarticulating above the glumes and between the florets; glumes subequal to unequal, broad or narrow, rounded on back or keeled, (2-) 5- to 7- (11)-nerved, usually glabrous or scabrous on the midnerve, awn-tipped or awnless; lemmas firm, lanceolate, (3-) 5- to 7-nerved, rounded on back or slightly keeled above, awnless or awned from the apex, the awn straight or divergent; palea membranous, subequal to the lemma; lodicules 2, usually ciliate on the margins and sometimes at the apex; stamens 3 with relatively large anthers; caryopsis oblong, furrowed in back, hairy at the summit, tending to adhere to the palea; x = 7.

A genus of about 100 species of cool and temp. regions of both hemispheres, with 14 native and 8 introduced species in N. Amer., most abundant in w. U.S. (Name from the Greek *agrios,* wild, and *puros,* wheat, the 2 original species being weeds in wheat fields.)

A few species of *Agropyron* are among our most important native range grasses, most notably *A. smithii, A. dasystachyum, A. spicatum* and *A. trachycaulum.*

Agropyron is a difficult genus because of a high degree of variability due primarily to allopolyploidy and the presence of naturally occurring interspecific and intergeneric hybridization with species of *Elymus* and *Sitanion.* Cytological studies have shown that a number of species assigned to *Agropyron* are more closely related to certain species of *Elymus* than to one another. In Eurasia the genus is commonly split into *Agropyron, Roegneria* and *Elytrigia.* We have chosen to treat *Agropyron* in its broad sense until the phylogenetic relationships among *Agropyron, Elymus, Sitanion* and related genera have become more clearly defined.

Sterile hybrids are relatively frequently encountered in our area. Two of them are intergeneric and one is interspecific. X *Agrositanion saxicola* and X *Agrositanion saundersii* are treated in the key and discussion following the tribe description (p. 295). *Agropyron* X *pseudorepens* is treated here, following the key.

References:

Bowden, W. M. Cytotaxonomy of the species and interspecific hybrids of the genus *Agropyron* in Canada and neighboring areas. Canad. J. Bot. 43: 1421–1448. 1965.

Boyle, W. S., and A. H. Holmgren. A cytogenetic study of natural and controlled hybrids between *Agropyron trachycaulum* and *Hordeum jubatum.* Genetics 40: 539–545. 1955.

Daubenmire, R. An experimental study of variation in the *Agropyron spicatum-A. inerme* complex. Bot. Gaz. 122: 104–108. 1960.

Dewey, D. R. Natural hybrids of *Agropyron trachycaulum* and *Agropyron scribneri.* Bull. Torrey Bot. Club 90: 111–122. 1963.

———. Natural and synthetic hybrids of *Agropyron spicatum* X *Sitanion hystrix.* Bull. Torrey Bot. Club 91: 396–405. 1964.

———. Genome relations between *Agropyron scribneri* and *Sitanion hystrix.* Bull. Torrey Bot. Club 94: 395–404. 1967.

HORDEUM

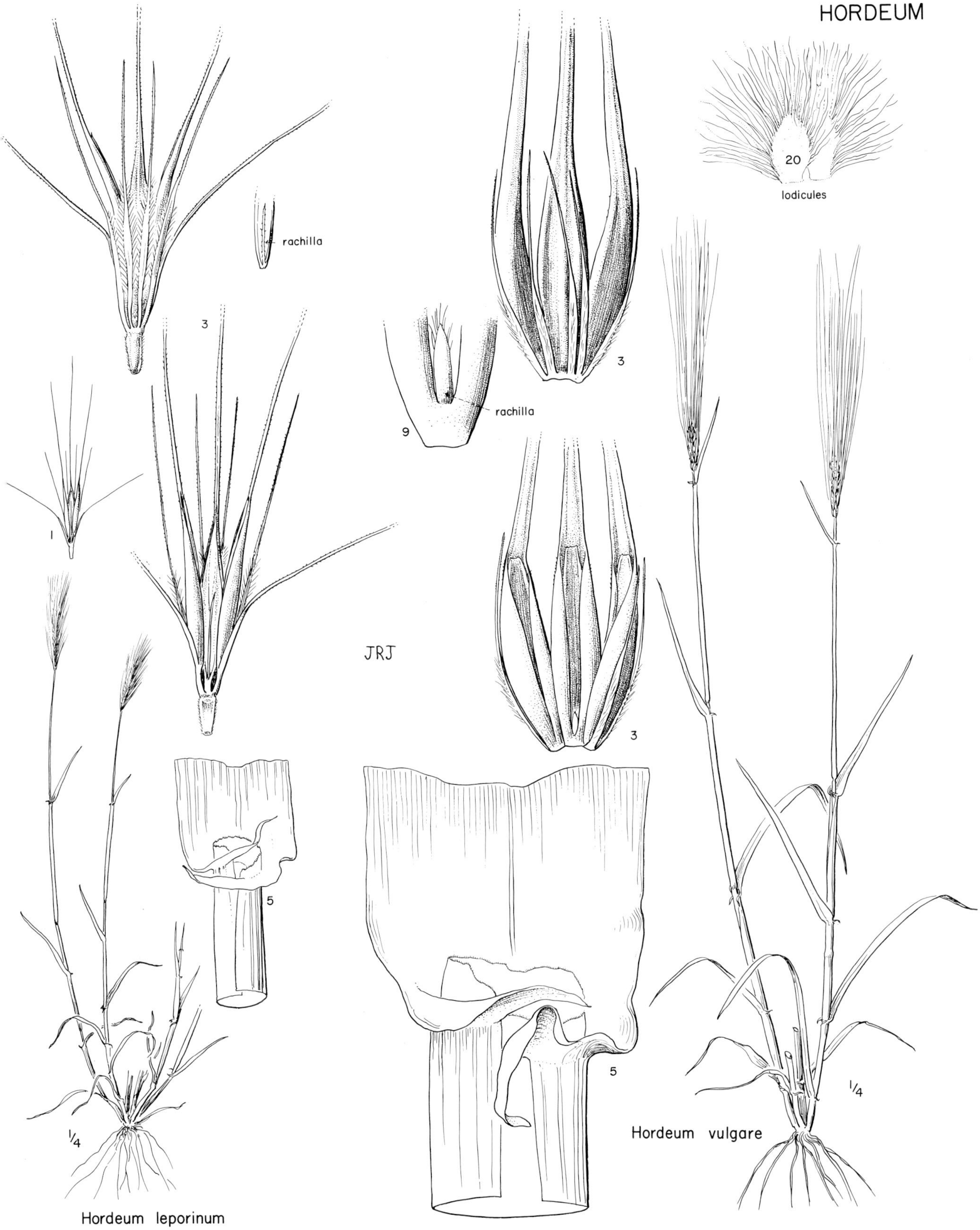
20
lodicules
rachilla
3
3
9
rachilla
1
3
JRJ
3
5
5
1/4
1/4
Hordeum vulgare
Hordeum leporinum

Dewey, D. R. The origin of *Agropyron albicans*. Amer. J. Bot. 57: 12–18. 1970.
Fernald, M. L. *Agropyron* § Goularda in eastern North America. Rhodora 35: 161–185. 1933.
Gillett, J. M., and H. A. Senn. Cytotaxonomy and infraspecific variation of *Agropyron smithii* Rydb. Canad. J. Bot. 38: 747–760. 1960.
Malte, M. O. The so-called *Agropyron caninum* (L.) Beauv. of North America. Annual Rep. Natl. Mus. Canad. 1930: 27–57. 1932.
Piper, C. V. *Agropyron tenerum* and its allies. Bull. Torrey Bot. Club 32: 543–547. 1905.
Pohl, R. W. *Agropyron* hybrids and the status of *Agropyron pseudorepens*. Rhodora 64: 143–147. 1962.
Sarkar, P. Crested wheatgrass complex. Canad. J. Bot. 34: 328–345. 1956.
Stebbins, G. L., Jr., and F. T. Pun. Artificial and natural hybrids in the Gramineae, tribe Hordeae. V. Diploid hybrids of *Agropyron*. Amer. J. Bot. 40: 444–449. 1953.
———, and R. Singh. Artificial and natural hybrids in the Gramineae, tribe Hordeae. IV. Two triploid hybrids of *Agropyron* and *Elymus*. Amer. J. Bot. 37: 388–393. 1950.
———, and L. A. Snyder. Artificial and natural hybrids in the Gramineae, tribe Hordeae. IX. Hybrids between western and eastern North American species. Amer. J. Bot. 43: 305–312. 1956.
———, J. I. Valencia, and R. M. Valencia. Artificial and natural hybrids in the Gramineae, tribe Hordeae. I. *Elymus, Sitanion* and *Agropyron*. Amer. J. Bot. 33: 338–351. 1946.

1 Spikelets spreading from the axis of the spike, much compressed and crowded, the rachis internodes only 0.3–2.2 (3) mm long in the middle of the spike .... 1. *A. cristatum*
1 Spikelets appressed to the rachis of the spike, the rachis internodes (2.5) 4–25 mm long in the middle of the spike (the short internodes found in *A. scribneri* and X *Agrositanion saundersii*).
  2 Plants with long, creeping rhizomes.
    3 Lemmas awnless or merely awn-tipped with awns no more than 5 mm long (up to 10 mm in some *A. repens*) and straight.
      4 Glumes and lemmas obtuse or rounded to acute, if awned the awn arising from a minutely bifid apex; blades usually lax and flat; plants usually green, rarely glaucous.
        5 Glumes and lemmas obtuse to rounded .... 2. *A. intermedium*
        5 Glumes acute.
          6 Glumes about 1/2 the length of the spikelet .... 4. *A. repens*
          6 Glumes 2/3 to nearly equaling the spikelet .... *A.* X *pseudorepens*
      4 Glumes acute to acuminate, gradually tapering into a point or awn tip; lemmas acute; blades firm, stiff, mostly involute; plants usually glaucous.
        7 Glumes narrow-lanceolate, acuminate, tapering from below the middle into an awn tip .... 5. *A. smithii*
        7 Glumes oblong-lanceolate, acute to acuminate, broadest at or above the middle .... 6. *A. dasystachyum*
    3 Lemmas with awns 5 mm or more long, the awns divergent at maturity .... 7. *A. albicans*
  2 Plants without creeping rhizomes (occasionally short rhizomes present in *A. spicatum* and *A. trachycaulum*).
    8 Rachis continuous; glumes awnless or shortly awn-tipped.
      9 Glumes no more than half as long as the spikelet; anthers 4–6 mm long; internodes of the rachis 9–25 mm long.
        10 Lemmas awnless; glumes thick, indurate, obtuse to truncate .... 3. *A. elongatum*
        10 Lemmas usually distinctly awned with awns up to 25 mm long; glumes not thick or indurate, acute to acuminate, sometimes obtuse .... 8. *A. spicatum*
      9 Glumes 2/3 to nearly as long as the spikelet; anthers small, 1–1.8 (2) mm long; internodes of the rachis 4–8 (11) mm long .... 9. *A. trachycaulum*
    8 Rachis tardily disarticulating; glumes long-awned.
      11 Culms prostrate-spreading, or at least decumbent at the base, mostly less than 3 dm tall; fertile subalpine or alpine species .... 10. *A. scribneri*
      11 Culms erect, mostly more than 4 dm tall; sterile hybrids of low to midmontane elevs. (see treatment following the tribe description).
        12 Awns of the lemma long and spreading, often recurved, mostly (14) 18–37 mm long; internodes of the rachis mostly 7–10 mm long .... X *Agrositanion saxicola*
        12 Awns of the lemma shorter, 4–17 (35) mm long, if very long, straight; internodes of the rachis 2.5–6 (7) mm long .... X *Agrositanion saundersii*

Agropyron X pseudorepens Scribn. & Smith (pro sp.)

Agropyron pseudorepens Scribn. & Smith, U.S.D.A. Div. Agrostol. Bull. 4: 34. 1897. *A. tenerum* var. *pseudorepens* M. E. Jones, Contr. W. Bot. 14: 19. 1912. *Zeia pseudorepens* Lunell, Amer. Midl. Naturalist 4: 226. 1915. *Elymus pauciflorus* subsp. *pseudorepens* Gould, Madroño 10: 94. 1949. *A. trachycaulum* var. *majus* f. *pseudorepens* A. A. Beetle, Rhodora 54: 196. 1952. (*Neally*, Texas, in 1889; lectotype by Piper.)

False quackgrass.

False quackgrass is relatively common and has been collected from s. Alta. to Mich., s. to Oregon, Nev., Ariz., and N.M. *Agropyron* X *pseudorepens* has been found to consist of sterile F_1 hybrids and partially fertile introgressant forms. These plants apparently all involve *A. trachycaulum*, probably var. *trachycaulum*, as one of the parents. The other parent is thought by Pohl (1962) to be *A. smithii*, or by Bowden (1965) and Dewey (Bot. Gaz. 136: 122–128. 1975) to be *A. dasystachyum*.

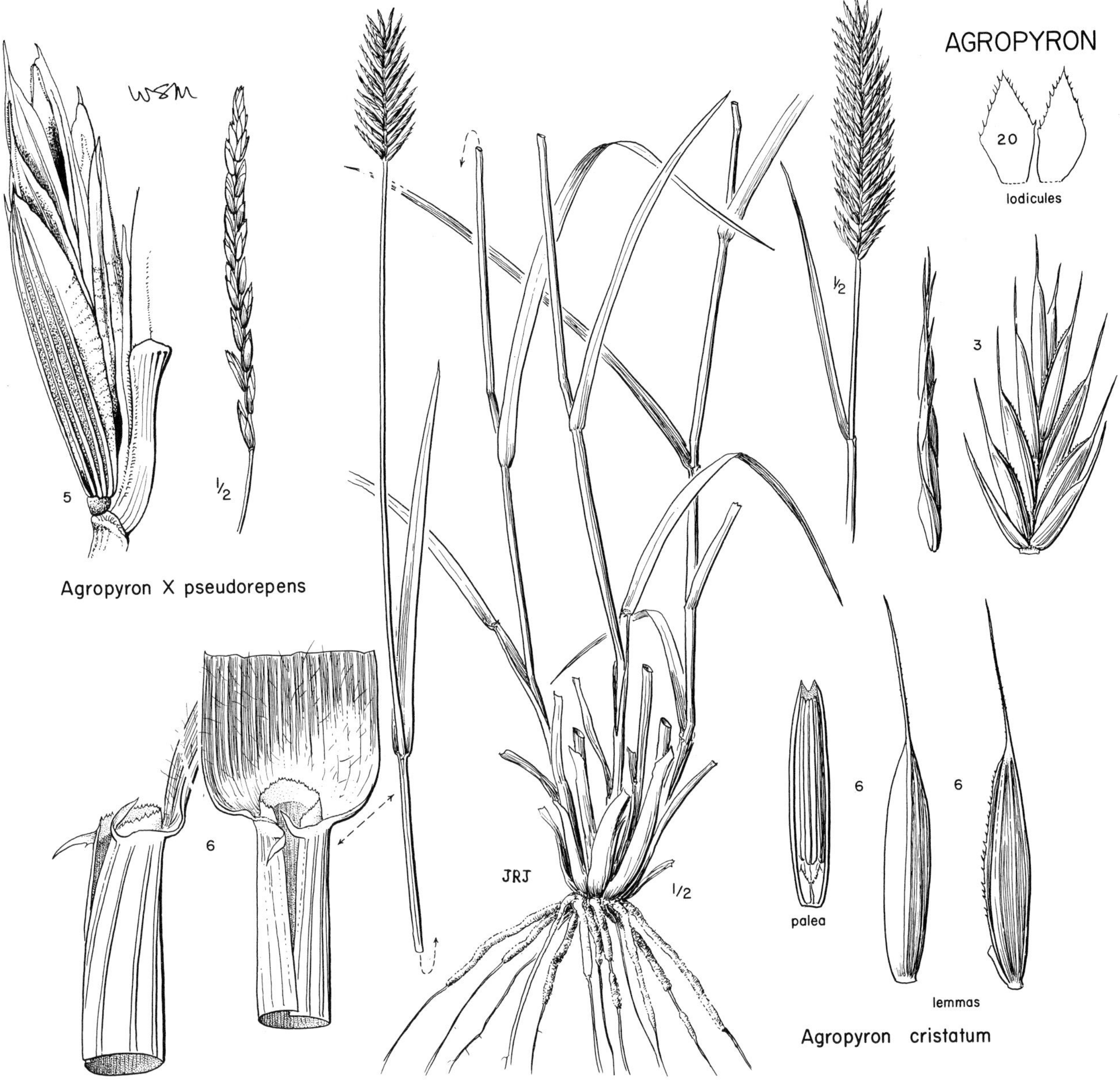

1. Agropyron cristatum (L.) Gaertn.

Bromus cristatus L. Sp. Pl. 78. 1753. *Triticum cristatum* Schreb. Beschreibung Gräser 2: 12, pl. 23, fig. 2. 1772. *Agropyron cristatum* Gaertn. Novi Comment. Acad. Sci. Imp. Petrop. 14: 540. 1770. *Avena cristata* Roemer & Schultes, Syst. Veg. 2: 758. 1817. *Costia cristata* Willk. Bot. Zeitung (Berlin) 16: 377. 1858. *Eremopyrum cristatum* Willk. & Lange, Prodr. Fl. Hispan. 1: 108. 1861. *Zeia cristata* Lunell, Amer. Midl. Naturalist 4: 226. 1915. ("Habitat in Sibiria, Tataria.")

Triticum pectinatum M. Bieb. Fl. Taur.-Caucas. 1: 87. 1808; not *Agropyron pectinatum* (Labill.) Beauv. 1812. *Agropyron pectiniforme* Roemer & Schultes, Syst. Veg. 2: 758. 1817. *Triticum pectiniforme* Roemer & Schultes ex Steudel, Nomencl. Bot. 855. 1821. *A. cristatum* var. *pectinatum* Roschev. ex Fedtsch. Izv. Imp. Bot. Sada Petra Velikago 14 (suppl. 2): 97. 1915. *A. cristatum* subsp. *pectinatum* Tzvelev, Spisok Rast. Gerb. Fl. SSSR. 18: 25. 1970. (". . . in Tauriae rupestribus rarius.")

Triticum sibiricum Willd. Enum. Pl. Hort. Regii Bot. Berol. 135. 1809. *A. sibiricum* Beauv. Essai Nouv. Agrostogr. 102, 146, 181. 1812. ("Habitat in Sibiria.")

Triticum desertorum Fisch. ex Link, Enum. Pl. Hort. Regii Bot. Berol. 1: 97. 1821. *A. desertorum* J. A. Schultes, Mantissa 2: 412. 1824. ("Hab. in deserto Cumano," along River Kuma, se. European Russia.)

Crested wheatgrass.

Tufted perennials, the rhizomes lacking; culms (2.5) 3.5–7 (10) dm tall, ascending to erect; sheaths glabrous or the lowermost sometimes pilose to villous; ligules very short, mostly less than 0.5 mm long, erose-ciliolate; blades flat, 1.5–6 (10) mm broad, scabrous to pilose on the upper surface, the auricles small to well developed and slender; spikes relatively short, 3–7 (10) cm long, broad and tapering at both ends, the spikelets closely imbricate and divergent from the rachis, the rachis internodes short, 0.3–2.2 (3) mm long in the middle of the spike, scabrous to pilose on the angles; spikelets short, (6) 7.5–10 (14) mm long, 3- to 5- (8)-flowered; glumes subequal, short, 3–5.2 mm long, firm, asymmetrical, 3-nerved, the prominent midnerve sometimes pilose-ciliate, often extending into an awn 1.5–3.5 (5) mm long; lemmas relatively small, 5–7 (9) mm long, firm, except for the scarious margins, 5-nerved, with an awn (0.2) 1–2.5 (3.5) mm long; palea with pectinate-ciliate nerves; lodicules rhombic-elliptic, ciliate, about 1 mm long; anthers 2–4 (5) mm long; $2n$ = 14, 28, 42.

Arid sagebrush deserts up to middle elevs. in the mts.; introduced from Russia; Alaska to N.S., s. throughout the w. U.S. and n. parts of the c. and e. U.S. June–Aug.

Several taxa of crested wheatgrass from the USSR and Turkey have been planted on our arid, winter-cold desert and foothill ranges for forage and soil stabilization. There has been a great deal of confusion with the taxonomy of these plants. Sarkar (1956) recognized 6 species in the complex, while Bowden (1965) reduced these to 3. There is enough intergradation between these taxa to obscure their distinctness. The following key, a modification from Bowden, reflects the distinctions usually employed to separate the different species in the complex.

1 Spikelets divaricate at broad angles, giving the spike an oblong-oval, flat shape, the spikelets crowded, separated by short rachis internodes (0.7) 1–1.9 (2.3) mm long.
  2 Spikes subglabrous to glabrate; glumes scabrous only on the keel .... *A. pectiniforme*
  2 Spikes densely hairy .... *A. cristatum*
1 Spikelets more ascending, giving the spike a narrow-oblong, subcylindrical shape, the spikelets separated by longer rachis internodes (1.5) 2.5–3.5 (5) mm long (including *A. desertorum*) .... *A. sibiricum*

2. Agropyron intermedium (Host) Beauv.

Triticum intermedium Host, Icon. & Descript. Gram. Austriacorum 3: 23. 1805, based on the description and plate for *Triticum junceum* of Host, ibid., 2: 18, tab. 22. 1802; not L. 1755, 1759. *Agropyron intermedium* Beauv. Essai Nouv. Agrostogr. 102, 146, 180. 1812. *A. glaucum* var. *intermedium* Beck, Wiss. Mitt. Bosner Herzegowina 9: 460. 1904. *Elytrigia intermedia* Nevski, Trudy Bot. Inst. Akad. Nauk SSSR, Ser. 1, Fl. Sist. Vysš. Rast. 1: 14. 1933. (Austria.)

Triticum trichophorum Link, Linnaea 17: 395. 1843. *A. trichophorum* K. Richter, Pl. Eur. 1: 124. 1890. *A. glaucum* var. *trichophorum* Beck, Wiss. Mitt. Bosner Herzegowina 9: 460. 1904. *A. intermedium* var. *trichophorum* Halac. Conspectus Fl. Graecae 3: 437. 1904. *Elytrigia trichophora* Nevski, Trudy Sredne-Aziatsk. Gosud. Univ., Ser. 8b, Bot. 17: 57. 1934. *Elytrigia intermedia* subsp. *trichophora* Tzvelev, Novit. Syst. Pl. Vasc. 10: 31. 1973. (s. Europe.)

Intermediate wheatgrass.

Rhizomatous, glaucous (occasionally green) perennials, the rhizomes sometimes poorly developed; culms (4) 7–10 dm tall; sheaths ciliate-margined, otherwise glabrous to scabrous; ligules very short, mostly about 0.5 mm long, ciliolate; blades flat to loosely involute, 2–5 (7.5) mm broad, scabrous to pilose on the upper surface and glabrous to scabrous underneath, the auricles usually well developed; spikes (5) 6–15 (20) cm long, slender, erect, stiff, the rachis internodes (6) 8–12 (15) mm long in the middle of the spike; spikelets (7) 10–18 mm long, (2-) 3- to 8-flowered; glumes elliptic-lanceolate, obtuse to rounded, rigid, glabrous or densely hirsute, awnless, the first glume (3.5) 5–8 mm long, 3- to 5-nerved, the second (5) 6.5–9 mm long, 5- to 7-nerved; lemmas (6.5) 7.5–10 (11) mm long, broadly lanceolate, obtuse to acute, glabrous or densely hirsute above and on the margins, faintly 3- to 5-nerved, awnless or sometimes short-awned up to 1 mm; palea also very blunt, the nerves ciliate; anthers (2.2) 3–5 mm long; $2n$ = 42 (in var. *intermedium*).

Growing in pastures and rangelands; introduced from Eurasia; var. *intermedium* near Laramie, Albany Co., Wyo. and near Boise, Elmore Co., Idaho; var. *trichophorum* on the Utah State University campus. June–Aug.

This species has been successfully introduced on cool ranges of western United States. Two varieties can be separated by the following characters:

1 Lemmas glabrous; spikelets 3- to 8-flowered ... var. *intermedium*
1 Lemmas hirsute; spikelets 2- to 3- (6)-flowered .... var. *trichophorum* (Link) Halac.

3. Agropyron elongatum (Host) Beauv.

Triticum elongatum Host, Icon. & Descript. Gram. Austriacorum 2: 18, tab. 23. 1802. *Agropyron elongatum* Beauv. Essai Nouv. Agrostogr. 102, 146, 180. 1812. *Elytrigia elongata* Nevski, Trudy Bot. Inst. Akad. Nauk SSSR, Ser. 1, Fl. Sist. Vysš. Rast. 1: 23. 1933. (Austria.)

Tall wheatgrass.

Tall, tufted, glaucous perennials; culms tall, 7–15 (20) dm, erect, stout, glabrous; sheaths ciliate margined, otherwise glabrous; ligules inconspicuous, with small, densely puberulent areas; blades flat to loosely involute, 2.5–5 (7) mm broad, stiff, thick-veined, glabrous or scabrous on the upper surface, sometimes sparsely pilose on both surfaces, the auricles 1–2 mm long, standing erect instead of clasping the culm; spikes long, 15–30 cm, often loose and open, especially on the lower part of the spike where the internodes may be twice as long as the spikelets, the rachis internodes very long, (9) 15–20 (25) mm in the middle of the spike, scabrous on the angles; spikelets long, (13) 16–22 mm, 6- to 12-flowered, appressed at first, becoming arcuate in age; glumes oblong, obtuse to truncate, indurate, 5- to 7- (9)-nerved, the first glume 6–9 mm long, the second 7.5–10 mm long; lemmas relatively large, 8.5–11 mm long, broadly lanceolate, obtuse to rounded, strongly 5-nerved near the apex with an especially thickened mid-nerve, awnless; anthers 4–5.5 mm long; $2n$ = 14, 28, 42, 56, 70.

Waste places along roadsides where it has escaped; introduced from the Medit. region; known from scattered localities in the Intermountain Region.

Agropyron elongatum is especially well suited to alkaline soils.

4. Agropyron repens (L.) Beauv.

Triticum repens L. Sp. Pl. 86. 1753. *Triticum infestum* Salisb. Prodr. Stirpium 27. 1796. *Agropyron repens* Beauv. Essai Nouv. Agrostogr. 102, 146, 180. 1812. *Elytrigia repens* Desvaux ex B. D. Jackson, Index Kew. 1: 836. 1895; Nevski, Trudy Bot. Inst. Akad. Nauk SSSR, Ser. 1, Fl. Sist. Vysš. Rast. 1: 14. 1933. *Zeia repens* Lunell, Amer. Midl. Naturalist 4: 227. 1915. *Elymus repens* Gould, Madroño 9: 127. 1947. ("Habitat in Europae cultis.")

Triticum repens var. *aristatum* Schumacher, Enum. Pl. Partibus Saellandiae Sept. & Orient. 1: 38. 1801. *A. repens* f. *aristatum* O. R. Holmberg, Skand. Fl. 2: 274. 1926. (*Herb. Schumacher*, without locality; lectotype by Bowden.)

Quackgrass.

Strongly rhizomatous perennials, the herbage usually green or occasionally glaucous; culms 5–10

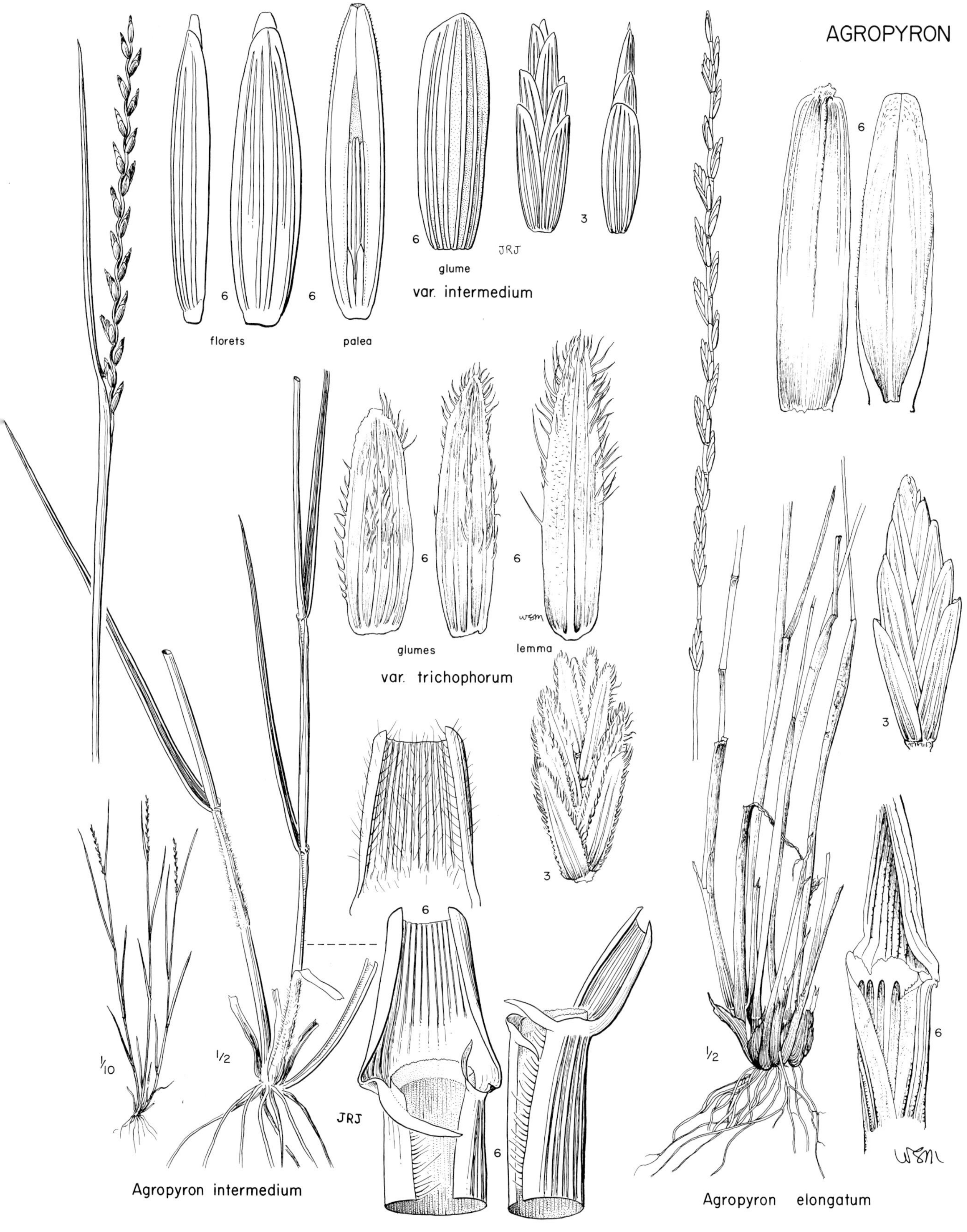
AGROPYRON
6
6
6
6
florets
palea
6
glume
JRJ
3
var. intermedium
6
6
glumes
lemma
var. trichophorum
3
6
1/10
1/2
JRJ
6
Agropyron intermedium
6
3
1/2
6
Agropyron elongatum

dm tall, erect to decumbent; sheaths glabrous or the lowermost soft-pilose; ligules very short, mostly about 0.5 mm long, erose-ciliolate; blades usually flat, (2) 3–6 (14) mm broad, scabrous, sometimes glabrous on the upper surface, the auricles often narrow; spikes (8) 10–15 (20) cm long, erect, the rachis joints 4–6.5 (8) mm long in the middle of the spike; spikelets (9) 12–16 mm long, 3- to 5- (8)-flowered, the florets closely crowded; glumes subequal, 6–10 mm long, rigid, lanceolate, acute and usually awned, 5- to 7-nerved, glabrous or the midnerve usually scabrous, the awn somewhat distinct from the apex of the body, 0.5–4 (5) mm long; lemmas (6) 7–10 (12) mm long, acute, tapering into a point or an awn, glabrous to scaberulous apically, 5-nerved, the awn 0.5–4 (10) mm long, straight; lodicules ovate, ciliolate apically, about 1 mm long; anthers 4–5.5 mm long; $2n = 42$.

Growing as a weed along ditch banks, roadsides and moderately moist waste places in valley bottoms to subalpine elevs.; native to Europe and Asia; Alaska to Newfl., s. to the w. states and across the c. and e. states to Kansas, Ark., Ky., and N.C.; also in Mex. June–Aug.

Quackgrass is an excellent soil binder and valuable forage grass which too often becomes a troublesome weed. Fernald (1933) recognized 8 forms in northeastern United States. The only one that is easily distinguishable here is f. *aristatum* (Schumacher) O. R. Holmberg, to which specimens with awned lemmas (2–9 mm long) could be referred.

5. **Agropyron smithii** Rydb.

Agropyron spicatum var. *palmeri* Scribn. & Smith, U.S.D.A. Div. Agrostol. Bull. 4: 33. 1897. *A. smithii* var. *palmeri* A. A. Heller, Cat. N. Amer. Pl. ed. 2. 3. 1900. *A. occidentale* var. *palmeri* Scribn. U.S.D.A. Div. Agrostol. Circ. 27: 9. 1900. *A. palmeri* Rydb. Agric. Exp. Sta. Agric. Coll. Colorado Bull. (Fl. Colorado) 100: 55. 1906. (*Palmer,* Ariz., in 1869; lectotype by A. S. Hitchcock.)

A. spicatum var. *molle* Scribn. & Smith, U.S.D.A. Div. Agrostol. Bull. 4: 33. 1897. *A. molle* Rydb. Mem. New York Bot. Gard. 1: 65. 1900. *A. occidentale* var. *molle* Scribn. U.S.D.A. Div. Agrostol. Circ. 27: 9. 1900. *A. smithii* var. *molle* M. E. Jones, Contr. W. Bot. 14: 18. 1912. *Zeia mollis* Lunell, Amer. Midl. Naturalist 4: 226. 1915. *A. smithii* f. *molle* Gillett & Senn, Canad. J. Bot. 38: 750. 1960. (*Rydberg 3193,* East Gallatin swamps, Mont., 24 July 1896; lectotype by A. S. Hitchcock.)

A. smithii Rydb. Mem. New York Bot. Gard. 1: 64. 1900. *Zeia smithii* Lunell, Amer. Midl. Naturalist 4: 227. 1915. *Elymus smithii* Gould, Madroño 9: 127. 1947. *Elytrigia smithii* A. Löve, Bot. Not. 1950: 31. 1950. (*Geyer,* "Valley of the Missouri," in 1839.)

Western wheatgrass, bluestem.

Strongly rhizomatous, usually glaucous perennials; culms 3–8 (9) dm tall; leaves few to many; sheaths glabrous to densely pubescent; ligules inconspicuous, mostly about 0.5 mm long, erose-ciliolate; blades mostly flat when fresh, becoming involute in drying, (1) 2–4.5 mm broad, firm, scabrous and sometimes pilose on the upper surface, the auricles prominent, often up to 2 mm long and clasping the culm; spikes (3) 5.5–15 (20) cm long, stiff, erect, the rachis scabrous on the angles, the rachis internodes (4.5) 5.5–10 (16) mm long in the middle of the spike, the spikelets mostly closely imbricate, sometimes 2 per node; spikelets usually long, 12–24 (26) mm long, (3-) 4- to 8- (10)-flowered; glumes rather rigid, linear-lanceolate to lanceolate, gradually tapering from below the middle and passing imperceptibly into an awn tip, asymmetrical, 3- to 5-nerved, glabrous to scabrous on the nerves, the first glume (6) 7–9 (12) mm long and the second (7) 9–10 (15) mm long; lemmas relatively large, 8–11 (14) mm long, acute, glabrous to glabrate, sometimes densely hirsute, 5-nerved, the nerves sometimes apparent only towards the apex, mostly awn-tipped or with an awn up to 5 mm long; lodicules ovate, often lobed, ciliolate, usually more than 1 mm long; anthers 2–4.4 mm long; $2n = 56$.

Dry sagebrush deserts and foothills, also along ditch banks and roadsides in sandy to heavy soils, often in alkaline soils in bottomlands; B.C. to Ont., s. to ne. Calif., ne. Nev., Ariz., N.M., Texas, Mo., Ill., Ind., Ky., and N.Y. June–Aug.

Western wheatgrass often forms large, pure types in hilly, sandy areas and is also characteristic of alkaline meadows in bottomlands. It is easily recognized by its glaucous foliage and often becomes a dominant roadside plant, especially in areas sprayed with 2,4-D. Western wheatgrass is a nutritious and important grass that withstands more grazing than most of our native wheatgrasses.

Plants along the Green River often have spikes no more than 3 cm long, much shorter than found throughout the remainder of our area. Gillett and Senn (1960) found *A. smithii* to be an octoploid throughout its range and suggested it to be of hybrid origin involving a species of *Elymus. Agropyron smithii* is a peculiar *Agropyron* with its narrow glumes and sometimes paired spikelets, which are suggestive of *Elymus. Agropyron dasystachyum* and *Elymus triticoides* are the putative parents (Dewey, Amer. J. Bot. 62: 524–530. 1975). Gillett and Senn found that the often recognized var. *molle,* a phase with pubescent lemmas, does not segregate into geographically distinct areas and should be regarded as no more than a form. The isotype of var. *molle* at NY, as a matter of fact, includes 2 plants, one that would key to "*molle*" and the other to the typical "*smithii.*" Variety *palmeri,* however, may be worthy of recognition and can be distinguished in the following key.

1 Sheaths glabrous; range of the species .... var. *smithii*
1 Sheaths finely puberulent, at least on the lower part of the plant; Utah to e. Colo. and e. to Ariz. and N.M. .... var. *palmeri* (Scribn. & Smith) A. A. Heller

6. **Agropyron dasystachyum** (Hook.) Scribn.

Triticum repens var. *dasystachyum* Hook. Fl. Boreali-Amer. 2: 254. 1840. *Triticum dasystachyum* A. Gray, Manual Bot. 602. 1848. *Agropyron dasystachyum* Scribn. Bull. Torrey Bot. Club 10: 78. 1883. *Zeia dasystachya* Lunell, Amer. Midl. Naturalist 4: 226. 1915. *Elytrigia dasystachya* A. Löve & D. Löve, Bull. Torrey Bot. Club 81: 33. 1954. (*Richardson,* "Carlton House Fort, on the Saskatchawan.")

Triticum repens var. *subvillosum* Hook. Fl. Boreali-Amer. 2: 254. 1840. *A. dasystachyum* var. *subvillosum* Scribn. & Smith, U.S.D.A. Div. Agrostol. Bull. 4: 33. 1897. *A. repens* zeta *aristatum* eta *subvillosum* P. Candargy, Tribu des Hordés 48. 1901. *A. subvillosum* E. Nels. Bot. Gaz. 38: 378. 1904. *Elymus subvillosus* Gould, Madroño 9: 127. 1947. (*Richardson,* "Fort Norman on the Mackenzie River.") = var. *dasystachyum.*

A. lanceolatum Scribn. & Smith, U.S.D.A. Div. Agrostol. Bull. 4: 34. 1897. *Elymus lanceolatus* Gould, Madroño 10: 94. 1949. (*Palmer 266,* Blackfoot, Idaho, in 1893; lectotype by A. S. Hitchcock.) = var. *dasystachyum.*

A. riparium Scribn. & Smith, U.S.D.A. Div. Agrostol. Bull. 4: 35. 1897. *A. smithii* var. *riparium* M. E. Jones, Contr. W. Bot. 14: 19. 1912. *Zeia riparia* Lunell, Amer. Midl. Naturalist 4: 227. 1915. *Elymus riparius* Gould, Madroño 9: 127. 1947; not Wieg. 1918. *Elymus rydbergii* Gould, Madroño 10: 94. 1949. *A. dasystachyum* var. *riparium* Bowden, Canad. J. Bot. 43: 1434. 1965. (*Rydberg 2127,* "river-bank," Garrison, Mont., 10 July 1895; lectotype by A. S. Hitchcock.)

Thickspike wheatgrass.

Strongly rhizomatous, usually glaucous perennials; culms 3.5–9 (13) dm tall; sheaths glabrous to hirsute;

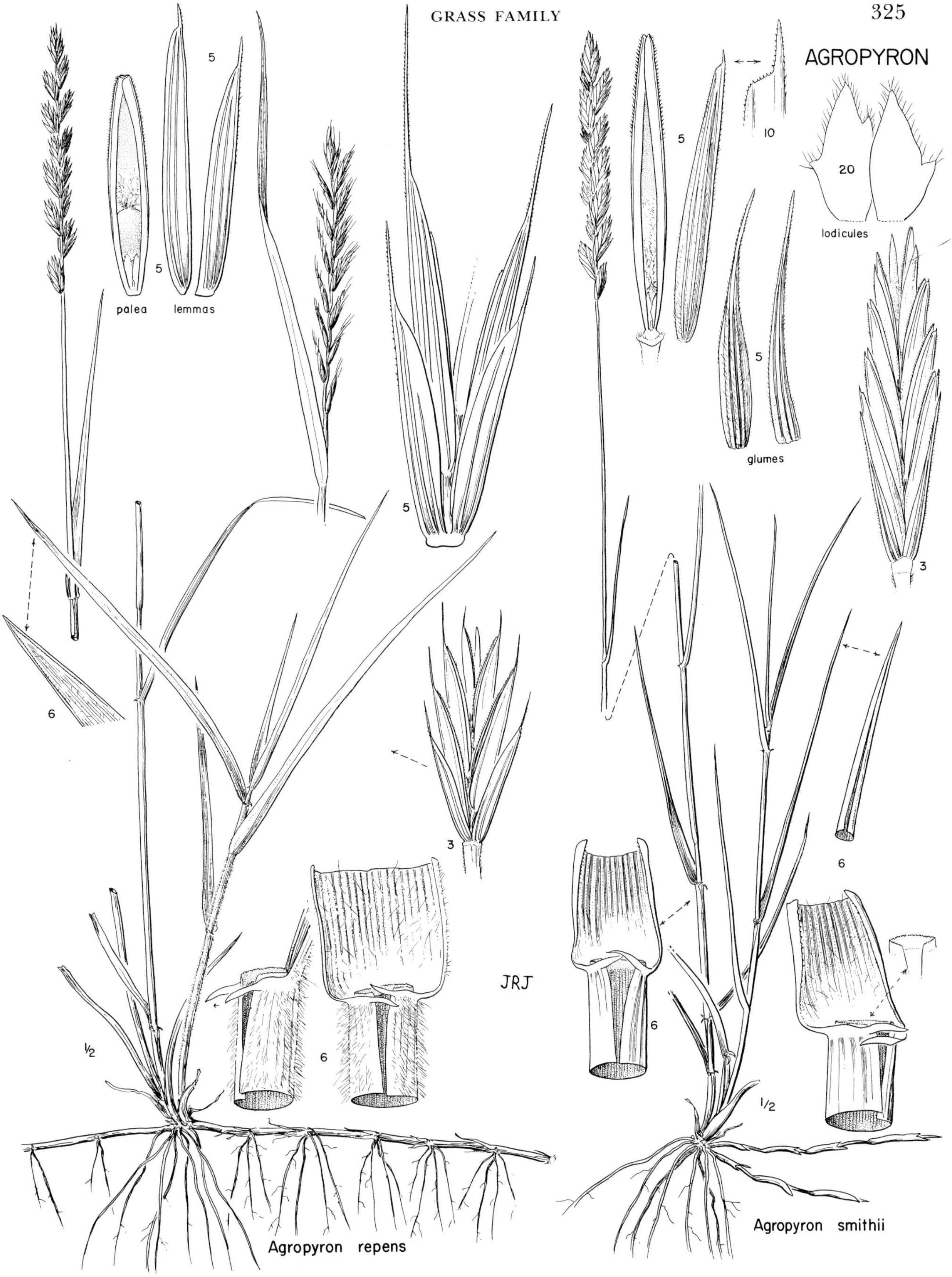
AGROPYRON
palea
lemmas
5
6
3
1/2
JRJ
Agropyron repens
10
20
lodicules
glumes
Agropyron smithii

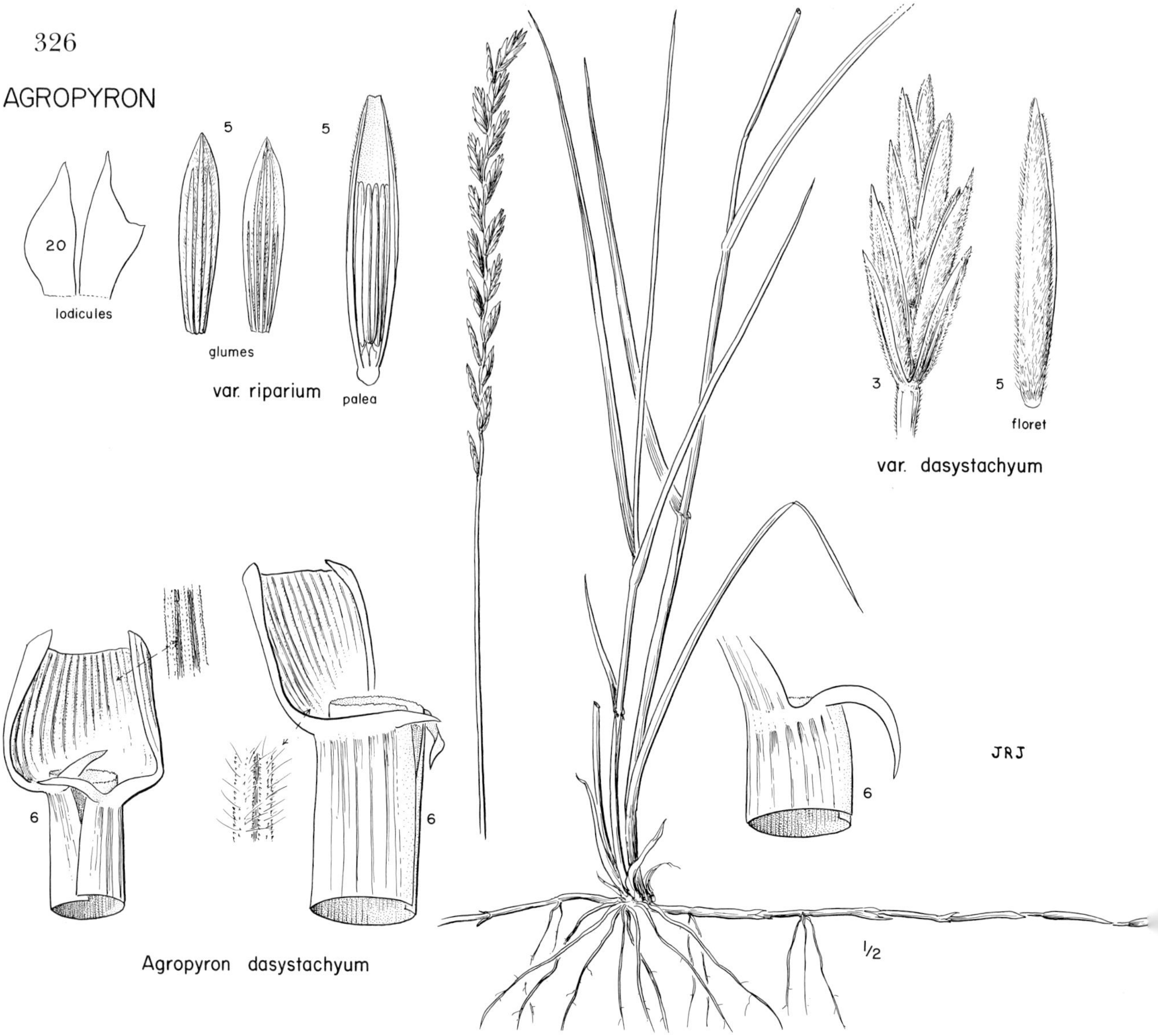

ligules very short, mostly about 0.5 mm long, erose-ciliolate; blades involute or sometimes flat, 1–3.5 (5) mm broad, firm and stiff, glabrous to scaberulous, sometimes pilose, the auricles up to 1.5 mm long; spikes sometimes long, 6–22 cm, stiff, slender, erect, the rachis internodes (5) 7–12.5 mm long in the middle of the spike, the spikelets sometimes 2 per node; spikelets 11–18 (24) mm long, 3- to 6- (11)-flowered; glumes oblong-lanceolate, acute to acuminate, broadest at or above the midlength, pubescent, rarely glabrous, inconspicuously 3- to 5-nerved, the first glume 4–8.5 mm long and the second 5.5–10 (11) mm long; lemmas 7–10 (11) mm long, villous or sometimes glabrous, acute, sometimes awn-tipped, 5-nerved apically; lodicules lanceolate, sometimes lobed, about 1 mm long; anthers (3.5) 4–5 mm long; $2n = 28$.

Sagebrush deserts and foothill woodlands, in sandy to heavy soils; Alaska to Ont., s. to Calif., Ariz., N.M., Nebr., Ill. and Mich. June–Aug.

Agropyron dasystachyum is closely related to *A. smithii* (see discussion above). Two morphologically, and to some extent ecologically, distinct varieties can be recognized as follows.

1 Lemmas scabrous to villous; plants of dry, usually sandy soils .... var. *dasystachyum*

1 Lemmas glabrous to scaberulous; plants of more mesic habitats, usually in heavy soils .... var. *riparium* (Scribn. & Smith) Bowden

7. **Agropyron albicans** Scribn. & Smith

Agropyron albicans Scribn. & Smith, U.S.D.A. Div. Agrostol. Bull. 4: 32. 1897. *Zeia albicans* Lunell, Amer. Midl. Naturalist 5: 94. 1917. *A. dasystachyum* subsp. *albicans* D. R. Dewey, Amer. J. Bot. 56: 669. 1969. (*Rydberg 3405,* Yogo Gulch, Mont., 22 Aug. 1896.)

A. griffithsi Scribn. & Smith ex Piper, Proc. Biol. Soc. Wash. 18: 148. 1905. *Zeia griffithsii* Lunell, Amer. Midl. Naturalist 5: 234. 1918. *A. albicans* var. *griffithsii* A. A. Beetle, Rhodora 54: 196. 1952. *A. dasystachyum* subsp. *albicans* var. *griffithsii* D. R. Dewey, Amer. J. Bot. 56: 669. 1969. (*Williams & Griffiths 140,* dry hills, North Fork of Clear River, Wyo., 10 Aug. 1898.)

Rhizomatous perennials; culms 4–8 (10) dm tall; sheaths glabrous; ligules very short, scarcely 0.5 mm long, ciliolate; blades narrow, involute, rarely flat,

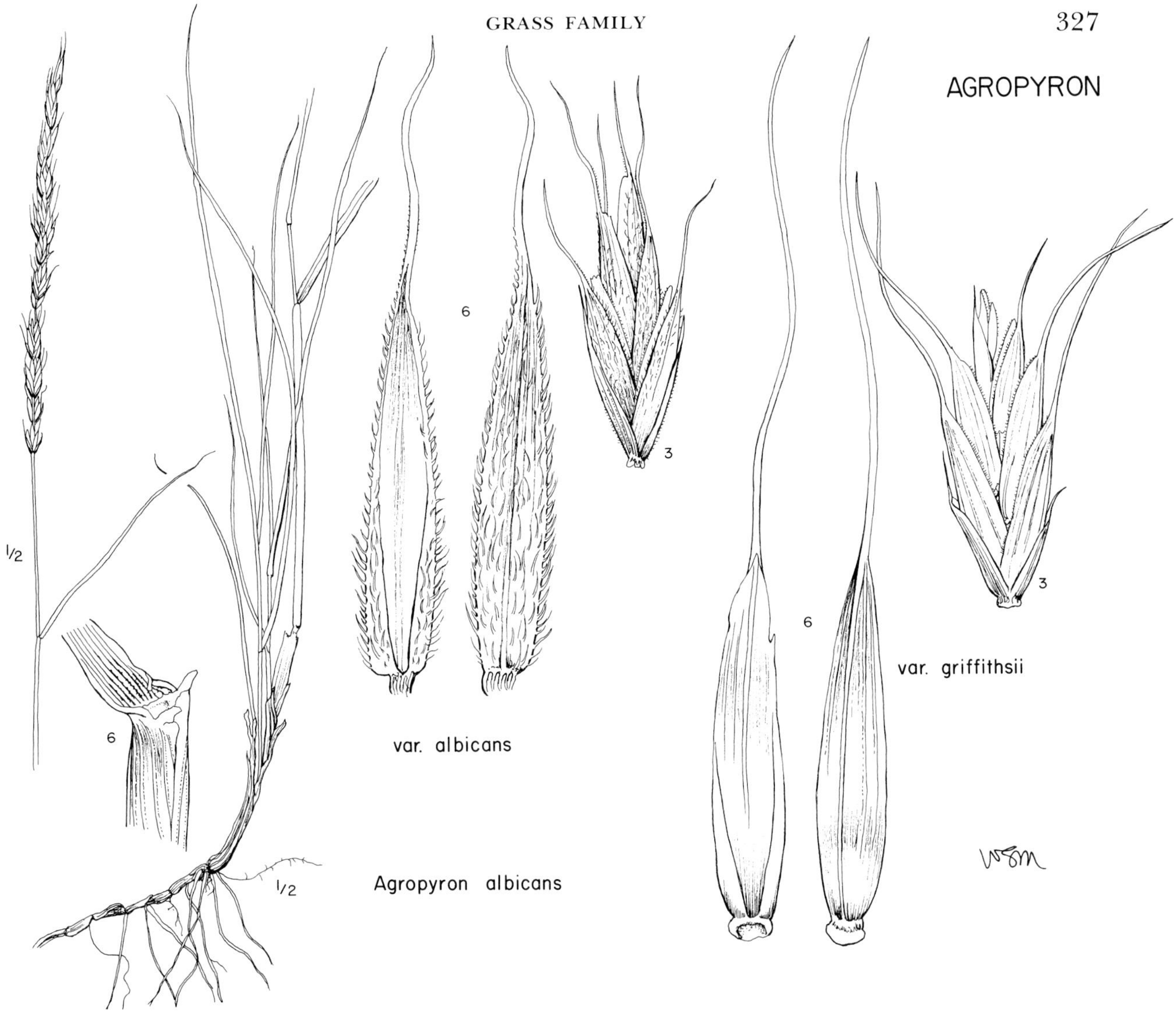

1–3 mm broad, firm, glabrous or sometimes scaberulous on the upper surface, the auricles usually present; spikes (4) 6–10 (14) cm long, slender, erect, the rachis internodes 6–10 mm long in the middle of the spike; spikelets 10–16 (18) mm long, 3- to 6- (7)-flowered; glumes oblong-lanceolate, acute or acuminate, usually with a short awn 0.5–3 (4) mm long, 3- to 5-nerved, scabrous or glabrous, the first glume (4) 4.5–6.5 (8) mm long and the second (4.5) 5.5–8 mm long; lemmas 7.5–9.5 mm long, glabrous to copiously pubescent, 5-nerved apically, the nerves sometimes obscured by pubescence, the awn (4) 5–12 mm long, arcuate, divergent at maturity; anthers 3–5 mm long; $2n = 28$.

Dry sagebrush hillsides and wooded slopes in the mts. in shallow rocky soil; B.C. and Sask., s. to Wash., Idaho, n. Utah, Colo. and S.D. Late June–Aug.

Agropyron albicans has two varieties which can be distinguished as follows.

1 Lemmas pubescent .... var. *albicans*
1 Lemmas glabrous . . var. *griffithsii* (Scribn. & Smith) A. A. Beetle

Dewey (1970) demonstrated that *A. albicans* and its varieties are hybrids involving *A. spicatum* var. *spicatum* and both varieties of *A. dasystachyum.* Such a parentage would account for the presence of the divergent awns (from *A. spicatum* var. *spicatum*) and for the pubescent and glabrous lemmas that separate the two varieties of *A. dasystachyum,* var. *dasystachyum* and var. *riparium.* Populations are known in which *A. albicans* appears to be reproducing itself.

8. **Agropyron spicatum** (Pursh) Scribn. & Smith

Festuca spicata Pursh, Fl. Amer. Sept. 83. 1814. *Schenodorus spicatus* Roemer & Schultes, Syst. Veg. 2: 707. 1817. *Agropyron spicatum* Scribn. & Smith, U.S.D.A. Div. Agrostol. Bull. 4: 33. 1897. *Zeia spicata* Lunell, Amer. Midl. Naturalist 4: 227. 1915. *Elymus spicatus* Gould, Madroño 9: 125. 1947. (*Lewis & Clark,* "On the plains of Columbia," between Camp Chopunniash and the Weippe prairie, likely from the mt. ridges between Clearwater Canyon and Lolo Creek Canyon n. of Kamiah, Idaho Co., Idaho, 10 June 1806; lectotype by A. S. Hitchcock.)

A. divergens var. *tenue* Vasey, Grasses U. S. 96. 1885, nomen nudum. *A. vaseyi* Scribn. & Smith, U.S.D.A. Div. Agrostol. Bull. 4: 27. 1897. *A. spicatum* var. *vaseyi* E. Nels. Bot. Gaz. 38: 378. 1904. (*Rydberg 2299,* Dillon, Mont., 2 Aug. 1895; lectotype by A. S. Hitchcock.) The *A. spicatum* phase.

A. divergens var. *tenuispicum* Scribn. & Smith, U.S.D.A. Div. Agrostol. Bull. 4: 27. 1897. *A spicatum* var. *tenuispicum* Rydb. Mem. New York Bot. Gard. 1: 61. 1900. (*T. Howell*

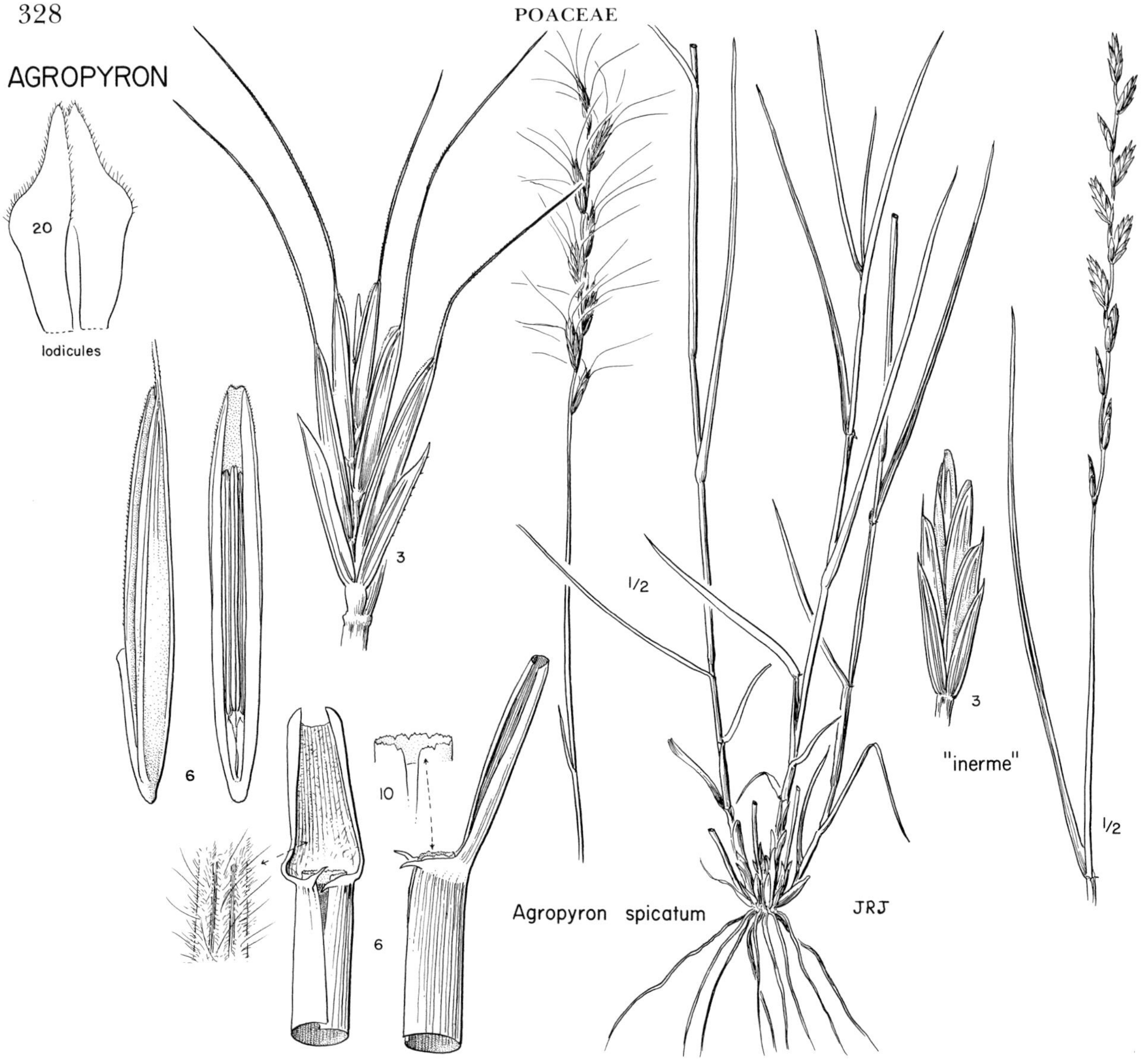

181, Oregon, in 1885; lectotype by A. S. Hitchcock.) A rhizomatous form of the *A. spicatum* phase.

A. divergens var. *inermis* Scribn. & Smith, U.S.D.A. Div. Agrostol. Bull. 4: 27. 1897. *A. spicatum* var. *inerme* A. A. Heller, Cat. N. Amer. Pl. ed. 2. 3. 1900. *A. inerme* Rydb. Bull. Torrey Bot. Club 36: 539. 1909. *A. spicatum* f. *inerme* A. A. Beetle, Leafl. W. Bot. 6: 162. 1951. (*Henderson 3058,* Idaho, in 1895; lectotype by A. S. Hitchcock.)

A. strigosum of authors; not (M. Bieb.) Boiss., 1884.

Bluebunch wheatgrass.

Tufted perennials, sometimes producing rhizomes as a result of introgression with *A. dasystachyum,* the bunches sometimes up to 1.5 dm across the base, the herbage green or glaucous; culms (3) 4–9 (10) dm tall, erect, glabrous or puberulent below the nodes; leaves numerous, mostly cauline; sheaths glabrous to retrorse-puberulent, the old sheaths persisting in large clumps; ligules less than 1 mm long, erosc-ciliolate; blades flat to loosely involute, (1) 2–3.5 (4.5) mm broad, usually pilose on the upper surface, rarely pubescent on both surfaces, the auricles well developed; spikes (5) 8–16 (20) cm long, slender, loose, open, the remote spikelets just exceeding the length of the internodes or somewhat shorter, the rachis internodes 9–17 mm long in the middle of the spike, scabrous on the angles; spikelets 12–16 (20) mm long, 4- to 6- (9)-flowered; glumes narrowly oblong to obovate, rounded to acute, rarely awn-tipped, 4- to 5-nerved, the margins scarious, glabrous or scabrous on the nerves, about 1/2 the length of the spikelet, the first glume 4.5–8 (10) mm long and the second 5.5–10 (11) mm long; lemmas (7) 8–10 (11) mm long, mostly glabrous, rarely scaberulous, 5-nerved apically, awnless or usually awned with divergent awns 9–15 (25) mm long; lodicules lanceolate, ciliolate, about 1.5 mm long; anthers 4–6 mm long, purplish; $2n = 14, 28$.

Dry mt. slopes at middle elevs. with sagebrush, Gambel oak and pinyon-juniper vegetation, sometimes at subalpine elevs.; Alaska to B.C. and Alta., s. to ne. Calif., Ariz., Texas and e. to Sask. and the Keweenaw Peninsula, Mich. June–early Aug.

Bluebunch wheatgrass often forms pure stands and is one of the most important bunchgrasses of our area. It is one of the co-dominants of the Palouse Prairie of eastern Washington, adjacent

Idaho and parts of eastern Oregon, southern Idaho and northern Utah, along with *Festuca idahoensis* and *Poa sandbergii.*

Agropyron spicatum has been recognized as having two varieties, var. *spicatum* and var. *inerme.* Variety *inerme* is sometimes treated as a distinct species. Stebbins and Pun (1953) presented cytogenetic evidence showing the two varieties to be completely interfertile with normal pairing in meiosis. When found growing together every conceivable form from one to the other may be found, as they hybridize freely and form fertile progeny. The *"inerme"* form may be distinguished from the *"spicatum"* form by its awnless or nearly awnless lemmas.

Daubenmire (1960) studied both the short rhizomatous and the cespitose forms of *A. spicatum,* crossing the two and experimentally testing the progeny for ecological preference. He concluded that they are ecotypes, with the cespitose phase showing preference for arid grasslands and semidesert and the rhizomatous phase for moderately mesic grassland.

Agropyron spicatum is the only native North American diploid species of the genus ($2n = 14$). Autotetraploid ($2n = 28$) races also occur. It has been found that the *"spicatum"* genome is similar to one of the genomes in several tetraploid species such as *A. trachycaulum, Elymus virginicus* and *Sitanion jubatum* (Stebbins and Snyder, 1956), *Elymus glaucus* (Stebbins and Singh, 1950) and *Sitanion hystrix* (Dewey, 1964). The hybrid *A. spicatum* X *Sitanion hystrix* has been named X *Agrositanion saxicola* (Scribn. & Smith) Bowden.

Agropyron arizonicum Scribn. & Smith of the Southwest (s. Ariz., s. N.M., Trans-Pecos Texas, and adj. Mex.) has been reported from the Toiyabe Range of Nevada (*McGowan 60,* North Twin River, 25 June 1917). *Agropyron arizonicum* has longer, flexuous spikes, 15–30 cm long (straight and 5–16 [20] cm in *A. spicatum*) and broader blades, 4–6 mm broad (1–3.5 [4.5] mm in *A. spicatum*). The flexuous spike of the McGowan specimen appears to be due to improper collecting techniques, and the spike size and blade width are those of a robust specimen of *A. spicatum.*

9. **Agropyron trachycaulum** (Link) Malte

Triticum pauciflorum Schwein. in Keating, Narr. Exped. St. Peter's River 2: 383. 1824. *Agropyron pauciflorum* A. S. Hitchc. ex Silveus, Texas Grasses 158. 1933; not Schur, 1859. *Roegneria pauciflora* Hylander, Uppsala Univ. Årsskr. 7: 36, 89. 1945. *Elymus pauciflorus* Gould, Madroño 9: 126. 1947; not Lam. 1791. (*Say,* Prairies of St. Peter, Minn., in 1823.) = var. *trachycaulum.*

Triticum trachycaulum Link, Hort. Regius Bot. Berol. 2: 189. 1833. *Crithopyrum trachycaulon* H. Prag. ex Steudel, Syn. Pl. Glum. 1: 344. 1854, as synonym. *A. trachycaulon* Hort. ex Steudel, Syn. Pl. Glum. 1: 344. 1854, as a synonym of the basionym. *A. trachycaulum* Malte, Annual Rep. Natl. Mus. Canad. 1930: 42. 1932. *Roegneria trachycaulon* Nevski in Komarov, Fl. USSR 2: 599. 1934. *Elymus trachycaulus* Gould ex Shinners, Rhodora 56: 28. 1954. (Grown from seed collected by *Richardson* in w. N. Amer.)

Triticum subsecundum Link, Hort. Regius Bot. Berol. 2: 190. 1833. *A. subsecundum* A. S. Hitchc. Amer. J. Bot. 21: 131. 1934. *Elymus pauciflorus* subsp. *subsecundus* Gould, Madroño 9: 126. 1947. *Elymus subsecundus* A. Löve & D. Löve, Taxon 13: 201. 1964. (Grown from seed collected by *Richardson* in w. Can.) = var. *unilaterale.*

Triticum richardsoni Schrader, Linnaea 12: 467. 1838. *A. richardsonii* Schrader, ibid., as a synonym. *Cryptopyrum richardsonii* Heynh. Nomencl. Bot. Hort. 2: 174. 1846. *A. caninum* var. *richardsoni* M. E. Jones, Contr. W. Bot. 14: 18. 1912. *Zeia richardsonii* Lunell, Amer. Midl. Naturalist 4: 227. 1915. *A trachycaulum* var. *richardsoni* Malte in H. F. Lewis, Canad. Field-Naturalist 45: 201. 1931. (*Richardson,* "America borealis arctica.") = var. *unilaterale.*

A. tenerum Vasey, Bot. Gaz. 10: 258. 1885. *A. repens* var. *tenerum* Beal, Grasses N. Amer. 2: 637. 1896. *A caninum* var. *tenerum* Pease & Moore, Rhodora 12: 71. 1910. *Zeia tenera* Lunell, Amer. Midl. Naturalist 4: 227. 1915. *A. trachycaulum* var. *tenerum* Malte, Annual Rep. Natl. Mus. Canad. 1930: 44. 1932. (*Vasey,* Fort Garland, Colo., in 1884; lectotype by Piper.) = var. *trachycaulum.*

A. unilaterale Cassidy, State Agric. Coll. Colorado Bull. 12: 63. 1890; not Beauv. 1812. *A. caninum* var. *unilaterale* Vasey, Contr. U. S. Natl. Herb. 1: 279. 1893. *A caninum* f. *violacescens* Ramaley, Minnesota Bot. Stud. 1: 107. 1894. *A. violacescens* Beal, Grasses N. Amer. 2: 635. 1896. *A. trachycaulum* var. *unilaterale* Malte, Annual Rep. Natl. Mus. Canad. 1930: 46. 1932. *A. caninum* subsp. *majus* var. *unilaterale* C. L. Hitchc. Univ. Wash. Publ. Biol. 17(1): 450. 1969. (Cassidy in publishing his later homonym did not cite a type, but Vasey in publishing the new name cites the following lectotype, *Scribner 422,* Hound Creek Valley, Mont., 1 Aug. 1883.)

A. violaceum var. *major* Vasey, Cont. U. S. Natl. Herb. 1: 280. 1893. *A. tenerum* subsp. *majus* Vasey ex Piper, Bull. Torrey Bot. Club 32: 543. 1905. *A. trachycaulum* var. *majus* Fern. Rhodora 35: 171. 1933. *A. pauciflorum* subsp. *majus* Melderis, Ark. Bot. II. 7(1): 20. 1968. *A. caninum* subsp. *majus* C. L. Hitchc. Univ. Wash. Publ. Biol. 17(1): 450. 1969. *Elymus trachycaulus* subsp. *major* Tzvelev, Novit. Syst. Pl. Vasc. 10: 24. 1973. (*Cusick 1134,* [possibly Union Co.] Oregon, in 1884.) = var. *trachycaulum.*

A. violaceum f. *caninoides* Ramaley, Minnesota Bot. Stud. 1: 108. 1894. *A. caninoides* Beal, Grasses N. Amer. 2: 640. 1896. *A. trachycaulum* f. *caninoides* B. Boivin, Naturaliste Canad. 94: 520. 1967. (*MacMillan & Sheldon 84,* Minn.) = var. *unilaterale.*

A. tenerum var. *ciliatum* Scribn. & Smith, U.S.D.A. Div. Agrostol. Bull. 4: 30. 1897. *A. tenerum* subsp. *trichocoleum* Piper, Bull. Torrey Bot. Club 32: 546. 1905. *A. caninum* var. *tenerum* f. *ciliatum* Pease & Moore, Rhodora 12: 72. 1910. *A. trachycaulum* var. *trichocoleum* Malte, Annual Rep. Natl. Mus. Canad. 1930: 45. 1932. *A. trachycaulum* var. *novae-angliae* f. *ciliatum* Carpenter in Dole, Fl. Vermont, ed. 3. 31. 1937. *A. trachycaulum* var. *ciliatum* Gleason, Phytologia 4(1): 21. 1952; not Malte, 1932. (*Vasey,* Duluth, Minn., in 1881; lectotype by Piper.) = var. *trachycaulum* or var. *latiglume.*

A. violaceum var. *latiglume* Scribn. & Smith, U.S.D.A. Div. Agrostol. Bull. 4: 30. 1897. *A. biflorum* subsp. *latiglume* Piper, Bull. Torrey Bot. Club 32: 547. 1905. *A. latiglume* Rydb. Bull. Torrey Bot. Club 36: 539. 1909. *A. caninum* var. *latiglume* Pease & Moore, Rhodora 12: 73. 1910. *Roegneria latiglumis* Nevski, Trudy Bot. Inst. Akad. Nauk SSSR, Ser. 1, Fl. Sist. Vysš. Rast. 2: 55. 1936, without basionym; Hylander, Uppsala Univ. Årsskr. 1945, no. 7: 36. 1945. *A. trachycaulum* var. *latiglume* A. A. Beetle, Rhodora 54: 196. 1952. *A. caninum* subsp. *majus* var. *latiglume* C. L. Hitchc. Univ. Wash. Publ. Biol. 17(1): 450. 1969. (*Tweedy 1011,* "Rocky subalpine slides," 9500 ft., Lone Mt., Gallatin Co., Mont., Aug. 1886; lectotype by A. S. Hitchcock.)

A. violaceum var. *andinum* Scribn. & Smith, U.S.D.A. Div. Agrostol. Bull. 4: 30. 1897. *A. biflorum* subsp. *andinum* Piper, Bull. Torrey Bot. Club 32: 547. 1905. *A. andinum* Rydb. Agric. Exp. Sta. Agric. Coll. Colorado Bull. (Fl. Colorado) 100: 54. 1906. *A. caninum* var. *andinum* Pease & Moore, Rhodora 12: 75. 1910. *A. latiglume* var. *andinum* Malte, Annual Rep. Natl. Mus. Canad. 1930: 36. 1932. *A. subsecundum* var. *andinum* A. S. Hitchc. Amer. J. Bot. 21: 132. 1934. *A. trachycaulum* var. *unilaterale* f. *andinum* A. A. Beetle, Rhodora 54: 196. 1952. *A. violaceum* subsp. *andinum* Melderis, Ark. Bot. II. 7(1): 19. 1968. *A. caninum* subsp. *majus* var. *andinum* C. L. Hitchc. Univ. Wash. Publ. Biol. 17(1): 450. 1969. (*Jones 720,* Grays Peak, Colo., 28 Aug. 1878; lectotype by A. S. Hitchcock.) According to Bowden (1965) the type is a hybrid between *A. violaceum* (Hornem.) Lange and *A. scribneri* Vasey (see following synonym).

A. brevifolium Scribn. U.S.D.A. Div. Agrostol. Bull. 11: 55, pl. 13. 1898. (*Elmer 676,* North Fork of Bridge Creek, Okanogan Co., Wash., 10 Sept. 1897.) Bowden (1965) designated *A.* X *brevifolium* (pro sp.) as a hybrid between *A. violaceum* (Hornem.) Lange (not in our region) and *A. scribneri* Vasey.

A. novae-angliae Scribn. in Brainerd, Jones & Eggleston, Fl. Vermont 9, 103. 1900. *A. tenerum* var. *novae-angliae* Farw. Annual Rep. Michigan Acad. Sci. 21: 355. 1920. *A. trachycaulum* var. *novae-angliae* Fern. Rhodora 35: 174. 1933. *Elymus donianus* subsp. *novae-angliae* D. Löve, Taxon 17: 89. 1968. *A. pauciflorum* subsp. *novae-angliae* Melderis, Ark. Bot. II. 7(1): 20. 1968. *Elymus trachycaulus* subsp. *novae-angliae* Tzvelev, Novit. Syst. Pl. Vasc. 10: 23. 1973. (*Grout & Eggleston,* Willoughby Mt., Westmore, Vt., 2 July 1894.) = var. *trachycaulum.*

AGROPYRON

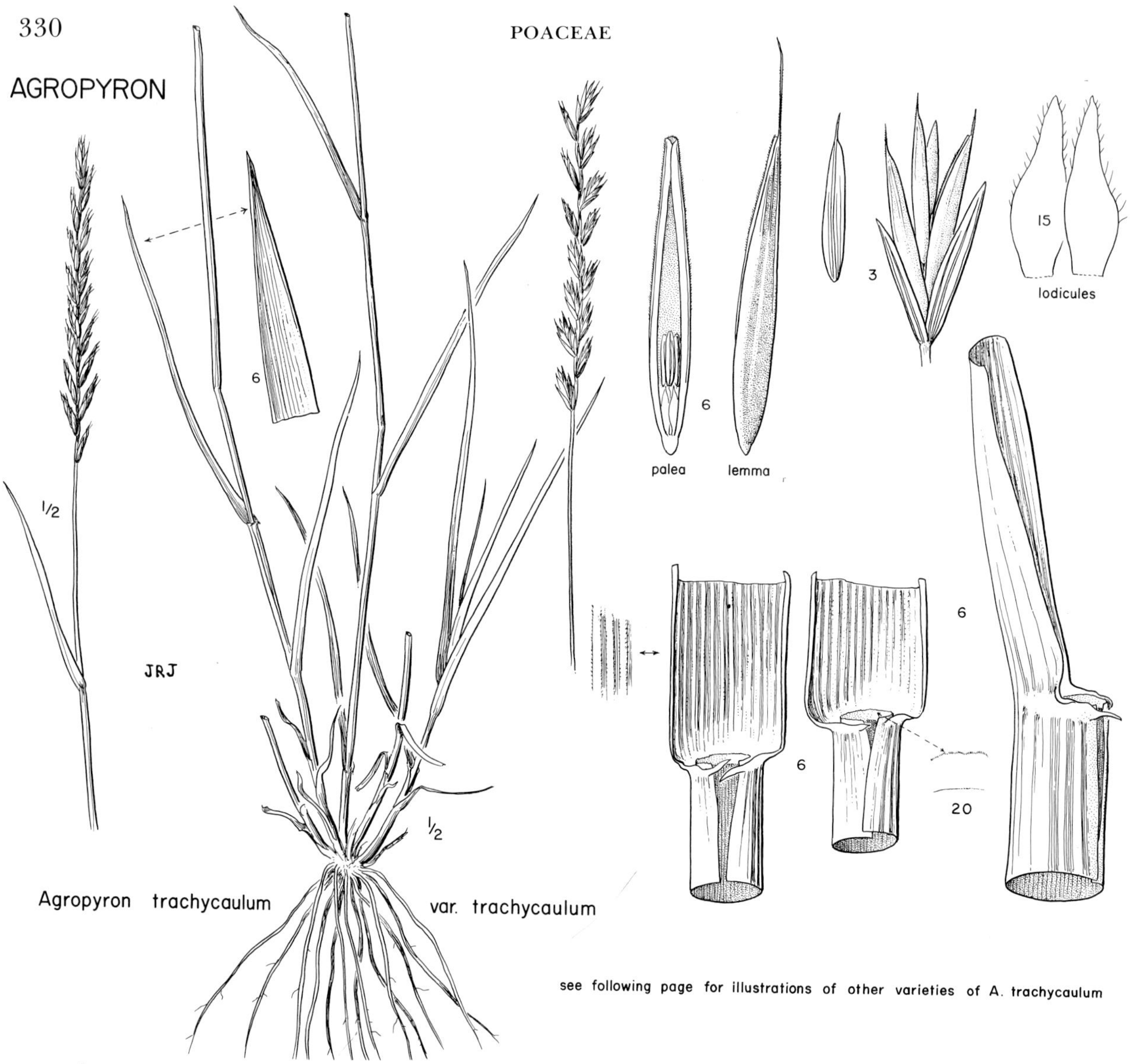

A. caninum f. *glaucum* Pease & Moore, Rhodora 12: 71. 1910. *A. trachycaulum* var. *glaucum* Malte, Annual Rep. Natl. Mus. Canad. 1930: 47. 1932. (*Fernald 1367,* Kelly Point, Pembroke, Me., 31 July 1909.)

A. caninum of authors; not (L.) Beauv. 1812. = var. *unilaterale.*

A. violaceum of authors; not (Hornem.) Lange, 1880. = var. *latiglume.*

Slender wheatgrass, bearded wheatgrass.

Strongly tufted perennials, rarely producing rhizomes; culms often tall, 3–10 (15) dm; sheaths glabrous or finely hispid to pilose; ligules very short, scarcely 0.5 mm long, ciliolate; blades usually flat, (1) 2–6 (8) mm broad, scabrous to pilose, at least on the upper surface, the auricles short or lacking; spikes sometimes very long, 4–15 (25) cm, compact, the spikelets overlapping, the rachis internodes (4) 5–8 (11) mm long in the middle of the spike; spikelets 9–16 (18) mm long, 3- to 5- (7)-flowered; glumes lanceolate to oblong-elliptic, acute to short-awned, the margins scarious, 5- to 7-nerved, the first glume 6–10 (11) mm long and the second 7–12 mm long; lemmas 7.5–10 (13) mm long, glabrous or scabrous towards the apex and sometimes short hirsute on the margins, awnless to awned with a straight awn 1–4 or 5–30 mm long; lodicules lanceolate, ciliolate, about 1 mm long; anthers 1–1.8 (2) mm long; $2n = 28$.

Dry to moderately moist roadsides, ditchbanks, streambanks, meadows and woodlands from the valley bottoms to subalpine and alpine elevs.; Alaska to Newfl., s. to Calif., Ariz., N.M., Kansas, Mo., Ky., and N.C. June–Aug.

Agropyron trachycaulum is a valuable forage grass, occasionally planted in reclaimed rangeland.

Agropyron caninum (L.) Beauv., an Old World species, is considered by many to be conspecific with *A. trachycaulum* of the New World. *Agropyron caninum* differs in having narrow, mostly 3- (5)-nerved, rigid glumes and statistically longer anthers which average 2 mm long (see Malte, 1932). *Agropyron trachycaulum* has broader, 5- to 7-nerved, more or less membranous margined glumes while the anthers average 1.5 mm in length.

There is considerable variation in *A. trachycaulum* with gradations from glabrous to pubescent herbage, from awnless to long-awned lemmas and from short to long spikes. A. S. Hitchcock's (Manual Grasses U.S., 1951) 4 taxa, *A. trachycaulum, A. latiglume, A. subsecundum* var. *subsecundum* and *A. subsecundum* var. *andinum* are regarded as varieties of a single species here. Even then they may be no more than altitudinal races of the awnless and awned phases. They are customarily distinguished in the following manner.

AGROPYRON

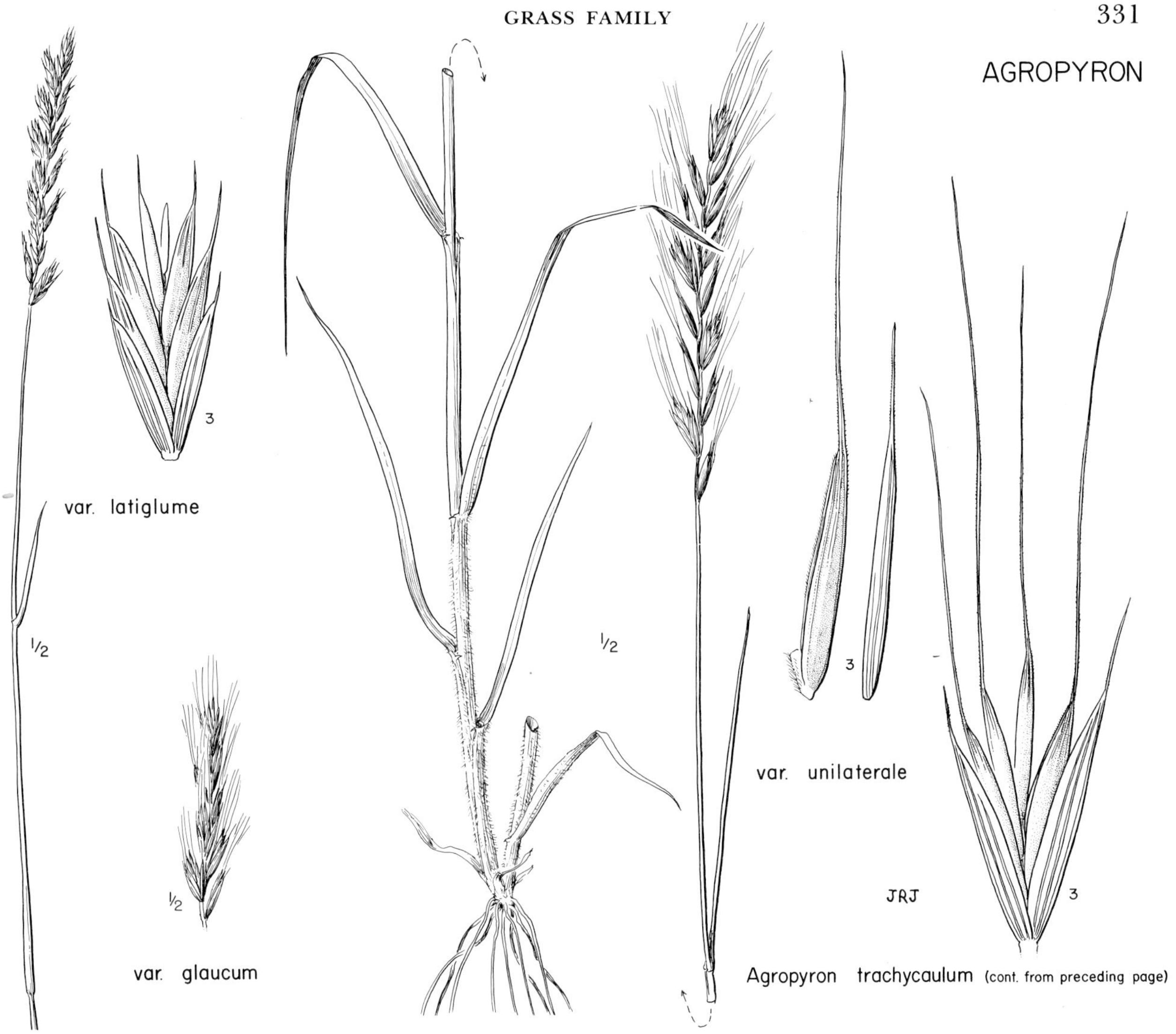

1 Lemmas awnless or at most with awns less than 6 mm long.
  2 Spikelets scarcely imbricated, the tips rarely reaching the bases of those above on the same side; spikes mostly more than 10 cm long; valleys to middle montane habitats . . . . var. *trachycaulum*
  2 Spikelets mostly closely imbricated; spikes mostly less than 10 cm long; subalpine and alpine habitats . . . . var. *latiglume* (Scribn. & Smith) A. A. Beetle
1 Lemmas awned, the awn (6) 10–30 mm long.
  3 Glumes 6–10 mm long; plants 2–3 dm tall, decumbent at the base; awn of the lemma (5) 7–16 (20) mm long; spikes slender, 3–10 mm thick; high mts. (including *A. subsecundum* var. *andinum*) . . . . var. *glaucum* (Pease & Moore) Malte
  3 Glumes 10–18 mm long; plants usually more than 3 dm tall, erect or ascending; awn of the lemma 17–40 mm long; spikes 6–13 mm thick; valleys and middle montane habitats (including *A. subsecundum*) . . . . var. *unilaterale* (Cassidy) Malte

Agropyron trachycaulum is the putative parent of two well known hybrids. When crossed with *Hordeum jubatum* a sterile hybrid comparable in every way to X *Agrohordeum macounii* (Vasey) Lepage is produced (Boyle and Holmgren, 1955), whereas with *Sitanion jubatum* the sterile hybrid X *Agrositanion saundersii* (Vasey) Bowden is produced (Stebbins et al., 1946). Bowden (1965) suspects that the *A. subsecundum* var. *andinum* type is a hybrid involving *A. scribneri* X *A. violaceum,* which he calls *A.* X *brevifolium* Scribn. (pro sp.).

10. Agropyron scribneri Vasey

Agropyron scribneri Vasey, Bull. Torrey Bot. Club 10: 128. 1883. *Elymus scribneri* M. E. Jones, Contr. W. Bot. 14: 20. 1912. (*Scribner 427,* "alpine," Mont., in 1883.)

Scribner wheatgrass.

Small tufted perennials, the rhizomes lacking; culms short, 1.5–3.2 (5.5) cm long, decumbent to spreading at the base; sheaths short-pilose; ligules very short, mostly about 0.5 mm long, ciliolate, entire to erose; blades narrow, involute or sometimes flat, 1.5–3.5 mm broad, pilose, the auricles usually present; spikes short, 3.5–6 (8) cm long, crowded, the rachis disarticulating at maturity, the internodes short, 2.5–5 mm long in the middle of the spike; spikelets 9–14 mm long, often purplish, 3- to 6-flowered; glumes narrow-lanceolate, tapering nearly imperceptibly into a scabrous awn, the body of the glumes 4–7 mm long, 2- to 3- (5)-nerved, the awns divergent, 12–20 mm long, rarely straight, sometimes split into a second awn; lemmas 7–9 (10) mm long, glabrous to scaberulous, faintly 5-nerved, attenuated into an arcuate, divergent, scabrous awn, the awn 15–25 mm long; palea acuminate, usually slightly longer than the body of the lemma; lodicules narrow, acuminate, about 0.7 mm long; anthers 1–1.6 mm long; $2n = 28$.

Rocky ridges and slopes in openings of subalpine forests and above timberline; Wash., sw. Alta. and Mont. s. to the White and Sweetwater mts. of Calif., n. Ariz. and n. N.M. July–Aug.

In its tardily disarticulating rachis, narrow, sometimes divided glumes and long, slender, divergent awns of the lemma, *A. scribneri* shows affinities with *Sitanion.* At the same time it shows close affinities to *A. trachycaulum* in its small anthers, self-fertility and anthesis in the morning. These are characters peculiar to these two species, with the remaining native *Agropyron* species having large anthers, self-sterility and anthesis in the late afternoon. In an experimental cross between *A. scribneri* and *A. trachycaulum,* Dewey (1963) found in the hybrid a high percentage of chromosome pairing in meiosis which also shows close relationship. More recently, Dewey (1967) crossed *A. scribneri* with *Sitanion hystrix* which also showed nearly normal pairing in meiosis. The separation of *Agropyron* and *Sitanion* is not as simple as using the number of spikelets per node, nor will a disarticulating versus continuous rachis be definitive.

34. EREMOPYRUM (Ledeb.) Juab. & Spach

Annuals; culms hollow; sheaths open; ligules membranous, short; blades usually flat and auriculate; inflorescence a short, oblong to ovoid, bilateral spike with solitary spikelets closely imbricated and set at nearly right angles to the rachis, the rachis disarticulating; spikelets 3- to 6-flowered; glumes linear to lanceolate (ours), awn-tipped or awnless, keeled, firm; lemmas awn-tipped; palea shorter than the lemma; lodicules 2, acute to somewhat truncate and toothed, often ciliate on the margins; $x = 7$.

A genus of 5 to 8 species of the Medit. region, c. Asia and nw. India. (Name from the Greek *erem,* desert, and *pyros,* wheat.)

1. Eremopyrum triticeum (Gaertn.) Nevski

Agropyron triticeum Gaertn. Novi Comment. Acad. Sci. Imp. Petrop. 14(1): 540. 1770. *Eremopyrum triticeum* Nevski in Komarov, Fl. USSR 2: 662. 1934. (Russia.)

Annual wheatgrass.

Annuals; culms geniculate at the base, 0.8–3.5 dm tall, glabrous except for retrorse hirsute below the spike; sheaths glabrous to puberulent, the upper sheaths inflated; ligules 0.2–1 mm long, erose to lacerate; blades flat to involute, 1–4 (6) mm broad, short, scabrous, the auricles small; spikes small, 0.8–2 cm long, ovate, bilateral, becoming long exserted from the sheaths, the spikelets densely crowded and divergent, the rachis tardily disarticulating; spikelets 3- to 6-flowered; glumes subequal, (4) 5–7.5 mm long, thick, cartilaginous except for the membranous margins, prominently 1-nerved, tapering into a small awn 0.8–1.7 mm of its length, somewhat gibbous at the base; lemmas (5) 6–7.5 mm long, lanceolate, acuminate into an awn tip, the awn 1–2 mm of its length, 3- to 5-nerved, the nerves faint below, scabrous; palea shorter than the lemma, bifid at the apex; anthers 0.8–1 mm long; $2n = 14$.

Disturbed areas; introduced from Russia; drier regions of w. N. Amer., known from our area in e. Oregon, sw. Idaho, n. Nev., and w. and c. Utah. May–July.

Eremopyrum triticeum has often been referred to the genus *Agropyron.* The annual habit, distinctive morphology and cytology readily distinguish it from any of the perennial agropyrons.

35. AEGILOPS L.

Annuals; culms hollow; sheaths open; ligules membranous, short; blades narrow, flat, usually pubescent, auriculate to non-auriculate; inflorescence usually a cylindric, compact spike with solitary spikelets, the spikelets closely appressed to the rachis, the uppermost spikelet sterile, the mature spike usually disarticulating at the base and falling entire or disarticulating into individual spikelets; spikelets 2- to 5-flowered, the upper florets often sterile; glumes flat or rounded on back, sometimes obscurely keeled, indurate, many-nerved, scabrous to pubescent, truncate, often toothed at the apex, the teeth sometimes extending into awns; lemmas rounded on back, the nerves not convergent at the summit, 1- to 3-toothed or awned; lodicules 2; stamens 3; caryopsis oblong-oval, furrowed in back, hairy at the summit, adhering to the lemma and palea (ours) or free; $x = 7$.

A genus of 5 or 6 species of the Medit. region and w. Asia. (Name from the Greek *aegilops,* a name used by Theophrastus for a kind of wild oat.)

1. Aegilops cylindrica Host

Aegilops cylindrica Host, Icon. & Descript. Gram. Austriacorum 2: 6, pl. 7. 1802. *Triticum cylindricum* Ces., Pass. & Gibbelli, Compend. Fl. Ital. 86. 1867. *A. caudata* subsp. *cylindrica* Hegi, Ill. Fl. Mittel-Eur. 1: 390. 1908. (Europe.)

Jointed goatgrass.

Weedy annuals; culms often geniculate at the base, 3–7 dm tall; sheaths glabrous with ciliate margins to hirsute or pilose; ligules short, 0.2–0.5 (0.8) mm long; blades flat, 1–4 (6) mm broad, glabrous or more often hirsute, the auricles present but not conspicuous; spikes 6–11 (15) cm long, narrow, terete, the spikelets closely appressed and sunken into the recesses of the rachis flush with the nodes; spikelets 2- to 5-flowered; glumes 7–10 mm long, cartilaginous, asymmetrical, unequally 9- to 13-nerved, scabrous to ascending hispid, tipped by a lobe in front (away from the rachis) and an awn in back, the awns (2) 3.5–8.5 mm long, the glumes of the terminal spikelet awned from between 2 short teeth; lemmas 8–11 mm long, asymmetrical, 5-nerved, awned from a bifid apex, the awn short, 1–3 (5) mm long or merely mucronate, the lemma of the terminal spikelet with a long awn; palea ciliate nerved, bifid; anthers about 2.8 mm long; $2n = 28$.

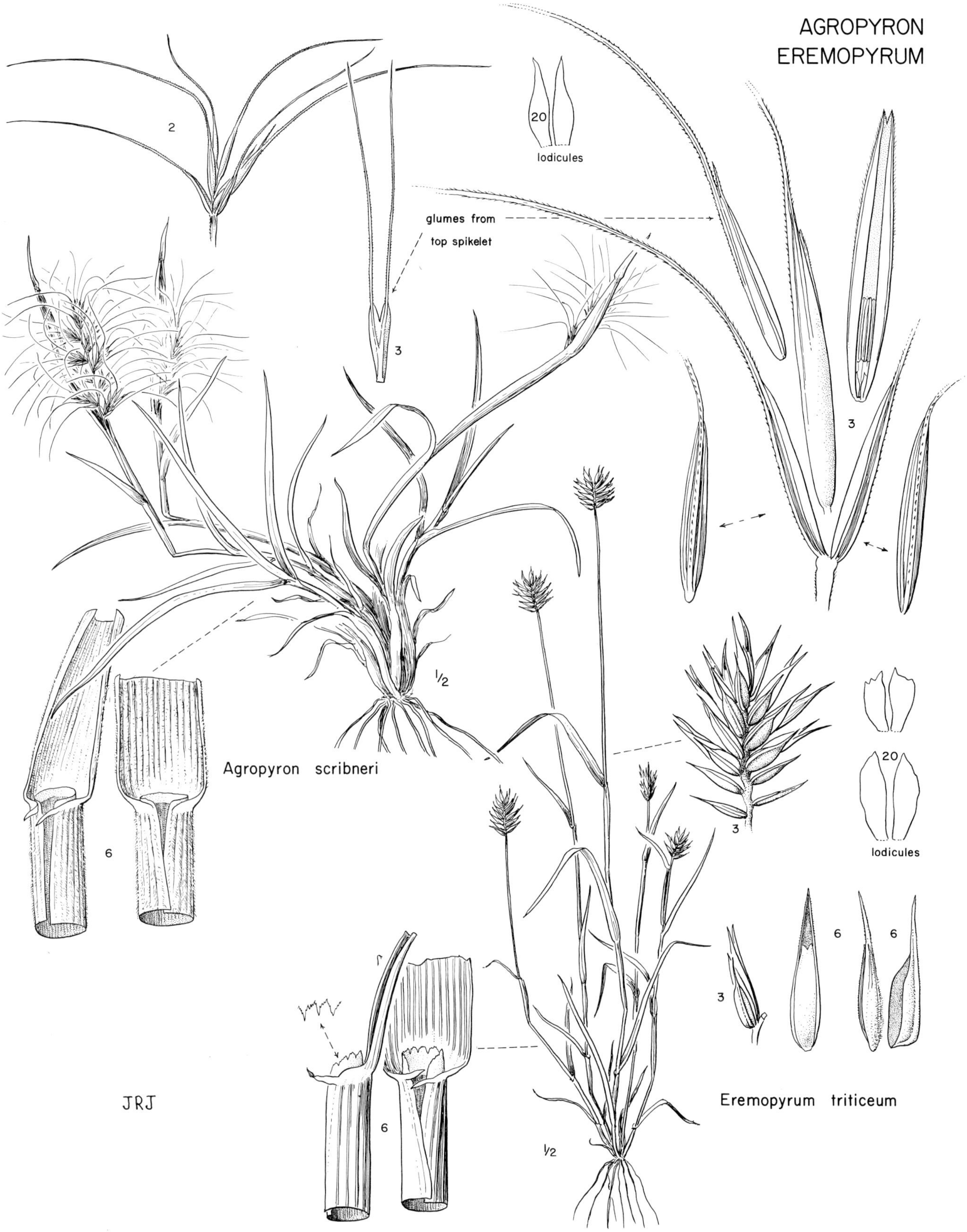
AGROPYRON
EREMOPYRUM
2
3
20
lodicules
glumes from
top spikelet
3
1/2
Agropyron scribneri
6
3
20
lodicules
3
6
6
JRJ
6
1/2
Eremopyrum triticeum

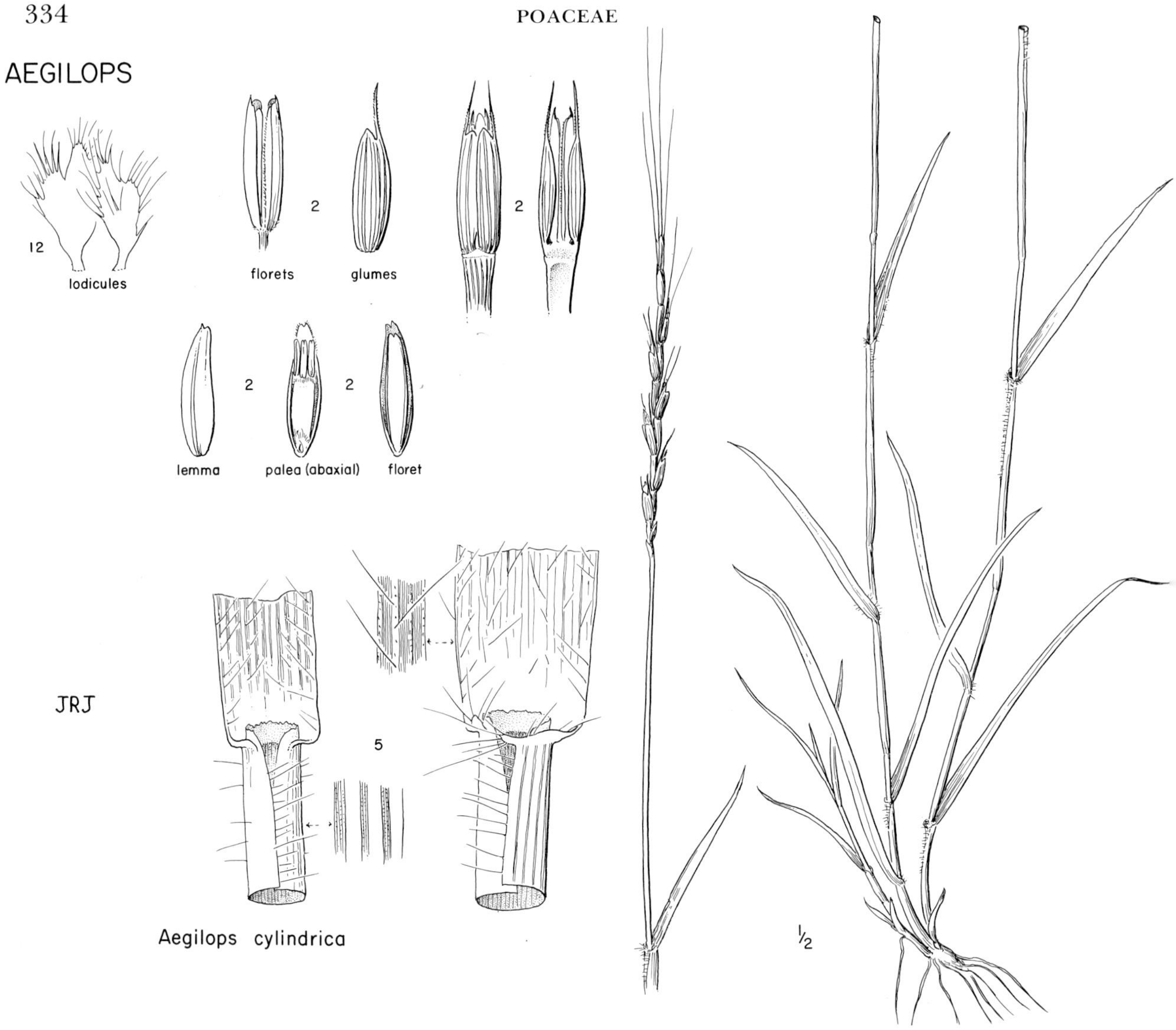

Dry waste places and wheatfields; introduced from s. Europe and c. Asia; widely scattered throughout w. U.S. Late May–June.

Jointed goatgrass often becomes a troublesome weed in wheatfields where it can hybridize with the wheat. Two other species, *A. ovata* L. and *A. triuncialis* L., have been introduced into western United States but have not as yet been found in our area.

36. TRITICUM L. Wheat

Annuals (biennials); culms hollow; sheaths open; ligules membranous, short; blades broad, flat, bearing well developed auricles; inflorescence a bilateral spike with solitary spikelets borne flatwise to the continuous or disarticulating rachis; spikelets 2- to 5-flowered, turgid but somewhat laterally flattened, disarticulating above the glumes and between the florets; glumes thick and firm, 3- to several-nerved, keeled, the apex toothed, mucronate or awned; lemmas broad, asymmetrically keeled, firm, 5- to 7-nerved, the nerves non-convergent, acute to awned; lodicules 2, ciliate on the margins; stamens 3; caryopsis rather plump, oval or oblong, deeply furrowed in back, slightly hairy at the summit; $x = 7$.

A genus of from 15 to 20 species of cool and temp. regions of Europe and w. Asia. (The classical Latin name for wheat.)

1. Triticum aestivum L.

Triticum aestivum L. Sp. Pl. 85. 1753. *T. cereale* var. *aestivum* Baumg. Enum. Stirp. Transsilv. 3: 266. 1816. *T. vulgare* var. *aestivum* Spenner, Fl. Friburg. 1: 163. 1825. *T. sativum* var. *aestivum* A. Wood, Cl.-Book Bot. ed. of 1847. 619. 1847. *T. vulgare* subsp. *eu-vulgare* f. *aestivum* A. Brand, Koch's Syn. Deutsch. & Schweiz. Fl. ed. 3. 3: 2792. 1907. *Zeia vulgaris* var. *aestiva* Lunell, Amer. Midl. Naturalist 4: 226. 1915. ("Habitat. . . .")

Common wheat.

Annuals, often winter annuals; culms erect, usually robust, 5–10 (12) dm tall, freely branching

TRITICUM
SECALE

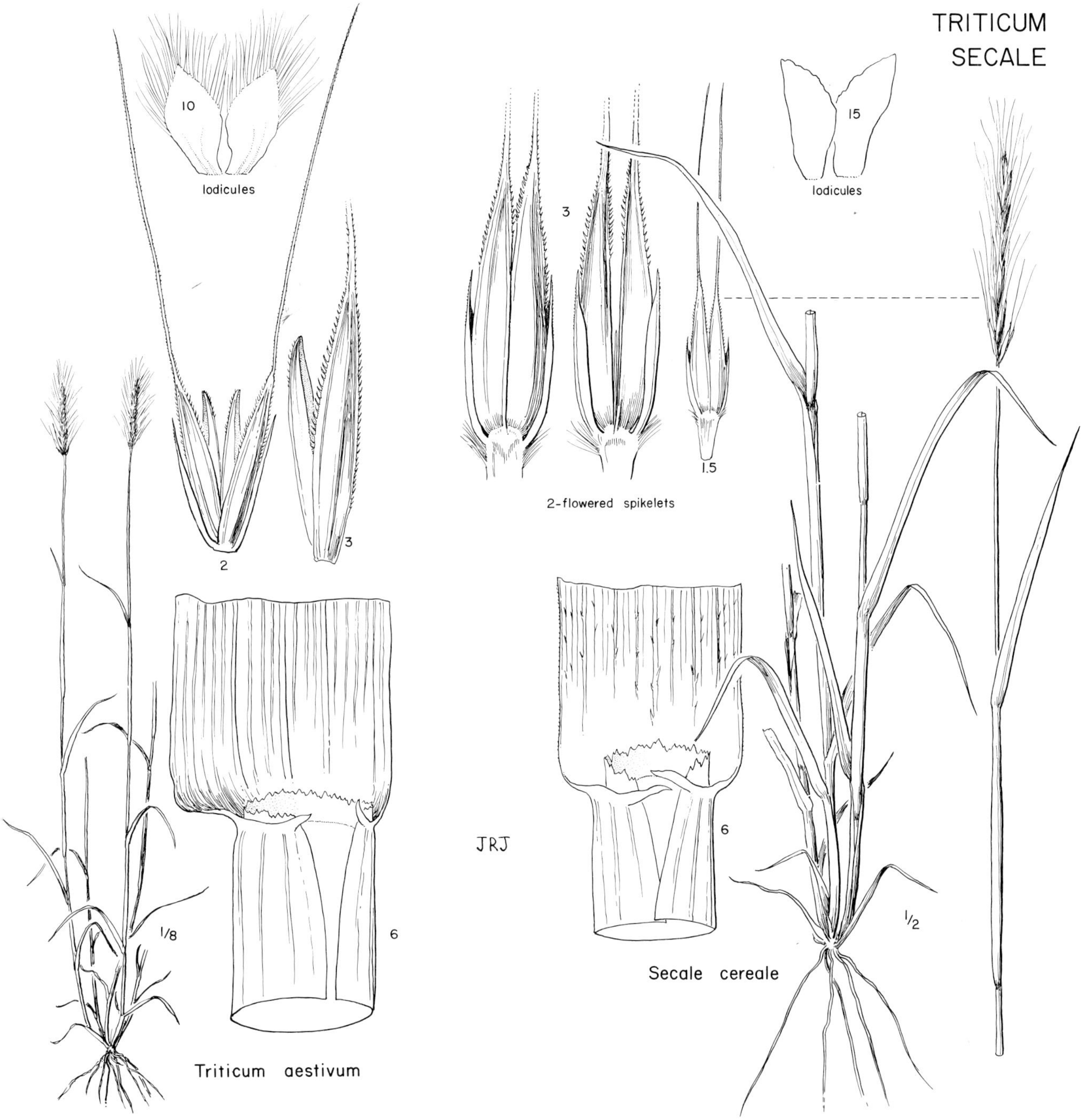

below; sheaths glabrous; ligules 0.5–1.5 mm long, entire; blades flat, sometimes very broad, 2–8 (20) mm, glabrous or scabrous above, the auricles prominent; spikes (3) 5–10 (12) cm long, the rachis continuous, flattened; spikelets (2-) 3- to 5-flowered; glumes subequal, 6–9 (11) mm long, indurate, asymmetrical, glabrous, scabrous or densely hispid, somewhat keeled, unequally (3-) 5- to 7-nerved, only the keel nerve and sometimes a pair of lateral nerves conspicuous, awned from between 2 broad lobes, the awns 1–4 (6) mm long; lemmas 7–12 mm long, similar to the glumes, asymmetrically keeled, indurate but with scarious margins, awned or mucronate from a broadly lobed apex, the awns short, 2–20 mm long, or long and then 47–70 mm long; anthers about 2.5 mm long; $2n = 42$.

Escaping cultivation along roadsides and in fallow fields; planted throughout cooler parts of the U.S. April–July.

There are many kinds of wheats of bearded and beardless types, too numerous to mention and describe here. Wheat probably had its origin in eastern Asia, where a number of wild species occur. It has been introduced throughout the world and perhaps occupies more acreage of farm land than any other crop plant.

37. SECALE L. Rye

Annuals (ours) and perennials; culms hollow; sheaths open; ligules membranous, short; blades flat, bearing prominent auricles; inflorescence a bilateral spike with solitary spikelets borne flatwise to the continuous or disarticulating rachis; spikelets 2-flowered, the rachilla prolonged behind the upper floret and sometimes bearing a vestige of a third floret, disarticulating above the glumes and between the florets (in ours); glumes narrow, rigid, keeled; lemmas broad, asymmetrically keeled, 5-nerved, shortly bilobed, the keel and margins strongly ciliate, tapering into a long, straight, scabrous awn; palea bidentate, the 2 nerves ciliate apically; lodicules 2, usually ciliate above; stamens 3; caryopsis oblong, somewhat laterally compressed, deeply furrowed in back, hairy at the summit; $x = 7$.

A genus of 5 species, widely distributed in temp. Europe and Asia, 1 species introduced to the U.S. for cultivation. (Classical Latin name used by Pliny for rye.)

1. Secale cereale L.

Secale cereale L. Sp. Pl. 84. 1753. *Triticum cereale* Salisb. Prodr. Stirpium 27. 1796; not Schrank, 1789. ("Habitat. . . .")

Robust annuals or sometimes biennials; culms 6–12.5 dm tall, glabrous except below the spike; sheaths glabrous; ligules 0.5–1.5 (2) mm long, erose-membranous; blades flat, often relatively broad, (1) 3–7 (10) mm, scabrous, at least above, the auricles about 1 mm long, clasping the culm; spikes large, (4) 8–14 (17) cm long, arcuate and somewhat nodding; spikelets 2-flowered, the glumes and florets attached nearly at the same level on a broadened, contracted rachilla, the rachilla then extending between the 2 florets as a narrow bristle; glumes equal, (6) 7.5–15 (17) mm long, narrow-lanceolate to linear, gradually acuminate, 1- (3)-nerved, strongly keeled, the keel ciliolate, scabrous; lemmas large, (10) 12–16 mm long, asymmetrical, 5-nerved, pectinate-ciliate on the keel and on the outer margins (away from the rachis), bearing scabrous awns 20–40 (70) mm long; palea subequal to the lemma; anthers large, 7.5–8.5 mm long; $2n = 14, 16, 27–29$.

Widely cultivated and often occurring in abundance along roadsides, railroad right of ways, and waste places. Late May–July.

Tribe 4. MELICEAE

Moderately tall perennials, often of wet to moist places; culms hollow; sheath margins connate, usually to the top; ligule membranous; blades non-auriculate; inflorescence a panicle (in ours) or rarely a raceme; spikelets 2- to several-flowered, rarely 1, disarticulating above or sometimes below the glumes; glumes often papery or membranous; lemmas 3- to several-nerved, awned to awnless; lodicules 2, thick, fleshy and truncate, often fused forming a collar around $^1/_2$ to $^2/_3$ of the ovary base (unique in the Pooideae); embryo pooid; $x = 8, 9$, or 10.

The absence of silica cells in the leaf epidermis, the thick, fleshy, truncate lodicules and basic chromosome number of $x = 8, 9$, or 10 will distinguish members of the Meliceae from those in the Poeae, where they have traditionally been placed.

1 Lemmas awnless; callus of the florets not bearded.
  2 Lemmas (5) 6–14 mm long; panicles 5–17 (20) cm long; glumes papery, the first glume 3–16 mm long; plants of dry to moderately moist places .... 38. *Melica*
  2 Lemmas 1.5–5 (6.5) mm long, if greater than 5 mm long the panicles 17–50 cm long; glumes not papery, except in some *Glyceria,* the first glume 0.5–2.8 (3.2) mm long; plants mostly of fresh water marshes or moist to wet places.
    3 Lemmas mostly 7-nerved, sometimes 5- to 9-nerved .... 39. *Glyceria*
    3 Lemmas 3-nerved .... 40. *Catabrosa*
1 Lemmas awned; callus of the floret bearded .... 41. *Schizachne*

38. MELICA L. Melic Grass, Oniongrass

Cespitose or rhizomatous perennials; culms usually simple, hollow, often from bulbous bases; sheaths closed nearly to the top; ligules membranous, erose, toothed or deeply lacerate, often decurrent, sometimes united in front; blades flat to involute, narrow to moderately broad, non-auriculate; inflorescence an open or contracted panicle or raceme, usually less than 20 cm long; spikelets relatively large, 2- to several-flowered and only the first 1 or 2 fertile, the uppermost florets represented by empty lemmas in a compact cluster called the rudiment, the callus glabrous, disarticulating above the glumes and between the florets or below the glumes with the spikelet falling intact; glumes broad, rounded on back, membranous or papery, unequal or subequal, large, the first (1-) 3- (7)-nerved, the second (3-) 5- (9)-nerved; lemmas obtuse to acuminate, firmer than the glumes but with scarious margins, rounded on back, (5-) 7- to 13- (15)-nerved, the nerves unequally developed or spaced, but parallel and not converging towards the apex, awnless (ours) or awned from a bifid or entire apex; palea with 2 submarginal nerves; lodicules often fused forming a collar around $^1/_2$ to $^2/_3$ of the ovary base, thick, fleshy, about 0.5 mm long; stamens 3; ovary glabrous with terminal styles; caryopsis oblong or ovate, free from the palea; $x = 9$.

MELICA

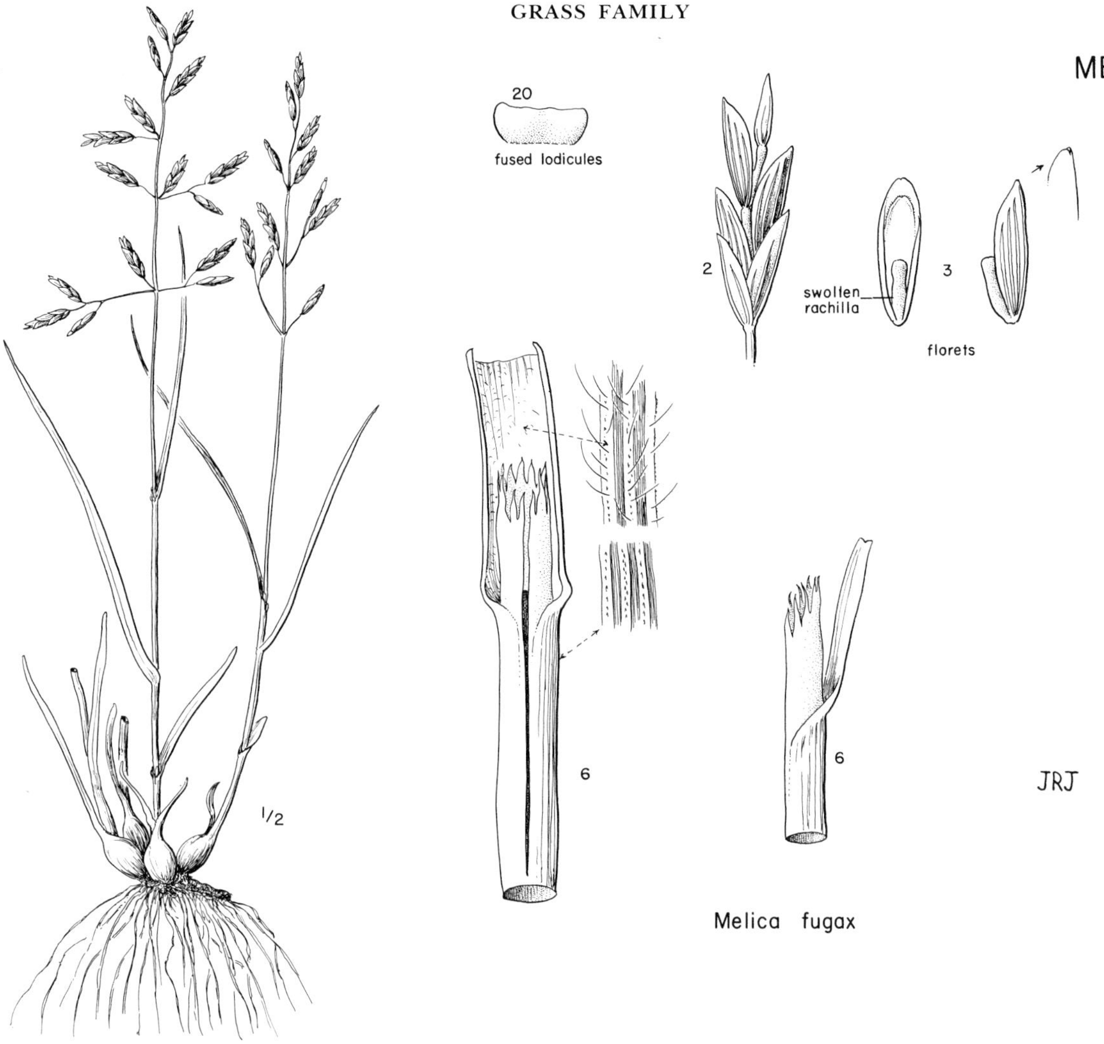

Melica fugax

A genus of about 60 species of temp. and trop. mts. of the world, 17 from the U.S., centering mostly in the west. (Classical name for some plant, possibly an Italian name for a sorghum with a sweet juice [Latin *mel,* honey].)

Melic grasses are palatable but do not occur in sufficient quantity to furnish much forage.

Reference: Boyle, W. S. A cyto-taxonomic study of the North American species of *Melica.* Madroño 8: 1–26. 1945.

1 Glumes unequal, the first glume 3–8 (9) mm long; articulation above the glumes and between the florets (Bromelica).
  2 Rachilla swollen, usually wrinkled in drying; lemmas 5–6.8 mm long .... 1. *M. fugax*
  2 Rachilla normal, not swollen; lemmas (6) 7–12 mm long.
    3 Bulb of the culm attached to the rhizome by a slender stem; first glume less than half as long as the spikelet, the glume 3.5–6 mm long .... 2. *M. spectabilis*
    3 Bulb of the culm attached directly to the rhizome (if rhizome present); first glume more than half as long as the spikelet, the glume (5) 6–8 (9) mm long .... 3. *M. bulbosa*
1 Glumes subequal, the first glume (9) 10–16 mm long; articulation below the glumes, each spikelet at maturity falling entire (Melica) .... 4. *M. stricta*

1. **Melica fugax** Boland.

Melica fugax Boland. Proc. Calif. Acad. Sci. 4: 104. 1870. (*Bolander & Kellogg,* Donner Lake, Calif.)

M. macbridei Rowland ex A. Nels. Bot. Gaz. 54: 404. 1912. *M. fugax* var. *macbridei* A. A. Beetle, Leafl. W. Bot. 4: 286. 1946. (*Macbride 948,* Silver City, Idaho, 20 June 1911.)

Little oniongrass.

Tufted perennials; culms 1.5–3.5 (6) dm tall, bulbous at the base, clustered on short rhizomes; sheaths glabrous to pilose, sometimes split open at the top; ligules 1–3 (4.5) mm long, erose-lacerate; blades flat to loosely involute, 1.5–3.5 mm broad, glabrous, sometimes pilose, at least above; panicles (5) 7–12 (20) cm long, relatively narrow, the branches short, spreading to appressed; spikelets 8–14 (17) mm long, loosely 2- to 4- (5)-flowered, the rachilla joints yellowish, inflated, wrinkling on drying, the rudiment narrow, 2–3.5 mm long, disarticulating above the glumes and between the florets; glumes oval, obtuse, membranous, glabrous or scaberulous on the nerves distally, often purplish, the first glume 3–4.8 mm long, 1- to 3- (5)-nerved, the second 4.5–7 mm long, 3- to 5- (9)-nerved; lemmas 5–6.8 mm long, ovate, sometimes broadly so, obtuse or sometimes bifid or emarginate, glabrous to scaberulous, usually with a purplish band within the scarious margins, 5- to 9-nerved, awnless; palea rather firm, the nerves ciliate distally; anthers 1–1.5 mm long; $2n = 18$.

Dry, open woods and sagebrush meadows at middle elevs. in the mts., usually in volcanic derived soils; Wash. to n. half of Calif., e. across n. Nev. to the Santa Rosa Mts. and sw. Idaho in the Owyhee Mts. June–Aug.

2. Melica spectabilis Scribn.

Melica spectabilis Scribn. Proc. Acad. Nat. Sci. Philadelphia 1885: 45, pl. 1, fig. 11–13. 1886. (*Scribner 385,* Crow Creek Mts., Lake Co., Mont., 29 June 1883.)

Purple oniongrass.

Rhizomatous perennials; culms 2.5–7.5 (12) dm tall, bulbous at the base, the bulbs not attached directly to the rhizome but connected by a slender stem and spaced at intervals along the rhizome; sheaths glabrous to pilose, sometimes open at the top; ligules (0.7) 1.2–3.2 mm long, erose-lacerate, acute to truncate; blades flat, folded or involute, 1–4 mm broad, glabrous to scabrous, at least above; panicles (5) 8–13 (20) cm long, narrow or sometimes somewhat open, the branches often flexuous; spikelets (8) 10–15 (20) mm long, 4- to 7-flowered, the rudiment rarely exserted, about 2.5 mm long, disarticulating above the glumes and below the florets; glumes obtuse to broadly ovate, membranous, scabrous on the nerves, the first glume (3.5) 4–6 mm long, 1- to 3-nerved, the second 5–7.5 mm long, 7- to 9-nerved; lemmas (6) 7–8.5 mm long, broadly ovate to obtuse, sometimes emarginate or bifid, scaberulous, often a purple-tinged band within the scarious margins and apex, 9- to 15-nerved, the strong nerves alternate with weak ones, awnless; palea with ciliate nerves; anthers 1.8–2.6 mm long.

Dry to moderately moist sagebrush–grass meadows at middle and subalpine elevs. in the mts.; se. B.C. to w. Mont., s. in the mts. to n. Calif., n. Nev., Utah and Colo. June–Aug.

3. Melica bulbosa Geyer ex Porter & Coulter

Melica bulbosa Geyer ex Hook. Hooker's J. Bot. Kew Gard. Misc. 8: 19. 1856, nomen nudum; Geyer ex A. Gray, Proc. Amer. Acad. Arts 8: 409. 1872, nomen nudum; Geyer ex Porter & Coulter, Syn. Fl. Colo. 149. 1874. *M. bella* Piper in Scribn. U.S.D.A. Div. Agrostol. Circ. 27: 10. 1900. (*Geyer 11,* "Rocky ravine, Upper Platte," near Independence Rock, Natrona Co., Wyo., July 1843.)

M. bulbosa var. *caespitosa* Cronq. Madroño 7: 77. 1943. (*Cronquist & Cronquist 2603,* w. side of Alturas Lake, Blaine Co., Idaho, 22 June 1941.)

Oniongrass.

Tufted perennials; culms 3–10 dm tall, the bulbous bases clustered on short rhizomes; sheaths glabrous to scaberulous, sometimes short-pilose, closed nearly full length; ligules 2–5 (7) mm long, often deeply lacerate; blades flat to involute, 2–4 (5) mm broad, glabrous to scabrous; panicles (7) 10–16 (20) cm long, narrow, elongate, the branches short and erect; spikelets (9) 10–13 (15) mm long, 3- to 6-flowered, the rudiment exserted, nearly 3.5 mm long, disarticulating above the glumes and between the florets; glumes ovate, obtuse, scaberulous, the nerves sometimes scabrous, the first glume (5) 6–8 (9) mm long, (1-) 3- (4)-nerved, the second (6.5) 7–9 (11) mm long, 5- (7)-nerved; lemmas (6) 8–9.5 (12) mm long, obtuse to rounded or sometimes emarginate, with a purple-tinged band within the scarious margins and apex, scaberulous, (5-) 9- to 11- (15)-nerved, awnless; palea with ciliate nerves; anthers 2.5–3.5 mm long; $2n = 18$.

Dry to moderately moist open sagebrush slopes and woodlands from mid to subalpine elevs. in the mts.; s. B.C. and s. Alta., s. in the mts. to n. and Sierran Calif., n. Nev., Utah and Colo. and reported to occur in Trans-Pecos Texas. Late May–Aug.

The nonbulbous extreme has been described as var. *caespitosa* Cronq., but it does not appear to be ecologically set off from the more common, bulbous form.

4. Melica stricta Boland.

Melica stricta Boland. Proc. Calif. Acad. Sci. 3: 4. 1863. (*Dunn,* Silver City, Nev.)

Rock melic.

Tufted perennials; culms 2–5.5 (7) dm long, not bulbous at the base; sheaths scaberulous, open near the top; ligules 3–5.5 mm long, erose-lacerate; blades flat, 1.5–3 (5) mm broad, scaberulous, sometimes hirsutulous on the upper surface; panicles 5–17 (20) cm long, narrow, the pedicels appressed, but curved at the tip making the spikelets secund, white hirsute just below the spikelet attachment; spikelets (11) 14–17 (20) mm long, (2-) 4- (5)-flowered, broadly V-shaped when mature, the rudiment narrow, 2–7 mm long, disarticulating below the glumes; glumes subequal, large, (9) 10–15 (16) mm long, oblanceolate, glabrous to scaberulous, 5- to 7-nerved, the second glume narrowed at the base; lemmas large, (9) 10–13 (14) mm long, scabrous, the scarious margins and apex extensive, 9-nerved, awnless; palea 1/2 to 2/3 as long as the lemma; anthers 1–2 mm long; $2n = 18$.

Rocky mt. slopes; se. Oregon and Calif., e. across the Great Basin through Nev. to n. Utah. July–Aug.

39. GLYCERIA R. Br. nom. conserv. Mannagrass

Tall, marsh or aquatic perennials, usually strongly rhizomatous; culms hollow, erect, sometimes decumbent at the base and freely rooting at the nodes; sheaths closed, sometimes to the top; ligule membranous, erose to deeply lacerate, sometimes united in front; blades flat to folded or loosely involute, usually broad, non-auriculate; inflorescence an open or contracted panicle, often with slender, flexuous branches, some-

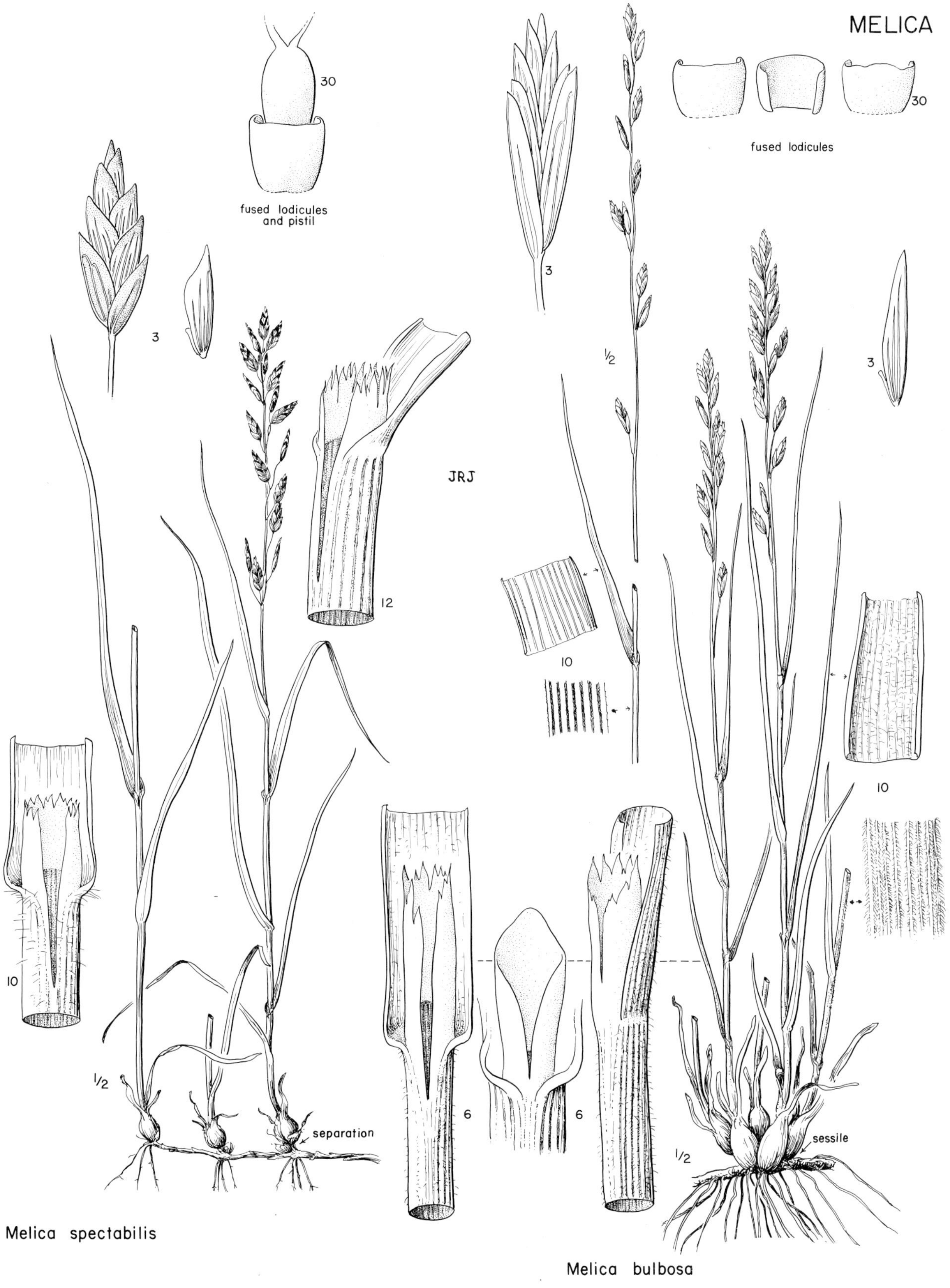
MELICA
30
fused lodicules
and pistil
3
3
JRJ
12
10
1/2
separation
6
6
3
1/2
30
fused lodicules
3
10
10
1/2
sessile
Melica spectabilis
Melica bulbosa

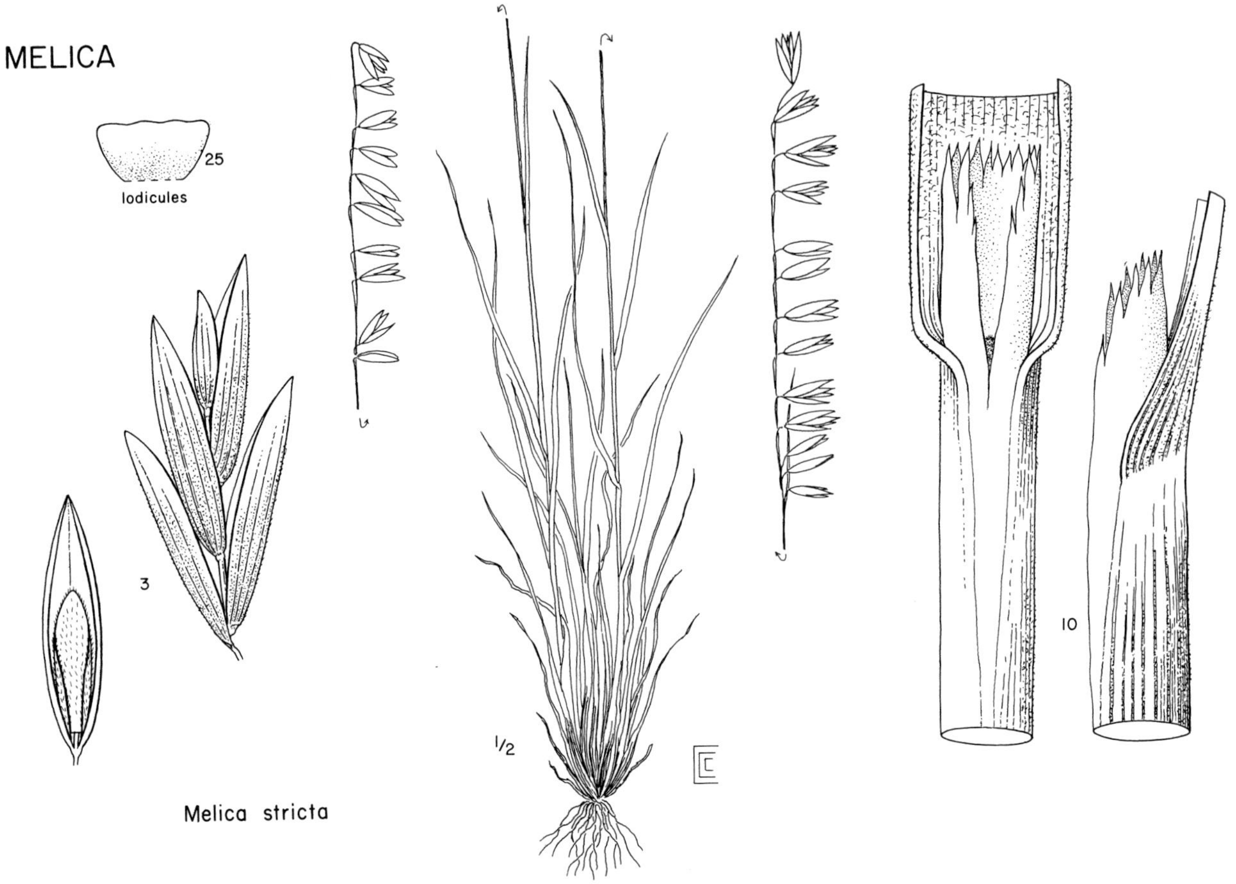

times reduced to a raceme; spikelets few- to many-flowered, linear to ovate, often purplish, the callus glabrous, disarticulating above the glumes and between the florets; glumes unequal, small, shorter than the first floret, broad, usually membranous and 1-nerved, awnless; lemmas broad, acute, obtuse or rounded, rounded on back, firm, usually with scarious margins and apex, often erose, prominently (5-) 7- (9)-nerved, the nerves strongly and uniformly developed, parallel, not converging towards the apex, awnless; palea large, broad, sometimes exceeding the lemmas, the 2 strong nerves submarginal; lodicules thick, fleshy, united or free, 0.2 mm long; stamens 2 or 3; ovary glabrous, the styles plumose in upper half only; $x = 10$.

A genus of about 30 species from temp. and subarctic regions of the N. Hemisphere with about 15 in the U.S. (Name from the Greek *glyceros*, sweet, possibly in reference to the taste of the grain of some of the species.)

Species of *Glyceria* provide food and cover for birds and small mammals in marshes and swamps. They are palatable to livestock but the wetland habitat is too restricted in our region to make them important as forage grasses.

Reference: Church, G. L. A cytotaxonomic study of *Glyceria* and *Puccinellia*. Amer. J. Bot. 36: 155–165. 1949.

1 Spikelets linear, nearly terete, 9–23 mm long, (5-) 8- to 12-flowered; lemmas 3.3–6.5 mm long; ligules 4–13 mm long (Glyceria).
  2 Lemmas glabrous between the slightly scaberulous nerves, 3.3–4 mm long; spikelets 9–15 (18) mm long, with short rachilla joints 0.7–1 mm long . . . . 1. *G. borealis*
  2 Lemmas scaberulous or hirtellous between the scabrous nerves, 4–6.5 mm long; spikelets 15–23 mm long, the rachilla joints (1) 1.4–2.1 mm long . . . . 2. *G. occidentalis*
1 Spikelets ovate or oblong, more or less compressed, 2.5–6.5 mm long, 3- to 6- (7)-flowered; lemmas 1.5–2.8 mm long; ligules 1–4 (7) mm long (Hydropoa).
  3 Spikelets 4–6.5 mm long, with longer rachilla joints, 0.6–0.8 mm long; first glume 1–1.8 mm long . . . . 3. *G. grandis*
  3 Spikelets 2.5–4 (5) mm long, the rachilla joints 0.3–0.5 (0.6) mm long; first glume 0.5–1 (1.2) mm long.
    4 Blades (4.5) 6–12 mm broad; first glume 0.8–1.2 mm long, mostly 1 mm; ligules (2) 2.5–6 mm long, puberulent, not as thin; plants (6) 10–15 dm tall . . . . 4. *G. elata*
    4 Blades 2–4.5 (6) mm broad; first glume 0.5–0.8 (1) mm long, mostly 0.7 mm; ligules 1–2.2 (3.5) mm long, glabrous, thin; plants 3–10 (13) dm tall . . . . 5. *G. striata*

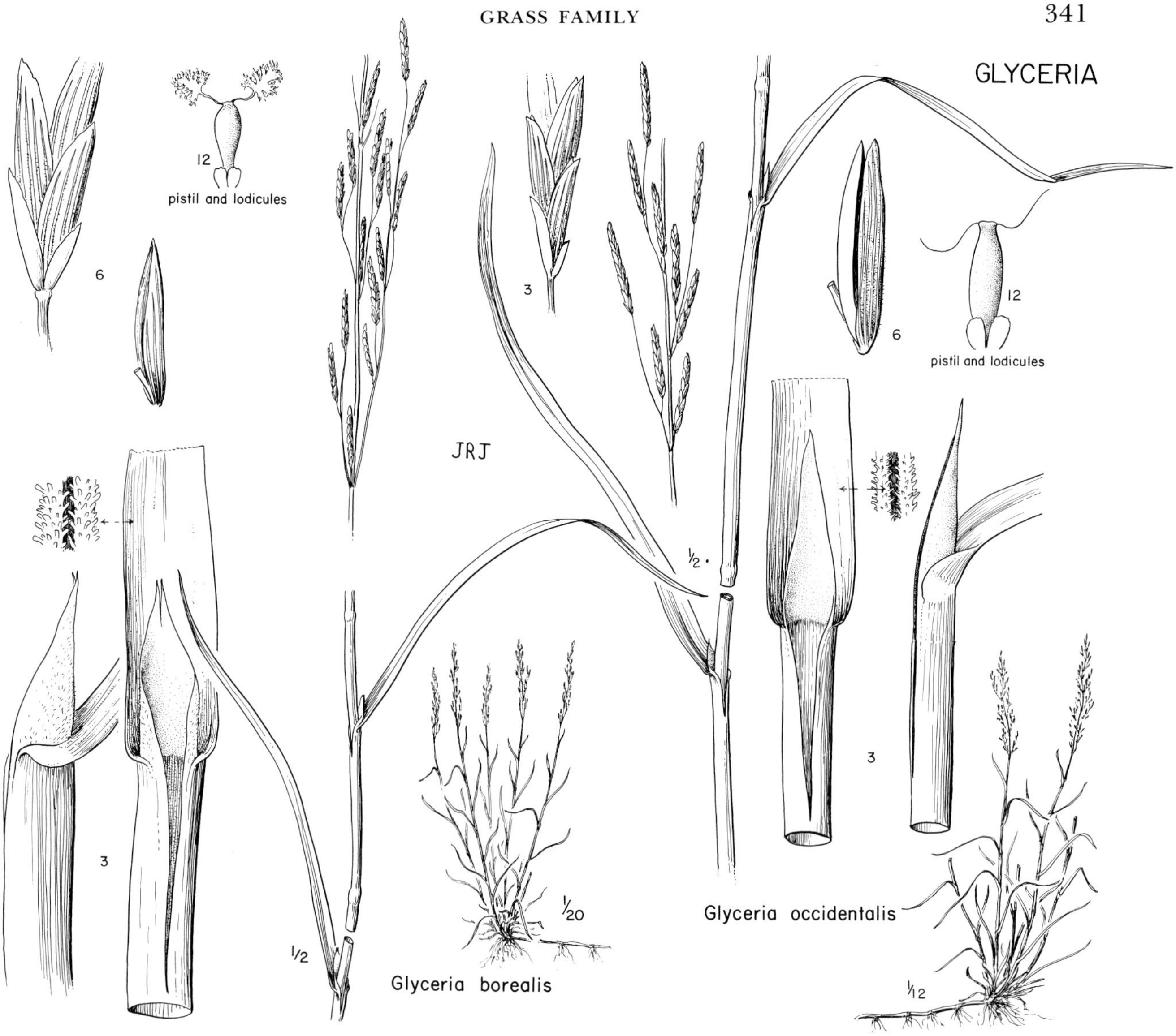

1. Glyceria borealis (Nash) Batchelder

Glyceria fluitans var. *angustata* Vasey ex Fern. Proc. Portland Soc. Nat. Hist. 2: 91. 1895; not *G. angustata* (R. Br.) T. M. Fries, 1869. *Panicularia borealis* Nash, Bull. Torrey Bot. Club 24: 348. 1897. *G. borealis* Batchelder, Proc. Manchester Inst. Sci. 1: 74. 1900. (*Fernald 193,* Violette Brook, Van Buren, Aroostook Co., Me., 25 July 1893.)

Northern mannagrass.

Rhizomatous perennials; culms 8–15 dm long, erect or decumbent at the base; sheaths glabrous, closed to within 1–4 cm of the top; ligules relatively large, 4–12 mm long, the shorter ones often broad and entire and the longer ones narrow and sometimes lacerate; blades flat or folded, 2–5 (7) mm broad and 9–23 cm long, glabrous or scaberulous on the upper surface; panicles relatively large, 18–40 (50) cm long, narrow, the branches erect to appressed, the base often enclosed in the upper sheath; spikelets relatively long, 9–15 (18) mm long, linear, subterete, 8- to 12-flowered, the rachilla joints 0.7–1 mm long; glumes glabrous, smooth, the first 1.2–2.2 mm long, lanceolate, elliptic or oblanceolate, nerveless, the second 2.2–3.5 mm long, broadly elliptic or obovate, rounded, 1-nerved; lemmas 3.3–4 mm long, glabrous, except sometimes scaberulous on the nerves, erose at the apex and margins, prominently (5-) 7-nerved, except on the broad and scarious margins, often short nerves between near the base; palea nearly as long as the lemma, the nerves winged; lodicules united; stamens 3, the anthers 0.5–0.8 mm long; $2n = 20$.

Shallow water and muddy shores of ponds and lakes and wet meadows at middle and subalpine elevs. in the mts. of our area; se. Alaska to Newfl., s. in the w. to n. Calif., Ariz. and N.M. and across the c. and e. states to S.D., Iowa, Ill., Ind., Ohio, Pa. and N.J. July–Aug.

2. Glyceria occidentalis (Piper) J. C. Nels.

Panicularia occidentalis Piper in Piper & Beattie, Fl. N.W. Coast 59. 1915. *Glyceria occidentalis* J. C. Nels. Torreya 19: 224. 1919. (*Piper 4905,* Vancouver, Wash., 5 June 1904.)

Rhizomatous perennials; culms 7–15 dm tall, erect or decumbent at the base; sheaths glabrous or minutely scaberulous, closed to within 1–4 cm of the top; ligules relatively large, 4–8 (13) mm long, acute to rounded, often lacerate, decurrent; blades flat or sometimes folded, 2–7 mm long, glabrous or sometimes scaberulous above; panicles relatively large, 17–40 (50) cm long, loose, somewhat narrow, the branches ascending to spreading at anthesis; spikelets long, 15–20 (23) mm long, linear, subterete, (5-) 8- to 12-flowered, the rachilla joints (1) 1.4–2.1 mm long; glumes erose-ciliolate, 1-nerved, the first glume 1.5–3.2 mm long, obovate, obtuse to rounded, the second 2.5–5 mm long, elliptic to lanceolate, acute to obtuse; lemmas relatively large, 4–5.5 (6.5) mm long, obtuse to broadly rounded, erose-ciliolate, sometimes toothed, prominently 7- (9)-nerved, scaberulous on the nerves and internerves; palea subequal to the lemma, the nerves winged; lodicules united; stamens 3, the anthers 0.8–1.5 mm long; $2n = 20, 40$.

Shallow water and muddy banks of ponds, wet marshes and quiet waters of streams; sw. B.C., Wash. and Idaho, s. to n. Calif. and n. Nev. Late June–Aug.

A population in Nevada (*Holmgren & Ballenger 7073*, 1/2 mile above Lee, Elko Co., Nev.) was found by Church (1949) to be a diploid which he identified as *G. declinata* Breb., a diploid European species. Morphologically the Nevada collection matches our material of *G. occidentalis*, which, in the opinion of Church, should all be tetraploid. The distinction Church used to separate the two species, besides ploidy level, is the depth of the apical cleft in the palea. Our material of *G. occidentalis* has a relatively wide range of depth variation within which the Nevada collection fits. According to Munz (Calif. Fl., 1959) the California material of *G. declinata* has culms 1–7 dm tall and blades 2–3 mm broad while the culms of *G. occidentalis* are 6–10 dm tall and the blades 3–12 mm broad. The Nevada collection has the broad blades of *G. occidentalis*, but the decumbent stems may have been less than 6 dm high. It is our opinion that this collection represents a diploid *G. occidentalis*.

3. Glyceria grandis S. Wats.

Poa aquatica var. *americana* Torr. Fl. N. & Middle Sect. U. S. 1: 108. 1823. *Panicularia americana* MacMillan, Metasp. Minnesota Valley 81. 1892. *Glyceria americana* Pammel, Rep. Iowa Geol. Surv. Suppl. 1903: 271. 1905. *G. maxima* var. *americana* B. Boivin, Naturaliste Canad. 94: 524. 1967. (*Cooley*, Deerfield, Mass.)

G. grandis S. Wats. Gray's Manual Bot. ed. 6. 667. 1890. *Panicularia grandis* Nash in Britton & Brown, Ill. Fl. ed. 2. 1: 265. 1913. *G. maxima* subsp. *grandis* Hultén, Acta Univ. Lund, II. (Fl. Alaska & Yukon 2:) 38: 229. 1942. *G. maxima* var. *grandis* Breit. Canad. Field-Naturalist 71: 45. 1957. (*Munro*, Quebec, in 1858.)

American mannagrass.

Rhizomatous perennials; culms (5) 9–15 (20) dm tall, erect or decumbent at the base; sheaths glabrous to minutely scaberulous, closed to the top or open for about 1 cm; ligules 2–4 (7) mm long, truncate or rounded, entire to somewhat lacerate; blades flat, folded or loosely involute, broad, 4.5–12 (15) mm, glabrous to scaberulous; panicles 16–32 (40) cm long, open, lax with spreading branches; spikelets 4–6.5 mm long, ovate, slightly flattened, 4- to 6- (7)-flowered, purplish, the rachilla joints 0.6–0.8 mm long; glumes somewhat erose and minutely ciliolate, 1-nerved, the first glume 1–1.8 mm long, elliptic or lanceolate, sometimes broadly so, acute, the second 1.5–2.5 mm long, ovate or obovate, obtuse; lemmas 2–2.8 mm long, ovate, truncate to obtuse, erose to toothed, prominently (5-) 7-nerved, the nerves finely scaberulous, the internerves glabrous; palea broad, as long to slightly longer than the lemma, the nerves not winged; lodicules free; stamens (2) 3, the anthers 0.5–0.9 mm long; caryopsis ovoid, blackish, shiny; $2n = 20, 28?$

In water of streams, ditches, ponds and wet meadows in the valleys and lower elevs. of the mts.; Alaska to N.S., s. throughout the mt. states and in the c. and e. states to Nebr., Iowa, Ill., Ky. and Va. Late June–Aug.

Glyceria grandis and the following 2 species appear to be closely related. They more or less represent a reduction series in panicle and spikelet size. Plants of *G. grandis* have been found with either 2 or 3 stamens whereas all plants studied of *G. elata* and *G. striata* have only 2.

4. Glyceria elata (Nash) M. E. Jones

Panicularia elata Nash in Rydb. Mem. New York Bot. Gard. 1: 54. 1900. *Panicularia nervata* subsp. *elata* Piper, Contr. U. S. Natl. Herb. 11: 140. 1906. *Glyceria elata* M. E. Jones, Bull. Montana State Univ., Biol. Ser. 15: 17. 1910. *G. nervata* var. *elata* J. K. Henry, Fl. S. Brit. Columb. & Vancouver Island 38. 1915. (*Flodman 176*, Sweet Grass Canyon, Crazy Mts., Mont., 8 Sept. 1896.)

Tall mannagrass.

Rhizomatous perennials; culms (6) 10–15 dm tall, erect or decumbent at the base; sheaths scaberulous, closed to near the top; ligules (2) 2.5–4 (6) mm long, truncate, erose to lacerate, finely puberulent; blades flat, (4.5) 6–10 (12) mm broad, the larger leaves of the plant usually over 6 mm broad; panicles 15–25 cm long, open, lax with spreading branches; spikelets small, 2.5–5 mm long, ovate, slightly flattened, 3- to 4- (6)-flowered, the rachilla joints 0.4–0.5 (0.6) mm long; glumes relatively small, erose-ciliolate, the first glume 0.8–1.2 mm long, mostly 1 mm, ovate to lanceolate, the second 1–1.5 mm long, ovate; lemmas small, 1.7–2.2 mm long, ovate to obovate or rounded, erose, prominently (5-) 7-nerved; palea subequal to the lemma, often slightly longer, the nerves not winged; lodicules free; stamens usually 2, the anthers 0.5–0.8 mm long; $2n = 20, 28$.

In water of streams, ditches, ponds, marshes and wet meadows in mts.; B.C., Wash. and Mont., s. to Calif., Ariz. and N.M. July–Aug.

5. Glyceria striata (Lam.) A. S. Hitchc.

Poa striata Lam. Tableau Encycl. & Méth. 1: 183. 1791. *Poa sulcata* Roemer & Schultes, Syst. Veg. 2: 550. 1817; not Lag. 1816. *Poa lamarckii* Kunth, Enum. Pl. 1: 362. 1833. *Glyceria striata* A. S. Hitchc. Proc. Biol. Soc. Wash. 41: 157. 1928. *Panicularia striata* A. S. Hitchc. in Small, Manual S.E. Fl. 132. 1933. (E. Virginia; Carolina.)

Poa nervata Willd. Sp. Pl. 1: 389. 1797. *G. nervata* Trin. Mém. Acad. Imp. Sci. St. Pétersbourg, Sér. 6, Sci. Math. 1: 365. 1830. *Panicularia nervata* Kuntze, Revisio Gen. Pl. 2: 783. 1891. *G. aquatica* subsp. *nervata* Douin in Bonnier, Fl. Compl. France, Suisse & Belgique 12: 24. 1932. ("Habitat in America boreali.")

Panicularia nervata stricta Scribn. in A. Nels. U.S.D.A. Div. Agrostol. Bull. 13: 44. 1898. *G. nervata* var. *stricta* Scribn. ex A. S. Hitchc. in Robinson & Fern. Gray's Manual Bot. ed. 7. 159. 1908. *Panicularia nervata* f. *stricta* H. D. House, New York State Mus. Bull. (Ferns & Fl. Pls. New York State) 254: 118. 1924. *G. striata* var. *stricta* Fern. Rhodora 31: 47. 1929. *G. striata* subsp. *stricta* Hultén, Acta Univ. Lund. II. (Fl. Alaska & Yukon 2:) 38: 232. 1942. *G. striata* f.

GLYCERIA
6
15
pistil and lodicules
3
1/2
JRJ
3
15
pistil and lodicules
6
floret
1/2
1/20
6
Glyceria grandis
Glyceria elata

stricta B. Boivin, Naturaliste Canad. 94: 524. 1967. (*A. Nelson 3818,* Beaver Basin, Colo.-Wyo. state line, 22 July 1897.)

Panicularia nervata var. *rigida* Nash in Rydb. Mem. New York Bot. Gard. 1: 54. 1900. *G. nervata* var. *rigida* Lunell, Amer. Midl. Naturalist 4: 223. 1915. *Panicularia rigida* Rydb. Fl. Rocky Mts. 83, 1060. 1917 [1918]. *G. rigida* Rydb. Fl. Prairies & Plains Centr. N. Amer. 122. 1932; not Smith, 1824. (*Rydberg 2068,* Lima, Mont., 29 June 1895.)

Fowl mannagrass.

Rhizomatous perennials; culms 3–10 (13) dm tall, erect or decumbent at the base; sheaths glabrous to scaberulous, usually closed to the top; ligules small, 1–2.2 (3.5) mm long, membranous, truncate, erose-lacerate; blades flat or folded, relatively narrow, 2–4.5 (6) mm broad, glabrous or sometimes scaberulous above; panicles relatively short, (6) 8–16 (25) cm long, lax and open with loosely divergent branches; spikelets small, 2.5–3.7 (4.5) mm long, ovate, slightly flattened, purplish, 3- to 6-flowered, the rachilla joints 0.3–0.4 mm long; glumes small, ovate or obovate, rounded to obtuse, somewhat erose on the margins, 1-nerved, the first glume 0.5–0.8 (1) mm long, mostly 0.7 mm, the second 0.7–1 (1.2) mm long; lemmas small, 1.5–2 mm long, ovate or obovate, rounded, erose, prominently 5- to 7-nerved; palea subequal to the lemma, the nerves not winged; lodicules free; stamens 2, the anthers 0.3–0.6 mm long; $2n = 20, 28?$

Along slow streams, often in aspen or coniferous woods or willow thickets at middle and subalpine elevs. in our area; Alaska to Newfl., s. throughout most of temp. U.S. July–Aug.

Our material belongs to var. *stricta* (Scribn.) Fern., which differs from the eastern var. *striata* in being a smaller plant with less closure of the sheaths, narrower blades, smaller panicles and longer lemmas with distinctly scarious margins.

40. CATABROSA Beauv. Brookgrass

Aquatic perennials, bearing rhizomes; culms decumbent and rooting freely at the lower nodes, hollow; sheaths closed, at least at the base; ligules membranous, entire to erose; blades flat, moderately broad, non-auriculate; inflorescence an open panicle; spikelets 1- to 5- (ours mostly 2)-flowered, the callus glabrous, disarticulating above the glumes and between the florets; glumes flat, membranous, irregularly toothed or erose at the broad apex and upper margins, obtuse, unequal, small, shorter than the first floret, nerveless or the second faintly nerved; lemmas broad, truncate, rounded on back, relatively firm, but scarious apically, erose, conspicuously 3-nerved, the nerves parallel, not converging towards the apex; palea broad, subequal to the lemma, the 2 strong nerves submarginal; lodicules membranous, flabellate and minutely lobed; stamens 3; ovary glabrous, the styles terminal, short; caryopsis flat, oval, brownish, loosely enclosed in the lemma and palea; $x = 10$.

A genus of 7 species, mostly in n. Eurasia, 1 species in the U. S. (Name from the Greek *catabrosis,* an eating or devouring, in reference to the nibbled appearance of the broadly erose glumes and lemma tips.)

1. Catabrosa aquatica (L.) Beauv.

Aira aquatica L. Sp. Pl. 64. 1753. *Molinia aquatica* Wibel, Primitiae Fl. Werthemensis 116. 1799. *Poa airoides* Koeler, Descript. Gram. 194. 1802. *Catabrosa aquatica* Beauv. Essai Nouv. Agrostogr. 97, 149, 157. 1812. *Glyceria aquatica* Presl, Fl. Čech. 25. 1819. *Diarrhena aquatica* Rasp. Ann. Sci. Nat. (Paris) 5: 447. 1825. *Glyceria airoides* Reichenb. in Mössler, Handb. Gewächsk. ed. 2. 3: 1827. 1828. *Melica aquatica* Loisel. Fl. Gallica, ed. 2. 1: 59. 1828. *Colpodium aquaticum* Trin. Mém. Acad. Imp. Sci. St.-Pétersbourg, Sér. 6, Sci. Math. 1: 395. 1830. *Glyceria catabrosa* Klett & Richter, Fl. Leipzig 96. 1830. *Catapodium aquaticum* Trin. ex Willk. & Lange, Prodr. Fl. Hispan. 1: 77. 1861, as a synonym. ("Habitat in Europae pascuis aquosis.")

Brookgrass.

Rhizomatous perennials; culms 2–5 (8) dm tall, decumbent at the base and erect above; sheaths closed at least half their length, usually glabrous; ligules 2–6 (8) mm long, entire to erose, acute to obtuse, decurrent; blades flat, 3–8 (12) mm broad, glabrous to scaberulous; panicles 7–20 (25) cm long, open, the branches divergent, sometimes finely scaberulous; spikelets small, (2) 2.5–3.5 (4) mm long, 1- to 2- (3)-flowered, the rachilla joints 7–12 mm long; glumes erose to irregularly toothed, flat, 1-nerved, the nerves sometimes weak, the first glume 0.6–1.5 (2) mm long, ovate to linear, the second 1.2–2.5 mm long, ovate, rounded to truncate; lemmas 2–3 mm long, broad, truncate, scarious, erose, 3-nerved; palea subequal to the lemma and similar in texture; lodicules about 0.3 mm long; anthers 1.3–1.8 mm long; $2n = 20, 42$.

Along streams, seepages and edges of ponds and small lakes in the mts.; Alta. to Newfl., s. through Idaho, e. Oregon, Nev. and Utah, e. to nw. N.M., Colo., Nebr. and Wisc.; also Europe, nw. Africa and temp. Asia. Late June–Sept.

41. SCHIZACHNE Hackel

Tufted perennials; culms erect, unbranched, hollow; sheaths closed; ligule membranous, usually united in front; blades flat to loosely involute, relatively narrow, non-auriculate; inflorescence a simple panicle or loose raceme of few spikelets; spikelets large, 3- to 7-flowered, the callus strongly bearded, disarticulating above the glumes and between the florets; glumes membranous, rounded on back, (1-) 2- to 5-nerved, awnless, unequal, large, but shorter than the first floret; lemmas lanceolate, rounded on back, 7- to 13-nerved, 7 of the nerves prominent and converging towards the apex, awned from below a bifid apex; palea with very hairy submarginal nerves, much shorter than the lemma; lodicules thick, fleshy, truncate, about 0.5 mm long; stamens 3; ovary glabrous with subterminal styles; caryopsis smooth and shiny, free from the palea; $x = 10$.

GLYCERIA
CATABROSA

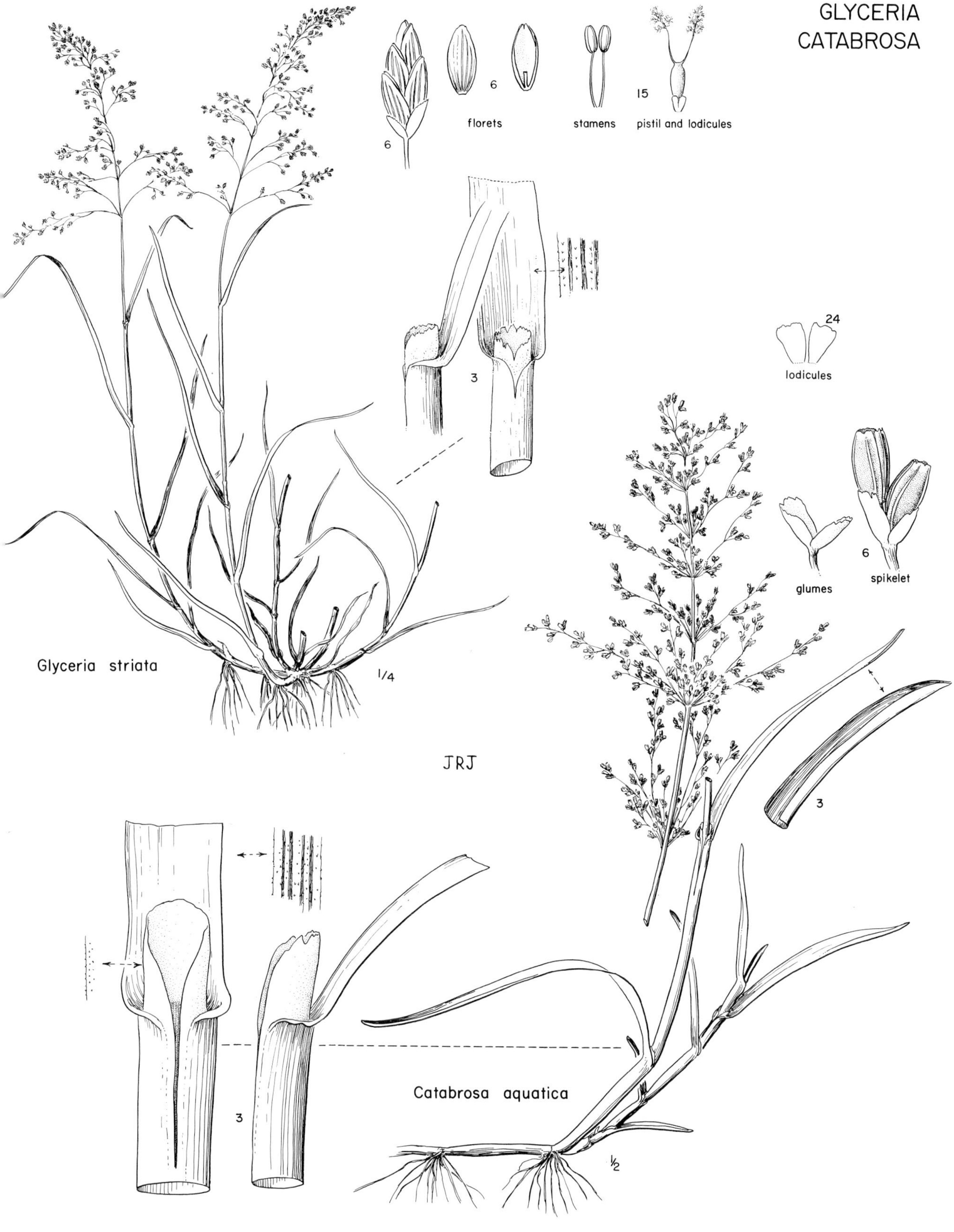

A monotypic genus of N. Amer. and e. Asia. (Name from the Greek *schizo*, to split, and *achne*, chaff, referring to the bifid apex of the lemma.)

References:
Boyle, W. S. Cytological evidence for the taxonomic position of *Schizachne purpurascens*. Madroño 7: 129–130. 1944.
Swallen, J. R. The grass genus *Schizachne*. J. Wash. Acad. Sci. 18: 203–206. 1928.

1. Schizachne purpurascens (Torr.) Swallen
Avena striata Michx. Fl. Boreali-Amer. 1: 73. 1803; not Lam. 1783. *Melica striata* A. S. Hitchc. Rhodora 8: 211. 1906. *Bromelica striata* Farw. Rhodora 21: 77. 1919. *Schizachne striata* Hultén, Svensk Bot. Tidskr. 30: 518. 1936. (*Michaux*, "a sinu Hudsonis per tractus montium ad Canadam," 1792.)
Trisetum purpurascens Torr. Fl. N. & Middle Sect. U. S. 1: 127. 1823. *Melica purpurascens* A. S. Hitchc. Contr. U. S. Natl. Herb. 12: 156. 1908. *Avena torreyi* Nash in Britton & Brown, Ill. Fl. N. U.S. Canada & Brit. Possessions, ed. 2. 1: 219. 1913. *S. purpurascens* Swallen, J. Wash. Acad. Sci. 18: 204, fig. 1. 1928. (*Dewey*, Williamstown, Mass.; *Bigelow*, near Boston, Mass.; *Eaton*, Catskill Mts.; *Paine*, near Montreal, Can.)

False melic.

Tufted glabrous perennials; culms (3) 5–8 (11) dm tall, erect, sometimes slightly decumbent at the base; sheaths closed to the top, sometimes split open above; ligules small, 0.5–1.5 (3) mm long, erose, united, usually longer in front forming an acute lobe but when split appearing as sheath auricles; blades flat, folded or loosely involute, 2–4 (5) mm broad, sometimes scaberulous above; panicles relatively short, 7–13 (17) cm long, sparsely flowered, the branches drooping, flexuous, the upper branches ascending and the lower divergent; spikelets 11.5–17 mm long, 3- to 6-flowered, the callus long-pilose with hairs 1–1.5 (2) mm long; glumes acute, glabrous, large, the first glume 4.2–6.2 mm long, faintly (1-) 3- to 5-nerved, the second 6–8.5 (9) mm long, faintly (3) 5-nerved; lemmas large, (8) 8.5–10.5 (12) mm long, scabrous, 7-nerved, sometimes with lesser nerves between (13-nerved), especially basally, the apex bifid with narrow teeth 1.2–2 mm long; awn arising from below the bifid apex, 8–15 mm long, diverging by a curve near the base; palea shorter than the lemma, the nerves strongly ciliate; anthers 1.4–2 mm long; $2n = 20$.

Moderately moist woods and open hillsides; Alaska to Newfl., s. in the Rocky Mts. to Mont., e. Utah and n. N.M. and in the midw. and e. to S.D., Minn., Ill., Ind., Ohio, Pa. and N.J.; also Siberia and Japan. Late June–Aug.

Tribe 5. STIPEAE

Cespitose perennials of dry places; culms hollow or sometimes solid (in some *Oryzopsis* and *Stipa webberi*); sheath margins free; ligule membranous or sometimes a fringe of hairs; blades non-auriculate; inflorescence a panicle, often very narrow; spikelets 1-flowered, disarticulating above the glumes; glumes papery or membranous, sometimes firm, faintly several-nerved; lemma indurate to membranous, obscurely nerved, awned from the tip, the awn becoming twisted and geniculate in *Stipa;* callus well-developed, often bearded; lodicules 2 or 3, large, narrowly-lanceolate to lanceolate or somewhat cuneate; caryopsis permanently enclosed by the lemma and palea; embryo pooid, except for the absence of an epiblast; $x = 7$ and 11.

The Stipeae are well represented in warm, dry, temp. regions of the N. and S. hemispheres. Traditionally *Stipa* and *Oryzopsis* were placed at the apex of the Agrostideae tribe along with the superficially similar *Aristida,* which has been moved to the Eragrostoideae subfamily in the new system. The presence of the primitive number of 3 lodicules in some species of the Stipeae is curious in a group with such a highly specialized spikelet.

1 Awn deciduous, mostly less than 10 mm long, not twisted, or only very slightly, sometimes bent, rarely more than 3 or 4 times as long as the plump lemma; callus short, usually obtuse.
  2 Awns 7–18 mm long with villous lemmas 2.7–5.5 mm long; sterile hybrids .... X *Stiporyzopsis*
  2 Awns mostly 2–7 mm long, up to 9.5 mm in *O. micrantha* which has glabrous lemmas that are 1.8–2.8 mm long; fertile species.
    3 Glumes 2.5–7.5 (8) mm long .... 42. *Oryzopsis*
    3 Glumes (7) 7.5–10 mm long .... 43. *Stipa webberi*
1 Awn usually persistent, (10) 14–200 mm long, twisted, several to many times longer than the narrow lemma; callus sharp pointed .... 43. *Stipa*

X STIPORYZOPSIS Johnson & Rogler

Reference: Johnson, B. L. Natural hybrids between *Oryzopsis hymenoides* and several species of *Stipa*. Amer. J. Bot. 32: 599–608. 1945.

X Stiporyzopsis bloomeri (Boland.) B. L. Johnson
Stipa bloomeri Boland. Proc. Calif. Acad. Sci. 4: 168. 1872. *Oryzopsis bloomeri* Ricker in Piper, Contr. U. S. Natl. Herb. 11: 109. 1906. X *Stiporyzopsis bloomeri* B. L. Johnson, Amer. J. Bot. 32: 602, figs. 14–18. 1945. (*Bolander* [*6116*], Bloody Canyon, near Mono Pass, Calif., Sept. 1866.)

Densely tufted perennials; culms 3–9.5 dm tall; leaves crowded at the base; sheaths glabrous to scaberulous, sometimes pubescent only at the throat; ligules 1–4 mm long, truncate, entire, puberulent; blades involute, firm, scabrous; panicles (7) 15–35 cm long, narrow, the branches erect or ascending; glumes subequal, (4.5) 5.5–8.5 (9.5) mm long, ovate-acuminate, sometimes forming an awn up to 1 (2) mm long, indistinctly 3- (5)-nerved, scabridulous, greenish below and anthocyanous or hyaline above; lemma (2.7) 3.5–5.5 mm long, elliptic, 5-nerved,

SCHIZACHNE
X STIPORYZOPSIS

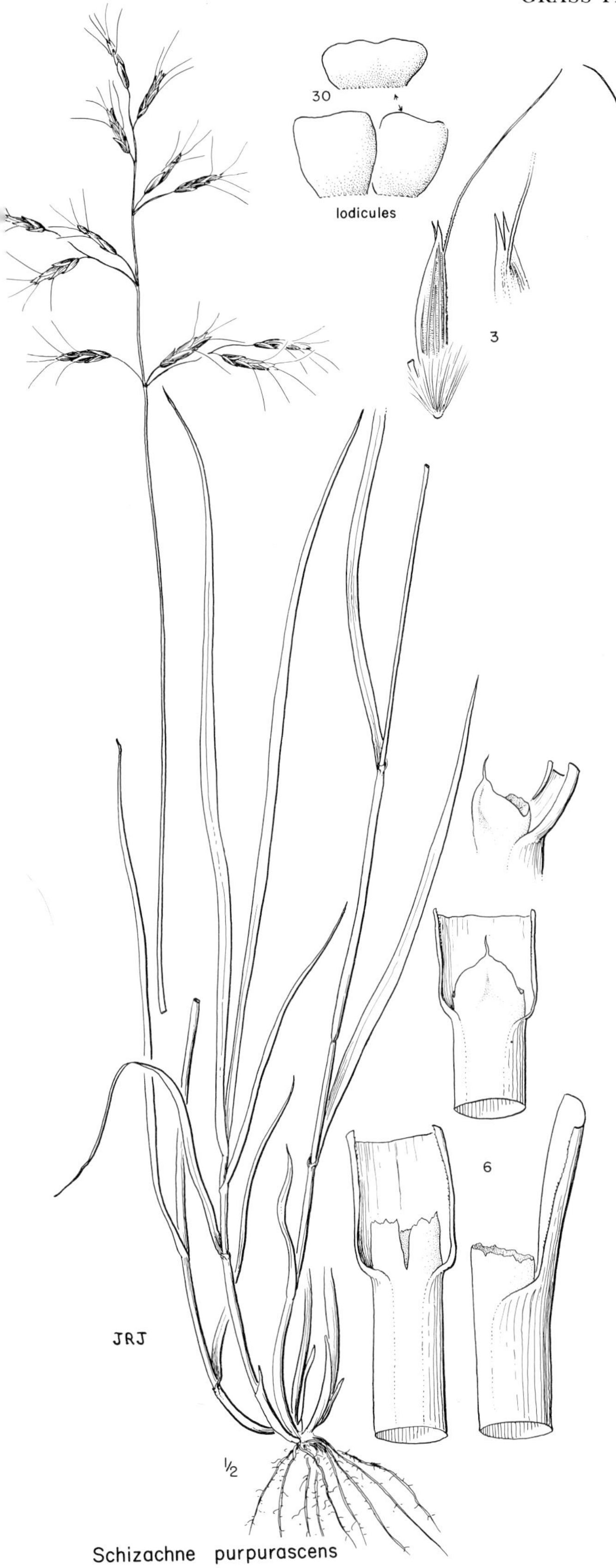

Schizachne purpurascens

X Stiporyzopsis bloomeri

densely long villous with hairs nearly 3 mm long, the callus somewhat pointed; awns 7–18 mm long, weakly once geniculate and twisted below, appressed pubescent, at least basally, only tardily deciduous; palea slightly shorter than the lemma; lodicules 2, about 1.8 mm long, elliptic-lanceolate; anthers 1.2–2.5 mm long with an apical tuft of hairs, the pollen sterile.

Sporadically occurring in a wide range of community types and soil types from the foothills to subalpine elevs.; e. Wash. to sc. Mont., s. to Calif., Nev. and Colo. Late June–Aug. (–Sept.).

X *Stiporyzopsis bloomeri* has been shown to consist of a series of sterile hybrids, all of which involve *O. hymenoides* as one parent and various species of *Stipa* as the other parent. These hybrids which collectively constitute this unnatural X *S. bloomeri* taxon are similar enough, regardless of the differences in the *Stipa* parent, to have been regarded as a species in the past. In our area the *Stipa* parentage may include *S. neomexicana* (Thurber) Scribn., *S. speciosa* Trin. & Rupr., *S. thurberiana* Piper, *S. occidentalis* Thurber, *S. columbiana* Macoun, *S. scribneri* Vasey, and *S. pinetorum* M. E. Jones. Careful observations in the field will reveal which species of *Stipa* has entered into the cross. Johnson (1945) intended the name X *Stiporyzopsis bloomeri* to apply only to the *O. hymenoides* X *S. occidentalis* cross.

42. **ORYZOPSIS** Michx. Ricegrass

Cespitose perennials; culms solid or hollow; sheaths open, mostly at ground level and often appearing partly buried; ligules membranous, truncate to elongate and acute; blades erect, flat or more often involute, non-auriculate; inflorescence a narrow or open, sometimes reduced panicle; spikelets 1-flowered, disarticulating above the glumes; glumes equal or nearly so, broad, obtuse to acuminate, membranous to firm, usually 3- to 5-nerved; lemma often the same length as the glumes or slightly shorter, broad, oval or oblong, nearly terete, indurate, appressed-puberulent to hirsute or glabrous, the callus short and blunt and often bearded with hairs about the same length as those on the lemma body; awn 2–8 (18) mm long, straight or curved, not more than 3 or 4 times longer than the body of the lemma, early deciduous, attached terminally or subterminally; palea subequal to the lemma; lodicules 2 or 3, large, narrowly lanceolate to elliptic-lanceolate or somewhat cuneate; stamens 3; caryopsis permanently enclosed by the lemma and palea; x = 11, 12, 14.

A genus of about 20 species distributed in the cool and temp. regions of both hemispheres. (From the Greek *oruza,* rice, and *opsis,* like, alluding to the similarity of the grains of some of the species to unpolished rice.)

The ricegrasses are all edible, but only *O. hymenoides* (Indian ricegrass) is of any importance. It is a dominant species in sandy desert areas where it cures well on the ground. Indian ricegrass becomes an important grass on winter ranges, and the seed has been used for food by the Indians.

Oryzopsis is closely related to the genus *Stipa. Oryzopsis webberi* (Thurber) Benth. ex Vasey is here treated as *Stipa webberi* (Thurber) B. L. Johnson because of its close affinities to *S. pinetorum.* Also many intergeneric hybrids have been reported between *O. hymenoides* and several species of *Stipa* (see discussions under *O. hymenoides and* X *Stiporyzopsis bloomeri*).

References:

Johnson, B. L. Cytotaxonomic studies in *Oryzopsis.* Bot. Gaz. 107: 1–32. 1945.

Kam, Y. K., and J. Maze. Studies in the relationships and evolution of supraspecific taxa utilizing developmental data. II. Relationships and evolution of *Oryzopsis hymenoides, O. virescens, O. kingii, O. micrantha,* and *O. asperifolia.* Bot. Gaz. 135: 227–247. 1974.

1 Lemmas usually smooth or nearly so, 1.8–3 mm long.
  2 Blades flat, 3 mm wide or more; plants tall, 6–15 dm; leaves mostly cauline; introduced species .... 1. *O. miliacea*
  2 Blades involute, or if flat less than 2 mm wide; plants mostly smaller, 2.5–7 (9) dm tall; leaves mostly basal; native species .... 2. *O. micrantha*
1 Lemmas copiously hairy, (2.5) 3–6 mm long.
  3 Pubescence of the lemma of short, appressed hairs; awns flexuous or geniculate; panicles short, 3.5–8 (11) mm long, narrow.
    4 Blades involute-filiform, less than 1.5 mm broad; glumes 3–5 (6) mm long; awn of the lemma 3–6.5 mm long, geniculate .... 3. *O. exigua*
    4 Blades flat or loosely involute, (3) 5–9 mm broad; glumes 5–7.5 mm long; awn of the lemma 9–14 mm long, flexuous .... 4. *O. asperifolia*
  3 Pubescence of the lemma of long, silky hairs; awns straight or weakly geniculate.
    5 Panicles (5) 7.5–35 cm long; ligules 1–7.5 mm long.
      6 Panicle branches and capillary pedicels divaricately spreading; awn of the lemma 3–5.5 mm long .... 5. *O. hymenoides*
      6 Panicle branches erect or ascending; awn of the lemma 7–18 mm long .... X *Stiporyzopsis bloomeri*
    5 Panicles short, 3–6 (7.5) cm long; ligules 0.2–0.5 (1) mm long; see species no. 1 of next genus .... *Stipa webberi*

1. Oryzopsis miliacea (L.) Aschers. & Schweinf.

Agrostis miliacea L. Sp. Pl. 61. 1753. *Achnatherum miliaceum* Beauv. Essai Nouv. Agrostogr. 20, 146, 148. 1812. *Urachne miliacea* C. Koch, Linnaea 21: 439. 1848. *Piptatherum miliaceum* Coss. Notes Pl. Crit. 129. 1851. *Oryzopsis miliacea* Benth. & Hook. ex Aschers. & Schweinf. Mém. Inst. Égypt. 2: 169. 1887. *Stipa miliacea* Hoover, Leafl. W. Bot. 10: 340. 1966. (Europe.)

Smilo grass.

Stout perennials; culms 6–15 dm tall, erect from a

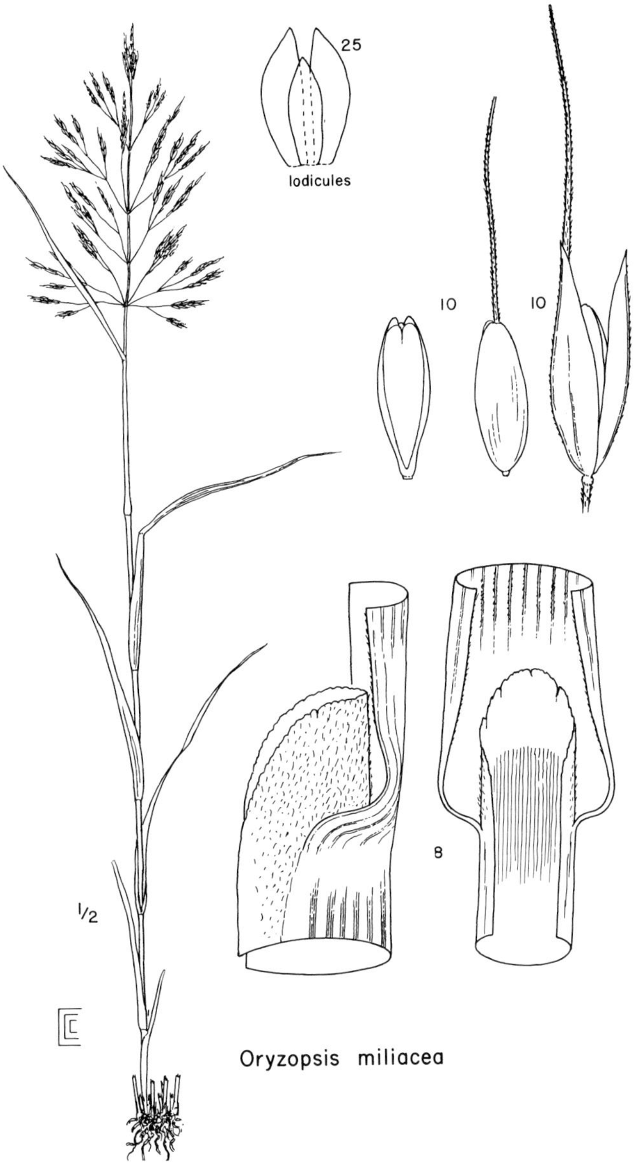

decumbent base, glabrous; leaves mostly cauline; sheaths glabrous, persistent, becoming papery; ligules 1.5–3 mm long, rounded to truncate, scaberulous; blades flat, 3–10 mm broad, glabrous, sometimes scabrous on the upper surface; panicles 15–30 cm long, loose, the branches ascending or spreading; glumes broadly lanceolate, acuminate, 3-nerved, minutely scaberulous to glabrous, the first glume 3–4 mm long, the second 2.5–3.5 mm long; lemma 2–3 mm long, oblong-elliptic, 3-nerved, smooth, shiny; awn 2–3.5 (4.2) mm long, straight, early deciduous; palea 2-nerved, same texture as the lemma, subequal; lodicules 3, the larger pair narrow ovate, the smaller one lanceolate; anthers about 1 mm long; $2n = 24$.

Dry to moderately moist disturbed places; introduced into Calif., Nev. (Elko Co.) and Utah from the Medit. region, possibly becoming established in our area.

2. Oryzopsis micrantha (Trin. & Rupr.) Thurber

Urachne micrantha Trin. & Rupr. Mém. Acad. Imp. Sci. Saint-Pétersbourg, Sér. 6, Sci. Math., Seconde Pt. Sci. Nat. 5(1): 16. 1842. *Oryzopsis micrantha* Thurber, Proc. Acad. Nat. Sci. Philadelphia 1863: 78. 1864. (Sask.)

Little ricegrass.

Densely tufted perennials; culms 2.5–7 (9) dm tall, numerous, hollow, glabrous; leaves mostly basal; sheaths smooth to scaberulous; ligules 0.2–1 (2) mm long, truncate, finely ciliolate; blades involute or if flat mostly less than 2 mm broad, scabridulous; panicle (4.5) 6–14 (20) cm long, narrow, the short branches ascending-appressed or slightly spreading; glumes subequal, 2.7–3.5 (4) mm long, ovate-acute, 3- to 5-nerved, glabrous to minutely scabridulous, greenish below, hyaline above; lemma 1.8–2.8 mm long, oblong-elliptic, usually glabrous, shiny, becoming brownish, indurate, the callus glabrous; awn (2) 4–8 (9.5) mm long, straight or somewhat flexuous, early deciduous; palea subequal to the lemma and the same texture; lodicules 3, the larger pair elliptic-oblanceolate, nearly 1 mm long, the third one smaller, lanceolate; anthers about 1 mm long; $2n$ = 22.

Dry open places from sagebrush foothills to coniferous woodlands at middle elevs., often in sandy soil or on rocky ridges; B.C. to Sask., s. to e. Calif. and Texas. June–early Aug.

3. **Oryzopsis exigua** Thurber

Oryzopsis exigua Thurber in Wilkes, U. S. Explor. Exped. Bot. 17: 481. 1874. (*Thurber 829,* "1st Camp beyond Mts to Chief's Place," Cascade Mts., Oregon.)

Little ricegrass.

Densely tufted perennials, sometimes forming large bunches; culms 1.5–3.5 (4) dm tall, stiff and hollow; sheaths persisting, becoming papery and finally fibrous, smooth to slightly scabrous; ligules 1.5–3.5 (4) mm long, acute, puberulent; blades involute-filiform, 5–10 cm long, stiffly erect, scabrous; panicle 3.5–8 (11) cm long, narrow, spike-like, the few short branches stiffly appressed-ascending; spikelets short pedicelled; glumes subequal, (3) 4–5 (6) mm long, broadly lanceolate, obtuse to abruptly pointed, faintly 3- to 5- (7)-nerved, scaberulous to glabrous, often anthocyanous, becoming hyaline above; lemma 3.5–5.5 mm long, elongate-elliptic, appressed-hispid; awn 3–6.5 mm long, stout, geniculate, the first segment sometimes twisted, attached subterminally behind a bifid apex; palea slightly shorter than the lemma; lodicules 2, narrowly oblanceolate, slightly longer than 1 mm; anthers 1.5–2 mm long; $2n$ = 22.

Open gravelly or rocky hillsides, ridges and dry meadows, sometimes in open woodlands at subalpine elevs. to near timberline; s. B.C. and s. Alta., s. to Oregon, Nev., Utah and n. Colo. Late June–Aug. (–Sept.).

4. **Oryzopsis asperifolia** Michx.

Oryzopsis asperifolia Michx. Fl. Boreali-Amer. 1: 51, pl. 9. 1803. *O. aspera* Michx. ex Muhl. Cat. Pl. N. Amer. 11. 1813. *Urachne asperifolia* Trin. Gram. Unifloris & Sesquifloris 174. 1824. (*Michaux,* "Hab. a sinu Hudsonis ad Quebec, per tractus montium.")

Tufted perennials; culms (2) 3–5 (6) dm tall, hollow; sheaths glabrous, sometimes auriculate at apex; ligules 0.2–0.7 mm long, truncate, ciliolate; blades flat to loosely involute, (3) 5–9 mm broad, erect, stiff, often exceeding the culms, scaberulous to hirsute beneath, glabrous above, the blades of the culms much reduced; panicle 3.5–7 (8) cm long, narrow, sometimes reduced to a raceme; glumes subequal, 5–7.5 mm long, broadly ovate, obtuse, mucronate, sometimes truncate and shallowly 3-toothed, 5- to 7-nerved, ciliate apically; lemma 5–7 mm long, broadly fusiform, turgid, shiny, hirsute with fine appressed hairs, the callus densely tufted in upper half; awn 9–14 mm long, flexuous, deciduous; palea slightly shorter than the lemma; lodicules 2, lanceolate, about 5 mm long; anthers 2.7–3.5 mm long; $2n$ = 46, 48.

Wooded slopes, usually under conifers; B.C. to Newfl., s. in the w. U.S. to n. Idaho, Mont., Wyo., ne. Utah (Uinta Mts., n. of Vernal, Uintah Co., *D. Austin s.n.*), Colo., and N.M., and e. to S.D., Minn., Wisc., Ill., Mich., N.Y. and New England. July–Aug.

5. **Oryzopsis hymenoides** (Roemer & Schultes) Ricker

Stipa membranacea Pursh, Fl. Amer. Sept. 728. 1814; not L. 1753. *Stipa hymenoides* Roemer & Schultes, Syst. Veg. 2: 339. 1817. *Oryzopsis membranacea* Vasey, U.S.D.A. Div. Bot. Bull. 12(2): pl. 10. 1891. *Eriocoma membranacea* Beal, Grasses N. Amer. 2: 232. 1896. *O. hymenoides* Ricker in Piper, Contr. U. S. Natl. Herb. 11: 109. 1906. *Eriocoma hymenoides* Rydb. Bull. Torrey Bot. Club 39: 102. 1912. (*Bradbury,* on the banks of the Missouri River, in 1811.)

Eriocoma cuspidata Nutt. Gen. N. Amer. Pls. 1: 40. 1818. *Milium cuspidatum* Sprengel, Syst. Veg. 1: 251. 1825. *O. cuspidata* Benth. ex Vasey, Grasses U. S. 23. 1883. (*Nuttall,* "On the grassy plains of the Missouri," in 1811.)

Indian ricegrass.

Densely tufted perennials; culms (2) 3–6 (8.5) dm tall, hollow with thick walls; sheaths glabrous to puberulent, partly buried in the sand, persisting and becoming papery and finally fibrous in old clumps; ligules 2.5–7.5 mm long, acuminate, entire, becoming lacerate; blades strongly involute, about 1 mm wide, nearly as long as the culms, numerous, smooth; panicle (5) 7–14 (19) cm long, the capillary branches in pairs, branchlets dichotomous, divaricately spreading, often flexuous; glumes ovate-acuminate or caudate, sometimes tapering into an awn up to 2 mm long, 3- (5)-nerved, the nerves prominent at the greenish base, becoming indistinct in the anthocyanous portion above, the margins hyaline, finely puberulent to nearly glabrous, the first glume (4.5) 5–7.5 (8) mm long, the second slightly shorter, 4.2–6.5 (7.5) mm long; lemma 2.5–4 (5) mm long, fusiform, turgid, maturing dark brown or nearly black, shiny, densely pilose-hirsute, the whitish hairs nearly extending the length of the glumes; awn 3–5.5 mm long, straight, readily deciduous; palea slightly shorter than the lemma; lodicules 3, nearly 2 mm long, the longer pair broadly cuneate, the third one rhombic and slightly shorter; anthers 0.8–1.2 (1.5) mm long with an apical tuft of hairs; $2n$ = 28 + 0 –8B, 48, 65, 130.

Mostly in deserts, sagebrush plains and pinyon-juniper woodlands, often dominating communities in sandy soils, less frequent in gravelly soils, from low valley bottoms to the foothills, but sometimes up to subalpine elevs.; B.C. to Man., s. to Calif., Ariz., N.M. and Texas. May–early July (–Aug.).

Indian ricegrass often forms dense grasslands where it appears to be a subclimax species. It is a nutritious grass that cures well on the ground because of thick-walled culms and numerous, long, involute leaf blades.

Sterile hybrids between *O. hymenoides* and several species of *Stipa* are often found. Johnson (Amer. J. Bot. 32: 599–608. 1945; 47: 736–742. 1960; 49: 540–546. 1962; 50: 228–234. 1963) discussed hybrids of Indian ricegrass and eleven species of *Stipa* that are similar enough to have been collectively included under X *Stiporyzopsis bloomeri*. Some of these sterile hybrids are beautiful grasses with hybrid vigor and promise of being favorable range grasses. They should be treated with colchicine in an attempt to induce chromosome doubling and subsequent fertility.

ORYZOPSIS

Oryzopsis exigua Oryzopsis asperifolia Oryzopsis hymenoides

43. STIPA L. Needlegrass, Feathergrass

Cespitose perennials; culms hollow or solid; leaves mostly basal; sheaths open, often persistent; ligules membranous, acute or truncate and often lacerate, sometimes consisting of a ring of hairs or a low crown; blades usually involute, long and narrow, usually disposed in a basal clump, non-auriculate; inflorescence a strongly contracted and spike-like to loose and open panicle; spikelets 1-flowered, large, disarticulating above the glumes; glumes subequal or the first longer than the second, both longer than the body of the lemma, narrow-acute to long-acuminate or infrequently aristate, mostly 1- to 5-nerved, membranous, often papery; lemma firm or indurate, relatively narrow, terete or angular, mostly faintly 5-nerved, the margins usually overlapping, the callus and rachilla long, usually sharp pointed from an oblique break and bearded with hairs usually longer than those on the lemma body; awn (5) 15–200 mm long, once or twice geniculate, usually several to many times as long as the body of the lemma (except in *S. webberi*), persistent or tardily deciduous, the lower segment(s) scabrous or pubescent, sometimes glabrous, twisted, attached terminally or subterminally, the junction of the lemma body and awn clearly evident, the terminal segment not twisted; palea shorter than the lemma, usually hidden by overlapping margins of the lemma, glabrous to pubescent; lodicules 2 or usually 3, slender and blunt, the lateral pair usually longer than the middle one; stamens 3; caryopsis tightly invested by the lemma and palea; $x = 7, 11$.

A genus of about 100 species, ranging from the tropics almost to the Arctic. Most of the U.S. species are distributed in the w. states. (Name from the Greek *stupe,* tow, in reference to the feathery appearance of the plumose awns of the type species, *S. pennata* L.)

The needlegrasses are generally valuable forage species. *Stipa comata* and *S. neomexicana* should be grazed only when the plants are young, because the sharp callus of mature plants is capable of causing mechanical injury to mouth parts of livestock. Most of the other species are important and palatable throughout the growing season and often occur in abundance on the range. Species of *Stipa* often dominate in prairie communities beyond the Intermountain Region borders. They are common within our area, but rarely form extensive communities.

Several species of *Stipa* are known to hybridize freely among themselves and with the closely related *Oryzopsis hymenoides* (Roemer & Schultes) Ricker. The resulting intergeneric hybrids show dominating characters of the *Oryzopsis* parent and are often referred to X *Stiporyzopsis bloomeri* (Boland.) B. L. Johnson since they all look so much alike. All are completely sterile. Euploidy and aneuploidy run rampant in *Stipa* with chromosome numbers ranging from $2n = 22$ to 70.

We have deviated from the widely used sequence of species established by A. S. Hitchcock (1925) in order to better reflect the possible phylogeny of species. Hitchcock began his sequence with the more specialized species such as *S. neomexicana, S. speciosa* and *S. coronata* and ended it with some of the *Oryzopsis*-like species such as *S. pinetorum.* It seems more logical to more or less reverse this sequence in order to place the *Oryzopsis*-like species at the point of divergence between the two obviously closely related genera. Cytologically this places the lower chromosome numbers at the beginning.

References:
Hitchcock, A. S. The North American species of *Stipa.* Contr. U. S. Natl. Herb. 24: 215–262. 1925.
Johnson, B. L. Natural hybrids between *Oryzopsis hymenoides* and several species of *Stipa.* Amer. J. Bot. 32: 599–608. 1945.
———. Natural hybrids between *Oryzopsis* and *Stipa.* I. *Oryzopsis hymenoides* X *Stipa speciosa.* ibid. 47: 736–742. 1960.
———. Amphiploidy and introgression in *Stipa.* ibid. 49: 253–262. 1962a.
———. Natural hybrids between *Oryzopsis* and *Stipa.* II. *Oryzopsis hymenoides* X *Stipa nevadensis.* ibid. 49: 540–546. 1962b.
———. Natural hybrids between *Oryzopsis* and *Stipa.* III. *Oryzopsis hymenoides* X *Stipa pinetorum.* ibid. 50: 228–234. 1963.

1 Terminal segment of the awn plumose; glumes (20) 30–60 mm long .... 13. *S. neomexicana*
1 Terminal segment of the awn not plumose; glumes less than 25 mm long, except in some *S. comata.*
  2 Lower segment of the once-geniculate awn strongly plumose, the hairs 4.5–8 mm long .. 10. *S. speciosa*
  2 Lower segment of the awn glabrous, scaberulous, or if plumose the hairs not more than 2 mm long.
    3 Lemma densely villous with white hairs about 2–5 mm long.
      4 Spikelets (8) 11–20 mm long; awns (10) 17–42 mm long; culms 2–20 dm tall .... 11. *S. coronata*
      4 Spikelets 6–11 (12) mm long; awns 5–17 (22) mm long; culms 1–3.5 (5) dm tall.
        5 Awn deciduous (sometimes tardily so), 5–7 mm long, straight or only slightly bent, not twisted .... 1. *S. webberi*
        5 Awn persistent, 11–22 mm long, twice geniculate .... 2. *S. pinetorum*
    3 Lemma often villous, but the hairs not more than 2 mm long (except the apical hairs in *S. scribneri*).
      6 Awn plumose on the lower two segments, the hairs 0.5–2 mm long.
        7 Ligules 2–7 mm long; glumes often purple .... 4. *S. thurberiana*
        7 Ligules 0.2–0.7 (2) mm long; glumes usually greenish.
          8 Pubescence of the lemma tip and the first awn segment similar, the hairs subequal in length; palea mostly less than half as long as the lemma .... 5. *S. occidentalis*
          8 Pubescence of the lemma tip and the first awn segment dissimilar, the hairs of the lemma longer; palea mostly more than half as long as the lemma .... 6. *S. nevadensis*
      6 Awn hirtellous to nearly glabrous, never plumose, the hairs, when present, less than 0.5 mm long.
        9 Lemma (8) 10–14 mm long; awn (60) 70–160 mm long; panicle (13) 18–40 cm long; glumes 15–35 mm long .... 12. *S. comata*

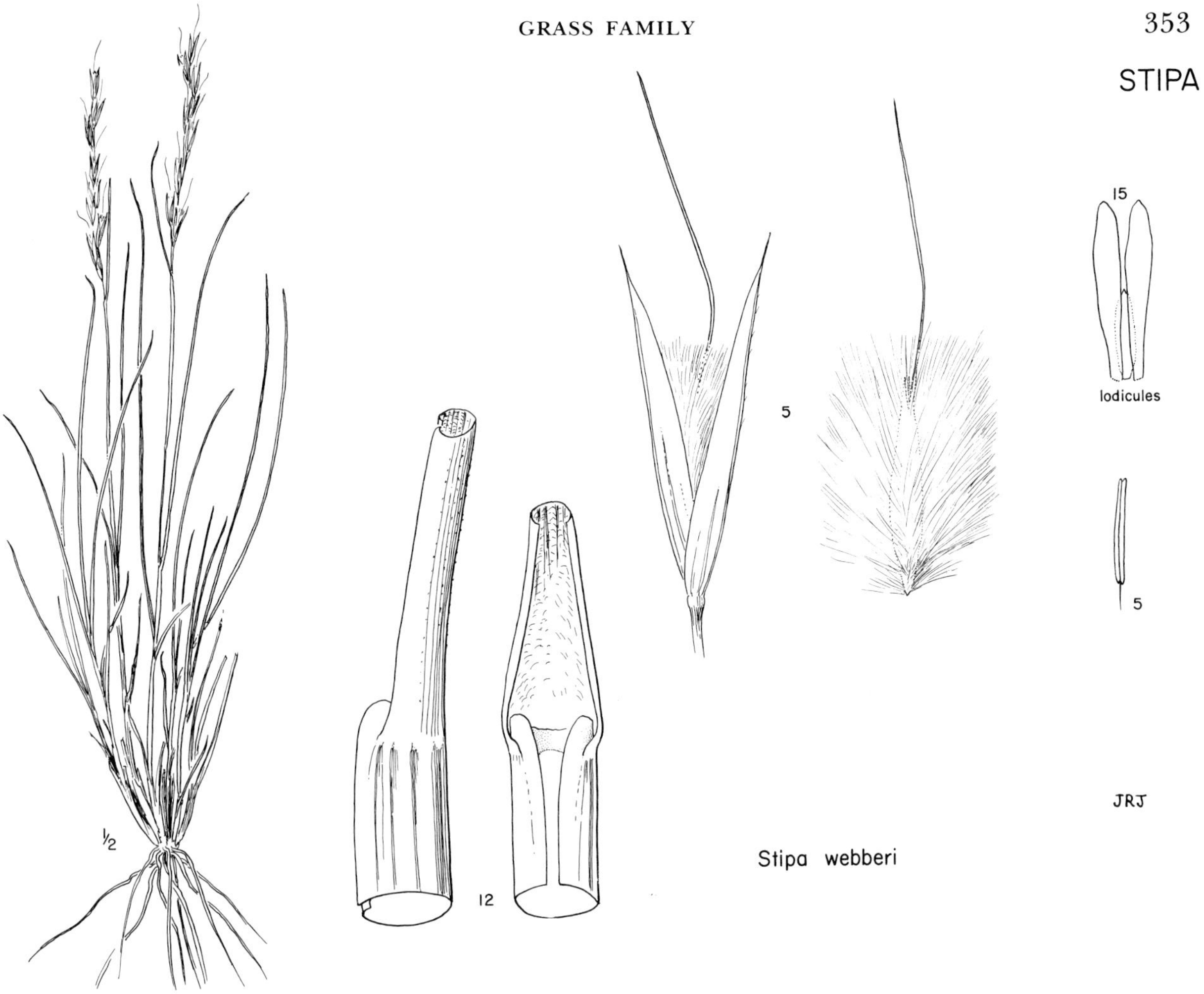

9 Lemma 3.5–8 mm long or less; awn 5–70 mm long, sometimes longer in *S. arida;* panicle 4–20 (30) cm long; glumes 5–15 (19) mm long.
- 10 Hairs on the lemma copious, 1.5–4 mm long at the summit.
  - 11 Lemma about 3.5–5 (6) mm long; sheaths glabrous at the throat .... 2. *S. pinetorum*
  - 11 Lemma (4.5) 6–8 mm long; sheath with a tuft of villous hairs at the throat ... . 3. *S. scribneri*
- 10 Hairs on the lemma not copious, rarely exceeding 1.5 mm at the summit.
  - 12 Glumes broad, abruptly acuminate; awn subterminal; lemma indurate; approaching our area on the west and possibly entering .... *S. lemmonii* (Vasey) Scribn.
  - 12 Glumes narrow, gradually acuminate; awn terminal; lemma usually not indurate.
    - 13 Awn 10–35 (50) mm long, twice geniculate and usually tightly twisted below, the terminal segment straight.
      - 14 Awn 20–50 mm long; lemma 5.5–7.2 mm long .... 7. *S. columbiana*
      - 14 Awn 10–22 mm long; lemma 4.5–5.7 (6.7) mm long .... 8. *S. lettermanii*
    - 13 Awn 40–75 mm long, obscurely geniculate and loosely twisted for 10–20 mm and flexuous beyond .... 9. *S. arida*

1. Stipa webberi (Thurber) B. L. Johnson

Eriocoma webberi Thurber in S. Wats. Bot. Calif. 2: 283. 1880. *Oryzopsis webberi* Benth. ex Vasey, Grasses U. S. 23. 1883. *Stipa webberi* B. L. Johnson, Bot. Gaz. 107: 25. 1945. (*Bolander,* Sierra Valley, Calif.)

Webber needlegrass.

Low, densely tufted perennials; culms 1–2.5 dm tall, solid, erect from a decumbent base and with numerous innovations; leaves in basal tufts, numerous, the tufts clothed in old leaf sheaths; sheaths usually shiny and smooth; ligules short, 0.2–0.5 (1) mm long, somewhat higher on the conspicuous edges than towards the back, subentire; blades strongly involute-filiform, 5–9 cm long, straight or often flexuous, glabrous to scaberulous below and puberulent above; panicles short, 3–6 (7.5) cm long, narrow, the few branches appressed and the base often enclosed in the upper sheath; glumes subequal, (7) 7.5–10 mm long, narrow-lanceolate, acuminate, glabrous, usually purplish, obscurely 3- to 5-nerved; lemma relatively small, (3.5) 4.5–6 mm long (including the callus), copiously spreading pilose, the whitish hairs 2–3.5 mm long and of about equal length throughout, the summit with 2 papery teeth, 0.5–1 mm long; callus 0.3 mm long, blunt, with shorter and more dense hairs than those on the lemma body; awn short, 5–7 mm long, straight or weakly once-geniculate, not twisted, glabrous, readily deciduous; lodicules 3, slender, the larger lateral pair about 1.5 mm long and the middle $^1/_2$ as long; anthers 1.4–2.4 mm long; $2n = 32$.

Dry, open sagebrush valleys and foothills; se. Oregon and s. Idaho, s. to e. Calif. and Nev., reported in w.-central Colo. Mid June–Aug.

Stipa webberi has been treated as an *Oryzopsis* by most botanists, but we prefer to follow B. L. Johnson and treat it as a *Stipa* that is very closely related to *S. pinetorum*. Both species have $2n = 32$ chromosomes, a number not known in any species of *Oryzopsis*. *Stipa pinetorum* is found at high elevations and *S. webberi* is the desert counterpart of that species. Only in awn characters is *S. webberi* like an *Oryzopsis;* it is definitely a *Stipa* in all other respects. *Oryzopsis swallenii* C. L. Hitchc. & Spellenberg, which is barely outside of our area in central Idaho, is very closely related to *S. webberi*. *Oryzopsis swallenii* has $2n = 34$ chromosomes instead of $2n = 32$. Spikelets of *O. swallenii* have the appearance of an *Oryzopsis,* with relatively broader glumes and lemmas. The inflorescence stands well above the numerous basal leaves, giving the plant an appearance of a reduced *O. exigua*. It is through these species that *Oryzopsis* and *Stipa* appear to merge and it is likely that a species like *S. webberi* has *Oryzopsis* parentage in its distant past.

2. **Stipa pinetorum** M. E. Jones

Stipa pinetorum M. E. Jones, Proc. Calif. Acad. Sci. II. 5: 724. 1895. (*Jones 6023p,* "open places among the pine forests," Panguitch Lake, Utah, 8 Sept. 1894.)

Pine needlegrass.

Low perennials, often forming large circular tufts as the center dies out in old bunches; culms (1) 1.5–3.5 (5) dm tall, often puberulent below the nodes, otherwise usually glabrous; leaves mostly basal; sheaths often shiny, smooth, persistent and clothing the base of the plant; ligules short, only 0.1–0.5 mm long, membranous, sometimes merely a crown; blades involute, capillary, less than 0.5 mm broad and 4–12 cm long, straight or sometimes flexuous, scaberulous or sometimes puberulent; panicles short, 4.5–11 (15) cm long, narrow with relatively few spikelets; glumes subequal, 7–11 (12) mm long, narrow-lanceolate, acuminate, glabrous, often deep-purple, 1- to 3-nerved, the lateral nerves obscure; lemma relatively small, 3.5–5 (6) mm long (including the callus), pale or brownish, densely villous, the whitish hairs nearly uniform in length from top to bottom, 1–4 mm long, the summit bearing 2 slender teeth at the base of the awn, the callus relatively short, 0.5 mm long with shorter and more dense hairs than those on the body of the lemma; awn relatively short, 11–17 (22) mm long, twice-geniculate, the lower segment loosely twisted, the terminal segment 5–12 mm long, not twisted, glabrous or nearly so, usually deep-purple; palea about $^1/_2$ as long as the lemma, villous above; lodicules 3, slender, the middle one $^2/_3$ the length of the lateral pair; anthers purple, 1.4–2.2 mm long; $2n = 32$.

High sagebrush slopes, open pine woods and rocky ridges, mostly from middle elevs. to near timberline; se. Oregon, s. Idaho and sw. Wyo., and s. to e. Calif. (mostly Inyo and Mono cos.), Nev., Utah and Colo. Late June–Aug.

3. **Stipa scribneri** Vasey

Stipa scribneri Vasey, Bull. Torrey Bot. Club 11: 125. 1884. (*Vasey,* "on dry hillsides at Sante Fe, New Mexico," Aug. 1884.)

Scribner needlegrass.

Tufted perennials; culms 3.5–8 dm tall, glabrous except at the finely puberulent, purplish nodes; sheaths smooth, strongly striate, persistent, becoming brown in age and clothing the base of the culms, the throat ciliate with long tawny hairs that diminish in size on down the margins; ligules 0.2–1 (2) mm long, irregularly truncate, finely ciliolate, becoming lacerate, sometimes merely a crown; blades involute, or if flat, 1.5–3.5 (5) mm broad, often involute towards the tip, 15–25 cm long, glabrous to scaberulous, the margins scabrous to ciliate; panicles 7–20 cm long, narrow with short, appressed branches; glumes distinctly unequal, linear-lanceolate, acuminate, glabrous to scaberulous, pale and often hyaline or somewhat purplish, prominently 3-nerved, the first glume (10) 12–15 (19) mm long, the second 9–12 (14) mm long; lemma (4.5) 6–8 mm long (including the callus), pale, villous, with hairs becoming more dense and longer at the tip (2–3.5 mm long) than below (0.5–1.5 mm long), the callus 0.7–1 mm long, densely pubescent; awn relatively short, 14–22 mm long, once- or rarely twice-geniculate, the lower segment(s) twisted, scaberulous to finely puberulent, the terminal segment 8–12 mm long, glabrous and untwisted; palea 2–3 mm long; lodicules 2, narrow-elliptic, about 1.5 mm long; anthers about 4 mm long; $2n = 40$.

Mesas and dry rocky mt. slopes; s. Utah, Colo., s. to n. Ariz. and N.M. Late June–early Sept.

4. **Stipa thurberiana** Piper

Stipa occidentalis Thurber in Wilkes, U.S. Explor. Exped. Bot. 17: 483. 1874; not Thurber, 1871. *S. thurberiana* Piper in Scribn. U.S.D.A. Div. Agrostol. Circ. 27: 10. 1900. (*Pickering & Brackenridge,* "N. branch of the Columbia, Washington.")

S. occidentalis var. *montana* Merr. & Davy, Univ. Calif. Publ. Bot. 1: 62. 1902. (*Watson 1296,* Pah-Ute Range [East Range], 5000 ft., Nev.)

Thurber needlegrass.

Densely tufted perennials; culms 3.5–5.5 (8) dm tall, puberulent, at least at the nodes; sheaths usually glabrous, sometimes puberulent below, striate; ligules exceptionally long, 2–5 (7) mm, acute, sometimes truncate, lacerate at the tip, decurrent, thin, hyaline, glabrous to sparsely puberulent; blades filiform-involute, scaberulous, 10–25 cm long; panicles 7–16 (24) cm long, narrow, the branches erect and relatively few-flowered; glumes subequal, (8) 10–16

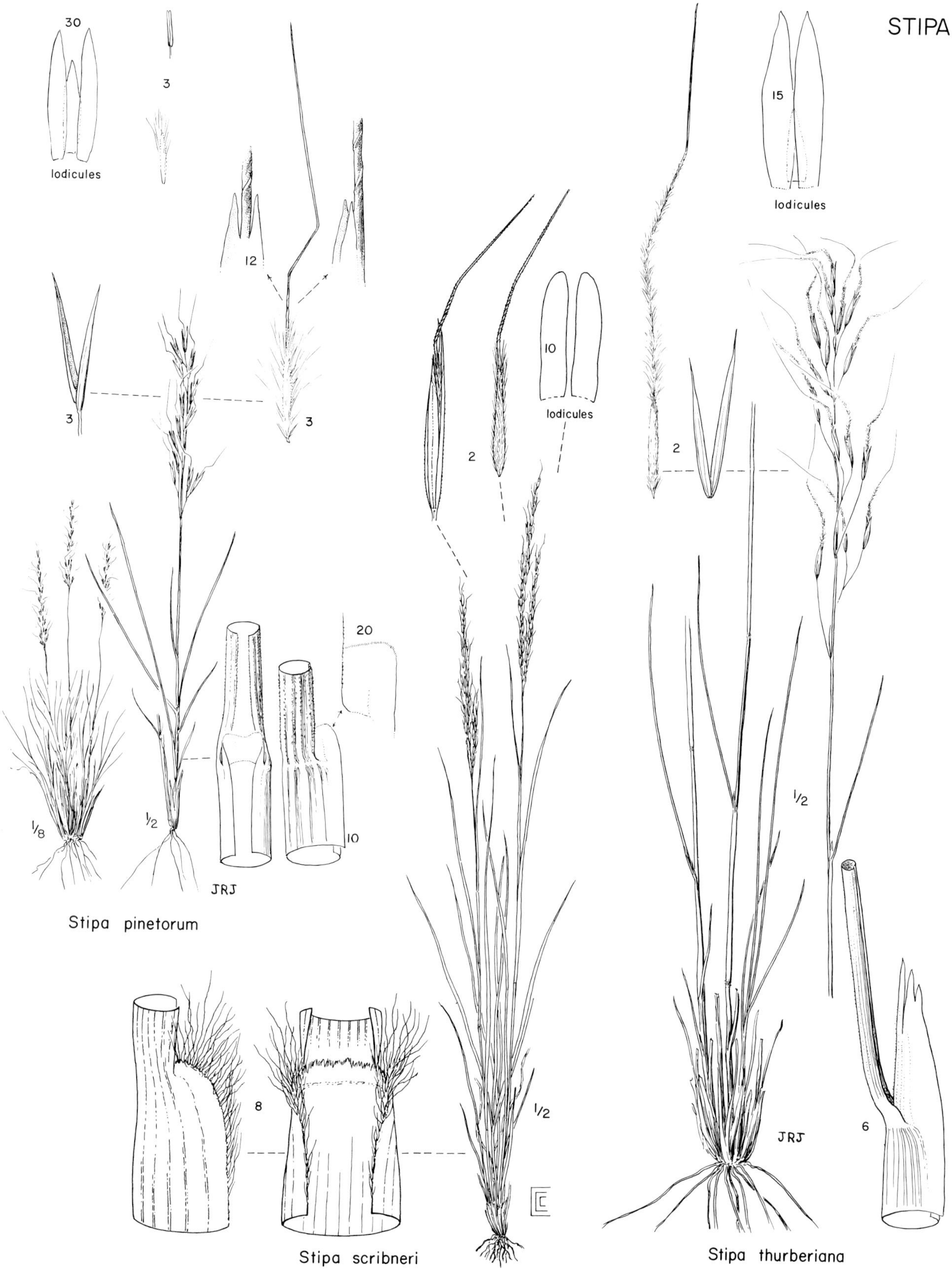
STIPA
lodicules
lodicules
lodicules
JRJ
JRJ
Stipa pinetorum
Stipa scribneri
Stipa thurberiana

mm long, the second slightly shorter, lanceolate-attenuate, 3- (5)-nerved, often purplish, hyaline above forming a soft point; lemma 6.2–8 (8.7) mm long (including the callus), sparsely appressed-pilose, the hairs 0.5–1 (2) mm long, the callus about 1 mm long, appressed-villous; awn 30–45 mm long, twice geniculate, the lower segments twisted, hirsute-plumose, the hairs 0.5–1.5 (2) mm long, the terminal segment 16–30 mm long, scaberulous to glabrous; lodicules 3, lanceolate, the lateral pair about 2 mm long and the middle one only about 1 mm long; anthers 3–4 mm long, purplish; $2n = 34$.

Sagebrush deserts, canyons and juniper wooded foothills; e. parts of Wash., Oregon and Calif., across n. Nev. to s. Idaho, sw. Mont. and ne. Wyo. Late May–July.

Stipa thurberiana is known to hybridize with *Oryzopsis hymenoides* (Roemer & Schultes) Ricker (Johnson, 1945). Thurber needlegrass is often abundant enough in parts of northern Nevada and southeastern Oregon to be an important range grass. The purplish glumes and plumose lower segments of the awn make this species a beautiful photographic subject with proper back-lighting.

5. **Stipa occidentalis** Thurber

Stipa occidentalis Thurber in King, Rep. Geol. Explor. 40th Parallel 5: 380. 1871. *S. stricta* var. *sparsiflora* Vasey, Contr. U. S. Natl. Herb. 3: 51. 1892. (*Bolander 5038,* Yosemite Trail, 8000 ft., Calif., in 1866.)

S. stricta Vasey, Bull. Torrey Bot. Club 10: 42. 1883; not Lam. 1791. *S. oregonensis* Scribn. U.S.D.A. Div. Agrostol. Bull. 17: 130, fig. 426. 1899. (*Suksdorf,* Wash. [erroneously given as Oregon by Vasey], in 1882.)

S. viridula var. *pubescens* Vasey, Contr. U. S. Natl. Herb. 3: 50. 1892; not *S. pubescens* R. Br. 1810. *S. elmeri* Piper & Brodie ex Scribn. U.S.D.A. Div. Agrostol. Bull. 11: 46. 1898. (*Suksdorf,* "Dry ground, Columbia River," Wash., in 1883.)

Western needlegrass.

Strongly tufted perennials; culms 2.5–4 (8) dm tall, glabrous to hispidulous, especially at the nodes; sheaths glabrous to hirsute; ligules mostly short, 0.2–0.7 (2) mm long, truncate, sometimes no more than a low crown, often projecting higher on the sides than in back, entire to slightly erose-ciliolate, sometimes flanked by tufts of hairs which extend down the margins of the sheath; blades filiform-involute, or rarely flat and up to 2 mm broad, 10–30 cm long, puberulent; panicles (5) 10–25 (35) cm long, narrow, sometimes rather loose and lax, the branches ascending; glumes subequal, 9–15 mm long, narrow-lanceolate, acuminate, 3- (5)-nerved, papery; lemma (5) 6–8 (8.5) mm long (including the callus), more or less indurate, soft pubescent, sometimes with longer hairs at the tip, the hairs 0.2–1.2 (1.7) mm long, yellowish to pale-brown, the callus 1–1.5 mm long, sharp, usually curved; awn 25–35 (50) mm long, twice geniculate, the lower segments twisted and plumose with hairs 0.5–1.2 mm long, the terminal segment glabrous to short plumose, not twisted; palea 2–3.2 mm long, pubescent; lodicules 2, narrow-elliptic, about 2 mm long; anthers 2.5–4 mm long; $2n = 36$.

In mts. on dry to moderately moist slopes and ridges in sagebrush–grass and coniferous woodland vegetation from middle to subalpine elevs., becoming abundant in western juniper woodlands, Wash., Idaho and Wyo., s. to Calif., Nev. and Utah. June–Aug.

Western needlegrass is an important bunchgrass of the palouse prairies. The species, as treated here, includes *S. elmeri. Stipa elmeri* has been distinguished as being more than 4 dm tall and having pubescent culms, while *S. occidentalis* should be less than 4 dm tall with glabrous culms, but these characters are so variable that many collections could be keyed to either species. Johnson's (1962) cytological and statistical study gives evidence for this. Johnson found that what has been called *S. californica* Merr. & Davy [*S. occidentalis* var. *californica* (Merr. & Davy) C. L. Hitchc.] in the Great Basin and east slopes of the Sierra Nevada turns out to be an amphiploid which he named *S. nevadensis.* See under that species for distinctions. *Stipa californica* has its greatest development on the west slope of the Sierra Nevada and apparently does not enter the Intermountain Region. The *Stipa occidentalis* complex is discussed by Johnson (1962) and Maze (Leafl. W. Bot. 10: 157–161. 1965).

6. **Stipa nevadensis** B. L. Johnson

Stipa nevadensis B. L. Johnson, Amer. J. Bot. 49: 257. 1962. (*Johnson 211,* Upper Twin Lake, near Bridgeport, Mono Co., Calif.)

S. californica of authors.

Tufted perennials; culms 2–7 (8.5) dm tall, puberulent below the nodes; sheaths glabrous to slightly scaberulous, the throat usually glabrous; ligules short, 0.2–0.7 mm long, truncate, often projecting higher on the sides than in back, erose-ciliolate; blades involute or flat, 1–3 mm broad, often involute towards the tip, 10–25 cm long, pubescent above, glabrous below; panicles 6–15 (20) cm long, narrow, the lower branches often enclosed by the upper sheath; glumes subequal, 8–14 mm long, narrow-lanceolate, acuminate, 3-nerved, papery; lemma 5.5–7 (8.5) mm long (including the callus), more or less indurate, soft pubescent, sometimes with longer hairs at the tip, these averaging about 1.5 mm long, the callus 0.7 mm long, sharp; awn 20–30 (35) mm long, twice geniculate, the lower segments twisted and plumose with hairs averaging about 0.8 mm long (0.5–1.2), the terminal segment glabrous to short plumose, not twisted; palea 2.8–4.2 (6) mm long; lodicules 2, narrow-elliptic, about 2 mm long; anthers 2.5–4 mm long; $2n = 68$.

Dry foothills and open woodlands up to 3048 m (10,000 ft.) elev.; from the e. slope of the Sierra Nevada, Calif., across n. Nev. to s. Idaho. June–Aug.

Stipa nevadensis has been confused with *S. californica,* differing from it in having longer paleas in relation to lemma length and slightly longer awn hairs. *Stipa californica,* which is best considered as a variety of *S. occidentalis,* remains to the west of our region. Johnson (1962) suggests that *S. nevadensis* with $2n = 68$ may be an amphiploid derivative of *S. occidentalis* with $2n = 36$ and *S. lettermanii* with $2n = 32$. *Stipa californica* may have had similar ancestry with *S. occidentalis* as one parent and *S. columbiana,* a close relative of *S. lettermanii,* as the other. Both of these parents are $2n = 36$. The idea of considerable backcrossing with the parents is consistent with the wide variation in *S. californica.* However small the characters are that separate *S. nevadensis* and *S. occidentalis,* they seem fairly consistent.

7. **Stipa columbiana** Macoun

Stipa columbiana Macoun, Cat. Canad. Pl. 2(4): 191. 1888. (*Macoun 28940,* "On rocks, Yale, B.C.," 17 May 1875.)

S. viridula var. *minor* Vasey, Contr. U. S. Natl. Herb. 3: 50. 1892. *S. minor* Scribn. U.S.D.A. Div. Agrostol. Bull. 11: 46. 1898. *S. occidentalis* var. *minor* C. L. Hitchc. Univ. Wash. Publ. Biol. 17(1): 714. 1969. (*Letterman 95,* "Kelso Mt., near Torrey's Peak, Colo.," 13 Aug. 1885.) = var. *columbiana.*

S. williamsii Scribn. U.S.D.A. Div. Agrostol. Bull. 11: 45. 1898. (*Williams 2804,* "West side of Big Horn Mts. near Monument Spring," Wyo., 3 Aug. 1897.) = var. *nelsonii.*

S. nelsoni Scribn. U.S.D.A. Div. Agrostol. Bull. 11: 46. 1898. *S. columbiana* subsp. *nelsoni* A. S. Hitchc. Contr. U. S. Natl.

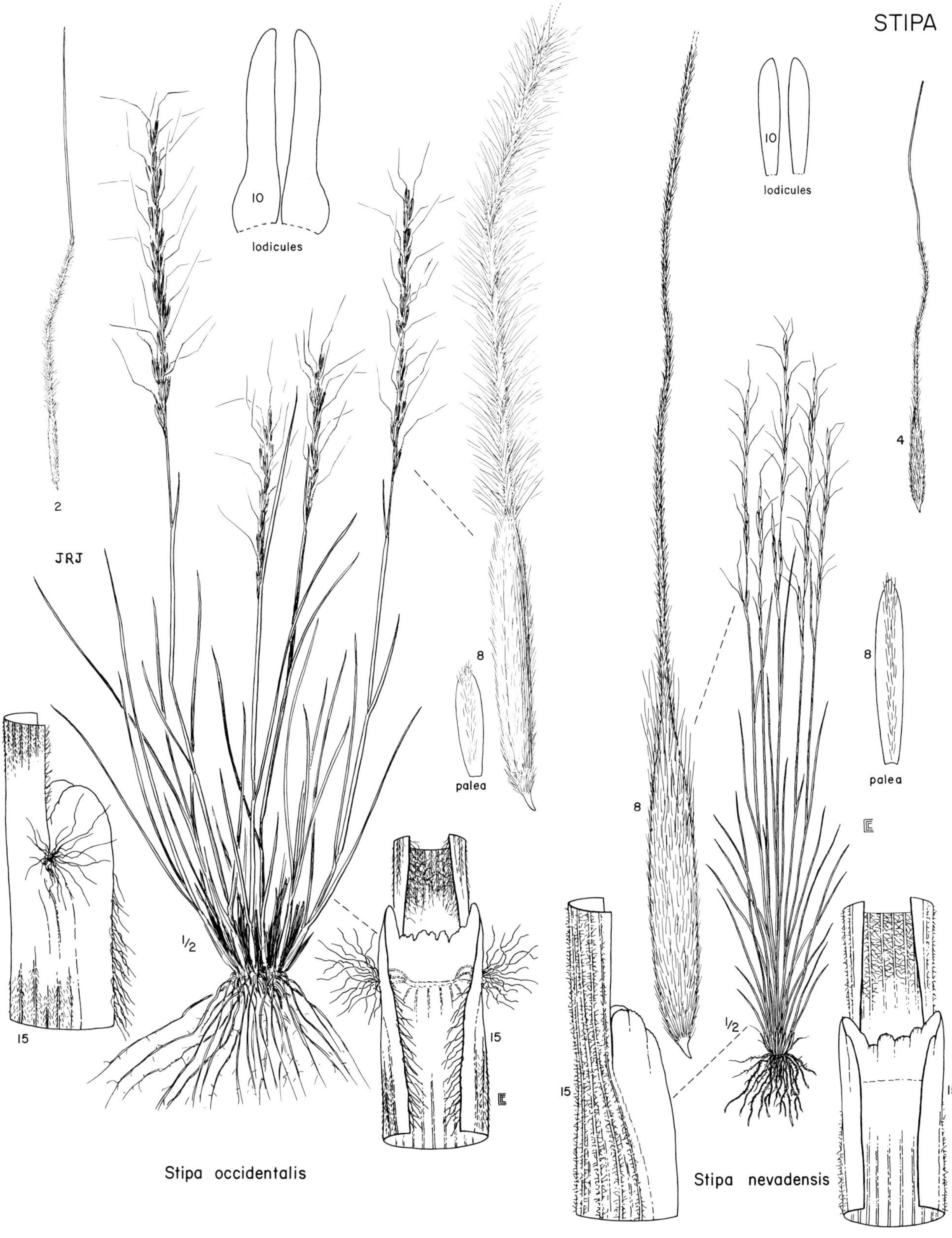
2
JRJ
10
lodicules
8
palea
1/2
15
15
Stipa occidentalis
10
lodicules
4
8
palea
8
1/2
15
15
Stipa nevadensis

Herb. 24: 254. 1925. *S. columbiana* var. *nelsoni* St. John, Fl. S.E. Wash. & Adj. Idaho 61. 1937. *S. occidentalis* var. *nelsonii* C. L. Hitchc. Univ. Wash. Publ. Biol. 17(1): 715. 1969. (*Nelson 3963*, Woods Landing, Woods Creek, 8700 ft., Albany Co., Wyo., 9 Aug. 1897.)

Columbia needlegrass.

Stout, tufted perennials; culms relatively tall, (3) 4–10 (12) dm, erect; sheaths glabrous to densely pubescent, strongly ribbed; ligules mostly short, 0.2–1 (2) mm long, truncate, rather firm, usually longest on the sides, decurrent; blades involute but sometimes flat and 2.5–5 mm broad, 10–20 cm long, glabrous or slightly scaberulous to densely pubescent, often strongly striate; panicles 12–25 (30) cm long, narrow, compact or rather loose with the lower branches separated from the remainder of the inflorescence, the branches short and appressed with numerous spikelets; glumes subequal, 7–11 (14) mm long, acuminate or awn-pointed, somewhat purplish, scaberulous, papery, becoming somewhat hyaline above, 3-nerved; lemma 5.5–7.2 mm long (including the callus), appressed pubescent, the hairs usually longer at the tip (0.7–1.5 mm long) than below (0.2–0.5 mm long), the callus short, 0.7–1 mm long; awn 20–30 (50) mm long, thin, twice geniculate, the lower segments twisted and scabrous, the terminal segment 10–16 mm long, not twisted, scaberulous to glabrous; palea shorter than the lemma, pubescent; lodicules 2, about 1.5 mm long, slender; anthers 2.2–3 mm long; $2n$ = 36, 44.

Dry sagebrush plains and slopes at middle and high elevs.; B.C. to w. S.D., and s. to n. and e. Calif., Ariz., N.M. and reported to occur in w. Texas. June–Sept.

C. L. Hitchcock [Univ. Wash. Publ. Biol. 17(1): 714. 1969] relegated *S. columbiana* under *S. occidentalis* to var. *minor* (Vasey) C. L. Hitchc. The two are certainly closely related, but it seems better to us to separate them. We considered treating *S. williamsii* as a variety here but the degree of pubescence of the culms, blades and sheaths seemed rather trivial as the plants are like var. *nelsonii* in all other respects. Two varieties are discernible and may be separated as follows:

1 Awns mostly less than 2.5 cm long; callus blunt; culms 3–6 dm tall . . . . var. *columbiana*
1 Awns 2.5 or more cm long, up to 5 cm; callus sharp; culms as much as 1 m tall . . . . var. *nelsonii* (Scribn.) St. John

8. Stipa lettermanii Vasey

Stipa lettermani Vasey, Bull. Torrey Bot. Club 13: 53. 1886. *S. viridula* var. *lettermani* Vasey, Contr. U. S. Natl. Herb. 3: 50. 1892. (*Letterman 102*, Snake River, Idaho, Aug. 1885.)

Letterman needlegrass.

Tufted perennials, often forming large clumps; culms 2.5–6 (8.5) dm tall, glabrous or minutely scaberulous, with numerous innovations; sheaths glabrous, sometimes scaberulous; ligules 0.2–1.2 (2) mm long, firm, rounded, truncate, decurrent; blades involute, filiform, rarely flat and up to 2 mm wide, 10–20 cm long, hispid above, glabrous to minutely scaberulous below; panicles 7–19 (24) cm long, narrow, the branches erect with relatively few spikelets; glumes subequal, relatively short, (5) 6.5–9 (10.5) mm long, acuminate, glabrous or sometimes scaberulous, often purple with hyaline margins and apex, 3-nerved; lemma 4.5–5.7 (6.7) mm long (including the callus), pale, only slightly indurate, sericeous with hairs longer at the summit (1–1.5 mm long) than below (0.2–0.5 mm long), the callus often with dense white hairs less than 1 mm long; awn relatively short, (10) 16–22 mm long, slender, twice-geniculate, the lower segments loosely twisted, scaberulous with hairs about 0.2–0.3 mm long, the terminal segment 7–12 mm long and glabrous; palea about 3.5 mm long, pubescent, sometimes exposed; lodicules 2, about 1.5 mm long, slender; anthers 1.7–2.5 mm long, purplish; $2n$ = 32, 66, 68.

Sagebrush slopes, rocky ridges and open coniferous or aspen woods at middle and subalpine elevs. in the mts.; Oregon to Mont. and s. to s. and e. Calif., Ariz. and N.M. July–Sept.

Specimens of *S. lettermanii* and *S. columbiana* are often misidentified as they approach one another in size. *Stipa lettermanii* has a slender palea which is about 2/3 the lemma length and is often exposed, and chromosome counts of $2n$ = 32, 66 and 68, whereas *S. columbiana* has a shorter palea, less than half as long as the lemma, and chromosome numbers of $2n$ = 36 and 44. Subalpine forms of *S. lettermanii* are distinctive and rarely mistaken for the robust *S. columbiana*. The difference in hair length between the summit and midlength of the lemma in *S. lettermanii* may not be easily discernible when the hairs are spreading at the base and somewhat appressed above.

9. Stipa arida M. E. Jones

Stipa arida M. E. Jones, Proc. Calif. Acad. Sci. II. 5: 725. 1895. (*Jones 5377*, "on very dry talus slopes in shingle," Marysvale, Utah, 4 June 1894.)

S. mormonum Mez, Repert. Spec. Nov. Regni Veg. Beih. 17: 209. 1921. (*Jones 2106*, Utah.)

Mormon needlegrass.

Densely tufted perennials; culms 4–7 (10) dm tall, scaberulous below the nodes; leaves mostly basal, scabrous, the persistent sheaths densely clothing the base of the plant; sheaths sometimes finely pubescent below the collar; ligules short, 0.2–0.8 mm long, variously lacerate-lobed with ciliolate edges, decurrent; blades mostly involute-filiform, the culm blades sometimes flat and 1–2 mm wide, 10–20 cm long, the margins scabrous; panicles short, 4–12 (16) cm long, narrow, compact with numerous, congested spikelets; glumes pale, linear-lanceolate, long-acuminate, papery, becoming hyaline above, glabrous or sometimes scaberulous, the first glume 9–10 (11.5) mm long, 3-nerved, the second 7–9 mm long, indistinctly 5-nerved; lemma relatively small, 4.5–5.7 mm long (including the callus), narrow, pale or light-brown, sparsely short appressed-hirsute below, glabrate above, the whitish hairs less than 0.7 mm long, the callus about 0.7 mm long, sharp, somewhat more pubescent than the lemma body; awn 40–75 mm long, arising from a distinct darker joint, capillary, weakly once- or twice-geniculate, glabrous, loosely twisted below, untwisted and flexuous above; palea about half as long as the lemma; lodicules 2, about 1.5 mm long, oblanceolate.

Rocky, shadscale and sagebrush deserts and foothills up to the pinyon-juniper woodland; se. Calif. and s. parts of Nev., Utah and sw. Colo., s. to n. Ariz., N.M. and reported to occur in w. Texas. May–July.

This is one of our most distinctive desert stipas, readily recognized by the short, narrow panicle, long awns and short, narrow lemma.

STIPA

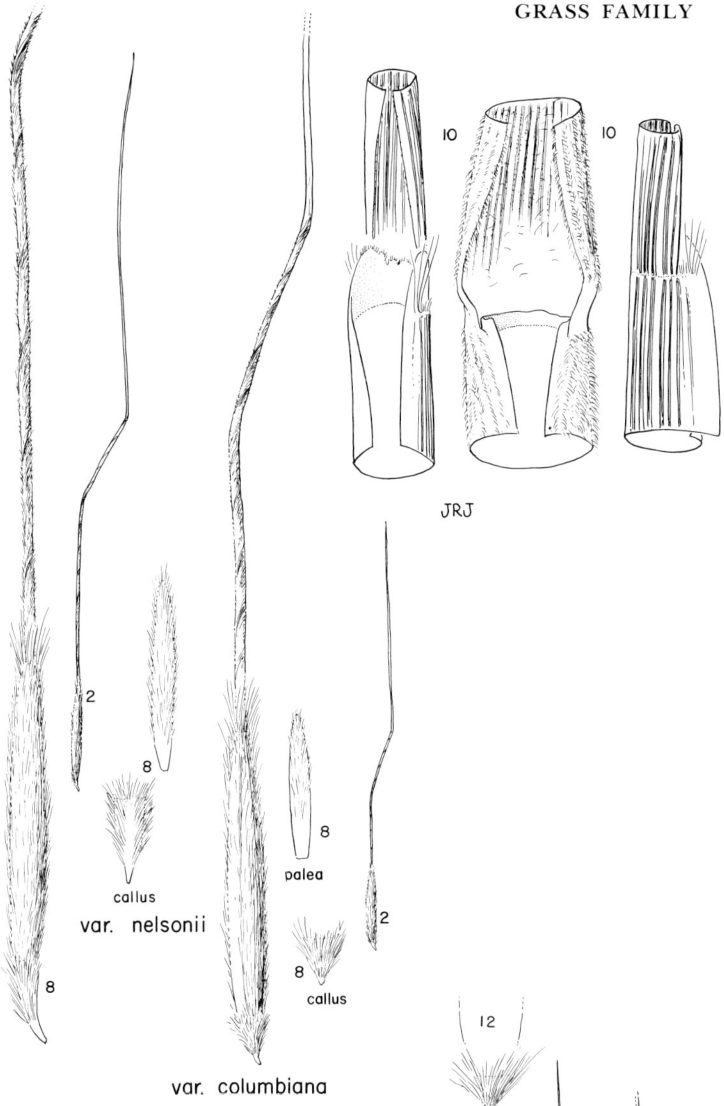

Stipa columbiana

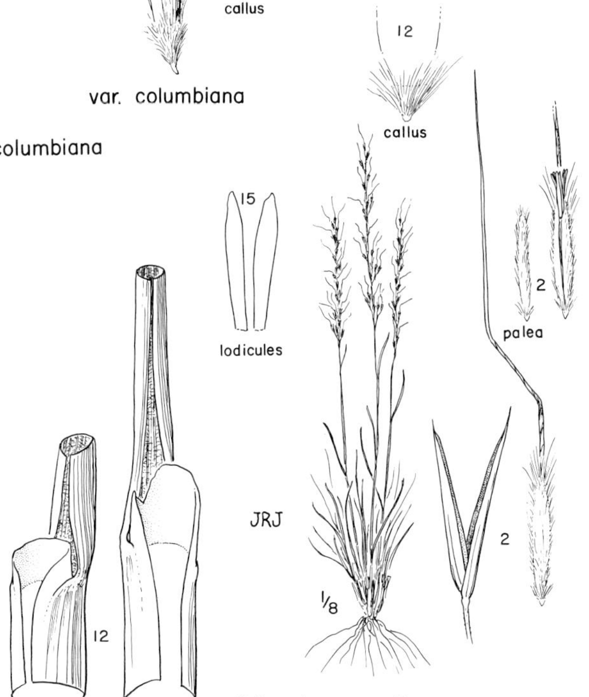

Stipa lettermanii

Stipa arida

10. Stipa speciosa Trin. & Rupr.

Stipa speciosa Trin. & Rupr. Mém. Acad. Imp. Sci. Saint-Pétersbourg, Sér. 6, Sci. Math., Seconde Pt. Sci. Nat. 5(1): 45. 1842. *S. humilis* var. *speciosa* Kuntze, Revisio Gen. Pl. 3(2): 371. 1898. (*Cuming,* Chile.) Fragments of the type at US have lemmas and awns that exceed ours in length.

S. californica Vasey, Proc. Amer. Acad. Arts 24: 80. 1889, nomen nudum; not Merr. & Davy, 1902. (*Palmer 505,* Los Angeles Bay, Baja Calif., in 1887.)

S. speciosa var. *minor* Vasey, Contr. U. S. Natl. Herb. 3: 52. 1892. (*Jones,* Empire City, Nev., in 1882.)

S. humilis var. *jonesiana* Kuntze, Revisio Gen. Pl. 3(2): 371. 1898. (*Jones 4111,* Empire City, Nev., 20 June 1882.)

Desert needlegrass.

Stout, densely tufted perennials; culms 3.5–6.5 dm tall, erect; sheaths firm, striate, shiny, the upper glabrous, the lowermost pilose, persistent and clothing the base of the plant; ligules very short, less than 0.5 mm long, densely ciliolate, a conspicuous tuft of hairs on each side of the throat, the hairs 0.2–0.8 (2) mm long; blades tightly involute, slender, 1 mm thick or less, 20–40 cm long, very firm, glaucous, glabrous on exposed surface with scaberulous margins, the tip pungent; panicles (7) 10–20 cm long, compact, narrow, scarcely exceeding the leaves, lower portion often included in the inflated sheath, the short, appressed branches spikelet-bearing nearly to the base; glumes long-acuminate, terminating in a fine point, scaberulous to glabrous, papery, pale or tawny, the first glume (12) 15–20 (25) mm long, 1- to 3-nerved, the second 11–18 (23) mm long, 3- to 5-nerved; lemma 7–10 mm long (including the callus), terete, indurate, short appressed-pilose, becoming somewhat glabrate towards the tip, the callus densely appressed-pilose, the hairs about 1.5–2.5 mm long; awn 30–43 (50) mm long, once geniculate, the lower segment tightly twisted and densely plumose, the hairs 4.5–6 (8) mm long, the terminal segment not twisted, 20–30 mm long, scaberulous; lodicules 3, narrow-oblanceolate, the lateral pair about 2 mm long, the middle one shorter; anthers about 4 mm long; $2n = 60, 64$.

Sagebrush deserts, canyons and pinyon-juniper wooded foothills in rocky or sandy soils; s. and e. Calif. across s. Nev. to s. and e. Utah, sw. Colo. and n. Ariz.; also in Baja Calif. and S. Amer. (April–) May–June.

Hybridization of *Stipa speciosa* with *Oryzopsis hymenoides* (Roemer & Schultes) Ricker has been studied by Johnson (1960). Desert needlegrass is a beautiful, plumy grass that should be used as an ornamental in desert planting of native species.

11. Stipa coronata Thurber

Stipa coronata Thurber in S. Wats. Bot. Calif. 2: 287. 1880. (*Bolander,* "in a cañon around springs on hillsides near Julian City," San Diego Co., Calif., April 1872.)

S. parishii Vasey, Bot. Gaz. 7: 33. 1882. *S. coronata* subsp. *parishii* A. S. Hitchc. Contr. U. S. Natl. Herb. 24: 227. 1925. (*Parish Bros. 1079,* Bear Valley, Mill Creek Falls, San Bernardino Mts., Calif., Aug. 1882.) = var. *depauperata.*

S. parishii var. *depauperata* M. E. Jones, Contr. W. Bot. 14: 11. 1912. *S. coronata* var. *depauperata* A. S. Hitchc. J. Wash. Acad. Sci. 24: 292. 1934. (*Jones,* Detroit, Juab Co., Utah, 25 May 1891.)

Stout, tufted perennials; culms 2–7 (20) dm tall, erect; sheaths glabrous or sometimes retrorsely puberulent, persistent, becoming smooth and brown and clothing the base of the plant; ligules 0.5–1 mm long, membranous and ciliate, flanked by tufts of whitish hairs, the hairs 1–3.5 (6) mm long and sometimes spreading down the margins of the sheath; blades usually flat, 2.5–5 (6) mm broad, 30–50 cm long, often glaucous, striated, becoming somewhat involute towards the long slender tips, glabrous to scaberulous on the surfaces and strongly scabrous on the margins; panicles (7) 14–23 (35) cm long, narrow and densely flowered, somewhat nodding; glumes linear-lanceolate, gradually tapering into a long, slender point, 3- (5)-nerved, often keeled, pale or purplish below, scarious to hyaline on the margins and upper portion, glabrous or minutely scaberulous, unequal, the first glume (10) 12–18 (20) mm long, the second (8) 11–15 (18) mm long; lemma 5–8 mm long (including the callus), obscurely 5-nerved, densely appressed-villous with long, whitish hairs, the longer apical hairs 3–5 mm long; awn (10) 17–35 (42) mm long, once or twice geniculate, the lower segment scabrous, sometimes hispid, the terminal segment scaberulous at most, not twisted, 10–15 mm long; palea shorter than the lemma, pubescent between the nerves; lodicules 3, the large lateral pair about 2.5 mm long, elliptic-lanceolate, the shorter middle one about 1 mm long, subulate-lanceolate; anthers 3.5–5 mm long; $2n = 40$.

Dry, rocky or sandy slopes and ridges, mostly in the pinyon-juniper zone; s. parts of Utah, Nev. and Calif., n. Ariz. and Baja Calif. June–July.

We have two varieties that may be separated by the following characters.

1 Culms 10–20 dm tall; panicles 20–35 cm long, with numerous spikelets; lemma mostly about 8 mm long; awn usually twice-geniculate, 25–35 (42) mm long; Monterey, Calif. to Baja Calif., e. to n. Ariz. and entering our area in s. Utah .... var. *coronata*

1 Culms shorter, 2–6 dm tall; panicles less than 20 cm long, with fewer spikelets; lemma about 5–7 mm long; awn usually once-geniculate, about (10) 17–25 mm long; Utah, Nev., Ariz. and s. Calif. .... var. *depauperata* (M. E. Jones) A. S. Hitchc.

The variety *depauperata* is recognized here with some misgivings as many intermediates occur, and the depauperate plants may be a mere modification due to the environment. *Stipa coronata* var. *coronata* is the largest of our needlegrasses and probably the largest in North America. It is common in the grasslands of southern California. It would make a beautiful ornamental that should lend itself well to cultivation in the native wild landscape scene across the southern part of the Intermountain Region.

12. Stipa comata Trin. & Rupr.

Stipa juncea sensu Pursh, Fl. Amer. Sept. 72. 1814; not L. 1753. (*Lewis,* "Valleys of the Missouri in the Rocky Mts.," 8 July 1806.)

S. capillata sensu Hook. Fl. Boreali-Amer. 2: 237. 1840; not L. 1762. (*Drummond,* "Carlton House Fort ad fl. Saskatchawan.")

S. comata Trin. & Rupr. Mém. Acad. Imp. Sci. Saint-Pétersbourg, Sér. 6, Sci. Math., Seconde Pt. Sci. Nat. 5(1): 75. 1842. (*Drummond,* Saskatchewan River.)

S. comata var. *intermedia* Scribn. & Tweedy, Bot. Gaz. 11: 171. 1886. *S. tweedyi* Scribn. U.S.D.A. Div. Agrostol. Bull. 11: 47. 1898. *S. spartea* var. *tweedyi* M. E. Jones, Contr. W. Bot. 14: 11. 1912. (*Tweedy 610,* Junction Butte, Yellowstone Park, July 1885.)

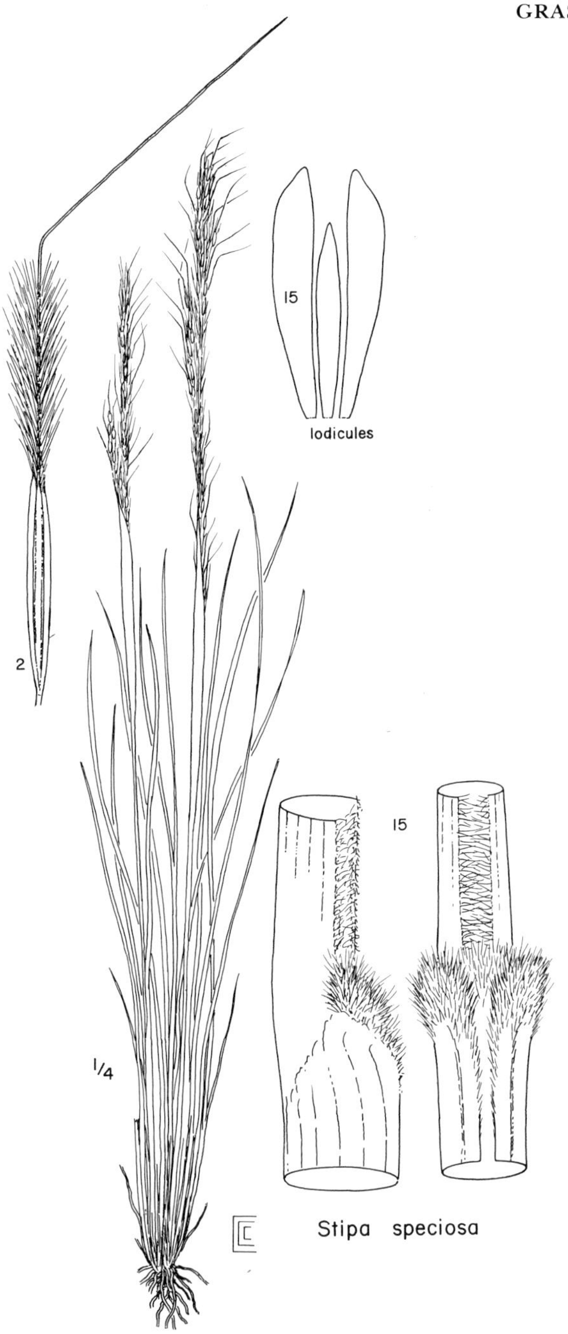

S. comata subsp. *intonsa* Piper, Contr. U. S. Natl. Herb. 11: 109. 1906. (*Suksdorf 1026,* "On dry, sandy ground near Rockland," Klickitat Co., Wash., 3 June 1890.) = var. *comata.*

S. comata var. *suksdorfii* St. John, Fl. S. E. Wash. & Adj. Idaho 61. 1937. (*Suksdorf 8990,* "In a grove southwest of Philleo Lake," Spokane Co., Wash., 3 Aug. 1916.) = var. *comata.*

Needle-and-thread grass.

Tufted perennials; culms 3–8 (11) dm tall, glabrous, often puberulent at the nodes; sheaths smooth to scabrous, strongly ribbed, usually longer than the internodes; ligules relatively long, 1–5 (6) mm, usually acute, becoming lacerate, decurrent, more or less puberulent; blades involute, or if flat up to 3 mm broad, 10–30 cm long, smooth to scaberulous beneath, scabrous on the upper surface; panicles relatively long, (13) 18–34 (40) cm, narrow, usually partly enclosed in an inflated sheath, the branches usually slender and ascending, the spikelets sometimes drooping at anthesis; glumes long and narrow, tapering to a fine point, 5-nerved, slightly convoluted, glabrous, papery, the margins and tip hyaline, the first glume 18–30 (35) mm long, the second 15–27 (34) mm long; lemma relatively long, (8) 10–12.5 (14) mm (including the callus), pale green to yellowish or brownish, sparsely appressed-pilose, often glabrate apically, the hairs 0.2–0.5 mm long, the callus very sharp, about 3 mm long, densely barbed with tawny hairs, the hairs 1–2 mm long; awn very long, (60) 70–160 mm, once or twice geniculate, the first joint distinct, the second indistinct or more often merely flexuous, the lower segment tightly twisted, scabrous, the terminal segment scabrous, not twisted, 40–90 (120) mm long; palea nearly as long as the lemma, pubescent between the nerves; lodicules 2 or 3, lanceolate, about 1.5 mm long; $2n$ = 44–46.

Dry hills and plains, in sandy (sand dunes) and stony soils, from the low sagebrush valleys and deserts to open subalpine coniferous woods; B.C. and Yukon to Ont., s. to Calif., Ariz., N.M. and Texas. May–July.

Stipa comata is an important component of the mixed prairies of the high plains and the palouse prairie of the Northwest. It is perhaps the most common and widespread of the western stipas. It can be a problem on rangelands when the caryopses are mature. With their sharp, retrorsely-barbed lemma bases they can work into mouthparts of livestock and cause injury.

We have two rather distinctive varieties that come close to being acceptable as species. They may be separated as follows:

1 Terminal segment of the awn flexuous, usually well over 50 mm long; lower branches of the panicle usually included in the sheath; B.C. to Ont., s. to Calif. and Texas .... var *comata*

1 Terminal segment of the awn nearly straight, rather firm, usually less than 50 mm long; panicle usually exserted from the sheath; e. Wash. to Mont., s. to e. Oregon and Wyo. .... var. *intermedia* Scribn. & Tweedy

Some accounts of early surveys reported *S. spartea* Trin. in the region, but they were probably dealing with robust forms of *S. comata.*

13. **Stipa neomexicana** (Thurber) Scribn.

Stipa pennata var. *neo-mexicana* Thurber in J. M. Coulter, Manual Bot. Rocky Mt. Region 408. 1885. *S. neo-mexicana* Scribn. U.S.D.A. Div. Agrostol. Bull. 17: 132. 1899. (*Thurber 269,* Rio Mimbres, N.M., April 1851.)

New Mexican feathergrass.

Densely tufted, glaucous perennials; culms 4–8 (10) dm tall, erect, glabrous; sheaths glabrous to pilosulous, often with a few villous hairs at the throat, persistent and clothing the base of the plant; ligules no more than a low ciliolate ring of hairs, 0.2–0.5 (2) mm long; blades strongly involute, filiform, slightly less than 1 mm broad when unrolled, abruptly narrowed at the junction of the sheath and blade, 10–30 cm long, firm, retrorsely hispidulous to glabrate or smooth; panicles (11) 15–26 (30) cm long, narrow, lower portion included in an inflated sheath; glumes very long and subequal, (20) 30–50 (60) mm long, linear-lanceolate, gradually tapering to a point, 4- to 7-nerved, papery, glabrous; lemma long, (10) 14–17 (19) mm (including the callus), 5-nerved, sparsely appressed-villous, becoming dark brown at maturity, the callus often 4–5 mm long, tapered, densely pilose; awn very long, 120–200 mm, sometimes readily deciduous, once or sometimes indistinctly twice geniculate, the lower segment twisted and appressed-villous, the long, untwisted terminal segment often more than 100 mm long, densely plumose, the pale or tawny hairs 2–3 mm long; lodicules 3, relatively long, about 3 mm, the lateral pair lanceolate, the middle one slightly narrower; $2n$ = 44.

Dry sandy or rocky sagebrush mesas, canyons and hillsides at middle elevs.; s. Utah (Marysvale and the Colorado River drainage) and Ariz., e. to s. Colo., N.M. and w. Texas and n. Mex. June.

Stipa neomexicana is an important component of the Southwest desert grasslands. Weber (Rhodora 59: 273–277. 1957) reported an intergeneric hybrid involving *S. neomexicana* and *Oryzopsis hymenoides* (Roemer & Schultes) Ricker on a hill 6 miles north of Boulder, Colorado, in which the length and plumose characters of the awn, so distinctive of *S. neomexicana,* were completely suppressed.

Subfamily II. **PANICOIDEAE**

Epidermal cells in the region of root hairs all alike and all capable of giving rise to root hairs; culm internodes mostly solid or nearly so; sheath margins free; ligule membranous, a fringe of hairs, or absent; blades non-auriculate; inflorescence a panicle, sometimes of spicate or racemose branches that are digitately or racemosely arranged on a central axis, the spikelets often in pairs, often along 2 sides of a midrib of a flattened rachis or along 2 sides of a 3-angled rachis; spikelets dorsally compressed, turgid to plano-convex, (1-) 2-flowered with 1 perfect floret above and 1 staminate or sterile floret below; disarticulation below the glumes, sometimes at the rachis node, the rachis segment falling with the spikelet pair; lemmas (1-) 3- to 5-(9)-nerved, indurate or membranous; lodicules mostly short, truncate, thick and with considerable vasculation; stamens usually 3; embryo relatively large, more than $^1/_3$ the length of the caryopsis, with the typical panicoid morphology with vascular traces to the scutellum and coleoptile separated by an internode, the epiblast absent, the lower part of the scutellum and coleorhiza separated by a cleft, and the embryonic leaf margins overlapping; x = 9 or 10, the chromosomes mostly small.

The Panicoideae subfamily is composed of genera centering in trop. or subtrop. regions and therefore is not well represented in our region. Slightly less than $^1/_{10}$ of our grass species belong to the Panicoideae and most of them enter the region in the southern parts. Eight of the 13 genera of the Panicoideae in our area contain weedy species, several of which are among our most troublesome grasses.

In the new system the Panicoideae remains nearly the same as in A. S. Hitchcock's treatment (Manual Grasses U.S., 1951), with some minor changes in the Andropogoneae tribe. *Zea mays* L. (cultivated corn), which was included in the Tripsaceae or Maydeae in older treatments is now placed with the Andropogoneae. Corn is treated only briefly here as it does not become established as a part of our flora.

Key to the Tribes of Panicoideae

1 Glumes and lemma of the reduced floret similar in texture, soft; lemma of the perfect floret indurate; spikelets usually not organized into pairs .... Tribe 6. Paniceae, p. 363

1 Glumes, at least the first, firm and indurate; lemma of the reduced and perfect florets membranous; spikelets in pairs of 1 sessile (or subsessile) and 1 pedicelled .. Tribe 7. Andropogoneae, p. 382

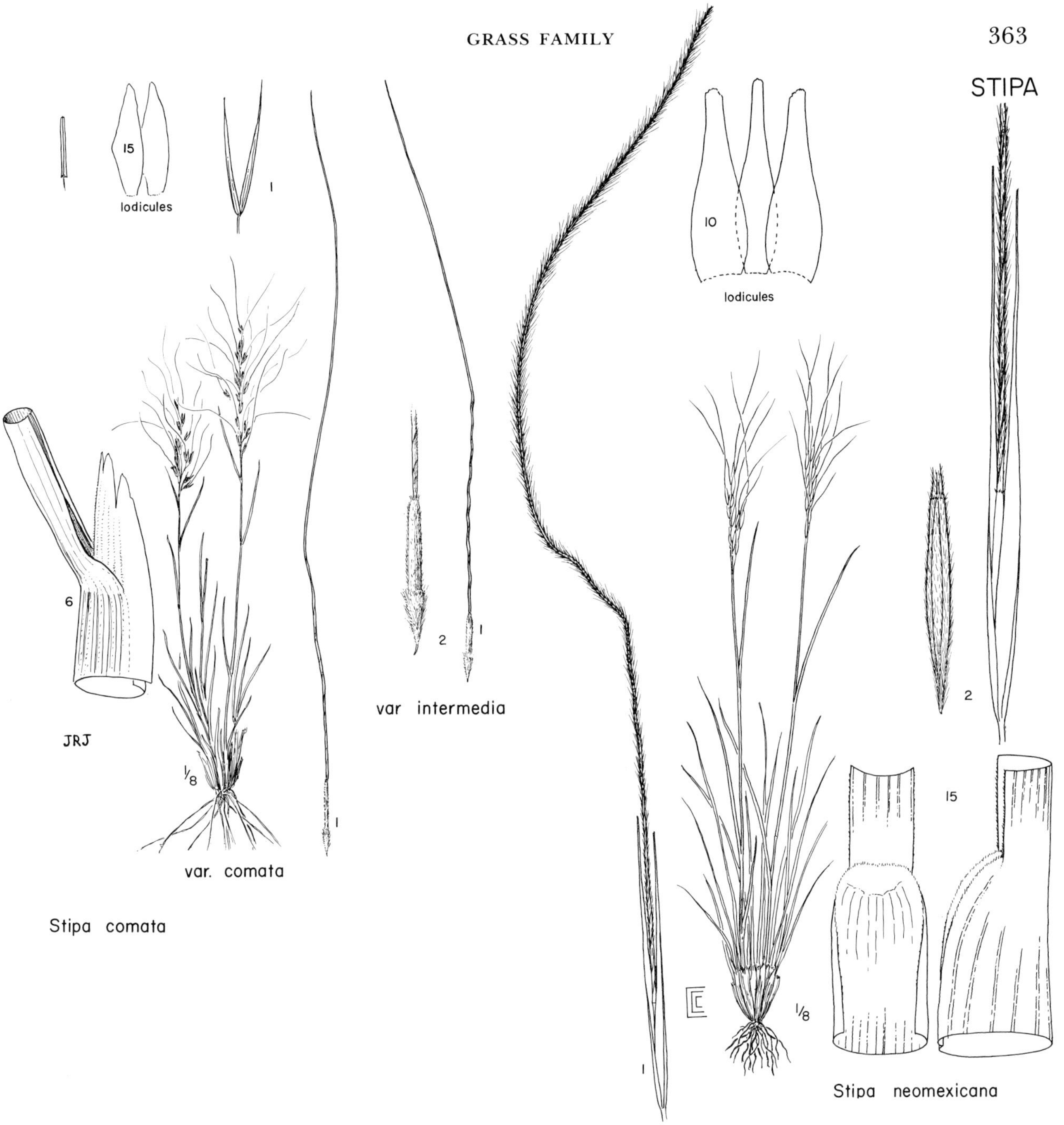

Tribe 6. **PANICEAE**

Annuals or perennials, cespitose or rhizomatous; culms solid to hollow, sometimes pith filled; sheath margins free; ligule membranous, a fringe of hairs, or absent; blades non-auriculate; inflorescence a diffuse panicle or a panicle of digitately or racemosely arranged spicate-racemes with the spikelets along 2 sides of a midrib of a flattened rachis, or along 2 sides of a 3-angled rachis, some of the inflorescence branches sometimes modified to form bristles that subtend the spikelets in *Setaria,* or a bur (involucre) that surrounds the spikelets in *Cenchrus;* spikelets dorsally compressed, 2-flowered, the lower floret staminate or sterile, disarticulating below the glumes; glumes usually membranous, soft, the first glume often much reduced or absent; lemma of the first floret similar in size and texture to the second glume, awnless or awned in *Echinochloa;* lemma of the fertile floret indurate, the margins overlapping the palea, usually smooth and shiny, sometimes rugose, indistinctly 3- (5)-nerved, awnless; palea of the first floret membranous, or often absent, the palea of the fertile floret indurate, smooth and shiny; lodicules 2, thick, truncate, vasculated; embryo panicoid; $x = 9$ or 10.

Rhynchelytrum repens (Willd.) C. E. Hubbard (ruby grass) is sometimes planted in our region. The inflorescence is rosy purple to pink with silky-villous spikelets and is used in dry bouquets. The spikelet is very characteristic with the similar second glume and sterile lemma separated from the small first glume by a stipe. The second glume and sterile lemma taper to an awned apex. *Pennisetum setaceum* (Forsk.) Chiov. (fountain grass) is commonly planted around fountains. The inflorescence is a purplish, spike-like panicle. The spikelets are subtended by long, unfused bristles that form an involucre and fall with the spikelet at maturity.

The genera of tribe Paniceae are recognized here in the same order presented by A. S. Hitchcock (Manual Grasses U.S., 1951), an arrangement that has withstood all modern reinterpretations.

1 Spikelets enclosed in a bur-like involucre . . . . 50. *Cenchrus*
1 Spikelets not enclosed by an involucre.
 2 Spikelets subtended or surrounded by 1 to many bristles . . . . 49. *Setaria*
 2 Spikelets not subtended or surrounded by bristles.
  3 Sterile lemma mucronate or awned . . . . 48. *Echinochloa*
  3 Sterile lemma awnless.
   4 Spikelets short-pedicellate on 1 side of the flattened panicle branches.
    5 First glume much reduced to absent, the second glume from $^1/_3$ to as long as the spikelet.
     6 Plants annual; culms hollow . . . . 44. *Digitaria*
     6 Plants perennial; culms solid . . . . 45. *Paspalum*
    5 Glumes subequal and both as long as the spikelet . . . . 46. *Panicum obtusum*
   4 Spikelets in open or somewhat contracted panicles.
    7 Second glume and sterile lemma glabrous, or sometimes scaberulous only on the nerves towards the apex; basal leaves similar to the culm leaves, not forming a winter rosette; spikelets all fertile . . . . 46. *Panicum*
    7 Second glume and sterile lemma short pilose all over (sometimes glabrous in *D. oligosanthes*); basal leaves distinctly different from the culm leaves, forming a winter rosette; plants passing from one growth phase to another, the so-called spring or vernal phase, characterized by simple stems and terminal panicles, becomes transformed to the autumnal phase with branched culms and axillary, often cleistogamous panicles . . . . 47. *Dichanthelium*

44. DIGITARIA Heister Crabgrass

Annuals and perennials, sometimes stoloniferous; culms erect to prostrate, hollow; sheaths open; ligules membranous; blades flat and short, non-auriculate; inflorescence a panicle of few to several, slender, spike-like, mostly unbranched racemes, the racemes digitately or approximately arranged on a short axis, the spikelets solitary, in pairs or in groups of 3 to 5, if in pairs (like ours) one spikelet subsessile and the other short-pedicellate, in 2 rows on 2 sides of the midrib of a flattened rachis or on 2 sides of a 3-angled rachis; spikelets dorsally compressed, lanceolate or elliptic, slightly plano-convex, 2-flowered, the first floret sterile and the second perfect, disarticulating below the glumes; glumes unequal, the first glume minute or absent, the second glume well developed, usually shorter than, but somewhat similar in texture to the sterile lemma, 3-nerved; sterile lemma the same size as the fertile lemma, 5-nerved, the nerves prominent and glabrous, puberulent or long-ciliate; fertile lemma dorsiventrally compressed, rounded on the back, more or less indurate, smooth and shiny, indistinctly 3-nerved, the margins hyaline, inrolled and clasping the palea, awnless; palea of the fertile floret similar to the lemma in texture; lodicules 2, oblong-obovate, slightly bilobed; stamens 3; $x = 9$.

A genus of about 100 species of temp. and trop. regions of the world, about 24 of which are native to the U.S. (From the Latin *digitus*, finger, referring to the finger-like or digitate arrangement of racemes.)

Our species are troublesome weeds in lawns. They form a good green mat during the summer but die in the fall, leaving brown areas.

References:

Ebinger, J. E. Validity of the grass species *Digitaria adscendens*. Brittonia 14: 248–253. 1962.
Gould, F. W. Cytotaxonomy of *Digitaria sanguinalis* and *D. adscendens*. Brittonia 15: 241–244. 1963.
Henrard, J. T. Monograph of the genus *Digitaria*. Universitare Pers, Leiden. 1950.

1 Second glume equaling the length of the spikelet; spikelets 1.8–2.3 mm long; sheaths glabrous, the upper margins sometimes ciliate . . . . 1. *D. ischaemum*
1 Second glume $^1/_3$ to $^4/_5$ the length of the spikelet; spikelets 2.5–3.6 mm long; sheaths papillose-pilose.
 2 Blades glabrous or only sparsely papillose-pilose on the upper surface near the throat; lateral nerves of the sterile lemma smooth; second glume $^1/_2$ to $^4/_5$ the length of the spikelet, (1) 1.5–2.5 mm long . . . . 2. *D. adscendens*
 2 Blades papillose-pilose, the upper surface often densely so; lateral nerves of the sterile lemma often minutely scaberulous; second glume $^1/_3$ to $^3/_5$ the length of the spikelet, 0.8–1.8 mm long . . . . 3. *D. sanguinalis*

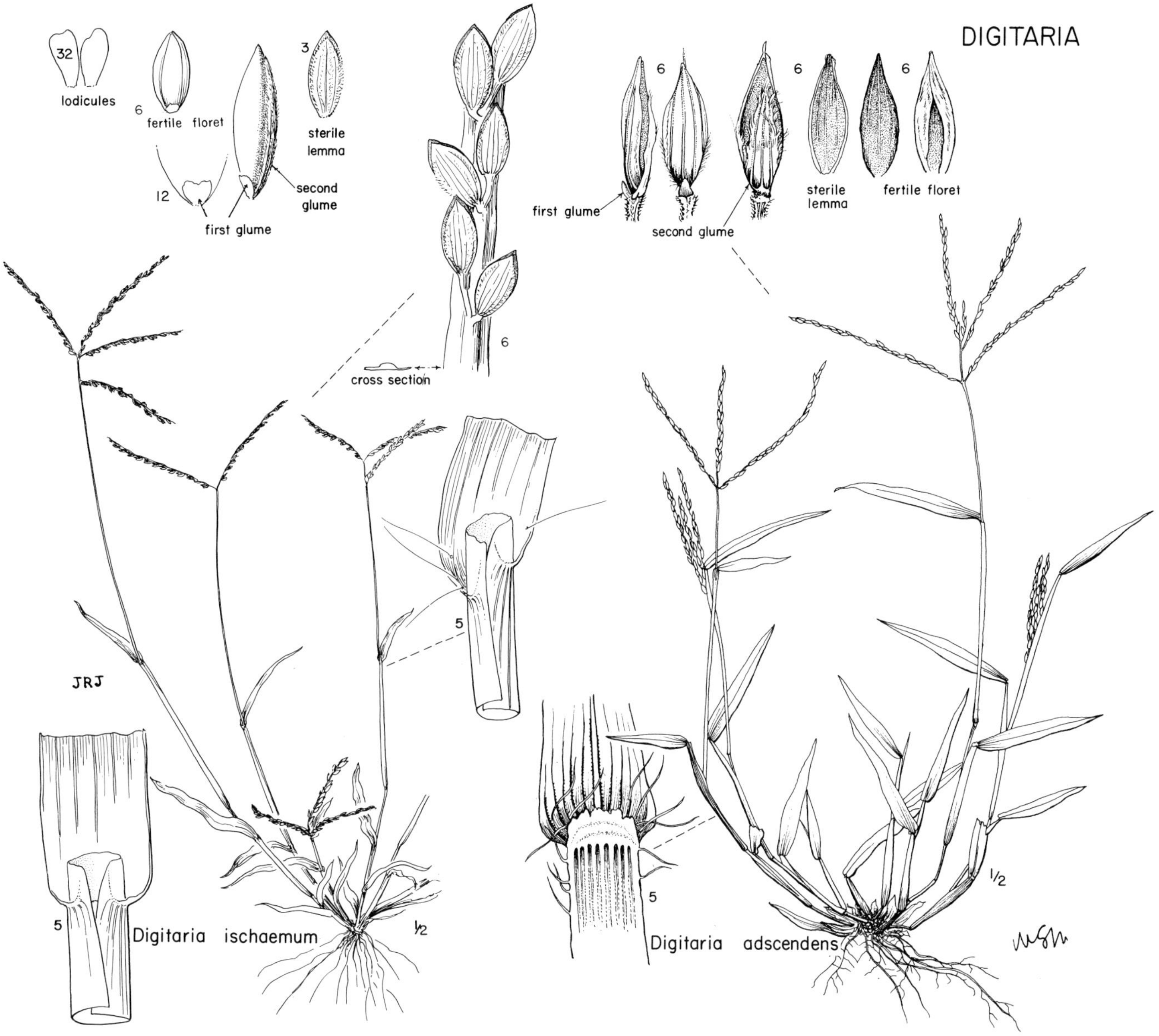

1. Digitaria ischaemum (Schreber) Schreber ex Muhl.

Panicum ischaemum Schreber in Schweigger, Sp. Fl. Erlangensis 16. 1805. *Digitaria ischaemum* Muhl. Cat. Pl. N. Amer. 9. 1813, name only; Schreber ex Muhl. Descript. Gram. 131. 1817. *Syntherisma ischaemum* Nash, N. Amer. Fl. 17: 151. 1912. (Germany.)

Smooth crabgrass.

Decumbent annuals, rooting at the nodes; culms short, 0.5–3 dm long; sheaths keeled, glabrous or often ciliate apically, the lower sheaths often hirsute at the base; ligules 1–2 mm long, truncate, entire to lacerate; blades flat, 2–4.5 (6) mm broad, glabrous to slightly scabrous, sometimes sparsely pilose near the collar; racemes 1–6, digitately arranged, 2–5 (12) cm long, the midrib of the rachis nearly flat on the side of the pedicel attachment and rounded on the opposite side; spikelets dorsally-compressed, 1.8–2.3 mm long; first glume 0.1–0.3 mm long, sometimes absent, broad, truncate, hyaline; second glume 1.7–2.2 mm long, ovate, 3-nerved, about equaling the lemmas; sterile lemma equaling the fertile floret, slightly shaggy-pubescent, the hairs minutely enlarged apically, prominently 5-nerved; fertile lemma about 2 mm long, indurate with hyaline margins, becoming dark-brown, glabrous; lodicules about 0.3 mm long; anthers about 0.5 mm long; $2n = 36, 45$.

A weed in lawns and moist waste places of lower, warmer valleys of the region; native of Europe; s. B.C. to Newfl., s. throughout most of the U.S. and throughout the Amer. tropics. Aug.–Sept.

2. Digitaria adscendens (H.B.K.) Henrard

Panicum adscendens H.B.K. Nova Gen. & Sp. Pl. 1: 97. 1815. *Digitaria adscendens* Henrard, Blumea 1: 92. 1934. (*Humboldt & Bonpland,* Venezuela, Peru and Mex.)

D. marginata of authors; not Link [*Syntherisma marginatum* (Link) Nash].

Southern crabgrass.

Annuals, sometimes producing stolons; culms 3–10 dm long, decumbent to ascending, often branched below; sheaths keeled, papillose-pilose, the upper margins often projecting as scarious margined auricles; ligules 1–2.5 (4) mm long, erose, sometimes lacerate; blades flat, 3.5–6.5 (9) mm broad, glabrous or sometimes sparsely papillose-pilose on the upper surface near the throat, the margins crisped and scabrous; racemes 2–7 (9), digitately or nearly so arranged, (4) 5–15 (18) cm long, the midrib of the rachis winged on the side of the pedicel attachment and merely rounded on the opposite side, the spikelet pairs alternating from 2 sides of the resulting 3-angled rachis; spikelets dorsally-compressed, 2.7–3.6 mm long; first glume 0.1–0.4 mm long, ovate, obtuse or acute; second glume (1) 1.5–2.5 mm long, $^2/_3$ to $^4/_5$ the length of the spikelet, lanceolate, 3-nerved, hirsute, ciliate; sterile lemma equaling the fertile floret, appressed-pubescent on the margins, glabrous on the back, prominently 5-nerved, the lateral nerves smooth, without hairs; fertile lemma 2.6–3.5 mm long, indurate with hyaline margins, glabrous, obscurely 3-nerved; $2n$ = 36, 54, 60, 70, 72.

A weed in gardens, lawns and moist waste places, mostly of trop. and subtrop. climates; s. Utah, Kansas, Mo., s. Ind. and Va., southward (more common in se. U.S.); Mex., C. Amer., W.I. and S. Amer. Aug.–Sept.

Without a close look, *D. adscendens* can pass for *D. sanguinalis.* A. S. Hitchcock (Manual Grasses U.S., 1951) considered them to be conspecific. Using the characters in the key above, Ebinger (1962) found that *D. adscendens* segregates from *D. sanguinalis* as a species, replacing it in southern latitudes. The two species overlap in a narrow area across southern United States, and some instances of hybridization are known. Gould (1963) added supportive evidence when he discovered that American specimens identifiable as *D. adscendens* are hexaploids with $2n$ = 54 while plants keying to *D. sanguinalis* are tetraploids with $2n$ = 36. It has been suggested that *D. adscendens* is a native to the New World tropics and subtropics while *D. sanguinalis* has been introduced from temperate regions of the Old World. Howell [Wasmann J. Biol. 29(1): 101. 1971] points out that N. L. Bor (in K. H. Rechinger, Fl. Iranica 70: 490. 1970) found that *D. adscendens* is the same as the nomenclaturally older *D. ciliaris* (Retz.) Koeler and should be referred to by that name. Without seeing the type, we choose to use the familiar name.

3. **Digitaria sanguinalis** (L.) Scop.

Panicum sanguinale L. Sp. Pl. 57. 1753. *Digitaria sanguinalis* Scop. Fl. Carniolica, ed. 2. 1: 52. 1772. *Dactilon sanguinale* Vill. Hist. Pl. Dauphiné 2: 69. 1787. *Paspalum sanguinale* Lam. Tableau Encycl. & Méth. 1: 176. 1791. *Syntherisma sanguinalis* Dulac, Fl. Dépt. Hautes-Pyrén. 77. 1867. ("Habitat in America, Europa australi.")

Crabgrass.

Prostrate to ascending annuals, rooting at the nodes and forming mats; culms 3–10 dm long; sheaths keeled, papillose-pilose with hairs 2–3 mm long, the upper margins projecting as scarious margined auricles; ligules 0.5–1.5 (2.5) mm long, truncate to obtuse, erose; blades flat, 2.5–8 (10) mm broad, papillose-pilose, the margins crisped; racemes few to several, digitately or nearly so arranged, 6–16 cm long, the midrib of the rachis winged on the side of the pedicel attachment and nearly flat on the opposite side, the spikelet pairs alternate from 2 sides of the resulting 3-angled rachis; spikelets dorsally-compressed, 2.5–3.2 mm long; first glume 0.1–0.5 mm long, ovate, acute; second glume 0.8–1.8 mm long, $^1/_3$ to $^3/_5$ the length of the spikelet, lanceolate, 3-nerved, hirsute, ciliate; sterile lemma equalling the fertile floret, the margins appressed-pubescent, prominently 5-nerved, the lateral nerves often scaberulous; fertile lemma 2.5–3.2 mm long, indurate with hyaline margins, glabrous, obscurely 1-nerved, sometimes with 1 or 2 pair of very faint, lateral nerves; lodicules about 0.4 mm long; $2n$ = 18, 28, 36–48, 76.

A weed in lawns and areas of cultivation in the warmer, lower valleys of the region; native of Europe; s. B.C. to N.S., s. throughout most of the U.S., and rare in Mex. July–Sept.

45. **PASPALUM** L.

Annuals and perennials, many with rhizomes or stolons; culms solid; sheaths open; ligules membranous; blades usually flat, often broad, non-auriculate; inflorescence a panicle of 1 to many unilateral spicate or racemose branches, these scattered, or in some paired at the apex, the spikelets subsessile or short-pedicelled, borne in pairs or singly in 2 rows on 1 side of a flattened, occasionally broadly-winged rachis; spikelets dorsally-compressed, plano-convex, 2-flowered, the first floret sterile, the second perfect, disarticulating below the glumes; glumes unequal, the first glume usually absent, the second glume and sterile lemma usually subequal, broad and rounded at the apex and slightly exceeding the fertile lemma; sterile lemma similar to the second glume; fertile lemma usually obtuse, rounded on the back, indurate, smooth and shiny, indistinctly nerved, the margins inrolled and clasping the palea, awnless; palea of the fertile floret broad, flat or slightly convex and of similar texture to the lemma; lodicules 2; stamens 3; x = 10.

A large genus of from 250 to 300 species distributed throughout the warmer regions of the world, about 45 of which are native to the U.S. plus a number of introduced species, commonest in the warm, humid Southeast. (From the Greek *paspalos,* for some kind of millet.)

1. **Paspalum distichum** L.

Paspalum distichum L. Syst. Nat. ed. 10. 2: 855. 1759. *Milium distichum* Muhl. Cat. Pl. N. Amer. 9. 1813, nomen; Descript. Gram. 78. 1817. *Digitaria disticha* Fiori in Fiori & Paol. Fl. Analitica Ital. 1: 50. 1896. (*Browne,* Jamaica.)

Knot grass.

Stoloniferous perennials; culms (2) 4–10 dm long, decumbent, creeping, often rooting at the lower nodes, glabrous or sometimes the nodes pubescent; sheaths somewhat keeled, pilose at the base and on the upper margins, sometimes pilose all over or sometimes mostly glabrous; ligules 1.2–2.5 mm long,

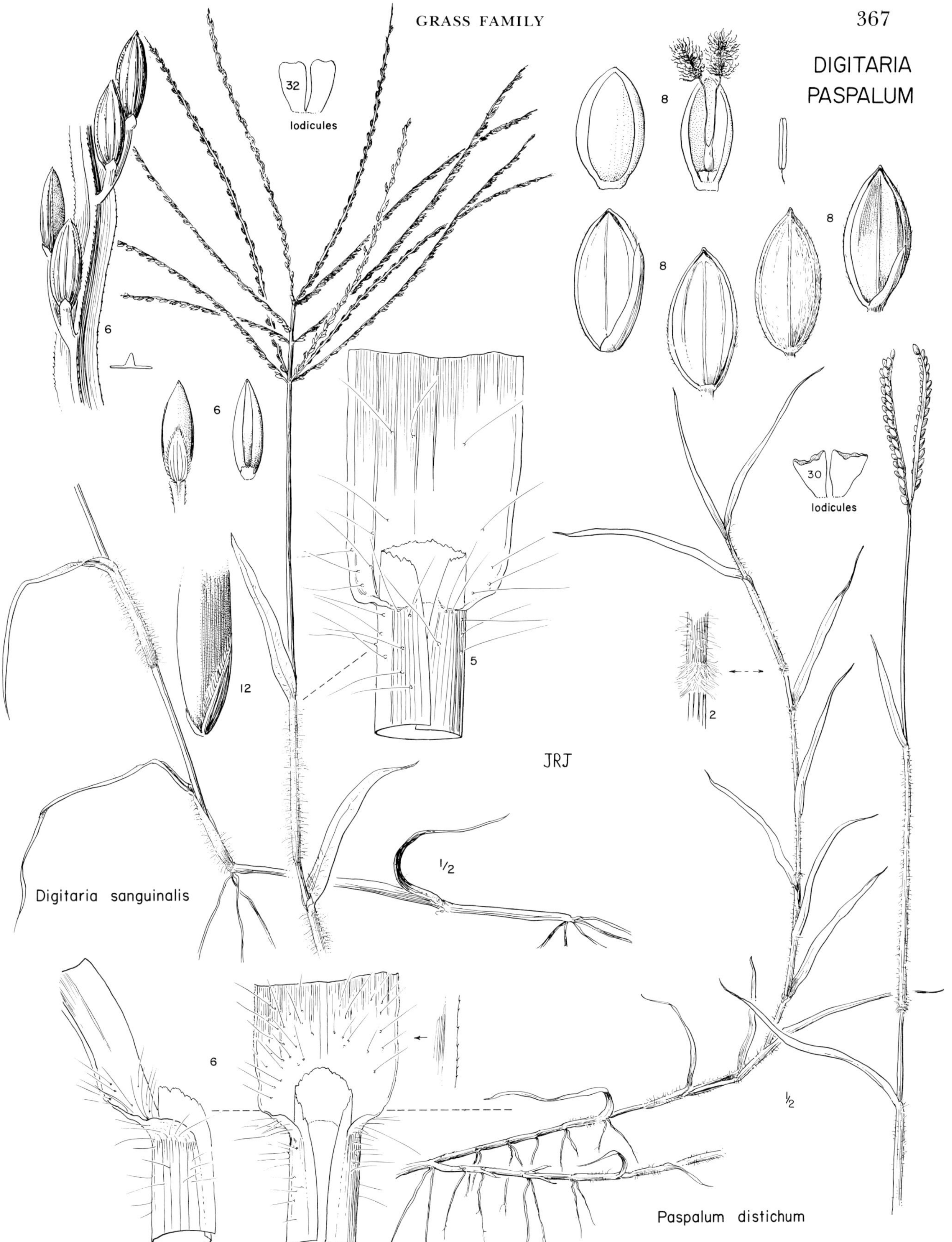
DIGITARIA
PASPALUM
32
lodicules
8
8
8
8
6
6
12
5
30
lodicules
2
JRJ
1/2
Digitaria sanguinalis
6
1/2
Paspalum distichum

membranous, slightly erose; blades flat or folded, (1.5) 3–6 mm broad, glabrous, the margins usually scabrous, sometimes pilose below; racemes (1) 2 (3) in number and (1.5) 3–7 cm long, often arcuate, pilose in the axils, the flattened rachis 1–1.5 mm broad, the spikelets borne singly, forming 2 rows on 2 sides of the midrib, pedicellate; spikelets dorsally-compressed; first glume minute, up to 1.5 mm long, truncate or sometimes lanceolate; second glume 2.5–3.2 mm long, broadly lanceolate, obtuse to somewhat pointed, pale-green, minutely puberulent, 3-nerved, the lateral nerves submarginal; sterile lemma similar to the second glume in size, shape and texture, although slightly smaller and glabrous; fertile lemma 2.2–2.8 mm long, ovate, indurate, smooth and shiny, glabrous, faintly 3-nerved; lodicules about 0.3 mm long; anthers 1–1.8 mm long, dark purple; stigmas dark purple; $2n = 20, 30, 40, 48, 60$.

Moist places along stream, canal and ditch banks at lower elevs.; throughout lower elevs. of w. U.S., excluding Mont. and Wyo., and e. across the s. part of the U.S. through Okla., Ark., Tenn. and n. to N.Y., s. in trop. Amer., and warm places of the Old World. Aug.–Sept.

Paspalum distichum is a stoloniferous species important as a soil binder along streams. However, knot grass becomes a problem when it chokes irrigation ditches.

46. PANICUM L. Panic Grass

Annuals and perennials; culms hollow; sheaths open; ligules mostly of hairs, sometimes membranous or absent; blades usually flat, non-auriculate; inflorescence an open to contracted panicle with rebranched primary branches, rarely reduced to racemes as in *Digitaria* and *Paspalum;* spikelets turgid to plano-convex or somewhat dorsiventrally-compressed, usually pedicellate, 2-flowered, disarticulating below the glumes; glumes unequal or rarely subequal, the first glume usually small and often clasping the base of the spikelet, the second subequal to and similar in texture to the sterile lemma; first floret sterile or sometimes staminate, the lemma several-nerved, membranous, usually longer than the fertile lemma, often bearing a membranous or hyaline palea; fertile floret more or less dorsally-compressed, the lemma mostly obtuse, rounded on back, indurate, smooth and shiny or rugose and dull, indistinctly nerved, the margins thick, inrolled and clasping the palea by its margins, awnless; palea of the fertile floret of similar texture to the lemma; lodicules 2, tubular-funnelform, the margins irregular; stamens 3; $x = 9, 10$.

With some 300 to 500 species, *Panicum* is the largest genus of the grass family; widely distributed throughout warmer parts of the world. (From the Latin name, *panus,* for some kind of millet.)

In 1974 Gould (Brittonia 26: 59–60. 1974) elevated the subgenus Dichanthelium to the generic level as *Dichanthelium,* which greatly reduced the number of species of *Panicum* in the United States. In the Hitchcock grass manual (Manual Grasses U.S., 1951) this represents the removal of 111 taxa, leaving only 61 in *Panicum.* See the discussion under *Dichanthelium* for a brief review of evidence to support Gould's conclusions.

Reference: Hitchcock, A. S., and A. Chase. The North American species of *Panicum.* Contr. U.S. Natl. Herb. 15: 1–396. 1910.

1 Plants perennial; second glume and sterile lemma completely glabrous.
  2 First glume subequal to the second glume and sterile lemma; spikelets short-pedicelled along 2 sides of 3-angled rachises, borne in 2–6 spike-like racemes attached to a central axis (Obtusa) . . . . 1. *P. obtusum*
  2 First glume 1/2 to 3/4 as long as the second glume and sterile lemma; spikelets in open or contracted, diffusely branched panicles.
    3 Culms swollen and bulbous at the base; fertile lemma finely transversely rugose, often dull (Maxima) . . . . 2. *P. bulbosum*
    3 Culms not swollen and bulbous at the base; fertile lemma smooth and shiny.
      4 Plants with conspicuous, creeping, scaly rhizomes; culms stout, 6–20 dm tall; spikelets (3.5) 4–5.2 mm long (Virgata) . . . . 3. *P. virgatum*
      4 Plants without rhizomes; culms 2–6 (7.5) dm tall; spikelets 3–4 (4.2) mm long (Diffusa) . . . . 4. *P. hallii*
1 Plants annual; second glume and sterile lemma often scaberulous at the apex (glabrous in *P. dichotomiflorum*).
  5 First glume short, 0.5–1 mm long, obtuse or rounded, 1/5 to 2/5 as long as the second glume (Dichotomiflora) . . . . 5. *P. dichotomiflorum*
  5 First glume 1–2 mm long, acute to acuminate, 1/2 to 3/4 as long as the second glume (Capillaria).
    6 Spikelets 2.5–3.7 mm long, the second glume and sterile floret 2–3.5 mm long; panicles neither densely floriferous nor nodding.
      7 Panicles narrow, usually less than 1/2 as broad as long, 1–5 (7) cm broad, the branches ascending; blades 1. 5–4 (7) mm broad . . . . 6. *P. flexile*
      7 Panicles about as broad as long, or nearly so, 6–30 cm broad, the branches divaricate; blades 4–15 mm broad . . . . 7. *P. capillare*
    6 Spikelets 4.5–6 mm long, the second glume and sterile floret 4–5.5 mm long; panicles densely floriferous and nodding . . . . 8. *P. miliaceum*

PANICUM

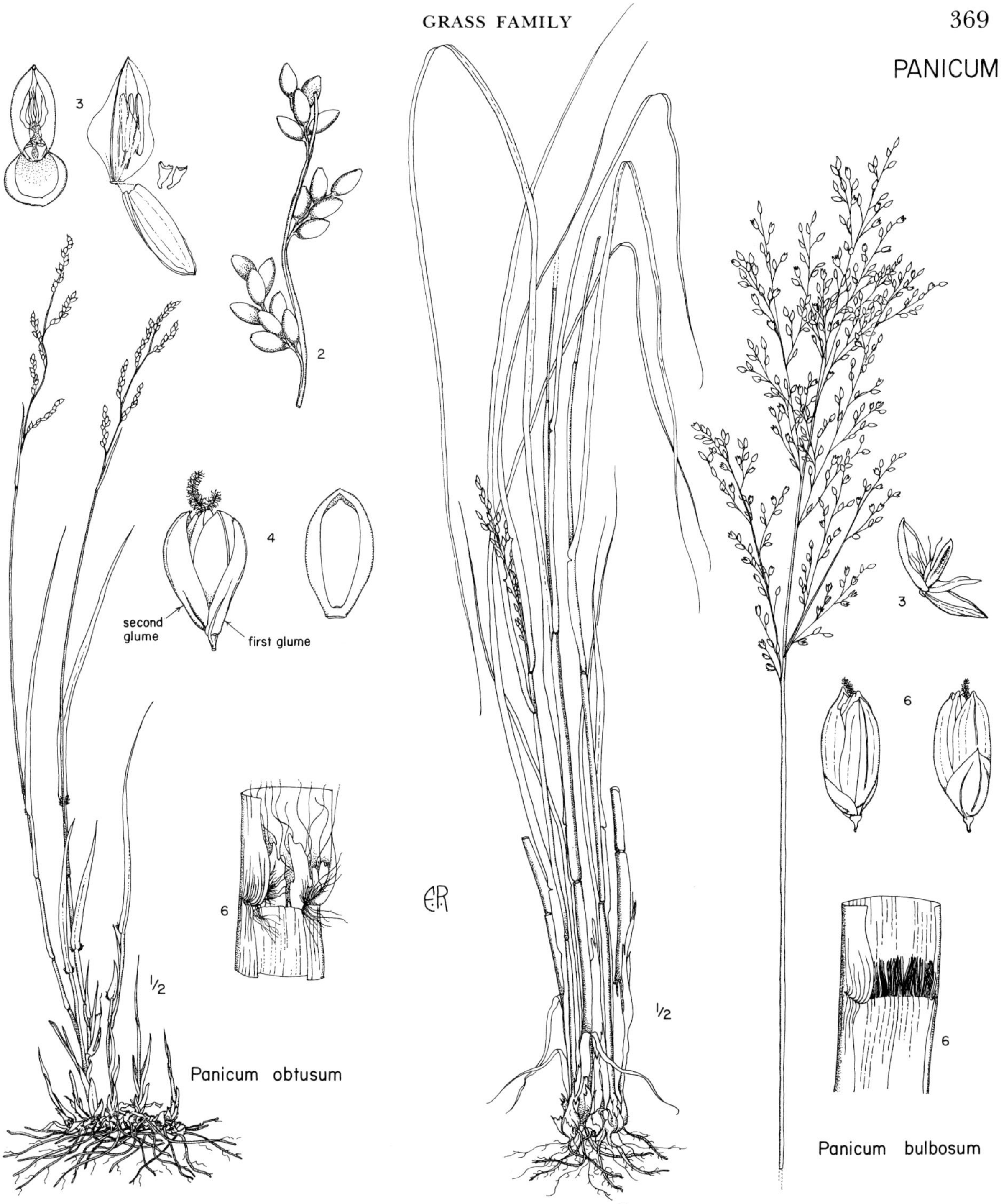

1. Panicum obtusum H.B.K.

Panicum obtusum H.B.K. Nova Gen. & Sp. Pl. 1: 98. 1815. *Oplismenus obtusum* Smyth, Trans. Kansas Acad. Sci. 16: 164. 1899. *Brachiaria obtusa* Nash in Britton, Manual Fl. N. States & Canad. 77. 1901. (*Humboldt & Bonpland,* "Crescit in planitie montana regni Mexicani prope Guanaxuato et Burras.")

Vine mesquite.

Stoloniferous perennials, sometimes bearing rhizomes, the stolons sometimes up to 2 m long with long internodes and villous nodes; culms (2) 3–6 (8) dm tall, erect or decumbent at the base, unbranched, glaucous, glabrous; sheaths glabrous, sometimes the lower sheaths villous towards the base; ligule a membrane, 0.2–1.8 mm long, entire to lacerate; blades flat, 2–5 (7) mm broad, becoming involute towards the tips, the margins sometimes scabrous, sometimes long-pilose near the throat on the upper surface, otherwise glabrous; panicles 5–10 (15) cm long, bearing (2) 3–5 (6) racemes, the spikelets single or in pairs on short pedicels on 2 sides of a 3-angled rachis as in *Digitaria* and *Paspalum;* spikelets 2.8–4 mm long, obovoid, obtuse to rounded, glabrous, brownish; glumes subequal, (2.2) 2.6–3.6 mm long, 5-nerved, ovate, glabrous; first floret sterile or sometimes staminate, the lemma similar to the glumes, ovate, 5- (7)-nerved, with a palea subequal in size; second floret fertile, the lemma (2.2) 2.8–3.2 mm long, elliptic or ovate, obtuse, indurate, smooth and shiny, the margins clasping the palea only towards the base; palea clasping the anthers; anthers about 2 mm long; $2n = 20, 36, 40$.

Moist situations, often associated with mesquite in the Southwest; s. Utah, Colo., Kansas and Mo., s. to Ariz., N.M., Texas, Ark. and c. Mex. Aug.–Sept.

By virtue of its stoloniferous habit, vine-mesquite often grows in dense stands and is sometimes planted in gullies to arrest erosion. *Panicum obtusum* is also palatable enough to be considered valuable forage, even though it is sometimes regarded as a weed.

2. Panicum bulbosum H.B.K.

Panicum bulbosum H.B.K. Nova Gen. & Sp. Pl. 1: 99. 1815. *P. maximum* var. *bulbosum* Vasey in Wheeler, Rep. U. S. Geogr. Surv. W. 100th Meridian 6: 295. 1879. (*Humboldt & Bonpland,* "Santa Rosa, Los Joares et Guanaxuato," Mex.)

P. bulbosum var. *minor* Vasey, U.S.D.A. Div. Bot. Bull. 8: 38. 1889. (*Rusby,* N.M., in 1880.)

Bulb panicum.

Stout, tufted, bulbous based perennials; culms tall, 5–10 (20) dm, erect, somewhat compressed, glabrous; sheaths keeled, glabrous or sometimes pilose near the collar; ligule a ciliate, lacerate membrane, 0.5–1 mm long; blades flat, 2–6 (10) mm broad, glabrous or the upper surface scabrous and sometimes pilose near the throat; panicles relatively large, 18–40 (50) cm long, long-exserted, diffuse, open, the branches ascending to somewhat spreading; spikelets (2.8) 3–3.8 (4) mm long, acuminate; first glume 1.3–2.2 (2.5) mm long, about $^2/_3$ as long as the second, ovate, acute, 5-nerved; second glume 2.5–3.4 mm long, lance-ovate, acute, 5-nerved, the nerves weak, glabrous; first floret sterile or sometimes staminate, the lemma similar to the second glume, with a well developed palea; second floret fertile, the lemma 2.8–3.2 mm long, often projecting beyond the first floret and second glume, narrowly ovate to lanceolate, acute, finely transversely rugose, not smooth, usually dull; $2n = 36, 54, 72$.

Moist meadows in canyons and on slopes in the mts., often in ponderosa pine woods; s. Utah and Ariz. to Texas, s. to c. Mex. July–Sept.

Bulb panicum is so called because of the large, corm-like swellings at the base of the culms. It is an important forage grass of the Southwest mountain ranges and is sometimes cut for hay.

3. Panicum virgatum L.

Panicum virgatum L. Sp. Pl. 59. 1753. *Chasea virgata* Nieuwl. Amer. Midl. Naturalist 2: 64. 1911. *Milium virgatum* Lunell, ibid. 4: 212. 1915. (*Clayton 578,* "Habitat in Virginia.")

Switchgrass.

Stout and strongly rhizomatous perennials, the scaly rhizomes often branching; culms tall, 6–15 (20) dm, erect, glabrous; sheaths glabrous, except for the ciliate upper margins; ligule a short, dense fringe of hairs, 0.5–1 mm long, sparsely interspersed with longer hairs up to 5 mm long; blades flat, 4–8 (15) mm broad, glabrous or sometimes pilose towards the base on the upper surface, the margins scabrous; panicles large, 15–50 cm long, long-exserted, open and diffuse, the branches ascending or spreading; spikelets (3.5) 4–5.2 mm long, glabrous, acuminate, strongly nerved; first glume large, 2–4 mm long, $^3/_4$ as long as the second, lance-ovate, acuminate, 3- (5)-nerved; second glume large, 3.5–5 mm long, lance-ovate to ovate, 7- (9)-nerved, the nerves becoming faint toward the margins; first floret sterile or sometimes staminate, the lemma similar to the second glume, its palea well developed, nearly as long as the lemma; second floret fertile, the lemma (2.4) 2.8–3.2 (3.5) mm long, lanceolate to narrowly ovate, acute, indurate, smooth and shiny, the margins clasping the palea only towards the base; anthers about 2 mm long, orange; $2n = 18, 21, 25, 30, 32, 36, 54–65, 70, 72, 108$.

Moist places along streams and ditchbanks, sand dunes; Sask. to N.S., s. through the Dakotas and Me. to all of the U.S. but the Pacific Northwest and Calif., in our area in the s. parts, common in the Great Plains and the e. states, but infrequent in the w. states; s. in e. Mex. and W.I. June–Sept.

Switchgrass, one of the well-known midwestern "tall grasses," is a good forage grass that is sometimes cut for hay.

4. Panicum hallii Vasey

Panicum hallii Vasey, Bull. Torrey Bot. Club 11: 61. 1884. (*Hall 816,* "Dry hills," Austin, Texas, 18 May 1872.)

Hall's panicum.

Tufted, glaucous perennials; culms 2–6 (7.5) dm tall, erect, sometimes branched towards the base, the nodes pubescent and the internodes glabrous; sheaths papillose-pilose; ligule a fringe of hairs, 0.3–1.5 mm long; blades flat, 1.5–5.5 (7) mm broad, glabrous at least above, sometimes sparsely papillose-pilose below and usually sparsely ciliate near the base, the old leaves becoming curled; panicles (6) 9–20 (25) cm long, usually long-exserted, open, the branches stiffly ascending; spikelets 3–4 (4.2) mm long, glabrous, acuminate; first glume 1.7–2.5 mm long, $^2/_3$ as long as the second, broadly ovate, acuminate, 3- (5)-nerved; second glume 2.5–3.5 (3.9) mm long, ovate to lance-ovate, acuminate, 9-nerved, glabrous; sterile lemma similar to the second glume, sometimes with a small hyaline palea; fertile lemma 2–2.6 mm long, ovate, obtuse, indurate, smooth and shiny; $2n = 18$.

Dry prairie communities, hills and canyons in rocky or gravelly soil; s. Utah (*Hutchings & Stachmann,* Red Knoll, e. side of Pine Valley, Beaver Co.) and Ariz. to Texas and c. Mex. Aug.–Sept.

Panicum lepidulum Hitchc. & Chase is a closely related, but weakly distinguished, species reported by A. S. Hitchcock to be in Utah (Manual Grasses U.S. 692. 1951). It is supposed to have larger spikelets, and the second glume and sterile lemma are supposed to be scabrous, rather than smooth, towards the apex.

5. Panicum dichotomiflorum Michx.

Panicum dichotomiflorum Michx. Fl. Boreali-Amer. 1: 48. 1803. *P. geniculatum* Muhl. Trans. Amer. Philos. Soc. 4: 235. 1799, nomen nudum; Cat. Pl. N. Amer. 9. 1813; not Lam. 1798. *Leptoloma dichotomiflora* Smyth, Trans. Kansas Acad.

6
3
6
6
6
30
lodicules
6
6
6
1/2
1/2
1/2
6
6
1/8
Panicum hallii
Panicum virgatum
Panicum dichotomiflorum
JRJ

Sci. 25: 86. 1913. (*Michaux,* "occidentalibus montium Alleghanis.")

Fall panicum.

Tufted annuals; culms 3–10 (20) dm tall, erect, ascending or decumbent at the base, sometimes sparingly branched at the lower nodes; sheaths glabrous; ligule of hairs, 1–1.5 mm long; blades flat, 3–15 (18) mm broad, glabrous or sometimes scaberulous above; panicles 15–25 cm long, diffuse, often included at the base; spikelets 2.5–3.5 mm long, acute; first glume very short, 0.5–1 mm long, $^1/_5$ to $^2/_5$ as long as the second glume, obtuse to rounded, 1- to 5-nerved, the lateral nerves usually weak; second glume 1.8–3.2 mm long, lanceolate or narrowly ovate, acute, 7- to 9-nerved; sterile lemma similar to the second glume, without a palea; fertile lemma 1.8–2.5 mm long, narrowly ovate to elliptic, indurate, smooth and shiny; $2n$ = 36, 54.

Moist streambanks, roadsides and meadows; common from e. Nebr. to e. Texas, e. throughout e. U.S.; sporadic in the w. states, in Calif., Utah, Idaho and N.M., known in the Intermountain Region only from *Arnow 3779,* collected in Salt Lake City. Aug.–Sept.

6. Panicum flexile (Gattinger) Scribn.

Panicum capillare var. *flexile* Gattinger, Tennessee Fl. 94. 1887. *P. flexile* Scribn. in Kearney, Bull. Torrey Bot. Club 20: 476. 1893. *Chasea flexilis* Nieuwl. Amer. Midl. Naturalist 2: 65. 1911. (*Gattinger,* "Cedar glades near Nashville," Tenn.)

Tufted annuals; culms 1.5–5 (7) dm tall, erect, branched from the base, glabrous to somewhat hispid, especially around the nodes; sheaths papillose-pilose; ligules of hairs, 0.5–1 mm long; blades flat, 1.5–4 (7) mm broad, papillose-hispid, sometimes hirsute to glabrous above; panicles (6) 10–17 (30) cm long, usually exserted, oblong, narrow, the branches ascending, 1–5 (7) cm broad, relatively few-flowered; spikelets 2.8–3.5 (3.7) mm long, glabrous, acuminate; first glume 1.2–2 mm long, 3- to 5-nerved, the mid-nerve prominent; second glume (2.2) 2.6–3.5 mm long, lanceolate or narrowly ovate, acuminate, 5- to 7-nerved, the nerves not prominent, scaberulous above and glabrous below; sterile lemma similar to the second glume, without a palea; fertile lemma 1.8–2.2 mm long, narrowly ovate to elliptic, acute to obtuse, indurate, smooth and shiny, faintly 3-nerved; anthers about 1.2 mm long; $2n$ = 18.

Meadows and open woods, often moist places; Ont., Que., s. through N.D. and N.Y. to e. Texas and Fla. and introd. in s.-central Utah. Aug.–Sept.

Panicum hirticaule Presl, if found in the region, would key here. See distinction and comments below in the *P. capillare* discussion.

7. Panicum capillare L.

Panicum capillare L. Sp. Pl. 58. 1753. *Milium capillare* Moench, Meth. Pl. Horti Bot. & Agri Marburg. 203. 1794; not Rottb. 1778. *Chasea capillaris* Nieuwl. Amer. Midl. Naturalist 2: 64. 1911. *Leptoloma capillaris* Smyth, Trans. Kansas Acad. Sci. 25: 86. 1913. (*Clayton 454,* "Habitat in Virginia.")

P. capillare var. *occidentale* Rydb. Contr. U. S. Natl. Herb. 3: 186. 1895. (*Rydberg 1788,* "west of Whitman in dry lake," Nebr., 19 Sept. 1893.)

P. capillare var. *brevifolium* Vasey ex Rydb. & Shear, U.S.D.A. Div. Agrostol. Bull. 5: 21. 1897; not *P. brevifolium* L. 1753. (*Shear 436,* "on a small sand bar in the Gallatin River," Manhattan, Gallatin Co., Mont., 19 July 1895.)

P. barbipulvinatum Nash in Rydb. Mem. New York Bot. Gard. 1: 21. 1900. *Leptoloma barbipulvinata* Smyth, Trans. Kansas Acad. Sci. 25: 86. 1913. *Milium barbipulvinatum* Lunell, Amer. Midl. Naturalist 4: 212. 1915. (*Rydberg & Bessey 3544,* Lower Geyser Basin, Yellowstone Park, Wyo., 4 Aug. 1897.)

Old witchgrass.

Tufted annuals; culms 1.5–4.5 (10) dm tall, erect, ascending or decumbent at the base and rooting at the lower nodes, sparingly branched at the lower nodes; sheaths papillose-pilose; ligules of hairs, 1–1.5 mm long; blades flat or folded, 4–12 (15) mm broad, sparsely papillose-pilose, sometimes only ciliate and nearly glabrous; panicles (8) 10–30 (50) cm long, the branches diffuse, divaricately spreading, nearly as broad as long and including half the length of the entire plant, 6–20 (30) cm broad, often partially included in the upper sheaths, relatively densely flowered; spikelets 2.5–3.2 (3.5) mm long, acuminate; first glume 1–1.7 (2) mm long, broadly ovate, acuminate, 3- (5)-nerved, the lateral nerves often weak; second glume 2–2.8 (3.4) mm long, lanceolate, acuminate, 7-nerved, the nerves scaberulous towards the apex, otherwise glabrous; sterile lemma similar to the second glume, without a palea; fertile lemma 1.5–2 mm long, broadly oblanceolate, obtuse, indurate, smooth and shiny; lodicules about 0.3 mm long, flabellate; anthers 0.8–1.2 mm long, purple; $2n$ = 18, ca. 54.

Moist, open places along ditchbanks, around springs (often thermal) and edges of alkali flats; s. Can. from B.C. to Que. and s. throughout the U.S. June–Sept.

Our plants have been referred to *Panicum capillare* var. *occidentale* Rydb. (*P. barbipulvinatum* Nash), but the characters used to distinguish this variety are not convincingly consistent. Variety *occidentale* is supposed to have a lower average height, shorter, less pubescent blades, more exserted and divaricate panicles and longer, narrower, more acuminate spikelets than the typical variety has.

Panicum hirticaule Presl, of the Southwest, has not been collected in our area but may eventually be discovered in the Grand Canyon Plateaus of northwestern Arizona. It would key to *P. flexile* with its narrower panicle and stiffly ascending branches. Smaller panicles (5–15 cm long) that include only $^1/_4$ to $^1/_3$ the height of the plant distinguish *P. hirticaule* from *P. flexile* and *P. capillare* which have panicles (6) 10–50 cm long that include more than $^1/_2$ the height of the plant.

8. Panicum miliaceum L.

Panicum miliaceum L. Sp. Pl. 58. 1753. *Milium panicum* Miller, Gard. Dict. ed. 8. (no pages, alphabetical order). 1768. *Milium esculentum* Moench, Meth. Pl. Horti Bot. & Agri Marburg. 203. 1794. *P. milium* Pers. Syn. Pl. 1: 83. 1805. *Leptoloma miliacea* Smyth, Trans. Kansas Acad. Sci. 25: 86. 1913. ("Habitat in India.")

Broom corn millet, hog millet.

Stout annuals; culms 2–7.5 (12) dm tall, erect or decumbent at the base, sometimes branching at the lower nodes, hispid; sheaths loose, papillose-pilose; ligules of relatively long hairs, 1–3 mm; blades flat, 6–14 (20) mm broad, papillose-pilose, rarely glabrate, the margins crisped; panicles 10–25 (30) cm long, relatively densely flowered, often nodding, usually more or less compact, the branches ascending, often included at the base, axillary panicles common; spikelets 4.5–6 mm long, strongly nerved, acu-

PANICUM

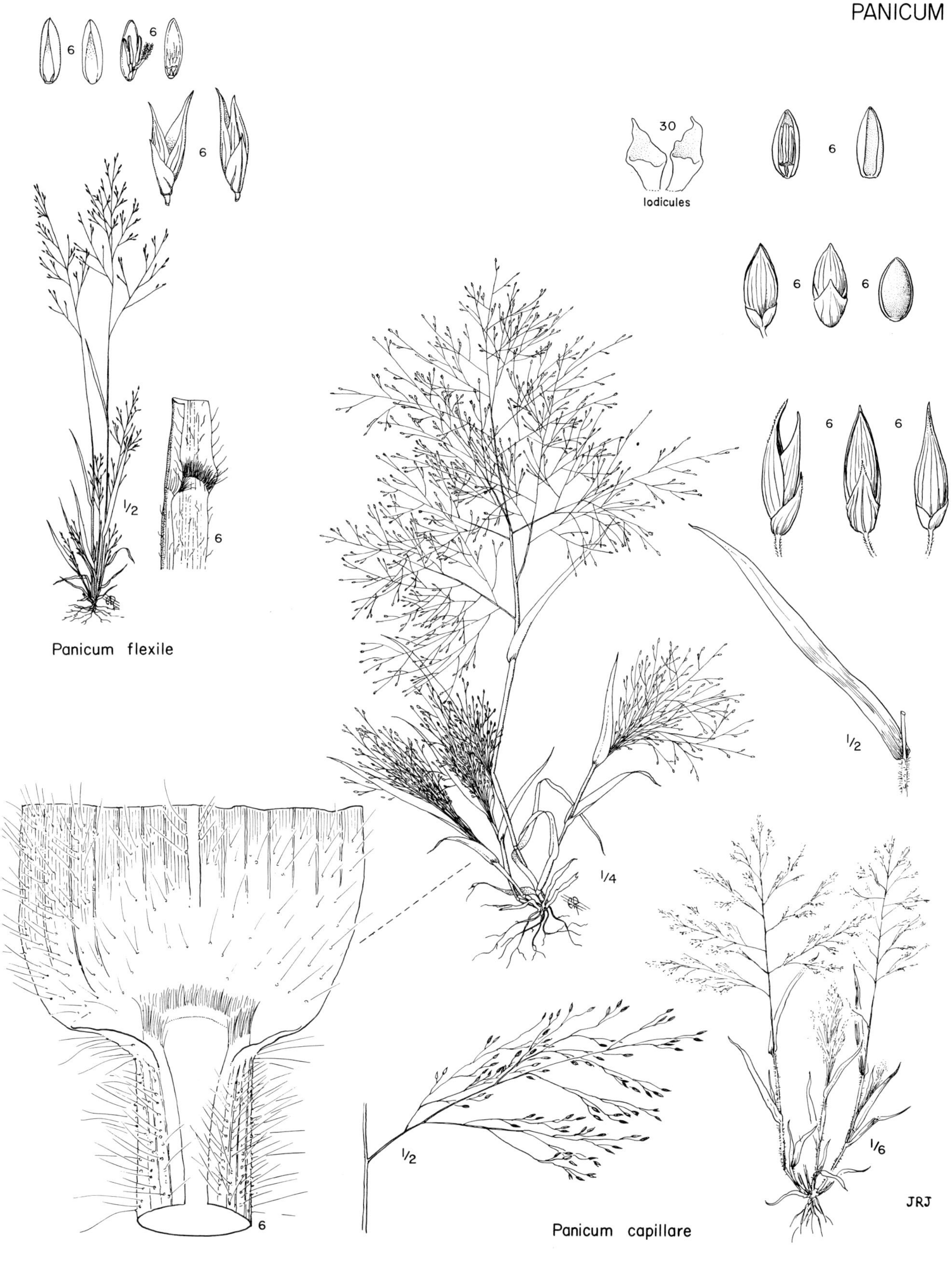

Panicum flexile

Panicum capillare

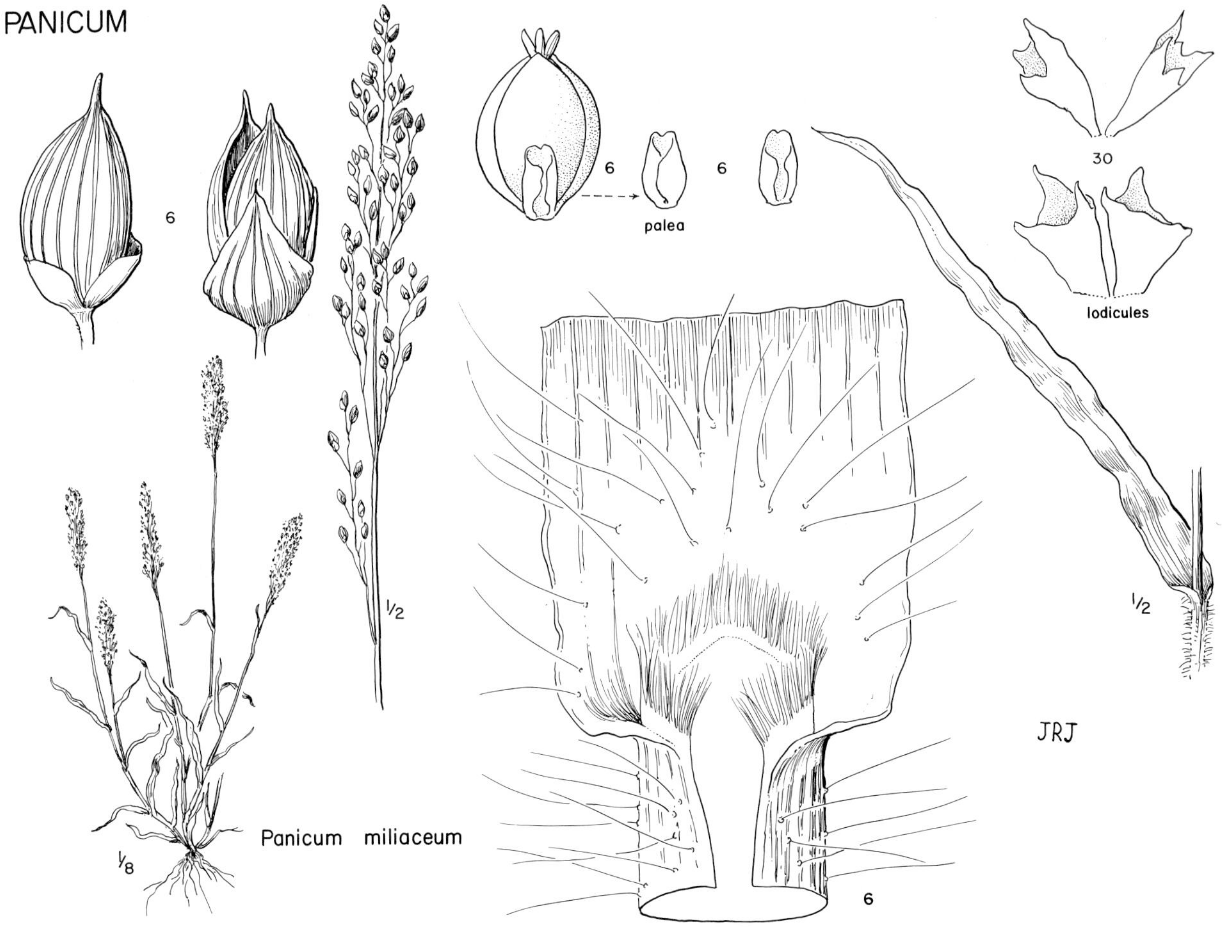

minate; first glume 2.5–3.5 (4.7) mm long, broadly ovate, acuminate, 5- to 7-nerved; second glume 4–5 (5.5) mm long, ovate, acuminate, 9- to 11- (13)-nerved, usually scaberulous on the nerves near the apex, otherwise glabrous; sterile lemma similar to the second glume, with a small hyaline palea; fertile lemma 3–3.5 (3.8) mm long, broadly ovate, obtuse, indurate, shiny and glabrous, sometimes faintly 3-nerved; lodicules about 0.5 mm long, semi-tubular; anthers about 1.5 mm long; $2n$ = 36, 40, 42, 49, 54, 72.

In waste places adjacent to cultivated fields; introduced from the temp. parts of the Old World; Ill. to Me., s. to N.C. and less common in scattered localities in se. states and in the w. states in Calif., Utah, Oregon, Wash., Idaho and Mont. July–Aug.

Panicum miliaceum is cultivated for forage and hog food in the United States, hence the name hog millet. In some regions of the Old World, it is cultivated for human food.

47. DICHANTHELIUM (Hitchc. & Chase) Gould Dichotomous Panicums

Perennials (and annuals?); plants passing through 3 seasonal phases, a winter phase as a rosette, a spring phase (vernal) with a simple culm terminating in an inflorescence, and a summer and fall phase (autumnal) with axillary fascicles and/or well developed branches bearing axillary panicles; culms hollow; sheaths open; ligules mostly of hairs, sometimes membranous or absent; blades usually flat, non-auriculate, short and broad in the winter rosette, long and narrow on the spring culm and much reduced in summer fascicles and branches; inflorescence of the spring phase a terminal, open to contracted primary panicle, and of the summer phase small secondary panicles, mostly enclosed in leaf sheaths; spikelets turgid to plano-convex or somewhat dorsiventrally compressed, pedicellate, 2-flowered, disarticulating below the glumes, chasmogamous, but easily aborting in the primary panicle, cleistogamous in secondary panicles; glumes unequal, the first glume usually small and often clasping the base of the spikelet, the second subequal and similar in texture to the sterile lemma; first floret sterile or sometimes staminate, the lemma several-nerved, membranous, longer than the fertile lemma, often bearing a membranous or hyaline palea; fertile floret more or less dorsally-compressed, the lemma mostly obtuse, rounded on back, indurate, smooth and shiny, the margins thick, inrolled and clasping the palea by its margins, awnless; palea of the fertile floret of similar texture to the lemma; lodicules 2, tubular-funnelform, the margins irregular; stamens 3; x = 9.

From 50 to 100 species, mostly confined to the Americas from s. Canada to n. S. Amer., most abundant in se. U.S., also a few spp. in the Hawaiian Islands. (The meaning of the name *Dichanthelium* is obscure, but perhaps it is in reference to the seasonal dimorphism of the plants.)

Until recently (Gould, 1974) *Dichanthelium* was considered a subgenus of *Panicum*. It appears to be a monophyletic group that can be readily distinguished by the 3 seasonal growth forms. All species of *Dichanthelium* studied anatomically and physiologically have the non-Kranz leaf anatomy and associated C_3 photosynthesis. The species remaining in the genus *Panicum* are all Kranz types with C_4 photosynthesis. A Kranz plant has a chlorenchymatous sheath of large thick-walled cells surrounding the vascular bundles in the leaves. In the grass family this bundle sheath is surrounded by radially arranged mesophyll cells where carbon is fixed into 4-carbon acids, a process of photosynthesis which minimizes or avoids photorespiration and can be efficient at relatively high temperatures. Smith and Brown (1973) have shown that the Kranz and non-Kranz conditions are conservative in the grass family. In Poaceae all but 3 tribes contain either all Kranz or non-Kranz species, and following the removal of *Dichanthelium* from *Panicum* only 1 genus, *Alloteropsis*, is known to contain both kinds of species. Clark and Gould (1975) discovered that in *Dichanthelium* the apical portion of the palea is covered with simple papillae arranged in regular rows whereas in *Panicum* the palea tip has compound or both compound and simple papillae arranged irregularly.

References:

Brown, W. V., and B. N. Smith. The genus *Dichanthelium* (Gramineae). Bull. Torrey Bot. Club 102: 10–13. 1975.

Clark, C. A., and F. W. Gould. Some epidermal characteristics of paleas of *Dichanthelium, Panicum,* and *Echinochloa*. Amer. J. Bot. 62: 743–748. 1975.

Gould, F. W. Nomenclatural changes in the Poaceae. Brittonia 26: 59–60. 1974.

Hitchcock, A. S., and A. Chase. The North American species of *Panicum*. Contr. U. S. Natl. Herb. 15: 1–396 (subgenus *Dichanthelium,* 142–320). 1910.

Smith, B. N., and W. V. Brown. The Kranz syndrome in the Gramineae as indicated by carbon isotopic ratios. Amer. J. Bot. 60: 505–513. 1973.

Spellenberg, R. Synthetic hybridization and taxonomy of western North American *Dichanthelium,* group Lanuginosa (Poaceae). Madroño 23: 134–153. 1975.

1 Spikelets 1.5–2 mm long, the second glume and sterile lemma 1.2–1.8 mm long; ligular hairs 1.5–5 mm long . . . . 1. *D. lanuginosum*

1 Spikelets 2.9–3.5 mm long, the second glume and sterile lemma 2.4–3 mm long; ligular hairs 0.5–1.4 mm long . . . . 2. *D. oligosanthes*

1. Dichanthelium lanuginosum (Elliott) Gould

Panicum lanuginosum Elliott, Sketch Bot. S. Carol. & Georgia 1: 123. 1816. *Panicum dichotomum* var. *lanuginosum* A. Wood, Cl.-Book Bot. ed. 3. 786. 1861. *Dichanthelium lanuginosum* Gould, Brittonia 26: 60. 1974. (*Baldwin,* Ga.)

Panicum dichotomum var. *fasciculatum* Torr. Fl. N. & Middle Sect. U.S. 145. 1824. *Panicum dichotomum* f. *fasciculatum* S. Wats. Gray's Manual Bot. ed. 6. 633. 1890. *Panicum huachucae* var. *fasciculatum* Hubbard, Rhodora 14: 171. 1912. *Panicum lindheimeri* var. *fasciculatum* Fern. Rhodora 23: 228. 1921. *Panicum lanuginosum* var. *fasciculatum* Fern. Rhodora 36: 77. 1934. *D. lanuginosum* var. *fasciculatum* Spellenberg, Madroño 23: 145. 1975. ("In sandy fields, New-Jersey.")

Panicum nitidum var. *pilosum* Torr. Fl. N. & Middle Sect. U.S. 146. 1824. *Panicum lindheimeri* var. *fasciculatum* subvar. *pilosum* Farw. Amer. Midl. Naturalist 11: 45. 1928. *Panicum lanuginosum* var. *fasciculatum* subvar. *pilosum* Farw. Pap. Michigan Acad. Sci. 26(1): 5. 1941. (N.Y.) = var. *fasciculatum.*

Panicum pubescens sensu Presl in C. Presl, Reliquiae Haenkeanae 1: 306. 1830; not Lam. 1798, or Michx. 1803. *Panicum occidentale* Scribn. Annual Rep. Missouri Bot. Gard. 10: 48. 1899. (*Haenke,* "Hab. in Nootka-Sund," Vancouver I., B.C., in 1791.) = var. *fasciculatum.*

Panicum huachucae Ashe, J. Elisha Mitchell Sci. Soc. 15: 51. 1898. *Panicum lanuginosum* var. *huachucae* A. S. Hitchc. Rhodora 8: 208. 1906. (*Lemmon,* Huachuca Mts., Ariz., in 1882.) = var. *fasciculatum.*

Panicum tennesseense Ashe, J. Elisha Mitchell Sci. Soc. 15: 52. 1898. *Panicum lindheimeri* var. *tennesseense* Farw. Amer. Midl. Naturalist 11: 45. 1928. *Panicum lanuginosum* var. *tennesseense* Gleason, Phytologia 4: 21. 1952. (*Biltmore Herbarium 7087,* Cedar glades, La Vergne Co., Tenn., 7 Aug. 1897.) = var. *fasciculatum.*

Panicum pacificum Hitchc. & Chase, Contr. U.S. Natl. Herb. 15: 229, fig. 241. 1910. (*A. S. Hitchcock 3070,* "one fourth mile east of hotel," Castle Crags, Shasta Co., Calif., 3 Aug. 1908.) = var. *fasciculatum.*

Panicum ferventicola Schmoll, Madroño 5: 92. 1939. (*Chase 137,* "growing in fine gravel in warm soil close to a boiling spring, the plants bathed in hot vapor," 5 mi. n. of Norris Geyser Basin, Yellowstone Natl. Park, Wyo., 10 Aug. 1908.) = var. *sericeum.*

Panicum ferventicola var. *sericeum* Schmoll, Madroño 5: 93. 1939. *D. lanuginosum* var. *sericeum* Spellenberg, Madroño 23: 150. 1975. (*A. & E. Nelson 6037,* "in wet ground about hot springs, Mammoth Hot Springs," Yellowstone Natl. Park, Wyo., 21 July 1899.)

Panicum ferventicola var. *papillosum* Schmoll, Madroño 5: 94. 1939. (*A. S. Hitchcock 220,* "on a declivity wet with water from hot spring," Banff, Alta., 31 July 1914.) = var. *sericeum.*

Panicum lassenianum Schmoll, Madroño 5: 95. 1939. (*Jepson 4082,* Devil's Kitchen, Hot Spring Valley, Plumas Co., Calif., 6 June 1910.) = var. *fasciculatum.*

Perennials with well developed winter rosettes; plants of the vernal phase with simple, or few branched, erect culms terminated by a small panicle; plants of the autumnal phase branched at the upper, lower or all nodes, the culms erect to decumbent or prostrate-spreading, the secondary leaves reduced, forming fascicles and the axillary panicles partially included in the sheaths; culms 0.4–6.5 (10) dm tall, the pubescence of the nodes retrorse; sheaths papillose-pilose to velvety-pubescent; ligules of relatively long hairs, 1.5–5 mm long; blades usually flat, (3) 5–8 (12) mm broad, variously pubescent, from velvety on both sides to sparsely pilose or glabrous on the upper surface; terminal panicle of the vernal phase short, 2–10 (16) cm long, open, the axillary panicles even smaller; spikelets 1.5–2 mm long, cupulate at the base and rounded to obtuse at the apex; first glume very small, 0.2–0.8 mm long, broadly ovate, glabrous, obscurely 3-nerved; second glume short, 1.2–1.8 mm long, ovate, rounded to obtuse, short-pilose, 7-nerved, the nerves relatively prominent; sterile lemma similar to the second glume, sometimes with a very thin, hyaline palea; fertile lemma 1.2–1.5 (1.8) mm long, obovate, obtuse, indurate, shiny and smooth; $2n = 18$.

Along river and stream banks, margins of ponds and lakes, marshy areas and around springs (often thermal), usually in moist ground but sometimes drier places in our area; B.C. to N.S. and s. throughout the U.S. July–Sept.

Spellenberg (Univ. Microfilms, Ann Arbor, Michigan, Diss. Abstr. 68: 7080. 1968) has conducted biosystematic studies on the western members of the *Panicum* [*Dichanthelium*] *lanuginosum* complex and has discovered that the plants are predominantly autogamous with occasional hybridization which contributes to the taxonomic complexity. The complex has received many different interpretations. As it now appears the taxa recognized by Hitchcock and Chase (1910) in this group are for the most part based on an arbitrary selection of specimens showing various recombinations of characters. Regional investigations have been conducted in the Northeast by Fernald (Rhodora 36: 61–87. 1934) and Pohl (Amer. Midl. Naturalist 38: 575–590. 1947); in the Midwest by Shinners (Amer. Midl. Naturalist 32: 164–180. 1944); and in the Northwest by Spellenberg (1968). All these studies treated the species as Panicums. Fernald recognized the eastern taxa of the complex as varieties under the oldest North American name, *Panicum lanuginosum*. Shinners and Pohl and more recently Voss (Michigan Fl. 1: 238. 1972) combine *P. tennesseense* and *P. huachucae* under *P. implicatum* Scribn., since the midwestern members appear distinct from the velvety-pubescent *P. lanuginosum* of the eastern and southeastern coastal plains. Spellenberg's biosystematic investigation included the western elements of the complex, *P. occidentale, P. pacificum,* and *P. thermale,* and more recently (Spellenberg, 1975) *P. tennesseense, P. huachucae, P. ferventicola,* and *P. lanuginosum.* Spellenberg concluded that Fernald's expanded *P. lanuginosum* has some merit and should be further expanded to include the western elements. Spellenberg has reinterpreted the western members of the complex to include 3 discrete, although somewhat intergrading varieties, 2 of which enter our region. We recognize the following 2 varieties of *Dichanthelium lanuginosum* in our area, which can be distinguished in the following manner.

1 Culms 3–8 dm long; vernal leaf blades long, mostly more than 8 times as long as broad; s. B.C. to N.S., s. to s. Calif. and the Gulf states ...... var. *fasciculatum* (Torr.) Spellenberg
1 Culms 0.4–3 (4) dm long; vernal leaf blades short and broad, usually less than 8 times as long as broad; Alta. and w. Mont. to nw. Wyo. and a garden specimen from Salt Lake City, Utah (usually around hot springs and geysers) .... var. *sericeum* (Schmoll) Spellenberg

Variety *fasciculatum* is an extremely variable taxon which in the past has been recognized as 4 species in our area. The following key shows how these "species" have traditionally been distinguished, but the characters completely intergrade and the geographical distributions have been found to be inaccurate.

1 Vernal blades more or less densely pilose on the upper surfaces.
  2 Vernal blades appressed-pubescent on the upper surface; autumnal culms stiffly erect or ascending; Ont. to Newfl., s. to N.C. and Texas and w. to Utah and s. Calif. .... *Panicum huachucae* Ashe
  2 Vernal blades spreading-pilose on the upper surface; autumnal culms decumbent or prostrate-spreading; s. B.C., nw. Mont., Wash. and Idaho, s. to Calif., Nev. and Ariz. .... *Panicum pacificum* Hitchc. & Chase
1 Vernal blades glabrous to sparsely pilose on the upper surfaces.
  3 Autumnal culms branching from the middle and upper nodes; Que. and N.S., s. throughout e. U.S. and w. to Utah and Ariz. .... *Panicum tennesseene* Ashe
  3 Autumnal culms branching from the middle and lower nodes; s. B.C., s. to Calif., Utah and Ariz. .... *Panicum occidentale* Scribn.

2. Dichanthelium oligosanthes (J. A. Schultes) Gould

Panicum pauciflorum Elliott, Sketch Bot. S. Carol. & Georgia 1: 120. 1816; not R. Br. 1810. *Panicum oligosanthes* J. A. Schultes, Mantissa 2: 256. 1824. *Panicum scoparium* var. *pauciflorum* Scribn. Tennessee Agric. Exp. Sta. Bull. 7: 48. 1894. *Dichanthelium oligosanthes* Gould, Brittonia 26: 60. 1974. (Ga.)

Panicum scoparium sensu S. Wats. Gray's Manual Bot. ed. 6. 632. 1890; not Lam. 1798. *Panicum scribnerianum* Nash, Bull. Torrey Bot. Club 22: 421. 1895. *Panicum oligosanthes* var. *scribnerianum* Fern. Rhodora 36: 80. 1934. *D. oligosanthes* var. *scribnerianum* Gould, Brittonia 26: 60. 1974. (*Carey,* Pa., July 1836.)

Perennials, the winter rosettes not as well developed as in the preceding species; plants of the vernal phase with mostly unbranched, erect or ascending culms terminated by a panicle; plants of the autumnal phase branching from the middle and upper nodes, the secondary leaves reduced, forming fascicles and the axillary panicles partially included in the sheaths; culms 1.5–4.5 (6) dm tall, glabrous or variously pubescent; sheaths papillose-pilose; ligules of hairs, 0.5–1.4 mm long; blades flat, 4–10 (12) mm broad, glabrous or hispid, the margins crisped, sometimes ciliate; terminal panicle of the vernal phase short, (3.5) 4.5–7 (9) cm long, open, the axillary panicles even; spikelets 2.9–3.5 mm long, broadly obtuse above and abruptly narrowed to a cupulate base; first glume 1–1.5 mm long, about 1/2 as long as the second, broadly ovate and obtuse, usually nerveless; second glume 2.4–3 mm long, broadly ovate, obtuse to rounded, glabrous to short-pilose, 7- to 11-nerved; sterile lemma similar to the second glume with a hyaline palea 1/2 its length; fertile lemma 2.5–3 mm long, obovate, obtuse, indurate, shiny and glabrous, sometimes faintly 3-nerved; lodicules 0.6–0.7 mm long, tubular-flabellate; $2n$ = 18.

Variety *scribnerianum* is found on banks and bars of creeks and rivers in sandy or gravelly soils; B.C. to Me., s. throughout most of the U.S. June–Sept.

The *D. oligosanthes* group is another complex that requires further study. The complexities, however, occur to the east of our region as the "*scribnerianum*" segregate is the only western representative. Variety *oligosanthes* ranges from Michigan to Massachusetts, south to eastern Texas and Florida. It differs from our var. *scribnerianum* (Nash) Gould with its larger spikelets, 3.4–4 mm long (ours 2.9–3.5 mm), and a pubescence on the sheaths of a mixture of papillose hairs and fine, non-papillose hairs (ours of papillose hairs only). Where the two come together, along with var. *helleri* (Nash) Fern., there is enough intergradation to give problems.

48. ECHINOCHLOA Beauv.

Coarse annuals and perennials; culms succulent, weak, rarely erect, often rooting at the lower nodes, pithy or hollow; sheaths more or less compressed, open; ligule a fringe of hairs or lacking (ours); blades flat, non-auriculate; inflorescence a contracted or open panicle with few to numerous, simple or rebranched, densely flowered, lateral racemes, the spikelets subsessile, in irregular fascicles or regular rows along 1 side of panicle branches; spikelets somewhat dorsally-compressed, plano-convex or turgid, 2-flowered, the first floret sterile or staminate, the second one fertile, disarticulating below the glumes; glumes unequal, usually with stout spicules and long or short hairs which are often papillose at the base, membranous, 3 (5)-nerved, the first glume often 1/2 as long as the second, acute, the second subequal to the sterile lemma, acute, short- to

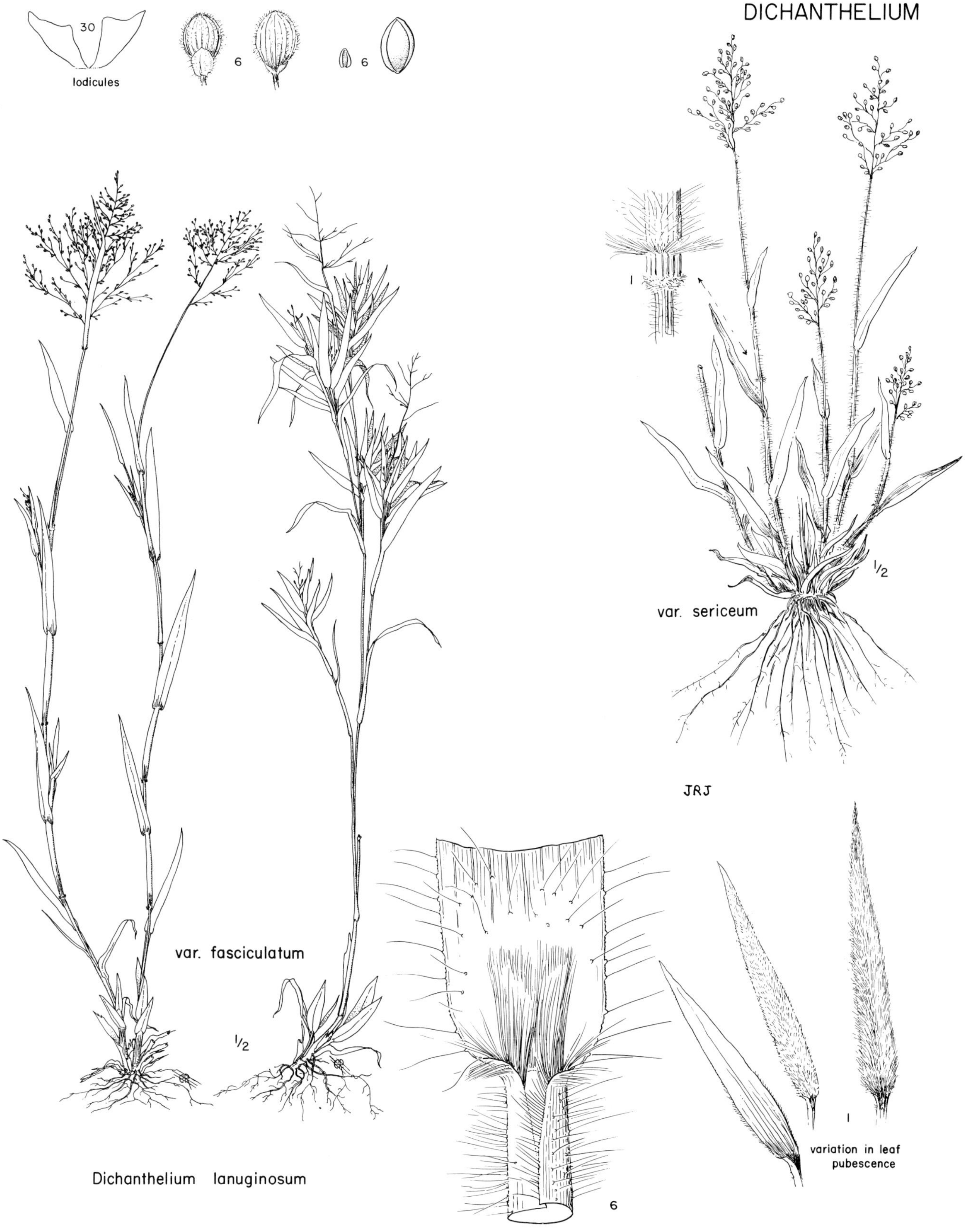
DICHANTHELIUM
30
lodicules
6
6
1
var. sericeum
1/2
JRJ
var. fasciculatum
1/2
6
1
variation in leaf pubescence
Dichanthelium lanuginosum

long-awned; sterile (or sometimes staminate) lemma similar to the glumes in texture, sometimes prominently awned; fertile lemma broadly elliptical, apiculate, awnless, firm or indurate, smooth and shiny, indistinctly nerved, the margins revolute and clasping the base of the palea of the fertile floret; palea similar to the lemma in texture, broad, but narrowing to a pointed tip that is free from the lemma margins; lodicules 2, narrowly flabellate; stamens 3; $x = 9$.

A genus of 15 to 20 species of warm temp., subtrop., and trop. regions of the world. (From the Greek *echinos,* hedgehog or sea urchin, and *chloa,* a grass, alluding to the appearance of the echinate spikelets.)

References:
Gould, F. W., M. A. Ali, and D. E. Fairbrothers. A revision of *Echinochloa* in the United States. Amer. Midl. Naturalist 87: 36–59. 1972.
Wiegand, K. M. The genus *Echinochloa* in North America. Rhodora 23: 49–65. 1921.

1. Echinochloa crusgalli (L.) Beauv.

Panicum crusgalli L. Sp. Pl. 56. 1753. *Milium crus galli* Moench, Meth. Pl. Horti Bot. & Agri Marburg. 202. 1794. *Echinochloa crus galli* Beauv. Essai Nouv. Agrostogr. 53, 161, 169. 1812. *Pennisetum crus galli* Baumg. Enum. Stirp. Magno Transsilv. Princ. 3: 277. 1816. *Oplismenus crus galli* Dumort. Observ. Gram. Fl. Belg. 138. 1823. *Orthopogon crus galli* Sprengel, Syst. Veg. 1: 307. 1825. ("Habitat in Europae, Virginiae cultis.")

Panicum crus galli var. *mite* Pursh, Fl. Amer. Sept. 66. 1814. *E. crus-galli* var. *mitis* Peterm. Fl. Lips. 82. 1838. *E. crus-galli* f. *mitis* Farw. Rep. Michigan Acad. Sci. 21: 349. 1920. (N. Amer.)

Oplismenus zelayensis H.B.K. Nova Gen. & Sp. Pl. 1: 108. 1815. *E. zelayensis* J. A. Schultes, Mantissa 2: 269. 1824. *Panicum zelayense* Steudel, Nomencl. Bot. ed. 2. 2: 265. 1841. *E. crusgalli* var. *zelayensis* A. S. Hitchc. U.S.D.A. Bull. (1915–23) 772: 238. 1920. *Panicum crusgalli zelayense* Stratman, Catholic Univ. Amer. Biol. Ser. 19: 60. 1935. *E. crusgalli* f. *zelayensis* Farw. Pap. Michigan Acad. Sci. 26(1): 4. 1941. *E. crusgalli* subsp. *zelayensis* Shinners, Rhodora 56: 33. 1954. (*Humboldt & Bonpland,* "prope Zelaya, Queretaro et Patzcuaro.")

E. muricata var. *microstachya* Wieg. Rhodora 23: 58. 1921. *E. microstachya* Rydb. Brittonia 1: 82. 1931. *E. pungens* var. *microstachya* Fern. & Griscom, Rhodora 37: 137. 1935. *E. crusgalli* var. *microstachya* Shinners, Rhodora 56: 34. 1954. (*Palmer 97,* Cayuga Lake Basin, Tompkins Co., N.Y.)

Barnyard grass.

Stout, tufted annuals; culms (1.5) 3–10 (14) dm tall, geniculate and decumbent below, ascending above, glabrous, leafy; sheaths glabrous; ligules lacking; blades flat, 4–12 (16) mm broad, a few papillae on the margins near the collar, these sometimes with hairs, the margins scabrous; panicles 5–21 (27) cm long, erect or nodding, consisting of 5–12 (20) racemosely arranged lateral branches; spikelets 2.8–3.5 mm long, somewhat turgid to more or less laterally compressed; first glume 1.2–1.6 mm long, broad, clasping, scaberulous, apiculate to caudate, sometimes with an awn up to 0.5 mm long, 3-nerved; second glume 2.5–3.2 mm long, 5-nerved, the lateral nerves pectinate near the margins and scabrous below, acuminate to awned, the awn 0.7–1.5 mm long; sterile lemma terminating in an awn from 0.5 mm to over 30 mm long, usually with a well developed palea; fertile lemma 2.2–3 (3.3) mm long, plano-convex, abruptly acute, indurate, smooth and shiny, 5- to 9-nerved, awned up to 0.5 mm long, its palea similar in texture; lodicules about 0.25 mm long; anthers 0.5–0.8 mm long; $2n = 36, 48, 54, 72$.

Seepages, banks of irrigation ditches and moist waste places; B.C. to N.S., and s. throughout the U.S.; also trop. and subtrop. regions of both hemispheres. Aug.–Sept.

Some forms of barnyard grass are cultivated in Asia and Africa, where the seeds are harvested for human food. Varieties *mitis* (Pursh) Peterm. and *zelayensis* (H.B.K.) A. S. Hitchc. have been recognized as awnless or mucronate forms of the species. Variety *mitis* has been distinguished from var. *zelayensis* by its denser, more spreading racemes and less strongly hispid spikelets. The presence of all degrees of intergradation and lack of geographical or ecological distinction between these forms suggest that they are not worthy of taxonomic recognition.

49. SETARIA Beauv. nom. conserv. Bristlegrass

Annuals and perennials; culms usually geniculate, decumbent to erect, often branching at the base, solid or hollow; sheaths often keeled, open; ligule a ciliate membrane or a fringe of hairs; blades flat or folded, non-auriculate; inflorescence a slender, compact, usually cylindrical, spike-like, bristly panicle; spikelets somewhat dorsiventrally compressed to subterete, subsessile, unawned, 2-flowered, the first floret sterile or staminate, the second fertile, subtended by 1 to several, persistent, scabrous or scabrous-barbellate bristles (reduced branchlets), disarticulating below the glumes and above the bristles; glumes acute or obtuse, glabrous, unequal, prominently nerved, the first glume usually small, (1-) 3- (5)-nerved, the second 1/2 to as long as the sterile lemma, 5- to 7- (9)-nerved; sterile (or staminate) lemma similar in texture to the second glume, several-nerved, usually not quite as long as the fertile lemma, its palea obsolete to well developed; fertile lemma firm or hard, indistinctly nerved, rounded on the back, rounded at the apex, usually faintly or coarsely transverse-rugose corrugated, rarely smooth and shiny, the margins revolute and clasping the palea; palea of the fertile floret similar to the lemma in texture; lodicules 2, cuneate-flabellate, the upper margin irregular; stamens 3; $x = 9$.

A genus of some 125 species of temp., subtrop., and trop. regions of the world, most abundant in trop. Africa. There are 43 species in N. Amer., 12 of which are introduced from the Old World and S. Amer. (From the Latin *seta,* a bristle, in reference to the bristle-like sterile branches of the inflorescence.)

Reference: Rominger, J. M. Taxonomy of *Setaria* (Gramineae) in North America. Illinois Biol. Monogr. 29: 1–132. 1962.

1 Second glume as long or nearly as long as the spikelet; spikelets 2–2.7 mm long, each subtended by 1–3 (4) bristles; sheath margins ciliate towards the summit.

DICHANTHELIUM
ECHINOCHLOA

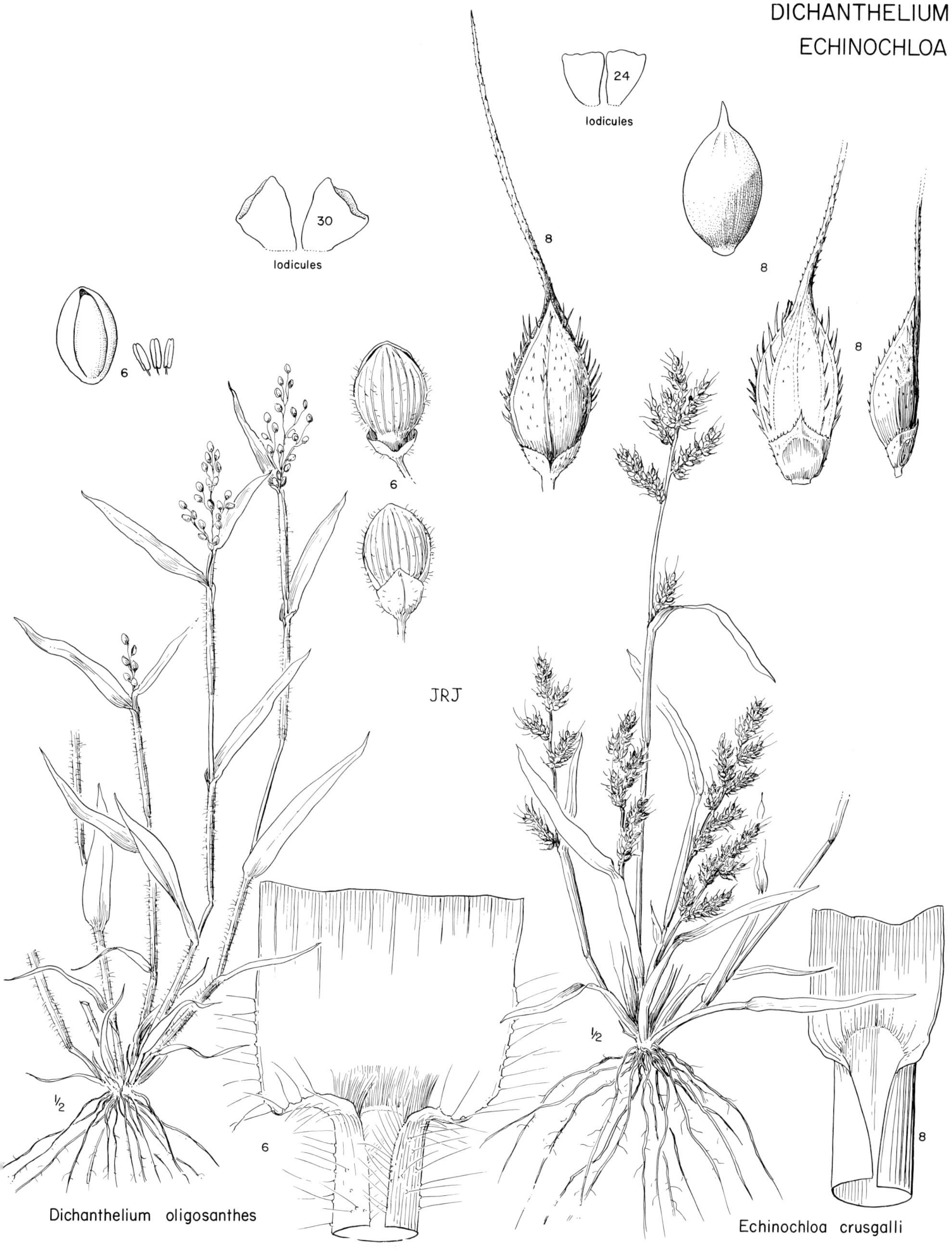

Dichanthelium oligosanthes

Echinochloa crusgalli

2 Bristles antrorsely scabrous, (4) 5.5–11 mm long; panicle branches not verticillate .... 1. *S. viridis*
2 Bristles retrorsely scabrous, 2–4.5 (6) mm long; panicle branches verticillate .... 2. *S. verticillata*
1 Second glume 1/2 the length of the spikelet; spikelets 2.8–3.4 mm long, each subtended by 5–12 bristles; sheath margins glabrous .... 3. *S. glauca*

1. Setaria viridis (L.) Beauv.

Panicum viride L. Syst. Nat. ed. 10. 2: 870. 1759. *Pennisetum viride* R. Br. Prodr. Fl. Novae Holl. & Insulae van-Diemen 1: 195. 1810. *Setaria viridis* Beauv. Essai Nouv. Agrostogr. 51, 171, 178. 1812. *Panicum italicum* var. *viride* Koern. in Koern. & Wern. Handb. Getriedebau 1: 277. 1885. *Chamaeraphis italica* var. *viridis* Kuntze, Revisio Gen. Pl. 2: 767. 1891. *Chamaeraphis viridis* Millsp. West Virginia Agric. Exp. Sta. Bull. 2: 466. 1892. *Ixophorus viridis* Nash, Bull. Torrey Bot. Club 22: 423. 1895. *Chaetochloa viridis* Scribn. U.S.D.A. Div. Agrostol. Bull. 4: 39. 1897. *S. italica* subsp. *viridis* Thell. Mém. Soc. Sci. Nat. Cherbourg 38: 85. 1912. *Setariopsis viridis* G. Samp. Herb. Portug. Seg. Apênd. Lista Espéc. 4. 1914. ("Habitat in Europa australi.")

Green bristlegrass.

Tufted annuals; culms solid, 1–5.5 (10) dm tall, decumbent and branched at the base and ascending above, glabrous, leafy; sheaths keeled, only slightly compressed, the upper margins densely ciliate, sometimes a line of hairs across the collar, otherwise glabrous; ligule a dense fringe of hairs, 1–2 mm long; blades flat or folded, (3) 4–10 (14) mm broad, scaberulous above and scabrous on the margins; panicles 2–7 (9) cm long, densely flowered, cylindric, greenish, the axis densely hispid, the branches short, each spikelet subtended by 1–3 (4) bristles, the bristles (4) 5.5–11 mm long, antrorsely scabrous; spikelets 2–2.7 mm long, plano-convex; first glume (0.7) 1–1.2 mm long, about 1/3 the length of the spikelet, ovate, obtuse, 1- to 3-nerved; second glume 2–2.4 mm long, subequal to the sterile lemma, ovate, rounded, glabrous, 5- to 7-nerved; sterile lemma 2–2.5 (2.7) mm long, ovate, the edges embracing the fertile floret, 5- to 7-nerved, with a small hyaline palea; fertile lemma 1.8–2.5 mm long, ovate, weakly 3-nerved, finely transverse-rugose, its palea similar in texture; lodicules about 0.2 mm long; anthers 0.5–0.7 mm long; $2n$ = 18, 35, 36.

Moist roadsides, irrigated fields and gardens; introduced from Europe; B.C. to Newfl. and throughout the U.S.; in temp. regions of the world. June–Sept.

2. Setaria verticillata (L.) Beauv.

Panicum verticillatum L. Sp. Pl. ed. 2. 1: 82. 1762. *Pennisetum verticillatum* R. Br. Prodr. Fl. Novae Holl. & Insulae van-Diemen 1: 195. 1810. *Setaria verticillata* Beauv. Essai Nouv. Agrostogr. 51, 171, 178. 1812. *Chamaeraphis italica* var. *verticillata* Kuntze, Revisio Gen. Pl. 2: 768. 1891. *Chamaeraphis verticillata* T. C. Porter, Bull. Torrey Bot. Club 20: 196. 1893. *Ixophorus verticillatus* Nash, Bull. Torrey Bot. Club 22: 422. 1895. *Chaetochloa verticillata* Scribn. U.S.D.A. Div. Agrostol. Bull. 4: 39. 1897. *Setariopsis verticillata* G. Samp. Herb. Portug. Seg. Apênd. List. Espéc. 4. 1914. ("Habitat in Europa australi & Oriente.")

Bur bristlegrass.

Tufted annuals; culms (1.5) 3–10 dm tall, the lower nodes often geniculate, rooting and often branched and ascending to erect above, glabrous except above near the panicle, leafy; sheaths compressed-keeled, the upper margins ciliate, otherwise glabrous; ligule a dense fringe of hairs, 0.7–2 mm long; blades flat, 5–10 (13) mm broad, glabrous to scaberulous, sometimes sparsely pilose near the base on the upper surface, scabrous on the margins; panicles (2) 4–10 (12) cm long, densely flowered, cylindric, yellowish-green to purplish, the axis densely retrorsely hispidulous, the branches short, verticillate, each spikelet subtended by 1 (2) bristle(s), the bristles 2–4.5 (6) mm long, retrorsely scabrous; spikelets 2–2.5 mm long, glabrous, plano-convex; first glume 0.7–1 (1.2) mm long, about 1/3 the length of the spikelet, broadly ovate, obtuse, sometimes mucronate tipped, 1- (3)-nerved; second glume 1.6–2 mm long, slightly shorter than the sterile lemma, ovate, rounded, 5- to 7-nerved; sterile lemma 1.9–2.4 mm long, 7-nerved, rounded, with a small membranous palea; fertile lemma 1.8–2.3 mm long, ovate, weakly 3-nerved, finely transverse-rugose, its palea similar in texture; lodicules about 0.25 mm long; anthers about 0.5 mm long; $2n$ = 18, 36, 54.

A weed in greenhouses, gardens and moist waste places; introduced from Europe; s. B.C. to Mass., s. nearly throughout the U.S. July–Sept.

3. Setaria glauca (L.) Beauv.

Panicum glaucum L. Sp. Pl. 56. 1753. *Pennisetum glaucum* R. Br. Prodr. Fl. Novae Holl. & Insulae van-Diemen 1: 195. 1810. *Setaria glauca* Beauv. Essai Nouv. Agrostogr. 51, 169, 178. 1812. *Chamaeraphis glauca* Kuntze, Revisio Gen. Pl. 2: 767. 1891. *Ixophorus glaucus* Nash, Bull. Torrey Bot. Club 22: 423. 1895. *Chaetochloa glauca* Scribn. U.S.D.A. Div. Agrostol. Bull. 4: 39. 1897. *Setariopsis glauca* G. Samp. Herb. Portug. Seg. Apênd. List. Espéc. 4. 1914. ("Habitat in India.")

Panicum lutescens Weigel, Observ. Bot. 20, tab. 2, fig. 1. 1772. *Chaetochloa lutescens* Stuntz, U.S.D.A. Bur. Pl. Industr. Invent. Seeds 31: 36, 86. 1914. *Setaria lutescens* Hubbard, Rhodora 18: 232. 1916. (Germany.)

Yellow bristlegrass.

Tufted annuals; culms 3–8 (12.5) dm tall, the lower nodes geniculate, often rooting and branching and ascending to prostrate above, leafy; sheaths compressed-keeled, glabrous, the margins scarious; ligule a densely ciliate membrane, 0.2–1.2 mm long; blades flat, 3–8 (10) mm broad, the upper surface glabrous or sometimes scabrous or pilose and usually papillose-pilose near the throat, the lower surface glabrous; panicles 3–10 (20) cm long, densely flowered, cylindric, yellowish, the axis and branches densely hirsute, each spikelet subtended by (5) 6–10 (12) bristles, the bristles irregular in length, 3.5–9 mm long, antrorsely scabrous; spikelets 2.8–3.4 mm long, plano-convex; first glume (0.7) 1–1.7 mm long, about 1/3 the length of the spikelet, broadly ovate, mucronate, 3 nerved; second glume (1.2) 1.4–2.2 mm long, about 1/2 the length of the spikelet, ovate, acute, apiculate, the margins scarious, 5- (9)-nerved; sterile lemma 2.2–3 mm long, ovate, apiculate, the margins embracing the fertile lemma, 5-nerved, sometimes staminate with a well-developed palea;

SETARIA

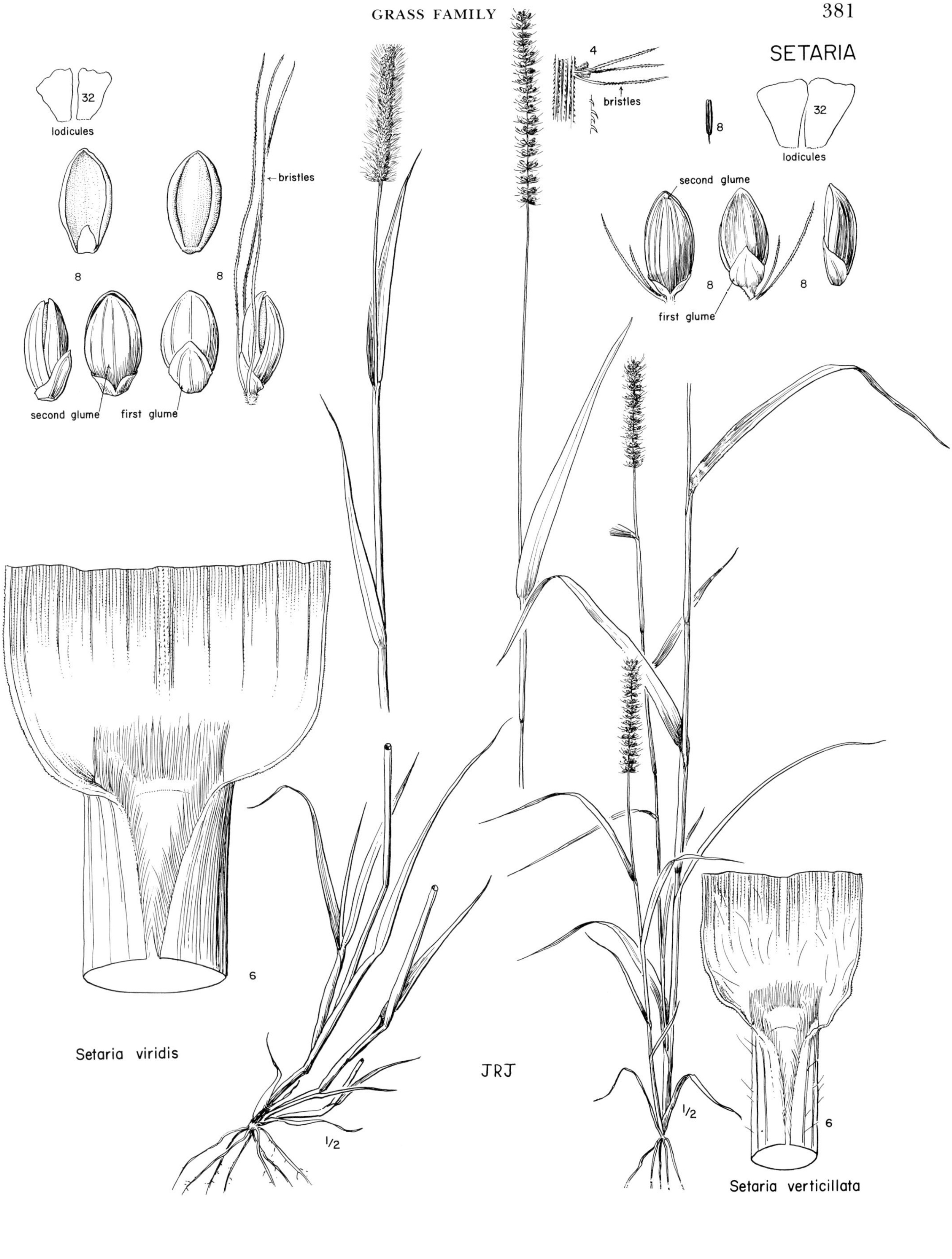

Setaria viridis

Setaria verticillata

fertile lemma 2.3–3.2 mm long, ovate, acute, sometimes with 3 faint ridges on the back terminating as 3 apical lumps, coarsely transversely corrugated, its palea similar in texture; lodicules about 0.25 mm long; anthers 0.5–1.5 mm long; caryopsis plano-convex; $2n = 18, 36, 72$.

A weed along ditchbanks, moist roadsides and cultivated areas in lower valleys; introduced from Europe; s. B.C. to N.S. and throughout the U.S. and common in temp. regions of the world, infrequent in trop. regions. July–Sept.

There are conflicting viewpoints as to which name should apply to yellow bristlegrass, *Setaria lutescens* (Weigel) Hubbard or *S. glauca* (L.) Beauv. Chase (Amer. J. Bot. 8: 41–49. 1921) suggested that Linnaeus' description for the older epithet "*glauca*" was based on what is now considered a species of *Pennisetum*. Linnaeus' somewhat ambiguous description in Species Plantarum referred to what are now regarded as about 5 different species. Reeder (Rhodora 53: 27–30. 1951) points out that Linnaeus, in his 10th edition of Systema Naturae, narrowed down his circumscription of the binomial to apply only to yellow bristlegrass. Reeder also points out that Weigel did not use the epithet "*lutescens*" in a combination with a genus name which makes the name invalid in accordance with article 33 of the International Code of Botanical Nomenclature.

50. CENCHRUS L. Sandbur

Annuals and perennials; culms usually geniculate, decumbent or sometimes erect, solid; sheaths compressed-keeled, open; ligule a fringe of hairs; blades flat or folded, non-auriculate; inflorescence a highly modified panicle, the spikelets borne singly or in clusters and partly enveloped by a subtending spiny involucre (bur) of basally fused bristles and/or flattened spines (modified branches) and disarticulating as a unit, the spines often retrorsely barbed, the burs sessile or subsessile on a short rachis; spikelets more or less turgid or somewhat dorsally compressed, 2-flowered, the first floret sterile or staminate, the second fertile; glumes usually unequal, the first glume small, 1- or 3-nerved, the second larger, 1- to 7-nerved; sterile (or staminate) lemma similar in size and texture to the second glume, 1- to 7-nerved, its palea subequal to the lemma; fertile lemma usually indurate, smooth, shiny or dull, 3- to 5-nerved, tapering to a slender, acuminate tip, awnless, rounded on the back, the margins revolute and clasping the palea; palea of the fertile floret similar to the lemma in texture; lodicules lacking; stamens 3; caryopsis elliptic to ovoid and dorsally flattened; $x = 9$.

A genus of 20 species, mostly American, but a few in s. Europe and warm regions of Africa, Asia, Australia and the S. Pacific. There are 6 species in the U.S., 1 of which is introduced. (From the Greek name, *kenchros,* for some millet.)

Reference: DeLisle, D. G. Taxonomy and distribution of the genus *Cenchrus*. Iowa State Coll. J. Sci. 37: 259–351. 1963.

1. Cenchrus longispinus (Hackel) Fern.

Cenchrus echinatus f. *longispinus* Hackel in Kneucker, Allg. Bot. Z. Syst. 9: 169. 1903. *C. longispinus* Fern. Rhodora 45: 388. 1943. *C. pauciflorus* var. *longispinus* Jansen & Wachter in Jansen, Ned. Kruidk. Arch. 56: 246. 1949. (*Kneucker 426,* Oxford, Conn., in 1903.)

C. pauciflorus of authors; not Benth. 1844.

C. carolinianus of authors; not Walter, 1788.

Sandbur.

Loosely tufted, often prostrate and mat forming annuals; culms 1–4 (9) dm long, branched at the base, sometimes rooting at the nodes, geniculate, glabrous, leafy; sheaths compressed-keeled, the collar sometimes puberulent, otherwise glabrous, the margins scarious, sometimes extended as membranous auricles, sometimes diverging from the culm; ligule a dense fringe of hairs, 0.5–1.5 mm long, the marginal hairs sometimes up to 2.5 mm long; blades flat to folded, 3–6 (8) mm broad, scabrous, at least above; panicle axis 3–6 (10) cm long, bearing 4–12 (25) burs, the rachis geniculate, flattened and angled, scabrous to glabrous; bur urceolate to globose, the body 3–5 (6) mm broad when pressed, enclosing (1) 2 (3) sessile spikelets and covered with 45–55 (75) spines, the upper spines large, flat, spreading to ascending, 3.5–7 mm long, the lower spines more numerous, smaller, pointing down or curved up, some of them retrorsely scabrous, the body of the bur and base of the spines hispid; spikelets 5.5–7.5 mm long, the glumes and lemmas glabrous, sometimes minutely scaberulous towards the apex, plano-convex; first glume 1.5–4.2 mm long, narrow-lanceolate, 1- (3)-nerved; second glume 4.2–5.8 mm long, broadly lanceolate, acuminate, 3- to 5-nerved; sterile lemma 4.7–6.8 mm long, broadly lanceolate, acuminate, 3- to 7-nerved, its palea subequal, narrow-lanceolate and puberulent; fertile lemma 5.2–7 (7.5) mm long, broadly lanceolate, acuminate, faintly 5-nerved, the median-lateral nerves prominent below and the mid and marginal-lateral nerves prominent above, smooth, dull, not shiny, its palea similar in texture; lodicules lacking; anthers 0.6–2 mm long; caryopsis 2.7–3.8 mm long; $2n = 34, 36$.

A weed in disturbed areas, often in sandy soil; Ont., s. throughout most of the U.S., but most common in the c. and e. parts; s. to Mex., C. Amer., W.I., S. Amer. and warm parts of the Old World. July–Sept.

The sharp, spiny burs make this grass an obnoxious weed. The retrorsely barbed spines are extremely sharp and are hard to extract after puncturing the skin. The end may break off and cause infection.

Tribe 7. ANDROPOGONEAE

Perennials or only a few annuals in our flora; culms solid, sometimes hollow, often branched above; sheath margins free, prolonged at the top into erect sheath auricles in *Sorghastrum;* ligule membranous, a fringe of hairs, or absent; blades non-auriculate; inflorescence a panicle, diffusely branched or of unbranched spicate-racemose branches digitately or racemosely arranged, the spikelets borne in pairs of 1 sessile and 1 pedicelled (2 pedicelled spikelets at the branch tips), the pedicelled spikelet usually reduced to

30
lodicules
bristles
8
8
8
first glume
second glume
6
Setaria glauca
1/2
JRJ
3
3
involucre removed
spikelets
3
involucre (bur)
6
1/2
Cenchrus longispinus

sometimes absent; spikelets dorsally compressed, (1-) 2-flowered, the lower floret usually sterile, in the monoecious genus *Zea* the staminate and pistillate spikelets strikingly different; disarticulation below the glumes or at the rachis nodes with the spikelet pair falling as a unit that includes the pedicel and rachis segment; glumes large, indurate (except in *Imperata*), enclosing the florets; lemmas of both sterile and fertile florets membranous, hyaline, often reduced, the lemma of the fertile floret sometimes consisting of no more than a geniculate awn; palea reduced or absent; lodicules 2, thick, truncate, vasculated; embryo panicoid; x = 5, 9 or 10.

The Andropogoneae has been expanded to include the Tripsaceae (*Zea*) of A. S. Hitchcock (Manual Grasses U.S., 1951), and the sections of *Andropogon* have been elevated to genera.

Zea mays L., the cultivated corn, does not survive in our area outside of cultivation. It is an important and familiar crop, however, and is treated with a species description and discussion following *Bothriochloa*. *Erianthus ravennae* (L.) Beauv., ravenna grass, is sometimes planted in the region for ornament. This native of southern Europe grows up to 4 m tall with silky panicles as long as 60 cm. It is characterized by having spikelets all alike, arranged in pairs, 1 sessile and 1 pedicelled and the pair disarticulating as a connected unit. The glumes have a tuft of long, spreading, silky hairs at the base. *Coix lacryma-jobi* L., Job's tears, a native of the tropics, is sometimes grown for dry bouquets. It is a coarse annual with peculiar monoecious inflorescences. The pistillate spikelet is enclosed by a hard, globular involucre formed from highly modified leaf sheaths. These white to bluish-gray involucres are often used as beads.

1 All spikelets unisexual, the staminate spikelets borne in terminal panicles (tassels) and the pistillate spikelets in highly modified lateral spikes (cobs); cultivated corn (treated following 56. *Bothriochloa*) . . . . *Zea mays* L.
1 At least some of the spikelets perfect.
  2 Fertile spikelets 2.4–3.5 mm long, awnless; pubescence of the panicle of long, silky hairs, the hairs 8–11 mm long . . . . 51. *Imperata*
  2 Fertile spikelets (3.8) 4–12 mm long, awned (sometimes awnless in *Andropogon hallii*); pubescence of the inflorescence (rachis joints and pedicels) of hairs 0.5–5 (9) mm long.
    3 Inflorescence an open or contracted panicle; hairs of the rachis joints and pedicels 0.5–1.2 mm long.
      4 Spikelets in 2s or 3s, 1 sessile, fertile and 1 or 2 pedicellate, sterile or staminate; upper margins of the sheath not auricled . . . . 52. *Sorghum*
      4 Spikelets single, sessile, with empty pedicels present; upper margins of the sheath projected into sheath auricles . . . . 53. *Sorghastrum*
    3 Inflorescence of 1 to several spicate-racemes (with both sessile and pedicellate spikelets present); hairs of the rachis joints and pedicels 1–9 mm long.
      5 Flowering culms branched into 2 or more short, leafy branchlets, each bearing 1–6 racemes; awns less than 20 mm long, sometimes absent; nodes glabrous to pubescent.
        6 Racemes 2–6 on each peduncle; spikelets (6) 7–12 mm long . . . . 54. *Andropogon*
        6 Racemes solitary on each peduncle; spikelets 4–6.5 (7.5) mm long . . . . 55. *Schizachyrium*
      5 Flowering culms unbranched, terminating as a panicle bearing more than 6 racemes; nodes distinctly bearded . . . . 56. *Bothriochloa*

51. IMPERATA Cyr.

Erect, slender, rhizomatous perennials; culms erect; sheaths open; ligules membranous; blades flat, nonauriculate; inflorescence a relatively long, slender, spike-like panicle with appressed branches, the spikelets unequally pedicellate on slender, continuous rachises, obscured by long silky hairs which impart a silvery appearance to the panicle; spikelets 1-flowered, dorsally compressed, all alike, disarticulating below the glumes, the spikelets falling individually from a persistent rachis; glumes subequal, thin, several-nerved, clasping and permanently enclosing the second glume and florets; lemma membranous and hyaline, usually greatly reduced or absent; palea of the fertile floret small, membranous; stamens 1 or 2; stigmas plumose on a long style; x = 10.

A genus of 7 to 10 species of warmer parts of the world. (Named in honor of Ferrante Imperate, a Neapolitan pharmacist who lived in the 16th century.)

1. Imperata brevifolia Vasey

Imperata arundinacea subsp. *hookeri* Rupr. ex Anderss. Öfvers. Förh. Kongl. Svenska Vetensk.-Akad. 12: 160. 1855. *I. hookeri* Rupr. ex Hackel in A. & C. DC. Monogr. Phanerog. 6: 97. 1889. (*Drummond 283*, Texas.)

I. brevifolia Vasey, Descript. Cat. Grasses U. S. 24. 1885, nomen nudum; Bull. Torrey Bot. Club 13: 26. 1886. (*S. B. & W. F. Parish 1031*, "wet soils, San Bernardino Valley," Calif., 15 August 1881.)

Satintail.

Slender perennials from hard, scaly rhizomes; culms erect, 7–15 dm tall, unbranched, smooth, terete, leafy; leaves rather rigid, 4 or 5, reduced upward; sheaths smooth, longer than the internodes; ligules membranous, about 1 mm long, truncate, ciliate; blades flat, (4) 6–9 mm broad and 10–25 cm long, glabrous, except for the long hairs at the base behind the ligule, the hairs 6–7 mm long, somewhat scabrous margined; panicles 12–25 cm long and 1.5–2.5 cm broad, the stigmas and anthers adhering to the velvety hairs imparting a brown, speckled ap-

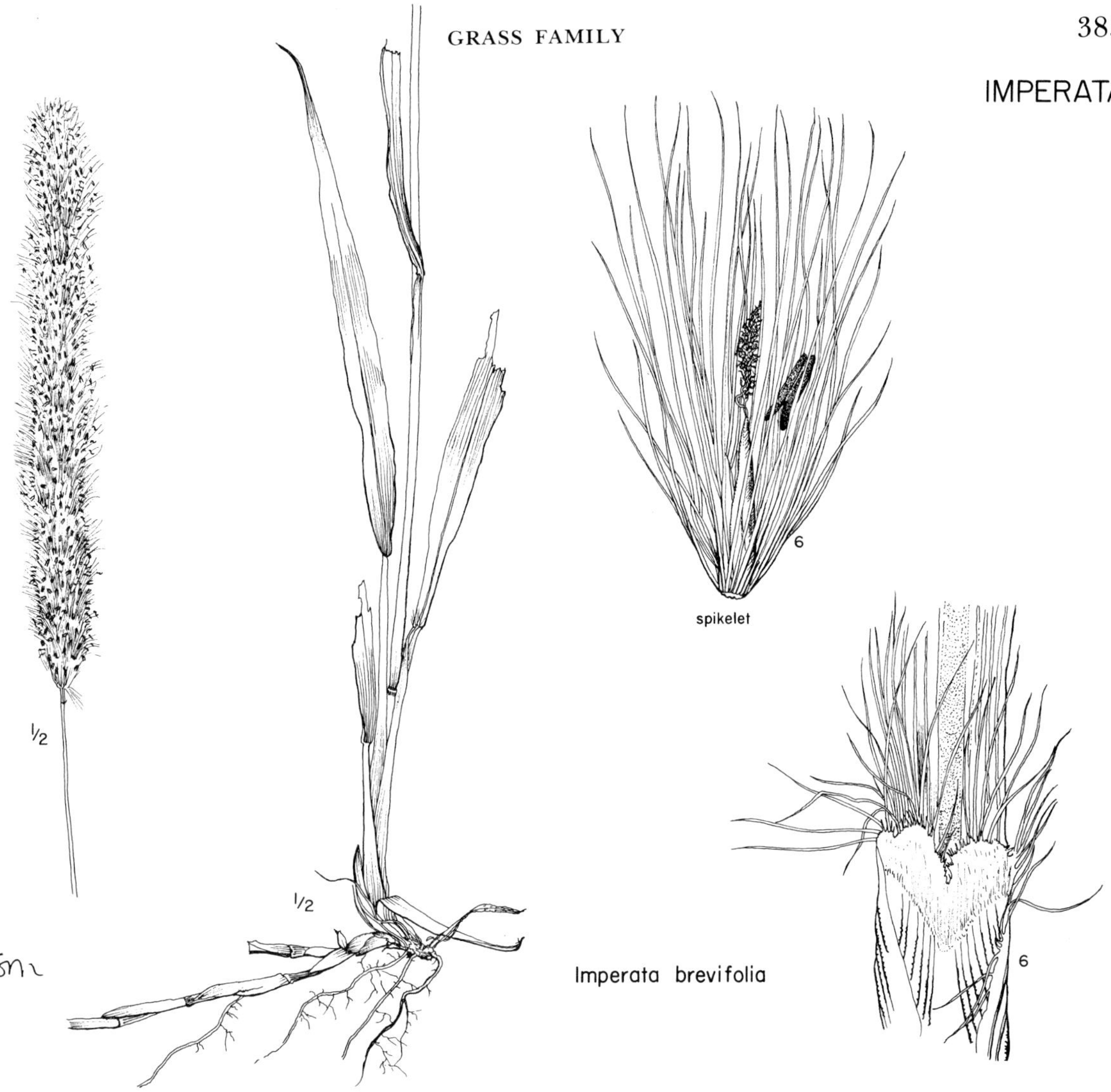

Imperata brevifolia

pearance to the silver pubescence, the branches in 3s or 4s; spikelets all alike, subdorsally compressed, 1-flowered, awnless, unequally pedicelled, silky pubescent with hairs 8–11 mm long; glumes unequal, the first glume 2.4–3 mm long, broadly lanceolate, often blunt and finely ciliate at the apex, more or less flat, the hairs mostly marginal and at the base, membranous on the margins and apex, the second glume 2.7–3.5 mm long, lanceolate, narrowly rounded, 5-nerved, keeled with long hairs on the keel; lemma 1.5–3 mm long, lanceolate, acute, hyaline, translucent, 1-nerved; palea irregularly dentate or lacerate; anthers 1.4–2 mm long; style elongate and long exserted; caryopsis obovate.

Moist to wet sandy places around springs or along watercourses in arid and hot places; s. Calif. (Tulare Co., s.), s. Nev., s. Utah, s. and e. to Ariz., N.M., Trans-Pecos Texas and Chihuahua. Aug.–Sept.

Satintail grass should be a beautiful ornamental in the southern part of our area, and the silvery inflorescence would be choice material for dry bouquets.

52. SORGHUM Moench, nom. conserv.

Robust annuals and perennials; culms usually stout and tall, solid or sometimes hollow below; sheaths open; ligules membranous and fimbriate; blades flat to folded, broad or narrow, long, non-auriculate; inflorescence a large, open or contracted panicle of numerous branches, the spikelet clusters on short racemose branchlets, in 3s at the tip with 1 sessile and 2 pedicelled and in pairs below with 1 sessile and 1 pedicelled; spikelets 2-flowered, somewhat dorsiventrally compressed, the sessile spikelets perfect and the pedicelled spikelets staminate, usually disarticulating as a unit consisting of the spikelet pair and a section of the rachis; glumes subequal, the first large, coriaceous, nerveless, shiny, clasping or enclosing the second glume, awnless; lemmas of the sterile and fertile florets membranous, hyaline, reduced, sometimes the fertile lemma with a geniculate and twisted, readily deciduous awn arising dorsally; palea of the fertile floret membranous; lodicules 2, cuneate, finely many-veined, delicately long-pilose from the upper corners; stamens 3; $x = 5$.

A genus of about 35 species, mostly of warmer parts of Africa, 2 species native to Mex. and C. Amer. (Possibly from the Italian vernacular name for the plant, *sorgho.*)

1 Rhizomatous perennials; leaf blades 4–15 (25) mm broad; sessile spikelet enclosing the mature caryopsis; terminal segment of the awn 4–7 mm long . . . . 1. *S. halepense*
1 Robust annuals; leaf blades (12) 20–52 mm broad; sessile spikelet usually spread apart by the caryopsis at maturity; terminal segment of the awn 2–4 mm long . . . . 2. *S. bicolor*

1. Sorghum halepense (L.) Pers.

Holcus halepensis L. Sp. Pl. 1047. 1753. *Blumenbachia halepensis* Koeler, Descript. Gram. Gallia & German. 29. 1802. *Milium halepense* Cav. Descripcion Pl. 306. 1802. *Andropogon halepensis* Brot. Fl. Lusit. 1: 89. 1804. *Sorghum halepense* Pers. Syn. Pl. 1: 101. 1805. *Sorghum halepense* var. *genuinum* Hackel, Martii Fl. Brasil. 2(3): 272. 1883. *Andropogon sorghum* subsp. *halepensis* Hackel in A. & C. DC. Monogr. Phanerog. 6: 501. 1889. *Sorgum saccharatum* var. *halepense* Kuntze, Revisio Gen. Pl. 3(2): 368. 1898. ("Habitat in Syria, Mauritania.")

Johnson grass.

Perennials from stout, scaly rhizomes; culms 5–15 dm tall, sometimes branched, leafy, the nodes sometimes puberulent; sheaths glabrous, sometimes puberulent across the collar; ligule a densely ciliate membrane, 1.5–3 (5) mm long, the cilia about 1/3 the length; blades flat, 4–15 (25) mm broad; panicles 15–35 (50) cm long, open, the pedicel of the pedicelled spikelet 1.5–3 (4) mm long with hairs 0.5–0.7 mm long; pedicellate spikelet staminate, usually slightly longer than the sessile spikelet, 4–6.5 (7) mm long, broadly lanceolate; glumes of the pedicellate spikelet subequal, the first glume 9-nerved, flat to sulcate on back, the lateral nerves scabrous, the second glume 7-nerved, somewhat keeled or rounded on back, glabrous or the nerves scabrous or appressed-hirsute; sessile spikelet 4–5.5 (6.2) mm long, ovate; glumes of the sessile spikelet subequal, the first glume smooth and shiny on back, hispidulous on the margins or sometimes all over, 5- to 7-nerved, the second glume glabrous, smooth and shiny below and hispidulous towards the apex, the margins inrolled and clasping the fertile lemma, 3-nerved; sterile lemma 4–4.5 mm long; fertile lemma much reduced but with a stout, twisted, geniculate awn, the nerves sometimes more prominent apically and marginally, the awn 9–12 (15) mm long with a terminal segment 4–7 mm long; lodicules about 0.5 mm long; anthers 2.2–3 mm long; $2n = 20, 33 + f, 39 + f, 40$.

Moist roadsides and ditchbanks; introduced from the Medit. region and naturalized across most of the U.S., but more common in warmer climates. July–Sept.

Although Johnson grass is widely used for forage, the plant produces a prussic acid under certain conditions of growth, which is poisonous to livestock. Johnson grass can also become a troublesome weed along irrigation ditches and cultivated fields.

2. Sorghum bicolor (L.) Moench

Holcus sorghum L. Sp. Pl. 1047. 1753. *Andropogon sorghum* Brot. Fl. Lusit. 1: 88. 1804. *Sorghum vulgare* Pers. Syn. Pl. 1: 101. 1805. *Andropogon vulgaris* Rasp. Ann. Sci. Nat. Bot. 5: 307. 1825. *Sorghum sorghum* Karsten, Deutsche Fl. 367, fig. 189. 1880. *Andropogon sorghum* subsp. *sativus* var. *vulgaris* Hackel in A. & C. DC. Monogr. Phanerog. 6: 515. 1889. *Sorgum saccharatum* var. *vulgare* Kuntze, Revisio Gen. Pl. 2: 793. 1891. ("Habitat in India.")

Holcus bicolor L. Mantissa Pl. 2: 301. 1771. *Sorghum bicolor* Moench, Meth. Pl. Horti Bot. & Agri Marburg. 207. 1794. *S. vulgare* var. *bicolor* Pers. Syn. Pl. 1: 101. 1805. *Andropogon bicolor* Roxb. Hort. Bengalensis 7. 1814. *Andropogon sorghum* var. *bicolor* Koern. & Wern. Handb. Getreidebau. 1: 313. 1885. *Andropogon sorghum* subsp. *sativus* var. *bicolor* Hackel in A. & C. DC. Monogr. Phanerog. 6: 519. 1889. *S. vulgare* subsp. *bicolor* Maire & Weiller, Fl. Afrique N. 1: 270. 1952. ("Habitat in Persia.")

Sorghum.

Robust annuals; culms 10–30 dm tall; sheaths glabrous; ligules membranous, 1.5–5.5 mm long; blades flat, (12) 20–40 (52) mm broad; panicles 10–25 (35) cm long, very dense and compact, the pedicel of the pedicellate spikelet 0.7–2 mm long with hairs 0.5–0.7 mm long; pedicellate spikelet usually sterile, rarely staminate, usually slightly shorter than the sessile spikelet, 3.8–5 (6) mm long, narrow-lanceolate; glumes of the pedicellate spikelet subequal, the first glume 9- to 11-nerved, flat to sulcate on back, clasping the second glume, the second glume narrow, 3-nerved, appressed-hirsute near the margins; sessile spikelet perfect, 4–6 (6.5) mm long, broadly ovate, acute; glumes of the sessile spikelet subequal, the first glume rounded across the back, becoming shiny black, usually appressed-hirsute, the 9 to 11 nerves prominent at the apex, the second glume narrower, slightly keeled, 3-nerved at the apex; sterile lemma 3.8–5 mm long; fertile lemma 3.9–4.3 mm long, bearing a geniculate awn 5–8.5 (11) mm long, the terminal segment short, 2–4 mm long, twisted below; caryopsis filling the spikelet and spreading the glumes apart at maturity; $2n = 20$.

Sometimes escaping cultivation in warm climates in moist waste places; introduced from Eurasia; warm parts of the U.S., found in our region in s. Utah, and cultivated as far n. as the Snake River Plains, Idaho. July–Sept.

The long-used species epithet, *S. vulgare* Pers., was found to be illegitimate according to the International Rules of Botanical Nomenclature by Shinners (Baileya 4: 141–142. 1956). Persoon based the name on *Holcus sorghum* L. while recognizing two named varieties based on older names at the rank of species. These older names are var. *bicolor* from *Holcus bicolor* of Linnaeus in 1771 and var. *rubens* from *Holcus rubens* of Gaertner in 1791. *Sorghum bicolor* (L.) Moench is the oldest epithet, and most botanists agree it is the same species as Linnaeus' *Holcus sorghum*.

53. SORGHASTRUM Nash

Tufted perennials, with or without rhizomes; culms tall; sheaths open, the upper margins projected into prominent sheath auricles; ligule a membranous lobe extending between the sheath auricles; blades flat, non-auriculate; inflorescence a narrow, contracted panicle, the spikelets in pairs, 1 sessile and 1 pedicelled,

SORGHUM

15
lodicules
4
4
4
4
in fruit
4
JRJ
6
Sorghum halepense
1/2
1/2
Sorghum bicolor
WSM

the pedicelled spikelet reduced, sometimes represented only by the slender, hairy pedicel, often in 3s at the tips of the branchlets with 1 sessile and 2 pedicelled; spikelets 2-flowered, dorsiventrally compressed, the sessile spikelet perfect and the pedicelled spikelets sterile, disarticulating as a unit consisting of the spikelet pair and a section of the rachis; glumes subequal, large, firm or hard, elliptical, hirsute, the first glume clasping or enclosing the second glume; lemmas of the sterile and fertile florets membranous, the lemma of the sterile floret absent or membranous and rudimentary, the lemma of the fertile floret membranous with a stout, geniculate and twisted awn; palea of the fertile floret hyaline; $x = 10$.

A genus of about 15 species of warmer regions of Africa and the Americas with 3 species in the U.S. (Named for its resemblance to *Sorghum.*)

1. Sorghastrum nutans (L.) Nash

Andropogon nutans L. Sp. Pl. 1045. 1753. *Sorghum nutans* A. Gray, Manual Bot. 617. 1848. *Chrysopogon nutans* Benth. J. Linn. Soc. Bot. 19: 73. 1881. *Sorghum nutans* subsp. *linnaeanum* Hackel, Martii Fl. Brasil. 2(3): 276. 1883. *Andropogon nutans* var. *linnaeanus* Hackel in A. & C. DC. Monogr. Phanerog. 6: 531. 1889. *Chrysopogon nutans* var. *linneanus* Mohr, Bull. Torrey Bot. Club 24: 21. 1897. *Andropogon linnaeanus* Scribn. & Kearney, in Scribn. & Ball, U.S.D.A. Div. Agrostol. Bull. 24: 40. 1901. *Sorghastrum nutans* Nash in Small, Fl. S.E. U. S. 66, 1326. 1903. *Sorghastrum linnaeanum* Nash in Small, ibid. 66, 1326. 1903, misapplied to *Sorghastrum elliottii* (Mohr) Nash. *Holcus nutans* Kuntze ex Stuck. Anales Mus. Nac. Hist. Nat. Buenos Aires 11: 48. 1904. *Chalcoelytrum nutans* Lunell, Amer. Midl. Naturalist 4: 212. 1915. *Trichachne nutans* Baum, Canad. J. Bot. 45: 1851. 1967. (*Kalm,* "Habitat in Virginia, Jamaica," probably e. Amer.)

Andropogon avenaceum Michx. Fl. Boreali-Amer. 1: 58. 1803. *Sorghum avenaceum* Chapman, Fl. S. U. S. 583. 1860; not Beauv. 1812. *Chrysopogon avenaceus* Benth. J. Linn. Soc. Bot. 19: 73. 1881. *Sorghum nutans* subsp. *avenaceum* Hackel, Martii Fl. Brasil. 2(3): 274. 1883. *Sorghum nutans* subsp. *avenaceum* var. *genuinum* Hackel, ibid. *Andropogon nutans* var. *avenaceus* Hackel in A. & C. DC. Monogr. Phanerog. 6: 530. 1889. *Chrysopogon nutans* var. *avenaceus* Brann. & Cov. Annual Rep. Geol. Surv. Arkansas 1888. 4: 234. 1891. *Sorghastrum avenaceum* Nash in Britton, Manual Fl. N. States and Canad. 71. 1901. *Holcus nutans* var. *avenaceus* Hackel ex Stuck. Anales Mus. Nac. Hist. Nat. Buenos Aires 11: 48. 1904. *Sorghum nutans* subvar. *avenaceum* Roberty, Boissiera 9: 309. 1960. (*Michaux,* "Hab. in vastissimis pratis Illinoensibus," in 1795.)

Indian grass.

Perennials with short, scaly rhizomes; culms 10–25 dm tall, the nodes ascending hirsute; sheaths glabrous to hirsute, the sheath auricles prominent; ligules 1.5–6.5 (9) mm long, more or less as long as the sheath auricles, erose, fairly stiff; blades flat, (3) 6–10 (13) mm broad, scabrous; panicles 15–30 (45) cm long, narrow, rather dense, the branchlets yellowish-villous, the pedicel empty (rarely with a vestigial spikelet), (2.5) 5–5.5 mm long, with hairs 0.5–1.2 mm long; spikelets usually single and sessile, associated with an empty pedicel, 2-flowered, lanceolate; glumes subequal, (5.5) 6–7 (8) mm long, the first glume flat to somewhat rounded across the back, acuminate to a narrowly truncate apex, 5- to 7-nerved above, smooth and somewhat shiny and nerveless below, sparsely scaberulous and sometimes sparsely hispid as well, the second glume somewhat keeled, acuminate, 3-nerved, glabrous or ciliate; sterile lemma 4–5 (6.5) mm long, lanceolate, truncate, ciliate, hyaline to transparent; fertile lemma 4.5–6.5 mm long, transparent-hyaline, ciliate, bifid at the apex and bearing a long geniculate awn (8) 12–15 (20) mm long, twisted below; anthers 2.5–3.5 mm long; $2n = 20, 40, 80$.

Prairies, open woods and dry slopes; Man. to Que., s. throughout the midw. and e. states and as far w. as the Dakotas, e. Wyo., e. and s. Colo., s. Utah and Ariz. and s. to c. Mex. Aug.–Sept.

Indian grass is one of the important components of the mid-western tall-grass prairie and is a good forage plant.

54. ANDROPOGON L. Bluestem

Robust perennials with or without rhizomes; culms stiffly erect, branched, solid; sheaths keeled, open; ligule membranous; blades flat or folded, non-auriculate; inflorescence a panicle, variously branched, the flowering culm terminating in a stiffly erect panicle of 2 to few, more or less digitately arranged spicate branches, or the flowering culms much-branched above and terminating in numerous leafy branchlets bearing 2 to several peduncled flower clusters, the rachis joints mostly pubescent, sometimes with a shallow groove on 1 side, the spikelets in pairs of 1 sessile and 1 pedicelled; spikelets 2-flowered, dorsiventrally compressed, the pedicelled spikelet well developed, reduced or represented by a pedicel only, the sessile spikelet fertile, disarticulating as a unit consisting of the spikelet pair and a section of the rachis; glumes subequal, large, narrow, coriaceous, the first glume flat or slightly concavo-convex, strongly 4-nerved, clasping the second, the second glume sometimes slightly keeled; lemmas of the sterile and fertile florets membranous, hyaline, sometimes greatly reduced, the lemma of the fertile floret narrow, entire or bifid and sometimes bearing a twisted awn from the apex or from between the lobes; palea of the fertile floret small or absent, hyaline; lodicules 2, cuneate-flabellate, finely many-nerved, often with a few long terminal hairs; stamens 1–3; $x = 9, 10$.

A large genus of warm and temp. regions of the world with 17 species in the U.S., most of which center in the southeast. (Name from the Greek *andros,* man, and *pogon,* beard, in reference to the hairy pedicellate spikelet that is usually staminate.)

Reference: Gould, F. W. The grass genus *Andropogon* in the United States. Brittonia 19: 70–76. 1967.

1 Hairs of the rachis joints and pedicels yellowish, (3) 4–5 mm long; awns mostly less than 5 mm long, sometimes absent; plants strongly rhizomatous .... 1. *A. hallii*
1 Hairs of the rachis joints and pedicels whitish or grayish, 1–3 (3.5) mm long; awns (8) 12–20 mm long; plants tufted, sometimes with short rhizomes .... 2. *A. gerardii*

SORGHASTRUM
ANDROPOGON

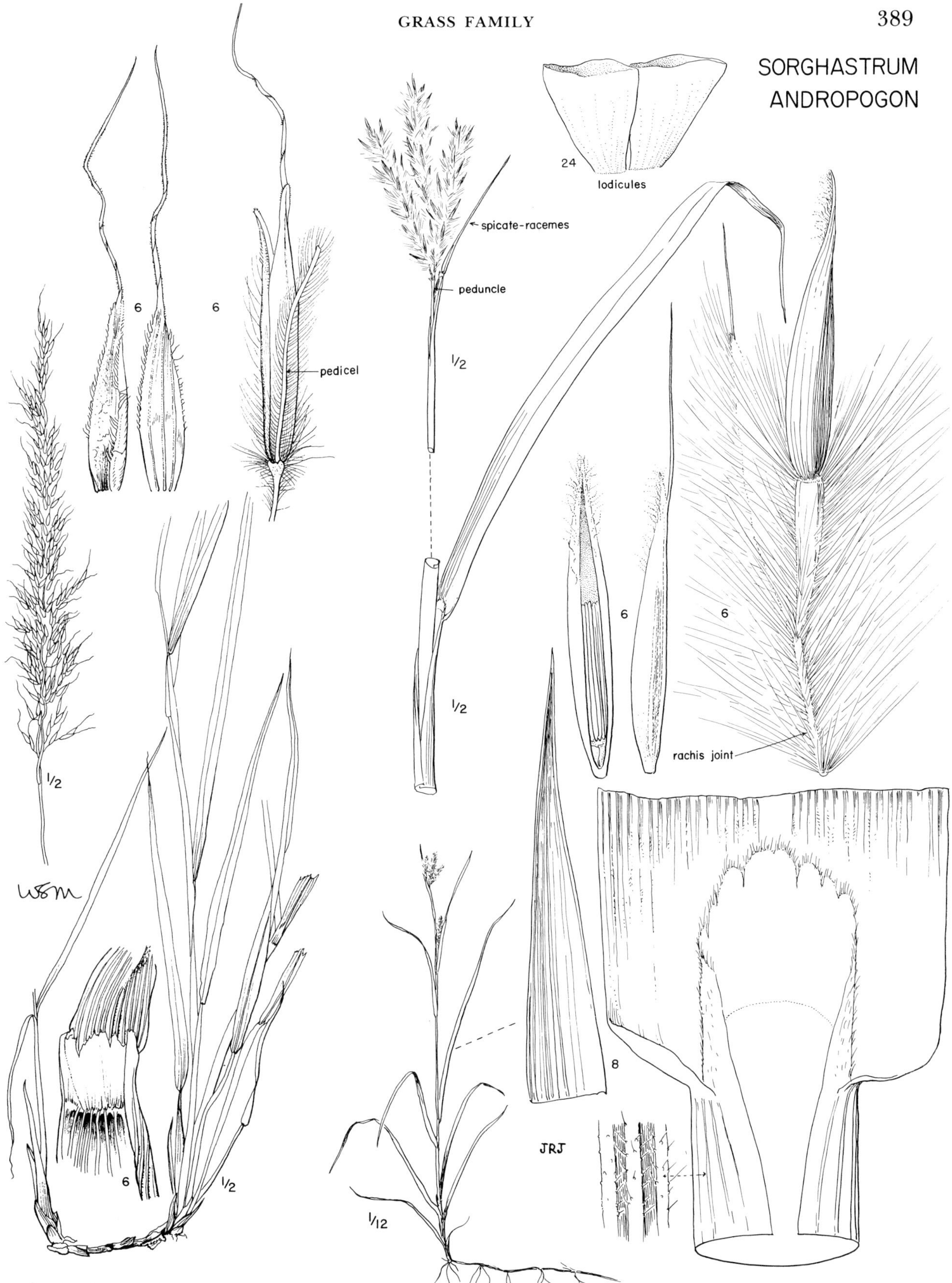

Sorghastrum nutans

Andropogon hallii

1. Andropogon hallii Hackel

Andropogon hallii Hackel, Sitzungsber. Kaiserl. Akad. Wiss., Math.-Naturwiss. Cl., Abt. 1. 89: 127. 1884. *Sorgum hallii* Kuntze, Revisio Gen. Pl. 2: 791. 1891. (*Hall & Harbour 651,* Nebr., in 1862.)

Sand bluestem, turkey-foot.

Stout, glaucous perennials from well developed, creeping rhizomes; culms (5) 10–20 dm tall, glabrous; sheaths smooth and glabrous, sometimes auricled above; ligule 2–3.5 mm long, puberulent, lacerate to sometimes nearly entire, ciliate, decurrent; blades flat or loosely rolled, 3–9 mm broad, glabrous, sometimes scabrous above; panicles of 2–5 approximate or digitate spicate branches, 4–8 (10) cm long, the pedicel of the pedicellate spikelet 3.5–6 mm long with yellowish hairs, the hairs (3) 4–5 mm long; pedicellate spikelet 7–10 (12) mm long, narrow-lanceolate, the glumes subequal with the first glume slightly longer, 5- to 9-nerved, the lateral nerves scabrous, the second glume 3-nerved, slightly keeled; sessile spikelet (7) 8–10 (11) mm long; glumes of the sessile spikelet subequal, the first glume 4-nerved, grooved, hirsute in the groove, the outermost pair of nerves ciliate apically, the second glume slightly shorter, keeled, 3-nerved, ciliate on the margins and keel nerves distally; sterile lemma 3- to 5-nerved, lanceolate, usually slightly longer than the fertile lemma; fertile lemma (5) 6–8 mm long, 1- to 3-nerved, lanceolate, awned, transparent, membranous, the awn mostly less than 5 mm long, sometimes absent; $2n = 60, 70, 100$.

Dry, usually sandy plains and hills or dunes; common in the Great Plains from e. Mont. and the Dakotas to Texas and e. N.M., less common w. to s. Utah and Ariz., s. to Chihuahua. July–Sept.

2. Andropogon gerardii Vitman

Andropogon sp. Gérard, Fl. Gallo-Prov. 107, fig. 4. 1761. *A. provincialis* Lam. Encycl. Méth. Bot. 1: 376. 1785; not Retz. 1783. *A. gerardi* Vitman, Summa Pl. 6: 16. 1792. *A. provincialis* subvar. *genuinus* Hackel in A. & C. DC. Monogr. Phanerog. 6: 442. 1889. *Sorgum provinciale* Kuntze, Revisio Gen. Pl. 2: 792. 1891. (Based on Gérard's diagnosis and figure and a plant in the Botanic Garden in Paris.)

A. furcatus Muhl. in Willd. Sp. Pl. 4: 919. 1806. *A. provincialis* subvar. *furcatus* Hackel in A. & C. DC. Monogr. Phanerog. 6: 442. 1889. ("Habitat in America boreali," probably Pa.)

A. hallii var. *grandiflorus* Scribn. in Rydb. & Shear, U.S.D.A. Div. Agrostol. Bull. 5: 21. 1897. (*Shear 747,* "Mountains near Golden," Colo., 30 Aug. 1895.)

Big bluestem.

Robust, tufted perennials, sometimes producing short rhizomes, often glaucous; culms 5–20 dm tall; sheaths usually glabrous, sometimes villous or papillose-pilose; ligule 1–2.5 mm long, membranous, lanceolate, glabrous, ciliate, erose; blades flat, 3–6 (10) mm broad, sometimes (papillose-) pilose near the throat, glabrous or scabrous; panicles of (2) 3–6 spicate branches, the branches (3.5) 5–10 cm long, the pedicel of the pedicellate spikelet 2.5–4.5 mm long with whitish or grayish hairs 1–3 (3.5) mm long; pedicellate spikelet 6–9 (11) mm long, narrow lanceolate, the glumes subequal, the first 9-nerved, flat to slightly keeled on the back and with inrolled margins, the second glume 3-nerved, rounded on back; sessile spikelet (6) 7–10 (11) mm long, narrow lanceolate; glumes of the sessile spikelet subequal, the first glume 4- to 6-nerved, the margins folded inward, clasping the edges of the second glume, scaberulous, grooved, the second glume acuminate, keeled or rounded, hispidulous on the nerves, 3-nerved; lemmas transparent-membranous, deeply cleft, 5–7 mm long, bearing a long, geniculate awn, twisted below and loosely twisted above, (8) 12–20 mm long; lodicules about 0.8 mm long, cuneate, flabellate; $2n = 20, 40, 60, 80–86$.

Gravelly or clay, less often sandy soils, grassland habitats, usually dry open hillsides in our area; Sask. to Que., in all states of the e. and midw., especially of the prairies and plains, extending w. to s. Utah and Ariz. where uncommon. July–Sept.

Big bluestem is one of the most important range grasses in the Great Plains where it is a major component of the tall-grass prairie.

55. SCHIZACHYRIUM Nees

Cespitose perennials, with or without rhizomes; culms usually erect; sheaths rounded or compressed and keeled, open; ligules membranous; blades flat or folded, non-auriculate; inflorescence a panicle, the flowering culms much-branched above, each leafy branch terminating in a single, pedunculate raceme, the rachis joint oblique and hollow at the summit, the spikelets in pairs of 1 sessile and 1 pedicelled, the pedicelled spikelet reduced but often fertile, the sessile spikelet fertile; spikelets somewhat dorsiventrally compressed, disarticulating as a unit consisting of the spikelet pair and a section of the rachis; glumes large, coriaceous, the first glume narrow, permanently clasping or enclosing the second glume, flat or somewhat concavo-convex, the second sometimes slightly keeled; lemmas of sterile and fertile florets hyaline, greatly reduced, the lemma of the fertile floret entire or bifid with a geniculate awn from the apex or between the lobes of a bifid apex; palea of the fertile floret small or absent, hyaline; lodicules 2, cuneate-flabellate, often hairy apically; ovary narrow and elongate; $x = 10$.

A relatively large genus of trop. and subtrop. regions of the world with 11 native species in the U.S. (From the Greek *schizo,* split or cleft, and *achuron,* chaff, referring to a cleft palea in the type species.)

1. Schizachyrium scoparium (Michx.) Nash

Andropogon scoparium Michx. Fl. Boreali-Amer. 1: 57. 1803. *Pollinia scoparia* Sprengel, Pl. Minus Cogn. Pugillus 2: 13. 1815. *Sorgum scoparium* Kuntze, Revisio Gen. Pl. 2: 792. 1891. *Schizachyrium scoparium* Nash in Small, Fl. S.E. U. S. 59, 1326. 1903. (*Michaux,* "Hab. in aridus sylvarum Carolinae.")

Andropogon neo-mexicanus Nash, Bull. Torrey Bot. Club 25: 83. 1898. *Schizachyrium neo-mexicanum* Nash, N. Amer. Fl. 17: 107. 1912. *Andropogon scoparius* var. *neomexicanus* A. S. Hitchc. Proc. Biol. Soc. Wash. 41: 163. 1928. *S. scoparium* var. *neomexicanum* Gould, Brittonia 19: 73. 1967. (*Wooton 583,* White Sands, Doña Ana Co., N.M., 26 Aug. 1897.)

Little bluestem.

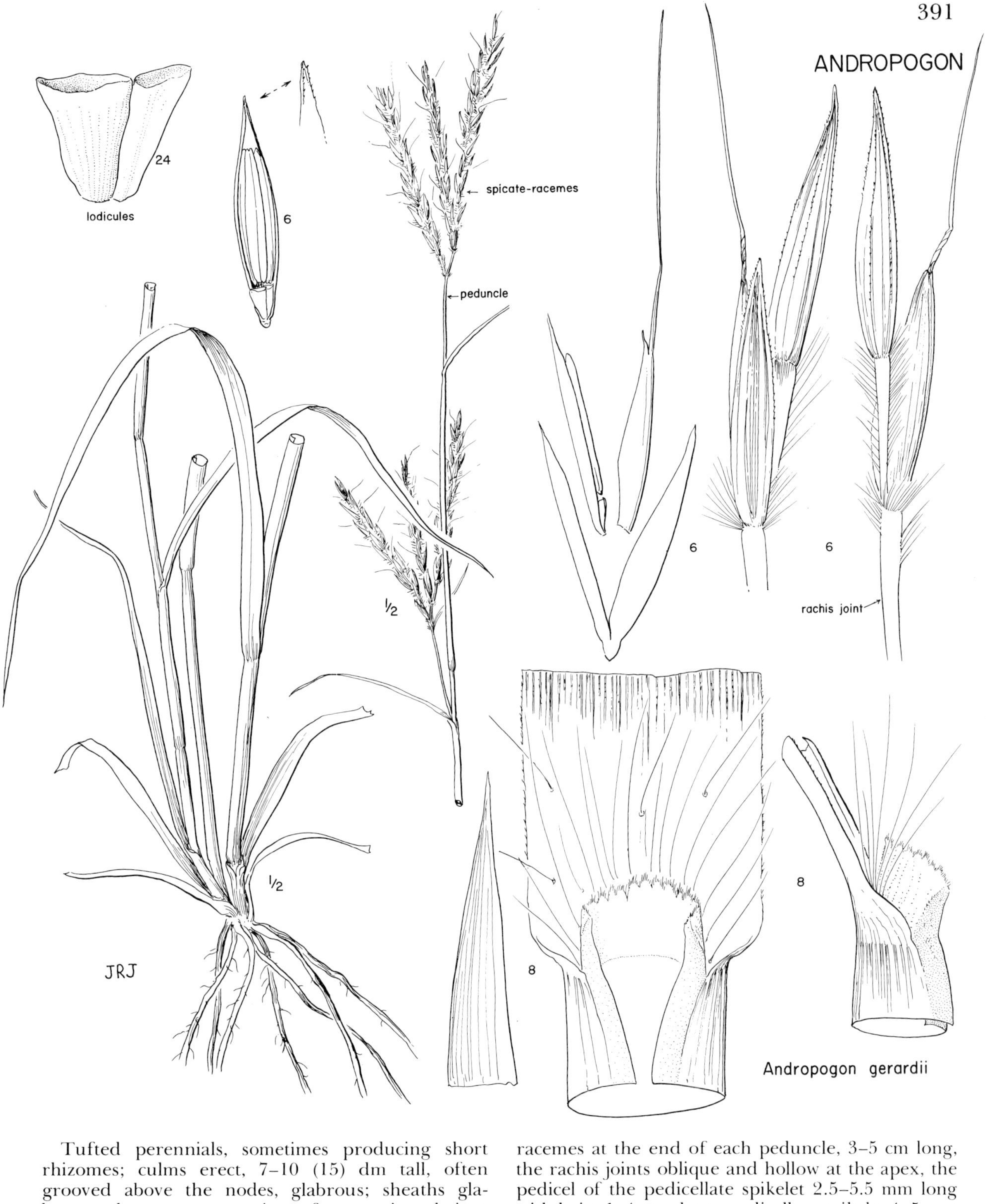

Tufted perennials, sometimes producing short rhizomes; culms erect, 7–10 (15) dm tall, often grooved above the nodes, glabrous; sheaths glabrous, the upper margins often projected into sheath auricles; ligules 0.5–2.7 mm long, membranous, erose, lacerate, ciliolate, minutely puberulent; blades flat or folded, 1.5–5 (6) mm broad, glabrous to scaberulous above, the margins scabrous and sometimes pilose near the throat; panicles of solitary racemes at the end of each peduncle, 3–5 cm long, the rachis joints oblique and hollow at the apex, the pedicel of the pedicellate spikelet 2.5–5.5 mm long with hairs 1–4 mm long; pedicellate spikelet 4–5 mm long, subulate; sessile spikelet (4) 5–6.5 (7.5) mm long, narrow-lanceolate, gradually acuminate; glumes of the sessile spikelet usually subequal, hirsute on the lower half or glabrous on all but the most prominent scabrous lateral nerves, the first glume 5-

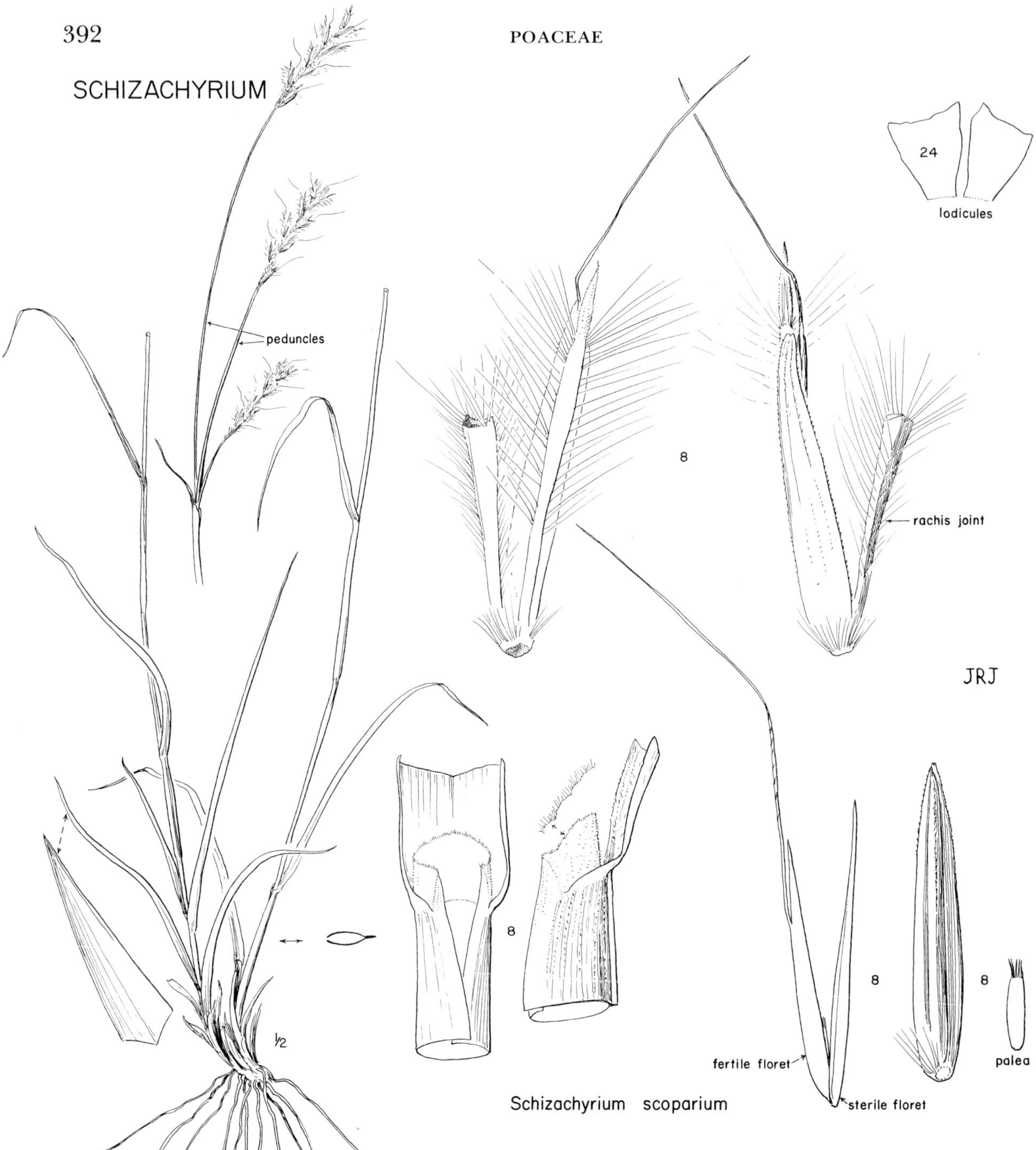

to 7-nerved, flat or sulcate across the back, the margins folded inward clasping the second glume, the second glume keeled, 3-nerved, the margins sometimes hyaline and ciliate; fertile lemma 4.2–6.5 mm long, bearing a geniculate awn 6–16 mm long, the awn usually twisted below; lodicules about 0.7 mm long; anthers 4–4.7 mm long; $2n = 40, 56$.

Dry places, canyons and rocky slopes; Alta. to Que. and in the U.S. from se. Idaho, Utah and Ariz., e. throughout the remaining states; also in Mex. July–Sept.

Little bluestem is a major component of the tall-grass prairie with *Andropogon gerardii* and *Sorghastrum nutans*. It is a highly variable species with several intergrading varieties. Variety *neomexicanum* enters our area in southern Utah and can be distinguished from var. *scoparium* by the following characters.

1 Rachis joints partially bald, often the basal 1/3 sparingly pubescent or glabrous, the hairs grayish; panicles little if at all exserted; range of the species but infrequent in the Southwest, in our area in se. Idaho . . . . var. *scoparium*

1 Rachis joints more conspicuously and completely bearded, the hairs whitish; panicles longer, much exserted; s. Utah, Ariz. to w. Texas and n. Mex. . . . . var. *neomexicanum* (Nash) Gould

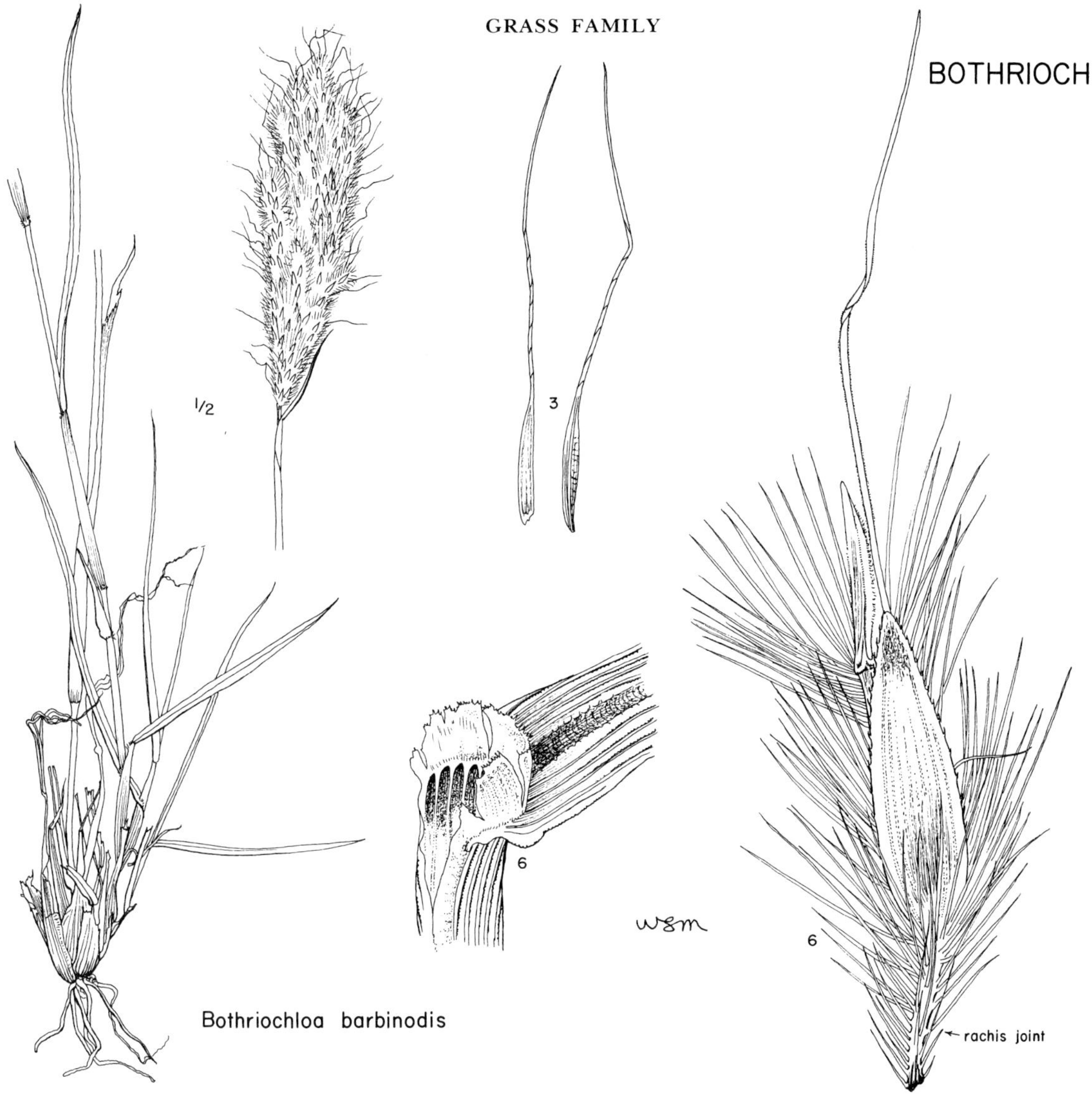

56. BOTHRIOCHLOA Kuntze

Cespitose perennials; culms erect or decumbent-spreading; sheaths rounded, open; ligules membranous; blades flat, non-auriculate; inflorescence a terminal panicle, the spicate branches large or small, sparingly rebranched in some, the upper pedicels and upper rachis joints with a central groove or membranous area, the spikelets in pairs of 1 sessile and 1 pedicelled, the pedicelled spikelet staminate or sterile, but relatively well-developed, awnless, the sessile spikelet fertile; spikelets disarticulating as a unit consisting of the spikelet pair and a section of the rachis; glumes large, firm or hard, the first glume dorsally flattened, clasping or enclosing the second glume, some with a depressed glandular pit in the upper portion, the second glume with a median keel; lemmas of both florets membranous, hyaline, the lemma of the fertile floret more or less triangular in outline, entire or bifid, bearing a bent or twisted awn from the apex or between apical lobes; palea of the fertile floret small or absent, hyaline; $x = 10$.

A genus of about 30 species of warm regions of the world with 11 in the U.S., 3 of which are introduced. (Name from the Greek *bothr,* pit, and *chloa,* grass, in reference to the glandular depression on the median line of the first glume of several of the species.)

References:

Gould, F. W. A cytotaxonomic study in the genus *Andropogon.* Amer. J. Bot. 40: 277–306. 1953.
———. New andropogons with a key to the native and naturalized species of section Amphilophis in the United States. Madroño 14: 18–29. 1957.

1. Bothriochloa barbinodis (Lag.) Herter

Andropogon barbinode Lag. Gen. & Sp. Pl. 3. 1816. *Andropogon saccharoides* var. *barbinodis* Hackel in A. & C. DC. Monogr. Phanerog. 6: 494. 1889. *Amphilophis barbinodis* Nash in Small, Fl. S.E. U. S. 65, 1326. 1903. *Holcus saccharodes* var. *barbinodis* Hackel ex Stuck. Anales Mus. Nac. Hist. Nat. Buenos Aires 11: 48. 1904. *Bothriochloa barbinodis* Herter, Revista Sudamer. Bot. 6: 135. 1940. (*Sessé,* Mex.)

Cane bluestem.

Tufted perennials; culms spreading to ascending, sometimes geniculate below, 5–10 (15) dm long, often branched below, the nodes densely villous; sheaths glabrous; ligules 1–2.5 mm long, membranous, erose-lacerate; blades flat, (2) 3–6 mm broad, scaberulous, long pilose at the base; panicles (4) 5–10 (16) cm long, of numerous, clustered racemes, the pedicel of the pedicellate spikelet 3–4 mm long, flat, the margins thick and the center thin, the rachis joints and pedicels long villous, the hairs (3.5) 6–9 mm long; pedicellate spikelet rudimentary, often deciduous; sessile spikelet (4) 5–6 mm long, lanceolate; glumes of the sessile spikelet subequal, the first glume 7- to 11-nerved, flat on back, acuminate to a narrowly truncate apex, villous below, scabrous above, the second glume somewhat keeled, 3-nerved, glabrous, shiny; fertile lemma about 4 mm long, membranous, bearing a long, geniculate awn 20–30 mm long, twisted below; anthers 0.5–1.2 mm long; stigmas on long styles; $2n = 70, 80, 120, 180$.

Open sandy or gravelly places and rocky slopes; s. Calif., across s. parts of Nev., Utah and Colo., s. through Ariz., N.M., Okla. and Texas to s. Mex. and recurring in s. S. Amer. April–Nov.

Zea mays L.

Zea mays L. Sp. Pl. 971. 1753. *Mays zea* Gaertn. Fructibus & Seminibus Pl. 1: 6, pl. 1, fig. 9. 1788. *Z. segetalis* Salisb. Prodr. Stirpium 28. 1796. *Mays americana* Baumg. Enum. Stirp. Magno Transsilv. Princ. 3: 281. 1816. *Mayzea cerealis* Raf. Med. Fl. 2: 241. 1830. ("Habitat in America.")

Corn, maize.

Robust, monoecious annuals; culms thick, succulent; blades flat, broad, 3 cm or more, with a well developed midrib; inflorescence of 2 strikingly different types, the staminate inflorescence a panicle (tassel) terminating the culm with pairs of unequally pedicelled spikelets arranged along 1 side of the panicle branches, the pistillate inflorescence a spike (ear) borne in the axils of the leaves, the spikelets in pairs in 8–many rows on a thickened rachis (cob) enclosed in numerous foliaceous bracts (husks); spikelets 2-flowered, in the pistillate spikelets the lower floret sterile or occasionally both fertile; glumes of the staminate spikelet membranous, broad, acute, several-nerved, those of the pistillate spikelets short, broad, thin, rounded to emarginate; lemmas of both pistillate and staminate spikelets hyaline; ovary with a single, long, filamentous style (silk); caryopsis much exceeding the glumes; $2n = 20–22, 24$.

Zea is a genus of 1 or 3 species, native of Mex. and C. Amer. (From the Greek *zea*, a kind of grain.)

Corn probably originated in central and southern Mexico. It is considered a cultigen (Galinat, 1971), bred by the early American Indians from a teosinte [*Zea mays* subsp. *mexicana* (Schrader) Iltis] ancestor. It was not known in the Old World before Columbus discovered America, and now is grown as a crop throughout the world. Thousands of strains have been developed, some of the familiar ones called dent, flint, flour, sweet and pop. Most of the corn grown in the United States is "hybrid corn" grown from seed resulting from the crossing of 2 inbred lines. Corn is unable to become established in the wild. The heavy cob remains intact and the close-packed germinating seeds crowd each other to death.

References:
Galinat, W. C. The origin of Maize. Annual Rev. Genet. 5: 447–478. 1971.
Manglesdorf, P. C., and R. G. Reeves. The origin of corn. Bot. Mus. Leafl. 18: 329–440. 1959.

Subfamily III. **ERAGROSTOIDEAE**

Epidermal cells in the region of root hairs all alike and all capable of giving rise to root hairs; culm internodes solid, hollow, or sometimes pith filled; sheath margins free; ligule usually a fringe of hairs, sometimes membranous or absent; blades non-auriculate; inflorescence usually a panicle, often of spicate or racemose branches that are digitately or racemosely arranged on a central axis, the spikelets on 2 sides of a midrib of a flattened rachis or 2 sides of a 3-angled rachis; spikelets usually laterally compressed, 1- to several-flowered; reduced florets, when present, usually above the perfect ones; unisexual spikelets not uncommon; disarticulation mostly above the glumes; lemmas mostly 3-nerved, but 1-nerved in *Calamovilfa, Sporobolus, Crypsis,* and *Spartina* and several-nerved in the small tribes Aeluropodeae and Pappophoreae; lodicules very small or rarely lacking, cuneate or obcuneate, rarely oblong; stamens usually 3; embryo relatively large, $1/3$ to $5/6$ the length of the caryopsis, with the typical eragrostoid morphology with vascular traces to the scutellum and coleoptile separated by an internode, the epiblast present (except absent in Tribe Aristideae), the lower part of the scutellum and coleorhiza separated by a cleft, and the embryonic leaf margins not overlapping; $x = 8, 9, 10$, the chromosomes small.

The Eragrostoideae is one of the 3 major subfamilies (besides Pooideae and Panicoideae), with its own leaf anatomy, epidermal features, embryos and chromosomes. The subfamily was proposed as recently as 1954 by Pilger (Bot. Jahrb. Syst. 76: 334. 1954). Genera grouped in this subfamily were included in the Festuceae (Poeae), Agrostideae (Aveneae, in part), Chlorideae, and Zoysieae by A. S. Hitchcock (Manual Grasses U.S., 1951). Even though the subfamily has reached its best development in the tropics and subtropics it is well represented in the Intermountain Region. As in the Pooideae, characters found in the pattern of the inflorescence, number of florets per spikelet and number of lemma nerves provide the distinctions that separate the tribes.

Key to the Tribes of Eragrostoideae

1 Lemmas 3-nerved or sometimes 1-nerved in *Sporobolus, Calamovilfa, Crypsis,* and *Spartina,* the nerves often obscure.
  2 Inflorescence an open or contracted panicle.
    3 Lemmas indurate, narrowed at the apex to an awn column bearing 3 awns, the callus well developed and usually minutely bearded; spikelets 1-flowered .... Tribe 12. Aristideae, p. 452

3 Lemmas usually not indurate, or if so, then not 3-awned as above; spikelets 1- to several-flowered .... Tribe 8. ERAGROSTEAE, p. 395
2 Inflorescence a unilateral spike or of a few to several unilateral, spicate primary branches .... Tribe 9. CHLORIDEAE, p. 432
1 Lemmas 7- or more-nerved.
4 Spikelets unisexual, the plants dioecious; nerves of the lemma not extending as awns; ligule a short membrane, sometimes tipped with stiff hairs .... Tribe 10. AELUROPODEAE, p. 450
4 Spikelets with perfect florets; nerves of the lemma extending as awns; ligule a ring of hairs .... Tribe 11. PAPPOPHOREAE, p. 452

Tribe 8. **ERAGROSTEAE**

Annuals or perennials, with or without rhizomes; culms hollow or solid, sometimes pith filled; sheath margins free; ligule membranous, a fringe of hairs or absent; blades non-auriculate; inflorescence an open or contracted panicle; spikelets 1- to several-flowered, unisexual in *Scleropogon;* disarticulation usually above the glumes, below in *Lycurus, Crypsis* and some species of *Muhlenbergia;* lemmas 3-nerved, or 1-nerved in *Calamovilfa, Sporobolus* and *Crypsis,* awnless or awned from the tip; lodicules 2 or lacking, usually broad, truncate, rounded or emarginate; fruit a caryopsis, except an achene in *Calamovilfa, Sporobolus* and *Crypsis* in that the seed coat remains free from the pericarp; embryo eragrostoid; $x = 8$ (*Erioneuron*), 9, and 10.

The tribe Eragrosteae centers in trop. and warm temp. regions of the world. The tribe includes genera that A. S. Hitchcock (Manual Grasses U.S., 1951) placed in the Festuceae (Poeae), Agrostideae (Aveneae, in part) and Chlorideae.

1 Spikelets 3- to many-flowered.
2 Plants dioecious or rarely monoecious, the sexes very dissimilar, the pistillate lemmas long-awned, the awns 70–140 mm long, the staminate lemmas only mucronate .... 62. *Scleropogon*
2 Plants with perfect flowers.
3 Inflorescence hidden among sharp-pointed leaves .... 60. *Munroa*
3 Inflorescence not concealed.
4 Callus densely hairy; panicles long and diffuse .... 61. *Redfieldia*
4 Callus not densely hairy; panicles various.
5 Glumes 0.5–2.5 (2.8) mm long; lemmas 1.5–2 (3) mm long, the lateral nerves glabrous, or at most scaberulous; most of our species annual .... 57. *Eragrostis*
5 Glumes 3–8.5 mm long; lemmas (2.5) 3–7.5 mm long, the lateral nerves hairy; perennials.
6 Lemmas not deeply 3-lobed; spikelets 5- to 18-flowered, all but sometimes the terminal floret fertile.
7 Panicles open or contracted, (4) 6–15 cm long, when less than 10 cm long, then the lemmas inconspicuously puberulent on the nerves .... 58. *Tridens*
7 Panicles 1.5–4 (8) cm long, contracted, ovoid or oblong; lemmas conspicuously long hairy on the nerves, at least below .... 59. *Erioneuron*
6 Lemma deeply 3-lobed; spikelets 4-flowered, only the third floret is fertile .... 63. *Blepharidachne*
1 Spikelets 1-flowered, rarely with some 2- or 3-flowered.
8 Callus bearded; lemma 6.5–8 mm long, 1-nerved .... 65. *Calamovilfa*
8 Callus not bearded, except in some *Muhlenbergia;* lemma 1–4.5 (5.5) mm long.
9 Spikelets disarticulating below the glumes.
10 Perennials; glumes awned, the first glume 2- (3)-awned; ligule a membrane, (2) 3–9 mm long .... 64. *Lycurus*
10 Annuals; glumes awnless; ligule a fringe of hairs, 0.2–1 mm long .... 69. *Crypsis*
9 Spikelets disarticulating above the glumes; glumes awnless or sometimes relatively short-awned in some *Muhlenbergia.*
11 Nerves of the lemma glabrous to scabrous.
12 Lemma 3-nerved, awned from the tip or mucronate; ligule membranous .... 66. *Muhlenbergia*
12 Lemma 1-nerved, awnless; ligule a short fringe of hairs .... 67. *Sporobolus*
11 Nerves of the lemma densely silky .... 68. *Blepharoneuron*

57. **ERAGROSTIS** Beauv. Lovegrass

Cespitose annuals and perennials, some with rhizomes; culms hollow; sheaths open with tufts of hairs at the apex (throat); ligules of short, dense, whitish hairs; blades flat to folded or involute, non-auriculate; inflorescence usually an open or sometimes contracted panicle; spikelets 3- to many-flowered, disarticulating in most with the lemma and palea falling separately allowing the caryopsis to fall free; glumes mostly unequal, shorter than the first floret, entire, rounded or keeled, 1- (3)-nerved; lemmas keeled or rounded on back, 3-nerved, the lateral nerves faint to prominent and non-convergent, awnless; palea as long or nearly as long as the lemma, hyaline, 2-keeled, often ciliolate on the keels, usually persistent; lodicules short, fleshy, truncate, 0.2 mm or less long; stamens 3; ovary oblong or subelliptic, glabrous, the styles terminal; caryopsis reddish-brown and translucent; $x = 10$.

About 250 species of trop. and temp. regions of the world, about 35 of which are native to the U.S., centering in the Southwest. There are about 15 species in the U.S. introduced from S. Africa and the Medit. region. (Name from the Greek *eros,* God of love, and *agrostis,* a grass.)

Eragrostis is a very difficult genus in the Southwest.

References:

Harvey, L. H. *Eragrostis. In:* F. W. Gould. Grasses S.W. U. S. Univ. of Ariz. Press. pp. 75–89. 1951.

Koch, S. D. The *Eragrostis pectinacea-pilosa* complex in North and Central America (Gramineae: Eragrostoideae). Illinois Biol. Monogr. 48: 1–74. 1974.

1 Perennials; rachilla disarticulating, the palea falling with the lemma; lemmas 2–3 mm long . . . . 1. *E. curvula*
1 Annuals; rachilla not disarticulating, the palea persistent; lemmas 1.5–2 (2.5) mm long.
  2 Culms prostrate, often rooting at the nodes, forming mats; panicles short, 1.5–3 (3.5) cm long . . . . 2. *E. hypnoides*
  2 Culms erect or sometimes decumbent, but not forming mats or rooting at the nodes; panicles 4–36 cm long.
    3 Plants not glandular on the panicle branches nor the lemmas.
      4 Spikelets mostly 1.5 mm broad or more (1.2–2 mm), ovate to narrow-lanceolate; plants short, 1–6 (7.5) dm tall . . . . 3. *E. pectinacea*
      4 Spikelets mostly less than 1 mm broad (0.6–1.2 [1.5] mm), linear, slender; plants tall, (3) 6–10 dm . . . . 4. *E. orcuttiana*
    3 Plants with minute glandular depressions on the panicle branches and sometimes on the keels and lemmas.
      5 Lemmas not glandular on the keel; spikelets 1.2–1.8 mm broad; panicles narrow throughout . . . . 5. *E. lutescens*
      5 Lemmas glandular on the keel; spikelets 1.5–3 mm broad; panicles broad at the base, tapered upward . . . . 6. *E. cilianensis*

1. Eragrostis curvula (Schrader) Nees

Poa curvula Schrader, Gött. Gel. Anz. 3: 2073. 1821. *Eragrostis curvula* Nees, Fl. Africae Australioris Ill. Monogr. 397. 1841. (Cape of Good Hope.)

Weeping lovegrass.

Tufted perennials; culms relatively tall, 6–15 dm, erect; sheaths keeled, the lower sheaths hispid, the upper more sparingly hispid to glabrous, pilose at the apex and upper margins; ligules a dense row of white hairs, scarcely 1 mm long; blades involute, stiff, attenuate to a fine point, arcuate, scaberulous above and glabrous below; panicles large, 20–34 cm long, open, oblong to ovoid, the main branches pilose in the axils; spikelets (5) 7–8.5 (9) mm long and 1–1.5 mm broad, 5- to 9- (11)-flowered, short-pedicellate, compressed, grayish-green, the rachilla disarticulating; glumes relatively large, lanceolate, 1-nerved, compressed-keeled, scabrous on the keel, the first glume 1.5–2 (2.5) mm long, the second 2–2.8 mm long; lemmas 2–3 mm long; palea hyaline, falling with the lemma; anthers about 0.5 mm long; $2n = 20, 40, 50, 60$.

Sandy roadsides; native of S. Africa, cultivated for ornament or range grass in the Southwest; sporadically distributed across the s. U.S., entering our region in sc. Utah. July.

2. Eragrostis hypnoides (Lam.) Britton, Sterns, & Poggenb.

Poa hypnoides Lam. Tableau Encycl. & Méth. 1: 185. 1791. *Megastachya hypnoides* Beauv. Essai Nouv. Agrostogr. 74, 167, 175. 1812. *Eragrostis hypnoides* Britton, Sterns, & Poggenb. Prelim. Cat. Antho. & Pterid. within 100 Mi. New York City. 69. 1888. *Neeragrostis hypnoides* Bush, Trans. Acad. Sci. St. Louis 13: 180. 1903. *Erosion hypnoides* Lunell, Amer. Midl. Naturalist 4: 221. 1915. (*Richards,* Trop. Amer.)

Creeping lovegrass.

Prostrate annuals, often rooting at the nodes and forming small mats; culms short, the internodes becoming longer than the sheaths; leaves sparingly pubescent; sheaths short, inflated, often pilose at the base and apex and sometimes ciliate margined; ligule a dense row of whitish hairs about 0.6 mm long; blades flat or folded, 0.7–2.2 (3) mm broad, short, pilose; panicles short, 1.5–3 (3.5) cm long, usually branching below only and racemose above, the branches and pedicels short; spikelets (3) 4.5–7 (15) mm long, 1.7–2.7 mm broad, 7- to 15- (40)-flowered, linear; glumes unequal, the first 0.5–1 mm long, narrow-lanceolate, 1-nerved, the second 1–1.5 mm long, lanceolate to lance-ovate, 1- to 3-nerved; lemmas 1.5–2 mm long, lanceolate to lance-ovate, strongly 3-nerved, the lateral nerves nearly parallel, keeled, glabrous; palea about 1 mm long, remaining on the rachilla after the lemma and caryopsis have dropped; anthers 0.2–0.3 mm long.

Muddy, sandy or gravelly banks of streams, ponds or lakes; Que. and Ont. and throughout the e. half of the U.S., and sporadic in the w., often becoming locally abundant in the n. half of our region. July–Aug.

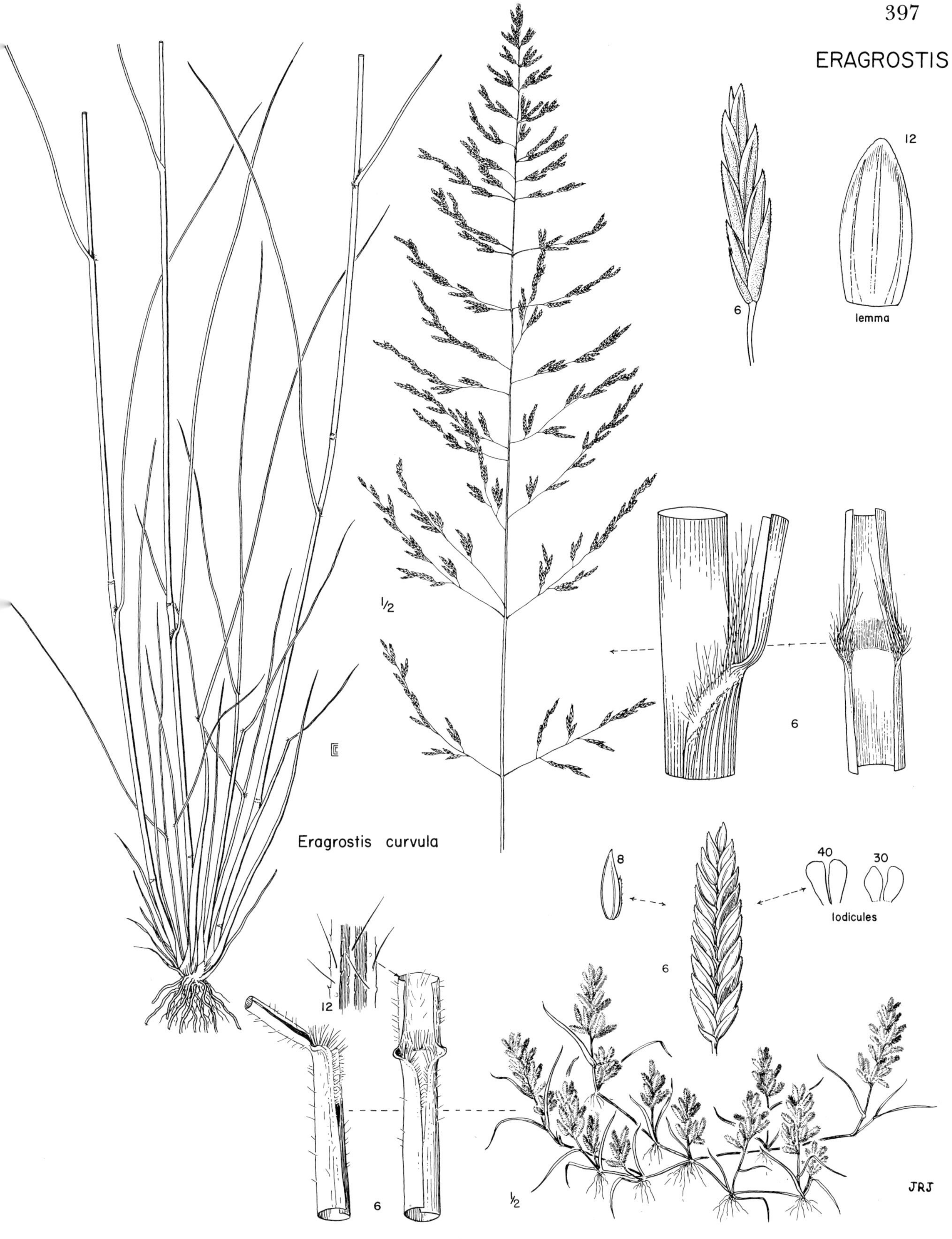

Eragrostis curvula

Eragrostis hypnoides

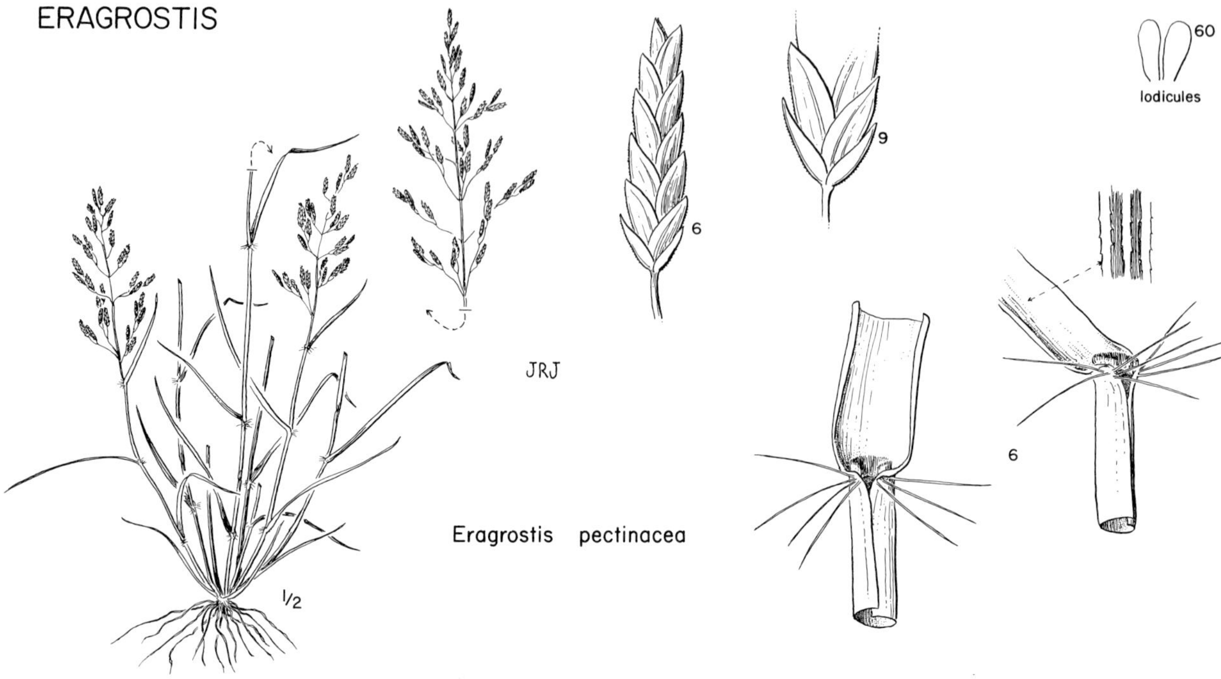

3. Eragrostis pectinacea (Michx.) Nees
Poa pectinacea Michx. Fl. Boreali-Amer. 1: 69. 1803. *Eragrostis pectinacea* Nees, Fl. Africae Australioris Ill. Monogr. 406. 1841. (*Michaux,* "in arvis Illinoensibus," in 1795.)
E. diffusa Buckley, Proc. Acad. Nat. Sci. Philadelphia 1862: 97. 1863. *E. purshii* var. *diffusa* Vasey ex J. M. Coulter, Contr. U. S. Natl. Herb. 1: 59. 1890. (*Buckley,* n. Texas.)
E. caroliniana sensu A. S. Hitchc. in Abrams, Ill. Fl. Pacific States 1: 187. 1923; not (Spreng.) Scribn. 1894.

Tufted annuals; culms (1) 1.5–6 (7.5) dm tall, erect or somewhat decumbent; sheaths smooth, except for the pilose tufts at the apex (throat) with hairs 1.5–3.5 (4) mm long; ligule a dense row of white hairs about 0.2–0.4 mm long; blades flat to involute, 1–2 (3.5) mm broad, glabrous below and scabrous above; panicles 5–20 (35) cm long, ovoid, the branches ascending to spreading; spikelets (3.7) 4.5–6.5 (7.5) mm long, 1.2–2 mm broad, 6- to 12-flowered, compressed; glumes acute to acuminate, keeled, 1- (3)-nerved, scabrid on the nerves, the first glume 0.9–1.5 mm long, the second 1.2–1.8 mm long; lemmas 1.5–2 mm long, ovate, acute, loosely imbricate, conspicuously 3-nerved, scabrid on the nerves, scaberulous between; palea 1–1.5 mm long, hyaline, persistent, convex, ciliate on the keels; anthers 0.2–0.4 mm long; $2n = 40, 60$.

Along stream courses and moist places, usually in sandy soil; Que. and Ont., and throughout the e. and c. U.S., sporadic in the w. states; also in Mex.; entering our area in the Snake River Plains, Idaho, and in the Bonneville Basin and se. Utah. July–Aug.

A robust form of this species with more branched and densely flowered panicles has been called *E. diffusa* Buckley. Rydberg (Fl. Rocky Mts. 1918) reported *E. neomexicana* Vasey as entering s. Utah from the south. We doubt this report. Koch (1974), in his distribution map for *E. tephrosanthos* Schultes, shows a dot in the Bonneville Basin of Utah. The only character Koch used to distinguish *E. tephrosanthos* is its spreading to divaricate pedicels as opposed to the appressed to ascending pedicels of *E. pectinacea.* Because the collection is not documented, we have been unable to verify the report.

4. Eragrostis orcuttiana Vasey
Eragrostis orcuttiana Vasey, Contr. U.S. Natl. Herb. 1: 269. 1893. (*Orcutt 1313,* Chollas Valley, San Diego Co., Calif., Aug. 1885.)

Orcutt lovegrass.

Annuals; culms (3) 6–10 dm tall, ascending from a more or less decumbent base; sheaths smooth, except for the pilose tufts at the apex (throat), the hairs 1.5–3 (4) mm long; ligule a dense row of whitish hairs 0.2–0.5 mm long; blades flat, (2) 2.5–4 (6) mm broad; panicles relatively large, 15–30 (36) cm long, pyramidal, the slender branches spreading to reflexed, flexuous; spikelets (4) 5–8 mm long, 0.6–1.2 (1.5) mm broad, linear, nearly terete, 6- to 10-flowered; glumes lanceolate, purplish, 1-nerved, scabrous keeled, the first glume 0.7–1.2 mm long, the second 1.2–1.7 mm long; lemmas 1.5–2 mm long, 3-nerved, scaberulous on the keel and nerves and sometimes towards the margins and apex, the lateral nerves relatively weak, somewhat loosely arranged, the rachilla often exposed; palea 1–1.7 mm long; anthers 0.3–0.5 mm long.

A weed in moist waste places and along streambanks; Calif., Nev., Utah, Colo. and Ariz., sporadic in Oregon, Idaho and Wash. July–Sept.

5. Eragrostis lutescens Scribn.
Eragrostis lutescens Scribn. U.S.D.A. Div. Agrostol. Circ. 9: 7. 1899. (*Piper 2624,* "sandy banks of Snake River at Almota, Wash.," 30 July 1897.)

Six-weeks lovegrass.

Small tufted annuals with depressed glands (appearing warty) on the leaves, culms and panicle branches; culms 0.6–1.5 (2) dm tall, erect or ascending, sometimes decumbent at the base; sheaths more

ERAGROSTIS

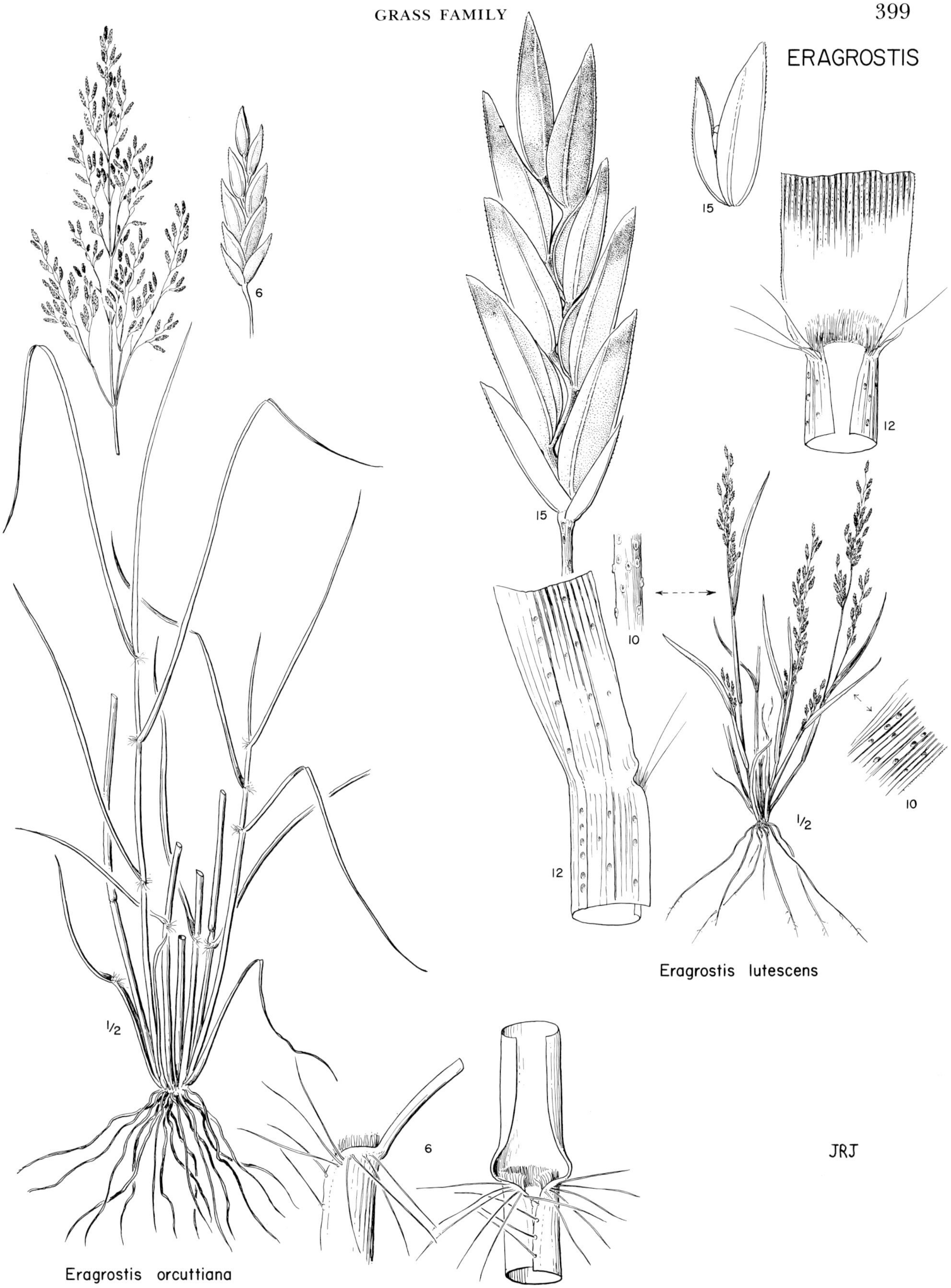

Eragrostis lutescens

Eragrostis orcuttiana

or less pilose at the apex (throat), the hairs 1.5–2 mm long; ligule a dense row of whitish hairs 0.2–0.5 mm long; blades flat, folded or involute, 1–2.5 mm broad, glabrous or scaberulous above; panicles short, 4–10 cm long, narrow and dense, the branches appressed; spikelets (3.5) 4.5–6.5 (8) mm long, 1.2–1.8 mm broad, ovate-lanceolate, compressed, 6- to 10-flowered; glumes lanceolate, 1-nerved, scaberulous on the compressed keel, the first glume 0.7–1.2 mm long, the second 1–1.8 mm long; lemmas 1.5–2.2 mm long, conspicuously 3-nerved, scaberulous, mostly on the nerves; paleas 1.2–2 mm long, hyaline, persistent; anthers 0.2–0.3 mm long.

Sandy banks of streams and lakes and moist alkaline flats; rare from Wash. and Idaho, s. to s. Calif. and Ariz.; in our area in w. Nev. July–Aug.

6. Eragrostis cilianensis (All.) Mosher

Briza eragrostis L. Sp. Pl. 70. 1753. *Poa megastachya* Koeler, Descript. Gram. Gallia & German. 181. 1802. *Megastachya eragrostis* Beauv. ex Roemer & Schultes, Syst. Veg. 2: 575, 584. 1817. *Eragrostis megastachya* Link, Hort. Regius Bot. Berol. 1: 187. 1827. *Briza megastachya* Steudel, Nomencl. Bot. ed. 2. 1: 225. 1840, as synonym of *Poa megastachya*. *E. vulgaris* var. *megastachya* Coss. & Germain, Fl. Descript. & Analytique Environs Paris 2: 641. 1845. *E. poaeoides* var. *megastachya* A. Gray, Manual Bot. ed. 2. 563. 1856. *E. eragrostis* MacMillan, Metasp. Minnesota Valley 75. 1892; not Beauv. 1812. *E. minor* var. *megastachya* Davy in Jepson, Fl. W. Middle Calif. 60. 1901. *E. eragrostis* var. *megastachya* Farw. Ann. Rep. Michigan Acad. Sci. 17: 182. 1916. *E. vulgaris* subsp. *megastachya* Douin in Bonnier, Fl. Compl. France, Suisse & Belgique 12: 32. 1933. *E. cilianensis* subsp. *major* f. *megastachya* Maire & Weiller, Fl. Afrique N. 2: 175. 1953. (Habitat in Europa.)

Poa cilianensis All. Fl. Pedemontana 2: 246. 1785. *E. megastachya* var. *cilianensis* Aschers. & Graebner, Syn. Mitteleur. Fl. 2: 371. 1900. *E. cilianensis* Lutati, Malpighia 18: 386. 1904, the combination suggested but not accepted by Lutati; Mosher, Univ. Illinois Agric. Exp. Sta. Bull. 205: 381. 1918, possibly the earliest author accepting the combination and validating it (see Shinners, Rhodora 56: 26–27. 1954, who attributed it to A. S. Hitchcock). *E. multiflora* var. *cilianensis* Maire, Bull. Soc. Hist. Nat. Afrique N. 30: 369. 1939. (*Bellardi*, "Ciliani" estate, Italy.)

Stink grass.

Tufted annuals with glandular pit-like depressions (appearing warty) on the culms, leaves, branches of the inflorescence, and keels of the glumes and lemmas; culms 1–5 (8) dm tall, erect to prostrate; sheaths pilose at the apex (throat), the hairs 1–3 mm long; ligule a dense row of hairs about 0.5 mm long; blades flat to folded, 2–5 (6.5) mm broad; panicles 4–13 (23) cm long, somewhat congested, ovoid to oblong, the branches ascending to spreading; spikelets 4–8 (14) mm long, 1.5–3 mm broad, ovate-oblong, flattened, (7-) 9- to 14- (35)-flowered; glumes keeled, the first glume 1–2 mm long, lanceolate, 1-(3)-nerved, the second 1.2–2.5 mm long, lance-ovate, 1- to 3-nerved; lemmas 1.5–2.5 mm long, the keels glandular pitted, conspicuously 3-nerved, scaberulous towards the margins, closely imbricate; paleas 1–2 mm long, hyaline, persistent; anthers about 0.4 mm long; $2n = 20, 40$.

Weedy roadsides, fields and along watercourses; introduced from the Old World, now established throughout the U.S., also in C. Amer., W.I. and s. to Argentina. July–Sept.

Much has been written about the validity of the name *Eragrostis cilianensis* based on Allioni's *Poa cilianensis*. Most recent botanists use the epithet *cilianensis*, suggesting that Bellardi's collection was merely an immature specimen. Henrard (Blumea 3: 420–422. 1940) rejects the basionym, however, since it is his interpretation that the type was a mixed collection of *Poa trivialis* L. and what he prefers to call *E. megastachya* and that the description was almost entirely taken from the *Poa* except for the description of the pilose throat of the sheath.

58. TRIDENS Roemer & Schultes Tridens

Tufted or sometimes shortly rhizomatous perennials; culms moderately tall; sheaths open; ligules of short white fringed scales; blades flat, non-auriculate; inflorescence an open or contracted panicle; spikelets several-flowered, disarticulating above the glumes and between the florets; glumes subequal to unequal, mostly thin and membranous, the first glume 1-nerved, the second 1- to 3- (5)-nerved; lemmas broad, thin, 3-nerved, short-hairy on the nerves below (ours), rounded on the back, mostly bidentate at the apex, the midnerve and often the lateral nerves extending as minute mucros; palea slightly shorter than the lemma, glabrous to short-hairy, smooth to ciliolate on the keels; anthers dark purple; caryopsis dark brown and opaque, the embryo about $^{2}/_{5}$ the caryopsis length; $x = 10$.

A N. Amer. genus of some 16 species (in the strict sense with *Erioneuron* removed), mostly in the e. and s. U.S. and n. Mex. (Name from the Latin *tria*, thrice, and *dens*, tooth, referring to the excurrent nerves of the lemma forming three minute mucros.)

The American species of *Tridens* and *Erioneuron* are treated under the name *Triodia* in earlier manuals. *Triodia* has since been interpreted as a genus restricted to Australia. Chase, in her revision of Hitchcock's Manual (Manual Grasses U.S., 1951), recognized this and included all United States species under *Tridens*. Tateoka (1961) split *Tridens* into two groups, one of which was transferred to the revived genus *Erioneuron*.

Reference: Tateoka, T. A biosystematic study of *Tridens* (Gramineae). Amer. J. Bot. 48: 565–573. 1961.

1. Tridens muticus (Torr.) Nash

Tricuspis mutica Torr. in Whipple, Explor. & Surv. Railroad Route from Mississippi River to Pacific Ocean 4(5): 156. 1857. *Triodia mutica* Scribn. Bull. Torrey Bot. Club 10: 30. 1883. *Sieglingia mutica* Kuntze, Revisio Gen. Pl. 2: 789. 1891. *Tridens muticus* Nash in Small, Fl. S.E. U.S. 143, 1327. 1903. (*Bigelow* [*793*], Laguna, Valencia Co., N.M., 22 Sept. 1853.)

Slim tridens.

Densely tufted perennials; culms slender, erect, 1.5–5 (6.5) dm tall; leaves smooth to scaberulous and often sparsely pilose, the hairs more concentrated around the collar; ligule a ciliate fringe around the back of the throat and sometimes with membranous lateral auricles; blades narrow, involute or sometimes flat, 1–3 mm broad when flat, often glaucous; panicles (4) 6–15 cm long, narrow, sometimes racemose, the spikelets rather distant; spikelets (7.5) 9–11 (14) mm long, 5- to 7- (9)-flowered, the florets

ERAGROSTIS
TRIDENS

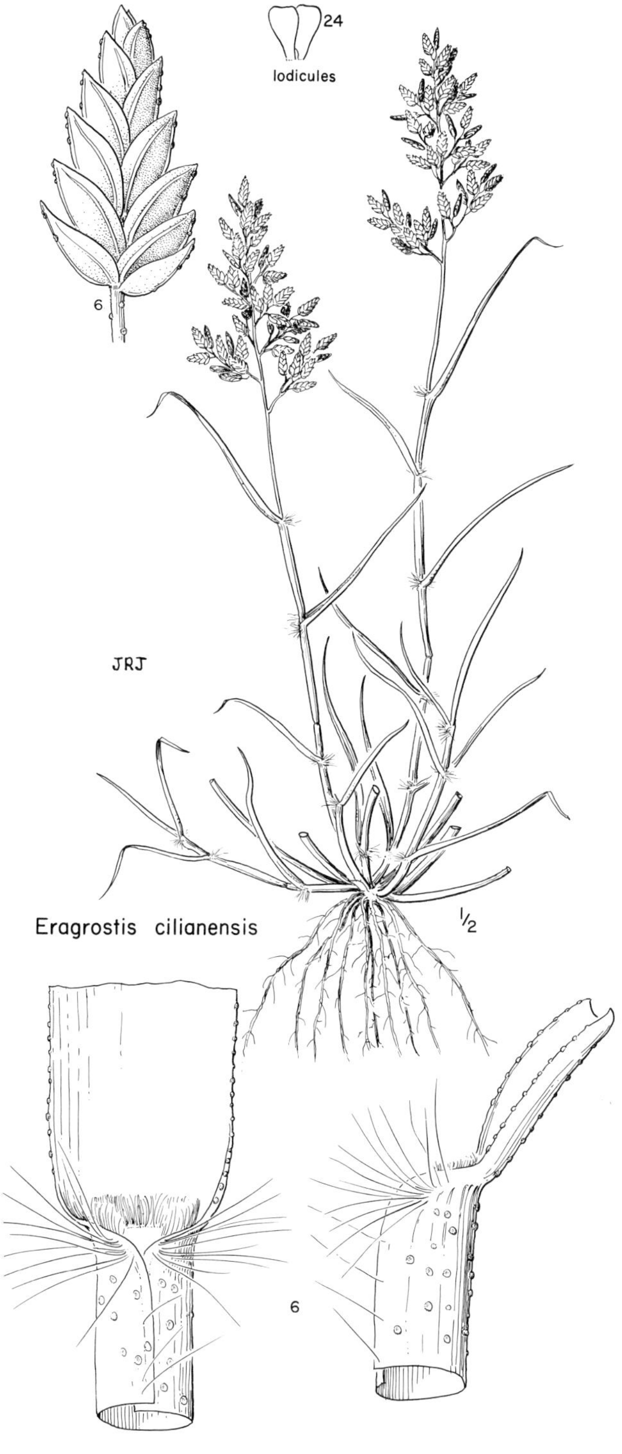

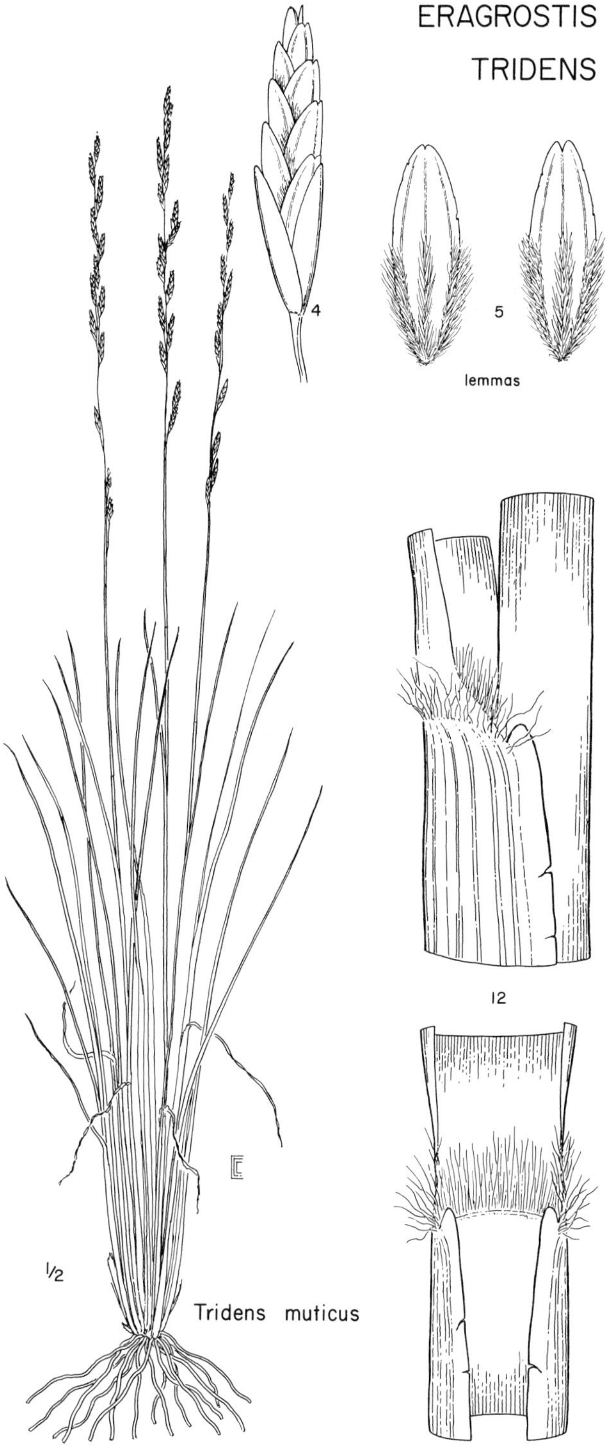

strongly imbricate, not flattened; glumes smooth, thin and hyaline, 1-nerved, the first glume 3–5 (6) mm long, ovate, the second 4–6 mm long, elliptic-ovate, rarely faintly 3-nerved; lemmas 4.5–5.5 (6) mm long, broadly obovate, broadly rounded to obtuse at the apex, this occasionally cleft or mucronate, sometimes splitting deeper, hyaline, 3-nerved, the lateral nerves nearly converging apically, pilose below, the hairs mostly on the nerves; palea 2–3.5 mm long, ovate, the keels ciliate towards the base; $2n = 32, 40$.

Dry rocky slopes, desert grasslands and oak woodlands; s. Calif., e. across the s. parts of Nev., Utah and Colo., s. to Ariz., N.M., Texas and n. Mex. May–Sept.

59. ERIONEURON Nash

Densely tufted, low, mostly stoloniferous perennials; culms numerous, erect or spreading; leaves mostly in a basal tuft; sheaths open; ligules of short, whitish hairs; blades narrow, often involute, white-cartilaginous margined, non-auriculate; inflorescence a short, usually capitate raceme or panicle; spikelets 4- to 18-flowered, on short pedicels, flattened, disarticulating above the glumes and between the florets; glumes subequal to unequal, deeply bifid at the apex, membranous, 1-nerved; lemmas broad, rounded on the back, 3-nerved with long conspicuous hairs on the nerves, at least below, bilobed at the apex (except in *E. pilosum*), the mid-nerve terminating in a short awn and each lateral nerve often somewhat projected as a short mucro; palea slightly shorter than the lemma, ciliate on the keels with long hairs between the keels basally; ovary with white stigmas; caryopsis oblong, glossy and translucent with a visible embryo that is more than half the caryopsis length; $x = 8$.

A genus of 5 species of arid regions of the sw. U.S. and Mex. (Name from the Greek *erion,* woolly, and *neuron,* nerve, alluding to the villous nerves of the lemma.)

The species of *Erioneuron* were included in *Tridens* or *Triodia* by A. S. Hitchcock and other authors. Tateoka (1961, see reference under *Tridens*) found that the differences between the two groups are numerous and striking. He found differences in basic chromosome number, shape and color of the caryopsis, color of the stigma, shape and pubescence of the lemma, pubescence of the palea, shape of the bicellular microhairs of the epidermis, leaf anatomy, general plant habit and nature of the panicle. Tateoka suggests that *Erioneuron* is more closely related to *Munroa* than to *Tridens.*

1 Panicles usually exceeded by a fascicle of leaves; lemmas 2.5–4 (5) mm long, deeply cleft; plants becoming stoloniferous .... 1. *E. pulchellum*
1 Panicles exserted; lemmas 4–7.5 mm long, not deeply cleft; plants not stoloniferous .... 2. *E. pilosum*

1. Erioneuron pulchellum (H.B.K.) Tateoka

Triodia pulchella H.B.K. Nova Gen. & Sp. Pl. 1: 155, tab. 47. 1816. *Koeleria pulchella* Sprengel, Syst. Veg. 1: 332. 1825. *Uralepis pulchella* Kunth, Révision Gram. 1: 108. 1829. *Dasyochloa pulchella* Willd. ex Steudel, Nomencl. Bot. ed. 2. 1: 484. 1840. *Tricuspis pulchella* Torr. in Whipple, Explor. & Surv. Railroad Route from Mississippi River to Pacific Ocean 4(5): 156. 1857. *Sieglingia pulchella* Kuntze, Revisio Gen. Pl. 2: 789. 1891. *Tridens pulchellus* A. S. Hitchc. in Jepson, Fl. Calif. 1: 141. 1912. *Erioneuron pulchellum* Tateoka, Amer. J. Bot. 48: 572. 1961. (*Humboldt & Bonpland,* Guanajuato, Mex.)

Fluffgrass.

Low, tufted, stoloniferous perennials; culms 0.5–1.5 dm tall, scabrous or puberulent, consisting of an elongated internode topped by a fascicle of leaves, the fascicle finally bending over, rooting and producing other culms; sheaths striate, the margins open and slightly scarious, pilose at the base and bearing a tuft of hairs at the apex (throat); ligule a low ciliate fringe; blades strongly involute, often curved, scaberulous below, scabrous above; panicles small and compact, 1.5–4 (8) cm long, capitate, usually not exceeding the blades of the fascicle, appearing white-woolly; spikelets (5.5) 6.5–10 mm long, 3–6 mm broad, 5- to 8- (14)-flowered; glumes subequal, (3.5) 6–8 mm long, glabrous, often tinged with purple, 1-nerved, acuminate, awn-tipped; lemmas 2.5–4 (5) mm long, deeply cleft, long-pilose in the lower half, often tinged with purple, the awns barely extended beyond the lobes, becoming divergent; palea 2–3 mm long, the keels ciliate above, long-pilose below; anthers about 0.5 mm long; caryopsis light-yellowish; $2n = 16$.

Desert foothills and mesas in rocky soils; s. Calif., across the s. parts of Nev., Utah and Colo., s. to Ariz., N.M., w. Texas, and Mex. April–Aug.

2. Erioneuron pilosum (Buckley) Nash

Uralepis pilosa Buckley, Proc. Acad. Nat. Sci. Philadelphia 1862: 94. 1863. *Sieglingia pilosa* Nash in Britton & Brown, Ill. Fl. N. U.S., Canada & Brit. Possessions 3: 504. 1898. *Tricuspis pilosa* A. A. Heller, Cat. N. Amer. Pl. ed. 2. 28. 1900. *Triodia pilosa* Merr. U.S.D.A. Div. Agrostol. Circ. 32: 9. 1901. *Erioneuron pilosum* Nash in Small, Fl. S.E. U.S. 144, 1327. 1903. *Tridens pilosus* A. S. Hitchc. Contr. U. S. Natl. Herb. 17: 357. 1913. (*Buckley,* middle Texas.)
Triodia acuminata Benth. ex Vasey, Dept. Agric. Special Rep. 63: 35. 1883, nomen nudum; Vasey, U.S.D.A. Div. Bot. Bull. 12(2): pl. 32. 1891. *Sieglingia acuminata* Kuntze, Revisio Gen. Pl. 2: 789. 1891. (*Hall 779,* hills, Austin, Texas, 18 May 1872.)

Hairy tridens.

Weakly rooted, tufted perennials; culms 1–2 (4.5) dm tall, usually 1 internode showing above the dense tuft of basal leaves; sheaths smooth, striated, strongly keeled, the margins open and slightly scarious, pilose at the apex (throat); ligule a low ciliate fringe; blades flat or more often folded, 1–1.5 mm broad and 1–6 cm long, long-pilose near the collar, sometimes sparsely pilose at the tip on both surfaces, the margins and midnerve white-cartilaginous, abruptly narrowed at the tip and somewhat mucronate; panicles 1.5–3 (4.5) cm long, exserted from the leaves, contracted, ovoid, sometimes racemose, consisting of 3–9 large, subsessile or short-pedicelled spikelets; spikelets 8.5–14 (20) mm long, 3–6 (8) mm broad, 6- to 12- (18)-flowered, strongly compressed, pale or purplish; glumes glabrous, acuminate, pointed or awn-tipped, 1-nerved, the first glume 3.5–6 (7) mm long, the second 4.5–7 (8.5) mm long; lemmas 4–7.5 mm long, lanceolate, acute, 3-nerved, the lateral pair near the margin, densely pilose on the keel nerve toward the base, pilose on the margins, often the full length, the mid-nerve produced into a short, straight awn, the lateral nerves forming mucros, the awn 1–2 mm long; palea 3–4 mm long, elliptic, pilose on the back and margins below; anthers 1–1.2 mm long; $2n = 16, 32$.

Dry rocky hills and mesas, often in oak and pinyon-juniper woodlands; s. parts of Nev., Utah, Colo. and Kansas, s. through Ariz., N.M. and Texas to c. Mex. May–July.

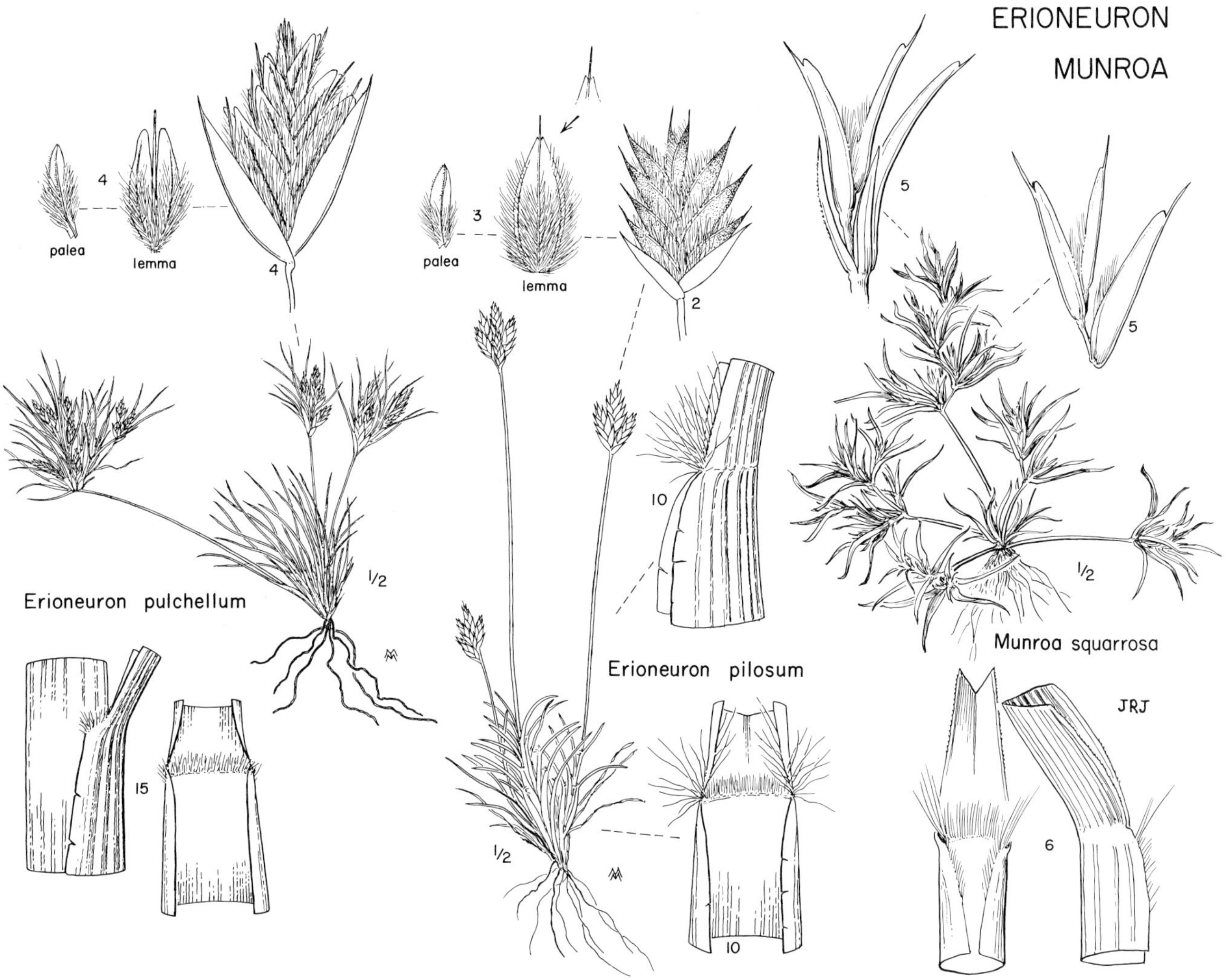

60. MUNROA Torr.

Low, spreading, much-branched annuals; culms hollow, with long, naked internodes and fascicles of leaves bearing inflorescences, sometimes rooting at the nodes; sheaths open; ligules of short hairs; blades stiff, pungent and flat or folded, white-margined, non-auriculate; inflorescence a reduced panicle of small subsessile clusters of spikelets, almost hidden in the fascicles of leaves at the branch tips; spikelets few-flowered, unlike in size, the lower ones smaller, 3- or 4-flowered, the upper 2- or 3-flowered, disarticulating above the glumes and between the florets; glumes shorter than the lowermost lemma, narrow, acute, 1-nerved, subequal in the lower spikelet, the upper spikelets with reduced or absent first glumes; lemmas lanceolate, prominently 3-nerved, the nerves often hairy, the hairs often in tufts, the nerves gradually narrowing above, abruptly mucronate or short-awned from a slender, spreading tip; palea narrow, membranous, strongly 2-nerved; lodicules lacking; stamens 3; caryopsis translucent, dorsiventrally compressed and oval in outline; $x = 7, 8$.

A genus of 3 species, *M. squarrosa* in N. Amer. and the other 2 in s. S. Amer. (Named for William Munro, 1818–1880, an English agrostologist.)

Munroa has in the past been referred to the tribe Chlorideae, but Tateoka (1961, see reference under *Tridens*) found good reason to place it here in the Eragrosteae near *Erioneuron.*

Reference: Parodi, L. R. Contribucion al estudio de las Gramineas del "*Munroa.*" Revista Mus. La Plata, secc. Bot. 34: 171–193. 1934.

1. **Munroa squarrosa** (Nutt.) Torr.

Crypsis squarrosa Nutt. Gen. N. Amer. Pls. 1: 49. 1818. *Munroa squarrosa* Torr. in Whipple, Explor. & Surv. Railroad Route from Mississippi River to Pacific Ocean 4(5): 158. 1857. (*Nuttall,* "On arid plains near the 'Grand Detour' of the Missouri" River, in S.D., in 1811.)

M. squarrosa var. *floccuosa* Vasey ex Beal, Grasses N. Amer. 2: 456. 1896. (*Jones 4088,* Peach Springs, Ariz., 27 Aug. 1884.) Described from a plant with white web-like remains of egg cases of a woolly aphid.

False buffalograss.

Low, matted annuals; culms 0.3–1.5 (3) dm long, the internodes scabrous to puberulent, naked; leaves in tight clusters; sheaths short, broad, the apical margins with stiff hairs 1–2 mm long, sometimes the lower sheaths of a cluster pilose at the base; ligule a fringe of hairs 0.5 (1) mm long; blades involute, about 2 mm broad and 10–25 mm long, stiff and pungent, white-margined; panicles much reduced, bearing only 2–4 spikelets, much exceeded by the leaf blades; spikelets 6–8 mm long, 3- to 5-flowered, irregular, the first solitary and sessile and the upper in a fascicle, pedicellate; glumes narrow, acute, 1-nerved, those of the lower spikelet subequal, 2.5–4.2 mm long, those of the upper spikelets unequal, the first reduced or even absent in the terminal spikelet; lemmas 3.5–5 mm long, lanceolate, scaberulous, 3-nerved, the lateral nerves with a tuft of hairs just below the midlength, excurrent at the tip, the midnerve forming an awn 0.5–2 mm long; palea nearly as long as the lemma, with 2 conspicuous nerves, elliptic-lanceolate; anthers 1–1.5 mm long; $2n = 16$.

Dry open areas, usually in sandy soil or disturbed roadsides; Great Plains from Sask. and Alta. to Texas, w. through s. and e. Utah and s. Nev. to Calif. and s. to n. Mex. June–Sept.

Munroa squarrosa may be locally abundant but is relatively unpalatable to livestock. The fascicles bearing inflorescences readily separate from the mat and are carried by the wind, disseminating the seeds in its path.

61. REDFIELDIA Vasey

Stout rhizomatous perennials; culms tall, firm, solitary and simple; sheaths open, firm; ligule a fringe of short hairs; blades long, firm, often involute, non-auriculate; inflorescence a large, open panicle with long, spreading, flexuous, capillary branches and pedicels; spikelets (1-) 2- to 6-flowered, disarticulating above the glumes and between the florets; glumes subequal, narrow, lanceolate, 1-nerved; lemmas lanceolate, distinctly 3-nerved, glabrous on the nerves, occasionally mucronate or 3-toothed at the apex; callus silky-pilose; palea nearly as long as the lemma; caryopsis oblong, free from the palea.

A monotypic genus centered in the prairie states of c. U.S. (Named in honor of John Howard Redfield, 1815–1895, a distinguished botanist at the Academy of Natural Sciences of Philadelphia.)

1. **Redfieldia flexuosa** (Thurber) Vasey

Graphephorum flexuosum Thurber in A. Gray, Proc. Acad. Nat. Sci. Philadelphia 1863: 78. 1864. *Redfieldia flexuosa* Vasey, Bull. Torrey Bot. Club 14: 133, pl. 70. 1887. (*Hall & Harbour 635,* "Colorado Territory," latitude 41° [probably Nebr.], in 1862.)

Blowout grass, sandgrass.

Tall, stout perennials with long creeping or vertical, tap-root like rhizomes; culms tall, (5) 7–12 dm long; sheaths overlapping, smooth; ligule a dense row of hairs about 1 mm long; blades loosely involute, glabrous or sometimes scaberulous above; panicles long, 22–35 (55) cm long, diffuse; spikelets 5–7 (8) mm long and 2.5–5 mm broad, (1-) 3- to 6-flowered; glumes subequal, 2.5–4 (4.5) mm long, glabrous, 1-nerved; lemmas 4.5–5.7 (6.5) mm long, lanceolate, chartaceous, compressed-keeled, 3-nerved, the lateral nerves nearly convergent, sometimes slightly excurrent, forming 3 minute teeth with the excurrent midnerve, the callus with long silky hairs 1–1.2 mm long; palea about as long as the lemma, stout, 2-nerved and folded in the middle; styles long with plumose stigmas.

Sandhills and dunes, most common in blowout areas; Great Plains from S.D. to Texas and w. through Colo. and N.M. to s. Utah and n. Ariz. (June–) July–Aug.

Blowout grass is present on sandy hills and dune formations. Being so wiry, it is of little value for forage but is important as a soil binder in areas of drifting sand.

62. SCLEROPOGON Phil.

Low, dioecious or monoecious (rarely with perfect flowers), stoloniferous, mat-forming perennials; culms usually bearing staminate or pistillate panicles; leaves basally disposed; sheaths open; ligules of short hairs; blades firm, flat below and folded above, non-auriculate; inflorescence a contracted panicle or spike-like raceme with few spikelets; pistillate spikelets long awned with 3–5 fertile florets and 1 to several rudimentary florets (reduced to awns); staminate spikelets awnless, the rachilla disarticulating above the glumes, the florets falling together; glumes of the pistillate spikelets unequal, acuminate, (1) 3-nerved or with additional, fine lateral nerves; glumes of the staminate spikelets thin, pale, lanceolate, 1-nerved or obscurely 3-(5)-nerved, awnless, the first and second separated by a short internode; lemmas of the pistillate spikelets firm, rounded on the back, 3-awned from the 3 nerves, the awns 5–10 (14) cm long; lemmas of the staminate spikelets similar to the glumes; callus not long hairy; paleas shorter than the lemmas, obtuse; $x = 10$.

A monotypic genus with 2 centers of distribution, one in the sw. U.S. and n. Mex. and the other in Chile and Argentina. (Name from the Greek *skleros,* hard, and *pogon,* bearded, alluding to the hard awns of the pistillate lemmas.)

1. **Scleropogon brevifolius** Phil.

Scleropogon brevifolius Phil. Anales Univ. Chile 36: 206. 1870. (Mendoza, Argentina.)

Lesourdia karwinskyana Fourn. Bull. Soc. Bot. France 27: 102, pl. 4, fig. 12. 1880. *S. karwinskyanus* Benth. ex S. Wats. Proc. Amer. Acad. Arts 18: 181. 1883. (*Karwinsky 992,* Mex.)

Burrograss.

Perennials with numerous stolons; plants dioecious or monoecious, the different sexes on different culms, sometimes in the same inflorescence or rarely in the same spikelet (the pistillate above); culms

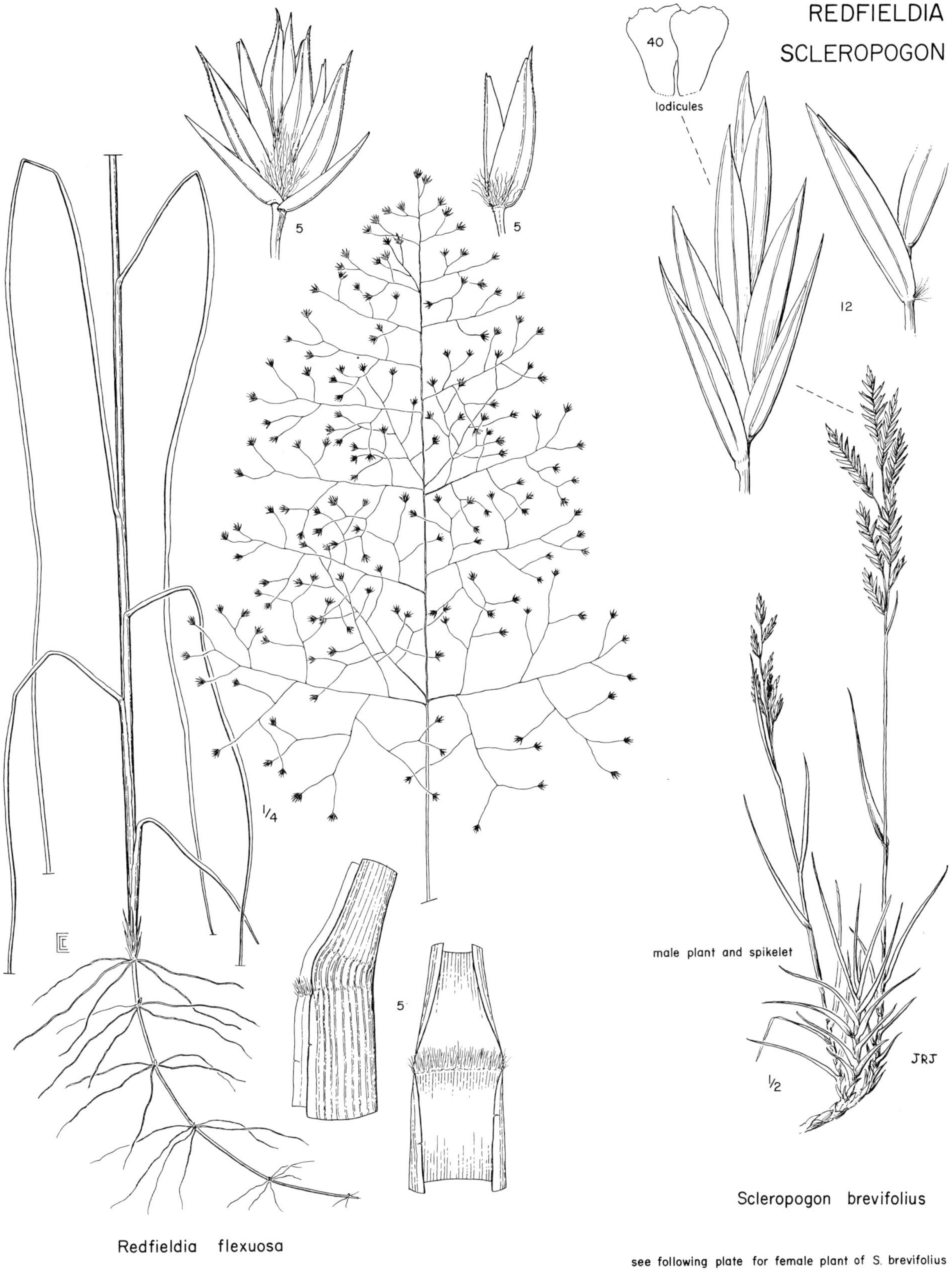
REDFIELDIA
SCLEROPOGON
40
lodicules
5
5
12
1/4
5
male plant and spikelet
JRJ
1/2
Scleropogon brevifolius
Redfieldia flexuosa
see following plate for female plant of S. brevifolius

1.5–3 (3.5) dm long, several, unbranched; leaves crowded near the base; sheaths smooth, the margins sometimes scarious; ligule a very short fringe of hairs, 0.2–0.5 mm long; blades 1.5–2.5 mm broad, flat and sharply bent away from the culm at the base, folded and arcuate above, sometimes sparsely pilose, especially at the collar, the margins sometimes scabrous; pistillate panicles 10–15 (18) cm long (measurement includes the awns), few-flowered; staminate panicles 3–7 cm long, loosely spike-like; pistillate spikelets 80–150 mm long (including the awns), 1- to 4-flowered, usually only the first one fertile, straw-colored to purplish; staminate spikelets 10–24 mm long, 6- to 10- (15)-flowered, slightly laterally compressed, straw-colored, with no abscissions, the rachilla internodes long and exposed, with a sparse tuft of hairs at each node; glumes of the pistillate spikelets subtended by a glume-like bract, about as long as the first glume, the first glume 10–20 mm long, 1- to 5-nerved, the second glume 16–30 mm long, 3-nerved; glumes of the staminate spikelets usually subequal, 3–7 (10) mm long, 1- to 5-nerved; lemmas of the pistillate spikelets narrow, the body 7–11 mm long, 3-awned, the awns 70–140 mm long, the callus bearded; lemmas of the staminate spikelets similar to the glumes, 3.5–7 (10) mm long, 3-nerved, attenuate into a short awn; palea of the staminate spikelets flat, nearly as long as the lemma; lodicules often absent on pistillate spikelets, those of the staminate spikelets truncate, erose, about 0.8 mm long; anthers 2.5–3 mm long; $2n = 28, 40$.

Dry, open hills and plains among grass or low shrub vegetation types; s. Nev. and Ariz., e. to Okla. and Texas, and s. to c. Mex. and recurring in Chile and Argentina. June–Sept.

Scleropogon brevifolius is a coarse grass of low palatability, often increasing on heavily grazed rangelands. After maturing it becomes a problem because the awns penetrate clothing and wool.

63. BLEPHARIDACHNE Hackel

Low, tufted perennials or annuals; culms spreading; leaves mostly in basal tufts; sheaths open; ligules of minute hairs; blades narrow-involute, arcuate and rigid, non-auriculate; inflorescence a short, congested, few-flowered panicle, usually surpassed by the subtending leaves; spikelets compressed, subsessile, 4-flowered, but only the third fertile, the fourth reduced to a 3-awned rudiment, disarticulating above the glumes but not between the florets; glumes more or less unequal, thin, 1-nerved, acute or acuminate; lemmas 3-nerved, 3-lobed and 3-awned; callus not long hairy; awns conspicuously plumose; palea nearly as long as the lemma; $x = 7$?

A genus of 3 species, 2 perennials in the U.S. and 1 annual in Argentina. *Blepharidachne kingii* is endemic to the Great Basin and *B. bigelovii* (S. Wats.) Hackel to Trans-Pecos Texas and adjacent Coahuila. (Name from the Greek *blepharis,* eyelash, and *achne,* chaff, alluding to the ciliate lemma.)

1. Blepharidachne kingii (S. Wats.) Hackel

Eremochloë kingii S. Wats. in King, Rep. Geol. Explor. 40th Parallel 5: 382, pl. 40, figs. 10–16. 1871. *Blepharidachne kingii* Hackel in A. & C. DC. Monogr. Phanerog. 6: 261. 1889. (*Watson 1300,* "dry barren foot-hills of the Trinity Mountains," Pershing Co., Nev., May 1868.)

Low, tufted perennials with crowded basal leaves; herbage and branches of the inflorescence scabrous; culms 0.3–0.8 (1.4) dm long; sheaths with broad hyaline margins, often becoming much flattened and up to 3 mm across, striate, those subtending the inflorescence often spathe-like, with a tuft of hairs at the apex (throat); ligule a very low, ciliate fringe of hairs; blades strongly involute, rigid, abruptly narrowed above the collar, less than 1 mm broad and 0.7–3 cm long, sharp-pointed, arcuate, more or less deciduous at the collar; panicles 1–2.5 cm long, subcapitate, usually exceeded by the upper blades; spikelets pale or purplish, 4-flowered, the florets congested and only the third fertile; glumes exceeding the florets, shiny, hyaline, scaberulous towards the base, 1-nerved, acuminate, more or less awn-tipped, the first glume 6–7.5 mm long, the second glume 6.8–8.5 mm long; lemmas 4, the first lemma 3.4–5.8 mm long, cleft nearly to the middle, 3-lobed, the lateral lobe 2.2–3 mm long, the middle lobe narrow and tapering quickly into the awn, only the margins and awns plumose on the fertile lemma, the awns of the lateral lobes 0.5–1.5 mm long and the awn of the middle lobe 3–5 mm long, the fourth lemma reduced to a rudiment of 3 plumose awns; callus villous; palea 3.5–4.5 mm long, the 2 nerves close to the middle, pilose above, 2-toothed at the apex, the paleas of the sterile lemmas linear and plumose; caryopsis compressed, about 2 mm long; anthers about 1.5 mm long.

Dry, gravelly shadscale deserts and sagebrush valleys; nw. edge of Utah, across c. Nev. to Mono and Inyo cos., Calif., all within the Great Basin. May–July.

A. S. Hitchcock originally reported this species as "apparently rather rare" but it is more widely distributed and common than previously thought. There are large areas in eastern Elko and White Pine counties of Nevada and adjacent parts of Utah where *Blepharidachne kingii* is nearly the dominant grass. The low stature and rigid, sharp-pointed blades make this species rather unpalatable to livestock even though it becomes locally abundant.

Blepharidachne kingii is often mistaken for *Erioneuron pulchellum* as the two species do superficially resemble one another. *Blepharidachne kingii* does not root at the upper nodes, and the broad hyaline sheath margins are somewhat more conspicuous, although they are scarious-margined to a slight extent in *E. pulchellum.* The surest difference, of course, will be found in the spikelet, where the glumes of *E. pulchellum* are usually shorter than the spikelet or at most only as long. Also, the number of florets will vary in *Erioneuron,* the number often exceeding 4, and the lemmas are shorter, about 4 mm long. The spikelets of *B. kingii* disarticulate above the glumes but not between the florets, while in *Erioneuron,* disarticulation occurs above the glumes and between the florets.

64. LYCURUS H.B.K. Wolftail

Low to moderately tall, tufted perennials; culms solid, erect; leaves mostly in a basal tuft, typically grayish-green in color; sheaths open, laterally compressed and sharply keeled; ligules conspicuous, pointed

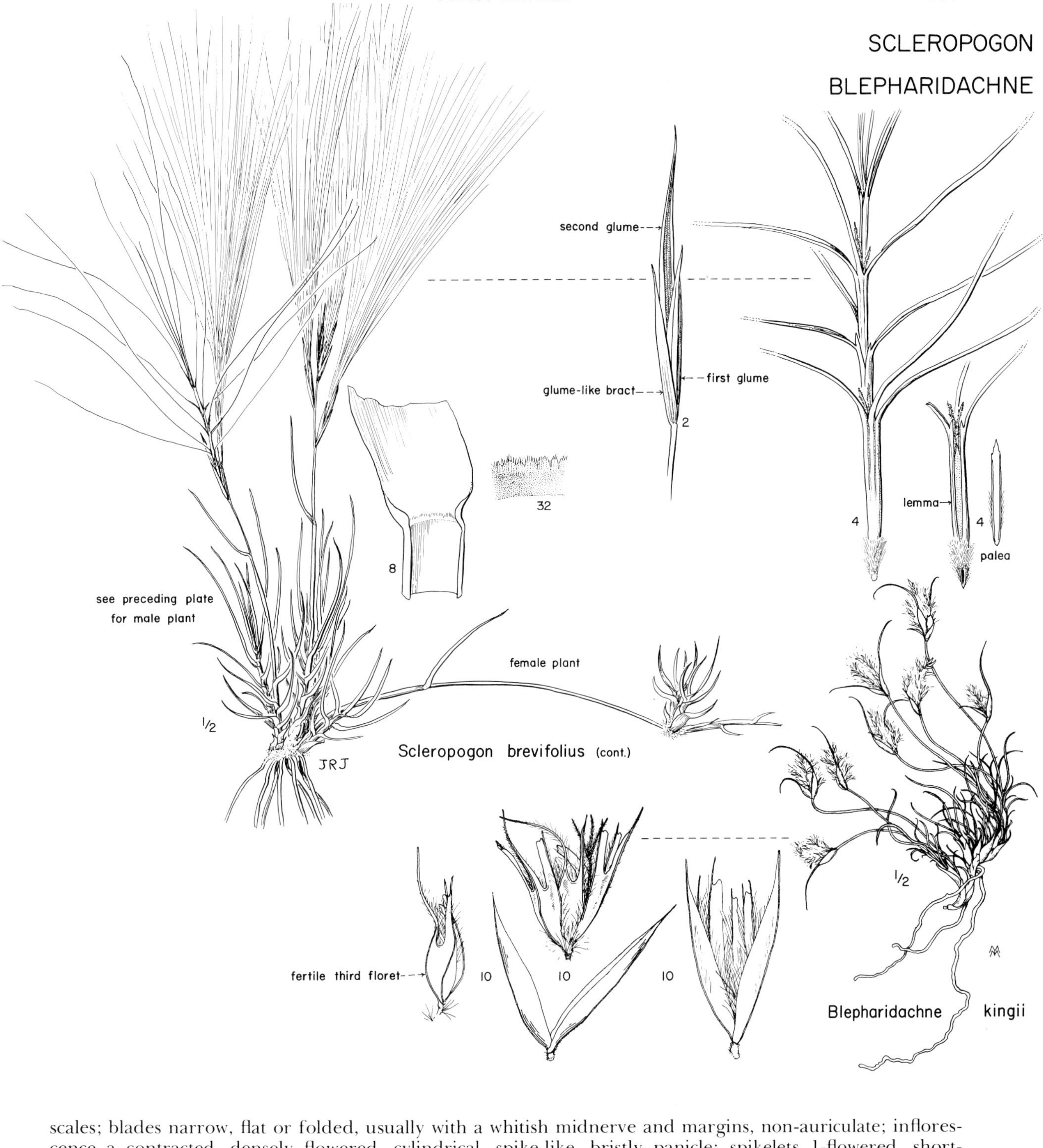

scales; blades narrow, flat or folded, usually with a whitish midnerve and margins, non-auriculate; inflorescence a contracted, densely flowered, cylindrical, spike-like, bristly panicle; spikelets 1-flowered, short-pedicelled, borne in pairs, the lower of the two sterile or staminate, the 2 falling together on short inflorescence branches; glumes unequal, shorter than the lemmas, the first glume usually 2- or 3-nerved and 2- or 3-awned, the second 1-nerved and 1-awned; lemma narrow, firm, tapering into an awn, 3-nerved; palea about as long as the lemma and similar in texture, enclosed by the lemma only at the base; stamens 3; styles short, distinct; $x = 10$.

A genus of 3 species of sw. U.S., n. Mex. and S. Amer. (Name from the Greek *lukos,* wolf, and *oura,* tail, alluding to the spike-like panicles.)

Lycurus, until recent years, was placed in the tribe Agrostideae.

1. Lycurus phleoides H.B.K.

Lycurus phleoides H.B.K. Nova Gen. & Sp. Pl. 1: 142, tab. 45. 1815. (*Humboldt & Bonpland,* Mex.)

Wolftail, Texas timothy.

Tufted perennials; culms low to moderately tall, (2) 3.5–4.5 (6) dm tall, erect or decumbent, solid, slightly flattened, finely puberulent on the ridges of the angled culms, more densely so at the nodes, the nodes sometimes geniculate; sheaths considerably shorter than the internodes, open, glabrous, usually strongly keeled; ligule a prominent membrane, (2) 3–6 (9) mm long, narrow, scarious, acute or erose-lacerate, decurrent with the sheath; blades flat or folded, 1–3 mm broad when unfolded and 4–10 cm long, midribs and margins whitish, glabrous or decidedly scarious beneath, scabrous above; panicles (3) 5–9 (13) cm long, dense and spike-like, 3–7 mm thick; spikelets borne in pairs on short pedicels, the lowermost often sterile or staminate, 1-flowered; glumes subequal, 1–1.8 mm long, the first glume 2-(3)-nerved with an awn 3.5–7 mm long, the second glume 1-nerved with an awn 3.1–6.5 mm long; lemma longer than the glumes, 2.7–3 (4.2) mm long, ciliate, 3-nerved, tapered into an awn 2–6 (8.5) mm long; palea 2–3 mm long, pubescent, enclosed by the lemma only at the base; lodicules about 0.3 mm long; anthers about 1 mm long; $2n = 28, 40$.

Dry, open foothills on gravelly or rocky soil; w. Okla., w. Texas, w. across s. Colo. to se. Utah and Ariz. and s. to n. Mex. June–Aug.

Lycurus phleoides is similar to *Muhlenbergia wrightii* Vasey in general appearance and is often confused with that species.

65. CALAMOVILFA Hackel Sandreed

Tough, wiry perennials with long or short rhizomes; culms erect and thick, semi-solid; leaves distributed up the culm; sheaths open; ligules of short hairs; blades firm, flat, involute, the collar and basal margin of the blades with long hairs, non-auriculate; inflorescence an open or loosely contracted panicle; spikelets 1-flowered, disarticulating above the glumes, the rachilla not prolonged behind the floret; glumes more or less unequal, the first glume shorter than the second, firm, 1-nerved, keeled, acute; lemma longer than the glumes, chartaceous, rounded on the back, glabrous or pubescent, 1-nerved (rare in Eragrosteae), awnless; callus bearing a beard of soft, white hairs; palea about the same length as the lemma; lodicules 2, broad, membranous; stamens 3; ovary glabrous, the styles terminal; fruit not a true caryopsis with its pericarp free from the seed coat (as in *Sporobolus*), elongate; $x = 10$.

A genus of 5 species, endemic to N. Amer. (Name from the Greek *kalamos,* a reed, and *Vilfa,* another genus in the Poaceae.)

Calamovilfa closely resembles *Calamagrostis* and *Ammophila* in gross morphological characters and was placed with them in tribe Agrostideae until Reeder and Ellington (1960) discovered closer affinities within tribe Eragrosteae in characters of the embryo, lodicules, leaf anatomy, epidermal microhairs, and chromosome size and number. *Calamovilfa* belongs close to *Sporobolus,* with which it shares many features, including the 1-nerved lemmas and the peculiar fruit character of having seed coats free from the pericarp.

References:

Reeder, J. R., and M. A. Ellington. *Calamovilfa:* a misplaced genus of Gramineae. Brittonia 12: 71–77. 1960.
Thieret, J. W. Synopsis of the genus *Calamovilfa* (Gramineae). Castanea 31: 145–152. 1966.

1. Calamovilfa gigantea (Nutt.) Scribn. & Merr.

Calamagrostis gigantea Nutt. Trans. Amer. Philos. Soc. II. 5: 143. 1837. *Calamovilfa gigantea* Scribn. & Merr. U.S.D.A. Div. Agrostol. Circ. 35: 2. 1901. *Toxeumia gigantea* Nutt. ex Scribn. & Merr. ibid., as a synonym. (*Nuttall,* "On the sandy banks of Great Salt river of the Arkansas," in 1819.)

Big sandreed.

Robust perennials from strong creeping rhizomes; culms 10–20 (22) dm tall, up to 7 mm thick at the base, usually solitary; sheaths glabrous, except sometimes villous at the throat, much overlapping; ligule a ciliate fringed membrane about 1 mm long; blades 5–10 mm broad at the base, gradually tapering to the tip, involute or sometimes flat for nearly half the length, glabrous, wiry; panicles 25–45 (60) cm long, open with stiffly spreading branches, the lowermost sometimes reflexed; spikelets yellowish, sometimes tinged with purple, 1-flowered; glumes keeled, glabrous, 1-nerved, the first glume (4.5) 5.5–6.5 mm long, the second 6.2–8 (8.5) mm long; lemma 6.5–8 mm long, rounded on back, 1-nerved, villous along the back, becoming glabrous towards the acute tip, the callus blunt and copiously villous with hairs about 4 mm long; palea about the same length as the lemma, villous between the nerves, often villous to the tip; lodicules flabellate, about 0.8 mm long; anthers 3–5.5 mm long, orange; $2n = 60$.

Sand dunes and loose sandy hills and riverbanks; s. Utah (sand dunes n. of Kanab, Kane Co.) and Ariz., e. to sw. Nebr., Kansas, Okla. and c. Texas. July–Sept.

Calamovilfa gigantea, a valuable sand binder, is closely related to *C. longifolia* (Hook.) Scribn. of the midwestern states. *Calamovilfa gigantea* is more robust and has pubescent lemmas and paleas whereas *C. longifolia* has glabrous lemmas and paleas.

66. MUHLENBERGIA Schreber Muhly

Low and delicate to moderately tall or sometimes robust perennials with or without rhizomes, or rarely annuals; culms simple to much-branched, solid or hollow; sheaths open; ligules usually prominent, acute or truncate, sometimes finely erose-ciliate, often longest on the sides; blades narrow, flat or more or less involute, sometimes pungent-tipped, non-auriculate; inflorescence an open to much-contracted panicle; spikelets 1- (rarely 2- or 3)-flowered, the florets terete, disarticulating above the glumes; glumes usually subequal in length, mostly shorter than the lemmas, sometimes as long or rarely exceeding them, 1- (3)-

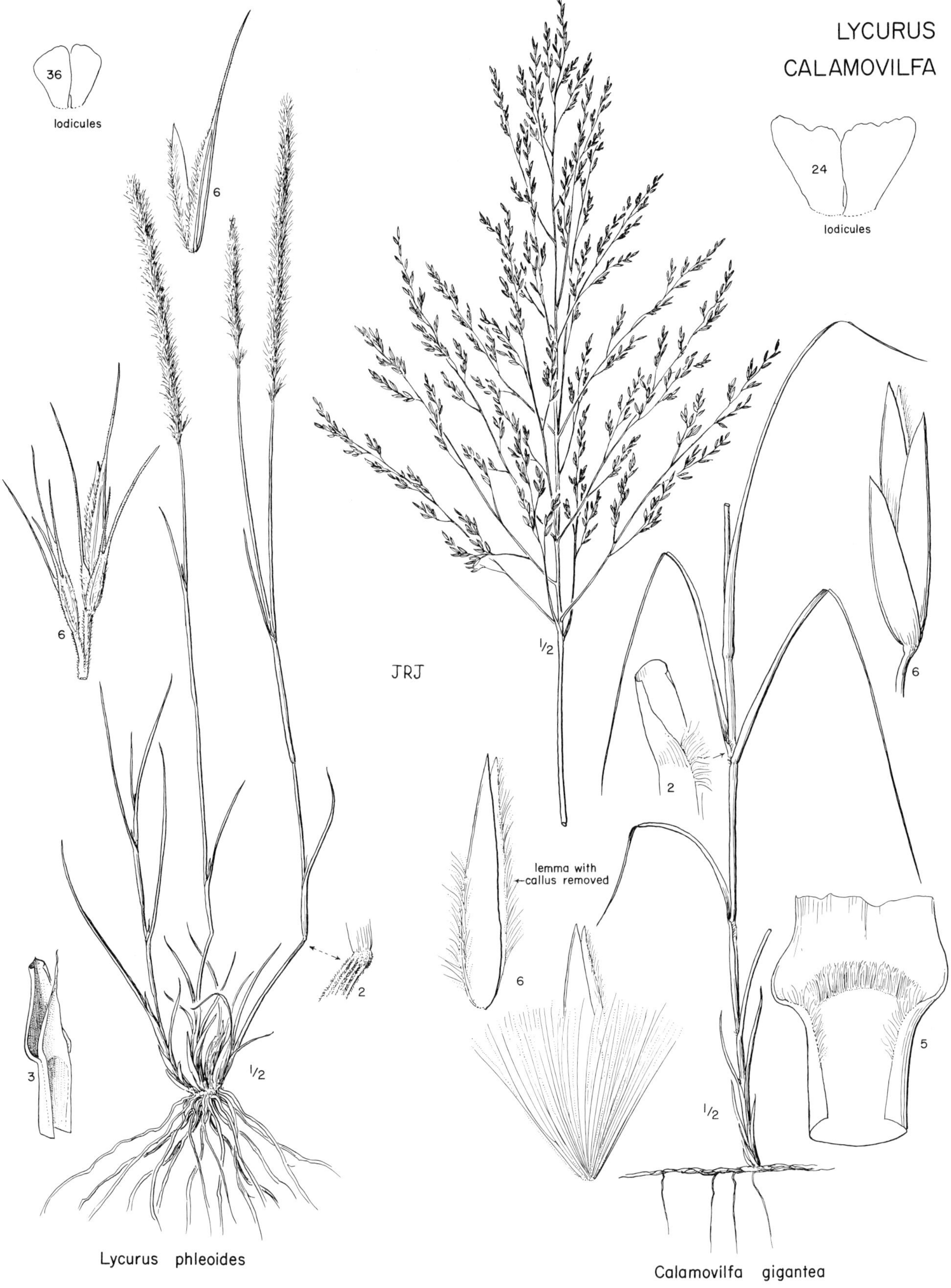
LYCURUS
CALAMOVILFA
36
lodicules
6
6
1/2
3
2
JRJ
Lycurus phleoides
24
lodicules
1/2
6
2
lemma with
←callus removed
6
5
1/2
Calamovilfa gigantea

nerved or occasionally nerveless, obtuse, acute, acuminate or sometimes short-awned; lemma(s) relatively thin, acute or acuminate into an awn-tip or terminating in a slender awn, as firm or firmer than the glumes, 3-nerved (sometimes obscurely so), the nerves glabrous or scabrous; awn present or absent, or only a mucro, the awn straight or flexuous, arising from an entire or minutely cleft apex; callus glabrous to sometimes long-pilose; rachilla not prolonged behind the palea; palea well developed, more or less as long as the lemma and similar in texture, 2-nerved, rounded on back; lodicules 2, short, rarely as much as 0.3 mm long, fleshy; stamens 3; caryopsis elongate, cylindrical or slightly dorsally compressed, usually not falling free from the lemma and palea; $x = 10$.

A large, diverse genus of from 80 to 120 species (perhaps constituting more than one genus), most of which are in the New World, a few in Japan, e. Asia and the Himalayas. (Named for Gotthilf Henry Ernest Muhlenberg, 1753–1815, a Lutheran minister in Pennsylvania who was a student of the Poaceae.)

Several species are well known by the common name of "Muhly" by range men. *Muhlenbergia* is best developed in the Southwest states. In Arizona, where *Muhlenbergia* is the largest genus of the family, there are about 45 species. The number of species diminishes northward, with 16 species in the Intermountain Region and only 9 in the Pacific Northwest. These more northern members are widespread species which are also represented in the Arizona flora and most of which are adapted to relatively mesic environments.

References:
Gould, F. W., and C. G. Reeder. *Muhlenbergia. In:* F. W. Gould. Grasses S.W. U. S. Univ. of Ariz. Press. pp. 186–220. 1951.
Mitchell, W. W., and R. W. Pohl. Variation and aneuploidy in *Muhlenbergia glomerata.* Amer. Midl. Naturalist 76: 211–221. 1966.
Pohl, R. W. *Muhlenbergia,* subgenus Muhlenbergia (Gramineae) in North America. Amer. Midl. Naturalist 82: 512–542. 1969.
———, and W. W. Mitchell. Cytogeography of the rhizomatous American species of *Muhlenbergia.* Brittonia 17: 107–112. 1965.

1 Plants annual, mostly weak rooted and completing a life cycle in six weeks or so.
  2 Panicles narrow and short, 0.5–3.5 (6) cm long; glumes glabrous; lemma 1.5–2.8 mm long .... 15. *M. filiformis*
  2 Panicles open and longer, (4) 10–20 cm long; glumes minutely pilose; lemma 1–1.2 (1.5) mm long .... 16. *M. minutissima*
1 Plants perennial.
  3 Plants with well-developed rhizomes, these usually scaly, creeping, and often branching.
    4 Panicles narrow, the spikelets on short pedicels.
      5 Blades 2 mm broad or less, usually short and involute; lemma about twice as long as the glumes, except in *M. thurberi;* anthers 1–2.7 mm long.
        6 Lemma short, 2–2.5 (3) mm long, at most mucronate, the mucro no longer than 0.3 mm long and the body of the lemma not at all pilose; glumes 1–1.3 (1.5) mm long; widespread .... 1. *M. richardsonis*
        6 Lemma (3) 3.5–5 mm long with a slender awn, the awn 0.7–20 mm long and the body of the lemma pilose at least on the lower half (except most of *M. pauciflora*); glumes 1.5–3.2 mm long; s. portion of the region.
          7 Awns 0.7–3 mm long or the lemmas merely mucronate; spikelets 2–3.5 (4) mm long; plants of moist habitats .... 2. *M. thurberi*
          7 Awns 5–20 mm long; spikelets (3) 3.5–5 mm long; plants of dry habitats.
            8 Culms loosely tufted, hard and wiry at the base, erect or decumbent; lemma sparsely pilose at the base; glumes acuminate, (2) 2.5–3 mm long .... 3. *M. arsenei*
            8 Culms more erect; lemma strigose, not pilose; glumes short-awned, the body 1.5–2.2 mm long .... 4. *M. pauciflora*
      5 Blades mostly more than 3 mm broad, flat; lemma as long as or shorter than the glumes; anthers 0.3–1 (1.5) mm long (Sect. Muhlenbergia).
        9 Hairs of the callus almost or fully as long as the lemma, the hairs 2–2.7 mm long; lemma with an awn 2.5–8 mm long .... 7. *M. andina*
        9 Hairs of the callus less than half as long as the lemma, the hairs 0.2–1.5 mm long; lemma mucronate, short-awned or with an awn to 4–6 mm long in some *M. mexicana.*
          10 Glumes (3) 4–7 mm long, tapering into an awn that exceeds the lemma in length; panicles contracted with tightly appressed branches.
            11 Internodes dull, puberulent; ligules 0.2–0.6 mm long; anthers 0.8–1.5 mm long; lemma pilose at the base and up the margins .... 8. *M. glomerata*
            11 Internodes smooth and polished, except sometimes near the apex; ligules 0.6–1.5 mm long; anthers 0.4–0.6 (0.8) mm long; lemma pilose at the base only .... 9. *M. racemosa*
          10 Glumes 1.8–3.5 (3.8) mm long, not as long awned, not exceeding the lemma in length; panicles with slightly spreading branches .... 10. *M. mexicana*
    4 Panicles open when fully extended from the sheath, the spikelets on slender, more or less elongate pedicels.
      12 Glumes awnless, 0.7–1.7 (2) mm long; lemma only slightly longer than the glume; blades flat or folded; wet meadow habitats .... 6. *M. asperifolia*

12 Glumes short-awned, 1.5–3 mm long (measurement including the awn); lemma nearly twice as long as the glumes; blades involute, sharp-pointed; dry, sandy habitats ..... 13. *M. pungens*
3 Plants without rhizomes, the culms tufted, erect or spreading.
13 Second glume 3-toothed, rarely only erose or entire.
14 Lemma (3) 3.2–4.5 mm long, the awn (3) 6–20 mm long; panicles 5–16 cm long; ligules prominent, 4–17 mm long ..... 11. *M. montana*
14 Lemma 2.2–3 (3.2) mm long, the awn 0.2–2 (3) mm long; panicles only 2–4.5 cm long; ligules 1.5–4.2 mm long ..... 12. *M. filiculmis*
13 Second glume not distinctly 3-toothed, usually acute or awned.
15 Panicles open, or at least loose, naked at the base ..... 14. *M. porteri*
15 Panicles narrow or spike-like, spikelet-bearing to the base.
16 Lemma 1.5–2.2 (2.8) mm long ..... 15. *M. filiformis*
16 Lemma (2.5) 3–5 mm long, exclusive of any awn.
17 Lemma with an awn 5 mm or more long.
18 Culms loosely tufted, hard and wiry at the base; glumes (2) 2.5–3 mm long, acuminate, awnless; lemma pilose below; plants sometimes appearing rhizomatous ..... 3. *M. arsenei*
18 Culms more erect, branching only at the base; body of the glumes 1.5–2.2 mm long, short-awned; lemma rarely pilose, more or less strigose ..... 4. *M. pauciflora*
17 Lemma awnless or short awn-tipped, the awn less than 1 mm long ..... 5. *M. wrightii*

1. Muhlenbergia richardsonis (Trin.) Rydb.

Vilfa richardsonis Trin. Mém. Acad. Imp. Sci. Saint-Pétersbourg, Sér. 6, Sci. Math., Seconde Pt. Sci. Nat. 4(1): 103. 1840. *Muhlenbergia aspericaulis* Nees ex Trin. ibid., as a synonym. *Sporobolus aspericaulis* Scribn. Bot. Gaz. 21: 15. 1896. *Sporobolus richardsonii* Merr. Rhodora 4: 46. 1902. *M. richardsonis* Rydb. Bull. Torrey Bot. Club 32: 600. 1905. *M. brevifolia* var. *richardsonis* M. E. Jones, Contr. W. Bot. 14: 12. 1912. (*Richardson,* N. Amer.)

Vilfa squarrosa Trin. Mém. Acad. Imp. Sci. Saint-Pétersbourg, Sér. 6, Sci. Math., Seconde Pt. Sci. Nat. 4(1): 100. 1840. *M. squarrosa* Rydb. Bull. Torrey Bot. Club 36: 531. 1909. (Menzies I., Columbia River, Wash.)

Vilfa depauperata Torr. in Hook. Fl. Boreali-Amer. 2: 257, tab. 236. (July) 1840; not *M. depauperata* Scribn. 1884. *Sporobolus depauperatus* Scribn. Bull. Torrey Bot. Club 9: 103. 1882. (*Douglas,* Columbia River, "from Menzies' Island upwards.")

Mat muhly.

Mat-forming perennials from hard, creeping, scaly rhizomes; culms short, (0.5) 1–2.5 (4.5) dm tall, erect or decumbent at the base, often geniculate at the nodes, becoming much-branched at the base, solid, slightly flattened and minutely nodulose-roughened (this character scarcely perceptible); sheaths glabrous, striate; collar usually brownish, somewhat cartilaginous towards the edges; ligules 1–2 (3) mm long, acute, strongly decurrent; blades often nearly flat or involute towards the tip or involute throughout, 1–1.5 (2) mm broad and 1–5 cm long, tapered to a point and often nearly pungent, puberulent above and scaberulous beneath; panicles 1–4 (7) cm long, narrow and spike-like, the branches short, appressed; spikelets 1-flowered, green to nearly black; glumes subequal, small, 1–1.3 (1.5) mm long, broad, ovate, the midnerves prominent, sometimes lightly scaberulous with scabrous nerves, the first glume 1-nerved, the second 3-nerved, the lateral nerves often weak; lemma 2–2.5 (3) mm long, about twice as long as the glumes, narrow, rounded on back, 3-nerved, glabrous to minutely scaberulous, usually mucronate or shortly awn-tipped, the awn up to 0.3 mm long; callus naked; palea about the same length as the lemma, the upper margins scarious; anthers 1.2–1.5 mm long; $2n = 40$ (for *M. squarrosa*).

Moist to moderately moist situations in meadows, along streams and margins of ponds and lakes, often in alkaline soils, mostly in the mts., sometimes up to timberline; B.C. to N.B., s. in the w. states to Calif., Nev., n. Ariz. and n. N.M., in the c. states to S.D., Minn., Wisc. and Mich. and in the e. states to Me.; also in Baja Calif. (June) July–Sept.

Mat muhly sometimes forms dense mats in the moist mountain situations where it is an important forage grass as well as a good soil binder. An occasional plant has florets modified into bulbils. The low phase of this species has been called *M. squarrosa,* but in spikelet characters the plants are rather uniform.

2. Muhlenbergia thurberi Rydb.

Sporobolus filiculmis Vasey, Descript. Cat. Grasses U. S. 44. 1885, name only. *Vilfa filiculmis* Thurber ex Vasey, ibid., as a synonym. *Sporobolus filiculmis* Vasey ex Beal, Grasses N. Amer. 2: 288. 1896; not L. H. Dewey, 1894. *Sporobolus thurberi* Scribn. U.S.D.A. Div. Agrostol. Bull. 11: 48, fig. 5. 1898. *Muhlenbergia thurberi* Rydb. Bull. Torrey Bot. Club 32: 601. 1905; based on *Sporobolus filiculmis* Vasey ex Beal without reference to *S. thurberi* Scribn. *M. filiculmis* M. E. Jones, Contr. W. Bot. 14: 12. 1912; not Vasey, 1893. (*Bigelow 778,* Plaza Larga, N.M., 21 Sept. 1853.)

M. curtifolia Scribn. Bull. Torrey Bot. Club 38: 328. 1911. (*Jones 6047j,* "between Kanab and Carmel, Utah," 18 Sept. 1894.)

M. curtifolia subsp. *griffithsii* Scribn. Bull. Torrey Bot. Club 38: 328. 1911. (*Griffiths 5837,* Du Chelly Canyon, Ariz., in 1903.)

Slender, wiry, glaucous perennials from long, creeping rhizomes; culms low, 1–3.5 dm tall, erect or decumbent at the base, loosely branched below or tufted from the branches of the rhizomes, solid, pubescent below the lower nodes or glabrous; leaves 2-ranked, 7–10 on each culm, sometimes crowded near the base or rather evenly spaced over 5–7 cm of the lower culms; sheaths more or less keeled, glabrous or with spreading pubescence, becoming slightly more dense below the glabrous collar; ligules short, 0.2–0.7 mm long, ciliate on the edge, decurrent; blades mostly flat towards the base, becoming

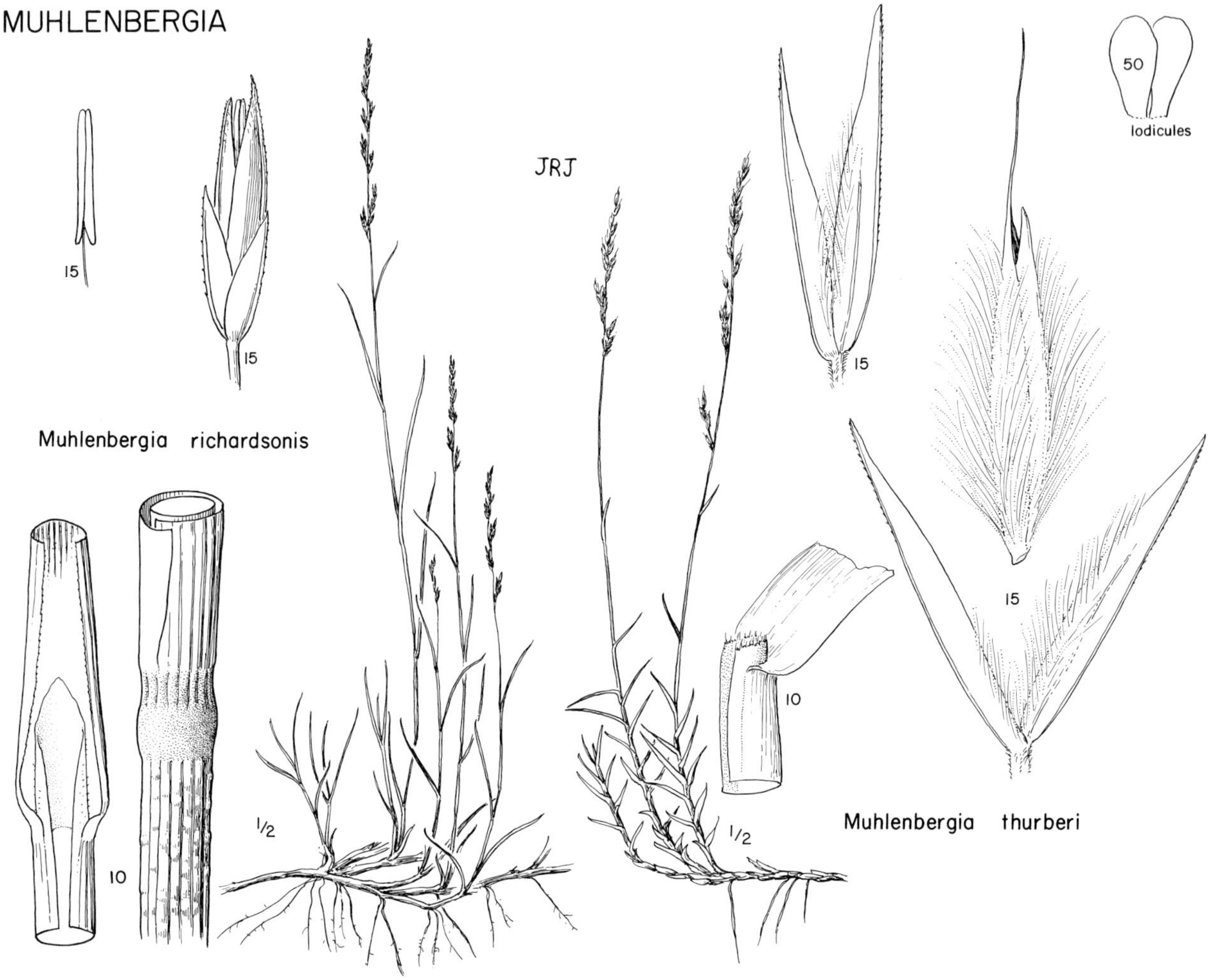

involute above, 1–1.2 (2) mm broad and 8–25 cm long, usually terminating in a sharp point, pubescent on both sides or scabrous below with spreading hairs above; panicles 2.5–8 (9) cm long, long exserted, narrow, often interrupted, the branches appressed, the slender and pubescent pedicels 3–4 mm long; spikelets 1-flowered, often tinged with purple; glumes subequal, 2.5–3.2 mm long, lanceolate, acute, sometimes scabrous along the keel towards the tip, often villous on the margins below, the first glume 1-nerved, the second 1- (3)-nerved; lemma 3–3.5 (4) mm long, barely exceeding the glumes, villous on the margins and lower half, 3-nerved, acuminate, sometimes mucronate or short-awned, the awn 0.7–2.5 (3) mm long; callus naked; palea as long as the lemma, villous between the nerves; anthers 1–2 mm long.

In hanging gardens or on rocky ledges and rocky slopes, often in moist places; s. Nev. (Charleston Mts.), s. Utah, n. Ariz. and sw. Colo. Late July–Sept.

Swallen included both *M. curtifolia* and *M. thurberi* in his Gramineae of Nevada (Contr. Toward Fl. Nevada 1: 69. 1940) but all the Nevada material we have seen was identified as *M. curtifolia.* The asserted differences between these taxa, such as culms being glabrous below the nodes, sheaths glabrous, ligule 1 mm long, lemma mucronate or short-awned in *M. thurberi* and culms strigose below the nodes, sheaths often strigose to hirsute, ligule 0.5–1 mm long, lemma with an awn 1–3 mm long in *M. curtifolia,* are not readily discernible, and for our purposes it seems desirable to combine the two. For a time we tried to recognize both of them and even considered the possibility that plants from the Charleston Mountains (not in our area) constituted yet a third and undescribed taxon.

3. **Muhlenbergia arsenei** A. S. Hitchc.

Muhlenbergia arsenei A. S. Hitchc. Proc. Biol. Soc. Wash. 41: 161. 1928. (*Arsène & Benedict 16405,* "Vicinity of Sulphur Springs, Sandoval Co.," N.M., 17 Aug. 1926.)

Loosely tufted, glaucous perennials; rhizomes lacking; culms (1) 2–3.5 (4) dm tall, erect or decumbent at the much-branched base and often appearing rhizomatous, solid, wiry; leaves somewhat crowded towards the base; sheaths glabrous, striate; collar brownish, somewhat cartilaginous; ligules up to 2 mm long towards the edges where nearly appearing to be acuminate from the corners of the sheaths, more or less decurrent, shorter around the back, becoming lacerate; blades filiform-involute, sometimes loosely so, 1–3 cm long, sharp-pointed; panicles 3–8 (12) cm long, long-exserted, narrow, often interrupted towards the base, the ascending branches floriferous to the base, the pedicels scabrous-pubescent; spikelets 1-flowered, often tinged

MUHLENBERGIA

with purple; glumes subequal, (2) 2.5–3 mm long, acuminate, somewhat hispidulous towards the acute tip, the prominent keel sometimes nearly ciliate; lemma (3.2) 4–5 mm long, nearly twice as long as the glumes, narrow, 3-nerved, slightly pubescent on the lower part, the long flexuous awn 5–10 (12) mm long; callus naked; palea about the same length as the lemma, slightly pubescent towards the base; anthers large, 1.6–2.7 mm long, yellowish-orange; caryopsis elongate, reddish.

Dry slopes and rocky ledges; s. Calif. (Clark Mt.), s. Utah and n. N.M. August.

The report of *M. polycaulis* Scribn. from southeastern Utah by Rydberg (Fl. Rocky Mts. 1918) and by Chase (in Tidestrom, Contr. U. S. Natl. Herb. 25: 75. 1925) was probably based on *Rydberg & Garrett 9498,* a paratype of Hitchcock's *M. arsenei.*

4. Muhlenbergia pauciflora Buckley

Muhlenbergia pauciflora Buckley, Proc. Acad. Nat. Sci. Philadelphia 1862: 91. 1863. (*Wright 732,* hillsides, w. Texas, in 1849.)

M. sylvatica var. *pringlii* Scribn. Bull. Torrey Bot. Club 9: 89. 1882. *M. pringlei* Scribn. in Vasey, Contr. U. S. Natl. Herb. 3: 71. 1892. (*Pringle 480,* "Dry Cliffs, Santa Rita Mts.," Ariz., 14 July 1881.)

M. neo-mexicana Vasey, Bot. Gaz. 11: 337. 1886. (*Vasey,* "Rocky hills and mountain sides," N.M. and Ariz.)

New Mexican muhly.

Glaucous perennials from very short, firm, knotty rhizomes; herbage smooth to scaberulous; culms (2) 3–5 (6.5) dm tall, erect, becoming sparingly branched at the lower nodes and often rooting, solid; sheaths glabrous, striate, the old sheaths often becoming flat and corkscrew-like in age; ligules 0.5–2.5 mm long, longest on the sides of the throat, sometimes these lateral extensions lanceolate to linear, lacerate-ciliate, appearing as extensions of the sheath; blades flat or involute, 0.5–1.5 mm broad and up to 8 cm long, scaberulous, somewhat deciduous from the sheath; panicles 5–12 (16) cm long, narrow, interrupted, rather densely flowered, the pedicels scabrous-pubescent; spikelets 1-flowered, usually tinged with purple; glumes subequal, 1.5–2.2 mm long, broad, acuminate, often erose apically, 1-nerved, the nerves scabrous, sometimes scaberulous across the back, often terminating in an awn 0.2–1 mm long; lemma (3) 3.5–4.5 mm long, about twice as long as the glumes, narrow, glabrous or usually strigose in short lines above the callus, 3-nerved, gradually tapering into a long awn (5) 8–20 mm long; callus glabrous or sometimes with a few hairs; palea more or less as long as the lemma, narrow, glabrous; anthers large, about 1.5 mm long, orangish; caryopsis elongate, reddish.

Rocky hills, steep ledges and canyons in the sw. U.S.; s. Utah, s. Colo., Ariz., N.M. and Trans-Pecos Texas, s. to Baja Calif., Sonora and Chihuahua. Aug.–Sept.

The short, firm, knotty rhizomes of this species are often not present in poorly prepared herbarium specimens. The rhizomes alone will separate *M. pauciflora* from *M. arsenei,* a species with which it is often confused.

5. Muhlenbergia wrightii Vasey ex J. M. Coulter

Muhlenbergia wrightii Vasey ex J. M. Coulter, Manual Bot. Rocky Mt. Region 409. 1885. (*Wright 1986,* N.M., in 1851–1852.)

M. wrightii var. *annulata* Vasey, Descript. Cat. Grasses U. S. 41. 1885, nomen nudum. (*Lemmon 3179,* Ariz.)

M. coloradensis Mez, Repert. Spec. Nov. Regni Veg. Beih. 17: 213. 1921. (*Jones 806,* "Chiann [Cheyenne] Canyon," Colo., 3 Oct. 1878.)

Spike muhly.

Closely tufted, non-rhizomatous perennials; culms 1.5–4.5 (6) dm tall, usually erect, rarely somewhat decumbent, often forming large, dense clumps from a hard crown and usually branching below, solid, wiry, the nodes puberulent, scabrous below the panicle; herbage scaberulous to shortly puberulent; sheaths compressed-keeled, glabrous, shorter than the internodes; ligules sometimes relatively prominent, 1–5 mm long, ciliolate, relatively broad, often truncate on the lower leaves and rounded to acute on the upper, usually decurrent; blades flat, sometimes folded, rarely somewhat involute, 1–4 mm broad and 6–15 cm long, scabrous above, glabrous beneath; panicles 2.5–12 (15) cm long, spike-like, interrupted below, densely flowered; spikelets 1- (2)-flowered; glumes subequal, 1.7–3.5 mm long, rather thin, 1-nerved, scabrous, tapering to an awn-point, the second glume with hyaline edges; lemma (2.5) 3.2–4.5 (5) mm long, 3-nerved, the lateral nerves sometimes indistinct, glabrous to scaberulous and with a blackish cast, acuminate, awn-tipped, the awns 0.2–1 mm long; callus naked; palea mostly 1.5–2 mm long; anthers large, 1.5–2 mm long; stigmas bright orange to chestnut brown.

Dry open hillsides, drying edges of meadows and open ponderosa or gamble oak woodlands in the sw. U.S.; s. Utah, Colo., w. Okla., s. through Ariz., N.M. and adj. Texas into n. Mex. July–Sept.

6. Muhlenbergia asperifolia (Nees & Meyen) L. R. Parodi

Vilfa asperifolia Meyen, Reise Erde 1: 408. 1834, name only; Nees & Meyen, Mém. Acad. Imp. Sci. Saint-Pétersbourg, Sér. 6, Sci. Math., Seconde Pt. Sci. Nat. 4(1): 95. 1840. *Sporobolus asperifolius* Nees, Nov. Actorum Acad. Caes. Leop.-Carol. German. Nat. Cur. 19: Sup. 1: 9. 1841. *Muhlenbergia asperifolia* L. R. Parodi, Revista Fac. Agron. Veterin. 6: 117, fig. 1. 1928. (*Meyen,* Rio Mayno and Copiapo, Chile.)

Scratchgrass muhly, alkali muhly.

Glaucous perennials with long, scaly rhizomes; culms 1–5 dm tall, slender or ascending, branching at the base, compressed, hollow; sheaths compressed-keeled, glabrous, overlapping, somewhat cartilaginous towards the edges with the blades abruptly widened immediately above the collar, finely pubescent on the margins; collar yellowish-brown; ligules 0.2–0.8 (1) mm long, truncate, finely erose-ciliolate, somewhat decurrent; blades flat or folded, 1–2.5 mm broad and about 2–5 cm long, crowded, scabrous, sometimes finely puberulent on the upper surface; panicles large, 6–14 (28) cm long and nearly as wide, open, diffuse, the lower branches often remaining enclosed by the upper sheaths, the capillary, scabrous branches widely spreading, often flexuous, the pedicels slightly enlarged below the spikelets, the panicle breaking away

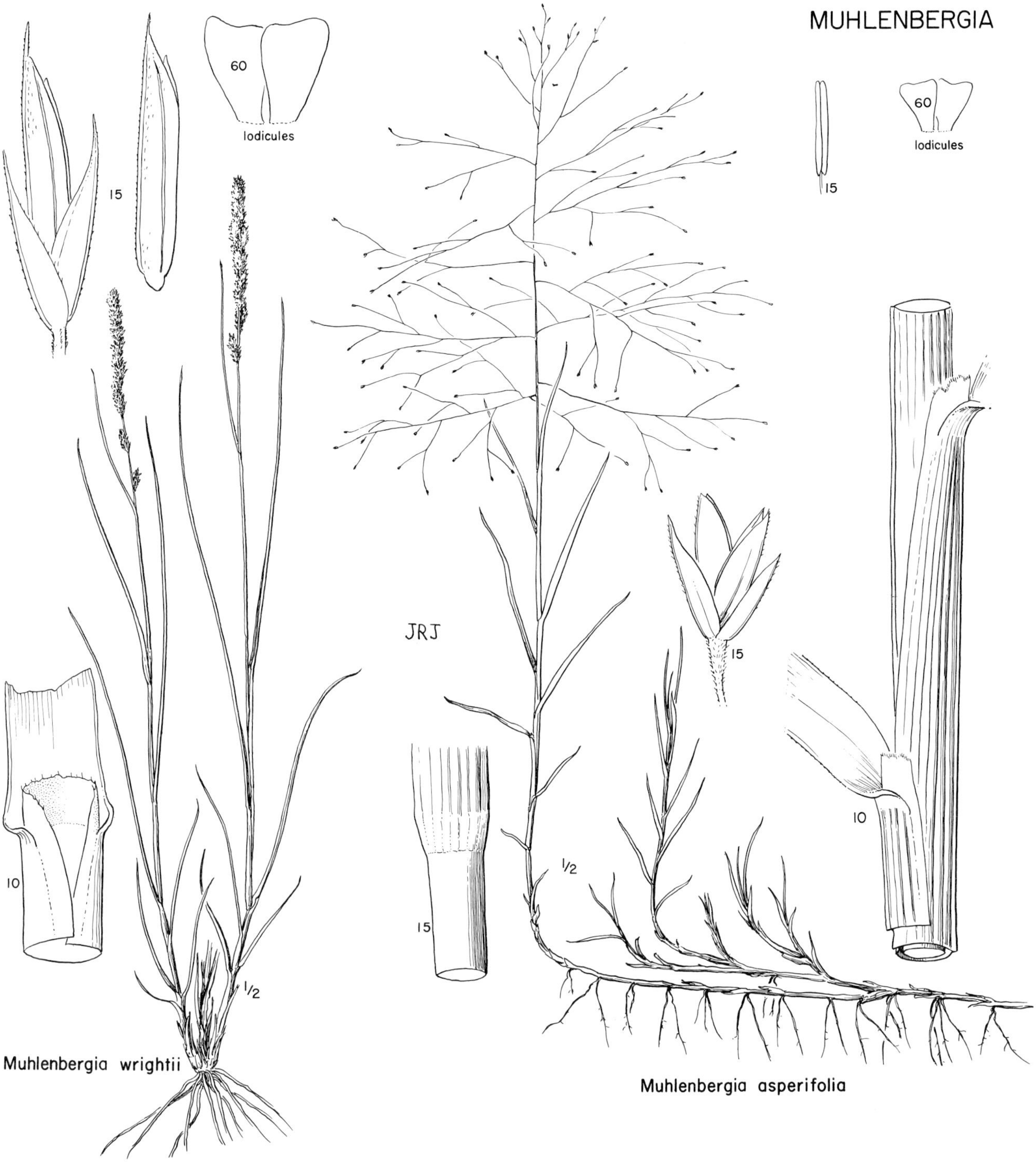

at maturity; spikelets 1- to 2- (3)-flowered, often deep-purple; glumes subequal, small, 0.7–1.7 (2) mm long, acute to nearly mucronate, puberulent-scabrous on the keel, sometimes with scarious margins; lemmas 1.2–1.6 mm long, only slightly longer than the glumes, 3-nerved, thin, broad, minutely mucronate, often with a blackish cast; palea as long or slightly longer than the lemma and about as broad; anthers about 1 mm long; $2n$ = 20, 22, 28.

Moist, usually in alkaline soil of meadows, margins of playas, streams (some from thermal springs) and margins of ponds; B.C. to Man., s. in the w. U.S. to Calif., Ariz., N.M. and w. Texas and in the c. states to Mo., Ill. and Ohio, on s. into Mex. and recurring in s. S. Amer. (June) July–Aug.

Scratchgrass, most noticeable in the spring when it is late greening up, is often a troublesome lawn weed. Competing vigorously with Kentucky bluegrass, a small, circular, glaucous (clearly marked on a dewy morning) patch soon increases in size. It probably gets a foothold in old established lawns, as the deciduous panicles become airborne during high winds. Scratchgrass often occurs in nearly pure stands in saline or alkaline bottomlands where the water table is high.

Specimens without spikelets can be identified by the keeled sheath, short, truncate, ciliolate ligule, and abruptly expanded blade immediately above the collar. Early floras treated *M. asperifolia* as a *Sporobolus*.

The caryopsis is often affected by the smut fungus, *Tilletia asperifolia* Ellis & Everh., which produces a large globular body. Infestations are often local and heavy in cool, wet years.

7. Muhlenbergia andina (Nutt.) A. S. Hitchc.

Calamagrostis andina Nutt. Proc. Acad. Nat. Sci. Philadelphia 4: 24. 1848. *Muhlenbergia andina* A. S. Hitchc. U.S.D.A. Bull. (1915–23) 772: 145. 1920. (*Gambel*, "In Upper California, on the Colorado of the West," probably from Utah, late Sept. or early Oct. 1841.)

Vaseya comata Thurber in A. Gray, Proc. Acad. Nat. Sci. Philadelphia 1863: 79. 1864. *M. comata* Thurber ex Benth. J. Linn. Soc., Bot. 19: 83. 1881. (*Hall & Harbour 685*, Nebr. [probably Wyo.].)

Foxtail muhly.

Erect perennials from moniliform, shiny, white rhizomes; culms relatively tall, 2.5–6 (8.5) dm, hollow, scabrous-puberulent around the nodes; sheaths keeled, glabrous or more often puberulent; ligules 0.5–1.2 mm long, truncate, erose-ciliolate; blades flat or subinvolute, 2–3.2 (5) mm broad and 5–12 cm long, scabrous, bright-green or sometimes glaucous; panicles (3.5) 5–10 (15) cm long, narrow and spike-like, interrupted or lobed; spikelets 1-flowered, shiny, often purple-tinged; glumes subequal, 2.5–3.6 (4) mm long, usually exceeding the lemma, linear to nearly filiform, acuminate, often bulged at the base, the keels ciliate-scabrous, mucronate or often short-awned; lemma 2.2–3.2 mm long, often with a blackish cast, 3-nerved, tapering into a capillary, straight or more often flexuous awn, the awn 2.5–8 mm long; callus bearded with dense, long, silky hairs, about as long as the body of the lemma or only slightly shorter, the hairs 2–2.7 mm long; palea about as long as the lemma; anthers small, 0.4–0.6 mm long; $2n = 10, 20$.

Meadows, marshes, streambanks and gravel bars at middle elevs. in the mts. of w. U.S.; e. Wash., Oregon and Calif., e. to the Rocky Mts. of Mont., Wyo., Colo., N.M. and Trans-Pecos Texas. July–Aug. (–Sept.).

8. Muhlenbergia glomerata (Willd.) Trin.

Polypogon glomeratus Willd. Enum. Pl. Hort. Regii Bot. Berol. 87. 1809. *Alopecurus glomeratus* Poiret in Lam. Encycl. Méth. Bot. Suppl. 5: 495. 1817. *Trichochloa glomerata* Trin. Fund. Agrostogr. 117. 1820. *Muhlenbergia glomerata* Trin. Gram. Unifloris & Sesquifloris 191, pl 5. fig. 10. 1824. *Podosemum glomeratum* Link, Hort. Regius Bot. Berol. 1: 84. 1827. *Cinna glomerata* Link, ibid., 2: 237. 1833; not Walter, 1788. ("Habitat in America boreali" [Pa.].)

Perennials from long, creeping, scaly rhizomes; culms tall, 3–12 dm, erect, unbranched or sparingly branched, hollow, slightly compressed, the internodes finely puberulent; sheaths slightly keeled, scaberulous to glabrous; ligules short, 0.2–0.6 mm long, membranous, erose-ciliate; blades flat, usually 2–6 mm broad and 6–15 cm long, scaberulous; panicles 2–11 cm long, narrow and spike-like, somewhat lobed with densely clustered spikelets; spikelets 1-flowered, green to purplish; glumes subequal, long, 3.2–5 (7) mm long, usually exceeding the lemma, narrow, tapering into a short awn (included in the measurement); lemma 2.2–3.7 mm long, 3-nerved, acuminate, pubescent on the nerves and margins, sometimes nearly to the tip, awnless or sometimes with a short awn-tip; callus bearded, the hairs about 1 mm long; palea about the same length as the lemma; anthers small, 0.8–1 (1.5) mm long; caryopsis 1–1.6 mm long, cylindrical; $2n = 20, 40$.

Streambanks, meadows, marshes and shores of ponds and lakes; B.C. across s. N.W.T. to Newfl., and s. in the w. states (sporadically) to Nev., Utah and Colo., in the c. and e. states to Nebr., Iowa, Ill., Ind., Ohio, W.Va. and N.C.; rare in the Intermountain Region. July–Aug.

Muhlenbergia glomerata closely resembles *M. racemosa* and has been merged with that species by some. Pohl (1969) separates them on the basis of culm and lemma pubescence and ligule and anther length as in the key above. The two species also have a different chromosome number and occupy different habitats. *Muhlenbergia glomerata*, a diploid, is confined to moister, non-disturbed habitats. It is a rare species that exists as isolated populations, perhaps relict colonies of a wide Pleistocene distribution. *Muhlenbergia racemosa*, on the other hand, is a tetraploid that is found in drier ground and often in disturbed sites.

9. Muhlenbergia racemosa (Michx.) Britton, Sterns, & Poggenb.

Agrostis racemosa Michx. Fl. Boreali-Amer. 1: 53. 1803. *Vilfa racemosa* Beauv. Essai Nouv. Agrostogr. 16, 148, 182. 1812. *Polypogon racemosum* Nutt. Gen. N. Amer. Pls. 1: 51. 1818. *Cinna racemosa* Kunth, Révision Gram. 1: 67. 1829. *Muhlenbergia racemosa* Britton, Sterns, & Poggenb. Prelim. Cat. Antho. & Pterid. within 100 Mi. New York City 67. 1888. (*Michaux*, "In ripas sabulosis inundatis fluminis, Mississippi" River, Ill., in 1795.)

M. glomerata var. *ramosa* Vasey, Descript. Cat. Grasses U. S. 40. 1885. *M. racemosa* var. *ramosa* Vasey ex Beal, Grasses N. Amer. 2: 253. 1896. (*Vasey*, "Dakota and Wisconsin.")

Perennials from long, creeping, scaly rhizomes; culms relatively tall, 2.5–6 (13) dm, stiffly erect, usually branching at about the midlength, hollow, terete or only slightly compressed, the internodes glabrous or slightly puberulent near the apex; sheaths keeled, smooth; ligules 0.6–1.5 mm long, erose-ciliate, truncate, more or less decurrent; blades flat, 2–5 (7) mm broad and 5–15 cm long, rather evenly distributed along the culm, scabrous; panicles 4–17.5 cm long, narrow and spike-like, often lobed with densely clustered spikelets; spikelets 1-flowered, usually light-green, often sessile; glumes subequal, (3) 4–6 (7) mm long, usually exceeding the lemma, narrow, attenuate, tapering into a short awn (included in the measurement), ciliate-scabrous on the keels, scaberulous across the back; lemma 2.2–3.7 (4.5) mm long, 3-nerved, pilose on the lower half, awnless or minutely awn-tipped; callus bearded, the hairs 0.7–1.5 mm long; palea about the same length as the lemma, ascending-pilose on the keels; anthers small, (0.4) 0.5–0.8 mm long; caryopsis 1.4–2.3 mm long, cylindrical, brown; $2n = 40$.

Drying meadows, rocky slopes, along irrigation ditches and often in cultivated areas; Alta. to Ont., s. in the mt. states from e. Wash.,

MUHLENBERGIA
45
lodicules
15
5
15
15
JRJ
15
8
1/2
1/2
1/2
10
10
Muhlenbergia andina
Muhlenbergia glomerata

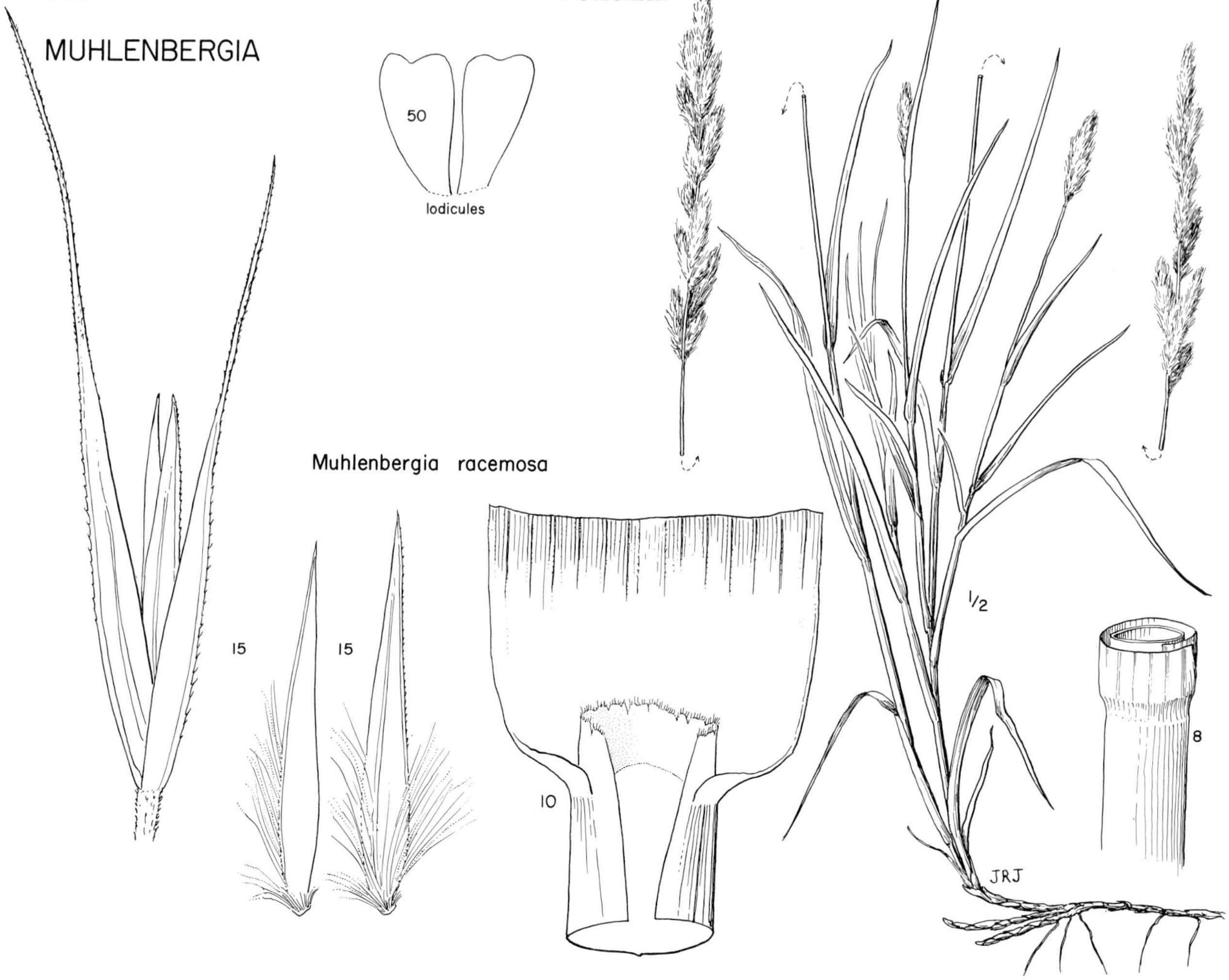

e. Idaho, Utah, n. Ariz., N.M. and Texas, and in the c. states to Mo., Ill. and Ind.; infrequent in the Intermountain Region. July–Sept.

Muhlenbergia racemosa is a tetraploid with a genome possibly from *M. glomerata* (Pohl and Mitchell, 1965; Pohl, 1969) and one from an extinct diploid that would have been without awns and with glabrous internodes, slightly hairy lemmas, greater leafiness, more branching, and more drought and heat resistance. *Muhlenbergia racemosa* is sometimes confused with *M. andina* in our area. The lemma characters alone will readily separate them as *M. andina* has long, copious hairs from the callus (about the same length as the lemma), and the lemma tapers into a capillary awn about 4–8 mm long. *Muhlenbergia racemosa* is found in drier ground and often in disturbed sites.

10. Muhlenbergia mexicana (L.) Trin.

Agrostis mexicana L. Mantissa Pl. 1: 31. 1767. *Vilfa mexicana* Beauv. Essai Nouv. Agrostogr. 16, 148, 181. 1812. *Cinna? mexicana* Beauv. ibid., 32, 148, 158. 1812. *Trichochloa mexicana* Trin. Fund. Agrostogr. 117. 1820. *Muhlenbergia mexicana* Trin. Gram. Unifloris & Sesquifloris 189. 1824. *Podosemum mexicanum* Link, Hort. Regius Bot. Berol. 1: 84. 1827. (Grown at Uppsala from seed sent by Jacquin, the source erroneously given as trop. Amer.)

M. ambigua Torr. in Nicollet, Rep. Upper Mississippi River 164. 1843. *M. foliosa* subsp. *ambigua* Scribn. Rhodora 9: 20. 1907. *M. foliosa* f. *ambigua* Wieg. Rhodora 26: 1. 1924. *M. mexicana* f. *ambigua* Fern. Rhodora 45: 236. 1943. (*Geyer,* "stony banks of Lake Okaman, Sioux Country," 20 Sept. 1838.) A form with awned lemmas.

M. sylvatica var. *setiglumis* S. Wats. in King, Rep. Geol. Explor. 40th Parallel 5: 378. 1871. *M. foliosa* subsp. *setiglumis* Scribn. Rhodora 9: 20. 1907. *M. setiglumis* Nels. & Macbr. Bot. Gaz. 61: 30. 1916. *M. mexicana* f. *setiglumis* Fern. Rhodora 45: 236. 1943. (*Watson 1288,* "Near warm springs in Humboldt Pass," Nev., Sept. 1868.) A form with both glumes and lemmas awned.

Agrostis foliosa of Amer. authors; not "Hortul." Roemer & Schultes, 1817.

Perennials from creeping, scaly rhizomes; culms relatively tall, (3) 4–8 (10) dm, hollow, the internodes dull and puberulent, especially just below the nodes; sheaths keeled, glabrous; ligules 0.5–1 (1.5) mm long, truncate to acute, erose-ciliolate; blades flat, 1.5–4 (5) mm broad and 5–15 cm long, lax, scaberulous; panicles (3) 5–12 (20) cm long, narrow, short-exserted or partly included, the several branches closely appressed, the secondary panicles arising from each of the upper 2 or 3 nodes; spikelets 1-flowered, purple-tinged; glumes subequal, 1.8–3.5 (3.8) mm long, very narrow, the keel scabrous, short-awned, the awn about 1.2 mm long; lemma 2.2–3.8 mm long, barely exceeding the glumes, 3-nerved, pilose below, acuminate, narrowed to an awn tip or with an awn 4–6 mm long; callus bearded, the hairs 0.2–1.5 mm long; palea about the same length as the lemma; anthers small, 0.3–0.5 mm long; $2n = 40$.

MUHLENBERGIA

Muhlenbergia mexicana

Muhlenbergia montana

Dry or moist habitats at middle elevs. in the mts.; B.C. to Que., s. sporadically in the w. states to Calif., Ariz. and N.M., in the c. and e. states to Texas, Mo., Ky. and N.C. Aug–Sept.

The length of the awn of the lemma has been used to distinguish f. *ambigua* (Torr.) Fern., and the presence of awns on the glumes as well distinguishes f. *setiglumis* (S. Wats.) Fern., but these trivial characters do not seem to represent important races. The specific epithet *"mexicana"* is a misnomer, since the species does not extend as far south as Mexico. Linnaeus must have been confused about the origin of the seed.

11. **Muhlenbergia montana** (Nutt.) A. S. Hitchc.

Calycodon montanum Nutt. J. Acad. Nat. Sci. Philadelphia, II. 1: 186. 1848. *Muhlenbergia montana* A. S. Hitchc. U.S.D.A. Bull. (1915–23) 772: 145, 147. 1920. (*Gambel,* "In the Rocky Mountains, near Sante Fé," N.M., in 1841, probably Sept.)

M. trifida Hackel, Repert. Spec. Nov. Regni Veg. 8: 518. 1910. (*Arsène 3217,* Michoacan, Mex.)

M. gracilis of Amer. authors; not (H.B.K.) Kunth, 1829.

Mountain muhly.

Densely tufted, non-rhizomatous perennial bunchgrasses; culms 1.5–5.5 (8) dm tall, erect, with numerous innovations, solid, glabrous; leaves in a dense basal cluster; sheaths minutely roughened, longer than the internodes; ligules prominent, 4–10 (17) mm long, acute, entire, hyaline or nearly so, somewhat decurrent; blades flat to involute, mostly 1–2 mm broad and 8–15 cm long, the culm blades somewhat broader, very scabrous, firm; panicles 5–12 (16) cm long, long-exserted, narrow, rather loosely flowered, floriferous to the base; spikelets 1-flowered; glumes strikingly different, only slightly keeled, the first glume 1.5–3 mm long, ovate, acute to obtuse and usually short-awned, scabrous on the keel and scaberulous across the back, 1-nerved, the second 2–3.2 mm long, 3-nerved, the nerves excurrent as awn tips or shortly 3-lobed, scaberulous; lemma (3) 3.2–4 (4.5) mm long, 3-nerved, short-pilose on the lower half or only towards the bottom, scaberulous above, often with black blotches, tapered to a slender, flexuous, long awn, the awn (3) 6–12 (20) mm long; callus naked; palea slightly shorter than the lemma; anthers large, 1.5–2.3 mm long; $2n = 40$.

Grassy parks, slopes and foothills, often in ponderosa pine and spruce-fir forests at middle and high elevs.; Wyo., c. Utah, n. and Sierra Nevadan Calif., s. through Ariz., N.M. and Trans-Pecos Texas to s. Mex.; apparently absent from Nev. July–Sept.

Mountain muhly is a robust species that often provides considerable amounts of forage in the eastern part of our area. On the Kaibab Plateau occurs a form with short-awned lemmas which, according to Goodding (Gould and Reeder, 1951), is very common. This form shows some intermediate characters with the following species and has been confused with it.

12. **Muhlenbergia filiculmis** Vasey

Muhlenbergia filiculmis Vasey, Contr. U. S. Natl. Herb. 1: 267. 1893. (*Sheldon 321,* "on sandy soil, Ute Pass, Green Mountain Falls," El Paso Co., Colo., 2 Aug. 1892.)

Slimstem muhly.

Low, densely tufted, non-rhizomatous perennials; culms low, 0.8–2 dm tall, erect, hollow, filiform, glabrous; leaves in a short, dense basal cluster, the old clumps clothed with persistent sheaths, becoming loose and more or less flattened and papery; sheaths glabrous, strongly nerved, the margins usually edged with the texture of the decurrent ligule; ligules relatively prominent, 1.5–4.2 mm long, hyaline, acute, entire, noticeably decurrent; blades filiform-involute, 3–6 cm long, rarely longer, sharp-tipped, very scabrous; panicles small, 2–4.5 cm long, slender, with erect branches; spikelets 1-flowered; glumes dissimilar and unequal, about 1/2 as long as the floret, the first glume 1–2 mm long, hyaline or scarious, except for the prominent keel nerve which terminates in a mucro or short awn, rarely erose, the second slightly longer and wider, 1.5–2.5 mm long, scarious, 3-nerved, sharply 3-toothed, rarely entire or erose only; lemma 2.2–3 (3.2) mm long, lanceolate-terete, pubescent on the lower half, scabrous above, tapering to an awn tip, the awn 0.2–2 (3) mm long; callus naked; palea slightly shorter than the lemma, pubescent below, glabrous above; anthers large, 1.5–1.8 mm long, deep purple.

Dry hillsides to open sandy or rocky places from middle elevs. to timberline; Utah, Wyo., Colo. and N.M. Aug.–Sept.

13. **Muhlenbergia pungens** Thurber

Muhlenbergia pungens Thurber in A. Gray, Proc. Acad. Nat. Sci. Philadelphia 1863: 78. 1864. *Podosemum pungens* Bush, Amer. Midl. Naturalist 7: 32. 1921; not (Schreb.) Link, 1827. (*Hall & Harbour 632,* Colo.)

Sandhill muhly.

Rhizomatous, glaucous perennials; culms 1.5–4.5 (5) dm tall, becoming much-branched and rooting at the decumbent base, then ascending, solid, terete, finely pubescent; sheaths glabrous, crowded towards the base and much overlapping, more or less scarious on the margins above; ligule a ciliate fringe around the back of the throat with membranous, ciliate lateral extensions, these 0.2–1 mm long; blades closely involute, 3–5 cm long, rigid, sometimes deciduous at the junction of the sheath, sharp-pointed, glabrous or minutely pubescent on both faces or scabrous when the rough margins are exposed; panicles large, (7) 9–15 cm long and about 2–6 (8) cm wide, remaining closed until entirely out of the sheath and then becoming long-exserted, open and diffuse, consisting of a main axis with 2–7 side branches, each paniculately branched and few-flowered, each side branch easily discernible in the young, only slightly opened, interrupted inflorescence and soon continuous in the diffuse state, the main branches usually puberulent, the pedicels capillary, pulvini-like at the base; spikelets 1- (2)-flowered, purplish to brownish; glumes subequal, 1.5–3 mm long, sometimes scaberulous, the first glume ovate, acute, entire or sometimes 3-toothed, often short-awned, stramineous, the second narrower with an awn about 1 mm long, often erose or toothed; lemma 2.5–4.8 (5.5) mm long, nearly twice as long as the glumes, narrow, terete, gradually tapering into a short awn about 1–1.5 mm of the length; callus naked; palea somewhat longer than the body of the lemma, minutely 2-toothed by excurrence of the 2 nerves; anthers large, 2–2.5 mm long; $2n = 42, 60$.

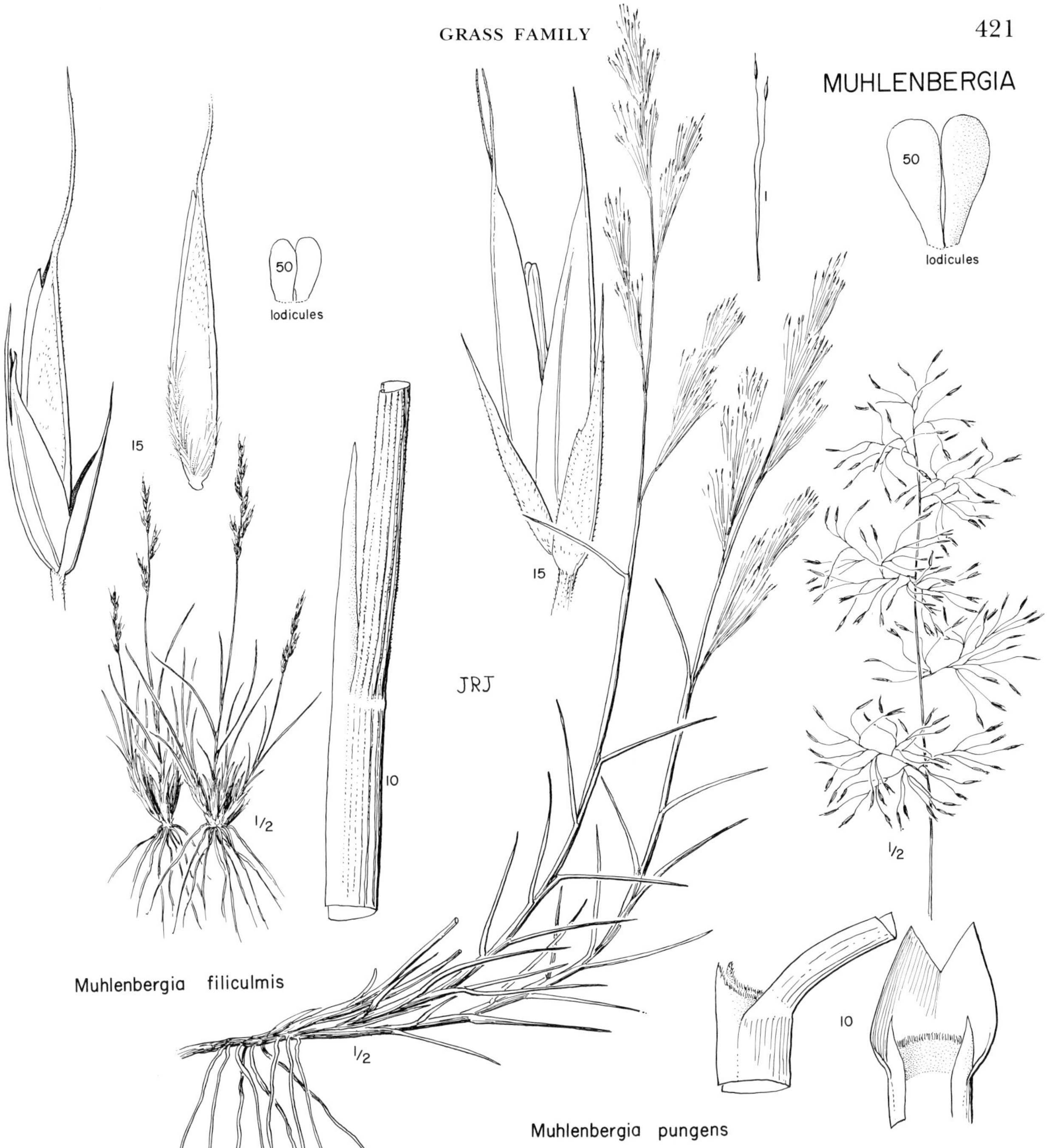

Muhlenbergia filiculmis

Muhlenbergia pungens

Loose, sandy soils, often in sand dune areas; Utah and Ariz., e. through Wyo., Colo., N.M. and w. Texas to S.D. and Nebr. July–Sept.

Sandhill muhly furnishes considerable amounts of forage in sandy areas of southeastern Utah.

14. Muhlenbergia porteri Scribn. ex Beal

Muhlenbergia texana Thurber ex Porter & Coulter, Syn. Fl. Colo. 144. 1874; not Buckley, 1863. *M. porteri* Scribn. ex Beal, Grasses N. Amer. 2: 259. 1896. *Podosemum porteri* Bush, Amer. Midl. Naturalist 7: 36. 1921. (*Bigelow,* Rio San Pedro, 5 Nov. 1850; *Parry,* Presidio del Norte, July 1852; *Wright 734,* "Western Texas to El Paso," in 1849; all in Texas.)

Bush muhly, mesquite grass.

Loosely cespitose, non-rhizomatous perennials from hard, knotty bases producing numerous culms; culms relatively tall, 3–10 dm or even more, becoming ascending or spreading through bushes, branching from the numerous nodes, geniculate, the nodes often swollen, solid, wiry, fragile, finely puberulent or scaberulous below, glabrous above; sheaths soon spreading from the culms revealing the prophylls; ligules 1–2 mm long, hyaline, lacerate-ciliate, decurrent, longer on the edges; blades flat, 1–1.5 (2) mm broad and 2–5 cm long, scabrous, early deciduous from the sheath; panicles 5–10 cm long, slightly longer than wide, open, diffuse, the branches brittle and easily broken, bearing few spikelets; spikelets 1-flowered, purplish; glumes subequal, 1.5–3 mm long, narrow, acuminate, glabrous, except the scabrous keel; lemma 3–4 mm long, acuminate, sparsely pubescent, purple, with a delicate awn 5–10 mm long; callus naked; palea about the same length as the lemma, glabrous; anthers large, 1.6–2.3 mm long; $2n = 20, 40$.

Dry rocky desert slopes and mesas among xerophytic shrubs; sw. U.S.; s. portions of Calif., Nev., Utah and Colo., s. through Ariz., N.M., and w. Texas to n. Mex. July–Sept.

Bush muhly gets its common name from its habit of branching and rebranching profusely at the upper nodes. In spite of its wiry culms it is highly palatable to livestock. Originally bush muhly existed in extensive stands throughout the southwest, but now it occurs in the protection of shrubs, especially shrubs armed by spines such as mesquite, hence the second name, mesquite grass.

15. **Muhlenbergia filiformis** (Thurber) Rydb.

Vilfa depauperata var. *filiformis* Thurber in King, Rep. Geol. Explor. 40th Parallel 5: 376. 1871. *Sporobolus filiformis* Rydb. Contr. U.S. Natl. Herb. 3: 189. 1895. *Muhlenbergia filiformis* Rydb. Bull. Torrey Bot. Club 32: 600. 1905. (*Bolander 6091,* Yosemite Valley, Calif.; *Torrey 565,* Donner Lake; *Watson 1280,* East Humboldt Mts., Nev.; *Watson 1281,* Uinta Mts., Utah.)

Sporobolus depauperatus var. *filiformis* Beal, Grasses N. Amer. 2: 296. 1896. (*Williams,* Yellowstone Natl. Park, Mont., 23 Aug. 1888; *Jones 1179,* Alta, Wasatch Mts., Utah, 4 Aug. 1879.)

Sporobolus simplex Scribn. U.S.D.A. Div. Agrostol. Bull. 11: 48, fig. 6. 1898. *M. simplex* Rydb. Bull. Torrey Bot. Club 32: 600. 1905; not Kunth, 1829. *M. filiformis* var. *fortis* Kelso, Rhodora 38: 298. 1936. (*Rydberg 2411,* Georgetown, Colo., 19 Aug. 1895.)

Sporobolus aristatus Rydb. Bull. Torrey Bot. Club 28: 266. 1901. *M. aristulata* Rydb. Bull. Torrey Bot. Club 32: 600. 1905. *M. aristata* Rydb. Agric. Exp. Sta. Agric. Coll. Colorado Bull. 100: 27. 1906; not Pers. 1805. (*Tweedy 2196,* Big Horn Mts., Sheridan Co., Wyo., Aug. 1899.)

Sporobolus simplex var. *thermale* Merr. Rhodora 4: 48. 1902. (*Griffiths 302a,* "margin of a stream of hot water," Lolo Hot Springs, Mont., 17 Sept. 1898.)

M. idahoensis St. John, Fl. S.E. Wash. & Adj. Idaho 50. 1937. (*St. John 9085,* "meadow along creek, Zaza, Craig Mts., Nez Perce Co.," Idaho, 9 Oct. 1927.)

Pull-up muhly.

Low, delicate annuals, sometimes perennating in areas near warm springs; culms low, 0.5–2 (4) dm tall, decumbent at the base and freely rooting at the nodes, freely branching at the lower nodes, solid, somewhat flattened, sometimes minutely nodulose-roughened; sheaths glabrous; ligules 1–3 mm long, obtuse to acute, decurrent, entire; blades flat, 1–2 mm broad and 2–3 (4) cm long, often puberulent on the upper surface; panicles small, 0.5–3.5 (6) cm long, narrow, few-flowered; spikelets 1-flowered, sometimes tinged with purple; glumes subequal, small, 0.7–1.2 (2) mm long, ovate, often scarious and erose; lemma 1.5–2.2 (2.8) mm long, twice as long as the glumes, lanceolate, acute, mucronate, minutely pubescent, becoming scaberulous towards the tip, often black tinged; callus naked; palea nearly as long as the lemma, glabrous or with a few hairs below; anthers 0.7–1.5 mm long; $2n = 18$.

Wet places in mt. meadows, sometimes forming dense stands at edges of pools and seeps (sometimes around thermal springs); B.C., s. to Calif. and e. to S.D., Nebr., Colo., and N.M. June–Aug.

16. **Muhlenbergia minutissima** (Steudel) Swallen

Agrostis minutissima Steudel, Syn. Pl. Glum. 1: 171. 1854. *Sporobolus minutissimus* A. S. Hitchc. Proc. Biol. Soc. Wash. 41: 161. 1928. *Muhlenbergia minutissima* Swallen, Contr. U.S. Natl. Herb. 29: 207. 1947. (*Fendler 986,* N.M., in 1847.)

Sporobolus confusus of early Amer. botanists; not (Fourn.) Vasey, which does not extend north of the Mex. border.

Sporobolus microspermus sensu A. S. Hitchc. J. Wash. Acad. Sci. 23: 453. 1933; not *Milium microspermum* Lag. 1816, on which the name was based.

Minutissima.

Delicate, weakly-rooted annuals; culms low, 1–2.5 (3) dm tall, geniculately branched near the base, erect or spreading, solid, puberulent below the nodes; sheaths strongly striate, shorter than the nodes; ligules 1–2 mm long, a hyaline scale soon becoming lacerate and somewhat decurrent; blades flat, folded or finally becoming involute, about 1–2 mm broad and 1.5–5 cm long, minutely pubescent; panicles (4) 10–15 (20) cm long and 2–4 cm broad, open and diffuse, often dominating the whole plant, the pedicels capillary, 3–5 mm long, flexuous; spikelets 1-flowered; glumes subequal, small, 0.6–1 (1.5) mm long, minutely pubescent, rarely nearly glabrous, usually finely ciliate, obtuse; lemma 1–1.2 (1.5) mm long, broadly elliptic, blunt, glabrous to finely pubescent; callus naked; palea slightly shorter than the lemma, glabrous; anthers slightly less than 1 mm long; $2n = 60, 80$.

Sandy soils of meadows, streambanks and shores of ponds and lakes, often forming dense stands along receding shores of reservoirs; Wash. to Mont., s. through Calif., Ariz., N.M. and w. Texas to n. Mex.; apparently absent from Utah. July–Sept.

Muhlenbergia minutissima is a distinctive annual with the whole plant seemingly consisting of a relatively large inflorescence.

67. **SPOROBOLUS** R. Br. Dropseed

Cespitose perennials or annuals, none of our species with rhizomes; culms more often solid and pithy than hollow, often forming large bunches; leaves mostly in a basal clump; sheaths open, often with tufts of hairs at the flanks of the collar and along the upper margins; ligules short, consisting of a short fringe of hairs or a short fringed membrane; blades flat, folded or involute, non-auriculate; inflorescence an open and diffuse or contracted, spike-like panicle; spikelets small, 1-flowered, the florets terete, the pedicels sometimes with an abscission zone below the glumes, but more often disarticulating above the glumes; glumes subequal to unequal, the second sometimes nearly as long as the lemma, sometimes deciduous before the floret, 1-nerved, awnless; lemma thin, membranous, usually acute, usually glabrous, 1-nerved (rare in Eragrosteae),

MUHLENBERGIA

Muhlenbergia porteri

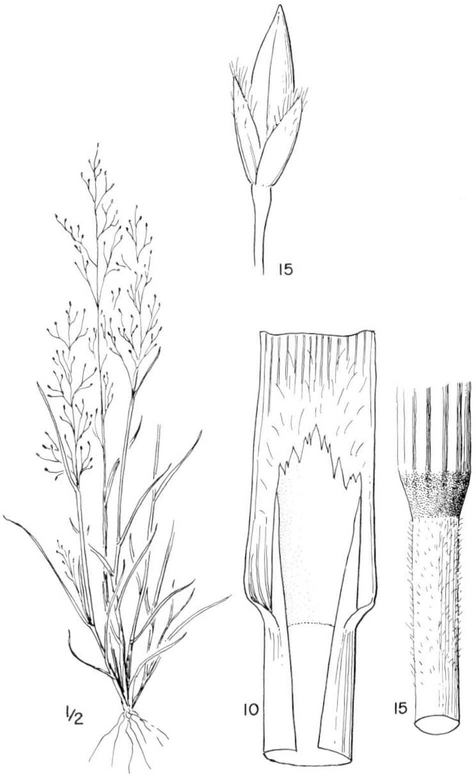

Muhlenbergia minutissima

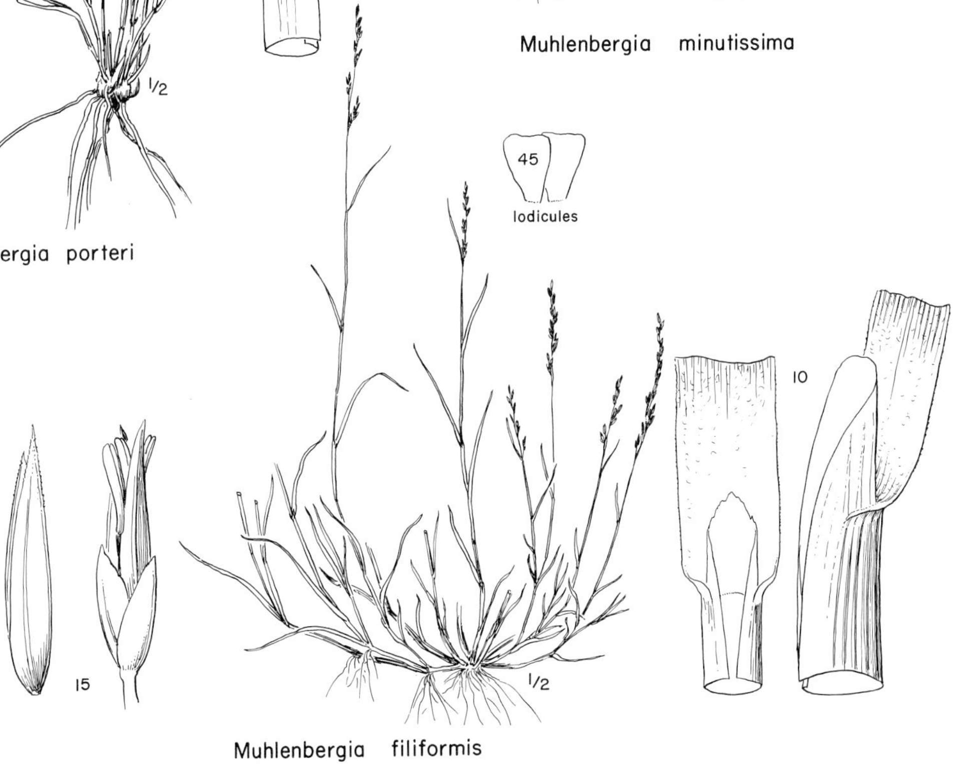

Muhlenbergia filiformis

awnless; callus not long hairy; rachilla not prolonged behind the palea; palea well developed, nearly as long as the lemma to slightly longer, 2-nerved, often splitting at maturity and appearing as an extra lemma; lodicules short, more or less fleshy, usually at least as broad as long; stamens 3; fruit with the seed coat (testa) remaining separate from the pericarp making it an achene or utricle rather than a true caryopsis (when the inner portion of the pericarp is moistened it becomes mucilaginous and then the seed is expelled from the pericarp upon drying), obovate, somewhat asymmetrical and flattened, usually falling free from the lemma and palea upon maturity, often reddish; $x = 6, 9$.

A genus of about 100 species, chiefly of warmer regions of the world, but most abundant in the Americas. (Name from the Greek *spora,* seed, and *ballein,* to throw, referring to the free seeds.)

Sporobolus differs from the closely related genus *Muhlenbergia* in having thin, 1-nerved, consistently awnless lemmas, obovate, flattened fruits and usually ciliate ligules flanked by tufts of hairs at the corners of the sheaths. *Muhlenbergia* has 3-nerved, often awned lemmas, elongate, usually cylindrical fruits and membranous ligules without tufts of hairs at the corners of the sheaths.

Sporobolus airoides forms hummocks in low-lying areas of saltgrass meadows where it is often a dominant species. Several species should be listed as fair forage grasses mostly because of their abundance rather than their palatability. *Sporobolus contractus* and *S. cryptandrus* are fair forage species on winter rangelands in western Utah and eastern Nevada with the latter being somewhat more palatable. *Sporobolus flexuosus* becomes abundant enough in areas of southeastern Utah to furnish some forage. The seeds of *S. cryptandrus* and *S. flexuosus* have been used for food by Indians.

Reference: Jones, E. K., and N. C. Fassett. Subspecific variation in *Sporobolus cryptandrus.* Rhodora 52: 125–126. 1950.

1 Plants low annuals, the culms less than 3 dm tall; first glume minute, 0.3–0.7 mm long, sometimes overlooked .... 7. *S. pulvinatus*
1 Plants perennials, often forming large tufts with considerable amounts of old plant remnants in bunches; culms 4–20 dm tall; first glume 0.7–4 mm long.
  2 Spikelets 3.5–6 mm long; anthers 1.5–2.5 mm long .... 1. *S. asper*
  2 Spikelets 1–3 (3.5) mm long; anthers 0.4–1.5 mm long.
    3 Sheaths with only a few spreading hairs at the summit, sometimes nearly glabrous, the old persisting sheaths at the base of the plant cream-colored, shiny, smooth; panicle branches with spikelets borne chiefly near the tips; spikelets usually falling entire; glumes usually glabrous; anthers 1.2–1.5 mm long .... 2. *S. airoides*
    3 Sheaths with a conspicuous tuft of white hairs at the summit; panicle branches usually spikelet-bearing nearly to the base; empty glumes usually left on the inflorescence; glumes scabrous on the keels or sometimes glabrous; anthers 0.4–1 mm long.
      4 Panicle open, oblong or pyramidal, at least the exserted portion (1.5) 2.5–12 cm broad, often the lower part enclosed in the sheath.
        5 Panicle oblong, 5–12 cm broad, the lower branches about the same length as the middle ones, flexuous, implicate (tangled and often nearly reflexed), the pedicels often widely spreading, the spikelets loosely arranged .... 3. *S. flexuosus*
        5 Panicle narrowly pyramidal or somewhat contracted, 1.5–4 (6) cm broad, the lower branches much exceeding the middle ones, the spikelets crowded on the branchlets .... 4. *S. cryptandrus*
      4 Panicle compact and spike-like, 0.3–1.6 cm broad, rarely up to 4 cm in some *S. giganteus.*
        6 Spikelets 2–2.5 mm long; panicles strongly contracted, 0.3–0.8 (1) cm broad; culms not exceeding 5 mm in thickness .... 5. *S. contractus*
        6 Spikelets 2.5–3.5 mm long; panicles usually thicker, 0.5–1.6 (4) cm broad; culms 4–7 (10) mm thick .... 6. *S. giganteus*

1. Sporobolus asper (Michx.) Kunth

Agrostis aspera Michx. Fl. Boreali-Amer. 1: 52. 1803; not Web. 1787. *Vilfa aspera* Beauv. Essai Nouv. Agrostogr. 16, 147, 181. 1812. *Sporobolus asper* Kunth, Révision Gram. 1: 68. 1829. *Muhlenbergia aspera* Trin. ex Kunth, Enum. Pl. 1: 210. 1833, in synonymy. (*Michaux,* "Habitat in collibus rupibusque regionis Illinoensis," in 1795.)

S. pilosus Vasey, Bot. Gaz. 16: 26. 1891. *S. asper* var. *pilosus* A. S. Hitchc. Proc. Biol. Soc. Wash. 41: 161. 1928. (*Smythe 217,* Coolidge, Hamilton Co., Kansas, in 1890.)

Tall dropseed.

Stout, tufted perennials; culms 5–12 dm tall, glabrous, sulcate, more or less solid; sheaths glabrous, overlapping, usually pilose at the throat and around the collar; ligules very short, less than 0.5 mm long, a thick, but low, truncate, ciliolate crown; blades flat, becoming involute, 1–4.5 (5.5) mm broad, elongate, tapering to a scabrous point, pilose near the base on the upper surface, occasionally pilose beneath at the base; panicles 10–17 (27) cm long, spike-like, partially or mostly enclosed in an inflated sheath; spikelets strongly flattened, glabrous, usually pale or green, sometimes purplish-tinged; glumes blunt, carinate-keeled, 1-nerved, scabrous on the keel, large, the first glume 2.5–3.5 (4) mm long, the second 3–4.5 (5.2) mm long; lemma longer than the second glume, (3.5) 4.5–5.5 mm long, blunt, the tips boat-shaped, carinate-keeled, 1-nerved, smooth but sometimes scabrous on the keel; palea about the same length as the lemma; lodicules about 0.5 mm long; anthers large, (1.5) 2–2.5 mm long; $2n = 54, 88, 108$.

Usually in dry, sandy soils of prairies and foothills; e. Wash., Oregon, Utah and Ariz. where rare, common e. of the Rocky Mts. to Que. and Vt. and s. to Texas, La., Ala., Tenn. and Va. Aug.–Sept.

A close relative of *S. asper* from the south is *S. interruptus* Vasey, which may enter our region on the Grand Canyon Plateaus of

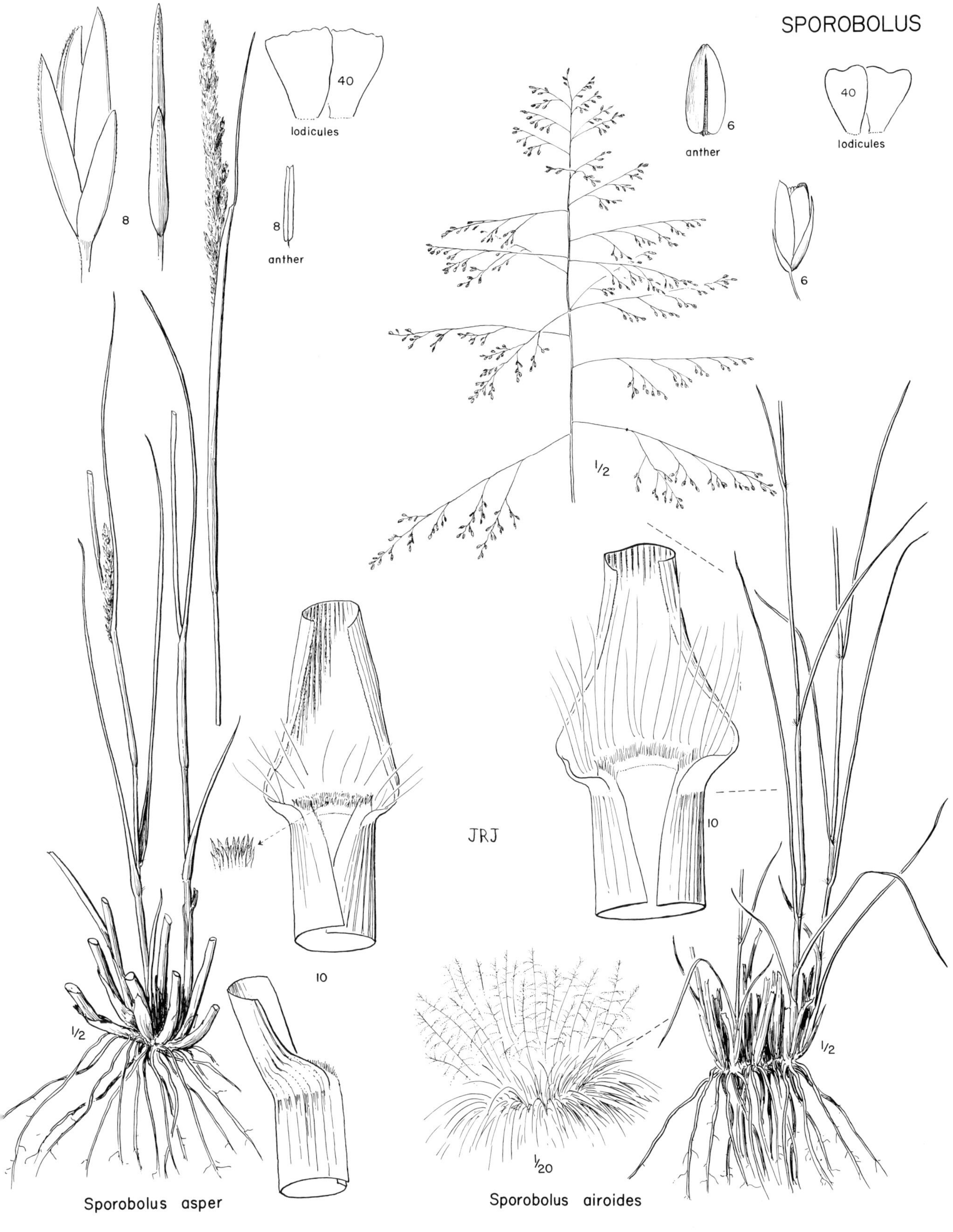
SPOROBOLUS
40
lodicules
8
8
anther
6
anther
40
lodicules
6
1/2
JRJ
10
10
1/2
1/20
1/2
Sporobolus asper
Sporobolus airoides

northwestern Arizona. *Sporobolus interruptus* differs from *S. asper* in having open, exserted panicles.

2. Sporobolus airoides (Torr.) Torr.

Agrostis airoides Torr. Ann. Lyceum Nat. Hist. New York 1: 151. 1824. *Vilfa airoides* Trin. ex Steudel, Nomencl. Bot. ed. 2. 2: 766. 1841. *Sporobolus airoides* Torr. in Parke, Explor. & Surv. Railroad Route from Mississippi River to Pacific Ocean 7(3): 21. 1856. (*James,* "On the branches of the Arkansas, near the Rocky Mountains," Colo., in 1820, probably late July.)

S. wrightii Munro ex Scribn. Bull. Torrey Bot. Club 9: 103. 1882. *S. airoides* var. *wrightii* Gould, Madroño 10: 94. 1949. (*Pringle,* Pantano, Santa Cruz Valley, Pima Co., Ariz., 28 June 1881.)

Alkali sacaton.

Stout, tufted perennials, the base of the large clumps densely clothed with slick and shiny, cream-colored sheaths; culms 4–9 (12) dm tall, glabrous, terete, usually hollow, but sometimes loosely pith-filled; sheaths glabrous or slightly pilose at the upper corners of the margins, otherwise smooth, the collar glabrous; ligules very short, mostly a dense band of hairs; blades flat to strongly involute, 2–4 mm broad, smooth on the lower surface, finely scaberulous above and sparsely to copiously hirsute near the throat; panicles 12–40 cm long, open and pyramidal, (4) 10–18 (20) mm broad, often fully exserted or the lower part remaining enclosed in a sheath, spikelet bearing mostly toward the tips of the branches; spikelets with glumes often deciduous before the floret; glumes acute, 1-nerved, scarious or often hyaline throughout, glabrous, the first glume 0.7–1.2 (2) mm long, the second about twice as long, 1.5–2 (3) mm; lemma usually longer than the second glume, 1.8–2.5 (3) mm long, acute; palea more or less as long as the lemma, usually splitting as the plump caryopsis ripens; lodicules usually less than 0.5 mm long; anthers relatively large, 1.2–1.5 mm long; $2n$ = 80, 90, 108, 126.

Usually in low-lying alkaline saltgrass meadows in our region where forming hummocks or sometimes in sandy soils of desert foothills or wash banks; e. Wash. to S.C. and Mo., s. to s. Calif., Ariz., N.M., Texas and Mex. June–Aug.

Alkali sacaton, a salt tolerant plant, is a characteristic bunchgrass of low-lying alkaline meadows in our region. It becomes an important forage grass in these habitats where higher quality grasses are not available. Upon becoming the dominant grass in some areas, the open, diffuse panicles give a smoky hue to the landscape that makes a beautiful sight.

Variety *wrightii* may enter the southern part of our area. It is a more robust plant and may be separated by the following key.

1 Panicles 1 to 2 times as long as wide, the branches not crowded; leaf blades mostly involute; range of the species . . . . var. *airoides*

1 Panicles mostly more than 3 times as long as wide, relatively dense; leaf blades mostly flat; s. Calif. to w. Texas and c. Mex. . . . . var. *wrightii* (Munro ex Scribn.) Gould

3. Sporobolus flexuosus (Thurber) Rydb.

Vilfa cryptandra var. *flexuosa* Thurber ex Vasey in Wheeler, Rep. U.S. Geogr. Surv. W. 100th Meridian 6: 282. 1878. *Sporobolus cryptandrus* var. *flexuosus* Thurber in S. Wats. Bot. Calif. 2: 269. 1880. *S. flexuosus* Rydb. Bull. Torrey Bot. Club 32: 601. 1905. (*Wheeler Exped.,* "Nevada and Arizona," in 1872.)

Mesa dropseed.

Tall, slender perennials, often with the appearance of an annual; culms 4–10 (12) dm tall, terete, solid; sheaths at the base of small clumps short, 1.5–3 cm long and persisting, strongly overlapping, glabrous, striate, with conspicuous tufts of long white hairs at the upper corners which extend down the margins for 1–2 cm or sometimes the full length, the edges sometimes with a narrow band of anthocyanin, the collar often glabrous; ligules consisting of a dense band of hairs on a low, thickened membrane; blades flat or involute, 1.5–4 (6) mm broad and 5–20 cm long, smooth beneath, sometimes scabrous above; panicles 10–25 (30) cm long, oblong, ovate, 5–12 cm broad, the main axis glabrous, the branches usually with swollen pulvinous-like, sparsely pilose bases, often reflexed, or widely spreading, flexuous and tangled, making the long inflorescence appear implicate; spikelets loosely arranged, often on widely divergent pedicels; glumes acute, 1-nerved, usually scabrous on the keel, otherwise glabrous, sometimes early deciduous, the first glume 0.7–1.2 (1.8) mm long, the second twice as long, 1.7–2.8 mm; lemma about as long as the second glume, 1.7–2.5 mm long, acute; palea about the same length as the lemma, acute; anthers 0.4–0.7 (1) mm long; $2n$ = 38.

Mesas, deserts in sandy and rocky soils, often in stands of blackbrush, also in marshes and seepages; s. Utah, s. Nev. and s. Calif., e. to s. Kansas, w. Texas and n. Mex. June–Sept.

Sporobolus flexuosus is similar to the following species, but the panicle is always open and the flexuous branches are somewhat drooping.

4. Sporobolus cryptandrus (Torr.) A. Gray

Agrostis cryptandra Torr. Ann. Lyceum Nat. Hist. New York 1: 151. 1824. *Vilfa cryptandra* Torr. ex Trin. Mém. Acad. Imp. Sci. Saint-Pétersbourg, Sér. 6, Sci. Math., Seconde Pt. Sci. Nat. 4(1): 69. 1840. *Sporobolus cryptandrus* A. Gray, Manual Bot. 576. 1848. (*James,* "On the Canadian river," Texas or Okla., Aug. 1820.)

Vilfa tenacissima var. *fuscicolor* Hook. Fl. Boreali-Amer. 2: 239. 1839. *S. cryptandrus* subsp. *fuscicolus* E. K. Jones & Fassett, Rhodora 52: 126. 1950. *S. cryptandrus* var. *fuscicolor* R. W. Pohl, Iowa State Coll. J. Sci. 40: 500. 1966. (*Douglas,* "Menzies Island and higher parts of Columbia" River, Wash.)

Vilfa triniana Steudel, Syn. Pl. Glum. 1: 156. 1854. ([British] Columbia.)

S. cryptandrus var. *occidentalis* E. K. Jones & Fassett, Rhodora 52: 125. 1950. (*Cusick 2222,* "Bars of Snake river, Ballard's Landing," Baker Co., Oregon, about 3 mi. ne. of Homestead, 8 July 1899.)

Sand dropseed.

Tufted perennials, sometimes appearing like annuals; culms 4–7 (9) dm tall, erect or decumbent below, glabrous, solid; sheaths strongly overlapping, glabrous, with pubescent margins and long pilose tufts at the upper corners of the margins, 2–3 mm long, sometimes pilose in a line across the collar, but more often glabrous; ligules about 0.5 mm long, composed of a line of dense hairs; blades usually flat, more or less involute in drying, becoming involute towards the tip, 1.5–4 (5) mm broad and 5–15 cm long, the margins cartilaginous and scabrous; panicles 10–20 (25) cm long, open, (1.5) 2.5–4 (6) cm wide, the lower part usually included in a sheath or

Sporobolus flexuosus

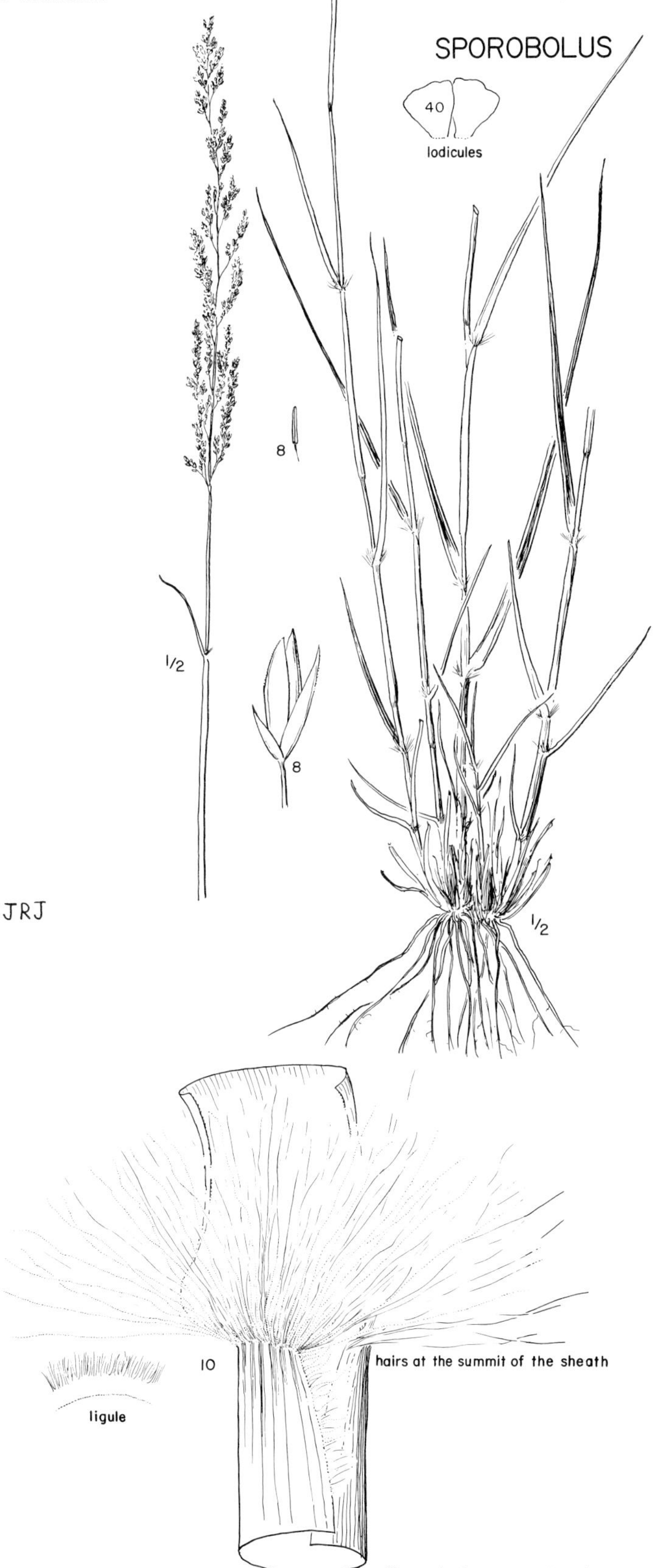

Sporobolus cryptandrus

sometimes entirely included, the branches spikelet-bearing to near the base, usually rather stiffly ascending, glabrous; spikelets pale or olivaceous, glabrous, the glumes occasionally deciduous at maturity; glumes acute, 1-nerved, the first 0.7–1 mm long, sometimes nearly subulate, hyaline, usually scabrous on the keel and otherwise glabrous or scabrous on the body near the keel, the second glume twice as long, 1.4–1.8 (2.2) mm long, broader; lemma about as long as the second glume, 1.5–2 (2.5) mm long, acute; palea about as long as the lemma; anthers about 0.5 mm long; $2n$ = 18, 36, 38, 72.

Sandy ground in sagebrush or in shadscale vegetation, often on disturbed sites; B.C. to Que. and Me., s. throughout the U.S., but only sporadic in the Southeast; n. Mex. June–Aug.

Sand dropseed has enlarged and inflated sheaths subtending and partially enclosing the inflorescence. The spikelets are cleistogamous. This and the following two species form a taxonomically difficult complex. *Sporobolus cryptandrus,* the smallest of the three, is the most widespread and common, especially in the Southwest. Sand dropseed, a good forage grass, is commonly used in reseeding work.

5. **Sporobolus contractus** A. S. Hitchc.

Sporobolus cryptandrus var. *stricta* Scribn. Bull. Torrey Bot. Club 9: 103. 1882. *S. strictus* Merr. U.S.D.A. Div. Agrostol. Circ. 32: 6. 1901; not Franch. 1893. *S. contractus* A. S. Hitchc. Amer. J. Bot. 2: 303. 1915. (*Pringle,* "Banks of the Rillita, near Camp Lowell," near Tucson, Ariz.)

S. cryptandrus var. *robustus* Vasey, Contr. U.S. Natl. Herb. 1: 56. 1890. (*Nealley 746,* Texas.) The type approaches *S. giganteus* in measurements.

Spike dropseed.

Large, tufted perennials, often with old sheaths and culm remnants at the base of the clumps; culms 4–10 (12) dm tall, erect, glabrous, not exceeding 5 mm in thickness; sheaths overlapping, glabrous with pubescent margins and long, white, pilose tufts at the upper corners of the margins, 3–4 mm long, usually pilose in a line across the collar; ligules 0.5–1 mm long, composed of a dense ciliate fringe; blades flat, becoming involute towards the tip, or involute throughout upon drying, 2–5 (6) mm broad and 4–30 cm long, the margins scabrous; panicles 15–35 (50) cm long, usually densely contracted and spike-like or somewhat interrupted on the lower half in robust plants, 0.3–0.8 (1) cm thick, usually included at the base, rarely entirely included in the sheath, the branches spikelet-bearing to the base; spikelets pale or olivaceous; glumes acute, 1-nerved, usually hyaline, glabrous, scabrous on the keel and over the back on the upper 1/3, narrow, the first glume 0.7–1.2 mm long, sometimes nearly subulate, the second twice as long, 1.5–2.5 mm, and broader; lemma about as long as the second glume, 2–2.5 mm long, acute, scaberulous above; palea slightly shorter than the lemma; anthers 0.5–1 mm long; $2n$ = 36.

Dry and gravelly and sandy soils from low deserts to pinyon-juniper foothills; se. Calif., Nev., Utah and Colo., s. to Ariz., N.M., w. Texas, and n. Mex. June–Sept.

Sporobolus contractus and the preceding species, when sympatric, are strikingly different, inasmuch as *S. contractus,* the larger of the two, has persistent old papery blades and sheaths that wind up in a corkscrew-like fashion.

6. **Sporobolus giganteus** Nash

Sporobolus giganteus Nash, Bull. Torrey Bot. Club 25: 88. 1898. *S. cryptandrus* var. *giganteus* M. E. Jones, Contr. W. Bot. 14: 11. 1912. (*Wooton 394,* "On the White Sands, Doña Ana Co.," N.M., 26 Aug. 1897.)

Giant dropseed.

Robust perennials; culms (6) 7–15 (20) dm tall, erect, glabrous, shiny, 4–7 (10) mm thick and solid; sheaths strongly overlapping, glabrous, with long, pilose hairs on the margins, becoming longer on the upper corners of the margins forming 2 long tufts of white hairs 3–5 mm long, often in a line across the collar; ligules 1–1.5 mm long, composed of a dense, ciliate fringe; blades flat, becoming involute towards the tip, 2.5–8 (10) mm broad and 8–30 cm long, often disarticulating at the junction of the sheath; panicles 20–65 cm long, contracted and spike-like or sometimes opening slightly, 0.5–1 (1.6) cm broad, rarely somewhat open and up to 4 cm across, often the lower part included in the inflated sheath, the branches strictly erect and spikelet-bearing to the base; spikelets pale or olivaceous; glumes 1-nerved, usually hyaline, scabrous on the keel, sometimes scaberulous above, otherwise glabrous, the first glume (0.7) 1–1.5 mm long, the second more than twice as long, 2.2–3.2 (3.5) mm; lemma about as long as the second glume, 2.5–3.5 mm long, acute; palea subequal to the lemma; anthers 0.5–0.8 mm long; $2n$ = 36.

Dry sandy soils, mostly around dunes; s. Utah, s. Colo. and w. Okla., s. to Ariz., N.M., w. Texas and n. Chihuahua. June–Sept.

Sporobolus giganteus is the largest of our dropseeds, with the culms often exceeding 1 m in height and often up to 1 cm in thickness. The larger, sometimes somewhat open inflorescence will help to identify this species, which occupies a somewhat more southerly range than the preceding species although the two are often sympatric. *Sporobolus giganteus* will be found in the valley bottoms below *S. contractus,* and more often in loose sandy soils on small shifting dunes.

7. **Sporobolus pulvinatus** Swallen

Sporobolus pulvinatus Swallen, J. Wash. Acad. Sci. 31: 351, fig. 4. 1941. (*Griffiths 5107,* Adamana, Ariz., 6–15 Aug. 1903.)

Six-weeks dropseed.

Low, tufted annuals or short-lived perennials; culms 0.4–1.5 (3) dm tall, erect or spreading, often strongly geniculate at the base, glabrous; sheaths keeled, usually shorter than the internodes, glabrous, the margins often scarious or hyaline with a sparse tuft of whitish hairs at the upper corners of the margins 1–2 mm long; ligules about 0.5 mm long, composed of a ciliate fringe; blades flat, 1–2.5 (3) mm broad and 1.5–4 cm long, often sparsely pilose on the upper surface; panicles small, 1.5–4 (8) cm long, pyramidal, many-flowered, the lower branches verticillate with erect or spreading branches, the branches naked at the base, densely flowered above; spikelets olivaceous; glumes 1-nerved, the first minute, 0.3–0.7 mm long, the second about three times as long, 1.3–1.9 mm long, somewhat obtuse and nearly cucullate; lemma small, 1.3–1.7 mm long, acute to subobtuse; palea broad, same length as the lemma; anthers small, 0.3–0.4

SPOROBOLUS

Sporobolus contractus

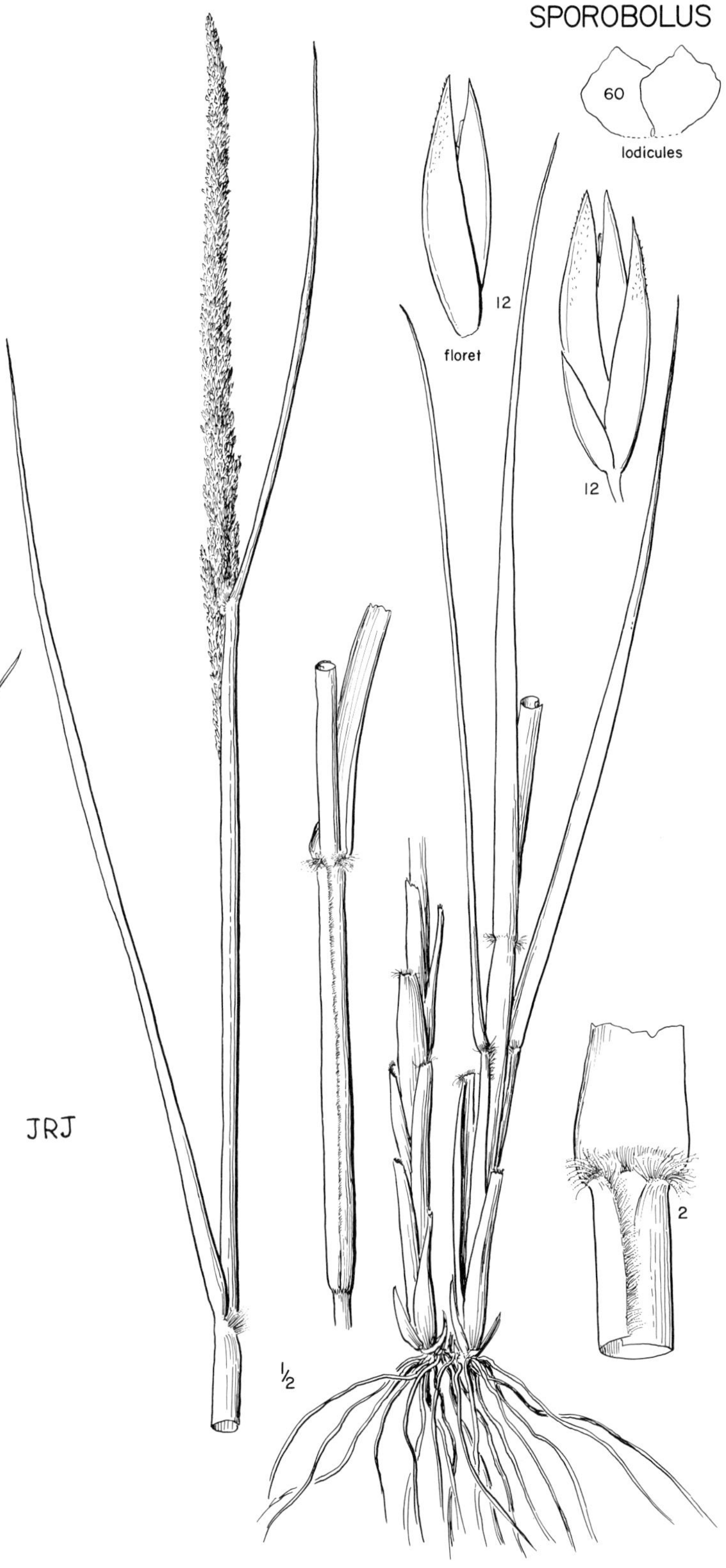

Sporobolus giganteus

mm long, pale or pinkish; caryopsis reddish; $2n$ = 24, 36, 54.

Sandy or clay soils; se. Utah (near the Four Corners), Ariz. and N.M. July–Sept.

68. BLEPHARONEURON Nash Hairy Dropseed

Cespitose perennials; culms simple, hollow, occasionally loosely pithy; leaves mostly in a basal tuft; sheaths open; ligules membranous, short, rounded, appearing as a continuation of the sheath margin; blades narrow, flat, but soon becoming involute, non-auriculate; inflorescence a loosely contracted panicle; spikelets 1-flowered, very little compressed, borne on slender pedicels, disarticulating above the glumes; florets terete; glumes subequal, broad, rounded dorsally and broadly acute or obtuse, faintly 1-nerved, awnless; lemma firmer than the glumes and slightly longer, rounded on the back, 3-nerved, the nerves usually densely silky-villous, the apex acute, occasionally mucronate; callus not long hairy; palea slightly exceeding the lemma, densely villous between the 2 nerves; lodicules club-shaped, about 0.2 mm long; stamens 3; styles not at all united at the base, the stigmas plumose; $x = 8$.

A monotypic genus of sw. U.S. and Mex. (Name from the Greek *blepharis,* eyelash, and *neuron,* nerve, in reference to the silky nerves of the lemma.)

1. Blepharoneuron tricholepis (Torr.) Nash

Vilfa tricholepis Torr. in Whipple, Explor. & Surv. Railroad Route from Mississippi River to Pacific Ocean 4(5): 155. 1857. *Sporobolus tricholepis* J. M. Coulter, Manual Bot. Rocky Mt. Region 411. 1885. *Blepharoneuron tricholepis* Nash, Bull. Torrey Bot. Club 25: 88. 1898. (*Bigelow,* Sandia Mts., N.M.)

Hairy dropseed, pine dropseed.

Slender, densely tufted perennials; culms 2.5–6.5 (8.5) dm tall, erect, glabrous; leaves mostly basal, often with 1 or 2 on the culm; sheaths crowded near the base, broad, glabrous on back and ciliate on the scabrous margins; ligules 0.5–2 mm long, erose-dentate or ciliolate, hyaline, noticeably decurrent; blades filiform-involute, 4–15 cm long, often strongly flexuous, scabrous; panicles 6–15 (22) cm long, narrow to open and often pyramidal, the branches glabrous, straight or flexuous and naked towards the base, the pedicels capillary; spikelets pale or olivaceous, 1-flowered; glumes 1-nerved, the first glume 1.5–2.2 (3) mm long, ovate, obtuse, the second 1.7–3.2 (3.7) mm long, broader and more blunt, sometimes rounded apically; lemma 2–3.2 (3.7) mm long, abruptly pointed, usually mucronate, 3-nerved, with silky hairs on the keel and marginal nerves, scarious towards the tip; palea more or less as long as the lemma, acute; anthers nearly 2 mm long, dark brown; $2n$ = 16.

Dry rocky, open slopes, open coniferous woods or grassy parks at middle to high elevs.; Uinta Mts. and Colorado Plateau region of Utah, Colo., s. to Ariz., N.M., Trans-Pecos Texas and n. Mex. Late July–Sept.

Blepharoneuron tricholepis is an important range grass in many parts of the Uinta Mountains and in the plateaus of eastern Utah where it is often the dominant grass species in open grassy parks.

69. CRYPSIS Aiton, nom. conserv.

Annuals; culms usually much branched below; sheaths open, often short and becoming inflated; ligules consisting of a fringe of hairs; blades flat to involute, short, non-auriculate; inflorescence a short, dense, spike-like or capitate panicle, terminating the main culms and often also in the axils of the upper leaves, often remaining partially enclosed by the spathe-like sheaths; spikelets small, 1-flowered, disarticulating below (in ours) or above the glumes; glumes acute, keeled, 1-nerved, awnless; lemma thin, acute, 1-nerved, keeled, awnless (in ours) to mucronate or short-awned; callus not long hairy; palea broad and slightly shorter than the lemma, 2-nerved and splitting between the nerves at maturity, sometimes 1-nerved; lodicules absent; stamens 3 (in ours) or 2; fruit with the seed coat (testa) remaining separate from the pericarp, the fruit then an achene or utricle rather than a true caryopsis, readily falling from the lemma and palea; $x = 9$.

A genus of about 7 species, these native to the Medit. region, 3 of which have been introduced in the U.S. (Name from the Greek *krupsis,* concealment, because of the partly hidden inflorescence.)

Reference: Lorch, J. A revision of *Crypsis* Ait. (Gramineae). Bull. Res. Council Israel, Sect. D, Bot. 11: 91–102. 1962.

1 Inflorescence narrow, 2.5–4.2 mm thick, usually extending beyond the uppermost leaf sheath; lemma 1.7–2.4 (2.8) mm long .... 1. *C. alopecuroides*
1 Inflorescence (4) 5–8.5 mm thick, remaining partly enclosed by the spathe-like leaf sheath; lemma 2.4–3 mm long .... 2. *C. schoenoides*

1. Crypsis alopecuroides (Piller & Mitterp.) Schrader

Phleum alopecuroides Piller & Mitterp. Iter Poseganum Sclavoniae Prov. 147, pl. 16. 1783. *Heleochloa alopecuroides* Host, Icon. & Descript. Gram. Austriacorum 1: 23, pl. 29. 1801. *Crypsis alopecuroides* Schrader, Fl. German. 1: 167. 1806. (Europe.)

Annuals; culms prostrate to somewhat erect, 0.3–2 (3.5) dm long, branching at the base, the nodes reddish brown; sheaths glabrous, the margins membranous; ligule a fringe of hairs 0.2–1 mm long, sometimes extended across the apex of the sheath; blades flat, folded or loosely involute, 1.2–2.5 mm broad, acuminate, glabrous to pilose on both surfaces; panicles 0.7–4.5 (6.5) cm long and 2.5–4.2 mm thick, cylindric, spike-like, the lower part enveloped in

SPOROBOLUS
BLEPHARONEURON
CRYPSIS

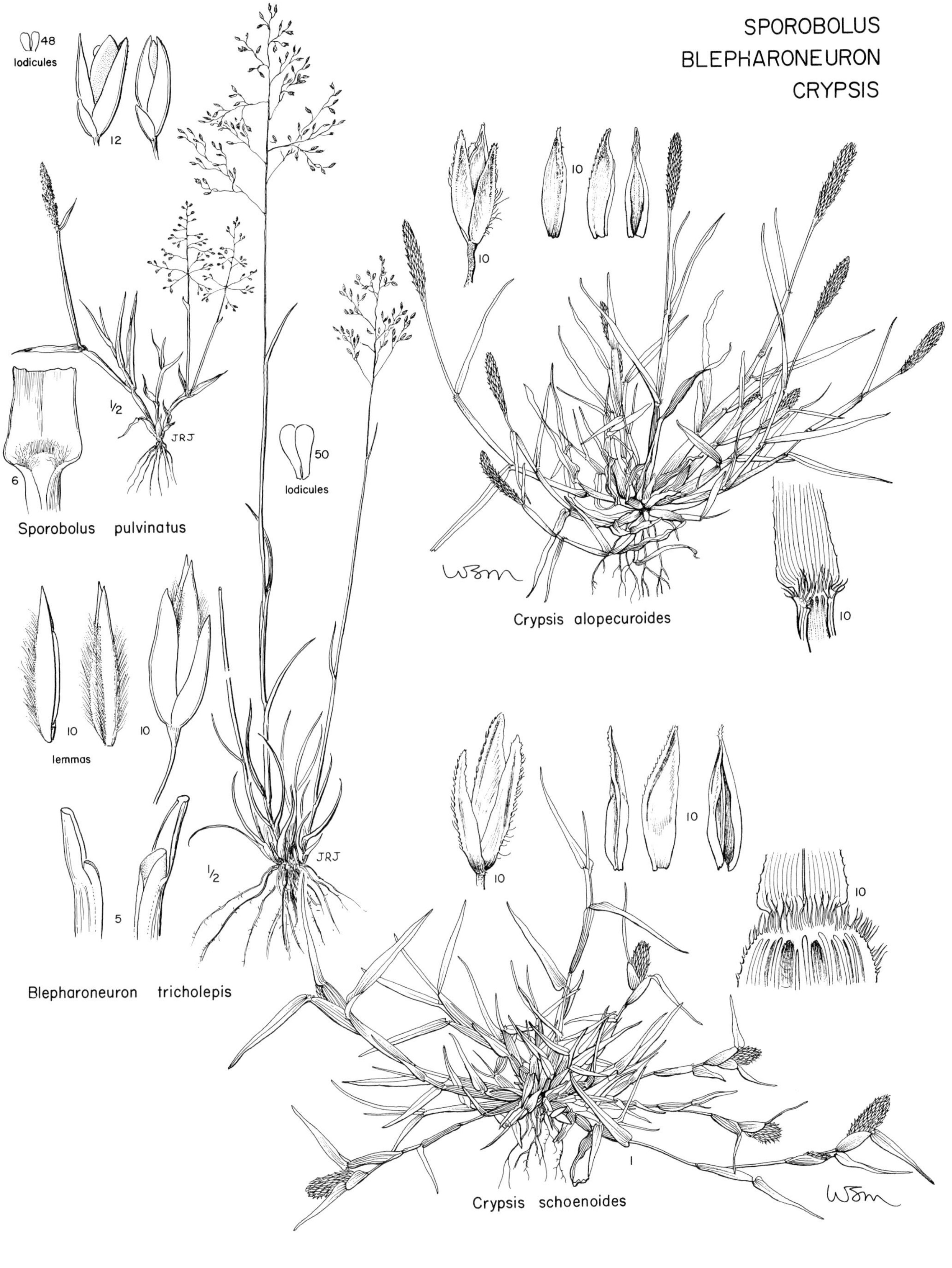

the upper sheath at anthesis, often dark-tinged; spikelets compressed, sessile to subsessile, 1-flowered; glumes 1-nerved, strongly keeled, the keel ciliate, the first glume 1.2–1.7 (2) mm long, narrow-lanceolate, the second 1.4–2 (2.4) mm long, lanceolate; lemma 1.7–2.4 (2.8) mm long, lanceolate, 1-nerved, glabrous; palea 1–1.5 mm long, ovate, faintly 2-nerved, hyaline; stamens 3, the anthers 0.5–1 mm long; caryopsis about 1 mm long, ovoid, brownish; $2n = 18$.

Wet sandy or muddy soil, often emersed in the spring and early summer; Europe and n. Africa to Iraq; introduced to the U.S. on ballast near Philadelphia, Pa., and Portland, Oregon, escaping up the Willamette and Columbia rivers and found in our region on the shores of American Falls Reservoir, Power Co., Idaho (*R. Baker 10*), and sw. shore of Washoe Lake, Washoe Co., Nev. [J. T. Howell, N. Nevada Native Pl. Soc. Newsletter 1(4): 2. 1975]. Aug.–Sept.

2. Crypsis schoenoides (L.) Lam.

Phleum schoenoides L. Sp. Pl. 60. 1753. *Crypsis schoenoides* Lam. Tableau Encycl. & Méth. 1: 166, pl. 42, fig. 1. 1791. *Heleochloa schoenoides* Host, Icon. & Descript. Gram. Austriacorum 1: 23, pl. 30. 1801. (*Loefling*, "Habitat in Italia, Smyrna, inque Hispania.")

Much branched, mat-forming annuals; culms prostrate to erect, geniculate, 0.5–4 (6) dm long, branching at the base, the nodes reddish; sheaths short, glabrous, inflated, especially those below the panicles, the margins membranous; ligule a fringe of hairs, 0.5–1 mm long, sometimes extended across the apex of the sheath; blades flat, folded or loosely involute, 2–4 (6) mm broad, scabrous to pilose above and glabrous beneath, readily deciduous from the sheaths; panicles 0.7–3 (4) cm long and (4) 5–8.5 mm thick, cylindric to oblong, spike-like, partly to mostly enveloped by the spathe-like upper sheaths, terminating the culm or in the axils of the upper leaves, pale; spikelets compressed, sessile to subsessile, 1-flowered; glumes 1-nerved, strongly keeled, the keels ciliate, the first glume 1.8–2.3 mm long, narrow-lanceolate and sometimes curved, the second 2.2–2.7 mm long, lanceolate; lemma 2.4–3 mm long, lanceolate, 1-nerved, glabrous, sometimes scabrous on the keel; palea subequal to the lemma, ovate, faintly 2-nerved, hyaline; stamens 3, the anthers 0.7–1 mm long; caryopsis ovoid, brownish; $2n = 36$.

Wet, waste places, often in sandy soil; s. Europe and Africa to India; introduced to the U.S. on ballast and escaping inland; scattered localities in the midw. and e. states, cismontane Calif. and in our region along shores at the Bear River Migratory Bird Refuge w. of Brigham City, Utah (*McDonald s.n.*). June–Sept.

Tribe 9. CHLORIDEAE

Annuals or perennials, with or without rhizomes; culms hollow or pith filled to solid; sheath margins free; ligule membranous or usually a fringe of hairs; blades non-auriculate; inflorescence a unilateral spike or a panicle of few to several digitately or racemosely arranged spicate branches with the spikelets sessile to short-pedicellate and placed along 2 rows on 2 sides of a 3-angled rachis, or a bilateral spike in *Hilaria* with the spikelets in groups of 3 per node on alternate sides of a simple, zig-zag rachis; spikelets 1- to several-flowered, sometimes with a single perfect floret and 1 or more reduced florets above, rarely unisexual, disarticulating above or below the glumes, or in *Schedonnardus* the whole panicle breaking off at the base, or in some *Bouteloua* the spicate branches fall entire, or in *Hilaria* the clusters of 3 spikelets break and fall intact; lemmas (1-) 3- (7)-nerved; lodicules 2, short-truncate, thick as in the Panicoideae, but with little or no vasculation; fruit a caryopsis, except an achene in *Eleusine* in that the seed coat remains separate from the pericarp; embryo eragrostoid; $x = 9$ or 10.

The tribe Chlorideae centers mainly in the tropical regions of the world, but is not uncommon in the warm temperate regions. In concept there are only a few changes from A. S. Hitchcock's (Manual Grasses U.S., 1951) treatment. *Beckmannia* has been moved to the Aveneae and *Munroa* to the Eragrosteae, and *Hilaria* has been transferred here from the Zoysieae. In our flora *Hilaria* seems anomalous in the Chlorideae, but when compared with the closely related *Aegopogon* and *Cathestecum* (both from regions south of our area), a connection to *Bouteloua* can be seen. *Spartina* shares some characteristics in common with the Aeluropodeae. Some agrostologists place *Spartina* in its own tribe, the Spartineae. Recently Gould (The Grasses of Texas, 1975) has placed *Eleusine* and *Leptochloa* in the Eragrosteae tribe.

Buchloë dactyloides (Nutt.) Engelm., Buffalo-grass, was collected near Fallon, Nevada (*G. Dalton s.n.*, RENO) in 1938, and has not been seen since. Because this area is being actively collected, we assume that *Buchloë* has not survived. A common species of the Great Plains, it differs from our other Chlorideae in being dioecious with very dissimilar sexes. The 2-flowered staminate spikelets are arranged on exserted spicate branches similar to *Bouteloua*, and pistillate spikelets are borne in clusters of 2–5 in bur-like structures partly hidden by the leaves.

1 Spikelets borne in clusters of 3 at each node of a zig-zag rachis, disarticulating as a whole cluster . . . . 77. *Hilaria*
1 Spikelets borne singly.
  2 Spikelets with 2 or more perfect florets.
    3 Spikes digitate with a terminal whorl and sometimes 1 or 2 spikes racemosely arranged below . . . . 70. *Eleusine*
    3 Spikes racemose . . . . 71. *Leptochloa*
  2 Spikelets with only 1 perfect floret, sometimes with additional imperfect florets above.
    4 Spikelets without additional modified florets above the perfect ones (an occasional vestigial floret in *Cynodon*).
      5 Lemmas 1.8–4.7 mm long; spikelets not falling entire, disarticulating above the glumes (the whole panicle often breaking off in *Schedonnardus*).
        6 Spikes racemose; lemmas 3–4.7 mm long; tufted perennials . . . . 72. *Schedonnardus*

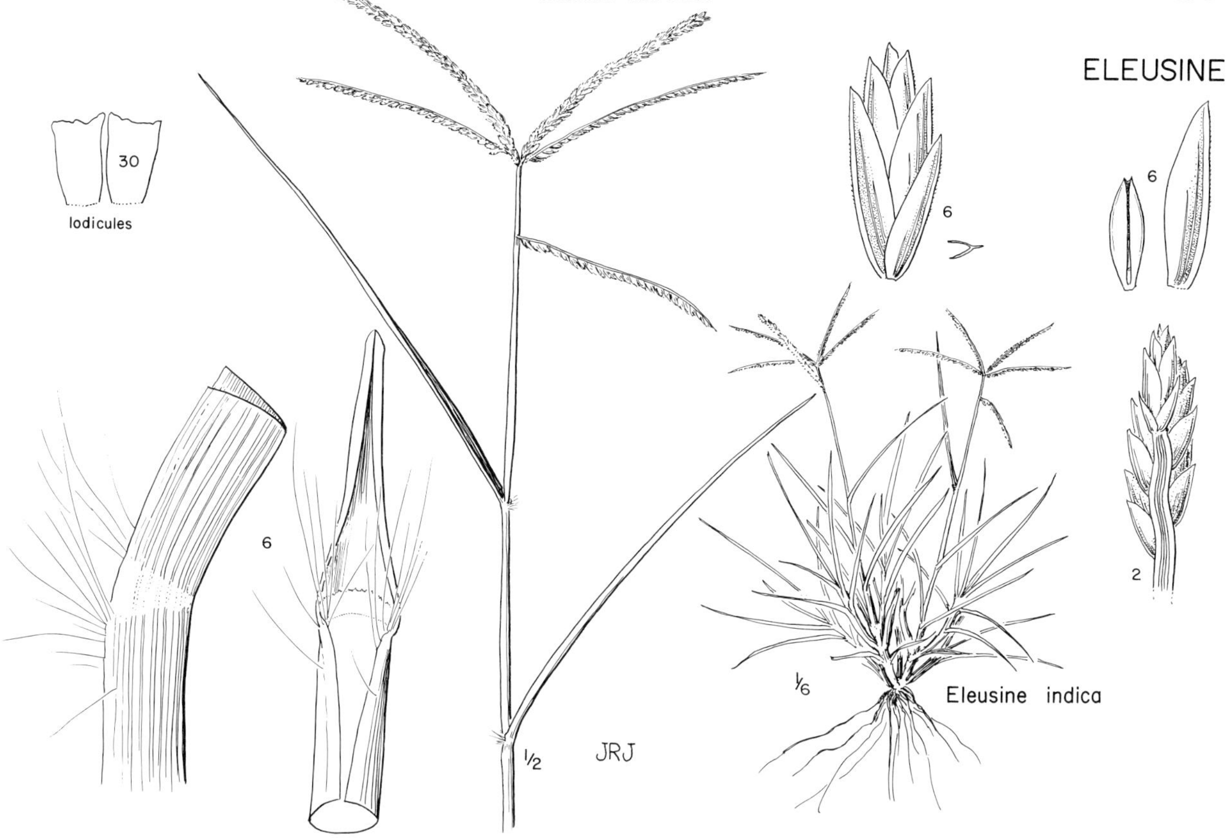

6 Spikes digitate; lemmas 1.8–2.5 mm long; rhizomatous or stoloniferous perennials . . . . 73. *Cynodon*

5 Lemmas 6.2–9.5 mm long; spikelets falling entire, disarticulating below the glumes . . . . 76. *Spartina*

4 Spikelets with one or more modified, staminate or sterile florets above the perfect one.

7 Spikes digitate, (3.5) 4.5–8 cm long . . . . 74. *Chloris*

7 Spikes racemose, sometimes only 1 terminal spike, (0.8) 1–4 (5) cm long . . . . 75. *Bouteloua*

70. ELEUSINE Gaertn.

Low, spreading annuals; culms hollow, thick and weak; sheaths open, usually with long hairs on the upper margins around the throat; ligules short, lacerate, membranous; blades flat or folded, soft, succulent, non-auriculate; inflorescence a panicle with (1) 2 to several digitately arranged spikes, sometimes with 1 or 2 spikes below the whorl, the sessile spikelets in 2 rows on 2 sides of the 3-angled rachis; spikelets 3- to several-flowered, disarticulating above the glumes and between the florets; glumes acute, awnless, firm, unequal, the first glume short, 1-nerved, the second 3- to 7-nerved; lemmas acute, 3- to 7-nerved, the lateral nerves very close to the midnerve, strongly keeled, awnless or mucronate; palea shorter than the lemma, 2-nerved; lodicules 2, wedge-shaped; stamens 3; grain not a true caryopsis since the seed coat remains separate from the pericarp as in *Sporobolus*, plump with a minutely transverse-rugose pericarp; $x = 9$.

A genus of about 6 species, mostly of warmer parts of the Old World, with 1 species native to S. Amer. and 1 adventive in the New World. (Named for Eleusis, an ancient Greek town where the goddess of harvests was worshiped.)

1. Eleusine indica (L.) Gaertn.

Cynosurus indicus L. Sp. Pl. 72. 1753. *Eleusine indica* Gaertn. Fructibus & Seminibus Pl. 1: 8. 1788. *E. gracilis* Salisb. Prodr. Stirpium 19. 1796. *E. domingensis* Sieber ex J. A. Schultes, Mantissa 2: 323. 1824; not Pers. 1805. *Cynodon indicus* Rasp. Ann. Sci. Nat. (Paris) 5: 303. 1825. ("Habitat in Indiis.")

Goose grass.

Annuals; culms 1–3 (6) dm tall, prostrate to ascending, much branched at the base; sheaths keeled, upper margins and throat sometimes pilose with hairs 2–6 mm long, the margins sometimes scabrous below; ligules membranous, 0.3–1.2 mm long, truncate, lacerate, finely ciliolate; blades flat or folded, (2) 3–5 (8) mm broad, sometimes pilose near the collar and the margins scabrous, otherwise glabrous; panicles of 2 to 4 digitately arranged spikes, sometimes with 1 or 2 racemosely arranged below, the spikes 3–7 (8) cm long; spikelets 5–6 (8) mm long and 5- to 7- (8)-flowered; glumes compressed, strongly keeled, scabrous on the keel and glabrous on the sides, the first glume 2–3.2 (7) mm long, only 1-nerved, the second 2.7–4 (8) mm long, ovate, 5- to 7-nerved, the lateral nerves close to the midnerve; lemmas 3.2–4 (5.5) mm long, glabrous, mucronate tipped, 5- to 7-nerved with a pair near the margins and a pair near the midnerve; palea 2–2.5 mm long, with 2 ciliolate nerves; lodicules about 0.4 mm long; anthers 0.5–0.7 mm long; $2n = 18, 36$.

A weed in lawns, roadsides and wasteplaces; introduced from the Old World, now common in warm regions throughout the world; sporadic in the w. states, in some lawns along the Wasatch Front, and in Washington Co., Utah, more common in the c. and e. states from S.D. to Mass. southward. July–Oct.

Eleusine and *Leptochloa* show close relationships with *Eragrostis* of tribe Eragrosteae, especially in having spikelets with more than 1 fertile floret. Some botanists have placed them in the same tribe.

71. LEPTOCHLOA Beauv. Sprangletop

Cespitose annuals and perennials; culms hollow; leaves distributed up the culms; sheaths open; ligules membranous; blades flat to involute, linear, non-auriculate; inflorescence a panicle with few to several spicate or subspicate branches distributed along a central axis, rarely more than 1 at each node, the spikelets sometimes closely spaced and overlapping, sessile to short-pedicellate in 2 rows on 2 sides of a 3-angled rachis; spikelets (1) 2- to several-flowered, disarticulating above the glumes and between the florets; glumes acute, awnless or mucronate, thin, usually unequal, the first glume 1-nerved, the second 1- to 3-nerved, keeled; lemmas rather broad, acute to obtuse or bifid at the apex, prominently 3-nerved, glabrous or puberulent on the nerves or at the base, awned, mucronate or awnless; palea well developed, occasionally puberulent on the nerves; lodicules 2, small, often bilobed; stamens 3; $x = 10$.

A genus of about 25 species in the warmer parts of the world, with 10 or 11 species native to the U.S. Most species are of moist or marshy, often alkaline or saline habitats. (Name from the Greek *leptos,* slender, and *chloa,* grass, in reference to the slender spikes.)

1 Spikelets 5- to 12-flowered; lemmas 3.5–6 mm long .... 1. *L. fascicularis*
1 Spikelets 1- to 3-flowered; lemmas 1–2.5 mm long .... 2. *L. filiformis*

1. Leptochloa fascicularis (Lam.) A. Gray

Festuca fascicularis Lam. Tableau Encycl. & Méth. 1: 189. 1791. *Diplachne fascicularis* Beauv. Essai Nouv. Agrostogr. 81, 160. 1812. *Cynodon fascicularis* Rasp. Ann. Sci. Nat. (Paris) 5: 303. 1825. *Leptochloa fascicularis* A. Gray, Manual Bot. 588. 1848. (*Richard,* "l'Amér. mérid.")

Diplachne tracyi Vasey, Bull. Torrey Bot. Club 15: 49. 1888. *L. tracyi* Beal, Grasses N. Amer. 2: 436. 1896. (*Tracy 216,* "growing in ditches at Reno, Nevada," in 1887.)

Sprangletop.

Loosely tufted annuals; culms 1–4 (7) dm tall, erect or decumbent, branched; sheaths smooth, somewhat inflated below and tapering above, sometimes diverging from the culms, strongly keeled; ligules membranous and long, 2–7 mm, narrow-lanceolate, acute to obtuse, entire to lacerate, often decurrent; blades mostly involute, if flat 1.5–3 (5.5) mm broad, glabrous to scaberulous, ascending and extending beyond the panicles; panicles 7–20 (40) cm long, usually partly enclosed by the upper sheath, the racemosely arranged racemes (3) 5–8 (10) cm long; spikelets (7) 9–12 mm long (measurement not including the awns), 5- to 9- (12)-flowered; glumes 1-nerved, scabrous on the nerves, otherwise glabrous, the first glume 2.2–4 mm long, narrow-lanceolate to linear, the second 3.7–5 (6) mm long, narrow-lanceolate, with an awn 0.5–1.5 mm long; lemmas (3.5) 4–5.2 (6) mm long, 3-nerved, the 2 lateral nerves near the margins and the 3 nerves appressed-pubescent, at least basally, awned from a bifid apex, the awn 2–3.5 mm long; palea subequal to the lemma with appressed-ascending hairs near the margins; lodicules about 0.3 mm long; anthers mostly only 0.2 mm long, ovoid, the flowers probably cleistogamous; $2n = 20$.

Shallow water of marshes, streams, reservoirs and ponds, often in alkaline situations; coastal marshes of both coasts and inland from Wash. to Mass. and s. throughout the remaining U.S. and trop. Amer. to Argentina. July–Sept.

Leptochloa uninervia (Presl) Hitchc. & Chase has been reported from Nevada and Utah, but no specimens have been seen for verification. *Leptochloa uninervia* differs from *L. fascicularis* in its shorter, broader and nearly awnless lemmas.

2. Leptochloa filiformis (Lam.) Beauv.

Festuca filiformis Lam. Tableau Encycl. & Méth. 1: 191. 1791; not Pourret, 1788. *Leptochloa filiformis* Beauv. Essai Nouv. Agrostogr. 71, 161, 166. 1812. *Oxydenia filiformis* Nutt. ex B. D. Jackson, Index Kew. 2: 392. 1894; merely suggested by Nutt. Gen. N. Amer. Pls. 1: 77. 1818. (*Richard,* "Amer. merid.")

Eleusine mucronata Michx. Fl. Boreali-Amer. 1: 65. 1803. *L. mucronata* Kunth, Révision Gram. 1: 91. 1829. (*Michaux,* "Hab. in cultis Illinoensibus," in 1795.)

Eleusine filiformis Pers. Syn. Pl. 1: 87. 1805. *Eleusine sparsa* Muhl. Cat. Pl. N. Amer. 12. 1813. *Leptostachys filiformis* G. Meyer, Primitiae Fl. Essequeboensis 74. 1818. ("Hab. in Americ. meridion.")

Red sprangletop.

Tufted annuals; culms (2) 3–8 (10) dm tall, sometimes decumbent at the base and rooting at the nodes, leafy; sheaths papillose-pilose; ligules membranous, 0.5–2 mm long, erose-lacerate, ciliate; blades flat, 1.5–6 (10) mm broad and 2–20 cm long, glabrous to scaberulous, sometimes papillose-pilose near the base, soft; panicles 10–30 (48) cm long, the racemosely arranged racemes 2–10 (15) cm long; spikelets 2–2.5 (3.5) mm long, (1-) 2- to 3-flowered, the florets barely overlapping, short pedicellate, appressed to the rachis; glumes small, narrow, acute, 1-nerved, the nerves scabrous, the first glume 1.2–2 (2.5) mm long, the second 1.6–2.2 (2.7) mm long; lemmas small, 1–1.5 (2.5) mm long, broadly ovate, blunt, awnless, 3-nerved, sometimes pubescent on the nerves; palea subequal to the lemma; caryopsis ovoid, reddish; $2n = 20$.

In seasonally moist to dry disturbed sites such as roadsides, ditchbanks and canyon slopes at low elevs.; se. Calif., se. Utah (*Higgins 823,* Beaver Dam Mts., Washington Co.), s. Colo., and e. through Kansas to Va. and s. throughout s. U.S. and trop. Amer. July–Sept.

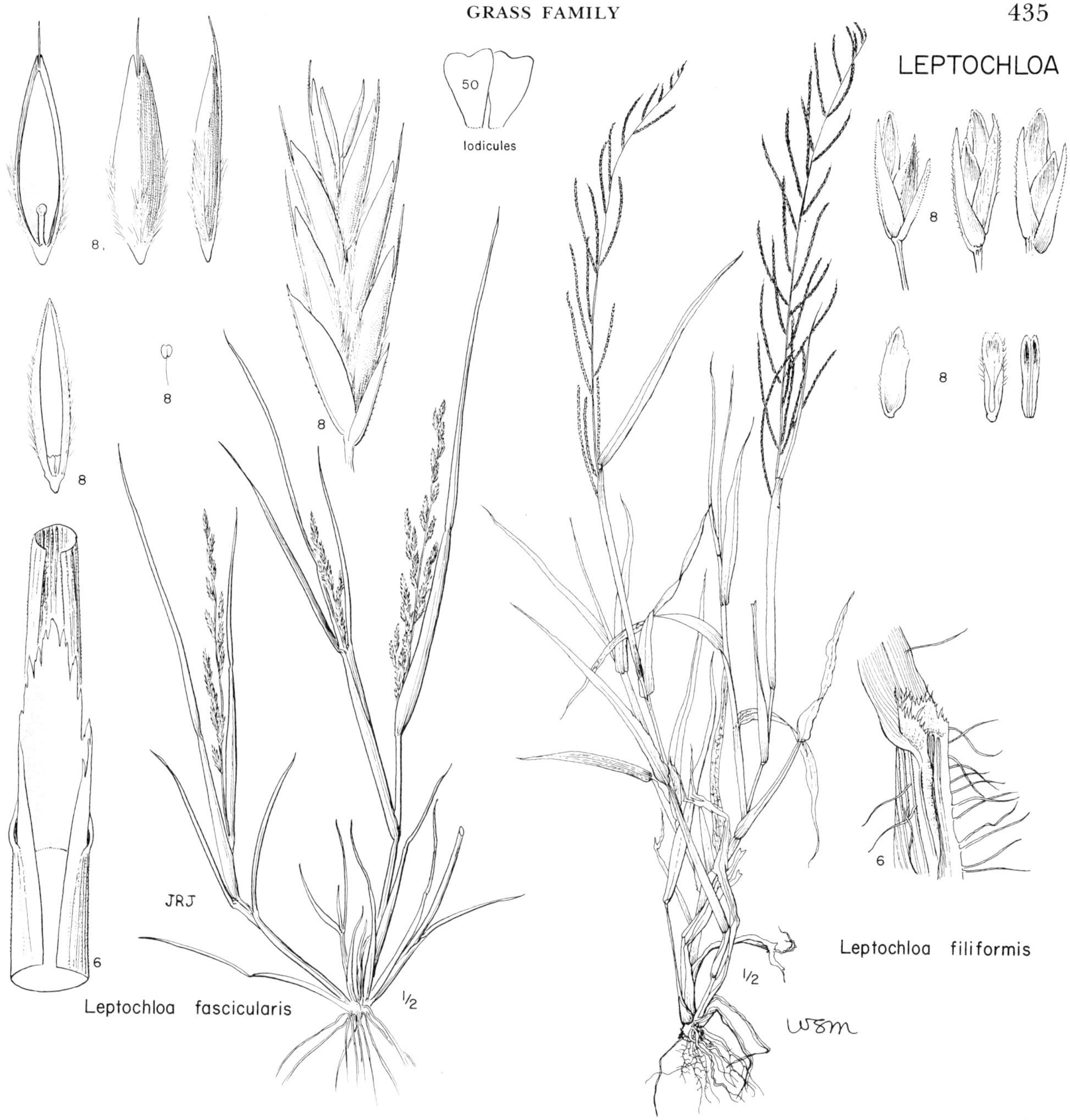

72. SCHEDONNARDUS Steudel

Low, tufted perennials; culms hollow or solid, slender, wiry, erect or decumbent, strongly angled above; leaves mostly basal; sheaths open; ligules membranous, decurrent, not ciliate; blades short, flat, usually spirally twisted and the crisped margins whitish, non-auriculate; inflorescence a panicle with few to several spikes distributed along a stiff, curved central axis, the panicle comprising $^1/_2$ to $^3/_4$ the height of the plant, the sessile spikelets widely spaced and appressed along the rachis and along the terminal segment of the central axis, in 2 rows on 2 sides of a 3-angled rachis and partially sunken in the rachis, at maturity the entire panicle breaking off at the base; spikelets 1-flowered, slender, disarticulating above the glumes; glumes narrow-lanceolate, acuminate and strongly 1-nerved, stiff, unequal, the first glume shorter, the second about as long as the lemma; lemma narrow, rigid, 3-nerved, awnless or with a minute awn-tip; palea similar to the lemma and about as long; lodicules 2, wedge-shaped; stamens 3; $x = 10$.

A monotypic genus of the Great Plains and adjacent prairies of the U.S. and recurring in Argentina. (From the Greek *schedon,* near, and *Nardus,* a genus of grass which Steudel took to be a close relative.)

1. Schedonnardus paniculatus (Nutt.) Trel.

Lepturus paniculatus Nutt. Gen. N. Amer. Pls. 1: 81. 1818. *Rottböllia paniculata* Sprengel, Syst. Veg. 1: 300. 1825. *Schedonnardus paniculatus* Trel. in Branner & Coville, Annual Rep. Geol. Surv. Arkansas 1888(4): 236. 1891. *Spirochloë paniculata* Lunell, Amer. Midl. Naturalist 4: 220. 1915. ("On dry saline plains, near Fort Mandan, on the Missouri," N.D., in 1811.)

Tumblegrass.

Tufted perennials, sometimes flowering the first year; culms relatively low, 1–4.5 dm tall, ascending, geniculate and branched below; sheaths compressed-keeled, glabrous with scarious margins; ligules membranous, relatively long, 1–3.5 mm, lanceolate, subentire, decurrent; blades firm, flat or usually folded, 0.7–2 mm broad and 1–10 cm long, becoming twisted in drying, the margins thick, whitish, often somewhat crisped; panicles 10–20 (30) cm long, the racemosely arranged, wiry spikes 4–10 cm long, spreading to divaricate; spikelets 3–4.8 mm long, 1-flowered, sessile, remote, appressed into slight depressions of the rachis; glumes, the first 1.5–3 mm long, sometimes filiform, acuminate, 1-nerved, the second 2.5–4 (5.5) mm long, linear-lanceolate; lemma 3–4.7 mm long, usually exceeding the glumes, narrow, acute, 3-nerved, enveloping the palea; palea subequal to the lemma; lodicules about 0.5 mm long; ovary short-stipitate; $2n = 20, 30$.

Usually in open, disturbed, relatively dry sites at lower elevs. in our area; Great Plains from se. Alta. and Man., s. through e. Mont., the Dakotas, Wisc. and Ill. to La., s. Utah and Ariz.; also in Argentina. June–Aug.

Although widespread and locally abundant, tumblegrass produces little herbage and is of little significance as a forage plant. It tends to increase in grama grass areas with overgrazing. The mature panicle breaks off at the base and functions as a tumbleweed in dispersing the seeds, hence its name.

73. CYNODON Rich. in Pers. nom. conserv.

Low, mostly mat-forming, stoloniferous and rhizomatous perennials; culms hollow or pith-filled, mostly with short internodes and much-branched; sheaths open; ligule a fringe of long, straight hairs; blades short, flat to involute, narrow, soft and succulent, non-auriculate; inflorescence a panicle with 2 to several digitately arranged spikes, the sessile spikelets in 2 rows on 2 sides of a 3-angled rachis; spikelets with 1 fertile floret, the rachilla prolonged behind the palea as a bristle and occasionally bearing a vestigial floret, disarticulating above the glumes; glumes lanceolate, 1-nerved, subequal, the second glume nearly as long as the lemma; lemma more or less indurate, laterally compressed, 3-nerved, ciliate-pubescent on the midnerve and the 2 submarginal nerves, awnless; palea as long as the lemma, narrow, 2-nerved; lodicules 2, more or less wedge-shaped; stamens 3; $x = 9$.

A genus of about 10 species, mostly in Africa and Australia, with 1 species, *C. dactylon,* widely distributed in warmer regions throughout the world. (Name from the Greek *kuon,* dog, and *odous,* tooth, probably in reference to the close rows of tooth-like spikelets.)

1. Cynodon dactylon (L.) Pers.

Panicum dactylon L. Sp. Pl. 58. 1753. *Digitaria dactylon* Scop. Fl. Carniolica, ed. 2. 1: 53. 1772. *Dactilon officinale* Vill. Hist. Pl. Dauphiné 2: 69. 1787. *Paspalum dactylon* Lam. Tableau Encycl. & Méth. 1: 176. 1791. *Digitaria littoralis* Salisb. Prodr. Stirpium 19. 1796. *Milium dactylon* Moench, Suppl. Meth. Pl. 67. 1802. *Fibichia umbellata* Koeler, Descript. Gram. Gallia & German. 308. 1802. *Cynodon dactylon* Pers. Syn. Pl. 1: 85. 1805. *Digitaria stolonifera* Schrader, Fl. German. 1: 165. 1806. *Chloris cynodon* Trin. Gram. Unifloris & Sesquifloris 229. 1824. *Capriola dactylon* Kuntze, Revisio Gen. Pl. 2: 764. 1891. *Fibichia dactylon* Beck, Wiss. Mitt. Bosner Herzegowina 9: 436. 1904. ("Habitat in Europa australi.")

Bermuda grass.

Strongly rhizomatous or stoloniferous perennials; culms 2–6 (8) dm long, decumbent and rooting at the lower nodes, leafy; sheaths smooth, the margins scarious, sometimes extending as lateral prolongations of the ligule, sometimes pilose towards the apex; ligule a short, fringed membrane, 0.2–0.5 mm long, sometimes with some of the longer hairs of the throat; blades flat, 1.5–3 (4) mm broad, lax, sometimes puberulent above, the margins scaberulous; panicles of 3–6 (9) digitately arranged spikes, the spikes 3–5 (6.5) cm long, the joints short-pilose; spikelets 2–2.8 mm long, sessile, 1-flowered, the rachilla prolonged behind the palea and sometimes bearing a vestigial floret; glumes small, compressed, 1-nerved and keeled, the keel scaberulous, the first glume 0.7–1.5 (2) mm long, lanceolate, curved, the second 1.2–2 (2.5) mm long, lanceolate, straight, tapered into a short awn tip; lemma 1.8–2.2 (2.5) mm long, broadly ovate and blunt, 3-nerved, the lateral nerves sub-marginal, glabrous or pubescent on the nerves; palea subequal to the lemma; lodicules about 0.3 mm long; anthers 0.6–1 mm long; stigmas dark-brown, plumose; $2n = 18, 27, 30, 36, 40, 54$.

Moist, disturbed sites such as ditchbanks, reservoir shores and around dwellings in the valleys; introduced from Africa and now widespread in warmer parts of the world; Oregon and Idaho, e. to Kansas, Mo., Ill., Ohio, Pa. and N.H. and s. throughout the remaining U.S. and trop. Amer. May–Sept.

Bermuda grass was introduced from Africa as a drought-resistant and alkali-tolerant species. It branches indefinitely with only a portion of the leafy stems terminating in inflorescences, making it an excellent lawn grass. Because it is able to withstand only moderate amounts of freezing temperatures, its use is limited to warm climates. In the south it is also an outstanding forage grass, being able to withstand heavy grazing pressure.

74. CHLORIS Swartz Fingergrass

Annuals and perennials, often stoloniferous; culms erect or decumbent, the lower portion often flattened; sheaths open, sharply keeled and laterally compressed; ligules membranous, short, blades flat or folded, non-auriculate; inflorescence a panicle with few to several spicate or subspicate branches, these digi-

SCHEDONNARDUS
CYNODON

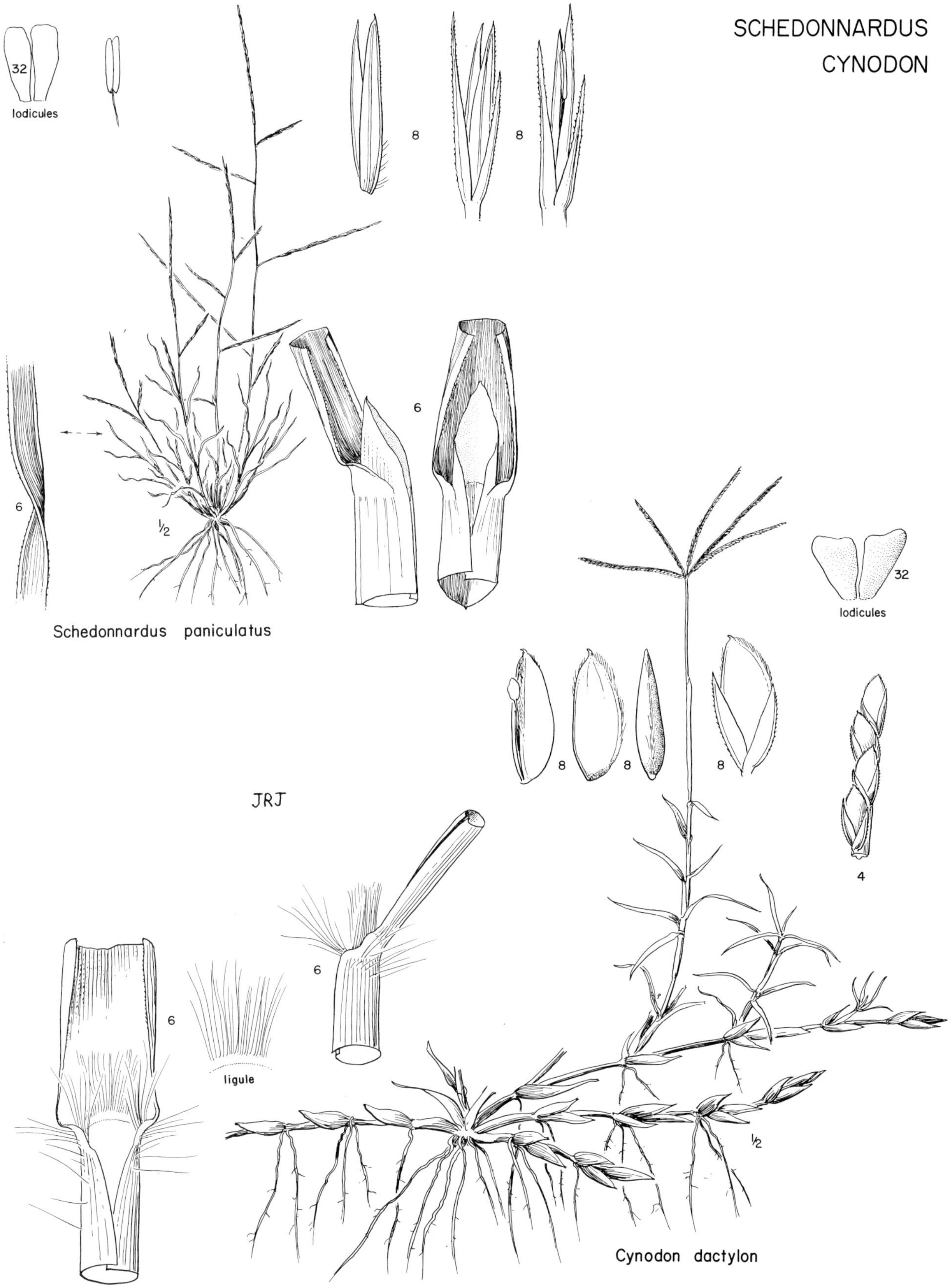

tately arranged or somewhat loosely clustered at the apex, but loosely distributed along the central axis in some, the sessile or short pedicellate spikelets in 2 rows on 2 sides of the 3-angled rachis; spikelets with a single, perfect floret and 1 to 3 staminate or rudimentary florets above, disarticulating above the glumes; glumes lanceolate, thin, 1-nerved, strongly unequal to subequal; fertile lemma broadly to narrowly ovate or oblong, 1- to 5-nerved, but usually with 3 strong nerves, glabrous to ciliate-pubescent on the nerves, awnless or more frequently awned from a minutely bifid apex; palea well developed, strongly 2-nerved; sterile lemmas shorter and often different in shape; lodicules 2; $x = 10$.

A genus of about 50 species distributed in warmer regions of the world, 13 of which are native to the U.S. where there are also about 10 introduced species. (Named for *Chloris,* who in Greek mythology was the mother of Nestor, goddess of flowers.)

Chloris is a difficult genus in the Southwest, where considerable introgression occurs.

1. Chloris virgata Swartz

Chloris virgata Swartz, Fl. Indiae Occid. 1: 203. 1797. *Rabdochloa virgata* Beauv. Essai Nouv. Agrostogr. 84, 158. 1812. (*Swartz,* Antigua.)

Feather fingergrass.

Tufted annuals; culms 2–9 dm tall, mostly ascending, sometimes decumbent below and rooting at the nodes, leafy; sheaths smooth, keeled; ligules membranous, 0.2–1 mm long, truncate, lacerate, ciliate; blades flat or folded, 2–6 (9) mm broad and 3–20 cm long, glabrous to scaberulous, sometimes sparsely pilose near the collar; panicles of (4) 6–14 digitately arranged spikes, the spikes (3.5) 4.5–8 cm long, ascending; spikelets 7.8–10 (13) mm long, with 1 fertile floret and 1 sterile one, short pedicellate, the pedicels often pilose; glumes narrowly-lanceolate, 1-nerved, the first glume 1.5–2.5 (3) mm long, sometimes aristate, the second 2.5–4 mm long, tapering into a short awn; fertile lemma 2.5–3.5 mm long, compressed, obovate with a prominent bulge on the keel, abruptly narrowed, 3-nerved, the keel appressed-pilose, the margins with long pilose tufts of hairs near the apex and shorter hairs down the sides, the lateral nerves depressed, sometimes lines of hairs between the lateral nerves and keel on the otherwise glabrous sides, awned from a bifid apex, the awn (5) 7–10 (12) mm long, the callus densely hairy; palea subequal to the lemma, narrow; sterile lemma somewhat smaller, truncate, with an awn 2–7 mm long; anthers 0.4–0.6 mm long; $2n = 14, 20, 26, 30$.

A weed in disturbed, moderately moist places such as roadsides and desert washes; Nebr. and Colo. to Texas and across the Southwest extending n. in suitably warm sites of Utah and Nev. and introduced in a few localities in the e. states, often around wool mills. Aug.–Sept.

Chloris virgata is not outstanding as a forage grass but does assume some importance when occasionally abundant on the southwest range.

75. BOUTELOUA Lag. Grama Grass

Annuals or more often perennials of diverse habit, some strongly tufted, but more often plants producing rhizomes or stolons; culms solid, pith-filled or hollow; leaves mostly basal; sheaths open; ligule usually a ring of straight hairs; blades flat, folded or involute, non-auriculate; inflorescence a panicle with 1 to several short spikes, closely or distantly spaced along the central axis and having 1 to several sessile spikelets in 2 rows on 2 sides of the 3-angled rachis; spikelets with 1 perfect floret and 1 to 3 staminate or vestigial florets above, disarticulating above the glumes or the spikes falling entire at maturity; glumes lanceolate, 1-nerved, awnless or short-awned, unequal to subequal; fertile lemma membranous, 3-nerved, the midnerve often extending into an awn, the lateral nerves occasionally short-awned; palea of the fertile lemma membranous, the 2 nerves occasionally awn-tipped; sterile lemma(s) small, with 3 strong awns, usually longer than those of the fertile lemma; lodicules 2, cuneate, the upper margin very slightly lobed; stamens 3; $x = 10$.

A genus of about 40 species, 17 of which occur in N. Amer., centering in the Southwest, and the rest in C. and S. Amer. (Named for Claudio Bouteloua, 1774–1842, a professor of agriculture in Madrid, and for his brother Esteban, 1776–1813, a gardener.)

Bouteloua curtipendula and *B. eriopoda* of medium and low altitudes and *B. gracilis* of medium and high altitudes are valuable forage species on the Great Plains and Southwest grasslands.

References:
Gould, F. W., and Z. J. Kapadia. Biosystematic studies in the *Bouteloua curtipendula* complex. II. Taxonomy. Brittonia 16: 182–208. 1964.
Griffiths, D. The grama grasses, *Bouteloua* and related genera. Contr. U.S. Natl. Herb. 14: 343–428. 1912.

1 Spikes 10–50 per culm, falling entire at maturity, each spike bearing only 1–5 (10) spikelets (Bouteloua).
  2 Spikes bearing (2) 3–10 spikelets; lemma 3.8–6 mm long .... 1. *B. curtipendula*
  2 Spikes bearing 1 (2) spikelets; lemma 6–7.5 mm long .... 2. *B. uniflora*
1 Spikes no more than 8 per culm, the florets falling from the persistent glumes, each spike bearing (8) 10–80 spikelets (Chondrosium).
  3 Plants perennial
    4 Culms white-lanate to arachnoid pubescent; rachis extending beyond the terminal spikelet as a needle .... 3. *B. eriopoda*
    4 Culms glabrous to scaberulous, not lanate or arachnoid pubescent.
      5 Spikes 3 to 7, rarely only 2 per culm; second glume 2–3 (4) mm long, awnless .... 4. *B. trifida*

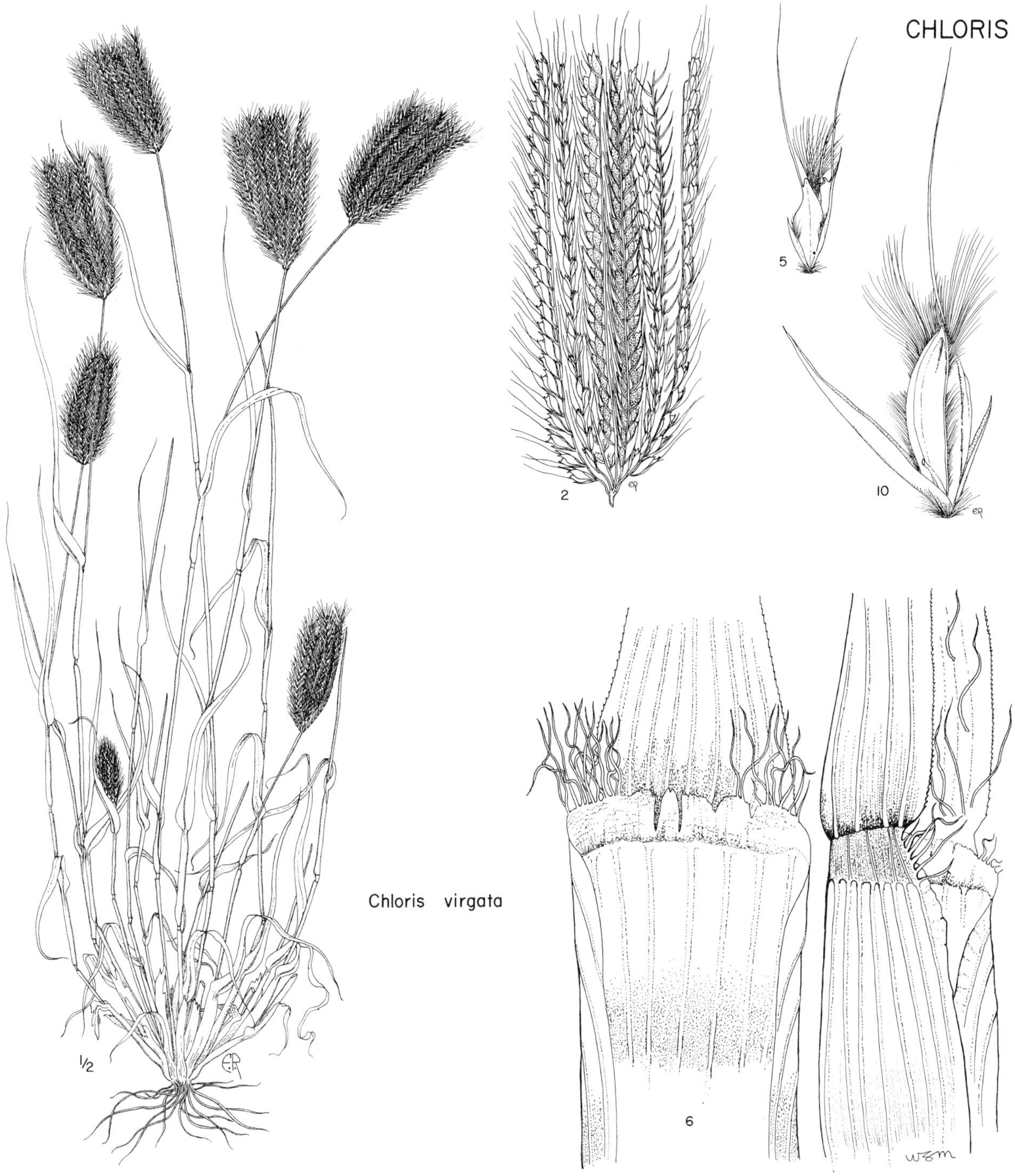

5 Spikes normally 2, rarely 1, 3 or 4 per culm; second glume (3.5) 4–6 (6.5) mm long with an awn tip up to 1.2 mm long.
 6 Rachis terminating in a spikelet that is sometimes sterile; callus of the sterile lemma densely bearded . . . . 5. *B. gracilis*
 6 Rachis extending beyond the terminal spikelet as a needle; callus of the sterile lemma naked . . . . 6. *B. hirsuta*
3 Plants annual.
 7 Spikes 3 to 6 per culm; second glume 1.5–2.5 mm long . . . . 7. *B. barbata*
 7 Spikes solitary; second glume 4–5.5 mm long . . . . 8. *B. simplex*

1. Bouteloua curtipendula (Michx.) A. Gray

Chloris curtipendula Michx. Fl. Boreali-Amer. 1: 59. 1803. *Dineba curtipendula* Beauv. Essai Nouv. Agrostogr. 98, 158, 160. 1812. *Eutriana curtipendula* Trin. Fund. Agrostogr. 161. 1820. *Cynodon curtipendula* Rasp. Ann. Sci. Nat. (Paris) 5: 303. 1825. *Andropogon curtipendulum* Sprengel ex Steudel, Nomencl. Bot. ed. 2. 1: 90. 1840, as synonym of *Eutriana curtipendula*. *Bouteloua curtipendula* Torr. in Emory, Notes Military Reconnaissance from . . . Missouri to . . . California 154. 1848, a provisional name. *B. curtipendula* A. Gray, Manual Bot. ed. 2. 553. 1856, with no mention of Torrey's provisional name. *Atheropogon curtipendulus* Fourn. Mex. Pl. 2: 138. 1886. (*Michaux*, "Hab. in aridis regionis Illinoensis ad Wabast et in rupibus ad prairie du rocher," in 1795.)

B. curtipendula var. *caespitosa* Gould & Kapadia, Brittonia 16: 203. 1964. (*Gould 10021*, 3 mi. e. of Bisbee, Cochise Co., Ariz., 10 Aug. 1962.)

Side-oats grama.

Tufted perennials, often developing rhizomes; culms (2.5) 4–7 (10) dm tall, erect, slender; leaves sometimes pustulose-pilose near the collar; sheaths glabrous to sparse-pilose, becoming flat and persistent below, often bright yellow and slick at the base; ligule a fringe of hairs, 0.2–0.6 (1) mm long; blades usually flat and somewhat involute at the tip, relatively broad, 2–3.5 (5) mm broad and 5–20 (35) cm long, scabrous above, glabrous below, glaucous; panicles long, (10) 14–20 (30) cm, of 30–40 (50) short, pendulous spikes, the spikes 0.8–1.5 (3) cm long with only 2–5 (10) spikelets alternating on a flattened rachis that extends beyond the distal spikelet as a needle, each spike deciduous as a unit; spikelets with a perfect floret and sometimes a vestigial spikelet; glumes 1-nerved, the first 2.5–4.5 (5.5) mm long, filiform, scarious, the second 4.5–7 (7.5) mm long, lanceolate, glabrous or scabrous, usually purplish-tinged; fertile lemma (3.8) 4.5–6 mm long, glabrous or scabrous-strigose, 3-nerved, the lateral nerves excurrent as short awns that equal the acute tip; palea of the fertile floret subequal to the lemma, relatively firm, strongly 2-nerved, the nerves sometimes terminating as 2 short bristles; sterile floret single or absent, 1–3 (4) mm long, its 3 awns 3–6 mm long; lodicules about 0.4 mm long; anthers 2.2–2.5 mm long, red, orange or yellow; $2n$ = 20, 28, 35, 40–66, 69–103.

Dry grassy, shrubby and open woodland hillsides, mesas and canyons at low elevs.; Sask. to Ont. and Mich., s. to se. Calif., Miss., Tenn. and S.C. and through Mex. to C. Amer. and Andean S. Amer. July–Sept.

There are rather well marked trends in variation that seem to justify infraspecific segregation. Two varieties occur in our area and can be separated by the following key.

1 Culms single or in small clusters from creeping rhizomes; anthers red or orange; $2n$ = 40, 41–64; Sask. and Man. to Mich., s. through se. Utah, ne. Ariz., N.M., and Texas . . . . var. *curtipendula*

1 Culms in large or small clumps, often from a hard, knotty base, without rhizomes; anthers yellow or orange; $2n$ = 58–103; se. Calif., s. Nev., s. Utah and s. Colo., s. through Ariz., N.M. and Texas to c. Mex. . . . . var. *caespitosa* Gould & Kapadia

2. Bouteloua uniflora Vasey

Bouteloua uniflora Vasey, Bot. Gaz. 16: 26. 1891. (*Nealley 222*, Crockett Co., Texas, in 1890.)

Tufted perennials, usually without rhizomes; culms 3.5–6 dm tall, erect, slender, glabrous, leafy; sheaths glabrous to scaberulous and sometimes sparsely pilose near the collar, becoming flat and persistent below; ligule a short fringe of hairs, 0.2–0.5 mm long; blades flat, becoming involute on drying, narrow, 1–2 mm broad and 12–16 cm long, scaberulous to hispidulous above and scabrous to papillose-ciliate on margins; panicles 5–10 (14) cm long, of 10–35 reflexed spikes, the spikes consisting of only 1, rarely 2, spikelets, the prolongation of the rachis about 4 mm long, needle-like, each spike deciduous as a unit; spikelets usually 1-flowered, sometimes with a minute, sterile floret above; glumes 1-nerved, the first 2.5–3.5 (4) mm long, narrow, scarious, the second 6.2–8 mm long, narrow-lanceolate, acute, usually scabrous on the nerve, otherwise glabrous; fertile lemma 6–7.5 mm long, glabrous, 3-nerved, narrow, acute, awnless; palea of the fertile floret subequal to the lemma; sterile floret, when present, minute with 1 or 2 short awns; anthers 2.5–3 mm long, bright lemon-yellow; $2n$ = 20.

Rocky hills and valleys; c. and sw. Texas and a disjunct station in sw. Utah (*Silveus 339*, Zion National Park, Utah).

3. Bouteloua eriopoda (Torr.) Torr.

Chondrosium eriopodum Torr. in Emory, Notes Military Reconnaissance from . . . Missouri to . . . California 153. 1848. *Bouteloua eriopoda* Torr. in Whipple, Explor. & Surv. Railroad Route from Mississippi River to Pacific Ocean 4(5): 155. 1857. (*Emory*, Del Norte [Rio Grande] River, N.M., 15 Oct. 1846.)

Black grama.

Tufted perennials from hard knotty bases, sometimes becoming stoloniferous as the lower nodes root and produce tufts of culms; culms 2.5–7 dm tall, sprawling to ascending, densely white-woolly; sheaths striate, glabrous; ligule a fringe of hairs, 0.2–7 mm long; blades flat, becoming twisted, involute towards the tips, 1–2.2 mm broad and 2–7 cm long, pustulose-pilose on the lower margins, glabrous below, scaberulous above; panicles (4) 7–12 (15) cm long, of 2–5 (8) loosely ascending spikes, the spikes 2–4 (5) cm long, each one bearing 9–20, uncrowded, appressed spikelets, the rachis bearded at the base, prolonged beyond the terminal spikelet as a needle; spikelets with 1 perfect floret and a sterile floret above; glumes 1-nerved, the first (2) 2.5–4.5 mm long, narrow, glabrous, sometimes scaberulous on the nerves, hyaline, the second 5–8 (9) mm long, linear-lanceolate, glabrous to minutely scaberulous, stramineous or sometimes purplish; fertile lemma 4.7–6.2 (7) mm long, linear, 3-nerved, narrowing into the terminal awn, purplish, the terminal awn 1.2–3 (4) mm long, the lateral awns shorter, 0.5–1.5 mm long, the callus densely hairy; palea of the fertile floret subequal to the lemma; sterile lemma 0.5–1.2 mm long, its 3 awns 3.5–7 (8.5) mm long, with a short, hairy callus; anthers 2.5–3.5 mm long, pale orange; $2n$ = 20, 21, 28.

Dry rocky or sandy hills and mesas at low elevs.; w. Okla. and w. Texas to s. Utah, se. Calif. and s. to n. Mex. June–Aug.

BOUTELOUA

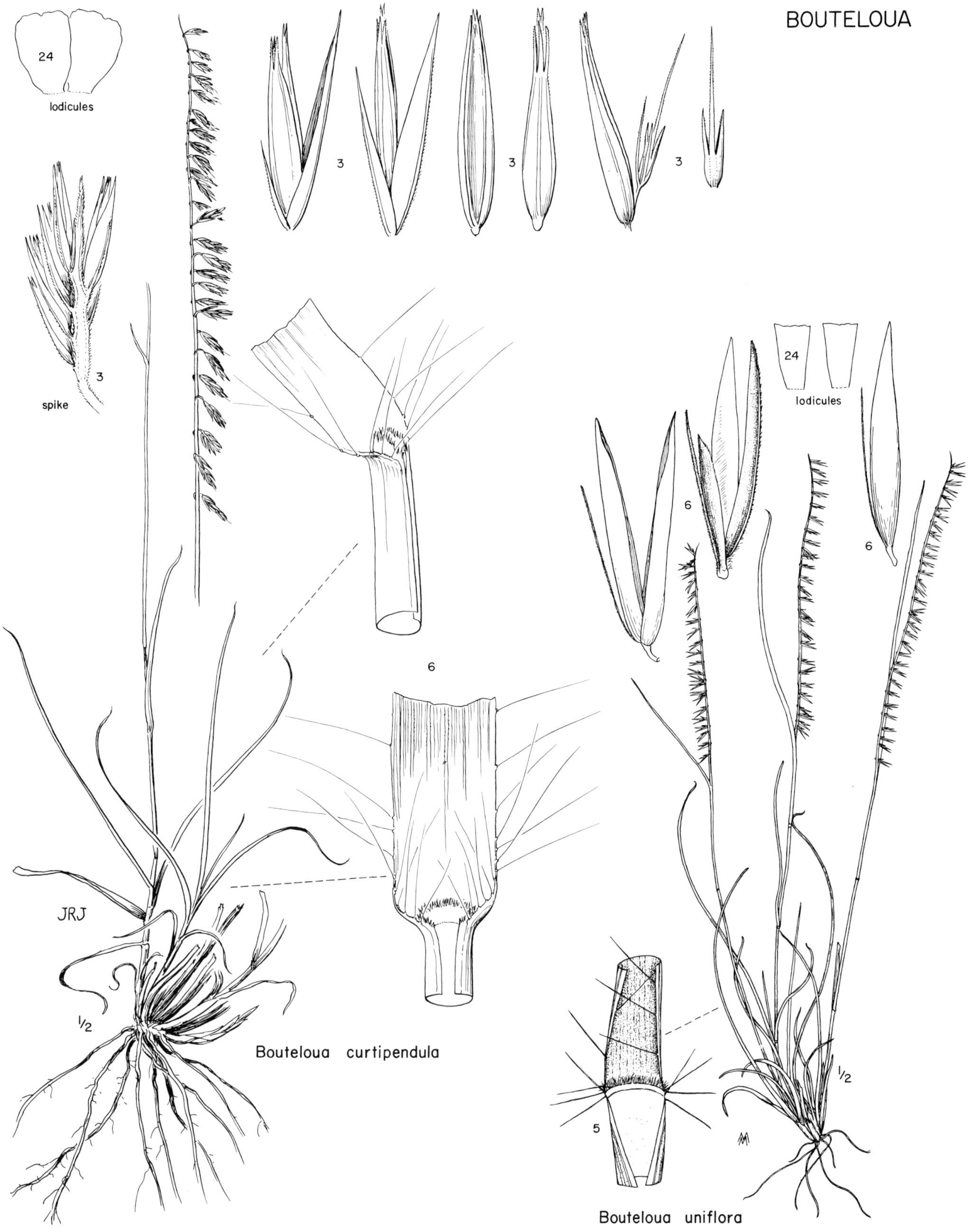
24
lodicules
spike
3
6
24
lodicules
JRJ
1/2
5
1/2
Bouteloua curtipendula
Bouteloua uniflora

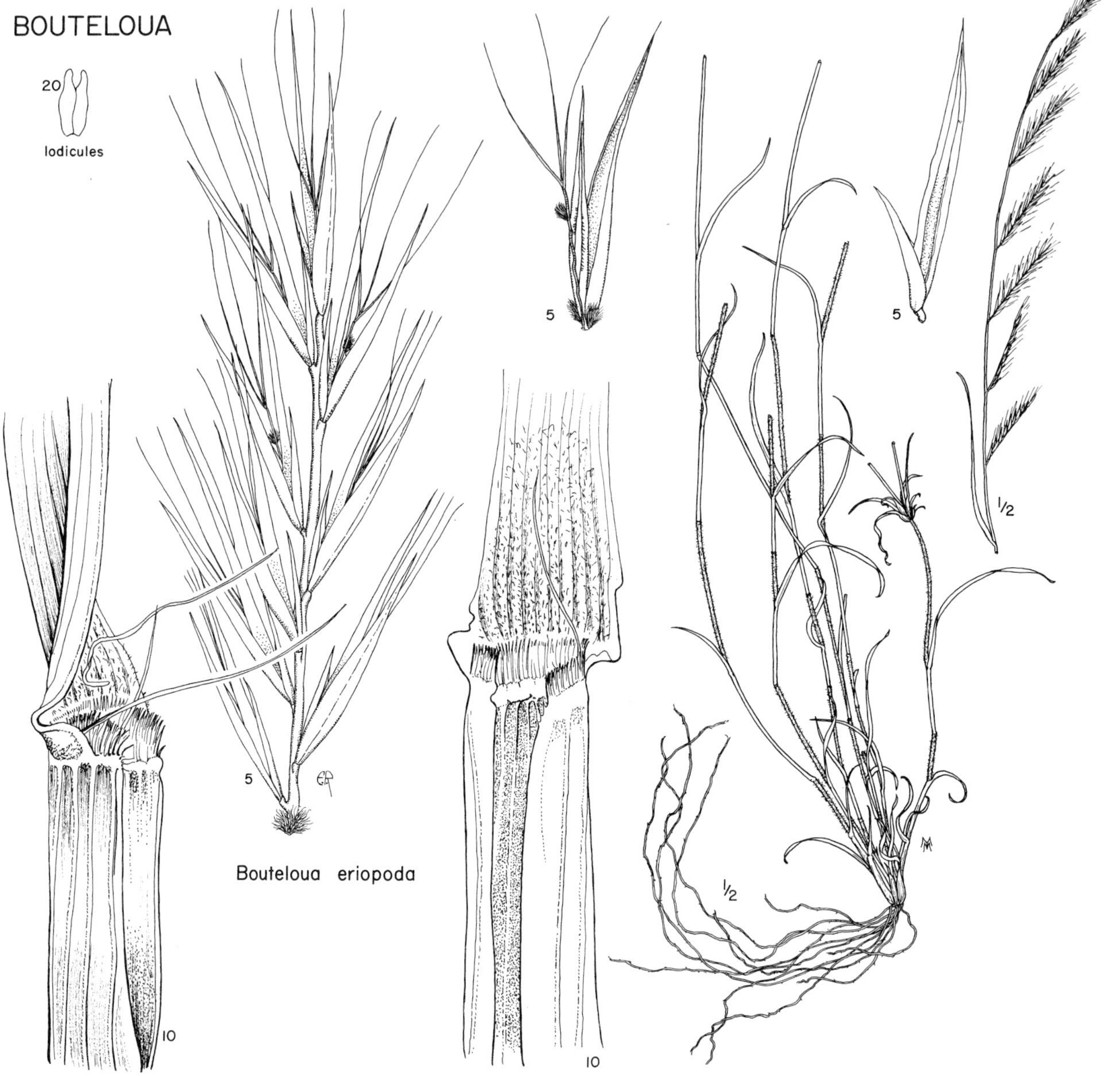

Black grama is a valuable range grass, but it does not hold up as well as *B. gracilis* and *B. curtipendula* under heavy grazing pressure. It is a very distinctive species with its white-woolly culms.

4. **Bouteloua trifida** Thurber

Bouteloua trifida Thurber in S. Wats. Proc. Amer. Acad. Arts 18: 177. 1883. (*Palmer 1355,* Monclova, Coahuila, Mex., Aug. 1880.)

B. burkei Scribn. in S. Wats. Proc. Amer. Acad. Arts 18: 179. 1883. *B. trifida* var. *burkei* Vasey ex L. H. Dewey in J. M. Coulter, Contr. U. S. Natl. Herb. 2: 532. 1894. (*Berlandier 167,* w. Texas and N.M.)

Chondrosium trinii Fourn. Mex. Pl. 2: 136. 1886. *B. trinii* Griffiths, Contr. U. S. Natl. Herb. 14: 387. 1912. (*Berlandier 1427,* Laredo, Texas.)

Red grama.

Delicate tufted perennials, sometimes very short rhizomatous on old plants; culms (0.5) 1.5–3 (4) dm tall, erect, often somewhat flattened, sometimes slightly geniculate at the lower nodes; leaves mostly basal; sheaths glabrous to scaberulous, the lower sheaths short and persistent, becoming flattened; ligule a short fringe of hairs, 0.2–0.5 mm long; blades flat, folded or involute, narrow, 0.5–1.2 (2) mm broad and 0.7–5 (8) cm long, scaberulous, sometimes sparsely pustulose-pilose, especially towards the collar, sometimes glabrous on back; panicles 3.5–8 (10) cm long, of (2) 3–7 spikes, the spikes 0.7–2.5 (4) cm long, ascending or appressed, straight or occasionally arcuate, bearing 8–23 (32) spikelets, the rachis terminating in a fertile spikelet; spikelets reddish-purple, with 1 perfect floret and a sterile floret; glumes usually subequal, (1.5) 2–3 (4) mm

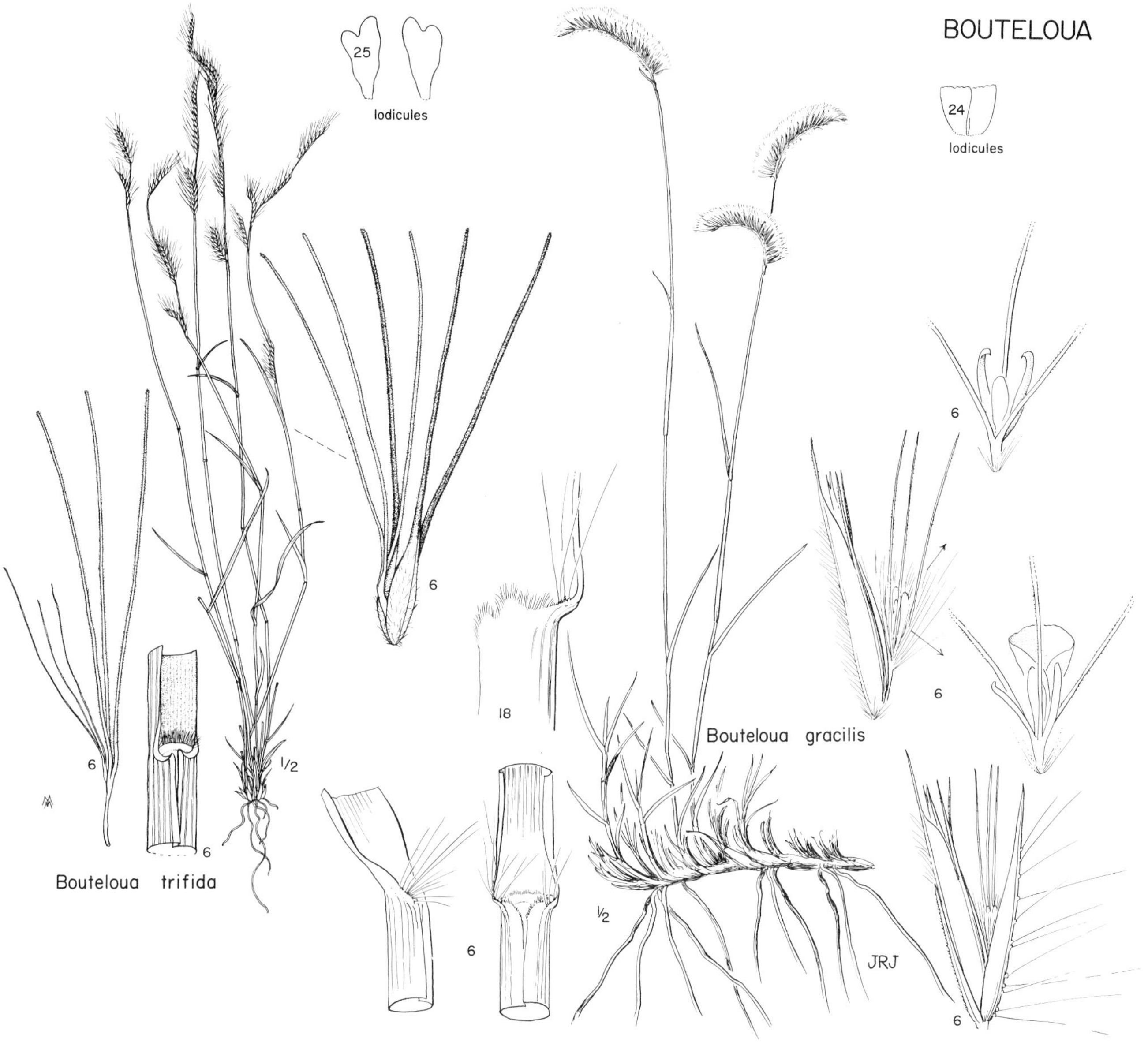

long, lanceolate, 1-nerved, glabrous to scaberulous, the first glume slightly shorter, hyaline to purplish, the second dark-purplish; fertile lemma 2.2–3 (4) mm long, glabrous to slightly pubescent basally, linear-lanceolate, gradually tapering into the terminal awn, 3-nerved, the terminal awn 1–4 mm long, the lateral awns 2–7 mm long, the lateral nerves appressed-pubescent, the callus hairy; palea of the fertile floret subequal to the lemma; sterile lemma 0.2–1 (2) mm long and its 3 awns 2.5–6 mm long, sometimes without a body; $2n = 20, 28$.

Dry open plains, shrubby hills and ravines at low elevs.; w. Texas, N.M., Ariz., sw. Utah, s. Nev., se. Calif. and n. Mex. April–Sept.

Red grama is sometimes misidentified in the field as an *Aristida* because of the 3 awns of the fertile lemma and the 3-awned rudiments. Our specimens have somewhat longer awns than in plants south of our area.

5. Bouteloua gracilis (H.B.K.) Lag. ex Steudel

Chondrosium gracile H.B.K. Nova Gen. & Sp. Pl. 1: 176, tab. 58. 1816. *Actinochloa gracilis* Willd. ex Roemer & Schultes, Syst. Veg. 2: 418. 1817. *Eutriana gracilis* Trin. Gram. Unifloris & Sequifloris 240. 1824. *Atheropogon gracilis* Sprengel, Syst. Veg. 1: 293. 1825. *Bouteloua gracilis* Lag. ex Steudel, Nomencl. Bot. ed. 2. 1: 219. 1840. (*Humboldt & Bonpland,* Guanajuato, Mex.)

Atheropogon oligostachyus Nutt. Gen. N. Amer. Pls. 1: 78. 1818. *Eutriana oligostachya* Kunth, Révision Gram. 1: 96. 1829. *Chondrosium oligostachyum* Torr. in Marcy, Explor. Red River Louisiana 300. 1852. *B. oligostachya* Torr. ex A. Gray, Manual Bot. ed. 2. 553. 1856. (*Nuttall,* "On the plains of the Missouri," in 1811.)

Blue grama.

Densely tufted perennials, the base often short rhizomatous and forming thick mats; culms 1.5–4 (5) dm tall, erect, ascending; leaves mostly basal and persistent, sometimes sparsely pilose around the collar, the hairs 1–2 mm long; sheaths glabrous to pilose; ligule a fringe of hairs, 0.2–0.7 mm long, sometimes membranous at the very base; blades flat or loosely involute, 1–2.2 mm broad and 5–15 cm long, often crisp or curly, scaberulous above, glabrous below; panicles bearing only (1) 2 (3) spikes, the spikes short-pedicellate, 1.7–4 (5) cm long, falcate-spreading at maturity, bearing 30–80 spikelets, the rachis terminating in a sterile spikelet; spikelets with 1 perfect floret and 1 or 2 sterile florets above; glumes 1-nerved, the first (1.5) 2–3.5 (4) mm long, linear to linear-lanceolate, sometimes aristate, hyaline, the second subequal to the lemma, (3.5) 4–6 (6.5) mm long, narrow-lanceolate, sometimes pustulose-pilose on the keel, glabrous to scaberulous on the sides, purplish; fertile lemma 4.2–5.7 (6.5) mm long, narrow-lanceolate, slightly keeled, 3-nerved, the terminal awn 0.5–1.5 mm long, arising between 2 narrow teeth, the lateral awns 1–2 mm long, ascending-hirsute on the nerves, the callus densely hairy; palea of the fertile floret slightly longer than the body of the lemma, narrow, the 2 nerves sometimes ending as minute bristles; sterile lemma with a body 1–2 mm long, buried in its callus hairs that are 2–3 mm long, its 3 awns 2.5–5 mm long, sometimes with an awnless, obovate second sterile lemma 1.5–1.7 mm long; anthers 2.5–3.5 (5) mm long; $2n$ = 20, 21, 28, 35, 40, 42, 60–62, 77, 84.

Grass or sagebrush plains and foothills, sometimes in pinyon-juniper and ponderosa pine woodlands at middle elevs. in the mts., in sandy loam or gravelly soils where often associated with *Hilaria jamesii;* Alta. and Man., s. to N.M., Texas and Mo., extending w. across s. halves of Utah and Nev. to s. Calif. and in scattered sites in the e., also s. into Mex. and introduced in S. Amer. July–Sept.

Blue grama, an important range grass throughout its distribution, is a major component of the short-grass prairie. In our area it is important in the Uinta Basin, the Bonneville Basin and Canyonlands. Where it forms a dense cover, it is one of our most important soil-building grasses. Blue grama is very resistant to drought and often remains dormant during long summer periods in what would normally be the growing season. Locally abundant in areas of the San Rafael Swell where the unknowing traveller would think of the grassy slopes as being beyond redemption during a dry season, blue grama quickly greens up as soon as moisture becomes available. The occurrence of tubercles or pustules on the second glume has been suggested as an indication of introgression with *B. hirsuta* Lag., but it occurs quite regularly on specimens outside the range of that species.

6. **Bouteloua hirsuta** Lag.

Bouteloua hirsuta Lag. Varied. Ci. 4: 141. 1805. *B. hirta* Lag., ibid., as a synonym. *Actinochloa hirsuta* Roemer & Schultes, Syst. Veg. 2: 419. 1817. *Eutriana hirta* Trin. Gram. Unifloris & Sesquifloris 240. 1824. *Chondrosium hirsutum* Sweet, Sweet's Hortus Brit. 1: 455. 1826, nomen. (Mex., probably collected by *Nees.*)

Hairy grama.

Densely tufted perennials; culms numerous, 2–4 (7.5) dm tall, erect, often slightly geniculate at the lower nodes; leaves mostly basal and persistent; sheaths glabrous to pilose; ligule a fringe of hairs mostly less than 0.3 mm long, sometimes membranous at the very base; blades flat, sometimes more or less involute towards the tips, 1–2.5 mm broad and 5–10 cm long with scattered, pilose hairs that become somewhat more dense on the lower margins, the hairs usually arising from tubercles; panicles 3–7 (15) cm long, of 1–4 spikes, the spikes 1–3.5 (5) cm long, ascending or spreading, straight or occasionally arcuate, bearing 18–50 spikelets, the rachis prolonged beyond the terminal floret as a long needle 5–20 mm long; spikelets with 1 perfect floret and a sterile floret above; glumes 1-nerved, the first 1.5–2.8 mm long, very narrow, sometimes awned up to 1 mm of its length, hyaline, the second 3.2–4 (5) mm long, broadly lanceolate, abruptly narrowed into an awn 0.5–1.2 mm of its length, purplish, tuberculate-hirsute with spreading hairs, the tubercles becoming black and visible to the naked eye; fertile lemma 3.2–4.5 mm long, deeply 3-cleft, 3-nerved, the nerves extending into short awns 0.2–2.5 mm long, the callus short-bearded; sterile lemma 0.5–1.2 mm long, its 3 awns 2–3 (4) mm long, the callus naked; second sterile lemma sometimes present, obdeltoid, awnless; $2n$ = 12, 20–22, 24, 28, 36, 37, 40, 42, 46, 50.

Dry sandy or sandy-loam soils, mostly at low elevs.; Great Plains from S.D. to Wisc., s. to N.M., Texas and La. and w. to s. Utah, s. Nev. and s. Calif., and on s. through Mex.; also in Fla. July–Aug.

7. **Bouteloua barbata** Lag.

Bouteloua barbata Lag. Varied. Ci. 4: 141. 1805. *Actinochloa barbata* Roemer & Schultes, Syst. Veg. 2: 420. 1817. *Eutriana barbata* Kunth, Révision Gram. 1: 96. 1829. (Mex., probably collected by *Nees.*)

Chondrosium polystachyum Benth. Bot. Voyage Sulphur 56. 1844. *B. polystachya* Torr. in Williamson, Explor. & Surv. Railroad Route from Mississippi River to Pacific Ocean 5(2): 366, pl. 10. 1857. (*Barclay,* Magdalena Bay, Baja Calif., in 1839.)

B. pumila Buckley, Proc. Acad. Nat. Sci. Philadelphia 1862: 93. 1863. (*Wright 754,* nw. Texas, in 1849.) A form with short awns.

Chondrosium microstachyum Fourn. Mex. Pl. 2: 138. 1886. *B. microstachya* L. H. Dewey in J. M. Coulter, Contr. U. S. Natl. Herb. 2: 531. 1894. (*Bourgeau 667,* Pâturages près Guadalupe, Mex., in 1866.)

B. arenosa Vasey in S. Wats. Proc. Amer. Acad. Arts 24: 81. 1889, nomen nudum; Vasey in S. Wats. U.S.D.A. Div. Bot. Bull. 12(1): pl. 34. 1890. (*Palmer 189,* "Plains and ridges about Guaymas," Mex., in 1887.) A form with long awns.

B. rothrockii of some authors; not Vasey, 1893.

Six-weeks grama.

Low, tufted annuals; culms numerous, (0.5) 1–2.5 (4) dm long, often somewhat flattened, geniculate, branching, sprawling or less often erect; leaves sometimes pilose around the margins of the throat; sheaths glabrous, the margins often scarious or hyaline; ligule a dense fringe of hairs, 0.5–1.2 mm long; blades mostly flat with a loosely involute tip, 1–2 (3) mm broad, scaberulous above, often with narrow whitish margins; panicles (3) 5–8 (10) cm long, of 3–5 (6) spikes, the spikes 1–2.5 cm long, often arcuate; spikelets closely crowded, each with 1 fertile floret and 1 or 2 sterile florets above; glumes small, glabrous to scaberulous, the first glume 1–2 mm long, hyaline, the second subequal to the lemma, 1.5–2.5 mm long, purplish with scarious to hyaline margins above, awned, the awn 0.2–0.7 mm long; fertile lemma small, 1.5–2.5 mm long, appressed-

BOUTELOUA

Bouteloua hirsuta

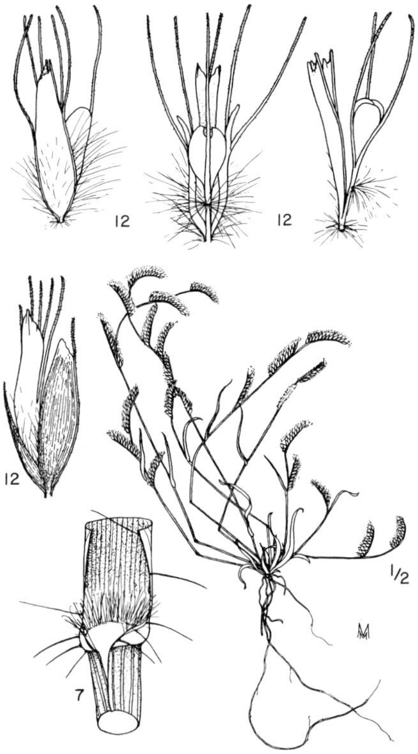

Bouteloua barbata

Bouteloua simplex

pilose, sometimes more so on the sides than along the back, the lower margins pilose, 3-nerved, the terminal awn 0.2–1.5 mm long, from a deeply bifid apex, the lobes erose-lacerate, the lateral awns 0.8–2.5 mm long, the callus densely hairy; palea of the fertile floret subequal to the lemma, erose-lacerate, the 2 nerves excurrent as short awns; sterile lemma 0.5–1 mm long, globose, empty, its 3 awns 1.5–3 mm long; anthers about 0.5 mm long; $2n$ = 20, 40.

Dry, open slopes of washes in deserts and bajadas, often in disturbed intermittent watercourses at low elevs.; s. parts of Calif., Nev., Utah and Colo., s. to Ariz., N.M., w. Texas and Mex. Aug.–Sept.

Robust, erect plants of *B. barbata* are sometimes misidentified as *B. rothrockii* Vasey, a related, perennial species of southern California, Arizona and northern Mexico. These species are very similar in appearance and difficult to distinguish on characters other than duration. Furthermore, Rothrock grama is often mistaken for an annual grass, because it is relatively short-lived and poorly rooted. The awns of *B. rothrockii* are longer than those of *B. barbata,* the stems are usually less branched and the glumes more noticeably scaberulous rather than glabrous and shiny as in *B. barbata.* We have seen no specimens from our area that we can call *B. rothrockii.*

8. Bouteloua simplex Lag.

Bouteloua simplex Lag. Varied. Ci. 4: 141. 1805. *Actinochloa simplex* Roemer & Schultes, Syst. Veg. 2: 418. 1817. *Chondrosium simplex* Kunth, Révision Gram. 1: 94. 1829. (Peru.)
Chloris procumbens Durande, Chloridis Sp. 16. 1808. *Chondrosium procumbens* Desvaux ex Beauv. Essai Nouv. Agrostogr. 41, 158. 1812. *Atheropogon procumbens* J. F. Jacq. Eclog. Pl. Rariorum 2: 16, pl. 12. 1813. *Actinochloa procumbens* Roemer & Schultes, Syst. Veg. 2: 417. 1817. *Cynodon procumbens* Rasp. Ann. Sci. Nat. (Paris) 5: 303. 1825. *B. procumbens* Griffiths, Contr. U. S. Natl. Herb. 14: 364. 1912. (Grown at Madrid from seed said to come from the Philippine Islands [collected by *Nees*] where the species is not known to occur, probably from S. Amer. or Mex., which regions Nees visited.)

Mat grama.

Low, densely tufted annuals; culms numerous, 0.5–2.5 (3) dm long, filiform, geniculate and somewhat branched at the decumbent base; sheaths sometimes pilose around the margins of the throat, otherwise glabrous; ligule a low fringe, 0.1–0.2 mm long; blades flat, becoming involute towards the tip, narrow, 0.5–1.5 mm broad and 2–4 (7) cm long; inflorescence a solitary spike, 1–2.5 cm long and arcuate, bearing (10) 20–40 spikelets, the rachis terminating in a sterile spikelet; spikelets with 1 fertile floret and 1 or 2 sterile florets above; glumes 1-nerved, lanceolate, glabrous, scaberulous on the nerves, the first glume small, 2–3 mm long, the second 4–5.5 mm long, usually purplish; fertile lemma 3–3.5 mm long, 3-nerved, appressed-pilose on both sides of the midnerve, at least basally, the hairs sometimes pustulose at the base, the terminal and lateral awns 1.5–2.5 mm long; palea of the fertile floret subequal to the lemma; sterile lemma 1–1.3 mm long, its 3 awns 3–3.7 (5) mm long, with a hairy callus, often an awnless, obovate second sterile lemma present; anthers 0.4–0.6 mm long; $2n$ = 20, 40.

Dry grass and shrub prairies and hills at low elevs.; s. Utah., s. Colo., s. to Ariz., N.M., Trans-Pecos Texas and c. Mex., also into Andean S. Amer. Aug.–Sept.

76. SPARTINA Schreber Cordgrass

Strongly rhizomatous perennials; culms hollow, slender or coarse, moderately tall to tall; leaves tough and firm; sheaths open, firm, smooth to somewhat striate; ligule a ring of fine, long or short hairs; blades long, flat or involute, non-auriculate; inflorescence a panicle with 2 to many spicate or subspicate branches racemosely arranged along a central axis, the short, congested branches erect-appressed to somewhat spreading, bearing closely placed, sessile spikelets in 2 rows on 2 sides of the 3-angled rachis, the rachis tip naked, often hidden by the terminal spikelet; spikelets 1-flowered, without any sterile florets, strongly compressed, disarticulating below the glumes; glumes usually 1-nerved or the second with 3 (5) closely placed nerves, sometimes the lateral nerves very obscure, strongly keeled, acute or the second glume short-awned, unequal; lemma firm, keeled, strongly 1- or 3-nerved, the lateral nerves obscure, often with additional indistinct lateral nerves, tapering to a narrow but usually rounded awnless tip; palea as long as or longer than the lemma, with broad, membranous margins on either side of the 2 closely placed nerves, delicate; stamens 3; x = 7, 10.

A genus of about 16 species, of the Americas, Europe and Africa, 9 of which occur in the U.S. with all but 2 distributed along the coasts. (Name from the Greek *spartine,* a cord, as was made from the bark of *Spartium junceum* L., a legume, and probably applied to *Spartina* because of the tough leaves.)

Reference: Mobberly, D. G. Taxonomy and distribution of the genus *Spartina.* Iowa State Coll. J. Sci. 30: 471–574. 1956.

1 Second glume 10–20 mm long, including the awn, greatly exceeding the lemma; spikes (4) 5–15 cm long; plants robust; (7) 10–30 dm tall; blades mostly more than 5 mm broad .... 1. *S. pectinata*
1 Second glume 7–8.5 (10) mm long, awnless or merely mucronate and only barely exceeding the lemma; spikes 2–4.5 (7) cm long; plants 3–7.5 (10) dm tall; blades rarely more than 5 mm broad .... 2. *S. gracilis*

1. Spartina pectinata Link

Spartina pectinata Link, Jahrb. Gewächsk. 1(3): 92. 1820. (*Bosc,* N. Amer., possibly at Wilmington, N.C.)
Trachynotia cynosuroides Michx. Fl. Boreali-Amer. 1: 64. 1803, based on the name *Dactylis cynosuroides* L., but misapplied. *S. michauxiana* A. S. Hitchc. Contr. U. S. Natl. Herb. 12: 153. 1908. *S. cynosuroides* var. *michauxiana* St.-Yves, Candollea 5: 58, pl. 3, fig. a-7. 1932. (*Michaux,* "in rupibus juxta sinum Hudsonis," probably in the vicinity of Lake Mistassini, s. of James Bay, Que., in 1792.)

Prairie cordgrass.

Coarse rhizomatous perennials, the rhizomes 4–8 mm thick with closely imbricate scales; culms usually

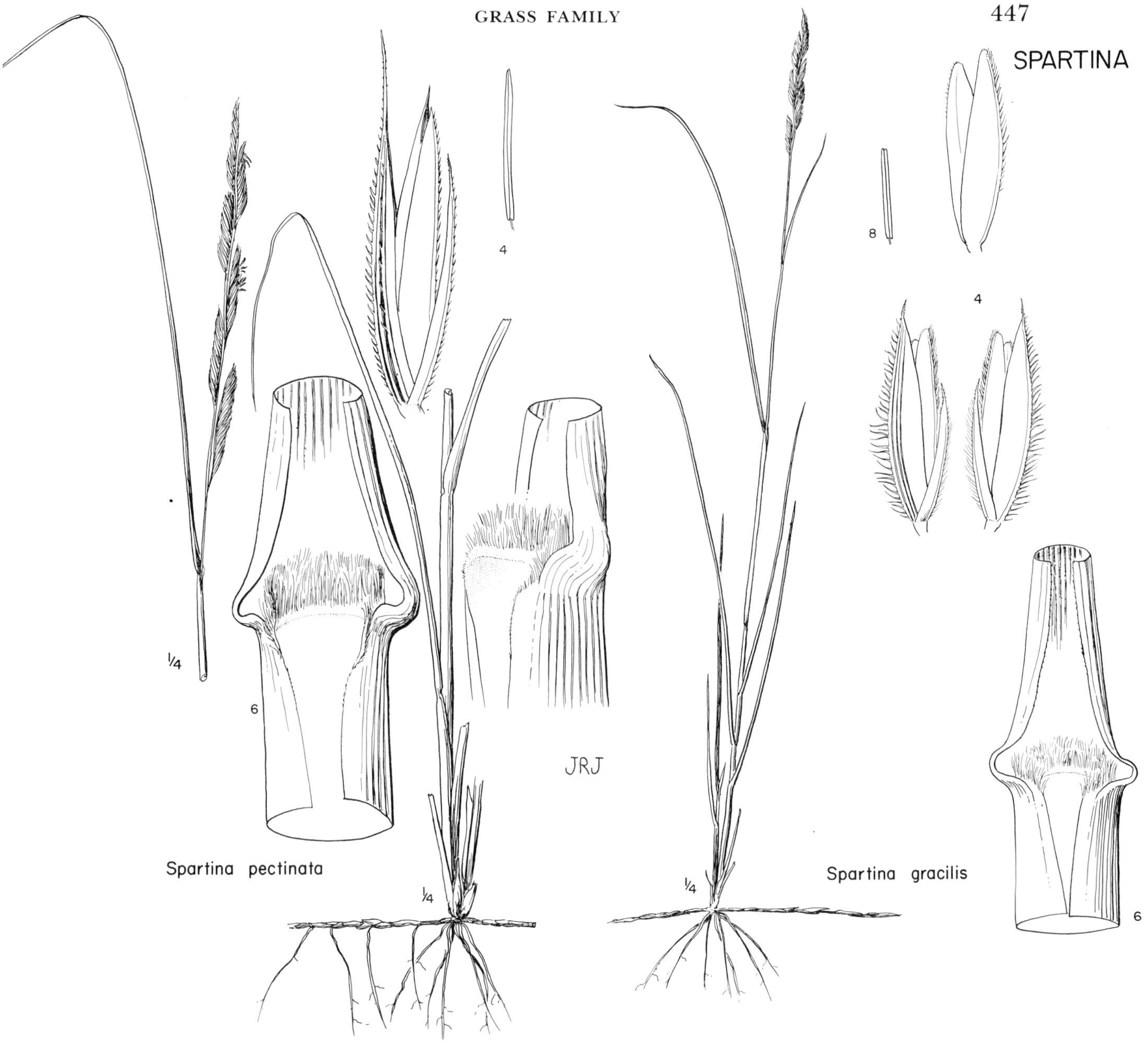

solitary, tall, (7) 10–20 (30) dm, erect, glabrous; sheaths smooth to slightly striate, often keeled above, glabrous, sometimes pilose at the upper margins; ligule a relatively long fringe of hairs, (1.5) 2–3 mm long; blades flat, becoming involute in drying, 6–11 mm broad and 20–60 cm long, glabrous on both sides, the margins scabrous; panicles (10) 20–32 (50) cm long, of 5–15 (50) appressed to somewhat spreading spikes, the spikes (4) 5–9 (15) cm long; spikelets densely crowded, 40–80 per spike, each 1-flowered; glumes large, glabrous to sparsely hispidulous, the keel pectinate, the first glume (5) 6–9.5 mm long, narrow, sometimes nearly subulate, 1-nerved, aristate, the second glume greatly exceeding the lemma, 10–20 mm long, including the awn, narrow-lanceolate, 3-nerved, the lateral nerves close to the midnerve, the awn 3–5 (8) mm long; lemma long, (6.5) 7.8–9.5 mm, lanceolate, 1-nerved, the nerve glabrous below and pectinate above, the margins hispidulous; palea thin, papery, exceeding the lemma by about 1 mm, the 2 nerves close together; anthers 4–6 mm long; $2n$ = 28, 40, 42, 70, 84.

Damp to wet meadows, marshes, streambanks, lake and pond shores, floodplains, in both salt and fresh water; B.C. to Newfl., s. in the w. states to Oregon, Idaho, n. Utah and n. N.M., in the c. states to Texas and Ark. and in the e. states to Tenn. and N.C. June–Sept.

Prairie cordgrass was not collected in Cache Valley until the early 1940s and since that time it has been increasing as a weed along roadsides and a worthless grass in pastures. The very scabrous leaf margins are capable of inflicting deep cuts on the unwary who may tug at a leaf blade in passing. The comb-like spikes get their characteristic look from the conspicuous awns of the second glumes.

2. Spartina gracilis Trin.

Spartina gracilis Trin. Mém. Acad. Imp. Sci. Saint-Pétersbourg, Sér. 6, Sci. Math., Seconde Pt. Sci. Nat. 4(1): 110. 1840. (*Hooker,* N. Amer., is cited, but probably collected by *Douglas.*)

Alkali cordgrass.

Strongly rhizomatous perennials, the rhizomes 3–5 mm thick with overlapping scales; culms solitary, 3–7.5 (10) dm tall, erect, glabrous; sheaths smooth to striate, glabrous; ligule a fringe of hairs, 0.5–1.5 mm long; blades flat, becoming involute in drying, 2.5–5 (8) mm broad at the base and 15–20 cm long, scabrous on the upper surface and margins, glabrous beneath; panicles (6) 8–16 (22) cm long of 2–6 (10) racemosely arranged, appressed spikes, the spikes 2–4.5 (7) cm long; spikelets closely crowded, 18–28 per spike, each 1-flowered, appressed; glumes glabrous except for the ciliate keel, the first 3.5–5 (7) mm long, linear, mucronate, the margins hispidulous, the second 7–8.5 (10) mm long, slightly exceeding the lemma, narrow-lanceolate, 5-nerved, the lateral nerves near the midnerve, the lateral nerves and or just the keel nerve scabrous to pectinate, awnless or mucronate; lemma long, 6.2–7.5 (8.5) mm, lanceolate, usually blunt tipped, 1-nerved, the keel ciliate, at least towards the apex, otherwise glabrous; palea large, subequal to the lemma, lanceolate, thin and papery, glabrous, the 2 nerves close together; anthers 2.5–5 mm long; $2n = 40, 42$.

Shores of lakes, streambanks, wet meadows and moist seeps, often alkaline situations; s. B.C. and Sask., s. through e. Wash., Idaho, Mont. and N.D. to e. Calif., Nev., c. Ariz., n. N.M. and Nebr. June–Sept.

77. HILARIA H.B.K. Galleta

Wiry, rhizomatous or stoloniferous perennials; culms stiff, solid; sheaths open; ligule a ciliate membrane or fringe of hairs; blades narrow, flat or involute, non-auriculate; inflorescence a slender, dense spike with clusters of 3 spikelets at each node of a zig-zag rachis; spikelet cluster with 2 lateral, 2- (3)-flowered, staminate spikelets and a central, 1-flowered, perfect spikelet, the combination of 6 glumes forming a false involucre, those of the central spikelet against the axis, disarticulating as a whole cluster from the persistent rachis; glumes firm, usually asymmetrical, some of them bearing 1 or more awns, each of the first glumes of the lateral spikelets with a prominent awn at one side from about the middle; lemmas thin and hyaline, 3-nerved, awned or awnless; palea subequal to the lemma and similar in texture; $x = 9$.

A genus of 7 species, 5 in arid and semiarid regions of the sw. U.S. and n. Mex., 1 species extending s. to Venezuela. (Named for Auguste St. Hilaire, a French naturalist.)

Hilaria was placed in the tribe Zoysieae by A. S. Hitchcock (Manual Grasses U.S., 1951). Species of *Hilaria* are important as range grasses in the Southwest due mostly to their abundance in dry habitats and perseverance under heavy grazing pressure rather than to palatability.

1 Culms and often the leaves lanate pubescent, (3.5) 5–8.5 (10) dm tall .... 1. *H. rigida*
1 Culms and leaves glabrous to hirsutulous, 1.5–4 (6.5) dm tall .... 2. *H. jamesii*

1. Hilaria rigida (Thurber) Benth. ex Scribn.

Pleuraphis rigida Thurber in S. Wats. Bot. Calif. 2: 293. 1880. *Hilaria rigida* Benth. ex Scribn. Bull. Torrey Bot. Club 9: 33, 86. 1882. (*Cooper 2230,* "Fort Mojave and Providence Mountains," Calif., 24 Feb. 1860 or 1861.)

Big galleta.

Coarse perennials, forming large clumps with short, woody rhizomes; culms tall, (3.5) 5–8.5 (10) dm long, spreading, much-branched, leafy, lanate; leaves lanate to glabrate, often scaberulous beneath the longer hairs; ligules 1–2 mm long, densely ciliate, usually hidden in the lanate pubescence around the collar; blades more or less involute, especially towards the tip, 2–5 mm broad and 2–5 cm long; spikes 4–8 (11) cm long; spikelet clusters densely long-villous at the base; glumes of the lateral spikelets subequal, 5–7 mm long, 4- to 5-nerved, ciliate, the first glume asymmetrical, awned from a displaced midnerve at one side, the awn 3–7 mm long, curved or geniculate, this glume often bearing another, shorter, straight awn also, the second glume with a short, straight awn 0.7–2 mm long; glumes of the central spikelet subequal, (4) 5–6.5 mm long, ciliate, flabellate, 5- to 7-nerved, each nerve extended into an irregular awn 2.5–7 mm long; lemmas of the lateral spikelets 5.7–6.5 mm long, glabrous below and ciliate on the scarious, rounded apex, 3-nerved, short-awned, the awn 0.7–1.8 mm long, the lemma of the central spikelet 6–7.2 mm long, also scarious and ciliate apically and glabrous basally, its awn longer, 2–5.5 mm; palea almost as long as the lemma; anthers 2.4–3.2 mm long; $2n = 18$.

Dry deserts to the juniper foothills; se. Utah, w. Ariz., s. Nev., se. Calif., Sonora and Baja Calif. March–June.

2. Hilaria jamesii (Torr.) Benth.

Pleuraphis jamesii Torr. Ann. Lyceum Nat. Hist. New York 1: 148, pl. 10. 1824. *Hilaria jamesii* Benth. J. Linn. Soc., Bot. 19: 62. 1881. (*James,* "at the sources of the Canadian River," [Texas or N.M.], late July or early Aug. 1820.)

H. sericea Benth. J. Linn. Soc., Bot. 19: 62. 1881, nomen nudum. *Pleuraphis sericea* Nutt. ex Benth. ibid., in synonymy. (*Nuttall,* Harris Fork of the Colorado; probably Ham's Fork of the Green River, Wyo., June 1834.)

Galleta.

Strongly rhizomatous or stoloniferous perennials; culms 1.5–4 (6.5) dm tall, erect, sometimes decumbent at the much-branched base, the nodes often puberulent; leaves mostly basal; sheaths glabrous to scaberulous, with a few villous hairs at the throat; ligules 1–2.5 mm long, long-ciliate, lacerate, decurrent; blades flat to folded, 1.5–3 mm broad and 2–5 cm long, becoming involute towards the tip, rigid,

HILARIA

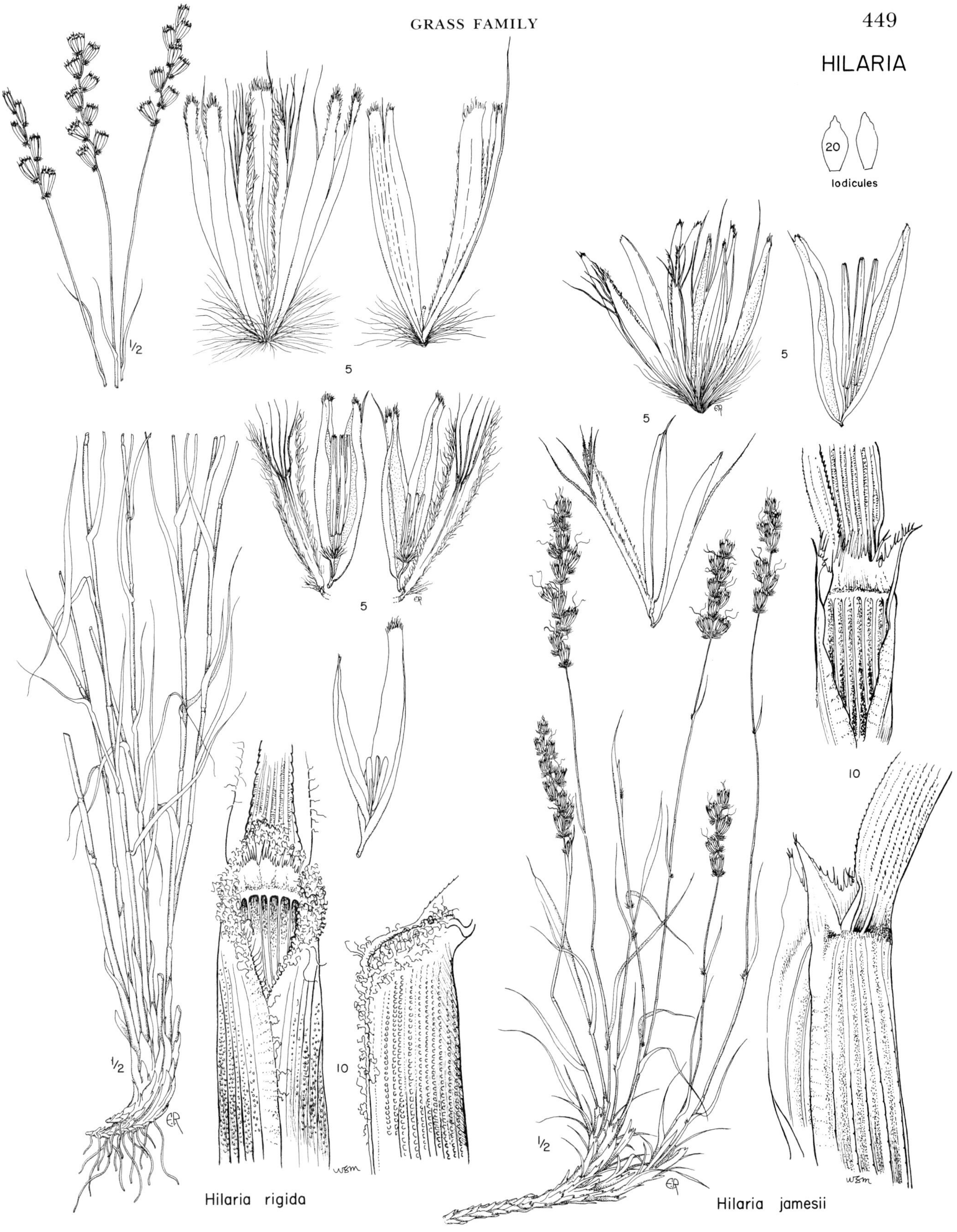

Hilaria rigida

Hilaria jamesii

often glaucous, glabrous to scaberulous, sometimes hispidulous above; spikes (2) 3–5 (7) cm long; spikelet clusters long-villous at the base; glumes of the lateral spikelets subequal, 5–7 (8.5) mm long, the apex erose, sometimes lacerate, scaberulous to hispidulous, the first glume asymmetrical, awned from a displaced midnerve at one side, the awn 3–5.5 mm long, the second glume awnless or very shortly awned; glumes of the central spikelet subequal, 4–5.5 (6.6) mm long, hispidulous, ciliate, flabellate, each of the 5 nerves extending into an irregular awn 1.5–5 (7) mm long; lemmas of the lateral and central spikelets (4.5) 6–7 (9) mm long, lanceolate, blunt-tipped, glabrous basally and scaberulous apically, 3-nerved, the single lemma of the central spikelet narrow-lanceolate, bearing a dorsal awn from below the bifid apex, the awn 1–2.5 mm long; palea shorter than the lemma, narrow; anthers 2.3–2.8 (3.5) mm long; stigmas long, reddish-brown; $2n = 36$.

From dry desert scrub of the lower valleys to the pinyon-juniper woodlands of the foothills; Utah (all but the Wasatch and Uinta divisions), s. Wyo. and w. Kansas, s. to se. Calif., s. half of Nev., Ariz., N.M. and possibly Trans-Pecos Texas. May–Aug.

Galleta often grows in association with blue grama in sagebrush areas where it may even become the dominant species over large areas. The production of an abundance of nutritious forage and ability to withstand heavy grazing and prolonged periods of drought make galleta an important range species.

Tribe 10. AELUROPODEAE

Stoloniferous or rhizomatous, dioecious (rarely monoecious) perennials of saline or alkaline places; culms solid; sheath margins free; ligule a short fringed membrane; blades usually short and stiff, non-auriculate; inflorescence a contracted panicle or sometimes a raceme; spikelets few- to many-flowered, mostly unisexual; disarticulation of the pistillate florets above the glumes and between the florets; lemmas 7- to several-nerved; lodicules 2, truncate, membranous; embryo eragrostoid; $x = 10$.

This small tribe of 4 mostly dioecious genera in the U.S. is represented only by *Distichlis* in our area. Superficially *Distichlis* resembles some members of the Poeae, and it was commonly referred to that tribe in older treatments. Members of the Aeluropodeae tribe usually inhabit alkaline or saline habitats.

78. DISTICHLIS Raf. Saltgrass

Low, dioecious (rarely monoecious), strongly rhizomatous or sometimes stoloniferous perennials; culms rigid, erect, solid, glabrous, usually completely clothed to the inflorescence by the distichous, closely overlapping leaves; leaves firm, glabrous; sheaths open; ligule a short membrane, more or less erose-ciliolate and sometimes tipped with long, stiff hairs; blades stiff and pungent, non-auriculate; inflorescence relatively few-flowered and of compact panicles or racemes; spikelets large, few- to many-flowered, laterally flattened, the pistillate spikelets disarticulating above the glumes and between the florets; glumes unequal, broad, acute, keeled, the first glume 3- to 5-nerved, the second 5- to 7- (9)-nerved, the lateral nerves sometimes indistinct; lemmas broad, faintly 7- to 13-nerved, awnless, laterally compressed, the lemma of the pistillate spikelets thicker than those of the staminate spikelets; palea more or less as long as the lemma, strongly 2-keeled, the keels more or less winged, the paleas of the pistillate florets enclosing the caryopsis; lodicules flabellate, membranous, erose-lobulate, 0.5–1 mm long; stamens 3; caryopsis brown, glabrous, the styles terminal; $x = 10$.

A genus of 4 species distributed in saline and alkaline marshes, moist flats and tidelands in N. and S. Amer. (From the Greek *disticous*, 2-ranked, in reference to the leaves.)

Desert saltgrass is not very palatable, but in some extensive alkaline flats it is the only forage available to livestock.

References:

Beetle, A. A. The North American variations of *Distichlis spicata*. Bull. Torrey Bot. Club 70: 638–650. 1943.
Reeder, J. R. State of *Distichlis dentata*. Bull. Torrey Bot. Club 70: 53–57. 1943.

1. Distichlis spicata (L.) Greene

Uniola spicata L. Sp. Pl. 71. 1753. *Briza spicata* Lam. Encycl. Méth. Bot. 1: 465. 1785; not Burm. f. 1768. *Distichlis maritima* Raf. J. Phys. Chim. Hist. Nat. Arts 89: 104. 1819. *Brizopyrum americanum* Link, Hort. Regius Bot. Berol. 1: 160. 1827. *Brizopyrum spicatum* Hook. & Arn. Bot. Beechey's Voyage 403. 1840. *D. spicata* Greene, Bull. Calif. Acad. Sci. 2: 415. 1887. ("Habitat in Americae borealis maritimis.")

Uniola stricta Torr. Ann. Lyceum Nat. Hist. New York 1: 155. 1824. *Brizopyrum spicatum* var. *strictum* A. Gray ex S. Wats. in King, Rep. Geol. Explor. 40th Parallel 5: 385. 1871. *D. maritima* var. *stricta* Thurber in S. Wats. Bot. Calif. 2: 306. 1880. *D. spicata* var. *stricta* Scribn. Mem. Torrey Bot. Club 5: 51. 1894; incorrectly attributed to A. A. Beetle, Bull. Torrey Bot. Club 70: 645. 1943. *D. stricta* Rydb. Bull. Torrey Bot. Club 32: 602. 1905. (*James*, "On the shores of the Canadian river," Okla., Aug. 1820.)

D. maritima var. *laxa* T. H. Holm, Bot. Gaz. 16: 277. 1891. *D. spicata* var. *laxa* Vasey ex Beal, Grasses N. Amer. 2: 519. 1896. *D. stricta* var. *laxa* Fassett & West in Munz, Manual S. Calif. Bot. 52, 597. 1935. (*Tracy*, Lake Park, Utah, in 1887.)

D. dentata Rydb. Bull. Torrey Bot. Club 36: 536. 1909. *D. stricta* var. *dentata* C. L. Hitchc. Univ. Wash. Publ. Biol. 17(1): 553. 1969. (*Sandberg & Leiberg 463*, Wash., 14 July 1893.) A pistillate plant.

Desert saltgrass, interior saltgrass.

Strongly rhizomatous, dioecious perennials; culms 1–4.5 (5) dm tall, usually decumbent at the base and becoming somewhat stolon-like; leaves sometimes pilose in the vicinity of the collar, especially at the upper margins of the sheath, the hairs 0.5–3 mm long; sheaths glabrous, the membranous margins sometimes extending apically as lateral auricles of the ligule; ligules 0.2–0.6 mm long, ciliate; blades

DISTICHLIS

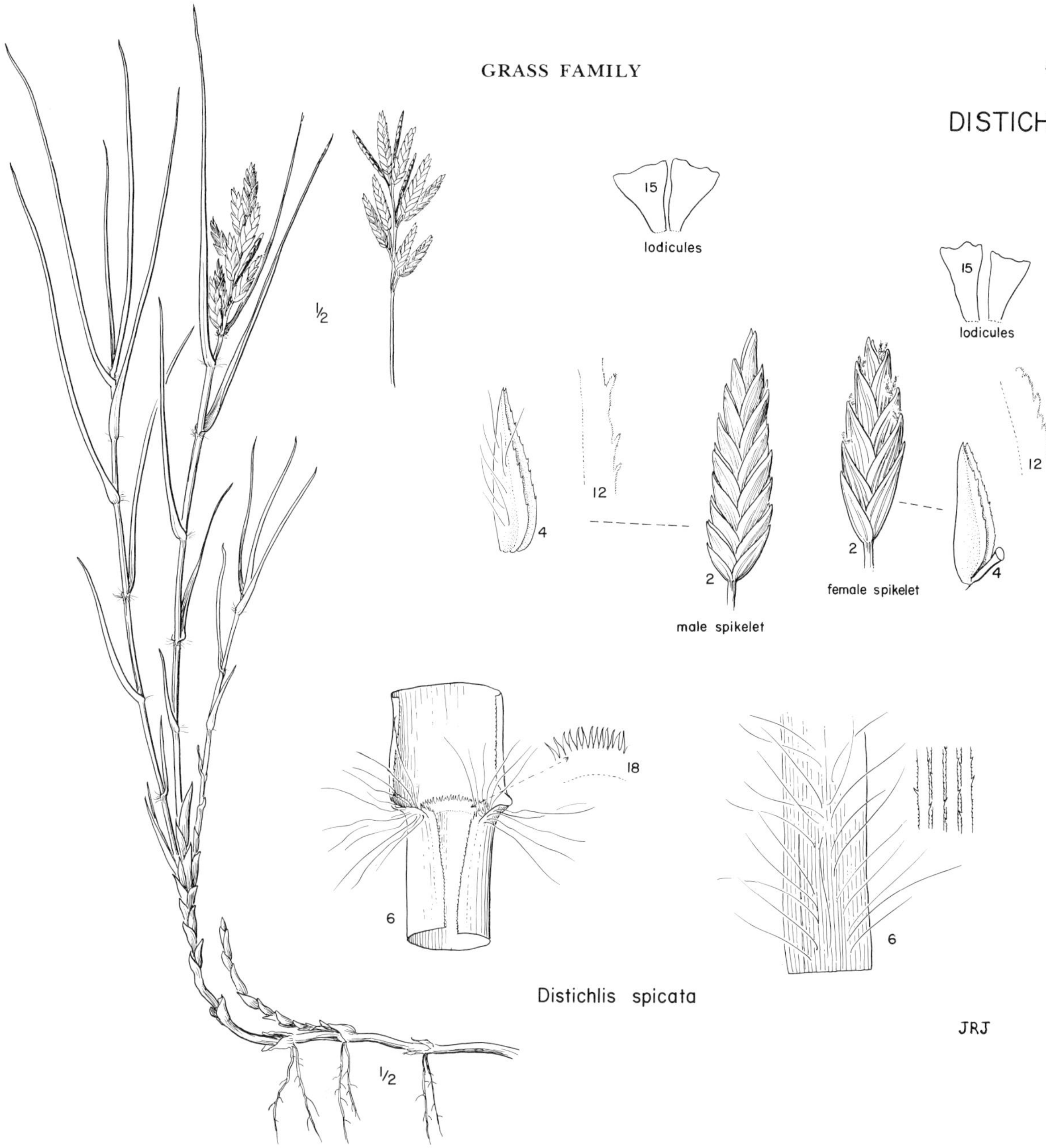

Distichlis spicata

loosely involute, 2–4 mm broad, the upper surface sometimes scabrous to pilose; panicles (3) 4–8 (10) cm long, with only a few, somewhat congested spikelets; spikelets of both sexes (7.5) 9–15 (20) mm long, strongly compressed, the pistillate spikelets 3- to 8-(12)-flowered, greenish, the staminate 6- to 11- (15)-flowered, yellowish; glumes lanceolate, glabrous, 5- to 7-nerved, the lateral nerves obscure, the pistillate glumes larger with the first glume 3–5 (6) mm long and the second (3.5) 4–6 (7) mm long, the staminate glumes with the first glume 2–4 mm long and the second 2.5–4.5 mm long; lemmas firmer than the glumes, 7- to 9-nerved, the lateral nerves obscure, usually glabrous, the pistillate lemmas 4.5–8 mm long and the staminate lemmas smaller, 3.5–5.5 mm long; paleas of the pistillate florets slightly shorter than the lemma, the winged keels with irregularly and shallowly serrate and minutely ciliolate margins, the palea of the staminate floret slightly longer than the lemma with ciliolate, less irregularly serrate, winged keels; lodicules about 1 mm long; anthers 2.3–3.5 mm long; $2n = 40$.

The interior var. grows in moderately moist to moist alkaline soils of flats and seepage areas, and along ditches; interior of s. N. Amer. from B.C. to Sask., s. through Wash., Idaho, Mont., N.D., and Minn. to Calif., Ariz., N.M., Texas and n. Mex. June–Sept.

The typical race, var. *spicata,* is called seashore saltgrass and is distributed along the Atlantic, Pacific and Gulf of Mexico coasts. Our material belongs to var. *stricta* (Torr.) Scribn., desert or interior saltgrass, which grows in alkaline flats. Desert saltgrass differs from the seashore variety in having more pilose pubescence in the vicinity of the leaf collars, looser panicles and broader keel wings on the pistillate paleas. C. L. Hitchcock revived the taxon *D. dentata* Rydb. as a variety of *D. stricta* (Torr.) Rydb. on the basis of less pubescent leaves and more dentate pistillate palea keels. Typical *D. stricta* as so defined occurs mostly in and east of the Rocky Mountains, whereas *D. dentata* occurs mainly west of the Rocky Mountains. We find the distinction to be so weak, however, that we prefer not to recognize *dentata* even at the varietal level.

Tribe 11. **PAPPOPHOREAE**

Cespitose perennials; sheath margins free; ligule a fringe of hairs; blades non-auriculate; inflorescence a panicle; spikelets 3- to several-flowered, disarticulating above the glumes, usually the florets falling together; lemmas 9- to several-nerved, the nerves extending into awns, the awns glabrous, scabrous, or plumose; embryo eragrostoid; $x = 10$.

Enneapogon and *Pappophorum* (not in our area) were included with the Festuceae (Poeae) by A. S. Hitchcock (Manual Grasses U.S., 1951) among such eragrostoid genera as *Blepharidachne.*

79. **ENNEAPOGON** Desvaux ex Beauv.

Low, cespitose perennials; culms numerous; sheaths open; ligule a fringe of hairs; blades flat to subinvolute, non-auriculate; inflorescence narrow, often spike-like panicles, grayish-green; spikelets 3- (ours) to several-flowered, the first floret fertile and the upper ones sterile and progressively smaller, disarticulating above the glumes and tardily between the florets; glumes subequal, lanceolate, 3- to numerous-nerved; lemmas much shorter than the glumes, firm, broad, truncate, rounded on the back, 9- or more-nerved, the nerves extending into 9, subequal, plumose awns; palea slightly longer than the body of the lemma, the 2 keels near the margins.

A genus of about 35 species, mostly of Australia, 10 species in Africa, 5 or 6 in Asia and only 1 in the Americas. (Name from the Greek *ennea,* nine, and *pogon,* beard, because of the 9 plumose awns of the lemma.)

Reference: Chase, A. *Enneapogon desvauxii* and *Pappophorum wrightii,* an agrostological detective story. Madroño 8: 187–189. 1946.

1. **Enneapogon desvauxii** Beauv.

Enneapogon desvauxii Beauv. Essai Nouv. Agrostogr. 82, 161, pl. 16, fig. 11. 1812. (*Nees,* locality given as "iles Manilles" but the genus is apparently not represented in the Philippines, probably from Argentina.)

Pappophorum wrightii S. Wats. Proc. Amer. Acad. Arts 18: 178. 1883. *E. wrightii* C. E. Hubbard, Hooker's Icon. Pl. 34: tab. 3337. 2. 1937. (*Wright 751,* Devils River, Texas; *2029,* N.M., in 1851–52.)

Spike pappusgrass.

Low, tufted perennials, sometimes flowering the first year; culms 1–2.5 (4.5) dm tall, geniculate below, ascending above, disarticulating at the lower nodes; culms and leaves finely pilose to villosulous, densely so at the nodes; sheaths with membranous margins that extend apically as lateral auricles of the ligule; ligules 0.7–1 mm long, consisting of a fringe of hairs; blades loosely involute, 1.5–2 mm broad and 3–12 cm long; panicles (2) 3–7 (9) cm long, spike-like, grayish; spikelets 3-flowered, only the first one fertile and the upper two sterile; cleistogamous spikelets sometimes produced within the lower sheaths; glumes strongly 3- to 7-nerved near the middle, the nerves sometimes unpaired and irregular in length and softly pilosulous, the first glume (2.5) 3.5–4.7 mm long, the second (3.5) 5–6 (7) mm long; lemma 1.5–2.5 mm long, strongly 9-nerved, pilose, the nerves extended into 9 subequal, plumose awns, the body ovate, the awns 2–4.5 mm long; palea about as long as the lemma, ciliate keeled; anthers 0.2–0.4 mm long; caryopsis ovoid, brownish; $2n = 20$.

Dry, rocky soils in desert valleys and slopes; s. Utah, se. Calif., Ariz., N.M. and w. Texas, s. to Oaxaca and in the s. Andes of S. Amer. Aug.–Sept.

Cleistogamous spikelets are produced in the lower sheaths. When mature they are dispersed with sections of the culms which disarticulate at the nodes.

Tribe 12. **ARISTIDEAE**

Strongly tufted annuals or perennials; culms hollow; sheath margins free; ligule a fringe of hairs or a short ciliate membrane; blades non-auriculate; inflorescence an open or contracted panicle; spikelets 1-flowered, disarticulating above the glumes; glumes large, thin; lemma 3-nerved, indurate, elongate, narrowed at the apex to an awn column bearing 3 awns, the lateral awns sometimes reduced or lacking; callus well developed and usually minutely bearded; lodicules 2, vasculated with membranous appendages similar to those of the Arundinoideae; caryopsis elongate; embryo eragrostoid, except for the absence of an epiblast; $x = 11$.

Aristida was included in the Agrostideae (Aveneae, in part) by A. S. Hitchcock (Manual Grasses U.S., 1951), next to *Stipa* and *Oryzopsis. Aristida* does bear a superficial resemblance to *Stipa* (now in the Stipeae), especially when the 2 lateral awns are suppressed. Based chiefly on characters of the embryo, Stebbins and Crampton (Recent Advances Bot. 1: 133–145. 1961) and Tateoka (J. Jap. Bot. 32: 275–287. 1957) placed *Aristida* in the Arundineae. This embryo type is unique in the Eragrostoideae.

80. **ARISTIDA** L. Three-awn

Strongly tufted annuals or perennials, without rhizomes or stolons; culms hollow; sheaths open; ligule a short fringe of hairs or a short, ciliate membrane; blades narrow, usually involute, non-auriculate; inflorescence an open or contracted panicle; spikelets large, 1-flowered, disarticulating obliquely above the glumes leaving a sharp callus; glumes equal or unequal, large, thin, lanceolate, acute to acuminate, sometimes awn-

ENNEAPOGON

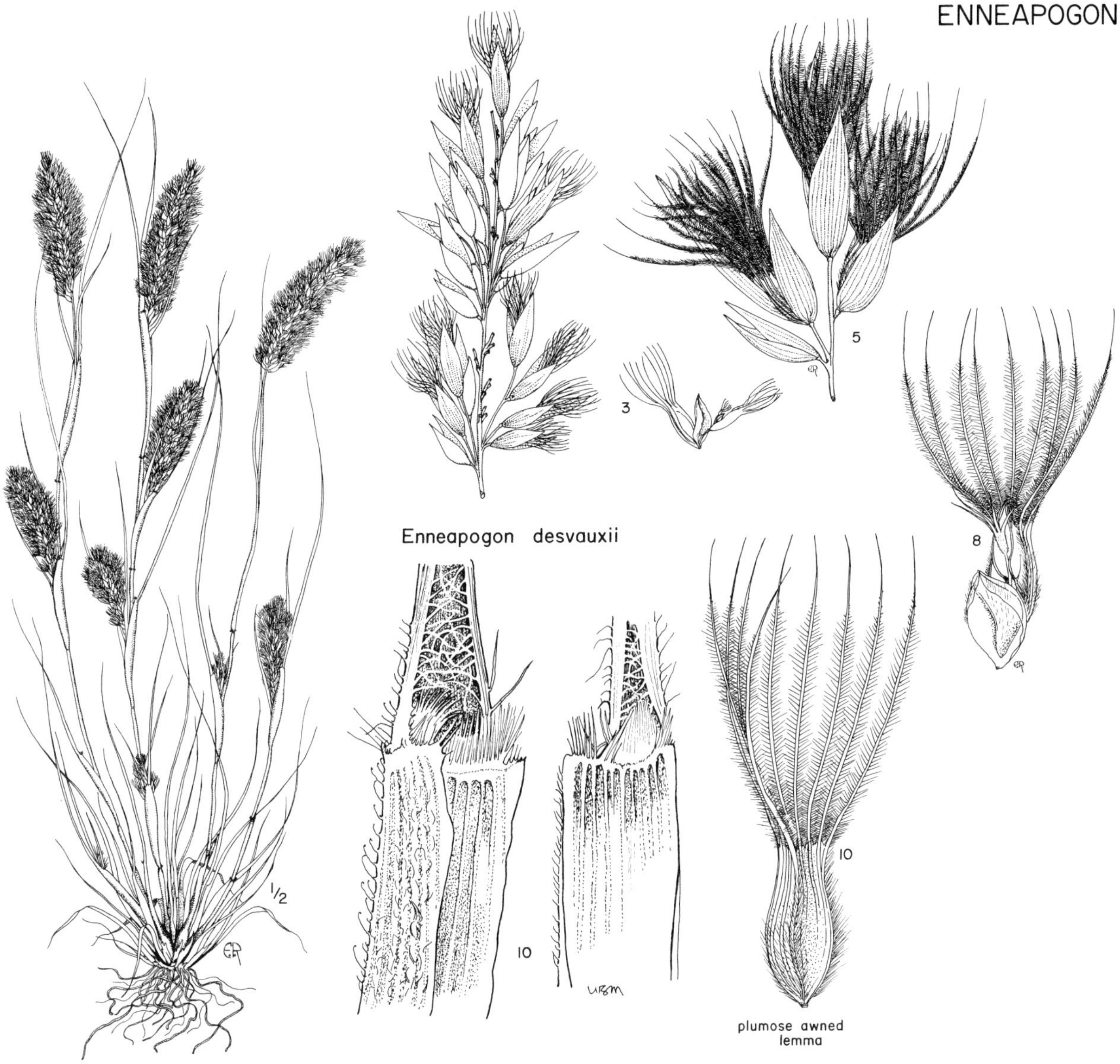

tipped, each with a strong central nerve and occasionally 2 lateral nerves; lemmas narrow, indurate, terete, 3-nerved, completely enclosing the palea, narrowing at the apex to an awn column (beak), bearing 3, usually stiff awns, the lateral awns sometimes short or rudimentary, the callus usually minutely bearded; palea small, faintly 2-nerved; lodicules 2, oblong, usually many-nerved, sometimes exceeding the palea; caryopsis elongate, tightly invested by the lemma and palea; $x = 11$.

A genus of about 200 species in the warmer parts of the world, especially important in Africa; according to A. S. Hitchcock (Manual Grasses U.S., 1951) there are 40 species in the U.S. in the semiarid Southwest and on sandy hills and pine barrens in the East and Southeast. (Name from the Latin *arista,* awn.)

The ripe awned fruits of *Aristida* are dispersed with ease in the strong winds characteristic of its desert environments. The sharp callus makes the fruits troublesome to livestock by becoming entangled in wool and causing injury to mouthparts.

Seven sections have been recognized in the genus, but all our members belong to sect. Chaetaria Trin. Taxonomically *Aristida* is a difficult genus with complexes of closely related, intergrading species. By A. S. Hitchcock's concept there are 5 perennial species entering the Intermountain Region. These are members of the "*purpurea* complex" which differ in characters that are either ambiguous or overlapping or are unreliable, being affected by differences in the environment. We have found it useful to combine them into one polytypic species, *A. purpurea* Nutt.

Aristida arizonica Vasey has been reported to occur in Utah and southern Nevada but we have not confirmed it. The subequal glumes would readily distinguish this species from *A. purpurea.*

1 Plants annual; second glume 6.2–10.5 mm long .... 1. *A. adscensionis*
1 Plants perennial; second glume (7) 10–25 mm long .... 2. *A. purpurea*

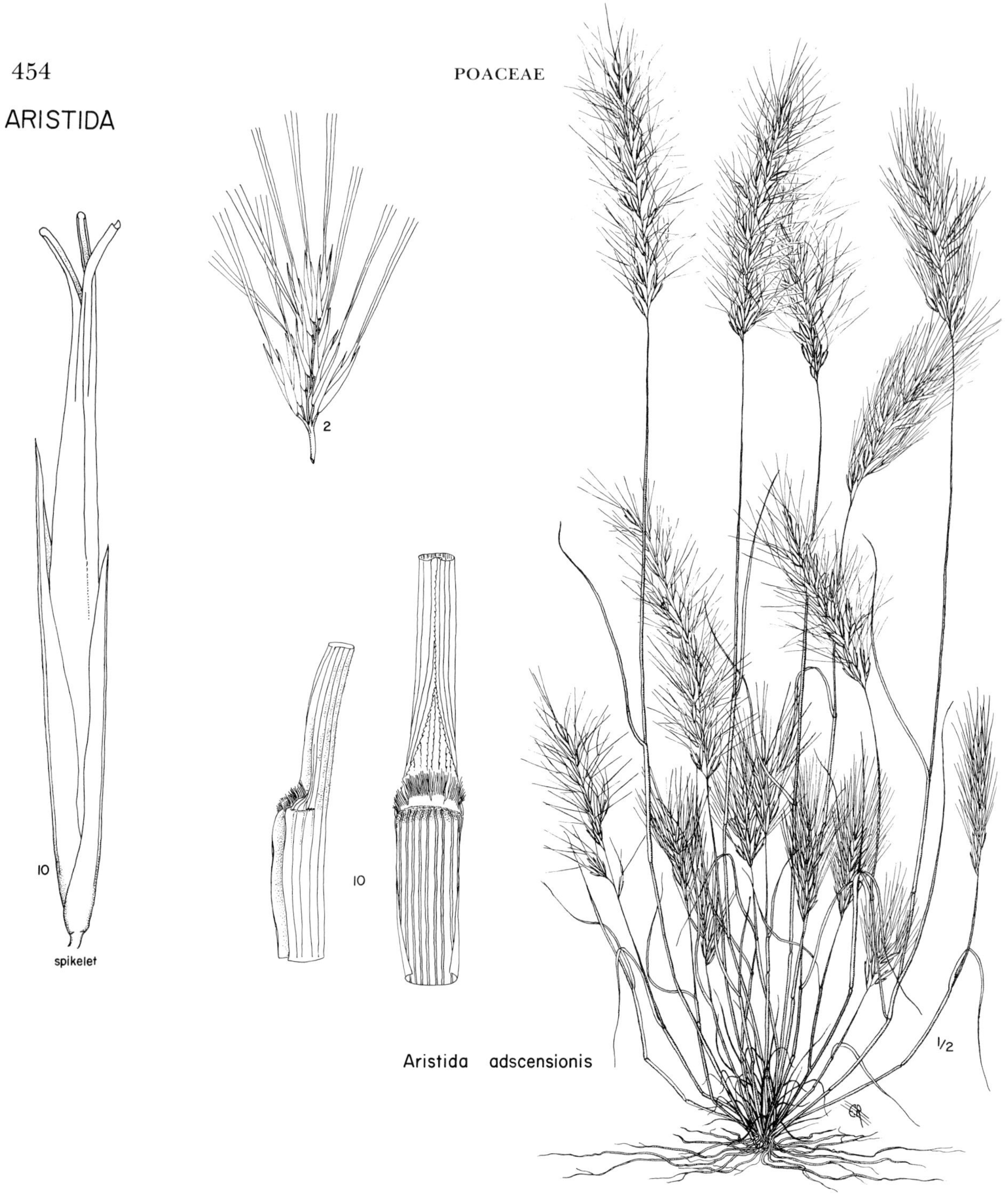

Aristida adscensionis

1. Aristida adscensionis L.

Aristida adscensionis L. Sp. Pl. 82. 1753. *Chaetaria ascensionis* [sic] Beauv. Essai Nouv. Agrostogr. 30, 151, 158. 1812. ("Habitat in Insula Adscensionis.")

A. bromoides H.B.K. Nova Gen. & Sp. Pl. 1: 122. 1815. *Chaetaria bromoides* Roemer & Schultes, Syst. Veg. 2: 396. 1817. *A. dispersa* var. *bromoides* Trin. & Rupr. Mém. Acad. Imp. Sci. Saint-Pétersbourg, Sér. 6, Sci. Math., Seconde Pt. Sci. Nat. 5(1): 130. 1842. *A. americana* var. *bromoides* Scribn. & Merr. U.S.D.A. Div. Agrostol. Circ. 32: 5. 1901. *A. adscensionis* var. *bromoides* Henrard, Meded. Rijks-Herb. 54: 62. 1926. (*Humboldt & Bonpland,* "in montanis regni Quitensis," Ecuador.)

A. fasciculata Torr. Ann. Lyceum Nat. Hist. New York 1: 154. 1824. *Chaetaria fasciculata* J. A. Schultes, Mantissa 3 (Addit. 1): 578. 1827. (*James,* "In forests of the Canadian river," Texas or Okla., Aug. 1820.)

A. oligantha of some authors; not Michx. 1803.

Six-weeks three-awn.

Tufted annuals; culms 2–5 (8) dm tall, often strongly geniculate and branched at the lower nodes and ascending above, glabrous; sheaths striate, the membranous margins sometimes extended apically as lateral auricles of the ligules; ligules about 0.5 mm long, consisting of a densely ciliate membrane; blades involute and short on small plants, more or less flat on larger plants, 2 mm broad and up to 10

ARISTIDA

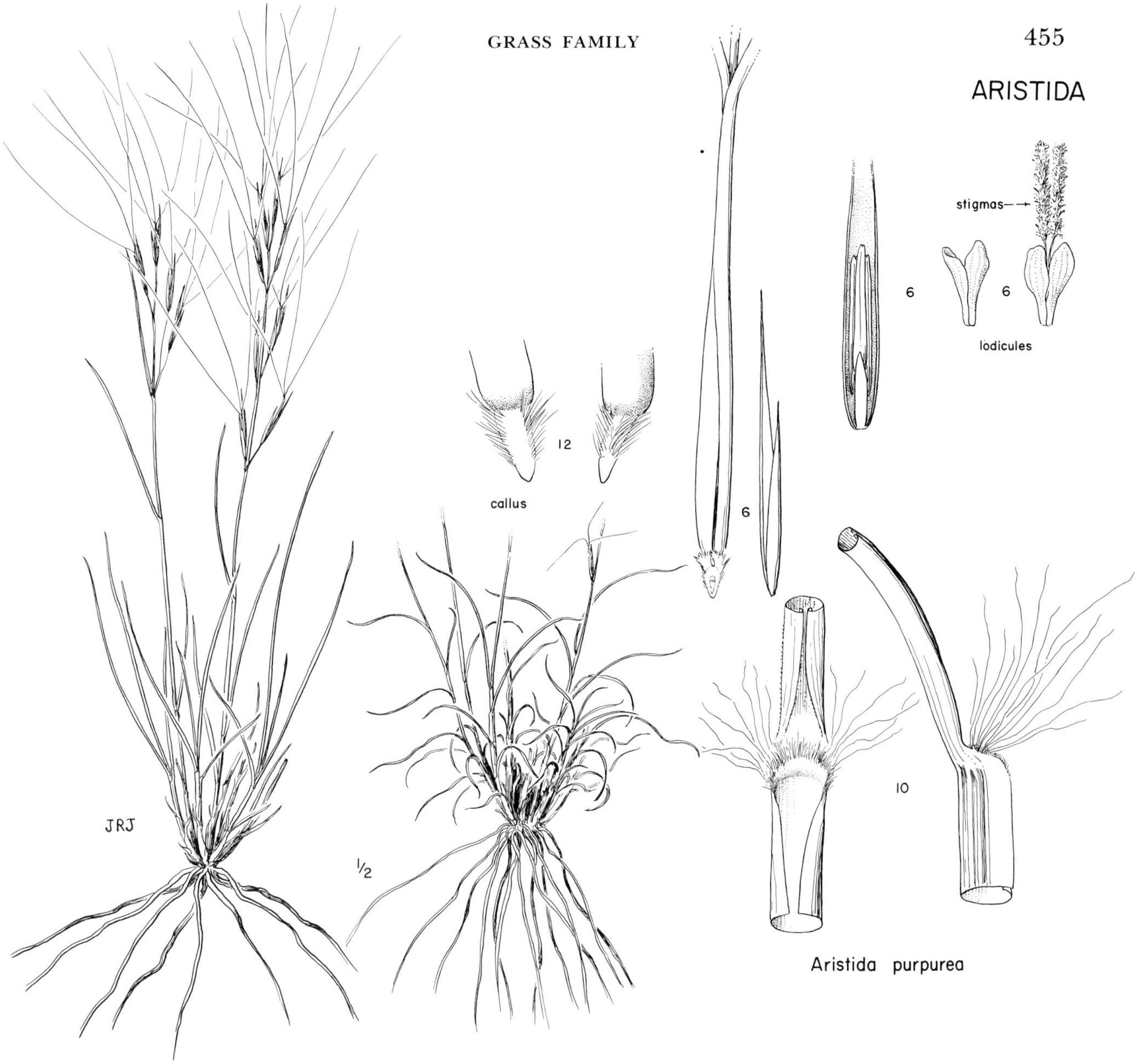

Aristida purpurea

cm long, scaberulous above, glabrous below; panicles 6–20 (30) cm long, narrow, congested to loose; glumes linear-lanceolate, glabrous, strongly 1-nerved, compressed-keeled, sometimes scabrous on the keel, the first glume (4) 4.8–7.7 mm long, acute, the second 6.2–10.5 mm long, less than 1.5 times as long as the first, acute, sometimes somewhat blunt with the nerve projected into a short mucro; lemma 5–8.5 (10.5) mm long (including the awn column), linear, 3-nerved, the lateral nerves obscure and submarginal, glabrous or often scaberulous above, the callus about 0.5 mm long, densely short-pilose, the 3 awns 9–20 (25) mm long, the lateral awns somewhat shorter to as long as the central awn, sometimes flattened below and terete above, strongly scabrous; palea much shorter than the lemma; anthers 0.7–1 mm long; $2n = 22$.

Dry creosote-bush desert valleys to the pinyon-juniper foothills and mesas; s. Calif., s. parts of Nev., Utah, Colo. to Mo. and s. to Ariz., N.M., Texas and on s. through trop. Amer. to Argentina and in warmer parts of the Old World. March–Sept.

The name *A. bromoides* H.B.K. has been used for plants only a few cm tall while *A. fasciculata* Torr. would refer to taller plants with loose, open panicles, but these are best interpreted as extremes in variation of one species. This annual weed responds to precipitation in such a way that accumulative rainfall can nearly be measured by the size of the plants. In years of drought the plants may be less than 10 cm high and a life cycle is completed in scarcely more than a month.

2. **Aristida purpurea** Nutt.

Aristida purpurea Nutt. Trans. Amer. Philos. Soc. II. 5: 145. 1837. (*Nuttall,* "On the grassy plains of Red river," Okla., late May or early June, 1819.)

Aristida purpurea var. **glauca** (Nees) A. Holmgren & N. Holmgren, hoc loc. *Chaetaria glauca* Nees in Nees & Schauer, Linnaea 19: 688. 1847. *A. glauca* Walp. Ann. Bot. Syst. 1: 925. 1849. (*Aschenborn 251,* Mex.)

A. longiseta Steudel, Syn. Pl. Glum. 1: 420. 1855. *A. purpurea* var. *longiseta* Vasey in Wheeler, Rep. U. S. Geogr. Surv. W.

100th Meridian 6: 286. 1878. *A. fasciculata* var. *nuttallii* Thurber ex Beal, Grasses N. Amer. 2: 208. 1896. (*Fendler 978,* N.M., in 1847.)

A. fendleriana Steudel, Syn. Pl. Glum. 1: 420. 1855. *A. purpurea* var. *fendleriana* Vasey, Grasses U.S. 21. 1883; Vasey, Contr. U. S. Natl. Herb. 3: 46. 1892. *A. fasciculata* var. *fendleriana* Vasey ex L. H. Dewey in J. M. Coulter, Contr. U. S. Natl. Herb. 2: 515. 1894. *A. longiseta* var. *fendleriana* Merr. U.S.D.A. Div. Agrostol. Circ. 34: 5. 1901. (*Fendler 973,* N.M., in 1847.) = var. *longiseta.*

Aristida purpurea var. **robusta** (Merr.) A. Holmgren & N. Holmgren, hoc loc. *A. longiseta* var. *robusta* Merr. U.S.D.A. Div. Agrostol. Circ. 34: 5. 1901. *A. purpurea* subsp. *robusta* Piper, Contr. U. S. Natl. Herb. 11: 107. 1906. (*Scribner 336,* Indian Creek, Mont., 5 July 1883.)

A. wrightii Nash in Small, Fl. S. E. U.S. 116, 1327. 1903. (*Reverchon 1061,* Dallas, Texas, in 1881.) = var. *glauca.*

Three-awn.

Tufted perennials; culms 1.5–6.5 (10) dm tall, erect, sometimes branched at the lower nodes, glabrous; sheaths glabrous, except for pilose tufts of white hairs at the apical margins, the hairs 2–3 mm long, with shorter hairs sometimes extending in a fine line across the collar, often a few hairs extending down the upper margins forming a ciliate shoulder; ligules scarcely 0.5 mm long, consisting of a fringe of hairs, somewhat higher on the edges than in the center; blades involute or rarely flat, 1–2 mm broad and (4) 5–10 (20) cm long, often curved or flexuous on small plants and straight on large plants, glabrous or sometimes scaberulous to hirsute above; panicles (5) 6–25 (35) cm long, narrow to somewhat open, the branches usually appressed and closely clustered, sometimes looser and divergent by curvature; glumes linear-lanceolate, 1-nerved, acuminate, sometimes awn-tipped, the nerves often scabrous, the first glume 5–10 (14) mm long, the second (7) 10–18 (25) mm long, 1.5 to twice as long as the first; lemma (7) 9–15 (17) mm long (including the awn column), linear, scaberulous in lines above, faintly 3-nerved, often gradually narrowing into a thin, often twisted awn column (beak), the callus 0.5–1 mm long, densely short-pilose, sharp, the 3 awns subequal, (10) 12–50 (100) mm long, the lateral awns divergent, slightly flattened and scabrous below, becoming terete and glabrous above; palea much shorter than the lemma; lodicules large, 2–2.5 mm long; anthers 1–2.5 mm long; $2n = 22, 44, 66$.

Dry desert valleys, hills and mesas; B.C. and Alta., s. through e. Wash., N.D., Iowa and Mo. to s. Calif., Ariz., N.M., Texas and Mex. March–Sept.

This is a complex that has been treated as five discrete "species," *A. wrightii* Nash, *A. purpurea* Nutt., *A. glauca* (Nees) Walp., *A. longiseta* Steudel and *A. fendleriana* Steudel. The species have been separated on such characters as beak (awn column) length, beak twisting, awn length, panicle length, panicle branching, lemma scabrosity, and leaf position. These characters can vary considerably in any population or appear different on the same plant in different years. The beak or awn column provides a poor character since it is often difficult to tell where the lemma body ends and the beak begins. The type specimen of *A. wrightii,* which is supposed to be nearly beakless, is as beaked as some carefully identified material of *A. glauca,* and many specimens of this latter "species" show characters associated with *A. purpurea.* The type specimen of *A. longiseta* is identical with material found in *A. fendleriana* folders, as the leaves are predominantly basal. Plants of *A. longiseta* often have loose panicles with branches becoming capillary and flexuous as in *A. purpurea.* Under extreme arid conditions plants of *A. longiseta* not only become reduced in size, but there is also a tendency for the leaves to be mostly basal and become more strongly involute and more curved or flexuous as in *A. fendleriana.* Plants collected as fully matured specimens are more likely to have lemmas with a beak, and as this beak dries it tends to twist. Because of the extensive intergradation we are including all these taxa in one species, *A. purpurea,* with the following varieties.

1 Panicle with slender, often nodding, flexuous or sigmoid branches, the pedicels capillary, the panicles usually deep purple throughout; s. Calif. to Utah, Ariz., N.M., Kansas, Texas and Mex. .... var. *purpurea*

1 Panicle branches stiff, usually erect (rarely some with open or slightly nodding panicles in var. *longiseta*), the panicles usually more reddish.

  2 Awns only 10–30 mm long; panicles 10–25 cm long; lemma gradually tapering into a slender, often twisted beak 5–6 mm long; Calif., Nev., Utah and s. to Ariz., N.M., Texas and Mex. (*A. wrightii*) .... var. *glauca* (Nees) A. Holmgren & N. Holmgren

  2 Awns (20) 25–100 mm long; panicles 5–11 cm long; lemma generally with a shorter beak or nearly beakless.

    3 Culms mostly 2–3 dm tall; leaves usually in a compact basal tuft; range of the species, but more common southward (*A. fendleriana*) ... var. *longiseta* (Steudel) Vasey

    3 Culms mostly 3–5 dm tall; leaves more loosely distributed over the lower half of the culm; range of the species, but more common northward .... var. *robusta* (Merr.) A. Holmgren & N. Holmgren

Subfamily IV. ORYZOIDEAE

Annuals or perennials (ours), often with rhizomes, usually of wet to moist places, some monoecious; epidermal cells in the region of root hairs not studied; culms hollow; sheath margins free, sometimes produced at the top into sheath auricles; ligule membranous; blades non-auriculate; inflorescence an open panicle; spikelets usually laterally compressed, 1-flowered, this often subtended by 2 scales or bristles that may represent reduced florets rather than vestigial glumes, often unisexual, disarticulating at the base of the spikelet; glumes absent; lemma strongly laterally compressed (in ours), 5- to several-nerved; palea 2- or usually 3-nerved, sometimes more; lodicules 2, ovate, entire or bilobed; stamens 6 (unusual in Poaceae), 3 (ours) or 1; ovary with 1 (also unusual in Poaceae) or 2 styles; embryo relatively small in respect to the caryopsis, the vascular traces to the scutellum and coleoptile diverging at approximately the same point (pooid), the epiblast present (pooid), the lower part of the scutellum fused in some and separated by a cleft in others, and the embryonic leaf margins overlapping (panicoid); $x = 12$ or 15, the chromosomes small.

The subfamily Oryzoideae is made up of a few small genera adapted to moist or wet habitats. Most species are tropical or subtropical in distribution, but *Leersia oryzoides* and *Zizania aquatica* (the latter not in our flora) are found as far north as Canada. The retention of 2 whorls of stamens in the flowers of some members is a primitive character, but the reduction of spikelets to 1 flower and the suppression of the glumes are advanced characters. Accompanying this unusual combination of characters in the inflorescence is a strange assortment of other characteristics such as pooid leaf anatomy, panicoid microhairs and an embryo that combines panicoid and pooid features.

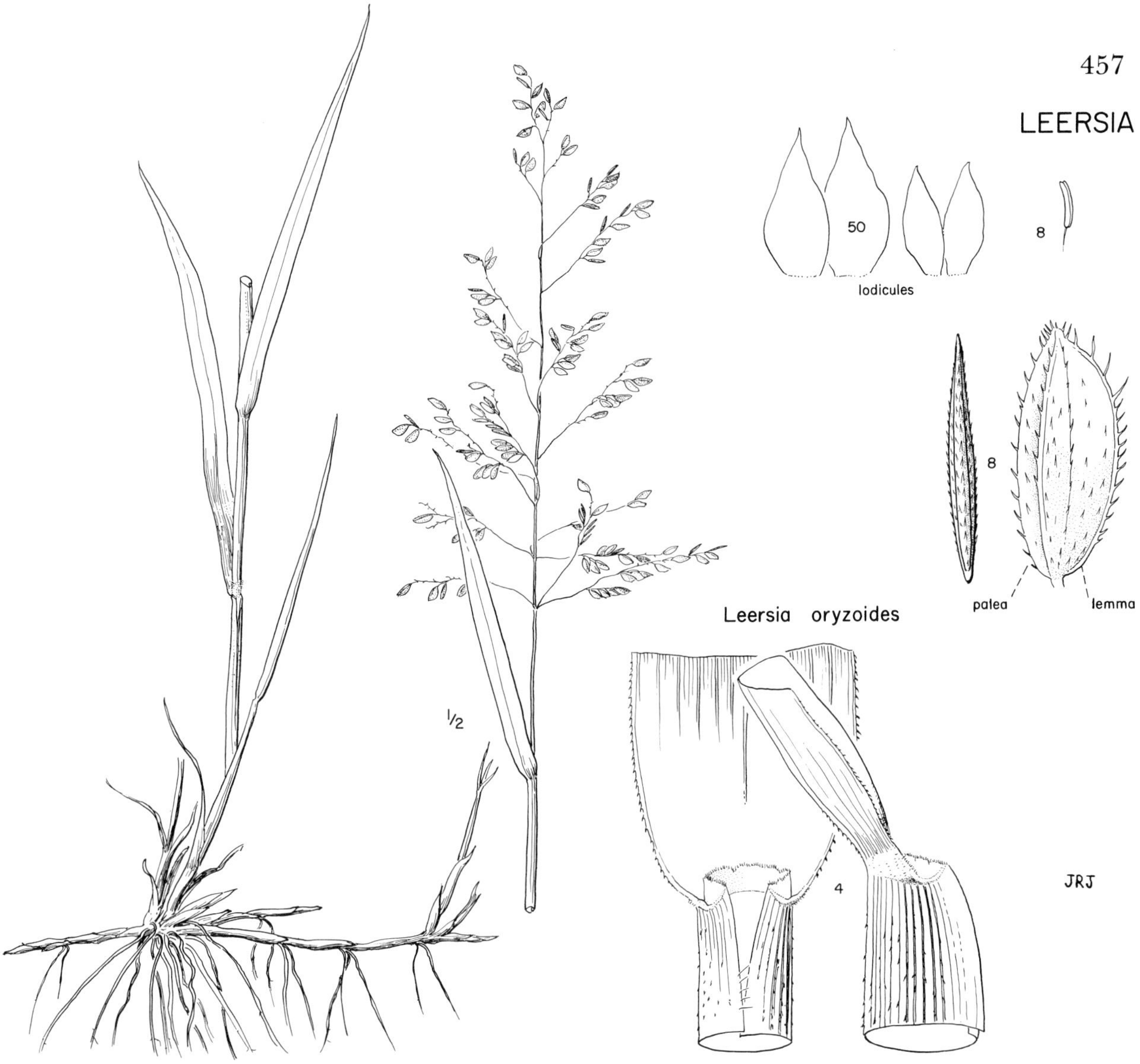

Tribe 13. ORYZEAE

Characters same as those of the subfamily.

81. LEERSIA Swartz, nom. conserv.

Rhizomatous perennials; culms hollow, leafy; sheaths open; ligule a short, firm membrane; blades flat to somewhat folded, scabrous, non-auriculate; inflorescence an open panicle with spikelets on the main axis as well as on the branches; spikelets 1-flowered, strongly laterally compressed, disarticulating below the spikelet and falling entire; glumes greatly reduced to lacking (what has been interpreted as rudimentary glumes may be reduced spikelets); lemmas broad, firm or indurate, boat-shaped, awnless, usually 5-nerved, the lateral-most pair marginal and hispid-ciliate as on the keel; palea narrow, subequal to the lemma, usually 3-nerved, the lateral nerves submarginal, the keel hispid-ciliate; lodicules 2, ovate, entire to lobed; stamens 1–6, ours 3; $x = 12$.

A genus of about 10 species of temp. and trop. regions of both hemispheres, 5 of which are present in the U.S., these mostly in the e. states. (Named in honor of Johann Daniel Leers, 1727–1774, a German botanist.)

The palea is well developed, possibly due to becoming exposed in a glumeless condition. Species of *Leersia* are adapted to marshy or aquatic habitats.

Reference: Pyrah, G. L. Taxonomic and distributional studies in *Leersia* (Gramineae). Iowa State Coll. J. Sci. 44: 215–270. 1969.

1. Leersia oryzoides (L.) Swartz

Phalaris oryzoides L. Sp. Pl. 55. 1753. *Homalocenchrus oryzoides* Pollich, Hist. Pl. Palat. Elect. 1: 52. 1776. *Ehrhartia clandestina* Weber, Primitiae Fl. Holsaticae 64. 1780. *Leersia oryzoides* Swartz, Nova Gen. & Sp. Pl. 21. 1788. *Asperella oryzoides* Lam. Tableau Encycl. & Méth. 1: 167. 1791. *Asprella oryzoides* Beauv. Essai Nouv. Agrostogr. 2, 153, 172. 1812. *Leersia asperrima* Willd. ex Trin. Mém. Acad. Imp. Sci. Saint-

Pétersbourg, Sér. 6, Sci. Math., Seconde Pt. Sci. Nat. 3(1): 171. 1840, as a synonym. *Oryza clandestina* A. Brown in Aschers. Fl. Prov. Brandenburg 799. 1864. *Laertia* [sic] *oryzoides* Gromow. in Trautv. Trudy Imp. S.-Peterburgsk. Bot. Sada 9: 354. 1884. *Oryza oryzoides* A. Brand, Koch's Syn. Deutsch. & Schweiz. Fl. ed. 3. 3: 2704. 1905. ("Habitat in Virginiae paludibus nemorosis.")

Rice cutgrass.

Slender perennials from long, creeping rhizomes; culms 5–15 dm tall, weakly decumbent at the base, pubescent at the nodes; sheaths glabrous to retrorsely scabrous, striate, the lateral margins extended at the apex; ligules about 1 mm long, truncate, minutely erose-ciliolate; blades flat, 6–10 mm broad, strongly retrorsely scabrous; panicles open, 10–20 (30) cm long, the lower branches flexuous and racemose; axillary panicles often present, enclosed by the upper sheaths and bearing cleistogamous spikelets; spikelets 1-flowered, lacking glumes; lemmas 4–5 mm long, strongly compressed laterally, semicircular, hispid-ciliate on the keel and marginal nerves, sparsely hispidulous between the nerves; palea slightly longer than the lemma, lanceolate; lodicules slightly less than 0.5 mm long; $2n = 48, 60$.

Wet, marshy places, growing in shallow to relatively deep water or muddy borders of streams, ponds and lakes; B.C., to Newfl., s. throughout all states of the U.S. except Nev., also in Europe, Japan and c. Asia; in Utah at the Bear River Bay and the e. side of the Great Salt Lake. July–Sept.

Rice cutgrass provides food for migratory fowl. The strongly retrorsely scabrous blade margins can easily inflict cuts on the unsuspecting person who tugs away at the blades; this gives the plant its common name, cutgrass.

Subfamily V. **ARUNDINOIDEAE**

Epidermal cells in the region of root hairs all alike and all capable of giving rise to root hairs; culm internodes hollow with small or large central cavities or solid; sheath margins free; ligule membranous or a fringe of hairs; blades non-auriculate; inflorescence a panicle, sometimes reduced to a raceme, or in *Danthonia unispicata,* reduced to a single spikelet; spikelets usually laterally compressed, 1- to several-flowered, rarely unisexual, disarticulating usually above the glumes and between the florets; lemmas 1- to several-nerved; lodicules 2; stamens 3; embryo never more than 1/2 as long as the caryopsis, with vascular traces to the scutellum and coleoptile separated by an internode (panicoid), the epiblast absent (in ours) or sometimes present, the lower part of the scutellum and coleorhiza separated by a cleft (panicoid), and the embryonic leaf margins not overlapping (in ours), or sometimes overlapping; $x = 6, 7,$ or 12.

Key to the Tribes of Arundinoideae

1 Rachilla with long silky hairs (in *Phragmites*); lemmas 1- to 5-nerved, not bifid at the apex .... Tribe 14. Arundineae, p. 458

1 Rachilla without long silky hairs; lemmas 7- to several-nerved, bifid or toothed at the apex .... Tribe 15. Danthonieae, p. 460

Tribe 14. **ARUNDINEAE**

Perennials, usually tall, robust reed grasses from stout rhizomes; culms hollow; sheath margins free; ligule a fringe of hairs (in ours) or a ciliate membrane or absent; blades usually flat and broad, non-auriculate; inflorescence a plumose panicle; spikelets (1) few- to several-flowered, perfect or unisexual, disarticulating above the glumes; glumes shorter to longer than the lowermost lemma; lemmas 1- to 5-nerved; lodicules 2, truncate; embryo panicoid except for the embryonic leaf margins not overlapping; $x = 12$.

The Arundineae tribe includes only a few genera of stout, perennial grasses which occur in both trop. and temp. regions of the world, mostly in aquatic habitats. These genera were placed in the Festuceae (Poeae) tribe by A. S. Hitchcock (Manual Grasses U.S., 1951).

Arundo donax L., giant reed, a native of the Old World tropics, is sometimes planted in the Intermountain Region for ornament. The culms grow 3 to 6 m high, and in warmer climates than ours, large plumose panicles, 20 to 40 cm or more long, are produced. The giant reed has villous lemmas and glabrous rachillas as opposed to the glabrous lemmas and villous rachillas of *Phragmites. Cortaderia selloana* (J. A. Schultes) Aschers. & Graebner, pampas grass, is a large dioecious grass introduced from southern South America as an ornamental. The pistillate plants are quite striking with their large plumose panicles. *Cortaderia* differs from *Arundo* and *Phragmites* in having non-rhizomatous densely clumped culms, densely clustered basal leaves and 1-nerved lemmas.

82. **PHRAGMITES** Trin. Reed

Tall, robust perennials with stout, creeping rhizomes or stolons, forming large colonies, the stolons sometimes up to 10 m long; culms hollow, leafy; sheaths open; ligule a short, fringed membrane or composed entirely of hairs; blades broad, flat, non-auriculate; inflorescence a large, open panicle; spikelets 1- to several-flowered, the lowest floret staminate or sterile, disarticulating above the glumes and between the florets; rachilla villous with long hairs giving the inflorescence a silky appearance; glumes thin, unequal, the first glume 1- to 3-nerved, the second much longer than the first but shorter than the florets, usually 3-nerved, lanceolate, acute and rounded on the back; lemmas narrow, acute or acuminate, rounded on back, (1-) 3- to 5-nerved, glabrous, the lemma of the first floret with only 1 (2) stamen(s) and the lemmas of the

PHRAGMITES

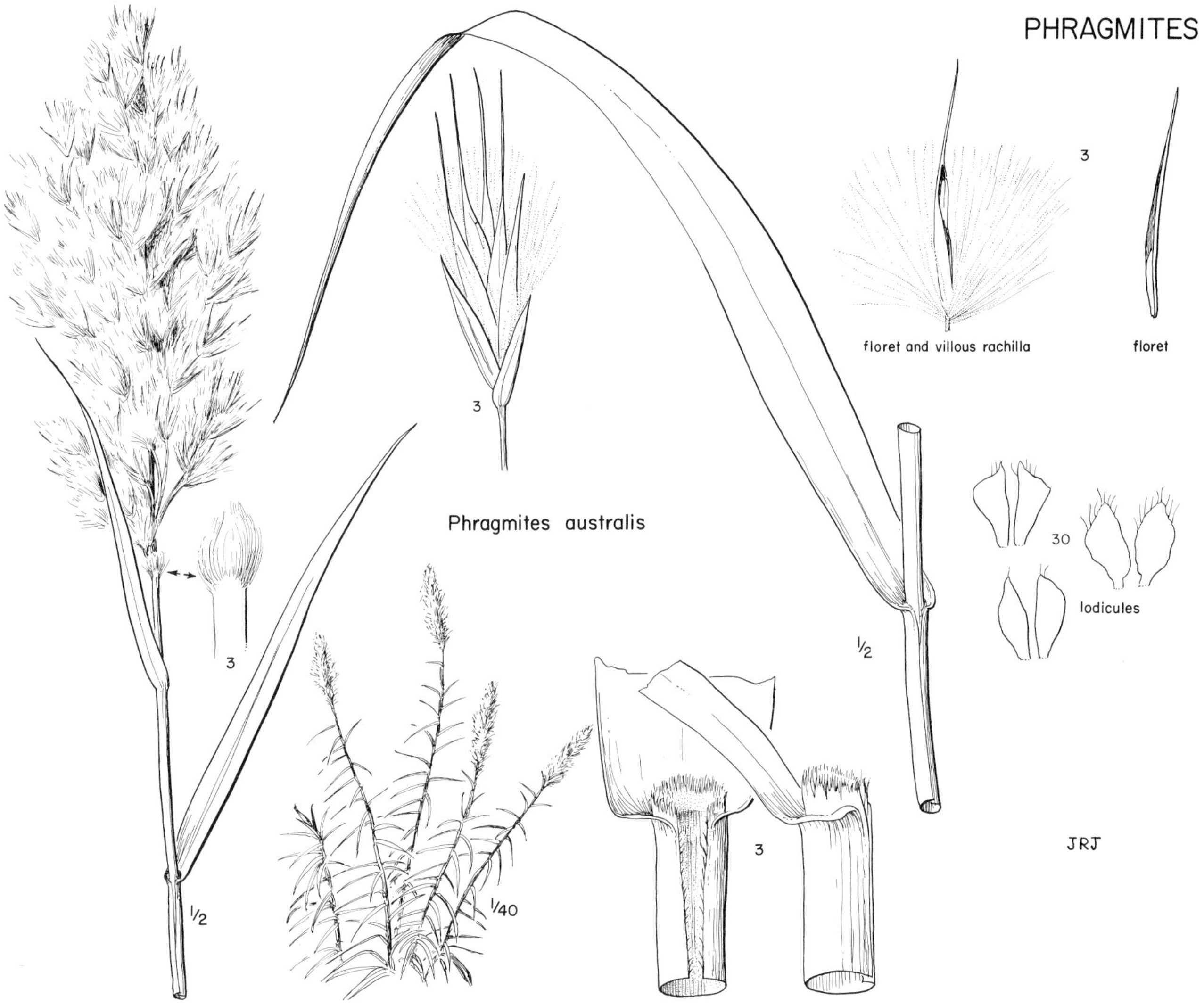

upper florets with 2 or 3 stamens each; palea much shorter than the lemma; lodicules 2, membranous, flabellate-deltoid and usually shallowly lobed; $x = 12$.

A genus of 3 species, 1 native to S. Amer., 1 to Asia and the third (ours) throughout the temp. and warm regions of the world. (The ancient Greek name meaning "growing in hedges," alluding to the appearance of the dense colonies fringing streams and ponds.)

Reference: Clayton, W. D. The correct name of the common reed. Taxon 17: 168–169. 1968.

1. Phragmites australis (Cav.) Trin. ex Steudel

Arundo phragmites L. Sp. Pl. 81. 1753. *Arundo vulgaris* Lam. Fl. Franç. 3: 615. 1778. *Arundo palustris* Salisb. Prodr. Stirpium 24. 1796. *Phragmites communis* Trin. Fund. Agrostogr. 134. 1820. *Cynodon phragmites* Rasp. Ann. Sci. Nat. (Paris) 5: 302. 1825. *P. vulgaris* Link ex Crépin, Manuel Fl. Belgique, ed. 2. 345. 1866. *P. phragmites* Karsten, Deutsche Fl. 379. 1883. *Trichoon phragmites* Rendle, Cat. African Pls. Collect. Welwitsch 2(1): 218. 1899. *Oxyanthe phragmites* Nieuwl. Amer. Midl. Naturalist 3: 332. 1914. ("Habitat in Europae lacubus fluviis.")

Arundo australis Cav. Anales Hist. Nat. 1: 100. 1799. *P. australis* Trin. ex Steudel, Nomencl. Bot. ed. 2. 2: 324. 1841. (Australia, near Port Jackson.)

P. berlandieri Fourn. Bull. Soc. Bot. France 24: 178. 1877. *P. communis* var. *berlandieri* Fern. Rhodora 34: 211. 1932. *P. maximus* var. *berlandieri* Moldenke, Torreya 36: 93. 1936. *P. communis* subsp. *berlandieri* A. Löve & D. Löve, Bull. Torrey Bot. Club 81: 33. 1954. (*Berlandier 1446*, Laredo, Texas, in 1828.)

Common reed.

Stout, erect perennials; culms large, 2–3 m tall; culms and leaves usually glabrous; ligules 0.5–2 (3) mm long, a ciliate membrane; blades flat and very broad, 10–30 (40) mm; panicles 15–35 cm long, often purplish, becoming stramineous; spikelets 10–15 mm long, (1-) 3- to 6-flowered, the rachilla with long, silky hairs, the hairs 7–9 mm long; glumes lanceolate, glabrous, the first glume 3.5–6 (7) mm long, 1- to 3-nerved, the lateral nerves sometimes not reaching the apex and joining with the mid-nerve, the second glume 6–9 (10) mm long, 3- to 5-nerved; lemmas (7) 9–14.5 mm long, linear, 5-

nerved, glabrous; palea much shorter than the lemma, 3–4 mm long, narrowly-elliptic; lodicules scarcely 1 mm long; $2n$ = 36, 48, 54, 72, 84, ca. 96.

Common in marshes and in water at edges of ponds, lakes and streams; usually in alkaline or saline habitats; cosmopolitan in temp. and many trop. parts of the world; in N. Amer. from B.C. to Newfl., s. throughout all of the U.S., except the Southeast. July–Sept.

Clayton (1968) gives good evidence that the correct name for the common reed is *Phragmites australis* (Cav.) Trin. ex Steudel, based on *Arundo australis* Cav. (1799) which predates the name *P. communis* Trin. (1820), a name long accepted for Linnaeus' *Arundo phragmites* to avoid tautonymy.

Tribe 15. DANTHONIEAE

Annuals or cespitose perennials; culms hollow; sheath margins free; ligule a dense fringe of hairs (in ours); blades non-auriculate; inflorescence a panicle, raceme, or in *Danthonia unispicata* reduced to a single spikelet; spikelets few- to several-flowered, disarticulating above the glumes and between the florets; glumes much longer than the lowermost lemma, often papery; lemmas 7- to several-nerved, bifid or toothed at the apex, awned or merely mucronate, the awn of *Danthonia* flat at the base, usually twisted; lodicules 2, truncate or sometimes lobed; embryo panicoid except the embryonic leaf margins not overlapping; x = 6, 7, 12.

The tribe Danthonieae is most common in the S. Hemisphere in warm and dry climates. *Danthonia* and *Schismus* were placed in the Aveneae by A. S. Hitchcock (Manual Grasses U.S., 1951), a disposition that may yet have some merit.

1 Awns conspicuous, flat, bent; spikelets 1 cm or more long; perennials .... 83. *Danthonia*
1 Awns minute or nearly obsolete; spikelets less than 1 cm long; annuals .... 84. *Schismus*

83. DANTHONIA DC. in Lam. & DC. nom. conserv. Oatgrass

Low to moderately tall, cespitose perennials; culms hollow, finally disarticulating at the lower nodes; sheaths open; ligule a dense fringe of short hairs; blades non-auriculate; inflorescence a small, often much-reduced panicle or raceme, sometimes reduced to a single spikelet in *D. unispicata;* spikelets several-flowered, short-pedicelled, disarticulating above the glumes and between the florets; glumes broad, acute, papery, subequal, the first glume usually slightly longer than the second and both exceeding the florets, rounded to keeled, (1) 3- to 5-nerved; lemmas broadly rounded on back, pilose on the margins, indistinctly several-nerved, the apex bifid, the lobes acute to acuminate, sometimes aristate; awn arising from between the apical lobes, usually flattened below, twisted and geniculate; callus well developed, bearded; palea broad, well developed but shorter than the lemma; lodicules 2, usually flabellate and strongly thickened to elongate and thin; stamens 3, sometimes aborted or vestigial; x = 6.

A genus of over 100 species distributed throughout temp. regions of the world, but most common in Australia, New Zealand and S. Africa; only 7 species are known in the U.S. (Named for Etienne Danthione, a French botanist of the early 19th century.)

Danthonia californica and *D. intermedia* are large and palatable enough to produce good forage, but in our area they are never abundant enough to be of importance. Many of the spikelets have abortive anthers, thus apomixis is undoubtedly common. All our species produce cleistogamous spikelets in the axils of the lower leaves. *Danthonia* was previously included in the tribe Aveneae because of its long glumes, but de Wet (1954, 1956) gives evidence to show their dissimilarities in embryology, leaf anatomy, cytology and structure of the epidermis, ligules and awns.

References:

de Wet, J. M. J. The genus *Danthonia* in grass phylogeny. Amer. J. Bot. 41: 204–211. 1954.
———. Leaf anatomy and phylogeny in the tribe Danthonieae. Amer. J. Bot. 43: 175–182. 1956.
Findlay, J. N., and B. R. Baum. The nomenclatural implications of the taxonomy of *Danthonia* in Canada. Canad. J. Bot. 52: 1573–1581. 1974.

1 Panicles narrow, the erect branches bearing (4) 5–13 spikelets; sheaths usually glabrous below the apex .... 1. *D. intermedia*
1 Panicles open with 5 or fewer spikelets, often reduced to a single spikelet; sheaths usually pilose, rarely glabrous.
  2 Spikelets 3–5, borne on widely spreading, flexuous branches; lobes of the lemma aristate, (2) 3–5.5 mm long; plants commonly over 5 dm tall .... 2. *D. californica*
  2 Spikelets usually 1, or if 2 or 3 the additional spikelets on stiffly erect branches; lobes of the lemma acuminate, not aristate, 1–3.5 mm long; plants seldom over 3 dm tall .... 3. *D. unispicata*

1. Danthonia intermedia Vasey

Danthonia intermedia Vasey, Bull. Torrey Bot. Club 10: 52. 1883. *Merathrepta intermedia* Piper, Contr. U. S. Natl. Herb. 11: 122. 1906. *Pentameris intermedia* Nels. & Macbr. Bot. Gaz. 56: 470. 1913. (*Allen,* "Rocky sides of ravine, Mt. Albert, Lower Canada," 26 July 1881; lectotype by Findlay & Baum.)

D. intermedia var. *cusickii* T. A. Williams in Scribn. U.S.D.A. Div. Agrostol. Circ. 30: 7. 1901. *Merathrepta intermedia* subsp. *cusickii* Piper, Contr. U. S. Natl. Herb. 11: 122. 1906. *D. cusickii* A. S. Hitchc. Amer. J. Bot. 2: 305. 1915. (*Cusick 2427,* Oregon, in 1899.)

Timber oatgrass.

Densely tufted perennials, the base of the plants often clothed with old, brown sheaths; culms 2–4.5 (7) dm tall, glabrous; leaves glabrous, sometimes pilose; sheaths pilose at the apical margins near the throat, the hairs up to 2 mm long; ligules 0.1–0.5 mm long, a low, ciliolate crown; blades flat to involute, 1–3.5 mm broad and 5–10 cm long, glabrous or sometimes pilose on the upper surface, sometimes

DANTHONIA

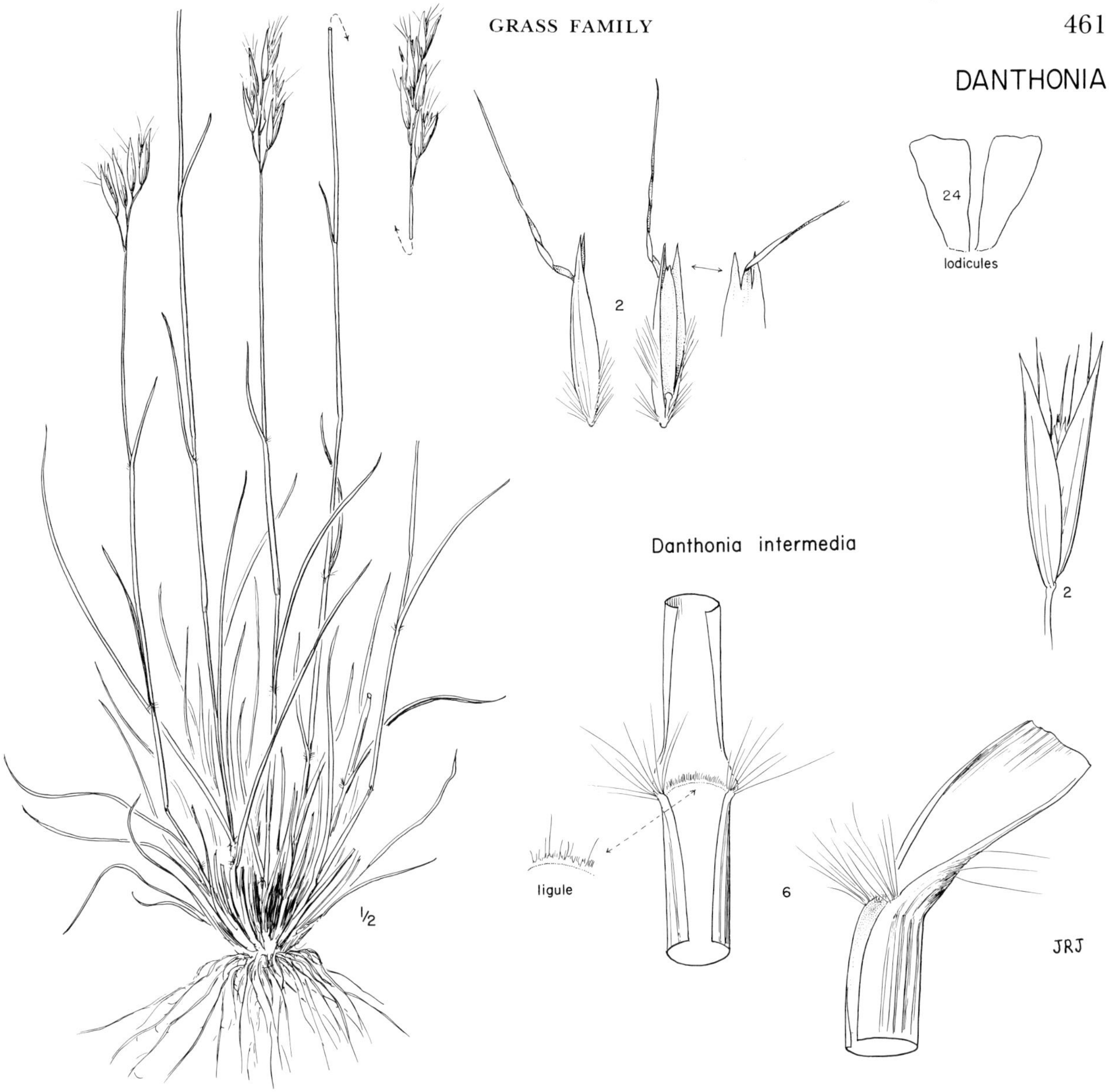

with a slight brownish cast; panicles relatively narrow, 3–6 (9) cm long, bearing 4–9 (13) spikelets, subracemiform, with erect, appressed branches or sometimes the lower-most branch spreading in larger panicles, often secund; spikelets 3- to 6-flowered, often deep purple; glumes subequal, (11) 12–14 (19) mm long, glabrous, broadly lanceolate to sparsely scaberulous, irregularly 3- to 5-nerved, the lateral nerves obscure; lemmas 6–9 (11) mm long, 7- to 9-nerved, the margins villous below, often the upper margins scabrous and the lower back shiny and glabrous, the apical lobes acuminate, 1.5–2.5 mm long, the callus sharp, villous-bearded; awns 7–10 (12.5) mm long, twisted and somewhat geniculate on the flattened lower part; palea broadly lanceolate, slightly shorter than the lemma, the finely ciliolate keels near the margins; lodicules less than 1 mm long, obovate-truncate, thick and hardened; anthers 1–3 mm long; $2n$ = 18, 36, ca. 96.

Dry to moderately moist meadows, aspen woods, rocky ridges and slopes up to timberline; Alaska to Newfl., and in the U.S. from Wash. to Mich. (Keweenaw Peninsula), s. through the mt. states to Calif., Ariz. and N.M. July–Aug.

Timber oatgrass appears to be apomictic throughout our area, as the anthers are mostly abortive.

2. **Danthonia californica** Boland.

Danthonia californica Boland. Proc. Calif. Acad. Sci. 2: 182. 1863. *Merathrepta californica* Piper, Contr. U. S. Natl. Herb. 11: 122. 1906. *Pentameris californica* Nels. & Macbr. Bot. Gaz. 56: 469. 1913. (*Bolander 10,* meadows, Oakland, Calif.; lectotype by Findlay & Baum.)

D. grandiflora Phil. Anales Univ. Chile 48: 568. 1873; not Hochst. 1851. *D. americana* Scribn. U.S.D.A. Div. Agrostol. Circ. 30: 5. 1901. *Merathrepta americana* Piper, Contr. U. S. Natl. Herb. 11: 123. 1906. *Pentameris americana* Nels. & Macbr. Bot. Gaz. 56: 469. 1913. *D. californica* var. *americana* A. S. Hitchc. Proc. Biol. Soc. Wash. 41: 160. 1928. (*Philippi 256,* Province of Nuble, Chile.)

California oatgrass.

Robust, perennial bunchgrasses, the base of the

plants clothed with broadened, old, brown sheaths; culms 3–7 (9) dm tall, glabrous, the nodes abruptly contracted; sheaths glabrous to long-pilose, bearing tufts of long hairs at the throat; ligules about 0.5 mm long, irregularly lacerate and ciliolate or a fringed crown; blades flat to loosely involute, 2–4 (6) mm broad and 10–25 cm long, often short-pilose or ciliolate on the margins, scaberulous on the surfaces, usually strongly striate; panicles 2–4.5 (7) cm long, bearing 3–5 widely spreading or reflexed spikelets, the branches and pedicels flexuous, usually strongly pubescent, often with pulvinis-like bases; spikelets 5- to 8-flowered, usually deep purple; glumes subequal, (10) 14–18 (20) mm long, exceeding the florets, glabrous or usually scaberulous, keeled, 1- to 7-nerved; lemmas 9.5–13 (15) mm long, glabrous except for the villous lower margins and sometimes finely scaberulous on the scarious lobes and finely ciliate on the margins, 9- to 11-nerved, the lateral nerves usually faint, the lobes (2) 3–5.5 mm long, narrow, becoming aristate, the callus villous; awns (5) 7–12 mm long, twisted and scabrous margined on the flattened lower portion, the awn in the terminal, reduced floret usually geniculate 1 mm above the body of the lemma; palea usually extending beyond the base of the lemma, cleft, ciliate-keeled; lodicules about 0.5 mm long, thickened; anthers 2.5–4 mm long; $2n = 36$.

Open sagebrush hills, basalt rimrock and mt. meadows at middle elevs.; B.C., N.W.T. and Alta., s. to Calif., Nev., Utah and n. N.M.; also in Chile. Mid June–Aug.

No attempt is made here to recognize var. *americana* (Scribn.) A. S. Hitchc., which is supposed to have shorter culms, more spreading tufts and spreading-pilose foliage. Plants exhibiting this entire range of variation are often found in a large mountain meadow. The meadow surrounding Dry Lake in Cache County, Utah, contains *D. californica* with plants that could be placed in either variety.

3. **Danthonia unispicata** (Thurber) Munro ex Macoun

Danthonia unispicata Munro ex Thurber in A. Gray, Proc. Acad. Nat. Sci. Philadelphia 1863: 78. 1864, name only. *D. californica* var. *unispicata* Thurber in S. Wats. Bot. Calif. 2: 294. 1880. *D. unispicata* Munro ex Macoun, Cat. Canad. Pl. 2(4): 215. 1888. *Merathrepta unispicata* Piper, Contr. U. S. Natl. Herb. 11: 123. 1906. *Pentameris unispicata* Nels. & Macbr. Bot. Gaz. 56: 470. 1913. ("San Diego to San Francisco," Calif., *Bolander, Parry, Lemmon.*)

One-spike oatgrass.

Low perennials forming dense, spreading tufts with centers finally dying out, the base of the plants clothed with old sheaths; culms (1) 1.5–3 (4) dm tall, glabrous, the lower nodes abruptly constricted and readily disarticulating; sheaths usually with spreading or reflexed, pilose hairs from somewhat pustular bases, rarely glabrous, the throat with dense tuft of spreading hairs 2–4 mm long, sometimes spreading across the collar, the collar light-yellow in contrast to the bright-green areas above and below; ligules usually less than 0.5 mm long, irregularly lacerate or fringed with ciliate hairs; blades flat to loosely involute, 1–2 mm broad and 3–8 cm long, spreading pilose, the hairs more dense on the upper surface, rarely glabrous, usually strongly striate; panicles usually reduced to a single spikelet, if 2 or 3 then the branches closely appressed and the lower spikelets smaller; spikelets 3- to 6-flowered, purple-tinged; glumes subequal, (12) 14–20 (24) mm long, the second somewhat shorter, narrow-lanceolate, glabrous, 5-nerved; lemmas (8) 10–14 mm long, glabrous to scaberulous with sparsely villous lower margins, 7-nerved, the apical lobes (1) 2–3.5 mm long, the callus bearded; awns (3) 5–9 mm long, sometimes twisted on the flattened lower portion or untwisted and geniculate; palea slightly shorter than the lemma, ciliate-keeled; lodicules about 0.5 mm long; anthers 2–3 mm long, usually aborted; $2n = 36$.

Open or rocky slopes and ridges, usually in dry places but occasionally in drying edges of meadows at middle and high elevs.; B.C. and Alta., s. to Calif., c. and n. Nev., n. Utah and nw. Colo. June–Aug.

Plants of this species are commonly covered with smut.

84. **SCHISMUS** Beauv.

Low, tufted annuals; culms hollow; sheaths open; ligule a dense fringe of short hairs from a short membrane; blades involute or flat, non-auriculate; inflorescence usually a small and contracted panicle; spikelets several-flowered, disarticulating above the glumes and between the florets; glumes subequal, large, lanceolate, slightly shorter than to about equalling the terminal floret, with white membranous margins, 5- to 7-nerved; lemmas several-nerved, broadly rounded on the back and bidentate at the apex with a short, delicate mucro between the lobes, hyaline at the glabrous apex and pilose on the lower margins; palea as long as the lemma or slightly shorter, hyaline, broad, 2-nerved, the nerves submarginal; stamens 3; $x = 6$.

A genus of about 5 species in warm parts of the Old World, native to Europe, Asia and Africa. (Name from the Greek *schismos,* a splitting, referring to the bidentate lemmas.)

Schismus has been variously placed in tribes Poeae and Aveneae, but anatomical and cytological data as well as spikelet structure indicate close affinities with *Danthonia* of tribe Danthonieae. *Schismus arabicus* Nees has been reported in southwestern Utah (*Higgins 331,* near the Utah-Arizona border along U.S. Highway 91). It can be separated from *S. barbatus* by its lemma with pubescence across the back and longer, acute lobes.

1. **Schismus barbatus** (L.) Thell.

Festuca barbata L. Amoen. Acad. 3: 400. 1756. *Schismus barbatus* Thell. Bull. Herb. Boissier, II. 7: 391. 1907. ("Habitat in Hispania.")

Low, erect or prostrate-spreading annuals; culms 0.5–1.5 (3.5) dm tall, glabrous, the nodes often deep-purple and usually geniculate, sheaths glabrous with a tuft of hairs at the throat, the hairs 2.2–3 mm long; ligules scarcely 1 mm long, irregularly lacerate or fringed, ciliate; blades filiform-involute, 5–10 cm long, scaberulous; panicles (1) 2–4 (7.5) cm long, dense, oval or narrow, sometimes hidden among the

DANTHONIA

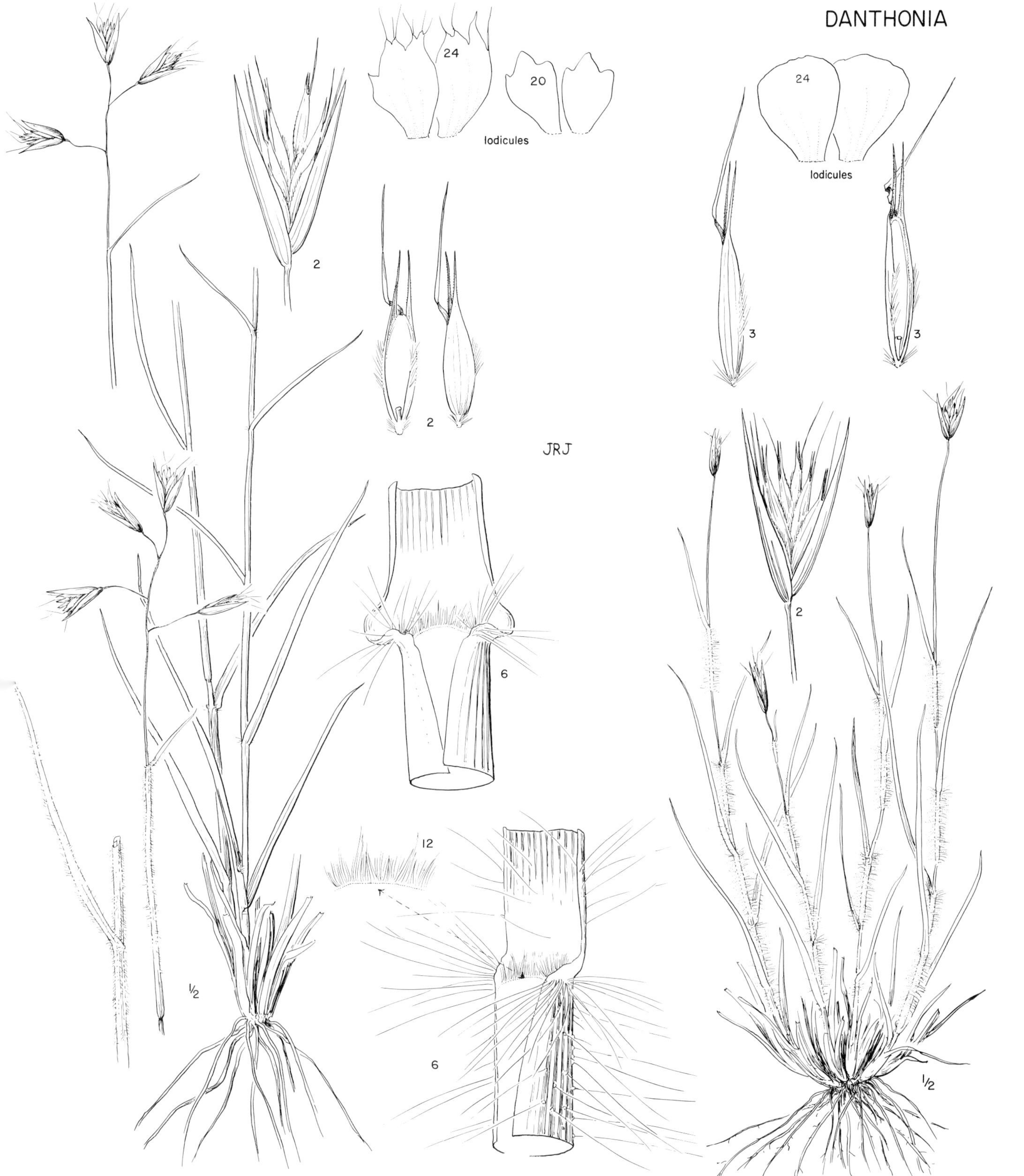

Danthonia californica

Danthonia unispicata

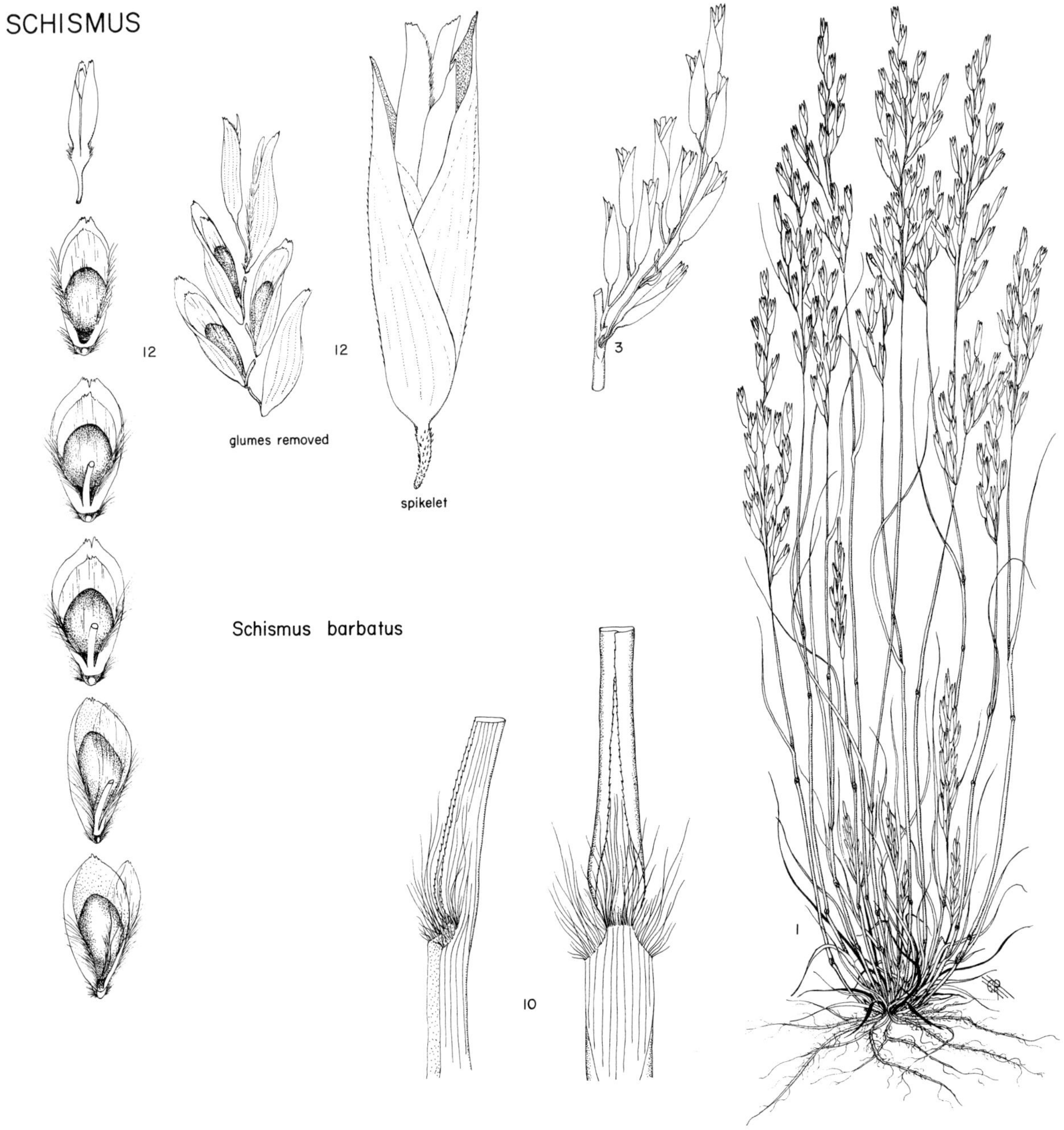

basal leaves; spikelets (4) 5–7.5 mm long; glumes and lemmas pale-green or sometimes tinged with red or purple; glumes subequal, 4–5 (6) mm long, shorter than the spikelets, lanceolate, acute, 7-nerved, the margins scarious, the nerves scaberulous; lemmas 1.5–2 mm long, obovate, keeled, usually with 9 conspicuous, parallel nerves, the lower margins and sometimes the keel with fine hairs, the apical lobes short, rounded, the keel nerve extending into a short, delicate awn from between the lobes, the apex scabrous to hyaline; palea as long or longer than the lemma, concave, hyaline, oblanceolate; anthers about 0.5 mm long; $2n = 12$.

Sandy places along roadsides, dry riverbeds and other weedy areas in creosote-bush desert country; introduced from the Medit. region; sw. Utah, s. Calif., Ariz. and Trans-Pecos Texas; also Argentina and Chile. April.

Order **TYPHALES**

Tall, wind pollinated, aquatic or semi-aquatic perennial herbs arising from an extensive rhizome, with alternating, linear, sheathing leaves; flowers imperfect, unisexual, borne in dense spikes or heads, the staminate ones uppermost; perianth apparently none or vestigial; stamens 2–5; carpels 1 (2), with a solitary ovule (or solitary in each locule), pendulous from the apex of the ovary, anatropous; pollen binucleate; seeds albuminous with the embryo embedded in the endosperm.

Two families, both of which occur in our area.

KEY TO THE FAMILIES OF TYPHALES

1 Inflorescences of globose heads; fruits large and nut-like achenes, sessile or nearly so, indehiscent .... SPARGANIACEAE, p. 465
1 Inflorescences of dense, cylindrical spikes; fruits small, buoyant, on long slender, hairy stipes, tardily dehiscent .... TYPHACEAE, p. 468

Family **SPARGANIACEAE**, the Bur-reed Family

Herbaceous perennials of aquatic places, with fibrous roots and an extensively creeping rhizome, with erect or floating, simple or branched stems bearing alternate, linear, erect or floating basally sheathing leaves; flowers imperfect, borne in capitate-globose clusters on the upper part of the main stem and often on the upper axillary branches, the staminate ones uppermost, sessile or peduncled, the lower pistillate ones with leaf-like bracts; staminate flowers of 3–5 stamens and 3–5 minute chaffy bracts, the slender filament much longer than the anther, greatly exceeding the bracts; pistillate flowers of a single ovary and 3–6 broad chaffy bracts, the ovary about equalling the bracts, the pistil simple or sometimes 2-carpellary with a simple or forked elongated style, the ovary 1 (2) celled and each locule 1-ovuled; fruits usually broadly fusiform to truncate-pyriform, hardened nut-like achenes with a strong beak and 1–2 albuminous seeds.

Only one genus.

1. **SPARGANIUM** L. Bur-reed

From 15–20 species of aquatic places in temp. and arctic regions of the N. Hemisphere and Australasia. (From the Greek *sparganion*, a swaddling band, apparently alluding to the long narrow leaves.)

Reference: Fernald, M. L. Notes on *Sparganium*. Rhodora 24: 26–34. 1922.

1 Stigmas 2, rarely one in an occasional flower; plants coarse, 5–20 dm high; mature achenes truncate-pyriform, narrowing abruptly to the beak; inflorescences usually branched ...... 1. *S. eurycarpum*
1 Stigma one, rarely 2 in an occasional flower; plants slender, mostly less than 10 dm high; mature achenes essentially fusiform, narrowing gradually to the beak; inflorescences usually simple.
  2 Staminate heads usually 2 or more, the anthers 0.8–1.5 mm long, less than half as wide as long; achene beak 1.5–5 mm long.
    3 Leaves (1) 2–6 (8) mm wide, rounded on the back, not scarious-margined, generally submerged or floating; mature pistillate heads 1–2 cm across; stigma 0.8–1 mm long; achene beak 1.5–2 mm long including the stigma .... 2. *S. angustifolium*
    3 Leaves (5) 6–12 (15) mm wide, flat or rounded above to slightly keeled below, scarious-margined, generally erect and emersed; mature pistillate heads 2–3 cm across; stigma 1–2 mm long; achene beak 3–4 (5) mm long including the stigma .... 3. *S. emersum*
  2 Staminate heads solitary (rarely 2), the anthers 0.3–0.6 (0.8) mm long, at least half as wide as long; achene beak 1–1.5 mm long including the stigma; leaves floating or erect and emersed, (2) 3–6 (8) mm wide .... 4. *S. minimum*

1. Sparganium eurycarpum Engelm.

Sparganium eurycarpum Engelm. in A. Gray, Manual Bot. ed. 2. 430. 1856. (*Engelmann*, St. Louis, Mo., June 1847; lectotype here selected, deposited at MO.)
S. californicum Greene, Bull. Calif. Acad. Sci. 1: 11. 1884. (*Greene*, swampy places in the western portion of Calistoga, Napa Co., Calif., June 1881.)
S. greenei Morong, Bull. Torrey Bot. Club 15: 77. 1888. *S. eurycarpum* var. *greenei* P. Graebner, Pflanzenr. IV. 10 (Heft 2): 13. 1900. (*Greene*, Olema, Marin Co., Calif., Sept. 1887.)
S. ramosum sensu Amer. authors; not Huds. 1778.

Tall erect herbaceous perennials 5–12 (20) dm tall, the stem stout, branching; leaves flat but slightly keeled below, linear, (2) 3–8 dm long, (7) 8–12 (15) mm wide, as long as or slightly shorter than the inflorescence; inflorescences branched, 1.5–4 (5) dm long, the upper branches with 5–12 staminate heads and 1–2 pistillate ones; perianth scales long, clawed, expanded and spatulate with irregularly lobed hyaline-margined apex, 3–7 mm long; anthers elliptic, 1–1.2 (1.5) mm long; style branches 2, rarely simple, filiform, 1.8–2 mm long; fruits sessile in bur-like heads 2–2.5 cm across, mostly 2-seeded, hard and thick at maturity, rounded to truncate-pyriform, (6) 8–10 mm long, narrowing abruptly to the stylar beak, the beak 2–3 mm long.

Wet meadows and marshes, mostly emersed; B.C. to Baja Calif., e. to Newfl. and Fla.; fairly common in our area from se. Oregon and s. Idaho s. into n. Nev., Utah and n. Ariz. May–Oct.

One of our more common species in the genus, and an important one for wild birds as food and cover.

SPARGANIUM

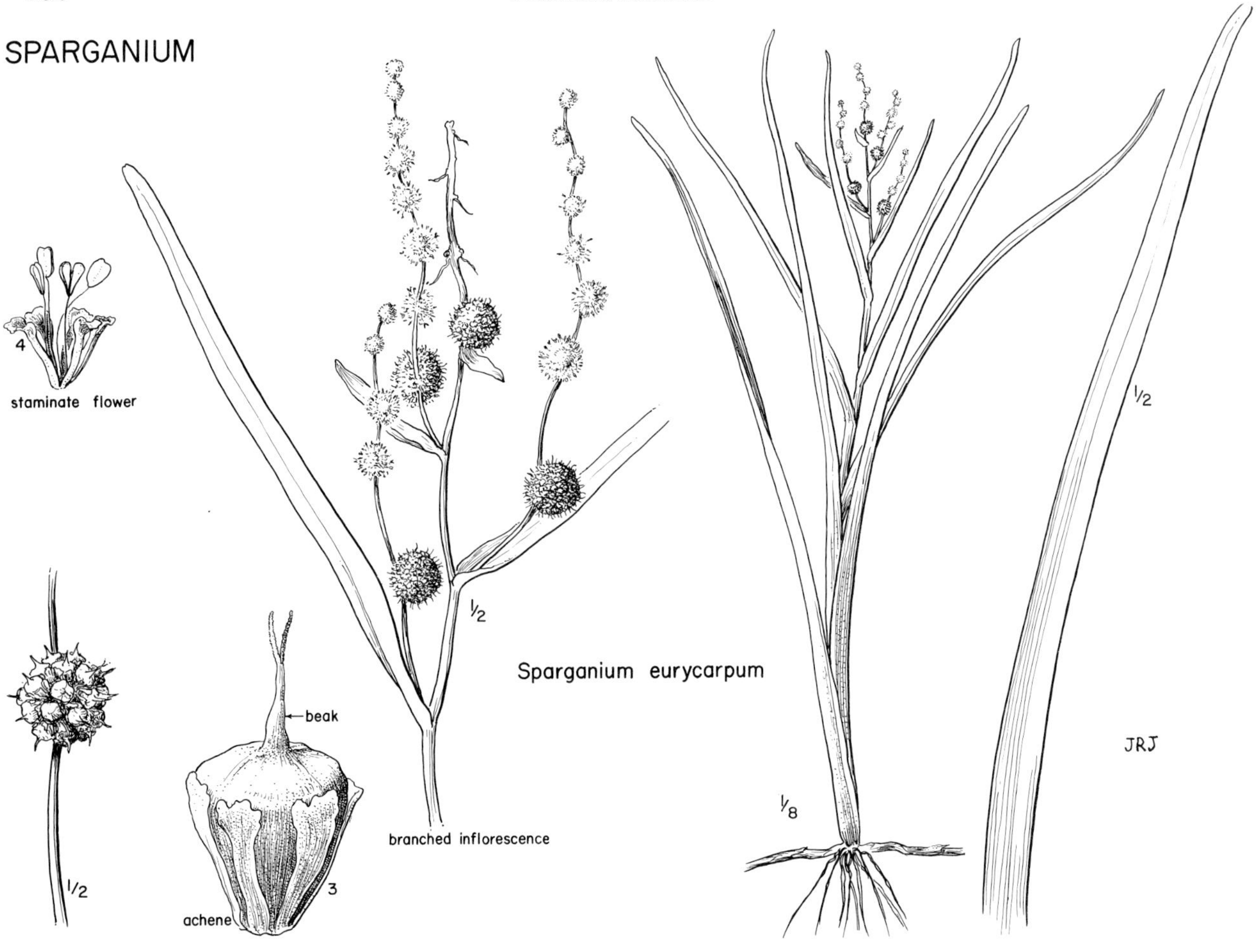

2. Sparganium angustifolium Michx.

Sparganium angustifolium Michx. Fl. Boreali-Amer. 2: 189. 1803. *S. natans* var. *angustifolium* Pursh, Fl. Amer. Sept. 34. 1814. *S. simplex* var. *angustifolium* Torr. Fl. State New York 2: 249. 1843. (*Michaux,* "in omnibus Canadae.")

Short floating or weakly erect herbaceous perennials, the stem slender and 3–10 dm long when floating, usually shorter and somewhat stouter when growing in shallow water, not obviously branched; leaves slightly rounded dorsally, linear, flaccid in most, mostly 3–10 dm long, (1) 2–6 (8) mm wide, the lower ones usually equalling the stem, the upper ones exceeding the inflorescence, usually dilated and sheathing the stem; inflorescences essentially simple with 2–5 staminate heads and 2–4 pistillate heads, or occasionally with one or two pistillate heads borne in the axils of the uppermost leaf-like bracts on peduncles 1–2 cm long; perianth scales short, clawed, only slightly expanded at the apex, 1–2 mm long; anthers broadly elliptic, 0.8–1 mm long; styles simple, the stigma 0.8–1 mm long; fruits sessile or more commonly stipitate on stipes 1–1.5 mm long, in bur-like heads 1–2 cm across, 1-seeded, fusiform, 4–5 mm long including the 1.5–2 mm long beak and stigma; $2n = 30$.

In shallow streams, ponds and lakes where submersed or partially floating, infrequently wholly emersed; Eurasia and N. Amer.; Alaska to s. Calif., e. to Colo. and N.M., then across s. Can. to Newfl. and Greenl., s. as far as Pa.; infrequent in the mts. of Utah (Aquarius Plateau, Wasatch, Uinta and LaSal mts.) and Nev. (Ruby Mts., Elko Co.). July–Aug.

3. Sparganium emersum Rehmann

Sparganium simplex sensu authors; not Huds. Fl. Anglica, ed. 2. 401. 1778, a superfluous name for *S. natans* L.

S. emersum Rehmann, Verh. Naturf. Vereins Brünn 10: 80. 1871. *S. simplex* var. *emersum* Aschers. & Graebner, Syn. Mitteleur. Fl. 1: 286. 1897. (*Rehmann,* "Im Flussbette des Row bei Bar in Podolien," Czech.)

S. simplex var. *multipedunculata* Morong, Bull. Torrey Bot. Club 15: 79. 1888. *S. multipedunculatum* Rydb. Bull. Torrey Bot. Club 32: 598. 1905. *S. emersum* var. *multipedunculatum* Reveal, Taxon 19: 797. 1970. (*R. S. Williams,* Great Falls, Mont.)

S. subvaginatum Meinsh. Bull. Acad. Imp. Sci. Saint-Pétersbourg 36(4): 34. 1893. *S. simplex* var. *subvaginatum* Aschers. & Graebner, Syn. Mitteleur. Fl. 1: 285. 1897, at least as applied to the Amer. syntype. (*Parry,* Rocky Mts. region.) = var. *multipedunculatum,* but as to the Alaska and Finland syntypes, the name would = var. *emersum.*

Tall erect herbaceous perennials (1.5) 2–5 (10) dm tall, the stem rather stout, simple; leaves flat or slightly keeled at the base (in ours) where expanded and scarious-margined, linear, (1.5) 2–5 (10) dm long, (5) 6–12 (15) mm wide, much longer than the inflorescence; inflorescences simple, rarely branched at the first node, up to 2 dm long, with 3–8 staminate heads and (2) 3–5 pistillate ones, the lowermost occasionally axillary or supra-axillary, or pedunculate on a slender peduncle up to 12 cm

SPARGANIUM

staminate heads
2
4
stamen
achene
4
1/2
Sparganium angustifolium
pistil
4
achene
4
JRJ
1/2
Sparganium emersum
4
stamen
1/2
pistillate head
1/2
1/4

long; perianth scales oblanceolate with a broadened erose apex, 2–4 mm long; anthers elliptic, 1–1.5 mm long; styles simple, the stigma (1) 1.5–2 mm long; fruits stipitate, the stipes 2.5–3.5 (6) mm long, the bur-like heads 2–3 cm across, 1-seeded, fusiform, 4–6 mm long, gradually to somewhat abruptly narrowing to a stylar beak 3–4 (5) mm long including the stigma; $2n = 30$.

Shallow ponds, lakes and marshes; Eurasia and N. Amer.; common from Alaska to c. Calif., e. to Colo. and N.M., and across Can. to N.S., then s. to N. Eng.; common throughout most of our area. June–Sept.

Until recently this species was known as *S. simplex,* but as I have noted (Reveal, Taxon 19: 796–797. 1970), this name is superfluous and thus the next oldest name, *S. emersum,* has been taken up. In North America the species is composed of two intergrading varieties, and our plants are only the var. *multipedunculatum* (Morong) Reveal. The var. *emersum* is characterized by its strongly keeled leaves and larger fruits (12–15 mm long including the stipe, body and beak), and is found along both coasts of North America and across Alaska and Canada. In some places the two may be found together.

4. Sparganium minimum Fries
Sparganium natans var. β L. Fl. Suec. ed. 2. 323. 1755. *S. minimum* Fries, Summa Veg. Scand. 2: 560. 1849. (Sweden.)

Short floating or weakly erect herbaceous perennials, the stem slender and 3–8 (10) dm long when floating, or usually 1–3 dm tall when emersed, not obviously branched; leaves flat, usually flaccid, 1–8 (10) dm long, (2) 3–6 (8) mm wide, about as long as the inflorescence; inflorescences simple, mostly of (2) 3–4 heads, the uppermost one staminate or with both staminate and pistillate flowers, the lowermost ones mixed or pistillate, on slender peduncles up to 6 cm long; perianth scales oblanceolate, 1–1.2 mm long, with essentially an entire apex; anthers broadly elliptic, 0.3–0.6 (0.8) mm long, at least half as wide; styles simple, the stigma 0.8–1 mm long; fruits sessile or stipitate, on stipes 0.7–0.9 (1) mm long, in bur-like heads 0.8–1.2 cm across, ellipsoid, 3.5–5 mm long including the 1–1.5 mm long beak and stigma; $2n = 30$.

In shallow ponds, streams and marshes; Eurasia and N. Amer.; Alaska to e.-central Calif., e. in Can. to Lab., s. in the U.S. to Utah, Mont., Tenn. and N.J.; infrequently found in our area in se. Oregon and n. Utah. July–Aug.

Family **TYPHACEAE,** the Cat-tail Family

Tall herbaceous perennials of wet and swampy places with extensive creeping rhizomes and erect, simple, cylindric, pithy stems bearing glabrous, linear, alternating sheathing leaves with tapering to scarious or membranous auricled shoulders on the open cylindric sheaths; flowers imperfect, borne in densely crowded, terminal, elongated cylindrical spike-like inflorescences, the staminate ones uppermost, sessile on the main axis, naked except for subtending capillary hairs; staminate flowers ephemeral, densely crowded in the spike and interspersed with numerous slender simple or forked hairs or with linear to linear-lanceolate bracts, composed of 2–5 stamens with free or united filaments and linear, basifixed anthers; pistillate flowers persistent, usually densely crowded on the spike, this composed of compound pedicels interspersed with sterile flowers and slender slightly clavate bracts attached to the base of the pedicel, the fertile flowers composed of a 1-carpellary pistil with the ovary 1-celled and 1-ovulate, the style linear and the stigma linear-lanceolate, the sterile flowers on elongated, hairy stipes which terminate in a swollen, aborted ovary and much longer than the functional ones; fruits dry, tardily dehiscent, buoyant, the seed albuminous.

Only one genus.

1. **TYPHA** L. Cat-tail

A genus of 9–10 species found widely distributed in temp. and trop. places. (From *typhe,* the ancient Greek name for the plant.)

References:

Hotchkiss, N., and H. L. Dozier. Taxonomy and distribution of N. American cat-tails. Amer. Midl. Naturalist 41: 237–254. 1949.
Lee, D. W., and D. E. Fairbrothers. A serological and disc electrophoretic study of North American *Typha.* Brittonia 21: 227–243. 1969.
Smith, S. G. Experimental and natural hybrids in North American *Typha* (Typhaceae). Amer. Midl. Naturalist 78: 257–287. 1967.

1 Staminate and pistillate parts of the spike usually contiguous (rarely slightly separated in Cache Valley, Cache Co., Utah), the pistillate spike dark brown, the subtending bracts of the pistillate flowers lacking, those of the staminate flowers simple and hair-like; leaves light green, 8–20 mm wide, flat on the back; pollen in tetrads, orange or yellow; stigma lanceolate to ovate-lanceolate; plants often more than 1.5 m tall; common .... 1. *T. latifolia*
1 Staminate and pistillate parts of the spike usually separated by an interval, the pistillate bracts present but occasionally minute; pollen in monads, or if more, then with many aborted; stigma linear to linear-lanceolate.
  2 Pistillate bracts broader than the linear stigma; spikes brown at anthesis when fresh; pollen in monads, all fertile.
    3 Uppermost or all cauline leaf sheaths lacking auricles, open at the throat, margins free, tapering to the blade, occasionally truncate; pistillate spikes buff-colored, the subtending

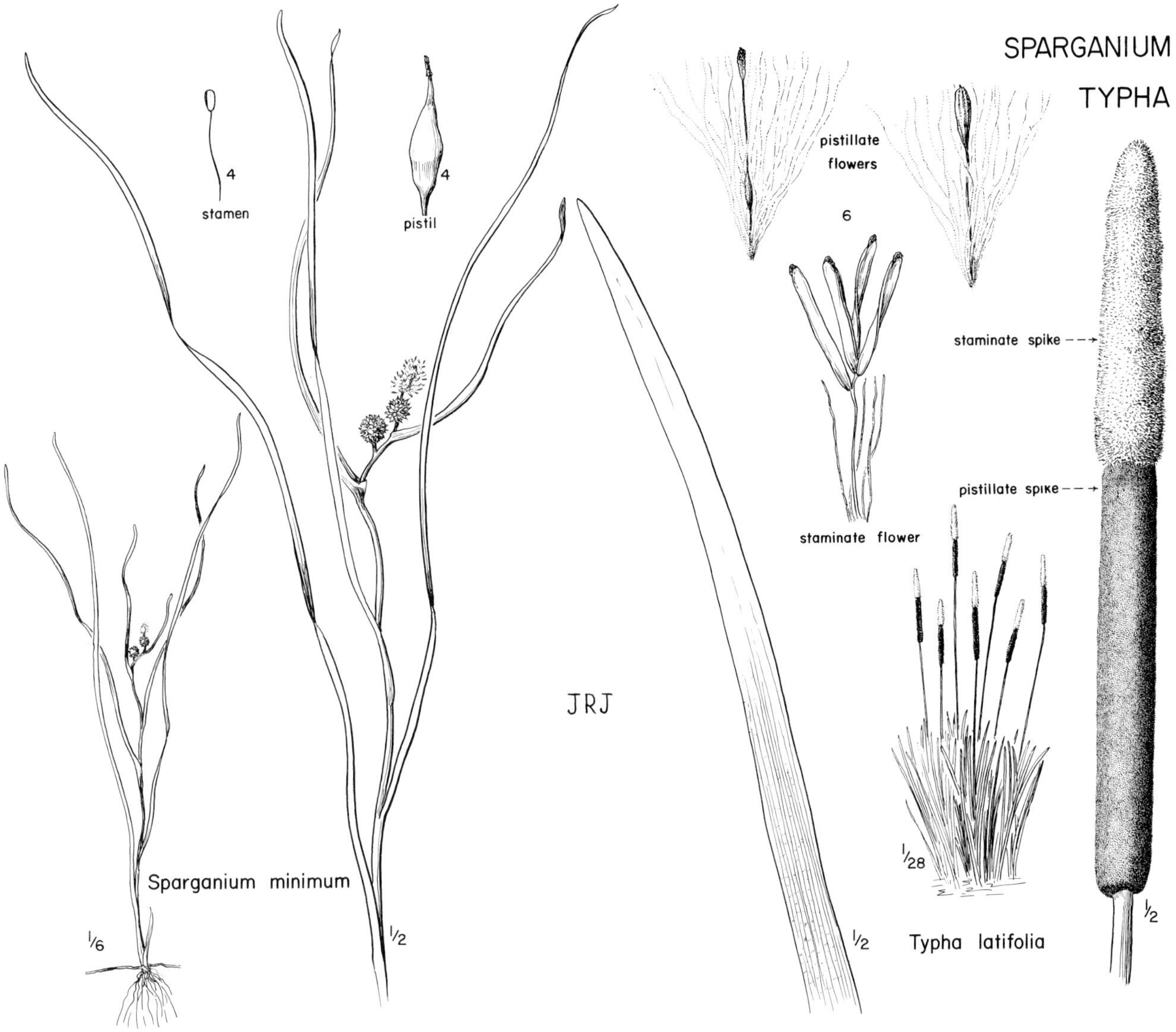

bracts paler than the stigma, acuminate, exceeding the hairs of the pedicel; leaves yellow-green, 7–15 mm wide, convex on the back; pollen golden-yellow; plants almost always more than 1.5 m high; common in e. Calif., Nev. and Utah .... 2. *T. domingensis*

3 Uppermost or all cauline leaf sheaths auricled, usually closed at the throat, margins free but parallel, rarely tapering at the throat; pistillate spikes dark brown to reddish-brown, the subtending bracts usually darker than the stigma, rounded to acute, equalling the hairs of the pedicel; leaves dark green and glaucous, (3) 5–10 (14) mm wide, somewhat convex on the back; pollen lemon-yellow; plants (0.7) 1–1.5 m tall; rare, se. Oregon .... 3. *T. angustifolia*

2 Pistillate bracts narrower than the linear-lanceolate stigma; spikes green at anthesis when fresh; pollen a mixture of monads, triads and tetrads, often with a high percent aborted, golden-yellow; leaf sheaths closed at the throat, auricled; leaves dark green, 6–10 (12) mm wide; plants 2–3.5 m tall; rare, Harney Co., Oregon, near Bishop, Inyo Co., Calif., and in Cache Valley, Cache Co., Utah .... 4. *T.* X *glauca*

1. Typha latifolia L.

Typha latifolia L. Sp. Pl. 971. 1753. *Massula latifolia* Dulac, Fl. Dépt. Hautes-Pyrén. 47. 1867. (Europe.)

Plants usually coarse and stout, the stem 1–3 m tall, arising from a stout spreading fleshy rhizome; leaves 12–16 per sterile stem, about equalling the inflorescence, light green, 8–20 mm wide, nearly flat, the leaf sheath cylindrical and open to the base, the scarious upper margin tapering or rarely truncate, not auricled; spike-bearing stem as long as or slightly longer than the leaves, the pistillate and staminate portions contiguous, rarely slightly separated; pistillate spikes dark brown, 10–18 cm long, 1.5–3 cm thick, the flowers without bracts or bract-like hairs but on slender, often hair-like, compound pedicels 1–2 mm long, the stigma medium brown to dark brown, lanceolate to lanceolate-ovate, the surface of the main stem axis appearing minutely pebbled; staminate spikes lighter brown, the stamens on branched filaments, the bracts lacking, the pollen orange or more commonly yellow, in tetrads; fruits ellipsoid, 1 mm long; $n = 15$.

In shallow, slow-moving waters of streams and marshy places; common throughout Eurasia, n. Africa and N. Amer.; e. Alaska and n. Can. s. to Mex.; common throughout most of our area. June–Aug.

2. **Typha domingensis** Pers.

Typha domingensis Pers. Syn. Pl. 2: 532. 1807. *T. angustifolia* var. *domingensis* Griseb. Fl. Brit. W. Indian Islands 512. 1864. (Santo Domingo.)

T. truxillensis H.B.K. Nova Gen. & Sp. Pl. 1: 83. 1816. (Truxillo, Peru.)

T. bracteata Greene, Bull. Calif. Acad. Sci. 2: 413. 1887. (*Greene,* marshy place near the sea on the south side of Santa Cruz Isl., above Laguna Cañon, Santa Barbara Co., Calif., 13 Aug. 1886.)

Plants coarse and stout, the stem 2.5–4 m tall, arising from a stout spreading rhizome; leaves 6–9 (12) per sterile stem, about equalling the inflorescence, light yellowish-green, 6–12 (15) mm wide, flat but moderately concave on the back, the leaf sheath cylindrical and open at the throat, with scarious margins tapering to the blade; spike-bearing stem about as long as the leaves, the pistillate and staminate portions separated by 5–8 cm; pistillate spikes light brown, 15–25 cm long, 1.5–2.5 cm thick, the flowers with obovate to oblanceolate, translucent to light brown bracts exceeding the gynophore hairs, arranged on a stout compound pedicel 0.5–0.8 mm long, these with numerous hair-like appendages, the stigma light brown, linear, the surface of the main stem axis minutely bristly; staminate spikes whitish-brown, the stamens on branched filaments, these interspersed with linear, brownish, simple bracts, the pollen golden-yellow, in monads; fruits ellipsoid, 0.8–1 mm long; $n = 15$.

Slow-moving streams, shallow ponds and marshy areas; sw. Oregon, n. Calif., Nev. and Utah, s. to Mex., e. across s. N.M., Texas and Okla. to the Gulf States of se. U.S., s. to the W.I., and in S. Amer.; common in the s. part of the Region, in e.-central Calif., s. and e. Nev., and rather common throughout most of Utah. July–Aug.

This large cat-tail is common in Utah and southern Nevada, but is not known to enter southeastern Oregon or southern Idaho. It is important as cover for many wild birds, and borders many springs, streams and ponds in our area. This species and *T. latifolia* often grow together, and some difficulties may be encountered in separating the two species and their hybrid, *T.* X *glauca,* especially in northern Utah.

3. **Typha angustifolia** L.

Typha angustifolia L. Sp. Pl. 971. 1753. *Massula angustifolia* Dulac, Fl. Dépt. Hautes-Pyrén. 47. 1867. (Europe.)

Plants slender, the stem (0.7) 1–1.5 m tall, arising from a slender creeping rhizome; leaves 6–8 per sterile stem, exceeding the inflorescence, dark green and glabrous, (3) 5–10 (14) mm wide, plano-concave or plano-convex to strongly convex on the back, the leaf sheath somewhat cylindrical and closed, with scarious margined auricles; spike-bearing stem shorter than the leaves, the pistillate and staminate portions separated by 1–5 (12) cm; pistillate spikes dark brown, 8–20 cm long, 1.5–2.5 cm thick, the flowers with oblanceolate to spathulate, dark brown bracts about equalling the gynophore hairs, arranged on a stout compound pedicel 0.5–0.7 mm long, these with numerous enlarged, brown appendages, the stigma dark brown, linear, the surface of the main stem axis minutely bristly; staminate spikes brown, the stamens on branched filaments, these interspersed with linear, brownish, often bifid bracts, the pollen lemon-yellow, in monads; fruits ellipsoid, 1 mm long; $n = 15$.

Shallow, quiet or slow-moving streams, lakes, ponds and marshy areas; Eurasia and N. Amer.; in scattered locations in Oregon and Calif., Wyo., s. Mont. and ne. Colo., hence e. across extreme s. Can. to the ne. U.S.; rare in our area and known only from the Lake Sect. of se. Oregon. July–Aug.

4. **Typha X glauca** Godron (pro sp.)

Typha glauca Godron, Fl. Lorraine 3: 20. 1844. (France.)

T. latifolia var. *elongata* Dudley, Cornell Univ. Sci. Bull. 2: 102. 1886. *T. elongata* Kronf. Verh. K.K. Zool.-Bot. Ges. Wien 39: 176. 1889. *T. angustifolia* var. *elongata* Wieg. Rhodora 26: 1. 1924. (*Dudley,* Canoga and Cayuga marshes, Cayuga Lake Basin, N.Y.)

Plants usually coarse and stout, the stem 2–3.5 m tall; leaves 8–12 per sterile stem, slightly exceeding the inflorescence, dark green, 6–10 (12) mm wide, the leaf sheath closed at the top, auricled; spikes separated; pistillate spikes dark brown, 18–50 cm long, 1.5–2 cm thick, the flowers bractless or nearly so, on pedicels 0.6–1.2 mm long, the stigma linear-lanceolate; staminate spikes light brown, the flowers bractless or nearly so, the pollen golden-yellow, in diads, triads or tetrads, often aborted; $n = 15$.

In shallow, slow-moving water in marshy areas; in widely scattered locations in Europe and N. Amer.; known in our area from Malheur Natl. Wildlife Refuge, Harney Co., Oregon, near Bishop, Inyo Co., Calif. and near Logan, Cache Co., Utah. July–Aug.

Under this name are two types of hybrids. From southeastern Oregon are those plants resulting from *T. latifolia* X *T. angustifolia,* while from the California and Utah locations come plants that are *T. latifolia* X *T. domingensis.* These hybrids differ morphologically from either parent and, according to Lee and Fairbrothers (1969), they also have chemical differences.

Subclass **ARECIDAE**

Monocotyledonous plants with mostly syncarpous flowers, the apocarpous forms borne on terrestrial, woody plants; flowers usually numerous, small, subtended by a prominent spathe or often aggregated into a spadix, with the petals and sepals much alike or reduced; plants often arborescent, but floating aquatics in ours; leaves in most tropical and subtropical species broad, petiolate and not with typical parallel venation, but in ours leafless; supporting cells typically 4, less often 3 or 2; pollen binucleate or trinucleate; ovary superior (sometimes sunken in the axis of the spadix); seeds mostly with nonstarchy endosperm or without endosperm.

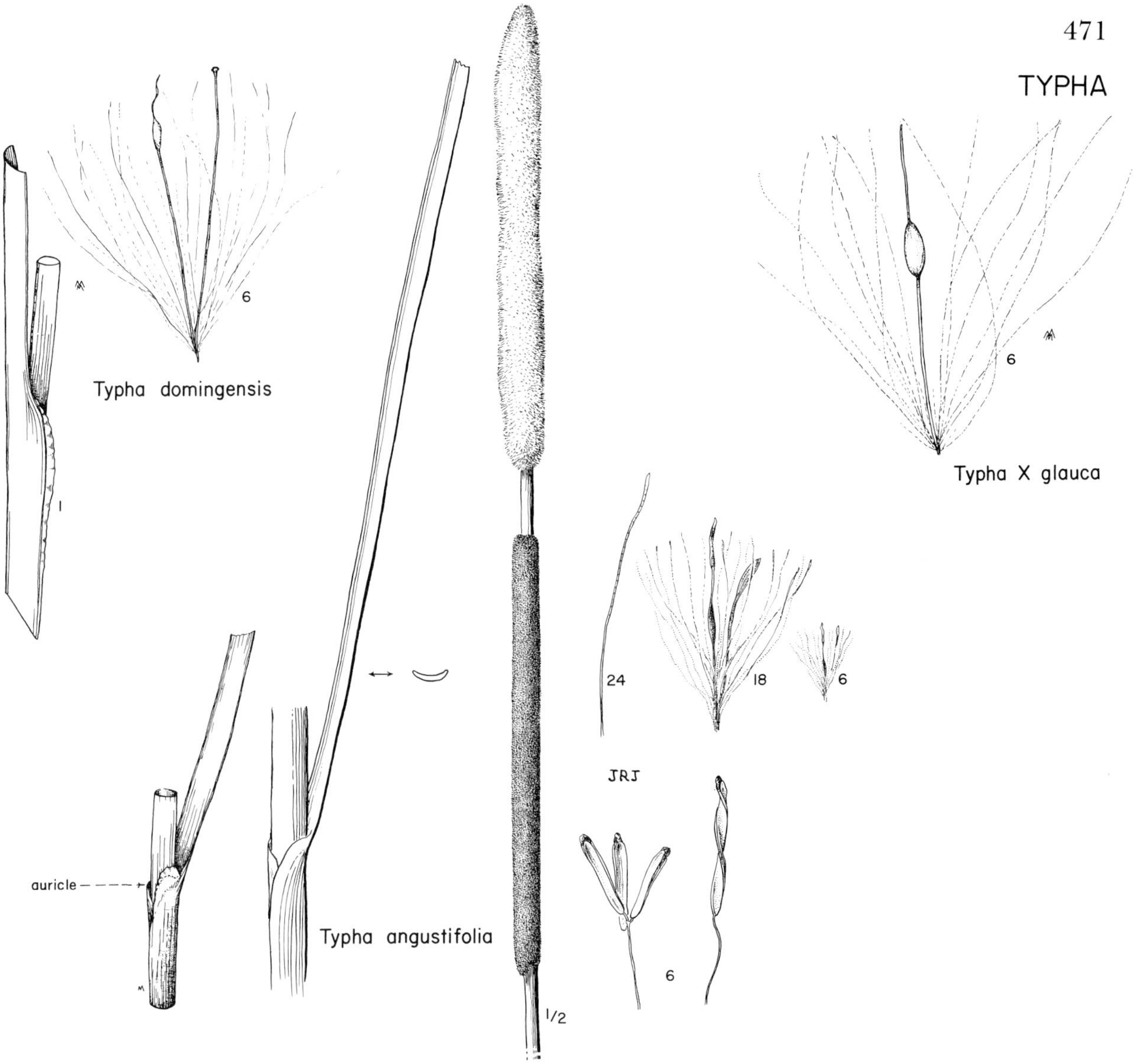

Four orders, only one of which (Arales) occurs in our area. The Arecales (Arecaceae), or the palms, are the largest order with some 2600 to 3500 species, and occur mainly in the warmer regions of the world; Cyclanthales (Cyclanthaceae) are a small order of about 180 species found in the Amer. tropics; and Pandanales (Pandanaceae), with nearly 900 species, are found mainly in the Old World tropics and especially on the Pacific Islands.

Various species of palms are cultivated in cities south of us, but they are rarely found in our area. Some potted plants are seen around universities, hotel lobbies and similar abodes, but these are rarely planted out-of-doors except in southernmost Utah. The California palm, *Washingtonia filifera* H. Wendl., may be occasionally planted in the St. George region.

Order **ARALES**

Flowers hypogynous or naked, dioecious or monoecious, perfect or imperfect, unisexual in some, the perianth, when present, in (1) 2 whorls of 2–3 (9) parts, not obviously differentiated into sepals and petals, borne on a spadix (Araceae) or in a small pouch or sheath (Lemnaceae); stamens 1–8; carpels 1–3 (9), connate; endosperm cellular; leaves, when present, often divided into a blade and petiole.

A large and complex order composed of two families, of which only the Lemnaceae occurs in our area. The remaining family, the Araceae, with some 1800–2000 species, is found mainly in the trop. and subtrop. regions of the world.

Family **LEMNACEAE,** the Duckweed Family

Minute to small free-floating to submersed, monoecious, fleshy perennials consisting of leafless, flat, disc-like or spheroid stems, rootless or with 1–several simple roots arising from the lower surface, often colonial and reproducing vegetatively; flowers (rarely seen) imperfect, usually (1) 2 staminate and 1 pistillate borne together in a small pouch or sheath, naked, the staminate one a single stamen with a 1- or 2-celled anther, the pistillate one of a single carpel with a 1-celled ovary and 1–several ovules arising from the base of the ovary; fruits utricular with 1–7 seeds, the seeds with little or no endosperm.

Six genera and 28–30 species in temp. and trop. waters.

References:
Daubs, E. H. A monograph of Lemnaceae. Illinois Biol. Monogr. 34: 1–118. 1965.
Den Hartog, C., and F. van der Plas. A synopsis of the Lemnaceae. Blumea 28: 355–368. 1970.
McClure, J. W., and R. E. Alston. A chemotaxonomic study of Lemnaceae. Amer. J. Bot. 53: 849–860. 1966.

1 Plants with one or more roots; thalli with 1–12 (15) nerves.
  2 Roots (4) 5–12 per frond; plants with 4–12 (15) nerves, 3–8 mm long, the undersurface usually reddish . . . . 1. *Spirodela*
  2 Roots solitary per frond; plants with 1–5 nerves, 1–6 (10) mm long, the undersurface greenish or purplish to red . . . . 2. *Lemna*
1 Plants rootless; thalli unnerved, 0.7–1.2 mm long, ellipsoidal . . . . 3. *Wolffia*

1. SPIRODELA Schleiden

Fronds floating on the surface of the water, 3–5 or more fronds connected by elongated stipes, reniform to obovate or orbicular in outline, each frond with 2–18 fascicled roots, the fronds usually reddish on the undersurface; flowers unisexual, usually 3 in each pouch, 2 (3) staminate ones and a solitary pistillate flower borne in 1 or 2 lateral pouches on either side of the basal end of the frond, enclosed in a spathe; anthers 2-celled; ovary with (1) 2 (4) ovules; fruits slightly winged, with longitudinally ribbed and transversely striated or infrequently smooth seeds.

A cosmopolitan genus of 4 or 5 species found in temp. and trop. waters. (From the Greek *speira,* cord, and *delos,* evident, referring to the fascicles of roots.)

1. Spirodela polyrhiza (L.) Schleiden

Lemna polyrhiza L. Sp. Pl. 970. 1753. *Lenticula polyrhiza* Lam. Fl. Franç. 2: 189. 1778. *Spirodela polyrhiza* Schleiden, Linnaea 13: 392. 1839. *Telmatophace polyrhiza* Godron, Fl. Lorraine 3: 18. 1844. (Europe.)

Fronds floating on the surface of the water, 2–5 remaining connected by elongated peltately attached stipes, obovate to rounded, (3) 4–8 mm long, 2–5.8 mm wide, dark green above, reddish below, with 5–11 (15) conspicuous nerves, each with a single vascular strand; roots (4) 5–12, fascicled; flowers borne in sac-like spathes on the sides of the frond, open only at the top, with 2 staminate flowers consisting of a solitary anther and a pistillate flower of a single ovary with 1–2 ovules; fruits slightly winged, the seeds compressed, smooth or slightly reticulated; reproducing chiefly by asexual budding, these overwintering buds (turions) dark brown and reniform, arising from the reproductive pouches and growing nearly to maturity before separating from the parent plant; $2n = 40$.

Quiet water of ponds, lakes, marshes and slow-moving streams; widely distributed throughout the world, from Eurasia and Africa to N. Amer. and Australia; known in our area only from scattered locations in Utah, but expected elsewhere.

2. LEMNA L. Duckweed

Fronds free floating on or just beneath the surface of the water, solitary or remaining attached, obovate to orbicular or much elongated in outline, each frond with a single peltately attached root, the fronds usually greenish or purplish to reddish on the undersurface; flowers unisexual, 3 in a pouch, 2 staminate and 1 pistillate, borne in an enclosed membranous sheath in a marginal cleft of the frond, opening along the side or nearly closed; anthers 2-celled; ovary 1-celled and with 1-several ovules; fruits utricles, the 1–6 seeds usually ribbed with minute transverse striations.

About 9 species, cosmopolitan in temp. and trop. waters. (A Greek name for a water plant, used by Theophrastus and possibly derived from *limne,* lake or swampy area, alluding to the general habitat of the species.)

Species of *Lemna* are among the smallest flowering plants in the world, and for that reason, they are extremely difficult to characterize. The size and shape of the frond changes as the plants mature and thus standard measurements are of limited value. Also, the effects of water temperature, salt concentration and other ecological factors are generally unknown, and what role they play in changing certain diagnostic features of the various species remains to be determined. The following key should separate most colonies, but as two or more entities often occur together, care must be taken when sampling a population and in the use of the key. A promising method of identification may be that outlined by McClure and Alston (1966) using chemotaxonomic procedures. Their method has proved to be highly successful in determining even the most closely related species, and may be used as a final means of identification.

Reference: de Lange, L., and A. H. Pieterse. A comparative study of the morphology of *Lemna gibba* L. and *Lemna minor* L. Acta Bot. Neerl. 22: 510–517. 1973.

1 Fronds frequently remaining attached in long-chain colonies by elongated stipes, the individual frond elliptic to lanceolate, often serrulate toward the apex, 6–10 (12) mm long, usually submerged; common . . . . 1. *L. trisulca*
1 Fronds soon separating, 2–5 (rarely more) temporarily joined by a very short stipe or completely

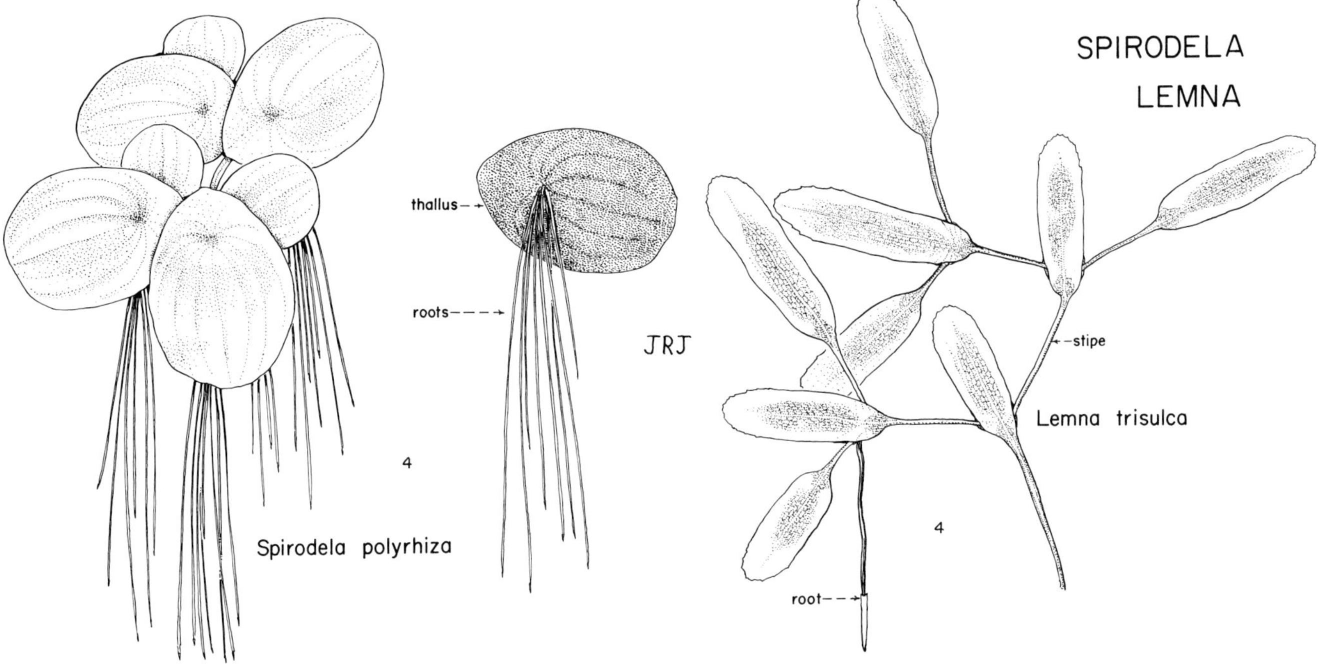

sessile, the individual frond oblong to obovate or orbicular, entire, less than 6 mm long, floating on the surface.

2 Dorsal surface with prominent protuberances, more or less evidently 3-nerved; fronds obovate to orbicular, (1) 1.5–5 mm wide.

3 Dorsal surface mottled yellowish-green, the frond (2.5) 3.5–6 mm long, 2.5–5 mm wide, obovate to orbicular, with an asymmetrical apex; air chambers within the frond conspicuous from below, often inflated so as to form a gibbous frond, the 3 nerves distinct; fruits winged, at least 2-seeded; common . . . . 2. *L. gibba*

3 Dorsal surface uniformly green; air chambers not prominent or inflated, or if so, then fronds 1.5–3.5 mm long and 1–2.5 mm wide, the 3 nerves obscure or indistinct; fruits not winged, usually 1-seeded.

4 Fronds 1.5–3.5 mm long, 1–2.5 mm wide, solitary or in groups of 2–3; air chambers usually prominent and slightly inflated; rare, c. Utah . . . . 3. *L. obscura*

4 Fronds 3–6 mm long, 1.5–4 mm wide, in groups of 2–5; air chambers prominent but not inflated; common . . . . 4. *L. minor*

2 Dorsal surface without prominent protuberances, flat, smooth, 1-nerved or nerveless; fronds mostly elliptic, 0.5–1.5 mm wide.

5 Fronds elliptic to obovate, 2.5–5 mm long, in colonies of (2) 8–10, with an asymmetrical base . . . . 5. *L. valdiviana*

5 Fronds ovoid to elliptic, 1–2.5 mm long, 0.7–1.5 mm wide, solitary or in 2s, asymmetrical at both ends . . . . 6. *L. minuta*

1. Lemna trisulca L.

Lemna trisulca L. Sp. Pl. 970. 1753. *Lenticula trisulca* Scop. Fl. Carniolica, ed. 2. 2: 213. 1772. *Staurogeton trisulcus* Schur, Verh. Mitth. Siebenbürg. Vereins Naturwiss. Hermannstadt 4: 70. 1853. *Hydrophace trisulca* Bubani, Fl. Pyrén. 4: 23. 1901. (Europe.)

Fronds remaining attached in large numbers, submersed except when flowering, the individual frond elongate-elliptic to lanceolate-elliptic, 6–10 (12) mm long, 2.5–5 mm wide, with the elongated stipe up to 10 mm long, symmetrical, flat on both surfaces, the air chambers obscure, faintly 3-nerved, the central one more prominent than the two lateral ones, the margin finely serrulate especially toward the apex at maturity; roots solitary or occasionally lacking on some fronds; flowers 2–3; fruits 1-seeded; $2n = 44$.

In slow-moving streams, ponds, ditches, springs or lakes; common throughout much of the temp. and subtrop. world, but apparently lacking from S. Amer.; found in our area from se. Oregon and e. Calif. across n. Nev. to Utah. June–Oct.

This common species is the most distinctive in the genus and may be quickly recognized by the long stipes and the numerous fronds in each colony. Reproduction is typically by asexual means, with the new fronds developing on the opposite sides of the parent plant and growing at nearly right angles to the base of the plant.

2. Lemna gibba L.

Lemna gibba L. Sp. Pl. 970. 1753. *Lenticula gibba* Moench, Meth. Pl. Horti Bot. & Agri Marburg. 319. 1794. *Telmatophace gibba* Schleiden, Linnaea 13: 391. 1839. (Europe.)

Fronds solitary or cohering in colonies of 2–4, floating, the individual frond obovate to orbicular, (2.5) 3.5–6 mm long, 2.5–5 mm wide, the apex asymmetrical, the base symmetrical, the upper surface flat or slightly convex and mottled yellowish-green, the lower surface flat to convex when immature, becoming strongly inflated and then globose at maturity, often reddish-purple, the air chambers large, mostly 3-nerved; fruits winged, with 1–3 (6) seeds; $2n = 64$.

In slow-moving streams, ponds and lakes; rather common throughout the temp. and trop. regions of the world; known in our area from se. Idaho and Utah, but to be expected elsewhere. June–Sept.

3. Lemna obscura (Austin) Daubs

Lemna minor var. *obscura* Austin in A. Gray, Manual Bot. ed. 5. 479. 1867. *L. obscura* Daubs, Illinois Biol. Monogr. 34: 20. 1965. (*Austin 8050,* New Durham Meadows, N.Y.)

Fronds solitary or cohering in colonies of 2–3, floating, the individual frond obovate to elliptic-orbicular, 1.5–3 (3.5) mm long, 1–2 (2.5) mm wide, slightly asymmetrical, the upper surface slightly convex and dark green, the lower surface convex to slightly inflated and reddish, the air chambers large, obscurely and indistinctly 3-nerved; fruits not winged, with a single seed.

In slow-moving streams and ponds, but possibly in lakes in our area; c. Calif., c. Utah, Ariz., N.M., Texas and the e. U.S., s. to Mex.; known in our area only from the Sevier R. drainage basin of c. Utah where it is fairly common. June–Sept.

This recently proposed species is rare in the western United States, but is to be expected throughout the southern half of the Intermountain Region. It is closely related to *Lemna minor* and *L. gibba* on the one hand, and *L. trisulca* on the other. According to the work by McClure and Alston (1966), these species may be easily recognized on chemotaxonomic bases. Den Hartog and van der Plas (1970) caution that *L. obscura* may only be a growth form of *L. gibba,* however.

4. Lemna minor L.

Lemna minor L. Sp. Pl. 970. 1753. *Lenticula minor* Scop. Fl. Carniolica, ed. 2. 2: 213. 1772. *Hydrophace minor* Bubani, Fl. Pyrén. 4: 23. 1901. (Europe.)

Fronds commonly cohering in colonies of 2–5, floating, the individual frond obovate to suborbicular, 3–6 mm long, 1.5–4 mm wide, essentially symmetrical, flat to slightly convex on both surfaces, green above, tinged with red below, not inflated and with obscure air chambers, obscurely 3-nerved; fruits not winged, with a single seed; $2n \doteq 40$.

In almost any kind of water habitat throughout our area and the world. June–Oct.

This is our most common species, and one of the most variable members of the genus. Careful observations are necessary to distinguish *L. minor* from immature, non-inflated plants of *L. gibba,* and the distinction between *L. minor* and *L. obscura* is obscure at best. All too often *L. minor* occurs in mixed populations with other species and genera of duckweeds, and care must be taken in making measurements not to include other species. For the most part, *L. gibba* can be distinguished by its mottled upper surface and three nerves, while *L. obscura* (at least in Utah) is a much darker green than *L. minor.* Two other species of *Lemna* may eventually be found in our area. These are *L. perpusilla* Torr. and *L. trinervis* (Austin) Small. Both have root sheaths with definite wings or appendages and are a light green in color; *L. trinervis* may be expected in northern Arizona and southern Utah.

Finally, recent studies by de Lange and Pieterse (1973) have shown that plants previously identified as *L. gibba* could be altered in culture to become *L. minor,* and concluded that the distinction between the two species is not always possible. I am reluctant to make the necessary taxonomic adjustments at this time, for this conclusion is by no means a new one. There has been considerable debate in the literature as to whether or not *L. gibba* represents a distinct genetically stable race, or is merely a minor environmental modification of *L. minor.* While currently most authors seem to agree that *L. gibba* does represent some element that is distinct from *L. minor,* the debate now is whether *L. gibba* is a good species or merely a variant of *L. minor.* If one were to reduce *L. gibba* to *L. minor,* then certainly *L. obscura* would also be reduced, and on a world-wide basis, other species would likely fall too. This would not solve the difficulties associated with *Lemna,* as a key to species or varieties would be equally difficult. I have opted not to alter the traditional taxonomy of the genus at the present time, but wish to note that further comparative studies on members of *Lemna* may substantially modify the taxonomy of the genus.

5. Lemna valdiviana Phil.

Lemna valdiviana Phil. Linnaea 33: 239. 1864. (*Philippi,* Valdivia, Chile.)

L. torreyi Austin in A. Gray, Manual Bot. ed. 5. 479. 1867. ("Pools, New Jersey (fertile) to Missouri and southward.")

Fronds cohering in colonies of (2) 8–10, floating or submersed, the individual frond oblong to elliptic or obovate, 2.5–5 mm long, 0.5–1.5 mm wide, the apex symmetrical, the base strongly asymmetrical, flat or nearly so on both surfaces, light green, the air chambers obscure, indistinctly 1-nerved; fruits not winged, with a single seed.

In slow-moving streams, ponds and lakes; N. and S. Amer.; e. Calif. s. to Mex. and e. to the e. U.S.; infrequent in our area in e. Calif., Nev., Utah, and n. Ariz. June–Sept.

6. Lemna minuta H.B.K.

Lemna minuta H.B.K. Nova Gen. & Sp. Pl. 1: 372. 1816. *L. minima* Kunth, Enum. Pl. 3: 5. 1841, in synonymy under *L. minor,* either a new name for, or misspelling of, *L. minuta;* not *L. minima* Thuill. ex Beauv. (1816), an actual synonym of *L. minor. Lenticula minima* Humb. & Kunth ex Miq. Fl. Ned. Indië 3: 222. 1856, in synonymy under *L. minor. Lemna minima* Humb. ex Phil. Linnaea 29: 45. 1857, and Linnaea 33: 239. 1864, nomen nudum. *L. minuscula* Herter, Revista Sudamer. Bot. 9: 185. 1954, a new name based on *L. minima* Humb. ex Phil. (1864). (*Bonpland & Humboldt,* "Innatans aquis stagnantibus prope Guaduas Novogranatensium," between Honda and Bogota, Colombia.)

L. valdiviana var. *minima* Hegelm. Lemnaceen 138. 1868. *L. minima* Hegelm. Bot. Jahrb. Syst. 21: 299. 1896. (*Philippi,* Santiago, Chile, is the first of several collections cited.)

Fronds solitary or rarely cohering in colonies of 2s, floating, the individual frond elliptic to ovoid, 1–2.5 mm long, 0.7–1.5 mm wide, asymmetrical, flat to only slightly convex on both surfaces, occasionally with a few papilla on the upper surface, green and often mottled, the air chambers obscure, indistinctly 1-nerved; fruits not winged, with a single seed.

In slow-moving streams, ponds or lakes; N. and S. Amer.; Oregon and Calif. e. to N.M. and Texas; infrequent from se. Oregon and e. Calif. across Nev. to Utah. June–Sept.

The nomenclature associated with this species is somewhat muddled, and is hopefully clarified here. When *Lemna minuta* was described in 1816, the name had already been used by Rafinesque (Med. Repos. II. 5: 353. 1808) as a nomen nudum. Kunth (1841) later altered the name to *L. minima,* either as an error (i.e., misspelling) or by design, when he placed the species in synonymy under *L. minor.* Philippi, in 1857 and 1864, used *L. minima* to contrast two new species of *Lemna,* citing "*L. minima* Humb." It was reasonable, at the time, for taxonomists to use the latest edition of a person's work, and thus he used *L. minima* (following Kunth) instead of the

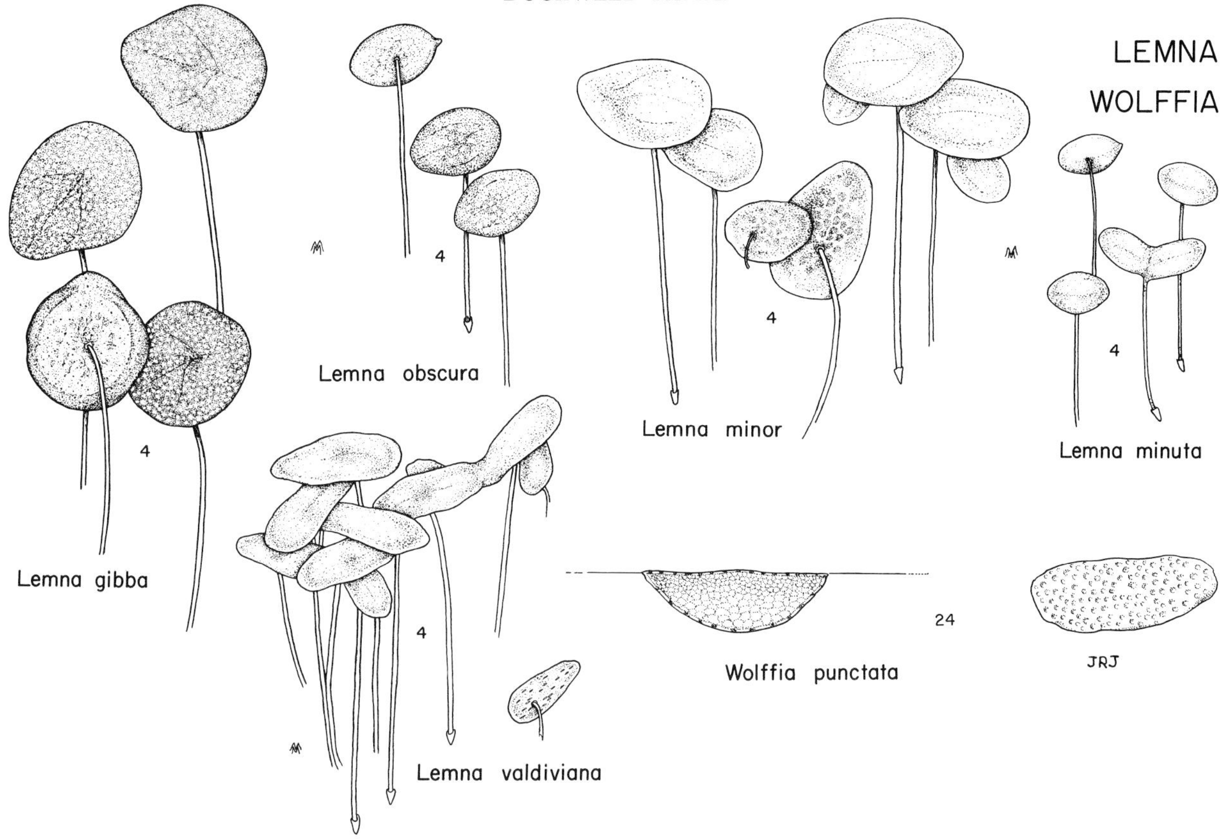

earlier *L. minuta* of 1816. At no time did Philippi describe a species of *Lemna* as *L. minima,* and I can not explain why the 1864 nomen nudum of Philippi's should be considered valid for publication purposes except that Hegelmaier cited this particular place of publication when he supposedly made a new combination of *L. minima* "Phil." under *L. valdiviana.* I have treated Hegelmaier's name as a new variety, rather than a new combination, in the list of synonyms cited above. Various sources have cited Philippi's specimen of this species from Santiago, Chile, as the type of *L. minima,* sensu Philippi, although no such collection was cited by Philippi in his publications. Herter (1954) proposed *L. minuscula,* as a new name for "*L. minima* Phil.," as he noted the name was already occupied by *L. minima* Chev. ex Schleiden (Linnaea 13: 391. 1839, as a synonym under *L. minor*); this too was an error as *L. minima* Chev. ex Schleiden, like *L. minima* Humb. ex Phil., was a nomen nudum!

In addition to the synonym, *Lemna minima* Thuill. ex Beauv. (J. Phys. Chim. Hist. Nat. Arts 82: 111. 1816) mentioned in the list of synonyms above, there is *L. minima* Steudel (Nomencl. Bot. 2: 469. 1823). Steudel incorrectly listed the name *Lemma minuta* Desr. in Lam. (Encycl. Méth. Bot. 3: 721. 1792) under *Lemna. Lemma,* an older Adanson name for *Marsilea,* was proposed by Desrousseaux as a new generic name for *Marsilea* (Marsileaceae), and thus this name is a synonym of *Marsilea minuta* L.

It seems reasonable to me that Philippi looked upon *Lemna minima,* not as his own species name, but as one proposed by Kunth (admittedly in error) in 1841. The final acceptance of the name and its attribution to Philippi must be laid to Hegelmaier who brought the name into vogue, reluctantly, in 1896, when he finally accepted the entity as a good species instead of merely a variant of *L. valdiviana.* As *L. minuta* has not been used in the botanical literature since 1816, perhaps after nearly 160 years of patience it deserves some award for merely surviving.

3. **WOLFFIA** Horkel ex Schleiden, nom. conserv. Water-meal

Tiny floating rootless plants rarely more than 1 mm wide, the fronds globose to ovoid, solitary, nerveless and lacking all vascular tissue except for a few cells of xylem in the staminal filament, usually greenish or punctated with brown pigment; flowers unisexual, 2 in a single cavity opening on the upper surface, 1 staminate and 1 pistillate, on the median line in most, not enclosed in a spathe; anthers 1-celled; ovary 1-celled with a single orthotropous ovule; fruits spherical utricles, the single globose to slightly compressed seed smooth.

A genus of about 7 species mostly in trop. or subtrop. water. (Named for Johann Friedrich Wolff, 1778–1806, a German physician and botanist who published "Commentatis de *Lemna*" in 1801.)

1. Wolffia punctata Grieseb.

Wolffia punctata Grieseb. Fl. Brit. W. Indian Islands 512. 1864.
Bruniera punctata Nieuwl. Amer. Midl. Naturalist 2: 306. 1912. (*Wullschlagel,* Jamaica.)

Fronds rootless, solitary, floating, ellipsoidal, 0.7–1.2 mm long and 0.4–0.7 mm wide when fresh, shrinking upon drying to 0.5–1 mm long and 0.3–0.5 mm wide, the upper surface flattened and bearing numerous stomata, the lower surface ovoid with large air chambers, profusely punctate on both surfaces with brown epidermal cells; fruits with a single seed; $2n$ = ca. 40.

In slow-moving water often in association with other Lemnaceae species; scattered in N. Amer. as in Calif., n. Oregon, s. Wash., n. Utah, but more common in the midw. and e. U.S., se. Can., n. and c. Mex. and the W.I.; known in our area only from n. Utah, but to be expected elsewhere. June–Sept.

So far as known this is the only species of *Wolffia* in our area. However, *W. columbiana* Karsten may be found. It is encountered just below the water's surface and has a globose frond 0.8–1.4 mm long and 0.4–0.8 mm wide when fresh; the upper surface has 1–6 stomata, but no punctate epidermal cells are present. At present, *W. columbiana* is known from northern Oregon, central California, and in the midwestern and eastern United States southward into Mexico and South America.

Subclass **LILIIDAE**

Monocotyledonous plants with the vessels commonly confined chiefly or wholly to the roots; inflorescence of various sorts, but not a spadix; flowers generally nectar-bearing, commonly with the sepals and petals both petaloid, seldom with the sepals more herbaceous and not petaloid; ovary superior or inferior, syncarpous; seeds either essentially without endosperm (Orchidales), or with an endosperm containing various sorts of food reserves, commonly cellulose or oils or protein, only seldom with much starch; embryo in Orchidales tiny and undifferentiated at maturity of the seed.

The subclass Liliidae consists of two orders (both represented in our area), 17 families (four in our area), and about 20,000 species.

Synoptical Arrangement of the Orders of Liliidae

1 Plants not obviously mycotrophic; seeds of ordinary number and structure, usually with a well differentiated embryo and well developed endosperm; flowers usually with septal nectaries, sometimes with other kinds of nectaries in addition or instead; flowers regular or nearly so (all our species) or less often distinctly irregular; ovary superior or inferior .... Order Liliales, p. 476

1 Plants strongly mycotrophic, sometimes even without chlorophyll; seeds very numerous and tiny, with undifferentiated embryo and very little or no endosperm; nectaries diverse, but generally not septal; flowers strongly irregular (including all our species) or less often regular; ovary inferior .... Order Orchidales, p. 546

Order **LILIALES**

Flowers mostly perfect, rarely imperfect or polygamo-dioecious, actinomorphic or zygomorphic, the perianth of equal or subequal petaloid tepals, or less often strongly differentiated into sepals and petals, mostly (2) 3 (4)-merous, the tepals mostly thin but occasionally thick and fleshy, the sepals often reduced, herbaceous and greenish, the petals mostly large, highly colored; stamens 3–6, sometimes abortive or reduced to staminodes, hypogynous or adnate to the corolla; ovary superior or inferior, mostly 3-carpellate, unilocular to trilocular, usually with septal nectaries, the styles 1–3; pollen essentially binucleate; seeds usually with a well differentiated embryo and well developed endosperm; most species essentially herbaceous perennials, but some shrubby to arborescent, mostly mesophytic to xerophytic, infrequently hydrophytic, from branching rhizomes or more often deep-seated bulbs or corms, with mostly basal or cauline, narrow to broad, alternate to opposite or whorled parallel-veined leaves.

An order, as here defined, of 13 families found throughout the world. We have 3 families in our flora.

Synoptical Key to the Families of Liliales

1 Habit liliaceous, i.e., plants strictly herbaceous, most commonly from a bulb or rhizome, the leaves neither pungent nor especially coarse and firm, and not persisting from one year to the next except in *Xerophyllum* (Liliaceae), which has a basal tuft of persistent, slender, grass-like leaves.

  2 Stamens 6; ovary superior in all our genera .... Liliaceae, p. 476

  2 Stamens 3; ovary inferior .... Iridaceae, p. 538

1 Habit agavaceous, i.e., plants coarse, either shrubby or arborescent, or with a persistent tuft of basal leaves, the leaves in any case perennial, coarse, firm, and usually (including all our spp.) strongly pungent .... Agavaceae, p. 526

Family **LILIACEAE,** the Lily Family

Leafy to scapose perennial herbs, or sometimes woody or climbing, from scaly or tunicated bulbs, solid corms or short to elongate rhizomes which may be somewhat tuberous, the stems simple to branched; leaves basal or cauline or both, mostly alternate, sometimes opposite or whorled, broad or more often narrow and even grasslike, parallel-veined; flowers perfect to polygamous or imperfect, small to large and often rather showy with colored parts, sometimes small and greenish but then usually many and as a group showy, 3-merous or rarely 2- or 4-merous; tepals usually 6 and distinct, mostly of 2 series of similar sets of corolla-like tepals, or less often clearly differentiated into a mostly greenish calyx and as often highly colored corolla, or the entire perianth sometimes gamophyllous and 6-lobed; stamens 6, infrequently 4, or 3, or sometimes 3 fertile and 3 staminodial, hypogynous or adnate to the perianth tube, the filaments slender to dilated or united into a crown or an expanded winglike base, the anthers cordate to oblong or linear to lanceolate, extrorse, versatile or introrse, or erect and laterally dehiscent, usually 2-celled, the pollen sacs sometimes confluent and the anthers opening across the top as well as up the sides; ovary superior to partially or wholly (though rarely) inferior, (1- to 2-) 3-celled with one to numerous ovules in each locule with axile placentation, the 1 or 3 styles terminated by a capitate to lobed stigma; fruits usually a loculicidal to septicidal, membranous to leathery capsule, or a fleshy berry; seeds 1–many, with abundant endosperm.

A rather large family of over 200 genera and some 4200 species widely distributed but most abundant in temp. and subtrop. regions of the world. Many of the species are used for food, fiber and horticultural purposes, and many ornamental plants found in the outdoor gardens of our area belong to this family.

As here defined, the Liliaceae includes the Alliaceae, Amaryllidaceae, Asparagaceae, and Trilliaceae, which are sometimes recognized as distinct. We are recognizing the Agavaceae, which in our area includes *Yucca, Nolina* and *Agave*.

1 Flowers solitary to racemose or paniculate, not in a scapose umbel subtended by spathelike bracts.
  2 Plants from rhizomes or thickened tuberous underground stems, not bulbous.
    3 Leaves all basal or the basal leaves much larger than the reduced cauline ones.
      4 Leaves long-lived and persistent on the plant, linear, (1.5) 3–7 (10) dm long; inflorescences densely flowered and up to 6 dm long; flowers 6–10 mm long, on pedicels 2.5–5 cm long; plants 3–15 (18) dm tall; se. Oregon .... 1. *Xerophyllum*, p. 478
      4 Leaves neither long-lived nor more than 4 dm long; inflorescences less than 3 dm long; plants smaller.
        5 Plants caulescent; inflorescences atop long, slender, leafy stems, rather densely flowered, racemose.
          6 Tepals 2.5–6.5 mm long, not united at the base; styles 3, (0.4) 0.5–2 (2.5) mm long; moist places, Mono Co., Calif. .... 2. *Tofieldia*, p. 479
          6 Tepals 10–20 mm long, united below and forming a tube up to 1 cm long; style 1, 3–4 mm long; sandy desert places, s. Utah and n. Ariz. .... 3. *Eremocrinum*, p. 480
        5 Plants acaulescent; inflorescence a sessile cluster atop a tuft of leaves, open; tepals 5–10 (12) cm long, united at the base, forming a tube 4–10 cm long; montane places ... . 4. *Leucocrinum*, p. 482
    3 Leaves well distributed on the stem, not all basal.
      7 Leaves large and greenish; flowers perfect, white to greenish-white.
        8 Inflorescence a raceme or panicle less than 2 dm long of few to many flowers; plants slender; fruits ovoid to globose berries 0.5–1.2 cm long.
          9 Flowers in terminal panicles or racemes, numerous, 1–7 mm long; seeds 1–2 (3) ... . 5. *Smilacina*, p. 482
          9 Flowers solitary or at most 2 (3), 8–15 mm long; seeds 3–15 (18).
            10 Flowers terminal, 1–2 (3) on stoutish, pendulous, pubescent pedicels; style 9–12 mm long; fruits 7–10 mm long .... 6. *Disporum*, p. 484
            10 Flowers axillary on slender, sharply geniculate and twisted, glabrous pedicels; style 4–5 mm long; fruits 10–12 mm long .... 7. *Streptopus*, p. 485
        8 Inflorescence a densely flowered panicle 2–8 dm long; plants stout; fruits narrowly ovoid capsules 2–3.5 cm long .... 9. *Veratrum*, p. 490
      7 Leaves scalelike and non-green, bearing in their axils several short, filiform, greenish branches that are often confused as fernlike leaves; plants functionally dioecious; flowers borne singly or in pairs along the stem; stems slender and wiry; fruits globose berries .... 10. *Asparagus*, p. 490
  2 Plants from tunicated or scaly bulbs, or corms, not from rhizomes.
    11 Flowers variously colored, not blue and globular-urceolate.
      12 Stems leafy, with 2–several leaves; flowers few to many in densely flowered racemes or panicles or few in open racemes or solitary and nodding, variously colored.
        13 Flowers in densely flowered racemes or panicles, 3–11 mm long, the tepals greenish to yellowish-white or white, distinct or adnate to the slightly inferior ovary, not united into a tube .... 8. *Zigadenus*, p. 486
        13 Flowers few in open racemes or solitary and nodding, 9–40 mm long or longer.
          14 Flowers 2–4 (5), nodding, pale yellow to golden, the tepals spreading and becoming reflexed, 2–4 cm long; leaves 2, seemingly opposite and nearly but not quite basal .... 11. *Erythronium*, p. 491

14 Flowers erect, or if nodding, then the plants with several leaves along the stem and the tepals not reflexed.

15 Tepals in 2 series of essentially similar segments, all glandless or with small, indistinct glands.

16 Flowers yellow to orange or red to purplish, (12) 15–40 mm long, somewhat nodding; leaves strictly cauline.

17 Anthers basifixed or subversatile; flowers campanulate, the tepals (1) 1.2–2 (2.5) cm long; common in our area .... 12. *Fritillaria,* p. 492

17 Anthers versatile; flowers broadly campanulate to more nearly funnelform, the tepals 3.5–4 cm long; Mono Co., Calif. .... 13. *Lilium,* p. 494

16 Flowers white with greenish or purplish veins, 9–12 (14) mm long, erect; leaves both basal and cauline; montane and rare, Uinta and Ruby mts. .... 14. *Lloydia,* p. 494

15 Tepals in 2 series of very dissimilar segments, either the sepals or petals or both with large conspicuous basal glands; leaves both basal and cauline .... 15. *Calochortus,* p. 496

12 Stems scapose; flowers numerous in dense to lax bracteate racemes and blue to violet, or few and the tepals white with broad longitudinal stripes.

18 Flowers blue to violet, numerous in congested racemes; native ....16. *Camassia,* p. 504

18 Flowers white, the tepals with longitudinal green stripes, few in an open, zig-zag raceme; introduced .... 17. *Ornithogalum,* p. 506

11 Flowers blue and globular-urceolate, 4–5 mm long; introduced ....18. *Muscari,* p. 506

1 Flowers several to many in a terminal umbel or head on a leafless scape, the inflorescence subtended by spathelike bracts.

19 Tepals distinct or nearly so, less than 1.5 cm long; anthers versatile.

20 Pedicels not subtended by bractlets within the main spathelike bracts; plants with onion-like odor and taste; widespread and common in our area .... 19. *Allium,* p. 508

20 Pedicels subtended by bractlets above the main spathelike bracts; plants lacking the onion-like odor and taste; rare, w. Nev. and e. Calif. .... 20. *Muilla,* p. 522

19 Tepals united into a definite basal tube more than 1.5 cm long (1 cm long in *Dichelostemma*); anthers basifixed or subversatile.

21 Filaments not united; stamens with anthers all alike on slender filaments; styles 2–3 (4) mm long; capsules short-stipitate; ne. Calif. and se. Oregon e. to w. Wyo. and n. Utah .... 21. *Triteleia,* p. 522

21 Filaments united into a tubular corona, or only partially so to form a tube with erect lobes between the anthers; styles 4–6 mm long; capsules sessile; s. part of the Intermountain Region.

22 Filaments united into a tubular corona with the 6 anthers in 2 series and dissimilar; plants (1) 3–6 (9) dm tall; leaves 2–5; capsules 4–6 mm long, the seeds 2.5–4 mm long; c. Nye Co., Nev. e. to sw. Utah .... 22. *Dichelostemma,* p. 524

22 Filaments partially united into a tube with the 6 anthers in a single series and all alike; plants 1–3 dm tall; leaves several, more than 5; capsules 10–15 mm long, the seeds 7–9 mm long; s. Nev. and s. Utah, s. to n. Ariz., hence n. along the Colorado River drainage system to Uintah Co., Utah .... 23. *Androstephium,* p. 526

1. XEROPHYLLUM Michx. Beargrass, Indianbasket Grass

Tall perennial herbs from a thick, tuberous rhizome, any one shoot not flowering for several years but eventually producing a simple, erect, conspicuously bracteate flowering stalk; leaves in dense clumps, linear and elongate, persistent, grasslike with needle-shaped apices; flowers small, whitish to cream-colored and long-pedicellate in a dense terminal raceme; tepals 6, alike, distinct, widely spreading, glandless and clawless, persistent but withering during post-anthesis, oblong to ovate, 5- to 7-nerved; stamens 6, hypogynous, the filaments subulate and as long as the tepals, the anthers basifixed, retrorse; ovary superior, 3-celled with few ovules, the 3 distinct elongate and filiform persistent styles stigmatic down the inner surface; fruits 3-lobed, loculicidal, globular to ovoid capsules; seeds oblong, 3-angled, 2–3 per locule.

A genus of 2 species restricted to N. Amer., the other species, *X. asphodeloides* (L.) Nutt., occurs to the e. of us. (From the Greek *xeros,* dry, and *phyllon,* leaf, alluding to the dry, hardened leaves.)

1. Xerophyllum tenax (Pursh) Nutt.
Helonias tenax Pursh, Fl. Amer. Sept. 243. 1814. *Xerophyllum tenax* Nutt. Gen. N. Amer. Pls. 1: 235. 1818. (*Lewis,* "On highlands near the Rocky-mountains," from just e. of Weippe Prairie, Clearwater Co., Idaho, 15 June 1806.)
X. douglasii S. Wats. Proc. Amer. Acad. Arts 14: 284. 1879.

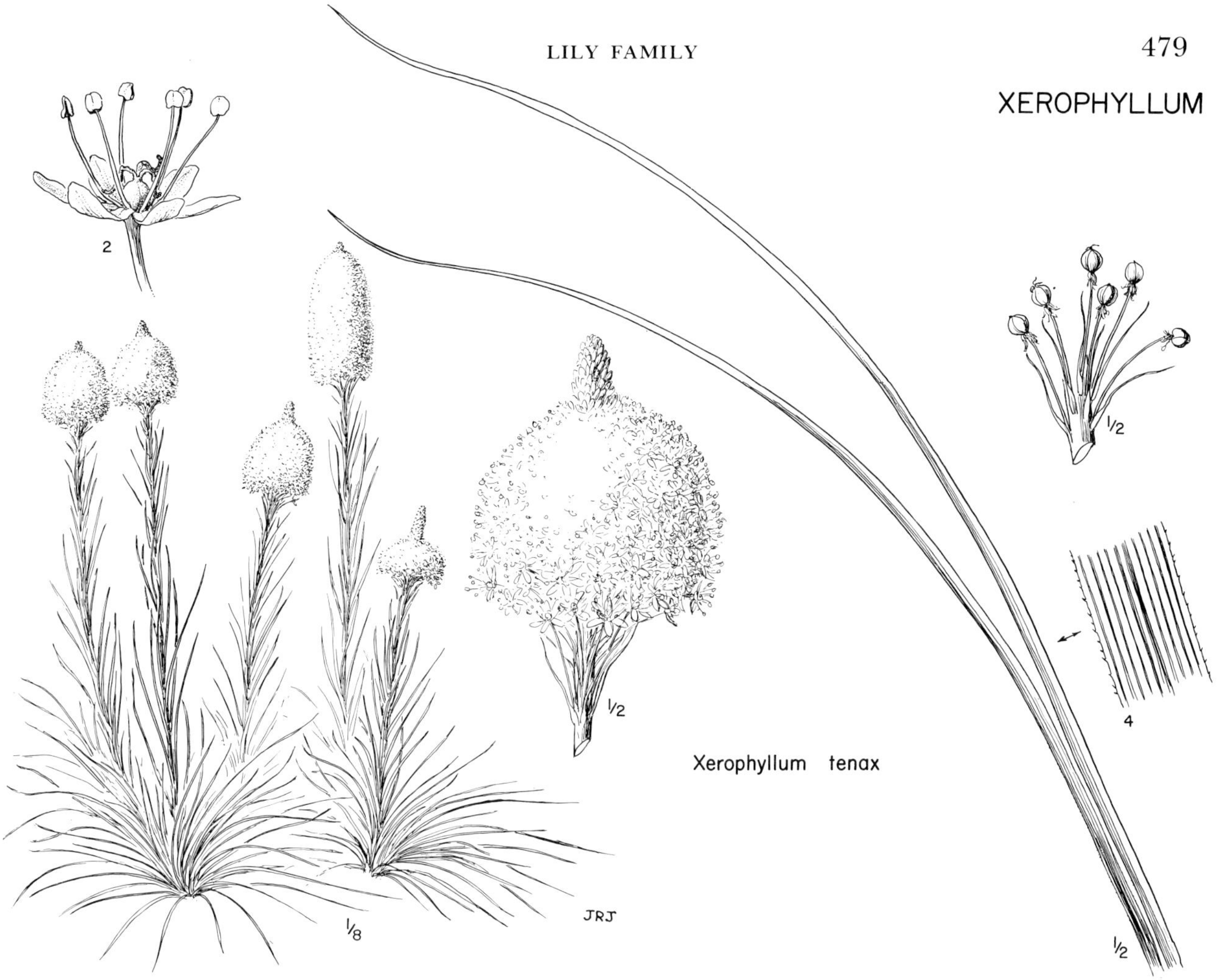

(*Douglas*, "probably in Oregon," *Hall 533*, near the Columbia River, and *Hayden*, mts. of Mont. on the headwaters of the Yellowstone R. are listed.)

Tall erect perennial herbs 3–15 (18) dm high from short, tuberous rhizomes 1–2 cm thick; leaves basal and cauline, the basal ones in large dense clumps, the blade (1.5) 3–7 (10) dm long, (1.5) 2–4 (6) mm wide, tough and wiry, strongly scabrous, roughened along the margin, the cauline leaves gradually passing into long, linear bracts; flowering stems stout, copiously leafy, up to 1.5 m long; inflorescences densely flowered, at first corymb-like, becoming elongate and up to 6 dm long; flowers semirotate, whitish, 6–10 mm long, on spreading to erect pedicels 2.5–5 cm long, the tepals mostly linear-oblong to oblong, (5) 6–8 (10) mm long; stamens equalling to somewhat exceeding the tepals, the filament subulate, the anther cream, ovate, ca. 0.5 mm long; ovary with a short style 1 mm long terminated by three stigmatic lobes 1.5–3 (4) mm long; capsules mostly ovoid, strongly 3-lobed, 5–7 mm long; seeds linear-oblong, 3–4 mm long; $n = 15$.

Dry open slopes and ridges mostly below 2150 m elev., from B.C. s. to c. Calif. and across Idaho into Mont., hence s. in the Rocky Mts. to nw. Wyo.; just entering our area in Crook and Deschutes cos., Oregon. May–Aug.

2. TOFIELDIA Huds.

Low perennial herbs with a solitary, erect and scapose flowering stem from a short, slender rhizome; leaves tufted, more or less 2-ranked, linear, grasslike, with 1–3 sheathing cauline leaves on the lower portion; flowers small, white to greenish-white, on short pedicels in a short, dense terminal raceme; tepals 6, alike, distinct, erect or somewhat spreading, glandless and clawless, persistent, oblong to obovate, 3-nerved; stamens 6, hypogynous, the filaments filiform and somewhat flattened below, about as long as the tepals, the anthers basifixed, introrse; ovary superior, 3-celled with numerous ovules, the 3 distinct short subulate styles stigmatic only at the apex; fruits 3-lobed, membranous, septicidal capsules; seeds ellipsoid, bearing a loose spongy testa with the free end tailed in ours.

A genus of about 15 species of the N. Hemisphere and in the Andes of S. Amer. (Named for Thomas Tofield, 1730–1779, an English botanist and correspondent of Hudson.)

Reference: Hitchcock, C. L. The *Tofieldia glutinosa* complex of western North America. Amer. Midl. Naturalist 31: 487–498. 1944.

1. Tofieldia glutinosa (Michx.) Pers.

Narthecium glutinosum Michx. Fl. Boreali-Amer. 1: 210. 1803. *Tofieldia glutinosa* Pers. Syn. Pl. 1: 399. 1805. *Triantha glutinosa* Baker, J. Linn. Soc., Bot. 17: 490. 1879. *Asphodeliris glutinosa* Kuntze, Revisio Gen. Pl. 2: 706. 1891. *Trianthella glutinosa* H. D. House, Amer. Midl. Naturalist 7: 127. 1921. *Tofieldia racemosa* var. *glutinosa* Ahles, J. Elisha Mitchell Sci. Soc. 80: 172. 1964. (*Michaux,* "Hab. a Quebec ad usque lacus Mistassins," in 1792.)

Tofieldia occidentalis S. Wats. Proc. Amer. Acad. Arts 14: 283. 1879. *Asphodeliris occidentalis* Kuntze, Revisio Gen. Pl. 2: 706. 1891. *Triantha occidentalis* Gates, J. Linn. Soc., Bot. 44: 137. 1918. *Tofieldia glutinosa* subsp. *occidentalis* C. L. Hitchc. Amer. Midl. Naturalist 31: 497. 1944. *Tofieldia glutinosa* var. *occidentalis* C. L. Hitchc. Univ. Wash. Publ. Biol. 17(1): 804. 1969. (*Kellogg & Harford 1022,* Mendocino Co., Calif.; lectotype selected by Hitchcock.)

Tofieldia intermedia Rydb. Bull. Torrey Bot. Club 27: 528. 1900. *Triantha intermedia* Gates, J. Linn. Soc., Bot. 44: 137. 1918. *Tofieldia occidentalis* var. *intermedia* M. E. Peck, Manual Higher Pls. Oregon 189. 1941. *Tofieldia glutinosa* var. *intermedia* B. Boivin, Naturaliste Canad. 94: 528. 1967. (*Gorman 78,* Sheh-shooh Lakes, Alaska, in 1895.)

Tofieldia glutinosa subsp. *brevistyla* C. L. Hitchc. Amer. Midl. Naturalist 31: 495. 1944. *T. glutinosa* var. *brevistyla* C. L. Hitchc. Univ. Wash. Publ. Biol. 17(1): 806. 1969. (*Hitchcock & Martin 7638,* Blackwater Lake, ca. 32 mi. n. of Golden, B.C., 11 July 1941.) = var. *intermedia.*

Tofieldia glutinosa subsp. *montana* C. L. Hitchc. Amer. Midl. Naturalist 31: 496. 1944. *T. glutinosa* var. *montana* R. J. Davis, Contr. Toward Fl. Idaho 13: 17. 1946. (*Barkley & Marsh 1731,* Glacier Natl. Park, Mont., 18 Sept. 1937.)

Slender, grasslike perennial herbs 1–5 (8) dm tall arising from slender, spreading rhizomes mostly less than 0.5 cm thick, with stems glabrous below or nearly so and copiously glandular-hairy above and in the subcapitate to short-racemose inflorescence, this 1–3 cm long in anthesis, often elongating in fruit and becoming 3–7 cm long; leaves mostly in basal tufts or with 1–3 sheathing up the stem a short distance, linear, 5–15 (20) cm long, 3–8 mm wide, mostly erect and somewhat rigid; flowers white to somewhat greenish, on thin, more or less erect 1–5 (10) mm long pedicels mostly arranged in 3s, with 3 separate, or more commonly (as in ours) almost completely connate, membranous involucral bracts at their tips, the tepals more or less spreading, oblong-obovate, (2.5) 3–5.5 (6.5) mm long, the inner tepals somewhat narrower and longer than the outer ones; stamens equaling or slightly exceeding the tepals, the filament filiform, the anther oblong, 0.3–0.5 mm long; ovary with distinct styles (0.4) 0.5–2 (2.5) mm long, these stigmatic only at the tip; capsules mostly ovoid to obovoid, strongly 3-lobed, 4–7 (9) mm long; seeds few to several, 0.6–1 (1.5) mm long, enclosed in a spongy testa, the testa attached to the body of the brown seed except at the end (in var. *glutinosa*) or (as in ours) white, inflated and free from the brown seed itself, the funiculus often as long as the seed.

Meadows and moist places, especially around lakes and along streambanks; Alaska and Can. s. to c. Calif., c. Idaho and w. Wyo., and in the e. U.S. from Me. to N.C.; just entering our area along our boundary in Mono Co., Calif., from 2000–2450 m elev. June–Aug.

This species is composed of five well-defined varieties. We have only one definitely known from our area, but two others closely approach our boundary and are therefore included in the following key.

1 Styles (1.2) 1.5–2 (2.5) mm long; tepals mostly 4–6 mm long; capsules 5–9 mm long; sw. Oregon and coastal nw. Calif., and in the Sierra Nevada from Lassen Co. s. to Inyo Co., entering our area in Mono Co. .... var. *occidentalis* (S. Wats.) C. L. Hitchc.

1 Styles 0.5–1 (1.5) mm long; tepals mostly (3) 3.5–4 mm long; capsules 4–6 mm long.

2 Hairs of the upper stem and inflorescence relatively thick, mostly 2 (3) times as long as thick, tapered and semi-papillose; Alaska s. to Oregon, approaching our area in se. Oregon .... var. *intermedia* (Rydb.) B. Boivin

2 Hairs of the upper stem and inflorescence relatively thin, mostly 3–4 times as long as thick, nearly uniform in diam.; s. Can. in the Rocky Mts. s. to Mont., most of Idaho and w. Wyo., approaching our area in s.-central Idaho and w.-central Wyo. .... var. *montana* (C. L. Hitchc.) R. J. Davis

3. EREMOCRINUM M. E. Jones Sand Lily

Low caulescent perennial herbs with erect, slender scapose stems from a short rhizome with fleshy roots; leaves tufted, narrowly linear, grasslike, each tuft surrounded at the base by fibrous, persistent sheaths; flowers small, rather showy, white to greenish-white, on a short pedicel subtended by a scarious bract, in a rather dense, somewhat elongate raceme; tepals 6, alike, united below into a tubular stipelike base, distinct above, erect or somewhat spreading, glandless and clawless, persistent, oblong, distinctly 3-nerved; stamens 6, hypogynous, the filaments linear, flattened, only about half as long as the tepals, the anthers basifixed, introrse; ovary superior, 3-celled with few ovules, the filiform and elongate style terminated by a discoid stigma; fruits somewhat bluntly 3-lobed, oblong, loculicidal capsules; seeds angled, black, few per locule.

A monotypic genus. (From the Greek *eremos,* desert, and *krinon,* lily.)

1. Eremocrinum albomarginatum (M. E. Jones) M. E. Jones

Hesperanthes albomarginata M. E. Jones, Zoe 2: 251. 1891. *Eremocrinum albomarginatum* M. E. Jones, Zoe 4: 53. 1893. (*M. E. Jones s.n.,* near Green River, Grand or Emery cos., Utah, 9 May 1890.)

Low slender caulescent perennial herbs 1.5–3.5 (4) dm tall with erect, slender, essentially leafless flowering stems (1) 1.5–3 dm long terminated by a short, dense raceme, from spreading rhizomes mostly less than 0.4 cm thick; leaves mostly in sheathing basal tufts, linear, 1–4 dm long, (1) 1.5–3 mm wide, erect or somewhat spreading, the thin, membranous linear sheaths (4) 5–8 (10) cm long, 5–9 mm wide, becoming fibrous; flowers white to greenish-white with 3 greenish to brown stripes on each tepal, 1–2 cm long, on short, erect or nearly peglike pedicels 1 mm long, these usually concealed by thin, membranaceous or sometimes chartaceous, triangular to awl-shaped bracts up to 2 cm long at the base of the inflorescence, but becoming much reduced and

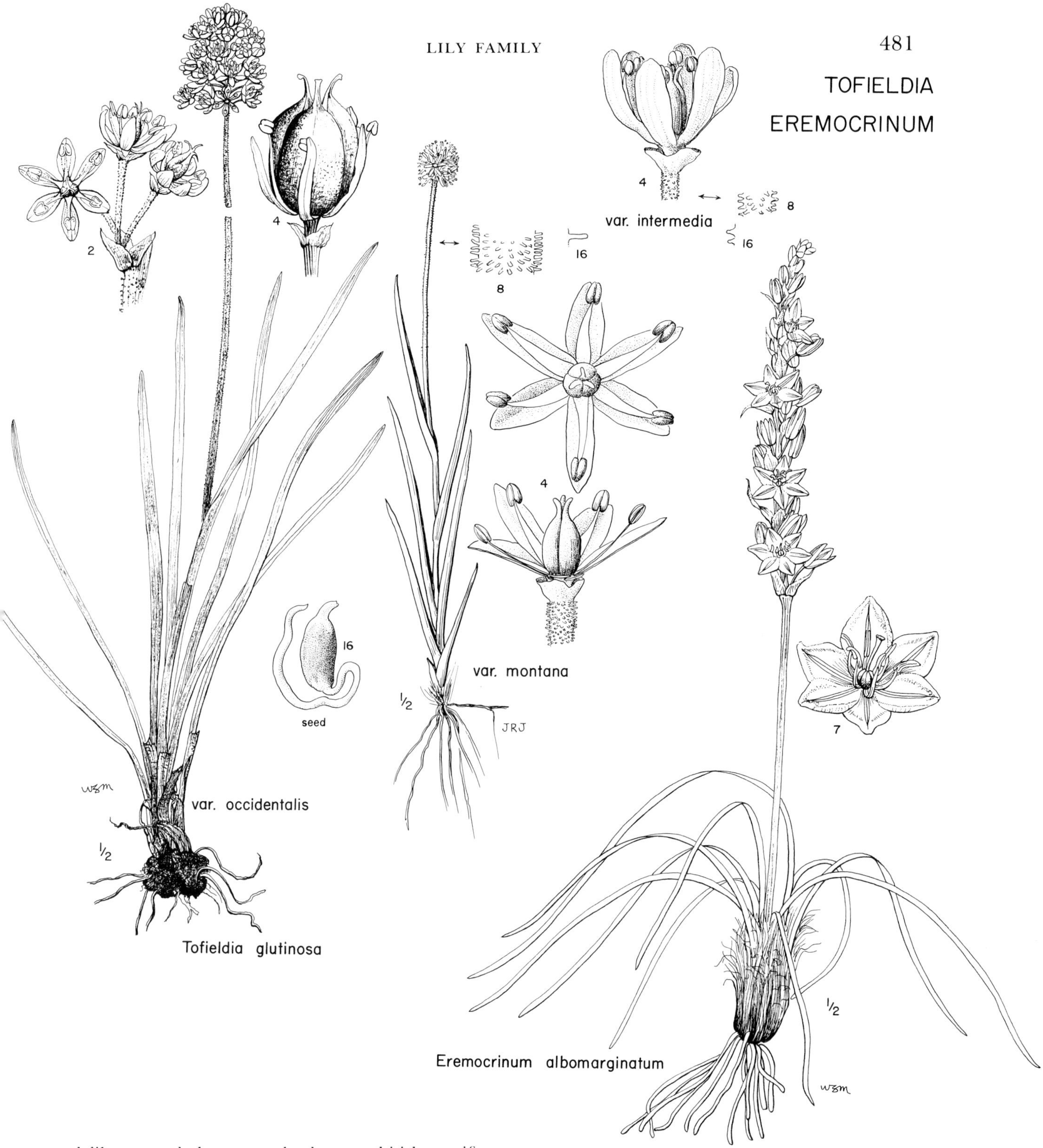

scalelike toward the apex, the bracts whitish, or if chartaceous, then greenish, the tepals distinct above with oblong, more or less spreading segments 6–10 mm long, forming a tube below with a stipelike base 4–8 (10) mm long; stamens shorter than the tepals, the filament linear, the anther lanceolate, 2.5–3 mm long; ovary with a 3–4 mm long style terminated by a discoid stigma; capsules mostly oblong, somewhat bluntly 3-lobed, 4–6 mm long; seeds few, black, angled, 2–2.5 mm long.

Deep sandy places usually in "blow" sand or on sand dunes from 1300 to 1750 m elev., essentially endemic to our area from Emery and Grand cos., Utah, s. into n. Navajo and Apache cos., Ariz. April–June.

4. LEUCOCRINUM Nutt. ex A. Gray Star Lily

Low, acaulescent, tufted perennial herbs from a short, deeply buried rootstalk; leaves tufted, narrowly linear, grasslike, with each tuft surrounded at the base by membranaceous sheathing bracts; flowers small, rather showy, white, few in a central cluster on the ground with underground, jointed pedicels arising directly from the rootstalk; tepals 6, alike, united below the middle into a long, slender tube, distinct and spreading above with narrowly oblong segments, glandless and clawless, persistent, 5- to 7-nerved; stamens 6, inserted near the apex of the floral tube, the filaments filiform, shorter than the tepals but usually well exserted, the anthers subversatile, introrse; ovary superior, 3-celled with several ovules, the filiform style with 3 short stigmatic lobes; fruits subterranean, obovoid, 3-angled loculicidal capsules; seeds few to several, black, angled.

A monotypic genus. (From the Greek *leucos,* white, and *krinon,* lily.)

1. Leucocrinum montanum Nutt. ex A. Gray

Leucocrinum montanum Nutt. ex A. Gray, Ann. Lyceum Nat. Hist. New York 4: 110. 1848. (*Nuttall,* plains of the upper Platte River, apparently somewhere in e. Wyo. or extreme w. Nebr., in 1834.)

L. montanum var. *major* Baker, J. Linn. Soc., Bot. 17: 453. 1879. (*Watson 1177,* near Oreana, Humboldt Valley, Pershing Co., Nev., June 1868.)

L. montanum var. *fibrosum* Kelso, Rhodora 35: 348. 1933. (*Kennedy 1047,* Paradise Valley, Humboldt Co., Nev., 30 April 1905.)

Low, tufted, acaulescent perennial herbs up to 1 dm high, with a central cluster of flowers atop a short, deep-seated vertical rootstalk; leaves in sheathing basal tufts, linear, 1–2 dm long, 2–6 (8) mm wide, mostly spreading, the sheathing bracts scarious, 3–8 cm long, the outermost occasionally fibrous; flowers white, 5–10 (12) cm long, on slender subterranean pedicels 5–30 mm long, the tepals distinct above with linear-lanceolate to narrowly oblong-elliptic, spreading segments 2–2.5 cm long, up to 7 mm wide, forming a long slender tube below, (4) 5–8 (10) cm long; stamens shorter than the tepals, the filament filiform, the anthers linear, 4–6 mm long; ovary with a long, slender style terminated by 3 short lobes; capsules underground, truncate, wrinkled in some, 5–7 (8) mm long; seeds few, 3–4 mm long; n = 11, 13, 14.

Sagebrush valley floors to montane coniferous forests but usually where moist; from c. Oregon s. to c. Calif., e. across Idaho and n. Nev. to Mont. and S.D., hence s. in the Rocky Mts. to s. Utah and N.M.; in scattered places in the northern valleys of our area and in the mts. throughout most of Utah and Nev. April–June.

5. SMILACINA Desf. nom. conserv. False Solomon's Seal

Erect, leafy perennial herbs with unbranched flowering stems arising from creeping rhizomes; leaves alternate, numerous, lanceolate to ovate, many-veined, sessile or short petioled; flowers small, white, arranged in terminal, open to dense racemes or panicles, on jointed pedicels; tepals 6, alike, distinct, withering-persistent; stamens 6, the filaments slender to broadly dilated, shorter to longer than the tepals, inserted at the base of the tepals, the anthers versatile, introrse; ovary superior, 3-celled with 2 ovules per locule, the short, columnar style obscurely 3-lobed; fruits globose, greenish to reddish berries; seeds subglobose, 1–2.

A complex genus of perhaps 25 species of N. Amer. and Asia. (The diminutive of *Smilax.*)

Reference: Galway, D. H. The North American species of *Smilacina.* Amer. Midl. Naturalist 33: 644–666. 1945.

1 Flowers paniculate, numerous, the tepals 1–2 mm long; stamens exceeding the tepals with broadly dilated filaments .... 1. *S. racemosa*

1 Flowers racemose, few, the tepals (3) 5–7 mm long; stamens not exceeding the tepals and with slender filaments .... 2. *S. stellata*

1. Smilacina racemosa (L.) Desf.

Convallaria racemosa L. Sp. Pl. 315. 1753. *Smilacina racemosa* Desf. Ann. Mus. Natl. Hist. Nat. 9: 51. 1807. *Maianthemum racemosum* Link, Enum. Pl. Hort. Regii Bot. Berol. 1: 343. 1821. *Unifolium racemosum* Britton, Trans. New York Acad. Sci. 8: 74. 1889. *Vagnera racemosa* Morong, Bull. Torrey Bot. Club 20: 480. 1893. ("Virginia, Canada.")

S. amplexicaulis Nutt. J. Acad. Nat. Sci. Philadelphia 7: 58. 1834. *S. racemosa* var. *amplexicaulis* S. Wats. in King, Rep. Geol. Explor. 40th Parallel 5: 345. 1871. *Unifolium amplexicaule* Greene, Bull. Torrey Bot. Club 15: 287. 1888. *Vagnera amplexicaulis* Greene, Manual Bot. Region San Francisco Bay 316. 1894. (*Wyeth,* "In the valleys of the Rocky Mountains, about the sources of the Columbia river," apparently from near the Salmon River, e. Idaho, June 1833.)

Vagnera pallescens Greene, Proc. Acad. Nat. Sci. Philadelphia 1895: 551. 1896. (Sierra Nevada of Calif., northward; no specimens cited.)

S. racemosa var. *brachystyla* L. F. Henderson, Bull. Torrey Bot. Club 27: 357. 1900. *Vagnera amplexicaulis* var. *brachystyla* A. A. Heller, Cat. N. Amer. Pl. ed. 2. 4. 1900. (*Henderson 4843,* Idaho.)

S. amplexicaulis var. *glabra* J. F. Macbride, Contr. Gray Herb. 56: 18. 1918. *Vagnera amplexicaulis* subsp. *glabra* Abrams, Ill. Fl. Pacific States 1: 453. 1923. *S. racemosa* var. *glabra* St. John, Res. Stud. State Coll. Wash. 1: 97. 1929. (*Culbertson 4252,* South Fork Kaweah River, Tulare Co., Calif., 22 July 1904.)

S. racemosa var. *cylindrata* Fern. Rhodora 40: 406. 1938. (*Fernald & Long 3859,* Little Neck, Princess Anne Co., Va., 8–9 Aug. 1934.)

Suberect to erect perennial herbs 3–9 (10) dm tall with glabrous or finely pubescent stems arising from well defined rhizomes; leaves alternate, flat, broadly ovate to ovate-oblong or elliptic, 7–20 cm long, (3) 4–8 (10) cm wide, the apex acuminate to acute, the

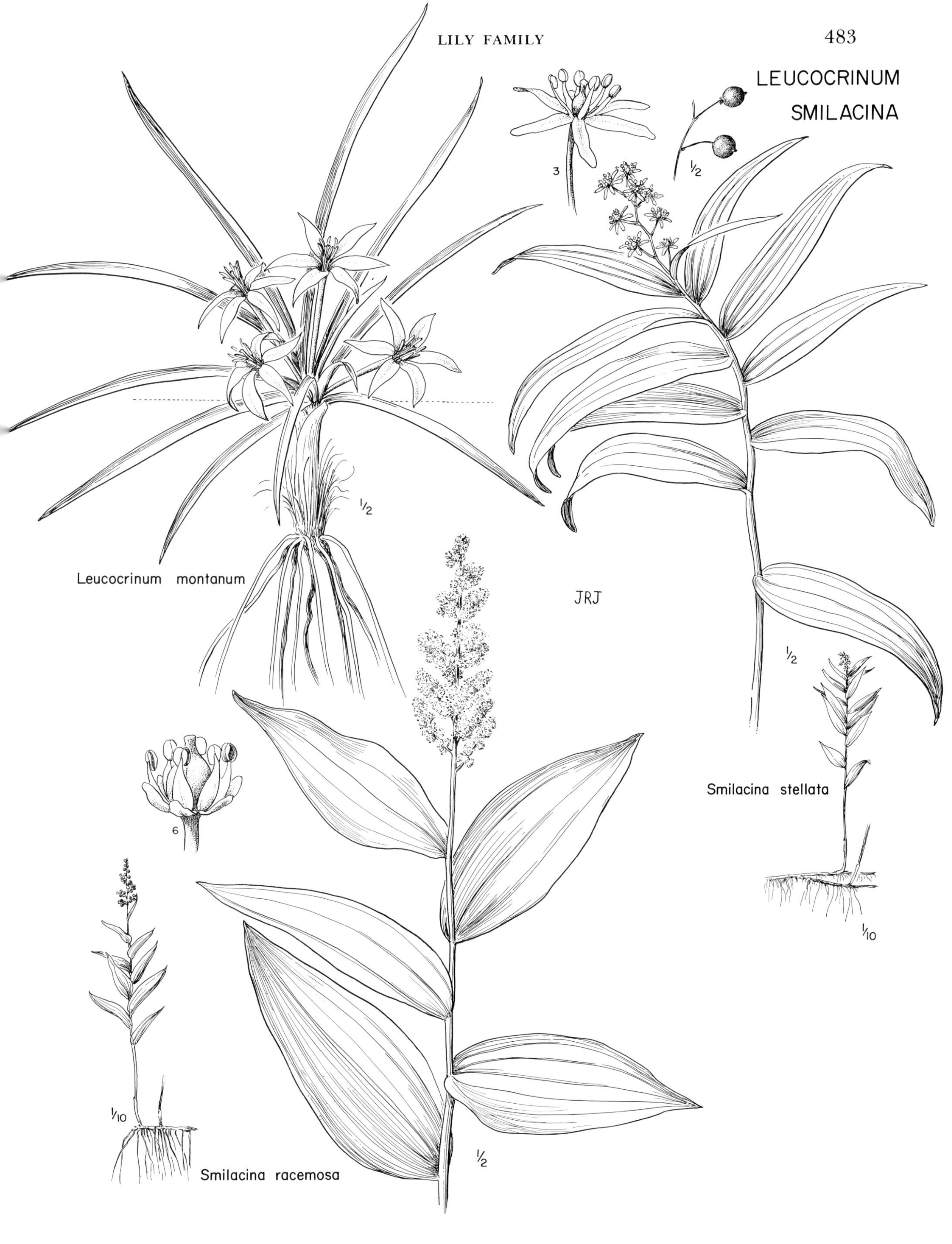
LEUCOCRINUM
SMILACINA
3
1/2
1/2
Leucocrinum montanum
JRJ
1/2
Smilacina stellata
6
1/10
1/10
Smilacina racemosa
1/2

blade short petiolate to sessile with a more or less clasping to strongly dilated clasping base; flowers numerous, on short stout pedicels 0.5–1 (2) mm long, arranged in dense, paniculate clusters, the panicle usually free branching, (3) 5–12 (18) cm long, rarely reduced to only a few subsessile flowers at each node, the tepals narrowly oblong, (1) 1.5–2 mm long; stamens erect, (1.5) 2.5–3 mm long, the filament broadly dilated at the base and often broader than the tepals; style 0.5–1 mm long; berries reddish, 5–7 mm long; seeds 1–2 (3), mostly 3.5–4 mm long; $n = 18$.

In shaded, often moist places throughout most of n. N. Amer.; common throughout most of our area except much more scattered in Nev., mostly below 2800 m elev. June–July.

In the past our material has been distinguished as var. *amplexicaulis* (Nutt.) S. Wats., differing—supposedly—from the more northern and eastern var. *racemosa* in having clasping and sessile, or at most short-petiolate, leaves with acute to short-acuminate apices. These differences are not consistent and it seems unwise to continue to perpetuate these minor extremes as distinct taxonomic entities. Plants which are glabrous occur randomly in the western United States, and these do not appear to represent a distinct element either.

2. Smilacina stellata (L.) Desf.

Convallaria stellata L. Sp. Pl. 316. 1753. *Smilacina stellata* Desf. Ann. Mus. Natl. Hist. Nat. 9: 52. 1807. *Maianthemum stellatum* Link, Enum. Pl. Hort. Regii Bot. Berol. 1: 343. 1821. *Tovaria stellata* Neck. ex Baker, J. Linn. Soc., Bot. 14: 565. 1875. *Unifolium stellatum* Greene, Bull. Torrey Bot. Club 15: 287. 1888. *Vagnera stellata* Morong, Mem. Torrey Bot. Club 5: 114. 1894. *Asteranthemum stellatum* Nieuwl. Amer. Midl. Naturalist 3: 109. 1913. ("Canada.")

Tovaria sessilifolia Baker, J. Linn. Soc., Bot. 14: 566. 1875. *S. sessilifolia* Nutt. ex S. Wats. Proc. Amer. Acad. Arts 14: 245. 1879. *Unifolium sessilifolium* Greene, Bull. Torrey Bot. Club 15: 287. 1888. *Vagnera sessilifolia* Greene, Manual Bot. Region San Francisco Bay 316. 1894. *S. stellata* var. *sessilifolia* L. F. Henderson, Bull. Torrey Bot. Club 27: 358. 1900. *Convallaria sessilifolia* Druce, Bot. Exch. Club & Soc. Brit. Isles 3: 408. 1914. ("America borealis occidentalis a Columbia britannica and Californiam et Mexicum Novum"; no specimens cited.)

Unifolium liliaceum Greene, Pittonia 1: 280. 1889. *Vagnera liliacea* Rydb. Mem. New York Bot. Gard. 1: 101. 1900. *S. liliacea* Wynd, Amer. Midl. Naturalist 17: 902. 1936. (Based on the description of *S. stellata* by S. Wats. in King, Rep. Geol. Explor. 40th Parallel 5: 345. 1871, and hereby lectotypified on *Bailey 1169,* Truckee Pass, Washoe Co., Nev., July 1867, deposited at GH.)

Erect and rather strict to flexuous perennial herbs (2) 3–6 (7) dm tall with glabrous or puberulent stems arising from well defined rhizomes; leaves distichous to spiral, flat to folded along the midvein, (4) 6–15 (17) cm long, 1.5–5 cm wide, lanceolate to oblong-lanceolate or narrowly ovate, the apex acuminate, the blade sessile or somewhat clasping at the base; flowers 5–10 (20), on ascending slender to stoutish pedicels 5–15 (20) mm long, arranged in open, loose racemes mostly 2–6 (10) cm long with the rachis straight or zigzag, the inflorescence infrequently compounded, the tepals lanceolate to narrowly oblong, (3) 5–7 mm long; stamens erect, 2–5 mm long, the filament slender; style 1 mm long; berries greenish-yellow, becoming reddish-purple to black, 6–10 mm long; seeds 2–3, 3–4 mm long; $n = 18$.

In shaded, moist places throughout most of n. N. Amer.; common throughout our area mostly below 2600 m elev. April–June.

6. DISPORUM Salisb. ex D. Don Fairy Bells

Sparsely branched, sparsely to strongly pubescent, leafy perennial herbs arising from spreading rhizomes; leaves alternate, linear to round or cordate, sessile, strongly veined, usually oblique and clasping the stem; flowers small, creamy-white (ours) to greenish or purplish, solitary or 2–several and more or less umbellate at the tips of branches, pendent on short pedicels, often concealed by the leaves; tepals 6, alike, distinct except at the very base where united into a short tube, deciduous, oblong-lanceolate to oblong-ellipsoid; stamens 6, the filaments slender (ours) or dilated below, borne on the base of the tepals, the anthers subversatile, introrse; ovary superior, 3-celled with 2–6 ovules per locule, the long slender style terminated by 3 shallow to deeply cleft stigmatic lobes; fruits yellowish to reddish (ours) or dark blue to black, ovoid to subcylindrical berries; seeds globose, 3–6 (18).

A genus of some 15 species, mainly of Asia, with 5 species in N. Amer. (From the Greek *dis,* double, and *spora,* seeds, alluding to the two-seeded berries of some species.)

Reference: Jones, Q. A cytotaxonomic study of the genus *Disporum* in North America. Contr. Gray Herb. 173: 1–39. 1951.

1. Disporum trachycarpum (S. Wats.) Benth. & Hook.

Uvularia lanuginosa var. *major* Hook. Fl. Boreali-Amer. 2: 174. 1838. *Disporum majus* Britton, Bull. Torrey Bot. Club 15: 188. 1888. (*Richardson,* "Between Norway House and Cumberland House Fort," Can.)

Prosartes trachycarpa S. Wats. in King, Rep. Geol. Explor. 40th Parallel 5: 344. 1871. *D. trachycarpum* Benth. & Hook. Gen. Pl. 3: 832. 1883. *Lethea trachycarpa* Farw. Annual Rep. Michigan Acad. Sci. 2: 20. 1923. (*Watson 1166,* Parley's Peak, Summit Co., Utah, June 1869; lectotype selected by Jones.)

D. trachycarpum var. *subglabrum* Kelso, Rhodora 39: 150. 1937. (*McDougal 64,* Weits Canyon, near Flagstaff, Coconino Co., Ariz., 2 May 1891.)

Fairy bells.

Suberect, thickly crisp-pubescent, sparsely branched perennial herbs 3–6 (8) dm tall from slender to thickish rhizomes; leaves alternate, 4–10 (12) cm long, 2–9 cm wide, ovate to ovate-oblong, often somewhat pubescent below, glabrate to glabrous above, the base semicordate to rounded, the apex acute to short-acuminate, the margin ciliate with spreading hairs; flowers 1–2 (3) on stoutish, pendulent, pubescent pedicels, creamy-white, (8) 10–15 mm long, narrowly campanulate, the tepals narrowly oblanceolate, spreading from near the narrowed base; stamens mostly exserted, the filiform filament as long to longer than the sepals, the anther white, 3–4 (5) mm long; ovary obovate, finely papillate especially at the apex, with 2–6 ovules per locule,

DISPORUM

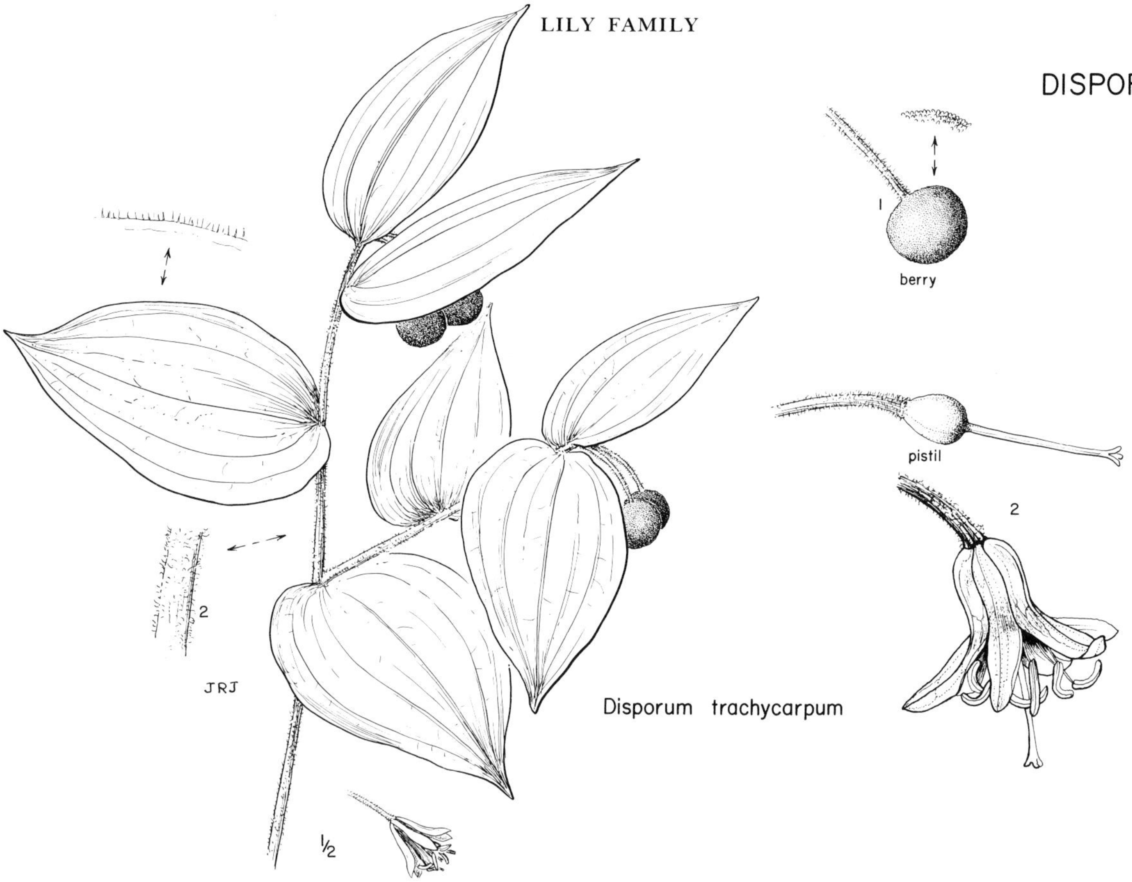

Disporum trachycarpum

the glabrous style 9–12 mm long with short stigmatic lobes 0.3–1 mm long; berries globose, orange to bright red, 7–10 mm long, strongly papillate; seeds 6–12 (18); $2n = 22$.

In shady, often wooded places, especially near streams, in the Rocky Mts. from Can. s. to s. Ariz. and N.M., w. across Idaho to c. Wash. and Oregon; in our area mainly in the mts. of se. Idaho and most of Utah, but also in Harney Co., Oregon, mostly below 2500 m elev. June–Aug.

7. STREPTOPUS Michx. Twisted Stalk

Simple to highly branched leafy perennial herbs from extensive horizontal rhizomes; leaves numerous, alternate, elliptic to ovate, with sessile to mostly clasping bases and acute to acuminate apices; flowers small, white to greenish-white or pink to rose-red, borne beneath the leaves with 1 or sometimes 2 together on a slender, bent or twisted peduncle usually adnate to the stem for one internode and then apparently arising from opposite the leaf axil above the point in which it had its origin; tepals 6, alike, forming a rotate to campanulate perianth, deciduous, mostly oblanceolate to oblong, erect to spreading or recurved, the outer tepals usually flat with the inner ones somewhat keeled and slightly narrower; stamens 6, hypogynous, the filaments short, broad and flat, the anthers basifixed, extrorse, sagittate, apiculate or aristate at the apex; ovary superior, 3-celled with many ovules, the slender to bulbous style with an entire or short 3-cleft stigma.

A genus of 7 species of temp. N. Amer. and Eurasia. (From the Greek *streptos,* twisted, and *pous,* foot, alluding to the bent or twisted peduncles.)

References:

Fassett, N. C. A study of *Streptopus.* Rhodora 37: 88–113. 1935.

Löve, D., and H. Harries. *Streptopus oreopolus* Fern., a hybrid taxon. Rhodora 65: 310–317. 1963.

1. Streptopus amplexifolius (L.) DC.

Uvularia amplexifolia L. Sp. Pl. 304. 1753. *Streptopus distortus* Michx. Fl. Boreali-Amer. 1: 200. 1803, nom. illeg. *S. amplexifolius* DC. in Lam. & DC. Fl. Franç. 3: 174. 1805. *Convallaria amplexifolia* E. H. L. Krause in Sturm, Deutschl. Fl. ed. 2. 1: 115. 1906. *Tortipes amplexifolius* Small, Manual S.E. Fl. 298. 1933. (Czechoslovakia.)

S. distortus sensu Torr. Fl. N. & Middle Sect. U.S. 353. 1824; not Michx. *S. amplexifolius* var. *americanus* J. A. Schultes in Roemer & Schultes, Syst. Veg. 7(1): 311. 1829. *S. amplexifolius* subsp. *americanus* A. Löve & D. Löve, Univ. Colorado Stud. Ser. Biol. 17: 18. 1965. (*Hadley,* near Fairfield, N.Y.)

S. amplexifolius var. *chalazatus* Fassett, Rhodora 37: 97. 1935. (*H. P. Hanson 1100,* Wallowa Lake, Wallowa Co., Oregon, 15 June 1933.)

S. amplexifolius var. *denticulatus* Fassett, Rhodora 37: 98. 1935. (*Fassett & Popko 16716,* Mellen, Ashland Co., Wisc., 7 June 1934.) = var. *americanus.*

S. amplexifolius var. *grandiflorus* Fassett, Rhodora 37: 99. 1935. (*J. C. Nelson 1143,* Falls City, Polk Co., Oregon, 12 May 1917.) = var. *americanus.*

Stout, erect perennial herbs (3) 5–8 (12) dm tall with glabrous or sparsely to moderately pubescent stems and branches with the thick rigid hairs often reddish near the base, the plants arising from stout, highly branched rhizomes; leaves lanceolate to ovate, 5–12 (15) cm long, 2.5–5 cm wide, the base cordate-clasping, the apex acuminate, the margin often with minute teeth; flowers few, white with a greenish or yellowish tinge, 9–15 mm long, narrowly campanulate, with arched peduncles (0.5) 1–3 cm long bearing a short-stalked gland at the tip, the sharply geniculate and twisted pedicel 1–2 cm long and bearing 1–2 flowers, the tepals narrowly oblong-lanceolate, spreading and somewhat recurved at the tip, the inner tepals somewhat clasping the stamens; stamens unequal, those of the outer series 0.8–1 mm long, those of the inner series 2–3 mm long, the filament only slightly less broad than the lanceolate anther, the anther 3–3.5 mm long with a narrowed, setose, simple or minutely bifid apex; ovary elliptic, with a thick style 4–5 mm long bearing a minutely lobed stigma; berries elliptic to oval, yellow to red, 10–12 mm long; seeds many, 2.5–3 mm long; $n = 8$ for var. *amplexifolius,* $n = 16$ for the N. Amer. varieties.

Moist places especially along streams and in forested places; Eurasia and N. Amer., from Alaska s. to Ariz. and e. across Can. and the U.S.; in our range mainly in the mts. of Utah, with isolated and scattered populations in s. Idaho. May–July.

The nomenclature of *Streptopus amplexifolius* is here modified from that presented by previous authors, and differs greatly from the Fassett treatment. In 1935, Fassett recognized seven varieties. The var. *amplexifolius* is found in the mountainous regions of southern Europe, with var. *papillatus* Ohwi restricted to Japan. In North America, Fassett proposed vars. *chalazatus, americanus, denticulatus* and *grandiflorus,* and reduced a Fernald species, *S. oreopolus,* to the varietal rank under *S. amplexifolius.* The last variant should actually be called *S.* X *oreopolus* Fern. (pro sp.) as Löve and Harries have shown this eastern North American element to be a sterile triploid hybrid formed by hybridization between *S. amplexifolius* and *S. roseus* Michx. As for the remaining variants, all are found in western North America.

The var. *chalazatus* is maintained as distinct. It can be recognized by its glabrous stems and entire leaf margins, although this latter feature is not entirely consistent. Hitchcock, in his treatment of the genus in 1969 for the Pacific Northwestern United States, reduced both var. *denticulatus* and var. *grandiflorus* to synonymy under var. *americanus.* This concept is followed here, with a slight nomenclatural adjustment. When Michaux proposed *S. distortus,* he included Linnaeus' earlier *Uvularia amplexifolia* in synonymy, thus creating a nomenclaturally superfluous name. In the past, it has been assumed that Schultes based his var. *americanus* on *S. distortus* of Michaux, but in fact, he based his name on a description of the taxon by Torrey. By following the current edition of the Code, I can maintain var. *americanus* for our plants by assuming "by implication" that Schultes was proposing a new name for the plant Torrey was discussing and not that described by Michaux. Under the previous edition of the Code, this would have been impossible.

We have two variants of the species in our area.

1 Stems coarsely pubescent below the first branch; leaves often with minute marginal teeth, not copiously papillate below; across s. Can. and n. U.S., extending s. in the mts. to n. Calif. and N.M., mostly in the mts. of n. Utah and s. Idaho in our area .... var. *americanus* J. A. Schultes

1 Stems glabrous; leaves with entire margins, coarsely papillate below; w. U.S. from s. B.C. e. to S.D., s. to s. Utah and N.M., mostly in the mts. of c. and s. Utah in our area ... . var. *chalazatus* Fassett

8. ZIGADENUS Michx. Death Camas

Erect, simple, leafy, glabrous perennial herbs from deeply-seated tunicated bulbs (ours) or rhizomes; leaves many, linear, grasslike, clasping, mostly basal, some cauline, but if so then reduced in both size and number; flowers small, greenish to yellowish-white or white, rotate to shallowly campanulate, perfect or unisexual, on erect to spreading pedicels subtended by long-acuminate bracts, arranged in terminal, erect to spreading racemes or panicles; tepals 6, alike, distinct or adnate to the lower part of the ovary, spreading to suberect, clawed at the base and with 1 (ours) or 2 yellow to greenish glands near the base, withering-persistent, lanceolate to ovate; stamens 6, hypogynous or perigynous, the filaments basally dilated, the anthers basifixed, extrorse, cordate-reniform; ovary superior or partially inferior, 3-celled with many ovules, the 3 styles distinct and stigmatic only at the tips; fruits 3-lobed loculicidal capsules; seeds subrhomboid, irregularly angled.

A genus of about 15 species of N. Amer. and Asia. (From the Greek *zugon,* yoke, and *aden,* gland, alluding to the yellow, paired glands of some species.)

The genus is commonly called death camas, and with good reason for there are several poisonous species. In our area, *Zigadenus* (occasionally spelled *Zygadenus*) *venenosus* is largely responsible for the majority of sheep losses in the mountains of Utah, as the leaves of this species are often confused with other grasslike plants and its presence is not often detected. Care should be taken in all attempts to gather liliaceous bulbs, and "pioneer stories" of surviving by eating bulbs of plants should be told to young children only with caution.

1 Tepals (6) 7–11 mm long, not clawed, the gland deeply obcordate; stamens perigynous; ovary partially inferior; common in the mts. of our area .... 1. *Z. elegans*

1 Tepals 3–6 mm long, those of the inner series clawed, the gland broadly ovate; stamens hypogynous; ovary superior; common in the foothills and in montane meadows.

2 Inflorescences usually racemose, rarely with a few branches near the base; flowers perfect; tepals of the outer whorl with a claw 0.3–1 mm long; gland with a well defined toothed upper margin; stamens as long as to just slightly longer than the tepals .... 2. *Z. venenosus*

2 Inflorescences paniculate; flowers mostly polygamous with those of the main axis perfect and those of the lower panicle branches staminate; tepals of the outer whorl scarcely or not clawed; gland with an indefinite margin; stamens 1–2 mm longer than the tepals .... 3. *Z. paniculatus*

STREPTOPUS
ZIGADENUS
1/2
1/2
JRJ
var. chalazatus
1/2
var. americanus
1/2
gland
2
3
1/6
1/8
Streptopus amplexifolius
Zigadenus elegans

1. Zigadenus elegans Pursh

Zigadenus elegans Pursh, Fl. Amer. Sept. 241. 1814. *Anticlea elegans* Rydb. Bull. Torrey Bot. Club 30: 273. 1903. (*Lewis,* "On the waters of Cokahlaishkit river, near the Rocky-mountains"; Lewis was along the Blackfoot River [called by him the Cokahlaishkit] from 3 July to 6 July, but on the 7th turned away from the River and headed for Lewis and Clark Pass where, just below the summit, the expedition collected several plants; as the collection is dated 7 July 1806, it is here proposed that the type came from just west of the Pass in Lewis and Clark Co., Mont.)

Z. chloranthus Richardson in Franklin, Narr. Journey Shores Polar Sea, Bot. Appendix 736. 1823. *Anticlea chlorantha* Rydb. Bull. Torrey Bot. Club 30: 273. 1903. (*Richardson,* "wooded country from latitude 54°–64° north.")

Z. chloranthus var. *major* Hook. Fl. Boreali-Amer. 2: 177. 1838. (*Drummond, Douglas,* "Rocky Mountains," probably of Can.)

Z. longus Greene, Pittonia 4: 240. 1901. *Anticlea longa* A. A. Heller, Muhlenbergia 6: 12. 1910. (*Cusick 2060,* mts. of Oregon, 28 July, no year given.)

Z. dilatatus Greene, Pl. Bakerianae 1: 51. 1901. (*Baker et al. 522,* La Plata Mts., Colo., 13 July 1898.)

Z. alpinus Blankinship, Montana Coll. Agric. Sci. Stud., Bot. 1: 44. 1905. *Anticlea alpina* A. A. Heller, Muhlenbergia 6: 12. 1910. (*Vogel,* Spanish Peaks, Madison Co., Mont., 20 July 1901, is the first collection cited.)

Anticlea vaginata Rydb. Bull. Torrey Bot. Club 39: 108. 1912. *Zigadenus vaginatus* J. F. Macbride, Contr. Gray Herb. 53: 4. 1918. (*Rydberg & Garrett 9407,* Armstrong Canyon, near the Natural Bridges, Utah, 4–6 August 1911.) A lowland phase, found in hanging gardens along the Colorado and Green rivers, blooming in August, sometimes robust and clustered, up to 1 m tall, but sometimes of more typical aspect; tepals relatively small, only 6–8 mm long. Monographic study may possibly support the recognition of this phase as a distinct variety, but no new combination is here proposed.

Erect perennial herbs 1.5–7 (10) dm tall with 3–several flowered racemose to freely branched terminal paniculate inflorescences up to 3.5 dm long from deep-seated, narrowly globose bulbs 1–2 (3) cm long; leaves basal, linear, 1–2.5 (3.5) dm long, 2–15 (20) mm wide, becoming reduced above when present along the stems; flowers rotate, greenish-white, on ascending pedicels 10–35 mm long, the tepals spreading, oblong-obovate to ovate, mostly 7–11 mm long, all about alike, not clawed at the base, the gland dark green, deeply obcordate; stamens perigynous, as long as the tepals, the filament broadly dilated basally, the anther reniform; ovary slightly inferior with the 3 distinct styles 2.5–3 mm long; capsules 15–20 mm long; seeds straw-colored, 4–5 mm long; $2n = 32$.

Forested and moist meadow areas or infrequently in grasslands mostly in montane or alpine situations from Alaska and B.C. s. through Wash. to e. Oregon and n. Nev., hence e. to Alta., the Dakotas, and s. to Ariz., N.M., Texas and n. Mex.; fairly frequent in our area across the n. half of the Region, but throughout most of the higher mts. in Utah from 1600 to 4000 m elev. June–Aug.

2. Zigadenus venenosus S. Wats.

Zigadenus venenosus S. Wats. Proc. Amer. Acad. Arts 14: 279. 1879. *Toxicoscordion venenosum* Rydb. Bull. Torrey Bot. Club 30: 272. 1903. ("California . . . to British Columbia and east to Utah and Idaho," no specimens cited; here lectotypified on *Watson 1163,* Parley's Peak, Summit Co., Utah, 6000 ft. elev., June 1869, deposited at GH.)

Z. gramineus Rydb. Bull. Torrey Bot. Club 27: 535. 1900. *Toxicoscordion gramineum* Rydb. Bull. Torrey Bot. Club 30: 272. 1903. *Z. venenosus* var. *gramineus* Walsh ex M. E. Peck, Manual Higher Pls. Oregon 190. 1941. (*Rydberg & Bessey 3848,* Spanish Basin, Gallatin Co., Mont., 28 June 1897.)

Z. intermedius Rydb. Bull. Torrey Bot. Club 27: 535. 1900. *Toxicoscordion intermedium* Rydb. Bull. Torrey Bot. Club 30: 272. 1903. (*Sandberg 10564,* Nez Perce Co., Idaho, May 1892.) = var. *gramineus.*

Toxicoscordion arenicola A. A. Heller, Muhlenbergia 2: 182. 1906. (*Heller 8321,* foothills w. of Bishop, Inyo Co., Calif., 23 May 1906.) = var. *venenosus.*

Z. salinus A. Nels. Bot. Gaz. 54: 406. 1912. *Toxicoscordion salinum* Gates, J. Linn. Soc., Bot. 44: 156. 1918. (*Macbride 889,* Emmett, Canyon Co., Idaho, 9 June 1911.) = var. *venenosus.*

Erect perennial herbs (1.5) 2–6 (7) dm tall with a densely flowered terminal raceme up to 2 dm long from deep-seated, oblong-ovoid bulbs 1.5–2.5 cm long; leaves essentially basal, linear, (1) 1.5–3 (4) dm long, (2) 3–6 (10) mm wide, becoming strongly reduced above; flowers narrowly campanulate, white to cream-colored, on strongly ascending pedicels 5–20 (30) mm long, the tepals suberect, those of the outer series ovate-lanceolate to ovate and (4) 4.5–5 mm long with a claw almost lacking to 1 mm long, those of the inner series (4.5) 5–6 mm long with a narrow claw 0.7–1.2 mm long, the gland yellowish-green, broader than long, with a thick, well defined toothed upper margin; stamens hypogynous, as long to only slightly longer than the tepals, the filament slender, the anther ovate-cordate; ovary superior with the distinct styles 2–3 mm long; capsules 8–15 mm long; seeds light brown, 5–6 mm long; $n = 11$.

Sagebrush slopes and montane forests or in coastal places from s. B.C. to Baja Calif., e. to Sask., the Dakotas, Nebr. and Colo.; common in the n. part of our area from se. Oregon and n. Nev. e. to Utah and Wyo., mostly below 2500 m elev. April–July.

This is a highly variable species that is currently defined to consist of three variants. Our common phase is var. *venenosus,* but just to the east of our line in Wyoming is var. *gramineus* (Rydb.) Walsh ex M. E. Peck. It is to be expected in the extreme northeastern corner of Utah. The two variants may be distinguished as follows:

1 Upper stem leaves not sheathing; outer tepals with claw 0.5–1 mm long .... var. *venenosus*

1 Upper stem leaves, except for the bracts of the inflorescence, all sheathing; outer tepals not clawed or with a claw less than 0.5 mm long .... var. *gramineus* (Rydb.) Walsh ex M. E. Peck

3. Zigadenus paniculatus (Nutt.) S. Wats.

Helonias paniculata Nutt. J. Acad. Nat. Sci. Philadelphia 7: 57. 1834. *Zigadenus paniculatus* S. Wats. in King, Rep. Geol. Explor. 40th Parallel 5: 343. 1871. *Toxicoscordion paniculatum* Rydb. Bull. Torrey Bot. Club 30: 272. 1903. (*Wyeth,* "In the Kamas prairie, near Flat-Head river," apparently near the Camas River, Jefferson Co., Idaho, on or about 20 June 1833.)

Erect perennial herbs (2) 3–5 (6) dm tall with a densely flowered terminal panicle 1–3 dm long from deep-seated, ovoid bulbs 3–4 cm long; leaves essentially basal but along the lower stem as well, linear, (1) 1.5–4 (5) dm long, (3) 5–15 (20) mm wide, becoming much reduced above; flowers broadly campanulate, yellowish-white, all perfect or the upper ones perfect and the lower flowers functionally staminate, on ascending pedicels (5) 10–20 (25) mm long, the tepals spreading, those of the outer series broadly ovate, (3) 3.5–4.5 mm long, scarcely or not clawed, those of the inner series nearly oblong-lanceolate, (3.5) 4–5 (5.5) mm long, with a narrow claw

ZIGADENUS

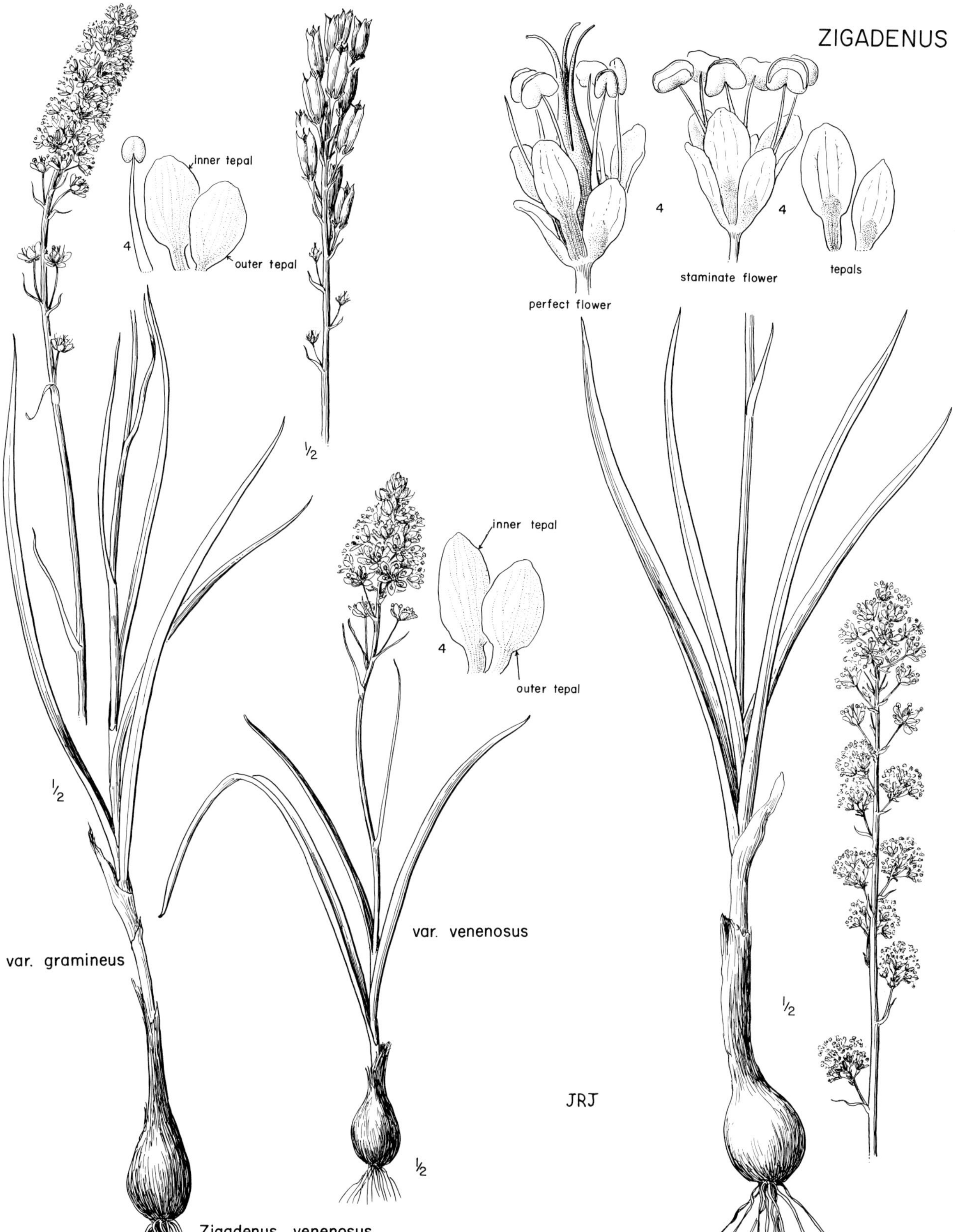

inner tepal
4
outer tepal
1/2
1/2
var. gramineus
inner tepal
4
outer tepal
var. venenosus
1/2
Zigadenus venenosus
perfect flower
4
staminate flower
4
tepals
1/2
JRJ
Zigadenus paniculatus

0.5–1 mm long, the gland yellowish-green, broadly ovate, with an indefinite margin; stamens hypogynous, usually 1–2 mm longer than the tepals, the filament slender, the anther reniform-cordate; ovary superior with distinct styles 2.5–3 mm long; capsules (10) 14–18 mm long; seeds light brown, 6–8 mm long.

Sagebrush slopes to coniferous montane forests from e. Wash. s. to c. Calif. and e. to c. Mont., w. Wyo. and Colo. s. into n. Ariz. and nw. N.M.; our most common species, found throughout the Intermountain Region except for the interior of Nev. where only widely scattered and very infrequent. March–June.

9. VERATRUM L. False Hellebore, Skunk Cabbage

Tall, stout and erect, leafy, unbranched perennial herbs from short, thick rhizomes; leaves numerous, narrow to broad, often plicated, strongly clasping in most; flowers small, greenish-white to yellow or purple, broadly campanulate, polygamous, the upper perfect and the lower staminate, borne on short pedicels in large terminal panicles or compound racemes; tepals 6, distinct or sometimes with a short hypanthium, spreading, glandless in most, sometimes clawed, deciduous, oblong-ovate, obtuse, adnate to the base of the ovary; stamens 6, opposite the tepals, free, almost as long as the tepals, the filaments slender, the anthers cordate-ovoid with confluent pollen sacs; ovary superior, 3-celled with several ovules, glabrous or woolly, the 3 short styles each terminated by an elongated stigma; fruits membranous, 3-lobed, septicidal, ovoid capsules tipped with persistent styles; seeds numerous, flat and completely surrounded by a broadly winged, loose testa.

A complex genus of perhaps 30 species found in the N. Hemisphere. (An ancient Latin name for hellebore, supposedly from *vere,* true, and *ater,* black, alluding to the black rhizomes of some species.)

References:

Kupchan, S. M., J. H. Zimmerman, and A. Afonso. The alkaloids and taxonomy of *Veratrum* and related genera. Lloydia 24: 1–26. 1961.

Loesener, O. Übersicht über die Arten der Gattung *Veratrum.* Repert. Spec. Nov. Regni Veg. 24: 61–72. 1927.

Zimmerman, J. H. A monograph of *Veratrum.* Unpublished doctoral dissertation, University of Wisconsin, Madison. 1958.

1. Veratrum californicum Dur.

Veratrum californicum Dur. J. Acad. Nat. Sci. Philadelphia, II. 3: 103. 1855. (*Pratten,* near Nevada City, Nevada Co., Calif., in 1854.)

V. eschcholtzii var. *watsoni* Baker, J. Linn. Soc., Bot. 17: 471. 1879. (Based on the description of *V. album* by S. Wats. in King, Rep. Geol. Explor. 40th Parallel 5: 344. 1871, and lectotypified here upon *Anderson 105,* near Carson City, Nev., deposited at GH.)

V. speciosum Rydb. Bull. Torrey Bot. Club 27: 531. 1900. (*Flodman 344,* near the pass, Bridger Mts., Mont., 28 July 1896.)

V. tenuipetalum A. A. Heller, Muhlenbergia 1: 39. 1905. (*Scoville,* Colo., in 1870.)

V. jonesii A. A. Heller, Muhlenbergia 1: 124. 1905. (*M. E. Jones 6584,* Middle Fork of the Weiser R., Washington Co., Idaho, 3 Aug. 1899.)

Tall erect, stout, leafy perennial herbs (1) 1.5–2 (2.5) m tall, usually glabrous below and becoming tomentose above in the freely branching, stiffly erect paniculate inflorescence, the panicles (2) 3–6 (8) dm long, the plants arising from thick, stout rhizomes; leaves numerous, oblong-lanceolate to broadly elliptic or ovate, 2.5–4 dm long, 1–2 dm wide, lightly pubescent especially below, sometimes moderately pubescent between the veins, plicated and strongly nerved, the base strongly sheathing, the apex acute, the uppermost leaves of the stem and in the inflorescence mostly lanceolate and shorter; flowers broadly campanulate, dull white to greenish-white, subsessile or on short pedicels 1–3 (6) mm long, these becoming gradually reduced upwardly, the tepals lanceolate to oblong-ovate or elliptic, (8) 10–14 (17) mm long, entire or slightly erose to somewhat denticulate in some, glabrous or slightly lanate, the margin green, the glands Y-shaped, two, separate or confluent, greenish; stamens shorter than the tepals, 6–8 mm long, the filament slender, the anther broader than long; ovary glabrous or with a few simple hairs in some; capsules narrowly ovoid, 2–3.5 mm long, glabrous; seeds straw-colored, 10–12 (15) mm long; $n = 16$.

Wet places along streams, lakes, and in meadows and thickets; w. Wash. s. to Calif., e. to Mont., Colo. and N.M.; infrequent and scattered in most of our area, especially in s. Nev., but more common (and locally frequent) in e. Calif., se. Oregon and most of montane Utah from about 600 to 3500 m elev. June–Aug.

Our plant is var. *californicum.*

This species is occasionally the cause of livestock poisoning. Reduction of cattle losses can be made simply by avoiding overgrazing montane meadows, as *V. californicum* tends to increase with mismanagement of range lands.

10. ASPARAGUS L. Asparagus

Tall, fernlike perennial herbs from thick matted rhizomes or tuberous roots with freely branched, erect to somewhat climbing stems with filiform to flattened branchlets which function as leaves borne in the axils of the leaves; leaves alternate, scale-like, scarious; flowers small, greenish-yellow, perfect or (ours) functionally dioecious, borne singly or in pairs along the stems on slender jointed pedicels in racemes or umbels; tepals 6, alike, divided nearly to the base (ours) or more or less connate; stamens 6, inserted on the tepals just above their base, the anthers introrse, versatile; ovary superior, 3-celled, 2–few ovules per locule, the short and slender style terminated by 3 small stigmas; fruits globose berries.

A large and exceedingly complex genus of from 100 to 300 species ranging from Siberia to S. Amer. Various species are grown for cultivation and food. (The ancient Greek name.)

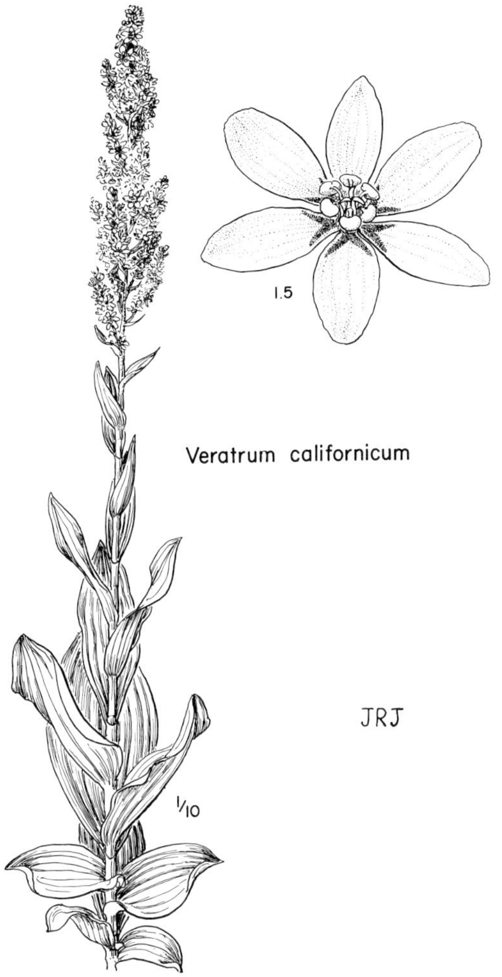

1. Asparagus officinalis L.
Asparagus officinalis L. Sp. Pl. 313. 1753. (Europe.)

Tall, usually much-branched, finely dissected perennial herbs 1–2 (3) m tall from thickly matted rhizomes, the new stems stout and simple with scale leaves up to 5 mm long, forming the edible portion of the plant; stems of mature plants freely branching, their leaves scarious, 1–2 mm long, with clusters of axillary, filiform to needlelike branchlets 1–2 cm long; flowers drooping on filiform, jointed pedicels up to 2 cm long, the tepals bell-shaped, greenish-white, 3–7 mm long; berries red, 6–8 mm in diam.; n = 10, 20.

An escaped weed in waste places throughout much of N. Amer.; especially common in the farming areas and along drainages in our area, infrequently found as a waif in isolated places in the deserts of w. Utah and Nev. July–Aug.

11. ERYTHRONIUM L. Dogtooth Violet

Erect or spreading perennial herbs with bractless scapes or with a reduced leaf at the base of the lowermost flower, arising from deep-seated, elongate corms; leaves in seemingly opposite, nearly basal pairs, petiolate to subsessile, green or mottled; flowers large and showy, pale to deep yellow (ours) or white to deep pink or violet, nodding, solitary or 2–5 in a loose raceme on a naked peduncle; tepals 6, alike (ours) or with the inner series broader, occasionally the members of one or both series with 2–4 inflated sac-like appendages near the base abutting against the filaments, lanceolate, acute or acuminate, spreading at anthesis and becoming eventually highly reflexed; stamens 6, in two slightly unequal sets, borne on the receptacle, the filaments slender and erect, tapering only slightly, if at all, from the base, or with a broadly dilated base and covering the ovary and thus closely surrounding the style, the anthers basifixed, extrorse, withering and shrinking to near half their length after dehiscence; ovary superior, 3-celled with several ovules, the slender style extending as far as the longest stamens with a shallow (ours) to deeply 3-lobed stigma, these lobes from as long as broad to slender and up to 6 mm long; fruits more or less obovoid to cylindric-clavate loculicidal capsules.

A genus of perhaps 12 species, all but one of which occur in N. Amer. (From the Greek *erythro,* red, in reference to the pink or red flowers seen in some species.)

Reference: Applegate, E. I. The genus *Erythronium:* A taxonomic and distributional study of the western North American species. Madroño 3: 58–113. 1935.

1. Erythronium grandiflorum Pursh

Erythronium grandiflorum Pursh, Fl. Amer. Sept. 231. 1814. *E. giganteum* Lindl. Edward's Bot. Reg. 21: 1786. 1835, a superfluous substitute. *E. grandiflorum* var. *minor* Hook. Fl. Boreali-Amer. 2: 182. 1838. *E. grandiflorum* var. *giganteum* Hook. Fl. Boreali-Amer. 2: 182. 1838. *E. grandiflorum* f. *giganteum* Voss, Vilmorin's Blumengärt. ed. 3. 1: 1115. 1895. (*Lewis,* "On the Kooskoosky," along the Clearwater River opposite Kamiah, Lewis Co., Idaho, 5 June 1806.)

E. nuttallianum Regel, Gartenflora 20: 227. 1871; not Roemer & Schultes, 1820. *E. grandiflorum* var. *parviflorum* S. Wats. Proc. Amer. Acad. Arts 26: 129. 1891. *E. parviflorum* Goodd. Bot. Gaz. 33: 67. 1902. (No specimens cited.)

E. obtusatum Goodd. Bot. Gaz. 33: 67. 1902. (*Nelson & Nelson 5606,* Glen Cr., Yellowstone Natl. Park, Wyo., in 1889.)

E. utahense Rydb. Fl. Rocky Mts. 165, 1061. 1917 [1918]. (*Stansbury,* Salt Lake City, Utah, in 1850.)

E. leptopetalum Rydb. Fl. Rocky Mts. 165, 1061. 1917 [1918]. (*Mulford,* near Boise, Ada Co., Idaho, June 1892.)

E. grandiflorum subsp. *chrysandrum* Appleg. Contr. Dudley Herb. 1: 190. 1933. (*Applegate 6760,* near Winchester, Nez Perce Co., Idaho, 3 May 1931.)

Dogtooth violet, trout lily, fawn lily, glacier lily, adder's tongue.

Tall, usually erect perennial herbs 1–3 (3.5) dm tall with slender scapes arising from a long slender corm; leaves green, not mottled, narrowly to broadly oblong-elliptic, 1–2 dm long, 1–5 cm wide, narrowing gradually to a broad petiole or subsessile, the apex acuminate; flowers 2–4 (5) on nodding, naked peduncles, pale yellow to golden (ours) or white, the base lighter within and streaked with green without, the tepals (2) 2.5–3.5 (4) cm long, 4–7 mm wide, lanceolate, spreading and becoming reflexed from near the base with the basal appendages well developed, globular but not strongly inflated; stamens long exserted from the reflexed tepals but still shorter than the tepals, the filament erect, slender, not concealing the ovary nor surrounding the style, the anther white, yellow, red or somewhat purplish, (8) 10–12 mm long before dehiscence; ovary with elongated styles projecting the 3 spreading, 1–2 (4) mm long stigmatic lobes beyond the dehiscing anthers; capsules cylindric-clavate, 3–6 cm long, tan to dark brown; $2n = 24$.

Dry sagebrush slopes to moist coniferous forest floors in the Rocky Mts. and Cascades from s. Can. s. to Utah and Colo., and to Calif.; common and mainly in the n. and e. part of the Intermountain Region. May–July.

Applegate proposed various subspecies of this species based on anther color. This feature has proved untenable and none of his infraspecific elements assigned to our plants is recognized.

This beautiful species does fairly well in some gardens, especially those along the Wasatch Front. However, there is more failure than success with the species in maintaining it for a long period of time. Much the same situation occurs here as in *Calochortus nuttallii:* it is much simpler to observe *Erythronium grandiflorum* locally in its native habitats than to attempt to grow it in a garden.

12. FRITILLARIA L. Fritillary

Erect, unbranched, glabrous perennial herbs with slender scapes or stems from small bulbs consisting of a few fleshy scales and usually numerous ricelike or grainlike offset bulblets; leaves few, alternate to whorled, sessile, linear to lanceolate; flowers showy, white, yellow, red or purplish and often mottled or striped with white, yellow, brown, red or purple, usually nodding, campanulate, solitary or 2–several in a loose, terminal raceme; tepals 6, much alike, distinct, spreading, each usually with a shallow gland or nectar-bearing area at the base, oblanceolate to rhombic, acute to rounded; stamens 6, borne on the receptacle, the filaments slender, the anthers extrorse, subversatile; ovary superior, 3-celled with numerous ovules, the slender style entire or distinct nearly to the base but stigmatic only at the tips of the branches; fruits membranaceous, 6-angled or winged loculicidal capsules; seeds flat, brownish, obovate.

A genus of perhaps 50 species in the N. Hemisphere, with 18 species found in N. Amer. (From the Latin *fritillus,* dice box, alluding to the shape of the capsules.)

This genus would be an excellent addition to gardens, but most attempts to transplant the species are met with limited success in our area. It would seem best to preserve and protect the limited number of native habitats near large intermountain cities where *Fritillaria* grows rather than grow it under artifical garden conditions.

Reference: Beetle, D. E. A monograph of the North American species of *Fritillaria.* Madroño 7: 133–159. 1944.

1 Flowers yellow or orange, fading red to purple; style one, the stigma discoid; n. half of our area . . . . 1. *F. pudica*

1 Flowers purplish, mottled; styles three, connate only at the base, the stigmas elongate.

  2 Stems slender in region of the leaves; bulblets lacking or few; leaves 7–14; capsule angled; n. Calif. and se. Oregon e. to the Dakotas . . . . 2. *F. atropurpurea*

  2 Stems fistulose in region of the leaves; bulblets numerous; leaves 12–20; capsule with curved hornlike processes at the base and top of each wing; Storey Co., Nev. s. to Mono Co., Calif. s. to s. Calif. . . . . 3. *F. pinetorum*

1. Fritillaria pudica (Pursh) Sprengel

Lilium pudicum Pursh, Fl. Amer. Sept. 228. 1814. *Amblirion pudicum* Raf. Amer. Monthly Mag. & Crit. Rev. 2: 265. 1818. *Fritillaria pudica* Sprengel, Syst. Veg. 2: 64. 1825. *Theresia pudica* Klatt, Hamburger Garten-Blumenzeitung 16: 439. 1860. *Ochrocodon pudicus* Rydb. Fl. Rocky Mts. 1061. 1917 [1918]. (*Lewis,* "On the head-waters of the Missouri, May," an error, actually obtained on the "Plains of Columbia near the Kooskooskee," above the Potlatch R., Nez Perce Co., Idaho, 8 May 1806.)

F. leucella Gandoger, Bull. Soc. Bot. France 66: 291. 1919. (*Linford,* Cache Co., Utah.)

F. dichroa Gandoger, Bull. Soc. Bot. France 66: 291. 1919. (*Linford,* Cache Co., Utah.)

F. washingtonensis Gandoger, Bull. Soc. Bot. France 66: 291. 1919. (*Sandberg,* Spokane, Spokane Co., Wash.)

F. utahensis Gandoger, Bull. Soc. Bot. France 66: 291. 1919. (*Linford,* Cache Co., Utah.)

F. oregonensis Gandoger, Bull. Soc. Bot. France 66: 291. 1919. (*Cusick 1831,* e. Oregon, 11 April 1898.)

F. oredoxa Gandoger, Bull. Soc. Bot. France 66: 291. 1919. (*Suksdorf 5917,* Bingen, Klickitat Co., Wash.)

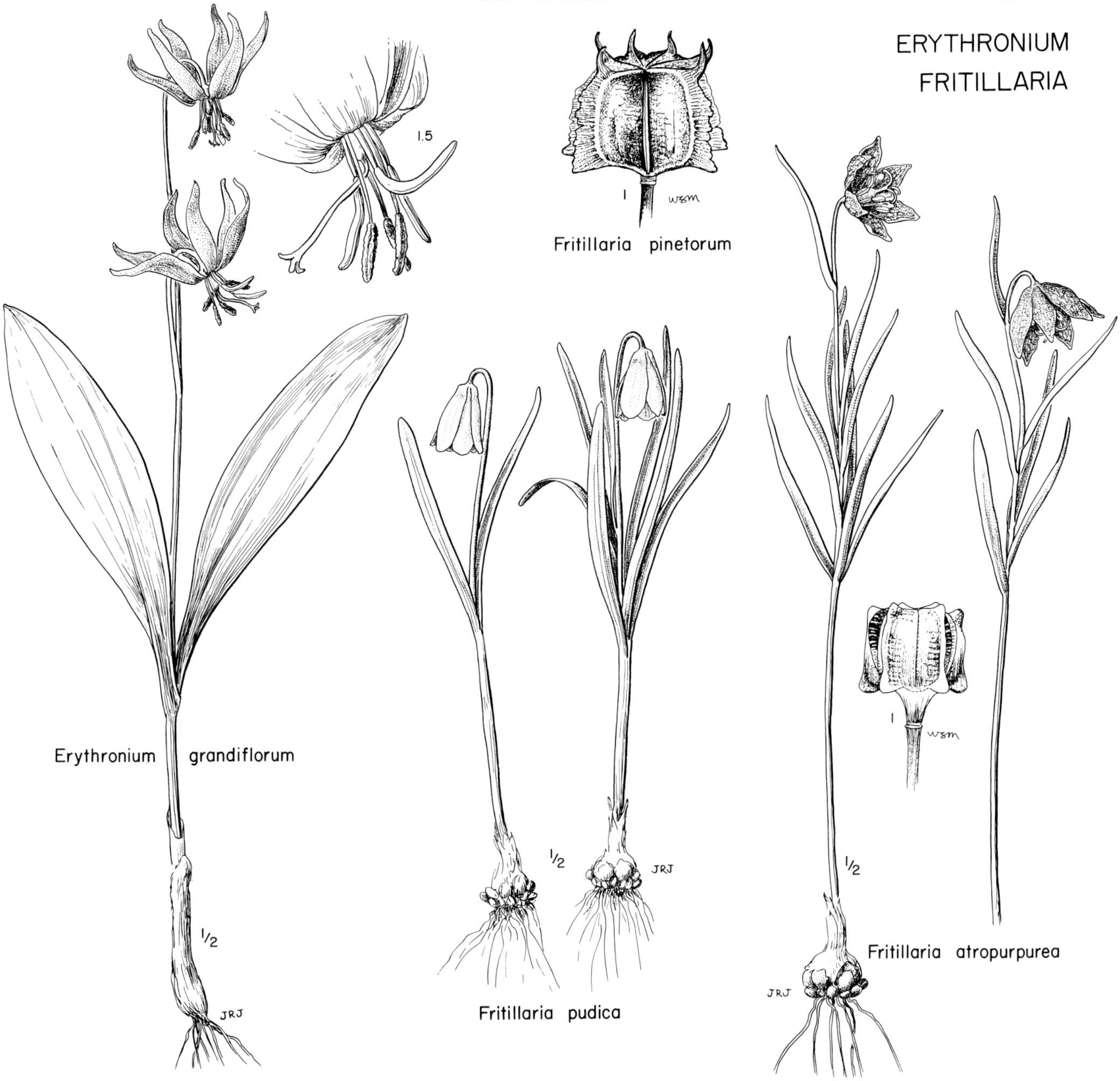

Yellow bell.

Erect, slender perennial herbs (0.5) 1–3 (3.5) dm tall with leafy stems arising from cream-colored bulbs 1–3 cm long with a few, small, thick fleshy scales and many bulblets; leaves 2 and alternate or 3–8 and semiwhorled, linear to linear-oblanceolate, (3) 6–15 (20) cm long, 2–12 mm wide; flowers usually solitary or occasionally 2–3, pendent, rather narrowly campanulate, yellow to orange and fading red to purple, often with brown or purple streaks without, the tepals oblanceolate to narrowly oblong, (12) 15–20 (25) mm long, 4–7 mm wide, the gland small; stamens 5–15 mm long, the filament slender, the anther 3–6 mm long; style one, shorter to slightly longer than the tepals, with a discoid (rarely scarcely lobed) stigma; capsules cylindric-obovoid to globose-obovoid, (15) 20–30 mm long; n = 12, 13, 18, 39.

Grasslands and sagebrush deserts to coniferous montane forests from B.C. s. to n. Calif. and e. to Alta., Mont., w. Wyo. and Utah; rather common in the foothills and mts. of w. Nev., se. Oregon, s. Idaho and n. Utah. March–June.

2. **Fritillaria atropurpurea** Nutt.

Fritillaria atropurpurea Nutt. J. Acad. Nat. Sci. Philadelphia 7: 54. 1834. (*Wyeth,* "With the preceding. Flowering early in June," referring to "On the borders of the Flat-Head river"; the two statements are contradictory, as Wyeth left the Flathead R. in Mont. by 18 May and was on the Snake R. Plains n. of Howe, Butte Co., Idaho, during the first week of June, 1833, when the specimens were most likely collected.)

F. gracillima Smiley, Univ. Calif. Publ. Bot. 9: 143. 1921. (*Hall & Babcock 3562,* Mt. Lyell, Madera Co., Calif., 18 July 1902.)

Leopard, checker or chocolate lily.

Erect, slender perennial herbs 1–6 dm tall with leafy stems arising from tannish bulbs 1–1.5 (2) cm long with a few, small, thin scales and few (if any) bulblets; leaves several, mostly 7–14, linear, (4) 6–10 (12) cm long, 2–4 (7) mm wide, alternate or whorled and scattered on the upper half of the stem; flowers commonly 1–4, spreading or nodding, broadly campanulate, greenish-brown to chocolate-brown with white or yellow spots or mottling, the tepals oblong to rhombic and tapering abruptly at the base, 10–20 mm long, 3–8 mm wide, the gland an indistinct brownish-yellow spot at the base; stamens 6–15 mm long, the filament slender, the anther 3–4 mm long; styles connate only 1–2 mm, the branches 6–9 mm long; capsules broadly obovoid, 10–15 (17) mm long, acutely angled; $n = 12$.

Grassy slopes and hills to coniferous montane forests from se. Oregon e. to the Dakotas, s. to n. Calif., Nev., Ariz. and N.M.; mostly in open woodlands throughout most of the Intermountain Region, but rather infrequent and widely scattered in Nev. as it is restricted to the higher mts. April–July.

3. **Fritillaria pinetorum** A. Davidson

Fritillaria pinetorum A. Davidson, Muhlenbergia 4: 67. 1908. *F. atropurpurea* var. *pinetorum* I. M. Johnston, Pl. World 22: 84. 1919. (*Hasse & Davidson 1739,* Mt. Cummings, Tehachapi Mts., Kern Co., Calif., 12 June 1907.)

Erect, fistulose perennial herbs 1–3 (5) dm tall with leafy stems arising from tannish bulbs 1–2 cm long with a few, small, thin scales and many bulblets; leaves several, mostly 12–20, linear, 5–15 cm long, 2–7 mm wide, whorled and scattered on the upper half of the fistulose stem; flowers commonly 3–9 (11), erect or nearly so, broadly campanulate, purplish and mottled with greenish-yellow, the tepals oblong, gradually tapering to the base, 14–19 mm long, 2–6 mm wide, the gland an indistinct brownish-purple spot at the base; stamens 9–15 mm long, the filament slender, the anther 2–3 mm long; styles connate only 1–1.5 mm, the branches 4–7 mm long; capsules broadly obovoid, 12–15 mm long, acutely angled with a short hornlike process at the base and at the summit of each angle; $n = 13$.

Exposed slopes in coniferous forests and open woodlands of Calif. and adjacent w. Nev. from Alpine Co., Calif., s. to San Bernardino Co.; known in our area from near Virginia City, Storey Co., Nev., and from Mono Co., Calif. May–July.

13. **LILIUM** L. Lily

Tall, erect, glabrous perennial herbs (ours) with unbranched leafy stems arising from fleshy-scaled bulbs or short, scaly rhizomes; leaves numerous, alternate or whorled, sessile, usually lanceolate; flowers large and showy, white or pink to orange or brick-red and often spotted with red to brown or purple, nodding or erect, funnelform to campanulate, solitary or 2–several in a loose, terminal, often leafy-bracteate raceme; tepals 6, alike, distinct, spreading to strongly recurved, often semiclawed at the base and each with a nectar-bearing gland near the base; stamens 6, borne on the receptacle, the filaments slender, the anthers extrorse or somewhat introrse, versatile; ovary superior, 3-celled with many ovules, the slender to stoutish style with 3 short lobes; fruits ovoid to cylindric loculicidal capsules; seeds many, flat.

A large genus of perhaps 75 species from temp. N. Amer. and Eurasia. (The Latinized form of the classical Greek, *leirion.*)

1. **Lilium parvum** Kellogg

Lilium parvum Kellogg, Proc. Calif. Acad. Sci. 2: 179. 1862. *L. canadense* var. *parvum* Baker, J. Linn. Soc., Bot. 14: 241. 1874. (*Stivers,* "from the [Sierra Nevada] mountains," and *Dorr,* "from Nevada Territory," are cited.)

Tall, erect perennial herbs 4–15 (20) dm tall with leafy, slender to stoutish stems from rhizomatous bulbs 2.5–3.5 cm long which bear 3-jointed scales 1–2 cm long; leaves linear to broadly lanceolate, 5–20 cm long, 5–30 mm wide, whorled and with a few, scattered alternate leaves below the inflorescence; flowers mostly 2–10 (15), racemosely arranged or occasionally in whorls, all ascending to erect, campanulate, orange to dark red with maroon spots, the tepals 3.5–4 cm long, the upper third of the lanceolate segments recurved; stamens long exserted, the filament slender, the anther 2.5–3 mm long; capsules ovoid, 12–16 mm long; $n = 12$.

In moist places near springs and along streams, and in grassy meadows, throughout the Sierra Nevada of Calif.; just entering our area in Mono Co., Calif., but to be expected from s. Washoe Co., Nev. s. to Lyon or Douglas cos., Nev. July–Sept.

The genus *Lilium* is not well represented in our area, but several species occur just outside our boundaries, especially along the foothills of the Sierra Nevada in California and Oregon. *Lilium parvum,* which is in the Intermountain Region, is found just north of Bridgeport, near Lee Vining, and at a spring east of Sand Flat north of Glass Mountain—all in Mono Co., California. It may also be found along the edge of the Sierra Nevada in our portion of western Nevada, but no specimens have been seen. *Lilium pardalinum* Kellogg, the tiger lily, is found in the Mammoth area of Mono Co., and it too may eventually be found within our area. It differs from *L. parvum* by having nodding yellow flowers with red to maroon spots.

14. **LLOYDIA** Salisb. ex Reichenb. nom. conserv.

Slender, erect perennial herbs from short rhizomes and persistent sheaths simulating a tunicated bulb; leaves 2–few, linear, becoming reduced and alternate above; flowers small, white with greenish to purplish veins, solitary or few in a loose, terminal leafy-bracteate raceme; tepals 6, alike, distinct, spreading, transversely corrugate with a small gland somewhat above the base; stamens 6, inserted on the base of the tepals, the filaments slender, the anthers basifixed, dehiscing laterally; ovary superior, 3-celled with many ovules, the short persistent style with 3 short stigmatic lobes; fruits obovoid to globose loculicidal capsules.

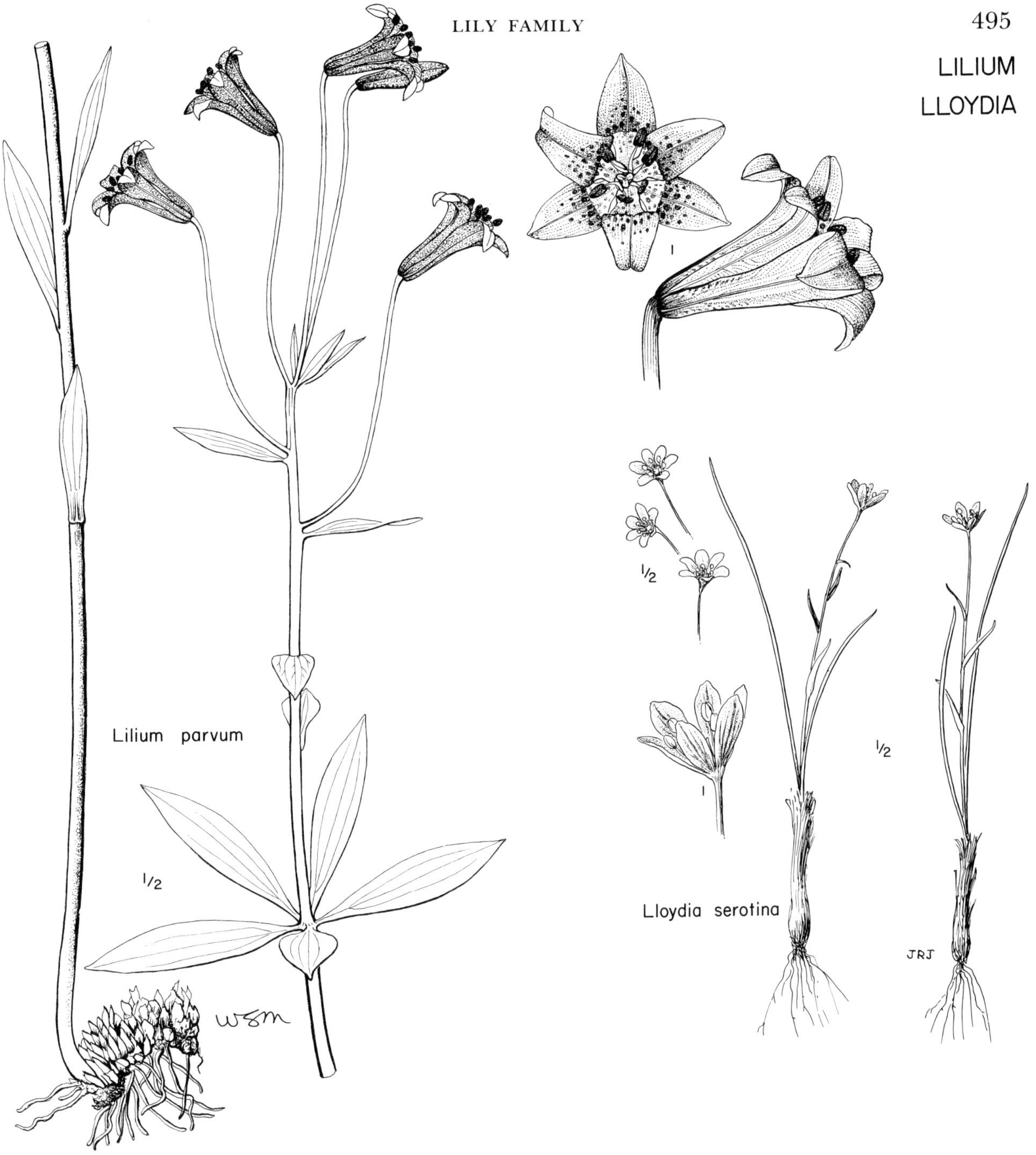

A genus of 10–12 species all of Eurasia except one. (Named for Edward Lloyd, 1660–1709, curator of the Oxford Museum who first found *Lloydia serotina* in the mts. of Wales.)

1. Lloydia serotina (L.) Salisb. ex Reichenb.

Bulbocodium serotinum L. Sp. Pl. 294. 1753. *Anthericum serotinum* L. Sp. Pl. ed. 2. 444. 1762. *Gagea serotina* Ker-Gawl. J. Sci. Arts (London) 1: 180. 1816. *Rhabdocrinum serotinum* Reichenb. Consp. Reg. Veg. 65. 1828. *Lloydia serotina* Salisb. ex Reichenb. Fl. German. Excursoria 102. (Jan–Apr) 1830. *L. serotina* Sweet, Sweet's Hortus Brit. ed. 2. 527. (Oct–Dec) 1830, a later publication of the same name often cited as the first place of publication. *Cronyxium serotinum* Raf. Fl. Telluriana 2: 28. 1837. ("Habitat in Alpibus Helvetiae, Angliae.")

Alpine lily.

Slender, erect perennial herbs from short, thick rhizomes, the stems and leaves of each year's growth surrounded by brownish, persistent bracts that partially wither away to ground level and resemble a tunicated bulb; leaves basal and cauline, the basal two linear, 4–10 (20) cm long, 0.8–1 mm wide, the 2–4 cauline leaves lanceolate, 1–5 cm long, 1–2 mm wide; flowers 1–2, erect, turbinate, white with greenish or more often purplish veins, the tepals oblong to broadly oblanceolate, 9–12 (14) mm long; stamens erect, the filament slender, 3–8 mm long, the anther 0.6–1 (1.2) mm long; capsules obovoid, 6–8 mm long; $2n = 24$.

Gravelly to rocky places or sometimes on cliffs or in rock crevices; Eurasia and N. Amer., from Alaska s. to w. Oregon in the Cascade Mts., and from e. Alta. s. to n. N.M. in the Rocky Mts., with scattered populations elsewhere; infrequent to rare in alpine places in the Uinta Mts. of ne. Utah and the Ruby Mts. of ne. Nev., mostly above 2700 m elev. June–Sept.

15. CALOCHORTUS Pursh Sego or Mariposa Lily

Glabrous perennial herbs with scapiform or leafy, simple or branched stems, bulbiferous in the axils of the lowest leaf or leaves in some, arising from deep-seated tunicate bulbs with membranous or fibrous-reticulate coats; leaves few, linear, the basal one solitary and large, the cauline ones reduced up the stem; flowers large and showy, white, yellow, red, lavender or purple and usually variously tinged, solitary or few together in a false umbel, erect or nodding; tepals 6, distinct and well differentiated, the three sepals ovate to lanceolate, usually naked and typically green or similarly colored as the petals, the three petals larger and broader, cuneate to clawed, distinctly colored and usually spotted or patterned with specks of contrasting color, more or less hairy within and with an often densely bearded, depressed gland or spot near the base; stamens 6, in two series, slightly adhering to the base of the petals, the filaments subulate and basally dilated, the anthers basifixed, dehiscing laterally; ovary superior, 3-celled with many ovules, the apex abruptly contracted or tapering to a trifid, sessile, persistent stigma; fruits orbicular to linear, 3-angled or -winged septicidal, erect or nodding capsules; seeds numerous, irregular or flattened.

About 60 species in w. N. Amer. from B.C. and the Dakotas s. to Guatemala with the greatest concentration of species in Calif. (From the Greek *kalo,* beautiful, and *chortos,* grass, alluding to the striking flowers atop the stems with otherwise narrow, grasslike leaves.)

The genus *Calochortus* is widely noted for its remarkable beauty and is therefore highly prized by gardeners. Alas, many of the species do not readily accept ordinary garden conditions and most attempts to introduce the genus into the horticultural trade have not been greatly successful. With care, sandy well-drained soil, and only natural rainfall, some species will flower in gardens in our area. If one remembers that most species of *Calochortus* are basically desert plants and treats them accordingly, some success may be expected.

Efforts have been made to divide *Calochortus* into three genera, *Calochortus, Mariposa* and *Cyclobothra.* The late Robert F. Hoover championed this concept beginning in 1944, but the idea has not been accepted by other workers.

References:
Beal, J. M., and M. Ownbey. Cytological studies in relation to the classification of the genus *Calochortus,* III. Bot. Gaz. 104: 553–562. 1943.
Ownbey, M. A monograph of the genus *Calochortus.* Ann. Missouri Bot. Gard. 27: 371–560. 1940.
———. *Calochortus.* Univ. Wash. Publ. Biol. 17(1): 765–779. 1969.

1 Ovary and fruit elliptic-oblong, distinctly 3-winged; flowers white to lavender with a large reddish-purple blotch in the middle of each petal; glands surrounded and partially covered by a margin; n. part of our area . . . . 1. *C. eurycarpus*
1 Ovary and fruit essentially linear, merely 3-angled; flowers various but not with a large blotch of color in the middle of each petal, merely with a band of color above the gland or no color at all; glands various.
  2 Glands not (or only slightly) depressed nor surrounded by a membrane.
    3 Stems decumbent to weakly erect, rarely bulbiferous; inflorescences monochasial; petals with a transverse band of yellow at the gland; deserts . . . . 2. *C. flexuosus*
    3 Stems strict and erect, bulbiferous; inflorescences subumbellate; petals not banded with yellow; montane places in e.-central Calif. and w. Nev. . . . . 3. *C. leichtlinii*
  2 Glands depressed and surrounded (at least in part) by a membrane.
    4 Petals with a longitudinal green stripe, this sometimes becoming obscured in old or poorly prepared specimens.
      5 Sepals usually longer than the petals; flowers purple; glands triangular-oblong; anthers linear to linear-lanceolate; ne. Calif. and se. Oregon e. across n. Nev. and s. Idaho . . . . 4. *C. macrocarpus*
      5 Sepals usually shorter than the petals; flowers white and often tinged with lilac; glands circular; anthers oblong; e. Calif. ne. across Nev. into s. Idaho . . . . 5. *C. bruneaunis*
    4 Petals without a longitudinal green stripe (rarely grayish in *C. ambiguus*).
      6 Glands circular and surrounded by a continuous margin; hairs on the petals simple, not branched or enlarged.
        7 Flowers white with a reddish-brown or purple band or spot above the gland, or lemon-yellow; basal leaf withering by late anthesis; dry places.
          8 Flowers white, tinged with lilac or infrequently magenta; e. Nev. and n. Ariz. n. through Utah to e. Idaho, common . . . . 6. *C. nuttallii*
          8 Flowers lemon-yellow; n. Ariz. and se. Utah, infrequent . . . . 7. *C. aureus*
        7 Flowers lavender to pink with a purple spot above the gland; basal leaf persistent; damp to wet meadows near Bishop, Inyo Co., Calif. . . . . 8. *C. excavatus*
      6 Glands transversely lunate to (infrequently) circular, discontinuously surrounded by a margin; hair on the petals branched or enlarged.

CALOCHORTUS

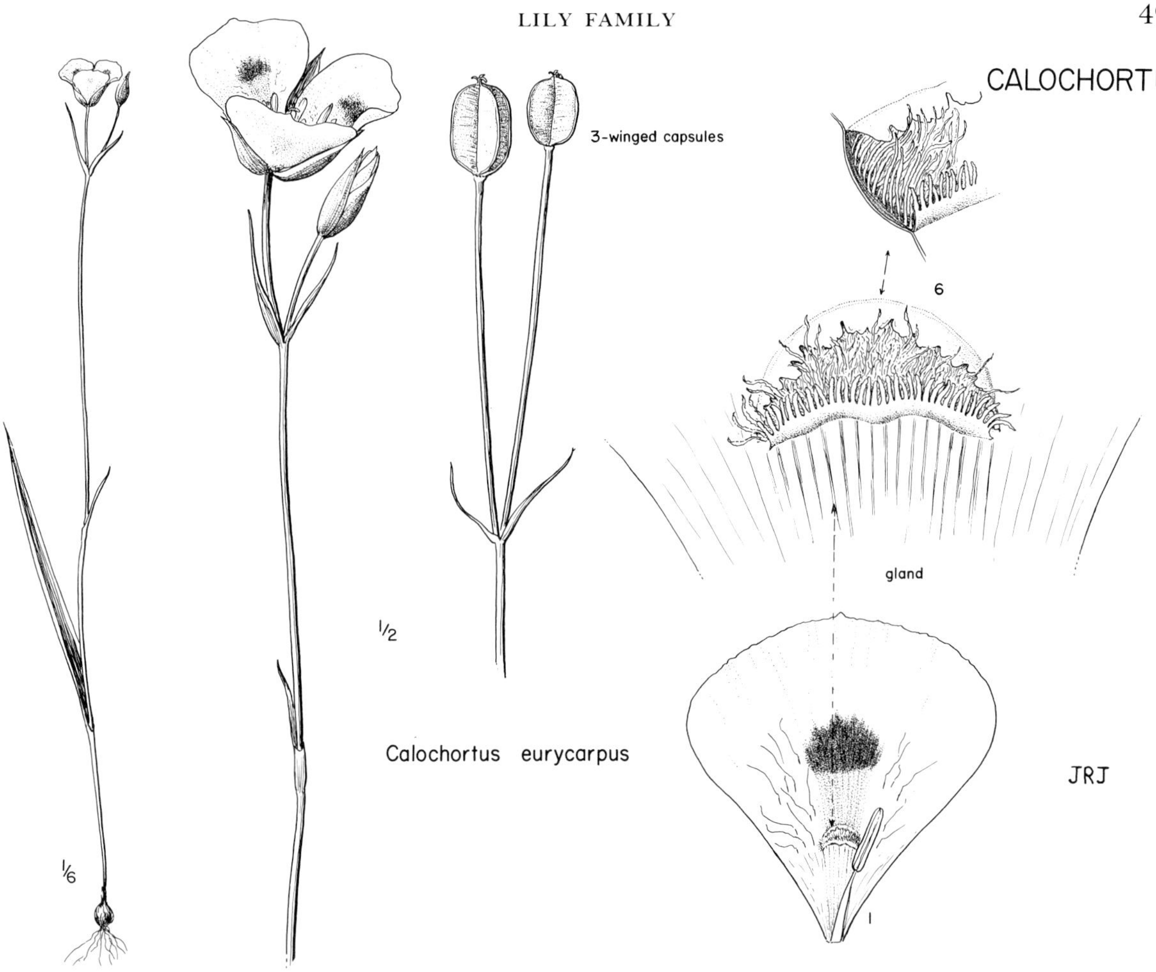

9 Glands transverse lunate to nearly circular, the margin of the gland continuous except at the base; plants typically bulbiferous; Pine Valley Mts., Utah, and Kaibab Plateau, Ariz. .... 9. *C. ambiguus*

9 Glands transversely oblong and elongate, often arched, the margin of the gland irregularly discontinuous; plants rarely bulbiferous; e. Utah from San Juan Co. n. to Grand and Carbon cos. .... 10. *C. gunnisonii*

1. Calochortus eurycarpus S. Wats.

Calochortus eurycarpus S. Wats. in King, Rep. Geol. Explor. 40th Parallel 5: 348. 1871. *C. nitidus* var. *eurycarpus* L. F. Henderson, Bull. Torrey Bot. Club 27: 356. 1900. (*Watson 1173,* East Humboldt [= Ruby] Mts., Elko Co., Nev., July 1868.)

C. parviflorus Nutt. ex Baker, J. Linn. Soc., Bot. 14: 306. 1874. (*Nuttall,* "Rocky Mts.," probably from s. Idaho, in 1834.)

C. umbellatus A. Nels. Bot. Gaz. 54: 405. 1912; not A. Wood, 1868. *C. euumbellatus* A. Nels. Bot. Gaz. 54: errata. 1912. (*Nelson & Macbride 1197,* Ketchum, Blaine Co., Idaho, 19 July 1911.)

Erect perennial herbs 1–5 dm tall with a single, bractlike leaf about midway and a simple, subumbellate, 1- to 5-flowered inflorescence from an ovoid bulb, the stem not bulbiferous; basal leaf flat, 1–3 dm long, 5–25 mm wide, involute and tapering at both ends, usually exceeding the inflorescence; bracts two or more, narrowly lanceolate to linear, 1–5 cm long, up to 3 mm wide, becoming reduced upwardly when along the stem, otherwise of unequal length at the base of the inflorescence; flowers campanulate, erect, white (drying yellowish) or rarely (in ours) lavender with a large reddish-purple blotch in the middle of each petal; sepals lanceolate to narrowly ovate, acute to acuminate, 1–3 cm long, shorter than the petals, glabrous; petals obovate, cuneate, rounded to acute at the apex, 2–4 cm long, with a few long (up to 5 mm), flexuous hairs with dark purple bases and whitish to yellowish tips near the gland; glands triangular-lunate, slightly depressed and bordered below by a narrow, deeply fringed membrane and with a narrower, crenate membrane above, the enclosed surface of the gland densely covered with long, yellowish processes, the membrane and surface often minutely papillose; anthers oblong, obtuse, nearly as long as the filaments; ovary 3-winged; capsules elliptic-oblong, 3-winged, 1.5–3 cm long, erect, light brown to tan; seeds straw-colored; $n = 10$.

Dry grassy slopes, margins of meadows, open grasslands and coniferous forests of sw. Mont. and nw. Wyo. w. across c. Idaho to e. Oregon, and disjunct in ne. Nev.; infrequent in our area, mostly on the foothills along the n. boundary of the Intermountain Region from se. Oregon e. into the Snake River Sect. from Ada Co. e. to Bonneville Co., Idaho, and with a disjunct location in the Ruby Mts., Elko Co., Nev., from 1200–2000 m elev. June–Aug.

This is one of our more infrequently collected species of *Calochortus* although it is not particularly rare. In our part of its range, habitats suitable for its survival are being reduced by increased demands for farm and grazing lands; it is, however, still common to the north of us. In Nevada, *C. eurycarpus* is restricted to the middle elevations in the Ruby Mountains, but in southern Idaho and southeastern Oregon, it is found at lower elevation, and near the Craters-of-the-Moon National Monument, it occurs on the floor of the valley. The majority of our plants have white petals, but some with lavender petals may be found in Oregon.

2. Calochortus flexuosus S. Wats.

Calochortus flexuosus S. Wats. Amer. Naturalist 7: 303. 1873. (*Thompson,* "southern Utah and Northern Arizona," from near Kanab, Kane Co., Utah, in 1872.)

Erect or more often more or less decumbent perennial herbs twining among other plants or straggling along the ground with the branched stems (1.5) 2–4 dm long, with scattered leaves and a monochasial, 1- to 4-flowered inflorescence, arising from an ovoid bulb, the stem rarely bulbiferous; leaves linear, somewhat involute, 1–2.5 dm long, 1–5 mm wide, attenuate and reduced upward, often exceeding the inflorescence; bracts two or more, linear, 1–3 cm long, up to 3 mm wide, unequal in length at the base of the inflorescence; flowers campanulate, erect, white with a lilac tinge, each petal with a transverse yellow band at the gland and usually a purple spot on the claw, the sepals similarly marked but usually with a more golden or orangish band; sepals lanceolate, obtuse to acute, 1–2.5 cm long, shorter than the petals, glabrous; petals obovate, cuneate, rounded to obtuse at the apex, 2–3.5 cm long, with only a few short, thick hairs near the gland or entirely glabrous; glands transverse-lunate to nearly circular, not depressed nor covered with a membrane, densely covered with rather short processes; anthers oblong, obtuse, about as long as the filaments; ovary linear, not winged; capsules lanceolate, acute, 3-angled, 2.5–3.5 cm long, erect, light tan; seeds light tan; $n = 7$.

Rather common on dry stony slopes and rocky mesas and flats from the Mojave Desert of se. Calif. e. across Nev. and Utah to sw. Colo. and Ariz.; along the s. boundary of our area from s. Esmeralda Co., Nev., se. into the s. tier of cos. in Utah, hence n. and e. to Beaver Co. in w. Utah, and Emery and Grand cos. in e. Utah, mostly 500 to 2000 m elev. April–June.

3. Calochortus leichtlinii Hook. f.

Calochortus leichtlinii Hook. f. Bot. Mag. 96: 5862. 1870. *C. nuttallii* var. *leichtlinii* Smiley, Univ. Calif. Publ. Bot. 9: 139. 1921. *Mariposa leichtlinii* Hoover, Leafl. W. Bot. 4: 2. 1944. (*Roezl,* "in the Sierra Nevada of California" was the original source of bulbs which flowered at the Royal Botanic Gardens, Kew, England, in 1870.)

C. nuttallii var. *subalpinus* M. E. Jones, Contr. W. Bot. 12: 78. 1908. (*M. E. Jones s.n.,* Summit, Placer Co., Calif., 26 July 1900; lectotype selected by Ownbey.)

Erect perennial herbs 2–4 (6) dm tall with erect usually unbranched stems bearing only a few scattered leaves and a subumbellate, (1-) 2- to 4- (5)-flowered inflorescence from an ovoid bulb, the stem bulbiferous with a solitary bulblet in a pouchlike sheath at the base of the lowest cauline leaf; leaves linear, flat, the lower ones 1–1.5 dm long, 2–3 mm wide, becoming reduced above, not exceeding the inflorescence; bracts mostly two, linear, 1–5 cm long, 1–3 mm wide, equal or only slightly unequal in length at the base of the inflorescence; flowers campanulate, erect, white or smoke-colored and often tinged with pink or lavender and with a dark red to nearly black spot just above the gland on each petal; sepals lanceolate, acute or acuminate, (1) 1.5–2 cm long, much shorter than the petals, glabrous; petals obovate, cuneate, usually rounded to obtuse at the apex, (1.5) 2–4 cm long, with only a few short hairs near the gland; glands subovate, slightly depressed but not covered by a membrane, densely covered with short hairlike processes; anthers linear-oblong, more or less sagittate at the base, longer than the filaments; ovary linear, not winged; capsules broadly lanceolate, acute, 3-angled, 1–3 cm long, erect, dark brown; seeds dark brown to nearly black with inflated coats; $n = 7$, $2n = 28$.

Open gravelly slopes and ridges in the Sierra Nevada of Calif. and adjacent ranges in w. Nev.; rare and infrequent in our area, occurring only in the Sweetwater Mts. of Mono Co., Calif., and Mineral Co., Nev., from 2500 to 3100 m elev., and along the edge of our boundary near Lee Vining, Mono Lake and Fales Hot Springs from 2000 to 2300 m elev. June–Aug.

4. Calochortus macrocarpus Douglas

Calochortus macrocarpus Douglas, Trans. Hort. Soc. London 7: 276. 1828. *Mariposa macrocarpa* Hoover, Leafl. W. Bot. 4: 2. 1944. (*Douglas,* Great Falls of the Columbia; lectotype selected by Ownbey.)

C. acuminatus Rydb. Bull. Torrey Bot. Club 24: 189. 1897. (*Rydberg 2600,* Lima, Beaverhead Co., Mont., 5 Aug. 1895.)

C. cyaneus A. Nels. Bot. Gaz. 53: 219. 1912. *C. macrocarpus* var. *cyaneus* J. F. Macbr. Contr. Gray Herb. 56: 14. 1918. (*Macbride 268,* Boise, Ada Co., Idaho, 18 June 1910.)

Erect perennial herbs 2–5 dm tall with stout, usually unbranched stems bearing several leaves and a subumbellate, 1- to 3-flowered inflorescence from an ovoid bulb, the stem often bulbiferous; leaves linear, strongly involute and curled at the tip, 0.5–1 dm long, 0.5–3 mm wide, becoming reduced above, not exceeding the inflorescence; bracts two or more, linear, unequal in length, the 2 longer ones 3–6 (10) cm long, the shorter ones, when present, up to 3 (5) cm long; flowers campanulate, erect, purple, with a median longitudinal green stripe and sometimes a transverse band of darker purple above the gland on each petal; sepals narrowly lanceolate, acuminate, 4–5 cm long, usually longer than the petals, glabrous; petals narrowly obovate, acuminate, 4–6 cm long, with slender to stoutish, yellowish to golden hairs up to 4 mm long at the edge and above the gland; glands triangular-oblong with a sagittate base, slightly depressed, surrounded by an essentially continuous, fringed, broad membrane, densely covered by slender, usually distally branched processes; anthers linear to linear-lanceolate, obtuse, longer than the filaments; ovary linear, not winged; cap-

CALOCHORTUS

Calochortus flexuosus

Calochortus leichtlinii

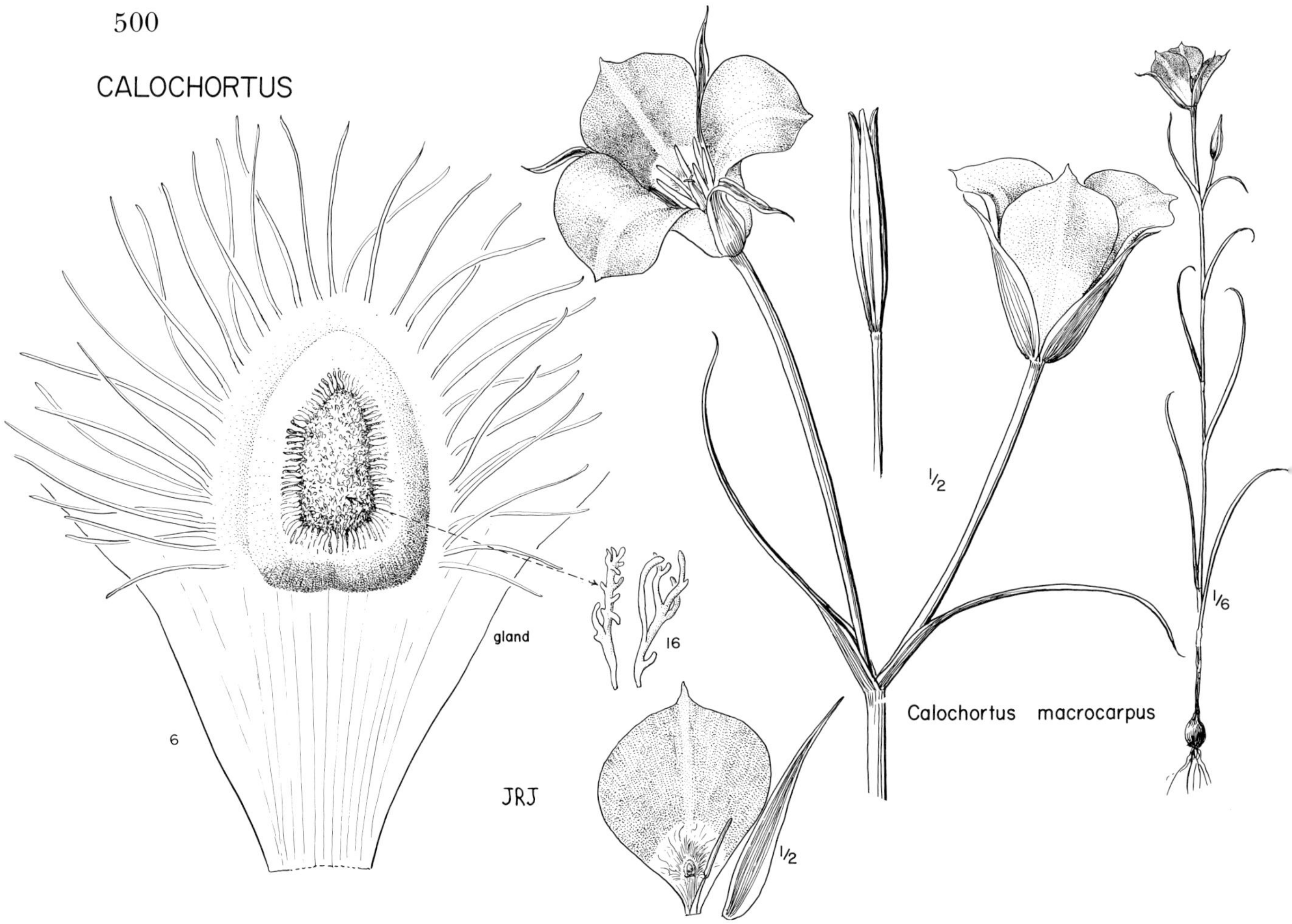

sules linear-lanceolate, acuminate, 3-angled, (3) 4–5 cm long, erect, tan to light brown; seeds straw-colored with inflated coats; $n = 7$.

Dry hills and flats of mostly volcanic soils from s. B.C. s. across Wash., Idaho and w. Mont. to ne. Calif. and n. Nev.; common across the nw. corner of our area from Lassen Co., Calif., ne. across se. Oregon and onto the ne. portion of the Snake River Sect. in Idaho, less frequent in n. Nev. but currently known from n. Washoe Co. e. to the Ruby Mts. of Elko Co. where it is found on limestone soils, mostly 850 to 2500 m elev. May–July.

This elegant *Calochortus* is slowly disappearing in the northwestern part of the Intermountain Region. It is highly palatable to livestock, and the few isolated populations in Oregon and Nevada are being seriously reduced by this grazing pressure. On the Snake River Plains of Idaho, and to a lesser degree also in Oregon, the increasing cultivation of desert lands is also leading to the gradual extinction of *C. macrocarpus* for this part of its range. Unless firm conservation efforts are made to protect and preserve "park lands" in the Snake River Plains, numerous plants will be lost forever from this part of Idaho.

Our plants belong to the var. *macrocarpus.* The more local and more northern variant, var. *maculosus* (Nels. & Macbr.) Nels. & Macbr. ex Macbr. has white petals with red-purple bands. It occurs in the Lewiston, Idaho, area.

5. Calochortus bruneaunis Nels. & Macbr.

Calochortus bruneaunis Nels. & Macbr. Bot. Gaz. 55: 372. 1913.
C. nuttallii var. *bruneaunis* Ownbey, Ann. Missouri Bot. Gard. 27: 492. 1940. (*Nelson & Macbride 1881,* high on the canyon sides of the "Hot Hole," East Fork of Bruneau Cr., Owyhee Co., Idaho, 2 July 1912.)
C. discolor A. Davidson, Bull. S. Calif. Acad. Sci. 14: 11. 1915. (*Davidson 2672,* Andrews Camp along Bishop Cr., w. of Bishop, Inyo Co., Calif., July 1911.)

Erect perennial herbs 2–4 dm tall with slender, usually simple stems bearing only a few leaves and a subumbellate, 1- to 4-flowered inflorescence from ovoid bulbs, the stem often bulbiferous; leaves linear, flat but becoming involute, the lower one 0.5–1.5 (2) dm long, (1) 2–4 (5) mm wide, becoming reduced above, not usually exceeding the inflorescence; bracts two or more, linear, (1) 2–4 (7) cm long, unequal in length at the base of the inflorescence; flowers campanulate, erect, white and often tinged with lilac, with a median longitudinal green stripe and a dark red or purple spot above the gland on each sepal and petal; sepals lanceolate, acuminate, 1.5–3.5 cm long, usually shorter than the petals, glabrous; petals narrowly obovate to oblanceolate, conspicuously acuminate, 2–4 cm long, glabrous or with a few short golden hairs near the gland; glands circular, depressed, surrounded by a conspicuously fringed membrane and densely covered with short, simple or distally branched processes; anthers oblong, obtuse, yellowish, bluish or reddish-brown, slightly longer than the filaments; ovary linear, not winged; capsules linear-lanceolate, acuminate, 3-angled, 3–6 (7) cm long, erect, light tan to brown; seeds straw-colored with loose-fitting coats; $n = 7$.

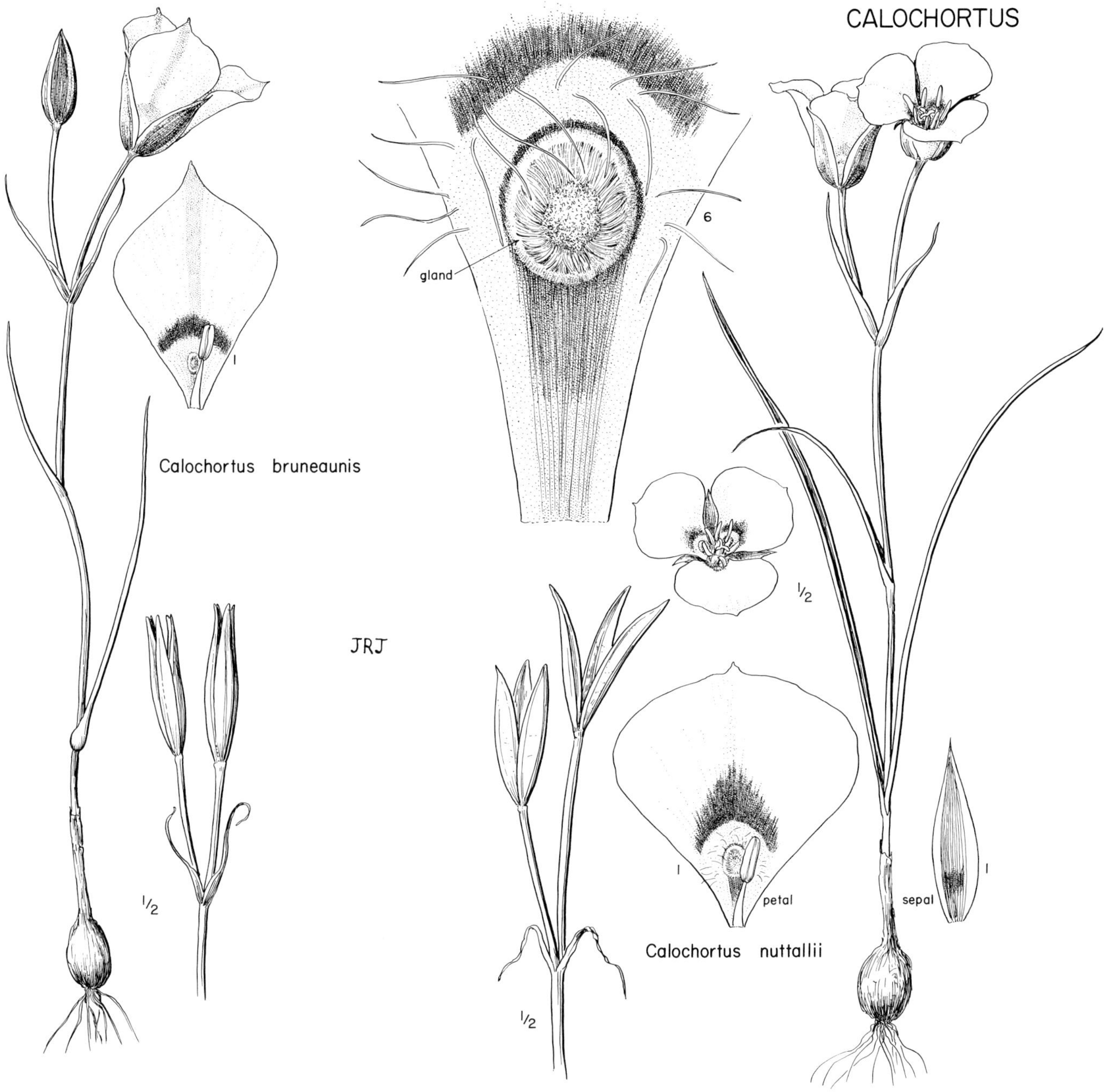

Dry sandy to rocky plains and hills from e. Calif. ne. across Nev. into s. Idaho and sw. Mont.; usually associated with sagebrush and pinyon-juniper woodlands in the w. and n. portions of our area, ranging from e. Calif. (Lassen Co. s. to Inyo Co.), ne. across Nev. and se. Oregon onto the Snake River Plains and just entering Utah in Box Elder Co., mostly 850 to 2200 m elev. May–July.

In the past, *C. bruneaunis* has been relegated to the varietal rank under *C. nuttallii.* Ownbey (1969) has suggested that this should not be the case and elevated this element to the species rank, supporting his conclusions on the basis of chromosome findings, various morphological criteria and the distinct geographical range. His opinion is followed here, and expanded to elevate *C. nuttallii* var. *panamintensis* Ownbey (Ann. Missouri Bot. Gard. 27: 493. 1940) to the species rank, **Calochortus panamintensis** (Ownbey) Reveal, hoc loc. The references to this species found in Lloyd and Mitchell (Fl. White Mts. Calif. Nev. 191. 1973) were based on specimens of *C. bruneaunis.* In general, *C. bruneaunis* occurs to the west of *C. nuttallii,* often in more rocky places, and almost always in association with *Artemisia. Calochortus nuttallii* occupies similar habitats, but several other niches as well, especially outside our floristic boundaries. *Calochortus panamintensis* is found outside our area, to the south of us in the Inyo and Panamint mountains of Inyo Co., California.

6. **Calochortus nuttallii** Torr. & Gray

Calochortus luteus Nutt. J. Acad. Nat. Sci. Philadelphia 7: 53. 1834; not Douglas, 1833. *C. nuttallii* Torr. & Gray in Beckwith, Explor. & Surv. Railroad Route from Mississippi River to Pacific Ocean 2: 124. 1854. *C. watsoni* M. E. Jones, Contr. W. Bot. 14: 26. 1912. (*Wyeth,* "Towards the sources of the Columbia," certainly an error; most likely the type came from near (or at) Pierre's Hole, Teton Basin, Teton Co., Idaho, or just e. of there on the w. slope of the Teton Range, Teton Co., Wyo., July 1833.)

C. rhodothecus Clokey, Bull. S. Calif. Acad. Sci. 37: 1. 1938. (*Clokey 7479,* Charleston Peak, Charleston Mts. [= Spring Mts.], Clark Co., Nev., 21 July 1937.)

Erect perennial herbs 2–4 (5) dm tall with slender, usually simple stems bearing only a few leaves and a subumbellate, 1- to 4-flowered inflorescence from ovoid bulbs, the stem often bulbiferous; leaves linear, flat but becoming involute, the lower one 1–2.5 (3) dm long, 2–4 mm wide, becoming reduced above, rarely exceeding the inflorescence; bracts two or more, broadly linear, 3–9 cm long, unequal at the base of the inflorescence; flowers campanulate, erect, white and tinged with lilac or infrequently magenta, yellow at the base of the claw and with a reddish-brown or purple band or spot above the gland on each sepal and petal; sepals lanceolate to broadly lanceolate, acuminate, 2–3 (3.5) cm long, shorter than the petals, glabrous; petals broadly obovate, cuneate, short-acuminate, 3–4 (5) cm long, with scattered yellowish to golden, slender hairs near the gland; glands circular, depressed, surrounded by a conspicuously fringed membrane and densely covered by short, simple or distally branched processes; anthers oblong, obtuse, yellowish or pinkish, about equalling the length of the filaments; ovary linear, not winged; capsules linear-lanceolate, acuminate, 3-angled, 3–5 cm long, erect, tan to brown; seeds straw-colored with loose-fitting coats; $n = 8$.

Dry slopes and flats from the n. Great Plains of sw. N.D. and e. Mont. s. to w. Nebr., hence w. across Wyo. to se. Idaho, e. and s. Nev., Utah, w. Colo. and as far s. as nw. N.M. and n. Ariz.; common in our area throughout most of Utah and onto the Kaibab Plateau in n. Ariz., but less frequent in e. Nev. and rare in c. Nev. (Nye Co.) and known only se. of the Snake River in se. Idaho and our portion of adjacent sw. Wyo., mostly 750–2500 m elev. June–July.

Calochortus nuttallii (see the frontispiece) is well known to the people of Utah as the state's official flower and as a source of food for the Indians and early pioneers. The plant has become somewhat sacred because of these associations, and hardly a year goes by without a newspaper story documenting the near demise of sego lily. Needless to say, the truth lies more closely toward the middle. True, the native habitats for *C. nuttallii* are being largely replaced by homes, pavement, wheatfields or orchards, especially along the Wasatch Front. Away from the population centers of Utah though, the species is still rather common and is only threatened by enthusiastic wildflower collectors or an occasional hungry cow. Without firm protection of native, undisturbed niches along the front ranges of the Wasatch, the state flower of Utah could be rendered exceedingly rare in this area. As noted in the discussion of the genus, it is somewhat difficult to successfully cultivate species of *Calochortus,* and this is particularly true of *C. nuttallii.* It is far better to observe this plant where it occurs naturally and preserve these places than dig up the bulbs in order to "save the species" by growing it in gardens.

As now defined, *C. nuttallii* is a plant of the Great Plains, Wyoming, western Colorado and Utah, with a smattering of isolated populations in adjacent areas. Ownbey (1969) has pointed out the need to redefine the species, and we are following his suggestion. I have associated *C. panamintensis* and *C. bruneaunis* (the latter is $n = 7$), and recognized *C. aureus* ($n = 16$) as a distinct species differing from *C. nuttallii* ($n = 8$) in ploidy level and flower color. Ownbey (1940, 1969) repeatedly stated that the *C. nuttallii* complex is composed of weakly defined elements as far as their general morphology is concerned, but that the various species are consistent in all attributes. No doubt the differing chromosome levels, distinct geographical ranges and varied habitat requirements have all played active roles in maintaining the species distinctions.

7. Calochortus aureus S. Wats.

Calochortus aureus S. Wats. Amer. Naturalist 7: 303. 1873. *C. nuttallii* var. *aureus* Ownbey, Ann. Missouri Bot. Gard. 27: 493. 1940. (*Thompson,* near Kanab, Kane Co., Utah, in 1872.)

Erect perennial herbs 1–3 dm tall with slender, usually simple stems bearing only a few leaves and a subumbellate, 1- to 4-flowered inflorescence from ovoid bulbs, the stem often bulbiferous; leaves linear, flat but becoming involute, the lower one 0.8–1.5 (2) dm long, 1–2.5 (3) mm wide, becoming reduced above, rarely exceeding the inflorescence; bracts two or more, linear, 1–6 cm long, unequal at the base of the inflorescence; flowers campanulate, erect, lemon-yellow with a maroon blotch above the gland on each sepal and petal; sepals broadly lanceolate, acuminate, 1.5–2.5 cm long, shorter than the petals, glabrous; petals broadly obovate, cuneate, short-acuminate, 3–3.5 cm long, glabrous or with only a few lemon-yellow hairs near the gland; glands circular, depressed, surrounded by a conspicuously fringed membrane and densely covered with short, simple or distally branched processes; anthers oblong, obtuse, yellowish to cream, about equalling the length of the filaments; ovary linear, not winged; capsules linear-lanceolate, acuminate, 3-angled, 2.5–5 cm long, light tan to light brown; seeds straw-colored with loose-fitting coats; $2n = 32$.

Dry sandy to clayey places in e.-central Ariz. and adjacent w.-central N.M., with disjunct populations from n. Ariz. n. into s. and e. Utah; rather infrequent in our area, mainly on clay soils from the Kanab area of Utah and Ariz. ne. into the San Rafael Desert as far n. as Emery Co., from about 1200–2000 m elev. May–June.

In our area, *C. aureus* is easily distinguished by the lemon-yellow flowers, our only species with such coloration. A Mojave Desert species with vermilion to orange flowers, *C. kennedyi* T. C. Porter, approaches our area in Esmeralda Co., Nevada. It further differs from *C. aureus* by having much longer and more pronounced hairs near the gland.

8. Calochortus excavatus Greene

Calochortus excavatus Greene, Pittonia 2: 71. 1890. (*Shockley 427,* Bishop Cr., w. of Bishop, Inyo Co., Calif., 30 May 1886.)

C. campestris A. Davidson, Bull. S. Calif. Acad. Sci. 14: 12. 1915. (*Davidson 2657,* meadows in Bishop, Inyo Co., Calif.)

Erect perennial herbs 1–3 dm tall with slender, usually simple stems bearing only a few leaves and a subumbellate, 1- to 4-flowered inflorescence from ovoid bulbs, the stem often bulbiferous; leaves linear, flat or involute, the lower one 1–2.5 dm long, 2–4 mm wide, only slightly reduced above, infrequently exceeding the length of the inflorescence; bracts two, linear, 3–5 cm long, occasionally with more hairlike bracts but these less than 1 cm long; flowers campanulate, erect, lavender to pinkish, usually with a purple spot above the gland on each sepal and petal; sepals lanceolate, acuminate, 2–2.5 cm long, shorter than the petals, glabrous; petals broadly obovate, cuneate, obtuse to rounded, 2.5–3.5 cm long, with scattered yellowish to golden hairs near the gland; glands circular, depressed, surrounded by a conspicuously fringed membrane and densely covered by short, simple or distally branched processes; anthers oblong, obtuse, reddish-brown,

CALOCHORTUS

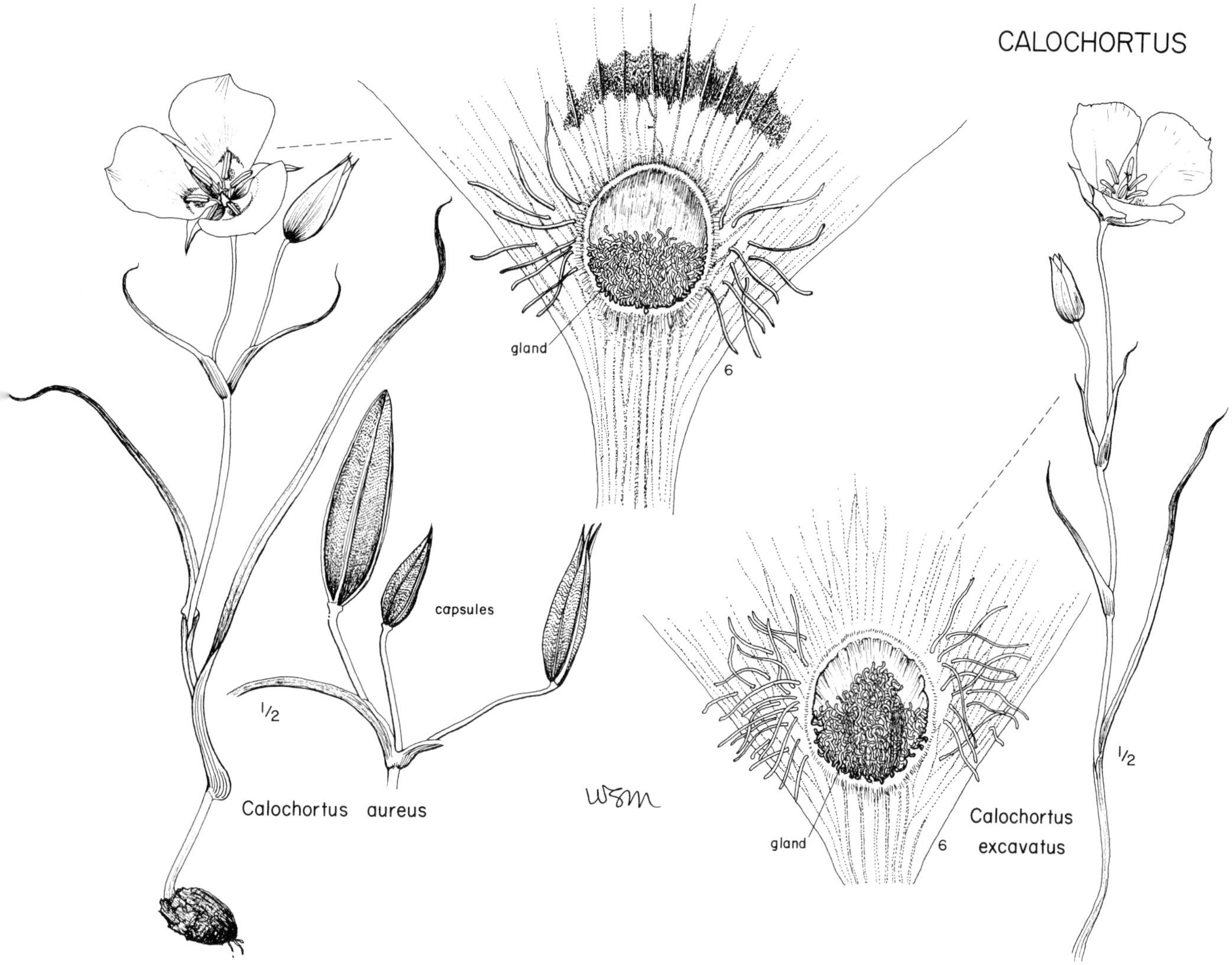

about equaling the length of the filaments; ovary linear, not winged; capsules linear-lanceolate, acute, 3-angled, 2–3 cm long, brown; seeds tan.

Infrequent in damp to wet grassy meadows in and near Bishop s. to Lone Pine, mostly just w. of our boundary on the lower foothills of the Sierra Nevada, Inyo Co., Calif., 1200–2000 m elev. April–May.

9. Calochortus ambiguus (M. E. Jones) Ownbey

Calochortus watsoni var. *ambiguus* M. E. Jones, Contr. W. Bot. 14: 27. 1912. *C. ambiguus* Ownbey, Ann. Missouri Bot. Gard. 27: 505. 1940. (*M. E. Jones s.n.,* Flagstaff, Coconino Co., Ariz., 5 Aug. 1884.)

Erect perennial herbs 1.5–5 dm tall with slender, usually simple stems bearing several leaves and a subumbellate, 1- to 4-flowered inflorescence from ovoid bulbs, the stem often bulbiferous; leaves linear, flat but becoming involute, 1–2.5 dm long, 1–3 mm wide, becoming only slightly reduced above, not exceeding the inflorescence; bracts two or more, linear, (3) 5–10 cm long, unequal at the base of the inflorescence; flowers campanulate, erect, pinkish to bluish-gray, sometimes marked with a curved band of purple above the gland and a purple spot on the claw, infrequently with a longitudinal stripe of gray on each sepal and petal; sepals lanceolate to broadly lanceolate, acuminate, (2) 2.5–3.5 cm long, shorter than the petals, glabrous; petals obovate, cuneate, rounded to obtuse or sharply acuminate, 3–4.5 cm long, with yellowish, slender to distally enlarged or branched hairs near the gland; glands transversely lunate to circular, depressed, continually surrounded with a narrow, deeply fringed membrane on all edges but the base, and densely covered by short, distally branched processes; anthers oblong-lanceolate to (rarely) acute, mostly pinkish to maroon, longer than the filaments; ovary linear, not winged; capsules linear-oblong, acute, 3-angled, 4–7 cm long, erect, usually dark tan to dark reddish-brown; seeds light brown with inflated coats; $n = 9$.

Dry slopes and hills throughout most of Ariz. with scattered populations ranging into sw. N.M. and sw. Utah; known from only two locations in our area, the Kaibab Plateau in Coconino and Mohave cos., Ariz., and in the Pine Valley Mts., Washington Co., Utah, from 2000–2500 m elev. May–July.

Calochortus ambiguus, which closely resembles *C. nuttallii,* often occurs with that species on the Kaibab Plateau, and separation of the two is rather difficult in this location. The margin of the gland in *C. nuttallii* is wider and more finely divided, and totally surrounds the gland. In *C. ambiguus*, the membrane is narrower and more deeply fringed with the individual segments less fine, and discontinuously surrounds the gland. The shape of the glands differs as well, with those of *C. nuttallii* being circular, and those of *C. ambiguus* usually—but unfortunately not always—transversely lunate. There is a purplish band above the gland in our specimens of *C. ambiguus*, but this feature is not consistent and a similar band occurs in some collections of *C. nuttallii*. The hairs on the petals are usually distally enlarged in *C. ambiguus*, but this is not the case in the majority of our specimens. Fortunately, the range of *C. gunnisonii*, with which *C. ambiguus* also is related, occurs to the northeast of that of *C. ambiguus*, and geographically the two do not overlap.

10. Calochortus gunnisonii S. Wats.

Calochortus gunnisoni S. Wats. in King, Rep. Geol. Explor. 40th Parallel 5: 348. 1871. (*Creuzefeldt*, "high mountains. Utah," certainly an error, probably obtained in c. Colo., in 1853; lectotype selected by Ownbey. See discussion below.)

Erect perennial herbs (1.5) 3–6 dm tall with slender, usually simple stems bearing several leaves and a subumbellate, 1- to 3-flowered inflorescence from ovoid bulbs, the stem rarely bulbiferous; leaves linear, flat but becoming involute, 1–3 (3.5) dm long, (2) 3–6 (8) mm wide, becoming only slightly reduced above, not exceeding the inflorescence; bracts two or more, linear, 3–15 cm long, unequal at the base of the inflorescence; flowers campanulate, erect, white to purple (ours) or yellow, often with a narrow, transverse purple band above the gland and a purple spot on the claw of each sepal and petal; sepals lanceolate to broadly lanceolate, acute, 2–3 (3.5) cm long, shorter than the petals, glabrous; petals obovate, cuneate, obtuse to rounded, (2.5) 3–4 cm long, with numerous golden (except for the dark basal portion), distally branched, gland-tipped hairs on the claw and near the gland; glands transversely oblong and more or less arched, depressed, discontinuously surrounded by a deeply fringed but indistinct membrane and densely covered with short, distally branched processes; anthers lanceolate, narrowly acute to strongly apiculate, mostly yellowish to cream, longer than the filaments; ovary linear, not winged; capsules linear-oblong, acute, 3-angled, (2.5) 3–5 (6) cm long, erect, usually light tan; seeds straw-colored with inflated coats; $n = 9$.

Dry to moist slopes primarily in the Rocky Mts. of Mont. s. through Wyo. and Colo. into n. N.M. with outlying populations in the Black Hills of S.D., se. Ariz. and adjacent w. N.M.; restricted in our area to the e. tier of cos. in Utah from Carbon and Grand cos. s. to San Juan Co., mostly in the higher mts., 2200–2800 m elev. June–Aug.

The type of *C. gunnisonii* was supposedly collected in the "high mountains" of Utah. The specimens were obtained by Creuzefeldt, the botanist for the Gunnison Expedition, but, as I have noted in the first volume of this flora (p. 51), Creuzefeldt was killed by Indians in western Utah on 21 October 1853, along with Gunnison and several others. It is likely that his death prevented the completely accurate labeling of his specimens, and this collection was erroneously labeled Utah.

This supposition is further supported in that the expedition did not even enter Utah until the last week of September, 1853, and was restricted to the clay hills and flats of the Grand Valley until it crossed the southern Wasatch Mountains in mid October. It is reasonable to assume that, as the species does not occur in these mountains, and as it flowers several months earlier, the type had to come from somewhere else. As a result, I suspect the type was obtained in the southern Rocky Mountains of southern Colorado, where the expedition was in late July and where the species is very common.

Our plants belong to var. *gunnisonii*. The var. *perpulcher* Cockerell has yellow petals and occurs in Mora and San Miguel counties of New Mexico. References to *C. gunnisonii* in northern Arizona (i.e., McDougall, Seed Pls. N. Ariz. 99. 1973) are without foundation.

16. CAMASSIA Lindl. nom. conserv. Camas

Tall, elegant scapose glabrous perennial herbs from deep-seated tunicate bulbs; leaves numerous, basal, long-linear, keeled; flowers large and showy, white to deep blue or violet, many and somewhat clustered in dense to lax bracteated racemes, ascending to spreading on jointed pedicels arising from the axils of lanceolate, scarious bracts; tepals 6, alike, distinct, regular or occasionally somewhat spreading with five tepals curving upward and the sixth downwardly turned; stamens 6, borne on the receptacle, the filaments slender, the anthers versatile, introrse; ovary superior, 3-celled with many ovules, the long and slender style terminated by a 3-lobed or -parted stigma; fruits ovoid to oblong or subglobose loculicidal capsules; seeds pyriform to ovoid-ellipsoid, lustrous black.

A genus of 5 species found in N. Amer. (From the Shoshone name, *camas* or *quamash*, as recorded by Lewis and Clark.)

Reference: Gould, F. W. A systematic treatment of the genus *Camassia* Lindl. Amer. Midl. Naturalist 28: 712–742. 1942.

1 Flowers regular, usually with 1–3 flowers in bloom at a time; tepals covering the capsule; capsules not appressed to the scape; e.-central Calif. . . . . 1. *C. leichtlinii*
1 Flowers slightly irregular, with 5 tepals curving upwardly and the sixth downward, usually with numerous flowers in bloom at one time; tepals separated by the maturing capsule; capsules closely appressed to the scape; ne. Calif. e. across se. Oregon and s. Idaho to n. Utah . . 2. *C. quamash*

1. Camassia leichtlinii (Baker) S. Wats.

Chlorogalum leichtlinii Baker, Gard. Chron. II. 1: 689. 1874. *Camassia esculenta* var. *leichtlinii* Baker, Bot. Mag. 103: 6287. 1877. *C. leichtlinii* S. Wats. Proc. Amer. Acad. Arts 20: 376. 1885. *Quamasia leichtlinii* Coville, Proc. Biol. Soc. Wash. 11: 63. 1897. (Grown from bulbs collected by *John Jeffrey* in Umpqua Valley, Douglas Co., Oregon, 3 May 1853.)

C. suksdorfii Greenman, Bot. Gaz. 34: 307. 1902. *Quamasia suksdorfii* Piper, Contr. U.S. Natl. Herb. 11: 191. 1906. *C. leichtlinii* subsp. *suksdorfii* Gould, Amer. Midl. Naturalist 28: 723. 1942. *C. leichtlinii* var. *suksdorfii* C. L. Hitchc. Univ. Wash. Publ. Biol. 17(1): 780. 1969. (*Suksdorf 251*, wet meadows of Falcon Valley, Klickitat Co., Wash., June–July 1883.)

Tall, erect perennial herbs (2) 3–10 (13) dm tall with slender scapes terminated by an elongated cluster of racemosely arranged flowers arising from ovate bulbs 2–4 cm long and 1.5–3 cm thick; leaves

CALOCHORTUS

Calochortus ambiguus

Calochortus gunnisonii

flat, 2–6 dm long, 0.5–2.5 cm wide, several from the base; flowers in close to loose many-flowered racemes 1–2 dm long, on ascending pedicels 1.5–5 cm long, the tepals cream-white to (in ours) blue-violet, the latter rarely producing albino forms, (2) 2.5–3.5 (4) cm long, 5–10 mm wide, narrowly lanceolate, narrowing gradually to the base, all regular, tending to twist together and covering the ovary as they wither; stamens 6, on a slender, usually included filament, the anther cream to blue, 4–7 mm long; capsules oblong to ovate, 1–3 cm long, tan to brown; $2n = 30$.

Grasslands to moist meadows from s. B.C. s. to Tulare Co., Calif.; just entering our area at Mono Lake, Mono Co., Calif. May–July.

Our plant is var. *suksdorfii* (Greenman) C. L. Hitchc. The var. *leichtlinii* is restricted to Douglas Co., Oregon.

2. Camassia quamash (Pursh) Greene

Phalangium quamash Pursh, Fl. Amer. Sept. 226. 1814. *Quamasia esculenta* Raf. Amer. Monthly Mag. & Crit. Rev. 2: 265. 1818. *Anthericum esculentum* Sprengel, Syst. Veg. 2: 84. 1825. *Camassia esculenta* Lindl. Edward's Bot. Reg. 18: 1486. 1832. *C. quamash* Greene, Manual Bot. Region San Francisco Bay 313. 1894. *Quamasia quamash* Coville, Proc. Biol. Soc. Wash. 11: 64. 1897. *C. leichtlinii* var. *watsoni* M. E. Jones, Contr. W. Bot. 14: 29. 1912. (*Lewis,* "near the foot of the Rocky Mountains on the Quamash flats," Weippe Prairie, Clearwater Co., Idaho, 23 June 1806.)

C. quamash subsp. *breviflora* Gould, Amer. Midl. Naturalist 28: 737. 1942. *C. quamash* var. *breviflora* C. L. Hitchc. Univ. Wash. Publ. Biol. 17(1): 780. 1969. (*Gould 436,* 7 mi. n. of Truckee, Sierra Co., Calif., 25 June 1938.)

C. quamash subsp. *utahensis* Gould, Amer. Midl. Naturalist 28: 740. 1942. *C. quamash* var. *utahensis* C. L. Hitchc. Univ. Wash. Publ. Biol. 17(1): 780. 1969. (*Maguire & Maguire 3265,* 12 mi. up Blacksmith Canyon, Cache Co., Utah, 5 June 1932.)

Tall erect perennial herbs (2) 3–7 dm tall with slender to stoutish scapes terminated by an elongated cluster of racemosely arranged flowers arising from ovoid bulbs 2–4 (5) cm long and 1–2.5 cm thick; leaves flat, 1–4 (5) dm long, 1–2.5 cm wide, several from the base; flowers in close to loose many-flowered racemes 1–2.5 dm long, on spreading or arching to ascending pedicels (0.7) 1–2 cm long, the tepals pale to deep blue or violet, rarely producing albino forms, narrowly lanceolate, 1.5–3 (4) cm long in ours, 2–6 mm wide, narrowed and more or less clawed at the base, slightly irregular with the lowermost tepal curved outward and downward from the others, all tending to twist together and covering the ovary as they wither, becoming separated by the maturing capsule; stamens 6, on a slender, usually exserted filament, the anther dark blue or yellow, 2.5–4 mm long; capsules ovoid, 1–2.5 cm long, tan to brown; $2n = 30$.

Moist areas, especially in grassy meadows or adjacent drier hillsides from B.C. s. to c. Calif., e. to Alta., Mont., nw. Wyo. and n. Utah; restricted in our area to the n. portions of the Intermountain Region. April–June (July).

We have only two of the eight variants of *C. quamash* in our area. The var. *breviflora* (Gould) C. L. Hitchc. has tepals 1.5–2 cm long that are bright blue to a deep blue-violet, and has bright yellow anthers. It occurs just inside our area in the Reno Section, but it becomes more frequent in the Lake Section of northeastern California and southeastern Oregon. The var. *utahensis* (Gould) C. L. Hitchc. has tepals that are 2–3 cm long, pale to a deep blue-violet in color, and anthers that are tinged with blue or brown. This variant ranges across the northern edge of the Snake River Section into the front ranges of the Wasatch Mountains, with a few outlying populations, as in Owyhee Co., Idaho.

This species was frequently used by the Indians, trappers, early explorers and pioneers as a major source of food. It is of some interest to visit portions of our area where, at one time, large concentrations of native Americans had established villages; today, *C. quamash* is often rather rare in these places. The same observation has been made regarding its frequency near the major population centers in Utah and Idaho today. In areas where man, grazing, or pollution have not had a detrimental effect upon *Camassia,* it forms large, colorful masses across the land, the blue to violet of the flowers harmonizing with the greens of the meadow grasses in a most pleasing manner, at least to this botanist.

17. ORNITHOGALUM L. Star-of-Bethlehem

Erect scapose perennial herbs with simple stems arising from tunicate bulbs; leaves few to several, basal, linear to lanceolate; flowers large and showy, few to many in open, bracteate racemes, white to greenish-white or yellow, on slender, non-jointed pedicels in the axils of narrow elongate bracts; tepals 6, subequal, distinct, spreading; stamens 6, the filaments broad, flattened, borne on the receptacle, the anthers versatile, introrse, oblong; ovary superior, 3-celled with few to several ovules, the short to long and slender style 3-lobed; fruits ovoid to subglobose, obtusely 3-angled loculicidal capsules; seeds few, dark, roundish.

An Old World genus of about 100 species with several species in cultivation. (A meaningless name derived from the Greek *ornis,* a bird, and *gala,* milk.)

1. Ornithogalum umbellatum L.

Ornithogalum umbellatum L. Sp. Pl. 307. 1753. ("Habitat in Germannia, Galiia.")

Erect perennial herbs 1–3 dm tall with stoutish stems terminated by a few-flowered corymb, arising from ovoid bulbs; leaves flat, 1–3 dm long, 4–8 mm wide; inflorescences open, 0.3–0.8 (1) dm long; flowers 1–2 cm long, on long, stoutish, ascending to spreading pedicels 3–5 cm long; tepals white with a broad green longitudinal stripe on the back of each tepal, oblong to oblong-lanceolate, slightly acuminate; stamens half as long as the tepals; capsules oblong-ovoid, 6-angled; $2n$ = from 18 to 90.

Native of Europe; escaping from cultivation in gardens, roadsides, weedy lots and grassy meadows; infrequent in our area. June.

18. MUSCARI Miller Grape Hyacinth

Erect slender scapose perennial herbs with simple stems arising from ovoid to subglobose tunicate bulbs; leaves few, basal, linear; flowers small, many in dense, ebracteate racemes, blue, azure to violet or rarely white to pinkish, on slender descending pedicels; tepals 6, alike, connate nearly their entire length, tubular

CAMASSIA

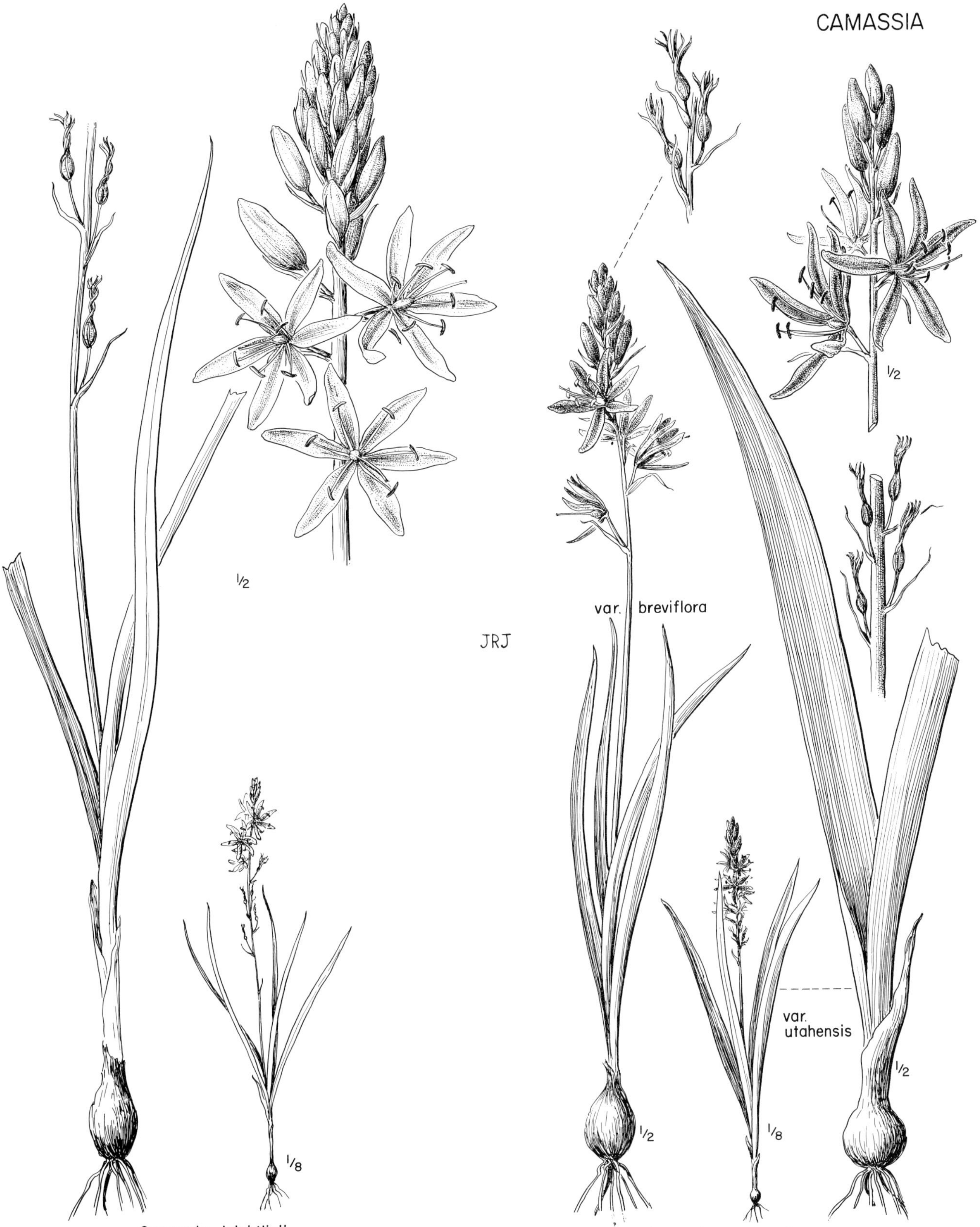

Camassia leichtlinii

Camassia quamash

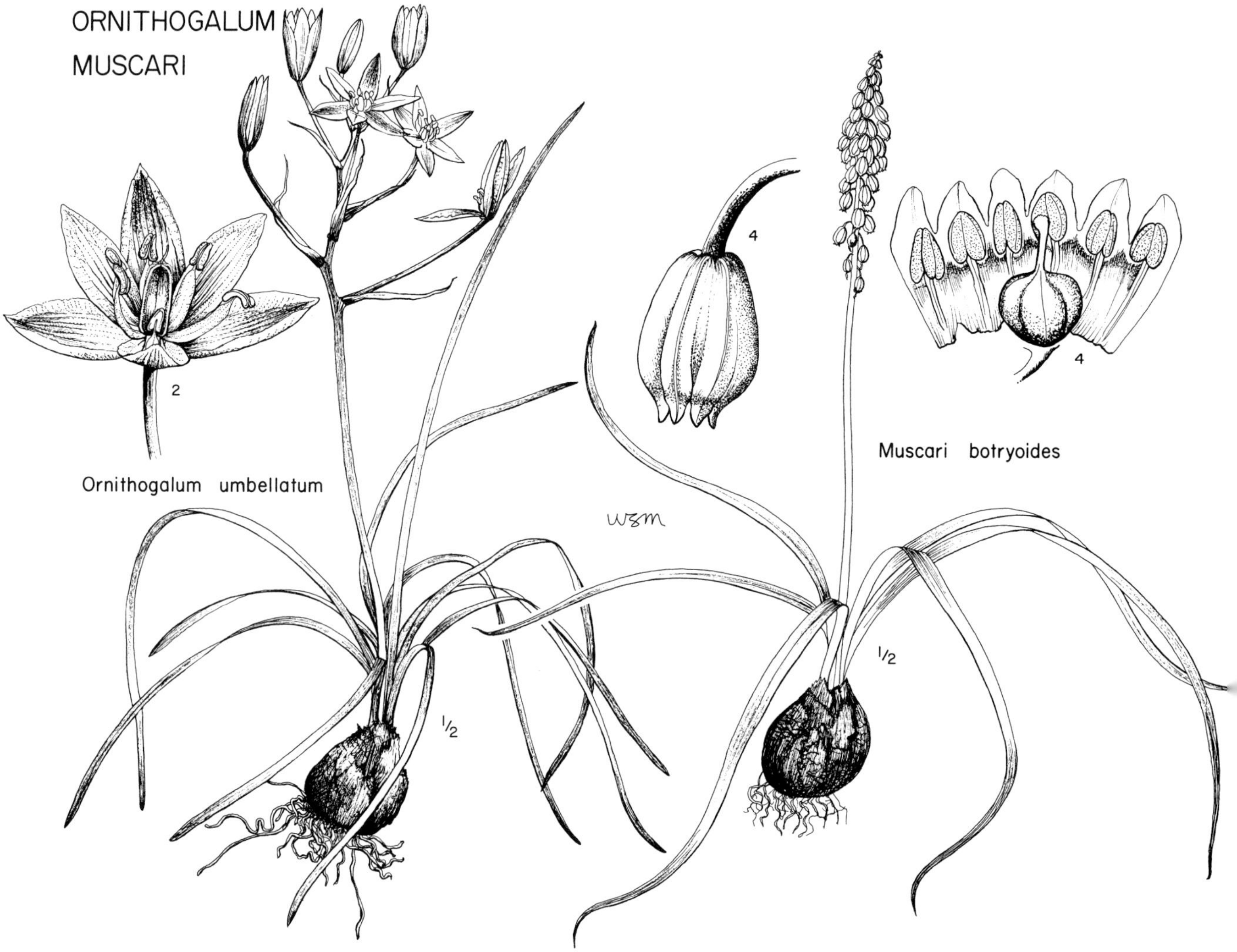

or urceolate; stamens 6, the filaments slender, inserted on the tepals, included, the anthers versatile, introrse, ovate; ovary superior, 3-celled with 2 ovules in each locule, with a short, slender style terminated by a capitate stigma; fruits distinctly 3-angled or -winged loculicidal capsules; seeds 2, black, angled.

A genus of perhaps 50 species native to the Old World with several species in cultivation. (From the Arabic *muskarimi,* alluding to the scent which is reminiscent of nutmeg or musk as applied to one species.)

1. Muscari botryoides (L.) Miller

Hyacinthus botryoides L. Sp. Pl. 318. 1753. *Muscari botryoides* Miller, Gard. Dict. ed. 8. unnumbered. 1768. *Botryanthus vulgaris* Kunth, Enum. Pl. 4: 311. 1843. *Botryphile botryoides* Salisb. Gen. Pls. 25. 1866. (Italy.)

Small erect perennial herbs 0.5–2 (3) dm tall with slender stems terminated by a densely flowered raceme, arising from ovoid bulbs; leaves flat, somewhat succulent, narrowly linear to linear-oblanceolate, (0.5) 1–2 (3) dm long, 3–8 (10) mm wide; inflorescences dense, racemose, 2–4 cm long at anthesis, elongating in fruit; flowers 4–5 mm long, dark blue, globular-urceolate, on cernuous pedicels 2–4 mm long; tepals united except for the slightly recurved acute tips; stamens small and included; capsules obovoid, 2–4 mm long; $2n = 18, 36, 48, 63$.

Native of the Medit. region, but widely cultivated and naturalized; mostly in grassy or weedy places in our area. April–June.

19. ALLIUM L. Wild Onion, Wild Garlic

Perennial herbs, ordinarily (including all our spp.) with an onion- or garlic-like odor, scapose from tunicated bulbs that are reformed annually, these sometimes terminal on primary or secondary rhizomes; leaves 1–several, usually linear, concave-convex or carinate to flat or terete, solid or hollow, often soon withering and deciduous; flowers perfect, several to many in a terminal, involucrate false umbel, or subcapitate, the involucral bracts conspicuous, commonly forming a spathe, often connate below, usually dry and membranous, 1- to several-nerved; tepals 6, essentially distinct, erect or spreading, all petaloid, often becoming dry and persistent in fruit; stamens 6, the filaments somewhat dilated, more or less connate at the base and adherent to the opposing tepals; ovary sessile, 3-lobed and sometimes crested, 3-celled with usually 2 (several) ovules in each cell; style 1, with a capitate to trifid stigma; capsule loculicidal; seeds in our species 6 or fewer, black, alveolate, often with a minute central pustule in each alveolus; some or all of the flowers sometimes replaced by bulblets.

Perhaps 500 species, widely distributed in the N. Hemisphere, about 80 in N. Amer. (The classical Latin name for garlic.)

The treatment here presented was prepared in collaboration with the late Dr. Marion Ownbey, making extensive use of his published and unpublished work and the unpublished theses of his students Louis Mingrone and Dale McNeal, Jr.

The cellular reticulations of the bulb coats of some species of *Allium* represent the persistent walls of the cells of the inner epidermis of the outer bulb coats. They become easily apparent only after the outer layers of the bulb scale have been removed by decay or otherwise. These outer bulb coats are frequently stripped from the bulbs at the time of collection.

Several Old World species of *Allium* are cultivated in the U.S. for flavoring, or as garden ornamentals. The most important of these are the common onion (*Allium cepa* L.), garlic (*A sativum* L.), leek (*A. porrum* L.), and chives (*A. schoenoprasum* L.). These species may persist after cultivation, but in our area they do not spread.

Allium schoenoprasum, mentioned above, is circumboreal in distribution, extending south in the Rocky Mts. to Wyo. and Colo., and closely approaching our range at Buffalo River, Fremont Co., Idaho. It may be sought in the Uinta Mts. of Utah, but its presence there remains to be demonstrated. It has cylindrical, hollow leaves and a subcapitate, centrifugally flowering inflorescence, whereas our other species have solid, flat to channeled or sometimes terete leaves and an umbelliform, centripetally flowering inflorescence. In *A. schoenoprasum* $n = 8$, whereas in all the following species $n = 7$ or a multiple of 7.

References:
McNeal, D. W., Jr. Comparative studies of the *Allium acuminatum* alliance. Ph.D. thesis, Washington State University. 1970.
Mingrone, L. V. A comparative study of the *Allium falcifolium* alliance. Ph.D. thesis, Washington State University. 1968.
Ownbey, M. The genus *Allium* in Arizona. Res. Stud. State Coll. Wash. 15: 211–232. 1947 [1949].
———. The genus *Allium* in Idaho. *ibid.* 18: 3–39. 1950.
———. The genus *Allium* in Texas. *ibid.* 18: 181–222. 1950 [1951].
———. *Allium,* pp. 739–760 in C. L. Hitchcock, A. Cronquist, M. Ownbey, and J. W. Thompson, Vasc. Pls. Pacific N.W. Part 1. 1969. Univ. Wash. Press. Seattle. [Univ. Wash. Publ. Biol. 17(1): 739–760.]
———, and H. C. Aase. Cytotaxonomic studies in *Allium.* I. The *Allium canadense* alliance. Res. Stud. State Coll. Wash. Monogr. Suppl. 1: 1–106. 1955 [1956].

1 Bulb coats persisting as conspicuous reticula of coarse, anastomosing fibers.
  2 Bracts of the involucre mostly 1-nerved; ovary inconspicuously low-crested, or more obviously crested in *A. textile.*
    3 Leaves flat, falcate, relatively broad, commonly 3–6 mm wide, 2 or 3 per scape; tepals pink; Box Elder Co., Utah . . . . 1. *A. passeyi*
    3 Leaves channeled or concave-convex to semiterete, not notably falcate, usually narrower, seldom over 3 (4) mm wide.
      4 Leaves 3 or more per scape; tepals pink, seldom white; alveoli on seeds pustuliferous; montane species . . . . 2. *A. geyeri*
      4 Leaves usually only 2 per scape; tepals white, seldom pink; alveoli on seeds not pustuliferous; desert species . . . . 3. *A. textile*
  2 Bracts of the involucre mostly 3- to 5-nerved; ovary usually strongly crested; leaves mostly 2 per scape . . . . 4. *A. macropetalum*
1 Bulb coats otherwise, either nonfibrous or with more or less parallel, nonreticulate fibers.
  5 Inflorescence nodding, at least by the time of anthesis; ovary strongly crested; stamens exserted . . . . 5. *A. cernuum*
  5 Inflorescence erect at anthesis and thereafter; ovary crested or not; stamens exserted or not.
    6 Bulb elongate, terminating a stout, *Iris*-like primary rhizome; outer bulb coats striate with elongate cells (several times as long as wide) in regular vertical rows.
      7 Stamens and style longer than the perianth; stigma capitate; capsule longer than broad, the seeds correspondingly long and slender; Cascade-Sierran sp., extending e. to ne. Nev., e. Oregon, and w. Idaho . . . . 6. *A. validum*
      7 Stamens and style half as long as the perianth; stigma trifid; capsule broader than long, the seeds short and thick; Rocky Mt. sp., reaching our range in e. Utah . . . . . . 7. *A. brevistylum*
    6 Bulb ovoid to subglobose; rhizomes, if present, slender and secondary; outer bulb coats various, often cellular-reticulate, but not with vertically elongate cells several times as long as wide in regular rows.
      8 Leaf solitary, terete or subterete.
        9 Outer bulb coats with obscure to distinct, contorted reticulations . . . . 8. *A. nevadense*
        9 Outer bulb coats nonreticulate . . . . 9. *A. atrorubens*
      8 Leaves usually 2 or more per scape, flat or channeled.
        10 Walls of the cellular reticulations of the outer bulb coats thin, minutely very sinuous; outer bulb coats striate above; ovary strongly crested.
          11 Leaves usually green at anthesis; widespread in s. Idaho, se. Oregon, Utah, Nev., and southward . . . . 10. *A. bisceptrum*

11 Leaves usually withering at anthesis; chiefly Sierran, Steens, and Blue Mt. sp. . . . . 11. *A. campanulatum*

10 Walls of the cellular reticulations of the outer bulb coats not at once thin and sinuous, sometimes (as in *A. acuminatum*) obscurely sinuous, but then thick; bulb coats not striate; ovary variously crested or crestless.

12 Leaves 2–4, narrowly linear, concave-convex or subterete, flexuous, shorter than the terete, persistent scape; bulb coats thick, with prominent (under a lens) thick-walled cellular reticulations.

13 Tepals lanceolate, acuminate, serrulate-denticulate (especially the inner set); bulb coat reticulations quadrate, regular . . . . 12. *A. acuminatum*

13 Tepals elliptic-lanceolate, acute to emarginate, entire; bulb coat reticulations more elongate, regularly arranged in a herringbone pattern . . . . 13. *A. amplectens*

12 Leaves 2, or rarely 1, commonly broad, flat, and falcate (if concave-convex then not flexuous), usually overtopping the often strongly flattened scape; bulb coat reticulations various, but usually not prominent.

14 Scapes terete or only slightly compressed; leaves linear, channeled, 1–4 (5) mm wide.

15 Leaves usually twice as long as the scape; tepals more or less serrulate, evidently longer than the pedicels; scape and leaves deciduous at maturity; cellular reticulations of the outer bulb coats obscure.

16 Tepals white with green midrib; ovary crested . . . . 14. *A. simillimum*

16 Tepals deep pink; ovary crestless . . . . 15. *A. aaseae*

15 Leaves less than twice as long as the scape; tepals entire, usually shorter than the pedicels; scape and leaves persistent; cellular reticulations of the outer bulb coats evident . . . . 16. *A. brandegei*

14 Scapes flattened or at least somewhat compressed; leaves flat, often falcate, (1) 3–15 mm wide; scape and leaves deciduous at maturity.

17 Tepals slender, linear to lance-linear or linear-oblong, mostly more than 4 times as long as wide.

18 Stamens exserted; leaves 8–15 mm wide . . . . 17. *A. platycaule*

18 Stamens equaling or generally shorter than the tepals; leaves 4–6 mm wide . . . . 18. *A. anceps*

17 Tepals broader, lanceolate to oblong or ovate, mostly less than 4 times as long as wide.

19 Tepals lanceolate or lance-ovate, acute or acuminate, the tip involute.

20 Stamens about as long as or longer than the tepals; leaves barely exceeding the scape . . . . 19. *A. lemmonii*

20 Stamens 1/2 to 3/4 as long as the tepals; leaves about twice as long as the scape . . . . 20. *A. tolmiei*

19 Tepals lance-oblong or oblong-obtuse, the tip not involute.

21 Outer bulb coats distinctly reticulate with transversely elongate meshes at 20X; scape merely flattened; c. and se. Oregon . . . . 21. *A. punctum*

21 Bulb coats without evident reticulations; scape winged as well as flattened; more widespread . . . . 22. *A. parvum*

1. **Allium passeyi** N. Holmgren & A. Holmgren

Allium passeyi N. Holmgren & A. Holmgren, Brittonia 26: 309. 1974. (*A. Holmgren, Passey & Hugie 13125,* Blue Creek Valley, Box Elder Co., Utah.)

Bulbs ovoid, the outer coats persisting as a set of fibrous reticula enclosing one or more bulbs; leaves 2 or 3 per scape, flat, falcate, nearly or quite equaling the scape, mostly 10–20 cm long and 3–6 mm wide; scape mostly 10–20 cm high, terete or somewhat wing-angled; spathe breaking into 2–4 ovate, acuminate to caudate, basally somewhat connate, mostly 1-nerved bracts; pedicels stiffly spreading or curved, mostly 12–18 mm long; tepals pinkish, mostly 7–9 mm long, persistent and permanently investing the capsule, the outer ones lance-ovate to lanceolate, acuminate, strongly keeled at the gibbous base, the inner ones narrowly lanceolate, acuminate; stamens shorter than the tepals; ovary inconspicuously low-crested; seeds shining, each alveolus with a minute central pustule.

Shallow, stony, lithosolic soil over dolomitic limestone; hilltops in Blue Creek Valley, Box Elder Co., Utah. June.

2. **Allium geyeri** S. Wats.

Allium geyeri S. Wats. Proc. Amer. Acad. Arts 14: 227. 1879. (*Geyer 226,* stony banks of the Kooskooskie [Clearwater] River, n. Idaho; lectotype [GH] by Ownbey.)

A. rubrum Osterh. Bull. Torrey Bot. Club 27: 506. 1900. (*Osterhout 1321,* North Park, along the Colorado-Wyoming line.)

A. geyeri var. *tenerum* M. E. Jones, Contr. W. Bot. 10: 28. 1902. *A. geyeri* subsp. *tenerum* Traub & Ownbey, Pl. Life 23: 110.

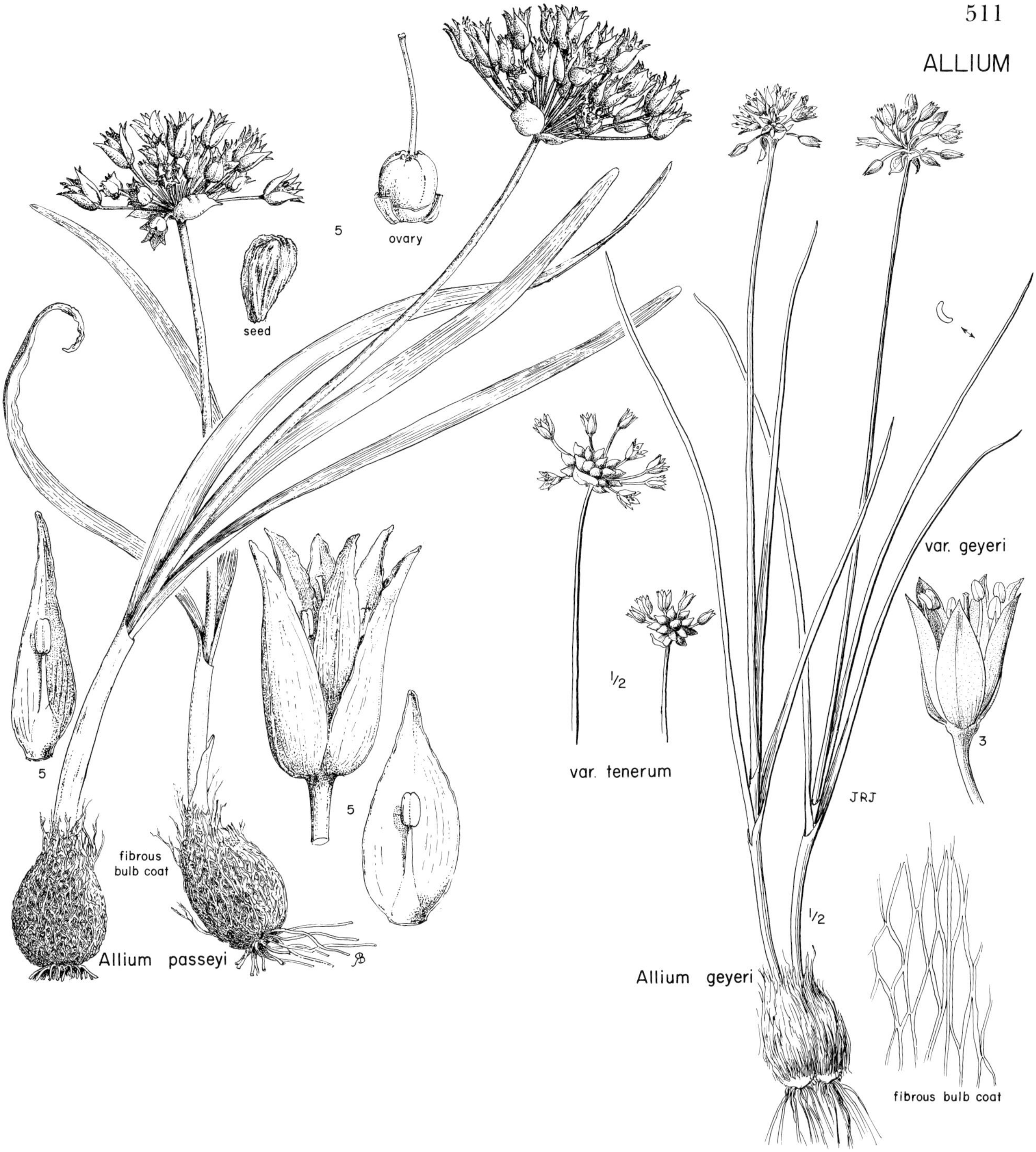

1967. *A. geyeri* f. *tenerum* B. Boivin, Naturaliste Canad. 94: 521. 1967. (*Jones 6597,* alpine meadows, Deer Lake, Washington Co., Idaho.)

A. geyeri var. *graniferum* L. F. Henderson, Rhodora 32: 22. 1930. (*Henderson 5397,* near Austin, Grant Co., Oregon.) = var. *tenerum.*

Bulbs ovoid or more elongate, usually clustered, the outer coats persisting as a set of coarse, fibrous reticula enclosing one or more bulbs; leaves generally 3 or more per scape, concave-convex, 1–3 (5) mm wide, usually shorter than the scape; scape mostly 15–50 cm tall, terete or somewhat angled; spathe breaking into 2 or 3 separate or partly connate involucral bracts, these ovate to lanceolate, acuminate, mostly 1-nerved; pedicels subequal, often less than twice as long as the perianth, becoming rigid and stiffly spreading in fruit (flowers and fruit mostly replaced by bulbils in var. *tenerum*); tepals pink (white), (4) 6–8 (10) mm long, ovate to acuminate, often obscurely toothed, erect, becoming callous-keeled and permanently investing the capsule;

stamens usually shorter than the perianth; ovary inconspicuously low-crested; stigma capitate, entire to obscurely 3-lobed; seeds shining, each alveolus with a minute central pustule; diploid, tetraploid, pentaploid and hexaploid on a base of $x = 7$.

Moist or wet meadows and along streams in the mts.; s. Alta. to N.M. and sw. Texas, w. irregularly to sw. B.C., c. Wash., c. and se. Oregon, ne. Nev., and c. Ariz. May–June.

The species consists of two almost equally widespread but irregularly and independently distributed varieties, frequently only one or the other in a given area. Var. *geyeri*, diploid and tetraploid, has normal flowers, usually 10–25 or more per umbel. Var. *tenerum* M. E. Jones, tetraploid, pentaploid, and hexaploid, has fewer than 10 normal flowers in an umbel, the others being replaced by bulbils.

3. Allium textile Nels. & Macbr.

Allium reticulatum Nutt. Fraser's Cat. Pls. Upper Louisiana 1813, nomen nudum, and J. Fraser ex G. Don, Mem. Wern. Nat. Hist. Soc. 6: 36. 1827; not J. & C. Presl, 1819. *A. textile* Nels. & Macbr. Bot. Gaz. 56: 470. 1913. (Cultivated specimens, from material collected by *Nuttall,* banks of the Missouri.)

A. reticulatum var. *playanum* M. E. Jones, Contr. W. Bot. 12: 79. 1908. (*Jones,* Pocatello, Idaho; lectotype [POM] by Ownbey.)

Bulbs ovoid, usually clustered, the outer coats persisting as a set of coarse, fibrous reticula enclosing one or more bulbs; leaves ordinarily 2, concave-convex, 1–3 (5) mm wide, usually equaling or surpassing the scape; scape 6–20 (30) cm tall, terete or nearly so; involucral bracts 3, ovate, acuminate, mostly 1-nerved; umbel with 15–30 or more flowers, the pedicels unequal, mostly 2–3 times as long as the perianth, flexuous, becoming rigid in fruit; tepals white (pink), with reddish-brown midrib, 5–7 mm long, becoming callous-keeled and permanently investing the capsule, the outer ones broadly lance-ovate, obtuse to acuminate, often obscurely toothed above, erect, the inner narrower and distinctly spreading at the tip; stamens shorter than the perianth; ovary more or less conspicuously crested, or crestless; stigma capitate, entire to obscurely 3-lobed; seeds shining, the alveoli not pustuliferous; $n = 7, 14$.

Dry plains and hills, n. Great Plains and dry intermontane valleys; extending w., in our range, to se. and sc. Idaho and e. Utah; an isolated station in Elko Co., Nev. May–June.

4. Allium macropetalum Rydb.

Allium reticulatum var. *deserticola* M. E. Jones, Contr. W. Bot. 10: 30. 1902. *A. deserticola* Wooton & Standley, Contr. U. S. Natl. Herb. 16: 114. 1913. (*Jones s.n.,* Cisco, Grand Co., Utah; lectotype [POM] by Ownbey.)

A. macropetalum Rydb. Bull. Torrey Bot. Club 31: 401. 1904. (*Crandall 4082,* Palisades, Mesa Co., Colo.)

Bulbs ovoid, usually clustered, the outer coats persisting as a set of coarse, fibrous reticula enclosing 1 or more bulbs; leaves ordinarily 2, semi-terete or concave-convex, 1–3 mm wide, surpassing the scape; scape 5–20 cm tall, terete or somewhat angled; spathe membranous, breaking into 2 or 3 separate or basally connate, ovate to lanceolate, mostly 3- to 5-nerved bracts; flowers mostly 10–20, erect, the pedicels unequal, 2–3 times as long as the perianth, becoming more or less flexuous and rigid in fruit; perianth broadly campanulate, the tepals pink, varying to nearly white except for the midrib, 8–12 mm long, lanceolate, obtuse to acuminate, entire, spreading above, becoming papery in fruit, the midrib darker but scarcely thickened; stamens shorter than the perianth; ovary strongly crested; stigma capitate, entire to obscurely lobed; seeds shining, the alveoli not pustuliferous; $n = 7$.

Desert plains and hills; Colorado Plateau of w. Colo. and e. Utah, s. to s. Ariz., s. N.M., and w. Texas. May–June.

5. Allium cernuum Roth

Allium cernuum Roth, Arch. Bot. (Leipzig) 1(3): 40. 1798. (No type given.)

Bulbs elongate, clustered, often stipitate (short-rhizomatous) at the base, the outer coats membranaceous, minutely striate with elongate cells in regular vertical rows; leaves several per bulb, concave-convex to nearly flat, 1–6 mm wide, shorter than the scape; scape 10–50 cm tall, terete or ridged, or distally flattened and narrowly winged, abruptly recurved near the tip, the inflorescence nodding; spathe usually caducous at anthesis; pedicels slender, 2–3 times as long as the perianth, in fruit becoming stout, elongating, and bending abruptly upward at the point of attachment; perianth campanulate, pink or white, the tepals 4–6 mm long, elliptic-ovate, obtuse or obtusish, entire or nearly so, withering in fruit, the midrib not thickened; stamens exserted; ovary conspicuously crested; style exserted, with capitate, entire stigma; seeds dull, the alveoli variously smooth, or roughened, or with a tiny central pustule; $n = 7$.

Widespread in moist soils in the mountainous and cool regions of N. Amer., s. in the Rocky Mt. region to Mex.; in our range known only in the Colorado Plateau region of e. Utah, but to be expected also in the Uinta Mts. June–July.

6. Allium validum S. Wats.

Allium validum S. Wats. in King, Rep. Geol. Explor. 40th Parallel 5: 350. 1871. (*Watson 1179,* East Humboldt and Clover mts., Nev., 7000–9000 ft. alt.)

Bulb elongate, 1–1.5 cm thick, terminating a thick, *Iris*-like rhizome, the outer coats minutely striate with elongate cells in regular vertical rows, not fibrous-reticulate but with coarse, persistent, parallel fibers; leaves several, flat, obtuse, entire, 4–15 mm wide, shorter than the scape; scape 30–70 cm tall, flattened and narrowly winged distally; bracts 2, basally connate, broadly ovate, acute, 5- to 7-nerved; umbel 15- to 30-flowered, the pedicels slender, about equaling the perianth at anthesis, elongating and becoming stout in fruit; tepals pink, 8–10 mm long, narrowly lanceolate, acuminate, entire, withering in fruit, the midrib scarcely thickened; stamens much exceeding the perianth; ovary crestless; style exserted, with capitate, entire stigma; capsules mostly longer than broad, with oblong, barely emarginate valves; seeds correspondingly long and slender, dull, the alveoli not pustuliferous; $n = 14, 28$.

Swampy meadows at middle and upper alt. in the mts.; chiefly Cascade-Sierran sp., also in the Coast ranges, and in the mts. of n. Nev., w. Idaho, and e. Oregon. June–Aug.

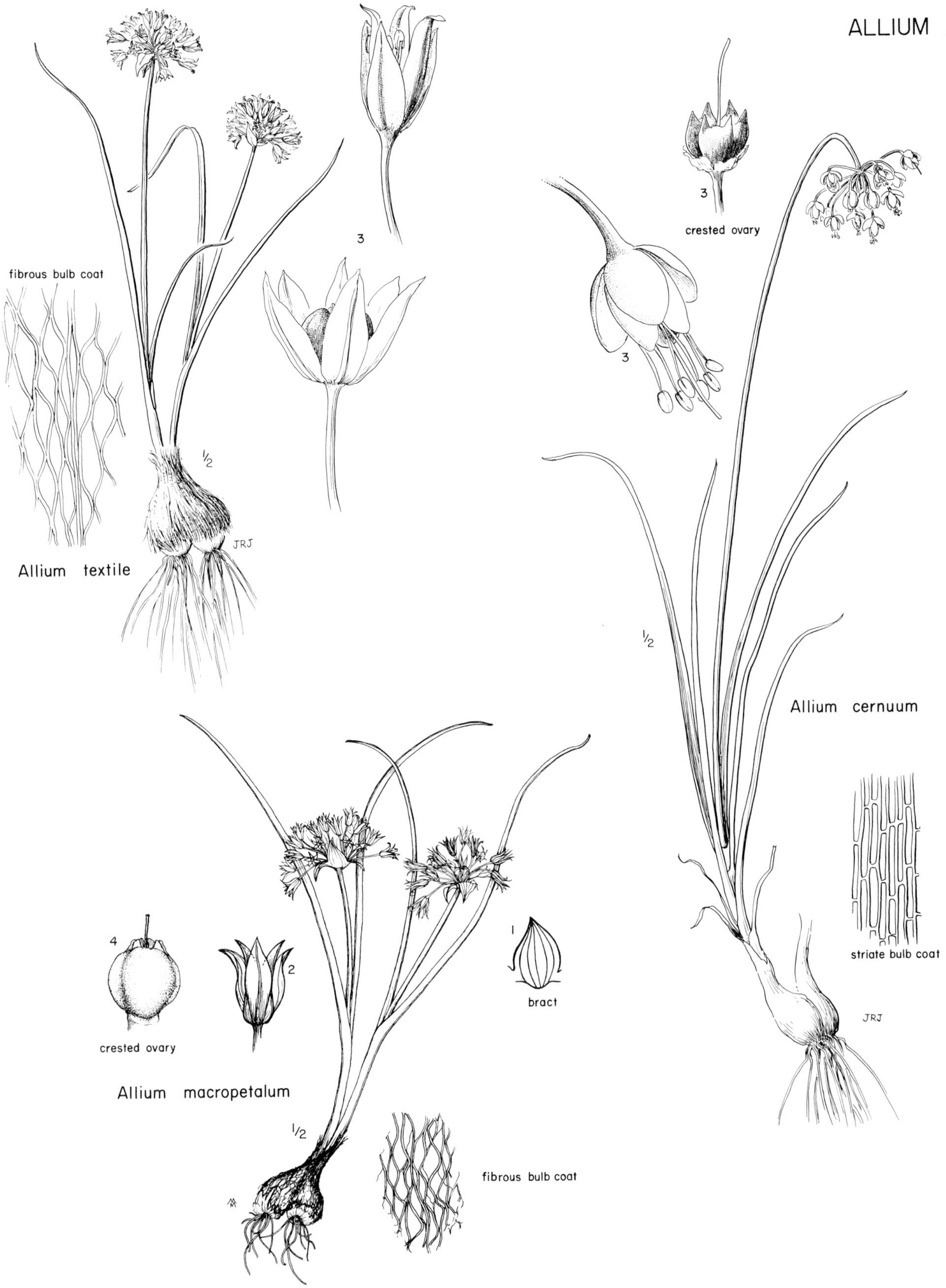
ALLIUM
3
fibrous bulb coat
1/2
JRJ
Allium textile
3
crested ovary
3
1/2
Allium cernuum
striate bulb coat
JRJ
1
bract
4
2
crested ovary
Allium macropetalum
1/2
fibrous bulb coat

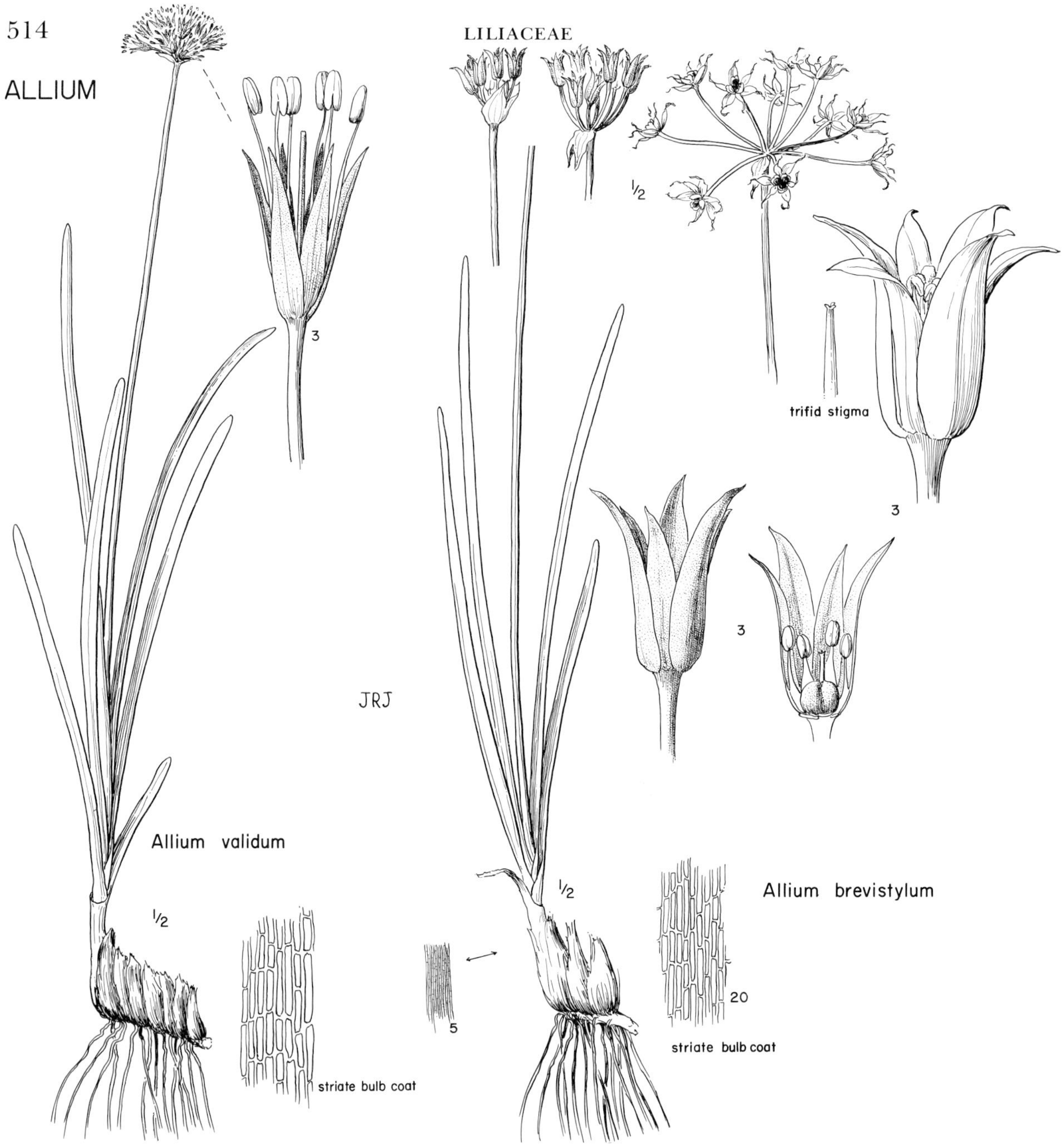

7. Allium brevistylum S. Wats.

Allium brevistylum S. Wats. in King, Rep. Geol. Explor. 40th Parallel 5: 350. 1871. (*Watson 1180,* shaded banks of Provo R., Uinta Mts., Utah.)

Bulb elongate, mostly less than 1 cm thick, terminating a thick, *Iris*-like rhizome, the outer coats membranous, minutely striate with elongate cells in regular vertical rows, not fibrous-reticulate but with persistent parallel fibers; leaves 2–several, flat, obtuse or obtusish, entire, 2–8 mm wide, shorter than the scape; scape 20–60 cm tall, flattened and narrowly winged distally; involucral bracts 2, connate at the base and often along one side, ovate, acute, 3- to 5-nerved; umbel mostly 7- to 15-flowered, the pedicels slender, about equaling the perianth at anthesis, elongating and becoming stout and curved in fruit; tepals pink, 10–13 mm long, lanceolate, acuminate, entire, withering in fruit, the midrib somewhat thickened; stamens about half as long as the perianth; ovary crestless; style about half as long as the perianth, with 3 cleft stigma; capsule broader than long, the valves obcordate, distinctly emarginate; seeds correspondingly short and thick, dull, the alveoli from distinct and pustuliferous to obscure and minutely roughened; $n = 7$.

Swampy meadows and along streams, rarely on wooded slopes; middle and upper alt. in the Rocky Mt. region from c. Mont. and ne. Idaho through Wyo. to e. Utah (Uinta Mts., LaSal Mts.) and Colo. June–Aug.

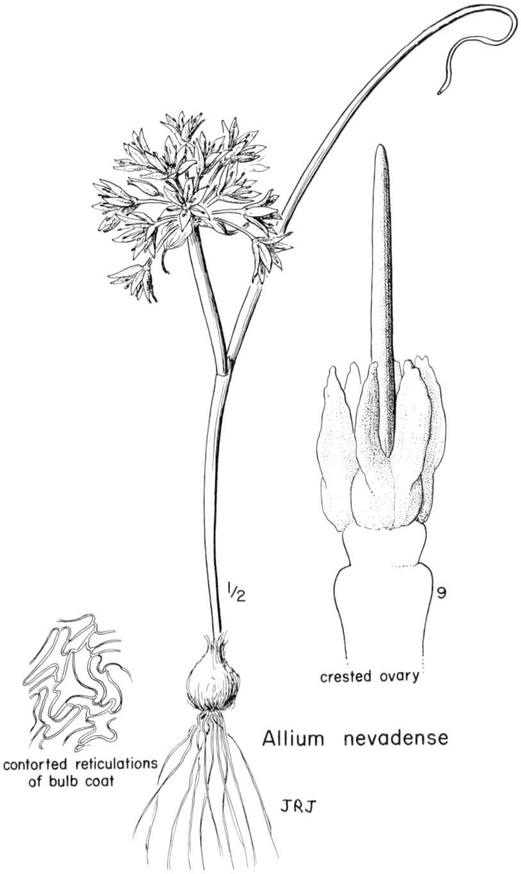

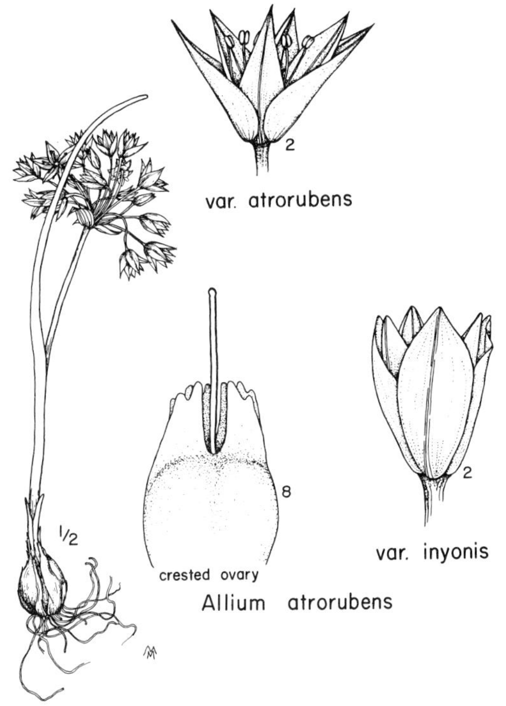

8. Allium nevadense S. Wats.

Allium nevadense S. Wats. in King, Rep. Geol. Explor. 40th Parallel 5: 351. 1871. (*Watson 1183* p.p., Trinity Mts., Pershing Co., Nev.; lectotype [GH] by Ownbey.)

A. cristatum S. Wats. Proc. Amer. Acad. Arts 14: 232. 1879. *A. nevadense* var. *cristatum* Ownbey, Res. Stud. State Coll. Wash. 15: 228. 1947 [1949]. *A. nevadense* subsp. *cristatum* Traub & Ownbey, Pl. Life 23: 110. 1967. (*Palmer 454,* St. George, Utah, in 1877.)

A. nevadense var. *macropetalum* M. E. Peck, Leafl. W. Bot. 4: 177. 1945. (*Peck 21704,* low sagebrush 17 mi. n. of McDermitt, Malheur Co., Oregon.) = var. *nevadense.*

Bulb ovoid, often proliferating by stalked basal bulblets, the outer coats (or some of them) with contorted reticulations; leaf solitary, terete or subterete, up to about twice as long as the scape, or even longer, the terminal part often coiled but frequently broken off; scape slender, 3–5 (10) cm tall; bracts 2 or 3, usually connate at the base, ovate to lanceolate, acuminate, spreading or reflexed, 3- to 7-nerved; umbel few- to many-flowered, the pedicels slender, up to twice as long as the perianth; tepals whitish or pinkish with deep pink midrib, 8–12 mm long, the inner and outer about equal, lanceolate, acute to acuminate, entire, spreading or somewhat recurved at the tip, becoming papery in fruit, the midrib not thickened; stamens 1/2 to 4/5 as long as the tepals; ovary conspicuously crested, with 6 distinct, narrow, thin, entire or toothed processes; stigma capitate, entire; seeds dull, the alveoli minutely roughened; n = 7, 14.

Sandy, gravelly, or occasionally clay soils, on the desert from the Snake River Plains of s. Idaho to se. Oregon, s. through Utah, w. Colo., and Nev. to n. Ariz. and se. Calif. Late April–July.

The species consists of two morphologically ill-defined but geographically significant varieties. Most of our material belongs to the fairly widespread var. *nevadense,* with fairly evident reticulations on the outer bulb coats. The more southwestern var. *cristatum* (S. Wats.) Ownbey, with the reticulations obscure, enters our range in sw. Utah (Washington and Kane cos.).

9. Allium atrorubens S. Wats.

Allium atrorubens S. Wats. in King, Rep. Geol. Explor. 40th Parallel 5: 352. 1871. (*Watson 1184,* near Unionville, West Humboldt Mts., Pershing Co., Nev.)

Allium atrorubens var. **inyonis** (M. E. Jones) Ownbey & Aase, hoc loc. *A. decipiens* M. E. Jones, Contr. W. Bot. 10: 16. 1902; not Fischer, 1812. *A. inyonis* M. E. Jones, Contr. W. Bot. 10: 86. 1902. *A. atrorubens* subsp. *inyonis* Traub, Pl. Life 28: 66. 1972. (*Jones,* Summit, Owens Valley, Calif.)

Much like *A. nevadense;* bulb coats not reticulate; scape a little taller, (3) 5–10 (15) cm tall; tepals not strongly recurved at the tip.

Rocky, gravelly, or clay soil in the desert hills and mts. of Nev. and adj. Calif. Late April–June.

The species consists of two varieties. Variety *atrorubens,* with deep reddish-purple (rarely white), lanceolate, attenuate-setaceous tepals, is well distributed throughout most of Nev. except Esmeralda Co. Variety *inyonis* (M. E. Jones) Ownbey & Aase, with broader, merely acute to acuminate, pale tepals with deep pink midrib, occurs in Esmeralda Co., Nev., w. to the eastern foothills of the Sierra Nevada in Calif.

10. Allium bisceptrum S. Wats.

Allium bisceptrum S. Wats. in King, Rep. Geol. Explor. 40th Parallel 5: 351. 1871. (*Watson 1182* p.p., Trinity Mts., Pershing Co., Nev.; lectotype [GH] by Ownbey.)

Allium bisceptrum var. **palmeri** (S. Wats.) Cronq. hoc loc.

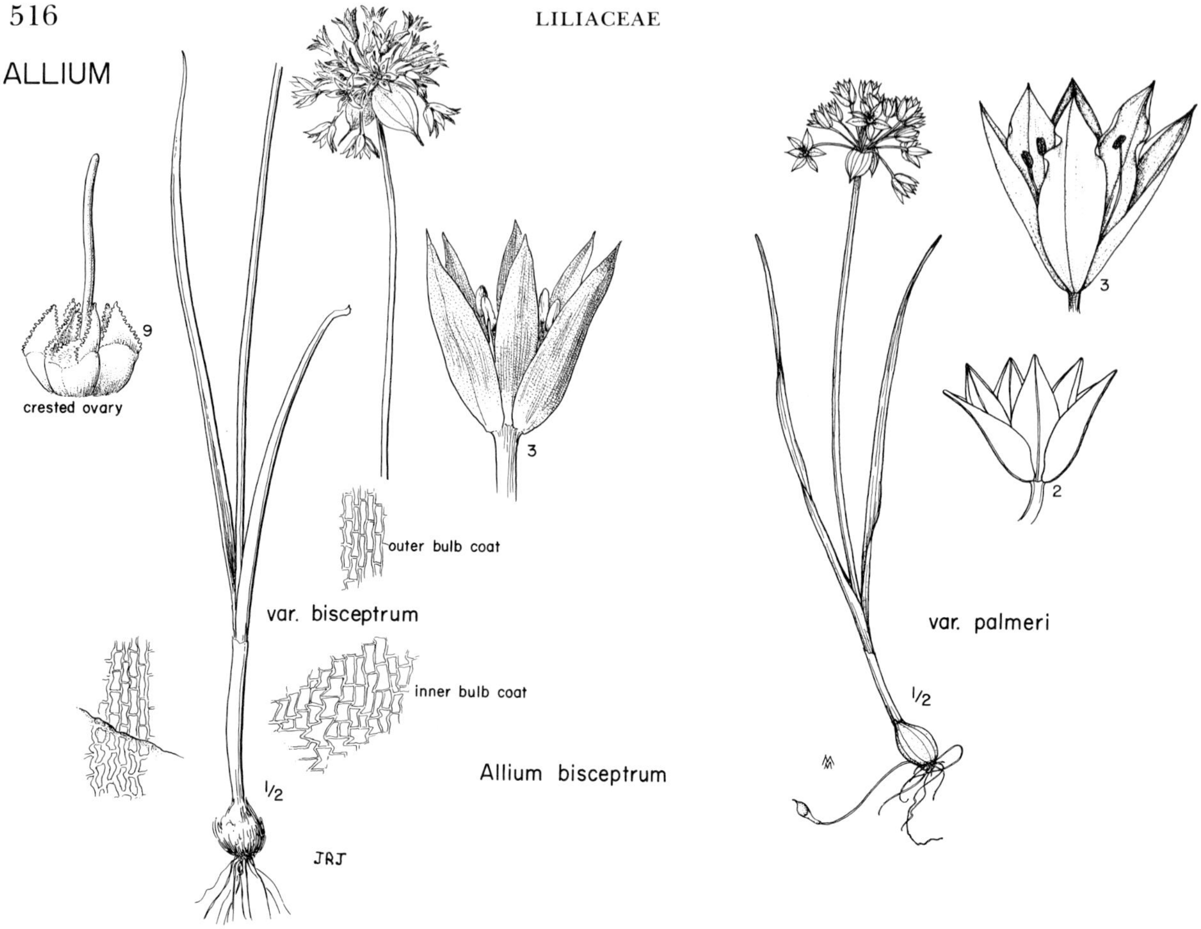

A. palmeri S. Wats. in King, Rep. Geol. Explor. 40th Parallel 5: 487. 1871. (*Palmer,* in 1869, possibly near Fort Wingate, N.M.)

A. bisceptrum var. *utahense* M. E. Jones, Contr. W. Bot. 10: 33. 1902. (*Jones 6647,* City Creek Canyon, Salt Lake City, Utah, 6 June 1898, the only collection in Jones' herbarium [POM] bearing the name in his hand.) = var. *bisceptrum.*

Bulb ovoid, commonly producing either a cluster of short-stalked basal bulblets or filiform rhizomes terminated by rather large bulblets, but the rhizomes generally lost when the specimen is collected; some of the outer bulb-coats cellular-reticulate, the meshes rectangular with minutely sinuous vertical walls, varying to irregular with all the walls sinuous; leaves 2 or more, broadly channeled, usually tapering from a broad base, 1–13 mm wide, usually shorter than the scape, green at anthesis; scape 10–30 (40) cm tall, terete; bracts 2, distinct, ovate to lanceolate, acuminate, 3- to 5-nerved; umbel with many to sometimes rather few flowers, the pedicels up to about 3 times as long as the perianth, often becoming flexuous and deflexed in fruit; tepals lilac or pale pink to white, 7–10 mm long, lanceolate, acuminate, entire, thin, becoming papery in fruit; stamens shorter than the perianth; ovary conspicuously crested, with 6 distinct, flattened, papillose-denticulate processes; style included; stigma capitate, entire; seeds wrinkled, and each alveolus with a minute central pustule; n = 7, 14.

Meadows and aspen groves, less commonly on open slopes in the mts.; se. Idaho to Klamath Co., Oregon, s. to the Sierra Nevada and the White Mts. of Calif., and to the mts. of Ariz. and w. N.M.; disjunct in n. Idaho. May–July.

Allium bisceptrum consists of a northern, diploid variety and a southern, tetraploid variety that also differ in a series of morphological tendencies. Where their ranges adjoin across c. Utah and c. Nev. they can as yet be securely distinguished only by the chromosome number.

1 Basal bulblets short-stalked; scape mostly (10) 20–30 (40) cm tall; leaves 1–13 mm wide; pedicels mostly 2–3 times as long as the perianth, usually slender, generally not becoming strongly flexuous and none but the outer deflexed in fruit; tepals developing neither a pronounced keel nor strongly involute margins; n = 7; from c. Utah, c. Nev., and the Sierra Nevada and the Sweetwater and White mts. of Calif., n. to s. Idaho and s. Oregon, and disjunct in n. Idaho . . . . var. *bisceptrum*

1 Basal bulblets terminating slender rhizomes; scape mostly 10–20 (30) cm tall; leaves 1–6 mm wide; pedicels up to about twice as long as the perianth, rather stout, becoming strongly flexuous and deflexed in fruit; tepals often becoming involute at the tip and sometimes developing a prominent keel; n = 14; mts. of Ariz. and w. N.M., n. to c. Utah and c. Nev., and sometimes to White Pine and s. Elko cos., Nev. . . . . var. *palmeri* (S. Wats.) Cronq.

11. **Allium campanulatum** S. Wats.

Allium campanulatum S. Wats. Proc. Amer. Acad. Arts 14: 231. 1879. (*Bolander 4943,* Mt. Bullion, Sierra Nevada, Calif.; lectotype [GH] by Ownbey.)

Bulb ovoid, commonly producing either a cluster of short-stalked basal bulblets or filiform rhizomes

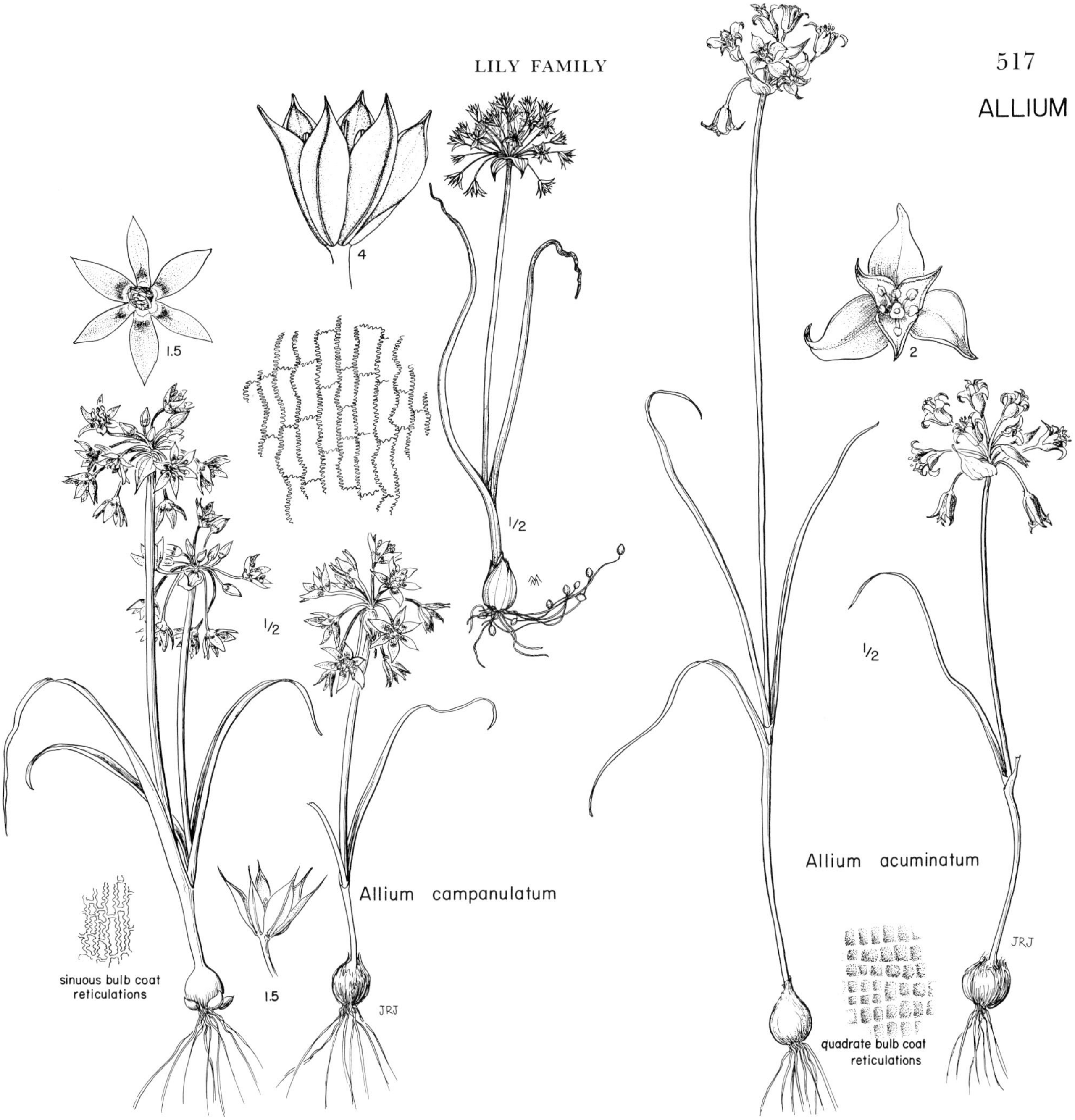

up to 1 dm long terminated by small bulblets; some of the outer bulb coats cellular-reticulate, the meshes transversely to vertically rectangular or irregular with at least the vertical walls very sinuous, often indistinct as though covered with a thick coat of varnish; leaves 2, 1–5 mm wide, about equaling the scape, generally withering at anthesis; scape 5–15 cm tall, terete, glaucous; bracts generally 2, distinct, ovate to lanceolate, acuminate, 3- to 5-nerved, rarely more; pedicels elongate, up to 4 times as long as the perianth, commonly becoming flexuous, usually strongly deflexed in fruit; tepals pinkish to reddish or purplish, commonly with a purple crescent near the base, or rarely white, (4) 6–8 (10) mm long, ovate to narrowly lanceolate, acuminate, entire, becoming rigid in fruit, the tip strongly involute, usually with a pronounced keel; stamens shorter than the perianth; ovary crested with 6 distinct, flattened, and conspicuous processes; style usually included, with capitate, entire stigma; seeds wrinkled, each alveolus with a minute central pustule; $n = 7, 14$.

Open or shaded places in the mts.; species chiefly of Calif., especially the Sierra Nevada, entering our range only in the Steens Mts. of Oregon. June–Aug.

12. Allium acuminatum Hook.

Allium acuminatum Hook. Fl. Boreali-Amer. 2: 184. 1838. (*Menzies,* Nootka, Vancouver I.)

Bulb ovoid to subglobose, the outer coat cellular-reticulate, the meshes square or polygonal with thick, obscurely sinuous walls; leaves 2–4, narrow, concave-convex, channeled, shorter than the scape; scape terete, mostly 10–35 cm tall; bracts 2, lanceolate to ovate, acuminate, 1–3 cm long; pedicels spreading, 3/4–2 times as long as the perianth; tepals serrulate-denticulate (the inner ones the more prominently so), white to more often pink or rose-purple, lanceolate to lance-ovate, acuminate, erect, becoming rigid and keeled in fruit, with somewhat involute margins, the outer 8–13 (16) mm long, with recurved or spreading tip, the inner shorter and narrower, with strongly recurved tip; stamens shorter than the tepals; ovary inconspicuously crested, with 3 minute, bilobed apical processes; stigma capitate, obscurely 3-lobed; seeds with minutely roughened alveoli; $n = 7$.

Dry hillsides and plains, sometimes becoming a weed in dryland wheat fields; s. B.C. to c. Calif., e. to Idaho, Wyo., Colo., and Ariz.; nearly throughout our range, except that in Nev. it is known only from Elko, White Pine, Humboldt, n. Eureka and nw. Washoe cos. May–July.

13. **Allium amplectens** Torr.

Allium amplectens Torr. in Whipple, Explor. & Surv. Railroad Route from Mississippi River to Pacific Ocean 4(5): 148. 1857. (*Bigelow s.n.*, hillsides, Sonoma, Calif.)

Bulbs ovoid to subglobose, often clustered, the outer coats cellular-reticulate with transversely elongate, wavy (not sharply V-shaped) meshes, forming a herringbone pattern; leaves 2–4, narrow, concave-convex or subterete, channeled, up to about as long as the scape; scape terete, slender, 15–45 cm tall; bracts 2 or more often 3, ovate, abruptly short-acuminate, 8–15 mm long; umbel compact, the pedicels 3/4–2 times as long as the perianth; pedicels and flowers deciduous as a unit at maturity; tepals entire, white or pale pink, spreading, lanceolate, acute to emarginate, scarcely involute, (5) 6–8 (9) mm long, becoming papery in fruit and connivent over the capsule; stamens a little shorter than the perianth; ovary 3-lobed, crested with 6 distinct, lateral processes; stigma capitate, scarcely thickened, entire or nearly entire; seeds with minutely roughened alveoli; $2n = 21, 28$.

Hillsides and open flats; species chiefly of Calif., n. to sw. B.C., entering our range only in the Steens Mts. of Oregon. May–June.

14. **Allium simillimum** L. F. Henderson

Allium simillimum L. F. Henderson, Bull. Torrey Bot. Club 27: 355. 1900. (*Henderson 3100*, Sesesh Peak, Idaho. Sesesh Peak is probably the present Squaw Meadow Lookout, on the line between Idaho and Valley cos. in S. 28, T. 22 N., R. 42 E.)

Bulb ovoid, the outer coats with more or less distinct, cellular, isodiametric to transversely elongate reticulations; leaves 2 per scape, linear, channeled, 1–2 mm wide, 2–3 times as long as the scape, deciduous at maturity; scape 1–5 cm tall, slightly flattened, but hardly if at all winged, deciduous at maturity; bracts 2, ovate, obtuse or acute; umbel 5- to 12-flowered, the pedicels much shorter than the perianth; tepals white with green midrib, 5–9 mm long, oblong to lanceolate, obtuse to acuminate, obscurely to distinctly serrulate-denticulate, not involute at the tip; stamens shorter than the tepals; ovary crested with 3 low processes; style included, the stigma capitate, entire; seeds dull or shining, the alveoli not pustuliferous; $n = 7$.

Coarse, sandy soils of granitic, calcareous, or basaltic origin, in the mts. of c. and sw. Idaho, and barely reaching Ravalli Co., Mont. May–July.

15. **Allium aaseae** Ownbey

Allium aaseae Ownbey, Res. Stud. State Coll. Wash. 18: 38. 1950. (*Ownbey & Ward 3095*, near Boise, Idaho.)

Bulb ovoid, the outer coats with or without obscure cellular reticulations; leaves 2 per scape, linear, channeled, concave-convex, 1–4 mm wide, entire or denticulate, twice as long as the scape, or longer, green at anthesis but deciduous at maturity; scape 1–5 cm long, slightly flattened, with or without narrow wings, deciduous at maturity; umbels 5- to 10-flowered, the pedicels much shorter than the perianth; tepals pink with darker midrib, fading to white, 6–9 mm long, oblong to lanceolate, obtuse to acuminate, obscurely to distinctly serrulate-denticulate, not involute at the tip; stamens shorter than the tepals; ovary crestless; style included; stigma capitate, entire; seeds shining, the alveoli not pustuliferous; $n = 7$.

Coarse, sandy soil and gravelly river benches in n. Ada and s. Gem cos., Idaho. April–May.

16. **Allium brandegei** S. Wats.

Allium brandegei S. Wats. Proc. Amer. Acad. Arts 17: 380. 1882. (*Brandegee s.n.*, Elk Mts., Colo., July 1881.)

A. tribracteatum var. *diehlii* M. E. Jones, Contr. W. Bot. 10: 18. 1902. *A. diehlii* M. E. Jones, Contr. W. Bot. 10: 86. 1902. (*Jones 6590*, Altus, Parley's Canyon, Utah.)

A. minimum M. E. Jones, Contr. W. Bot. 10: 19. 1902. (*Jones 6595*, Cuddy Mts., Washington Co., Idaho.)

Bulb ovoid, the outer coats with usually distinct, isodiametric or transversely elongate cellular reticulations; leaves 2 per scape, narrowly linear, channeled, 1–3 (5) mm wide, entire, longer than the scape, but less than twice as long, persistent at maturity; bracts 2, ovate, acuminate; umbel (5) 8- to 25 (35)-flowered, the pedicels usually less than twice as long as the perianth; tepals white with prominent, green to red midrib, 5–8 mm long, lanceolate, obtuse to acuminate, entire, becoming involute at the tip; stamens shorter than the tepals; style included, the stigma capitate, entire; ovary crestless; seeds shining, the alveoli not pustuliferous; $n = 7$.

Sandy or rocky soil in the mts. of w. Colo. and Wyo. to sw. Mont., w. across Idaho and n. Utah to e. Oregon (Grant and Harney cos.) and ne. Nev. (Elko Co.). June–July.

17. **Allium platycaule** S. Wats.

Allium platycaule S. Wats. Proc. Amer. Acad. Arts 14: 234. 1879. (*Lemmon s.n.*, Sierra Valley, Sierra Co., Calif., in 1873; lectotype [GH] by Ownbey, hoc loc.)

Bulb ovoid, the coats without evident reticulations; leaves 2 per scape, flat, thick, broad and strongly falcate, 8–15 mm wide, longer than the scape, deciduous at maturity; scape 10–20 cm tall, stout, strongly compressed and winged, deciduous at maturity; in-

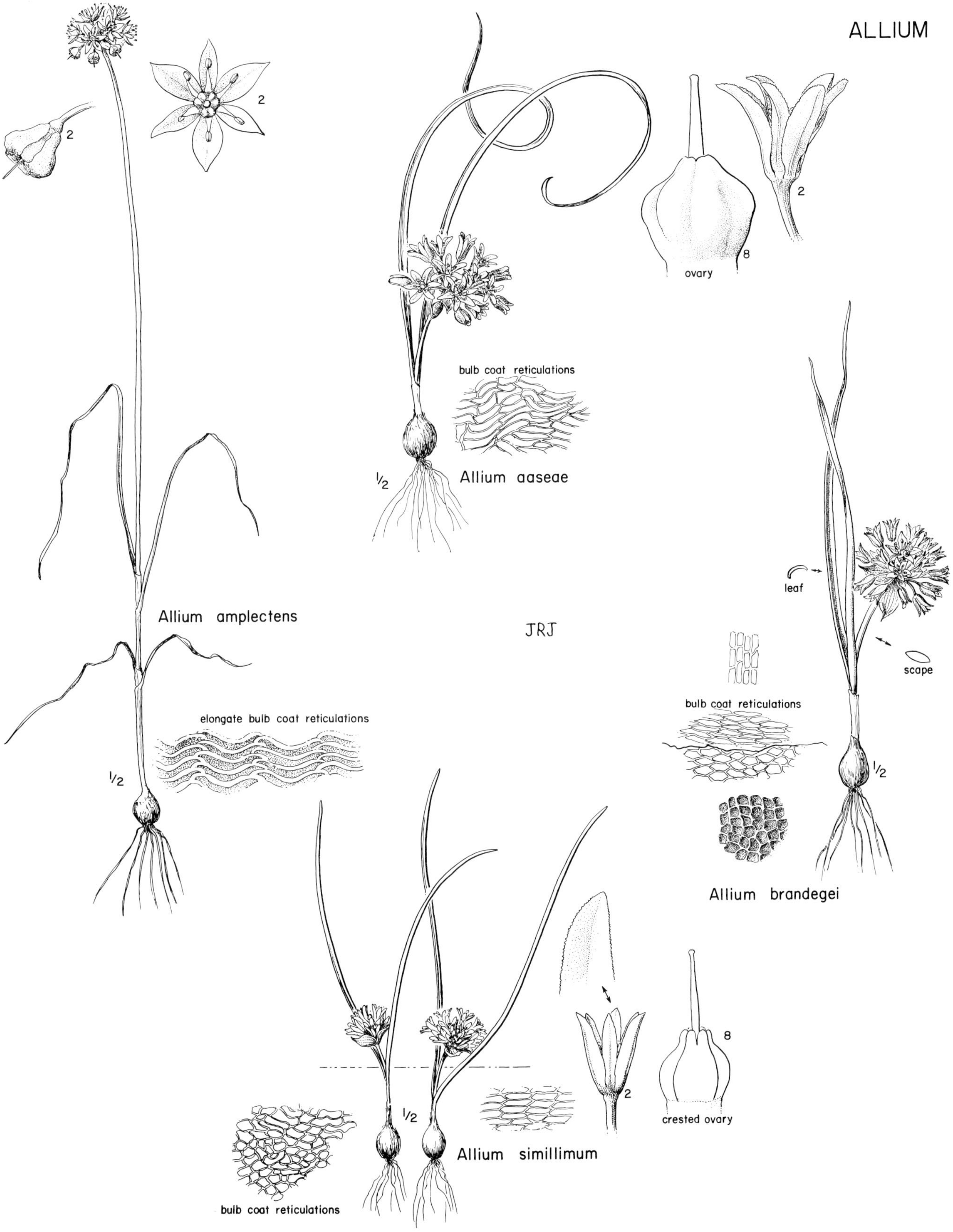
ALLIUM
2
2
Allium amplectens
elongate bulb coat reticulations
1/2
ovary
8
2
bulb coat reticulations
1/2
Allium aaseae
JRJ
leaf
scape
bulb coat reticulations
1/2
Allium brandegei
8
2
crested ovary
1/2
Allium simillimum
bulb coat reticulations

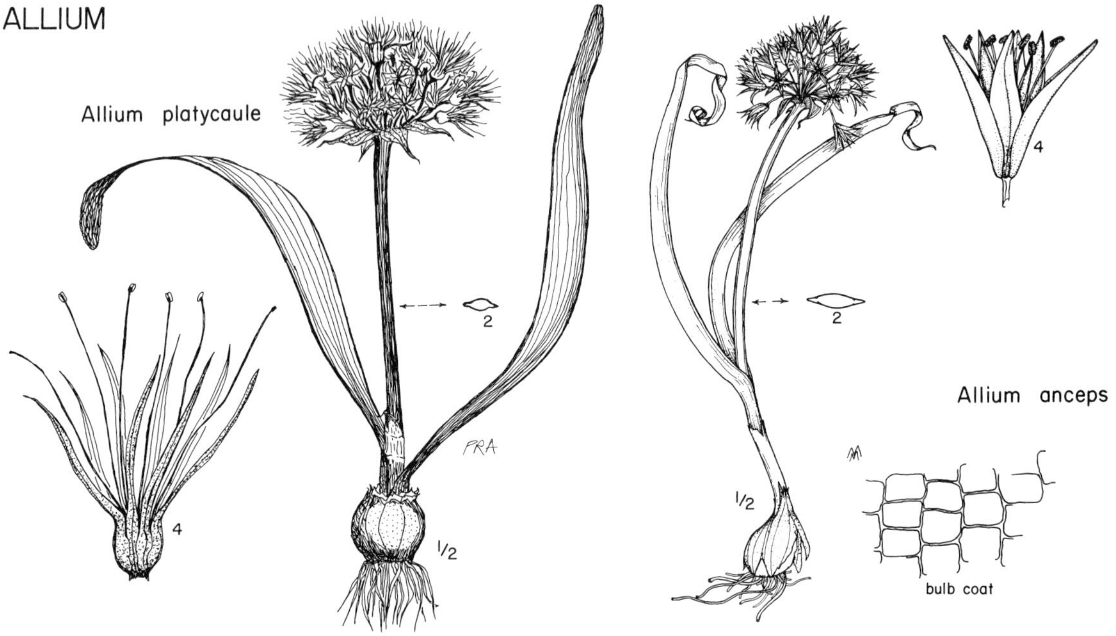

volucral bracts 3–5, acuminate; umbel 30- to 50- or more-flowered; pedicels twice as long as the perianth; tepals pink with conspicuously darker midrib, 8–15 mm long, lanceolate, very narrow, with long-acuminate, infolded tip; stamens exserted; ovary crestless; seeds dull or shining, the alveoli not pustuliferous; *n* = 7.

Rocky or sandy soil; n. Sierra Nevada of Calif., n. to s. Lake Co., Oregon, and e. to n. Humboldt Co., Nev. May–July.

18. **Allium anceps** Kellogg

Allium anceps Kellogg, Proc. Calif. Acad. Sci. 2: 109. 1863. (*Veatch s.n.,* Washoe, Nev.)

Bulb broadly ovoid, the coats thick and yellow, with thick cellular reticulations; leaves 2 per scape, slightly convex, striate, with scabrous margins, somewhat falcate, 4–6 mm wide, twice as long as the scape, deciduous at maturity; scape 10–15 cm tall, flattened, with scaberulous, translucent margins, deciduous at maturity; involucral bracts 2, ovate, acute; umbel 15- to 25-flowered, the pedicels 3 times as long as the perianth, or longer; tepals pinkish with diffuse green midrib, 6–10 (12) mm long, lance-linear, acute; stamens as long as the tepals or a little shorter; ovary with a 6-parted crest; style half as long as the stamens, with pointed, entire stigma; seeds dull, the alveoli not pustuliferous; *n* = 7.

Heavy barren soils on flats and slopes in the foothills and lowlands; widespread in Nev., extending into se. Oregon and ne. Calif. May–July.

19. **Allium lemmonii** S. Wats.

Allium lemmonii S. Wats. Proc. Amer. Acad. Arts 14: 234. 1879. *A. anceps* var. *lemmonii* Jepson, Fl. Calif. 1(6): 279. 1922. (*Lemmon s.n.,* Sierra Valley, Sierra Co., Calif. in 1874.)

A. incisum Nels. & Macbr. Bot. Gaz. 56: 470. 1913; not Fomin, 1909. *A. scissum* Nels. & Macbr. Bot. Gaz. 65: 58. 1918. (*Nelson & Macbride 1790,* House Creek, Owyhee Co., Idaho [actually in extreme sw. Twin Falls Co., fide Ownbey].)

Bulb ovoid, the outer coats with faint transverse cellular reticulations; leaves 2 per scape, thick, broadly linear, flat and somewhat falcate, 3–5 mm wide, a little longer than the scape, deciduous at maturity; scape 5–8 cm tall, stout, strongly flattened and winged, deciduous at maturity; bracts 2 or more, ovate or broadly lanceolate, gradually long-acuminate; umbel 10- to 30- or more-flowered, the pedicels 2–3 times as long as the perianth; tepals white to pale rose with slightly thickened green midrib, 6–7 mm long, lance-linear, acute to acuminate, entire; stamens up to about as long as or longer than the tepals; ovary with 6 prominent ridges giving the appearance of crests; style about equaling the perianth, with capitate, entire stigma; capsule crestless; seeds shining, the alveoli not pustuliferous; *n* = 7.

Heavy soils; n. Sierra Nevada of Calif., n. and e. to se. Oregon, sw. Idaho, and ne. Nev. May–July.

20. **Allium tolmiei** Baker ex S. Wats.

Allium tolmiei Baker ex S. Wats. Proc. Amer. Acad. Arts 14: 234. 1879. *A. douglasii* var. *tolmiei* Traub, Herbertia 12 (1945): 68. 1947. *A. idahoense* Traub, Herbertia 12 (1945): 69. 1947. ("A friend of Mr. Tolmie" [*John McLeod*], between Burnt and Malheur rivers, probably in ne. Malheur Co., Oregon.)

A. anceps var. *aberrans* M. E. Jones, Contr. W. Bot. 10: 10. 1902. (*Jones 6593,* Weiser, Washington Co., Idaho.)

Bulb ovoid, its coats without definite reticulations; leaves 2 per scape, thick, flat, commonly strongly falcate, 1–8 mm wide, up to twice as long as the scape, deciduous at maturity; scape 5–15 cm tall, strongly flattened and winged, varying to merely compressed, deciduous at maturity; bracts 2, ovate to lanceolate, acuminate, 5- to 11-nerved; umbel 10- to 40- or more-flowered, the pedicels 2–3 times as long as the

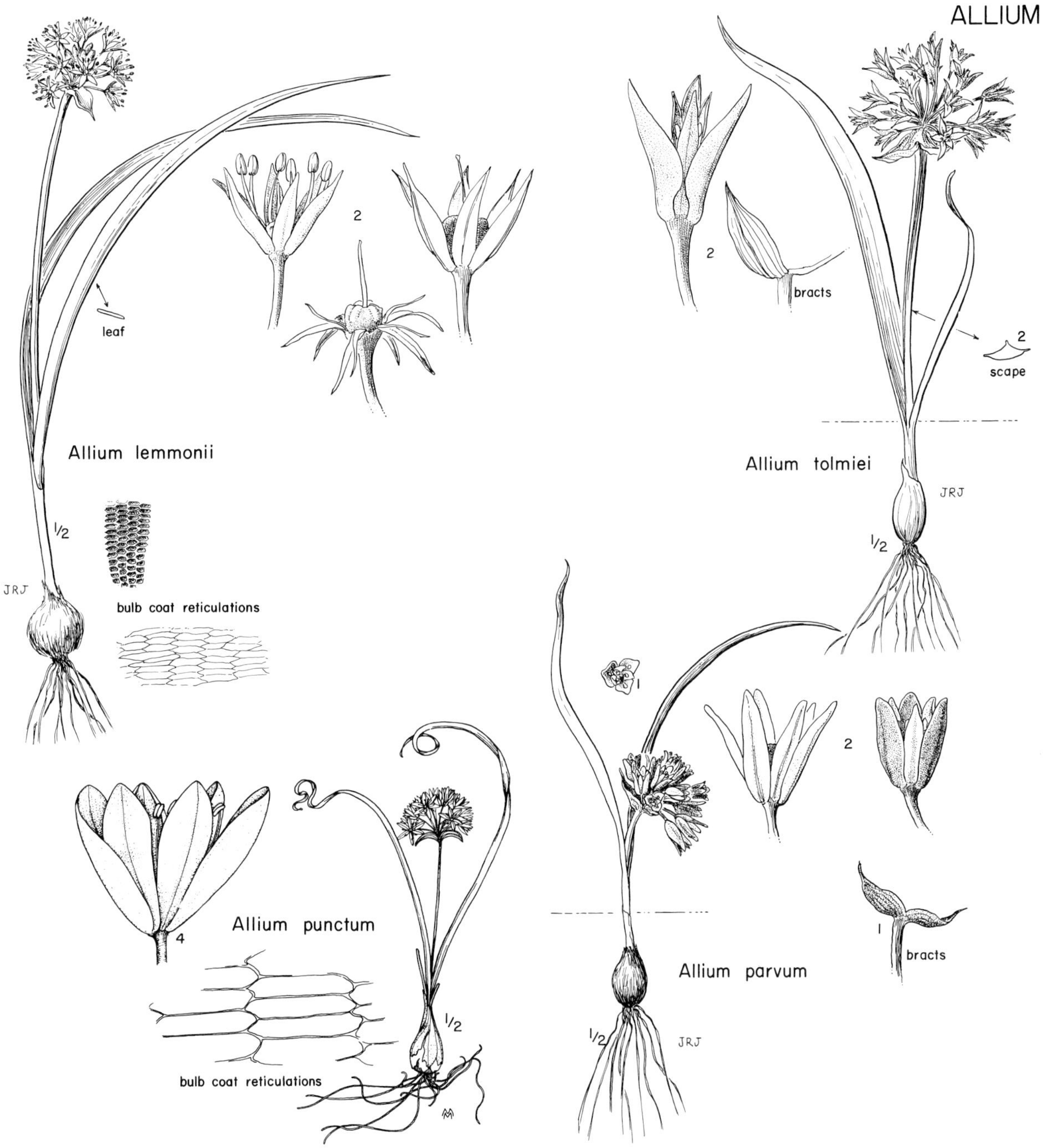

perianth; tepals pink to almost white, with darker midrib, 6–10 mm long, erect, lanceolate, acute, entire, involute at the tip; stamens 1/2 to 3/4 as long as the tepals; ovary crestless, or obscurely crested with (3) 6 low processes; style included; stigma capitate, entire or obscurely lobed; seeds dull, the alveoli not pustuliferous; $n = 7$.

Rocky, gravelly, or clay soils; widespread in e. Oregon, extending into adj. se. Wash., w. Idaho, and ne. Calif.; not yet found in Nev. April–July.

Our plants are var. *tolmiei.* Two other vars. occur to the n. of our range.

21. Allium punctum L. F. Henderson

Allium punctum L. F. Henderson, Rhodora 32: 23. 1930. (*Henderson 8813,* near the Donner-and-Blitzen River, 8 to 10 mi. n. of Frenchglen, Harney Co., Oregon.)

Much like the next species, but the bulb coats distinctly reticulate with transversely elongate meshes as seen at 20X; scape merely flattened, not winged.

Sandy, gravelly, or clay soils; Wheeler, Lake, Harney, and Malheur cos., in c. and se. Oregon. May–June.

22. Allium parvum Kellogg

Allium parvum Kellogg, Proc. Calif. Acad. Sci. 3: 54. 1863. *A.*

tribracteatum var. *parvum* Jepson, Fl. Calif. 1(6): 276. 1922. (Cultivated plants from bulbs taken by Veatch in Washoe Co., Nev.)

A. tribracteatum var. *andersoni* S. Wats. in King, Rep. Geol. Explor. 40th Parallel 5: 353. 1871. (*Anderson 63,* sandy ridges, Carson City, Nev.; lectotype [GH] by Ownbey, hoc loc.)

A. pleianthum var. *particolor* M. E. Jones, Contr. W. Bot. 10: 31. 1902. (*Jones 6656,* Centreville, Spring Valley, Nev.)

Bulb ovoid, its coats without evident reticulations; leaves (1) 2 or more per scape, thick, flat, strongly falcate, commonly serrulate-denticulate, 2–4 mm wide, 2–3 times as long as the scape, deciduous at maturity; scape 3–8 cm tall, flattened and winged, deciduous at maturity; bracts 2, ovate to lanceolate, acuminate; umbel 5- to 15-flowered, the pedicels about as long as the perianth; tepals rose to purple (varying to white), with conspicuous darker midrib, 7–10 mm long, the outer larger and broader than the inner, erect, entire, oblong, obtusish, the inner lance-linear, acutish; stamens shorter than the tepals; ovary crested, with 3 (6) rounded processes; style about equaling the tepals, the stigma capitate, entire or nearly so; seeds dull, the alveoli not pustuliferous; $n = 7$.

Gravelly and stony clay slopes; throughout most of e. Oregon, s. to the Sierra Nevada of Calif., w. into w. Idaho (and at the Craters of the Moon) and across Nev. to w. Utah (Tooele Co.). April–June.

20. MUILLA S. Wats. Muilla

Erect scapose glabrous perennial herbs from deep-seated, fibrous-coated corms; leaves 2–10, basal and surrounded by a fibrous sheath, narrowly linear and much elongated, flattened but neither keeled nor channelled; flowers small, white to greenish-white or bluish on slender non-jointed pedicels in open umbels subtended by several white or greenish-white scarious bracts; tepals 6, subequal, distinct almost to the base and there only slightly united, subrotate, lanceolate-oblong, spreading; stamens 6, inserted near the base of the tepals, the filaments filiform to petaloid, the anthers versatile, introrse; ovary superior, 3-celled with few to several ovules, the clavate style persistent but finally splitting, stigmatic only at the tips of the 3 segments; fruits globose, slightly lobed, loculicidal capsules; seeds black, angled.

As here defined, a genus of 3 species of Calif., Nev. and Baja Calif. (The generic name is an anagram of *Allium.*)

Reference: Ingram, J. A monograph of the genera *Bloomeria* and *Muilla* (Liliaceae). Madroño 12: 19–27. 1953.

1. Muilla transmontana Greene

Muilla transmontana Greene, Pittonia 1: 73. 1887. *Bloomeria transmontana* J. F. Macbr. Contr. Gray Herb. 56: 8. 1918. (*Pease,* Reno, Washoe Co., Nev., the date of the collection is unknown [but probably the spring of 1887] and the material upon which the name was based, which was a collection of fresh specimens, was apparently never preserved; a neotype is therefore selected here as *Sonne,* Reno, Washoe Co., Nev., May 1888, deposited at NDG.)

Tall erect perennial herbs 1.5–5 dm tall with slender scapes usually fusiform-enlarged at the base and terminated by open umbels, arising from thick corms 1–2 cm long and 2–3 cm wide; leaves 3–5, flat, not keeled, 1.5–6 dm long, 0.5–1 (1.5) mm wide; flowers several on pedicels (1) 1.5–2 (2.5) cm long, the tepals white and often tinged with lilac, largely distinct but forming a short tube 1–1.5 mm long, mostly lanceolate, 6–8 mm long, 1.5–2 mm wide, spreading; stamens 6, the filament 2.5–3 mm long, dilated at the base with the bases united to form a shallow basal cup around the ovary, the anther yellow, 1–1.5 mm long; styles short, 1.5–2 mm long; capsules globose, 8–10 mm long; seeds 4–5 mm long.

In coniferous forests in montane situations, but infrequently found on sagebrush flats and slopes at much lower elev. and even (though rarely) on heavy clay soils, mainly in the Sierra Nevada of Calif. and w. Nev. from Lassen Co., Calif. s. to Mono Co., and in Nev. from Washoe Co. s. to Mineral Co.; in our area known from near Reno, Nev., s. to Bridgeport, Calif., mostly 1400–2500 m elev. May–July.

21. TRITELEIA Douglas ex Lindl.

Erect scapose glabrous to scabrous or hairy at the base perennial herbs from deep-seated, scaly, straw-colored corms; leaves 1–2, arising directly from the corm, narrowly linear and much elongated, flattened, keeled beneath, double channelled above; flowers small, blue to lilac or white to straw-colored to yellow or golden-yellow on jointed ascending pedicels arranged in open umbels, subtended by an involucre of distinct, usually scarious or green acuminate bracts; tepals 6, alike, jointed below and forming a tube of varying lengths, funnelform to campanulate, the spreading segments mostly lanceolate to oblong; stamens 6, in two series, all functional, the filaments distinct, slender, either all alike and of unequal length or unequally inserted on the tepals, the anthers versatile, introrse; ovary superior, 3-celled with few ovules, short-stipitate, the style terminated by 3 minute stigmatic lobes; fruits thin-walled, 3-angled loculicidal capsules; seeds black, rounded, minutely pitted.

A genus of perhaps 14 species found in w. N. Amer. (From the Greek *trios,* three, and *teleios,* complete, apparently alluding to the minutely lobed stigma which seems entire on dried specimens.)

Reference: Hoover, R. F. A systematic study of *Triteleia.* Amer. Midl. Naturalist 25: 73–100. 1941.

1 Anthers attached at two different levels; flowers turbinate-campanulate with the tepals concealing the ovary, light to dark blue; se. Oregon e. to w. Wyo. and n. Utah . . . . 1. *T. grandiflora*

1 Anthers all attached at the same level; flowers broadly campanulate with the tepals spreading, white to bluish-tinged; se. Oregon s. to w. Nev. . . . . 2. *T. hyacinthina*

Muilla transmontana

Triteleia grandiflora

1. Triteleia grandiflora Lindl.

Triteleia grandiflora Lindl. Edward's Bot. Reg. 15: 1293. 1830. *Milla grandiflora* Baker, J. Linn. Soc., Bot. 11: 380. 1871. *Brodiaea douglasii* S. Wats. Proc. Amer. Acad. Arts 14: 237. 1879. *Hookera grandiflora* Kuntze, Revisio Gen. Pl. 2: 712. 1891. *Hookera douglasii* Piper, Contr. U.S. Natl. Herb. 11: 190. 1906. *Brodiaea grandiflora* J. F. Macbr. Contr. Gray Herb. 56: 9. 1918; not J. E. Smith, 1811. ("Found in North-west America by Mr. Douglas. It is growing in the Garden of the Horticultural Society, where it flowers in July.")

Tall erect perennial herbs 2–7 dm tall with slender scapes terminated by open umbels 1–7 cm long, arising from globose corms 1–2.5 cm long and 1–3 cm wide; leaves 1–2, flat but keeled beneath, 2.5–5 dm long, 3–10 mm wide, persistent; flowers several on pedicels 1–5 cm long, light to dark blue, the tepals united into a narrowly turbinate-campanulate tube 9–12 (14) mm long with the lobes 10–14 (16) mm long and in two series, the outer series oblong-lanceolate, plane and spreading, the inner series oblong with strongly undulate margins and erect for 2–3 mm, with these margins contracted at the point of the flare and closing the orifice of the tube; stamens 6, unequally inserted, the filaments about as long as wide, the adnate portion strongly keeled and touching the ovary, narrowing at the point of attachment with the anther, the outer series of 3 filaments scarcely half the length of the anthers and inserted 2–3 mm below the inner series of 3 filaments, these 1–1.5 times as long as the anthers, the anthers blue,

2–3 mm long; styles 2–3 (4) mm long; ovary subterete, 6-grooved, on a stipe 3–5 mm long; capsules cylindro-ovoid, 6–12 mm long; seeds ca. 2 mm long; $2n = 32$.

Grasslands or sagebrush and pinyon-juniper woodlands to pine forest slopes and hills from e. B.C. s. to e. Oregon and e. across Idaho to sw. Mont., w. Wyo. and n. Utah; across the n. part of our area where infrequent in se. Oregon, rare in s. Idaho, but rather common along the Wasatch Front in n. Utah. May–July.

Our plants are var. *grandiflora*.

2. Triteleia hyacinthina (Lindl.) Greene

Hesperoscordum hyacinthinum Lindl. Edward's Bot. Reg. 15: 1293. 1830. *Hesperoscordon lewisii* Hook. Fl. Boreali-Amer. 2: 185. 1838. *Milla hyacinthina* Baker, J. Linn. Soc., Bot. 11: 385. 1871. *Triteleia hyacinthina* Greene, Bull. Calif. Acad. Sci. 2: 142. 1886. *Hookera hyacinthina* Kuntze, Revisio Gen. Pl. 2: 712. 1891. *Brodiaea hyacinthina* Baker, Gard. Chron. III. 20: 459. 1896. *Calliprora hyacinthina* Gorman, Muhlenbergia 2: 375. 1916. (*Lewis*, "On the plains of the Columbia and Missouri," actually from near Celilo Falls, Klickitat Co., Wash., 20 April 1806.)

Hesperoscordum lacteum Lindl. Edward's Bot. Reg. 15: 1293. 1830. *Allium lacteum* Benth. Pl. Hartwegianus 339. 1857. *Milla hyacinthina* var. *lactea* Baker, J. Linn. Soc., Bot. 11: 385. 1871. *Brodiaea lactea* S. Wats. Proc. Amer. Acad. Arts 14: 238. 1879. *Brodiaea hyacinthina* var. *lactea* Jepson, Fl. W. Middle Calif. 117. 1901. *T. lactea* (attrib. to S. Wats. by) Davidson & Moxley, Fl. S. Calif. 87. 1923. (*Douglas*, Calif.)

Tall, erect perennial herbs 2.5–7 dm tall with slender scapes terminated by modestly open umbels 1–5 cm long, arising from ovoid bulbs 1–2 cm long and wide; leaves 1–2, flat but keeled beneath, 1.5–4 dm long, 3–10 (15) mm wide, persistent; flowers several on pedicels 1–4 cm long, white and usually bluish-tinged to light blue, the tepals united into a broadly campanulate to semirotate tube 9–16 mm long, with the lobes 7–12 mm long, broadly lanceolate to oblong-obovate, spreading, not constricted over the orifice of the tube; stamens 6, the filaments 2–4 mm long and all inserted at the same level, greatly expanded at the base and confluent, the anthers white or rarely blue, 1–2 mm long; styles 2.5–3 mm long; ovary terete, 6-grooved, on a stipe 0.5–1 mm long; capsules ovoid, 4–6 mm long; seeds ca. 1.5 mm long; $n = 14, 28, 35$, $2n = 28, 84$.

Grassy places to rocky, open flats from s. B.C. s. to s. Calif., e. into Idaho, se. Oregon and w. Nev.; infrequent in our area, mainly in grassy places in the Lake and Reno sects. below 2200 m elev. May–July (Aug.).

22. DICHELOSTEMMA Kunth

Erect scapose, glabrous to scabrous perennial herbs from deep-seated, fibrous, brown corms; leaves 2–5, arising directly from the corm, narrowly linear and much elongated, flattened, keeled beneath, double channelled above; flowers small but showy, violet or lilac to light pink or rose to bright red, on ascending to flexuous or spreading to recurved jointed pedicels, arranged in congested to capitate or rarely open (ours) umbels subtended by an enlarged outer bract forming an involucre and smaller inner bracts; tepals 6, alike, jointed below and forming a tube at least half their length or longer, the tube inflated or campanulate, the segments only slightly spreading; stamens 3 or 6 (ours), in two series, those opposite the outer tepals with reduced anthers or completely transformed into staminodia, those opposite the inner tepals with functional anthers, all equally adnate nearly their entire length to the tepals and broad, the anthers basifixed, introrse, sagittate; ovary superior, 3-celled with few ovules, sessile or short-stipitate, the slender style enlarged upwardly with 3 short stigmatic lobes; fruits firm-walled, 3-angled loculicidal capsules; seeds black, elongate, not pitted.

A small genus of 6 species of w. N. Amer. (From the Greek *dicha,* two, and *stema,* stamen, alluding to the two different types of stamens in some species.)

Reference: Hoover, R. F. The genus *Dichelostemma.* Amer. Midl. Naturalist 24: 463–476. 1940.

1. Dichelostemma pulchellum (Salisb.) A. A. Heller

Hookera pulchella Salisb. Paradisus Lond. 2: 117. 1808. *Brodiaea pulchella* Greene, Bull. Calif. Acad. Sci. 2: 133. 1886. *Dichelostemma pulchellum* A. A. Heller, Muhlenbergia 1: 132. 1906. *Dipterostemon pulchellus* Rydb. Bull. Torrey Bot. Club 39: 111. 1912. (*Menzies*, Calif.)

Brodiaea capitata Benth. Pl. Hartwegianus 339. 1857. *Dichelostemma capitata* (attrib. to Benth. by) A. Wood, Proc. Acad. Nat. Sci. Philadelphia 1868: 173. 1868. *Milla capitata* Baker, J. Linn. Soc., Bot. 11: 381. 1871. *Hookera capitata* Kuntze, Revisio Gen. Pl. 1: 712. 1891. *Dipterostemon capitatus* Rydb. Bull. Torrey Bot. Club 39: 111. 1912. (*Hartweg 58,* near Monterey, Monterey Co., Calif., in 1846 or 1847.)

Brodiaea capitata var. *pauciflora* Torr. in Emory, Rep. U.S. & Mex. Boundary Surv. 2(1): 218. 1859. *Milla capitata* var. *pauciflora* Baker, J. Linn. Soc., Bot. 11: 381. 1871. *Dichelostemma pauciflorum* Standley, Contr. U.S. Natl. Herb. 13: 179. 1910. *Dipterostemon pauciflorus* Rydb. Bull. Torrey Bot. Club 39: 111. 1912. *Hookera pauciflora* Tidestrom, Contr. U.S. Natl. Herb. 25: 123. 1925. *Dichelostemma pulchellum* var. *pauciflorum* Hoover, Amer. Midl. Naturalist 24: 471. 1940. (*Bigelow*, near the Santa Rita "Copper Mines," Grant Co., N.M., April 1851; lectotype selected by Hoover.)

Brodiaea insularis Greene, Bull. Calif. Acad. Sci. 2: 134. 1886. *Dichelostemma insulare* Burnham, Muhlenbergia 3: 74. 1907. *Dipterostemon insularis* Rydb. Bull. Torrey Bot. Club 39: 111. 1912. *Brodiaea capitata* var. *insularis* J. F. Macbr. Contr. Gray Herb. 56: 9. 1918. (*Greene*, Guadalupe Isl., Mex.)

Tall erect perennial herbs (1) 3–6 (9) dm tall with slender scapes terminated by open umbels 1–3 cm long, arising from ovoid bulbs 1–2.5 cm long and 1–2 cm wide; leaves 2–5, flat but keeled beneath, 1.5–4 dm long, 5–12 mm wide; flowers several on ascending pedicels 1–15 mm long, violet or occasionally varying to white, subtended by ovate and purple or lanceolate to linear-lanceolate and white-scarious bracts, the tepals united into a cylindro-campanulate tube (3) 4–8 mm long with the lobes 7–11 mm long, mostly oblong; stamens 6, of two series, the filaments opposite the outer tepals dilated and 2 mm long with anthers 2–3 mm long, those opposite the inner tepals nearly fully adnate to the tepals and bearing lateral, narrowly triangular to lanceolate

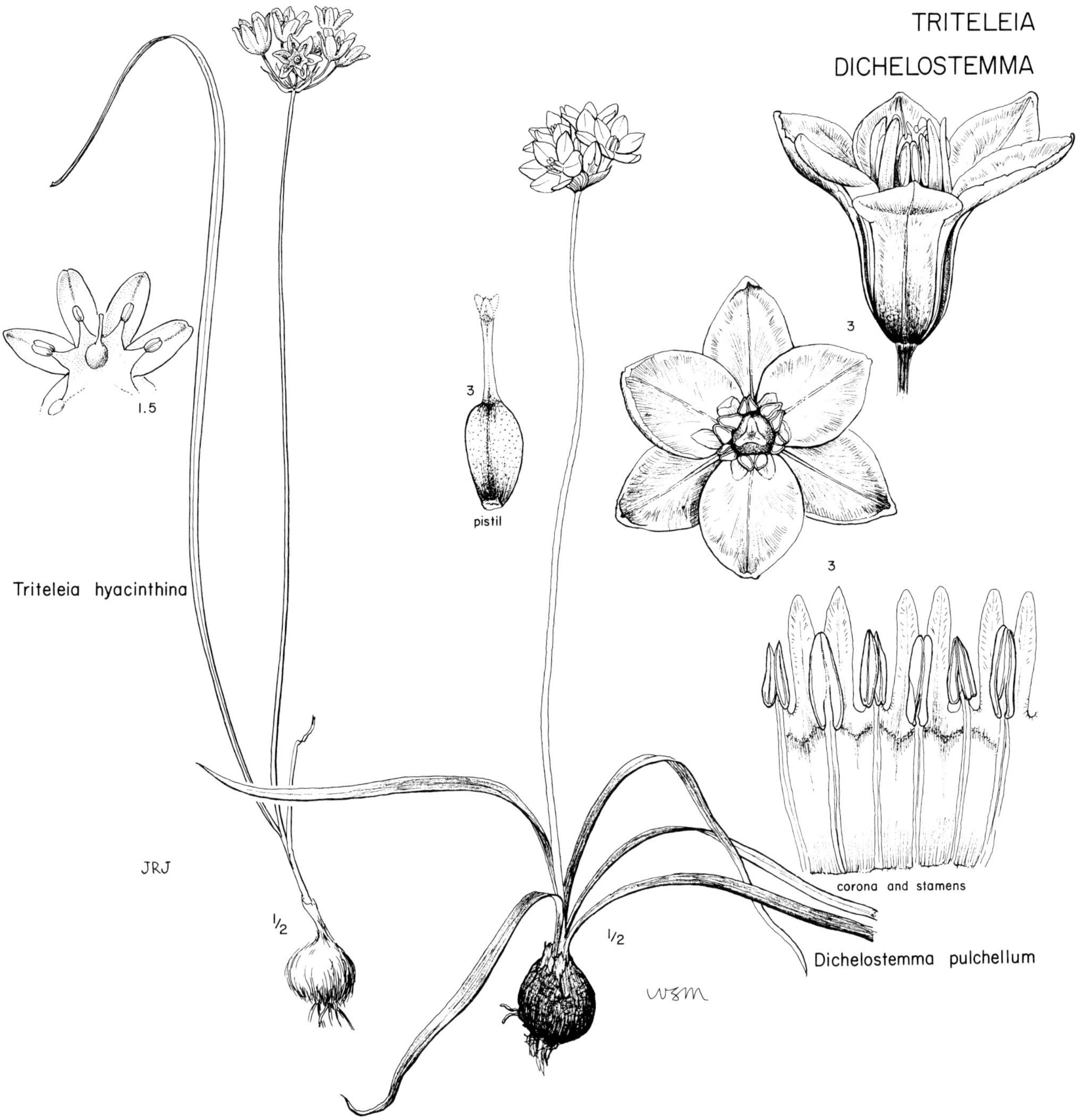

wings 4–6 mm long on either side of the anthers, these anthers 3.5–4.5 (5) mm long; styles 4–6 mm long; capsules ovoid, 4–6 mm long; seeds 2.5–4 mm long; n = 9, 18, 36, $2n$ = 18, 27, 36, 45, 54, 72.

Plains and hills in coniferous forests or open woodlands to sagebrush flats from s. Oregon s. throughout most of Calif. to Baja Calif., hence e. across s. Nev. to sw. Utah, Ariz. and sw. N.M.; infrequent in our area, where known only from c. Nye Co., Nev., and Washington Co., Utah. March–May.

I have continued to follow Hoover in using the specific name "*pulchellum*" for our plant, and have not taken up the later "*capitata*" as recently suggested by some California workers. The illustration given by Salisbury seems conclusive, a point agreed to by Lee Lenz, a noted authority on this and related genera, and there appears no reason to regard the Salisbury name as a nomen ambiguum.

Hoover divided *D. pulchellum* into two weakly defined varieties. The var. *pulchellum* was characterized by its ovate, purplish bract, and ranged from southwestern Utah westward. The var. *pauciflorum* (Torr.) Hoover was denoted by its linear-lanceolate to lanceolate white-scarious bracts, and more easterly range, being found from Washington Co., Utah, and adjacent Mohave Co., Arizona, southeastwardly to southwestern New Mexico. Lenz suggests that these differences proposed by Hoover are not consistent and that there is no sound basis for attempting to distinguish the two phases. The specimens from the Intermountain Region are particularly difficult to place, and I am following Lenz in reducing the var. *pauciflorum* to synonymy.

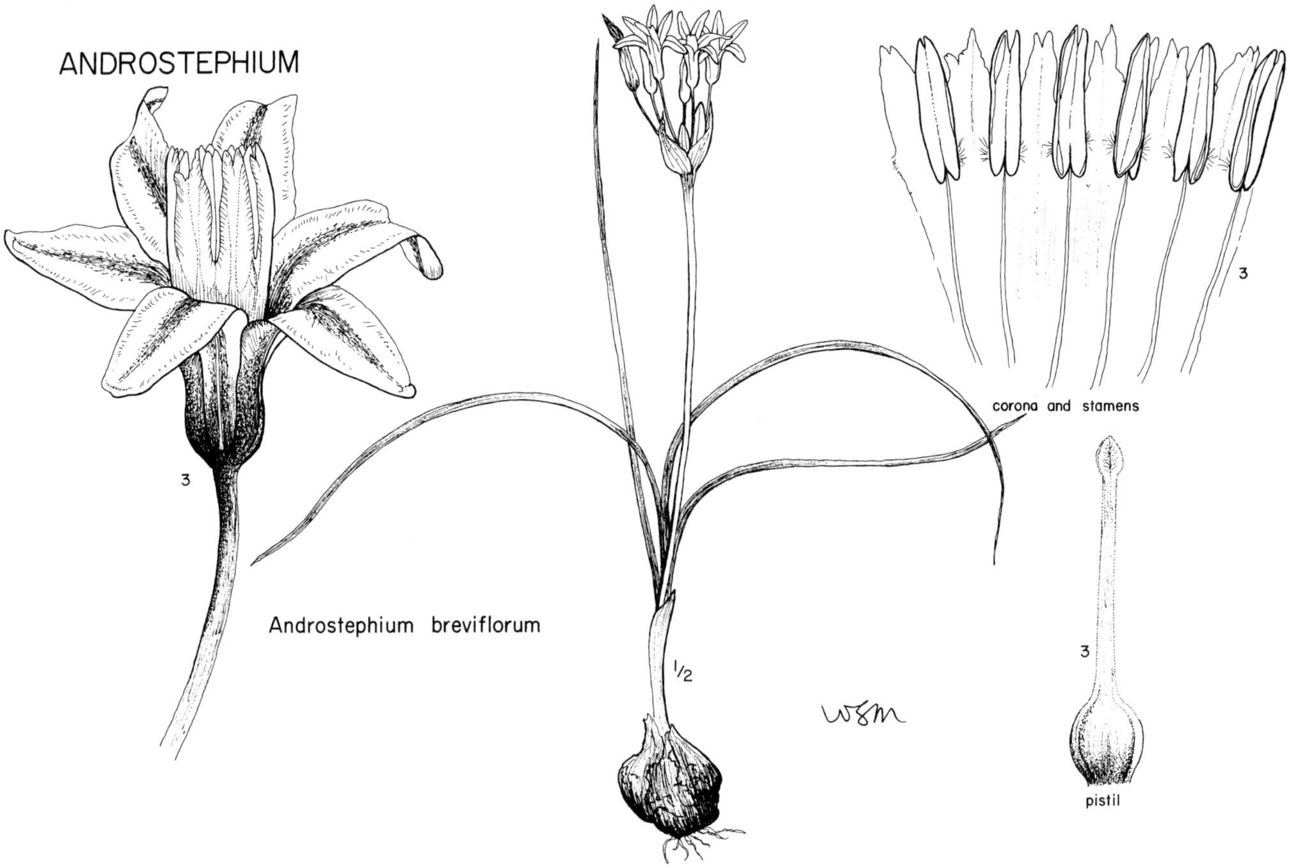

23. ANDROSTEPHIUM Torr.

Erect glabrous scapose perennial herbs from fibrous-coated corms; leaves several, linear, channelled; flowers small but rather showy, white to violet-purple, on slender non-jointed pedicels arranged in open umbels and subtended by 3 lanceolate to broadly lanceolate bracts; tepals 6, alike, jointed below and forming a funnelform tube slightly less than half the overall length of the tepals, the segments narrowly-oblong, spreading; stamens 6, in one row in the throat, the filaments partially united into a tube with erect bifid lobes between the introrse, basifixed, mostly acute anthers; ovary superior, 3-celled with several ovules, sessile, the long, slender, persistent style terminated by three small stigmas; fruits subglobose, obtusely 3-angled, loculicidal capsules; seeds black, flat.

A genus of two species in the sw. U.S. (From the Greek *andros,* stamens, and *stephanos,* crown, referring to the united filaments.)

1. Androstephium breviflorum S. Wats.

Androstephium breviflorum S. Wats. Amer. Naturalist 7: 303. 1873. *Brodiaea breviflora* J. F. Macbr. Contr. Gray Herb. 56: 9. 1918. *Bessera breviflora* Jepson, Fl. Calif. 1: 291. 1921. (*Thompson,* near Kanab, Kane Co., Utah, in 1872.)

Brodiaea paysonii A. Nels. Bot. Gaz. 56: 63. 1913. (*Payson 33,* Montrose, Montrose Co., Colo., 18 May 1912.)

Erect perennial herbs 1–3 dm high with slender scapes terminated by open umbels 1–5 cm long, arising from ovoid corms 0.5–2 cm long and 1–2.5 cm wide; leaves several, flat, linear, 1–3 dm long, 1.5–2.5 mm wide, enclosed in tan, underground, membranous sheaths 2–5 cm long; flowers 3–8 (12), on ascending pedicels (1) 1.5–2 (3) cm long, whitish to light violet-purple, often drying brownish-yellow with dark brownish longitudinal stripes on each tepal, subtended by lanceolate to broadly lanceolate, whitish bracts 8–15 mm long, the tepals united into a tube 4–8 mm long, the free lobes 5–12 (15) mm long; stamens 6, the filaments united and longer than the yellow to bluish anthers, the anther 3–3.5 mm long; capsules subglobose, 10–15 mm long, 3-lobed, dark brown; seeds large, 4–6 per capsule, black, 7–9 mm long, flat.

Dry sandy to rocky soil from s. Nev. (and supposedly adjacent se. Calif.) e. across n. Ariz. and s. Utah to w. Colo., hence n. to Uintah Co., Utah, mostly from 550–2300 m elev. April–May.

Family AGAVACEAE, the Agave Family

Succulent to subsucculent or xerophytic monocarpic to polycarpic perennials (ours) or multiannuals with long-lived leaves, acaulescent to caulescent or arborescent, frequently suckering at the base and forming large colonies or occasionally bulbiferous in the inflorescences, or small, slender erect perennial herbs from small perennial rosettes, the stems erect to decumbent or procumbent, often thick, simple or branched, terminated by elongated racemose or paniculate inflorescences; leaves basal or cauline or both, succulent to dry, often large and spine-tipped, the margin often armed in some or with curly fibers, mostly in whorls, obscurely parallel-veined; flowers perfect to polygamous or rarely imperfect, small to large and often rather showy, mostly white to yellow, orange or greenish, 3-merous; tepals of 6 distinct segments or united at the base into a funnelform to salverform, erect or more often curved tube, the segments mostly of 2 series of similar or subequal sets of corolla-like tepals, not differentiated into a distinct calyx or corolla, often fleshy; stamens 6, included or long exserted, sometimes abortive in fertile flowers, hypogynous or adnate to the perianth tube, the filaments slender to broadly dilated or infrequently united at the base, the anthers mostly linear to lanceolate, usually dorsifixed, introrse; ovary superior or inferior, 3-celled (but often seemingly 6-celled) with numerous to few (rarely 1–2) ovules per locule with axile or rarely parietal placentation, the slender and often very short style terminated by a capitate or short, 3-lobed stigma; fruit a berry or loculicidal capsule, membranous to leathery; seeds 1–many, with endosperm.

A family of some 450 species and about 20 genera widely distributed in the more xeric regions of the world. Members of this family are often placed in the Liliaceae or Amaryllidaceae, according to whether the ovary is superior or inferior.

1 Ovary superior; tepals white to creamy or rarely yellowish, free or nearly so; anthers included.
  2 Flowers large, 3 cm or more long, usually succulent, campanulate to globose; leaves with a terminal spine .... 1. *Yucca*
  2 Flowers small, less than 1 cm long, not succulent, campanulate; leaves without a terminal spine .... 2. *Nolina*
1 Ovary inferior; tepals yellow, united at the base into a tube; anthers long exserted .... 3. *Agave*

1. YUCCA L. Yucca, Spanish Bayonet

Arborescent to shrubby perennials with stout, short to rather large, simple or branched caulescent to subacaulescent, semi-woody caudices and stems, and with larger, terminal, annual, many-bracteated flowering stems or peduncles; leaves narrow, usually stiff, linear to lanceolate, thin to thick and fleshy, with expanded bases and terminal spinose tips; flowers large and showy, campanulate to globose, mostly drooping, cream to white, numerous, in large dense elongate terminal panicles or racemes; tepals 6, alike, distinct or somewhat united at the base, lanceolate to ovate, the tips tending to curve inward; stamens 6, hypogynous, the filaments broad and flat, often fleshy, the anthers sagittate, introrse; ovary superior, 3-locular or often falsely 6-locular by the intrusion of a new partition from the midrib of the carpel to the placental axis between the two rows of ovules, the ovules numerous, the style short and thick with a subglobose or 3-lobed stigma; fruits hardened loculicidal or septicidal capsules, or, in some, soft and fleshy or somewhat spongy; seeds obovoid to compressed, blackish.

A genus of about 30 species of the drier desert areas of N. Amer. (A Haitian name for manihot, an entirely different plant.)

References:

McKelvey, S. D. Yuccas of the southwestern United States. Part I, pp. 1–150. 1938. Part II, pp. 1–192. 1947. The Arnold Arboretum of Harvard University, Jamaica Plain, Mass.
Powell, J. A., and R. A. Mackie. Biological interrelationships of moths and *Yucca whipplei*. Univ. Calif. Publ. Entomol. 42: 1–59. 1966.
Webber, J. M. Yuccas of the southwest. Agric. Monogr. U.S.D.A. 17: 1–97. 1953.

Pollination in *Yucca* is by members of the moth genus *Tegeticula* (including *Pronuba*) which visit the plants usually at night. The ovary is pierced by a long ovipositor, and the eggs laid among the ovules. The female moth then packs the openings of the stigma with a mass of sticky pollen gathered from other flowers. Moth larvae and seeds of *Yucca* then develop together, with the larvae feeding on the immature seeds. When fully developed, the larvae chew their way through the ovary wall and escape. Other genera of moths also visit *Yucca*, but these play no role in pollination (Powell & Mackie, 1966). Many species of *Yucca* have been used by Indians to produce fibers, soap and other products. The fleshy fruits and seeds of some species are eaten by cattle, burros and wild horses, and the fruits and young flowering stalks were sometimes used as food by the Indians and early pioneers. One of the more interesting sights is watching cattle eat the young inflorescences of *Y. baccata;* there is simply no polite way to describe this. Some of our species do well in the garden, and one, *Y. brevifolia,* is sometimes transplanted outside its natural range, especially in Nevada and southern Utah. In addition, other species of *Yucca* may be cultivated in parks and on university campuses, but these are usually from sources outside the Intermountain Region.

Taxonomically, the genus is one of the most difficult in our area. The treatment presented here is not designed to apply strictly to our flora, as it has been necessary to review much of the genus in an extensive manner in order to treat our members. The descriptions given are for the species—as now defined—wherever they occur, for in many cases we have only a minor part of the overall range of variation in the plants found in the Intermountain Region.

It has proved difficult to prepare an adequate key that will faithfully distinguish all herbarium specimens. Therefore, the keys given here will prove more successful with material actually obtained by the user. In order to assist future students of the genus, it would be helpful to make the following observations and record them with the location data on herbarium labels:

1. Note if the plant is acaulescent, subcaulescent with procumbent or prostrate stems, or caulescent to arborescent. If caulescent or arborescent, record the length of the trunk to the first branch as well as the height of the entire plant.
2. Note the habit of the inflorescence and its length. Care should be given to distinguish between the height of the flowering stem or peduncle (that portion of the axis which extends from the base of the leaves to the lowermost flowers) and the inflorescence. If the inflorescence is actually included within the leaves, this fact should be recorded, and the length of overlap noted.

3. Flower color is helpful as is the length of fresh flowers. The fresh pistil should be measured and the length and color of the style recorded.
4. Mature fruits are often difficult to collect and curate. For those species with fleshy fruits, a cross-section will usually suffice, but for those with dried, dehiscent capsules, it is often best to collect the entire fruit, noting the degree—if any—of constriction.
5. Finally the leaves themselves. The majority of our species can be easily handled as the leaves are usually thin and of moderate length. If an entire leaf—including the base—can be collected, it is best. However, if this is impossible, a portion of the base, center and the apex may be obtained, or at least a cross-section of the leaf should be prepared.

With these items in mind, it will be much easier to discover the proper name for each collection of *Yucca.* The genus is rather variable, noted for its promiscuous nature, and the subject of little attention by most taxonomists. Hopefully, this attitude may alter in the future.

1 Fruits indehiscent, fleshy or dry and spongy; plants large and arborescent with small, thin, nonfibrous leaves, or cespitose perennials with large, rigid, broad, swordlike leaves with a long terminal spine.
  2 Plants cespitose or nearly so; leaves 5–7.5 dm long, concavo-convex, strong and rigid; fruits fleshy; s. Nev. and s. Utah .... 1. *Y. baccata*
  2 Plants arborescent; leaves (1) 1.5–3.5 dm long, plano-convex, thin; fruits spongy and dry; sw. Nev., se. Nev. and extreme sw. Utah .... 2. *Y. brevifolia*
1 Fruits dehiscent, loculicidal or (ours) septicidal; plants acaulescent to caulescent with long, narrow, flexible, grasslike leaves, or if broader and rigid, then less than 5 dm long.
  3 Leaves concavo-convex, 1–5 dm long, lanceolate; inflorescence racemose or nearly so, 3.5–7 dm long on a short flowering stalk 1–4 dm long; style bright green, 9–13 mm long; fruits constricted; throughout most of Utah and portions of Nev. .... 3. *Y. harrimaniae*
  3 Leaves plano-convex, mostly linear; style white to pale green; fruits mostly not constricted, but if so, then inflorescences and flowering stems much longer.
    4 Inflorescences simply racemose or at most with a few abortive branchlets, not paniculate; plants mostly acaulescent to slightly caulescent.
      5 Flowering stems short, less than 1 dm long; inflorescence within or just barely exceeding the leaves, in ours up to 0.5 m in length; fruits pendulous, not constricted; San Juan and e. Kane cos., Utah .... 4. *Y. baileyi*
      5 Flowering stems long, 3–25 dm long; inflorescence well beyond the leaves, 0.3–1.5 (2) m in length; fruits erect, constricted but sometimes only slightly so; c. Utah southward .... 6. *Y. angustissima*
    4 Inflorescences panicles with only the uppermost flowers racemose; plants caulescent or rarely acaulescent; capsules only slightly, if at all, constricted; Washington Co., Utah, s. into Mohave Co., Ariz. .... 5. *Y. elata*

1. Yucca baccata Torr.

Yucca baccata Torr. in Emory, Rep. U.S. & Mex. Boundary Surv. 2(1): 221. 1859. (*Bigelow s.n.,* Camp 52, Cedar Woods, Guadalupe Co., N.M., 24 Sept. 1853; lectotypification of McKelvey rejected and replaced with the above, deposited at NY—see discussion below.)

Y. baccata var. *vespertina* McKelvey, Yuccas S.W. U.S. 1: 45. 1938. (*McKelvey 2167,* e. of Peach Springs, Mohave Co., Ariz., 12 May 1931.)

Plants mostly simple or clumped, acaulescent or rarely caulescent with 2–6 short, procumbent stems, the clumps 1–5 m in diam; leaves in rather open, asymmetrical heads (5) 6–7.5 dm long and 13–15 dm across, those at the base spreading with the central leaves more erect, the blade straight or incurved, falcate or occasionally twisted, deeply concavo-convex, rigid, (3) 5–7 (7.5) dm long, 2.5–4 cm wide, or if flattened, 3–6 cm wide, dark green or blue-green, glabrous or glaucous, the margin separating on the upper half of the blade to form broad to wiry, coarse to fine, recurved to curly marginal fibers, the apex thickened, mostly acuminate with a stout, stiff spine 1.5–7 mm long; flowering stems erect, short, hidden by the leaves, up to 1.3 dm long; inflorescences heavy, narrow panicles with the uppermost clusters of flowers in racemes, (4) 6–7.5 dm long, usually as long as to just exceeding the leaves, the stalk glabrous to slightly scabrous, green to reddish-purple, the branchlets few, 5–20 cm long, spreading to ascending; bracts thick and rigid, the lowermost lanceolate, 1–2 dm long, 4–5 cm wide, becoming reduced above to 10 cm long, 5 cm wide and triangular-ovate; bractlets papery, lanceolate, 2.5–5 cm long, white or reddish-purple; flowers campanulate, 6–13 cm long, red-brown without, creamy-white within, tinged with reddish-purple in some, pendent on terete to slightly flattened pedicels 0.7–4 cm long, the tepals united at the base for 7–12 mm, lanceolate and 4–10 cm long above, the inner tepals slightly wider than the outer ones, fleshy pubescent; stamens included, the filament 3–4.5 cm long, united at the base, the anther 5–7 mm long; ovary (3) 5–7 (7.5) cm long, the style 4–7 (10) mm long; capsules ellipsoid, fleshy, (10) 15–17 cm long, (3) 5–6.5 cm in diam, the upper third constricted and lobed; seeds obovoid to compressed, dull black, thick and somewhat ridged, 7–11 mm long.

Dry slopes across the sw. U.S. from se. Calif. e. across s. Nev. and Ariz. to Colo., Texas and nw. Mex.; occasional in our area from s. Nev. (Lincoln and c. Nye cos. as far n. as the Quinn Canyon Range) across s. Utah in the Dixie Corridor Sect. into the Canyon Lands Sect. as far n. as Grand Co., and s. onto the n. edge of the Kaibab Plateau in n. Ariz., from 850–2000 m elev. April–June.

The type of *Y. baccata* was cited by McKelvey (1938) as Bigelow's

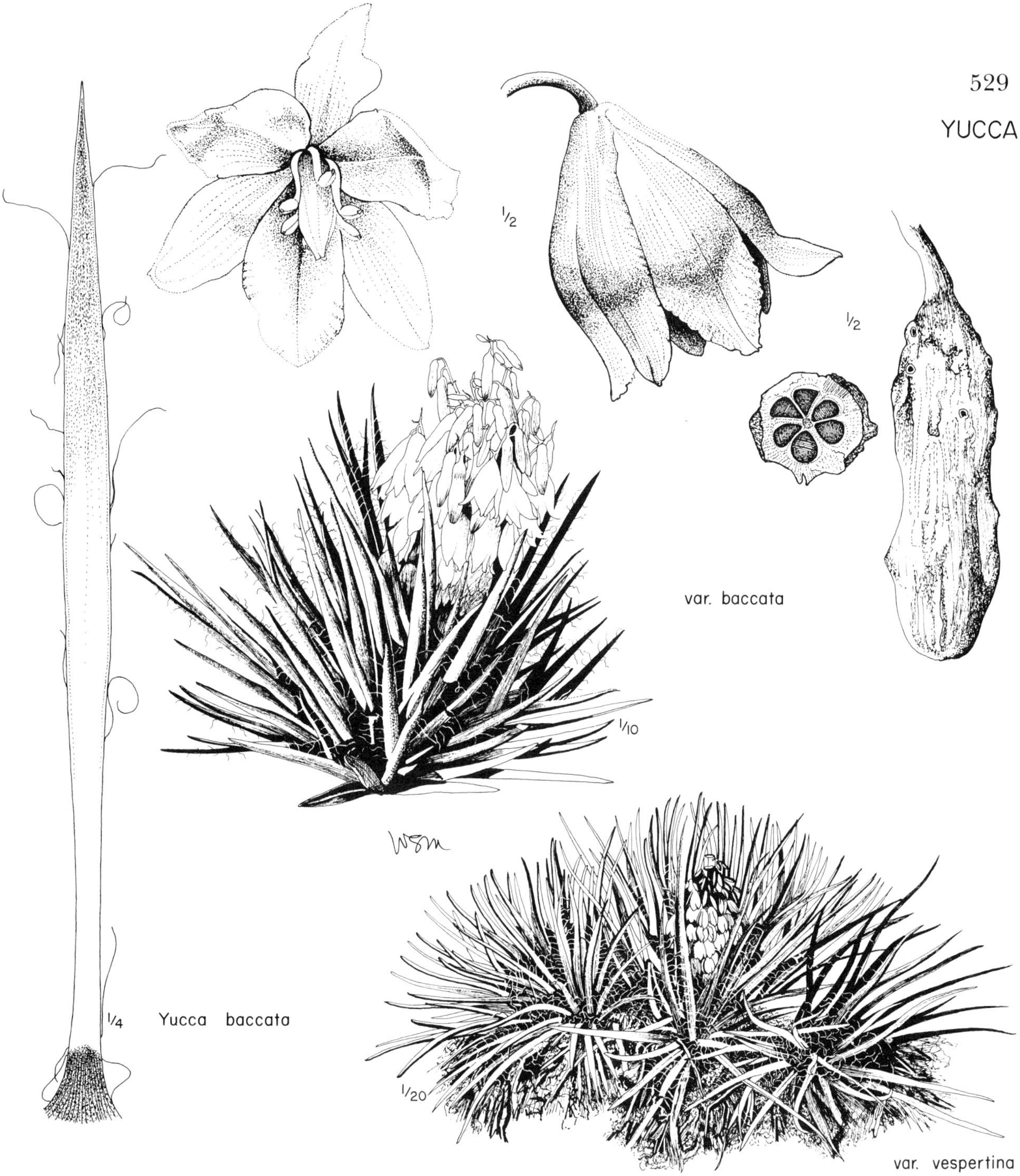

collection from Hurrah Creek, Guadalupe Co., New Mexico, gathered on 25 September 1853, and she gives the holotype as being located at the Smithsonian Institution. Two important facts stand in the way of this indirect lectotypification. First, Torrey did not select a type for *Y. baccata,* but discussed three collections directly and a fourth indirectly. Secondly, the sheet in Torrey's own herbarium is representative of this species, as is the one sheet in the Smithsonian. However, only rarely did Torrey give away his original types, and usually only when the type was represented by a single collection (he was pressured to give up his type material by the government when the U.S. National Herbarium was formed, but to a great extent, Torrey kept his types). McKelvey's selection of the Hurrah Creek collection over that of Cedar Woods must be overturned because the former collection does not represent the original description as precisely as the latter collection, and in this case, the Washington collection must be rejected in favor of the New York collection regardless. As a result, McKelvey's indirect lectotypification is overturned.

The species is composed of two weakly defined varieties, both of which occur in our area.

1 Leaves dark green, glabrous, the blade straight with the marginal fibers coarse and broad; inflorescence of ca. 15 branches, as long as to longer than the leaves, the scape and stalk mostly green; se. Calif. e. to Texas and Mex., our common form .... var. *baccata*

1 Leaves blue-green, glaucous, the blade falcate with the marginal fibers fine and wiry; inflorescence shorter to just slightly longer than the leaves, the scape and stem mostly reddish-purple; se. Calif. and s. Nev. e. to Ariz. and Utah, restricted to Washington and w. Kane cos., Utah in our area .... var. *vespertina* McKelvey

2. Yucca brevifolia Engelm.

Yucca draconis var. *arborescens* Torr. in Whipple, Explor. & Surv. Railroad Route from Mississippi River to Pacific Ocean 4(5): 147. 1857. *Y. brevifolia* Engelm. in King, Rep. Geol. Explor. 40th Parallel 5: 496. 1871. *Y. arborescens* Trelease, Annual Rep. Missouri Bot. Gard. 3: 163. 1892, prov. name. *Y. arborescens* Trelease ex Merriam, N. Amer. Fauna 7(2): 353. 1893. *Clistoyucca arborescens* Trelease, Rep. Annual Missouri Bot. Gard. 13: 41. 1902. *Clistoyucca brevifolia* Rydb. Fl. Rocky Mts. 1061. 1917 [1918]. (*Bigelow,* "Sandy and gravelly plains west of the Colorado [River]," along the north bank of the Mojave River near Barstow, San Bernardino Co., Calif., 15 March 1854.)

Y. brevifolia var. *jaegeriana* McKelvey, J. Arnold Arbor. 16: 269. 1935. (*McKelvey 2732,* near Shadow Mts., San Bernardino Co., Calif., 30 April 1932.)

Joshua tree.

Plants erect, often single stemmed trees (3) 5–12 (15) m tall, becoming branched usually (0.7) 1–3 m above ground with erratically spreading to ascending branches, the trunk stout, reddish-brown to gray, the bark checkered into squarish plates, the primary branches mostly erect but the secondary ones essentially diffuse, often covered, at least in part, with an appressed or shaggy thatch of dead, reflexed-spreading leaves below the crowns of green living leaves; leaves in crowns at first dense, eventually open, broad flat- or round-topped clusters 0.3–1 (1.5 or more in juvenile plants) m long, 3–5 dm in diam, the blade straight, plano-convex or triquetrous, rigid, (1) 1.5–3.5 dm long, 0.7–1.5 cm wide, green, the base whitish, 2.5–4 cm long and 4–5 cm wide at the point of attachment, with a narrow crimson to brown line marking the point of demarcation between the blade and base, the margin entire, thin, horny, minutely denticulate, the apex acuminate, dark red-brown, mostly with a stiff spine 7–12 mm long; flowering stems erect, short, 1–1.5 dm long; inflorescences heavy, broad, densely flowered panicles (1.5) 2–4 dm long, the stalk commonly glabrous, 5–15 cm long, green or rarely tinged with lilac, the branches numerous, 2.5–7.5 cm long, spreading to ascending; bracts leaf-like, the lowermost 2–3 dm long and linear-lanceolate with a broad laminar base as long to longer, the uppermost reduced, 10–12 cm long with the free tip 4–5 cm long; bractlets thin and papery, 3–5 cm long, broadly triangular, white but tinged with rose or lilac; flowers ellipsoid to globose, (3) 4–7 cm long, greenish-white to cream, on stout, stiff, erect or spreading, terete (below) to flattened (near the apex) pedicels 0.7–1.2 (2.5) cm long, the tepals united at least half their length, lanceolate to oblong, fleshy, waxen; stamens included, the filament 1–1.2 cm long, not united at the base, the anther 3–3.2 mm long; ovary 2.5–3 cm long, the style 1–1.5 mm long; capsules ellipsoid, dry and rather spongy, 6–8.5 (10) cm long, 3–5 (6.5) cm in diam, commonly deeply furrowed; seeds flat and thin, dull black, smooth, 8–11 mm long.

Dry slopes and mesas of se. Calif. e. across s. Nev. to sw. Utah and w. Ariz.; widely scattered in two disjunct locations in our area, most commonly near the borderline of the sagebrush and creosote communities and seldom extending very far into either, the first being in the Tonopah Sect. from near Lida Junction w. to the se. base of the White Mts. in Inyo Co., Calif., and e. to the Cactus Range of Nye Co., Nev., n. to Goldfield, and with a few planted specimens near the Tonopah airport e. of Tonopah, and the second area being in the Dixie Corridor Sect. along the w. slope of the Beaver Dam Mts. of Washington Co., Utah, and adjacent Mohave Co., Ariz., from 850–2200 m elev. April–May.

In preparation for this treatment of *Yucca,* types of most of the species were reviewed, including those of the Joshua tree, *Y. brevifolia.* The type was collected by John Bigelow along the north bank of the Mojave River just southwest of Barstow in San Bernardino Co., California, on 15 March 1854. The plants were not in flower and only leaves were obtained; none exceeds 9.5 cm in length. The collection was divided into three specimens (insofar as I have seen), with one sheet remaining in Torrey's collection now housed at the New York Botanical Garden, and with duplicates sent to the Academy of Natural Sciences in Philadelphia and the Smithsonian Institution. All but the specimen at Philadelphia have leaves, the exception consists only of bracts. These collections were annotated by Torrey as *Y. draconis* var. *arborescens.* In 1871, Engelmann proposed the specific name, *Y. brevifolia,* citing in synonymy Torrey's earlier name. In her treatment of the species, McKelvey (1938) selected the Smithsonian specimen of var. *arborescens* as the holotype, wholly without justification. Her typification is here ignored and the specimen deposited in Torrey's own collection considered the holotype.

At the same time, McKelvey lectotypified Engelmann's *Y. brevifolia* on a Palmer collection from Date Creek, Yuma Co., Arizona. This was subsequently followed by Webber (1953), but as *Y. brevifolia* is clearly based on var. *arborescens,* it must have the same type, and McKelvey's proposed lectotypification here too must be rejected.

For some unexplained reason, McKelvey could not place the type of var. *arborescens* from a specific area of California, although the New York collection is dated, and the Whipple report of 1857 gives the expedition's exact location on 15 March 1854. She placed the variety within her concept of var. *brevifolia* and noted that the leaves of the type of var. *arborescens* "suggests in size those of the dwarf variety." This dwarf variant was her var. *jaegeriana.* As she defined var. *jaegeriana,* it is a small tree 3–6 m tall with leaves 1–2 dm long, and ranges from eastern San Bernardino Co., California, northeastwardly to Washington Co., Utah. In reviewing herbarium specimens annotated by McKelvey, one finds that much of the western Mojave Desert material from San Bernardino County has short leaves, and as these plants occur outside the supposed range of var. *jaegeriana,* they were annotated "*Y. brevifolia* but with leaves like var. *jaegeriana.*" The type of var. *arborescens* falls into this group—having leaves like var. *jaegeriana* but occurring outside the geographical range of that variant.

As a result of these studies, it seems best to consider all of these elements as representing only one kind of plant, and call them *Y. brevifolia.* Those plants from the mountain ranges surrounding the Mojave Desert do have longer leaves (2–3.5 dm long), and may deserve some formal recognition. However, such a treatment would require a new varietal description, and I do not wish to make such a proposal until further, detailed work on *Y. brevifolia* can be done. For the record, however, the range of the small-leaved form of *Y. brevifolia* ranges from the Mojave Desert of California into Nevada, southern Utah and northern Arizona, and just entering our area in northern Inyo Co., California and western Nye Co., Nevada, and in Washington and Mohave counties of Utah and Arizona. The large-leaved form ranges from the mountains along the western edge of the Mojave Desert eastward into central and western Arizona, with scattered individuals found across southernmost Nevada, and just entering Utah on the western foothills of the Beaver Dam Mountains in Washington Co.

3. Yucca harrimaniae Trelease

Yucca harrimaniae Trelease, Annual Rep. Missouri Bot. Gard. 13: 59. 1902. (*Trelease,* Helper, Carbon Co., Utah, 7 Sept. 1901.)

Y. harrimaniae var. *gilbertiana* Trelease, Annual Rep. Missouri Bot. Gard. 18: 225. 1907. *Y. gilbertiana* Rydb. Fl. Rocky Mts. 1061. 1917 [1918]. (*Gilbert,* n. end of Fish Springs Range, Juab Co., Utah, 4 Aug. 1901.) = var. *harrimaniae.*

Yucca harrimaniae var. **neomexicana** (Wooton & Standley) Reveal, hoc loc. *Y. neomexicana* Wooton & Standley, Contr. U.S. Natl. Herb. 16: 115. 1913. (*Standley 6208,* 0.5 mi. n. of Des Moines, Union Co., N.M., 20 June 1911.)

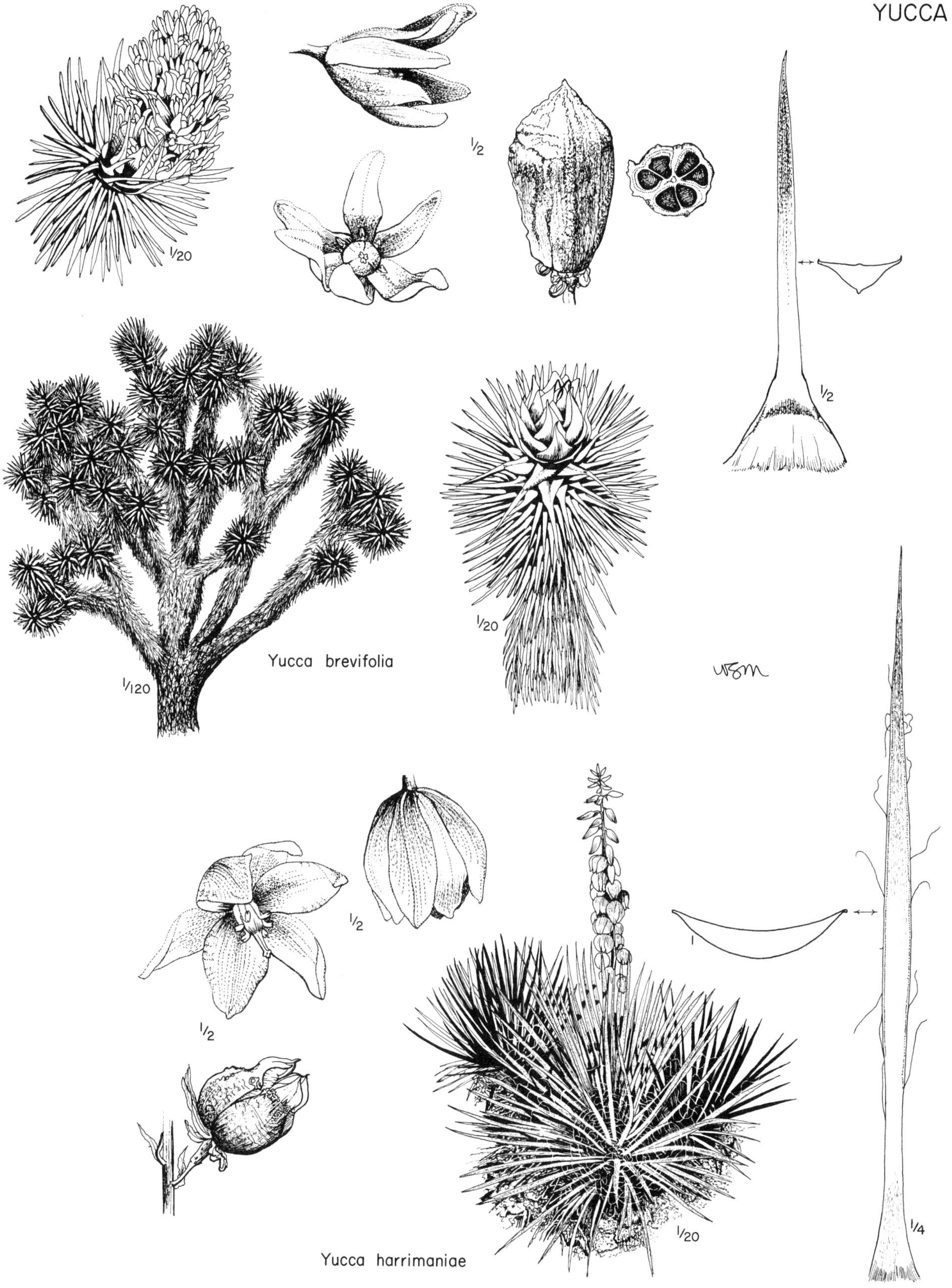
YUCCA
1/20
1/2
1/2
1/20
Yucca brevifolia
1/120
1/2
1/2
1
Yucca harrimaniae
1/20
1/4

Plants acaulescent forming dense to open clumps of 3–20 heads of leaves, the clumps 0.3–0.8 m in diam; leaves in small hemispherical, symmetrical or asymmetrical heads, the blade linear to lanceolate or spatulate-lanceolate, concavo-convex, 1–5 dm long, 0.7–4 cm wide, gray-green to blue-green or green to deep-green, rigid and stiff, rather glaucous, occasionally papillose, the base broad, thin, whitish, 1–2.5 (3) cm long, 3–4 (4.5) cm wide, the margin at first papery but soon separating into long, fine to coarse, curly fibers in some, the apex tapering gradually to a short, stiff ivory-colored spine; flowering stems erect, 1–4 dm long; inflorescences uniformly racemose or rarely with a few short branches, 3.5–7 dm long, well above the leaves, the stalk glabrous, pale green and tinged with reddish-purple, the branchlets, when present, 1–5 cm long, spreading; bracts thin, rigid, the lowermost spatulate, 1.5–2 dm long, becoming reduced above and 5–10 cm long; bractlets papery at least along their margins, triangular, 2–7 cm long, white and often with a dark reddish-purple center; flowers broadly campanulate, 4–6 (6.5) cm long, white to pale green or yellowish to greenish-yellow, commonly tinged with purple, pendent on pedicels 1–2 cm long, the tepals united only at the base for 2–4 mm, ovate, the inner tepals broader than the outer ones, fleshy; stamens included, the filament 1.5–3 cm long, hirsute, not united at the base, the anther 3–3.5 mm long; ovary 1.5–2 cm long, the style bright green, 9–13 mm long; capsules cylindrical, dry, (2.2) 4–5 (6) cm long, 2–3 cm in diam, deeply constricted toward the center and flaring open on drying; seeds thin, dull black, narrowly winged, 5–6 mm long.

Desert slopes and foothills of extreme e.-central Nev. e. across Utah to w. and s. Colo., ne. Ariz. and n. N.M.; mostly on limestone or volcanic slopes in the Bonneville Basin Sect. from Tooele Co., Utah, s. to Iron Co., Utah, and adjacent Lincoln Co., Nev., and in the Utah Plateaus and Canyon Lands sects. from Garfield Co. n. to Carbon Co., e. to Grand and San Juan cos., and in the Uinta Basin Sect. in Duchesne and Uintah cos., and also in the Quinn Canyon Range, Nev., mostly 1000–2500 m elev. April–July.

As here defined, *Y. harrimaniae* ranges from central Nevada to southeastern Colorado and adjacent northeastern New Mexico. The plants of the desert of western Utah were distinguished by Trelease as var. *gilbertiana.* This phase has generally short, narrowly spatulate leaves and a racemose inflorescence positioned well above the rosette of leaves. The flowers are about 5 cm long and the leaves are more obviously papillose than the other forms of the species. The basic aspect of this phase occurs mainly in the low, desert ranges of western Utah where it is often found in the *Atriplex* communities. In places of slightly higher elevation, and where the habitat is somewhat more mesic, the aspect is altered to a plant of greater stature, with much longer leaves. This type of plant is found in Nevada, and in Utah in the Deep Creek Range and along the western foothills of the southern Wasatch.

The type of *Y. harrimaniae* was gathered at Helper, Utah. This phase of the species is considerably larger than that found in western Utah, and has—generally—linear to linear-lanceolate leaves with the bottom portion of the inflorescence actually surpassed by the leaves. It occurs from Carbon Co. southward through Grand and San Juan cos. of Utah into Apache Co., Arizona, and from Mesa Co., Colorado southward in that state into Ouray Co. A series of isolated populations occur in the Uinta Basin of Utah, and these plants resemble more closely the plants of western Utah than those of eastern Utah found south of the Basin.

Taken together, these elements seem best maintained as a single variant of the species, called var. *harrimaniae,* but with the note that there are at least three separate, but not particularly distinct, geographical elements within the variety.

To the east of us, in southeastern Colorado and adjacent New Mexico, is var. *neomexicana* (Wooton & Standley) Reveal. It has even more narrowly linear leaves (7–20 mm wide instead of 15–45 mm wide as in most of ours) and white flowers instead of the yellowish to yellowish-green flowers of var. *harrimaniae.*

The treatment of this species presented here differs from that of both McKelvey (1947) and Webber (1953). McKelvey defined *Y. harrimaniae* much as here, but she did not recognize the more eastern var. *neomexicana.* Webber distinguished *Y. gilbertiana* and *Y. neomexicana,* but ignored the oldest available name, *Y. harrimaniae,* calling it a hybrid between these two extremes. I have seen no evidence to support Webber's contention that *Y. harrimaniae* is a hybrid between the *"gilbertiana"* phase and the *"neomexicana"* phase, and have returned the name to use.

4. **Yucca baileyi** Wooton & Standley

Yucca baileyi Wooton & Standley, Contr. U.S. Natl. Herb. 16: 114. 1913. (*Standley 7638,* Tunitcha Mts., San Juan Co., N.M., 8 Aug. 1911.)

Y. navajoa Webber, Madroño 8: 105. 1945. *Y. baileyi* var. *navajoa* Webber, Agric. Monogr. U.S.D.A. 17: 51. 1953. (*Webber s.n.,* 5 mi. n. of Tohatchi, McKinley Co., N.M., 28 May 1944.)

Y. standleyi McKelvey, Yuccas S.W. U.S. 2: 108. 1947. (*McKelvey 4609,* w. of Toadlena and Newcomb, Tunitcha Mts., San Juan Co., N.M., 24 May 1934.) = var. *baileyi.*

Yucca baileyi var. **intermedia** (McKelvey) Reveal, hoc loc. *Y. intermedia* McKelvey, Yuccas S.W. U.S. 2: 116. 1947. (*McKelvey 4902,* just w. of the Pecos River, San Miguel Co., N.M., 4 June 1934.)

Plants acaulescent or rarely with short, erect or suberect stems, solitary or 3–15 heads in clumps 0.5–2 m in diam; leaves somewhat crowded but in well-defined symmetrical, nearly hemispherical heads, the blade linear, plano-convex, (2) 2.5–6 (10) dm long, 0.3–0.8 cm wide, pale green or yellow-green, divergently spreading, somewhat rigid (ours) to flexible, the base broad, thin, whitish, 2–3 cm long, 2–4 cm wide, the margin entire and white, becoming separate and forming conspicuous, fine, curly fibers, the apex tapering gradually to a short spine 3–5 mm long; flowering stems erect, short, 0.1–1 dm long; inflorescences racemose, simple, 3–15 dm long, included to well beyond the leaves, the stalk glabrous, green or tinged with rose; bracts thin, often rigid, the lowermost linear, 0.5–2 (3) dm long, becoming reduced above and 5–10 cm long; bractlets thin, papery, triangular to ovate, 2–4 cm long, white with a purplish to greenish center; flowers campanulate to globose, 5–6.5 cm long, greenish-white, usually deeply tinged with purple especially without, pendent on pedicels 1–2 cm long, the tepals united at the base for 3–7 mm, ovate to obovate, the inner tepals slightly broader than the outer ones, rather thin and not fleshy; stamens included, the filament 2–2.5 cm long, hirsute, usually not united at the base, the anther 3–5 mm long; ovary 2–2.5 cm long, the style white to pale greenish-white, 7–9 mm long; capsules oblong-cylindric, dry, 4–7 cm long, 2.5–5 cm in diam, not constricted or only slightly so; seeds thin, black, with or without a narrowly winged margin, 6–10 mm long.

Dry forest floors to grasslands, or infrequently on exposed sandstone rims, extreme se. Utah and ne. Ariz. e. to e. N.M. and n. along the Front Ranges of Colo. to near the Wyo. line; known only

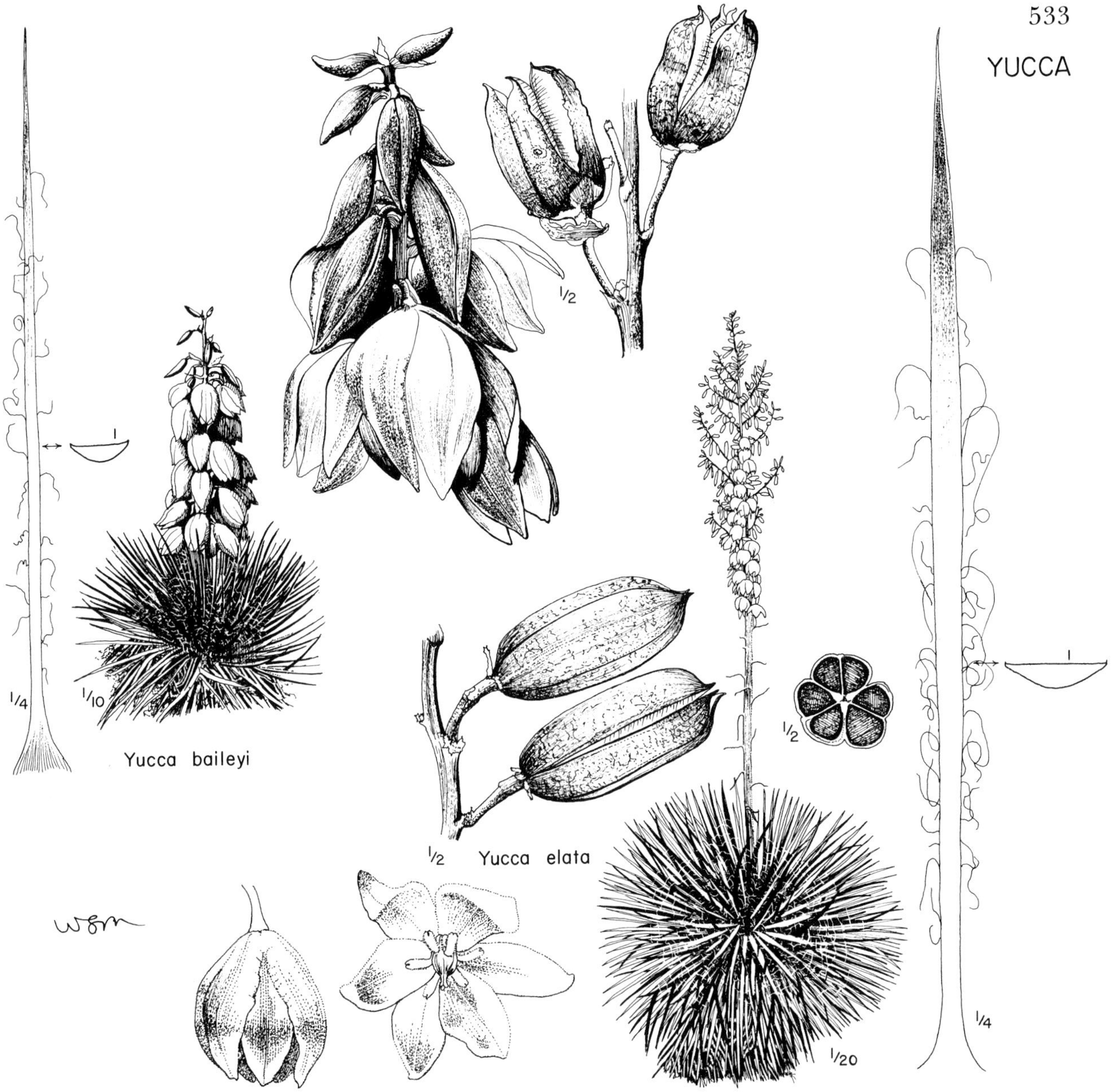

from San Juan and e. Kane cos., Utah, in our area, from 1200–2200 m elev. April–June.

Our plant is var. *baileyi*. It may be characterized by its acaulescent habit and short inflorescence (up to 5 dm long) which is included or just barely exceeds the leaves.

Once again the treatment presented here differs from that of McKelvey (1947) and Webber (1953).

McKelvey regarded *Y. baileyi* to be a hybrid between *Y. baccata* and what she called *Y. standleyi.* Webber could not support her conclusion and neither can I. The Standley type of *Y. baileyi* is typical of the species, and even McKelvey annotated the holotype "*Y. standleyi* n.sp!"

In 1945, Webber proposed *Y. navajoa* for a series of highly reduced plants found in McKinley Co., New Mexico. He later reduced the species to a variety of *Y. baileyi,* and expanded the range of the taxon to include similar plants in Coconino Co., Arizona. At the same time, he (Webber, 1953) excluded consideration of McKelvey's newly proposed species, *Y. intermedia,* calling it a hybrid between *Y. glauca* Nutt. and *Y. angustissima* or *Y. baileyi.* This concept is not supported by the high degree of viable pollen set and the mature fruits observed in *Y. intermedia,* nor do any of the proposed parents occur in the part of New Mexico where the supposed hybrids exist. I can not maintain *Y. intermedia* as a distinct species, however, and have therefore proposed var. *intermedia* (McKelvey) Reveal for those plants of central New Mexico with short erect stems.

5. **Yucca elata** Engelm.

Yucca angustifolia var. *radiosa* Engelm. in King, Rep. Geol. Explor. 40th Parallel 5: 496. 1871. *Y. angustifolia* var. *elata* Engelm. Trans. Acad. Sci. St. Louis 3: 50. 1873, a superfluous substitute. *Y. radiosa* Trelease, Annual Rep. Missouri Bot. Gard. 13: 56. 1902. ("Central Arizona, and northward to the borders of Utah," *Palmer 201,* "Arizona," near Navajo Springs, Apache Co., Ariz., 24 June 1869; lectotype selected here and deposited at MO.)

Y. elata Engelm. Bot. Gaz. 7: 17. 1882. (*Rothrock 382,* Camp Grant, Graham Co., Ariz., July 1874; lectotype selected by McKelvey.)

Yucca elata var. **utahensis** (McKelvey) Reveal, hoc loc. *Y. utahensis* McKelvey, Yuccas S.W. U.S. 2: 94. 1947. (*McKelvey 4167,* 7–9 mi. nw. of St. George in the Santa Clara Valley, Washington Co., Utah, 6 May 1934.)

Yucca elata var. **verdiensis** (McKelvey) Reveal, hoc loc. *Y. ver-*

diensis McKelvey, Yuccas S.W. U.S. 2: 98. 1947. (*McKelvey 2752,* between Clarkdale and Cottonwood, Yavapai Co., Ariz., 7 May 1933.)

Soaptree yucca.

Plants acaulescent or more often caulescent to nearly arborescent, solitary or in large clumps of several heads, or if arborescent, then with one to several stems 0.3–4.5 m tall and erect and trunklike, covered with leaves and often branched above, rarely with short, procumbent stems 0.6–1.3 m long (ours), the clumps of leaves few to several, 0.5–4 m in diam; leaves in rather loose and open to dense and crowded, symmetrical to asymmetrical (especially in age) heads, the blade linear, plano-convex or plano-keeled, (1.5) 3–7 (9) dm long, (0.3) 0.7–1.5 (2) cm wide, pale green to yellow-green, spreading, flexible, the base broad, white to cream, 3–5 (6) cm long, 3–6 (7) cm wide, the margin entire, white or greenish-white, becoming separated and forming fine, often dense, curly fibers, the apex tapering and long-acuminate with a short spine; flowering stems erect, 0.5–1.5 (2) m long; inflorescences paniculate, closed to open, only with the uppermost flowers arranged racemosely, 0.5–1.5 (3) m long, well beyond the leaves, the stalk glabrous, green to reddish or yellowish, the branchlets many to several, (0.5) 1–3.5 cm long, spreading to erect-ascending or horizontal; bracts linear-lanceolate, 1.5–2.5 dm long, 1–3 cm wide, thin, becoming reduced above to 3–5 (10) cm long, 1–2 cm wide and narrowly triangular; bractlets papery, ovate, 0.5–2.5 cm long, 0.5–2 cm wide, white to greenish or tinged with rose; flowers campanulate to somewhat globose, 4–6 cm long, white to cream, tinged with green or pink in some, pendent on slender to stout pedicels (0.7) 1–2.5 cm long, the tepals united at the base for 2–7 mm, ovate to obovate or broadly elliptic, the inner tepals broader than the outer ones, thin and not fleshy; stamens included, the filament 1.5–3 cm long, hirsute, not united at the base, the anther 2–5 mm long; ovary 1.5–2 cm long, the style white to pale green, 5–10 mm long; capsules oblong-cylindric, 4–7 (8) cm long, 3.5–6 cm in diam, dry, not constricted or only slightly so; seeds thin, dull black, with or without a marginal wing, 7–10 mm long; $n = 30$.

Desert places, often in sandy soil, from sw. Utah s. to s. Ariz., and hence e. into s. and c. N.M., w. Texas and n. Mex.; known in our area only from Washington Co., Utah, and adjacent Mohave Co., Ariz., from 850–1000 m elev. April–June.

As here defined, *Y. elata* is a wide ranging species with three distinct variants differing mainly in growth form. Our phase is var. *utahensis* (McKelvey) Reveal which occurs in southwestern Utah, southward into northern Arizona as far as the Grand Canyon. This plant is acaulescent or more often caulescent with procumbent stems 0.6–1.3 m long. It has capsules 5–6 cm long, leaves 2–7 dm long and 0.7–2 cm wide, and a pistil 2.5–3.5 cm in length. The var. *utahensis* flowers from late April to early June.

To the south of us are two additional varieties. The var. *verdiensis* (McKelvey) Reveal ranges mainly from the southern foothills of the Mogollon Rim southward to northern Gila Co., Arizona, with a few populations up on the upper slopes of the Rim in southern Coconino and Navajo cos. This plant, which is caulescent with a distinct but short trunk, has capsules only 4–4.5 cm long, leaves that are 2.5–4.5 dm long and 0.4–0.7 (1.3) cm wide, and a pistil 2–2.5 cm long. These plants flower from May to mid June.

The typical phase of the species, var. *elata,* occurs from southern Arizona to western Texas and northern Mexico. It is truly arborescent and has much larger capsules (5–8 cm long), with leaves 3–9 dm long and 0.5–0.7 (1.3) cm wide.

In defining *Y. elata* in this fashion, I have placed together those plants of the West which have paniculate inflorescences, and only the uppermost flowers arranged in racemes. These plants of *Yucca* are almost always caulescent, although an infrequent individual of var. *utahensis* may be acaulescent, and many are arborescent. Those plants which are generally acaulescent and only infrequently caulescent, and which have inflorescences which are almost always strictly racemose, I am placing in *Y. angustissima.*

The contention by Webber (1953) that var. *utahensis* and var. *verdiensis* represent hybrid populations between the Great Plains species *Y. glauca* Nutt. and the southern southwestern United States *Y. elata* is in my opinion wholly without justification.

6. **Yucca angustissima** Engelm. ex Trelease

Yucca angustissima Engelm. ex Trelease, Annual Rep. Missouri Bot. Gard. 13: 58. 1902. (*Bigelow,* deserts of the Colorado River area, apparently from near the San Francisco Mts., near Picacho, Mohave Co., Ariz., in 1854.)

Yucca angustissima var. **kanabensis** (McKelvey) Reveal, hoc loc. *Y. kanabensis* McKelvey, Yuccas S.W. U.S. 2: 122. 1947. (*McKelvey 4347A,* 9 mi. n. of Kanab, Kane Co., Utah, 14 May 1934.)

Yucca angustissima var. **toftiae** (S. L. Welsh) Reveal, hoc loc. *Y. toftiae* S. L. Welsh, Great Basin Naturalist 34: 308. 1975. (*Welsh 11935a,* at Three Gardens, 1 mi. n. of the confluences of the San Juan and Colorado rivers, Lake Powell, San Juan Co., Utah, 4 June 1973.)

Yucca angustissima var. **avia** Reveal, var. nov. (A var. *angustissimo* floribus brevioribus 3.5–4.5 cm longis, folii longioribus 4–6 dm longis differt. Typus, *M. E. Jones 5639a,* Loa Pass, Piute Co., Utah, 2550 m elev., 17 July 1898, holotype deposited at US.)

Plants acaulescent to short caulescent with short procumbent stems 1–4 dm long, or caulescent and erect with the stem up to 1 m tall, solitary or in small to large clumps of numerous heads, up to 3 m in diam; leaves in compact, usually symmetrical, hemispherical heads, the blade linear, plano-convex to plano-keeled, rarely concavo-convex, 2.5–6 (7.5 or rarely up to 15 in var. *kanabensis*) dm long, 0.4–1.5 cm wide, pale-yellow to blue-green, spreading, rigidly-spreading, flexible, the base broad, white to cream, 2–5 cm long, 2.5–6 cm wide, the margin entire, cream to tan or reddish-brown, becoming separated and forming fine, few, slightly curled fibers, the apex tapering and long-acuminate with a short, acicular spine 3–7 mm long; flowering stems erect, short, 0.2–2.5 m long; inflorescences racemose, simple, occasionally with a few abortive branchlets near the base, 0.2–1.5 (2) m long, well above the leaves, the stalk glabrous or finely pubescent, greenish and often tinged with reddish-purple, the abortive branchlets, when present, few, 5–10 cm long, erect-ascending; bracts linear, 1–2 dm long, 1–2 cm wide, thin, becoming reduced above to 3–8 cm long, 1–4 cm wide, narrowly triangular, greenish; bractlets papery, ovate, 1–3 cm long and wide, white to greenish or tinged with red or pink; flowers campanulate to globose, 3–6.5 cm long, white to cream or greenish-white and often tinged with rose or rose-purple, pendent on slender, terete pedicels 1–2.5 (4) cm long, the tepals united for 3–7 mm, elliptic to ovate, the inner tepals broader than the outer ones, thin and not fleshy; stamens included, the filament (0.7) 1–2.5 (2.8) cm long, hirsute, not united at the base,

YUCCA

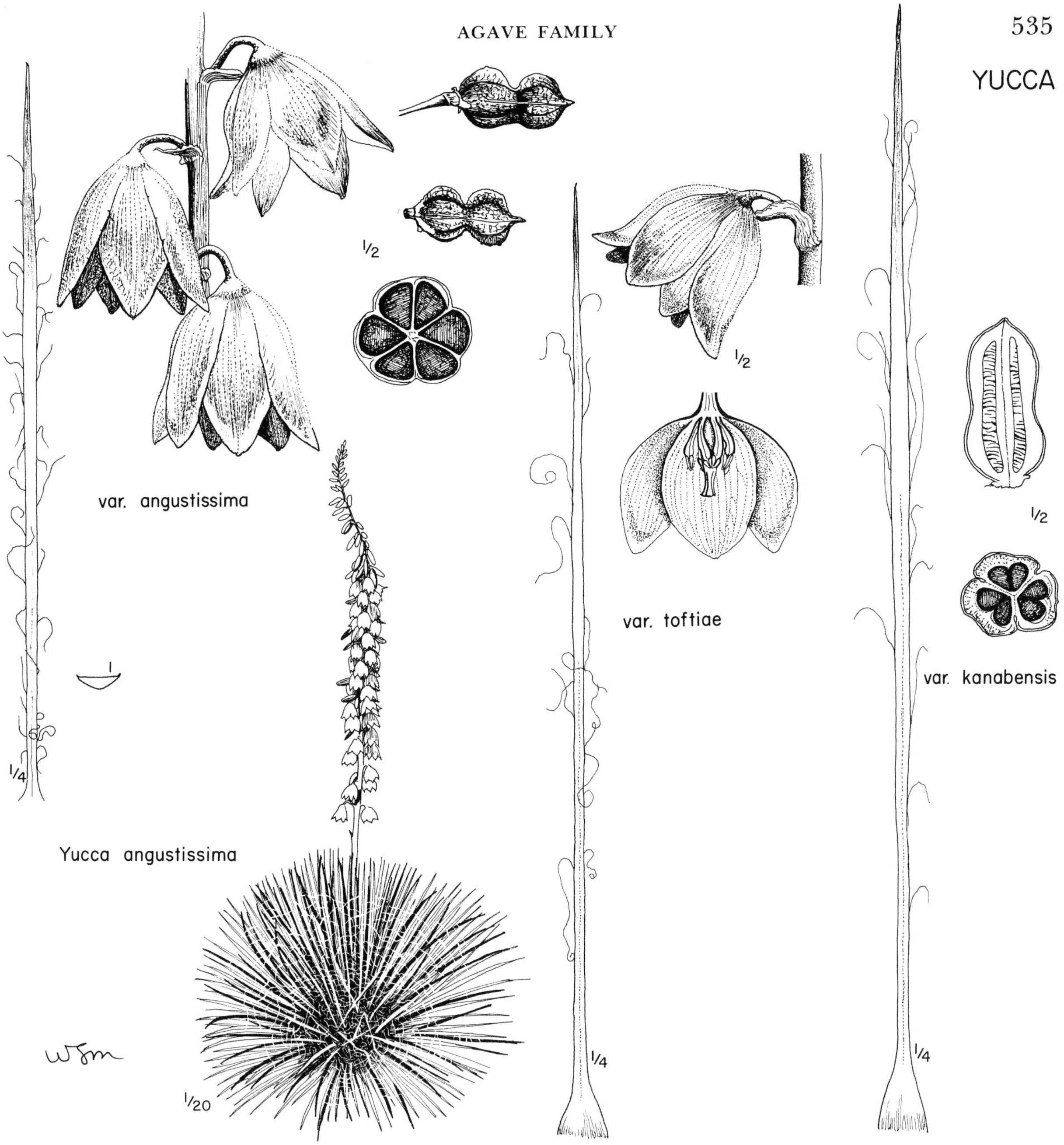

the anther 3–6 mm long; ovary (0.7) 1–2.5 cm long, the style white to pale green, 3–13 (15) mm long; capsules oblong-cylindric, 3.5–7.5 cm long, 2–3 cm in diam, dry, commonly with a deep central constriction; seeds thin, dull black, narrowly winged, 5–7 mm long; $n = 30$.

Desert flats or mesas, often in sandy places or on sandstone outcrops, from s. Utah s. to c. Ariz., hence e. to w. N.M.; isolated in Piute, Wayne and Sevier cos. of c. Utah, and across the s. tier of cos. of Utah, s. into n. Mohave and Coconino cos., Ariz., from 1050–2550 m elev. May–July.

As here defined, *Y. angustissima* includes those specimens of *Yucca* with tall, erect racemose inflorescences and narrow leaves. This is the concept of the species given to it by Trelease and others before the efforts of McKelvey (1947) and Webber (1953). As now understood, *Y. angustissima* consists of four closely associated variables, with var. *angustissima* and var. *avia* (from the Latin *avius,* out of the way, alluding to its geographical isolation from the rest of the species) being more grossly similar to each other, and var. *kanabensis* and var. *toftiae* forming a second phalanx. To a great degree, each element is well isolated from the other, but even so, morphologically, they overlap in many features. This is especially true of var. *angustissima* in central Coconino and Mohave cos., Arizona, where the capsules of this variant are only moderately to even slightly constricted and therefore somewhat similar to var. *kanabensis* in this one aspect. Nonetheless, based on the totality of the features, these Arizona plants seem better associated with var. *angustissima,* and the range of var. *kanabensis* restricted to the Dixie Corridor Section and the southernmost tip of the Utah Plateaus Section.

McKelvey's concept of her *Y. kanabensis* is restricted here to only those plants directly related to the type collection. The recently proposed *Y. toftiae,* which was known to McKelvey and associated by her with some reluctance with *Y. kanabensis,* is recognized here as a distinct variant of *Y. angustissima,* next to var. *kanabensis.* The var. *toftiae* occurs further up the Colorado River than var. *kanabensis,* and is known even on the lower foothills of the Abajo Mountains.

The most isolated variant of *Y. angustissima* is var. *avia.* It is restricted to the central portion of Utah, and occurs mainly on loamy, but rocky, soils rather than sandy soils like the other members of *Y. angustissima.* It too was included by McKelvey in her concept of *Y. kanabensis,* but the plants of var. *avia* are clearly more closely related to var. *angustissima* than to var. *kanabensis.*

Webber once again regarded the variation within *Y. angustissima* as due to hybridization between *Y. angustissima* and *Y. glauca* Nutt., *Y. baileyi* or *Y. elata.* I have seen no evidence to support this thesis.

The variants within *Y. angustissima* may be distinguished as follows:

1 Capsules small, 3.5–5 cm long, deeply constricted; flowering stems and inflorescences 0.5–1.5 (1.8) m tall.
  2 Flowers 4.5–5.5 cm long; leaves 2.5–4.5 dm long; style 10–13 mm long; plants of sw. Utah se. through n. Ariz. to w. N.M. .... var. *angustissima*
  2 Flowers 3.5–4.5 cm long; leaves 4–6 dm long; style 7–10 mm long; plants of c. Utah .... var. *avia* Reveal
1 Capsules large, 4.5–7.5 cm long, moderately constricted; flowering stems and inflorescences 2–4.5 m tall.
  3 Flowers 3–4.5 (5.2) cm long; leaves 2.5–6 (7) dm long; pistil 1.5–2.5 cm long with the style 3–10 mm long; flowering stems 1.2–2.5 m long with the inflorescences 0.2–2 m long; sandstone outcrops and mesas, Lake Powell region e. to the Abajo Mts., San Juan, Kane, and Garfield cos., Utah .... var. *toftiae* (S. L. Welsh) Reveal
  3 Flowers 5.5–6.5 cm long; leaves 4.5–7.5 (15) dm long; pistil 3–3.5 cm long with the style 5–8 mm long; flowering stems 1–1.5 m long with the inflorescences 1–2 m long; sandy places of Kane and Washington cos., Utah, s. into n. Coconino and Mohave cos., Ariz. .... var. *kanabensis* (McKelvey) Reveal

2. NOLINA Michx. Nolina

Erect perennial herbs or shrubs with a large woody underground caudex or a distinct above-ground trunk; leaves numerous, narrowly linear, often with a greatly distended base, entire or shredding to serrate along the margin, the apex tapering gradually to a short-pointed tip; flowers minute to small, both perfect and imperfect, campanulate, whitish to yellowish-white, on slender, jointed pedicels subtended by scarious bracts, borne in nearly naked, elongate, congested, compound racemose panicles, the main branches subtended by foliaceous to scarious, long-attenuated bracts; tepals 6, alike, distinct, oblanceolate to obovate, the tips curving outward; stamens 6, often abortive, hypogynous, the filaments short and slender to stoutish, the anthers sagittate, versatile; ovary superior, 3-celled with only 2 ovules per locule, the short style terminated by 3 short stigmas; fruits broadly 3-winged, membranaceous, indehiscent or irregularly dehiscent capsules; seeds subglobose to ovoid, light brown to brown, 1–2.

A genus of about 30 species largely of the deserts of sw. U.S. and Mex. (Named in honor of P. C. Nolin, apparently a friend of Michaux.)

Reference: Trelease, W. The desert group Nolineae. Proc. Amer. Philos. Soc. 50: 404–443. 1911.

1. Nolina microcarpa S. Wats.

Nolina microcarpa S. Wats. Proc. Amer. Acad. Arts 14: 247. 1879. *Beaucarnea microcarpa* Baker, J. Linn. Soc., Bot. 18: 236. 1880. (*Rothrock 278,* "Rocky Cañon," Ariz., July 1874.)

Sacahuista.

Erect to spreading, acaulescent or cespitose perennial shrubs; leaves narrowly linear, 6–12 dm long, 6–12 (15) mm wide, concavo-convex, thick and keeled, the margin unevenly serrulate-scabrous to strongly serrulate with single teeth or (ours) with the teeth irregularly clumped together, the tips splitting and fibrous-lacerate in some (not ours), these tufts of fibers 5–12 cm long; flowering stems erect, stout, up to 18 dm long; inflorescences paniculate, up to 1 m long, narrow, the primary branches 1.5–5 dm long and subtended by narrowly lanceolate bracts 8–15 cm long, 5–12 mm wide, the secondary and ascending branchlets (2) 5–7.5 (15) cm long, these subtended by broadly lanceolate to obovate or cordate, fimbriate bractlets 1–3 mm long; flowers campanulate, whitish, (1.5) 2–3.5 mm long, the tepals oblong to ovate; stamens with yellow anthers 0.9–1.3 (1.6) mm long; capsules 3-winged, subglobose, deeply notched at the apex, rounded to shallowly notched at the base, 4–5 mm high, 6–8 mm wide, papery, cells ruptured only at late maturity, on pedicels 4–6 mm long; seeds obovate, (2.5) 3–3.5 mm in diam, light brown; $n = 18$.

On rocky slopes and hills, especially in canyons, from sw. Utah and n. Ariz. s. to ne. Sonora, Mex., and e. through N.M. into (apparently) w. Texas and Chihuahua; known currently in our area only from Toroweap, Mohave Co., Ariz., where infrequent to rare at about 1200 m elev., the Zion Canyon population of Washington Co., Utah, has been extinct since 1925. May–July.

References to *N. parryi* S. Wats. are based on misidentifications of *N. microcarpa* and *N. bigelovii* (Torr.) S. Wats., the latter a species found to the south of us regardless of reports to the contrary.

3. AGAVE L. Century Plant

Acaulescent or sometimes short-stemmed monocarpic perennials forming succulent rosettes of long-lived leaves, frequently suckering at the base or bulbiferous in the inflorescence, arising from a thick fibrous-rooted crown and flowering only once after 8–20 or more years; leaves large, lanceolate to ovate, with distended bases, with or without spinous margins and usually tipped with a long, sharp-pointed spine; flowers relatively large and as a group showy, perfect, funnelform to tubular, greenish-yellow to yellow or orange, borne in terminal, open to dense, elongate, spicate to racemose or paniculate inflorescences with the flowers in umbellate clusters subtended by bracts; tepals 6, subequal, forming a short to long tube, the segments linear to oblong or deltoid, the tips ascending to curving outward; stamens 6, usually long exserted, attached to the tepals at the throat or in the tube, the filaments filiform, the anthers versatile, linear; ovary inferior, 3-celled with many ovules, the style awl-shaped with a capitate, 3-lobed stigma; fruits oblong, loculicidal capsules; seeds obovate, flat, black.

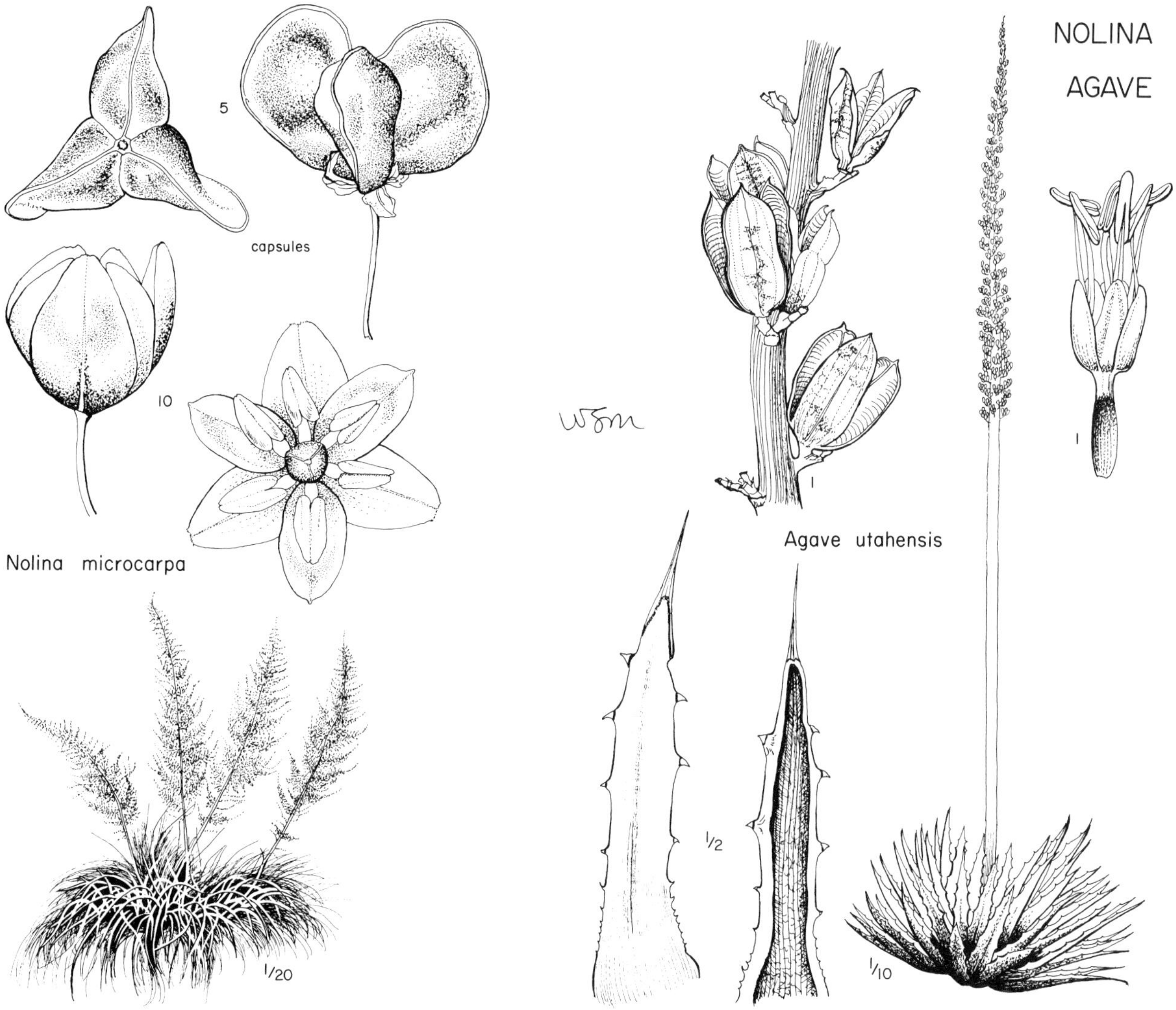

A large and exceedingly complex genus of perhaps 250 species found throughout the warmer parts of the New World but now widely cultivated throughout the world. The numerous cultivated forms are not included here. (From the Greek *agave*, noble or admirable.)

McDougall (Seed Pls. N. Ariz. 105. 1973) reports that *Agave deserti* Engelm. may occur on the North Rim of the Grand Canyon, but I have no indication that this species has ever been found in our area.

References:

Breitung, A. J. Cultivated and native agaves in the southwestern United States. Part 6. Cact. Succ. J. (Los Angeles) 32: 20–23. 1960.

Gentry, H. S. The agave family in Sonora. Agric. Handb. U.S.D.A. 399: 1–195. 1972.—While this reference does not apply directly to our flora, it is an excellent review of the morphology, ecology and biology of *Agave*.

1. Agave utahensis Engelm.

Agave utahensis Engelm. in King, Rep. Geol. Explor. 40th Parallel 5: 497. 1871. *A. haynaldii* var. *utahensis* Terracino, Pr. Contr. Monogr. Agave 28. 1875. (*Palmer s.n.*, near St. George, Washington Co., Utah, May 1870, sent to Engelmann by Parry; lectotype selected here and deposited at MO.)

A. newberryi Engelm. Trans. Acad. Sci. St. Louis 3: 309. 1875. (*Newberry*, Peacock Springs, Mohave Co., Ariz., 31 March 1858.) = var. *utahensis*.

A. scaphoidea Greenman & Roush, Ann. Missouri Bot. Gard. 16: 391. 1929. *A. utahensis* var. *scaphoidea* M. E. Jones, Contr. W. Bot. 17: 19. 1930. (*Palmer s.n.*, St. George, Washington Co., Utah, May 1877; lectotype selected here and deposited at MO.) = var. *utahensis*.

A. utahensis var. *discreta* M. E. Jones, Contr. W. Bot. 17: 19. 1930. (*M. E. Jones 25167*, Oatman, Yuma Co., Ariz., 14 June 1930; excluding the material of *A. mckelveyae* Gentry.) = var. *utahensis*.

A. kaibabensis McKelvey, J. Arnold Arbor. 30: 277. 1949. *A. utahensis* var. *kaibabensis* Breit. Cact. Succ. J. (Los Angeles) 32: 21. 1960. (*McKelvey 4381*, Kaibab Plateau, Coconino Co., Ariz., 15 May 1934.)

Solitary cespitose perennials occasionally with several off-sets; leaves in subglobose rosettes 2–14 dm in diam, the blade mostly narrowly lanceolate, 1–3.5 dm long, 2–6.5 cm wide, glaucous, curving inward at the tip, the margin straight or sinuate, the teeth few, 1–3.5 (7) mm long, 1–3 cm apart, the terminal spine 1–3 (ours) or up to 20 cm long; flowering stems erect, stoutish, 1.5–4.5 m high; inflorescences subracemose panicles, narrow, 0.5–2 m long, bearing clusters of 2–4 flowers at the tips of peduncles 2–5 cm long; flowers funnelform, yellow, (2.2) 2.5–3 (3.5) cm long, the tube 4–5 mm long, the tepals spreading, 1–1.5 cm long; stamens inserted near the middle of the tube, the filament 1.5–2 cm long, the anther 1–1.2 cm long; capsules cylindrical to oblong, 1–2 (3) cm long, 1–1.2 cm in diam; seeds 3.5–4 mm long; $n = 30$ (for var. *nevadensis*).

Dry slopes, mesas and hillsides from s. Utah s. to n. Ariz. and w. to se. Calif.; from the Dixie Corridor Sect. of Kane and Washington cos., Utah, s. to the Grand Canyon Plateaus Sect. of n. Ariz., from 800–2300 m elev. June–July (Aug.).

This species is composed of four variants, two of which occur in our area. I have reduced *A. newberryi* and *A. utahensis* var. *discreta* to synonymy under var. *utahensis*, and thereby this treatment differs somewhat from that of Breitung. The var. *eborispina* (Hester) Breit. occurs on the Sheep Range, Sunshine Mountain, and other desert ranges in Clark Co. and extreme southern Nye Co., Nevada, while the more western var. *nevadensis* Engelm. ex Greenman & Roush occurs only in the desert ranges of southeastern California, in spite of its name. Our two variants may be keyed out as follows:

1 Rosettes 2–3 dm in diam; leaves 1.2–2 dm long, 2–3 cm wide; teeth 1.5–2 mm long; flowering stems 1.5–2.5 m tall; s. Utah and nw. Ariz. .... var. *utahensis*

1 Rosettes 6–10 (14) dm in diam; leaves 3–3.5 dm long, 5–6.5 cm wide; teeth 1.5–3.5 mm long; flowering stems 2–4.5 m tall; Kaibab Plateau of n. Ariz. s. onto the South Rim and along the cliffs above the Little Colorado R. .... var. *kaibabensis* (McKelvey) Breit.

Family **IRIDACEAE**, the Iris Family

Perennial or annual herbs from rhizomes, bulbs or corms; stems 1–several, sometimes none; leaves mostly basal, usually long and narrow, 2-ranked, equitant (our native species); inflorescence generally racemose or paniculate, a flower or cluster of flowers subtended by 2 spathelike bracts; flowers perfect, regular (ours) to sometimes irregular, usually showy; perianth of 6 petaloid segments (tepals), in 2 whorls, these from similar to distinct and often dissimilar, usually connate basally in a perianth tube; stamens 3, the filaments partially adnate to the outer tepals, the anthers 2-celled; style usually 3-branched, the branches often flattened, opposite the stamens and outer tepals; ovary inferior, 3-loculed, with axile placentation; fruit a loculicidally dehiscent, many-seeded capsule; seeds membranous, usually with fleshy endosperm, sometimes arillate.

About 60 genera and 1500 species, widely distributed in temp. and trop. regions of both hemispheres.

The Iridaceae family is best known for its ornamentals. Species and hybrids of *Iris*, *Crocus*, and *Gladiolus* are common in gardens of the region. A key to the most commonly grown species of *Iris* is provided under that genus. *Crocus*, a stemless, early spring-flowering perennial, is characterized by having leaves and flowers arising directly from an underground corm. Its showy funnelform flowers consist of a long tube and 6 subequal tepals. *Gladiolus* is a summer-flowering, corm-bearing perennial. The showy flowers are usually in simple, secund spikes, each subtended by a simple spathe bract. The funnelform tube is curved, and the 6 tepals are unequal. The upper 3 (2 outer and 1 inner) tepals are arched over the throat, and the lower 3 (1 outer and 2 inner) tepals are usually smaller and recurved.

Hybrids are common throughout the Iridaceae family, and species are often difficult to recognize.

1 Perianth segments (tepals) alike, not more than 3 cm long, all spreading; style branches not petal-like; filaments united at least half their length .... 1. *Sisyrinchium*

1 Perianth segments (tepals) dissimilar, more than 3 cm long, differentiated in a distinct outer and inner whorl, the outer tepals usually spreading to reflexed, the inner tepals usually erect; style branches petal-like, opposite to and overarching the stamens; filaments distinct .... 2. *Iris*

1. **SISYRINCHIUM** L. Blue-eyed Grass

Cespitose perennial (ours) or annual herbs with clusters of fibrous roots or rhizomes; stems slender, compressed, often conspicuously 2-winged, simple or branched, with smooth to minutely toothed margins; leaves linear, grass-like, usually $^1/_2$–$^3/_4$ the length of the stem, equitant, flattened, sheathing at the base, lax to erect, with entire to finely toothed margins; flowers borne in 1- to few-flowered terminal umbellate inflorescence subtended by a pair of leaf-like spathe bracts; perianth yellow, pink, or commonly blue, deep blue-violet or red-violet, rarely white, the 6 spreading segments (tepals) all alike or nearly so, variously acute, rounded, truncate, emarginate, or retuse and often aristulate, slightly connate at base, generally the 3 outer tepals with 5 veins and the 3 inner tepals with 3 veins; stamens 3, united by their filaments from $^1/_2$ their length to near the apex and forming a column around the style; ovary inferior, usually stipitate-glandular, 3-celled, each cavity several-ovuled, with axile placentation; style branches filiform, opposite the stamens; fruit a subglobose or obovoid, loculicidally dehiscent capsule; seeds subglobose or ovoid, smooth or pitted, 3–35 (ours) per capsule.

Distributed throughout N. Amer., S. Amer., and W.I., with perhaps 75–100 species. (Name used by Theophrastus for some *Iris*-like plant.)

The taxonomy of our sisyrinchiums has been confused since the time when E. P. Bicknell and E. L. Greene named numerous western species from herbarium material. Characters used to separate our blue-flowered species were the relative length of the spathe bracts and the length of connation of the outer one, the relative width of the leaves, the degree of branching and the width of the stems. All of these characters are useful, since they are relatively well preserved in herbarium material. However, extensive morphological study has demonstrated the intergradation of characters used to define some of these taxa. The characteristics of the flowers are seldom evident in dried material nor are they described sufficiently in field notes. Therefore, no matter how specifically distinct the flower characters may appear in the field, they are virtually useless in dried material unless good notes appear on the label.

Using modern biosystematic methods, Douglass Henderson (1972, 1976) has studied the Pacific Northwest sisyrinchiums in depth. He is now expanding his study to include all western North American taxa. We maintain a somewhat conservative treatment of the genus in this Flora in order not to anticipate Henderson's much-needed taxonomic revision.

Three species, *S. bellum* S. Wats., *S. sarmentosum* Suksd., and *S. septentrionale* E. P. Bicknell, listed in the "Checklist of the Vascular Plants of the Intermountain Region" have not been found within the boundaries. *Sisyrinchium bellum* occurs to the west of the Region in cismontane California and southern Oregon, while *S. sarmentosum* is endemic to south-central Washington. *Sisyrinchium septentrionale* can be found in non-glaciated areas of northeastern Washington and Canada from British Columbia to Manitoba. Another species, *S. montanum* Greene, listed in the "Checklist" is a Rocky Mountain species that may be expected in the Uinta Mountains. We have seen no specimens from Utah clearly identifiable as *S. montanum* (see discussion under *S. idahoense*).

References:
Henderson, D. M. A biosystematic study of the genus *Sisyrinchium* (Iridaceae) in the Pacific Northwest. Ph.D. thesis. Univ. of Washington, Seattle. 185 pp. 1972.
———. A biosystematic study of Pacific Northwestern blue-eyed grasses (*Sisyrinchium*, Iridaceae). Brittonia 28: 149–176. 1976.
Mosquin, T. Chromosome numbers and a proposal for classification in *Sisyrinchium* (Iridaceae). Madroño 20: 269–275. 1970.

1 Filaments connate their full length; perianth light to dark blue; anthers (0.8) 1–2 (2.5) mm long; capsules (2) 2.2–6 (7.1) mm long.
  2 Spathe bracts subequal, the inner bract (10) 12–20 (24) mm long, the outer bract 13–24 (26) mm long; capsules 2–4.5 (6.2) mm long; outer tepals broadly oblanceolate to cuneate, often truncate, apiculate to aristulate, rarely emarginate.
    3 Stems predominantly branched with a leaf-like bract subtending two or more pedunculate spathes, the peduncles up to 14 cm long; s. tip of Nev. to s. and c. Utah in our area .... 1. *S. demissum*
    3 Stems predominantly unbranched; e. Calif. throughout Nev. except the s. tip ..... 2. *S. halophilum*
  2 Spathe bracts unequal, the inner bract 13–35 mm long, the outer bract (15) 20–75 mm long; capsules (2.5) 4–6 (7.1) mm long.
    4 Outer bract generally clasping at the base for more than 4 mm, not twice as long as the inner bract; inner bract not abruptly protruding from the outer bract; tepals narrowly elliptic to oblanceolate, rounded or more often truncate and emarginate; widespread in our region in e. Calif., n. Nev., throughout Idaho and Utah .... 3. *S. idahoense*
    4 Outer bract generally clasping at the base for less than 4 mm, nearly twice the length of the inner bract, somewhat broadened above the apex of the inner bract before tapering to a slender, acute apex; inner bract often abruptly protruding from the outer bract; tepals generally broadly elliptic and deeply emarginate or retuse, aristulate; probably not entering our range but perhaps expected in the Uinta, Wasatch, and se. Utah mts. .... *S. montanum* Greene
1 Filaments connate less than half their length; perianth magenta, pink, deep reddish-purple, or occasionally white; anthers 3–7 (9) mm long; capsules (4.5) 5.5–9 (11) mm long .... 4. *S. douglasii*

1. Sisyrinchium demissum Greene
Sisyrinchium demissum Greene, Pittonia 2: 69. 1890. (*Greene*, in moist meadows at the base of Bill Williams Mountain, Arizona, and also near Flagstaff, July 1889.)
S. radicatum E. P. Bicknell, Bull. Torrey Bot. Club 28: 576. 1901. (*Palmer 456*, St. George, Washington Co., Utah, in 1877.)

Blue-eyed grass.

Perennial herbs; roots fibrous; stems numerous, erect, (0.7) 1–2.5 (3.7) mm wide and up to 4.8 dm tall, usually branched, with a leaf-like bract subtending two or more pedunculate spathes, the margins entire to finely denticulate; leaves erect, shorter than the stems, 0.7–3 (5) mm wide, the margins entire to finely denticulate; spathe mostly pedunculate, the peduncles up to 14 cm long, the bracts equal to subequal; outer bract 11–21 (25) mm long, the inner margins broadly hyaline below, narrowly hyaline above, connate at the base (2.2) 3–6.5 (8.2) mm; inner bract 12–20 (24) mm long, the inner margins hyaline to the tip; flowers 1–7 (8) on slender, glabrous to glandular-puberulent pedicels equal to or frequently exceeding the outer bract; perianth light to dark blue, with a yellow eye, the tepals broadly oblanceolate to cuneate, truncate, apiculate to aristulate, sometimes emarginate, (6) 9–10 (12) mm long; filament column (3.5) 4.5–5.5 mm; anthers 1–1.5 (1.8) mm long; ovary at anthesis 1–2 mm long, sparsely to densely glandular-puberulent; capsules globose to obovoid, (2.4) 3–6 (6.2) mm long; $n = 16$.

Meadows, streambanks, and damp soils, especially where alkaline, from 610–2600 m (2000–8500 ft.) elev.; s. tip of Nev. to s. and c. Utah, Ariz., N.M., and Texas; also n. Mex. April–Aug.

Sisyrinchium demissum can be separated from the following closely related species, *S. halophilum*, on the basis of its branched stems, with a leaf-like bract subtending two or more pedunculate spathes. It is not uncommon to find both branched and unbranched stems in many populations although one type is always much more frequent than the other. Henderson (1972) has observed branched plants often occupying somewhat drier microsites within a predominantly single-stemmed population. *Sisyrinchium demissum* replaces *S. halophilum* geographically in southern Nevada.

2. Sisyrinchium halophilum Greene
Sisyrinchium halophilum Greene, Pittonia 4: 34. 1899. (*Greene s.n.*, Humboldt Wells, Nev., 25 July 1893.)
S. leptocaulon E. P. Bicknell, Bull. Torrey Bot. Club 26: 451. 1899. (*Greene s.n.*, Deeth, Elko Co., Nev., 14 July 1896; lectotype selected by D. Henderson.)

Nevada blue-eyed grass.

Perennial herbs; roots fibrous; stems numerous, erect, 0.7–2 mm wide and up to 4.2 dm tall, usually simple, the margins entire; leaves erect, shorter than the stems, 0.6–2.5 mm wide, the margins entire; spathe usually terminal on the stem, the bracts subequal; outer bract 13–26 mm long, tapering gradually to a blunt or acute tip, the inner margins broadly hyaline below, narrowly hyaline nearly to the tip, connate at the base (1.5) 2.2–6 mm; inner bract 10–20 mm long, with broadly hyaline inner margins extending to the tip; flowers 2–7 (9) on slender, erect, glandular-puberulent pedicels, equal to or often exceeding the outer bract; perianth light to dark blue, with a yellow eye, glandular-stipitate abaxially, the tepals broadly oblanceolate to cuneate, truncate, apiculate to aristulate, rarely emarginate, often with prominent dark blue veins on the adaxial surface, (6) 9–10 (12) mm long; filament column (3) 4–5 (6) mm long; anthers (0.8) 1–1.6 mm long; ovary at anthesis 1–1.8 (2.1) mm long, densely stipitate-glandular; capsules globose to obovoid, (2) 2.2–4.5 (5.5) mm long; seeds 1–1.5 (1.7) mm in diam, finely pitted; $n = 16$.

Meadows bordering springs and running water, especially in alkaline soils, at 1340–2440 m (4400–8000 ft.) elev.; e. Calif. (perhaps as far w. as Kern Co.) and throughout all of Nev. except the s. tip. May–Aug.

Sisyrinchium halophilum is the commonly encountered sisyrinchium throughout Nevada except for the southern tip, where it is replaced by the branched *S. demissum*, and the northern part, where *S. idahoense* also occurs. *Sisyrinchium halophilum* can be separated from *S. idahoense* by its wide, cuneate tepals with dark blue veins, narrow stems and leaves, and subequal spathe bracts. According to Henderson (1972), *S. halophilum* is one of the species that can be recognized by the juxtaposition of prominent, dark blue veins on the pale blue outer surface of the tepals. Unfortunately the character is rarely discernible in dried material.

Sisyrinchium demissum and *S. halophilum* are closely related morphologically to *S. bellum* S. Wats. of cismontane California and southern Oregon, a species with branched stems. The three taxa more or less replace each other geographically. Greenhouse hybridization studies by Henderson have demonstrated a close genetic relationship between *S. halophilum* and *S. bellum*. Both species are tetraploid obligate outcrossers. His study did not include *S. demissum*.

Sisyrinchium halophilum normally inhabits highly alkaline areas. When Henderson grew this species and *S. bellum* together in Seattle, *S. halophilum* tended, after two years, to assume some of the features of plants common to meadow soils of lower pH, i.e., somewhat wider and less glaucous stems and more vigorous growth.

Henderson is currently studying the *S. bellum* complex, which includes *S. demissum, S. halophilum*, and *S. funereum* E. P. Bicknell (a Death Valley–Mojave Desert species). The difficulty of separating these taxa morphologically plus the close genetic relationship indicated by Henderson's study may suggest that these taxa be merged, perhaps recognizing the "*halophilum*" phase at some infraspecific level. In order not to anticipate Henderson's publication, we retain the taxa at the specific level.

3. **Sisyrinchium idahoense** E. P. Bicknell

Sisyrinchium idahoense E. P. Bicknell, Bull. Torrey Bot. Club 26: 445. 1899. (*Sandberg, MacDougal, & Heller 317*, valley of Big Potlatch River, Nez Perce Co., Idaho, 6 June 1892.)

S. occidentale E. P. Bicknell, Bull. Torrey Bot. Club 26: 447. 1899. *S. idahoense* var. *occidentale* D. Henderson, Brittonia 28: 174. 1976. (*Palmer s.n.*, Arco, Idaho, 18 June 1893; lectotype selected by D. Henderson.)

S. segetum E. P. Bicknell, Bull. Torrey Bot. Club 26: 449. 1899. *S. idahoense* var. *segetum* D. Henderson, Brittonia 28: 174. 1976. (*Piper s.n.*, Seattle, Wash., May 1892; lectotype selected by D. Henderson.)

S. oreophilum E. P. Bicknell, Bull. Torrey Bot. Club 31: 381. 1904. (*Bioletti s.n.*, Yosemite Valley, Calif., May–June 1900.)

S. angustifolium sensu western authors; not Miller, 1768.

S. montanum sensu western authors; not Greene, 1899.

Idaho blue-eyed grass.

Cespitose perennial herbs; roots fibrous; stems usually simple, (0.9) 1–2.2 (4) mm wide, up to 5.5 dm tall, the margins entire to finely denticulate; leaves mainly shorter than the stems, erect and firm to ascending and lax, (0.5) 1.3–2.6 (4.5) mm wide, the margins entire to finely denticulate; spathe usually terminal on the stem, rarely pedunculate, the bracts usually very unequal in length, rarely subequal; outer bract (15) 20–40 (65) mm long, the inner margins hyaline at least 1/2 the length, connate at the base (2) 3.5–7 (8) mm; inner bract 13–28 (33) mm long, the inner margins narrowly to broadly hyaline nearly the entire length of the bract; flowers (1) 2–6 (7) on slender, glabrous to glandular pedicels shorter than, equal to, or occasionally exceeding the outer bract; perianth blue-violet to deep-violet purple, commonly with a yellow eye, the tepals narrowly elliptic to oblanceolate, rounded or more often truncate and emarginate, 6–15 (20) mm long; filament column 3–6.5 mm long; anthers 1–2 (2.5) mm long; ovary at anthesis (1.4) 1.8–2.5 (3) mm long, usually densely glandular-puberulent; capsules (2.5) 4–6 (6.5) mm long, globose to obovoid; seeds 1–1.8 mm in diam, finely pitted; $n = 32$, 48.

Meadows and moist grassy places, streambanks, in areas where moist in the spring, up to 3300 m (10,800 ft.) elev.; B.C. to Calif. (as far s. as the White Mts.), e. across n. Nev., Idaho, Utah, and w. of the continental divide in Colo., Wyo., and Mont. May–Aug.

Sisyrinchium idahoense is the most variable and widely distributed species of the genus in the western states. Henderson (1972, 1976) has shown the *S. idahoense* complex, including *S. occidentale* and *S. segetum*, to consist entirely of interfertile octoploid populations. Chemical data and morphological similarity further support the recognition of one highly variable species in this complex.

Although the two may not actually overlap in range, *S. idahoense* can be separated from *S. halophilum* in northern Nevada on the basis of its unequal spathe bracts and emarginate tepals.

Authors of western floras have often called these plants *S. montanum* or *S. angustifolium* (the latter here recognized as an eastern species). We follow Henderson's interpretation of *S. montanum* as a morphologically, chemically, and genetically distinct species of the Rocky Mountains, occurring in Colorado, Wyoming, Montana, and the Canadian Rockies, mostly east of the continental divide. *Sisyrinchium montanum* can be recognized by its greatly elongated outer bract that is often slightly widened above the level of the inner bract and shortly connate (for less than 4 mm) at the base, deeply emarginate tepals, stiffly erect leaves and stems, and coarsely pitted seeds. While the characters separating *S. idahoense* and *S. montanum* appear to hold throughout much of the range, the distinction between the two taxa seems to break down somewhat in the Uinta, Wasatch, LaSal and Abajo mts. and in Kane Co., Utah. Here the connation of the outer bract is that of *S. montanum* while the length of the outer bract is much like that of *S. idahoense*. We consider these Utah plants to be members of the variable *S. idahoense* until such time as the Utah populations can be studied chemically and genetically.

4. **Sisyrinchium douglasii** A. Dietr.

Sisyrinchium grandiflorum Douglas ex Lindl. Edward's Bot. Reg. 16: pl. 1364. 1830; not Cav. 1790. *S. douglasii* A. Dietr. Sp. Pl. 2: 504. 1833. *Olsynium grandiflorum* Raf. New Fl. N. Amer. 1: 72. 1836. *Bermudiana grandiflora* Kuntze, Revisio Gen. Pl. 2: 700. 1891. *Olsynium douglasii* E. P. Bicknell, Bull. Torrey Bot. Club 27: 237. 1900. (*Douglas s.n.*, "near the great falls of the river Columbia," in 1826.)

S. inalatum A. Nels. Bot. Gaz. 54: 136. 1912. (*Macbride 909*, Silver City, Owyhee Co., Idaho, 17 June 1911.) = var. *inflatum*.

Sisyrinchium douglasii var. inflatum (Suksd.) P. Holmgren, hoc loc. *Olsynium inflatum* Suksd. Werdenda 1: 8. 1923. *S. inflatum* St. John, Rep. Prov. Mus. Nat. Hist. 1930: 12, pl. 10, fig. 2. 1931. (*Suksdorf 8507*, Spangle, Wash., 24 April 1916.)

SISYRINCHIUM

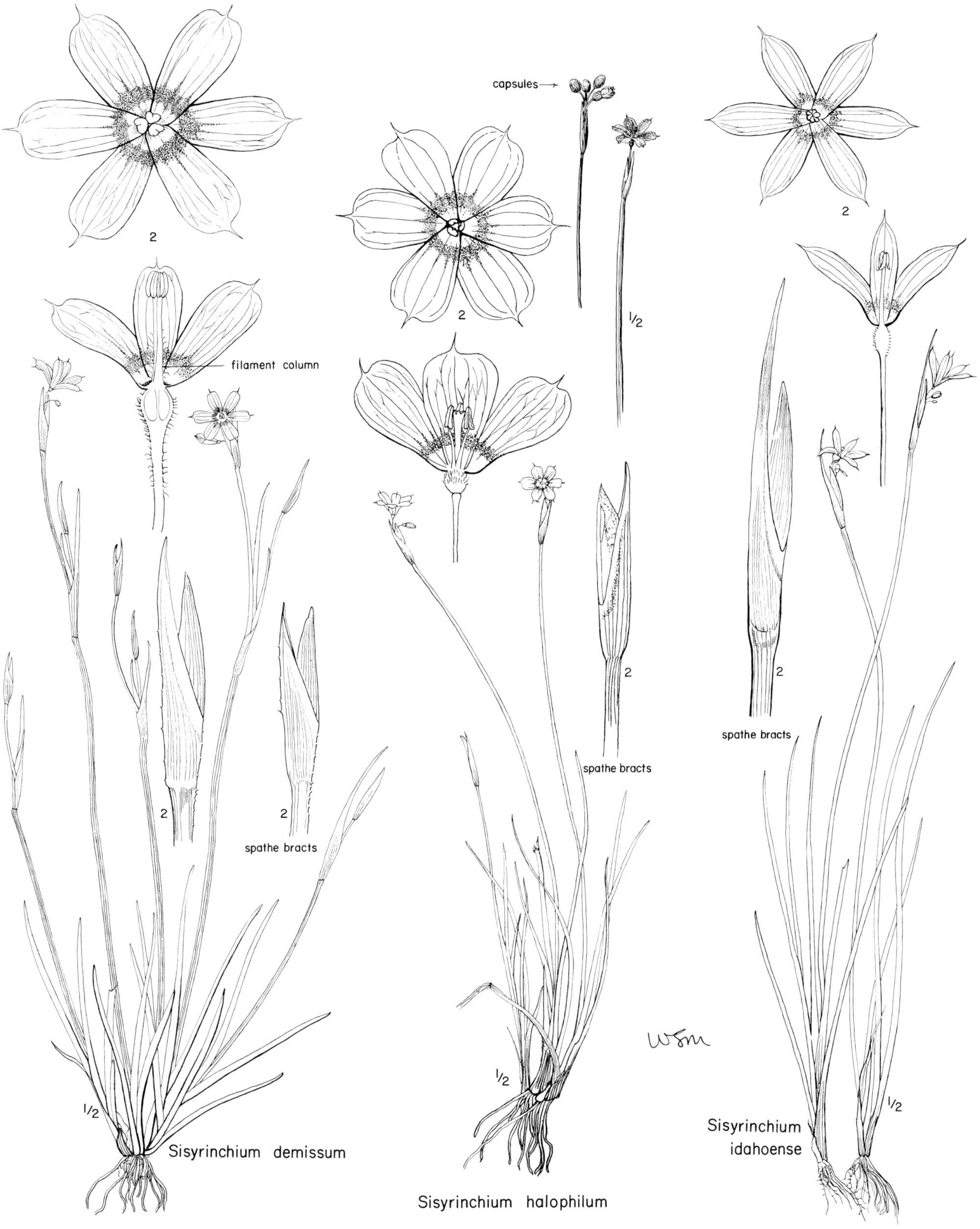

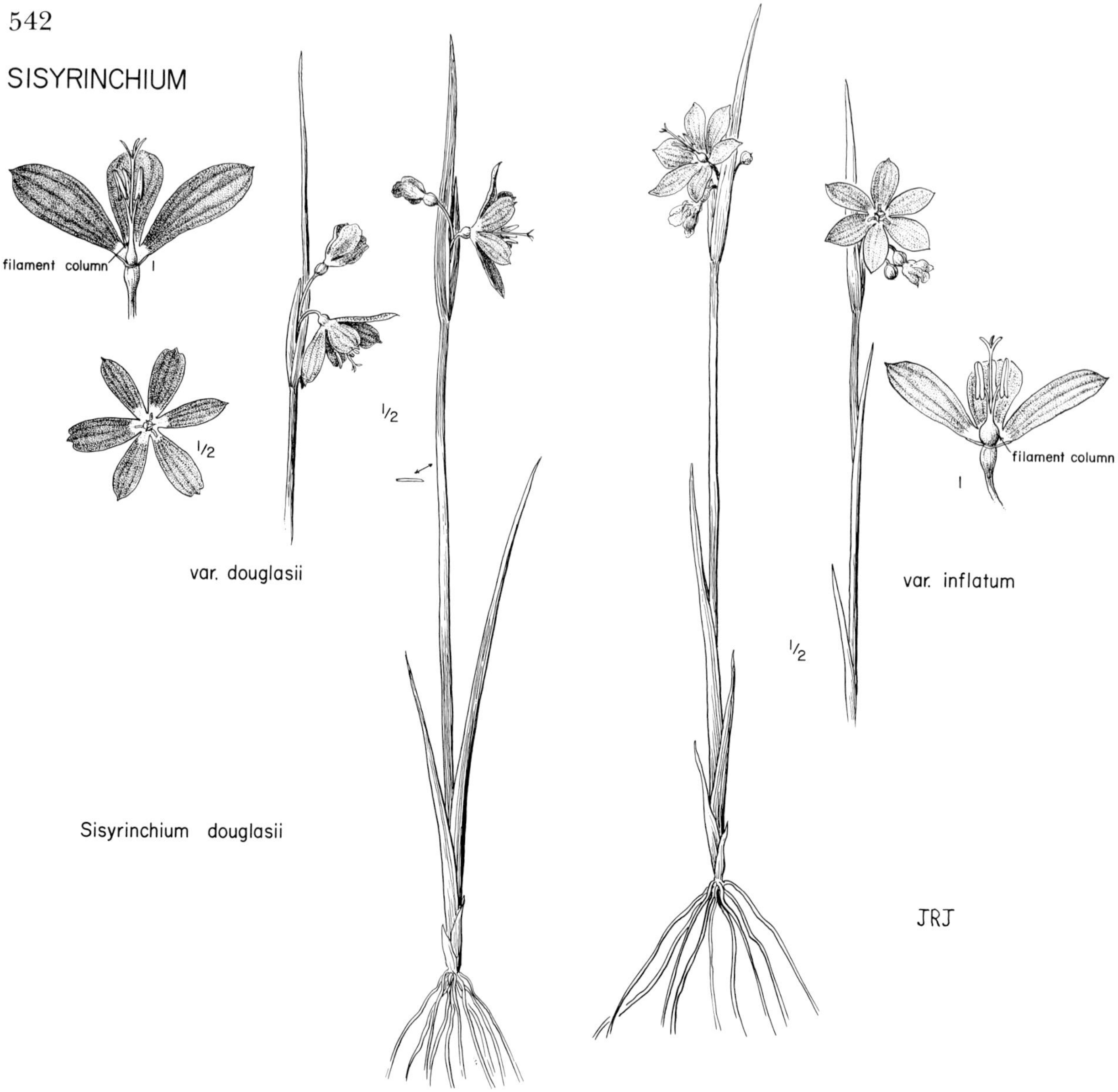

Sisyrinchium douglasii

S. inflatum f. *alba* St. John, Rep. Prov. Mus. Nat. Hist. 1930: 13. 1931. (*St. John, Jones, Moore, & Warren 9785,* Glade Creek, 2 mi. n. of Bluelight, Yakima Co., Wash., 12 April 1929.) = var. *inflatum.*

Purple-eyed grass, grass-widows.

Cespitose perennials; stems erect, simple, scabrous to smooth, 1–3.2 (4.5) mm broad, up to 3.8 dm tall; basal leaves bractlike; cauline leaves clasping the stem, the blade 1.8–3 (4.3) mm broad, scabrous to smooth with entire margins; spathes terminal on the stem, the bracts unequal, the inner 2.2–3.7 (6) cm long, hyaline-margined its entire length, the outer (3.5) 4.3–12 cm long, hyaline-margined ca. 3/4 its length, surpassing the flowers; flowers 1–3 (4) on glabrous pedicels 2.1–3.8 cm long, the pedicels subequal to the inner spathe bract; perianth deep reddish-purple to magenta, pink or white, the tepals 13–25 (30) mm long, oblanceolate to obovate, truncate to rounded or acute with an apiculate or short-attenuate tip; filaments connate up to 1/2 their length, the column abruptly inflated to slightly or not at all inflated immediately above the base; anthers 3–7 (9) mm long; style slender, with 3 stigmas; ovary glabrous to glabrate, at anthesis 2–5 mm long; capsules spherical to broadly obovoid, (4.5) 5.5–9 (11) mm long; seeds strongly angled, coarsely reticulate, 1.5–2.5 mm long; $n = 32$.

In dry areas where moisture is adequate during the flowering stage, in grasslands, damp prairies, and juniper woodland to exposed, rocky areas up to 2150 m (7000 ft.) elev.; s. B.C. to n. Calif., e. to Idaho, n. Nev., and Utah (Tooele Co.). March–June.

Because of its large, pink, magenta, reddish-purple flowers, *Sisyrinchium douglasii* is our most easily recognized sisyrinchium. It is rather uncommon in the Intermountain Region.

Past workers have considered *S. douglasii* and *S. inflatum* as separate species although they have all noted that the two are very similar. The main basis for separation has been the inflated base of the filament column (*S. inflatum*) versus the non-inflated or only slightly inflated base of the filament column (*S. douglasii*). Henderson (1972) has found a nearly complete morphological and chemical gradation in populations examined in Washington. We understand that on the basis of cytological, chemical, and morphological data, Henderson intends to merge the two species and recognize *S. inflatum* at the subspecific rank. We agree with his decision to merge the two but prefer to recognize the "*inflatum*" phase at the varietal level.

The two varieties can be separated as follows:

1 Filament column only slightly or not at all inflated; w. Wash., Columbia River Gorge, and e. slope of Cascades in Wash. .... var. *douglasii*

1 Filament column abruptly and broadly inflated above the base; e. of the Cascades in Wash., s. to n. Calif., e. to w. Idaho, n. Nev., and Utah (Tooele Co.) .... var. *inflatum* (Suksd.) P. Holmgren

2. IRIS L. Iris, Flag, Fleur-de-lis

Rhizomatous or bulbous perennials, sometimes spreading by stolons; stems erect, simple or branched; leaves mostly basal, sometimes cauline, equitant, linear to ensiform, flat, terete, or quadrangular, many-nerved; inflorescence a spike, raceme, or panicle, bearing 1–several flowers in clusters (spathes or heads) from paired spathe bracts; flowers showy, often brightly colored, short-lived; tepals epigynous, differentiated into an outer and inner whorl of 3 each, united below into a long or short perianth tube; outer tepals (falls or sepals) spreading, recurved or reflexed, sometimes erect and arching, in some species bearded on the upper surface, narrowed basally into a claw (haft); inner tepals (standards or petals) erect or arching, sometimes spreading or reflexed, narrowed basally into a claw; stamens 3, alternating with the inner tepals, nestled against the style branch, the anthers 2-celled, basifixed; ovary inferior, 3- or 6-angled, 3-loculed; style divided into 3 petaloid style branches, these usually arched over the falls and together simulating a tube for the pollinator, the stigmatic surface on the adaxial side behind a short lip and terminated by a bifid, often lacerate style crest; fruit a loculicidal capsule, the seeds numerous, in (1) 2 rows in each locule.

A genus of about 150 species in the N. Temp. Zone, mostly of Eurasia. (Name from the Greek *iris,* meaning rainbow, in reference to the many bright colors of flowers.)

Several species and many hybrids of *Iris* are well-known garden plants. The following key is to the more common species planted in flower gardens in our area. We have only one native species, *I. missouriensis,* and only two of the cultivated species, *I. germanica* and *I. pseudacorus,* are known to escape. For easy identification of these three species, see the second key.

References:

Dykes, W. R. The genus *Iris.* Cambridge Univ. Press, London. pp. 1–245. 1913.

Foster, R. C. A cyto-taxonomic survey of the North American species of *Iris.* Contr. Gray Herb. 119: 3–82. 1937.

Lawrence, G. H. M. A reclassification of the genus *Iris.* Gentes Herb. 8: 346–371. 1953.

———, and L. F. Randolph. The classification of irises. *In:* L. F. Randolph (ed.). Garden irises. Cayuga Press, Ithaca, N.Y. pp. 133–160. 1959.

Key to the Common Garden Irises
(as represented in our region)

1 Rootstock a bulb.
  2 Standards (inner tepals) small and spreading or reflexed between the larger, erect falls (outer tepals) (Juno irises).
    3 Plants dwarf, the stems very short or obsolete; claw (haft) of the fall winged; Persian Iris .... *I. persica* L.
    3 Plants larger, the stems 3–4.5 dm long; claw of the fall not winged .... *I. bucharica* Foster
  2 Standards larger and erect; falls spreading or recurved (Xiphium irises).
    4 Plants dwarf, the stems very short or obsolete; perianth tube 5–7.5 cm long; leaves quadrangular .... *I. reticulata* M. Bieb.
    4 Plants larger, the stems 3 dm or more in length; perianth tube less than 1.5 cm long; leaves flat.
      5 Fall with a fiddle-shaped base, with a constriction separating the blade from the oblong-oval claw; includes Spanish and Dutch irises .... *I. xiphium* L.
      5 Fall with a cuneate base, i.e., the blade gradually narrowing into the claw; English Iris .... *I. xiphioides* Ehrh.
1 Rootstock a rhizome.
  6 Falls (outer tepals) bearded or with fringed crests.
    7 Falls with longitudinal fringed crests running up the claw and lower part of the blade, not at all pubescent (Evansia irises).
      8 Plants dwarf, the stems absent or never more than 1 dm long; Dwarf Crested Iris .... *I. cristata* Soland.
      8 Plants taller, the stems evident, 3–5 dm long, equalling or exceeding the leaves .... *I. tectorum* Maxim.
    7 Falls bearded, the beard of multicellular hairs.
      9 Beard diffuse on the claw and lower blade; seeds with a conspicuous white aril (Oncocyclus irises).
        10 Falls cream-white to sky blue with fine purple veins and purplish dots; standards pale purple or yellowish .... *I. gatesii* Foster

10 Falls brownish-purple with dark brown spots; standards brownish-white; Mourning Iris . . . . *I. susiana* L.
9 Beard confined to a narrow line.
11 Falls and standards bearded; seeds with a large, cream-colored aril (Regelia irises) . . . . *I. hoogiana* Dykes
11 Falls bearded; standards not bearded; seeds without a conspicuous aril (Pogoniris irises).
12 Plants dwarf, the stems obsolete or very short . . . . *I. pumila* L.
12 Plants taller, the stems more than 3 dm long.
13 Inflorescence equalling the leaves; flowers yellow and reddish-brown . . . *I. variegata* L.
13 Inflorescence exceeding the leaves; flowers violet to white.
14 Spathe bracts silvery-scarious . . . . *I. pallida* Lam.
14 Spathe bracts green or only partly scarious; an introduced species sometimes escaping cultivation (see treatment below) . . . . 1. *I. germanica* L.
6 Falls without a beard or crest.
15 Inflorescence a raceme; seeds conspicuously winged with a circular, cream-colored aril (Pardanthopsis irises) . . . . *I. dichotoma* Pallas
15 Inflorescence a panicle; seeds not conspicuously winged (Apogon irises).
16 Leaves linear, generally less than 1 cm broad.
17 Stems hollow; leaf-sheaths splitting into fibers; garden plants . . . . *I. sibirica* L.
17 Stems solid; leaf-sheaths not splitting into fibers; native species (see treatment below) . . . . 3. *I. missouriensis* Nutt.
16 Leaves ensiform, usually more than 1.2 cm broad.
18 Stems bearing several long leaves; flowers yellow; an introduced species often escaping cultivation along water courses (see treatment below) . . . . 2. *I. pseudacorus* L.
18 Stems bearing only 2–4 reduced leaves; flowers variously colored; garden plants not known to escape cultivation.
19 Flowers yellowish to white.
20 Flowers pale yellow . . . . *I. monnieri* DC.
20 Flowers white with golden-yellow blotches on the falls . . . . *I. orientalis* Miller
19 Flowers blue or purple . . . . *I. spuria* L.

Key to the Native or Naturalized Irises

1 Leaves broadly ensiform, 12–40 mm broad; introduced species often escaping cultivation.
2 Flowers predominantly purplish or bluish; falls (outer tepals) bearded; standards (inner tepals) about the same size as the falls; plants of relatively dry habitats . . . . 1. *I. germanica*
2 Flowers yellow; falls not bearded; standards much smaller and narrower than the falls; plants often growing in water . . . . 2. *I. pseudacorus*
1 Leaves linear, 2.5–8 (9.5) mm broad; native plants . . . . 3. *I. missouriensis*

1. Iris germanica L.

Iris germanica L. Sp. Pl. 38. 1753. ("Habitat in Germaniae editis.")

Fleur-de-lis, German iris.

Stout rhizomatous perennials, the rhizomes up to 5 cm thick; stems 5–10 dm tall, often branched from the middle, the branches in axils of inflated bracts; leaves broadly ensiform, 15–40 mm broad, shorter than the flowering stems, glaucous; flowering stems terminating in a 2- to 3-flowered spathe, often with 1 or 2, usually 1-flowered branches below; spathe bracts mostly 4–5 cm long, oblong-lanceolate, scarious in the upper half; perianth tube 20–50 mm long, longer than the ovary; falls (outer tepals) 6–8 cm long and 3–5 cm broad, broadly obovate, recurved, purplish, varying to white with yellow, white and brown veins, the median line yellow bearded, the hairs multicellular-clavate; standards (inner tepals) about the same size as the falls, obovate, erect-arching, usually lighter colored than the falls; stamens white or light violet; style branches spatulate, usually the same color as the standards; capsules 4–5 cm long, ovoid, trigonous; $2n = 44, 48$.

Gravelly hillsides and waste places around habitations, probably where discarded; native of the Medit. region, widely cultivated in temp. regions of the world. April–May.

Iris germanica has yielded many garden forms, often as a result of hybridization and selection.

2. Iris pseudacorus L.

Iris pseudacorus L. Sp. Pl. 38. 1753. ("Habitat in Europa ad ripas paludum fossarum.")

Yellow water iris, yellow flag.

Rhizomatous perennials forming large clumps up to 1 m broad, the rhizomes up to 4 cm thick; stems 4–15 dm tall, leafy, branched; leaves broadly ensiform, 12–25 (33) mm broad, glaucous; flowering stems often branched and bearing several spathes, the terminal spathe 2- to 3-flowered, the lateral spathes 1- (2)-flowered; spathe bracts 4.5–7 cm long; flowers bright yellow; perianth tube short, about 12 mm long; falls (outer tepals) 5–6.5 cm long, ovate-

IRIS
standard
beard
falls
1/2
Iris germanica
wsm
style branch
1/2
Iris pseudacorus
JRJ
1/12
1/4
JRJ
Iris missouriensis
1/2

spatulate, rounded, not bearded, spreading, recurved, the claw (haft) narrow; standards (inner tepals) 1–3 cm long, narrow-elliptic to linear, obtuse, erect; anthers 12–16 mm long; style branches broader and longer than the standards, 3–5 cm long, oblanceolate, lacerate, arching outward; capsules 3.5–7 cm long, elliptic, apiculate, trigonal, with 6 faint grooves; seeds roundish or deltoid, compressed, D-shaped, shiny brown; $2n = 24, 30, 32–34$.

Along streams, often in water 0.5 m deep; native of Europe and the Medit. region; widely cultivated in Amer., escaping along watercourses, common in the lower Snake River Plains and marshes on east side of Great Salt Lake. May–June.

3. **Iris missouriensis** Nutt.

Iris missouriensis Nutt. J. Acad. Nat. Sci. Philadelphia 7: 58. 1834. (*Wyeth,* "towards the sources of the Missouri.")

I. haematophylla var. *valametica* Herbert ex Hook. Fl. Boreali-Amer. 2: 206. 1839. *I. tolmieana* Herbert in Hook. & Arn. Bot. Beechey's Voyage 396. 1840. ("Hab. River Walamet, N.W. America. *Tolmie.*")

I. missouriensis albiflora Cockerell, W. Amer. Sci. 6: 135. 1889, nomen nudum. (Wet Mountain Valley, Colo.)

I. longipetala var. *montana* Baker, Handb. Irideae 10. 1892. ("Widely spread in California, first gathered by *Douglas* in 1833.")

I. pelogonus Goodd. Bot. Gaz. 33: 68. 1902. *I. missouriensis* var. *pelogonus* R. C. Foster, Contr. Gray Herb. 119: 66. 1937. (*A. Nelson 7102,* Bush Ranch, Sweetwater Co., Wyo.)

I. montana Nutt. ex Dykes, Gen. Iris 91, pl. 22. 1913. (From a plant grown in Dykes' garden in England, sent to him by A. Nelson from near Laramie in 1909.) A superfluous name with *I. pelogonus* Goodd. cited in synonymy.

I. missouriensis f. *angustispatha* R. C. Foster, Contr. Gray Herb. 119: 67. 1937. (*H. J. Rust 530,* Brady Place, Coeur d'Alene, Idaho, May 1914.)

I. missouriensis f. *alba* St. John, Leafl. W. Bot. 9: 96. 1960. (*Pickett s.n.,* between Pullman and Wawawai, Wash., 3 June 1916.)

Western iris, western blue-flag.

Perennials spreading by thick rhizomes, the rhizomes 1–1.5 cm thick; stems 2–6 (10) dm tall; leaves linear, 2.5–8 (9.5) mm broad; flowering stems barely surpassing the leaves, terminating in a 2- (4)-flowered spathe; spathe bracts (3.5) 4.5–6 (9) cm long; perianth tube mostly 5–8 mm long; falls (outer tepals) 4.5–6 (8) cm long, obovate to oblanceolate, rounded to emarginate, recurved, the blue to violet nerves on a light blue, lavender or white background, often with a patch of yellow in the center, the claw (haft) long, expanding gradually into the blade; standards (inner tepals) about as long as the outer tepals, oblanceolate, emarginate, obtuse or acute, erect, pale blue, pale purple to white; anthers 13–17 mm long; style branches 2–4 cm long, dentate, same color as the standards; capsules 3–5 (6) cm long, short-cylindrical to subglobose, apiculate, 6-ridged; $2n = 38, 48$.

Marshes, moist meadows, streambanks, shores and aspen woods, often in soils that dry out by summer's end, in high sagebrush valleys and foothills to subalpine elevs.; B.C. and Wash., to Mont. and w. S.D., s. to s. Calif., Nev., Ariz., N.M. and Coahuila; in the Pacific states mostly e. of the Cascade-Sierran divide. May–July.

Order **ORCHIDALES**

Mycotrophic (and sometimes nongreen) monocotyledonous herbs with very numerous, tiny seeds that have an undifferentiated embryo and very little or no endosperm; sepals and petals generally both petaloid, but often dissimilar; ovary inferior, tricarpellate.

The order consists of four families. Three of these (Burmanniaceae, Corsiaceae, and Geosiridaceae) are small groups (less than 150 spp. in all) that do not occur in our range. The fourth family, Orchidaceae, occurs throughout the world and is one of the largest families of plants.

Family **ORCHIDACEAE,** the Orchid Family

Perennial herbs (in ours) to vines, often epiphytic, but ours all terrestrial; plants sometimes lacking chlorophyll and with much reduced leaves, being then dependent on their mycorrhizal symbiont for food; leaves mostly green, often sheathing, 1 to several, linear to nearly orbicular, often coarsely ribbed; plants frequently becoming rhizomatous, often with somewhat fleshy fibrous roots to much enlarged coral-like clusters of roots; flowers usually complete, rarely imperfect, solitary, spicate or in simple racemes in ours, minute to showy and often nearly grotesquely zygomorphic, epigynous; as the flower develops, the ovary and pedicel twist a half turn (resupinate) so the flower is turned 180° (some species, not ours, are non-resupinate so that the lip remains uppermost); perianth composed of six segments, 2 or more of them sometimes fused, the outer series (the sepals) sometimes somewhat similar to the petals but more often greenish, brownish and unlike the inner in color and especially in form; corolla usually highly colored but sometimes dull white to brown or even deep green, the two lateral petals often more or less similar to the sepals and occasionally somewhat adnate with the upper sepal and forming a hood, the third and lowermost petal highly specialized and referred to as the lip or labellum, larger and unlike the other perianth segments, variously grooved or concave and leading to a basal nectary or into a sac or spur; stamens more or less completely fused with the style and stigmas to form a very prominent structure, the column; fertile anthers 2 in the diandrous genus *Cypripedium* but 1 in all the remaining (monandrous) genera, the third stamen sometimes sterile and represented by a staminodium, the pollen sacs two, often separated by a broad connective, the pollen of each sac adhering in 1–4 masses called pollinia, or seldom the pollen pulpy or granular and somewhat viscid; pollinia often with a stalk-like base called the caudicle; stigmas 3, usually forming a sticky concave to convex surface on the lower side near the tip of the column, the three all sometimes fertile and alike, or only two functional and the third modified and forming part or most of a broadened to elongate and beaklike structure, the rostellum; rostellum separating the stigmatic surface from the anther, and usually attached to the caudicle of the 1–4 pollinia of each pollen sac by a small knob-like, sticky portion, the viscidium, which tends to adhere to any foreign object, thus facilitating pollen transfer; ovary inferior, 1-celled with 3 parietal placentae or 3-celled; fruit a capsule, dry in ours; seeds very numerous, minute, lacking endosperm, the embryo not differentiated into cotyledon, epicotyl, and hypocotyl.

One of the largest families of flowering plants, with estimates of numbers of genera ranging from about 450 to 800 and species from 8,000 to approximately 30,000. If the higher figure is near being correct, this family would account for about 10 percent of all flowering plants, but more realistic figures would place the numbers at about 15,000 species. It is possible that only Asteraceae have more species.

Five subfamilies may be recognized. Our eight genera are placed in four of them as follows:

Cypripedioideae	*Cypripedium*
Neottioideae	*Epipactis, Listera, Spiranthes, Goodyera*
Orchidoideae	*Habenaria*
Epidendroideae	*Corallorhiza, Calypso*

The vegetative parts of our orchids are relatively simple as compared to those of tropical rain forests where orchids attain their greatest development. In tropical rain forests, leaves are of myriad forms. Many persist as storage organs for water and nutrients.

The lip is a passageway into which insects are guided. An insect first passes by the stigma, then by the anther or anthers. The guiding of the insect is elaborate. The stamens are so reduced that pollination must be efficient.

The pollinium of the Orchidaceae is not quite unique, as it is simulated in the Asclepiadaceae where we find several instances of parallel evolution with striking analogous structures.

So far as we know today, all orchids live in symbiosis, at least in the young stage, with a root fungus.

Most of our orchids are found in moist to wet boggy places, but several grow in relatively dry soil. Our three species of *Corallorhiza, Goodyera oblongifolia, Habenaria unalascensis* are all found in relatively dry soils.

All our orchids are of rather wide distribution, with none of them being narrow endemics. Our two species of *Cypripedium* are our rarest species.

References:
Correll, D. S. Native orchids of North America north of Mexico. Chronica Botanica Co., Waltham, Mass. 399 pp. 1950.
Dressler, R. L. The structure of the orchid flower. Missouri Bot. Gard. Bull. 49: 60–69. 1961.
———, and C. H. Dodson. Classification and phylogeny in the Orchidaceae. Ann. Missouri Bot. Gard. 47: 25–68. 1960.
Luer, C. A. The native orchids of Florida. New York Botanical Garden. 293 pp. 1972.
———. The native orchids of the United States and Canada excluding Florida. New York Botanical Garden. 361 pp. 1975.
Williams, L. O. The Orchidaceae of the Rocky Mountains. Amer. Midl. Naturalist 18: 830–841. 1937.

1 Plants without chlorophyll; leaves reduced to brownish sheaths; roots coralloid .... 7. *Corallorhiza*
1 Plants with green leaves, often withering at flowering time; roots not coralloid.
  2 Fertile stamens 2; lip an inflated, sac-like pouch .... 1. *Cypripedium*
  2 Fertile stamen 1; lip not an inflated sac-like pouch.
    3 Lip prolonged at the base into a slender to saccate or scrotiform spur .... 6. *Habenaria*
    3 Lip not forming a spur.
      4 Flowers 1 or sometimes 2; plant with a subglobose corm and a single leaf .... 8. *Calypso*
      4 Flowers usually several to many; leaves 2 or more.
        5 Lip wedge-shaped; leaves 2, opposite or subopposite, borne near the middle of the stem .... 3. *Listera*
        5 Lip saccate or concave; leaves not opposite or subopposite.
          6 Roots from a creeping rhizome; leaves evergreen, usually mottled; growing in dry to damp places in forested area ....5. *Goodyera*
          6 Roots fleshy and fascicled; leaves not persisting, not at all mottled; plants of wet places at base of cliffs or in meadows.
            7 Flowers brownish-purple, rather showy, in a loose, bracted raceme; leaves lanceolate to ovate; lip saccate near the base ....2. *Epipactis*
            7 Flowers white or sordid in ours, in a tight, spirally twisted spike; leaves linear to linear-lanceolate; lip concave near the base ....4. *Spiranthes*

1. **CYPRIPEDIUM** L. Lady's Slipper

Caulescent herbs with short rhizomes and fibrous roots; leaves sessile, flat, plaited, often sheathing; stem and leaves more or less pubescent and usually glandular; flowers 1 to several in a terminal raceme, leafy bracted, usually showy; sepals 3, the two laterals partially to nearly completely united under the lip, the free tips the only indication two sepals are involved, the third sepal remaining in a dorsal position above the lip; the 2 lateral petals narrow, spreading, similar in color and texture to the sepals, but usually slightly longer; lip pouch-like and much inflated, more or less puckered and inrolled around the orifice; column bent over the orifice, glabrous, bearing a fertile anther on each side, the third represented by an enlarged and usually petaloid staminodium which exceeds the column; rostellum lacking; pollen pulpy or granular, somewhat viscid, not forming pollinia; stigma terminal, broadest at the base; capsule ribbed.

A genus of about 25 species of temp. N. Amer. and Eurasia. (Name from the Greek *Kypris* meaning Venus, and *pes,* foot, alluding to the moccasin-like lip.)

The diandrous genus, *Cypripedium,* is a member of the Cypripedioideae subfamily.

1 Leaves 2, opposite or subopposite; flowers 2–4, in a tight cluster immediately above the leaves; lip purplish .... 1. *C. fasciculatum*
1 Leaves several, alternate; flowers usually solitary, or 2 in robust plants; lip yellow, often purplish-dotted around the orifice .... 2. *C. calceolus*

1. Cypripedium fasciculatum Kellogg ex S. Wats.

Cypripedium fasciculatum Kellogg ex S. Wats. Proc. Amer. Acad. Arts 17: 380. 1882. *Cypripedilum fasciculatum* Pfitzer, Bot. Jahrb. Syst. 19: 39. 1894. (*Suksdorf*, "On the White Salmon River above the falls," Wash., May 1880, is the first of three collections cited.)

Cypripedium pusillum Rolfe, Bull. Misc. Inform. [Kew Bull.] 1892: 211. 1892. *Cypripedium fasciculatum* var. *pusillum* Hook. f. Bot. Mag. 119: pl. 7275. 1893. (Described from a cultivated plant of uncertain origin.)

Cypripedium knightae A. Nels. Bot. Gaz. 42: 48. 1906. (*Knight*, Cooper Hill, Medicine Bow Mts., Wyo., July 1905.)

Brownie lady's slipper, clustered lady's slipper.

A low herb about 0.4–1.2 dm high with a pair of leaves about halfway up on the villous-viscid stem, a sheathing bract near ground level; leaves opposite or subopposite, sessile, broadly elliptic to nearly orbicular, 5–8 cm long, glabrous or nearly so, prominently nerved, usually with 2 or 3 lanceolate bracts near the inflorescence, the bracts somewhat longer than the ovary and often enveloping it; flowers 2–4 in a contracted, bracted spike only slightly exceeding the leaves, with villous-viscid hairs extending into the inflorescence; sepals lanceolate, acuminate, 1.2–2 cm long, greenish-brown to greenish-purple and somewhat mottled or lined, the lower pair mostly connate throughout or sometimes with free tips; petals similar in color and texture to the sepals but somewhat broader; lip depressed-ovoid, shorter than the sepals, about 1 cm long, greenish-yellow, with deep purple margins; staminodium 2.5–3.0 mm long, oblong, obtuse, about the same length as the longest lobe of the stigma; ovary with villous-viscid hairs; $2n = 20$.

Somewhat dry to moist places in spruce forest at 2900 m (9500 ft.) on the s. side of Lake Blanche in the Wasatch Mts., where it was found by Dr. Walter P. Cottam in 1940 (we returned in 1972 and found two rather large patches in the dense spruce forest); also from the Uinta Mts.; a very rare orchid throughout its range, as it is known only from a few widely scattered localities; B.C. to Calif., e. to Idaho, Mont., Wyo., Utah, and Colo. June–July.

2. Cypripedium calceolus L.

Cypripedium calceolus L. Sp. Pl. 951. 1753. (Europe.)

C. parviflorum Salisb. Trans. Linn. Soc. London 1: 77, pl. 2, fig. 2. 1791. *C. luteum* var. *parviflorum* Raf. Med. Fl. 1: 142. 1828. *Cypripedilum parviflorum* Stein, Orchideenbuch 179. 1892. *Cypripedium hirsutum* var. *parviflorum* Rolfe, Orchid Rev. 15: 184. 1907. *Cypripedium bulbosum* var. *parviflorum* Farw. Rep. Michigan Acad. Sci. 15: 170. 1913. *Calceolus parviflorus* Nieuwl. Amer. Midl. Naturalist 3: 118. 1913. *Cypripedium calceolus* var. *parviflorum* Fern. Rhodora 48: 4. 1946. *Cypripedium calceolus* subsp. *parviflorum* Hultén, Ark. Bot. II, 7(1): 34. 1968. (Virginia.)

Yellow lady's slipper.

Often forming clumps with several flowering stems rising from the base; stems 3–4 dm tall, sparsely pubescent and more or less glandular, with about 2 sheathing bracts below the 3 or 4 alternate, broadly lanceolate to broadly elliptic, somewhat acuminate leaves, these 5–14 cm long, puberulent and usually glandular, prominently plaited; flower usually solitary, but 2 on robust plants, usually exceeded by an erect, leaf-like bract; sepals and petals greenish-yellow or more often purplish-brown in ours, with wavy margins, the dorsal sepal twisted, lanceolate, about 3–4 cm long and 1 cm wide, the lower pair completely connate or with only a slight notch at the tip; petals narrower and longer than the sepals and often with as many as 3 complete twists; lip 2–3 cm long, strongly pouched and of a firm, nearly coriaceous texture, yellow, often with purplish dots around the orifice; staminodium triangular, more or less lobed or auriculate at the base, up to 10 mm long, yellow with purplish dots; ovary ribbed; $2n = 20$.

Known in our area from canal banks near mouth of Logan Canyon in Cache Co. and river bottoms w. of Provo, in Utah Co., Utah; B.C. to Wash. and Oregon to Idaho, Wyo., Utah, Ariz., Colo., N.Y. and e. Can. and e. U.S.; also in Eurasia. May–June.

This circumboreal species is morphologically and ecologically diversified, especially in eastern North America. Our plants belong to the var. *parviflorum* (Salisb.) Fern., which is widespread in North America, especially toward the more northern or western part of the range of the species. The more strictly eastern var. *pubescens* (Willd.) Correll has larger, less odorous flowers and often grows in mesic woods instead of in wet low places. No sharp lines can be drawn between these two varieties, nor between them and the European var. *calceolus*.

When a housing development threatened the Logan Canyon stand of yellow lady's slipper, a few plants were brought into cultivation. Not only has the original clump thrived, it has been divided several times without a loss so there are now more than 30 plants. Several seedlings are coming along and (1974) are nearly ready to flower. Plants under cultivation produce 2 flowers on each stem more often than the solitary flower found in native stands.

2. EPIPACTIS Swartz Helleborine

Tall perennials from a rootstock, often forming clumps with several leafy stems and conspicuously bracted inflorescences; flowers few, in loose, bracted racemes; sepals and lateral petals essentially similar, separate, the lower two sepals spreading, the upper one and the two petals subequal; lip saccate at the base with the upper margins rounded at the apex, becoming abruptly constricted to a triangular blade, inconspicuously crested; column short, the stamen very slightly stipitate, attached on the upper side of the column near the tip; pollinia 4; stigmatic surface large, usually minutely lobed around the edges, the side nearest the lip with 2 short spur-like horns; capsule pendent.

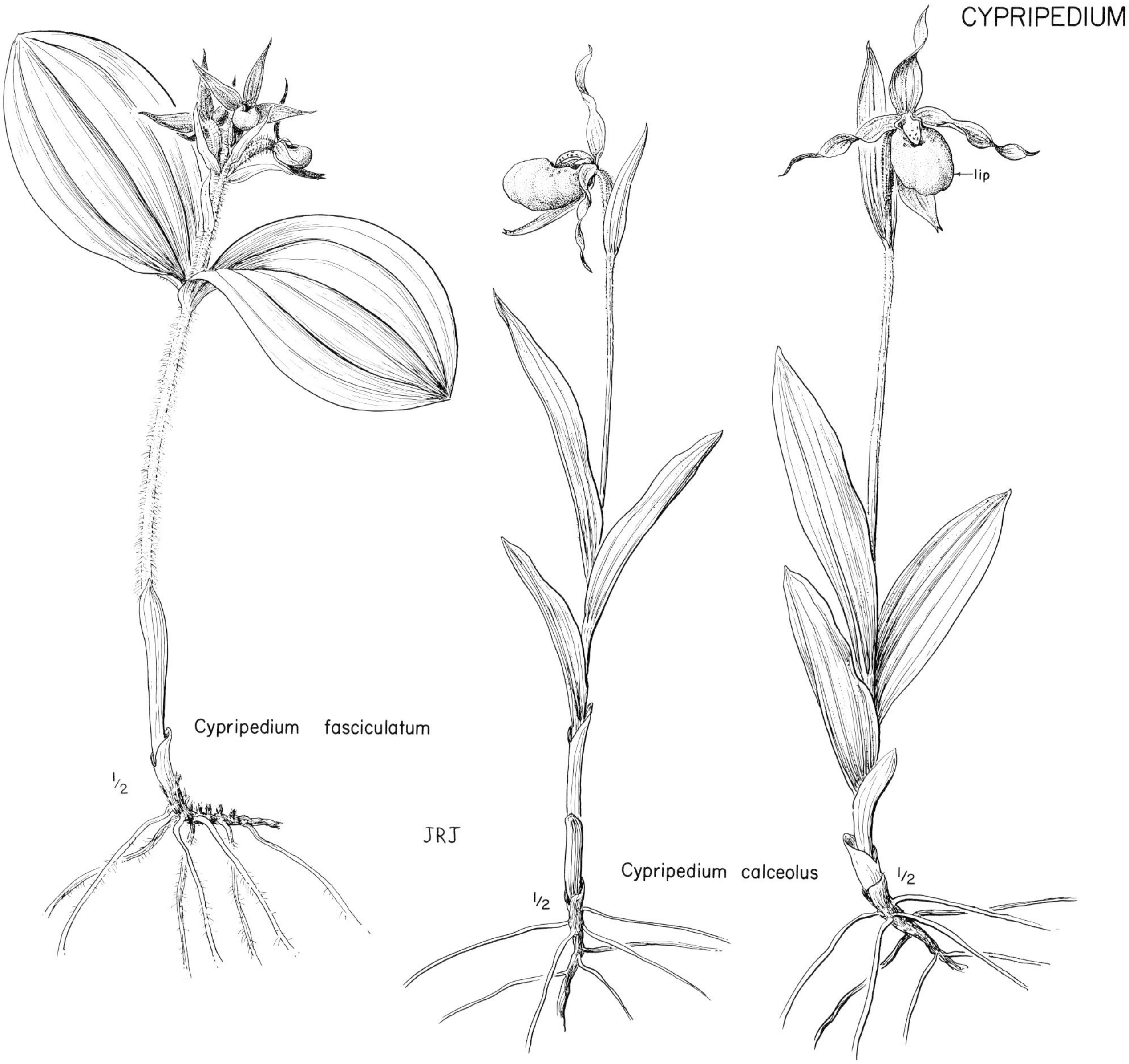

About a dozen species in temp. areas of N. Amer. and Eurasia. (Name probably from the Greek *epipaktis,* the ancient Greek name for hellebore.)

1. Epipactis gigantea Douglas ex Hook.

Epipactis gigantea Douglas ex Hook. Fl. Boreali-Amer. 2: 202, pl. 202. 1839. *Limodorum giganteum* Kuntze, Revisio Gen. Pl. 2: 672. 1891. *Peramium giganteum* J. M. Coulter, Contr. U.S. Natl. Herb. 2: 424. 1894. *Serapias gigantea* A. A. Eat. Proc. Biol. Soc. Wash. 21: 67. 1908. *Helleborine gigantea* Druce, Bull. Torrey Bot. Club 36: 547. 1909. *Amesia gigantea* Nels. & Macbr. Bot. Gaz. 56: 472. 1913. (*Douglas,* "Hab. N.W. America. On the subalpine regions of the Blue and Rocky Mountains.")

Giant helleborine.

Stems 1 to several, from creeping rhizomes, 3–7 (10) dm high, glabrous or nearly so and becoming pubescent in the inflorescence; lower leaves ovate, sessile, the upper becoming narrower, lanceolate to linear-lanceolate; flowers 3–9 (12), rather showy, the raceme usually secund, the long bracts becoming reduced above, the terminal one often exceeding the ovary; sepals with a greenish sheen, with brownish veins, 12–15 mm long; petals similar to the sepals but somewhat thinner, more brownish-purple; lip 15–18 mm long, the sac with raised purplish lines leading to the base, 3-lobed, with prominent outer lobes, the blade or central lobe about the same length as the basal lobes, slightly curved downward, the flattened tip with uprolled margins, greenish-yellow, the basal portion much thickened, yellow, with several crests leading into the sac; column 6–8 mm long, broadened above; capsule 2–2.5 cm long, reflexed, ovoid to ellipsoid, dark brown on the ridges and otherwise yellowish.

Wet places on seepage slopes and wet places at base of cliff along rivers or often wet places in desert areas; B.C. to Baja Calif. and most of w. N. Amer. to the Rocky Mts. and S.D. to n. Mex. April–early Aug.

Giant helleborine is easy to cultivate when a wet habitat can be furnished.

3. LISTERA R. Br. Twayblade

Small, rhizomatous herbs from fibrous roots, with two broad, sessile, opposite or subopposite leaves at about mid-point on the stem; stem glabrous or nearly so below, but usually glandular-pubescent above the leaves; flowers few to several in a bracted terminal raceme, rather small and usually greenish in ours; sepals and lateral petals distinct, almost alike, reflexed or spreading, 1-nerved; lip much longer than the sepals, notched or two-cleft at the apex, often with a pair of teeth or auricles near the base and sometimes clawed, without a spur, usually with a small nectary under the stigma; column nearly terete, wingless; stigmatic surface broader than long, the rostellum broad and flat, usually curving over the stigmatic surface; anther dorsal on the column and hinged to it, the pollinia 2, powdery, without caudicles, the viscidium circular; capsule with relatively few large seeds.

A genus of about 25 species, widely distributed in cool-temp. to subarctic regions of the N. Hemisphere. (Named for Dr. Martin Lister, 1638–1711, an English naturalist.)

The twayblades are named for the pair of leaves midway on the stem. These small, inconspicuous plants do not answer to our usual idea of orchids. The sensitive rostellum forcefully ejects the liquid glue at the slightest nudge from the back of a pollinator. The droplet dries in a few seconds, fixing the pollinia to the pollinator, which is frightened away to another flower.

1 Leaves not cordate; lip rounded to retuse or only shallowly lobed, the lobes broad.
  2 Leaves lanceolate to ovate-elliptic; lip shallowly bilobed, not especially narrowed toward the auriculate base . . . . 1. *L. borealis*
  2 Leaves broader, often suborbicular; lip retuse, abruptly narrowed to a short claw at the base . . . . 2. *L. convallarioides*
1 Leaves subcordate to cordate; lip cleft about halfway into 2 linear-lanceolate divergent lobes, with a pair of teeth near the base . . . . 3. *L. cordata*

1. Listera borealis Morong

Listera borealis Morong, Bull. Torrey Bot. Club 20: 31. 1893. *Ophrys borealis* Rydb. Bull. Torrey Bot. Club 32: 610. 1905. (*Miss E. Taylor*, Fort Smith, Slave River, Hudson Bay Territory, Can., 28 June 1892.)

Northern twayblade.

Plant 7–15 cm high, glabrous below the leaves, pubescent and more or less glandular above, with 1 or 2 sheathing bracts slightly below ground level; leaves opposite to subopposite, slightly above midlength of stem; blade 1.5–3 cm long, lanceolate to ovate-elliptic, rounded to acute, glabrous; raceme 2- to 8-flowered; flowers about 1 cm long, the rachis and pedicels glandular-pubescent, the bracts 1–2.5 mm long; pedicels 3–6 mm long, ascending; flowers yellowish-green; sepals 1-nerved, reflexed, about 5 mm long, oblong-lanceolate, the lateral pair slightly longer than the dorsal one; petals narrower and shorter than the sepals, also reflexed, 1-nerved; lip pointed downward at about a 45° angle, oblong-ovate, broadest at the tip, often somewhat puberulent on the dorsal surface, finely ciliolate, 7–12 mm long, 4.5–6.5 mm wide, with 2 obtuse lobes at the apex, with rounded auricles and 2 parallel, deep green callosities leading to the triangular nectary near the base, 3-nerved, the nerves purplish; column 3–4 mm long, pointing forward and more or less arcuate over the lip; capsule 6–8 mm long, oblanceolate, erect, more or less ribbed.

Moist woods and mossy places along small streams high in the mts.; Alaska to Hudson Bay, s. to Wash., Utah, Mont. and Wyo. June–July.

Probably more common than our scant material would indicate, as the plants often grow in moist thickets where, because of their small size, they escape detection.

The erect fruits of this species will identify our northern twayblade. The remaining two species in our area have fruits spreading at maturity.

2. Listera convallarioides (Swartz) Torr.

Epipactis convallarioides Swartz, Kongl. Vetensk. Acad. Nya Handl. 21: 232. 1800. *Listera convallarioides* Torr. Compend. Fl. N. Middle States 320. 1826. *Diphryllum convallarioides* Kuntze, Revisio Gen. Pl. 2: 659. 1891. *Ophrys convallarioides* Wight, Bull. Torrey Bot. Club 32: 380. 1905. *Bifolium convallarioides* Nieuwl. Amer. Midl. Naturalist 3: 129. 1913. ("E. Terra Nova Amer. sept.")

Broad-leaved twayblade.

Plants 6–25 cm high, the stem glabrous below the leaves and becoming glandular-pubescent above, with a sheathing bract slightly below ground level; leaves opposite or subopposite, glabrous, broadly ovate to suborbicular, rounded to abruptly acute, with a reticulate pattern between the more or less prominent veins, 2–6 cm long, up to about 5 cm wide; raceme loosely 5- to 10- (15)-flowered, the floral bracts 3–6 mm long; pedicels slender, 4–7 mm long; flowers yellowish-green; sepals and petals ovate-lanceolate to linear, 4–5 mm long, 2 mm wide or slightly less; petals slightly shorter and narrower than the sepals, strongly recurved, 1-nerved; lip 8–12 mm long, 5–7 mm wide near the apex, projecting forward and almost horizontal, cuneate, shallowly notched at the apex and often appearing subcordate, usually with a tiny darker green tooth in the sinus, abruptly contracted to a narrow basal claw about 1.5 mm long; column slender, with the anther

EPIPACTIS

LISTERA

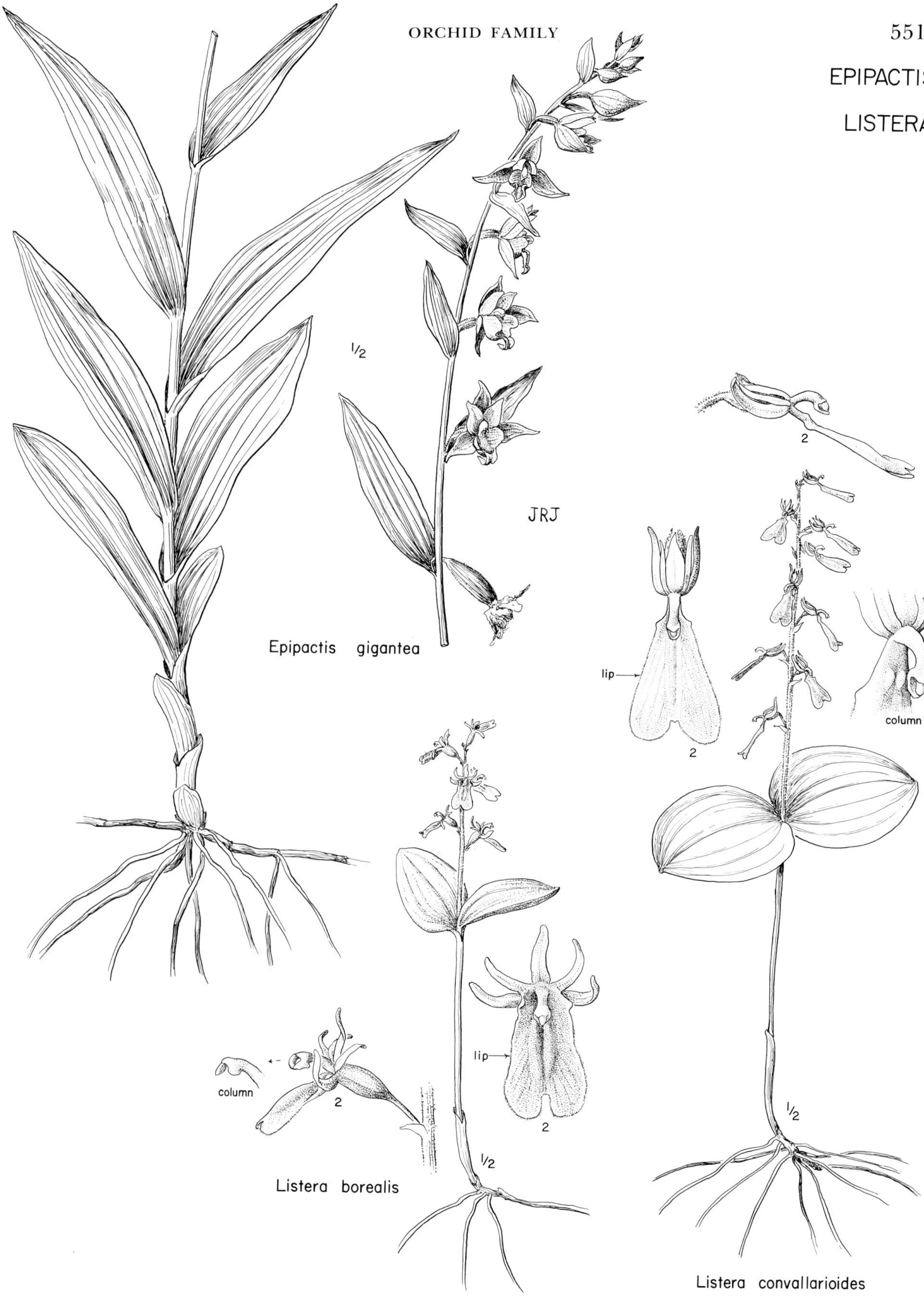

about 3 mm long; ovary slightly glandular-pubescent; capsule 6–8 mm long, narrowly elliptical, usually spreading at about 45°, only slightly ribbed; $2n = 36$.

Moist places in leaf mold in shaded areas, or in wet places along small streams high in the mts.; Alaska to Newfl., s. to Calif., Utah, Ariz., Colo., Mich., N.Y.; Asia. June–July.

The spreading narrowly elliptical fruit will separate this species from the following one, which has broader capsules.

3. Listera cordata (L.) R. Br.

Ophrys cordata L. Sp. Pl. 946. 1753. *Listera cordata* R. Br. in Aiton, Hort. Kew. ed. 2. 5: 201. 1813. *Distomaea cordata* Spenner, Fl. Friburg. 247. 1825. *Pollinirhiza cordata* Dulac, Fl. Dept. Hautes-Pyrén. 120. 1867. *Diphryllum cordatum* Kuntze, Revisio Gen. Pl. 2: 659. 1891. *Bifolium cordatum* Nieuwl. Amer. Midl. Naturalist 3: 129. 1913. (Europe.)

L. nephrophylla Rydb. Mem. New York Bot. Gard. 1: 108. 1900. *Ophrys nephrophylla* Rydb. Bull. Torrey Bot. Club 32: 610. 1905. *L. cordata* var. *nephrophylla* Hultén, Fl. Aleutian Islands 145. 1937. *L. cordata* subsp. *nephrophylla* A. Löve & D. Löve, Univ. Colorado Stud. Ser. Biol. 17: 19. 1965. (*Flodman 365,* Spanish Basin, Mont., in 1895, was the first of many specimens cited.)

Heart-leaved twayblade.

Plant 6–20 cm high, mostly glabrous throughout or the stem somewhat glandular-pubescent immediately above the leaves, with a sheathing bract below ground level; leaves opposite to subopposite at about midlength of the stem or slightly above, broadly ovate-cordate to reniform, short-mucronate at the apex, about 1–3.5 cm long and nearly as wide or somewhat wider; raceme 5- to 13-flowered, with bracts scarcely exceeding 1 mm in length; pedicels 1–3 mm long; flowers yellowish-green or purplish-green; sepals 2.5–4 mm long, about 1 mm wide, 1-nerved, ovate-oblong to linear-oblong, spreading forward over the column, the lower pair somewhat narrower; petals narrowly oblong, scarcely longer than the sepals; lip 8–10 mm long, linear-oblong, cleft one-half to two-thirds of the distance to the base into two linear-lanceolate lobes, with a subulate transverse tooth on each side near the base; column short and thick, about 1.5 mm long (including the anther); capsule broadly elliptical to nearly spherical, spreading at about 45°, often more striated than ribbed; $2n = 36, 38, 40, 42$.

In damp mossy places or in somewhat drier places than in two preceding species; Alaska to Greenl., s. to Calif., Nev., Utah, N.M., Minn., N.Y., and N.C.; Eurasia. June–July.

4. SPIRANTHES Rich. Ladies' Tresses

Small to fairly large herbs with fleshy-fascicled roots and narrow, basal or cauline leaves, reduced above to sheathing bracts; flowers in a congested terminal spike, usually in spirally twisted longitudinal rows in several ranks, mostly white, cream or greenish in ours but soon appearing sordid; sepals in ours connivent with the 2 petals and forming a hood enclosing the column and most of the lip, or the lateral sepals free and not forming part of the hood; lip sessile or short-clawed, concave and strongly grooved near the base and partly enclosing the column, simple or lobed, spreading or recurved distally; column short in ours, bearing the single anther on the back, tipped by a 2-toothed viscid rostellum; pollinia 2, with a common viscidium; capsule ellipsoid to ovoid.

A polymorphic genus as here defined (following Correll), with some 300 species found throughout trop. and temp. regions of the world. (Name from the Greek *speira,* coil, and *anthos,* flower, referring to the coil or twisted spike of flowers.)

The name "ladies' tresses" was used nearly two centuries ago, as the twisted spike has a fancied resemblance to braided hair.

1 Flowers upcurved or outcurved; sepals and lateral petals all more or less connivent to form a hood opposite the sharply deflexed lip; boreal and cordilleran species, widespread in the mts. of our range .... 1. *S. romanzoffiana*

1 Flowers widely spreading or somewhat downcurved; only the lateral petals more or less connivent with the upper sepal, the lateral sepals looser and more spreading; mainly more eastern species, rarely reaching our range .... 2. *S. cernua*

1. Spiranthes romanzoffiana Cham.

Spiranthes romanzoffiana Cham. Linnaea 3: 32. 1828. *Gyrostachys romanzowiana* MacMillan, Metasp. Minnesota Valley 171. 1892. *Orchiastrum romanzoffianum* Greene, Manual Bot. Region San Francisco Bay 306. 1894. *Ibidium romanzoffianum* H. D. House, Muhlenbergia 1: 129. 1906. *Triorchis romanzoffiana* Nieuwl. Amer. Midl. Naturalist 3: 123. 1913. (Unalaska.)

S. porrifolia Lindl. Gen. & Sp. Orchidaceous Pls. 467. 1840. *Gyrostachys porrifolia* Kuntze, Revisio Gen. Pl. 2: 664. 1891. *Orchiastrum porrifolium* Greene, Manual Bot. Region San Francisco Bay 306. 1894. *Ibidium porrifolium* Rydb. Bull. Torrey Bot. Club 32: 610. 1905. *S. romanzoffiana* var. *porrifolia* Ames & Correll, Bot. Mus. Leafl. 11: 1. 1943. (*Douglas,* "Hab. in America boreali-occidentali.")

Gyrostachys stricta Rydb. Mem. New York Bot. Gard. 1: 107. 1900. *Ibidium strictum* H. D. House, Bull. Torrey Bot. Club 32: 381. 1905. *S. stricta* A. Nels. in Coulter & Nels. New Manual Bot. Centr. Rocky Mts. 125. 1909. *Ibidium romanzoffianum* var. *strictum* F. P. Daniels. Fl. Boulder, Colo., & Vicinity 96. 1911. *Triorchis stricta* Nieuwl. Amer. Midl. Naturalist 3: 123. 1913. *S. romanzoffiana* var. *stricta* Druce, Brit. Pl. List, ed. 2. 108. 1928. (Typification recondite.)

Hooded ladies' tresses.

Plant glabrous below, somewhat glandular-pubescent above, or glabrous throughout, 0.8–2 (3) dm high, the roots fascicled, enlarged and fleshy; leaves usually several, near the base of the stem (stem rarely leafy above), linear-lanceolate to occasionally oblanceolate, to obovate in small plants, 5–10 cm long, about 0.5–1.2 cm wide, abruptly reduced above to short, lanceolate, sheathing bracts; spike dense, 3–6 (10) cm long; flowers white or cream to greenish-white, becoming sordid, aligned in 2 to 4 ranks variously spiralled to the right or left; floral bracts 10–20 (25) mm long; sepals viscid-pubescent, all connivent with the 2 lateral petals to form an

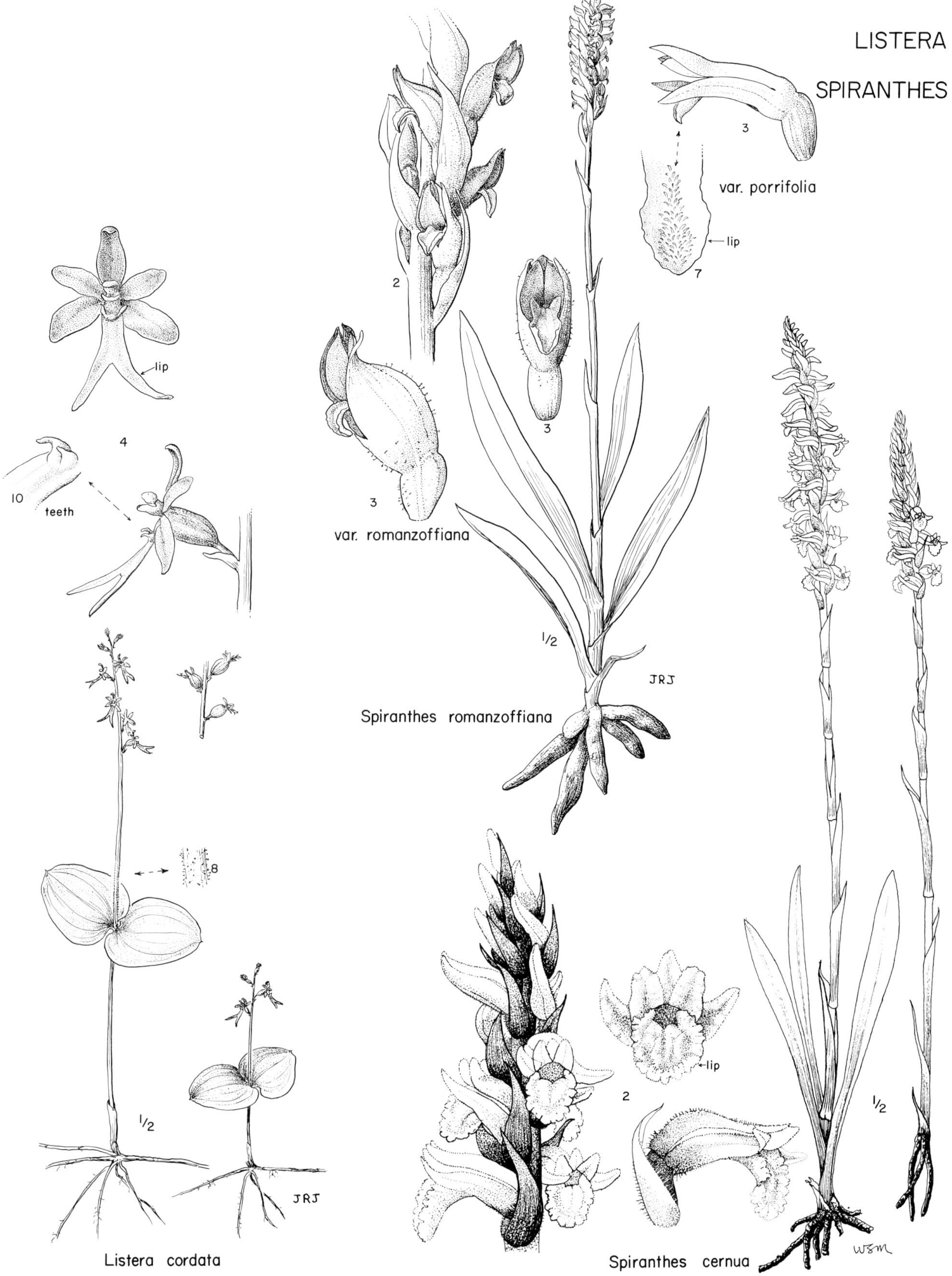
LISTERA
SPIRANTHES
3
var. porrifolia
lip
7
2
3
3
var. romanzoffiana
1/2
JRJ
Spiranthes romanzoffiana
lip
4
10
teeth
8
1/2
JRJ
Listera cordata
lip
2
1/2
Spiranthes cernua

open-tubular hood 7–12 mm long, the lateral sepals often with reflexed tip; lip sharply deflexed and slightly recurved, about as long as the sepals, the basal half concave with erect margins, abruptly tapering to a rounded tip or evidently constricted above the concavity, the tip erose; column 2–4 mm long, with a slender, forked terminal beak, about as long as the brownish viscidium, the anther shorter than the beak; $2n = 60$.

Usually in wet mt. meadows or boggy places along small mt. streams; Alaska to Newfl., s. to Calif., Nev., Utah, Ariz., and N.M.; Nebr., Iowa and N.Y. July–early Sept.

Our two varieties are rather distinctive and are sometimes treated as species.

1 Flowers white to cream; lip contricted below the tip, without prominent basal callosities or terminal puberulence; range of the species .... var. *romanzoffiana*

1 Flowers cream to greenish-white; lip nearly triangular, not constricted below the tip or only slightly so, slightly puberulent at the tip, the base with prominent callosities; Wash. to Oregon, Calif., and Utah .... var. *porrifolia* (Lindl.) Ames & Correll

Pollinating insects begin at the bottom of the flowering spike and work upward. A bee is treated to nectar at the bottom; further up the bee encounters freshly opened flowers with closed nectaries and hidden stigmas. The bee becomes laden with pollinia in his search, and flies to the bottom of another spike, where the elevated and sticky stigma of an older flower is waiting.

2. Spiranthes cernua (L.) Rich.

Ophrys cernua L. Sp. Pl. 946. 1753. *Spiranthes cernua* Rich. Mém. Mus. Hist. Nat. 4: 59. 1818. *Gyrostachys cernua* Kuntze, Revisio Gen. Pl. 2: 664. 1891. *Ibidium cernuum* H. D. House, Bull. Torrey Bot. Club 32: 381. 1905. *Triorchis cernua* Nieuwl. Amer. Midl. Naturalist 3: 122. 1913. ("Virginia, Canada.")

Nodding ladies' tresses.

Similar to *S. romanzoffiana* in general aspect, but averaging taller and more slender, up to 5 dm tall, and often with more slender roots; leaves up to 25 cm long and 2 cm wide; inflorescence not so dense and averaging more elongate, up to 15 cm long, its bracts averaging smaller, 8–15 mm long; flowers widespreading or somewhat downcurved so as to appear nodding; only the upper sepal and the lateral petals connivent, the lateral sepals loose and more spreading; lip arcuate-recurved, mostly ovate-oblong, neither strongly constricted near the middle nor abruptly tapering from the middle; $2n = 30, 60$.

Moist or wet places at lower elevs.; species principally of e. and c. U.S. and adj. Can., rarely and irregularly w. to Utah (Salt Lake Co.) and N.M. July–Aug.

5. GOODYERA R. Br.

Rhizomatous, scapose herbs from fleshy fibrous roots, with alternate and basal leaves often forming rosettes; leaves usually somewhat mottled; flowers small, dull white to greenish to pink, somewhat secund and spiralled in a slender bracteate raceme; upper sepal and lateral petals connivent, forming a hood pointing forward and covering the lip; lateral sepals free, not joined below the lip, the lip sessile, without lobes in ours, deeply saccate, with erect sides, tip pointed and slightly reflexed; column short, with a slender beak; rostellum 2-pronged, stigmatic area almost basal and circular; single anther short-stalked, attached near the base of the column, broadly concave and nearly petaloid; pollinia 2, attached to a common narrow viscidium between the prongs of the rostellum; capsule erect, ovoid to ellipsoid.

A genus of about 25 species widely distributed in temp. to subarctic regions of the N. Hemisphere and southward into the tropics. (Named in honor of the English botanist John Goodyear, 1592–1664.)

1. Goodyera oblongifolia Raf.

Goodyera oblongifolia Raf. Herb. Rafinesquianum 76. 1833. ("Oregon mountains.")

Spiranthes decipiens Hook. Fl. Boreali-Amer. 2: 203. 1839. *Orchiodes decipiens* Kuntze, Revisio Gen. Pl. 2: 675. 1891. *Peramium decipiens* Piper, Contr. U.S. Natl. Herb. 11: 208. 1906. *Epipactis decipiens* Ames, Orchidaceae 2: 261. 1908. *G. decipiens* Ames, Orchid. U.S. & Canad. 36. 1924. ("Hab. Lake Huron, *Dr. Todd.* Lake Simcoe, and near the sources of the Columbia River. *Drummond.* Fort Vancouver and Straits of De Fuca. *Dr. Scouler.*")

G. menziesii Lindl. Gen. & Sp. Orchidaceous Pls. 492. 1840. *Orchiodes menziesii* Kuntze, Revisio Gen. Pl. 2: 675. 1891. *Peramium menziesii* Morong, Mem. Torrey Bot. Club 5: 124. 1894. (*Menzies, Douglas,* "Hab. in Americae septentrionalis orâ occidentali.")

G. oblongifolia var. *reticulata* B. Boivin, Canad. Field-Naturalist 65: 20. 1951. (*F. Fyles,* Victoria, B.C., 20 Aug. 1914.)

Rattlesnake plantain.

Plant scapose, densely glandular-pubescent above, 2–3.5 dm tall, with rhizomes; leaves in basal rosettes, with broad, winged petioles, the blades subcoriaceous, ovate-lanceolate to elliptic-lanceolate, 3–8 cm long, up to 3.5 cm wide, the midrib usually white, or white lines on each side of the midrib, sometimes with white mottling or stripes throughout in a reticulated pattern; scape with 2–6 inconspicuous, nongreen, membranous, glandular-pilose sheathing bracts; raceme lax to closely flowered, (4–10) 14 cm long, the flowers mostly secund below and somewhat spiralled above, subsessile, mostly pale greenish-white, segments 1-veined, pubescent, subtended by ovate-lanceolate bracts scarcely longer than the ovary; dorsal sepal connivent with the fused lateral petals, forming a forward-pointing hood up to 10 mm long; lateral sepals free, somewhat shorter than the hood, covering and concealing the lip; lip shorter than the lateral sepals, swollen and deeply concave at the base, the tip slightly recurved, with a beak up to 8 mm long; column 3.5–5 mm long, nearly concealed by the lip, the stigmatic area somewhat circular; anther 2–2.5 mm long, abruptly short-acuminate, longer than the filament; capsule about 1 cm long, pubescent; $2n = 22$.

In dry or drying mossy places, usually in deep wooded areas, late flowering or early fruiting plants often in dry crusted humus; Alaska to N.S., s. to most of w. U.S. to Mex., and to Mich., Wisc., Minn., and Me. July–Aug.

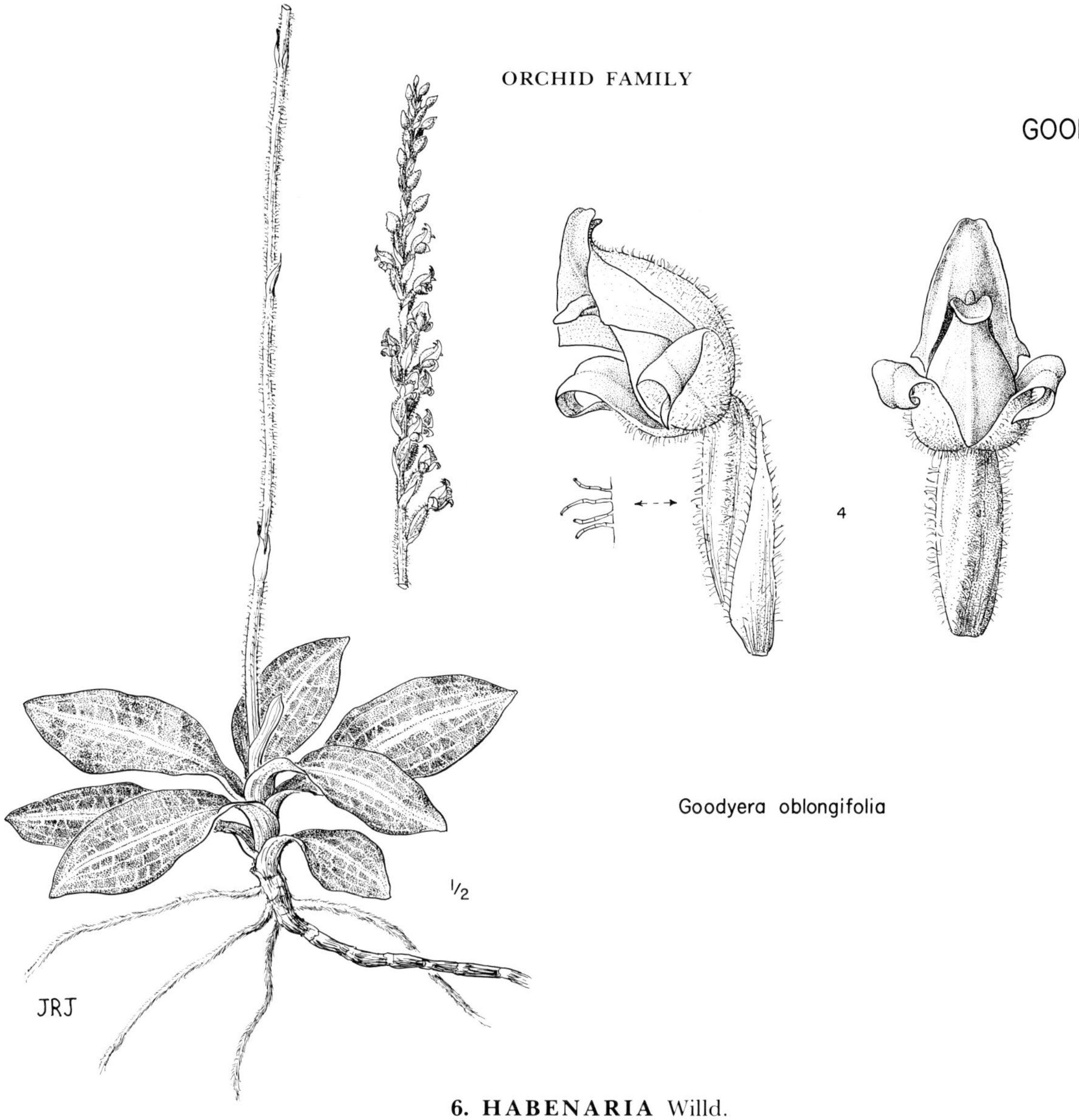

Goodyera oblongifolia

6. HABENARIA Willd.

Glabrous perennials from rootstocks or more often from tuberous or fleshy roots, the leaves sometimes 1 or 2 and basal but usually several and cauline and gradually reduced above, mostly with a sheathing base, often withering by anthesis on plants in dry places; flowers few to many in loose to congested, bracteate spikes or spikelike racemes, white, yellowish-green or green, sometimes purplish-tinged; upper sepal usually erect or concave and somewhat hooded, often connivent with the petals; lateral sepals spreading to reflexed; lip pendent to upcurved, entire to lobed, or even fringed (not in ours); column short, the pollen sacs separated by a broad connective, the anther sacs sometimes opening nearly laterally or in front; pollen powdery and loose to tightly massed in 2 pollinia; stigma concave; spur conspicuous and well developed, from saccate and didymous (scrotiform) to cylindric, straight to curved, sometimes twice as long as the lip, usually nearly translucent toward the base.

A genus of perhaps more than 500 species of terrestrial orchids widely distributed in temp. to trop. regions. (Name from the Latin *habena,* reins, in reference to rein-like appendages of the petals and the lip of some species.)

Most botanists today treat the genus as it is here. Our species have been treated under *Coeloglossum, Lysiella, Limnorchis, Orchis, Piperia,* and *Platanthera.*

The species in the section *Limnorchis* (*H. dilatata, H. hyperborea, H. saccata,* and *H. sparsiflora*) apparently hybridize freely. This creates the perplexing array of hybrid variants which makes them so difficult for the taxonomist.

Pollination is probably carried out by insects provided with a proboscis.

Reference: Correll, D. S. *Habenaria* in western North America. Leafl. W. Bot. 3: 233–247. 1943.

1 Plants more or less distinctly scapose, the leaves basal or on the lower third of the stem; flowers greenish or nearly white.
  2 Plants with a solitary leaf or rarely 2, not withering at anthesis; sepals 3- to many-nerved; flowers with pedicels; plants usually in moist areas . . . . 2. *H. obtusata*

2 Plants with 2–4 leaves, withering by anthesis; sepals 1-nerved; flowers subsessile, without obvious pedicels; plants of dry areas .... 7. *H. unalascensis*

1 Plants leafy-stemmed, the leaves all cauline, distributed along at least the lower half of the stem; flowers often white.

3 Lip unequally tridentate at the apex, pendent; bracts subtending flowers mostly more than 2 cm long, green and often leaflike .... 1. *H. viridis*

3 Lip without lobes or teeth at the tip, sometimes with shallow lobes behind the tip; bracts shorter, usually rather inconspicuous, or if prominent then less than 2 cm long.

4 Spur scrotiform, somewhat didymous near the tip; lip not at all enlarged toward the base; raceme usually open .... 6. *H. saccata*

4 Spur not scrotiform; lip usually widened toward the base, sometimes only slightly so; raceme mostly congested.

5 Flowers mostly white, rarely pale green; lip rhombic-lanceolate, prominently and rather abruptly dilated at the base .... 3. *H. dilatata*

5 Flowers greenish; lip linear to broadly lanceolate, not prominently dilated at the base.

6 Raceme usually elongated and laxly flowered; lip linear to linear-elliptic, often with a fleshy ridge in the center below the middle .... 4. *H. sparsiflora*

6 Raceme usually short and densely flowered; lip linear lanceolate to lanceolate or slightly broader, without a callosity .... 5. *H. hyperborea*

1. Habenaria viridis (L.) R. Br.

Satyrium viride L. Sp. Pl. 944. 1753. *Habenaria viridis* R. Br. in Aiton, Hort. Kew. ed. 2. 5: 192. 1813. *Orchis viridis* Pursh, Fl. Amer. Sept. 587. 1814. *Gymnadenia viridis* Sprengel, Mém. Mus. Hist. Nat. 4: 57. 1818. *Platanthera viridis* Finet, Bull. Soc. Bot. France 47: 284. 1900. (Europe.)

Orchis bracteata Muhl. ex Willd. Sp. Pl. 4: 34. 1805. *Satyrium bracteatum* Pers. Syn. Pl. 2: 507. 1807; not Lindl. 1813. *Satyrium bracteale* Salisb. Trans. Hort. Soc. London 1: 290. 1812. *H. bracteata* R. Br. in Aiton, Hort. Kew. ed. 2. 5: 192. 1813. *Gymnadenia bracteata* Presl, Reliquiae Haenkeanae 92. 1830. *Peristylus bracteatus* Lindl. Gen. & Sp. Orchidaceous Pls. 298. 1835. *Platanthera bracteata* Torr. Fl. State New York 2: 279. 1843. *Coeloglossum bracteatum* Parl. Fl. Ital. 3: 409. 1850. *H. viridis* var. *bracteata* Reichenb. ex A. Gray, Manual Bot. ed. 5. 500. 1867. *Coeloglossum viride* var. *bracteatum* K. Richter, Pl. Eur. 1: 278. 1890. *Peristylus viridis* var. *bracteatus* Kurtz, Bot. Jahrb. Syst. 19: 408. 1894. *Coeloglossum viride* subsp. *bracteatum* Hultén, Fl. Alaska & Yukon 3: 472. 1943. *H. viridis* subsp. *bracteata* R. T. Clausen, Cornell Univ. Agric. Exp. Sta. Mem. 291: 11. 1949. (Pa.)

Orchis virescens Muhl. ex Willd. Sp. Pl. 4: 37. 1805. *Satyrium virescens* Pers. Syn. Pl. 2: 507. 1807. ?*Orchis flava* var. *virescens* Jacob Green, Cat. Pl. N.Y. 60. (extract from Trans. Soc. Promot. Useful Arts 3). 1814, nomen nudum. *Habenaria virescens* Sprengel, Syst. Veg. 3: 688. 1826. *Peristylus virescens* Lindl. Gen. & Sp. Orchidaceous Pls. 298. 1835. *Perularia virescens* A. Gray, Bot. Gaz. 5: 63. 1880. *Habenaria flava* var. *virescens* Fern. Rhodora 23: 148. 1921. *Coeloglossum viride* var. *virescens* Luer, Native Orchids U.S. & Canada 172. 1975. (Pa.)

Satyr rein orchid.

Plant glabrous, leafy-stemmed, 2–5 dm tall, with fleshy roots; leaves sheathing, becoming reduced and narrowed above, basal ones without blades, the lower cauline leaves with rounded, oblong-lanceolate blades 3–7 cm long, the middle cauline blades larger, up to about 10 cm long, usually acute; racemes 8- to many-flowered, with conspicuous green, lanceolate bracts, the lower ones often leaflike, 2–4 cm long, reduced above, but the uppermost usually longer than the subtended flower; flowers subsessile, greenish; upper sepal broadly ovate to ovate-lanceolate, 4–6 mm long, concave, the lateral sepals broadly lanceolate, concave, often slightly longer than the upper ones, rounded or obtuse, usually turned upward, connivent with the petals and upper sepal and forming a hood over the lip, the petals, column and base of the lip sometimes purplish tinged; lip oblong, pendent, 7–10 mm long, about 3 mm broad, with a slight median keel, tridentate at the tip with the middle tooth or lobe the shortest and sometimes nearly wanting; spur scrotiform, up to half as long as the lip, pale and translucent; column not quite 1 mm long; stamen conspicuous, the pollen sacs about 1.5 mm long, the connective less than 1 mm wide; pollinium 1 per sac, the pollen powdery; capsule slightly more than 1 cm long, glabrous, ovate; $2n = 40$.

Moist, often drying places in woods and open mt. slopes, occasionally in mt. meadows; Alaska to Newfl., s. to Wash., Utah, Wyo., Colo., S.D., Ill., W. Va., and in the mts. to N.C.; Eurasia. June–early Aug.

The North American plants, var. *bracteata* (Muhl. ex Willd.) Reichenb. ex A. Gray, have longer bracts than the Old World var. *viridis*.

2. Habenaria obtusata (Banks ex Pursh) Richardson

Orchis obtusata Banks ex Pursh, Fl. Amer. Sept. 588. 1814. *Habenaria obtusata* Richardson in Franklin, Narr. Journey Shores Polar Sea, Bot. Appendix 750. 1823. *Platanthera obtusata* Lindl. Gen. & Sp. Orchidaceous Pls. 284. 1835. *Lysiella obtusata* Rydb. Mem. New York Bot. Gard. 1: 104. 1900. (*Hutchinson,* "On Hudson's Bay, near Ft. Albany.")

Northern small bog orchid.

A glabrous scapose plant usually producing a single leaf (rarely 2), 1 or 2 sheathing bracts below, 0.8–2.5 dm tall; leaf obovate to oblanceolate, gradually tapering to the sheathing base, 4–15 cm long, up to about 4 cm wide; scape usually leafless or sometimes with a sheathing bract near the middle; raceme loosely 3- to 12-flowered with pedicels 1–5 mm long, bracts greenish to yellowish, more or less prominent below and up to nearly 1.5 cm long and becoming much reduced above to less than 2 mm long; flowers greenish-white; sepals and petals mostly broadly ovate to lanceolate, 3.5–5.5 mm long, 1.5–4 mm wide, the upper sepal triangular-reniform to suborbicular, nearly erect, the petals connivent with the upper sepal, outlining the entrance to the orifice; lip

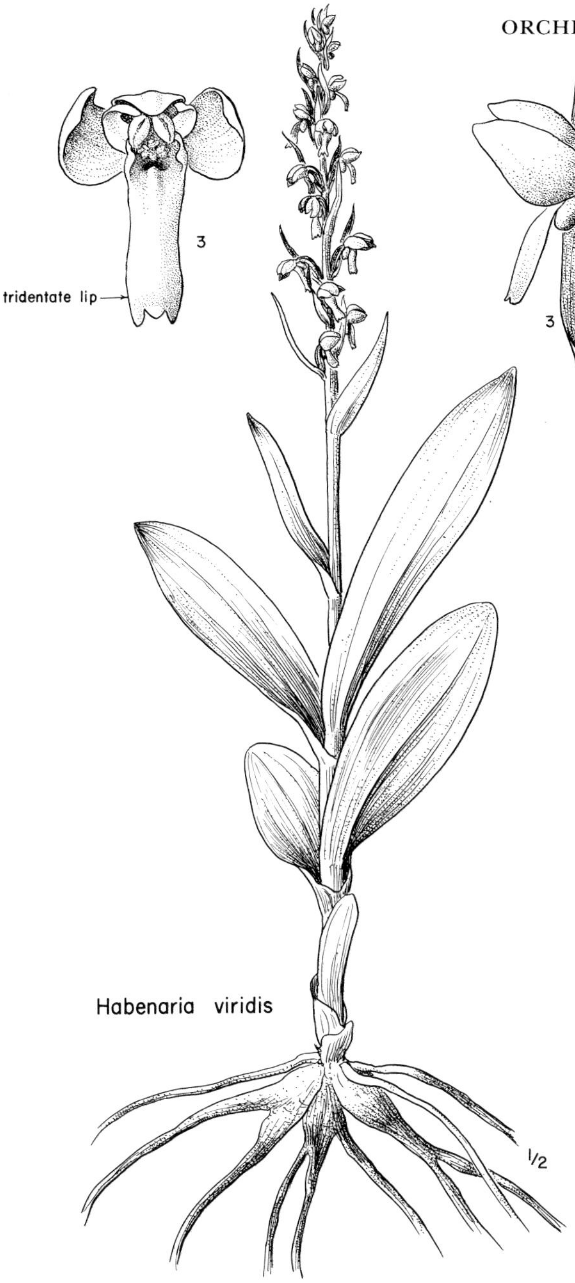

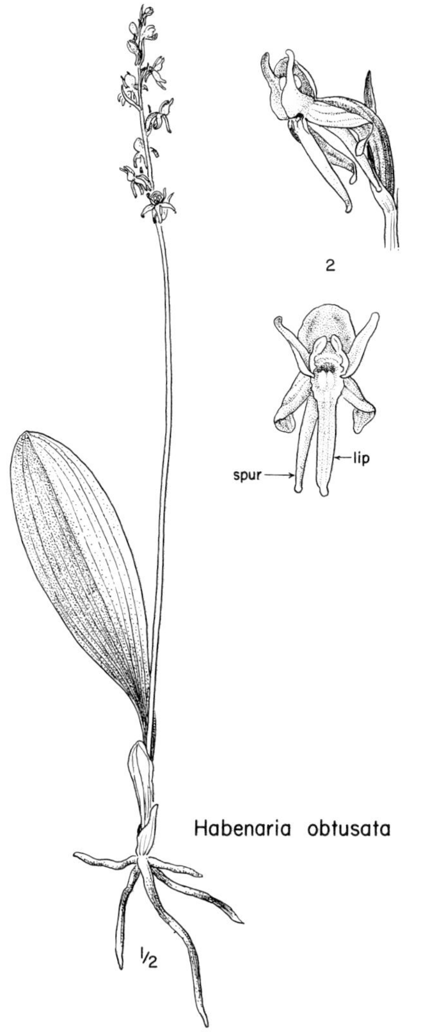

narrowly ligulate, not exceeding 1.5 mm across, 5–9 mm long, pendent, tip becoming slightly upturned, with 2 small calluses near the base; spur tapered from the broad base, 3–8 mm long, often cream to nearly white; column with triangular-lanceolate processes 2–3.5 mm long, projecting upward, connective about 1 mm wide; pollinia 1 per anther sac, the pollen loosely aggregated, the viscidium discoid; capsule about 1 cm long, glabrous, lanceolate to ovate; $2n = 42$.

In damp or wet places in forests, marshes, bogs, meadows and along streambanks; Alaska to Newfl., s. to B.C., ne. Oregon, Utah, Mont., Colo., Minn., Wisc. and N.Y.; Europe. June–July.

3. Habenaria dilatata (Pursh) Hook.

Orchis dilatata Pursh, Fl. Amer. Sept. 588. 1814. *Habenaria dilatata* Hook. Exot. Fl. 2: pl. 95. 1825. *H. borealis* var. *dilatata* Cham. Linnaea 3: 28. 1828. *Platanthera dilatata* L. C. Beck, Bot. N. & Middle States 347. 1833. *Platanthera hyperborea* var. *dilatata* Kraenzlin, Gen. & Sp. Orchidaceous Pls. 640. 1899. *Limnorchis dilatata* Rydb. in Britton, Manual Fl. N. States & Canada 294. 1901. (*Colmaster,* "In Labrador.")

H. borealis Cham. Linnaea 3: 28. 1828. *Limnorchis borealis* Rydb. Bull. Torrey Bot. Club 28: 621. 1901. *H. dilatata* var. *borealis* Muenscher, Fl. Whatcom County, Washington 76. 1941. (*Chamisso,* "Unalaschca.") = var. *dilatata.*

H. borealis var. *albiflora* Cham. Linnaea 3: 28. 1828. *Platanthera dilatata* var. *albiflora* Ledeb. Fl. Ross. 4: 71. 1853. *H. dilatata* var. *albiflora* Correll, Leafl. W. Bot. 3: 238. 1943. *Limnorchis dilatata* subsp. *albiflora* A. Löve & Simon, Southw. Naturalist 13: 339. 1968. (Unalaska.)

Platanthera gracilis Lindl. Gen. & Sp. Orchidaceous Pls. 288. 1835. *H. gracilis* S. Wats. Proc. Amer. Acad. Arts 12: 277. 1877; not Hook. 1825. *H. dilatatiformis* Rydb. Bull. Torrey Bot. Club 24: 189. 1897. *Limnorchis dilatatiformis* Rydb.

Mem. New York Bot. Gard. 1: 105. 1900. *Limnorchis gracilis* Rydb. Bull. Torrey Bot. Club 28: 627. 1901. *Platanthera stricta* var. *gracilis* Hultén, Fl. Alaska & Yukon 3: 483. 1943. *H. saccata* var. *gracilis* B. Boivin, Naturaliste Canad. 94: 636. 1967. *Platanthera hyperborea* var. *gracilis* Luer, Native Orchids U.S. & Canada 232. 1975. (*Menzies,* "Hab. in ora occidentali Americae septentrionalis.") = var. *dilatata.*

Platanthera leucostachys Lindl. Gen. & Sp. Orchidaceous Pls. 288. 1835. *H. leucostachys* S. Wats. Bot. Calif. 2: 134. 1880. *Platanthera hyperborea* var. *leucostachys* Kraenzlin, Orchidacearum Gen. & Sp. 640. 1899. *Limnorchis leucostachys* Rydb. Mem. New York Bot. Gard. 1: 106. 1900. *H. dilatata* var. *leucostachys* Ames, Orchidaceae 4: 71. 1910. *Limnorchis dilitata* var. *leucostachys* (attr. to Ames by) J. P. Anders. Iowa State Coll. J. Sci. 19: 189. 1945. *Platanthera dilatata* var. *leucostachys* Luer, Native Orchids U.S. & Canada 225. 1975. (*Douglas,* "Hab. in ora occidentali Americae septentrionalis.")

Platanthera graminea Lindl. Gen. & Sp. Orchidaceous Pls. 289. 1835; not *H. graminea* Sprengel, 1826. *Limnorchis graminifolia* Rydb. Bull. Torrey Bot. Club 28: 627. 1901. *H. graminifolia* J. K. Henry, Fl. S. Brit. Columb. & Vancouver Island 92. 1915. (*Menzies,* "Hab. in ora occidentali Americae septentrionalis.") A linear-leaved, coastal phase of var. *leucostachys.*

Limnorchis leptoceratitis Rydb. Bull. New York Bot. Gard. 2: 162. 1901. *H. leptoceratitis* J. K. Henry, Fl. S. Brit. Columb. & Vancouver Island 92. 1915. (*R. S. Williams,* Bennett City, Yukon Territory, 25 Aug. 1899.) = var. *dilatata.*

Limnorchis leucostachys var. *robusta* Rydb. Bull. Torrey Bot. Club 28: 626. 1901. *H. leucostachys* var. *robusta* J. K. Henry, Fl. S. Brit. Columb. & Vancouver Island 93. 1915. (*R. S. Vasey 77,* Washington, 1889.) = var. *dilatata.*

White bog orchid.

A tall leafy-stemmed plant, glabrous throughout, 1.5–8 (10) dm tall; leaves several, gradually reduced above, sheathing, the lower ones oblong-elliptic to lanceolate or occasionally oblanceolate, up to 20 cm long and 5.5 cm wide, the upper leaves more acute to acuminate and somewhat longer than the lower ones; raceme spiciform, 5–30 cm long, from lax to commonly dense and cylindrical; floral bracts from somewhat shorter than the flower to several times longer; flowers usually white or yellowish-white, occasionally tinged with green, with a spicy fragrance of cinnamon; upper sepal ovate, elliptic, or ovate-lanceolate, somewhat hooded and connivent with the petals, mostly 3-nerved, 4–6 mm long, the lower sepals elliptic-lanceolate, 3-nerved, 5–8 mm long, spreading, and usually twisted to present the lower edge to the front; petals 1- or 2-nerved, subequal to the upper sepal and connivent with it, forming a hood; lip pendent, 5–10 mm long, 2–5 mm across the base, with 5 to 7 prominent nerves and indistinct ridges between the nerves, margins finely serrulate, broadened above the base; spur cylindric to clavate, usually curved, varying from half as long to nearly twice as long as the lip; column about 2 mm long, the base nearly surrounding the orifice of the spur; pollen sacs scarcely separated by the connective, with 2 pollinia per sac, with an oblong viscidium; $2n = 42$.

Wet places in bogs, marshes in open or shaded woods, meadows, about mt. springs and along streams and lakes; Alaska to Greenl., s. to Calif., Nev., Utah, N.M., S.D., Mich., Pa., and N.Y. June–Aug.

This is one of the most polymorphic species in the genus in our area. Three varieties are recognized in our area as follows:

1 Spur usually one-half to two-thirds the length of the lip; plants usually less than 6 dm tall; raceme dense, flowers large; Alaska, B.C. to Alta., nw. U.S. to Utah, Wyo., and Colo. .... var. *albiflora* (Cham.) Correll

1 Spur from about as long to much longer than the lip; plants up to 10 dm tall.

2 Spur strongly curved, usually about one and one-half times as long as the lip or longer; usually found at high elevs.; Alaska to Calif., e. to Mont., Idaho and Utah; the most common variety in our area . . var. *leucostachys* (Lindl.) Ames

2 Spur not strongly curved, about same length as the lip; usually found at middle elevs.; widespread, but infrequent in our area .... var. *dilatata*

Many specimens are difficult to assign to any of the above varieties because of the variations in the length of the spur. This species probably crosses with *H. saccata* and *H. hyperborea.*

4. Habenaria sparsiflora S. Wats.

Habenaria sparsiflora S. Wats. Proc. Amer. Acad. Arts 12: 276. 1877. *Platanthera sparsiflora* Schlechter, Bull. Herb. Boissier 7: 538. 1899. *Limnorchis sparsiflora* Rydb. Bull. Torrey Bot. Club 28: 631. 1901. (*Bolander,* near Mariposa Grove, Calif.; lectotype by Rydberg.)

Limnorchis laxiflora Rydb. Bull. Torrey Bot. Club 28: 630. 1901. *H. laxiflora* Parish, Pl. World 20: 209. 1917. *H. sparsiflora* var. *laxiflora* Correll, Leafl. W. Bot. 3: 245. 1943. (*T. Howell,* Coast Mts., Oregon, in 1884.)

H. aggregata Howell, Fl. N.W. Amer. 628. 1902. *Limnorchis aggregata* Frye & Rigg, N. W. Fl. 114. 1912. ("Coast Mountains of southern Oregon.")

Canyon habenaria.

A glabrous plant 3–6 dm tall from a fascicle of more or less fleshy roots; leaves several, sheathing, with 1 or 2 sheathing bracts below, mostly on the lower half of the stem, narrowly lanceolate to lanceolate, often as long as 25 cm and mostly 1–3 cm broad, sometimes bluntly obtuse, gradually becoming reduced above; raceme laxly flowered, elongate, 7–15 (20) cm long; lower bracts sometimes longer than the flowers, becoming shorter above; flowers greenish; upper sepal broadly ovate to nearly orbicular, 6–8 mm long, mostly 3-nerved, concave and connivent with the upper petals to form a hood, the lateral sepals spreading, oblong-lanceolate, 6–10 mm long, 3-nerved; lip pendent with a fleshy thickening near the base, linear to linear-lanceolate, 6–10 mm long; spur cylindric to slightly clavate, becoming abruptly narrowed at the tip, about same length as the lip, usually slightly curved; column prominent, more than half as long as the upper sepal, the pollen sacs about 2 mm long, well separated by the connective.

Moist or wet soil in mt. meadows, marshes, bogs in open or dense woods, often about mt. springs or seepage slopes; Wash., Oregon, to Baja Calif., e. to Nev., Utah, Ariz., and N.M. June–July.

We have two varieties, both widespread in our area, separated as follows:

1 Column large, with a broad connective; lip about 8 mm long; spur about 10 mm long, cylindric; middle as well as upper elevs. in the mts. .... var. *sparsiflora*

1 Column small, usually without a broad connective; lip about 6 mm long; spur about 8 mm long, slightly clavate; at high elevs., up to 3350 m (11,000 ft.) .... var. *laxiflora* (Rydb.) Correll

5. Habenaria hyperborea (L.) R. Br.

Orchis hyperborea L. Mantissa Pl. 121. 1767. *Habenaria hyper-*

HABENARIA

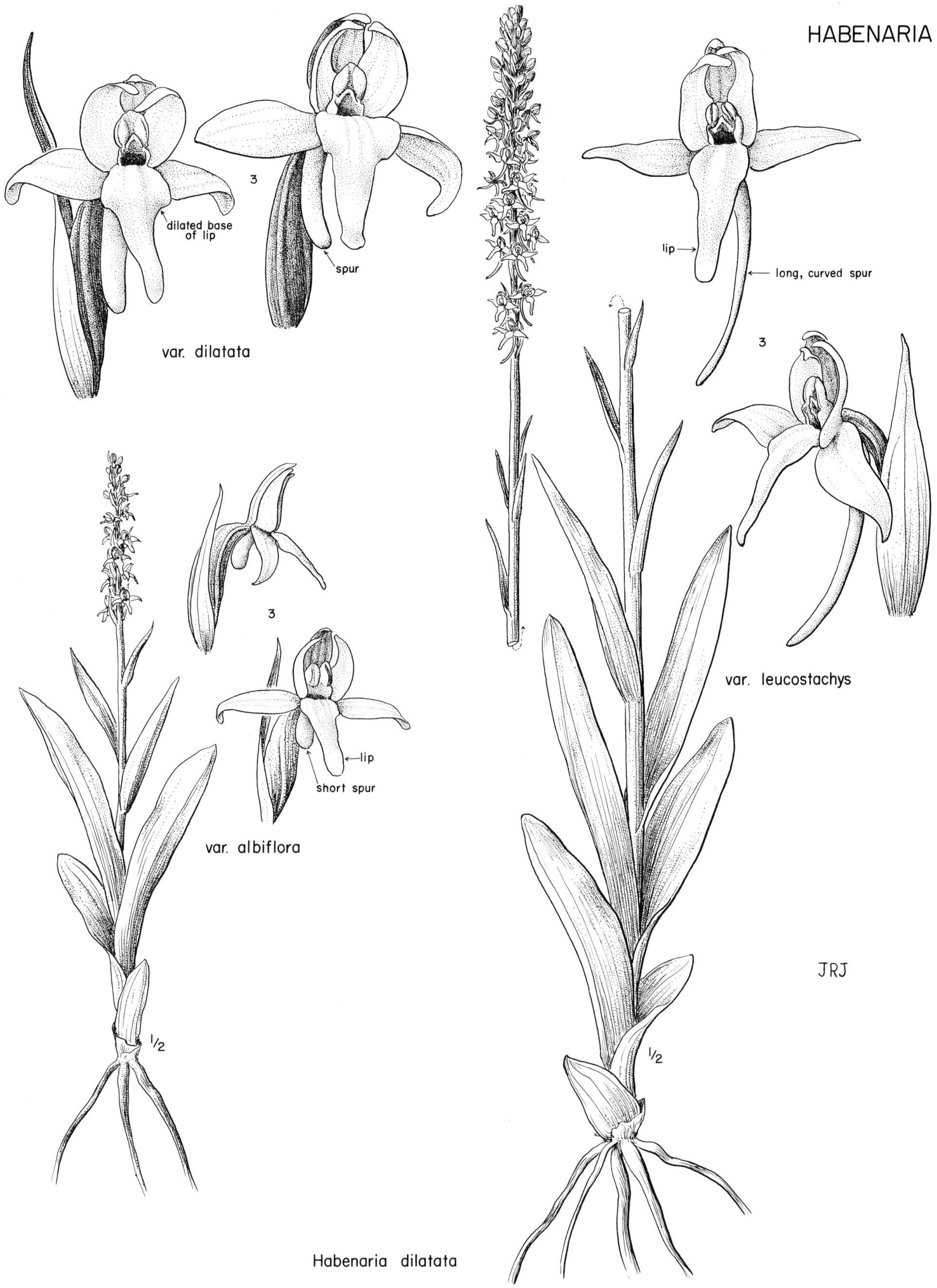

Habenaria dilatata

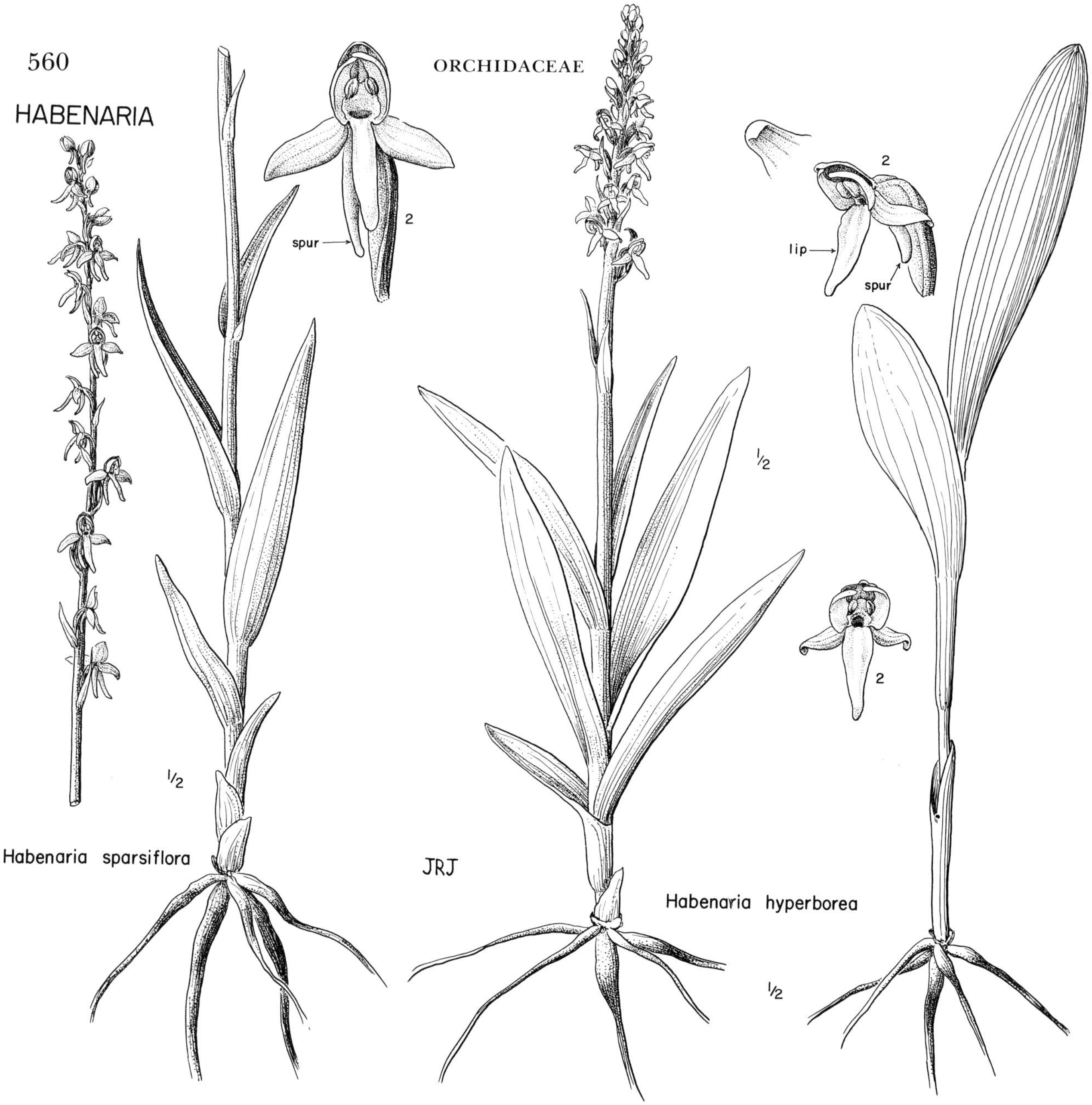

borea R. Br. in Aiton, Hort. Kew. ed. 2. 5: 193. 1813. *Platanthera hyperborea* Lindl. Gen. & Sp. Orchidaceous Pls. 287. 1835. *Limnorchis hyperborea* Rydb. Mem. New York Bot. Gard. 1: 104. 1900. (*Koenig*, Iceland.)

H. borealis var. *viridiflora* Cham. Linnaea 3: 28. 1828. *Platanthera dilatata* var. *viridiflora* Ledeb. Fl. Ross. 4: 71. 1853. *Limnorchis viridiflora* Rydb. Bull. Torrey Bot. Club 28: 616. 1901. *H. viridiflora* (incorrectly attributed to Rydb. by) J. K. Henry, Fl. S. Brit. Columb. & Vancouver Island 92. 1915; not Span. 1841. *H. septentrionalis* Tidestrom in Tidestrom & Kittell, Fl. Arizona & New Mex. 731. 1941. *Platanthera hyperborea* var. *viridiflora* Luer, Native Orchids U.S. & Canada 233. 1975. (*Chamisso*, "Unalaschca.")

Northern green habenaria.

Plant slender to stout, glabrous throughout, from a fascicle of fleshy roots, 1.5–10 dm tall; leaves several, mostly on the lower half of the stem, linear-lanceolate to oblanceolate, up to 30 cm long and about 4.5 cm wide, blunt to slenderly acute; flowers mostly numerous in a lax or dense spicate raceme 10–20 cm long, green or yellowish-green, rarely whitish green, often somewhat purplish-tinged or veined, the lower floral bracts often 2–3 times longer than the flowers, becoming reduced above to bracts shorter than the flowers; upper sepal ovate-elliptic to nearly orbicular, 4–6 mm long, 3-nerved, connivent with the upper petals to form a hood, the lower sepals spreading, 3-nerved, ovate-lanceolate, 4–6 mm long, the sepals and petals about equal; lip narrowly lanceolate to somewhat narrowly rhombic-elliptic, somewhat fleshy, usually upcurved or projecting forward, 4–7 mm long; spur cylindric to somewhat clavate, usually somewhat curved, usually shorter to barely longer than the lip; column nearly half as long as the upper sepal, the pollen sacs 1–1.5 mm long, well separated by the connective; pollen powdery, loose, in 1 pollinium per sac, the viscidium oblong-oval; capsule about 1.5 cm long; $2n = 42, 84$.

HABENARIA

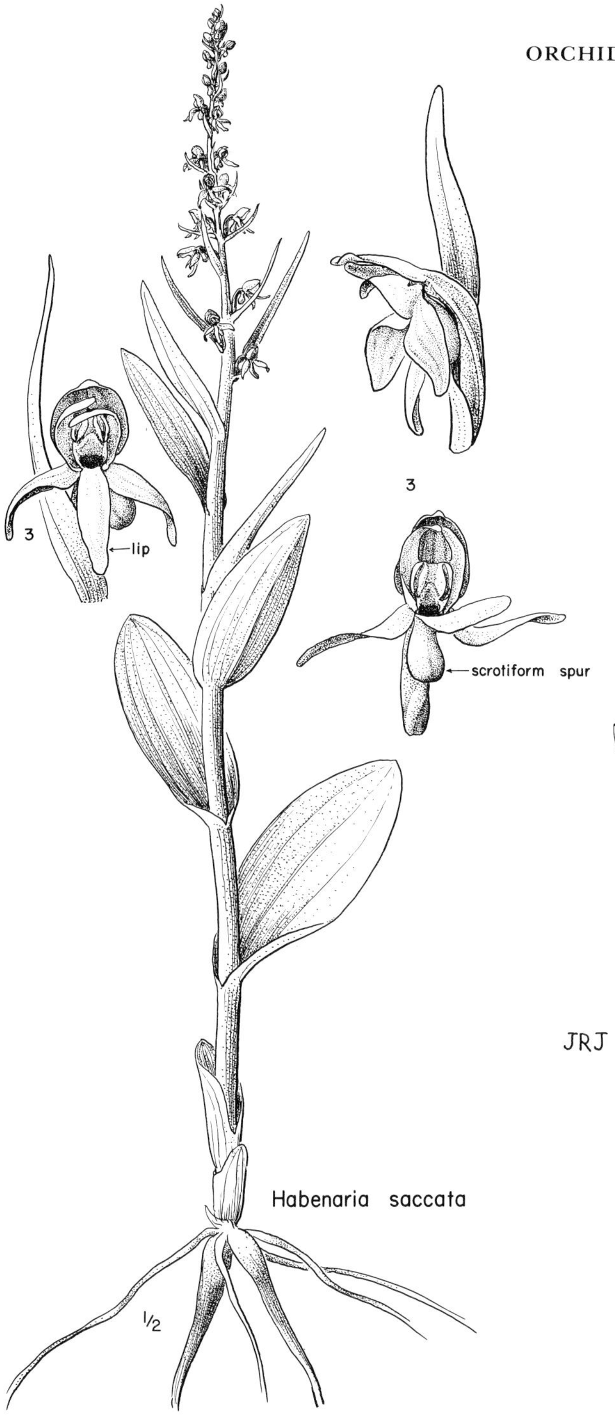

Moist or wet soil in meadows, bogs, thickets, coniferous forests, along streams, usually at high elevs.; Alaska to Newfl., Greenl., Iceland, s. to Calif., Nev., Utah, Ariz., N.M., Nebr., Minn., Ind., Pa. and N.Y. June–Aug.

This species is probably even more polymorphic than *H. dilatata*. Mention has already been made of hybrids within the *Limnorchis* section of the genus.

6. **Habenaria saccata** Greene

Habenaria saccata Greene, Erythea 3: 49. 1895. *Platanthera saccata* Hultén, Madroño 19: 223. 1968. *Limnorchis saccata* A. Löve & Simon, Southw. Naturalist 13: 339. 1968. (*Mrs. Austin,* Lassen Creek, Modoc Co., Calif., in 1894.)

Platanthera stricta Lindl. Gen. & Sp. Orchidaceous Pls. 288. 1835. *H. stricta* Rydb. Bull. Torrey Bot. Club 24: 189. 1897; not Ridley, 1885. *Limnorchis stricta* Rydb. Mem. New York Bot. Gard. 1: 105. 1900. *H. neomexicana* Tidestrom in Tidestrom & Kittell, Fl. Arizona & New Mex. 731. 1941. (*Douglas,* "Hab. in America boreali-occidentali.")

Limnorchis purpurascens Rydb. Bull. Torrey Bot. Club 28: 269. 1901. *H. hyperborea* var. *purpurascens* Ames, Orchidaceae 4: 90. 1910. *H. purpurascens* Tidestrom in Tidestrom & Kittell, Fl. Arizona & New Mex. 731. 1941. *Platanthera hyperborea* var. *purpurascens* Luer, Native Orchids U.S. & Canada 234. 1975. (*Rydberg & Vreeland 6414,* Iron Mountain, Colo., in 1900.)

Limnorchis brachypetala Rydb. Bull. New York Bot. Gard. 2: 161. 1901. (*R. S. Williams,* Bennett City, Yukon, 25 Aug. 1899.)

Slender bog orchid.

Plant glabrous throughout, 2–8 dm tall, from a fascicle of fleshy roots; 2 or 3 bladeless, sheathing bracts below the first leaf; lower two leaves usually oblong-elliptic and rounded to obtuse, 3–10 cm long and about 2–3.5 cm wide, the upper leaves becoming narrower and usually definitely acute to acuminate, gradually reduced to the cellular-serrulate bracts subtending the flowers; raceme spicate, open and lax, often secund, 5–15 (20) cm long with nearly sessile numerous flowers; flowers green, sometimes tinged with purplish-brown, with a slight fragrance; upper sepal broadly ovate, somewhat hooded, 3.5 mm long, the lower sepals usually 3-nerved, broadly lanceolate, about 5 mm long; petals more or less fleshy, about same length as the upper sepal, connivent with the upper sepal and forming a hood; lip linear to lanceolate or broader, 5–7 mm long, 1–2 mm wide, not at all dilated near the base, the orifice to the spur oblong-oval; spur scrotiform, greenish to purplish-brown, translucent, about half as long as the lip or slightly longer, rarely two-thirds as long; column small, usually slightly less than 2 mm long, the pollen sacs well separated by the connective; pollinia 2 per pollen sac, the viscidium nearly orbicular; capsule 1–1.5 cm long; $2n = 42$.

Moist to wet soil in meadows, bogs, thickets, or along mt. streams; Alaska, the Aleutian Islands, Can. and w. U.S. from Wash. to Calif. and e. to Mont., Colo. and N.M. June–Aug.

Sometimes very similar to *H. hyperborea* and differing mainly in the scrotiform spur, lax inflorescence, and rounded lower leaves. It is tempting to treat this as a mere variety of *H. hyperborea,* but intermediates in the *Limnorchis* section are many and Correll has preferred to treat *H. saccata* as a good species as did C. L. Hitchcock in the "Vascular Plants of the Pacific Northwest."

7. Habenaria unalascensis (Sprengel) S. Wats.

Spiranthes unalascensis Sprengel, Syst. Veg. 3: 708. 1826. *Herminium unalaschcense* Reichenb. Icon. Fl. German. & Helv. 13: 107. 1851. *Habenaria unalascensis* S. Wats. Proc. Amer. Acad. Arts 12: 277. 1877 (as *unalaschcensis*). *Platanthera unalaschcensis* Kurtz, Bot. Jahrb. Syst. 19: 408. 1894. *Montolivaea unalaschensis* Rydb. Mem. New York Bot. Gard. 1: 107. 1900. *Piperia unalaschensis* Rydb. Bull. Torrey Bot. Club 28: 270. 1901. ("Ins. Aleut.")

Alaska rein orchid.

Slender scapose plant, 3–5 dm high, from rounded ellipsoid or turbinate fleshy tubers 1.5–3 cm long and about 1 cm thick; leaves on the lower part of the stem, 2–4 that begin to wither by early anthesis, from narrowly lanceolate to broadly oblanceolate, 8–12 cm long, acute to rounded; stem with 1 or 2 sheathing bracts below the leaves; scape clothed with inconspicuous bracts; bracts of inflorescence much reduced; raceme spikelike, or often a spike as the flowers are sessile to subsessile, spirally arranged, 10–30 cm long, lax to densely flowered; flowers numerous, small, 8–10 mm long, greenish; petals and sepals 2–4 mm long; sepals 1-nerved, thin, the lower pair narrower but slightly longer than the upper one; petals fleshy and thicker than the sepals, usually greenish but sometimes nearly white or occasionally purplish-tinged; lip narrowly lanceolate to triangular-ovate, 2.5–4 mm long, somewhat contracted near the base, fused with the lateral sepals and the base of the column; spur narrowly cylindric to slightly saccate, 3–5 mm long, the orifice oblong-oval; column short, the anther about 1 mm long, the pollen sacs nearly contiguous; pollinia 2 per sac, the viscidium oblong.

Usually in dry wooded areas or open mountainsides but occasionally along gravelly streambanks; Alaska to Que., s. to Baja Calif., Nev., Utah, Colo., S.D. and Ont. June–early Aug.

New tubers are formed as offsets by the tubers producing the current flowering stem; the new tubers then produce the flowering stems of the next year.

The Alaska rein orchid is a distinctive species with scapose stems and leaves that usually begin to wither by early anthesis, especially on plants growing in rather dry sites.

7. CORALLORHIZA Chatel. Coralroot

Non-green, mycorrhizal plants from warty, short or elongate rhizomes which superficially resemble coral; roots none; stems scapose, yellowish to brownish-red, clothed with several membranous sheathing bracts, but without green leaves; flowers few to many from short pedicels in an inconspicuously bracted, lax to congested raceme, yellow to reddish-brown or purple; perianth usually cupped forward; sepals equal, the lower pair united at the base and forming a short mentum which is more or less adnate to the ovary; lip simple to 3-lobed, slightly adherent to the base of the column, the column curved upon the lip, the stigma circular to triangular; anther terminal on the column, deciduous; pollinia 4, waxy, without appendages; capsule ovoid to ellipsoid, pendent.

A genus of 12 species of terrestrial non-green orchids of N. Amer. and C. Amer. (From the Greek *korallion,* coral, and *riza,* root.)

Old desiccated flowering stems persisting singly or in clusters, often in extensive colonies, remain upright long after flowering and the dissemination of the tiny seeds. The fragile coralloid rhizomes remain imbedded in the rotting humus of the forest floor where they multiply and gather nutrients in conjunction with the mycelium of the fungus associated with them.

1 Sepals and petals prominently striped with reddish-brown or purple; spur lacking .... 1. *C. striata*
1 Sepals and petals not prominently striped, sometimes very lightly veined; spur small but usually present, adnate to the ovary.
  2 Sepals more than 6 mm long, often reddish, usually 3-nerved; lip usually strongly spotted.
    3 Lip with an evident though rather small lateral lobe on each side near the base; common species .... 2. *C. maculata*
    3 Lip without lateral lobes; species rare with us .... 3. *C. wisteriana*
  2 Sepals less than 6 mm long, yellowish to nearly white, 1-nerved; lip without spots or sometimes lightly red-spotted .... 4. *C. trifida*

CORALLORHIZA

Corallorhiza striata Corallorhiza maculata Corallorhiza wisteriana Corallorhiza trifida

1. Corallorhiza striata Lindl.

Corallorhiza striata Lindl. Gen. & Sp. Orchidaceous Pls. 534. 1840. *Neottia striata* Kuntze, Revisio Gen. Pl. 2: 674. 1891. (*Douglas,* "Hab. in America boreali-occidentali.")

C. ochroleuca Rydb. Bull. Torrey Bot. Club 31: 402. 1904. (*T. A. Williams 48,* War Bonnet Canyon, Nebr., in 1890.)

Striped coralroot.

Plant reddish-purple to brownish-purple, 1.5–4.5 dm tall; stems with many scales; flowers 7–25 on a bracteate raceme up to 2 dm long, the pedicels rather stout, 1–3 mm long; flowers pinkish yellow to whitish; sepals and petals with 3–5 purplish to reddish brown stripes, the sepals oblong-lanceolate, 10–15 mm long, 2–4 mm wide, the lateral pair horizontal; petals somewhat shorter and broader than the sepals, up to 7 mm wide; lip broadly elliptical, entire, 8–12 mm long, 6–8 mm wide when spread out, whitish, strongly striate-veined with reddish purple, the stripes merging at the apex, concave with thickened upturned margins and two median callosities near the base; column 4–7 mm long, about half as long as the petals, spotted with purple near the base; capsule ellipsoid, becoming reflexed, 1.5–2 cm long; $2n = 42$.

Dryish places in aspen and coniferous forests; B.C. to Que., and s. through most of w. U.S. in Nev., Utah, Ariz., N.M., Texas to n. Mex. and n. Mich. June–July.

Albino plants of a pale yellowish color of this species and those that follow may be found in most patches of any size.

2. Corallorhiza maculata Raf.

Corallorhiza maculata Raf. Amer. Monthly Mag. & Crit. Rev. 2: 119. 1817, as *Coralorhiza* ("Shady woods of Long Island, near Flatbush, Flushing, Oyster-Bay, etc.")

Cladorhiza maculata Raf. Amer. Monthly Mag. & Crit. Rev. 1: 429. 1817. (No collectors or locality cited.)

Corallorhiza multiflora Nutt. J. Acad. Nat. Sci. Philadelphia 3: 138, pl. 7. 1823. *Neottia multiflora* Kuntze, Revisio Gen. Pl. 2: 674. 1891. ("Hab. from New England to Carolina.")

Corallorhiza multiflora var. *occidentalis* Lindl. Gen. & Sp. Orchidaceous Pls. 534. 1840. *Corallorrhiza maculata* subsp. *occidentalis* Cockerell, Torreya 16: 232. 1916. (*Douglas,* in Calif.) The phase with the spur gibbosity rather prominent.

Corallorhiza multiflora var. *flavida* Peck, Annual Rep. New York State Mus. 50(1): 126. 1897. *Corallorrhiza maculata* var. *flavida* Cockerell, Torreya 16: 232. 1916. *Corallorrhiza maculata* f. *flavida* Farw. Amer. Midl. Naturalist 10: 208. 1927. (Menands, Albany Co., N.Y., no collector cited.) Plants pale yellow—the equivalent of an albino phase?

Corallorrhiza multiflora var. *sulphurea* Suksd. Allg. Bot. Z. Syst. 12: 42. 1906. (*Suksdorf 2810,* near Bingen, Klickitat Co., Wash., 4 Aug. 1896 and 19 July 1897.) A pale yellow phase.

Corallorrhiza leimbachiana Suksd. Allg. Bot. Z. Syst. 12: 42. 1906. (*Suksdorf 2811,* near Bingen, Klickitat Co., Wash., 2 July and 4 Aug. 1896.)

Coralliorrhiza hortensis Suksd. Werdenda 1: 18. 1927. (*Suksdorf 11647,* Bingen, Klickitat Co., Wash., 30 April 1924.)

Corallorhiza maculata var. *immaculata* M. E. Peck, Leafl. W. Bot. 7: 177. 1954. *C. maculata* f. *immaculata* J. T. Howell, Marin Fl. ed. 2. 363. 1970. (*Howard Taylor,* Mosby Creek, 10 mi. e. of Cottage Grove, Linn Co., Oregon.)

Spotted coralroot.

Plant reddish-purple to brownish-purple, 2–5.5 dm tall; stems with many scales; flowers 10–30 in a bracteate raceme, erect in the bud, widely spreading at anthesis and then becoming reflexed, the pedicels slender, 1–2 mm long; sepals and petals reddish-purple, linear to oblong-lanceolate to oblanceolate, 6–8 mm long, 3.5–5 mm wide; petals slightly shorter than the sepals, often with a few dark red spots, 3-nerved, the lateral pair extending under the base of the lip; lip white, usually with dark red spots, with 2 prominent lateral lobes about 1.3 mm long and curved, the larger mid-lobe obovate to nearly orbicular, more or less crenate; column thick and stout, yellow, with magenta spots on the ventral surface, 4–5 mm long; capsule 1.5–2 cm long, ellipsoid, reflexed; $2n = 42$.

Moist or usually somewhat dry places in decaying humus of aspen or coniferous forests; Alaska to B.C., e. to N.S. and s. to Calif., Nev., Utah, Ariz., N.M., Texas, and N.C.; also in Mex. and Guatemala. June–July.

Albino plants lack the purplish pigments and are pale yellowish throughout. The corolla is white and without spots. The two color forms often grow together, with the pigmented plants being more common.

3. Corallorhiza wisteriana Conrad

Corallorhiza wisteriana Conrad, J. Acad. Nat. Sci. Philadelphia 6: 145. 1829. (*Charles J. Wister,* Schuylkill River, between the Falls and the mouth of the Wissahickon Creek, Pa., in 1828.)

Spring coralroot.

Much like *C. maculata,* but averaging smaller and softer, and blooming earlier when growing in habitats that so permit; sepals and petals 1- or 3-nerved; capsule ca. 1 cm long; differing sharply from *C. maculata* in its merely undulate or erose-denticulate lip, without lateral lobes.

Along streams or in other moist areas, usually in rather deep woods; best developed in c. and s. (especially se.) U.S., but extending s. into Mex. and w. to the e. side of the Rocky Mts. in Mont., Wyo., and Colo., with outlying stations in Utah and Ariz.; in our range known only from two stations in Utah, both well up in the mts.: Uinta Mts. (Summit Co.) and Pine Valley Mts. (Washington Co.).

4. Corallorhiza trifida Chatel.

Corallorhiza trifida Chatel. Specimen Inaugurale *Corallorhiza* 8. 1760. (No locality or collector cited.)

Ophrys corallorhiza L. Sp. Pl. 945. 1753. *Epipactis corallorhiza* Crantz, Stirpium Austriacarum, ed. 2. 464. 1769. *Helleborine corallorrhiza* F. W. Schmidt, Fl. Boëmica 79. 1794. *Cymbidium corallorrhiza* Swartz, Kongl. Vetensk. Acad. Nya Handl. 21: 238. 1800. *Corallorhiza corallorhiza* Karsten, Deutsche Fl. 448. 1882. *Neottia corallorhiza* Kuntze, Revisio Gen. Pl. 2: 674. 1891. ("Habitat in Europae borealis desertis.")

Corallorhiza verna Nutt. J. Acad. Nat. Sci. Philadelphia 3: 136. 1823. *C. odontorhiza* var. *verna* A. Wood, Cl.-Book. Bot. ed. 2. 531. 1847. *C. trifida* var. *verna* Fern. Rhodora 48: 196. 1946. (*Muhlenberg,* Pa., is the first of two collections cited.)

Corallorrhiza innata var. *virescens* Farr, Contr. Bot. Lab. Morris Arbor. Univ. Pennsylvania 2: 425. 1904. *Corallorrhiza trifida* var. *virescens* Farw. Pap. Michigan Acad. Sci. 26: 9. 1941. (*Edith M. Farr,* Banff, Alta., 14 June 1904, and Field, B.C., 11 June 1904.)

Corallorrhiza corallorrhiza subsp. *coloradensis* Cockerell, Torreya 16: 231. 1916. *Corallorhiza corallorhiza* var. *coloradensis* Ames, Orchid. U.S. & Canad. 24. 1924. (*Cockerell,* near Long's Peak Inn, Larimer Co., Colo., 21 June 1916.)

Corallorrhiza wyomingensis Hellmayr & Hellmayr, Rhodora 33: 133. 1931. (*C. E. & K. J. Hellmayr 72,* near Libby Lodge, Medicine Bow Range, Wyo., 17 July 1929.)

CALYPSO

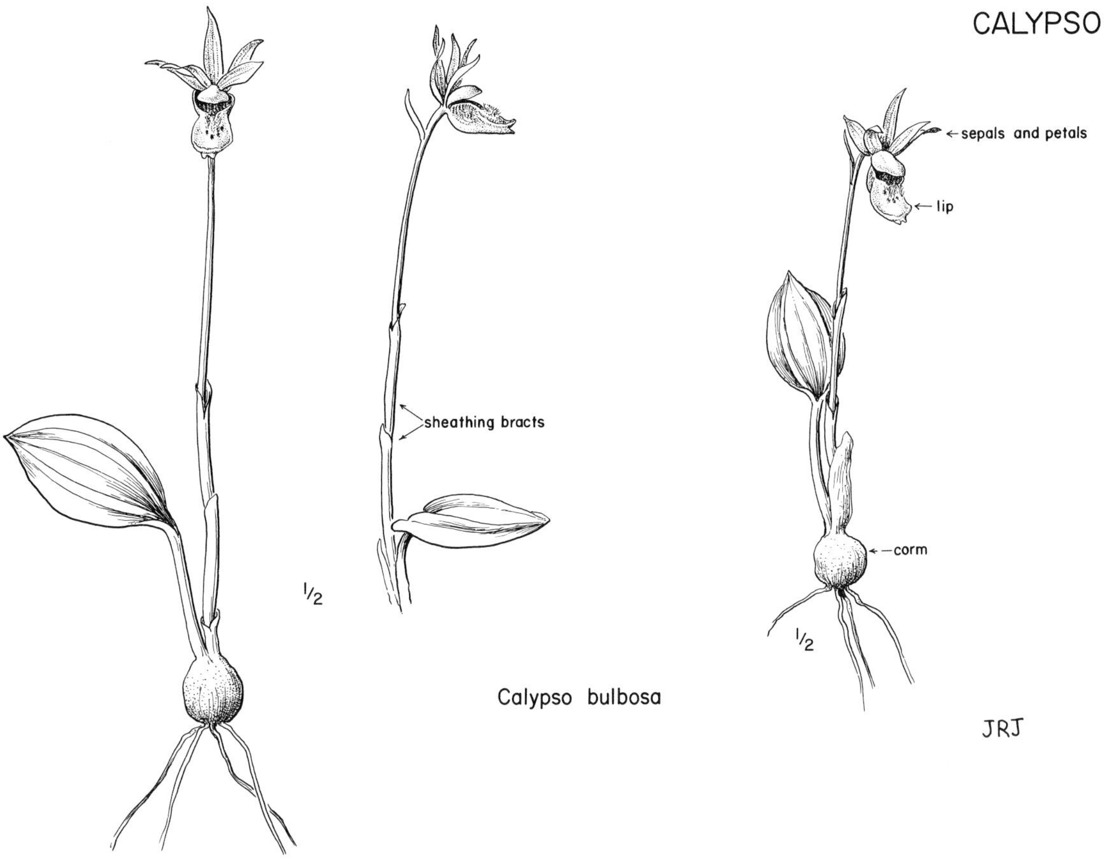

Northern coralroot.

Plant usually pale yellow or sometimes golden yellow or greenish yellow, 0.5–2.5 dm tall; stems with two or three sheathing bracts; flowers (3) 5–15, erect at first and then spreading on somewhat stout pedicels about 1 mm long, yellowish white to pale purple or greenish; sepals and petals linear-oblong to linear oblanceolate, about 5 mm long, slightly less than 2 mm wide, 1-nerved; sepals pointing forward, the lateral pair extending under the base of the lip; petals somewhat shorter than the sepals and slightly wider, extending forward; lip white or nearly so, sometimes with a few reddish or purplish spots, 3.5–4.5 mm long, 2–3 mm wide, oblong, usually with a small triangular upcurved tooth on each side near the base, 3-nerved, usually with callosities over the lateral nerves; column broad, 2.5–6 mm long; capsule about 1 cm long, strongly reflexed; $2n = 42$.

In moist places in coniferous woods, often in lodgepole pine forests; Alaska to Labrador, s. to Wash., Wallowa Mts. in Oregon, Idaho, Utah in the Uinta Mts., and near Teasdale in Wayne Co.; Colo., Ind. and N.J.; also Eurasia. June–early Aug.

This is the smallest of our species, often going undetected in deep woods of lodgepole pine in the Uinta Mountains where the pale plants may be no more than 5 cm tall and bear only three little flowers.

8. CALYPSO Salisb.

A low plant, with a single leaf and a short scape with 2 or 3 sheathing bracts, from a subglobose corm with fleshy roots or coralloid rhizome or both; scape with a solitary, showy blossom; sepals and petals similar, spreading; lip the largest and showiest part of the flower, deeply saccate, bicornute at the apex of the sac, dilated at the anterior margin of the orifice, forming an apron with 3 short rows of yellow or white hairs leading into the orifice; column concave, broadly winged and petaloid, stigmatic area near the base, bearing the single anther and 2 pollinia; capsule erect, ellipsoid-cylindric.

A genus with a single species widely distributed in N. Amer. and Eurasia. (Named for the Greek sea nymph, *Kalypso,* of Homer's Odyssey, the word meaning covered or hidden; probably alluding to the tiny plant which often goes undetected.)

1. Calypso bulbosa (L.) Oakes

Cypripedium bulbosum L. Sp. Pl. 951. 1753. *Calypso bulbosa* Oakes in Z. Thompson, Hist. Vermont 1: 200. 1842. *Cypripedilum bulbosum* Stein, Orchideenbuch 176. 1892. *Cytherea bulbosa* H. D. House, Bull. Torrey Bot. Club 32: 382. 1905. ("Hab. in Lapponia, Russia, Sibiria.")
Calypso borealis Salisb. Paradisus Lond. pl. 89. 1808. *Cytherea borealis* Salisb. Trans. Hort. Soc. London 1: 301. 1812.

(*Menzies,* Nova Scotia, is the first of three collections cited.) *Calypso bulbosa* f. *occidentalis* Holzinger, Contr. U.S. Natl. Herb. 3: 251. 1895. *Calypso occidentalis* A. A. Heller, Bull. Torrey Bot. Club 25: 193. 1898. *Cytherea occidentalis* A. A. Heller, Muhlenbergia 1: 132. 1906. *Cytherea bulbosa* var. *occidentalis* Cockerell, Bot. Gaz. 59: 331. 1915. *Calypso bulbosa* subsp. *occidentalis* Calder & Taylor, Canad. J. Bot. 43: 1393. 1965. *Calypso bulbosa* var. *occidentalis* B. Boivin, Naturaliste Canad. 94: 522. 1967. (*Sandberg, MacDougal & Heller 211,* Craig Mts. at Lake Waha, Nez Perce Co., Idaho, 21 May 1892.)

Fairy slipper, Venus' slipper, calypso.

Plant glabrous, 5–15 (20) cm tall, from a fleshy, globose to ellipsoid corm; a single leaf produced from the summit of the corm late in the season and persisting through the winter, withering in the summer, the leaf petiolate with the blade and petiole about the same length, the blade broadly ovate-elliptic to oblong-ovate or cordate, 3–6 cm long; flower usually solitary, on a scape with 2 or 3 sheathing bracts; sepals and petals similar, purplish, rarely white with deeper colored veins, narrowly-lanceolate to linear-oblong, erect to ascending, 15–23 mm long; lip pendent, ovate-oblong, usually slightly longer than the other perianth segments, about 10 mm wide, bearded, whitish to yellowish to pale reddish-brown in age, variously spotted on the flanged appendage; column petaloid, suborbicular, 7–12 mm long and about as wide, reddish-purple, nearly horizontal, concave, hooded over the opening to the lip; capsule erect, about 2 cm long; $2n = 28$.

In mossy places on cool moist coniferous forests, bogs, and along streams; Alaska to Labrador, s. to Calif., Utah, Ariz., Colo., Minn., Mich., N.Y. to Me.; also in Eurasia. May–July.

This beautiful little orchid occurs in widely scattered places but is usually abundant in large patches when found. In the Uinta Mountains it occurs at Soapstone, Moon Lake and Brown's Lake. It also grows in Engelmann spruce forest on Mt. Tukuhnikivatz in the LaSal Mountains in Grand County and west of Monticello in San Juan County.

List of Nomenclatural Innovations

Potamogeton filiformis var. *latifolius* (J. W. Robbins) Reveal, p. 26
Juncus ensifolius var. *brunnescens* (Rydb.) Cronq., p. 56
Scirpus pungens var. *longispicatus* (Britton) Cronq., p. 74
Carex scirpoidea var. *curatorum* (Stacey) Cronq., p. 113
Aristida purpurea var. *glauca* (Nees) A. Holmgren & N. Holmgren, p. 455
Aristida purpurea var. *robusta* (Merr.) A. Holmgren & N. Holmgren, p. 456
Calochortus panamintensis (Ownbey) Reveal, p. 501
Allium atrorubens var. *inyonis* (M. E. Jones) Ownbey & Aase, p. 515
Allium bisceptrum var. *palmeri* (S. Wats.) Cronq., p. 515
Yucca harrimaniae var. *neomexicana* (Wooton & Standley) Reveal, p. 530
Yucca baileyi var. *intermedia* (McKelvey) Reveal, p. 532
Yucca elata var. *utahensis* (McKelvey) Reveal, p. 533
Yucca elata var. *verdiensis* (McKelvey) Reveal, p. 533
Yucca angustissima var. *kanabensis* (McKelvey) Reveal, p. 534
Yucca angustissima var. *toftiae* (S. L. Welsh) Reveal, p. 534
Yucca angustissima var. *avia* Reveal, p. 534
Sisyrinchium douglasii var. *inflatum* (Suksd.) P. Holmgren, p. 540

INDEX

Page numbers in bold face indicate where the taxon is described. Names in *italics* indicate synonyms. This index is only for volume six. The cumulative index for all six volumes is scheduled to appear in volume two.

NOTES

NOTES

NOTES

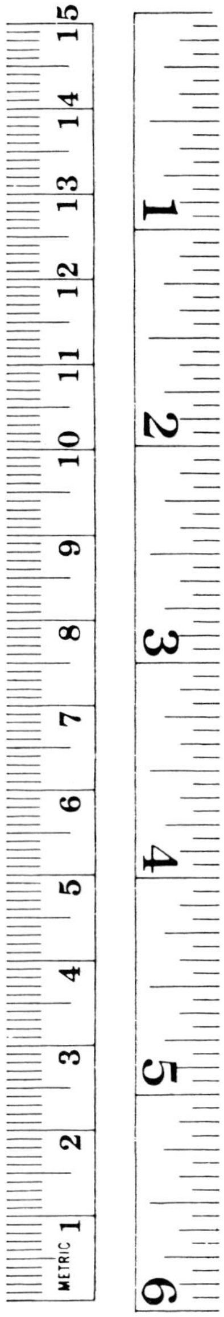
METRIC
1 2 3 4 5 6 7 8 9 10 11 12 13 14 15
1 2 3 4 5 6

OREGON
IDAHO
WYOMING
NEVADA
UTAH
CALIFORNIA
ARIZONA
CROOK
GRANT
DESCHUTES
LAKE
HARNEY
MALHEUR
KLAMATH
Burns
WASHINGTON
Weiser
VALLEY
PAYETTE
GEM
CANYON
Boise
BOISE
ADA
ELMORE
CUSTER
LEMHI
Kilgore
CLARK
FREMONT
MADISON
TETON
JEFFERSON
Arco
BUTTE
BONNEVILLE
CAMAS
BLAINE
GOODING
LINCOLN
BINGHAM
JEROME
MINIDOKA
POWER
BANNOCK
CARIBOU
Afton
SUBLETTE
OWYHEE
TWIN FALLS
CASSIA
ONEIDA
FRANKLIN
BEAR LAKE
MODOC
HUMBOLDT
ELKO
BOX ELDER
CACHE
RICH
WEBER
Evanston
UINTA
SWEETWATER
WASHOE
LASSEN
Susanville
PERSHING
DAVIS
MORGAN
SUMMIT
DAGGETT
SALT LAKE
TOOELE
WASATCH
DUCHESNE
UINTAH
PLUMAS
SIERRA
Reno
STOREY
CHURCHILL
LANDER
EUREKA
WHITE PINE
JUAB
CARBON
ORMSBY
LYON
DOUGLAS
MILLARD
SAN PETE
EMERY
GRAND
ALPINE
MINERAL
NYE
SEVIER
BEAVER
PIUTE
WAYNE
MONO
ESMERALDA
IRON
GARFIELD
SAN JUAN
Bishop
Big Pine
INYO
LINCOLN
Alamo
Motoqua
WASHINGTON
KANE
CLARK
MOHAVE
COCONINO
NAVAJO
APACHE
Neal